3.1–3.2 RULES FOR EXPONENTS	For all rational numbers m and n
	Product rule $a^m \cdot a^n = a^{m+n}$
	Quotient rule $\dfrac{a^m}{a^n} = a^{m-n} \quad (a \neq 0)$
	Power rules $(a^m)^n = a^{mn} \quad (ab)^m = a^m b^m \quad \left(\dfrac{a}{b}\right)^m = \dfrac{a^m}{b^m} \quad (b \neq 0)$
NEGATIVE EXPONENT	$a^{-n} = \dfrac{1}{a^n} = \left(\dfrac{1}{a}\right)^n \quad \dfrac{a^{-n}}{b^{-m}} = \dfrac{b^m}{a^n} \quad \left(\dfrac{a}{b}\right)^{-n} = \left(\dfrac{b}{a}\right)^n \quad (a, b \neq 0)$
ZERO EXPONENT	If a is any nonzero real number, then $a^0 = 1$.

MW01267852

3.7 SPECIAL FACTORIZATIONS	**Difference of two squares** $x^2 - y^2 = (x + y)(x - y)$
	Perfect square trinomial $x^2 + 2xy + y^2 = (x + y)^2 \quad x^2 - 2xy + y^2 = (x - y)^2$
	Difference of two cubes $x^3 - y^3 = (x - y)(x^2 + xy + y^2)$
	Sum of two cubes $x^3 + y^3 = (x + y)(x^2 - xy + y^2)$

5.1 RATIONAL EXPONENTS AND RADICALS

Let a be a real number and n be a positive integer.

$a^{1/n} = b$ means $b^n = a$.

$a^{1/n}$ is the principal nth root of a if a is positive and n is even.

$a^{1/n} = \sqrt[n]{a}$.

$\sqrt[n]{a^n} = |a|$ if n is even.

$\sqrt[n]{a^n} = a$ if n is odd.

If m and n are positive integers with m/n in lowest terms, then

$$a^{m/n} = (a^{1/n})^m = (a^m)^{1/n},$$

provided $a^{1/n}$ is a real number.

5.3 SIMPLIFYING RADICALS

If $\sqrt[n]{a}$ and $\sqrt[n]{b}$ are real numbers, and n is a natural number, then

$$\sqrt[n]{a} \cdot \sqrt[n]{b} = \sqrt[n]{ab} \quad \text{and} \quad \frac{\sqrt[n]{a}}{\sqrt[n]{b}} = \sqrt[n]{\frac{a}{b}} \quad (b \neq 0).$$

A simplified radical has

1. no factor in the radicand raised to a power greater than or equal to the index;
2. no fractions under the radicand;
3. no radical denominators;
4. exponents in the radicand and the index of the radical with no common factors (except 1).

ALGEBRA FOR COLLEGE STUDENTS

SECOND EDITION

MARGARET L. LIAL
American River College

CHARLES D. MILLER

E. JOHN HORNSBY, JR.
University of New Orleans

TO THE STUDENT

If you need further help with algebra, you may want to obtain a copy of the *Student's Solutions Manual* that goes with this book. It contains solutions to all the odd-numbered exercises and all the chapter test exercises. You also may want the *Student's Study Guide*. It has extra examples and exercises to complete, corresponding to each learning objective of the book. In addition, there is a practice test for each chapter. Your college bookstore either has these books or can order them for you.

On the cover: "Thy belly is like a heap of wheat set about with lilies"—Solomon's song of his beloved might also have celebrated the sensual curves of the grain-giving earth in California, where wheat and barley are grown on the crowns of rolling hills and in long narrow valleys at the northern approach to the Carrizo Plain. In May the pale gold winter wheat has just been harvested; these fields will be fallowed for eighteen months before being resown.

Sponsoring Editor: Anne Kelly
Developmental Editor: Adam Bryer
Project Editor: Cathy Wacaser
Assistant Art Director: Julie Anderson
Text and Cover Design: Lucy Lesiak Design, Inc.: Lucy Lesiak
Cover Photo: Comstock/George Gerster
Photo Researchers: Kelly Mountain and Sandy Schneider
Production: Jeanie Berke, Linda Murray, and Helen Driller
Compositor: York Graphic Services
Printer and Binder: R. R. Donnelley & Sons Company
Cover Printer: Lehigh Press Lithographers

Algebra for College Students, Second Edition

Library of Congress Cataloging-in-Publication Data
Lial, Margaret L.
Algebra for college students/Margaret L. Lial, Charles D. Miller, E. John Hornsby, Jr.—2nd ed.
p. cm.
Includes index.
ISBN 0-673-46469-5 (Free Copy: ISBN 0-673-46637-X)
1. Algebra. I. Miller, Charles David. II. Hornsby, E. John. III. Title.
QA154.2.L495 1992
512.9—dc20 91-27432
CIP

91 92 93 94 9 8 7 6 5 4 3 2 1

PREFACE

In the second edition of *Algebra for College Students,* we have maintained the strengths of the past edition, while enhancing the pedagogy, readability, and attractiveness of the text. Many new features have been added to make the text easier and more enjoyable for students and teachers to use, including new exercises, Quick Reviews, and the use of full color. We continue to provide an extensive supplemental package. For students, we offer a solutions manual, a study guide, interactive tutorial software, videotapes, and audiotapes. For instructors, we provide alternative forms of tests, additional exercises, computer-generated tests, complete solutions to all exercises, transparencies, and videos.

All the successful features of the previous edition are carried over in the new edition: learning objectives for each section, careful exposition, fully developed examples with comments printed at the side (more than 900 examples), and carefully graded section, chapter review, and chapter test exercises (more than 7200 exercises, with more than 50 percent of the applications exercises new to this edition). Screened boxes set off important definitions, formulas, rules, and procedures to further aid in learning and reviewing the course material.

NEW FEATURES

Conceptual and Writing Exercises

To complement the drill exercises, several exercises requiring an understanding of the concepts introduced in a section are included in almost every exercise set. There are more than 400 of these conceptual exercises. Furthermore, more than 180 exercises require the student to respond by writing a few sentences. (Note that some of these writing exercises are also labeled as conceptual exercises.) Directions to conceptual and writing exercises often include references to specific learning objectives to help students achieve a broader perspective should they need to turn back to the explanations and examples.

Focus on Problem Solving An application from the exercise set will be featured at the beginning of selected sections. Each application is chosen from among the most interesting in the exercise set and is intended to help motivate the study of the section.

FOCUS ON PROBLEM SOLVING

Erin is a biology student. She has heard that the number of times a cricket chirps in one minute can be used to find the temperature. In an experiment, she finds that a cricket chirps 40 times per minute when the temperature is 50° F, and 80 times per minute when the temperature is 60° F. What is the temperature when the cricket chirps 20 times per minute?

Many real-world situations can be described by straight-line graphs. By finding an equation that relates chirps to temperature, we can solve the problem stated above. It is Exercise 75 in the exercises for this section. After working through this section, you should be able to solve this problem.

Problem-Solving Strategies In special paragraphs clearly distinguished by the heading "Problem Solving," we have expanded our discussion of strategies to include connections to techniques learned earlier.

PROBLEM SOLVING

A chart was used in Example 4 to organize the information. Some students like to use "box diagrams" instead. Either method works well if you are careful while filling in the information. The next example uses box diagrams. ■

Design Use of full color and changes in format help create a fresh look. We have both enhanced the book's appeal and increased its usefulness.

Cautionary Remarks Common student errors and difficulties are now highlighted graphically and identified with the heading "Caution." **Notes** have a similar graphic treatment.

CAUTION The symbol $f(x)$ *does not* indicate "f times x," but represents the y-value for the indicated x-value. As shown above, $f(2)$ is the y-value that corresponds to the x-value 2.

NOTE If the procedure of Example 4 leads to an equation of the form $(x - h)^2 + (y - k)^2 = 0$, the graph is the single point (h, k). If the constant on the right side is negative, the equation has no graph.

SECOND EDITION

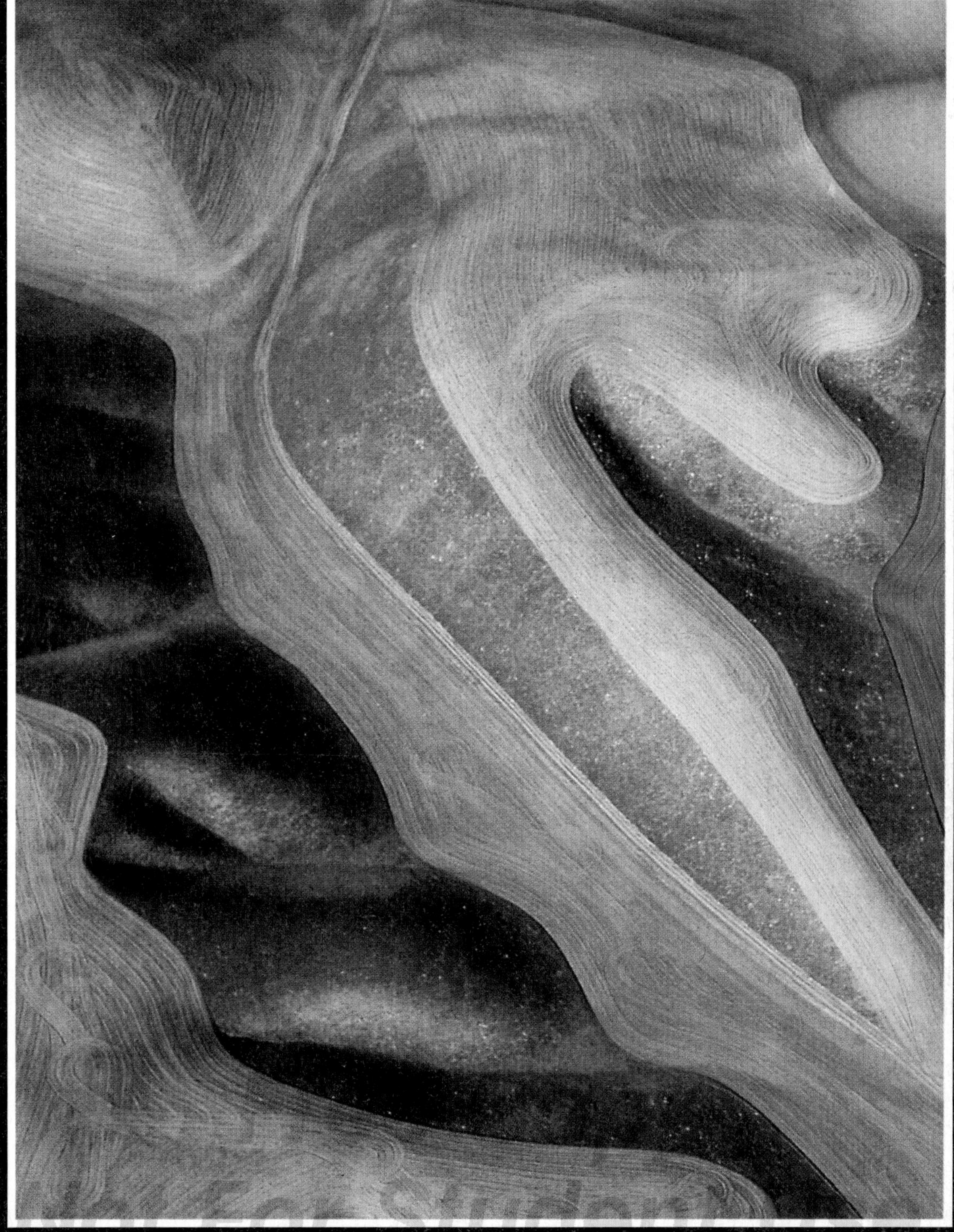

ALGEBRA FOR COLLEGE STUDENTS

Quick Review A Quick Review at the end of each chapter provides a capsule summary of key ideas and is set in tabular form to enable students to find the important concepts easily and review them more effectively. In addition, worked-out examples accompany each section-referenced key idea.

CHAPTER 8 QUICK REVIEW

SECTION	CONCEPTS	EXAMPLES
8.1 THE ALGEBRA OF FUNCTIONS	**Operations on Functions** $(f+g)(x) = f(x) + g(x)$ $(f-g)(x) = f(x) - g(x)$ $(fg)(x) = f(x) \cdot g(x)$ $\left(\frac{f}{g}\right)(x) = \frac{f(x)}{g(x)}, \quad g(x) \neq 0$ **Composition of Functions** If f and g are functions, then the composite function of g and f is $(g \circ f)(x) = g[f(x)].$	If $f(x) = 3x^2 + 2$ and $g(x) = \sqrt{x}$, then $(f+g)(x) = 3x^2 + 2 + \sqrt{x}$ $(f-g)(x) = 3x^2 + 2 - \sqrt{x}$ $(fg)(x) = (3x^2 + 2)(\sqrt{x})$ $\left(\frac{f}{g}\right)(x) = \frac{3x^2 + 2}{\sqrt{x}}, \ x > 0.$ If $g(x) = \sqrt{x}$ and $f(x) = x^2 - 1$, then the composite function of g and f is $(g \circ f)(x) = \sqrt{x^2 - 1}$
8.2 QUADRATIC FUNCTIONS; PARABOLAS	**1.** The graph of the quadratic function $f(x) = a(x - h)^2 + k, \ a \neq 0$ is a parabola with vertex at (h, k) and the vertical line $x = h$ as axis. **2.** The graph opens upward if a is positive and downward if a is negative. **3.** The graph is wider than $f(x) = x^2$ if $0 < \|a\| < 1$ and narrower if $\|a\| > 1$.	Graph: $f(x) = -(x + 3)^2 + 1.$ 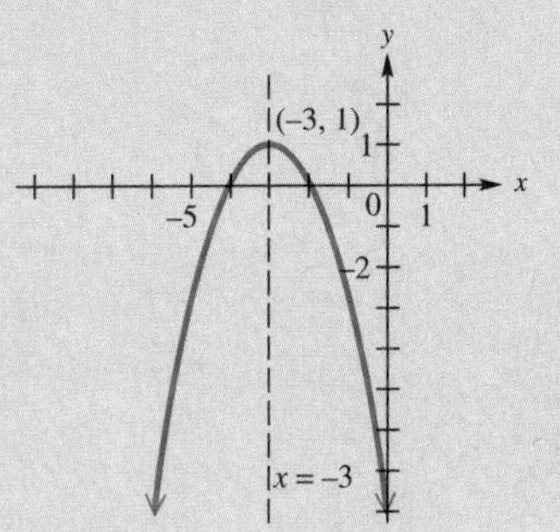

Geometry Exercises Review of geometry is a thread that runs through the text. We have increased the number of exercises that relate geometric concepts to the new algebraic concepts. A brief review of geometry is included in Appendix A, expanded from the first edition to include a greater number of basic definitions and relationships.

Applications These exercises have been extensively rewritten and updated to more closely reflect the student's world. More than 50 percent are new to the second edition.

Glossary A glossary of key terms, followed by a description of new symbols, is provided at the end of each chapter. A comprehensive glossary is placed at the end of the book. Each term in the glossary is defined and then cross-referenced to the appropriate section, where students may find a more detailed explanation of the term.

CHAPTER 11 GLOSSARY

KEY TERMS

11.1 system of equations Two or more equations that are to be solved at the same time form a system of equations.

linear system A linear system is a system of equations that contains only linear equations.

inconsistent system A system is inconsistent if it has no solution.

dependent equations Dependent equations are equations whose graphs are the same line.

elimination (or addition) method The elimination (or addition) method of solving a system of equations involves the elimination of a variable by adding the two equations.

Example Titles Each example now has a title to help students see the purpose of the example. The titles also facilitate working the exercises and studying for examinations.

EXAMPLE 2 FINDING THE SLOPE OF A LINE

Find the slope of the line $4x - y = 8$.

The intercepts can be used as the two different points needed to find the slope. Let $y = 0$ to find that the x-intercept is $(2, 0)$. Then let $x = 0$ to find that the y-intercept is $(0, -8)$. Use these two points in the slope formula. The slope is

$$m = \frac{-8 - 0}{0 - 2} = \frac{-8}{-2} = 4. \quad ■$$

Calculator Usage The use of a scientific calculator is explained and referred to at appropriate points in the text. An Introduction to Scientific Calculators, included just prior to Chapter 1, gives additional information on keystrokes and techniques.

Graphing Calculators At the end of each of Chapters 8–11, there is a brief discussion on how a graphing calculator can enhance the study of the relations covered in the chapter. Students are encouraged to explore new and previously graphed examples. (Answers in the answer section will not be provided for these explorations.)

Success in Algebra This foreword to the student provides additional support by offering suggestions for studying the course material.

Preview Exercises Formerly called Review Exercises, these are intended to sharpen the basic skills needed to do the work in the next section. Students need to review material presented earlier, even though it seems very basic. These exercises also help to show how earlier material connects with and is needed for later topics.

In the chapter review exercises, mixed review exercises are included for all of the chapters except Chapters 7, 10, 12, and 13; the material in those chapters does not lend itself to a mixed review with general instructions.

NEW CONTENT HIGHLIGHTS

In Chapter 1, the properties of real numbers are now presented after operations with real numbers. This section has been completely rewritten to present the use of the properties in algebra to combine terms, factor, and reduce or build up rational expressions, for example.

The discussion on solving applied problems in Chapter 2 has been expanded to two sections and reorganized to help students over this rough spot. New topics from geometry have been included. The slower presentation and additional applied problems will help students to acquire this important skill. Absolute value equations and inequalities, now combined into one section, are developed with a parallel treatment.

The first two sections of Chapter 5 have been reorganized. The first section introduces the basic ideas of rational exponents and the connection with radicals. In the second section we discuss the use of the rules of exponents with rational exponents and show how to approximate roots with a calculator. Rationalizing the denominator has been moved to Section 5.5; all discussion of rationalizing of denominators is now included in that section.

Functions are introduced at the end of Chapter 7 so that the function concept can be used in discussing parabolas later in Chapter 8. The distinction between relation and function is made clearer by using the conic sections as examples.

Most of the material in Chapters 8 and 9 of the first edition has been integrated into Chapter 8 of the new edition. Chapter 8 now begins with algebra of functions; these ideas are then used when new types of functions are introduced in the rest of Chapter 8 and in Chapters 9 and 10. Chapter 8 continues with two sections on the parabola. The first section concentrates on the vertex, vertical and horizontal shifts, and the ways in which the coefficient of the squared term affects the shape of the parabola. Functional notation is used in this section. The second section concentrates on general methods of graphing and applications of quadratic functions.

Chapter 9 now covers polynomial functions and their zeros. We have given more emphasis to the relationships among intercepts of a graph, zeros of a function, and solutions of an equation. The section on rational functions has been expanded and reorganized to clarify the presentation.

Chapter 10, on exponential and logarithmic functions, now opens with a discussion of inverse functions to help students understand the relationship between exponential and logarithmic functions. The section on using logarithms for calculations has been replaced with a section on using the calculator to evaluate common logarithms, natural logarithms, and logarithms to other bases. The use of tables of logarithms is explained in Appendix C. The section on logarithmic and exponential equations has been extensively rewritten to make it more accessible to students.

The discussion of systems of equations and inequalities in Chapter 11 has been reordered. We now present systems of equations in two variables first, followed by systems of equations in three variables, and then a section on applications, including some with two variables and some with three variables.

Chapter 13 has also been rearranged. The first two sections now discuss sequences, with both arithmetic and geometric series covered in the third section, and sums of infinite geometric sequences discussed in Section 13.4. This expansion of the first two sections gives students a better opportunity to absorb the many new ideas presented here. The material on sequences is followed by material on the binomial

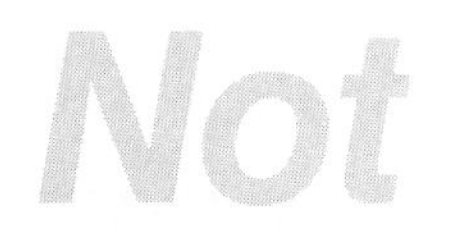

theorem and mathematical induction. The final sections cover permutations, combinations, and probability. This ordering of topics collects the material most instructors consider important in the earlier sections of the chapter.

Throughout the book, summaries of procedures have been moved earlier in the section, so students can refer to the summary while working through the examples.

SUPPLEMENTS

Our extensive supplemental package includes testing materials, solutions, software, videotapes, and audiotapes.

Instructor's Test Manual Included are two versions of a pretest—one open-response and one multiple-choice; six versions of a chapter test for each chapter—four open-response and two multiple-choice; three versions of a final examination—two open-response and one multiple-choice; and an extensive set of additional exercises, providing ten to twenty exercises for each textbook objective, which can be used as an additional source of questions for tests, quizzes, or student review of difficult topics. Answers to all tests and additional exercises are provided.

Instructor's Solutions Manual This manual includes solutions to all exercises and a list of all conceptual, writing, and challenging exercises.

Instructor's Answer Manual This includes answers to all exercises and a list of all conceptual, writing, and challenging exercises.

HarperCollins Test Generator for Mathematics The *HarperCollins Test Generator* is one of the top testing programs on the market for IBM and Macintosh computers. It enables instructors to select questions for any section in the text or to use a ready-made test for each chapter. Instructors may generate tests in multiple-choice or open-response formats, scramble the order of questions while printing, and produce twenty-five versions of each test. The system features printed graphics and accurate mathematical symbols. The program also allows instructors to choose problems randomly from a section or problem type or to choose questions manually while viewing them on the screen, with the option to regenerate variables if desired. The editing feature allows instructors to customize the chapter data disks by adding their own problems.

Transparencies Nearly 150 two-color overhead transparencies of figures, examples, definitions, procedures, properties, and problem-solving methods will help instructors present important points during lecture.

Student's Solutions Manual Solutions are given for odd-numbered exercises and all chapter test questions.

Student's Study Guide Written in a semiprogrammed format, the *Study Guide* provides additional practice problems and reinforcement for each learning objective in the textbook. A self-test is given at the end of each chapter.

Computer-Assisted Tutorials The tutorials offer self-paced, interactive review in IBM, Apple, and Macintosh formats. They include examples and practice problems keyed to objectives. Special features new in this edition are a Quick Review for each section; page references for each practice problem; more emphasis on new terms and on cautions and hints; solutions to practice problems that students answer correctly; the ability to move from practice problems to the tutorial and back without losing one's place in the problems; an online calculator.

Videotapes A new series has been developed to accompany *Algebra for College Students,* Second Edition. Text-specific videos follow the outline of Chapters 1–7, incorporating all sections and objectives. Topics for the advanced chapters present selected applied topics.

Audiotapes A set of audiotapes, one tape per chapter, guides students through each topic, allowing individualized study and additional practice for troublesome areas. These tapes are especially helpful for visually impaired students.

ACKNOWLEDGMENTS

We wish to thank the many users of the first edition for their insightful suggestions on improvements for this book. We also wish to thank our reviewers for their contributions:

Jean Airington, *Navarro College*
Brian Hayes, *Triton College*
Bruce Hoelter, *Raritan Valley Community College*
Daniel A. Hogan, *Hinds Community College*
Irma Holm, *Long Beach City College*
Mary Leeseberg, *Manatee Community College*
Carol Jean Martin, *Dodge City Community College*
Jerry V. McBee, *Tallahassee Community College*
Stuart E. Mills, *Louisiana State University in Shreveport*
Linda D. Norlien, *Winona State University*
Diane Porter, *Troy State University*
Norlin Rober, *Marshalltown Community College*
Radha Shrinivas, *Saint Louis Community College at Forest Park*
Bessie L. Tucker, *Jackson State University*

Paul Eldersveld, College of DuPage, deserves our gratitude for an excellent job coordinating all of the print ancillaries for us, an enormous and time-consuming task. Careful work by Barbara Wheeler, of California State University–Sacramento, has helped ensure the accuracy of the answers. Jim Walker and Paul Van Erden, of American River College, have done a superb job creating the indexes. We also want to thank Tommy Thompson of Seminole Community College for his suggestions for the feature "To the Student: Success in Algebra" that follows this preface.

Special thanks go to those among the staff at HarperCollins whose assistance and contributions have been very important: Jack Pritchard, Anne Kelly, Linda Youngman, Adam Bryer, Liz Lee, Ellen Keith, Julie Anderson, Janet Tilden, and Cathy Wacaser.

Margaret L. Lial
E. John Hornsby, Jr.

CONTENTS

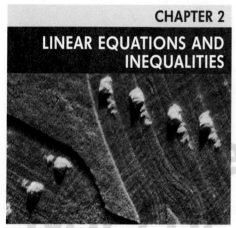

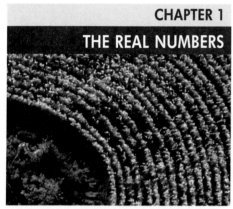

CHAPTER 3 EXPONENTS AND POLYNOMIALS

CHAPTER 4 RATIONAL EXPRESSIONS

CHAPTER 5 ROOTS AND RADICALS

CHAPTER 6
QUADRATIC EQUATIONS AND INEQUALITIES

CHAPTER 7
THE STRAIGHT LINE

CHAPTER 8
GRAPHING RELATIONS AND FUNCTIONS

CHAPTER 9
POLYNOMIAL AND RATIONAL FUNCTIONS AND THEIR ZEROS

CHAPTER 10
EXPONENTIAL AND LOGARITHMIC FUNCTIONS

CHAPTER 11
SYSTEMS OF EQUATIONS AND INEQUALITIES

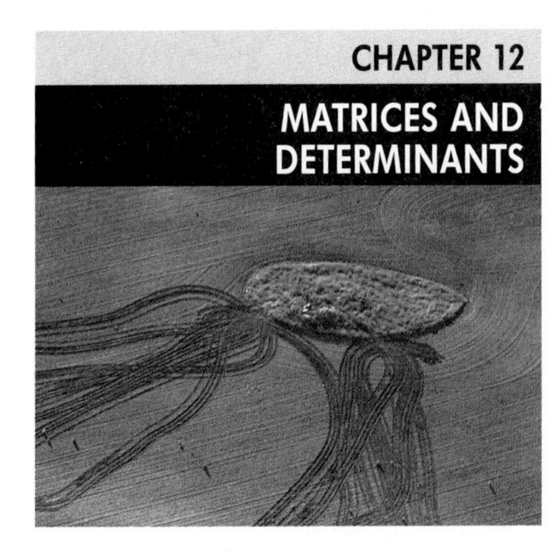

CHAPTER 12

MATRICES AND DETERMINANTS

CHAPTER 13

FURTHER TOPICS IN ALGEBRA

TO THE STUDENT: SUCCESS IN ALGEBRA

The main reason students have difficulty with mathematics is that they don't know how to study it. Studying mathematics *is* different from studying subjects like English or history. The key to success is regular practice.

This should not be surprising. After all, can you learn to play the piano or to ski well without a lot of regular practice? The same thing is true for learning mathematics. Working problems nearly every day is the key to becoming successful. Here is a list of things you can do to help you succeed in studying algebra.

1. *Attend class regularly*. Pay attention in class to what your teacher says and does, and make careful notes. In particular, note the problems the teacher works on the board and copy the complete solutions. Keep these notes separate from your homework to avoid confusion when you read them over later.
2. Don't hesitate to ask questions in class. It is not a sign of weakness, but of strength. There are always other students with the same question who are too shy to ask.
3. *Read your text carefully*. Many students read only enough to get by, usually only the examples. Reading the complete section will help you to be successful with the homework problems. Most exercises are keyed to specific Examples or Objectives that will explain the procedures for working them.
4. Before you start on your homework assignment, rework the problems the teacher worked in class. This will reinforce what you have learned. Many students say, "I understand it perfectly when you do it, but I get stuck when I try to work the problem myself."
5. Do your homework assignment only *after* reading the text and reviewing your notes from class. Check your work with the answers in the back of the book. If you get a problem wrong and are unable to see why, mark that problem and ask your instructor about it. Then practice working additional problems of the same type to reinforce what you have learned.
6. Work as neatly as you can. Write your symbols clearly, and make sure the problems are clearly separated from each other. Working neatly will help you to think clearly and also make it easier to review the homework before a test.

7. After you have completed a homework assignment, look over the text again. Try to decide what the main ideas are in the lesson. Often they are clearly highlighted or boxed in the text.
8. Use the chapter test at the end of each chapter as a practice test. Work through the problems under test conditions, without referring to the text or the answers until you are finished. You may want to time yourself to see how long it takes you. When you have finished, check your answers against those in the back of the book and study those problems that you missed. Answers are keyed to the appropriate sections of the text.
9. Keep any quizzes and tests that are returned to you and use them when you study for future tests and the final exam. These quizzes and tests indicate what your instructor considers most important. Be sure to correct any problems on these tests that you missed, so you will have the corrected work to study.
10. Don't worry if you do not understand a new topic right away. As you read more about it and work through the problems, you will gain understanding. Each time you look back at a topic you will understand it a little better. No one understands each topic completely right from the start.

AN INTRODUCTION TO SCIENTIFIC CALCULATORS

There is little doubt that the appearance of hand-held calculators twenty years ago and the later development of scientific calculators have changed the methods of learning and studying mathematics forever. Where the study of computations with tables of logarithms and slide rules made up an important part of mathematics courses prior to 1970, today the widespread availability of calculators make their study a topic only of historical significance.

Most consumer models of calculators are inexpensive. At first, however, they were more costly. One of the first consumer models available was the Texas Instruments SR-10, which sold for about $150 in 1973. It could perform the four operations of arithmetic and take square roots, but could do very little more.

In the past two decades, the hand-held calculator has become an integral part of our everyday existence. Today calculators come in a large array of different types, sizes, and prices. For the course for which this textbook is intended, the most appropriate type is the *scientific calculator,* which costs between ten and twenty dollars. While some scientific calculators have advanced features such as programmability and graphing capability, these two features are not essential for the study of the material in this text.

In this introduction, we explain some of the features of scientific calculators. However, remember that calculators vary among manufacturers and models, and that while the methods explained here apply to many of them, they may not apply to your specific calculator. For this reason, it is important to remember that *this is only a guide, and is not intended to take the place of your owner's manual.* Always refer to the manual in the event you need an explanation of how to perform a particular operation.

FEATURES AND FUNCTIONS OF MOST SCIENTIFIC CALCULATORS

Most scientific calculators use *algebraic logic*. (Models sold by Texas Instruments, Sharp, Casio, and Radio Shack, for example, use algebraic logic.) A notable exception is Hewlett Packard, a company whose calculators use *Reverse Polish Notation* (RPN). In this introduction, we explain the use of calculators with algebraic logic.

Arithmetic Operations

To perform an operation of arithmetic, simply enter the first number, touch the operation key ([+], [−], [×], or [÷]), enter the second number, and then touch the [=] key. For example, to add 4 and 3, use the following keystrokes.

(The final answer is displayed in color.)

Change Sign Key

The key marked [±] allows you to change the sign of a display. This is particularly useful when you wish to enter a negative number. For example, to enter −3, use the following keystrokes.

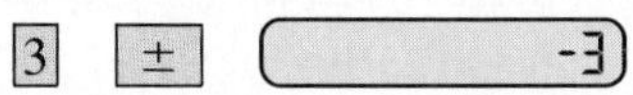

Memory Key

Scientific calculators can hold a number in memory for later use. The label of the memory key varies among models; two of these are [M] and [STO]. [M+] and [M−] allow you to add to or subtract from the value currently in memory. The memory recall key, labeled [MR], [RM], or [RCL], allows you to retrieve the value stored in memory.

Suppose that you wish to store the number 5 in memory. Enter 5, then touch the key for memory. You can then perform other calculations. When you need to retrieve the 5, touch the key for memory recall.

If a calculator has a constant memory feature, the value in memory will be retained even after the power is turned off. Some advanced calculators have more than one memory. It is best to read the owner's manual for your model to see exactly how memory is activated.

Clearing/Clear Entry Keys

These keys allow you to clear the display or clear the last entry entered into the display. They are usually marked [C] and [CE]. In some models, touching the [C] key once will clear the last entry, while touching it twice will clear the entire operation in progress.

Second Function Key

This key is used in conjunction with another key to activate a function that is printed *above* an operation key (and not on the key itself). It is usually marked [2nd]. For example, suppose you wish to find the square of a number, and the squaring function (explained in more detail later) is printed above another key. You would need to touch [2nd] before the desired squaring function can be activated.

Square Root Key

Touching the square root key, [$\sqrt{x}$], will give the square root (or an approximation of the square root) of the number in the display. For example, to find the square root of 36, use the following keystrokes.

The square root of 2 is an example of an irrational number (Chapter 5). The calculator will give an approximation of its value, since the decimal for $\sqrt{2}$ never terminates and never repeats. The number of digits shown will vary among models. To find an approximation of $\sqrt{2}$, use the following keystrokes.

An approximation

Squaring Key

This key, [x^2], allows you to square the entry in the display. For example, to square 35.7, use the following keystrokes.

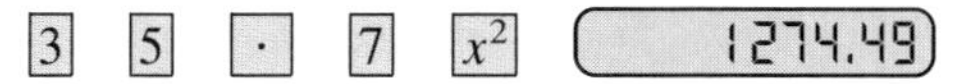

The squaring key and the square root key are often found on the same key, with one of them being a second function (that is, activated by the second function key, described above).

Reciprocal Key

The key marked [$1/x$] is the reciprocal key. (When two numbers have a product of 1, they are called *reciprocals*.) Suppose that you wish to find the reciprocal of 5. Use the following keystrokes.

Inverse Key

Some calculators have an inverse key, marked [INV]. Inverse operations are operations that "undo" each other. For example, the operations of squaring and taking the square root are inverse operations. The use of the [INV] key varies among different models of calculators, so read your owner's manual carefully.

Exponential Key

This key, marked [x^y] or [y^x], allows you to raise a number to a power. For example, if you wish to raise 4 to the fifth power (that is, find 4^5), use the following keystrokes.

Root Key

Some calculators have this key specifically marked [$\sqrt[y]{x}$] or [$\sqrt[x]{y}$]; with others, the operation of taking roots is accomplished by using the inverse key in conjunction with the exponential key. Suppose, for example, your calculator is of the latter type and you wish to find the fifth root of 1024. Use the following keystrokes.

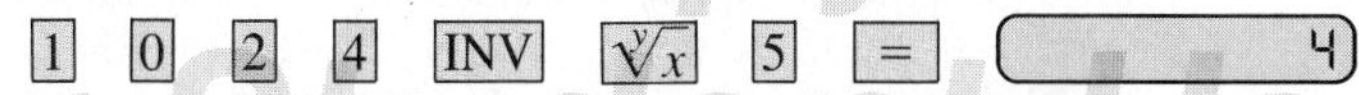

Notice how this "undoes" the operation explained in the exponential key discussion earlier.

Pi Key The number π is an important number in mathematics. It occurs, for example, in the area and circumference formulas for a circle. By touching the [π] key, you can get in the display the first few digits of π. (Because π is irrational, the display shows only an approximation.) One popular model gives the following display when the [π] key is activated: (3.1415927).

log and ln Keys Many students taking this course have never studied logarithms. Logarithms are covered in Chapter 10 in this book. In order to find the common logarithm (base ten logarithm) of a number, enter the number and touch the [log] key. To find the natural logarithm, enter the number and touch the [ln] key. For example, to find these logarithms of 10, use the following keystrokes.

Common logarithm: [1] [0] [log] (1)

Natural logarithm: [1] [0] [ln] (2.3025851) An approximation

10^x and e^x Keys These keys are special exponential keys, and are inverses of the log and ln keys. (On some calculators, they are second functions.) The number e is an irrational number and is the base of the natural logarithm function. Its value is approximately 2.71828. To use these keys, enter the number to which 10 or e is to be raised, and then touch the [10^x] or [e^x] key. For example, to raise 10 or e to the 2.5 power, use the following keystrokes.

Base is 10: [2] [·] [5] [10^x] (316.22777) An approximation

Base is e: [2] [·] [5] [e^x] (12.182494) An approximation

(Note: If no [10^x] key is specifically shown, touching [INV] followed by [log] accomplishes raising 10 to the power x. Similarly, if no [e^x] key is specifically shown, touching [INV] followed by [ln] accomplishes raising e to the power x.)

Factorial Key The factorial key, [$x!$], evaluates the factorial of any nonnegative integer within the limits of the calculator. Factorials are covered in Chapter 11 in this book. For example, $5! = 1 \cdot 2 \cdot 3 \cdot 4 \cdot 5$. To use the factorial key, just enter the number and touch [$x!$]. To evaluate 5! on a calculator, use the following keystrokes.

[5] [$x!$] (120)

OTHER FEATURES OF SCIENTIFIC CALCULATORS

When decimal approximations are shown on scientific calculators, they are either *truncated* or *rounded*. To see which of these a particular model is programmed to do, evaluate 1/18 as an example. If the display shows .0555555 (last digit 5), it truncates the display. If it shows .0555556 (last digit 6), it rounds off the display.

When very large or very small numbers are obtained as answers, scientific calculators often express these numbers in scientific notation (Section 3.2). For example, if you multiply 6,265,804 by 8,980,591, the display might look like this:

(5.6270623 13)

The "13" at the far right means that the number on the left is multiplied by 10^{13}. This means that the decimal point must be moved 13 places to the right if the answer is to be expressed in its usual form. Even then, the value obtained will only be an approximation: 56,270,623,000,000.

Advanced Features

Two features of advanced scientific calculators are programmability and graphing capability. A programmable calculator has the capability of running small programs, much like a mini-computer. A graphing calculator can be used to plot graphs of functions on a small screen. One of the issues in mathematics education today deals with how graphing calculators should be incorporated into the curriculum. Their availability in the 1990s parallels the availability of scientific calculators in the 1980s, and they will no doubt be a major component in mathematics education as we move into the twenty-first century.

CHAPTER ONE

THE REAL NUMBERS

Algebra is a language. Learning algebra is very similar to learning any new language. It is necessary to learn new symbols, rules, and patterns. This chapter reviews some of the basic symbols and rules of algebra that are studied in elementary algebra. These must be mastered before proceeding to learn more in the succeeding chapters about the language of algebra.

1.1 BASIC TERMS

OBJECTIVES

1. WRITE SETS.
2. DECIDE IF ONE SET IS A SUBSET OF ANOTHER.
3. USE NUMBER LINES.
4. FIND ADDITIVE INVERSES.
5. USE ABSOLUTE VALUE.
6. KNOW THE COMMON SETS OF NUMBERS.

Algebra depends on symbols, and many of the symbols used in this book are introduced in this first chapter. A list of the symbols used in each chapter is given in the summary for that chapter.

1 A basic term used in algebra is **set,** a collection of objects, numbers, or ideas. The objects in a set are the **elements** or **members** of the set. In algebra, the elements in a set are usually numbers. **Set braces,** $\{\ \}$, are used to enclose the elements. For example, 2 is an element of the set $\{1, 2, 3\}$.

A set can be defined either by listing or by describing its elements. For example,

$$S = \{\text{Oregon, Ohio, Oklahoma}\}$$

defines the set S by *listing* its elements. The same set might be *described* by saying that set S is the set of all states in the United States whose names begin with the letter "O."

Set S above has a finite number of elements. Some sets contain an infinite number of elements, such as

$$N = \{1, 2, 3, 4, 5, 6, \ldots\},$$

where the three dots show that the list continues in the same pattern. Set N is called the set of **natural numbers,** or **counting numbers.** A set containing no elements, such as the set of natural numbers less than 1, is called the **empty set,** or **null set,** usually written $\emptyset$. The empty set may also be written as $\{\ \}$.

CAUTION It is not correct to write $\{\emptyset\}$ for the empty set. $\{\emptyset\}$ represents a set with one element (the empty set) and thus is not an empty set. Also, the number 0 is not the same as the empty set. Always use only the notations $\emptyset$ or $\{\ \}$ for the empty set.

To write the fact that 2 is an element of the set {0, 1, 2, 3}, we use the symbol $\in$ (read "is an element of"):

$$2 \in \{0, 1, 2, 3\}.$$

The number 2 is also an element of set N above, so we may write

$$2 \in N.$$

To show that 0 is *not* an element of set N, we draw a slash through the symbol $\in$:

$$0 \notin N.$$

Two sets are **equal** if they contain exactly the same elements. For example,

$$\{1, 2\} = \{2, 1\},$$

because the sets contain the same elements. (The order doesn't matter.) On the other hand, $\{1, 2\} \neq \{0, 1, 2\}$ ($\neq$ means "is not equal to") since one set contains the element 0 while the other does not.

In algebra, letters called **variables** are often used to represent numbers. Variables also can be used to define sets of numbers. For example,

$$\{x \mid x \text{ is a natural number between 3 and 15}\}$$

(read "the set of all elements x such that x is a natural number between 3 and 15") defines the set

$$\{4, 5, 6, 7, \ldots, 14\}.$$

The notation $\{x \mid x$ is a natural number between 3 and 15$\}$ is an example of **set-builder notation.**

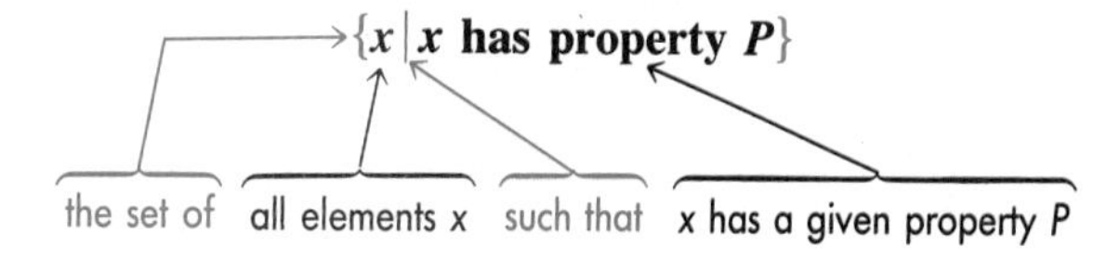

EXAMPLE 1
LISTING THE ELEMENTS IN A SET

List the elements in each set.

(a) $\{x \mid x$ is a natural number less than 4$\}$
The natural numbers less than 4 are 1, 2, and 3. The given set is

$$\{1, 2, 3\}.$$

(b) $\{y \mid y$ is one of the first five even natural numbers$\} = \{2, 4, 6, 8, 10\}$

(c) $\{z \mid z$ is a natural number at least 7$\}$
The set of natural numbers at least 7 is an infinite set; write it with three dots as

$$\{7, 8, 9, 10, \ldots\}. \quad \blacksquare$$

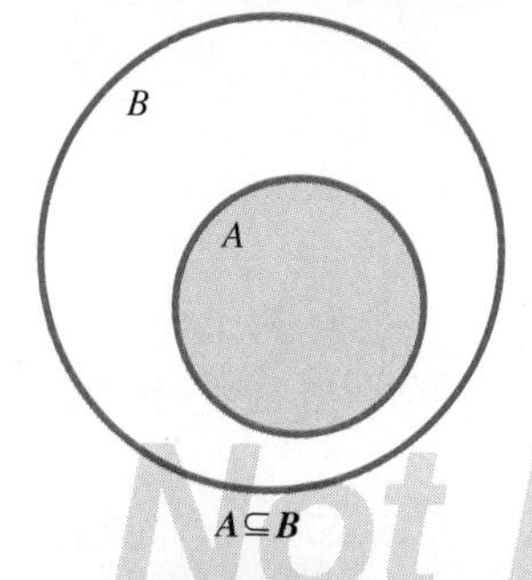

FIGURE 1.1

2 Set A is a **subset** of set B if every element of A is also an element of B. The symbol $\subseteq$ is used for subset, so $A \subseteq B$ says that A is a subset of B. The symbol $\not\subseteq$ indicates "is not a subset of." A sketch of a set A that is a subset of a set B is shown in Figure 1.1.

EXAMPLE 2
USING THE DEFINITION OF SUBSET

Decide whether each statement is *true* or *false*.

(a) $\{1, 2, 3\} \subseteq \{1, 2, 3, 4, 5\}$

Every element of the set $\{1, 2, 3\}$ is also an element of the set $\{1, 2, 3, 4, 5\}$. Because of this, $\{1, 2, 3\}$ is a subset of $\{1, 2, 3, 4, 5\}$ and the given statement is true.

(b) $\{0, 1, 2\} \subseteq \{1, 2, 3, 4, 5\}$

The number 0 is an element of $\{0, 1, 2\}$ but not of $\{1, 2, 3, 4, 5\}$, so that

$$\{0, 1, 2\} \not\subseteq \{1, 2, 3, 4, 5\}.$$

The statement is false.

(c) $\{0, 1, 2\} \subseteq \{0, 1, 2\}$

Every element of the first set is an element of the second set, so the given statement is true. ■

CAUTION Be careful not to confuse the symbol $\in$, "is an element of," with the symbol $\subseteq$, "is a subset of." The symbol $\in$ is used only between an element and a set, while $\subseteq$ is used only between two sets. For example,

$$5 \in \{3, 4, 5, 7, 9\},$$

but

$$\{5\} \subseteq \{3, 4, 5, 7, 9\}.$$

By the definition of subset, set A is a subset of set B if every element of A is also an element of B. Since every element of a set A is an element of the same set A itself, every set is a subset of itself. (See Example 2(c) above.) Rephrasing the definition of a subset, set A is a subset of set B if there are no elements of A that are not in B. Since there are no elements in the empty set, there are none that are not in set B. For this reason, *the empty set is a subset of every set*.

SPECIAL SUBSET RELATIONSHIPS

For any set A, $\quad A \subseteq A \quad$ and $\quad \emptyset \subseteq A$.

3 A good way to get a picture of sets of numbers is to use a diagram called a **number line.** To construct a number line, choose any point on a horizontal line and label it 0. Then choose a point to the right of 0 and label it 1. The distance from 0 to 1 establishes a scale that can be used to locate more points, with positive numbers to the right of 0 and negative numbers to the left of 0. A number line is shown in Figure 1.2.

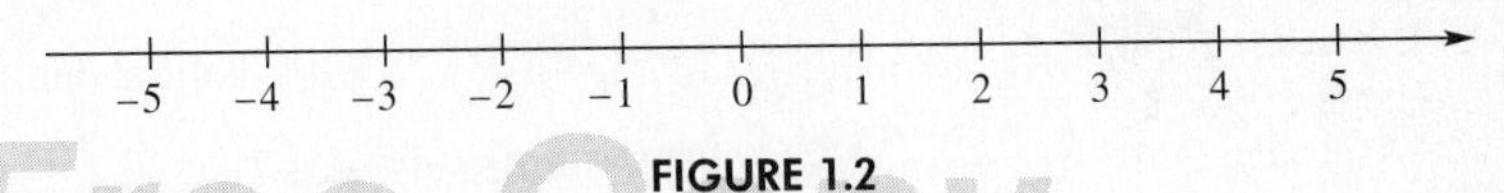

FIGURE 1.2

Each number is called the **coordinate** of the point that it labels, while the point is the **graph** of the number. A number line with several selected points graphed on it is shown in Figure 1.3.

FIGURE 1.3

4 Two numbers that are the same distance from 0 on the number line but on opposite sides of 0 are called **additive inverses,** or **opposites,** of each other. For example, 5 is the additive inverse of -5, and -5 is the additive inverse of 5.

ADDITIVE INVERSE

For any number a, the number $-a$ is the **additive inverse** of a.

The number 0 is its own additive inverse. The sum of a number and its additive inverse is always zero.

The symbol "$-$" can be used to indicate any of the following three things:

1. a negative number, such as -9 or -15;
2. the additive inverse of a number, as in "-4 is the additive inverse of 4";
3. subtraction, as in $12 - 3$.

In writing the number $-(-5)$, the symbol "$-$" is being used in two ways: the first $-$ indicates the additive inverse of -5, and the second indicates a negative number, -5. Since the additive inverse of -5 is 5, then $-(-5) = 5$. This example suggests the following property.

DOUBLE NEGATIVE

For any number a, $\quad -(-a) = a.$

Numbers written with a positive or negative sign, such as $+4$, $+8$, -9, -5, and so on, are called **signed numbers.** Positive numbers can be called signed numbers even if the positive sign is left off.

EXAMPLE 3 FINDING ADDITIVE INVERSES

The following list shows several signed numbers and the additive inverse of each.

NUMBER	ADDITIVE INVERSE
6	-6
-4	$-(-4)$, or 4
2/3	$-2/3$
0	-0, or 0 ■

5 The **absolute value** of a number a, written $|a|$, gives only the magnitude of a and so is never negative. The formal definition of absolute value is as follows.

ABSOLUTE VALUE

$$|a| = \begin{cases} a & \text{if } a \text{ is positive or } 0 \\ -a & \text{if } a \text{ is negative} \end{cases}$$

NOTE The second part of this definition, $|a| = -a$ if a is negative, is tricky. If a is a *negative* number, then $-a$, the additive inverse or opposite of a, is a positive number, so that $|a|$ is positive.

EXAMPLE 4 **EVALUATING ABSOLUTE VALUE EXPRESSIONS**

Find the value of each.

(a) $|13| = 13$ **(b)** $|-2| = -(-2) = 2$ **(c)** $|0| = 0$

(d) $-|8|$

Evaluate the absolute value first. Then find the additive inverse.

$$-|8| = -(8) = -8$$

(e) $-|-8|$

Work as in (d): $|-8| = 8$, so $-|-8| = -(8) = -8$.

(f) $|-2| + |5|$

Evaluate each absolute value first, then add.

$$|-2| + |5| = 2 + 5 = 7 \quad \blacksquare$$

One application of absolute value is describing distance. The distance on the number line from 0 to a is $|a|$. For example, the absolute value of 5 is the same as the absolute value of -5, since each number lies five units from 0. See Figure 1.4. That is, both

$$|5| = 5 \quad \text{and} \quad |-5| = 5.$$

Distance is 5, so $|-5| = 5$. Distance is 5, so $|5| = 5$.

-5 0 5

FIGURE 1.4

NOTE Since absolute value represents distance, and since distance is never negative, **the absolute value of a number is never negative.**

6 The following sets of numbers will be used throughout the book.

SETS OF NUMBERS

Real numbers	$\{x \mid x$ is a coordinate of a point on a number line$\}$*
Natural numbers or counting numbers	$\{1, 2, 3, 4, 5, 6, 7, 8, \ldots\}$
Whole numbers	$\{0, 1, 2, 3, 4, 5, 6, \ldots\}$
Integers	$\{\ldots, -3, -2, -1, 0, 1, 2, 3, \ldots\}$
Rational numbers	$\left\{\frac{p}{q} \middle\vert p \text{ and } q \text{ are integers, with } q \neq 0\right\}$ or all terminating or repeating decimals
Irrational numbers	$\{x \mid x$ is a real number that is not rational$\}$ or all nonterminating and nonrepeating decimals

*An example of a number that is not a coordinate of a point on a number line is $\sqrt{-1}$. This number, which is not a real number, is discussed in Chapter 5.

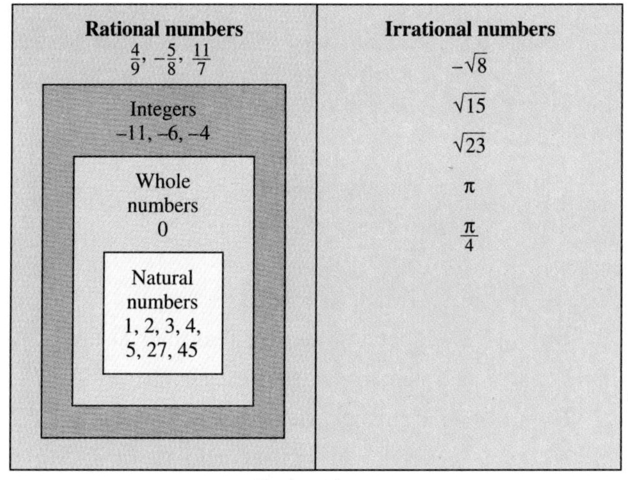

Real numbers.

(a)

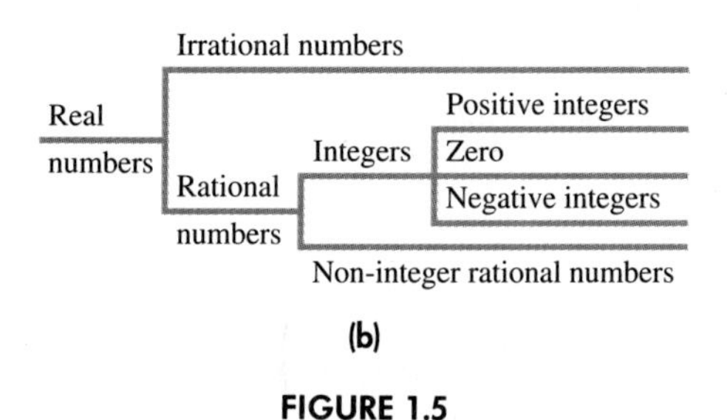

(b)

FIGURE 1.5

Examples of irrational numbers include most square roots, such as $\sqrt{7}$, $\sqrt{11}$, $\sqrt{2}$, and $-\sqrt{5}$. (Some square roots *are* rational: $\sqrt{16} = 4$, $\sqrt{100} = 10$, and so on.) Another irrational number is π, the ratio of the circumference of a circle to its diameter. All irrational numbers are real numbers.

Real numbers can also be defined in terms of decimals. By repeated subdivisions, any decimal can be located (at least in theory) as a point on a number line. Because of this, the set of real numbers can be defined as the set of all decimals. Also, the set of rational numbers can be shown to be the set of all repeating or terminating decimals. (A bar over the series of numerals that repeat is used to indicate a repeating decimal.) For example,

$$.\overline{6} = \frac{2}{3}$$ 6 repeats.

$$.25 = \frac{1}{4}$$

$$.2 = \frac{1}{5}$$

$$.\overline{142857} = \frac{1}{7}$$ The block of digits 142857 repeats.

$$.4\overline{3} = \frac{13}{30}$$ 3 repeats.

and so on. The set of irrational numbers is the set of decimals that do not repeat and do not terminate.

The relationships among these various sets of numbers are shown in Figure 1.5; in particular, the figure shows that the set of real numbers includes both the rational and the irrational numbers.

EXAMPLE 5
IDENTIFYING EXAMPLES OF NUMBER SETS

(a) 0, 2/3, −9/64, 28/7 (or 4), 2.45, and $1.\overline{37}$ are rational numbers.

(b) $\sqrt{3}$, π, $-\sqrt{2}$, and $\sqrt{7} + \sqrt{3}$ are irrational numbers.

(c) −8, 12/2, −3/1, and 75/5 are integers.

(d) All the numbers in parts (a), (b), and (c) above are real numbers.

(e) 4/0 is undefined, since the definition requires the denominator of a rational number to be nonzero. (However 0/4 = 0, which is a real number.) ■

1.1 EXERCISES

Graph each set on a number line.

1. $\{2, 3, 4, 5\}$

2. $\{0, 2, 4, 6, 8\}$

3. $\{-4, -3, -2, -1, 0, 1\}$

4. $\{-6, -5, -4, -3, -2\}$

5. $\left\{-\frac{1}{2}, \frac{3}{4}, \frac{5}{3}, \frac{7}{2}\right\}$

6. $\left\{-\frac{3}{5}, -\frac{1}{10}, \frac{9}{8}, \frac{12}{5}, \frac{13}{4}\right\}$

List the elements in each set. See Example 1.

7. $\{x \mid x$ is a natural number less than $7\}$

8. $\{m \mid m$ is a whole number less than $9\}$

9. $\{a \mid a$ is an even integer greater than $10\}$

10. $\{k \mid k$ is a counting number less than $1\}$

11. $\{x \mid x$ is an irrational number that is also rational$\}$

12. $\{r \mid r$ is a negative integer greater than $-1\}$

13. $\{p \mid p$ is a number whose absolute value is $3\}$

14. $\{w \mid w$ is a number whose absolute value is $12\}$

15. $\{z \mid z$ is a whole number multiple of $5\}$

16. $\{n \mid n$ is a counting number multiple of $3\}$

Write each set in set builder notation. (More than one correct answer is possible.) For example, $\{2, 4, 6, 8, \ldots\}$ can be written as "$\{x \mid x$ is a positive even integer$\}$." See Example 1.

17. $\{3, 6, 9, 12, \ldots\}$

18. $\{\ldots, -4, -2, 0, 2, 4, \ldots\}$

19. {January, February, March, April, May, June}

20. {Canada, United States, Mexico}

21. $\{1, 3, 5, 7, 9\}$

22. $\{3, 4, 5\}$

Find the value of each quantity. See Example 4.

23. $|-9|$

24. $|12|$

25. $-|6|$

26. $-|15|$

27. $-|-3|$

28. $-|-7|$

29. $|-2| + |3|$

30. $|-12| + |2|$

31. $|-8| - |3|$

32. $|-10| - |5|$

33. $|-6| - |-2|$

34. $|-15| + |-10|$

Give the additive inverse of each. See Examples 3 and 4.

35. 8

36. 15

37. -9

38. -12

39. 0

40. -0

41. $-(-3)$

42. $-(-5)$

43. $|10|$

44. $|17|$

45. $-|8|$

46. $-|21|$

47. $-|-1|$

48. $-|-9|$

49. $|-2| + |-3|$

50. $|-9| + |-4|$

Which elements of the set

$$S = \left\{-6, -\sqrt{3}, -\frac{1}{2}, -.1\overline{45}, 0, \frac{2}{3}, .75, \sqrt{2}, 2, 3, \pi, \frac{10}{2}\right\}$$

are elements of the following sets? See Example 5.

51. Natural numbers

52. Whole numbers

53. Integers

54. Rational numbers

55. Irrational numbers

56. Real numbers

Tell whether each statement is true *or* false. *(See Objective 6.)**

57. Every rational number is an integer.

58. Every natural number is an integer.

59. Every integer is a rational number.

60. Every whole number is a real number.

61. Some rational numbers are irrational.

62. Some natural numbers are whole numbers.

63. Some rational numbers are integers.

64. Some real numbers are integers.

---

65. Explain the difference between the rational numbers .36 and $.\overline{36}$. (See Objective 6.)

66. Explain why it is incorrect to write $\{\emptyset\}$ for the empty set. (See Objective 1.)

Let $N = \{x \mid x \text{ is a natural number}\}$,
$W = \{x \mid x \text{ is a whole number}\}$,
$I = \{x \mid x \text{ is an integer}\}$,
$Q = \{x \mid x \text{ is a rational number}\}$,
$H = \{x \mid x \text{ is an irrational number}\}$,
$R = \{x \mid x \text{ is a real number}\}$.

Place $\subseteq$ *or* $\not\subseteq$ *in each blank to make a true statement.*

67. $\{1, 2, 9, 14\}$ ____ N

68. $\{0, 3, 6, 9, 12\}$ ____ N

69. $\{5, 1, -3, -9, -11\}$ ____ I

70. $\{-1, -3, -5, -7, \ldots\}$ ____ I

71. $\{1, 3, 5, 7, 9, 11, 13, \ldots\}$ ____ W

72. $\{-1, 1\}$ ____ W

73. N ____ I

74. N ____ Q

75. W ____ N

76. W ____ I

77. I ____ Q

78. Q ____ H

79. Q ____ R

80. H ____ R

81. Q ____ Q

82. $\emptyset$ ____ H

Place $\in$, $\notin$, $\subseteq$, *or* $\not\subseteq$ *in each blank to make the statement true.*

83. 4 ____ $\{-2, 0, 2, 4, 6\}$

84. -1 ____ $\{-3, -2, -1, 0, 4\}$

85. -9 ____ $\{-10, -8, -6, -4\}$

86. 3 ____ $\{-5, -4, -3, -2, -1\}$

87. $\{5\}$ ____ $\{2, 3, 4, 5, 6\}$

88. $\{7\}$ ____ $\{2, 4, 6, 7, 8, 9\}$

89. Which of the following is a correct notation for the empty set? (There may be more than one correct answer.) (See Objective 1.)
(a) $\{\emptyset\}$ **(b)** $\emptyset$ **(c)** $\{0\}$ **(d)** $\{\ \}$ **(e)** 0

Under what conditions on sets A and B are the following statements true? (See Objective 2.)

90. $A \subseteq B$ and $B \subseteq A$

91. $A = B$ and $B \subseteq A$

92. $A \neq B$ and $A \subseteq B$

93. Suppose that $A = B$ and $B \subseteq C$. Is $A \subseteq C$?

Under what conditions on the real number x are the following statements true? (See Objective 5.)

94. $|-x| = |x|$

95. $|x| = |x + 2|$

96. $|x| = -|x|$

97. $|x - 1| = |x + 1|$

---

*For help with these exercises, read the discussion of Objective 6 in this section.

1.2 EQUALITY AND INEQUALITY

OBJECTIVES

1. USE THE PROPERTIES OF EQUALITY.
2. USE INEQUALITY SYMBOLS.
3. GRAPH SETS OF REAL NUMBERS.
4. USE THE PROPERTIES OF INEQUALITY.

The study of any object is simplified by finding the properties of the object. The property that water boils when heated to 100°C helps us predict and understand the behavior of water. The study of algebra is no different: it is made easier by writing down the basic properties of the real number system.

1 The first properties we shall consider are those of equality.

PROPERTIES OF EQUALITY

For all real numbers a, b, and c:

$a = a$	Reflexive property
If $a = b$, then $b = a$.	Symmetric property
If $a = b$ and $b = c$, then $a = c$.	Transitive property
If $a = b$, then a may replace b in any sentence without changing the truth of the sentence.	Substitution property

EXAMPLE 1 IDENTIFYING PROPERTIES OF EQUALITY

(a) $x + 2 = x + 2$. Reflexive property

(b) If $x + 5 = y$, then $y = x + 5$. Symmetric property

(c) If $p + q = r$, and $r = 3m$, then $p + q = 3m$. Transitive property or substitution property

(d) If $4k - 3 = x$, and $x + k = 0$, then $4k - 3 + k = 0$. Substitution property ■

CAUTION The transitive property and the substitution property often cause confusion. This is perhaps because the transitive property is a special case of the substitution property; whenever the transitive property can be used, so can the substitution property. The transitive property is studied separately since it also applies to inequalities.

2 A statement that two numbers are *not* equal is called an **inequality.** For example, the numbers 4/2 and 3 are not equal. Write this inequality as

$$\frac{4}{2} \neq 3,$$

where the slash through the equals sign is read "does not equal."

When two numbers are not equal, one must be smaller than the other. The symbol $<$ means "is smaller than" or "is less than." For example, $8 < 9$, $-6 < 15$, and

$0 < 4/3$. Similarly, "is greater than" is written with the symbol $>$. For example, $12 > 5$, $9 > -2$, and $6/5 > 0$. The number line in Figure 1.6 shows the graphs of the numbers 4 and 9. The graphs show that $4 < 9$. Starting at 4, add the positive number 5 to get 9. As this example suggests, $a < b$ means that there exists a positive number c such that $a + c = b$.

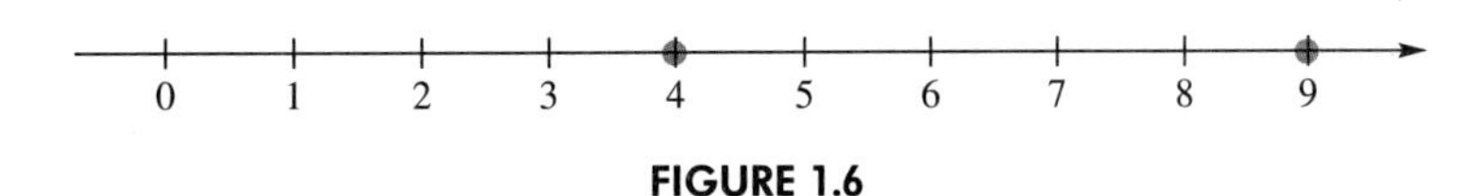

FIGURE 1.6

On the number line, the smaller of two given numbers is always located to the left of the other. Also, if a is less than b, then b is greater than a. The geometric definitions of $<$ and $>$ are as follows.

DEFINITIONS OF < AND >

On the number line,

$a < b$ if a is to the left of b; $\quad b > a$ if b is to the right of a.

EXAMPLE 2
USING A NUMBER LINE TO DETERMINE ORDER

(a) As shown on the number line in Figure 1.7, -6 is located to the left of 1. For this reason, $-6 < 1$. Also, $1 > -6$.

(b) From the same number line, $-5 < -2$, or $-2 > -5$. ■

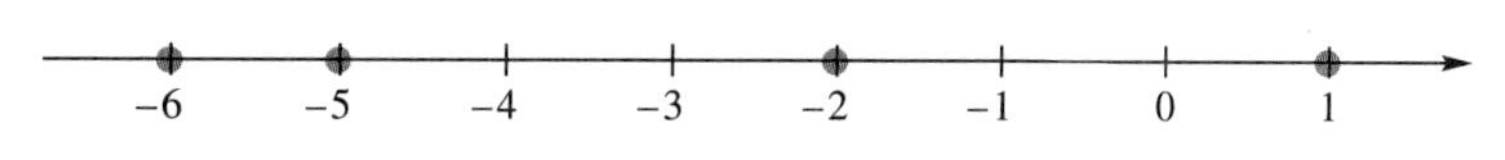

FIGURE 1.7

The following box summarizes results about positive and negative numbers. The same statement is given in both words and symbols.

Words	*Symbols*
Every negative number is smaller than 0.	If a is negative, then $a < 0$.
Every positive number is larger than 0.	If a is positive, then $a > 0$.

0 is neither positive nor negative.

In addition to the symbols $<$ and $>$, the following symbols often are used. (To complete the list, the symbols $<$ and $>$ are also included.)

SYMBOLS OF INEQUALITY

	Meaning	*Example*
$<$	is less than	$-4 < -1$
$>$	is greater than	$3 > -2$
$\leq$	is less than or equal to	$6 \leq 6$
$\geq$	is greater than or equal to	$-8 \geq -10$
$\nless$	is not less than	$5 \nless 2$
$\ngtr$	is not greater than	$-3 \ngtr -1$

EXAMPLE 3
INTERPRETING INEQUALITY SYMBOLS

The following table shows several statements and the reason that each is true.

STATEMENT	REASON
$6 \leq 8$	$6 < 8$
$-2 \leq -2$	$-2 = -2$
$-9 \geq -12$	$-9 > -12$
$-3 \geq -3$	$-3 = -3$
$8 \ngtr 10$	$8 < 10$
$6 \cdot 4 \leq 5(5)$	$24 < 25$

In the last line, the dot in $6 \cdot 4$ indicates the product 6×4, or 24. Also, $5(5)$ means 5×5, or 25. The statement is thus $24 \leq 25$, which is true. ■

3 Inequality symbols and variables can be used to write sets of real numbers. For example, the set $\{x \mid x > -2\}$ is made up of all the real numbers greater than -2. On a number line, we show the elements of this set (the set of all real numbers to the right of -2) by drawing a line from -2 to the right. We use a parenthesis at -2 since -2 is not an element of the given set. The result, shown in Figure 1.8, is called the **graph** of the set $\{x \mid x > -2\}$.

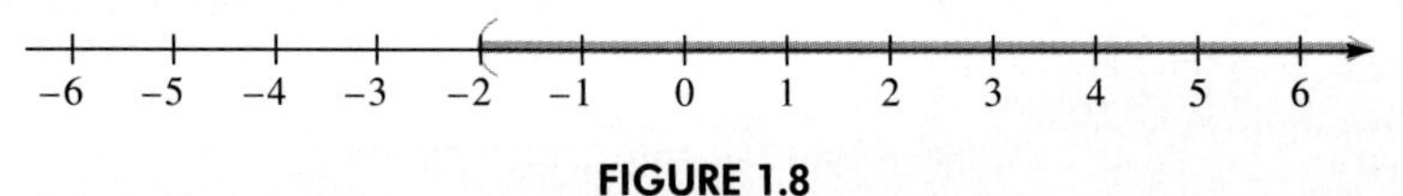

FIGURE 1.8

The set of numbers greater than -2 is an example of an **interval** on the number line. A simplified notation, called **interval notation,** is used for writing intervals. For example, using this notation, the interval of all numbers greater than -2 is written as $(-2, \infty)$. The infinity symbol ∞ does not indicate a number; it is used to show that the interval includes all real numbers greater than -2. The left parenthesis indicates that -2 is not included. A parenthesis is always used next to the infinity symbol in interval notation. The set of all real numbers is written in interval notation as $(-\infty, \infty)$.

EXAMPLE 4
GRAPHING AN INEQUALITY WRITTEN IN INTERVAL NOTATION

Write $\{x \mid x < 4\}$ in interval notation and graph the interval.

The interval is written as $(-\infty, 4)$. The graph is shown in Figure 1.9. Since the elements of the set are all the real numbers *less* than 4, the graph extends to the left. ■

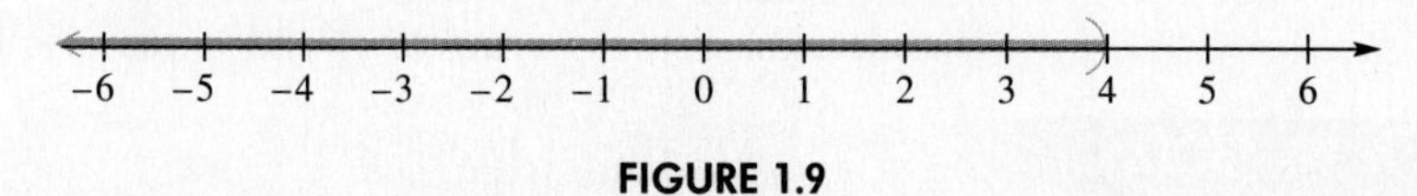

FIGURE 1.9

The set $\{x \mid x \leq -6\}$ contains all the real numbers less than or equal to -6. To show that -6 itself is part of the set, a *square bracket* is used at -6, as shown in Figure 1.10. In interval notation, this set is written as $(-\infty, -6]$.

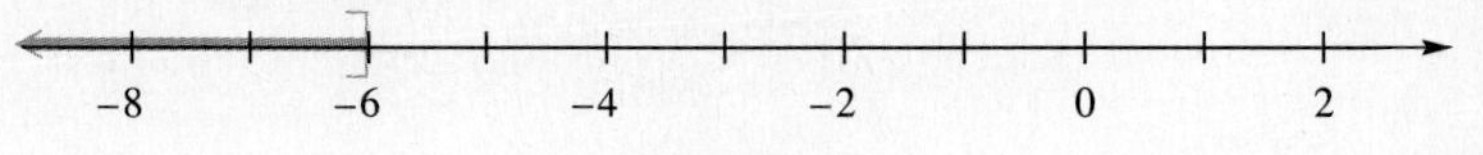

FIGURE 1.10

EXAMPLE 5
GRAPHING AN INEQUALITY WRITTEN IN INTERVAL NOTATION

Write $\{x \mid x \geq -4\}$ in interval notation and graph the interval.

This set is written in interval notation as $[-4, \infty)$. The graph is shown in Figure 1.11. A square bracket is used at -4 since -4 is part of the set. ■

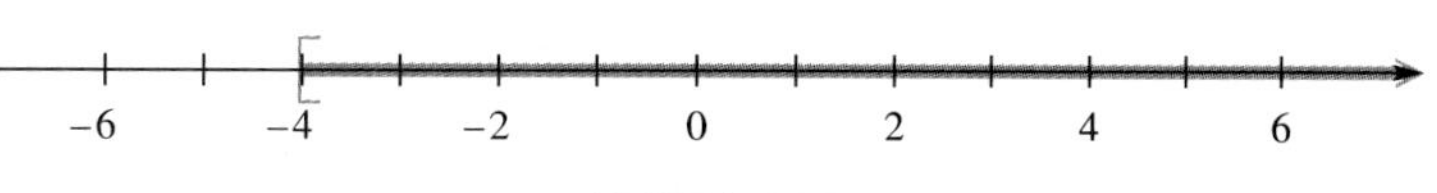

FIGURE 1.11

It is common to graph sets of numbers that are *between* two given numbers. For example, the set $\{x \mid -2 < x < 4\}$ is made up of all those real numbers between -2 and 4, but not the numbers -2 and 4 themselves. This set is written in interval notation as $(-2, 4)$. The graph has a heavy line between -2 and 4 with parentheses at -2 and 4. See Figure 1.12. The inequality $-2 < x < 4$ is read "x is greater than -2 and less than 4," or "x is between -2 and 4."

FIGURE 1.12

EXAMPLE 6
GRAPHING A THREE-PART INEQUALITY

Write in interval notation and graph $\{x \mid 3 < x \leq 10\}$.

Use a parenthesis at 3 and a square bracket at 10 to get $(3, 10]$ in interval notation. The graph is shown in Figure 1.13. Read the inequality $3 < x \leq 10$ as "3 is less than x and x is less than or equal to 10," or "x is between 3 and 10, excluding 3 and including 10." ■

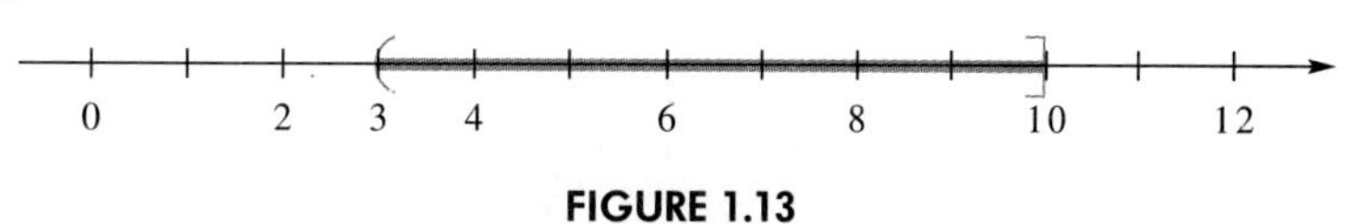

FIGURE 1.13

4 Two basic properties of inequalities are given next.

PROPERTIES OF INEQUALITY

For all real numbers a, b, and c:

Either $a < b$, $a = b$, or $a > b$. — Trichotomy property

If $a < b$ and $b < c$, then $a < c$. — Transitive property

EXAMPLE 7
IDENTIFYING PROPERTIES OF INEQUALITY

(a) By the trichotomy property, given the two real numbers m and 4, then

$$\text{either} \quad m < 4, \quad m = 4, \quad \text{or} \quad m > 4.$$

(b) If k, 6, and p are real numbers with

$$k < 6 \quad \text{and} \quad 6 < p,$$

then by the transitive property,

$$k < p. \quad ■$$

1.2 EXERCISES

Use inequality symbols to rewrite each of the following.

1. r is not equal to 4.

2. -6 is not equal to a.

3. $2p + 1$ is less than 9.

4. -5 is less than $a - 6$.

5. 3 is greater than or equal to $7y$.

6. 12 is greater than or equal to $8z + 3$.

7. -6 is less than or equal to -6.

8. -8 is greater than or equal to -8.

9. x is between 5 and 9.

10. x is between -4 and 3.

11. $3k$ is between -8 and 4, inclusive.

12. $7y$ is between 9 and 12, inclusive.

13. a is between 1 and 5, including 1 and excluding 5.

14. $4m$ is between -4 and 0, excluding -4 and including 0.

---

15. Use inequality symbols to write the lengths of the sides of the triangle in the figure in increasing order. (See Objective 2.)

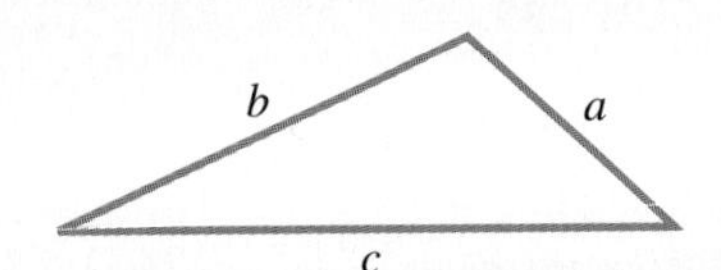

16. Use inequality symbols to write the angle measures in the figure in decreasing order. (See Objective 2.)

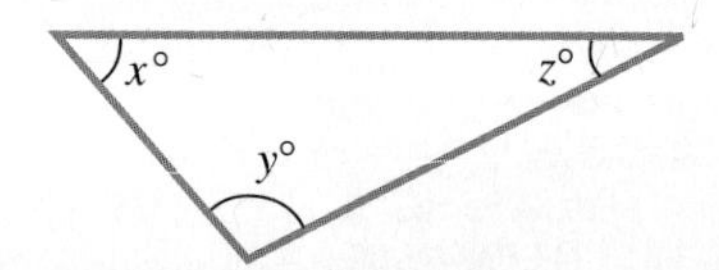

Identify each inequality as true *or* false. *See Example 3.*

17. $6 < -12$

18. $2 \geq -5$

19. $4 \leq 4$

20. $-8 \geq -8$

21. $4 \not< 12$

22. $8 \not> 3$

23. $-4 < 2 < 6$

24. $3 < 5 < 1$

Rewrite each statement with $<$ *so that it uses* $>$ *instead; rewrite each statement with* $>$ *so that it uses* $<$. *See Example 2.*

25. $-9 < 4$

26. $3 > -1$

27. $6 > 2$

28. $-1 > -6$

29. $x < 4$

30. $x > -3$

31. $5 < x < 6$

32. $9 > x > 1$

Using your knowledge of arithmetic, identify each of the following inequalities as true *or* false. *Do all work inside the parentheses before any other work. Recall that ab or $a \cdot b$ means "a times b."*

33. $9 \neq 5 + 4$

34. $11 \neq 15 - 4$

35. $3(8 + 2) > 6 - 5$

36. $2(1 + 4) \geq 8 + 2$

37. $(2 \cdot 4) + 7 \leq 4 \cdot 5$

38. $(3 \cdot 1) + 2 < 8 - 2$

39. $4 + |-4| \not> |4|$

40. $-|-6| < 0$

41. $9(2 + 4) < (9 \cdot 2) + (9 \cdot 4)$

42. $3(8 + 1) \geq (3 \cdot 8) + (3 \cdot 1)$

Write each set in interval notation and then graph it. See Examples 4–6.

43. $\{x \mid x > -1\}$

44. $\{x \mid x < 3\}$

45. $\{x \mid x \leq 4\}$

46. $\{x \mid x \geq -2\}$

47. $\{x \mid 0 < x < 3\}$

48. $\{x \mid -4 < x < 4\}$

49. $\{x \mid 2 \leq x \leq 5\}$

50. $\{x \mid -3 \leq x \leq -1\}$

51. $\{x \mid -4 < x \leq 2\}$

52. $\{x \mid 3 \leq x < 5\}$

Name the properties in Exercises 53–64 that justify each of the following. See Examples 1 and 7.

53. If $2 < 3$ and $3 < 7$, then $2 < 7$.

54. If $2 \cdot 5 = 10$ and $10 = 9 + 1$, then $2 \cdot 5 = 9 + 1$.

55. If $6 = 5 + 1$, then $5 + 1 = 6$.

56. If $x + 5 = 2x - 1$, then $2x - 1 = x + 5$.

57. If $x = 4$ and $4 = y$, then $x = y$.

58. If $x < 5$ and $5 < y$, then $x < y$.

59. If $x + y = 3$ and $5(x + y) = w$, then $5(3) = w$.

60. If $x < 5$ and $5 < 12$, then $x < 12$.

61. If $3x = 6$ and $6 = 5x + 2$, then $3x = 5x + 2$.

62. If $|-6| = 6$ and $6 = |6|$, then $|-6| = |6|$.

63. If $x = 5$ and $5 < 8$, then $x < 8$.

64. If x is a real number, then $x = 5$, $x < 5$, or $x > 5$.

Answer the following.

65. Suppose $x^2 \leq 16$. Is it always true then that $x \leq 4$? y

66. If $x^2 \geq 16$, must $x \geq 4$? y

67. Under what conditions on x is $\frac{1}{x} < x$? Assume $x \neq 0$.

68. Under what conditions on x is $x < x^2$?

1.3 OPERATIONS ON REAL NUMBERS

OBJECTIVES

1. ADD AND SUBTRACT REAL NUMBERS.
2. FIND THE DISTANCE BETWEEN TWO POINTS.
3. MULTIPLY AND DIVIDE REAL NUMBERS.
4. USE EXPONENTS AND ROOTS.
5. LEARN THE ORDER OF OPERATIONS.
6. EVALUATE EXPRESSIONS FOR GIVEN VALUES OF VARIABLES.

1 The answer to an addition problem is called the **sum.** The rules for addition of real numbers are given below.

ADDING REAL NUMBERS

Like Signs
Add two numbers with the *same* sign by adding their absolute values. The sign of the answer (either + or −) is the same as the sign of the two numbers.

Unlike Signs
Add two numbers with *different* signs by subtracting the absolute values of the numbers. The answer is positive if the positive number has the larger absolute value. The answer is negative if the negative number has the larger absolute value.

For example, to add -12 and -8, first find their absolute values:

$$|-12| = 12 \quad \text{and} \quad |-8| = 8.$$

Since these numbers have the *same* sign, add their absolute values: $12 + 8 = 20$. Give the sum the sign of the two numbers. Since both numbers are negative, the sign is negative and

$$-12 + (-8) = -20.$$

Find $-17 + 11$ by subtracting the absolute values, since these numbers have different signs.

$$|-17| = 17 \quad \text{and} \quad |11| = 11$$
$$17 - 11 = 6$$

Give the result the sign of the number with the larger absolute value.

$$-17 + 11 = -6$$

Negative since $|-17| > |11|$

It is important to be able to work these problems mentally and quickly.

EXAMPLE 1 ADDING REAL NUMBERS

(a) $(-6) + (-3) = -(6 + 3) = -9$

(b) $(-12) + (-4) = -(12 + 4) = -16$

(c) $4 + (-1) = 3$

(d) $-9 + 16 = 7$

(e) $-\frac{1}{4} + \frac{2}{3} = -\frac{3}{12} + \frac{8}{12} = \frac{5}{12}$

(f) $-16 + 12 = -4$

(g) $-\frac{7}{8} + \frac{1}{3} = -\frac{21}{24} + \frac{8}{24} = -\frac{13}{24}$ ■

We now turn our attention to subtraction of real numbers. The result of subtraction is called the **difference.** Thus, the difference between 7 and 5 is 2. To see how subtraction should be defined, compare the two statements below.

$$7 - 5 = 2$$
$$7 + (-5) = 2$$

In a similar way,

$$9 - 3 = 9 + (-3).$$

That is, to subtract 3 from 9, add the additive inverse of 3 to 9. These examples suggest the following rule for subtraction.

DEFINITION OF SUBTRACTION

For all real numbers a and b,

$$a - b = a + (-b).$$

(Change the sign of the second number and add.)

This method of observing patterns and similarities and generalizing from them as we did above is used often in mathematics. Looking at many examples strengthens our confidence in such generalizations. If possible, though, mathematicians prefer to prove the results using previously established facts.

EXAMPLE 2
SUBTRACTING REAL NUMBERS

Change to addition.
Change sign of second number.

(a) $6 - 8 = 6 + (-8) = -2$

Changed to addition.
Sign changed.

(b) $-12 - 4 = -12 + (-4) = -16$

(c) $-10 - (-7) = -10 + [-(-7)]$ This step is often omitted.

$= -10 + 7$

$= -3$

(d) $8 - (-5) = 8 + [-(-5)]$

$= 8 + 5$

$= 13$ ■

When a problem with both addition and subtraction is being worked, perform the additions and subtractions in order from the left, as in the following example. Do not forget to work inside the parentheses or brackets first.

EXAMPLE 3
ADDING AND SUBTRACTING REAL NUMBERS

(a) $15 - (-3) - 5 - 12 = (15 + 3) - 5 - 12$

$= 18 - 5 - 12$

$= 13 - 12$

$= 1$

(b) $-9 - [-8 - (-4)] + 6 = -9 - [-8 + 4] + 6$

$= -9 - [-4] + 6$

$= -9 + 4 + 6$

$= -5 + 6$

$= 1$ ■

2 One application of subtraction is finding the distance between two points on a number line. The number line in Figure 1.14 shows several points. Find the distance between the points 4 and 7 by subtracting the numbers: $7 - 4 = 3$. Since distance is never negative, we must be careful to subtract in such a way that the answer is not negative. Or, to avoid this problem altogether, take the absolute value of the difference. Then the distance between 4 and 7 is either

$$|7 - 4| = 3 \qquad \text{or} \qquad |4 - 7| = 3.$$

FIGURE 1.14

DISTANCE

The **distance** between two points on a number line is the absolute value of the difference between the numbers.

EXAMPLE 4

FINDING DISTANCE BETWEEN POINTS ON THE NUMBER LINE

Find the distance between the following pairs of points from Figure 1.14.

(a) 8 and -4

Find the absolute value of the difference of the numbers, taken in either order.

$$|8 - (-4)| = 12 \qquad \text{or} \qquad |-4 - 8| = 12$$

(b) -4 and -6

$$|-4 - (-6)| = 2 \qquad \text{or} \qquad |-6 - (-4)| = 2 \quad ■$$

3 A **product** is the answer to a multiplication problem. For example, 24 is the product of 8 and 3. The rules for products of real numbers are given below.

MULTIPLYING REAL NUMBERS

Like Signs The product of two numbers with the *same* sign is positive.

Unlike Signs The product of two numbers with *different* signs is negative.

EXAMPLE 5

MULTIPLYING REAL NUMBERS

(a) $(-3)(-9) = 27$

(b) $\left(-\frac{3}{4}\right)\left(-\frac{5}{3}\right) = \frac{5}{4}$

(c) $7 \cdot 9 = 63$

(d) $(-5)3 = -15$

(e) $\frac{2}{3}(-3) = -2$

(f) $-\frac{5}{8}\left(\frac{12}{13}\right) = -\frac{15}{26}$ ■

Addition and multiplication are the basic operations on real numbers. Subtraction was defined in terms of addition, and similarly, division is defined in terms of multiplication. The result of dividing two numbers is called the **quotient.** The quotient of the real numbers a and b $(b \neq 0)$ is the real number q such that $a = bq$. That is,

$$\frac{a}{b} = q \qquad \textbf{only if} \qquad a = bq.$$

For example, $\frac{36}{9} = 4$ since $36 = 9 \cdot 4$.

Also, $\frac{-12}{-2} = 6$ since $-12 = -2(6)$.

This definition of division is the reason division by 0 is undefined. To see why, suppose that the quotient of a nonzero real number a and 0 is the real number q, or

$$\frac{a}{0} = q.$$

By the definition of division, this means that

$$a = 0 \cdot q.$$

However, $0 \cdot q = 0$ for *every* number q, so no such quotient q is possible. On the other hand, if $a = 0$, we have

$$\frac{0}{0} = q, \qquad \text{or} \qquad 0 = q \cdot 0.$$

This statement is true for all values of q, so there is no single quotient $0/0$.

CAUTION Remember that division by 0 is always undefined.

The definition of division given above can be restated as follows.

DEFINITION OF DIVISION

If a and b are real numbers and $b \neq 0$, then

$$\frac{a}{b} = a \cdot \frac{1}{b}.$$

If $b \neq 0$, $1/b$ is the **reciprocal** of b. This definition of division is the reason we "multiply by the reciprocal of the denominator" or "invert the denominator and multiply" when a division problem involves fractions.

EXAMPLE 6

DIVIDING REAL NUMBERS

Find each quotient.

(a) $\dfrac{24}{-6} = 24\left(-\dfrac{1}{6}\right) = -4$

(b) $\dfrac{-\frac{2}{3}}{-\frac{1}{2}} = -\dfrac{2}{3}\left(-\dfrac{2}{1}\right) = \dfrac{4}{3}$ ■

NOTE Since division is defined as multiplication by the reciprocal, the rules for the sign of the quotient are the same as for the sign of the product.

The rules for multiplication and division suggest the results given below.

The fractions $\dfrac{-x}{y}$, $-\dfrac{x}{y}$, and $\dfrac{x}{-y}$ are equal.

Also, the fractions $\dfrac{x}{y}$ and $\dfrac{-x}{-y}$ are equal. (Assume $y \neq 0$.)

Generally, a positive fraction is written $\dfrac{x}{y}$ and a negative fraction as $\dfrac{-x}{y}$.

CAUTION Every fraction has three signs: the sign of the numerator, the sign of the denominator, and the sign of the fraction itself. As shown above, changing any two of these three signs does not change the value of the fraction. Changing only one sign, or changing all three, *does* change the value.

4 A **factor** of a given number is any number that divides evenly (without remainder) into the given number. For example, 2 and 6 are factors of 12 since $2 \cdot 6 = 12$. Other factors of 12 include 4 and 3, 12 and 1, -4 and -3, -12 and -1, and -6 and -2. A number is in **factored form** if it is expressed as a product of two or more numbers.

In algebra, exponents are used as a way of writing the products of repeated factors. For example, the product $2 \cdot 2 \cdot 2 \cdot 2 \cdot 2$ is written

$$2 \cdot 2 \cdot 2 \cdot 2 \cdot 2 = 2^5.$$

The number 5 shows that 2 appears as a factor five times. The number 5 is the **exponent,** 2 is the **base,** and 2^5 is an **exponential** or **power.** Multiplying out the five 2's gives

$$2^5 = 2 \cdot 2 \cdot 2 \cdot 2 \cdot 2 = 32.$$

DEFINITION OF EXPONENT

If a is a real number and n is a natural number,

$$a^n = \underbrace{a \cdot a \cdot a \ldots a}_{n \text{ factors of } a}.$$

EXAMPLE 7 EVALUATING AN EXPONENTIAL

(a) $5^2 = 5 \cdot 5 = 25$
Read 5^2 as "5 squared."

(b) $9^3 = 9 \cdot 9 \cdot 9 = 729$
Read 9^3 as "9 cubed."

(c) $2^6 = 2 \cdot 2 \cdot 2 \cdot 2 \cdot 2 \cdot 2 = 64$
Read 2^6 as "2 to the sixth power" or just "2 to the sixth." ■

We need to be careful when evaluating an exponential with a negative sign. Compare the results in the next example.

EXAMPLE 8 EVALUATING EXPONENTIALS WITH NEGATIVE SIGNS

(a) $(-3)^5 = (-3)(-3)(-3)(-3)(-3) = -243$

(b) $(-2)^6 = (-2)(-2)(-2)(-2)(-2)(-2) = 64$

(c) $-2^6 = -(2 \cdot 2 \cdot 2 \cdot 2 \cdot 2 \cdot 2) = -64$ ■

Example 8 suggests the following generalizations.

The product of an odd number of negative factors is negative.
The product of an even number of negative factors is positive.

CAUTION As shown by Example 8(b) and (c), it is important to be careful to distinguish between $-a^n$ and $(-a)^n$.

$$-a^n = -1(\underbrace{a \cdot a \cdot a \cdot \cdots \cdot a}_{n \text{ factors of } a})$$

$$(-a)^n = \underbrace{(-a)(-a)\cdots(-a)}_{n \text{ factors of } -a}$$

As shown above, $5^2 = 5 \cdot 5 = 25$, so that 5 squared is 25. The opposite of squaring a number is called taking its **square root.** For example, a square root of 25 is 5. Another square root of 25 is -5, since $(-5)^2 = 25$. Thus, 25 has two square roots, 5 and -5. The positive square root of a number is written with the symbol $\sqrt{}$. For example, the positive square root of 25 is written $\sqrt{25} = 5$. The negative square root of 25 is written $-\sqrt{25}$. Since the square of any nonzero real number is positive, a number like $\sqrt{-4}$ is not a real number.

EXAMPLE 9 FINDING SQUARE ROOTS

(a) $\sqrt{36} = 6$ since 6 is positive and $6^2 = 36$.

(b) $\sqrt{144} = 12$ since $12^2 = 144$.

(c) $\sqrt{0} = 0$ since $0^2 = 0$.

(d) $\sqrt{\dfrac{9}{16}} = \dfrac{3}{4}$

(e) $-\sqrt{100} = -10$

(f) $\sqrt{-16}$ is not a real number. ■

CAUTION The symbol $\sqrt{}$ is used only for the *positive* square root, except that $\sqrt{0} = 0$.

Since 6 cubed is $6^3 = 6 \cdot 6 \cdot 6 = 216$, the **cube root** of 216 is 6. We write this as

$$\sqrt[3]{216} = 6.$$

In the same way, the **fourth root** of 81 is 3, written

$$\sqrt[4]{81} = 3.$$

The number -3 is also a fourth root of 81, but the symbol $\sqrt[4]{}$ is reserved for roots that are not negative. Negative roots are discussed in Chapter 5.

EXAMPLE 10 FINDING HIGHER ROOTS

(a) $\sqrt[3]{27} = 3$ since $3^3 = 27$.

(b) $\sqrt[3]{125} = 5$ since $5^3 = 125$.

(c) $\sqrt[4]{16} = 2$ since $2^4 = 16$.

(d) $\sqrt[5]{32} = 2$ since $2^5 = 32$.

(e) $\sqrt[7]{128} = 2$ since $2^7 = 128$. ■

5 Given a problem such as $5 + 2 \cdot 3$, should 5 and 2 be added first or should 2 and 3 be multiplied first? When a problem involves more than one operation, we use the following order of operations. (This is the order used by computers and many calculators.)

ORDER OF OPERATIONS

If parentheses or square brackets are present:

Step 1 Work separately above and below any fraction bar.

Step 2 Use the rules below within each set of parentheses or square brackets. Start with the innermost and work outward.

If no parentheses or brackets are present:

Step 1 Simplify all powers and roots.

Step 2 Do any multiplications or divisions in the order in which they occur, working from left to right.

Step 3 Do any additions or subtractions in the order in which they occur, working from left to right.

EXAMPLE 11 USING ORDER OF OPERATIONS

Simplify $5 + 2 \cdot 3$.

To do this, first multiply, and then add.

$$\begin{aligned} 5 + 2 \cdot 3 &= 5 + 6 && \text{Multiply.} \\ &= 11 && \text{Add.} \end{aligned}$$ ■

EXAMPLE 12 USING ORDER OF OPERATIONS

Simplify $4 \cdot 3^2 + 7 - (2 + 8)$.

Work inside the parentheses.

$$4 \cdot 3^2 + 7 - (2 + 8) = 4 \cdot 3^2 + 7 - 10$$

Simplify powers and roots. Since $3^2 = 3 \cdot 3 = 9$,

$$4 \cdot 3^2 + 7 - 10 = 4 \cdot 9 + 7 - 10.$$

Do all multiplications or divisions, working from left to right.

$$4 \cdot 9 + 7 - 10 = 36 + 7 - 10$$

Finally, do all additions or subtractions, working from left to right.

$$\begin{aligned} 36 + 7 - 10 &= 43 - 10 \\ &= 33 \end{aligned}$$ ■

EXAMPLE 13 USING ORDER OF OPERATIONS

Simplify $\frac{1}{2} \cdot 4 + (6 \div 3 \cdot 7)$.

Work inside the parentheses first, doing the division before the multiplication.

$$\begin{aligned} \frac{1}{2} \cdot 4 + (6 \div 3 \cdot 7) &= \frac{1}{2} \cdot 4 + (2 \cdot 7) && \text{Divide.} \\ &= 2 + (14) && \text{Multiply.} \\ &= 16 && \text{Add.} \end{aligned}$$ ■

EXAMPLE 14 USING ORDER OF OPERATIONS

Simplify $\dfrac{5 + (-2^3)(2)}{6 \cdot \sqrt{9} - 9 \cdot 2}$.

$$\frac{5 + (-2^3)(2)}{6 \cdot \sqrt{9} - 9 \cdot 2} = \frac{5 + (-8)(2)}{6 \cdot 3 - 9 \cdot 2} \quad \text{Evaluate powers and roots.}$$

$$= \frac{5 - 16}{18 - 18} \quad \text{Multiply.}$$

$$= \frac{-11}{0} \quad \text{Subtract.}$$

Since division by zero is not possible, the given expression is undefined. ■

6 In many problems, an expression is given, along with the values of the variables in the expression. The expression can be **evaluated** by substituting the numerical values for the variables.

EXAMPLE 15 EVALUATING EXPRESSIONS

Evaluate each expression when $m = -4$, $n = 5$, and $p = -6$.

(a) $5m - 9n$

Replace m with -4 and n with 5.

$$5m - 9n = 5(-4) - 9(5) = -20 - 45 = -65$$

(b) $\dfrac{m + 2n}{4p} = \dfrac{-4 + 2(5)}{4(-6)} = \dfrac{-4 + 10}{-24} = \dfrac{6}{-24} = -\dfrac{1}{4}$

(c) $-3m^2 + n^3$

Replace m with -4 and n with 5.

$$-3m^2 + n^3 = -3(-4)^2 + 5^3$$
$$= -3(16) + 125$$
$$= -48 + 125 = 77 \quad ■$$

CAUTION When evaluating expressions, it is a good idea to use parentheses around any negative numbers that are substituted for variables. Notice the placement of the parentheses in Example 15(c) assures that -4 is squared, giving a positive result. Writing (-4^2) would lead to -16, which would be incorrect.

1.3 EXERCISES

Add or subtract as indicated. See Examples 1–3.

1. $-12 + (-8)$

2. $-5 + (-2)$

3. $-\dfrac{7}{3} + \dfrac{3}{4}$

4. $\dfrac{5}{8} + \left(-\dfrac{2}{3}\right)$

5. $13 - 15$

6. $-6 - 17$

7. $-12 - (-1)$

8. $-3 - (-14)$

9. $-5 + 11 + 3$

10. $-9 + 16 + 5$

11. $12 - (-3) - (-5)$

12. $15 - (-6) - (-8)$

13. $-9 - (-11) - (4 - 6)$

14. $-4 - (-13) + (-5 + 10)$

15. $-\frac{3}{4}-\left(\frac{1}{2}-\frac{3}{8}\right)$ **16.** $\frac{7}{5}-\left(\frac{9}{10}-\frac{3}{2}\right)$ **17.** $\left(\frac{1}{2}+\frac{2}{3}\right)-\frac{3}{2}$ **18.** $-\frac{5}{6}-\left(\frac{4}{3}-\frac{2}{5}\right)$

19. $\left(-\frac{5}{3}-\frac{1}{3}\right)+\frac{3}{4}$ **20.** $\left(-\frac{2}{7}+\frac{4}{3}\right)-\frac{1}{2}$ **21.** $-\sqrt{4}+|-3|+\sqrt{25}$

22. $|-5|+2^3-(-2^2)$ **23.** $|-11|+|5|-(|7|+|-2|)$ **24.** $|-40|-(|-3|+|18|)-(-16)$

The sketch shows a number line with several points labeled. Find the distance between the given pairs of points. See Example 4.

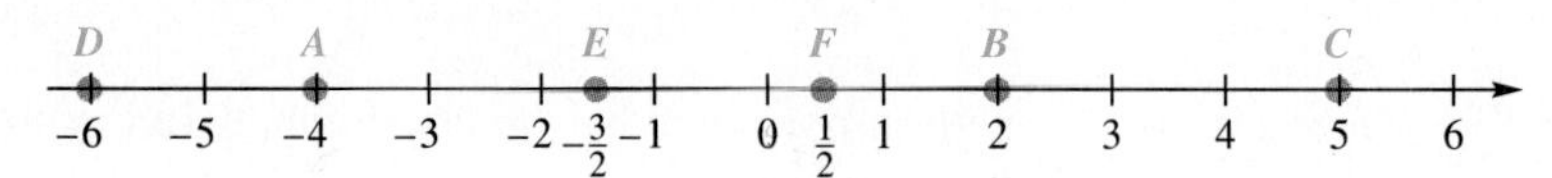

25. A and B **26.** A and C **27.** D and F **28.** E and C

---

29. Use absolute value to write an expression for the distance between x and a. (See Objective 2.)

Use operations with signed numbers to work each of the following problems.

30. The temperature at 4 A.M. was $-19°$. It then increased $39°$ by noon. Find the temperature at noon.

31. A diver goes to 84 feet below the surface. Another diver goes 18 feet lower than the first diver. How far is the second diver from the surface?

32. Telescope Peak, altitude 11,049 feet, is next to Death Valley, 282 feet below sea level. Find the difference between these altitudes.

33. A stunt pilot is flying 80 feet above the surface of the Dead Sea (altitude 1299 feet below sea level). She wants to clear a 3852-foot pass by 225 feet. How much altitude must she gain?

Photo for Exercise 31

Multiply or divide. See Examples 5 and 6.

34. $(-12)(-2)$ **35.** $(-3)(-5)$ **36.** $9(-12)(-4)(-1)3$ **37.** $-5(-17)(2)(-2)4$

38. $-\frac{5}{2}\left(-\frac{12}{25}\right)$ **39.** $-\frac{9}{7}\left(-\frac{35}{36}\right)$ **40.** $\frac{-18}{-3}$ **41.** $\frac{-100}{-50}$

42. $\frac{9}{0}$ **43.** $\frac{-15}{0}$ **44.** $-\frac{10}{17}\div\left(-\frac{12}{5}\right)$ **45.** $-\frac{22}{23}\div\left(-\frac{33}{4}\right)$

46. $\dfrac{\frac{7}{10}}{-\frac{8}{15}}$ **47.** $\dfrac{\frac{9}{16}}{-\frac{12}{7}}$

48. Explain in your own words why division by zero is undefined. (See Objective 3.)

Evaluate. See Examples 7–10.

49. 3^5 **50.** 7^3 **51.** $\left(-\frac{6}{7}\right)^3$ **52.** $\left(-\frac{2}{9}\right)^2$ **53.** -5^3 **54.** -4^4 **55.** $\sqrt{256}$

56. $\sqrt{529}$ **57.** $-\sqrt{361}$ **58.** $-\sqrt{841}$ **59.** $\sqrt[3]{27}$ **60.** $\sqrt[3]{343}$ **61.** $\sqrt[5]{243}$ **62.** $\sqrt[4]{625}$

63. Why is it incorrect to say that $\sqrt{16}$ equals 4 or -4? (See Objective 4.)

Find the following square roots or fourth roots on a calculator with a square root key. (Hint: For fourth roots, press the square root key twice.) Round to the nearest thousandth.

64. $\sqrt{18{,}499}$ **65.** $\sqrt{476.91}$ **66.** $\sqrt[4]{.08312}$ **67.** $\sqrt[4]{.90649}$

Perform the following operations wherever possible. Use the order of operations given in the text. See Examples 11–14.

68. $-6[2 + (-5)]$ **69.** $-8[4 + (-2^2 \cdot 3)]$ **70.** $-2^2(3)(-2) - (-5)$

71. $-7(-3) - (-2^3)$ **72.** $-4 - 3(-2) + 5^2$ **73.** $-6 - 5(-8) + 3^2$

74. $(-6 - 3)(-2 - 3)$ **75.** $(-8 - 5)(-2 - 1)$ **76.** $\dfrac{(-10 + 4) \cdot (-3)}{-7 - 2}$

77. $\dfrac{(-6 + 3) \cdot (-4)}{-5 - 1}$ **78.** $\dfrac{2(-5 + 3)}{-2^2} - \dfrac{(-3^2 + 2)3}{3 - (-4)}$ **79.** $\dfrac{3(-5 - 2^2)}{3^2} + \dfrac{(3^2 \cdot 2 - 7)(5)}{5 \cdot 2 - (-1)}$

80. $2\sqrt{100} - 10 \div 5$ **81.** $-6\sqrt{9} + 8\sqrt{25} + 3^2$

82. $\dfrac{2(-5) + (-3)(-2^2)}{-6 + 5 + 1}$ **83.** $\dfrac{3(-4) + (-5)(-2)}{2^3 - 2 + (-6)}$

84. $-\dfrac{4}{5}[6(-4) + (-5)(-5)]$ **85.** $-\dfrac{1}{4}[3(-5) + 7(-5) + 1(-2)]$

86. $\dfrac{5 - 3\left(\dfrac{-5 - 9}{-7}\right) - 6}{-9 - 11 + 3 \cdot 7}$ **87.** $\dfrac{-4\left(\dfrac{12 - (-8)}{3 \cdot 2 + 4}\right) - 5(-1 - 7)}{-9 - (-7) - [-5 - (-8)]}$

Evaluate each expression when $a = -4$, $b = 6$, and $c = -7$. See Example 15.

88. $-2(a + 4c)$ **89.** $-6(-b + 3a)$ **90.** $\dfrac{2c + a}{4b + 6a}$

91. $\dfrac{a - 6b}{7a - 4c}$ **92.** $2c^3 - a^2$ **93.** $-a^2 + b^3$

Replace a, b, and c with various integers to decide whether the statements in Exercises 94–97 are true or false. For each false statement, give an example showing it is false.

94. $a - b = b - a$ **95.** $a - (b - c) = (a - b) - c$

96. $|a + b| = |a| + |b|$ **97.** $|a - b| = |a| - |b|$

---

98. Do *any* numbers satisfy the statement $a - b = b - a$? If your answer is *yes*, give an example. (See Objective 1.)

99. Are there *any* numbers a and b for which $a/b = b/a$? If your answer is *yes*, give an example. (See Objective 3.)

1.4 PROPERTIES OF REAL NUMBERS

OBJECTIVES

1. USE THE DISTRIBUTIVE PROPERTY.
2. USE THE INVERSE PROPERTIES.
3. USE THE IDENTITY PROPERTIES.
4. USE THE COMMUTATIVE AND ASSOCIATIVE PROPERTIES.
5. USE THE MULTIPLICATION PROPERTY OF ZERO.
6. USE THE ADDITION AND MULTIPLICATION PROPERTIES OF EQUALITY.

In Section 1.2 several properties for equality and inequality were presented. In this section the list is expanded to include many of the basic properties of the real numbers. These properties are results that have been observed to occur consistently in work with numbers, so they have been generalized to apply to expressions with variables as well.

1 The properties we discuss in this section are used in simplifying algebraic expressions. For example, notice that

$$2(3 + 5) = 2 \cdot 8 = 16$$

and

$$2 \cdot 3 + 2 \cdot 5 = 6 + 10 = 16$$

so that

$$2(3 + 5) = 2 \cdot 3 + 2 \cdot 5.$$

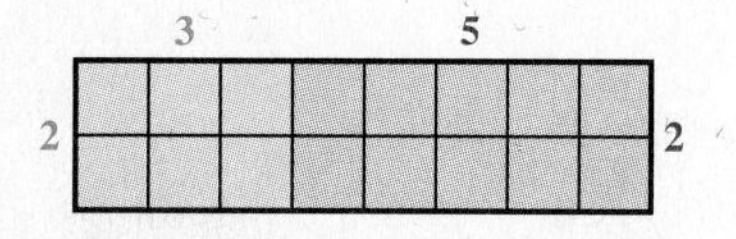

Area of left part is $2 \cdot 3 = 6$.
Area of right part is $2 \cdot 5 = 10$.
Area of total is $2(3 + 5) = 16$.

FIGURE 1.15

This idea is illustrated by the divided rectangle in Figure 1.15. Similarly,

$$-4[5 + (-3)] = -4(2) = -8$$

and

$$-4(5) + (-4)(-3) = -20 + 12 = -8$$

so

$$-4[5 + (-3)] = -4(5) + (-4)(-3).$$

These examples suggest the **distributive property.**

DISTRIBUTIVE PROPERTY

For any real numbers a, b, and c,

$$a(b + c) = ab + ac \quad \text{and} \quad (b + c)a = ba + ca.$$

The distributive property can also be written

$$ab + ac = a(b + c).$$

EXAMPLE 1 USING THE DISTRIBUTIVE PROPERTY

Use the distributive property to rewrite each expression.

(a) $3(x + y)$

In the statement of the property, let $a = 3$, $b = x$ and $c = y$. Then

$$3(x + y) = 3x + 3y.$$

(b) $-2(5 + k) = -2(5) + (-2)(k)$
$= -10 - 2k$

(c) $4x + 8x$
Use the second form of the property.

$$4x + 8x = (4 + 8)x = 12x$$

(d) $3r - 7r = 3r + (-7)r$ Definition of subtraction
$= [3 + (-7)]r$ Distributive property
$= -4r$

(e) $5p + 7q$
Since there is no common factor here, we cannot use the distributive property to simplify the expression. ■

As illustrated in Example 1(d), the distributive property can also be used for subtraction, so that

$$a(b - c) = ab - ac.$$

2 In Section 1.1 we saw that the additive inverse of a number a is $-a$. For example, 3 and -3 are additive inverses, as are -8 and 8. The number 0 is its own additive inverse. In Section 1.3, we saw that two numbers with a product of 1 are reciprocals. Another name for a reciprocal is **multiplicative inverse.** This is similar to the idea of an additive inverse. Thus, 4 and 1/4 are multiplicative inverses, and so are $-2/3$ and $-3/2$. (Note again that a pair of reciprocals has the same sign.) These properties are called the **inverse properties** of addition and multiplication.

INVERSE PROPERTIES

For any real number a, there is a single real number $-a$, such that

$$a + (-a) = 0 \quad \text{and} \quad -a + a = 0.$$

For any nonzero real number a, there is a single real number $1/a$ such that

$$a \cdot \frac{1}{a} = 1 \quad \text{and} \quad \frac{1}{a} \cdot a = 1.$$

Examples showing how these properties are used are given later in this section.

3 The numbers 0 and 1 each have a special property. Zero is the only number that can be added to any number to get that number. That is, adding 0 leaves the identity of a number unchanged. For this reason, 0 is called the **identity element for addition.** In a similar way, multiplying by 1 leaves the identity of any number unchanged, so 1 is the **identity element for multiplication.** The following **identity properties** summarize this discussion.

IDENTITY PROPERTIES

For any real number a,

$$a + 0 = 0 + a = a$$

and

$$a \cdot 1 = 1 \cdot a = a.$$

The identity property for 1 is especially useful in simplifying algebraic expressions.

EXAMPLE 2 USING THE IDENTITY PROPERTY $1 \cdot a = a$

Simplify each expression.

(a) $12m + m$

$$\begin{aligned} 12m + m &= 12m + 1m && \text{Identity property} \\ &= (12 + 1)m && \text{Distributive property} \\ &= 13m \end{aligned}$$

(b)
$$\begin{aligned} y + y &= 1y + 1y && \text{Identity property} \\ &= (1 + 1)y && \text{Distributive property} \\ &= 2y \end{aligned}$$

(c)
$$\begin{aligned} -(m - 5n) &= -1(m - 5n) && \text{Identity property} \\ &= -1 \cdot m + (-1)(-5n) && \text{Distributive property} \\ &= -m + 5n \quad \blacksquare \end{aligned}$$

Expressions such as $12m$ and $5n$ from Example 2 are examples of *terms*. A **term** is a number or the product of a number and one or more variables raised to powers. Terms with exactly the same variables raised to exactly the same powers are called **like terms.** The number in the product is called the **numerical coefficient** or just the **coefficient.** For example, in the term $5p$, the coefficient is 5.

4 Simplifying expressions as in Examples 2(a) and (b) is called **combining like terms.** Only like terms may be combined. To combine like terms in an expression such as

$$-2m + 5m + 3 - 6m + 8$$

we need two more properties. We are familiar with the fact that

$$3 + 9 = 12 \quad \text{and} \quad 9 + 3 = 12.$$

Also,

$$3 \cdot 9 = 27 \quad \text{and} \quad 9 \cdot 3 = 27.$$

The next properties generalize these statements about addition and multiplication.

COMMUTATIVE PROPERTIES

For any real numbers a and b,

$$a + b = b + a$$
$$ab = ba.$$

Furthermore, notice that

$$(5 + 7) + (-2) = 12 + (-2) = 10$$

and

$$5 + [7 + (-2)] = 5 + 5 = 10.$$

Also,

$$(5 \cdot 7)(-2) = 35(-2) = -70$$

and

$$(5)[7(-2)] = 5(-14) = -70.$$

These observations suggest the following properties.

ASSOCIATIVE PROPERTIES

For any real numbers a, b, and c,

$$a + (b + c) = (a + b) + c$$
$$a(bc) = (ab)c.$$

NOTE The associative properties are used to *regroup* the terms of an expression. The commutative properties are used to change the *order* of the terms in an expression.

EXAMPLE 3 USING THE COMMUTATIVE AND ASSOCIATIVE PROPERTIES

Simplify $-2m + 5m + 3 - 6m + 8$.

$$-2m + 5m + 3 - 6m + 8$$
$$= (-2m + 5m) + 3 - 6m + 8 \qquad \text{Order of operations}$$
$$= 3m + 3 - 6m + 8 \qquad \text{Distributive property}$$

By the order of operations, the next step would be to add $3m$ and 3, but they are unlike terms. To get $3m$ and $-6m$ together, use the associative and commutative properties. Begin by putting in parentheses and brackets, as shown.

$$[(3m + 3) - 6m] + 8$$
$$= [3m + (3 - 6m)] + 8 \qquad \text{Associative property}$$
$$= [3m + (-6m + 3)] + 8 \qquad \text{Commutative property}$$
$$= [(3m + [-6m]) + 3] + 8 \qquad \text{Associative property}$$
$$= (-3m + 3) + 8 \qquad \text{Combine like terms.}$$
$$= -3m + (3 + 8) \qquad \text{Associative property}$$
$$= -3m + 11 \qquad \text{Add.}$$

In practice, many of the steps are not written down, but you should realize that the commutative and associative properties are used whenever the terms in an expression are rearranged in order to combine like terms. ■

EXAMPLE 4 USING THE PROPERTIES OF REAL NUMBERS

Simplify each expression.

(a) $5y^2 - 8y^2 - 6y^2 + 11y^2$

$$5y^2 - 8y^2 - 6y^2 + 11y^2 = (5 - 8 - 6 + 11)y^2 = 2y^2$$

(b) $-2(m - 3)$

$$-2(m - 3) = -2(m) - (-2)(3) = -2m + 6$$

(c) $3x^3 + 4 - 5(x^3 + 1) - 8$

First use the distributive property to eliminate the parentheses.

$$3x^3 + 4 - 5(x^3 + 1) - 8 = 3x^3 + 4 - 5x^3 - 5 - 8$$

Next use the commutative and associative properties to rearrange the terms; then combine like terms.

$$= 3x^3 - 5x^3 + 4 - 5 - 8$$
$$= -2x^3 - 9$$

(d) $8 - (3m + 2)$

Think of $8 - (3m + 2)$ as $8 - 1 \cdot (3m + 2)$.

$$8 - 1 \cdot (3m + 2) = 8 - 3m - 2 = 6 - 3m$$

(e) $(3x)(5)(y) = [(3x)(5)]y$ Order of operations

$= [3(x \cdot 5)]y$ Associative property

$= [3(5x)]y$ Commutative property

$= [(3 \cdot 5)x]y$ Associative property

$= (15x)y$

$= 15(xy)$ Associative property

$= 15xy$

As mentioned earlier, many of these steps usually are not written out. ■

5 The additive identity property gives a special property of zero, namely that $a + 0 = a$ for any real number a. The **multiplication property of zero** gives a special property of zero that involves multiplication: The product of any real number and zero is zero.

MULTIPLICATION PROPERTY OF ZERO

For all real numbers a,

$$a \cdot 0 = 0 \quad \text{and} \quad 0 \cdot a = 0.$$

This property just extends to all real numbers what is true for positive numbers multiplied by 0. The multiplication property of zero will be used in Section 3.9 to solve equations.

6 The addition and multiplication properties of equality, given next, are used to solve equations.

ADDITION AND MULTIPLICATION PROPERTIES OF EQUALITY

Addition Property of Equality

If a, b, and c are real numbers, and

$$\text{if } a = b, \text{ then } a + c = b + c.$$

Multiplication Property of Equality

If a, b, and c are real numbers, and

$$\text{if } a = b, \text{ then } ac = bc.$$

In the next example, notice how several of the properties are used in the solution.

EXAMPLE 5 USING THE PROPERTIES TO SOLVE AN EQUATION

Solve $2x + 3 = 9$.

$2x + 3 = 9$	Given
$2x + 3 + (-3) = 9 + (-3)$	Addition property of equality
$2x + 0 = 6$	Inverse property of addition
$2x = 6$	Identity property of addition
$\frac{1}{2} \cdot 2x = \frac{1}{2} \cdot 6$	Multiplication property of equality
$1x = 3$	Inverse property of multiplication
$x = 3$	Identity property of multiplication ■

Solving equations as in Example 5 is covered in detail in Chapter 2. Other types of equations will be solved in later chapters.

1.4 EXERCISES

Use the properties of real numbers to simplify each expression. See Examples 1 and 2.

1. $5k + 3k$ **2.** $6a + 5a$ **3.** $9r + 7r$

4. $-4n + 6n$ **5.** $-8z + 3w$ **6.** $-12k + 2h$

7. $-a + 7a$ **8.** $7t + t$ **9.** $14c + c$

10. $2(m + p)$ **11.** $3(a + b)$ **12.** $12(x - y)$

13. $-5(2d + f)$ **14.** $-2(3m + n)$ **15.** $-4(5k - 3)$

16. $-(8p - 1)$ **17.** $-(3k + 7)$ **18.** $-(p - 3q)$

Simplify each of the following expressions by removing parentheses and combining terms. See Examples 1–4.

19. $4x + 3x + 7 + 9$

20. $5m + 9m + 8 + 4$

21. $-12y + 4y - 3 + 2y$

22. $5r - 9r + 8r - 5$

23. $-6p + 11p - 4p + 6 - 5$

24. $8x - 5x + 3x - 12 + 9$

25. $3(k + 2) - 5k + 6 - 3$

26. $5(r - 3) + 6r - 2r + 1$

27. $-2(m + 1) + 3(m - 4)$

28. $6(a - 5) - 4(a + 6)$

29. $3(2y + 3) - 4(3y - 5)$

30. $-2(5r - 2) + 4(5r - 1)$

31. $-(2p + 5) + 3(2p + 4) + p$

32. $-(7m - 12) - 2(4m + 7) - 3m$

33. $2 + 3(2z - 5) - 3(4z + 6) - 8$

34. $-4 + 4(4k - 3) - 6(2k + 8) + 7$

35. $m + 2(-m + 4) - 3(m + 6)$

36. $-k + 2(k - 5) + 5(2k - 1)$

37. $4(-2x - 5) + 6(2 - 3x) - (4 - 2x)$

38. $3(-4z + 5) - 2(4 - 5z) - (2 - 9z)$

For Exercises 39–58, use the indicated property or theorem to complete an equality. Simplify your answers.

39. $2x + 3x =$ ____; distributive property

40. $-4 \cdot 1 =$ ____; multiplicative identity property

41. $2(4x) =$ ____; associative property

42. $-3 + 3 =$ ____; commutative property

43. $-3 + 3 =$ ____; additive inverse property

44. $4y + 4z =$ ____; distributive property

45. $0 + 7 =$ ____; identity property

46. $13 \cdot \frac{1}{13} =$ ____; inverse property

47. $3a + 5a + 6a =$ ____; distributive property

48. $0 = 9($____$)$; multiplication property of zero

49. $0 = 2 + ($____$)$; inverse property

50. $6 + (2 + x) =$ ____; associative property

51. $8(2 + 3) = ($____$)2 + ($____$)3$; distributive property

52. $3(2 \cdot 5) = 3($____$)$; commutative property

53. If $x = 0$, then $x + 8 =$ ____; addition property of equality

54. If $y + 3 = 12$, then $y + 3 + (-3) = 12 +$ ____; addition property of equality

55. If $k = 4$, then $3k =$ ____; multiplication property of equality

56. If $5s = 60$, then $\frac{1}{5} \cdot 5s =$ ____; multiplication property of equality

57. $-(-4) =$ ____; double negative property

58. $-(-y) =$ ____; double negative property

59. Replace x with 5 to show that $2 + 6x \neq 8x$. (See Objective 1.)

60. Replace x with 5 to show that $4x - x \neq 4$. (See Objective 1.)

Use the distributive property to calculate the following values in your head. (See Objective 1.)

61. $96 \cdot 19 + 4 \cdot 19$

62. $27 \cdot 60 + 27 \cdot 40$

63. $58 \cdot \frac{3}{2} - 8 \cdot \frac{3}{2}$

64. $8.75(15) - 8.75(5)$

65. $4.31(69) + 4.31(31)$

66. $\frac{8}{5}(17) + \frac{8}{5}(13)$

67. By the distributive property, $a(b + c) = ab + ac$. This property is more completely named the **distributive property of multiplication over addition.** Is there a distributive property of addition over multiplication? That is, does

$$a + (b \cdot c) = (a + b)(a + c)$$

for all real numbers, a, b, and c? To find out, try various sample values of a, b, and c.

68. Identify the correct properties from Section 1.2 that justify the following proof of the addition property of equality.

$a + c = a + c$ ____________

$a = b$ Given

$a + c = b + c$ ____________

CHAPTER 1 GLOSSARY

KEY TERMS

1.1 set A set is a collection of objects.

elements The elements (or **members**) of a set are the objects in the set.

empty set The set with no elements is called the empty set (or **null** set).

variable A variable is a letter used to represent a number or a set of numbers.

set-builder notation Set-builder notation is used to describe a set of numbers without listing them.

subset Set A is a subset of set B if every element of A is also an element of B.

number line A number line is a line with a scale to indicate the set of real numbers.

coordinate The number that corresponds to a point on the number line is its coordinate.

graph The point on the number line that corresponds to a number is its graph.

additive inverse The additive inverse (**negative, opposite**) of a number a is $-a$.

signed numbers Positive and negative numbers are signed numbers.

absolute value The absolute value of a number is its distance from 0 or its magnitude without regard to sign.

1.2 inequality An inequality is a mathematical statement that two quantities are not equal.

interval An interval is a portion of a number line.

interval notation Interval notation uses symbols to describe an interval on the number line.

1.3 sum The result of addition is called the sum.

difference The result of subtraction is called the difference.

product The result of multiplication is called the product.

quotient The result of division is called the quotient.

reciprocal The reciprocal of a nonzero number b is $1/b$.

factors A factor of a given number is any number that divides evenly into the given number.

factored form A number is in factored form if it is expressed as a product of two or more numbers.

exponent An exponent is a number that shows how many times a factor is repeated in a product.

base The base is the repeated factor in an exponential expression.

exponential A base with an exponent is called an exponential or a **power.**

square root A square root of a number r is a number that can be squared to get r.

cube root The cube root of a number r is the number that can be cubed to get r.

1.4 multiplicative inverse The multiplicative inverse (reciprocal) of a nonzero number b is $1/b$.

identity element for addition The identity element for addition is 0.

identity element for multiplication The identity element for multiplication is 1.

term A term is a number or the product of a number and one or more variables raised to powers.

like terms Like terms are terms with the same variables raised to the same powers.

coefficient A coefficient (**numerical coefficient)** is the numerical factor of a term.

combining like terms Combining like terms is a method of adding or subtracting like terms by using the properties of real numbers.

NEW SYMBOLS

$\{a, b\}$	set containing the elements a and b
$\emptyset$	the empty set
$\in$	is an element of (a set)
$\notin$	is not an element of
$\neq$	is not equal to
$\{x\mid x \text{ has property } P\}$	the set of all x, such that x has property P
$\subseteq$	is a subset of
$\not\subseteq$	is not a subset of
$\lvert x\rvert$	the absolute value of x
$<$	is less than
$\leq$	is less than or equal to
$>$	is greater than
$\geq$	is greater than or equal to
$\nless$	is not less than
$\ngtr$	is not greater than
(a, ∞)	the interval $\{x\mid x > a\}$
$(-\infty, a)$	the interval $\{x\mid x < a\}$
$(a, b]$	the interval $\{x\mid a < x \leq b\}$
a^m	m factors of a
$\sqrt{a}$	the square root of a
$\sqrt[n]{a}$	the nth root of a

CHAPTER 1 QUICK REVIEW

SECTION	CONCEPTS	EXAMPLES
1.1 BASIC TERMS	**Sets of Numbers**	
	Real Numbers $\{x\mid x$ is a coordinate of a point on a number line$\}$	$-3, .7, \pi, -\frac{2}{3}$
	Natural Numbers $\{1, 2, 3, 4, \ldots\}$	10, 25, 143
	Whole Numbers $\{0, 1, 2, 3, 4, \ldots\}$	0, 8, 47
	Integers $\{\ldots, -2, -1, 0, 1, 2, \ldots\}$	$-22, -7, 0, 4, 9$
	Rational Numbers $\left\{\frac{p}{q}\middle\vert p \text{ and } q \text{ are integers}, q \neq 0\right\}$, or all terminating or repeating decimals	$-\frac{2}{3}, -.14, 0, 6, \frac{5}{8}, .\overline{3}$
	Irrational Numbers $\{x\mid x$ is a real number that is not rational$\}$ or all nonterminating, nonrepeating decimals	$\pi, .125469\ldots, \sqrt{3}, -\sqrt{22}$

SECTION	CONCEPTS	EXAMPLES
1.2 EQUALITY AND INEQUALITY	**Properties of Equality** For all real numbers a, b, and c:	
	Reflexive Property $a = a$	$m + 2 = m + 2$
	Symmetric Property If $a = b$, then $b = a$.	If $x = 3$, then $3 = x$.
	Transitive Property If $a = b$ and $b = c$, then $a = c$.	If $k = 5$, and $5 = r$, then $k = r$.
	Substitution Property If $a = b$, then a may replace b in any sentence without changing the truth of the sentence.	If $x = -4$, then $2x + y = 2(-4) + y$.
	Properties of Inequality For all real numbers a, b, and c:	
	Trichotomy Property Either $a < b$, $a = b$, or $a > b$.	Either $x < 2$, $x = 2$, or $x > 2$.
	Transitive Property If $a < b$ and $b < c$, then $a < c$.	If $z < -1$ and $-1 < 0$, then $z < 0$.
1.3 OPERATIONS ON REAL NUMBERS	**Addition** *Same sign:* Add the absolute values. The sum has the same sign as the numbers.	Add: $-2 + (-7)$. $-2 + (-7) = -(2 + 7)$ $= -9$
	Different signs: Subtract the absolute values. The answer has the sign of the number with the larger absolute value.	$-5 + 8 = 8 - 5 = 3$ $-12 + 4 = -(12 - 4) = -8$
	Subtraction Change the sign of the second number and add.	Subtract: $-5 - (-3)$. $-5 - (-3) = -5 + 3 = -2$
	Multiplication *Same sign:* The product is positive. *Different signs:* The product is negative.	Multiply: $(-3)(-8) = 24$. Multiply: $(-7)(5) = -35$.

SECTION	CONCEPTS	EXAMPLES
	Division *Same sign:* The quotient is positive.	Divide: $\frac{-15}{-5} = 3$.
	Different signs: The quotient is negative.	Divide: $\frac{-24}{12} = -2$.
	The product of an even number of negative factors is positive. The product of an odd number of negative factors is negative.	$(-5)^6$ is positive. $(-5)^7$ is negative.
	Order of Operations **1.** Work separately above and below any fraction bar. **2.** Use the rules below within each set of parentheses or square brackets. Start with the innermost set and work outward. **3.** Evaluate all powers and roots. **4.** Do any multiplications or divisions in the order in which they occur, working from left to right. **5.** Do any additions or subtractions in the order in which they occur, working from left to right.	Perform the indicated operations. $\frac{12 + 3}{5 \cdot 2} = \frac{15}{10} = \frac{3}{2}$ $(-6)[2^2 - (3 + 4)] + 3$ $= (-6)[2^2 - 7] + 3$ $= (-6)[4 - 7] + 3$ $= (-6)[-3] + 3$ $= 18 + 3$ $= 21$
1.4 PROPERTIES OF REAL NUMBERS	For any real numbers, a, b, and c:	
	Distributive Property $a(b + c) = ab + ac$	$12(4 + 2) = 12 \cdot 4 + 12 \cdot 2$
	Inverse Properties $a + (-a) = 0$ and $-a + a = 0$	$5 + (-5) = 0$
	$a \cdot \frac{1}{a} = 1$ and $\frac{1}{a} \cdot a = 1$ $(a \neq 0)$	$-\frac{1}{3}(-3) = 1$

SECTION	CONCEPTS	EXAMPLES
	Identity Properties $a + 0 = a$ and $0 + a = a$ $a \cdot 1 = a$ and $1 \cdot a = a$	$-32 + 0 = -32$ $(17.5)(1) = 17.5$
	Associative Properties $a + (b + c) = (a + b) + c$ $a(bc) = (ab)c$	$7 + (5 + 3) = (7 + 5) + 3$ $-4(6 \cdot 3) = (-4 \cdot 6)3$
	Commutative Properties $a + b = b + a$ $ab = ba$	$9 + (-3) = -3 + 9$ $6(-4) = (-4)6$
	Addition and Multiplication Properties of Equality If $a = b$, then $a + c = b + c$ and $ac = bc$.	If $3m + 2 = 5$, then $3m + 2 + (-2) = 5 + (-2)$ $3m = 3$ $\frac{1}{3} \cdot 3m = \frac{1}{3} \cdot 3$ $m = 1$.
	Multiplication Property of Zero $a \cdot 0 = 0$ and $0 \cdot a = 0$	$47 \cdot 0 = 0$; $0 \cdot -18 = 0$

CHAPTER 1 REVIEW EXERCISES

If you need help with any of these review exercises, look in the section indicated in brackets.

[1.1] *Graph each set on a number line.*

1. $\{-3, -5, 1, 1/4, 9/4\}$

2. $\{-6, -11/4, -1, 0, 3, 12/5\}$

Find the value of each expression.

3. $|-12|$

4. $|31|$

5. $-|4|$

6. $-|-7| + |-2|$

Give the additive inverse of each number.

7. 15

8. -16

9. $-(-9)$

10. $|-18|$

Let $S = \{-9, -1.\overline{3}, -.8, -\sqrt{10}, 0, 5/3, \sqrt{7}, \pi, 12/3\}$. *Simplify the elements of S as necessary, and list the elements that belong to the following sets.*

11. Natural numbers

12. Integers

13. Rational numbers

14. Irrational numbers

[1.2] *Use inequality symbols to rewrite each of the following.*

15. 3 is less than or equal to 10.

16. 3 is not equal to -2.

17. x is between -2 and 1.

18. x is between 0 and 3, excluding 0 and including 3.

Write true *or* false *for each inequality.*

19. $2 \geq -5$ **20.** $-3 < -3$ **21.** $6 < |-2|$

22. $3 \cdot 2 \leq |12 - 4|$ **23.** $2 + |-2| \geq 3$ **24.** $4(3 + 7) > -|-41|$

Write in interval notation and graph each set.

25. $\{x|x < -4\}$ **26.** $\{x|x \geq 1\}$ **27.** $\{x|-2 < x \leq 5\}$ **28.** $\{x|4 \leq x \leq 9\}$

[1.3] *Add or subtract, as indicated.*

29. $-\frac{5}{8} - \left(-\frac{7}{3}\right)$ **30.** $-\frac{4}{5} - \left(-\frac{3}{10}\right)$ **31.** $-5 + (-11) + 20$

32. $-9.42 + 1.83 - 7.6 - 1.9$ **33.** $-15 + (-13) + (-11)$ **34.** $-1 - 3 - (-10)$

35. $\frac{3}{4} - \left(\frac{1}{2} - \frac{9}{10}\right)$ **36.** $-\frac{2}{3} - \left(\frac{1}{6} - \frac{5}{9}\right)$ **37.** $-|-12| - |-9| + (-4)$

Find each product.

38. $2(-5)(-3)$ **39.** $-\frac{3}{7}\left(-\frac{14}{9}\right)$ **40.** $-4.6(-3.9)$

Divide.

41. $\frac{-15}{-3}$ **42.** $\frac{-7}{0}$ **43.** $-\frac{10}{21} \div \frac{5}{14}$

Evaluate each of the following.

44. 3^2 **45.** 10^4 **46.** $\left(\frac{3}{7}\right)^3$ **47.** $(-5)^3$ **48.** -5^3 **49.** 1.7^2

Find each root.

50. $\sqrt{400}$ **51.** $\sqrt[3]{27}$ **52.** $\sqrt[3]{343}$ **53.** $\sqrt[4]{81}$ **54.** $\sqrt[6]{64}$

Use the order of operations to simplify.

55. $-14\left(\frac{3}{7}\right) + 6 \div 2$ **56.** $-\frac{2}{3}[5(-2) + 8 - 4^2]$

57. $\frac{-4(\sqrt{25}) - (-3)(-5)}{3 + (-6)(\sqrt{9})}$ **58.** $\frac{-5(3^2) + 9(\sqrt{4}) - 5}{6 - 5\sqrt[3]{-8}}$

Evaluate. Assume that $k = -2$, $m = 3$, *and* $n = 16$.

59. $4k - 7m$ **60.** $-3(\sqrt{16}) + m + 5k$ **61.** $-2(3k^2 + 5m)$

62. $\frac{4m^3 - 3n}{7k^2 - 10}$ **63.** $\frac{4k - 3k(\sqrt{n} + 3)}{k - m}$ **64.** $\frac{k + \sqrt{n} - m(2 + k)}{m - 5 - 2k}$

[1.4] *Use the properties of real numbers to simplify each expression.*

65. $2q + 9q$ **66.** $3z - 17z$ **67.** $-m + 6m$

68. $5p - p$ **69.** $-2(k - 3)$ **70.** $6(r + 3)$

71. $9(2m + 3n)$ **72.** $-(3k - 4h)$ **73.** $-(-p + 6q)$

74. $2x^2 + 5 - 4x^2 + 1$ **75.** $-3y^3 + 6 - 5 - 4y^3$ **76.** $2a + 3 - a - 1$

77. $2(k^2 - 1) + 3k^2 - k^2$ **78.** $-3(4m^3 - 2) + 2(3m^3 - 1)$

79. $-(5p + 2) - 3p - 8 + p$ **80.** $-k + 3(k - 1) - 2(4k + 3)$

For each of the following, use the indicated property to complete the equality. Simplify your answers.

81. $7p + 7r =$ _____; distributive property

82. $6(mn) =$ _____; associative property

83. $0 = -11 + ($_____$)$; additive inverse property

84. If $p = 7$, then $p + 2 = ($_____$)$; addition property of equality

85. $-(-22) =$ _____; double negative property

86. $(9 + 2) + 7 = 7 + ($_____$)$; commutative property

MIXED REVIEW EXERCISES*

Perform the indicated operations.

87. $\left(-\frac{4}{5}\right)^4$

88. $-\frac{5}{8}(-32)$

89. $-25\left(-\frac{4}{5}\right) + 3^3 - 32 \div \sqrt{64}$

90. $-8 + |-14| + |-3|$

91. $\frac{6 \cdot \sqrt{4} - 3 \cdot \sqrt{16}}{-2 \cdot 5 + 7(-3) - 1}$

92. $\sqrt[5]{32}$

93. $\frac{-7}{0}$

94. $\frac{12}{7} - \frac{28}{5}$

95. $18 - (-9) - [6 - (-11)]$

96. -3^2

97. $\frac{3^3 + (-5)}{7}$

98. $\frac{3(-2^2) + 5(-2)}{7 + (-1)}$

99. $\frac{13}{14}\left(\frac{28}{9}\right) - 5\left(-\frac{9}{5}\right)$

100. $-36 - (-2)(3)$

CHAPTER 1 TEST

Let $A = \{-\sqrt{7}, -2, -.5, 0/2, 1.\overline{34}, 15/2, 3, 24/2, \sqrt{11}\}$. *List the elements from set A that belong to the following sets.*

1. Real numbers **2.** Irrational numbers **3.** Rational numbers **4.** Integers

Write in interval notation and graph each set.

5. $\{x \mid x < 2\}$ **6.** $\{x \mid -2 \le x < 5\}$

---

*The order of exercises in this final group does not correspond to the order in which topics occur in the chapter. This random ordering should help you prepare for the chapter test in yet another way.

Evaluate each of the following.

7. $-6-(-15)+(-3^2)-3$

8. $\left(-\frac{1}{3}\right)\left(-\frac{9}{5}\right)$

9. $-2+(-3)-|-5|+|-12+\sqrt{4}|$

10. $6+(-3)^2+5(-2^2)-\frac{2^3}{-2}$

11. $6-3^3+2(5)+(-4)^2$

12. $\frac{10-24+(-6)}{4(-5)}$

13. $\frac{-2[3-(-1-2)+2]}{3(-3)-(-2)}$

14. $\frac{8\cdot 4-3\cdot\sqrt{25}-2(-1)-(-1)}{-3\cdot\sqrt{4}+1}$

15. Name the coefficient, exponent, and base in $-8y^4$.

Find each of the following.

16. $(-2)^3$

17. $\left(\frac{3}{2}\right)^4$

18. $(-12)^2$

19. $-4^2+(-4)^2$

20. $\sqrt[3]{64}$

21. $\sqrt{64}$

Evaluate each expression if $k=-4$, $m=-3$, and $r=5$.

22. $8r^2+m^3$

23. $\frac{5k+6r}{k+r-1}$

24. $-6(2m-3k)$

25. $\frac{8k+2m^2}{r-2}$

Use the properties of real numbers to simplify each of the following.

26. $3(2x-y)$

27. $4k-2+k-8-3k$

28. $5p-3(p+2)+7$

Complete each statement using the indicated property.

29. $\frac{2}{3}\cdot$ ____ $=1$ (multiplicative inverse)

30. $32+$ ____ $=32$ (additive identity)

CHAPTER TWO

LINEAR EQUATIONS AND INEQUALITIES

A great deal of work in algebra involves the solving of equations and inequalities, which in turn allow us to solve applied problems. The word *algebra* comes from the title of the work *Hisâb al-jabr w'al muquâbalah,* a ninth century treatise by the Arab Muhammed ibn Mûsâ al-Khowârizmî. The title translates as "the science of reunion and reduction," or more generally, "the science of transposition and cancellation." These basic procedures for solving equations are introduced in this chapter with more modern terminology.

A balance scale provides a means of comparing the weights of two quantities. Equations and inequalities provide a way of comparing numerical quantities, one or both of which may contain unknowns, or variables. In this chapter we learn how to solve several different types of equations and inequalities, and how to apply them to problem solving.

2.1 LINEAR EQUATIONS IN ONE VARIABLE

OBJECTIVES

1 DEFINE LINEAR EQUATIONS.

2 SOLVE LINEAR EQUATIONS USING THE ADDITION AND MULTIPLICATION PROPERTIES OF EQUALITY.

3 SOLVE LINEAR EQUATIONS USING THE DISTRIBUTIVE PROPERTY.

4 SOLVE LINEAR EQUATIONS WITH FRACTIONS AND DECIMALS.

5 IDENTIFY CONDITIONAL EQUATIONS, CONTRADICTIONS, AND IDENTITIES.

FOCUS ON PROBLEM SOLVING

By studying the winning times in the 500-meter speed-skating event at Olympic games back to 1900, it was found that the winning times for men were given by the equation

$$y_m = 46.338 - .097x,$$

where y_m is the time in seconds needed to win for men, with x the Olympic year and $x = 0$ corresponding to 1900. (For example, 1992 winning times would be estimated by replacing x with $1992 - 1900 = 92$.) The corresponding equation for women is

$$y_w = 57.484 - .196x.$$

Find the year in which **(a)** $y_m = 42.458$; **(b)** $y_w = 44.940$.

Predicting the outcome of a particular phenomenon can sometimes be done by using an equation, or equations, such as the ones above. These are examples of linear equations, since the variable x appears to the first power. This section introduces the concepts of equation solving, and later sections will extend these concepts to more involved equations. This problem is Exercise 65 in the exercises for this section. After working through this section, you should be able to solve this problem.

An **algebraic expression** is the result of performing the basic operations of addition, subtraction, multiplication, and division (except by 0), or extraction of roots on any collection of variables and numbers. Some examples of algebraic expressions include

$$8x + 9, \qquad \sqrt{y} + 4, \qquad \text{and} \qquad \frac{x^3y^8}{z}.$$

1 Applications of mathematics often lead to **equations,** statements that two algebraic expressions are equal. A *linear equation* in one variable involves only real numbers and one variable. Examples include

$$x + 1 = -2, \qquad y - 3 = 5, \qquad \text{and} \qquad 2k + 5 = 10.$$

LINEAR EQUATION

An equation is **linear** if it can be written in the form

$$ax + b = c,$$

where a, b, and c are real numbers, with $a \neq 0$.

A linear equation is also called a **first-degree** equation, since the highest power on the variable is one.

If the variable in an equation is replaced by a real number that makes the statement true, then that number is a **solution** of the equation. For example, 8 is a solution of the equation $y - 3 = 5$, since replacing y with 8 gives a true statement. An equation is **solved** by finding its **solution set,** the set of all solutions. The solution set of the equation $y - 3 = 5$ is $\{8\}$.

2 **Equivalent equations** are equations with the same solution set. Equations are generally solved by starting with a given equation and producing a series of simpler equivalent equations. For example,

$$8x + 1 = 17, \qquad 8x = 16, \qquad \text{and} \qquad x = 2$$

are all equivalent equations since each has the same solution set, $\{2\}$. We use the addition and multiplication properties of equality to produce equivalent equations.

ADDITION AND MULTIPLICATION PROPERTIES OF EQUALITY

Addition Property of Equality

For all real numbers a, b, and c, the equations

$$a = b \qquad \text{and} \qquad a + c = b + c$$

are equivalent. (The same number may be added to both sides of an equation without changing the solution set.)

Multiplication Property of Equality

For all real numbers a, b, and c, where $c \neq 0$, the equations

$$a = b \qquad \text{and} \qquad ac = bc$$

are equivalent. (Both sides of an equation may be multiplied by the same nonzero number without changing the solution set.)

By the addition property, the same number may be added to both sides of an equation without affecting the solution set. By the multiplication property, both sides of an equation may be multiplied by the same nonzero number to produce an equivalent equation. Because subtraction and division are defined in terms of addition and multiplication, respectively, these properties can be extended: The same number may be subtracted from both sides of an equation, and both sides may be divided by the same nonzero number.

EXAMPLE 1
SOLVING A LINEAR EQUATION

Solve $4y - 2y - 5 = 4 + 6y + 3$.

First, combine terms separately on both sides of the equation to get

$$2y - 5 = 7 + 6y.$$

Next, use the addition property to get the terms with y on the same side of the equation and the remaining terms (the numbers) on the other side. One way to do this is first to add 5 to both sides.

$$2y - 5 + 5 = 7 + 6y + 5 \qquad \text{Add 5.}$$
$$2y = 12 + 6y$$

Now subtract $6y$ from both sides.

$$2y - 6y = 12 + 6y - 6y \qquad \text{Subtract } 6y.$$
$$-4y = 12$$

Finally, divide both sides by -4 to get just y on the left.

$$\frac{-4y}{-4} = \frac{12}{-4} \qquad \text{Divide by } -4.$$
$$y = -3$$

To be sure that -3 is the solution, check by substituting back into the *original* equation (not an intermediate one).

$$4y - 2y - 5 = 4 + 6y + 3 \qquad \text{Given equation}$$
$$4(-3) - 2(-3) - 5 = 4 + 6(-3) + 3 \quad ? \qquad \text{Let } y = -3.$$
$$-12 + 6 - 5 = 4 - 18 + 3 \quad ? \qquad \text{Multiply.}$$
$$-11 = -11 \qquad \text{True}$$

Since a true statement is obtained, -3 is the solution. The solution set is $\{-3\}$. ■

The steps used to solve a linear equation in one variable are as follows. (Not all equations require all of these steps.)

SOLVING A LINEAR EQUATION IN ONE VARIABLE

Step 1 **Clear fractions.** Eliminate any fractions by multiplying both sides by a common denominator.

Step 2 **Simplify each side separately.** Simplify each side of the equation as much as possible by using the distributive property to clear parentheses and by combining like terms as needed.

Step 3 **Isolate the variable terms on one side.** Use the addition property to get all terms with variables on one side of the equation and all numbers on the other.

Step 4 **Isolate the variable.** Use the multiplication property to get an equation with just the variable (with coefficient 1) on one side.

Step 5 **Check.** Check by substituting back into the original equation.

3 In Example 1 we did not use Step 1 and the distributive property in Step 2 as given above. Many other equations, however, will require one or both of these steps, as shown in the next examples.

EXAMPLE 2 **USING THE DISTRIBUTIVE PROPERTY TO SOLVE A LINEAR EQUATION**

Solve $2(k - 5) + 3k = k + 6$.

Since there are no fractions in this equation, Step 1 does not apply. Begin by using the distributive property to simplify and combine terms on the left side of the equation (Step 2).

$$2(k - 5) + 3k = k + 6$$

$$2k - 10 + 3k = k + 6 \qquad \text{Distributive property}$$

$$5k - 10 = k + 6 \qquad \text{Combine like terms.}$$

Next, add 10 to both sides (Step 3).

$$5k - 10 + 10 = k + 6 + 10 \qquad \text{Add 10.}$$

$$5k = k + 16$$

Now subtract k from both sides.

$$5k - k = k + 16 - k \qquad \text{Subtract } k.$$

$$4k = 16 \qquad \text{Combine like terms.}$$

The multiplication property of equality is used to get just k on the left. Divide both sides by 4 (Step 4).

$$\frac{4k}{4} = \frac{16}{4} \qquad \text{Divide by 4.}$$

$$k = 4$$

Check that the solution set is $\{4\}$ by substituting 4 for k in the original equation (Step 5). ■

In the rest of the examples in this section, we will not identify the steps by number.

4 When fractions or decimals appear as coefficients in equations, our work can be made easier if we multiply both sides of the equation by the least common denominator of all the fractions. This is an application of the multiplication property of equality, and it produces an equivalent equation with integer coefficients. The next examples illustrate this idea.

EXAMPLE 3 SOLVING A LINEAR EQUATION WITH FRACTIONS

Solve $\frac{y}{2} - \frac{3y}{5} = -1$.

Eliminate fractions by multiplying both sides of the equation by 10, since 10 is the least common denominator for 2 and 5.

$$10\left(\frac{y}{2} - \frac{3y}{5}\right) = 10(-1) \qquad \text{Multiply by 10.}$$

Use the distributive property on the left.

$$10\left(\frac{y}{2}\right) - 10\left(\frac{3y}{5}\right) = -10$$

$$5y - 6y = -10$$

$$-y = -10 \qquad \text{Combine like terms.}$$

Multiply both sides by -1 to get

$$-1(-y) = -1(-10)$$

$$y = 10.$$

(With this equation, it was not necessary to use the addition property.) Check by substituting 10 for y in the original equation.

$$\frac{y}{2} - \frac{3y}{5} = -1 \qquad \text{Original equation}$$

$$\frac{10}{2} - \frac{3(10)}{5} = -1 \quad ? \qquad \text{Let } y = 10.$$

$$5 - 6 = -1 \quad ?$$

$$-1 = -1 \qquad \text{True}$$

The solution checks, so the solution set is $\{10\}$. ■

EXAMPLE 4 SOLVING A LINEAR EQUATION WITH FRACTIONS

Solve $\frac{x+7}{6} + \frac{2x-8}{2} = -4$.

Start by eliminating the fractions. Multiply both sides by 6.

$$6\left[\frac{x+7}{6} + \frac{2x-8}{2}\right] = 6 \cdot (-4)$$

$$6\left(\frac{x+7}{6}\right) + 6\left(\frac{2x-8}{2}\right) = 6(-4)$$

$$x + 7 + 3(2x - 8) = -24$$

$$x + 7 + 6x - 24 = -24 \qquad \text{Distributive property}$$
$$7x - 17 = -24 \qquad \text{Combine terms.}$$
$$7x - 17 + 17 = -24 + 17 \qquad \text{Add 17.}$$
$$7x = -7$$
$$\frac{7x}{7} = \frac{-7}{7} \qquad \text{Divide by 7.}$$
$$x = -1$$

Check to see that $\{-1\}$ is the solution set. ■

PROBLEM SOLVING

In later sections we will solve problems involving interest rates and concentrations of solutions. These problems involve percents that are converted to decimal numbers. The equations that are used to solve such problems involve decimal coefficients. We can clear these decimals by multiplying by the largest power of 10 necessary to obtain integer coefficients. The next example shows how this is done. ■

EXAMPLE 5 **SOLVING A LINEAR EQUATION WITH DECIMALS**

Solve $.06x + .09(15 - x) = .07(15)$.

Since each decimal number is given in hundredths, multiply both sides of the equation by 100. (This is done by moving the decimal points two places to the right.)

$$.06x + .09(15 - x) = .07(15)$$
$$6x + 9(15 - x) = 7(15) \qquad \text{Multiply by 100.}$$
$$6x + 9(15) - 9x = 105 \qquad \text{Distributive property}$$
$$-3x + 135 = 105 \qquad \text{Combine like terms.}$$
$$-3x + 135 - 135 = 105 - 135 \qquad \text{Subtract 135.}$$
$$-3x = -30$$
$$\frac{-3x}{-3} = \frac{-30}{-3} \qquad \text{Divide by } -3.$$
$$x = 10$$

Check to verify that the solution set is $\{10\}$. ■

CAUTION When multiplying the term $.09(15 - x)$ by 100 in Example 5, do not multiply both .09 and $15 - x$ by 100. This step is not an application of the distributive property, but of the associative property. The correct procedure is

$$100[.09(15 - x)] = [100(.09)](15 - x) \qquad \text{Associative property}$$
$$= 9(15 - x). \qquad \text{Multiply.}$$

5 All the equations above had a solution set containing one element; for example, $2x + 1 = 13$ has solution set $\{6\}$, containing only the single number 6. An equation that has a finite number of elements in its solution set is a **conditional equation.** Sometimes an equation has no solutions. Such an equation is a **contradiction** and its solution set is $\emptyset$. It is also possible for an equation to have an infinite number of solutions. An equation that is satisfied by every number for which both sides are defined is called an **identity.** The next example shows how to recognize these types of equations.

EXAMPLE 6

IDENTIFYING CONDITIONAL EQUATIONS, IDENTITIES, AND CONTRADICTIONS

Solve each equation. Decide whether it is a conditional equation, an identity, or a contradiction.

(a) $5x - 9 = 4(x - 3)$

Work as in the previous examples.

$$5x - 9 = 4(x - 3)$$

$$5x - 9 = 4x - 12 \quad \text{Distributive property}$$

$$5x - 9 - 4x = 4x - 12 - 4x \quad \text{Subtract } 4x.$$

$$x - 9 = -12 \quad \text{Combine like terms.}$$

$$x - 9 + 9 = -12 + 9 \quad \text{Add 9.}$$

$$x = -3$$

The solution set, $\{-3\}$, has one element, so $5x - 9 = 4(x - 3)$ is a *conditional equation*.

(b) $5x - 15 = 5(x - 3)$

Use the distributive property on the right side.

$$5x - 15 = 5x - 15$$

Both sides of the equation are *exactly the same,* so any real number would make the equation true. For this reason, the solution set is the set of all real numbers, and the equation $5x - 15 = 5(x - 3)$ is an *identity*.

(c) $5x - 15 = 5(x - 4)$

Use the distributive property.

$$5x - 15 = 5x - 20 \quad \text{Distributive property}$$

$$5x - 15 - 5x = 5x - 20 - 5x \quad \text{Subtract } 5x.$$

$$-15 = -20 \quad \text{False}$$

Since the result, $-15 = -20$, is *false,* the equation has no solution. The solution set is $\emptyset$. The equation $5x - 15 = 5(x - 4)$ is a *contradiction*. ■

2.1 EXERCISES

Solve each of the following equations. See Examples 1 and 2.

1. $2k + 6 = 12$ **2.** $5m - 4 = 16$ **3.** $5 - 8k = -19$ **4.** $4 - 2m = 10$

5. $3 - 2r = 9$ **6.** $-5z + 2 = 7$ **7.** $-9y - 4 = 14$ **8.** $-7p - 3 = -17$

9. $-3 + 4z = -11$

10. $-1 + 2m = -11$

11. $-4 - 3p = 7$

12. $2x + 7 - x = 4x - 2$

13. $7z - 5z + 3 = z - 4$

14. $12a + 7 = 7a - 2$

15. $7m - 2m + 4 - 5 = 3m - 5 + 6$

16. $12z - 15z - 8 + 6 = 4z + 6 - 1$

17. $11p - 9 + 8p - 7 + 14p = 12p + 9p + 4$

18. $3(x + 5) = 2x - 1$

19. $2(k - 4) = 5k + 2$

20. $4(r - 1) + 2(r + 3) = 6$

21. $3(2t + 1) - 2(t - 2) = 5$

22. $6s - 3(5s + 2) = 4 - 5s$

23. $2y + 3(y - 4) = 2(y - 3)$

24. $6x - 4(3 - 2x) = 5(x - 4) - 10$

25. $4k - 3(4 - 2k) = 2(k - 3) + 6k + 2$

26. $-[z - (4z + 2)] = 2 + (2z + 7)$

27. $5y - (8 - y) = 2[-4 - (3 + 5y) - 13]$

28. $-9m - (4 + 3m) = -(2m - 1) - 5$

---

29. Explain in your own words the steps used to solve a linear equation. (See Objective 2.)*

30. Suppose that in solving the equation

$$\frac{1}{3}y + \frac{1}{2}y = \frac{1}{6}y,$$

you begin by multiplying both sides by 12, rather than the *least* common denominator, 6. Should you get the correct solution anyway? Explain. (See Objective 4.)

Solve the following equations having fractions or decimals as coefficients. See Examples 3–5.

31. $-\frac{5}{9}k = 2$

32. $\frac{3}{11}z = -5$

33. $\frac{6}{5}x = -1$

34. $-\frac{7}{8}r = 6$

35. $\frac{m}{2} + \frac{m}{3} = 5$

36. $\frac{y}{5} - \frac{y}{4} = 1$

37. $\frac{3k}{5} - \frac{2k}{3} = 1$

38. $\frac{8r}{3} - \frac{6r}{5} = 22$

39. $\frac{m - 2}{3} + \frac{m}{4} = \frac{1}{2}$

40. $\frac{y - 8}{5} + \frac{y}{3} = -\frac{8}{5}$

41. $.05y + .12(y + 5000) = 940$

42. $.09k + .13(k + 300) = 61$

43. $.02(50) + .08r = .04(50 + r)$

44. $.20(14{,}000) + .14t = .18(14{,}000 + t)$

45. $.05x + .10(200 - x) = .45x$

46. $.08x + .12(260 - x) = .48x$

---

47. The equation $x^2 + 2 = x^2 + 2$ is called a(n) ____________, because it has infinitely many solutions. The equation $x + 1 = x + 2$ is called a(n) ____________, because it has no solutions. (See Objective 5.)

48. Which one of the following is a conditional equation? (See Objective 5.)
(a) $2x + 1 = 3$ **(b)** $x = 3x - 2x$
(c) $2(x + 2) = 2x + 2$
(d) $5x - 3 = 4x + x - 5 + 2$

Decide whether the following equations are conditional, identities, *or* contradictions. *Give the solution set of each. See Example 6.*

49. $9k + 4 - 3k = 2(3k + 4) - 4$

50. $-7m + 8 + 4m = -3(m - 3) - 1$

51. $-2p + 5p - 9 = 3(p - 4) - 5$

52. $-6k + 2k - 11 = -2(2k - 3) + 4$

53. $-11m + 4(m - 3) + 6m = 4m - 12$

54. $3p - 5(p + 4) + 9 = -11 + 15p$

---

*For help with exercises, read the discussion of Objective 2 in this section.

55. $7[2 - (3 + 4r)] - 2r = -9 + 2(1 - 15r)$

56. $4[6 - (1 + 2m)] + 10m = 2(10 - 3m) + 8m$

57. $-4[6 - (-2 + 3q)] = 3(7 + 4q)$

58. $-3[-5 - (-9 + 2r)] = 2(3r - 1)$

Find a value of k so that the following pairs of equations are equivalent equations.

59. $6x - 5 = k$ and $2x + 1 = 15$

60. $3y + k = 11$ and $5y - 8 = 22$

Find a value of k so that each of the following equations has solution set $\{-4\}$.

61. $3x + 6 = k + 8$

62. $8x - 9 = k - 7$

63. $\dfrac{2x + 1}{3x - 4} = k$

64. $\dfrac{9 - 7x}{3 + 2x} = k$

By studying the winning times in the 500-meter speed-skating event at Olympic games back to 1900, it was found that the winning times for men were given by the equation

$$y_m = 46.338 - .097x,$$

where y_m is the time in seconds needed to win for men, with x the Olympic year and x = 0 corresponding to 1900. (For example, 1992 winning times would be estimated by replacing x with 1992 − 1900 = 92.) The corresponding equation for women is

$$y_w = 57.484 - .196x.$$

65. Find the year in which **(a)** $y_m = 42.458$; **(b)** $y_w = 44.940$.

66. Let $y_m = y_w$ and estimate the Olympic year in which women's times will surpass men's times.

Decide which of the following pairs of equations are equivalent equations.

67. $5x = 10$ and $\dfrac{5x}{x + 2} = \dfrac{10}{x + 2}$

68. $x + 1 = 9$ and $\dfrac{x + 1}{8} = \dfrac{9}{8}$

69. $y = -3$ and $\dfrac{y}{y + 3} = \dfrac{-3}{y + 3}$

70. $m = 1$ and $\dfrac{m + 1}{m - 1} = \dfrac{2}{m - 1}$

71. $k = 4$ and $k^2 = 16$

72. $p^2 = 36$ and $p = 6$

PREVIEW EXERCISES

Most of the exercise sets in the rest of this book end with a few "preview exercises." These exercises are designed to help you review ideas needed for the next section in the chapter. If you need help with these preview exercises, look in the section indicated each time.

Evaluate each expression, using the given values. See Section 1.3.

73. $2L + 2W$; $L = 10$, $W = 6$

74. rt; $r = 15$, $t = 9$

75. $\dfrac{1}{3}Bh$; $B = 27$, $h = 10$

76. prt; $p = 8000$, $r = .12$, $t = 2$

77. $\dfrac{5}{9}(F - 32)$; $F = 212$

78. $\dfrac{9}{5}C + 32$; $C = 100$

79. $\dfrac{1}{2}(B + b)h$; $B = 9$, $b = 4$, $h = 2$

80. $\dfrac{1}{2}bh$; $b = 21$, $h = 5$

2.2 FORMULAS AND TOPICS FROM GEOMETRY

OBJECTIVES

1 SOLVE A FORMULA FOR A SPECIFIED VARIABLE.

2 SOLVE APPLIED PROBLEMS USING FORMULAS.

3 SOLVE PROBLEMS ABOUT ANGLE MEASURES.

FOCUS ON PROBLEM SOLVING

The formula $A = \dfrac{24f}{b(p + 1)}$ gives the approximate annual interest rate for a consumer loan paid off with monthly payments. Here f is the finance charge on the loan, p is the number of payments, and b is the original amount of the loan.

Find the approximate annual interest rate for an installment loan to be repaid in 24 monthly installments. The finance charge on the loan is \$200 and the original loan balance is \$1920.

Formulas are an important part of applied mathematics, since they provide relationships among several quantities in a particular setting. Formulas are found in geometry, the mathematics of finance, branches of science, and so on. This problem is Exercise 49 in the exercises for this section. After working through this section, you should be able to solve this problem.

1 The solution of a problem in algebra often depends on the use of a mathematical statement or **formula** in which more than one letter is used to express a relationship. Examples of formulas are

$$d = rt, \qquad I = prt, \qquad \text{and} \qquad P = 2L + 2W.$$

A list of the formulas used in this book is given in Appendices A and B. Metric units are used in some of the exercises, although a working knowledge of the metric system is not needed to work the problems.

PROBLEM SOLVING

In some applications, the necessary formula is solved for one of its variables, which may not be the unknown number that must be found. The following examples show how to solve a formula for any one of its variables. This process is called **solving for a specified variable.** Notice how the steps used in these examples are very similar to those used in solving a linear equation. Keep in mind that when you are solving for a specified variable, treat that variable as if it were the only one, and treat all other variables as if they were numbers. ■

EXAMPLE 1 SOLVING FOR A SPECIFIED VARIABLE

Solve the formula $I = prt$ for t.

This formula gives the amount of simple interest, I, in terms of the principal (the amount of money deposited), p, the yearly rate of interest, r, and time in years, t. Solve this formula for t by assuming that I, p, and r are constants (having a fixed value) and that t is the variable. Then use the properties of the previous section as follows.

$$I = prt$$

$$I = (pr)t \qquad \text{Associative property}$$

$$\frac{I}{pr} = \frac{(pr)t}{pr} \qquad \text{Divide by } pr.$$

$$\frac{I}{pr} = t$$

The result is a formula for t, time in years. ■

PROBLEM SOLVING

While the process of solving for a specified variable uses the same steps as solving a linear equation from Section 2.1, the following additional suggestions may be helpful.

SOLVING FOR A SPECIFIED VARIABLE

Step 1 Use the addition or multiplication properties as necessary to get all terms containing the specified variable on one side of the equation.

Step 2 All terms not containing the specified variable should be on the other side of the equation.

Step 3 If necessary, use the distributive property to write the side with the specified variable as the product of that variable and a sum of terms.

In general, follow the steps used for solving linear equations given in Section 2.1. ■

EXAMPLE 2 **SOLVING FOR A SPECIFIED VARIABLE**

Solve the formula $P = 2L + 2W$ for W.

This formula gives the relationship between the perimeter of a rectangle, P, the length of the rectangle, L, and the width of the rectangle, W. See Figure 2.1.

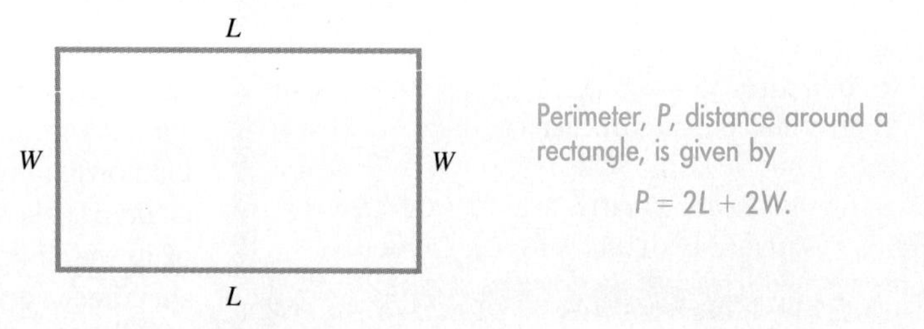

FIGURE 2.1

Solve the formula for W by getting W alone on one side of the equals sign. To begin, subtract $2L$ from both sides.

$$P = 2L + 2W$$

$$P - 2L = 2L + 2W - 2L \quad \text{Subtract } 2L.$$

$$P - 2L = 2W$$

Divide both sides by 2.

$$\frac{P - 2L}{2} = \frac{2W}{2}$$

$$\frac{P - 2L}{2} = W \quad ■$$

CAUTION Do not try to express the result in Example 2 by eliminating the 2 in the numerator and denominator. The 2 in the numerator is not a factor of the entire numerator. Expressing fractions in lowest terms will be explained in Chapter 4.

EXAMPLE 3
SOLVING FOR A SPECIFIED VARIABLE

Given the surface area, height, and width of a rectangular solid, write a formula for the length.

The formula for the surface area of a rectangular solid is found in Appendix A. It is

$$A = 2HW + 2LW + 2LH,$$

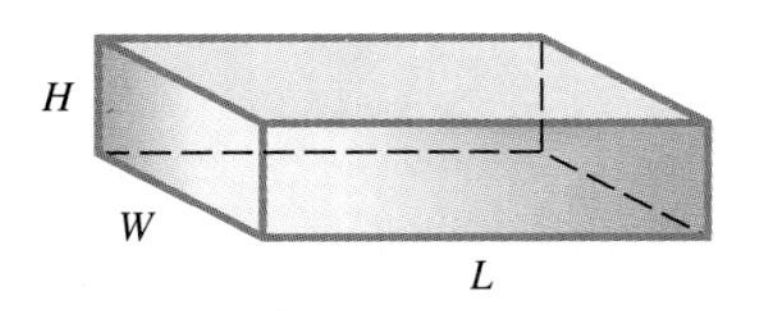

FIGURE 2.2

where A represents the surface area, H represents the height, W represents the width, and L represents the length. See Figure 2.2. To solve for L, write the equation so that only the terms involving L appear on the right side.

$$A = 2HW + 2LW + 2LH$$

$$A - 2HW = 2LW + 2LH \qquad \text{Subtract } \mathbf{2HW}.$$

Next, use the distributive property on the right to write $2LW + 2LH$ so that L, the variable for which we are solving, is a factor.

$$A - 2HW = L(2W + 2H) \qquad \text{Distributive property}$$

Finally, divide both sides by the coefficient of L, which is $2W + 2H$.

$$\frac{A - 2HW}{2W + 2H} = L \quad \text{or} \quad L = \frac{A - 2HW}{2W + 2H} \qquad ■$$

CAUTION The most common error in working a problem like Example 3 is not using the distributive property correctly. We must write the expression so that the variable for which we are solving is a *factor,* so that we can divide by its coefficient in the final step.

2 The next examples show how applications can be solved using formulas.

PROBLEM SOLVING

In general, when a formula applies to a problem, substitute the given values into the formula, so that only one variable remains, and then solve for that variable. In some cases it may be convenient to first use the methods of Examples 1–3 to solve for the particular unknown variable before substituting the given values. ■

EXAMPLE 4
FINDING AVERAGE SPEED

Janet Branson found that on the average it took her 3/4 hour each day to drive a distance of 15 miles to work. What was her average speed?

The formula needed for this problem is the distance formula, $d = rt$, where d represents distance traveled, r the rate, and t the time. Find the speed r, by solving $d = rt$ for r.

$$d = rt$$

$$\frac{d}{t} = \frac{rt}{t} \qquad \text{Divide by } t.$$

$$\frac{d}{t} = r$$

Now find r by substituting the given values of d and t into this formula.

$$r = \frac{d}{t}$$

$$r = \frac{15}{\frac{3}{4}} \qquad \text{Let } d = 15,\ t = \frac{3}{4}.$$

$$r = 15 \div \frac{3}{4} = 15 \cdot \frac{4}{3}$$

$$r = 20$$

Her average speed was 20 miles per hour. ■

When a consumer loan is paid off ahead of schedule, the finance charge is smaller than if the loan were paid off over its scheduled life. By one method, called *the rule of 78,* the amount of *unearned interest* (finance charge that need not be paid) is given by

$$u = f \cdot \frac{k(k+1)}{n(n+1)},$$

where u is the amount of unearned interest (money saved) when a loan scheduled to run n payments is paid off k payments ahead of schedule. The total scheduled finance charge is f.

Actually, if uniform monthly payments are made, the rule results in more interest in the early months of the loan than is strictly proper. Thus any refund for early repayment would be less by this rule than it really should be. However, the method is still often used, and its accuracy is acceptable for fairly short-term consumer loans. (With tables and calculators readily available, very accurate interest and principal allocations of loan payments are easily determined.)

The next example illustrates the use of the rule of 78.

EXAMPLE 5
USING THE RULE OF 78

Juan Ortega is scheduled to pay off a loan in 36 monthly payments. If the total finance charge is \$360, and the loan is paid off after 24 payments (that is, 12 payments ahead of schedule), find the unearned interest, u.

A calculator is helpful in this problem. Here $f = 360$, $n = 36$, and $k = 12$. Use the formula above.

$$u = f \cdot \frac{k(k+1)}{n(n+1)} = 360 \cdot \frac{12(13)}{36(37)} = 42.16$$

A total of \$42.16 of the \$360 finance charge need not be paid. The total finance charge for this loan is \$360 − \$42.16 = \$317.84. ■

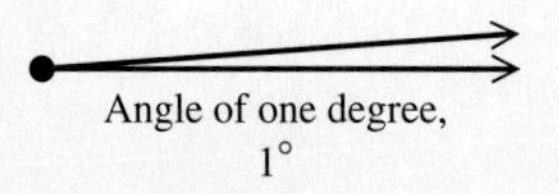

Angle of one degree, 1°

FIGURE 2.3

3 In this section and the next, we will look at some properties of angles and their measures that lend themselves to algebraic interpretation. Recall that a basic unit of measure of angles is the **degree;** an angle that measures one degree (1°) is 1/360 of a complete revolution. See Figure 2.3.

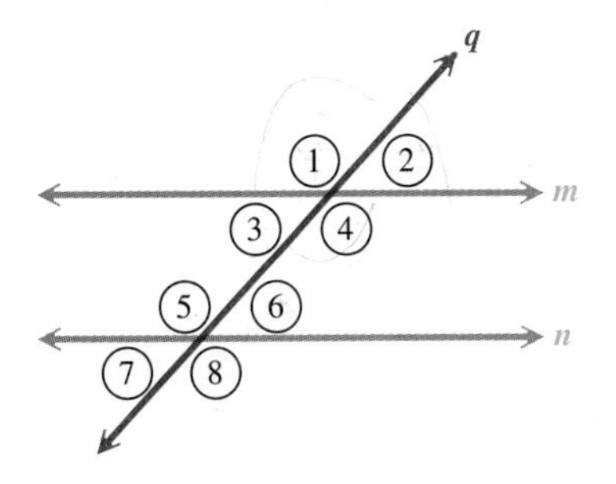

FIGURE 2.4

Parallel lines are lines that lie in the same plane and do not intersect. Figure 2.4 shows parallel lines m and n. When a line q intersects the two parallel lines, q is called a **transversal.** In Figure 2.4, the transversal intersecting the parallel lines forms eight angles, indicated by the circled numbers.

Pairs of angles like ① and ④, ② and ③, ⑤ and ⑧, and ⑥ and ⑦, are called **vertical angles.** They lie "opposite" each other. It is shown in geometry that *vertical angles have the same measure.*

Angles ① through ⑧ possess some special properties regarding their degree measures. The following chart gives their names with respect to each other, and the rules regarding their measures.

Name	Sketch	Rules
Alternate interior angles	⑤ ④ (also ③ and ⑥)	Angle measures are equal.
Alternate exterior angles	① ⑧ (also ② and ⑦)	Angle measures are equal.
Interior angles on same side of transversal	④ ⑥ (also ③ and ⑤)	Angle measures add to 180°.
Corresponding angles	② ⑥ (also ① and ⑤, ③ and ⑦, ④ and ⑧)	Angle measures are equal.

Use the information in the chart to solve the next problem.

EXAMPLE 6
SOLVING A PROBLEM ABOUT ANGLE MEASURES

Find x, given that lines m and n are parallel. Then find the measure of each marked angle.

In Figure 2.5, the marked angles are alternate exterior angles, which are equal. This gives

$$3x + 2 = 5x - 40$$
$$42 = 2x$$
$$21 = x.$$

One angle has a measure of $3x + 2 = 3 \cdot 21 + 2 = 65$ degrees, and the other has a measure of $5x - 40 = 5 \cdot 21 - 40 = 65$ degrees. ■

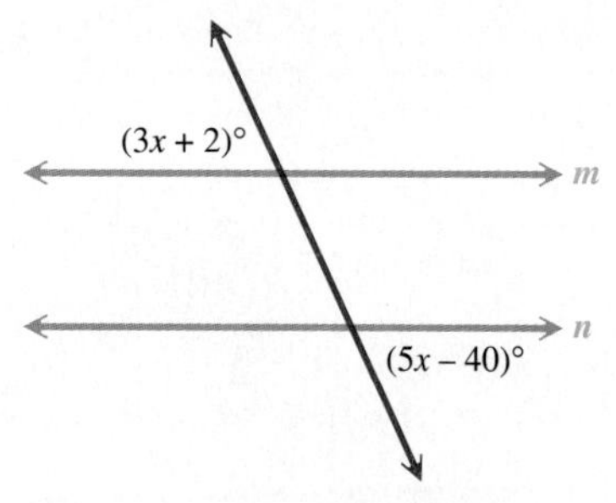

FIGURE 2.5

2.2 EXERCISES

Solve each formula for the specified variable. See Examples 1–3.

1. $d = rt$; for r (distance)
2. $I = prt$; for r (simple interest)
3. $A = bh$; for b (area of a parallelogram)
4. $P = 2L + 2W$; for L (perimeter of a rectangle)
5. $P = a + b + c$; for a (perimeter of a triangle)
6. $V = LWH$; for W (volume of a rectangular solid)
7. $A = \frac{1}{2}bh$; for h (area of a triangle)
8. $C = 2\pi r$; for r (circumference of a circle)
9. $S = 2\pi rh + 2\pi r^2$; for h (surface area of a right circular cylinder)
10. $A = \frac{1}{2}(B + b)h$; for B (area of a trapezoid)
11. $\mathrm{C} = \frac{5}{9}(\mathrm{F} - 32)$; for F (Fahrenheit to Celsius)
12. $\mathrm{F} = \frac{9}{5}\mathrm{C} + 32$; for C (Celsius to Fahrenheit)
13. $A = 2HW + 2LW + 2LH$; for H (surface area of a rectangular solid)
14. $V = \frac{1}{3}Bh$; for h (volume of a right pyramid)

---

15. Refer to Example 3. Suppose that the formula is "solved for L" as follows:

$$A = 2HW + 2LW + 2LH$$
$$A - 2LW - 2HW = 2LH$$
$$\frac{A - 2LW - 2HW}{2H} = L.$$

While there are no algebraic errors here, what is wrong with the final line, if we are interested in solving for L? (See Objective 1.)

16. When a formula is solved for a particular variable, several different equivalent forms may be possible. If we solve $A = (1/2)bh$ for h, one possible correct answer is

$$h = \frac{2A}{b}.$$

Which one of the following is *not* equivalent to this?

(a) $h = 2\left(\frac{A}{b}\right)$ **(b)** $h = 2A\left(\frac{1}{b}\right)$

(c) $h = \frac{A}{\frac{1}{2}b}$ **(d)** $h = \frac{\frac{1}{2}A}{b}$

17. Write an explanation, in step-by-step form, of the procedure you would use to solve $P = 2L + 2W$ for L. (See Objective 1.)

18. The formula for the area of a trapezoid is $A = (1/2)(B + b)h$. Suppose that you know the values of B, b, and h. Write *in words* the procedure you would use to find A. It may begin as follows: ''To find the area of this trapezoid, I would first add. . . .'' (See Objective 1.)

For each of the following, find the appropriate formula in Appendix A. Then solve the formula for the required variable. See Example 3.

19. If the measures of the area and height of a triangle are known, find a formula for the base.

20. Given the volume, length, and height of a box, write a formula for the width.

21. If rate, time, and simple interest of an investment are known, find a formula for the principal.

22. If two sides and the perimeter of a triangle are known, write a formula for the third side.

23. If the surface area and radius of a right circular cylinder are known, write a formula for the height.

24. The height, area, and short base of a trapezoid are known. Write a formula for the long base.

For each of the following,

(a) select the appropriate formula;
(b) solve the formula for the required variable;
(c) find the value of the required variable.

(In Exercises 33–36, part (b) is not necessary.) See Example 4.

25. Faye Korn traveled from Kansas City to Louisville, a distance of 520 miles, in 10 hours. Find her rate in miles per hour.

26. The distance from Melbourne to London is 10,500 miles. If a jet averages 500 miles per hour between the two cities, what is its travel time in hours?

27. The perimeter of a rectangular swimming pool is 36 meters, and the width is 3.5 meters. Find the length of the pool.

28. Two sides of a triangle measure 4.6 and 12.2 centimeters. The perimeter is 28.4 centimeters. Find the measure of the third side.

29. A newspaper recycling collection bin is in the shape of a box, 1.5 feet wide and 5 feet long. If the volume of the bin is 75 cubic feet, find the height.

30. A cord of wood contains 128 cubic feet of wood. A stack of wood is 4 feet wide and 4 feet high. How long must it be if it contains one cord?

31. If Kyung Ho deposits \$1200 at 12% (or .12) simple annual interest, how long will it take for the deposit to earn \$504?

32. Find the simple interest rate that Francesco Castellucio is getting, if a principal of \$30,000 earns \$7800 interest in 4 years.

33. On March 12, 1990, the high temperature in Morganza, Louisiana, was 41 degrees Fahrenheit. Find the corresponding Celsius temperature.

34. Water boils at 212 degrees Fahrenheit. Find the corresponding Celsius temperature.

35. If a child has a fever of 40 degrees Celsius, what is the child's temperature in Fahrenheit?

36. The lowest temperature ever recorded in Arizona was -40 degrees Celsius, on January 7, 1971. What was the corresponding Fahrenheit temperature?

37. A circle has a circumference of 30π meters. Find the radius of the circle.

38. The circumference of a circle is 120π meters. Find the radius of the circle.

39. The surface area of a can is 32π square inches. The radius of the can is 2 inches. Find the height of the can.

40. A can has a surface area of 12π square inches. The radius of the can is 1 inch. Find the height of the can.

For Exercises 41–44 assume 365 *days in a year.*

41. Find the interest on \$1574.35 at 12.64% (or .1264) for 179 days.

42. Find the amount of principal if an interest rate of 13.94% (or .1394) produces \$174.69 interest in 84 days.

43. Find the number of days (rounding to the nearest day) if \$5496.11 produces \$273.13 interest at 11.48% (or .1148).

44. Find the rate of interest (to the nearest hundredth of a percent) if \$11,428.54 produces \$1103.56 interest in 276 days.

Refer to the formula for the rule of 78 to solve each of the following. See Example 5.

45. Rhonda Alessi bought a new Ford and agreed to pay it off in 36 monthly payments. The total finance charge is \$700. Find the unearned interest if she pays the loan off 4 payments ahead of schedule.

46. Paul Lorio bought a car and agreed to pay in 36 monthly payments. The total finance charge on the loan was \$600. With 12 payments remaining, Paul decided to pay the loan in full. Find the amount of unearned interest.

47. The finance charge on a loan taken out by Vic Denicola is \$380.50. If there were 24 equal monthly installments needed to repay the loan, and the loan is paid in full with 8 months remaining, find the amount of unearned interest.

48. Adrian Ortega is scheduled to repay a loan in 24 equal monthly installments. The total finance charge on the loan is \$450. With 9 payments remaining he decides to repay the loan in full. Find the amount of unearned interest.

The formula

$$A = \frac{24f}{b(p + 1)}$$

gives the approximate annual interest rate for a consumer loan paid off with monthly payments. Here f is the finance charge on the loan, p is the number of payments, and b is the original amount of the loan.

Solve the following problems using the formula above.

49. Find the approximate annual interest rate for an installment loan to be repaid in 24 monthly installments. The finance charge on the loan is \$200, and the original loan balance is \$1920.

50. Find the approximate annual interest rate for an automobile loan to be repaid in 36 monthly installments. The finance charge on the loan is \$740 and the amount financed is \$3600.

In each of the following exercises, find the value of x, and then find the measure of each marked angle. In Exercises 53–56, assume that lines m and n are parallel. See Example 6.

51.

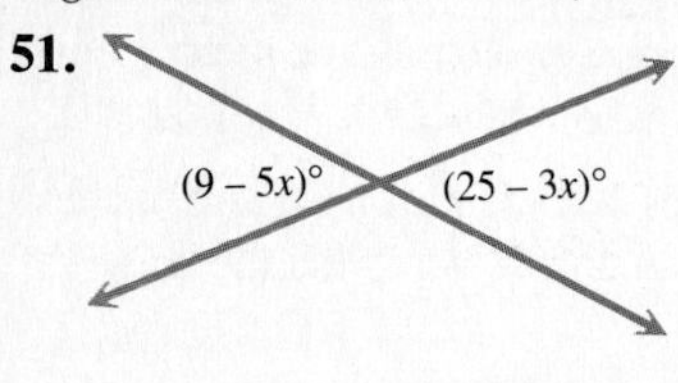

52.

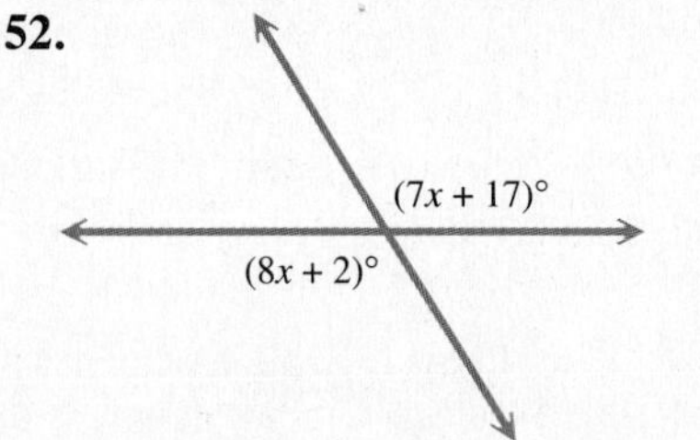

53.

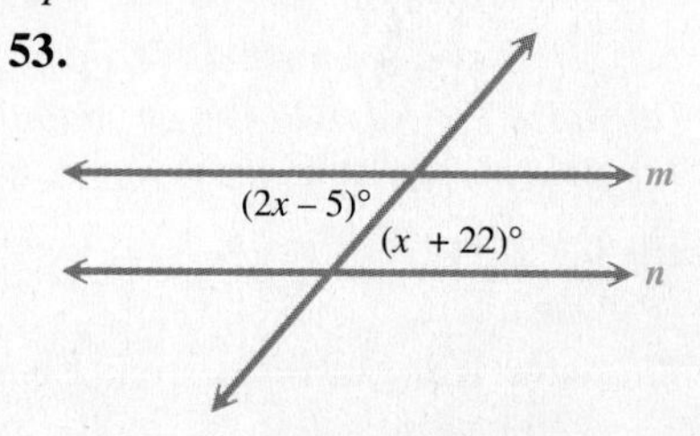

54.

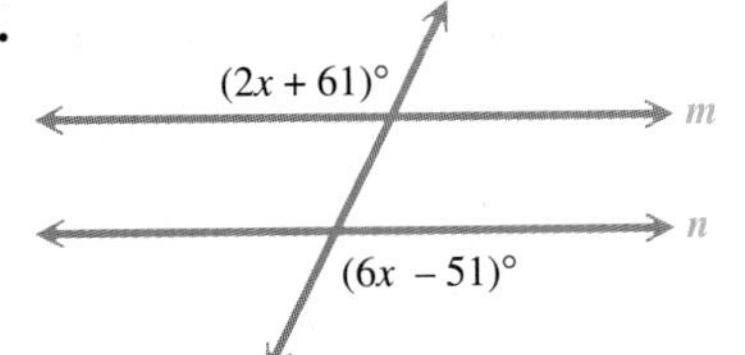

55.

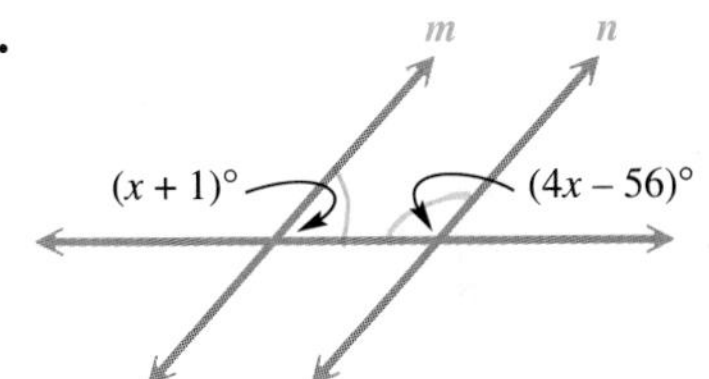

56.

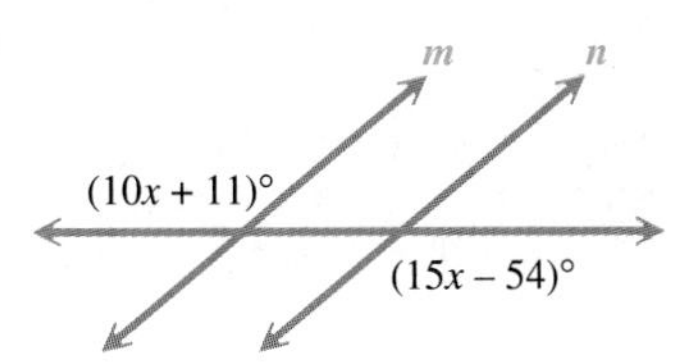

PREVIEW EXERCISES

Use the order of operations to evaluate the following. See Section 1.3.

57. 9 is added to the product of 8 and -2.

58. The product of -1 and 10 is divided by -5.

59. Twice 11 is subtracted from -3.

60. Half of -16 is added to the reciprocal of 1/2.

61. The product of -2 and 4 is added to the product of -9 and -3.

62. The sum of 6 and -9 is divided by the product of 2 and -3.

2.3 APPLICATIONS OF LINEAR EQUATIONS

OBJECTIVES

1 TRANSLATE FROM WORD EXPRESSIONS TO MATHEMATICAL EXPRESSIONS.

2 WRITE EQUATIONS FROM GIVEN INFORMATION.

3 SOLVE PROBLEMS ABOUT UNKNOWN NUMERICAL QUANTITIES.

4 SOLVE PROBLEMS ABOUT PERCENTS, SIMPLE INTEREST, AND MIXTURE.

5 SOLVE PROBLEMS ABOUT ANGLES: SUPPLEMENTARY, COMPLEMENTARY, AND SUMS OF ANGLES IN TRIANGLES.

FOCUS ON PROBLEM SOLVING

A medicated first aid spray on the market is 78% alcohol by volume. If the manufacturer has 50 liters of the spray containing 70% alcohol, how much pure alcohol should be added so that the final mixture is the required 78% alcohol?

This problem is one of a group of typical problems that can be solved using linear equations. Remember that multiplying the rate of concentration by the number of units of the solution gives the number of units of pure substance (in this case, alcohol). Keeping in mind that the amount of pure alcohol in the two concentrations will equal the amount of pure alcohol in the final mixture, a linear equation can be written so that the problem can be solved. This problem is Exercise 33 in the exercises for this section. After working through this section, you should be able to solve this problem.

1 When algebra is used to solve practical applications, it is necessary to translate the verbal statements of the problems into mathematical statements.

PROBLEM SOLVING

Usually there are key words and phrases in the verbal problem that translate into mathematical expressions involving the operations of addition, subtraction, multiplication, and division. Translations of some commonly used expressions are listed next.

TRANSLATION FROM WORDS TO MATHEMATICAL EXPRESSIONS

Verbal expression	*Mathematical expression*
Addition	
The sum of a number and 7	$x + 7$
6 more than a number	$x + 6$
3 plus 8	$3 + 8$
24 added to a number	$x + 24$
A number increased by 5	$x + 5$
The sum of two numbers	$x + y$
Subtraction	
2 less than a number	$x - 2$
12 minus a number	$12 - x$
A number decreased by 12	$x - 12$
The difference between two numbers	$x - y$
A number subtracted from 10	$10 - x$
Multiplication	
16 times a number	$16x$
Some number multiplied by 6	$6x$
2/3 of some number (used only with fractions and percent)	$\frac{2}{3}x$
Twice (2 times) a number	$2x$
The product of two numbers	xy
Division	
The quotient of 8 and some number	$\frac{8}{x}$
A number divided by 13	$\frac{x}{13}$
The ratio of two numbers or the quotient of two numbers	$\frac{x}{y}$ ■

CAUTION Because subtraction and division are not commutative operations, it is important to correctly translate expressions involving them. For example, "2 less than a number" is translated as $x - 2$, *not* $2 - x$. "A number subtracted from 10" is expressed as $10 - x$, not $x - 10$. For division, it is understood that the number doing the dividing is the denominator, and the number that is divided is the numerator. For example, "a number divided by 13" and "13 divided into x" both translate as $x/13$. Similarly, "the quotient of x and y" is translated as x/y.

2 The symbol for equality, =, is often indicated by the word "is." In fact, since equal mathematical expressions represent different names for the same number, any words that indicate the idea of "sameness" indicate translation to =.

EXAMPLE 1
TRANSLATING SENTENCES INTO EQUATIONS

Translate the following verbal sentences into equations.

VERBAL SENTENCE	EQUATION
Twice a number, decreased by 3, is 42.	$2x - 3 = 42$
If the product of a number and 12 is decreased by 7, the result is 105.	$12x - 7 = 105$
A number divided by the sum of 4 and the number is 28.	$\dfrac{x}{4 + x} = 28$
The quotient of a number and 4, plus the number, is 10.	$\dfrac{x}{4} + x = 10$ ■

PROBLEM SOLVING

Throughout this book we will be examining different types of applications. While there is no one method that will allow us to solve all types of applied problems, there are some guidelines that are helpful. They are listed below.

SOLVING AN APPLIED PROBLEM

Step 1 **Determine what you are asked to find.**
Read the problem carefully. Decide what is given and what must be found. Choose a variable and write down exactly what it represents.

Step 2 **Write down any other pertinent information.**
If there are other unknown quantities, express them using the variable. Draw figures or diagrams if they apply.

Step 3 **Write an equation.**
Write an equation expressing the relationships among the quantities given in the problem.

Step 4 **Solve the equation.**
Use the methods of the earlier sections to solve the equation.

Step 5 **Answer the question(s) of the problem.**
Re-read the problem and make sure that you answer the question or questions posed. In some cases, you will need to give more than just the solution of the equation.

Step 6 **Check.**
Check your solution by using the original words of the problem. Be sure that your answer makes sense. ■

3 The next example illustrates the use of the six steps outlined for problem solving.

EXAMPLE 2
FINDING UNKNOWN NUMERICAL QUANTITIES

The Perry brothers, Jim and Gaylord, were two outstanding pitchers in the major leagues during the past few decades. Together, they won 529 games. Gaylord won 99 games more than Jim. How many games did each brother win?

Step 1 We are asked to find the number of games each brother won. We must choose a variable to represent the number of wins of one of the men.

Let j = the number of wins for Jim.

Step 2 We must also find the number of wins for Gaylord. Since he won 99 games more than Jim,

let $j + 99$ = Gaylord's number of wins.

Step 3 The sum of the numbers of wins is 529, so we can now write an equation.

$$\underbrace{j}_{\text{Jim's wins}} + \underbrace{(j + 99)}_{\text{Gaylord's wins}} = \underbrace{529}_{529}$$

Step 4 Solve the equation.

$$j + (j + 99) = 529$$
$$2j + 99 = 529 \quad \text{Combine like terms.}$$
$$2j = 430 \quad \text{Subtract 99.}$$
$$j = 215 \quad \text{Divide by 2.}$$

Step 5 Since j represents the number of Jim's wins, Jim won 215 games. Gaylord won $j + 99 = 215 + 99 = 314$ games.

Step 6 314 is 99 more than 215, and the sum of 314 and 215 is 529.

The words of the problem are satisfied, and our solution checks. ■

CAUTION A common error in solving applied problems is forgetting to answer all the questions asked in the problem. In Example 2, we were asked for the number of wins for *each* brother, so there was an extra step at the end in order to find Gaylord's number.

4 The next few examples in this section involve percent, which means per one hundred. For example, 17% means 17/100, or .17, and 109% means 1.09.

(We will not number the steps in the following examples; see if you can identify them.)

EXAMPLE 3
FINDING THE AMOUNT OF INCOME TAX

This year Karen Harr's income tax of $2127.50 is 15% more than she paid last year. Find last year's tax.

Let x = the amount of last year's tax.

This year, the tax has increased by 15% of last year's tax. The increase is 15% of x, or $.15x$. Now write an equation.

Last year's tax	+	15% increase	=	this year's tax
↓	↓	↓	↓	↓
x	+	$.15x$	=	2127.50

Solve this equation. Since $x = 1x$, the sum on the left is $x + .15x = 1x + .15x = 1.15x$, giving $1.15x = 2127.50$. Divide both sides by 1.15 to get $x = 1850$. The tax last year was $1850. ■

CAUTION Watch out for two common errors that occur in solving problems like the one in Example 3. First, do not try to find 15% of this year's tax and subtract that amount. The 15% increase is in relationship to *last* year's tax. Second, avoid writing the equation incorrectly. It would be *wrong* to write the equation as

$$x + .15 = 2127.50.$$

.15 must be multiplied by x, as shown in the example.

The next example shows how to solve a mixture problem involving different concentrations. Notice how a sketch and a table are used.

EXAMPLE 4
SOLVING A MIXTURE PROBLEM

A chemist must mix 8 liters of a 40% solution of potassium chloride with some 70% solution to get a mixture that is a 50% solution. How much of the 70% solution should be used?

The information in the problem is illustrated in Figure 2.6.

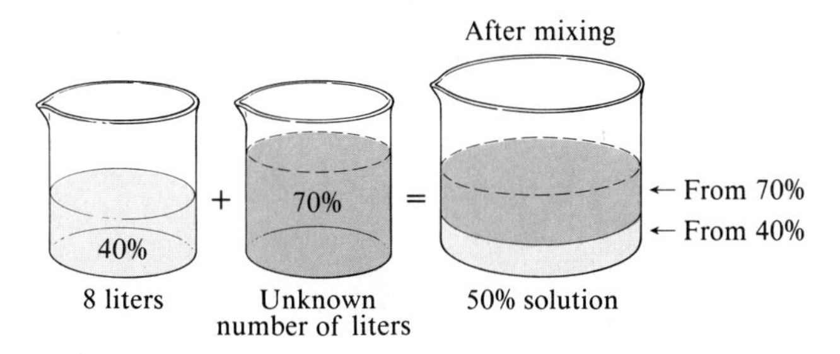

FIGURE 2.6

Let x = the number of liters of the 70% solution that should be used. The information in the problem is used to get the following table.

STRENGTH	LITERS OF SOLUTION	LITERS OF PURE POTASSIUM CHLORIDE
40%	8	$.40(8) = 3.2$
70%	x	$.70x$
50%	$8 + x$	$.50(8 + x)$

Sum must equal

The numbers in the right-hand column were found by multiplying the strengths and the numbers of liters. The number of liters of pure potassium chloride in the 40% solution plus the number of liters in the 70% solution must equal the number of liters in the 50% solution.

$$3.2 + .70x = .50(8 + x)$$

$$3.2 + .70x = 4 + .50x \quad \text{Distributive property}$$

$$.20x = .8 \quad \text{Subtract 3.2 and .50x.}$$

$$x = 4 \quad \text{Divide by .20.}$$

The chemist must use 4 liters of the 70% solution. ■

PROBLEM SOLVING

A chart was used in Example 4 to organize the information. Some students like to use "box diagrams" instead. Either method works well if you are careful while filling in the information. The next example uses box diagrams. ■

EXAMPLE 5
SOLVING AN INVESTMENT PROBLEM

After winning the state lottery, Grey Thornton had \$40,000 to invest. He put part of the money in an account paying 8% simple interest, with the remainder going into stocks paying 12% simple interest. The total annual income from the investments is \$4080. Find the amount invested at each rate.

Let x = the amount invested at 8%

$40{,}000 - x$ = the amount invested at 12%.

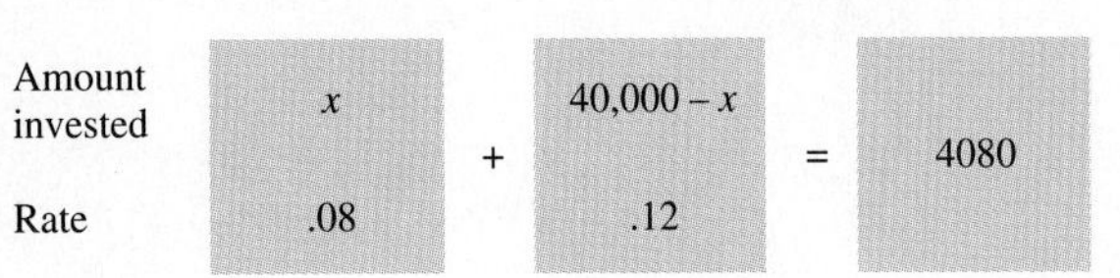

FIGURE 2.7

In the box diagram in Figure 2.7, fill in the important information. The formula for interest is $I = prt$. The interest earned in one year at 8% is

$$x(.08)(1) = .08x.$$

The interest at 12% is

$$(40{,}000 - x)(.12)(1) = .12(40{,}000 - x).$$

Since the total interest is \$4080,

$$\underset{\text{interest at 8\%}}{.08x} \; + \; \underset{\text{interest at 12\%}}{.12(40{,}000 - x)} \; = \; \underset{\text{total interest}}{4080.}$$

Solve the equation.

$$.08x + 4800 - .12x = 4080 \qquad \text{Distributive property}$$
$$-.04x + 4800 = 4080 \qquad \text{Combine terms.}$$
$$-.04x = -720 \qquad \text{Subtract 4800.}$$
$$x = 18{,}000 \qquad \text{Divide by } -.04.$$

Thornton has \$18,000 invested at 8% and \$40,000 − \$18,000 = \$22,000 at 12%. Check by finding the annual interest at each rate; they should add up to \$4080. ■

NOTE Refer to Example 5. We chose the variable to represent the amount invested at 8%. Students often ask "Can I let the variable represent the other unknown?" The answer is yes. The equation will be different, but in the end the two answers will be the same. You might wish to work Example 5, letting the variable equal the amount invested at 12%.

You may notice that the investment problems in this chapter deal with simple interest. In most "real-world" applications, compound interest is used. However,

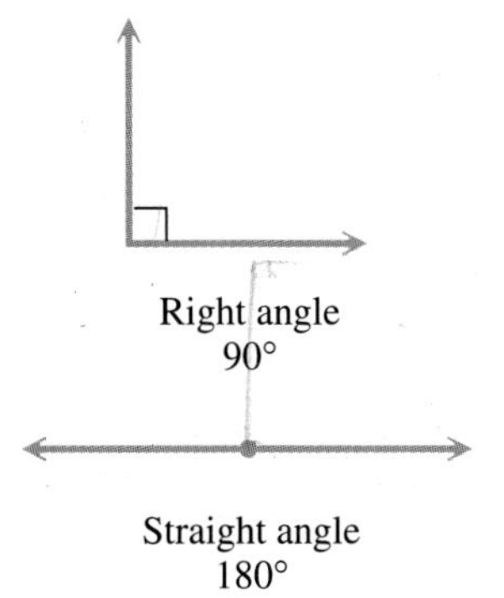

FIGURE 2.8

more advanced methods (covered in Chapter 10) are needed for compound interest problems, so we will deal only with simple interest until then.

5 In the last section we saw some angle relationships with regard to vertical angles and angles formed by transversals intersecting parallel lines. Here, we examine some other concepts involving angle measures.

Figure 2.8 shows some special angles, with their names. A **right angle** measures 90°; a **straight angle** measures 180°. If the sum of the measures of two angles is 90°, the angles are **complementary angles,** and they are called **complements** of each other. If the sum of the measures of two angles is 180°, the angles are **supplementary angles,** and they are called **supplements** of each other.

One of the important results of Euclidean geometry (the geometry of the Greek mathematician Euclid) is that the sum of the angle measures of any triangle is 180°. This property, along with the others mentioned above, is used in the next example.

EXAMPLE 6
SOLVING PROBLEMS ABOUT ANGLE MEASURES

(a) Find the value of x, and determine the measure of each angle in Figure 2.9.

Since the three marked angles are angles of a triangle, their sum must be 180°. Write the equation indicating this, and then solve.

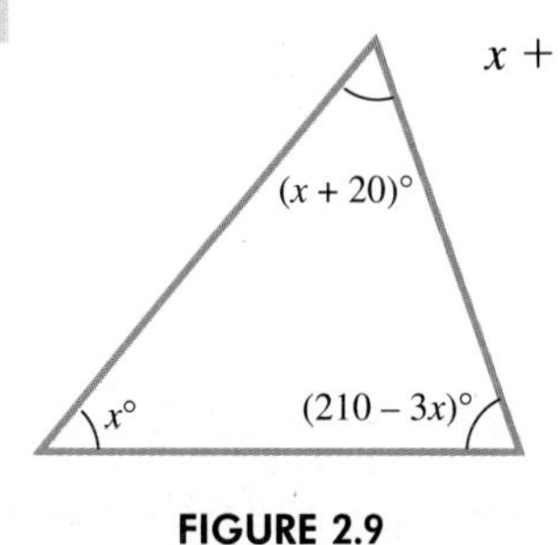

FIGURE 2.9

$$x + (x + 20) + (210 - 3x) = 180$$
$$-x + 230 = 180 \qquad \text{Combine like terms.}$$
$$-x = 180 - 230 \qquad \text{Subtract 230.}$$
$$-x = -50$$
$$x = 50 \qquad \text{Divide by } -1.$$

One angle measures 50°, another measures $x + 20 = 50 + 20 = 70°$, and the third measures $210 - 3x = 210 - 3(50) = 60°$. Since $50° + 70° + 60° = 180°$, the answers are correct.

(b) The supplement of an angle measures 10° more than three times its complement. Find the measure of the angle.

Let x = the degree measure of the angle. Then,
$180 - x$ = the degree measure of its supplement, and
$90 - x$ = the degree measure of its complement.

Now use the words of the problem to write the equation.

$$\underbrace{180 - x}_{\text{supplement}} \quad \underbrace{=}_{\text{measures}} \quad \underbrace{10 +}_{\text{10 more than}} \quad \underbrace{3(90 - x)}_{\text{three times its complement}}$$

Solve the equation.

$$180 - x = 10 + 270 - 3x \quad \text{Distributive property}$$
$$180 - x = 280 - 3x$$
$$2x = 100 \quad \text{Add } 3x\text{; subtract 180.}$$
$$x = 50 \quad \text{Divide by 2.}$$

The angle measures 50°. Since its supplement (130°) is 10° more than three times its complement (40°), that is

$$130 = 10 + 3(40) \quad \text{is true,}$$

the answer checks. ■

2.3 EXERCISES

Translate the following verbal phrases into mathematical expressions. Use x to represent the unknown. See Example 1.

1. A number decreased by 4

2. A number increased by 8

3. 11 increased by a number

4. 9 decreased by a number

5. The product of -8 and a number

6. The product of a number and 5

7. 8 less than a number

8. 6 more than a number

9. -3 increased by 4 times a number

10. 12 decreased by twice a number

11. The quotient of -1 and a number

12. The quotient of a number and 2

13. The ratio of a number and -4

14. The ratio of a number and 6

15. 5/9 of a number

16. 3/2 of a number

---

17. Explain why $9 - x$ is *not* the correct translation of "9 less than a number." (See Objective 1.)

18. Which one of the following is *not* a valid translation of "20% of a number"? (See Objective 1.)

(a) $.20x$ **(b)** $.2x$ **(c)** $\frac{x}{5}$ **(d)** $20x$

Solve each of the following problems. See Example 2.

19. Two of the highest paid business executives in a recent year were Mike Eisner, chairman of Disney, and Ed Horrigan, vice chairman of RJR Nabisco. Together their salaries totaled 61.8 million dollars. Eisner earned 18.4 million dollars more than Horrigan. What was the salary for each executive?

20. In a recent year, the two U.S. industrial corporations with the highest sales were General Motors and Ford Motor. Their sales together totaled 213.5 billion dollars. Ford Motor sales were 28.7 billion dollars less than General Motors. What were the sales for each corporation?

21. Babe Ruth and Rogers Hornsby were two great hitters. Together they got 5803 base hits in their careers. Hornsby got 57 more hits than Ruth. How many base hits did each get?

22. In the 1984 presidential election, Ronald Reagan and Walter Mondale together received 538 electoral votes. Reagan received 512 more votes than Mondale in the landslide. How many votes did each man receive?

23. The product of 5 and the sum of a number and 8 is 10 less than thirty times the number. Find the number.

24. When a number is multiplied by 6, and this product is decreased by 4, the result is 28. Find the number.

Solve each of the following problems. See Example 3.

25. A number is decreased by 12% of the number, giving 44. Find the number.

26. After a number is increased by 15% of the number, the result is 69. Find the number.

27. At the end of a day, Jeff found that the total cash register receipts at the motel where he works were $286.20. This included sales tax of 8%. Find the amount of the tax. (*Hint:* Let $x =$ the total due the motel, so $.08x =$ the amount of the tax.)

28. Dongming Wei wants to sell his house. He would like to receive $136,000 after deducting the 8% commission due the real estate agent. What should be the selling price for the house? (Round the answer to the nearest dollar.)

Solve each of the following problems. See Examples 4 and 5.

29. Five liters of a 4% acid solution must be mixed with a 10% solution to get a 6% solution. How many liters of the 10% solution are needed?

30. How many liters of a 14% alcohol solution must be mixed with 20 liters of a 50% solution to get a 20% solution?

31. In a chemistry class, 6 liters of a 12% alcohol solution must be mixed with a 20% solution to get a 14% solution. How many liters of the 20% solution are needed?

32. How many liters of a 10% alcohol solution must be mixed with 40 liters of a 50% solution to get a 20% solution?

33. A medicated first aid spray on the market is 78% alcohol by volume. If the manufacturer has 50 liters of the spray containing 70% alcohol, how much pure alcohol should be added so that the final mixture is the required 78% alcohol? (*Hint:* Pure alcohol is 100% alcohol.)

34. How much water must be added to 3 gallons of a 4% insecticide solution to reduce the concentration to 3%? (*Hint:* Water is 0% insecticide.)

35. It is necessary to have a 40% antifreeze solution in the radiator of a certain car. The radiator now holds 20 liters of 20% solution. How many liters of this should be drained and replaced with 100% antifreeze to get the desired strength? (*Hint:* The number of liters drained is equal to the number of liters replaced.)

36. A tank holds 80 liters of a chemical solution. Currently, the solution has a strength of 30%. How much of this should be drained and replaced with a 70% solution to get a final strength of 40%?

37. Mark Fong earned $12,000 last year by giving golf lessons. He invested part at 8% simple interest and the rest at 9%. He made a total of $1050 in interest. How much did he invest at each rate?

38. Kim Falgout won $60,000 on a slot machine in Las Vegas. She invested part at 8% simple interest and the rest at 12%. She earned a total of $6200 in interest. How much was invested at each rate?

39. Evelyn Pourciau invested some money at 8% simple interest and $1000 less than twice this amount at 14%. Her total annual income from the interest was $580. How much was invested at each rate?

40. Mabel Johnston invested some money at 9% simple interest, and $5000 more than 3 times this amount at 10%. She earned $2840 in interest. How much did she invest at each rate?

41. Jack Pritchard has $29,000 invested in bonds paying 11%. How much additional money should he invest in mortgages paying 15% so that the average return on the two investments is 14%?

42. Tommy Kreamer placed $14,000 in an account paying 12%. How much additional money should he deposit at 16% so that the average return on the two investments is 15%?

43. Prem Kythe invests $3000 at 8% annual interest. He would like to earn $740 per year in interest. How much should he invest in a second account paying 10% interest in order to accomplish this?

44. Rita Lieux invested some money at 7% interest and the same amount at 10%. Her total interest for the year was \$150 less than one-tenth of the total amount she invested. How much did she invest at each rate?

Solve the following problems. See Example 6.

45. The supplement of an angle measures 25° more than twice its complement. Find the measure of the angle.

46. The complement of an angle measures 10° less than one-fifth of its supplement. Find the measure of the angle.

47. Find the measure of each angle in the triangle.

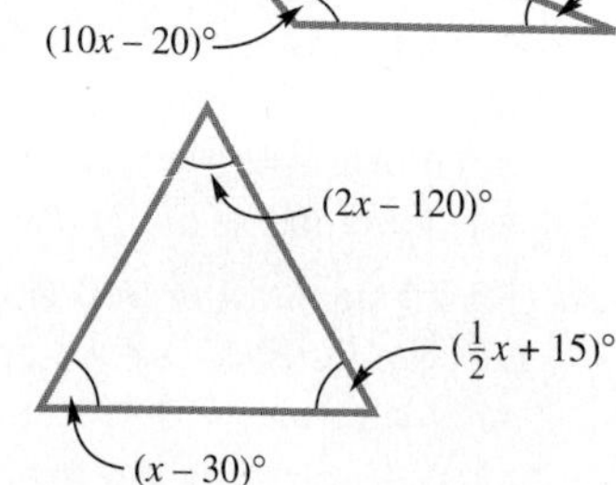

48. Find the measure of each angle in the triangle.

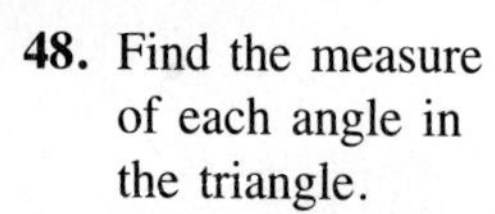

49. Find the measure of each marked angle.

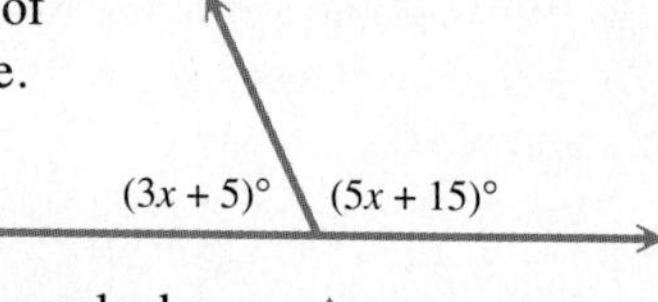

50. Together, the two marked angles form a right angle. Find the measure of each angle.

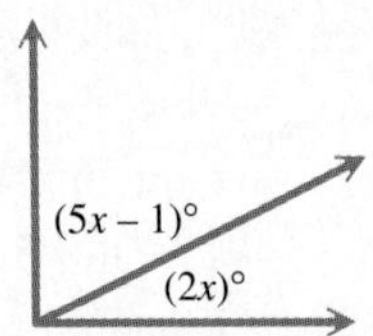

51. Use the sketch to find the measure of each numbered angle. Assume that m and n are parallel, and use the ideas about angle measures covered in Section 2.2 and this section.

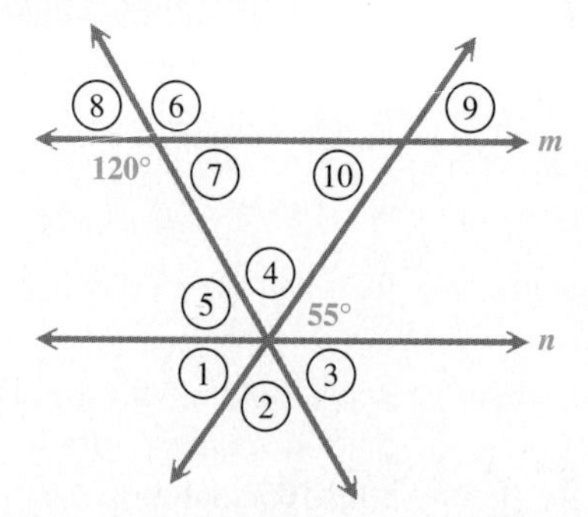

52. What angle is its own supplement? What angle is its own complement? (See Objective 5.)

PREVIEW EXERCISES

Solve each of the following. See Section 2.2.

53. Use $d = rt$ to find d if $r = 50$ and $t = 3$.

54. Use $d = rt$ to find r if $d = 75$ and $t = 5$.

55. Use $P = 2L + 2W$ to find P if $L = 10$ and $W = 4$.

56. Use $P = 2L + 2W$ to find L if $P = 80$ and $W = 6$.

57. Use $P = a + b + c$ to find a if $b = 13$, $c = 14$, and $P = 44$.

58. Use $A = \frac{1}{2}(B + b)h$ to find h if $A = 100$, $b = 12$, and $B = 13$.

2.4 MORE APPLICATIONS OF LINEAR EQUATIONS

OBJECTIVES

1 SOLVE PROBLEMS ABOUT DIFFERENT DENOMINATIONS OF MONEY.
2 SOLVE PROBLEMS ABOUT UNIFORM MOTION.
3 SOLVE PROBLEMS ABOUT GEOMETRIC FIGURES.

FOCUS ON PROBLEM SOLVING

In the nineteenth century, the United States minted two-cent and three-cent pieces. Charles Steib has three times as many three-cent pieces as two-cent pieces in his collection, and the face value of these coins is \$1.10. How many of each denomination does he have?

Very often we are able to use previous methods to solve problems which may seem to be different from others we have studied, but are actually quite similar. The method used to solve this problem is the same method used to solve investment and mixture problems like the ones seen in Section 2.3. This problem is Exercise 7 in the exercises for this section. After working through this section, you should be able to solve this problem.

There are three common applications of linear equations that we did not discuss in Section 2.3: problems about different denominations of money, uniform motion problems, and problems about geometric figures. The first type is very similar to the simple interest type problems seen in Section 2.3. Uniform motion problems use the formula $d = rt$. Problems about geometric figures usually require the use of the formulas studied in Section 2.2.

1 The first example shows how to solve problems about different denominations of money.

PROBLEM SOLVING

In problems involving money, use the basic fact that

$$\begin{bmatrix}\text{Number of monetary} \\ \text{units of the same kind}\end{bmatrix} \times [\text{Denomination}] = \begin{bmatrix}\text{Total monetary} \\ \text{value.}\end{bmatrix}$$

For example, 30 dimes have a monetary value of $30(.10) = 3.00$ dollars. Fifteen five-dollar bills have a value of $15(5) = 75$ dollars. ■

EXAMPLE 1 SOLVING A PROBLEM ABOUT DENOMINATIONS OF MONEY

For a bill totaling \$5.65, a cashier received 25 coins consisting of nickels and quarters. How many of each type of coin did the cashier receive?

Let x = the number of nickels received;

$25 - x$ = the number of quarters received.

The information for this problem may be arranged as shown in Figure 2.10. Multiply the numbers of coins by the denominations, and add the results to get 5.65.

Number of coins	x	+	$25 - x$	=	\$5.65
Denomination	\$.05		\$.25		
	Value of nickels, $.05x$	+	Value of quarters, $.25(25 - x)$	=	Total value, 5.65

FIGURE 2.10

$$.05x + .25(25 - x) = 5.65$$

$$5x + 25(25 - x) = 565 \quad \text{Multiply by 100.}$$

$$5x + 625 - 25x = 565 \quad \text{Distributive property}$$

$$-20x = -60 \quad \text{Combine terms; subtract 625.}$$

$$x = 3 \quad \text{Divide by } -20.$$

The cashier has 3 nickels and $25 - 3 = 22$ quarters. Check to see that the total value of these coins is \$5.65. ■

CAUTION Be sure that your answer is reasonable when working problems like Example 1. Since you are dealing with a number of coins, an answer can neither be negative nor a fraction.

2 The next examples involve uniform motion.

PROBLEM SOLVING

Uniform motion problems use the formula Distance = Rate × Time. In this formula, when rate (or speed) is given in miles per hour, time must be given in hours. When solving such problems, draw a sketch to illustrate what is happening in the problem, and make a chart to summarize the information of the problem. ■

EXAMPLE 2
SOLVING A MOTION PROBLEM

Two cars leave the same place at the same time, one going east and the other west. The eastbound car averages 40 miles per hour, while the westbound car averages 50 miles per hour. In how many hours will they be 300 miles apart?

A sketch shows what is happening in the problem: The cars are going in *opposite* directions. See Figure 2.11.

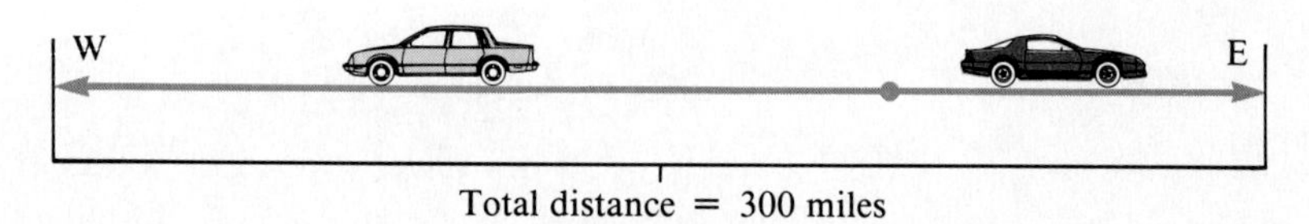

FIGURE 2.11

Let x represent the time traveled by each car. Summarize the information of the problem in a chart.

	RATE	TIME	DISTANCE
Eastbound car	40	x	$40x$
Westbound car	50	x	$50x$

The distances traveled by the cars, $40x$ and $50x$, are obtained from the formula $d = rt$. When the expressions for rate and time are entered in the chart, *fill in the distance expression by multiplying rate by time.*

From the sketch in Figure 2.11, the sum of the two distances is 300.

$$40x + 50x = 300$$

$$90x = 300 \qquad \text{Combine terms.}$$

$$x = \frac{300}{90} \qquad \text{Divide by 90.}$$

$$x = \frac{10}{3} \qquad \text{Lowest terms}$$

The cars travel $\frac{10}{3} = 3\frac{1}{3}$ hours, or 3 hours and 20 minutes. ■

CAUTION It is a common error to write 300 as the distance for each car in Example 2. Three hundred miles is the *total* distance traveled.

Example 2 involved motion in opposite directions. The next example deals with motion in the same direction.

EXAMPLE 3 SOLVING A MOTION PROBLEM

Train A leaves a town at 8:00 A.M., heading east. At 9:00 A.M., Train B leaves on a parallel track, also heading east. The rate of Train A is 80 kilometers per hour, and the rate of Train B is 100 kilometers per hour. How long will it take for Train B to catch up with Train A?

Begin by drawing a sketch that illustrates what is happening in the problem. See Figure 2.12.

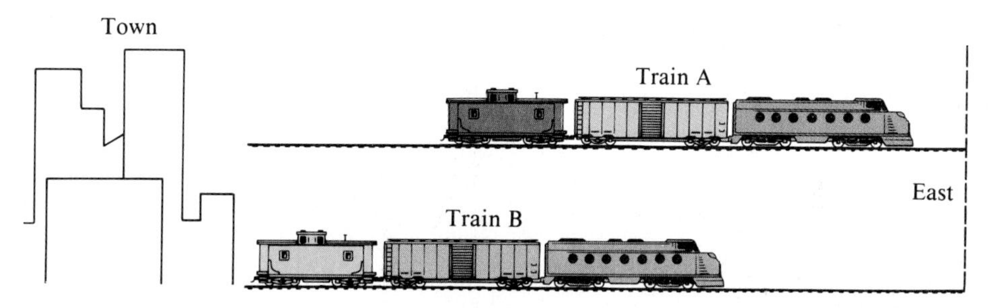

FIGURE 2.12

Let x = the time Train B will travel before it catches up with Train A (in hours);

then $x + 1$ = the time that Train A will travel (in hours),

since Train A left one hour earlier. The information is summarized in the following chart. Use $d = rt$ to get the expressions in the last column.

	RATE	TIME	DISTANCE
Train A	80	$x + 1$	$80(x + 1)$
Train B	100	x	$100x$

When Train B catches up with Train A, the distances traveled will be the same. Therefore, to get an equation to solve the problem, set the distance expressions equal to each other.

$$80(x + 1) = 100x$$

Now solve the equation.

$$80x + 80 = 100x \quad \text{Distributive property}$$
$$80 = 20x \quad \text{Subtract } 80x.$$
$$x = 4 \quad \text{Divide by 20.}$$

Train B will travel 4 hours before it catches up with Train A. (This will happen at 1:00 P.M. Do you see why?) ■

CAUTION Difficulties often arise in uniform motion problems because the step of drawing a sketch is neglected. *The sketch will tell you how to set up the equation.*

EXAMPLE 4
SOLVING A MOTION PROBLEM

Jeff Bezzone can bike to work in 3/4 hour. When he takes the bus, the trip takes 1/4 hour. If the bus travels 20 miles per hour faster than Jeff rides his bike, how far is it to his workplace?

Although the problem asks for a distance, it is easier here to let x be Jeff's speed when he rides his bike to work. Then the speed of the bus is $x + 20$. By the distance formula, for the trip by bike,

$$d = rt = x \cdot \frac{3}{4} = \frac{3}{4}x,$$

and by bus,

$$d = rt = (x + 20) \cdot \frac{1}{4} = \frac{1}{4}(x + 20).$$

Summarize the information of the problem in a chart.

	RATE	TIME	DISTANCE
Bike	x	$\frac{3}{4}$	$\frac{3}{4}x$
Bus	$x + 20$	$\frac{1}{4}$	$\frac{1}{4}(x + 20)$

The key to setting up the correct equation is to understand that the distance in each case is the same. See Figure 2.13.

FIGURE 2.13

Since the distance is the same in both cases, $\frac{3}{4}x = \frac{1}{4}(x + 20)$.

Solve this equation. First multiply on both sides by 4.

$$4\left(\frac{3}{4}x\right) = 4\left(\frac{1}{4}\right)(x + 20)$$

$3x = x + 20$ Multiply; distributive property

$2x = 20$ Subtract x.

$x = 10$ Divide by 2.

Now answer the question in the problem. The required distance is given by

$$d = \frac{3}{4}x = \frac{3}{4}(10) = \frac{30}{4} = 7.5.$$

Check by finding the distance using

$$d = \frac{1}{4}(x + 20) = \frac{1}{4}(10 + 20) = \frac{30}{4} = 7.5,$$

the same result. The required distance is 7.5 miles. ■

NOTE As mentioned in Example 4, it was easier to let the variable represent a quantity other than the one that we were asked to find. This is the case in some problems. It takes practice to learn when this approach is the best, and practice means working lots of problems!

3 In Section 2.2 we saw how formulas can be used to find areas and perimeters of geometric figures.

PROBLEM SOLVING

When applied problems deal with geometric figures, it is often necessary to use the appropriate formula so that the equation for the problem can be written. The next example illustrates how a figure and a formula are used in an application involving geometry. ■

EXAMPLE 5 SOLVING A PROBLEM ABOUT A GEOMETRIC FIGURE

A label is in the shape of a rectangle. The length of the rectangle is 1 centimeter more than twice the width. The perimeter is 110 centimeters. Find the length and the width. See Figure 2.14.

Let W = the width

$1 + 2W$ = the length (one more than twice the width).

The perimeter of a rectangle is given by the formula $P = 2L + 2W$. Replace P in this formula with 110 and L with $1 + 2W$, giving

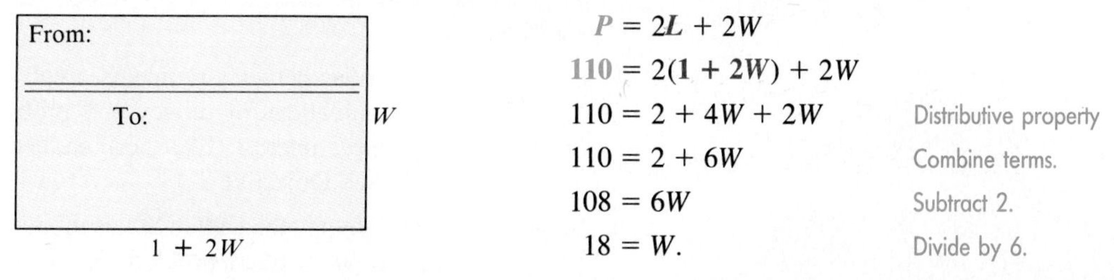

$$P = 2L + 2W$$
$$110 = 2(1 + 2W) + 2W$$
$$110 = 2 + 4W + 2W \quad \text{Distributive property}$$
$$110 = 2 + 6W \quad \text{Combine terms.}$$
$$108 = 6W \quad \text{Subtract 2.}$$
$$18 = W. \quad \text{Divide by 6.}$$

FIGURE 2.14

The width of the rectangle is 18 centimeters, and the length is $1 + 2W = 1 + 2(18) = 1 + 36 = 37$ centimeters. ■

2.4 EXERCISES

Solve each of the following problems. See Example 1.

1. Charlotte Lewis has 30 coins in her change purse, consisting of pennies and nickels. The total value of the money is $.94. How many of each type of coin does she have?

2. Meredith Many has 28 coins in her pocket, consisting of nickels and dimes. The total value of the money is $2.70. How many of each type of coin does she have?

3. Leslie Cobar's piggy bank has 45 coins. Some are quarters and the rest are half-dollars. If the total value of the coins is $17.50, how many of each type does she have?

4. Charles Rees has a jar in his office that contains 39 coins. Some are pennies and the rest are dimes. If the total value of the coins is $2.55, how many of each type does he have?

5. Gary Gundersen has a box of coins that he uses when playing poker with his friends. The box currently contains 40 coins, consisting of pennies, dimes, and quarters. The number of pennies is equal to the number of dimes, and the total value is $8.05. How many of each type of coin does he have in the box?

6. Carroll Blakemore found some coins while looking under his sofa pillows. There were equal numbers of nickels and quarters, and twice as many half-dollars as quarters. If he found $19.50 in all, how many of each type of coin did he find?

7. In the nineteenth century, the United States minted two-cent and three-cent pieces. Charles Steib has three times as many three-cent pieces as two-cent pieces, and the face value of these coins is $1.10. How many of each type does he have?

8. Gerry Moore collects U.S. gold coins. He has a collection of 30 coins. Some are $10 coins and the rest are $20 coins. If the face value of the coins is $540, how many of each type does he have?

9. The school production of *The Music Man* was a big success. For opening night, 300 tickets were sold. Students paid $1.50 each, while non-students paid $3.50 each. If a total of $810 was collected, how many students and how many non-students attended?

10. A total of 550 people attended a Frankie Valli and the Four Seasons concert last night. Floor tickets cost $20 each, while balcony tickets cost $14 each. If a total of $10,400 was collected, how many of each type of ticket were sold?

11. The matinee showing of *The Wizard of Oz* was attended by 54 people. Children three years old or younger were admitted free, and children between four and twelve paid $2. Anyone else paid $3. If the same number of children three or younger and children between four and twelve attended, and $82 was collected, how many of each type of ticket were sold?

12. On one day, 660 people visited the local zoo. Senior citizens attended free, children paid $3, and non-senior adults paid $5. If twice as many children visited as senior citizens, and $2292 was collected, how many of each age group visited the zoo that day?

13. Explain the similarities between problems involving different denominations of money and problems involving simple interest (like those studied in Section 2.3). (See Objective 1.)

14. In the nineteenth century, the United States minted half-cent coins. If an applied problem involved half-cent coins, what decimal number would represent this denomination? (See Objective 1.)

Solve each of the following problems. See Examples 2–4.

15. Two cars leave a town at the same time. One travels east at 50 miles per hour, and the other travels west at 55 miles per hour. In how many hours will they be 315 miles apart?

16. Two planes leave O'Hare Airport in Chicago at the same time. One travels east at 550 miles per hour, and the other travels west at 500 miles per hour. How long will it take for the planes to be 2100 miles apart?

17. A train leaves Coon Rapids, Minnesota, and travels north at 85 kilometers per hour. Another train leaves at the same time and travels south at 95 kilometers per hour. How long will it take before they are 990 kilometers apart?

18. Two steamers leave a port on a river at the same time, traveling in opposite directions. Each is traveling 12 miles per hour. How long will it take for them to be 66 miles part?

19. A pleasure boat on the Mississippi River traveled from Baton Rouge to New Orleans with a stop at White Castle. On the first part of the trip, the boat traveled at an average speed of 10 miles per hour. From White Castle to New Orleans the average speed was 15 miles per hour. The entire trip covered 100 miles. How long did the entire trip take if the two parts each took the same number of hours?

20. A jet airliner flew across the United States at an average speed of 500 miles per hour. It then continued on across the Atlantic at an average speed of 530 miles per hour. If the entire trip covered 14,420 miles and both parts of the trip took the same number of hours, how many hours did the whole trip take?

21. Carl leaves his house on his bicycle at 9:30 A.M. and averages 5 miles per hour. His wife, Karen, leaves at 10:00 A.M., following the same path and averaging 8 miles per hour. How long will it take for Karen to catch up with Carl?

22. Joey and Liz commute to work, traveling in opposite directions. Joey leaves the house at 7:00 A.M. and averages 35 miles per hour. Liz leaves at 7:15 A.M. and averages 40 miles per hour. At what time will they be 65 miles apart?

23. Maria Gutierrez can get to school in 1/4 hour if she rides her bike. It takes her 3/4 hour if she walks. Her speed when walking is 10 miles per hour slower than her speed when riding. What is her speed when she rides?

24. On an automobile trip, Susan Blohm maintained a steady speed for the first two hours. Rush hour traffic slowed her speed by 25 miles per hour for the last part of the trip. The entire trip, a distance of 150 miles, took 2 1/2 hours. What was her speed during the first part of the trip?

25. When Heath drives his car to work, the trip takes 1/2 hour. When he rides the bus, it takes 3/4 hour. The average speed of the bus is 12 miles per hour less than his speed when driving. Find the distance he travels to work.

26. On a 100-mile trip to the mountains, the Kwan family traveled at a steady speed for the first hour. Their speed was 16 miles per hour slower during the second hour of the trip. Find their speed during the first hour.

27. Steve leaves Nashville to visit his cousin David in Napa, 80 miles away. He travels at an average speed of 50 miles per hour. One-half hour later David leaves to visit Steve, traveling at an average speed of 60 miles per hour. How long after David leaves will it be before they meet?

28. In a run for charity Janet runs at a speed of 5 miles per hour. Paula leaves 10 minutes after Janet and runs at 6 miles per hour. How long will it take for Paula to catch up with Janet? (*Hint:* Change minutes to hours.)

29. When solving problems about uniform motion, you must be careful that rate is given in appropriate units. Suppose that distance is given in miles and rate is given in feet per second. How would you change feet per second to miles per hour? (*Hint:* There are 5280 feet in one mile, and 3600 seconds in one hour.)

30. Read over Example 4 in this section. The solution of the equation is 10. Why is 10 miles per hour not the answer to the problem? (See Objective 2.)

Solve each of the following problems. See Example 5.

31. A sign is in the shape of an isosceles triangle. (It has two sides of equal length.) The third side measures 3 feet less than twice the length of each of the equal sides. Find the lengths of the sides of the sign if its perimeter is 61 feet.

32. One side of a triangle is 5 meters longer than twice the shortest side. The third side is three times as long as the shortest side. Find the lengths of the three sides, if the perimeter of the triangle is 71 meters.

33. The perimeter of a rectangle is 43 inches more than the length of the rectangle. The width is 10 inches. Find the length of the rectangle.

34. The length of a rectangular billboard is 2 feet more than three times its width. The perimeter of the billboard is 68 feet. Find the length and the width of the billboard.

35. A farmer wishes to enclose a rectangular region with 105 meters of fencing in such a way that the length is twice the width and the area is cut into two equal parts, as shown in the figure. What length and width should be used?

Length

Width

36. Marjorie Jensen has a sheet of tin 12 centimeters by 16 centimeters. She plans to make a box by cutting squares out of each of the four corners and folding up the remaining edges. How large a square should she cut so that the finished box will have a length that is three times its width?

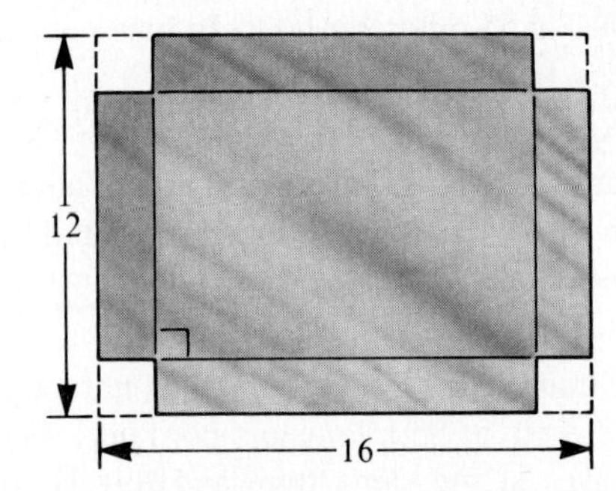

The following problems are a bit different from the ones studied in this section. Solve the problems, using the six-step method described in Section 2.3.

37. Jaime is twice as old as his brother, Ricardo. In five years, Ricardo will be the same age as his brother was ten years ago. What are their present ages?

38. Mr. Silvester is five years older than his wife. Five years ago his age was 4/3 her age. What are their ages now?

39. A grocer buys lettuce for \$10.40 a crate. Of the lettuce he buys, 10% cannot be sold. If he charges 40¢ for each head he sells and makes a profit of 20¢ on each head he buys, how many heads of lettuce are in the crate?

40. The monthly phone bill includes a monthly charge of \$10 for local calls plus an additional charge of 50¢ for each toll call in a certain area. A federal tax of 5% is added to the total bill. If all calls were local or within the 50¢ area and the total bill was \$17.85, find the number of toll calls made.

PREVIEW EXERCISES

Graph each of the following. See Section 1.2.

41. $(-\infty, -2]$

42. $(4, \infty)$

43. $(3, \infty)$

44. $(-\infty, -1]$

45. $\{y \mid -2 < y < 5\}$

46. $\{k \mid 0 \leq k \leq 3\}$

2.5 LINEAR INEQUALITIES IN ONE VARIABLE

OBJECTIVES

1. SOLVE LINEAR INEQUALITIES USING THE ADDITION PROPERTY.
2. SOLVE LINEAR INEQUALITIES USING THE MULTIPLICATION PROPERTY.
3. SOLVE LINEAR INEQUALITIES HAVING THREE PARTS.
4. SOLVE APPLIED PROBLEMS USING LINEAR INEQUALITIES.

FOCUS ON PROBLEM SOLVING

A couple wishes to rent a car for one day while on vacation. Ford Automobile Rental wants $15.00 per day and 14¢ per mile, while Chevrolet-For-A-Day wants $14.00 per day and 16¢ per mile. After how many miles would the price to rent the Chevrolet exceed the price to rent a Ford?

The word ''exceed'' in this problem indicates inequality. In earlier applications we have written equations to solve the problems. Now we will learn how to solve linear *inequalities*, and how they can be applied to problems like this one. This problem is Exercise 49 in the exercises for this section. After working through this section, you should be able to solve this problem.

An **inequality** says that two expressions are *not* equal. A **linear inequality in one variable** is an inequality such as

$$x + 5 < 2, \qquad y - 3 \geq 5, \qquad \text{or} \qquad 2k + 5 \leq 10.$$

LINEAR INEQUALITY

A linear inequality in one variable can be written in the form

$$ax + b < c,$$

where a, b, and c are real numbers, with $a \neq 0$.

(Throughout this section we give the definitions and rules only for $<$, but they are also valid for $>$, $\leq$, and $\geq$.)

1 An inequality is solved by finding all numbers that make the inequality true. Usually, an inequality has an infinite number of solutions. These solutions, like the solutions of equations, are found by producing a series of simpler equivalent inequalities. **Equivalent inequalities** are inequalities with the same solution set. The inequalities in this chain of equivalent inequalities are found with the addition and multiplication properties of inequality.

ADDITION PROPERTY OF INEQUALITY

For all real numbers a, b, and c, the inequalities

$$a < b \qquad \text{and} \qquad a + c < b + c$$

are equivalent. (The same number may be added to both sides of an inequality without changing the solution set.)

As with equations, the addition property can be used to *subtract* the same number from both sides of an inequality.

EXAMPLE 1
USING THE ADDITION PROPERTY OF INEQUALITY

Solve $x - 7 < -12$.

Add 7 to both sides. $\quad x - 7 + 7 < -12 + 7$

$$x < -5$$

The solution set, $(-\infty, -5)$, is graphed in Figure 2.15. ■

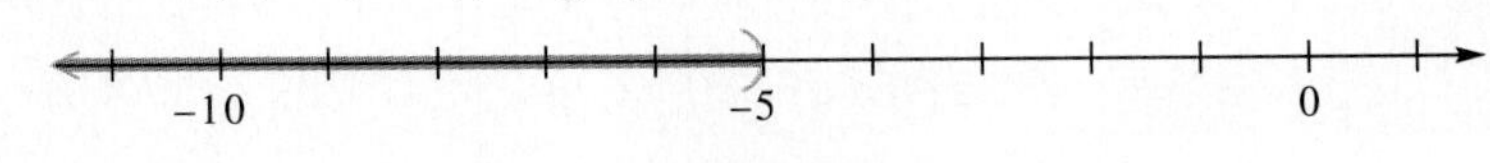

FIGURE 2.15

Interval notation was discussed briefly in Section 1.2. Because interval notation is very convenient in writing solution sets of inequalities, we will now discuss this type of notation in more detail. The solution set in Example 1 is an example of an **interval** on the number line. A simplified notation, called **interval notation,** is used for writing intervals. For example, using this notation, the interval $\{x | x < 5\}$ is written as $(-\infty, 5)$. The symbol $-\infty$ does not indicate a number; it is used to show that the interval includes all real numbers less than 5. The parenthesis indicates that 5 is not included. The interval $(-\infty, 5)$ is an example of an **open interval** since the endpoints are not included. Intervals that include the endpoints, as in the example $\{x | 0 \leq x \leq 5\}$, are **closed intervals.** Closed intervals are indicated with square brackets. The interval $\{x | 0 \leq x \leq 5\}$ is written as $[0, 5]$. An interval like $(2, 5]$, that is open on one end and closed on the other, is a **half-open interval.** Examples of other sets written in interval notation are shown in the following chart. In these intervals, assume that $a < b$. Note that a parenthesis is always used with the symbols $-\infty$ or ∞.

TYPE OF INTERVAL	SET	INTERVAL NOTATION	GRAPH
	$\{x \| a < x\}$	(a, ∞)	a
Open interval	$\{x \| a < x < b\}$	(a, b)	a b
	$\{x \| x < b\}$	$(-\infty, b)$	b
	$\{x \| a \leq x\}$	$[a, \infty)$	a
Half-open interval	$\{x \| a < x \leq b\}$	$(a, b]$	a b
	$\{x \| a \leq x < b\}$	$[a, b)$	a b
	$\{x \| x \leq b\}$	$(-\infty, b]$	b
Closed interval	$\{x \| a \leq x \leq b\}$	$[a, b]$	a b

EXAMPLE 2
USING THE ADDITION PROPERTY OF INEQUALITY

Solve the inequality $14 + 2m \le 3m$, and graph the solution set.

First, subtract $2m$ from both sides.

$$14 + 2m - 2m \le 3m - 2m \quad \text{Subtract } 2m.$$
$$14 \le m \quad \text{Combine like terms.}$$

The inequality $14 \le m$ (14 is less than or equal to m) can also be written $m \ge 14$ (m is greater than or equal to 14). Notice that in each case, the inequality symbol points to the smaller expression, 14. The solution set in interval notation is $[14, \infty)$. The graph is shown in Figure 2.16. ■

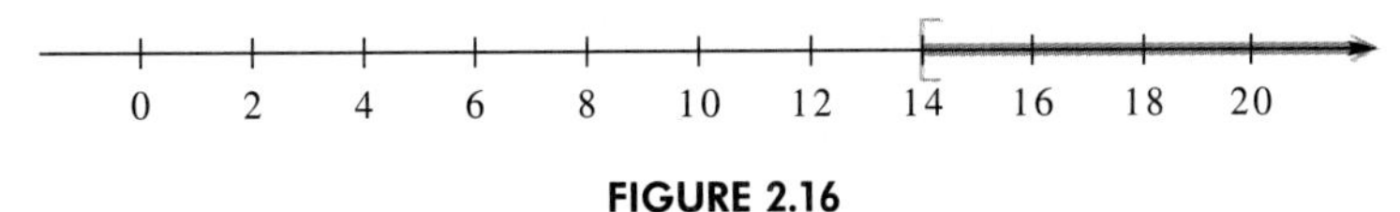

FIGURE 2.16

CAUTION Errors often occur in graphing inequalities where the variable term is on the right side. (This is probably due to the fact that we read from left to right.) To guard against such errors, it is a good idea to rewrite these inequalities so that the variable is on the left, as shown in Example 2.

2 An inequality such as $3x \le 15$ can be solved by dividing both sides by 3. This is done with the multiplication property of inequality, which is a little more involved than the corresponding property for equations. To see how this property works, start with the true statement

$$-2 < 5.$$

Multiply both sides by, say, 8.

$$-2(8) < 5(8)$$
$$-16 < 40 \quad \text{True}$$

This gives a true statement. Start again with $-2 < 5$, and this time multiply both sides by -8.

$$-2(-8) < 5(-8)$$
$$16 < -40 \quad \text{False}$$

The result, $16 < -40$, is false. To make it true, change the direction of the inequality symbol to get

$$16 > -40.$$

As these examples suggest, multiplying both sides of an inequality by a *negative* number forces the direction of the inequality symbol to be reversed. The same is true for dividing by a negative number, since division is defined in terms of multiplication.

CAUTION It is a common error to forget to reverse the direction of the inequality symbol when multiplying or dividing by a negative number.

MULTIPLICATION PROPERTY OF INEQUALITY

For all real numbers a, b, and c, with $c \neq 0$,

(a) the inequalities

$$a < b \quad \text{and} \quad ac < bc$$

are equivalent if $c > 0$;

(b) the inequalities

$$a < b \quad \text{and} \quad ac > bc$$

are equivalent if $c < 0$. (Both sides of an inequality may be multiplied by a *positive* number without changing the direction of the inequality symbol. Multiplying or dividing by a *negative* number requires that the inequality symbol be reversed.)

EXAMPLE 3 **USING THE MULTIPLICATION PROPERTY OF INEQUALITY**

Solve each inequality, then graph each solution.

(a) $5m \leq -30$

Use the multiplication property to divide both sides by 5. Since $5 > 0$, do *not* reverse the inequality symbol.

$$5m \leq -30$$

$$\frac{5m}{5} \leq \frac{-30}{5} \qquad \text{Divide by 5.}$$

$$m \leq -6$$

The solution set, graphed in Figure 2.17, is the interval $(-\infty, -6]$.

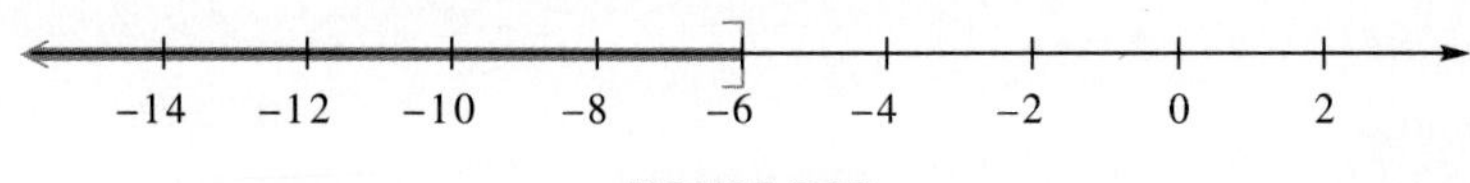

FIGURE 2.17

(b) $-4k \leq 32$

Divide both sides by -4. Since $-4 < 0$, the inequality symbol must be reversed.

$$-4k \leq 32$$

$$\frac{-4k}{-4} \geq \frac{32}{-4} \qquad \text{Divide by } -4 \text{ and reverse symbol.}$$

$$k \geq -8$$

Figure 2.18 shows the graph of the solution set, $[-8, \infty)$. ■

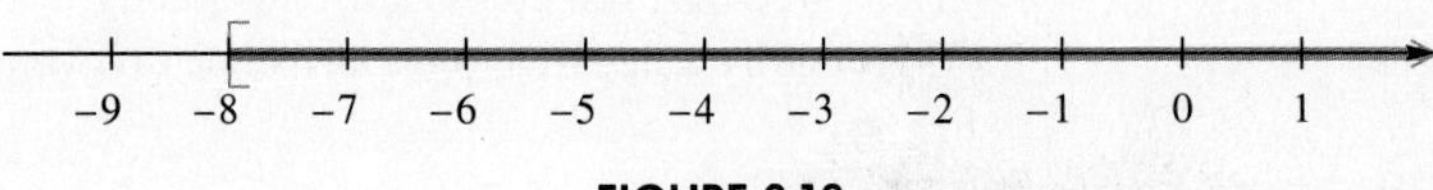

FIGURE 2.18

EXAMPLE 4
SOLVING A LINEAR INEQUALITY

Solve $-3(x+4)+2 \geq 8-x$.

Begin by using the distributive property on the left.

$$-3x - 12 + 2 \geq 8 - x \qquad \text{Distributive property}$$
$$-3x - 10 \geq 8 - x$$

Next, the addition property is used. First add x to both sides, and then add 10 to both sides.

$$-3x - 10 + x \geq 8 - x + x \qquad \text{Add } x.$$
$$-2x - 10 \geq 8$$
$$-2x - 10 + 10 \geq 8 + 10 \qquad \text{Add 10.}$$
$$-2x \geq 18$$

Finally, use the multiplication property and divide both sides of the inequality by -2. Dividing by a negative number requires changing $\geq$ to $\leq$.

$$\frac{-2x}{-2} \leq \frac{18}{-2} \qquad \text{Divide by } -2\text{; reverse inequality symbol.}$$
$$x \leq -9$$

Figure 2.19 shows the graph of the solution set, $(-\infty, -9]$. ■

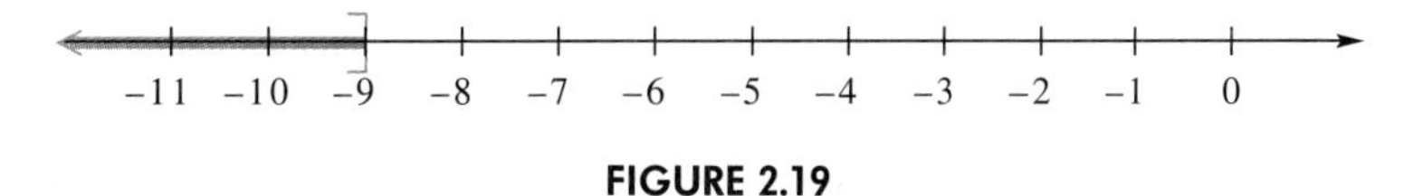

FIGURE 2.19

Most linear inequalities, like the one in Example 4, require the use of both the addition and multiplication properties. The steps used in solving a linear inequality are summarized below.

SOLVING A LINEAR INEQUALITY

Step 1 Simplify each side of the inequality as much as possible by using the distributive property to clear parentheses and by combining like terms as needed.

Step 2 Use the addition property of inequality to change the inequality so that all terms with variables are on one side and all terms without variables are on the other side.

Step 3 Use the multiplication property to change the inequality to the form $x < k$ or $x > k$.

Remember: Reverse the direction of the inequality symbol **only** when multiplying or dividing both sides of an inequality by a **negative** number, and never otherwise.

EXAMPLE 5
SOLVING A LINEAR INEQUALITY

Solve $2 - 4(r - 3) < 3(5 - r) + 5$, and graph the solution set.

Step 1 $\quad 2 - 4(r - 3) < 3(5 - r) + 5$

$2 - 4r + 12 < 15 - 3r + 5$ Distributive property

$14 - 4r < 20 - 3r$

Step 2 $\quad 14 - 4r + 3r < 20 - 3r + 3r$ Add $3r$ to both sides.

$14 - r < 20$

$14 - r - 14 < 20 - 14$ Subtract 14.

$-r < 6$

Solve for r by multiplying both sides of the inequality by -1. Since -1 is negative, change the direction of the inequality symbol.

Step 3 $\quad (-1)(-r) > (-1)(6)$ Multiply by -1, change $<$ to $>$.

$r > -6$

The solution set, $(-6, \infty)$, is graphed in Figure 2.20. ■

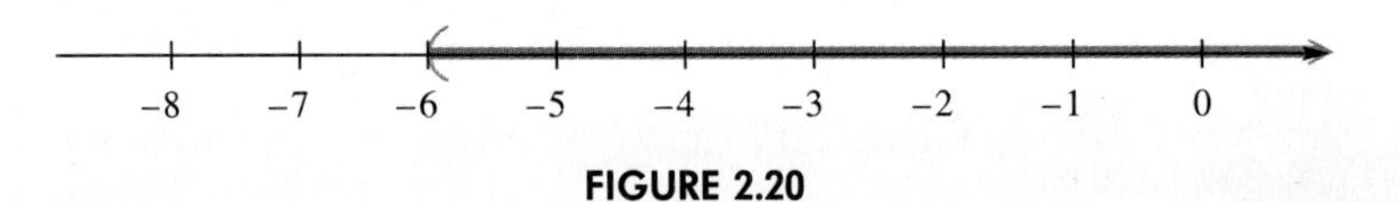

FIGURE 2.20

3 For some applications, it is necessary to work with an inequality such as

$$3 < x + 2 < 8,$$

where $x + 2$ is *between* 3 and 8. To solve this inequality, subtract 2 from each of the three parts of the inequality, giving

$$3 - 2 < x + 2 - 2 < 8 - 2$$

$$1 < x < 6.$$

The solution set, $(1, 6)$, is graphed in Figure 2.21.

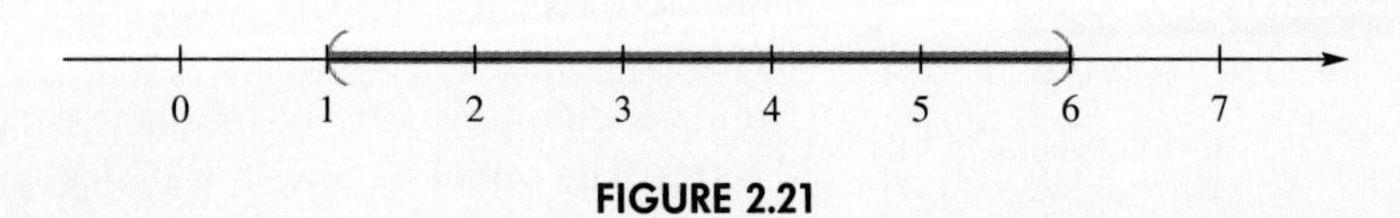

FIGURE 2.21

CAUTION When using inequalities with three parts like the one above, it is important to have the numbers in the correct positions. It would be *wrong* to write the inequality as $8 < x + 2 < 3$, since this would imply that $8 < 3$, a false statement. In general, three-part inequalities are written so that the symbols point in the same direction, and they both point toward the smaller number.

EXAMPLE 6
SOLVING A THREE-PART INEQUALITY

Solve the inequality $-2 \le 3k - 1 \le 5$ and graph the solution set.

To begin, add 1 to each of the three parts.

$$-2 + 1 \le 3k - 1 + 1 \le 5 + 1 \qquad \text{Add 1.}$$
$$-1 \le 3k \le 6$$

Now divide each of the three parts by the positive number 3.

$$\frac{-1}{3} \le \frac{3k}{3} \le \frac{6}{3} \qquad \text{Divide by 3.}$$
$$-\frac{1}{3} \le k \le 2$$

A graph of the solution set, $\left[-\frac{1}{3}, 2\right]$, is shown in Figure 2.22. ■

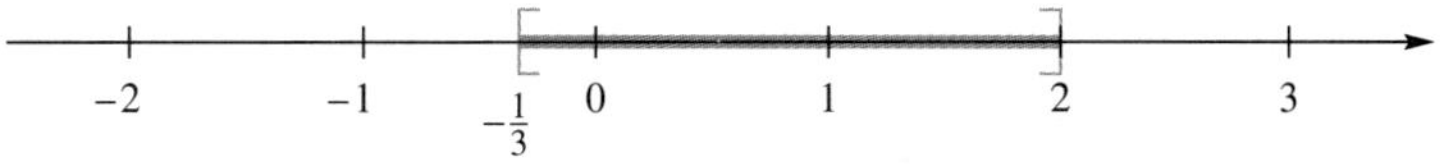

FIGURE 2.22

The types of solutions to linear equations or linear inequalities follow.

SOLUTION SETS OF LINEAR EQUATIONS AND INEQUALITIES

Equation or inequality	*Typical solution set*	*Graph of solution set*
Linear equation $ax + b = c$	$\{p\}$	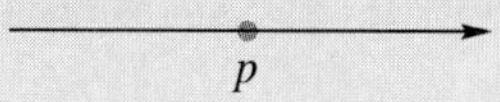
Linear inequality $ax + b < c$	$(-\infty, p)$ or (p, ∞)	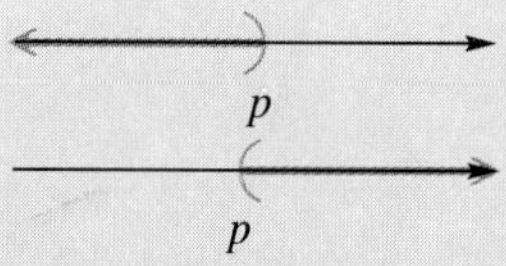
Three-part inequality $c < ax + b < d$	(p, q)	

4 There are several phrases that denote inequality. Some of them were discussed in Chapter 1. In addition to the familiar "is less than" and "is greater than" (which are examples of **strict** inequalities), the expressions "is no more than," "is at least," and others also denote inequalities. (These are called **nonstrict.**)

PROBLEM SOLVING

Expressions for nonstrict inequalities sometimes appear in applied problems that are solved using inequalities. The chart below shows how these expressions are interpreted.

WORD EXPRESSION	INTERPRETATION	WORD EXPRESSION	INTERPRETATION
a is at least b	$a \ge b$	a is at most b	$a \le b$
a is no less than b	$a \ge b$	a is no more than b	$a \le b$ ■

The final example shows how an applied problem can be solved using a linear inequality.

EXAMPLE 7
SOLVING AN APPLICATION USING A LINEAR INEQUALITY

A rental company charges \$15.00 to rent a chain saw, plus \$2.00 per hour. Pratap Puri can spend no more than \$35.00 to clear some logs from his yard. What is the maximum amount of time he can use the rented saw?

Let h = the number of hours he can rent the saw. He must pay \$15.00, plus \$2.00h, to rent the saw for h hours, and this amount must be *no more than* \$35.00.

cost of renting	is no more than	35 dollars	
$15 + 2h$	$\leq$	35	
$15 + 2h - 15$	$\leq$	$35 - 15$	Subtract 15.
$2h$	$\leq$	20	
h	$\leq$	10	Divide by 2.

He can use the saw for a maximum of 10 hours. (Of course, he may use it for less time, as indicated by the inequality $h \leq 10$.) ■

2.5 EXERCISES

Solve each inequality, then graph the solution. See Examples 1–6.

1. $4x > 8$
2. $6y > 18$
3. $2m \leq -6$
4. $5k \leq -15$
5. $3r + 1 \geq 16$
6. $2m - 5 \geq 15$
7. $\dfrac{3k - 1}{4} > 2$
8. $\dfrac{5z - 6}{8} < 3$
9. $-\dfrac{3}{4}r \geq 21$
10. $-\dfrac{2}{3}y < -10$
11. $-\dfrac{3}{2}y \leq -\dfrac{9}{2}$
12. $-\dfrac{2}{5}x \geq -4$
13. $-1.3m > 3.9$
14. $-2.5y \leq -7.5$
15. $\dfrac{2k - 5}{-4} > 1$
16. $\dfrac{3z - 2}{-5} \leq 4$
17. $-r \leq -7$
18. $-m > -12$
19. $-4x + 3 < 15$
20. $-6p - 2 \geq 16$
21. $-3 < x - 5 < 6$
22. $-6 < x + 1 < 8$
23. $-6 \leq 2z + 4 \leq 12$
24. $-15 < 3p + 6 < -9$
25. $-19 \leq 3x - 5 \leq -9$
26. $-16 < 3t + 2 < -11$
27. $-4 \leq \dfrac{2x - 5}{6} \leq 5$
28. $-8 \leq \dfrac{3m + 1}{4} \leq 3$
29. $3(x + 2) - 5x < x$
30. $2(3k - 5) + 7 > k + 12$
31. $y + 4(2y - 1) \geq 5y$
32. $-2(m - 4) \leq -3(m + 1)$
33. $-(4 + r) + 2 - 3r < -10$
34. $-(9 + k) - 5 + 4k \geq 1$
35. $-3(z - 6) > 2z - 5$
36. $-2(y + 4) \leq 6y + 8$
37. $\dfrac{2}{3}(3k - 1) \geq \dfrac{1}{2}(2k - 3)$
38. $\dfrac{7}{5}(10m - 1) < \dfrac{2}{3}(6m - 5)$
39. $-\dfrac{1}{4}(p + 6) + \dfrac{3}{2}(2p - 5) < 0$
40. $\dfrac{3}{5}(k - 2) - \dfrac{1}{4}(2k - 7) < 3$

41. Explain how to decide whether parentheses or brackets are used in graphing an inequality. (See Objective 1.)

42. Explain the distinction between a strict inequality and a nonstrict inequality. (See Objective 4.)

Solve each of the following problems. See Example 7.

43. Louise Siedelmann got scores of 88 and 78 on her first two tests. What score must she make on her third test to keep an average of 80 or greater?

44. Simon Cohen scored 94 and 82 on his first two tests. What score must he make on his third test to keep an average of 90 or greater?

45. A nurse must make sure that Ms. Carlson receives at least 30 units of a certain drug each day. This drug comes from red pills or green pills, each of which provides 3 units of the drug. The patient must have twice as many red pills as green pills. Find the smallest number of green pills that will satisfy the requirement.

46. A Christmas tree has twice as many green lights as red lights, and it has at least 15 lights. At least how many red lights does it have?

47. Samantha has a total of 815 points so far in her algebra class. At the end of the course she must have 82% of the 1100 points possible in order to get a B. What is the lowest score she can earn on the 100-point final to get a B in the class?

48. A nearby pharmacy school charges a tuition of \$6440 annually. Tom makes no more than \$1610 per year in his summer job. What is the least number of summers that he must work in order to make enough for one year's tuition?

49. A couple wishes to rent a car for one day while on vacation. Ford Automobile Rental wants \$15.00 per day and 14¢ per mile, while Chevrolet-For-A-Day wants \$14.00 per day and 16¢ per mile. After how many miles would the price to rent the Chevrolet exceed the price to rent a Ford?

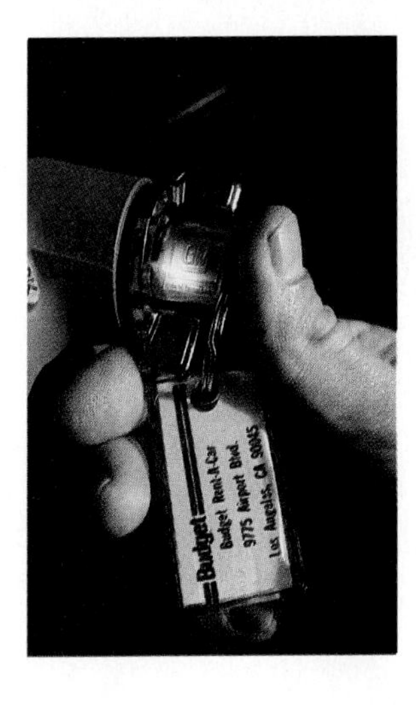

50. Jane and Terry Brandsma went to Long Island for a week. They needed to rent a car, so they checked out two rental firms. Avis wanted \$28 per day, with no mileage fee. Downtown Toyota wanted \$108 per week and 14¢ per mile. How many miles would they have to drive before the Avis price is less than the Toyota price?

A product will produce a profit only when the revenue R from selling the product exceeds the cost C of producing it. In the following exercises, find the smallest number of units, x, that must be sold for each of the following to show a profit.

51. The cost to produce x units of wire is $C = 50x + 5000$, while the revenue is $R = 60x$.

52. The cost to produce x units of squash is $C = 100x + 6000$, while the revenue is $R = 500x$.

53. $C = 85x + 900;\ R = 105x$

54. $C = 70x + 500;\ R = 60x$

Find the solution set of each of the following inequalities.

55. $3(2x - 4) - 4x < 2x + 3$

56. $7(4 - x) + 5x < 2(16 - x)$

57. $8\left(\frac{1}{2}x + 3\right) < 8\left(\frac{1}{2}x - 1\right)$

58. $10x + 2(x - 4) < 12x - 10$

Find the unknown numbers in each of the following.

59. Six times a number is between -12 and 12.

60. Half a number is between -3 and 2.

61. When 1 is added to twice a number, the result is greater than or equal to 7.

62. If 8 is subtracted from a number, the result is at least 5.

63. One third of a number is added to 6, giving a result of at least 3.

64. Three times a number, minus 5, is no more than 7.

---

65. Suppose that $-1 < x < 5$. What can you say about $-x$?

66. Suppose that $4 < y < 1$. What can you say about y?

67. Write an explanation of what is wrong with the following argument.

Let a and b be numbers, with $a > b$. Certainly, $2 > 1$. Multiply both sides of the inequality by $b - a$.

$$2(b - a) > 1(b - a)$$
$$2b - 2a > b - a$$
$$2b - b > 2a - a$$
$$b > a$$

But the final inequality is impossible, since we know that $a > b$ from the given information. (See Objective 2.)

68. Assume that $0 < a < b$, and go through the following steps.

$$a < b$$
$$ab < b^2$$
$$ab - b^2 < 0$$
$$b(a - b) < 0$$

Divide both sides by $a - b$ to get $b < 0$. This implies that b is negative. We originally assumed that b is positive. What is wrong with this argument? (See Objective 2.)

PREVIEW EXERCISES

Find all values of x satisfying both *of the following conditions* at the same time. *See Sections 1.2 and 2.5.*

69. $x \leq 2, \quad x \geq -1$ **70.** $x \leq -3, \quad x \geq -7$ **71.** $x > -1, \quad x < 8$ **72.** $x > 0, \quad x < 6$

Fill in the blank with the correct response. See Section 1.1.

73. A collection of objects is called a ____________.

74. The notation $\{x \mid x$ is a natural number greater than $6\}$ is an example of ____________ notation.

2.6 SET OPERATIONS AND COMPOUND INEQUALITIES

OBJECTIVES

1 FIND THE INTERSECTION OF TWO SETS.

2 SOLVE COMPOUND INEQUALITIES WITH THE WORD *AND*.

3 FIND THE UNION OF TWO SETS.

4 SOLVE COMPOUND INEQUALITIES WITH THE WORD *OR*.

The words *and* and *or* are very important in interpreting certain kinds of equations and inequalities in algebra. They are also used when studying sets. In this section we discuss the use of these two words as they relate to sets and inequalities.

1 We start by looking at the use of the word "and" with sets. The intersection of sets is defined below.

INTERSECTION OF SETS

For any two sets A and B, the **intersection** of A and B, symbolized $A \cap B$, is defined as follows:

$$A \cap B = \{x \mid x \text{ is an element of } A \textbf{ and } x \text{ is an element of } B\}.$$

EXAMPLE 1 FINDING THE INTERSECTION OF TWO SETS

Let $A = \{1, 2, 3, 4\}$ and $B = \{2, 4, 6\}$. Find $A \cap B$.

The set $A \cap B$ is made up of those elements that belong to both A *and* B at the same time: the numbers 2 and 4. Therefore,

$$A \cap B = \{1, 2, 3, 4\} \cap \{2, 4, 6\} = \{2, 4\}. \quad \blacksquare$$

2 A **compound inequality** consists of two inequalities linked by a connective word such as *and* or *or*. Examples of compound inequalities are

$$x + 1 \leq 9 \quad \text{and} \quad x - 2 \geq 3$$

and

$$2x > 4 \quad \text{or} \quad 3x - 6 < 5.$$

To solve a compound inequality with the word *and*, we use the following steps.

SOLVING INEQUALITIES WITH AND

Step 1 Solve each inequality in the compound inequality individually.

Step 2 Since the inequalities are joined with *and*, the solution will include all numbers that satisfy both solutions in Step 1 at the same time (the intersection of the solution sets).

The next example shows how a compound inequality with *and* is solved.

EXAMPLE 2 SOLVING A COMPOUND INEQUALITY WITH AND

Solve the compound inequality

$$x + 1 \leq 9 \quad \text{and} \quad x - 2 \geq 3.$$

Step 1 directs that we solve each inequality in the compound inequality individually.

$$x + 1 \leq 9 \quad \text{and} \quad x - 2 \geq 3$$

$$x \leq 8 \quad \text{and} \quad x \geq 5$$

Now we apply Step 2. Since the inequalities are joined with the word *and*, the solution set will include all numbers that satisfy both solutions in Step 1 at the same time. Thus, the compound inequality is true whenever $x \leq 8$ and $x \geq 5$ are both true. The top graph in Figure 2.23 shows $x \leq 8$ and the bottom graph shows $x \geq 5$.

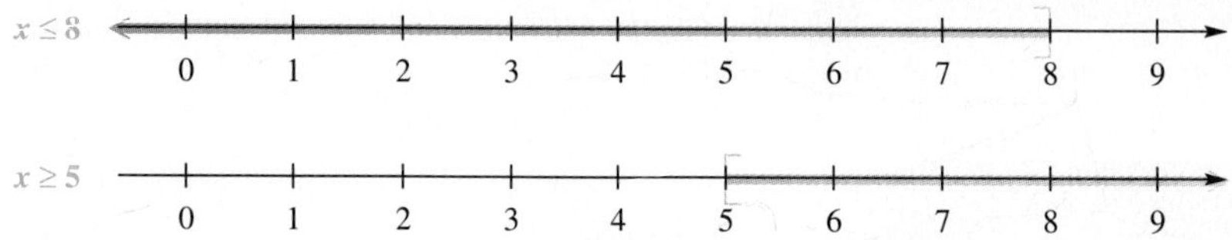

FIGURE 2.23

Now, find the intersection of the two graphs in Figure 2.23 to get the solution set of the compound inequality. The solution set consists of all numbers between 5 and 8, including both 5 and 8. This is the intersection of the two graphs, written in interval notation as [5, 8]. See Figure 2.24. ■

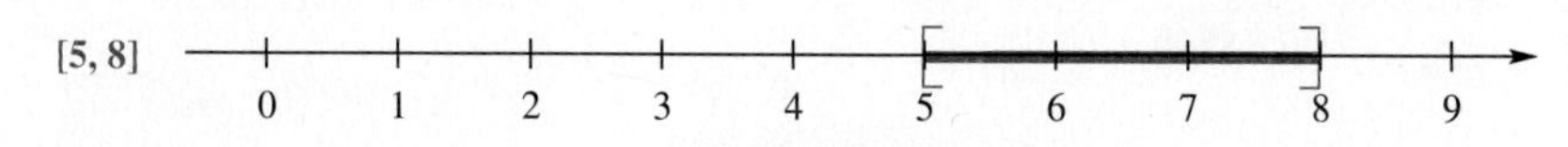

FIGURE 2.24

EXAMPLE 3 **SOLVING A COMPOUND INEQUALITY WITH AND**

Solve the compound inequality

$$-3x - 2 > 4 \quad \text{and} \quad 5x - 1 \leq -21.$$

Begin by solving $-3x - 2 > 4$ and $5x - 1 \leq -21$ separately.

$$\begin{aligned} -3x - 2 &> 4 & \text{and} \qquad 5x - 1 &\leq -21 \\ -3x &> 6 & \text{and} \qquad 5x &\leq -20 \\ x &< -2 & \text{and} \qquad x &\leq -4 \end{aligned}$$

The graphs of $x < -2$ and $x \leq -4$ are shown in Figure 2.25.

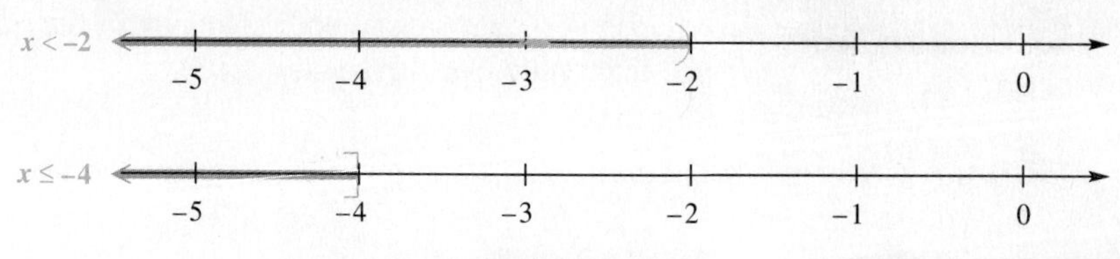

FIGURE 2.25

Now find all values of x that satisfy both conditions; that is, the real numbers that are less than -2 and also less than or equal to -4. As shown by the graph in Figure 2.26, the solution set is $(-\infty, -4]$. ■

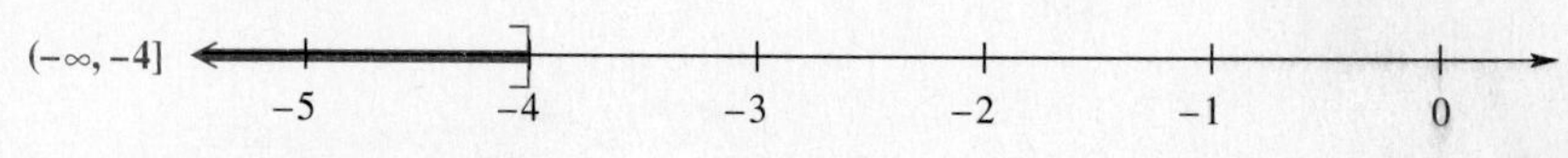

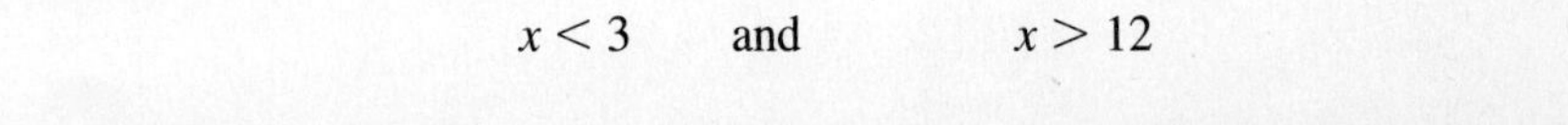

FIGURE 2.26

EXAMPLE 4 **SOLVING A COMPOUND INEQUALITY WITH AND**

Solve $x + 2 < 5$ and $x - 10 > 2$.

First solve each inequality separately.

$$\begin{aligned} x + 2 &< 5 & \text{and} \qquad x - 10 &> 2 \\ x &< 3 & \text{and} \qquad x &> 12 \end{aligned}$$

The graphs of $x < 3$ and $x > 12$ are shown in Figure 2.27.

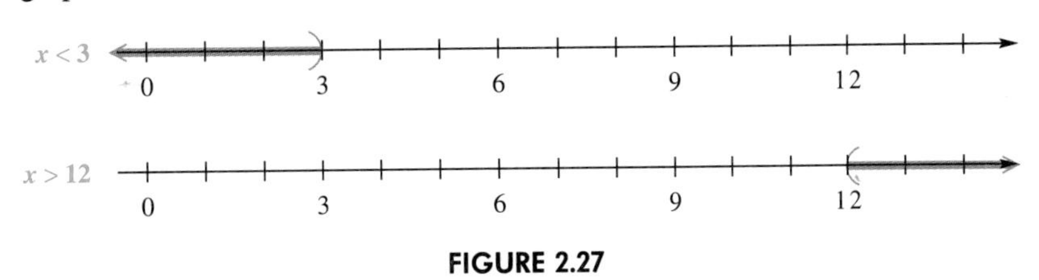

FIGURE 2.27

There is no number that is both less than 3 *and* greater than 12, so the given compound inequality has no solution. The solution set is $\emptyset$. See Figure 2.28. ■

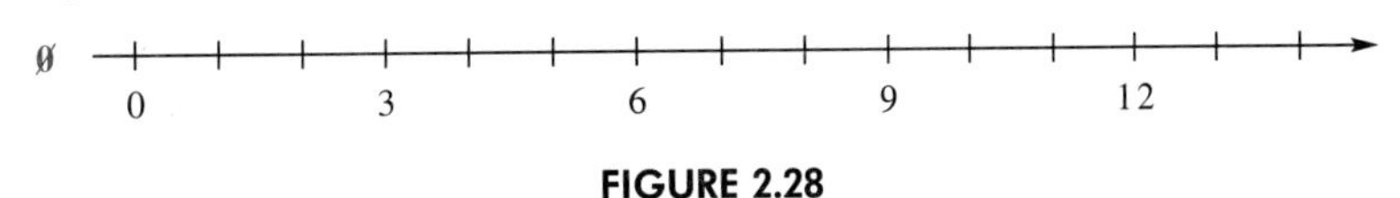

FIGURE 2.28

3 We now discuss the union of two sets, which involves the use of the word "or."

UNION OF SETS

For any two sets A and B, the **union** of A and B, symbolized $A \cup B$, is defined as follows:

$$A \cup B = \{x \mid x \text{ is an element of } A \textbf{ or } x \text{ is an element of } B\}.$$

EXAMPLE 5
FINDING THE UNION OF TWO SETS

Find the union of the sets $A = \{1, 2, 3, 4\}$ and $B = \{2, 4, 6\}$.

Begin by listing all the elements of set A: 1, 2, 3, 4. Then list any additional elements from set B. In this case the elements 2 and 4 are already listed, so the only additional element is 6. Therefore,

$$A \cup B = \{1, 2, 3, 4\} \cup \{2, 4, 6\} = \{1, 2, 3, 4, 6\}.$$

The union consists of all elements in either A *or* B (or both). ■

Notice in Example 5, that even though the elements 2 and 4 appeared in both sets A and B, they are only written once in $A \cup B$. It is not necessary to write them more than once in the union.

4 To solve compound inequalities with the word *or,* we use the following steps.

SOLVING INEQUALITIES WITH OR

Step 1 Solve each inequality in the compound inequality individually.

Step 2 Since the inequalities are joined with *or,* the solution will include all numbers that satisfy either one or both of the solutions in Step 1 (the union of the solution sets.)

The next examples show how to solve a compound inequality with *or*.

EXAMPLE 6
SOLVING A COMPOUND INEQUALITY WITH OR

Solve $6x - 4 < 2x$ or $-3x \leq -9$.

Solve each inequality separately (Step 1).

$$6x - 4 < 2x \quad \text{or} \quad -3x \leq -9$$
$$4x < 4$$
$$x < 1 \quad \text{or} \quad x \geq 3$$

The graphs of these two inequalities are shown in Figure 2.29.

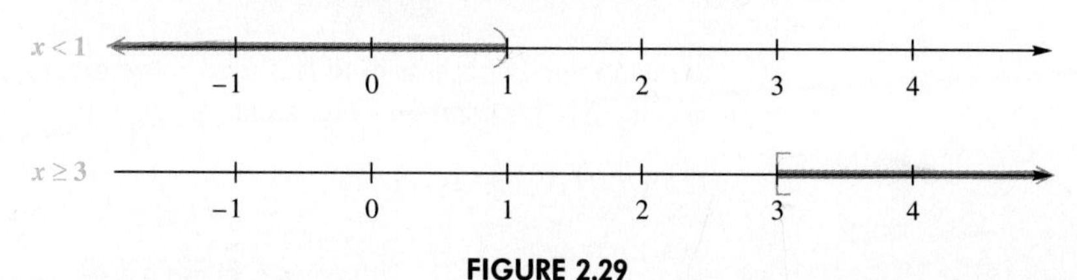

FIGURE 2.29

Since the inequalities are joined with *or,* find the union of the two sets (Step 2). The union is shown in Figure 2.30, and is written

$$(-\infty, 1) \cup [3, \infty). \quad \blacksquare$$

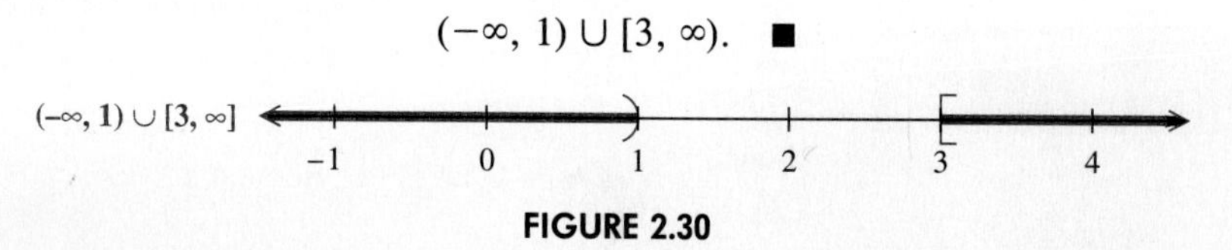

FIGURE 2.30

CAUTION When interval notation is used to write the solution of Example 6, it *must* be written as

$$(-\infty, 1) \cup [3, \infty).$$

There is no short-cut way to write the solution of a union.

EXAMPLE 7
SOLVING A COMPOUND INEQUALITY WITH OR

Solve $-4x + 1 \geq 9$ or $5x + 3 \geq -12$.

Solve each inequality separately.

$$-4x + 1 \geq 9 \quad \text{or} \quad 5x + 3 \geq -12$$
$$-4x \geq 8 \quad \text{or} \quad 5x \geq -15$$
$$x \leq -2 \quad \text{or} \quad x \geq -3$$

The graphs of these two inequalities are shown in Figure 2.31.

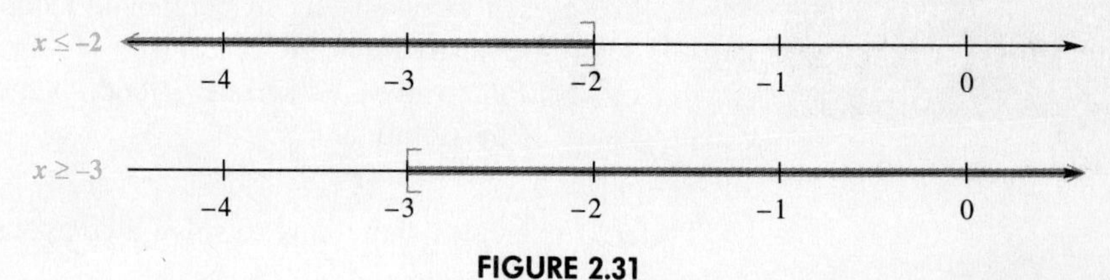

FIGURE 2.31

By taking the union, we obtain every real number as a solution, since every real number satisfies at least one of the two inequalities. The set of all real numbers is written in interval notation as

$$(-\infty, \infty).$$

The graph of the solution set is shown in Figure 2.32. ■

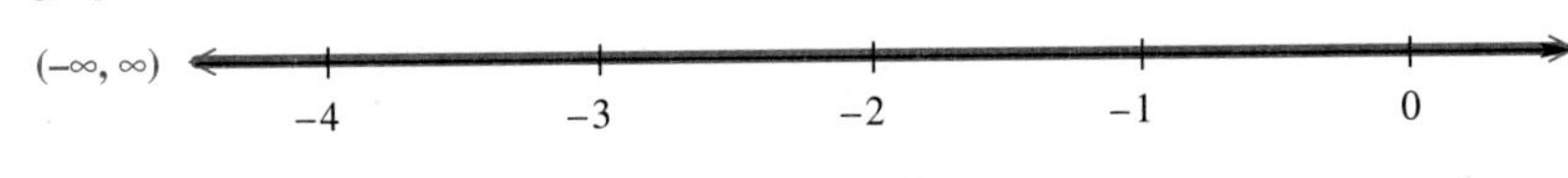

FIGURE 2.32

2.6 EXERCISES

Let $A = \{a, b, c, d, e, f\}$, $B = \{a, c, e\}$, $C = \{a, f\}$, *and* $D = \{d\}$. *List the elements in each of the following sets. See Examples 1 and 5.*

1. $A \cap B$ **2.** $B \cap A$ **3.** $A \cap D$ **4.** $B \cap D$

5. $B \cap C$ **6.** $A \cap \emptyset$ **7.** $A \cup B$ **8.** $B \cup D$

9. $B \cup C$ **10.** $C \cup B$ **11.** $C \cup D$ **12.** $D \cup C$

13. $B \cup \emptyset$ **14.** $A \cap (C \cap D)$ **15.** $A \cap (B \cap C)$ **16.** $(A \cap B) \cap C$

17. $B \cup (C \cup D)$ **18.** $(B \cup C) \cup D$ **19.** $(A \cap B) \cup C$ **20.** $(B \cup C) \cap A$

21. Compare the answers to Exercises 9 and 10. Notice that $B \cup C = C \cup B$. It turns out that for any sets B and C, this equality holds true. Recall the properties of real numbers studied in Section 1.4. What is the name of the property that says that the order of the operation makes no difference?

22. Compare the answers to Exercises 15 and 16. Notice that $A \cap (B \cap C) = (A \cap B) \cap C$. It turns out that for any sets A, B, and C, this equality holds true. Recall the properties of real numbers studied in Section 1.4. What is the name of the property that says that the grouping used in the operation makes no difference?

23. A compound inequality uses one of the words *and* or *or*. Explain how it is determined whether to use *intersection* or *union* when graphing the solution set. (See Objectives 2 and 4.)

24. Suppose that k is some real number. Explain why the solution set of

$$x < k \quad \text{and} \quad x > k$$

is $\emptyset$. (See Objective 2.)

Solve each of the following compound inequalities, and then graph each solution. See Examples 2, 3, 4, 6, and 7.

25. $x > 3$ and $x < 7$ **26.** $x < 2$ and $x > -1$ **27.** $x < 4$ and $x > 0$

28. $x \le 2$ or $x \le 6$ **29.** $x \le -8$ or $x \le -12$ **30.** $x \le 3$ and $x \le 5$

31. $x \ge 3$ and $x \ge 5$ **32.** $x \le 3$ or $x \ge 5$ **33.** $x \ge 3$ or $x \le 5$

34. $x \ge 3$ or $x \ge 5$ **35.** $x - 3 \le 5$ and $x + 2 \ge 6$ **36.** $x + 5 \le 9$ and $x - 3 \ge -2$

37. $3x < -3$ and $x + 2 > 0$ **38.** $3x > -3$ and $x + 2 < 2$ **39.** $2x - 1 < 3$ and $2x - 8 > -4$

40. $5x + 2 > 2$ and $5x + 2 \le 7$ **41.** $6x - 8 \le 16$ and $4x - 1 \le 15$ **42.** $7x + 6 < 48$ and $-2x \ge -6$

43. $x + 2 > 6$ or $x - 1 < -5$

44. $x - 3 > 2$ or $x + 4 < 3$

45. $2x + 3 > 7$ or $4x - 1 < 3$

46. $3x < x + 12$ or $x - 1 > 5$

47. $2x - 5 > 10$ or $4x > 4$

48. $3x - 6 > 15$ or $2x > 15$

49. $-5x + 2 \leq 17$ or $2x + 1 \leq 9$

50. $-3x - 4 > 9$ or $4x + 5 > -13$

Express each of the following in the simplest interval form.

51. $(-\infty, -1] \cap [-4, \infty)$

52. $[-1, \infty) \cap (-\infty, 9]$

53. $[4, \infty) \cap (-\infty, 12]$

54. $(-\infty, -6] \cap [-9, \infty)$

55. $(-\infty, 3) \cup (-\infty, -2)$

56. $(-\infty, 5) \cup (0, \infty)$

57. $(5, 11] \cap [6, \infty)$

58. $[-9, 1] \cap (-\infty, -3)$

59. $(-1, 4) \cap (2, 7)$

60. $[3, 6] \cap (4, 9)$

61. $[-1, 2] \cup (0, 5)$

62. $[3, 8] \cup (5, 11)$

In Exercises 63 and 64, let

$$A = \{1, 2, 3, 4, 5, 6\} \qquad B = \{2, 4, 5, 6\} \qquad \text{and} \qquad C = \{1, 2, 5\}.$$

It can be shown using advanced ideas from set theory that each of the following distributive properties holds for any sets A, B, and C. Use the given sets to illustrate these properties.

63. $A \cup (B \cap C) = (A \cup B) \cap (A \cup C)$

64. $A \cap (B \cup C) = (A \cap B) \cup (A \cap C)$

See the discussion of subsets in Section 1.1 to answer Exercises 65 and 66.

65. If $A \subseteq B$, must $A \cap B = A$?

66. If $A \subseteq B$, must $A \cup B = B$?

Under what conditions on sets A and B do the following hold?

67. $A \cap B = A \cup B$

68. $A \cap A = \emptyset$

69. $B \cup B = B$

70. $B \cap \emptyset = \emptyset$

PREVIEW EXERCISES

Solve each inequality. See Section 2.5.

71. $2y - 4 \leq 3y + 2$

72. $5t - 8 < 6t - 7$

73. $-5 < 2r + 1 < 5$

74. $-7 \leq 3w - 2 < 7$

Evaluate each of the following. See Sections 1.1 and 1.3.

75. $-|6| - |-11| + (-4)$

76. $(-5) - |-9| + |5 - 4|$

2.7 ABSOLUTE VALUE EQUATIONS AND INEQUALITIES

OBJECTIVES

1. USE THE DISTANCE DEFINITION OF ABSOLUTE VALUE.
2. SOLVE EQUATIONS OF THE FORM $|ax + b| = k$, FOR $k > 0$.
3. SOLVE INEQUALITIES OF THE FORM $|ax + b| < k$ AND OF THE FORM $|ax + b| > k$, FOR $k > 0$.
4. SOLVE ABSOLUTE VALUE EQUATIONS THAT INVOLVE REWRITING.
5. SOLVE ABSOLUTE VALUE EQUATIONS OF THE FORM $|ax + b| = |cx + d|$.
6. SOLVE ABSOLUTE VALUE EQUATIONS AND INEQUALITIES THAT HAVE ONLY NONPOSITIVE CONSTANTS ON ONE SIDE.

1 In Chapter 1 it was shown that the absolute value of a number x, written $|x|$, represents the distance from x to 0 on the number line. For example, the solutions of $|x| = 4$ are 4 and -4, as shown in Figure 2.33.

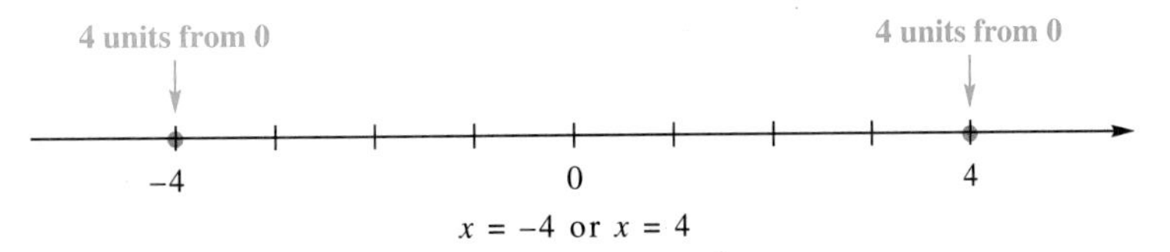

FIGURE 2.33

Since absolute value represents distance from 0, it is reasonable to interpret the solutions of $|x| > 4$ to be all numbers that are *more* than 4 units from 0. The set $(-\infty, -4) \cup (4, \infty)$ fits this description. Figure 2.34 shows the graph of the solution set of $|x| > 4$.

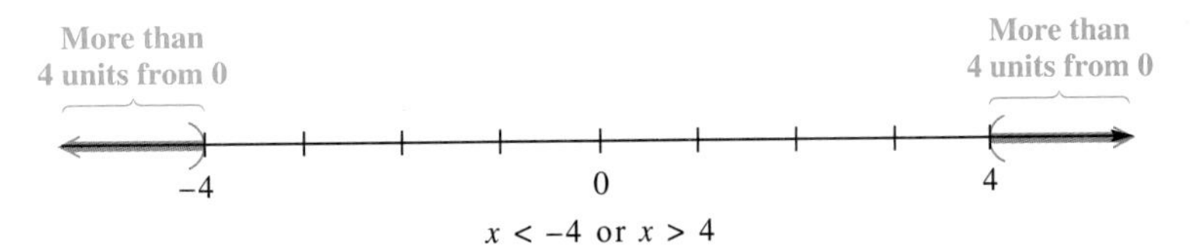

FIGURE 2.34

The solution set of $|x| < 4$ consists of all numbers that are *less* than 4 units from 0 on the number line. Another way of thinking of this is to think of all numbers between -4 and 4. This set of numbers is given by $(-4, 4)$, as shown in Figure 2.35.

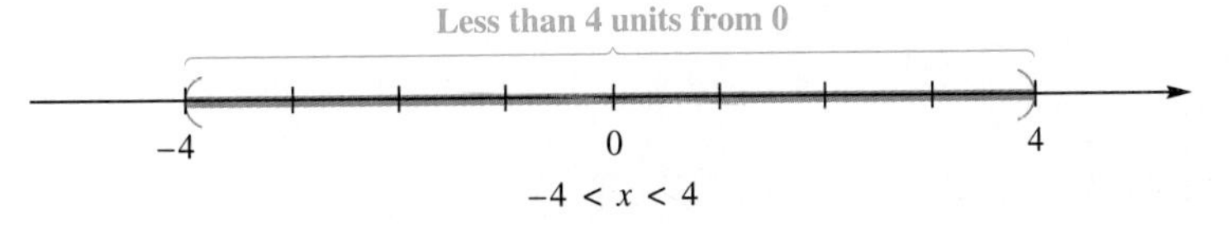

FIGURE 2.35

The equation and inequalities just described are examples of **absolute value equations and inequalities.** These are equations and inequalities that involve the absolute value of a variable expression. These generally take the form

$$|ax + b| = k, \qquad |ax + b| > k, \qquad \text{or} \qquad |ax + b| < k$$

where k is a positive number. We may solve them by rewriting them as compound equations or inequalities. The various methods of solving them are explained in the following examples.

2 The first example shows how to solve a typical absolute value equation. Remember that since absolute value refers to distance from the origin, an absolute value equation will have two cases.

EXAMPLE 1
SOLVING AN ABSOLUTE VALUE EQUATION

Solve $|2x + 1| = 7$.

For $|2x + 1|$ to equal 7, $2x + 1$ must be 7 units from 0 on the number line. This can happen only when $2x + 1 = 7$ or $2x + 1 = -7$. Solve this compound equation as follows.

$$\begin{aligned} 2x + 1 &= 7 \quad \text{or} \quad 2x + 1 = -7 \\ 2x &= 6 \quad \text{or} \quad 2x = -8 \\ x &= 3 \quad \text{or} \quad x = -4 \end{aligned}$$

The solution set is $\{-4, 3\}$. Its graph is shown in Figure 2.36. ■

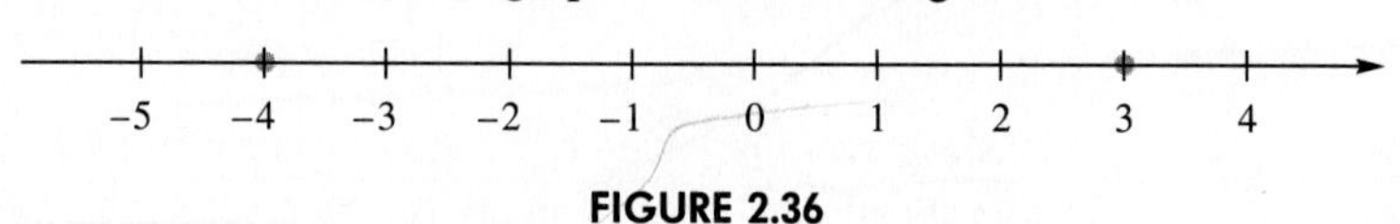

FIGURE 2.36

3 We now discuss how to solve absolute value inequalities.

EXAMPLE 2
SOLVING AN ABSOLUTE VALUE INEQUALITY WITH >

Solve $|2x + 1| > 7$.

This absolute value inequality must be rewritten as

$$2x + 1 > 7 \quad \text{or} \quad 2x + 1 < -7,$$

because $2x + 1$ must represent a number that is *more* than 7 units from 0 on either side of the number line. Now solve the compound inequality.

$$\begin{aligned} 2x + 1 &> 7 \quad \text{or} \quad 2x + 1 < -7 \\ 2x &> 6 \quad \text{or} \quad 2x < -8 \\ x &> 3 \quad \text{or} \quad x < -4 \end{aligned}$$

The solution set, $(-\infty, -4) \cup (3, \infty)$, is graphed in Figure 2.37. Notice that the graph consists of two intervals. ■

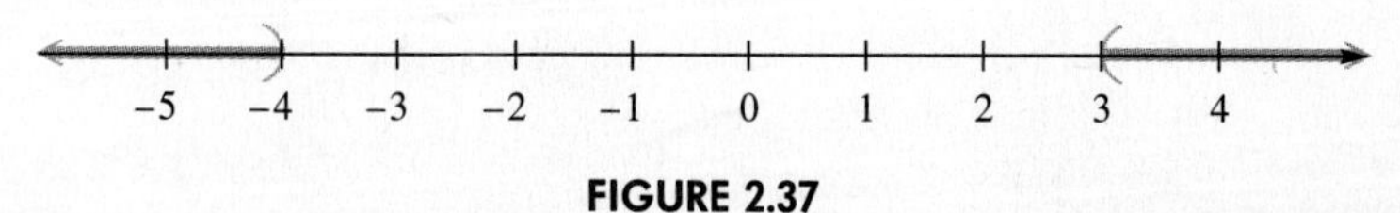

FIGURE 2.37

EXAMPLE 3
SOLVING AN ABSOLUTE VALUE INEQUALITY WITH <

Solve $|2x + 1| < 7$.

Here the expression $2x + 1$ must represent a number that is less than 7 units from 0 on the number line. Another way of thinking of this is to realize that $2x + 1$ must be between -7 and 7. This is written as the three-part inequality

$$-7 < 2x + 1 < 7.$$

We solved such inequalities in Section 2.5 by working with all three parts at the same time.

$$\begin{aligned} -7 &< 2x + 1 < 7 \\ -8 &< 2x < 6 && \text{Subtract 1 from each part.} \\ -4 &< x < 3 && \text{Divide each part by 2.} \end{aligned}$$

The solution set is $(-4, 3)$, and the graph consists of a single interval, as shown in Figure 2.38. ■

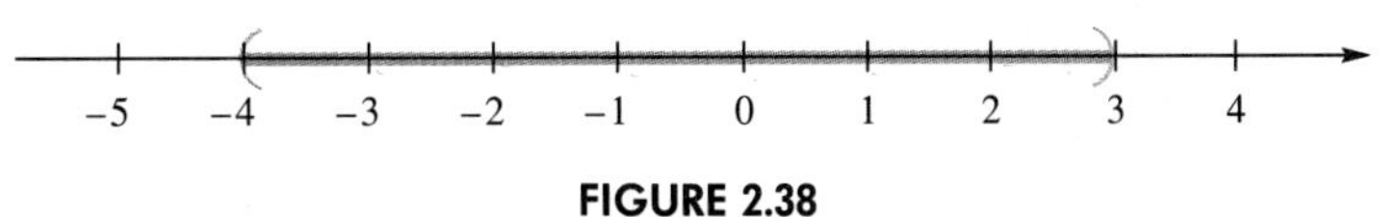

FIGURE 2.38

Look back at Figures 2.36, 2.37, and 2.38. These are the graphs of $|2x + 1| = 7$, $|2x + 1| > 7$, and $|2x + 1| < 7$. If we find the union of the three sets, we get the set of all real numbers. This is because for any value of x, $|2x + 1|$ will satisfy one and only one of the following: it is equal to 7, greater than 7, or less than 7.

CAUTION When solving absolute value equations and inequalities of the types in Examples 1, 2, and 3, be sure to remember the following.

1. The methods described apply when the constant is alone on one side of the equation or inequality and is *positive*.
2. Absolute value equations and absolute value inequalities written in the form $|ax + b| > k$ translate into ''or'' compound statements.
3. Absolute value inequalities in the form $|ax + b| < k$ translate into ''and'' compound statements, which may be written as three-part inequalities.
4. An ''or'' statement *cannot* be written in three parts. It would be *incorrect* to use

$$-7 > 2x + 1 > 7$$

in Example 2, since this would imply that $-7 > 7$, which is false.

4 Sometimes an absolute value equation or inequality is given in a form that requires some rewriting before it can be set up as a compound statement. The next example illustrates this process for an absolute value equation.

EXAMPLE 4
SOLVING AN ABSOLUTE VALUE EQUATION REQUIRING REWRITING

Solve the equation $|x + 3| + 5 = 12$.

First get the absolute value alone on one side of the equals sign. Do this by subtracting 5 on each side.

$$|x + 3| + 5 - 5 = 12 - 5 \quad \text{Subtract 5.}$$
$$|x + 3| = 7$$

Then use the method shown in Example 1.

$$x + 3 = 7 \quad \text{or} \quad x + 3 = -7$$
$$x = 4 \quad \text{or} \quad x = -10$$

Check that the solution set is $\{4, -10\}$ by substituting in the original equation. ■

Solving an absolute value *inequality* requiring rewriting is done in a similar manner.

The methods used in Examples 1, 2, and 3 are summarized on the following page.

SOLVING ABSOLUTE VALUE EQUATIONS AND INEQUALITIES

Let k be a positive number, and p and q be two numbers.

1. To solve $|ax + b| = k$, solve the compound equation

$$ax + b = k \quad \text{or} \quad ax + b = -k.$$

The solution set is usually of the form $\{p, q\}$, with two numbers.

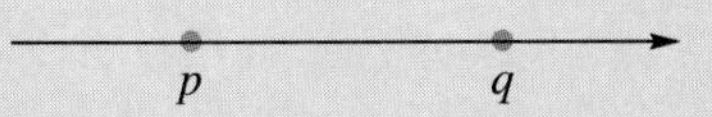

2. To solve $|ax + b| > k$, solve the compound inequality

$$ax + b > k \quad \text{or} \quad ax + b < -k.$$

The solution set is of the form $(-\infty, p) \cup (q, \infty)$, which consists of two separate intervals.

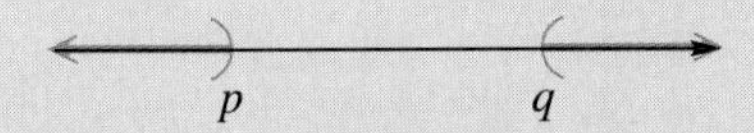

3. To solve $|ax + b| < k$, solve the compound inequality

$$-k < ax + b < k.$$

The solution set is of the form (p, q), a single interval.

5 The next example shows how to solve an absolute value equation with two absolute value expressions. For two expressions to have the same absolute value, they must either be equal or be negatives of each other.

SOLVING $|ax + b| = |cx + d|$

To solve an absolute value equation of the form

$$|ax + b| = |cx + d|,$$

solve the compound equation

$$ax + b = cx + d \quad \text{or} \quad ax + b = -(cx + d).$$

EXAMPLE 5 **SOLVING AN EQUATION WITH TWO ABSOLUTE VALUES**

Solve the equation $|z + 6| = |2z - 3|$.

This equation is satisfied either if $z + 6$ and $2z - 3$ are equal to each other, or if $z + 6$ and $2z - 3$ are negatives of each other. Thus, we have

$$z + 6 = 2z - 3 \quad \text{or} \quad z + 6 = -(2z - 3).$$

Solve each equation.

$$6 + 3 = 2z - z \quad \text{or} \quad z + 6 = -2z + 3$$
$$3z = -3$$
$$9 = z \quad \text{or} \quad z = -1$$

The solution set is $\{9, -1\}$. ■

6 When a typical absolute value equation or inequality involves a *negative* constant or *zero* alone on one side, simply use the properties of absolute value to solve. Keep in mind the following.

1. The absolute value of an expression can never be negative: $|a| \geq 0$ for all real numbers a.
2. The absolute value of an expression equals 0 only when the expression is equal to 0.

The next two examples illustrate these special cases.

EXAMPLE 6
SOLVING SPECIAL CASES OF ABSOLUTE VALUE EQUATIONS

Solve each equation.

(a) $|5r - 3| = -4$.

Since the absolute value of an expression can never be negative, there are no solutions for this equation. The solution set is $\emptyset$.

(b) $|7x - 3| = 0$

The expression $7x - 3$ will equal 0 *only* for the solution of the equation

$$7x - 3 = 0.$$

The solution of this equation is 3/7. The solution set is $\{3/7\}$. It consists of only one element. ■

EXAMPLE 7
SOLVING SPECIAL CASES OF ABSOLUTE VALUE INEQUALITIES

Solve each of the following inequalities.

(a) $|x| \geq -4$

The absolute value of a number is never negative. For this reason, $|x| \geq -4$ is true for *all* real numbers. The solution set is $(-\infty, \infty)$.

(b) $|k + 6| < -2$

There is no number whose absolute value is less than -2, so this inequality has no solution. The solution set is $\emptyset$.

(c) $|m - 7| \leq 0$

The value of $|m - 7|$ will never be less than 0. However, $|m - 7|$ will equal 0 when $m = 7$. Therefore, the solution set is $\{7\}$. ■

2.7 EXERCISES

Solve each equation. See Example 1.

1. $|x| = 8$ **2.** $|k| = 11$ **3.** $|4x| = 12$

4. $|5x| = 10$ **5.** $|y - 3| = 8$ **6.** $|p - 5| = 11$

7. $|2x + 1| = 9$ **8.** $|2y + 3| = 17$ **9.** $|4r - 5| = 13$

10. $|5t - 1| = 26$ **11.** $|2y + 5| = 12$ **12.** $|2x - 9| = 16$

13. $\left|\frac{1}{2}x + 3\right| = 4$ **14.** $\left|\frac{2}{3}q - 1\right| = 2$ **15.** $\left|1 - \frac{3}{4}k\right| = 3$

16. $\left|-4 + \frac{5}{2}m\right| = 14$ **17.** $|2(3x + 1)| = 4$ **18.** $|-3(x - 1)| = 7$

Solve each inequality. See Example 2.

19. $|x| > 2$ **20.** $|y| > 4$

21. $|k| \geq 3$ **22.** $|r| \geq 5$

23. $|t + 2| > 8$ **24.** $|r + 5| > 15$

25. $|3x - 1| \geq 5$ **26.** $|4x + 1| \geq 17$

27. $|3 - x| > 4$ **28.** $|5 - x| > 9$

Solve each inequality. See Example 3. (Hint: Compare your answers to those in Exercises 19–28.)

29. $|x| \leq 2$ **30.** $|y| \leq 4$ **31.** $|k| < 3$

32. $|r| < 5$ **33.** $|t + 2| \leq 8$ **34.** $|r + 5| \leq 15$

35. $|3x - 1| < 5$ **36.** $|4x + 1| < 17$ **37.** $|3 - x| \leq 4$ **38.** $|5 - x| \leq 9$

Solve each equation or inequality. See Examples 1, 2, and 3.

39. $|4 + k| > 9$ **40.** $|-3 + t| > 5$ **41.** $|7 + 2z| = 3$

42. $|9 - 3p| = 6$ **43.** $|3r - 1| \leq 8$ **44.** $|2s - 6| \leq 4$

45. $|6(x + 1)| < 1$ **46.** $|-2(x + 3)| < 5$

Solve each equation or inequality. See Example 4.

47. $|x| - 1 = 7$ **48.** $|y| + 3 = 9$ **49.** $|x + 4| + 1 = 3$

50. $|y + 5| - 2 = 8$ **51.** $|2x + 1| + 3 > 5$ **52.** $|6x - 1| - 2 > 1$

53. $|x + 5| - 6 \leq -2$ **54.** $|r - 2| - 3 \leq 6$

Solve each equation. See Example 5.

55. $|3x + 1| = |2x - 7|$ **56.** $|7x + 12| = |x - 4|$

57. $\left|m - \frac{1}{2}\right| = \left|\frac{1}{2}m - 2\right|$ **58.** $\left|\frac{2}{3}r - 2\right| = \left|\frac{1}{3}r + 3\right|$

59. $|6x| = |9x + 5|$ **60.** $|13y| = |2y - 1|$

61. $|2p - 6| = |2p + 5|$ **62.** $|3x - 1| = |3x + 4|$

Solve each equation or inequality. See Examples 6 and 7.

63. $|2x - 7| = -4$ **64.** $|8x + 1| = -9$ **65.** $|12t - 3| = -4$ **66.** $|13w + 1| = -\frac{1}{4}$

67. $|4x - 1| = 0$ **68.** $|6r + 2| = 0$ **69.** $|2q - 1| < -5$ **70.** $|8n + 4| < -3$

71. $|x + 5| > -8$ **72.** $|x + 9| > -4$ **73.** $|7x - 3| \leq 0$ **74.** $|4x + 1| \leq 0$

75. $|5x + 2| \geq 0$ **76.** $|4 - 7x| \geq 0$ **77.** $|10z + 9| > 0$ **78.** $|-8w + 1| > 0$

Use the properties of absolute value in Exercises 79–86.

79. Solve the equation $|x - 3| = |3 - x|$. (See Objective 5.)

80. Based on your solution in Exercise 79, find the solution set of $|x - a| = |a - x|$, for any real number a.

81. The only way that $|x + 6| = x + 6$ can be true is if $x + 6 \geq 0$. Use this fact to solve the equation $|x + 6| = x + 6$.

82. Solve $|x - 3| = -(x - 3)$.

83. Solve $|2x + 1| = 3x - 1$.

84. Using the result of Exercise 81, solve $|x + a| = x + a$, for any real number a.

85. Use the result of Exercise 82 to solve $|x + a| = -(x + a)$, for any real number a.

86. For what values of a and b is $|a - b| = |b - a|$?

By considering whether x is zero, positive, or negative, solve each of the following.

87. $x + |x| = 2$ **88.** $|x| + x = 10$ **89.** $|x| - x = 4$ **90.** $x - |x| = 8$

Solve each of the following.

91. $|x - 4| < 2x$ **92.** $|x + 1| \leq 3x$ **93.** $|x + 3| \geq 4x$ **94.** $|x + 5| > -x$

PREVIEW EXERCISES

Evaluate each of the following. See Section 1.3.

95. 2^3 **96.** 3^2 **97.** 5^2 **98.** 8^3 **99.** $(-6)^3$

100. -4^3 **101.** -6^3 **102.** $(-4)^3$ **103.** $\left(\frac{2}{3}\right)^3$ **104.** $\left(\frac{3}{2}\right)^2$

SUMMARY ON SOLVING LINEAR EQUATIONS AND INEQUALITIES

Students often have difficulty keeping the various equations and inequalities studied in this chapter separate. This section of miscellaneous equations and inequalities should help with this difficulty. Solve each of these, and give a graph of the solution of each inequality. (See the boxes in this chapter that summarize these equations and inequalities.)

1. $4z + 1 = 53$ **2.** $9y - 6 = 11$ **3.** $|m - 1| = 4$

4. $6q - 9 = 12 - q$ **5.** $3p + 7 = 9 - 8p$ **6.** $|a + 3| = 7$

7. $2m - 1 \leq m$ **8.** $8r + 2 \geq 7r$ **9.** $4(a - 11) + 3a = 2a + 11$

10. $2q - 1 = -5$ **11.** $|3q - 7| = 4$ **12.** $6z - 5 \leq 3z + 7$

13. $|x| < 3$ **14.** $|5z - 8| = 9$ **15.** $5a - 1 = 7$

16. $9y - 3(y + 1) = 8y - 3$ **17.** $|r - 2| = |4r + 1|$ **18.** $|y| \geq 4$

19. $9y - 5 \geq 4y + 3$

20. $13p - 5 > 8p + 1$

21. $|q| > 1$

22. $4z - 1 = 12 + 3z$

23. $\frac{2}{3}y - 8 = \frac{1}{4}y$

24. $-\frac{5}{8}y \geq 10$

25. $\frac{1}{4}p < -3$

26. $|y + 2| \geq 3$

27. $7z - 3 + 2z = 9z + 5$

28. $\frac{3}{5}q - \frac{1}{10} = 1$

29. $|r - 1| < 5$

30. $3r + 9 + 5r = 4(3 + 2r) - 3$

31. $11(q + 2) + 4q = (10 - 2q) + 12$

32. $6 - 3(2 - p) < 2(1 + p) + 1$

33. $|2p - 3| > 1$

34. $6q - 9 = 2(2q - 1) + q$

35. $\frac{x}{4} - \frac{2x}{3} = -5$

36. $|5a + 1| < 6$

37. $5z - (3 + z) \geq 6z + 2$

38. $|2q - 5| = 17$

39. $8(r - 1) + 2 = 7r - 3$

40. $-2 \leq 3x - 1 \leq 5$

41. $-1 \leq 4a + 3 \leq 5$

42. $2r - 6r + 4 = 3(1 - 3r) + 5r$

43. $|7z - 1| = |5z + 1|$

44. $|p + 2| = |p + 8|$

45. $|3r - 1| \geq 4$

46. $\frac{1}{2} \leq \frac{2}{3}r \leq \frac{3}{4}$

47. $-(m - 4) + 2 = 3m - 8$

48. $\frac{p}{6} - \frac{3p}{5} = -13$

49. $-6 \leq \frac{3}{2} - x \leq 2$

50. $|5 - y| < 3$

51. $|y - 1| \geq -2$

52. $|2r - 5| = |r + 3|$

53. $8q - (2 - q) = 3(1 + 3q) - 5$

54. $8y - (y + 3) = -(2y + 1) - 3$

55. $|r - 5| = |r + 7|$

56. $|r + 2| < -3$

57. $|2 + 5q| < 3$

58. $-1 \leq \frac{x}{4} + 2 \leq 1$

CHAPTER 2 GLOSSARY

KEY TERMS

2.1 algebraic expression An algebraic expression is an expression indicating any combination of the following operations: addition, subtraction, multiplication, division (except by 0) and taking roots on any collection of variables and numbers.

equation An equation is a statement that two algebraic expressions are equal.

linear equation or first-degree equation in one variable An equation is linear or first-degree in the variable x if it can be written in the form $ax + b = c$, where a, b, and c are real numbers, with $a \neq 0$.

solution A solution of an equation is a number that makes the equation true when substituted for the variable.

solution set The solution set of an equation is the set of all its solutions.

equivalent equations Equivalent equations are equations that have the same solution set.

addition and multiplication properties of equality These properties state that the same number may be added to (or subtracted from) both sides of an equation to obtain an equivalent equation; and the same nonzero number may be multiplied by or divided into both sides of an equation to obtain an equivalent equation.

conditional equation An equation that has a finite number of elements in its solution set is called a conditional equation.

contradiction An equation that has no solutions (that is, its solution set is $\emptyset$) is called a contradiction.

identity An equation that is satisfied by every number for which both sides are defined is called an identity.

2.2 **formula** A formula is a mathematical statement in which more than one letter is used to express a relationship.

degree An angle that measures one degree is 1/360 of a complete revolution.

parallel lines Lines that lie in the same plane and do not intersect are called parallel lines.

transversal A line intersecting parallel lines is called a transversal.

vertical angles Vertical angles are formed by intersecting lines, and have the same measure. (See, for example, angles ① and ④ in Figure 2.4.)

2.3 **right angle** An angle that measures 90° is called a right angle.

straight angle An angle that measures 180° is called a straight angle.

complementary angles (complements) If the sum of the measures of two angles is 90°, the angles are complementary angles. They are complements of each other.

supplementary angles (supplements) If the sum of the measures of two angles is 180°, the angles are supplementary angles. They are supplements of each other.

2.5 **linear inequality in one variable** An inequality is linear in the variable x if it can be written in the form $ax + b < c$, $ax + b \leq c$, $ax + b > c$, or $ax + b \geq c$, where a, b, and c are real numbers, with $a \neq 0$.

equivalent inequalities Equivalent inequalities are inequalities with the same solution set.

addition and multiplication properties of inequality The same number may be added to (or subtracted from) both sides of an inequality to obtain an equivalent inequality. Both sides of an inequality may be multiplied or divided by the same positive number. If both sides are multiplied by or divided by a negative number, the inequality symbol must be reversed.

strict inequality An inequality that involves $>$ or $<$ is called a strict inequality.

nonstrict inequality An inequality that involves $\geq$ or $\leq$ is called a nonstrict inequality.

2.6 **intersection** The intersection of two sets A and B is the set of elements that belong to both A and B.

union The union of two sets A and B is the set of elements that belong to either A or B (or both).

compound inequality A compound inequality is formed by joining two inequalities with a connective word, such as *and* or *or*.

2.7 **absolute value equation; absolute value inequality** Absolute value equations and inequalities are equations and inequalities that involve the absolute value of a variable expression.

NEW SYMBOLS

1° one degree $\cap$ set intersection $\cup$ set union

CHAPTER 2 QUICK REVIEW

SECTION	CONCEPTS	EXAMPLES
2.1 LINEAR EQUATIONS IN ONE VARIABLE	**Solving a Linear Equation** If necessary, eliminate fractions by multiplying both sides by the LCD. Simplify each side, and then use the addition property of equality to get the variables on one side and the numbers on the other. Combine terms if possible, and then use the multiplication property of equality to make the coefficient of the variable equal to 1. Check by substituting into the original equation.	Solve the equation $4(8 - 3t) = 32 - 8(t + 2)$. $32 - 12t = 32 - 8t - 16$ $32 - 12t = 16 - 8t$ $32 - 12t + 12t = 16 - 8t + 12t$ $32 = 16 + 4t$ $32 - 16 = 16 + 4t - 16$ $16 = 4t$ $\frac{16}{4} = \frac{4t}{4}$ $4 = t$ The solution set is $\{4\}$. This can be checked by substituting 4 for t in the original equation.
2.2 FORMULAS AND TOPICS FROM GEOMETRY	**Solving for a Specified Variable** Use the addition or multiplication properties as necessary to get all terms with the specified variable on one side of the equals sign, and all other terms on the other side. If necessary, use the distributive property to write the terms with the specified variable as the product of that variable and a sum of terms. Complete the solution.	Solve for h: $A = \frac{1}{2}bh$. $A = \frac{1}{2}bh$ $2A = 2\left(\frac{1}{2}bh\right)$ $2A = bh$ $\frac{2A}{b} = h$
2.3 APPLICATIONS OF LINEAR EQUATIONS	**Solving an Applied Problem** *Step 1* Determine what you are asked to find. *Step 2* Write down any other pertinent information. *Step 3* Write an equation. *Step 4* Solve the equation. *Step 5* Answer the question(s) of the problem. *Step 6* Check.	How many liters of a 30% alcohol solution and 80% alcohol solution must be mixed to obtain 100 liters of a 50% alcohol solution? Let x = number of liters of 30% solution needed; $100 - x$ = number of liters of 80% solution needed. The information of the problem is summarized in the chart on the following page.

SECTION	CONCEPTS	EXAMPLES

LITERS	CONCENTRATION	LITERS OF PURE ALCOHOL
x	.30	$.30x$
$100 - x$	.80	$.80(100 - x)$
100	.50	$.50(100)$

The equation is

$$.30x + .80(100 - x) = .50(100).$$

Solving this equation gives $x = 60$. 60 liters of 30% alcohol and $100 - 60 = 40$ liters of 80% alcohol should be used. Check this result.

2.4 MORE APPLICATIONS OF LINEAR EQUATIONS

To solve a uniform motion problem, draw a sketch and make a chart. Use the formula $d = rt$.

Two cars start from towns 400 miles apart and travel toward each other. They meet after 4 hours. Find the speed of each car if one travels 20 miles per hour faster than the other.

Let x = speed of the slower car in miles per hour;

$x + 20$ = speed of the faster car.

Use the information in the problem, and $d = rt$ to complete the chart.

	r	t	d
Slower car	x	4	$4x$
Faster car	$x + 20$	4	$4(x + 20)$

A sketch shows that the sum of the distances, $4x$ and $4(x + 20)$, must be 400.

$4x$ $4(x + 20)$

400

The equation is

$$4x + 4(x + 20) = 400.$$

Solving this equation gives $x = 40$. The slower car travels 40 miles per hour and the faster car travels $40 + 20 = 60$ miles per hour.

SECTION	CONCEPTS	EXAMPLES
2.5 LINEAR INEQUALITIES IN ONE VARIABLE	**Solving a Linear Inequality** Simplify each side separately, combining like terms and removing parentheses. Use the addition property of inequality to get the variables on one side of the inequality sign and the numbers on the other. Combine like terms, and then use the multiplication property to change the inequality to the form $x < k$ or $x > k$. If an inequality is multiplied or divided by a *negative* number, the inequality symbol *must be reversed*.	Solve $3(x + 2) - 5x \leq 12$. $3x + 6 - 5x \leq 12$ $-2x + 6 \leq 12$ $-2x \leq 6$ $\frac{-2x}{-2} \geq \frac{6}{-2}$ $x \geq -3$ The solution set is $[-3, \infty)$ and is graphed below. −3 −2 −1 0 1 2 3
2.6 SET OPERATIONS AND COMPOUND INEQUALITIES	**Solving a Compound Inequality** Solve each inequality in the compound inequality individually. If the inequalities are joined with *and*, the solution is the intersection of the two individual solutions. If the inequalities are joined with *or*, the solution is the union of the two individual solutions.	Solve $x + 1 > 2$ and $2x < 6$. $x + 1 > 2$ and $2x < 6$ $x > 1$ and $x < 3$ The solution set is $(1, 3)$. −1 0 1 2 3 4 Solve $x \geq 4$ or $x \leq 0$. The solution set is $(-\infty, 0] \cup [4, \infty)$. −1 0 1 2 3 4 5
2.7 ABSOLUTE VALUE EQUATIONS AND INEQUALITIES	Let k be a positive number. To solve $\|ax + b\| = k$, solve the compound equation $ax + b = k$ or $ax + b = -k$.	Solve $\|x - 7\| = 3$. $x - 7 = 3$ or $x - 7 = -3$ $x = 10$ $x = 4$ The solution set is $\{4, 10\}$. 0 2 4 6 8 10

SECTION	CONCEPTS	EXAMPLES
	To solve $\|ax + b\| > k$, solve the compound inequality $ax + b > k$ or $ax + b < -k$.	Solve $\|x - 7\| > 3$. $x - 7 > 3$ or $x - 7 < -3$ $x > 10$ or $x < 4$ The solution set is $(-\infty, 4) \cup (10, \infty)$. 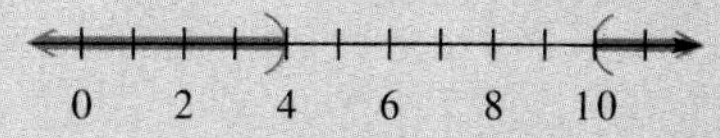
	To solve $\|ax + b\| < k$, solve the compound inequality $-k < ax + b < k$.	Solve $\|x - 7\| < 3$. $-3 < x - 7 < 3$ $4 < x < 10$ Add 7. The solution set is $(4, 10)$. 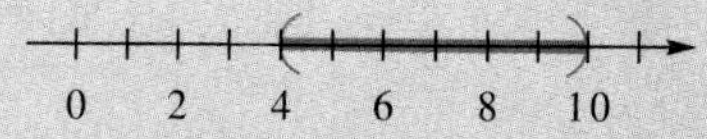
	To solve an absolute value equation of the form $\|ax + b\| = \|cx + d\|$ solve the compound equation $ax + b = cx + d$ or $ax + b = -(cx + d)$.	Solve $\|x + 2\| = \|2x - 6\|$. $x + 2 = 2x - 6$ or $x + 2 = -(2x - 6)$ $x = 8$ or $x = \frac{4}{3}$ The solution set is $\left\{\frac{4}{3}, 8\right\}$.

CHAPTER 2 REVIEW EXERCISES

[2.1] *Solve each equation.*

1. $7z + 9 = -5$

2. $3k - 6 = -4$

3. $-r + 9 = 4r - 7$

4. $5y - 9 + 4y = 4y + 1$

5. $-6r + 2r - 5 = r$

6. $7z - 3z + 12 - 5 = 8z + 7$

7. $5(k - 2) + 8k = 16$

8. $-(8 + 3y) + 5 = 2y + 3$

9. $-(r + 5) - (2 + 7r) + 8r = 3r - 5$

10. $\frac{y}{4} - \frac{y}{6} = -1$

11. $\frac{r}{6} + \frac{r}{2} = 2$

12. $\frac{2a}{3} + \frac{a}{4} = 6$

13. $\frac{m - 2}{4} + \frac{m + 1}{2} = 1$

14. $\frac{2q + 1}{3} + \frac{q - 1}{4} = \frac{13}{2}$

15. $.40x + .60(100 - x) = .45(100)$

16. $1.30q + .90(50 - q) = 1.00(50)$

Decide whether each equation is conditional, *an* identity, *or a* contradiction. *Give the solution set of each.*

17. $7r - 3(2r - 5) + 5 + 3r = 4r + 20$

18. $8p - 4p - (p - 7) + 9p + 6 = 12p - 7$

19. $-2r + 6(r - 1) + 3r - (4 - r) = -(r + 5) - 5$

20. $-(2y - 5) - y + 7(y + 2) = 6y + 14 - (2y - 5)$

[2.2] *Solve each formula for the specified variable.*

21. $V = LWH$; for H

22. $A = \frac{1}{2}(B + b)h$; for b

23. $C = \pi d$; for d

Find the unknown value.

24. If a vehicle travels 150 kilometers in 5 hours, find the rate of the vehicle.

25. A triangle has a perimeter of 120 meters. Two equal sides of the triangle each have a length of 50 meters. Find the length of the third side.

26. How much principal must be deposited for 6 years at 8% per year simple interest to earn $720 in interest?

27. A box has a volume of 180 cubic feet. Its length is 6 feet and its width is 5 feet. Find its height.

28. A triangle has an area of 42 square meters. The base is 14 meters long. Find the height of the triangle.

29. A circle has a circumference of 36π millimeters. Find its radius.

30. A loan has a finance charge of $450. The loan was scheduled to run for 24 months. Find the unearned interest if the loan is paid off with 5 payments left.

31. The Fahrenheit temperature is 68°. Find the Celsius temperature.

32. The Celsius temperature is 90°. Find the Fahrenheit temperature.

33. Find the measure of each marked angle.

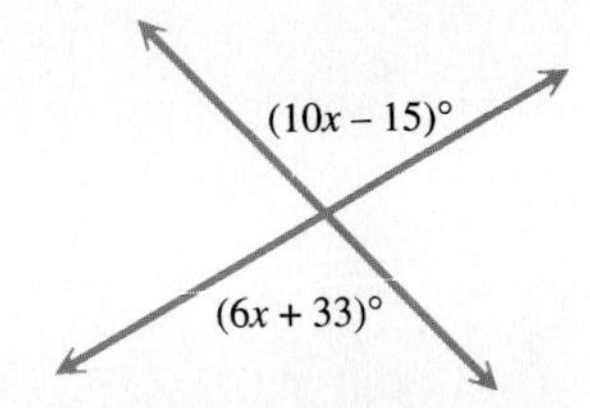

34. Find the measure of each marked angle, given that lines m and n are parallel.

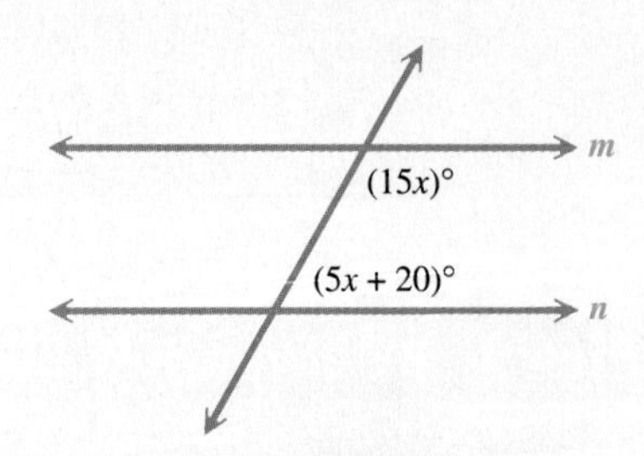

[2.3] *Write the following verbal phrases as mathematical expressions. Use x as the variable.*

35. Three times a number

36. The difference between 9 and twice a number

37. Half a number, added to 5

38. The product of 4 and a number, subtracted from 8

Solve each of the following.

39. When 9 is added to three times a number, the result is 15. Find the number.

40. The product of 6, and a number decreased by 2, is 18. Find the number.

41. If a number if multiplied by 2, the difference between this product and 3 is 11. Find the number.

42. After a number is increased by 20%, the result is 36. Find the number.

43. A number is decreased by 35%, giving 260. Find the number.

44. A candy clerk has three times as many kilograms of chocolate creams as peanut clusters. The clerk has 48 kilograms of the two candies altogether. How many kilograms of peanut clusters does the clerk have?

45. How many liters of a 20% solution of a chemical should be mixed with 15 liters of a 50% solution to get a mixture that is 30% chemical?

46. Chico Ruiz invested some money at 12% simple interest and \$4000 less than this amount at 14%. Find the amount invested at each rate if his total annual interest income is \$4120.

47. Carmella Santiago earned \$42,000 in book royalties. She invested part at 15% simple interest and the rest at 12%, earning \$5790 per year in interest. How much did she invest at each rate?

48. The supplement of an angle measures 10 times the measure of its complement. Find the measure of the angle.

49. Find the measure of each angle in the triangle.

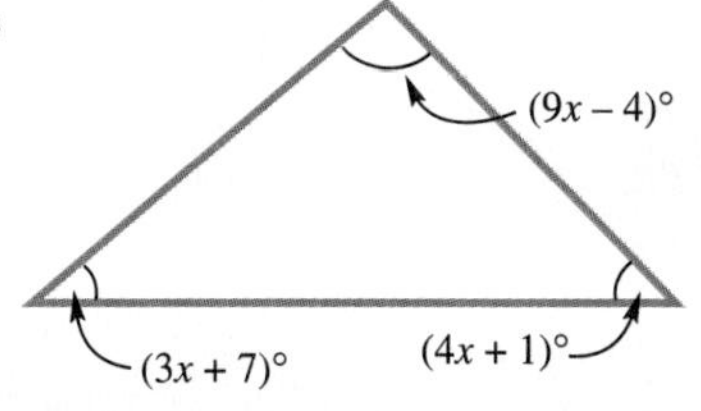

50. Find the measure of each marked angle.

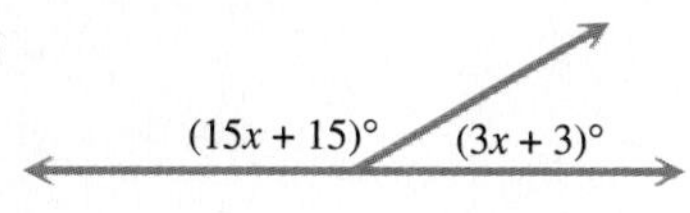

[2.4] *Solve each of the following.*

51. A total of 1096 people attended the Beach Boys concert yesterday. Reserved seats cost \$25 each and general admission seats cost \$20 each. If \$26,170 was collected, how many of each type of seat were sold?

52. There were 311 tickets sold for a soccer game, some for students and some for non-students. Student tickets cost 25¢ each and non-student tickets cost 75¢ each. The total receipts were \$108.75. How many of each type of ticket were sold?

53. Hazel Miller wishes to mix 30 pounds of candy worth \$6 per pound with candy worth \$3 per pound to get a mixture worth \$5 per pound. How much of the \$3 candy should be used?

54. A passenger train and a freight train leave a town at the same time and go in opposite directions. They travel at 60 miles per hour and 75 miles per hour, respectively. How long will it take for them to be 297 miles apart?

55. Two cars leave towns 230 kilometers apart at the same time, traveling directly toward one another. One car travels 15 kilometers per hour slower than the other. They pass one another 2 hours later. What are their speeds?

56. An automobile averaged 45 miles per hour for the first part of a trip and 50 miles per hour for the second part. If the entire trip took 4 hours and covered 195 miles, for how long was the rate of speed 45 miles per hour?

57. An 85-mile trip to the beach took the Rodriguez family 2 hours. During the second hour, a rainstorm caused them to average 7 miles per hour less than they traveled during the first hour. Find their average speed for the first hour.

58. The perimeter of a triangle is 34 inches. The middle side is twice as long as the shortest side. The longest side is 2 inches less than three times the shortest side. Find the lengths of the three sides.

59. The length of a rectangle is 3 meters less than twice the width. The perimeter of the rectangle is 42 meters. Find the length and the width of the rectangle.

60. A square is such that if each side were increased by 4 inches, the perimeter would be 8 inches less than twice the perimeter of the original square. Find the length of a side of the original square.

[2.5] *Solve each inequality.*

61. $3m < 24$

62. $5a \geq -15$

63. $2y + 7 > -21$

64. $-\frac{2}{3}k < 4$

65. $-5x - 4 \geq 10$

66. $-y < 4$

67. $2(p + 2) - 4p \geq -3p - 2$

68. $-2.4y \leq 14.4$

69. $-(4 + 2m) - 3 \leq 5m + 2$

70. $\frac{4m - 1}{3} > 5$

71. $\frac{2p + 1}{5} < -1$

72. $\frac{6a + 3}{-4} < -2$

73. $\frac{9y + 5}{-3} > 2$

74. $-6 \leq 2k \leq 12$

75. $8 \leq 3y - 1 < 14$

76. $-4 < 2k - 3 < 9$

77. $\frac{5}{3}(m - 2) + \frac{2}{5}(m + 1) > 1$

78. $\frac{3}{4}(a - 2) - \frac{1}{3}(5 - 2a) < -2$

---

79. To pass algebra, a student must have an average of at least 70% on five tests. On the first four tests, a student has grades of 75%, 79%, 64%, and 71%. What possible grades on the fifth test would guarantee a passing grade in the class?

80. A dietician must include three foods, A, B, and C, in a diet. He must include twice as many grams of food A as food C, and 5 grams of food B. The three foods must total at most 24 grams. What is the largest amount of food C that the dietician can use?

[2.6] *Let* $A = \{a, b, c, d\}$, $B = \{a, c, e, f\}$, *and* $C = \{a, e, f, g\}$. *Find each set.*

81. $A \cap B$

82. $A \cap C$

83. $B \cup C$

84. $A \cup C$

Solve each compound inequality. Graph each solution.

85. $x \leq 4$ and $x < 3$

86. $x > 6$ and $x < 8$

87. $x + 4 > 12$ and $x - 2 < 1$

88. $x > 5$ or $x \leq -1$

89. $x - 4 > 6$ or $x + 3 \leq 18$

90. $-5x + 1 \geq 11$ or $3x + 5 \geq 26$

Express each of the following in the simplest interval form.

91. $(-3, \infty) \cap (-\infty, 4)$

92. $(-\infty, 6) \cap (-\infty, 2)$

93. $(4, \infty) \cup (9, \infty)$

94. $(1, 2) \cup (1, \infty)$

[2.7] *Solve each absolute value equation.*

95. $|x| = 4$

96. $|y + 2| = 6$

97. $|-5 + 2x| = 1$

98. $\left|\frac{1}{2}p + 2\right| = 3$

99. $|2k - 7| + 4 = 9$

100. $|z - 4| = -4$

101. $|4a + 2| + 7 = 3$

102. $\left|\frac{1}{2}r - 2\right| = \left|\frac{1}{4}r - 1\right|$

103. $\left|\frac{2}{3}a + 2\right| = \left|\frac{3}{4}a - 5\right|$

104. $|3p + 1| = |p + 7|$

Solve each absolute value inequality. Graph each solution.

105. $|p| < 4$

106. $|-y + 6| \leq 5$

107. $|2p + 5| \leq 3$

108. $|m| > 2$

109. $|x + 1| \geq 9$

110. $|3k + 9| \geq 0$

111. $|2 - a| > 5$

112. $|1 - q| < 3$

MIXED REVIEW EXERCISES*

Solve.

113. $5 - (6 - 4k) > 2k - 5$

114. $S = 2HW + 2LW + 2LH;$ for L

115. $x < 3$ and $x \geq -2$

116. $-4(3 + 2m) - m = -3m$

117. A rectangle has a perimeter of 46 centimeters. The width is 8 centimeters. Find the length.

118. $-6z \leq 72$

119. $|5r - 1| > 14$

120. $x \geq -2$ or $x < 4$

121. How many liters of a 20% solution of a chemical should be mixed with 10 liters of a 50% solution to get a 40% mixture?

122. $|m - 1| = |2m + 3|$

123. $\frac{3y}{5} - \frac{y}{2} = 3$

124. $|m + 3| \leq 1$

125. $|3k - 7| = 4$

126. In an election, one candidate received 135 more votes than the other. The total number of votes cast in the election was 1215. Find the number of votes received by each candidate.

127. $5(2x - 7) = 2(5x + 3)$

128. Find the measure of each marked angle.

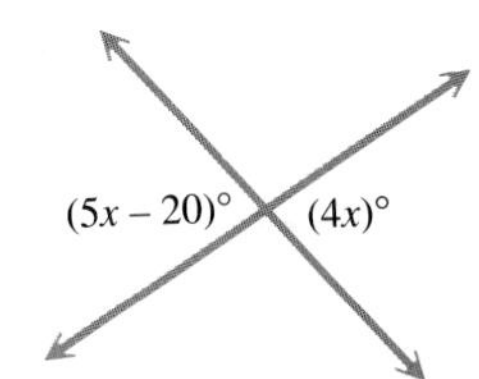

CHAPTER 2 TEST

Solve each equation.

1. $5r - 3 + 2r = 3(r - 2) + 11$

2. $9 - (3 + 4z) - 5z = 4(z - 5) - 9$

3. $\frac{2p - 1}{3} + \frac{p + 1}{4} = \frac{7}{4}$

4. Decide whether the equation

$$3p - (2 - p) + 4p = 7p - 2 - (-p)$$

is *conditional*, an *identity*, or a *contradiction*. Give its solution set.

5. Solve for v: $S = vt - 16t^2$.

---

6. Find the period of time for which \$2000 must be invested at 7% simple interest to produce \$560 interest.

7. A 90% acid solution is to be mixed with a 75% solution to make 20 liters of a 78% solution. How many liters of 90% and 75% solutions should be used?

*The order of exercises in this final group does not correspond to the order in which topics occur in the chapter. This random ordering should help you prepare for the chapter test.

8. Cesar Aparicio invested some money at 7% simple interest and three times as much at 6%. His total annual interest was \$375. How much did he invest at each rate?

9. Find the measure of each angle.

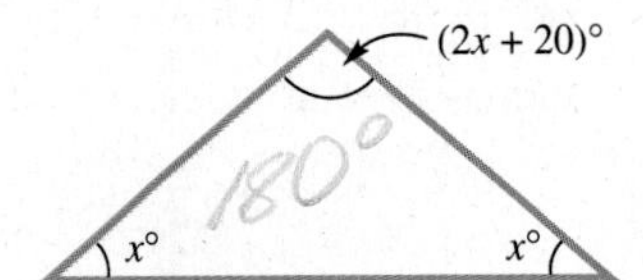

10. Two cars leave the same point at the same time, traveling in opposite directions. One travels 15 miles per hour faster than the other. After 3 hours they are 315 miles apart. Find the speed of each car.

11. The perimeter of a rectangular window is 28 feet. The length is 2 feet less than 3 times the width. Find the length and the width of the window.

Solve each inequality, then graph each solution.

12. $-\frac{3}{4}r > -6$

13. $2 - 3(p - 1) < 5p$

14. $6z - (4 + 5z) \leq 8 - 3z$

15. $-1 < \frac{2}{3}a - 2 < 2$

---

16. Which one of the following inequalities is equivalent to $x < -3$?
(a) $-3x < 9$ (b) $-3x > -9$ (c) $-3x > 9$
(d) $-3x < -9$

17. A student must have an average grade of at least 80% on the four tests in a course to get a grade of B. The student had grades of 83%, 76%, and 79% on the first three tests. What minimum grade on the fourth test would guarantee a B in the course?

18. Let $A = \{1, 2, 5, 7\}$ and $B = \{1, 5, 9, 12\}$. Find **(a)** $A \cap B$ and **(b)** $A \cup B$.

Solve each compound inequality.

19. $3x - 2 < 10$ and $-2x < 10$

20. $-4x \leq -20$ or $4x - 2 < 10$

Solve each absolute value equation or inequality. Graph the solution set of each inequality.

21. $|2k - 3| = 7$

22. $|m - 1| = |2m + 5|$

23. $|2k + 3| \leq 11$

24. $|3r + 2| \geq 6$

25. If $k < 0$, what is the solution set of $|-3x + 7| < k$?

CHAPTER THREE

EXPONENTS AND POLYNOMIALS

The expression $-.006x^4 + .14x^3 - .05x^2 + 2x$ can be used to determine the concentration of dye injected into a cardiac patient's bloodstream x seconds after injection. The expression itself is called a *polynomial*. Polynomials can be used in applications of medicine, biology, physics, and so on. In this chapter we study properties of exponents and begin our work with polynomials. The concept of factoring polynomials is crucial in many of the later chapters in this book, and the ability to factor is essential in more advanced mathematics courses.

3.1 INTEGER EXPONENTS

OBJECTIVES

1. IDENTIFY EXPONENTS AND BASES.
2. USE THE PRODUCT RULE FOR EXPONENTS.
3. DEFINE NEGATIVE EXPONENTS.
4. USE THE QUOTIENT RULE FOR EXPONENTS.
5. DEFINE A ZERO EXPONENT.

In Chapter 1 it was shown that two integers whose product is a third number are *factors* of that number. For example, 2 and 5 are factors of 10, since $2 \cdot 5 = 10$. Other factors of 10 include 1 and 10, -1 and -10, and -2 and -5.

1 *Exponents* are used to write products of repeated factors. For example, the product $3 \cdot 3 \cdot 3 \cdot 3$ is written

$$3 \cdot 3 \cdot 3 \cdot 3 = 3^4.$$

The number 4 shows that 3 appears as a factor four times. The number 4 is the *exponent* and 3 is the *base*. The quantity 3^4 is called an *exponential expession*. Read 3^4 as "3 to the fourth power," or "3 to the fourth." Multiplying out the four 3's gives

$$3^4 = 3 \cdot 3 \cdot 3 \cdot 3 = 81.$$

EXPONENTS If a is a real number and n is a natural number,

$$a^n = \underbrace{a \cdot a \cdot a \ldots a}_{n \text{ factors of } a}.$$

EXAMPLE 1 **EVALUATING EXPONENTIAL EXPRESSIONS**

Evaluate each exponential expression.

(a) $7^2 = 7 \cdot 7 = 49$ Read 7^2 as "7 squared."

(b) $5^3 = 5 \cdot 5 \cdot 5 = 125$ Read 5^3 as "5 cubed."

(c) $(-2)^4 = (-2)(-2)(-2)(-2) = 16$

(d) $(-2)^5 = (-2)(-2)(-2)(-2)(-2) = -32$

(e) $5^1 = 5$ ■

EXAMPLE 2
IDENTIFYING BASES AND EXPONENTS

Identify the base and the exponent.

(a) 3^6
The base is 3 and the exponent is 6.

(b) 5^4
Base, 5; exponent, 4.

(c) $(-2)^6$
The exponent of 6 refers to the number -2, so the base is -2. Evaluating $(-2)^6$ gives

$$(-2)^6 = (-2)(-2)(-2)(-2)(-2)(-2) = 64.$$

(d) $3z^7$
The exponent is 7. The base is z (*not* $3z$).

(e) $(-9y)^3$
Because of the parentheses, the exponent 3 refers to the base $-9y$.

(f) $-(9y)^3$
The base is $9y$ and the exponent is 3. ■

Look back at Example 2(d). Notice that in $3z^7$, the exponent 7 applies only to the quantity closest to it. Therefore,

$$3z^7 = 3 \cdot z \cdot z \cdot z \cdot z \cdot z \cdot z \cdot z,$$

while

$$(3z)^7 = (3z)(3z)(3z)(3z)(3z)(3z)(3z).$$

In Example 2(c) we evaluated $(-2)^6$. Notice the careful use of parentheses to indicate that the base was -2. Now, let us consider the exponential expression -2^6. In this expression, the base is 2, and *not* -2. The $-$ sign tells us to find the negative, or additive inverse, of 2^6. It acts as a symbol for the factor -1. We evaluate -2^6 as follows.

$$-2^6 = -(2 \cdot 2 \cdot 2 \cdot 2 \cdot 2 \cdot 2) = -64$$

The results shown in Example 2(c) and here show us that

$$(-2)^6 \neq -2^6,$$

since $64 \neq -64$. We can generalize as follows.

$(-a)^n$ and $-a^n$ do not mean the same thing. $(-a)^n$ means that $(-a)$ is used as a factor n times. $-a^n$ means that a is used as a factor n times, and then its negative is found.

CAUTION Be very careful when evaluating exponential expressions like

$$-4^2, \qquad -8^4, \qquad \text{and} \qquad -2^4.$$

As discussed above, a common error is to use the − sign as part of the base. In these expressions, the bases are 4, 8, and 2.

EXAMPLE 3 EVALUATING EXPONENTIAL EXPRESSIONS WITH LEADING − SIGNS

Evaluate each exponential expression.

(a) $-4^2 = -(4 \cdot 4) = -16$

(b) $-8^4 = -(8 \cdot 8 \cdot 8 \cdot 8) = -4096$

(c) $-2^4 = -(2 \cdot 2 \cdot 2 \cdot 2) = -16$ ■

2 There are several useful rules that simplify work with exponents. For example, the product $2^5 \cdot 2^3$ can be simplified as follows.

$$2^5 \cdot 2^3 = (2 \cdot 2 \cdot 2 \cdot 2 \cdot 2)(2 \cdot 2 \cdot 2) = 2^8 \qquad (5 + 3 = 8)$$

This result, that products of exponential expressions with the same base are found by adding exponents, is generalized as the **product rule for exponents.**

PRODUCT RULE FOR EXPONENTS

If m and n are natural numbers and a is any real number, then

$$a^m \cdot a^n = a^{m+n}.$$

To see that the product rule is true, use the definition of an exponent as follows.

$$a^m = \underbrace{a \cdot a \cdot a \ldots a}_{a \text{ appears as a factor } m \text{ times}} \qquad a^n = \underbrace{a \cdot a \cdot a \ldots a}_{a \text{ appears as a factor } n \text{ times}}$$

From this,

$$a^m \cdot a^n = \underbrace{a \cdot a \cdot a \ldots a}_{m \text{ factors}} \cdot \underbrace{a \cdot a \cdot a \ldots a}_{n \text{ factors}}$$

$$= \underbrace{a \cdot a \cdot a \ldots a}_{(m+n) \text{ factors}}$$

$$a^m \cdot a^n = a^{m+n}.$$

EXAMPLE 4 USING THE PRODUCT RULE FOR EXPONENTS

Apply the product rule for exponents in each case.

(a) $3^4 \cdot 3^7 = 3^{4+7} = 3^{11}$

(b) $5^3 \cdot 5 = 5^3 \cdot 5^1 = 5^{3+1} = 5^4$

(c) $y^3 \cdot y^8 \cdot y^2 = y^{3+8+2} = y^{13}$ ■

CAUTION A common error occurs in problems like Example 4(a), when students multiply the bases. Notice that $3^4 \cdot 3^7 \neq 9^{11}$. Remember, when applying the product rule for exponents, keep the *same* base and add the exponents.

EXAMPLE 5 USING THE PRODUCT RULE FOR EXPONENTS

Find each product.

(a) $(5y^2)(-3y^4)$

Use the associative and commutative properties as necessary to multiply the numbers and multiply the variables.

$$\begin{aligned}(5y^2)(-3y^4) &= 5(-3)y^2y^4 \\ &= -15y^{2+4} \\ &= -15y^6\end{aligned}$$

(b) $(7p^3q)(2p^5q^2) = 7(2)p^3p^5qq^2 = 14p^8q^3$ ■

3 A quotient, such as $\frac{a^8}{a^3}$, can be simplified in much the same way as a product. (In all quotients of this type, assume that the denominator is not zero.) Using the definition of an exponent,

$$\begin{aligned}\frac{a^8}{a^3} &= \frac{a \cdot a \cdot a \cdot a \cdot a \cdot a \cdot a \cdot a}{a \cdot a \cdot a} \\ &= a \cdot a \cdot a \cdot a \cdot a \\ &= a^5.\end{aligned}$$

Notice that $8 - 3 = 5$. In the same way,

$$\frac{a^3}{a^8} = \frac{a \cdot a \cdot a}{a \cdot a \cdot a \cdot a \cdot a \cdot a \cdot a \cdot a}$$

$$\frac{a^3}{a^8} = \frac{1}{a^5}.$$

Here again, $8 - 3 = 5$. In the first example, $\frac{a^8}{a^3}$, the exponent in the denominator was subtracted from the one in the numerator; the reverse was true in the second example, $\frac{a^3}{a^8}$. The order of subtracting the exponents depends on the larger exponent. In the second example, subtracting $3 - 8$ gives an exponent of -5. To simplify the rule for quotients, we first define a **negative exponent.**

NEGATIVE EXPONENT

For any natural number n and any nonzero real number a,

$$a^{-n} = \frac{1}{a^n}.$$

With this definition, the expression a^n is meaningful for any nonzero integer exponent n and any nonzero real number a.

EXAMPLE 6
USING NEGATIVE EXPONENTS

Write the following expressions with only positive exponents.

(a) $2^{-3} = \frac{1}{2^3} = \frac{1}{8}$ **(b)** $3^{-2} = \frac{1}{3^2} = \frac{1}{9}$

(c) $6^{-1} = \frac{1}{6^1} = \frac{1}{6}$ **(d)** $(5z)^{-3} = \frac{1}{(5z)^3}, \quad z \neq 0$

(e) $5z^{-3} = 5\left(\frac{1}{z^3}\right) = \frac{5}{z^3}, \quad z \neq 0$ **(f)** $(5z^2)^{-3} = \frac{1}{(5z^2)^3}, \quad z \neq 0$

(g) $-m^{-2} = -\frac{1}{m^2}, \quad m \neq 0$ **(h)** $(-m)^{-4} = \frac{1}{(-m)^4}, \quad m \neq 0$ ■

CAUTION A negative exponent does not necessarily lead to a negative number; negative exponents lead to reciprocals, as shown below.

EXPRESSION	EXAMPLE	
a^{-m}	$3^{-2} = \frac{1}{3^2} = \frac{1}{9}$	Not negative
$-a^{-m}$	$-3^{-2} = -\frac{1}{3^2} = -\frac{1}{9}$	Negative

EXAMPLE 7
USING NEGATIVE EXPONENTS

Evaluate each of the following expressions.

(a) $3^{-1} + 4^{-1}$

Since $3^{-1} = \frac{1}{3}$ and $4^{-1} = \frac{1}{4}$,

$$3^{-1} + 4^{-1} = \frac{1}{3} + \frac{1}{4} = \frac{4}{12} + \frac{3}{12} = \frac{7}{12}.$$

(b) $5^{-1} - 2^{-1} = \frac{1}{5} - \frac{1}{2} = \frac{2}{10} - \frac{5}{10} = -\frac{3}{10}$ ■

CAUTION In Example 7, note that

$$3^{-1} + 4^{-1} \neq (3 + 4)^{-1}$$

because $3^{-1} + 4^{-1} = \frac{7}{12}$ but $(3 + 4)^{-1} = 7^{-1} = \frac{1}{7}$.

Also,

$$5^{-1} - 2^{-1} \neq (5 - 2)^{-1}$$

because $5^{-1} - 2^{-1} = -\frac{3}{10}$ but $(5 - 2)^{-1} = 3^{-1} = \frac{1}{3}$.

EXAMPLE 8
USING NEGATIVE EXPONENTS

Evaluate each expression.

(a) $\frac{1}{2^{-3}} = \frac{1}{\frac{1}{2^3}} = 1 \div \frac{1}{2^3} = 1 \cdot \frac{2^3}{1} = 2^3 = 8$

(b) $\frac{1}{3^{-2}} = \frac{1}{\frac{1}{3^2}} = 3^2 = 9$

(c) $\frac{2^{-3}}{3^{-2}} = \frac{\frac{1}{2^3}}{\frac{1}{3^2}} = \frac{1}{2^3} \cdot \frac{3^2}{1} = \frac{3^2}{2^3} = \frac{9}{8}$ ■

Example 8 suggests the following generalizations.

SPECIAL RULES FOR NEGATIVE EXPONENTS

If $a \neq 0$ and $b \neq 0$, $\frac{1}{a^{-n}} = a^n$ and $\frac{a^{-n}}{b^{-m}} = \frac{b^m}{a^n}$.

4 When multiplying expressions such as a^m and a^n where the base is the same, *add* the exponents; when dividing, *subtract* the exponents, as stated in the following **quotient rule for exponents.**

QUOTIENT RULE FOR EXPONENTS

If a is any nonzero real number and m and n are nonzero integers, then

$$\frac{a^m}{a^n} = a^{m-n}.$$

EXAMPLE 9
USING THE QUOTIENT RULE FOR EXPONENTS

Apply the quotient rule for exponents in each case.

(a) $\frac{3^7}{3^2} = 3^{7-2} = 3^5$

(Numerator exponent ↓ 7; Denominator exponent ↓ 2; ↑ Minus sign)

(b) $\frac{p^6}{p^2} = p^{6-2} = p^4 \quad (p \neq 0)$

(c) $\frac{12^{10}}{12^9} = 12^{10-9} = 12^1 = 12$

(d) $\frac{7^4}{7^6} = 7^{4-6} = 7^{-2} = \frac{1}{7^2}$

(e) $\frac{k^7}{k^{12}} = k^{7-12} = k^{-5} = \frac{1}{k^5} \quad (k \neq 0)$ ■

EXAMPLE 10
USING THE QUOTIENT RULE FOR EXPONENTS

Write each quotient using only positive exponents.

(a) $\frac{2^7}{2^{-3}} = 2^{7-(-3)}$

Since $7 - (-3) = 10$, $\quad \frac{2^7}{2^{-3}} = 2^{10}$.

(b) $\frac{8^{-2}}{8^5} = 8^{-2-5} = 8^{-7} = \frac{1}{8^7}$

(c) $\frac{6^{-5}}{6^{-2}} = 6^{-5-(-2)} = 6^{-3} = \frac{1}{6^3}$

(d) $\frac{6}{6^{-1}} = \frac{6^1}{6^{-1}} = 6^{1-(-1)} = 6^2$

(e) $\frac{z^{-5}}{z^{-8}} = z^{-5-(-8)} = z^3 \quad (z \neq 0)$

(f) $\frac{q^{5r}}{q^{3r}} = q^{5r-3r} = q^{2r}$, if r is a nonzero integer and $q \neq 0$ ■

CAUTION As seen in Example 10, we must be very careful when working with quotients that involve negative exponents in the denominator. Always be sure to write the numerator exponent, then a minus sign, and then the denominator exponent.

As suggested by Example 10(f), the rules for exponents also apply with variable exponents, as long as the variables used as exponents represent integers. The product and quotient rules for exponents are summarized below.

SUMMARY OF PRODUCT AND QUOTIENT RULES

When multiplying expressions such as a^m and a^n where the base is the same, keep the same base and add the exponents. When dividing such expressions, keep the same base and subtract the exponent of the denominator from the exponent of the numerator.

5 By the quotient rule, $\frac{a^m}{a^n} = a^{m-n}$. Suppose that the exponents in the numerator and the denominator are the same. Then, for example, $\frac{8^4}{8^4} = 8^{4-4} = 8^0$. Since any nonzero number divided by itself is 1, $\frac{8^4}{8^4} = 1$. Therefore,

$$1 = \frac{8^4}{8^4} = 8^0.$$

Based on this, for consistency with past rules of exponents, the expression a^0 is defined to equal 1 for any nonzero number a.

ZERO EXPONENT If a is any nonzero real number, then

$$a^0 = 1.$$

The symbol 0^0 is undefined.

EXAMPLE 11 **USING ZERO AS AN EXPONENT**

Evaluate each expression.

(a) $12^0 = 1$

(b) $(-6)^0 = 1$

(c) $-6^0 = -(6^0) = -1$

(d) $5^0 + 12^0 = 1 + 1 = 2$

(e) $(8k)^0 = 1 \quad (k \neq 0)$ ■

3.1 EXERCISES

Identify the exponent and the base. Do not evaluate. See Example 2.

1. 5^7 **2.** 9^4 **3.** 12^5 **4.** 7^8

5. $(-9)^4$ **6.** $(-2)^{11}$ **7.** -9^4 **8.** -2^{11}

9. p^{-7} **10.** m^{-9} **11.** $5y^4$ **12.** $8x^{12}$

13. $-3q^{-4}$ **14.** $-9m^{-6}$ **15.** $(-m + z)^3$ **16.** $-(6a - 5b)^5$

Evaluate. See Examples 1, 3, 6, 7, 8, and 11.

17. 5^4 **18.** 10^3 **19.** $\left(\frac{5}{3}\right)^2$ **20.** $\left(\frac{3}{5}\right)^3$ **21.** $(-2)^5$

22. $(-5)^4$ **23.** -2^3 **24.** -3^2 **25.** $-(-3)^4$ **26.** $-(-5)^2$

27. 7^{-2} **28.** 4^{-1} **29.** -7^{-2} **30.** -4^{-1} **31.** $\frac{2}{(-4)^{-3}}$

32. $\frac{-3}{(-5)^{-2}}$ **33.** $\frac{2^{-3}}{3^{-2}}$ **34.** $\frac{5^{-1}}{4^{-2}}$ **35.** $\left(\frac{1}{2}\right)^{-3}$ **36.** $\left(\frac{1}{5}\right)^{-3}$

37. $\left(\frac{2}{3}\right)^{-2}$ **38.** $\left(\frac{4}{5}\right)^{-2}$ **39.** $3^{-1} + 2^{-1}$ **40.** $4^{-1} + 5^{-1}$ **41.** 8^0

42. 12^0 **43.** $(-23)^0$ **44.** $(-4)^0$ **45.** -2^0 **46.** -7^0

47. $3^0 - 4^0$ **48.** $-8^0 - 7^0$ **49.** $(-6)^0 + (-9)^0$ **50.** $(-21)^0 - 10^0$

---

51. In *some* cases, $-a^n$ and $(-a)^n$ do give the same result. Using $a = 2$, and $n = 2, 3, 4,$ and 5, draw a conclusion as to when they are equal and when they are opposites. (See Objective 1.)

52. Which one of the following is equal to 1 $(a \neq 0)$? (See Objective 5.)
(a) $3a^0$ **(b)** $-3a^0$ **(c)** $(3a)^0$ **(d)** $3(-a)^0$

Use the product rule or the quotient rule to simplify. Write all answers with only positive exponents. Assume that all variables represent nonzero real numbers. See Examples 4, 5, 9, and 10.

53. $2^6 \cdot 2^{10}$

54. $3^5 \cdot 3^7$

55. $7^{12} \cdot 7^{-8}$

56. $10^{-4} \cdot 10^5$

57. $\frac{3^5}{3^2}$

58. $\frac{5^7}{5^2}$

59. $\frac{6^7}{6^9}$

60. $\frac{7^5}{7^9}$

61. $\frac{3^{-5}}{3^{-2}}$

62. $\frac{2^{-4}}{2^{-3}}$

63. $\frac{9^{-1}}{9}$

64. $\frac{12}{12^{-1}}$

65. t^5t^{-12}

66. p^5p^{-6}

67. r^4r

68. $k \cdot k^6$

69. $a^{-3}a^2a^{-4}$

70. $k^{-5}k^{-3}k^4$

71. $\frac{x^7}{x^{-4}}$

72. $\frac{p^{-3}}{p^5}$

73. $\frac{r^3r^{-4}}{r^{-2}r^{-5}}$

74. $\frac{z^{-4}z^{-2}}{z^3z^{-1}}$

75. $7k^2(-2k)(4k^{-5})$

76. $3a^2(-5a^{-6})(-2a)$

77. $-4(2x^3)(3xy^2)$

78. $6(5z^3)(2zw^2)$

Use the product rule or the quotient rule to simplify. Write all answers with only positive exponents. Assume that all variables used as exponents represent nonzero integers and that all other variables represent nonzero real numbers.

79. $p^q \cdot p^{7q}$

80. $z^{m+1} \cdot z^{m+3}$

81. $a^r \cdot a^{r+1} \cdot a^{r+2}$

82. $k^{3+y} \cdot k^y \cdot k^{5y}$

83. $\frac{m^{2a} \cdot m^{3a}}{m^{4a}}$

84. $\frac{b^{5q} \cdot b^q}{b}$

Many students believe that each pair of expressions in Exercises 85–87 represents the same quantity. This is wrong. Show that each expression in the pair represents a different quantity by replacing x with 2 and y with 3.

85. $(x + y)^{-1}$; $x^{-1} + y^{-1}$

86. $(x^{-1} + y^{-1})^{-1}$; $x + y$

87. $(x + y)^2$; $x^2 + y^2$

88. Which one of the following does not represent the reciprocal of x ($x \neq 0$)?

(a) x^{-1} **(b)** $\frac{1}{x}$ **(c)** $\left(\frac{1}{x^{-1}}\right)^{-1}$ **(d)** $-x$

PREVIEW EXERCISES

Simplify each expression, writing answers in exponential form. See Section 1.3.

89. $5^2 \cdot 5^2 \cdot 5^2$

90. $(-2)^3 \cdot (-2)^3$

91. $-(2^3) \cdot (2^3)$

92. $\left(\frac{3}{4}\right)^2 \cdot \left(\frac{3}{4}\right)^2$

3.2 FURTHER PROPERTIES OF EXPONENTS

OBJECTIVES

1 USE THE POWER RULES FOR EXPONENTS.

2 SIMPLIFY EXPONENTIAL EXPRESSIONS.

3 USE THE RULES FOR EXPONENTS WITH SCIENTIFIC NOTATION.

FOCUS ON PROBLEM SOLVING

The planet Mercury has a mean distance from the sun of 3.6×10^7 miles, while the mean distance of Venus to the sun is 6.7×10^7 miles. How long would it take a spacecraft traveling at 1.55×10^3 miles per hour to travel from Venus to Mercury?

Scientists often use scientific notation to express numbers that are very large or very small. Scientific notation is an important use of exponents. In this section we look at exponents and properties of exponents that are used when performing the operations of arithmetic. This problem is Exercise 81 in the exercises for this section. After working through this section, you should be able to solve this problem.

In the previous section we learned some important rules for working with exponents. In this section we look at some additional rules.

1 The expression $(3^4)^2$ can be simplified as $(3^4)^2 = 3^4 \cdot 3^4 = 3^{4+4} = 3^8$, where $4 \cdot 2 = 8$. This example suggests the first of the **power rules for exponents;** the other two parts can be demonstrated with similar examples.

POWER RULES FOR EXPONENTS

If a is a real number and m and n are integers, then

$$(a^m)^n = a^{mn}$$

$$(ab)^m = a^m b^m$$

$$\left(\frac{a}{b}\right)^m = \frac{a^m}{b^m} \qquad (b \neq 0).$$

The parts of this rule can be proved in the same way as the product rule. In the statements of rules for exponents, we always assume that zero never appears to a negative power.

EXAMPLE 1 USING THE POWER RULES FOR EXPONENTS

Use a power rule in each case.

(a) $(p^8)^3 = p^{8 \cdot 3} = p^{24}$

(b) $\left(\frac{2}{3}\right)^4 = \frac{2^4}{3^4} = \frac{16}{81}$

(c) $(3y)^4 = 3^4y^4 = 81y^4$

(d) $(6p^7)^2 = 6^2p^{7 \cdot 2} = 6^2p^{14} = 36p^{14}$

(e) $\left(\frac{-2m^5}{z}\right)^3 = \frac{(-2)^3m^{5 \cdot 3}}{z^3} = \frac{(-2)^3m^{15}}{z^3} = \frac{-8m^{15}}{z^3} \quad (z \neq 0)$ ■

The reciprocal of a^n is $\frac{1}{a^n} = \left(\frac{1}{a}\right)^n$. Also, by definition, a^n and a^{-n} are reciprocals since

$$a^n \cdot a^{-n} = a^n \cdot \frac{1}{a^n} = 1.$$

Thus, since both are reciprocals of a^n, $\quad a^{-n} = \left(\frac{1}{a}\right)^n.$

Some examples of this result are $\quad 6^{-3} = \left(\frac{1}{6}\right)^3 \quad$ and $\quad \left(\frac{1}{3}\right)^{-2} = 3^2.$

The discussion above can be generalized as follows.

SPECIAL RULES FOR NEGATIVE EXPONENTS

Any nonzero number raised to the negative nth power is equal to the reciprocal of that number raised to the nth power. That is, if $a \neq 0$ and $b \neq 0$,

$$a^{-n} = \left(\frac{1}{a}\right)^n \quad \text{and} \quad \left(\frac{a}{b}\right)^{-n} = \left(\frac{b}{a}\right)^n.$$

EXAMPLE 2 USING NEGATIVE EXPONENTS WITH FRACTIONS

Write the following expressions with only positive exponents and then evaluate.

(a) $\left(\frac{3}{7}\right)^{-2} = \left(\frac{7}{3}\right)^2 = \frac{49}{9}$ **(b)** $\left(\frac{4}{5}\right)^{-3} = \left(\frac{5}{4}\right)^3 = \frac{125}{64}$ ■

This chapter began by restricting exponents to nonzero integers. The idea of exponent has now been expanded to include *all* integers, positive, negative, or zero. This has been done in such a way that all past rules for exponents are still valid. These definitions and rules are summarized below.

DEFINITIONS AND RULES FOR EXPONENTS

For all integers m and n and all real numbers a and b:

Product rule	$a^m \cdot a^n = a^{m+n}$	
Quotient rule	$\frac{a^m}{a^n} = a^{m-n}$	$(a \neq 0)$
Zero exponent	$a^0 = 1$	$(a \neq 0)$
Negative exponent	$a^{-n} = \frac{1}{a^n}$	$(a \neq 0)$
Power rules	$(a^m)^n = a^{mn}$	
	$(ab)^m = a^m b^m$	
	$\left(\frac{a}{b}\right)^m = \frac{a^m}{b^m}$	$(b \neq 0)$
	$a^{-n} = \left(\frac{1}{a}\right)^n$	$(a \neq 0)$
	$\left(\frac{a}{b}\right)^{-n} = \left(\frac{b}{a}\right)^n$	$(a, b \neq 0)$

2 The next three examples show how the definitions and rules for exponents are used to write expressions in equivalent forms.

EXAMPLE 3 USING THE DEFINITIONS AND RULES FOR EXPONENTS

Simplify each expression so that no negative exponents appear in the final result.

(a) $3^2 \cdot 3^{-5} = 3^{2+(-5)} = 3^{-3} = \dfrac{1}{3^3}$ or $\dfrac{1}{27}$

(b) $x^{-3} \cdot x^{-4} \cdot x^2 = x^{-3+(-4)+2} = x^{-5} = \dfrac{1}{x^5} \quad (x \neq 0)$

(c) $(2^5)^{-3} = 2^{5(-3)} = 2^{-15} = \dfrac{1}{2^{15}}$

(d) $(4^{-2})^{-5} = 4^{(-2)(-5)} = 4^{10}$

(e) $(x^{-4})^6 = x^{(-4)6} = x^{-24} = \dfrac{1}{x^{24}} \quad (x \neq 0)$

(f) $(xy)^3 = x^3y^3$ ■

CAUTION As shown in Example 3(f), $(xy)^3 = x^3y^3$, so that $(xy)^3 \neq xy^3$. Remember that ab^m is *not* the same as $(ab)^m$.

EXAMPLE 4 USING THE DEFINITIONS AND RULES FOR EXPONENTS

Simplify. Assume that all variables represent nonzero real numbers.

(a)
$$\frac{x^{-4}y^2}{x^2y^{-5}} = \frac{x^{-4}}{x^2} \cdot \frac{y^2}{y^{-5}} = x^{-4-2} \cdot y^{2-(-5)} = x^{-6}y^7 = \frac{y^7}{x^6}$$

(b)
$$(2^3x^{-2})^{-2} = (2^3)^{-2} \cdot (x^{-2})^{-2} = 2^{-6}x^4 = \frac{x^4}{2^6} \text{ or } \frac{x^4}{64}$$ ■

Recall that the rules for exponents can also be used with variable exponents, as shown in the next example. Assume that all variables used as exponents represent integers.

EXAMPLE 5 WORKING WITH VARIABLE EXPONENTS

Apply the rules for exponents in each case.

(a) $m^k \cdot m^{5-k} = m^{k+(5-k)} = m^5 \quad (m \neq 0)$

(b) $\dfrac{r^{2k}}{r^{5k}} = r^{2k-5k} = r^{-3k} = \dfrac{1}{r^{3k}} \quad (r \neq 0)$ ■

With the work of the first two sections of this chapter, a^n is meaningful for all integer exponents n. In Chapter 5, this will be extended to include all *rational* values of n.

3 Many of the numbers that occur in science are very large, such as the number of one-celled organisms that will sustain a whale for a few hours, 400,000,000,000,000; other numbers are very small, such as the shortest wavelength of visible light, about .0000004 meters. Writing these numbers is simplified by using *scientific notation.*

SCIENTIFIC NOTATION

A number is written in **scientific notation** when it is expressed in the form

$$a \times 10^n,$$

where $1 \leq |a| < 10$, and n is an integer.

As stated in the definition, scientific notation requires that the number be written as a product of a number between 1 and 10 (or -1 and -10) and some integer power of 10. (1 and -1 are allowed as values of a.) For example, since

$$8000 = 8 \cdot 1000 = 8 \cdot 10^3,$$

the number 8000 is written in scientific notation as

$$8000 = 8 \times 10^3.$$

(When using scientific notation, it is customary to use $\times$ instead of a dot to show multiplication.)

The steps involved in writing a number in scientific notation are given below. (If the number is negative, ignore the minus sign, go through these steps, and then attach a minus sign to the result.)

CONVERTING TO SCIENTIFIC NOTATION

Step 1 Place a caret, $_\wedge$, to the right of the first nonzero digit.

Step 2 Count the number of digits from the caret to the decimal point. This number gives the absolute value of the exponent on ten.

Step 3 Decide whether multiplying by 10^n should make the number larger or smaller. The exponent should be positive to make the product larger; it should be negative to make the product smaller.

EXAMPLE 6

WRITING A NUMBER IN SCIENTIFIC NOTATION

Write each number in scientific notation.

(a) 820,000

Place a caret to the right of the 8 (the first nonzero digit).

$$8_\wedge 20{,}000$$

Count from the caret to the decimal point, which is understood to be after the last 0.

$$8_\wedge 20{,}000. \leftarrow \text{Decimal point}$$

Count 5 places

Since the number 8.2 is to be made larger, the exponent on 10 is positive.

$$820{,}000 = 8.2 \times 10^5$$

(b) .000072
Count from right to left.

.00007 2
5 places

Since the number 7.2 is to be made smaller, the exponent on 10 is negative.

$$.000072 = 7.2 \times 10^{-5} \quad \blacksquare$$

To convert a number written in scientific notation to standard notation, just work in reverse.

CONVERTING FROM SCIENTIFIC NOTATION

Multiplying a number by a positive power of 10 makes the number larger, so move the decimal point to the right if n is positive in 10^n. Multiplying by a negative power of 10 makes a number smaller, so move the decimal point to the left if n is negative. If n is zero, leave the decimal point where it is.

EXAMPLE 7 **CONVERTING FROM SCIENTIFIC NOTATION TO STANDARD NOTATION**

Write the following numbers without scientific notation.

(a) 6.93×10^5

6.93000
5 places

The decimal point was moved 5 places to the right. (It was necessary to attach 3 zeros.)

$$6.93 \times 10^5 = 693{,}000$$

(b) $3.52 \times 10^7 = 35{,}200{,}000$

(c) 4.7×10^{-6}

000004.7
6 places

The decimal point was moved 6 places to the left.

$$4.7 \times 10^{-6} = .0000047$$

(d) $1.083 \times 10^0 = 1.083 \quad \blacksquare$

The next example shows how scientific notation and the rules for exponents can be used to simplify calculations.

EXAMPLE 8 **USING SCIENTIFIC NOTATION IN COMPUTATION**

Find $\dfrac{1{,}920{,}000 \times .0015}{.000032 \times 45{,}000}$.

First, express all numbers in scientific notation.

$$\frac{1{,}920{,}000 \times .0015}{.000032 \times 45{,}000} = \frac{1.92 \times 10^6 \times 1.5 \times 10^{-3}}{3.2 \times 10^{-5} \times 4.5 \times 10^4}$$

Next, use the commutative and associative properties and the rules for exponents to simplify the expression.

$$\frac{1{,}920{,}000 \times .0015}{.000032 \times 45{,}000} = \frac{1.92 \times 1.5 \times 10^6 \times 10^{-3}}{3.2 \times 4.5 \times 10^{-5} \times 10^4}$$
$$= \frac{1.92 \times 1.5}{3.2 \times 4.5} \times 10^4$$
$$= .2 \times 10^4$$
$$= (2 \times 10^{-1}) \times 10^4$$
$$= 2 \times 10^3$$

The expression is equal to 2×10^3, or 2000. ■

PROBLEM SOLVING

When problems require operations with numbers that are very large and/or very small, it is often advantageous to write the numbers in scientific notation first, and then perform the calculations using the rules for exponents. The next example illustrates this. ■

EXAMPLE 9
USING SCIENTIFIC NOTATION TO SOLVE A PROBLEM

A certain computer can perform an algorithm in .00000000036 second. How long would it take the computer to perform one billion of these algorithms? (One billion = 1,000,000,000)

In order to solve this problem, we must multiply the time per algorithm by the number of algorithms.

$$.00000000036 \times 1{,}000{,}000{,}000 = \text{total time}$$

Write each number in scientific notation, and then use rules for exponents.

$.00000000036 \times 1{,}000{,}000{,}000$

$= (3.6 \times 10^{-10}) \times (1 \times 10^9)$

$= (3.6 \times 1)(10^{-10} \times 10^9)$ — Commutative and associative properties

$= 3.6 \times 10^{-1}$ — Product rule

$= .36$ — Convert to standard notation.

It would take the computer .36 second. ■

3.2 EXERCISES

Simplify each expression. Write all answers with only positive exponents. Assume that all variables represent nonzero real numbers. See Examples 1–4.

1. $(4^2)^5$ **2.** $(3^3)^2$ **3.** $(5^2)^2$ **4.** $(4^3)^3$

5. $(6^{-3})^{-2}$ **6.** $(2^{-5})^{-4}$ **7.** $\left(\frac{3}{4}\right)^{-2}$ **8.** $\left(\frac{2}{5}\right)^{-3}$

9. $\left(\frac{6}{5}\right)^{-1}$ **10.** $\left(\frac{8}{3}\right)^{-2}$ **11.** $(3^{-2}\cdot 4^{-1})^3$ **12.** $(5^{-3}\cdot 6^{-2})^2$

13. $(4^{-3}\cdot 7^{-5})^{-2}$ **14.** $(9^{-3}\cdot 8^{-2})^{-2}$ **15.** $(z^3)^{-2}z^2$ **16.** $(p^{-1})^3p^{-4}$

17. $-3r^{-1}(r^{-3})^2$ **18.** $2(y^{-3})^4(y^6)$ **19.** $(3a^{-2})^3(a^3)^{-4}$ **20.** $(m^5)^{-2}(3m^{-2})^3$

21. $(x^{-5}y^2)^{-1}$ **22.** $(a^{-3}b^{-5})^2$ **23.** $(2p^2q^{-3})^2(4p^{-3}q)^2$ **24.** $(-5y^2z^{-4})^2(2yz^5)^{-3}$

25. $\frac{(p^{-2})^3}{5p^4}$ **26.** $\frac{(m^4)^{-1}}{9m^3}$ **27.** $\frac{4a^5(a^{-1})^3}{(a^{-2})^{-2}}$ **28.** $\frac{12k^{-2}(k^{-3})^{-4}}{6k^5}$

29. $\frac{(-y^{-4})^2}{6(y^{-5})^{-1}}$ **30.** $\frac{2(-m^{-1})^{-4}}{9(m^{-3})^2}$ **31.** $\frac{(2k)^2m^{-5}}{(km)^{-3}}$ **32.** $\frac{(3rs)^{-2}}{3^2r^2s^{-4}}$

33. $\frac{2^{-1}y^{-3}z^2}{2y^2z^{-3}}$ **34.** $\frac{2^{-2}m^{-5}p^{-3}}{2^{-4}m^3p^2}$ **35.** $\frac{(2k)^2k^3}{k^{-1}k^{-5}}(5k^{-2})^{-3}$ **36.** $\frac{(3r^2)^2r^{-5}}{r^{-2}r^3}(2r^{-6})^2$

37. $\left(\frac{3k^{-2}}{k^4}\right)^{-1}\cdot\frac{2}{k}$ **38.** $\left(\frac{7m^{-2}}{m^{-3}}\right)^{-2}\cdot\frac{m^3}{4}$ **39.** $\left(\frac{2p}{q^2}\right)^3\left(\frac{3p^4}{q^{-4}}\right)^{-1}$ **40.** $\left(\frac{5z^3}{2a^2}\right)^{-3}\left(\frac{8a^{-1}}{15z^{-2}}\right)^{-3}$

41. $\frac{2^2y^4(y^{-3})^{-1}}{2^5y^{-2}}$ **42.** $\frac{3^{-1}m^4(m^2)^{-1}}{3^2m^{-2}}$

---

43. In order to raise a fraction to a negative power, we may change the fraction to its ________ and change the exponent to the ________ of the original exponent. (See Objective 1.)

44. Explain in your own words how we raise a power to a power. (See Objective 1.)

45. Which one of the following is correct? (See Objective 1.)

(a) $-\frac{3}{4}=\left(\frac{3}{4}\right)^{-1}$ **(b)** $\frac{3^{-1}}{4^{-1}}=\left(\frac{4}{3}\right)^{-1}$

(c) $\frac{3^{-1}}{4}=\frac{3}{4^{-1}}$ **(d)** $\frac{3^{-1}}{4^{-1}}=\left(\frac{3}{4}\right)^{-1}$

46. Which one of the following is incorrect? (See Objective 1.)

(a) $(3r)^{-2}=3^{-2}r^{-2}$

(b) $3r^{-2}=(3r)^{-2}$

(c) $(3r)^{-2}=\frac{1}{(3r)^2}$

(d) $(3r)^{-2}=\frac{r^{-2}}{9}$

Write each number in scientific notation. See Example 6.

47. 230 **48.** 46,500 **49.** .02

50. .0051 **51.** .00000327 **52.** .00000898

53. 93,000,000 **54.** 176,000,000

Write each number without scientific notation. See Example 7.

55. 6.5×10^3 **56.** 2.317×10^5 **57.** 1.52×10^{-2}

58. 1.63×10^{-4} **59.** 5×10^{-3} **60.** 8×10^7

61. 4.69×10^{-2} **62.** 3.7×10^6

Use scientific notation to compute. In Exercises 73–76 round the numerical portion of the answer to the nearest thousandth. See Example 8.

63. $\dfrac{9 \times 10^2}{3 \times 10^6}$

64. $\dfrac{8 \times 10^6}{4 \times 10^5}$

65. $\dfrac{6 \times 10^{-3}}{2 \times 10^2}$

66. $\dfrac{5 \times 10^{-2}}{1 \times 10^{-4}}$

67. $\dfrac{.002 \times 3900}{.000013}$

68. $\dfrac{.009 \times 600}{.02}$

69. $\dfrac{.0004 \times 56{,}000}{.000112}$

70. $\dfrac{.018 \times 20{,}000}{300 \times .0004}$

71. $\dfrac{840{,}000 \times .03}{.00021 \times 600}$

72. $\dfrac{28 \times .0045}{140 \times 1500}$

73. $\dfrac{4.6 \times 85{,}000}{3570 \times .000064}$

74. $\dfrac{250 \times 2300}{542{,}000 \times .0091}$

75. $\dfrac{6863 \times 142{,}000}{.00965 \times .0843}$

76. $\dfrac{54.72 \times 136{,}000}{.0974 \times .0813}$

Use scientific notation to work the following exercises. See Example 9.

77. The distance to the sun is 9.3×10^7 miles. How long would it take a rocket, traveling at 2.9×10^3 miles per hour, to reach the sun?

78. A *light-year* is the distance that light travels in one year. Find the number of miles in a light-year if light travels 1.86×10^5 miles per second.

79. Use the information given in the previous two exercises to find the number of minutes necessary for light from the sun to reach the Earth.

80. A computer can do one addition in 1.4×10^{-7} seconds. How long would it take the computer to do a trillion (10^{12}) calculations? Give the answer in seconds and then in hours.

81. The planet Mercury has a mean distance from the sun of 3.6×10^7 miles, while the mean distance of Venus to the sun is 6.7×10^7 miles. How long would it take a spacecraft traveling at 1.55×10^3 miles per hour to travel from Venus to Mercury? (Give your answer in hours, without scientific notation.)

82. Use the information from the previous exercise to find the number of days it would take the spacecraft to travel from Venus to Mercury. Round your answer to the nearest whole number of days.

83. According to Bode's law, the distance d of the nth planet from the sun is

$$d = \frac{3(2^{n-2}) + 4}{10},$$

in astronomical units. Find the distance of each of the following planets from the sun.

(a) Venus $(n = 2)$
(b) Earth $(n = 3)$
(c) Mars $(n = 4)$

Photo for Exercise 81

84. When the distance between the centers of the Moon and the Earth is 4.60×10^8 meters, an object on the line joining the centers of the Moon and the Earth exerts the same gravitational force on each when it is 4.14×10^8 meters from the center of the Earth. How far is the object from the center of the Moon at that point?

85. Assume that the volume of the Earth is 5×10^{14} cubic meters and that the volume of a bacterium is 2.5×10^{-16} cubic meters. If the Earth could be packed full of bacteria, how many would it contain?

86. Our galaxy is approximately 1.2×10^{17} kilometers across. Suppose a spaceship could travel at 1.5×10^5 kilometers per second (half the speed of light). Find the approximate number of years needed for the spaceship to cross the galaxy.

Simplify each expression. Assume that all variables used as exponents represent integers and that all other variables represent nonzero real numbers. See Example 5.

87. $x^{-m} \cdot x^{3+4m}(x^{-2})^m$

88. $a^{-3k}(a^{-2})^k(a^{5-2k})$

89. $\dfrac{p^y p^{4y+2}}{(p^y)^3}$

90. $\dfrac{(y^2)^m(3y^{m+1})}{(y^{-1})^{-3}}$

91. $\dfrac{(a^{4+k}a^{3k})^{-1}}{(a^2)^{-k}}$

92. $\dfrac{(m^{2-p} \cdot m^{3p})^2}{(m^{-2})^p}$

93. $\dfrac{r^{-p}(r^{p+2})^p}{(r^{p-3})^{-2}}$

94. $\dfrac{z^{-3q}(z^{q-5})^q}{(z^{q+1})^{-q}}$

Simplify each expression. Write answers with only positive exponents. Assume that all variables represent nonzero real numbers.

95. $\dfrac{(2m^2p^3)^2(4m^2p)^{-2}}{(-3mp^4)^{-1}(2m^3p^4)^3}$

96. $\dfrac{(-5y^3z^4)^2(2yz^5)^{-2}}{10(y^4z)^3(3y^3z^2)^{-1}}$

97. $\dfrac{(-3y^3x^3)(-4y^4x^2)(x^2)^{-4}}{18x^3y^2(y^3)^3(x^3)^{-2}}$

98. $\dfrac{(2m^3x^2)^{-1}(3m^4x)^{-3}}{(m^2x^3)^3(m^2x)^{-5}}$

99. $\left(\dfrac{p^2q^{-1}}{2p^{-2}}\right)^2 \cdot \left(\dfrac{p^3 \cdot 4q^{-2}}{3q^{-5}}\right)^{-1} \cdot \left(\dfrac{pq^{-5}}{q^{-2}}\right)^3$

100. $\left(\dfrac{a^6b^{-2}}{2a^{-2}}\right)^{-1} \cdot \left(\dfrac{6a^{-2}}{5b^{-4}}\right)^2 \cdot \left(\dfrac{2b^{-1}a^2}{3b^{-2}}\right)^{-1}$

PREVIEW EXERCISES

Combine like terms. Remove all parentheses first, if necessary. See Section 1.4.

101. $9x + 5x - x$

102. $3p + 2p - 8p$

103. $6 - 4(3 - y)$

104. $2 + 3(5 + q)$

105. $7x - (5 + 5x) + 3$

106. $12q - (6 - 8q) - 4$

3.3 THE BASIC IDEAS OF POLYNOMIALS

OBJECTIVES

1 KNOW THE BASIC DEFINITIONS FOR POLYNOMIALS.
2 IDENTIFY MONOMIALS, BINOMIALS, AND TRINOMIALS.
3 FIND THE DEGREE OF A POLYNOMIAL.
4 ADD AND SUBTRACT POLYNOMIALS.
5 USE $P(x)$ NOTATION.

Just as whole numbers are the basis of arithmetic, *polynomials* are fundamental in algebra. A **term of a polynomial** is a number or the product of a number and one or more variables raised to a nonnegative power. Examples of terms include

$$4x, \quad \frac{1}{2}m^5, \quad -7z^9, \quad 6x^2z, \quad \text{and} \quad 9.$$

The number in the product is called the **numerical coefficient,** or just the **coefficient.** In the term $8x^3$, the coefficient is 8. In the term $-4p^5$, it is -4. The coefficient of the term k is understood to be 1. The coefficient of $-r$ is -1. More generally, any factor in a term is the coefficient of the product of the remaining factors. For example, $3x^2$ is the coefficient of y in the term $3x^2y$, and $3y$ is the coefficient of x^2 in $3x^2y$.

1 As mentioned earlier, any combination of variables or constants joined by the basic operations of addition, subtraction, multiplication, division (except by zero), or extraction of roots is called an **algebraic expression.** The simplest kind of algebraic expression is a **polynomial.**

POLYNOMIAL

A **polynomial** is a term or a finite sum of terms in which all variables have whole number exponents and no variables appear in denominators.

Examples of polynomials include

$$3x - 5, \quad 4m^3 - 5m^2p + 8, \quad \text{and} \quad -5t^2s^3.$$

Even though the expression $3x - 5$ involves subtraction, it is still called a sum of terms, since it could be written as $3x + (-5)$. Also, $-5t^2s^3$ can be thought of as a sum of terms by writing it as $0 + (-5t^2s^3)$.

Most of the polynomials used in this book contain only one variable. A polynomial containing only the variable x is called a **polynomial in *x*.** A polynomial in one variable is written in **descending powers** of the variable if the exponents on the terms of the polynomial decrease from left to right. For example,

$$x^5 - 6x^2 + 12x - 5$$

is a polynomial in descending powers of x.

NOTE The term -5 in the polynomial above can be thought of as $-5x^0$, since $-5x^0 = -5(1) = -5$.

EXAMPLE 1 WRITING POLYNOMIALS IN DESCENDING POWERS

Write each of the following in descending powers of the variable.

(a) $y - 6y^3 + 8y^5 - 9y^4 + 12$

Write the polynomial as

$$8y^5 - 9y^4 - 6y^3 + y + 12.$$

(b) $-2 + m + 6m^2 - 4m^3$ would be written as

$$-4m^3 + 6m^2 + m - 2. \quad ■$$

2 Polynomials with a certain number of terms are so common that they are given special names. A polynomial of exactly three terms is a **trinomial,** and a polynomial with exactly two terms is a **binomial.** A single-term polynomial is a **monomial.**

EXAMPLE 2 IDENTIFYING TYPES OF POLYNOMIALS

The list below gives examples of monomials, binomials, and trinomials, as well as polynomials that are none of these.

Monomials	$5x$, $7m^9$, -8
Binomials	$3x^2 - 6$, $11y + 8$, $5k + 15$
Trinomials	$y^2 + 11y + 6$, $8p^3 - 7p + 2m$, $-3 + 2k^5 + 9z^4$
None of these	$p^3 - 5p^2 + 2p - 5$, $-9z^3 + 5c^3 + 2m^5 + 11r^2 - 7r$ ■

3 The **degree of a term** with one variable is the exponent on the variable. For example, the degree of $2x^3$ is 3, the degree of $-x^4$ is 4, and the degree of $17x$ is 1. The degree of a term in more than one variable is defined to be the sum of the exponents of the variables. For example, the degree of $5x^3y^7$ is 10, because $3 + 7 = 10$.

The largest degree of any of the terms in a polynomial is called the **degree of the polynomial.** In most cases, we will be interested in finding the degree of a polynomial in one variable. For example, the degree of $4x^3 - 2x^2 - 3x + 7$ is 3, because the largest degree of any term is 3 (the degree of $4x^3$).

EXAMPLE 3 FINDING THE DEGREE OF A POLYNOMIAL

Find the degree of each polynomial.

Largest exponent is 2.

(a) $9x^2 - 5x + 8$

The largest exponent is 2, so the polynomial is of degree 2.

(b) $17m^9 + 8m^{14} - 9m^3$

This polynomial is of degree 14.

(c) $5x$

The degree is 1, since $5x = 5x^1$.

(d) -2

A constant term, other than zero, is of degree zero.

(e) $5a^2b^5$

The degree is the sum of the exponents, $2 + 5 = 7$.

(f) $x^3y^9 + 12xy^4 + 7xy$

The degrees of the terms are 12, 5, and 2. Therefore, the degree of the polynomial is 12, which is the largest degree of any term in the polynomial. ■

NOTE The number 0 has no degree, since 0 times a variable to any power is 0.

4 The distributive property can be used to simplify some polynomials by combining terms. For example, to simplify $x^3 + 4x^2 + 5x^2 - 1$, use the distributive property as follows.

$$\begin{aligned} x^3 + 4x^2 + 5x^2 - 1 &= x^3 + (4 + 5)x^2 - 1 \\ &= x^3 + 9x^2 - 1 \end{aligned}$$

On the other hand, it is not possible to combine the terms in the polynomial $4x + 5x^2$. As these examples suggest, only terms containing exactly the same variables to the same powers may be combined. As mentioned in Chapter 1, such terms are called *like terms*.

EXAMPLE 4 **COMBINING LIKE TERMS**

Combine terms.

(a) $-5y^3 + 8y^3 - y^3$

These like terms may be combined by the distributive property.

$$-5y^3 + 8y^3 - y^3 = (-5 + 8 - 1)y^3 = 2y^3$$

(b) $6x + 5y - 9x + 2y$

Use the associative and commutative properties to rewrite the expression with all the x's together and all the y's together.

$$6x + 5y - 9x + 2y = 6x - 9x + 5y + 2y$$

Now combine like terms.

$$= -3x + 7y$$

Since $-3x$ and $7y$ are unlike terms, no further simplification is possible.

(c) $$\begin{aligned} 5x^2y - 6xy^2 + 9x^2y + 13xy^2 &= 5x^2y + 9x^2y - 6xy^2 + 13xy^2 \\ &= 14x^2y + 7xy^2 \end{aligned}$$ ■

CAUTION Remember that only like terms can be combined.

Use the following rule to add two polynomials.

ADDING POLYNOMIALS

Add two polynomials by combining like terms.

EXAMPLE 5 ADDING POLYNOMIALS HORIZONTALLY

Add $4k^2 - 5k + 2$ and $-9k^2 + 3k - 7$.

Use the commutative and associative properties to rearrange the polynomials so that like terms are together. Then use the distributive property to combine like terms.

$$\begin{aligned}(4k^2 - 5k + 2) + (-9k^2 + 3k - 7) &= 4k^2 - 9k^2 - 5k + 3k + 2 - 7\\ &= -5k^2 - 2k - 5 \quad \blacksquare\end{aligned}$$

EXAMPLE 6 ADDING POLYNOMIALS VERTICALLY

The two polynomials in Example 5 can also be added vertically by lining up like terms in columns. Then add by columns.

$$\begin{array}{r} 4k^2 - 5k + 2\\ -9k^2 + 3k - 7\\ \hline -5k^2 - 2k - 5 \end{array} \quad \blacksquare$$

EXAMPLE 7 ADDING POLYNOMIALS

Add $3a^5 - 9a^3 + 4a^2$ and $-8a^5 + 8a^3 + 2$.

$$\begin{aligned}&(3a^5 - 9a^3 + 4a^2) + (-8a^5 + 8a^3 + 2)\\ &= 3a^5 - 8a^5 - 9a^3 + 8a^3 + 4a^2 + 2\\ &= -5a^5 - a^3 + 4a^2 + 2 \qquad \text{Combine like terms.}\end{aligned}$$

These same two polynomials can be added by placing them in columns, with like terms in the same columns.

$$\begin{array}{rrrr} 3a^5 & - 9a^3 & + 4a^2 & \\ -8a^5 & + 8a^3 & & + 2\\ \hline -5a^5 & - a^3 & + 4a^2 & + 2 \end{array}$$

For many people, there is less chance of error with vertical addition. ■

In Chapter 1, subtraction of real numbers was defined as

$$a - b = a + (-b).$$

That is, the first number and the opposite of the second are added. A similar definition for subtraction of polynomials can be given by defining the **negative of a polynomial** as that polynomial with every sign changed.

SUBTRACTING POLYNOMIALS

Subtract two polynomials by adding the first polynomial and the negative of the *second* polynomial.

EXAMPLE 8 SUBTRACTING POLYNOMIALS HORIZONTALLY

Subtract: $(-6m^2 - 8m + 5) - (-5m^2 + 7m - 8)$.

Change every sign in the second polynomial and add.

$$\begin{aligned}&(-6m^2 - 8m + 5) - (-5m^2 + 7m - 8)\\ &= -6m^2 - 8m + 5 + 5m^2 - 7m + 8\end{aligned}$$

Now add by combining like terms.

$$= -6m^2 + 5m^2 - 8m - 7m + 5 + 8$$
$$= -m^2 - 15m + 13$$

Check by adding the sum, $-m^2 - 15m + 13$, to the second polynomial. The result should be the first polynomial. ■

EXAMPLE 9
SUBTRACTING POLYNOMIALS VERTICALLY

Use the same polynomials as in Example 8 and subtract in columns.

Write the first polynomial above the second, lining up like terms in columns.

$$\begin{array}{r} -6m^2 - 8m + 5 \\ \underline{-5m^2 + 7m - 8} \end{array}$$

Change all the signs in the second polynomial, and add.

$$\begin{array}{rl} -6m^2 - 8m + 5 & \\ \underline{+5m^2 - 7m + 8} & \text{All signs changed} \\ -m^2 - 15m + 13 & \text{Add in columns.} \end{array}$$ ■

5 Sometimes one problem will involve several polynomials. To keep track of these polynomials, capital letters can be used to name the polynomials. For example, $P(x)$, read "P of x," or, "the value of P at x," can be used to represent the polynomial $3x^2 - 5x + 7$, so that

$$P(x) = 3x^2 - 5x + 7.$$

The x in $P(x)$ is used to show that x is the variable in the polynomial.

If $x = -2$, then $P(x) = 3x^2 - 5x + 7$ takes on the value

$$\begin{aligned} P(-2) &= 3(-2)^2 - 5(-2) + 7 \qquad \text{Let } x = -2. \\ &= 3 \cdot 4 + 10 + 7 \\ &= 29. \end{aligned}$$

EXAMPLE 10
USING $P(x)$ NOTATION

Let $P(x) = 4x^3 - x^2 + 5$. Find each of the following.

(a) $P(3)$

First, substitute 3 for x.

$$P(x) = 4x^3 - x^2 + 5$$
$$P(3) = 4 \cdot 3^3 - 3^2 + 5$$

Now use the order of operations from Chapter 1.

$$= 4 \cdot 27 - 9 + 5$$
$$= 108 - 9 + 5$$
$$= 104$$

(b) $P(-4) = 4 \cdot (-4)^3 - (-4)^2 + 5 \qquad \text{Let } x = -4.$

$$= 4 \cdot (-64) - 16 + 5$$
$$= -267$$ ■

3.3 EXERCISES

Give the coefficient and the degree of each term. See Example 3.

1. $9k$ **2.** $-14z$ **3.** $-11p^7$ **4.** $9y^{12}$ **5.** πk^3

6. $\sqrt{6}m$ **7.** y **8.** a **9.** $-x^4y^4$ **10.** $-p^3q^2$

11. The exponent in the expression 4^5 is 5. Explain why the degree of 4^5 is not 5. What is its degree? (See Objective 3.)

12. A polynomial with *exactly* two terms is called a binomial. Why is the word *exactly* important in this definition? (See Objective 2.)

Write each polynomial in descending powers of the variable. See Example 1.

13. $8x + 9x^2 + 3x^3$

14. $5m - 3m^4 + 8m^6$

15. $-6y^3 + 8y^2 + 4y^4 - 3$

16. $9 - 4z^3 + 8z - 12z^2$

17. $13y^3 - 9y^4 + 8y^2 - 7y + 2$

18. $-5a^4 - 3a^2 + 12a^3 + 8a^5$

Identify each polynomial as a monomial, binomial, trinomial, *or* none of these. *Give the degree of each. See Examples 2 and 3.*

19. $4k - 9$

20. $12z^5$

21. $-8a^2 - 9a + 2$

22. $7y^5 - 6y^{12}$

23. $a^3 - a^2 + a^4$

24. $-m^9 + m^7 - m^6 + m^2$

25. -2

26. $13x^4y^5$

27. $9a^3b^4 - 3ab^4 + 19a^5b^3 + 9$

Find the degree of each polynomial. See Example 3.

28. $-7y^6 + 11y^8$

29. $12k^5 - 9k^2$

30. $-5m^3 + 6m^2 - 9m$

31. $4z^2 - 11z + 2$

32. $8y^5 + 6y^4 - 9y^2 + 3$

33. 8

34. $6p + 11p^3 - 9p^4$

35. $3p^4q^5 - 5pq + 7p$

36. $7x^5y + 3x^3y^4 + 3x$

Add or subtract as indicated. See Examples 6 and 9.

37. Add.
$$\begin{array}{r} 21p - 8 \\ \underline{-9p + 4} \end{array}$$

38. Add.
$$\begin{array}{r} 15m - 9 \\ \underline{4m + 12} \end{array}$$

39. Add.
$$\begin{array}{r} 5k + 8 \\ \underline{-6k - 9} \end{array}$$

40. Add.
$$\begin{array}{r} -12p^2 + 4p - 1 \\ \underline{3p^2 + 7p - 8} \end{array}$$

41. Add.
$$\begin{array}{r} -6y^3 + 8y + 5 \\ \underline{9y^3 + 4y - 6} \end{array}$$

42. Subtract.
$$\begin{array}{r} 6m^2 - 11m + 5 \\ \underline{-8m^2 + 2m - 1} \end{array}$$

43. Subtract.
$$\begin{array}{r} -4z^2 + 2z - 1 \\ \underline{3z^2 - 5z + 2} \end{array}$$

44. Subtract.
$$\begin{array}{r} 5q^3 - 5q + 2 \\ \underline{-3q^3 + 2q - 9} \end{array}$$

45. Add.
$$\begin{array}{r} 6y^3 - 9y^2 \quad\quad + 8 \\ \underline{4y^3 + 2y^2 + 5y \quad\quad} \end{array}$$

Simplify each of the following polynomials by combining like terms. See Examples 4, 5, 7, and 8.

46. $6x - 5x + x^2$

47. $-3m^3 + 4m^3 - 5m^3$

48. $2a^2 + 3a - 5a + 6a^2$

49. $6p^2 - 11p + 8p^2 - 21p$

50. $3p^2 + 5p + 6p^2 - 9p$

51. $-m^4 + m^2 - m^3 + 5m^2 + 2m^4$

52. $3x^2 - 8x^3 - 9x^2 + 11x^3 - 4x + 8$

53. $6z^2 - 12z + 8z^3 - 5z^2 - 9$

54. $[4 - (2 + 3m)] + (6m + 9)$

55. $[8a - (3a + 4)] - (5a - 3)$

56. $[(6 + 3p) - (2p + 1)] - (2p + 9)$

57. $[(4x - 8) - (-1 + x)] - (11x + 5)$

58. $(3p^2 + 2p - 5) + (7p^2 - 4p^3 + 3p)$

59. $(y^3 + 3y + 2) + (4y^3 - 3y^2 + 2y - 1)$

60. $(2x^5 - 2x^4 + x^3 - 1) + (x^4 - 3x^3 + 2)$

61. $(y^2 + 3y) + [2y - (5y^2 + 3y + 4)]$

62. $(9a - 5a) - [2a - (4a + 3)]$

63. $(6r - 6r^2) - [(2r - 5r^2) - (3r + r^2)]$

64. $[(5m^2 + 2m^3) - (3m^2 - 5m^3)] + (7m^3 + 2m^2)$

65. $(8a + 3a^2) - (2a - a^2) - (4a + a^2)$

---

66. Which one of the following is a trinomial in descending powers, having degree 6? (See Objectives 1, 2, and 3.)
(a) $5x^6 - 4x^5 + 12$ **(b)** $6x^5 - x^6 + 4$
(c) $2x + 4x^2 - x^6$ **(d)** $4x^6 - 6x^4 + 9x^2 - 8$

67. Give an example of a polynomial of four terms in the variable x, having degree 5, written in descending powers, lacking a fourth degree term. (See Objectives 1, 2, and 3.)

For each polynomial, find **(a)** $P(-1)$ *and* **(b)** $P(2)$. *See Example 10.*

68. $P(x) = 3x + 3$

69. $P(x) = x - 10$

70. $P(x) = -2x^2 + 4x - 1$

71. $P(x) = 2x^2 - 5x + 2$

72. $P(x) = x^4 - 3x^2 + 2x$

73. $P(x) = 2x^4 + 3x^2 - 5x$

74. $P(x) = -x^2 + 2x^3 - 8$

75. $P(x) = -x^2 - x^3 + 11x$

76. $P(x) = 2x - x^2 + x^3$

Perform the following operations.

77. Find the sum of $-5y + 3a$ and $2a - 3y$ and $4y + 6a$.

78. Subtract $6mp - 4m^2$ from $3mp + 8m^2$.

79. Subtract $4y^2 - 2y + 3$ from $7y^2 - 6y + 5$.

80. Subtract $-(-4x + 2z^2 + 3m)$ from $[(2z^2 - 3x + m) + (z^2 - 2m)]$.

81. Subtract $(3m^2 - 5n^2 + 2n) + [-(3m^2 - 4n^2)]$ from $-4m^2 + 3n^2 - 5n$.

82. $[-(4m^2 - 8m + 4m^3) - (3m^2 + 2m + 5m^3)] + m^2$

83. $[-(y^4 - y^2 + 1) - (y^4 + 2y^2 + 1)] + (3y^4 - 3y^2 - 2)$

84. $(8rt + 6r - 4t) + (3rt - 4r + t) - (-2rt - 5r + 2t)$

85. $[(6xy - 10z^2) + (3z^2 - 2xy)] - [(5xy + 3z^2) - (-2xy - z^2)]$

86. $(2p - [3p - 6]) - [(5p - (8 - 9p)) + 4p]$

87. $-(3z^2 + 5z - [2z^2 - 6z]) + [(8z^2 - [5z - z^2]) + 2z^2]$

88. $5k - (5k - [2k - (4k - 8k)]) + 11k - (9k - 12k)$

89. $(3a^{2x} + a^x - 4) + (5a^{3x} - a^{2x} + 2a^x + 6)$

90. $(2m^{3y} - 4m^{2y} + m) - (4m^{3y} + 8m^{2y} + m^y - 3m)$

91. $(-3k^{4p} - 6k^{2p} + 10) - (5k^{4p} + 3k^{2p} + 8)$

92. $(7r^{3x} - 5r^{2x} + 6r^{x} + 1) + (-8r^{3x} + 3r^{2x} - 5r^{x} - 3)$

93. $(5p^{2k} + 3p^{k} - 4) + [(6p^{k} - 3) - (2p^{2k} + p^{k})]$

94. $[(-z^{3m} - 2z^{2m} + z^{m}) - 4z^{3m}] - (2z^{2m} + 9z^{m})$

PREVIEW EXERCISES

Find each product. See Section 3.1.

95. $9p^2(4p^3)$ **96.** $8r^2(6r^4)$ **97.** $10z(5z^4)$ **98.** $3y(9y^5)$

99. $6xy^3(2xy^4)$ **100.** $5p^3q(2p^3q^4)$ **101.** $-2a^4(3a^5b^2)$ **102.** $-9m(-4mn^4)$

3.4 MULTIPLICATION OF POLYNOMIALS

OBJECTIVES

1. MULTIPLY TERMS.
2. MULTIPLY ANY TWO POLYNOMIALS.
3. MULTIPLY BINOMIALS.
4. FIND THE PRODUCT OF THE SUM AND DIFFERENCE OF TWO TERMS.
5. FIND THE SQUARE OF A BINOMIAL.

The previous section showed how polynomials are added and subtracted. In this section polynomial multiplication is discussed.

1 Recall that the product of the two terms $3x^4$ and $5x^3$ is found by using the commutative and associative properties, along with the rules for exponents.

$$\begin{aligned}(3x^4)(5x^3) &= 3 \cdot 5 \cdot x^4 \cdot x^3 \\ &= 15x^{4+3} \\ &= 15x^7\end{aligned}$$

EXAMPLE 1 MULTIPLYING MONOMIALS

Find the following products.

(a) $(-4a^3)(3a^5) = (-4)(3)a^3 \cdot a^5 = -12a^8$

(b) $(2m^2z^4)(8m^3z^2) = (2)(8)m^2 \cdot m^3 \cdot z^4 \cdot z^2 = 16m^5z^6$ ■

2 The distributive property can be used to extend this process to find the product of any two polynomials.

EXAMPLE 2 MULTIPLYING A MONOMIAL AND A POLYNOMIAL

(a) Find the product of -2 and $8x^3 - 9x^2$.
Use the distributive property.

$$\begin{aligned}-2(8x^3 - 9x^2) &= -2(8x^3) - 2(-9x^2) \\ &= -16x^3 + 18x^2\end{aligned}$$

(b) Find the product of $5x^2$ and $-4x^2 + 3x - 2$.

$$5x^2(-4x^2 + 3x - 2) = 5x^2(-4x^2) + 5x^2(3x) + 5x^2(-2)$$
$$= -20x^4 + 15x^3 - 10x^2 \quad \blacksquare$$

EXAMPLE 3
MULTIPLYING TWO POLYNOMIALS

Find the product of $3x - 4$ and $2x^2 + x$.

Use the distributive property to multiply each term of $2x^2 + x$ by $3x - 4$.

$$(3x - 4)(2x^2 + x) = (3x - 4)(2x^2) + (3x - 4)(x)$$

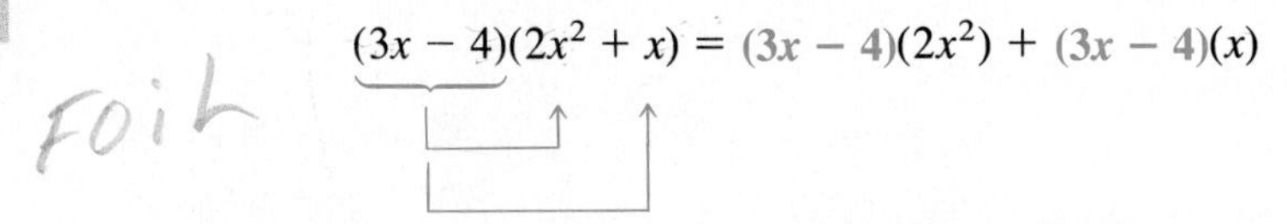

Here $3x - 4$ has been treated as a single expression so that the distributive property could be used. Now use the distributive property twice again.

$$= 3x(2x^2) + (-4)(2x^2) + (3x)(x) + (-4)(x)$$
$$= 6x^3 - 8x^2 + 3x^2 - 4x$$
$$= 6x^3 - 5x^2 - 4x \quad \blacksquare$$

It is often easier to multiply polynomials by writing them vertically. To find the product from Example 3, $(3x - 4)(2x^2 + x)$, vertically, proceed as follows. (Notice how this process is similar to that of finding the product of two numbers, such as 24×78.)

1. Multiply x and $3x - 4$.

$$\begin{array}{rr} & 3x - 4 \\ & 2x^2 + x \\ \hline x(3x - 4) \rightarrow & 3x^2 - 4x \end{array}$$

2. Multiply $2x^2$ and $3x - 4$. Line up like terms of the products in columns.

3. Combine like terms.

$$\begin{array}{rrrr} & & 3x & -4 \\ & & 2x^2 & +x \\ \hline & & 3x^2 & -4x \\ 2x^2(3x - 4) \rightarrow & 6x^3 & -8x^2 & \\ \hline & 6x^3 & -5x^2 & -4x \end{array}$$

EXAMPLE 4
MULTIPLYING POLYNOMIALS VERTICALLY

Find the product of $5a - 2b$ and $3a + b$.

$$\begin{array}{rrrl} & 5a & -2b & \\ & 3a & +b & \\ \hline & 5ab & -2b^2 & \leftarrow b(5a - 2b) \\ 15a^2 & -6ab & & \leftarrow 3a(5a - 2b) \\ \hline 15a^2 & -ab & -2b^2 & \blacksquare \end{array}$$

EXAMPLE 5
MULTIPLYING POLYNOMIALS VERTICALLY

Find the product of $3m^3 - 2m^2 + 4$ and $3m - 5$.

$$\begin{array}{rrrrrl} & 3m^3 & -2m^2 & & +4 & \\ & & & 3m & -5 & \\ \hline & -15m^3 & +10m^2 & & -20 & -5 \text{ times } 3m^3 - 2m^2 + 4 \\ 9m^4 & -6m^3 & & +12m & & 3m \text{ times } 3m^3 - 2m^2 + 4 \\ \hline 9m^4 & -21m^3 & +10m^2 & +12m & -20 & \text{Combine like terms.} \quad \blacksquare \end{array}$$

3 In work with polynomials, a special kind of product comes up repeatedly—the product of two binomials. A shortcut method for finding these products is discussed next.

You will recall that a binomial is a polynomial with just two terms, such as $3x - 4$ or $2x + 3$. The product of the binomials $3x - 4$ and $2x + 3$ can be found with the distributive property as follows.

$$\begin{aligned}(3x - 4)(2x + 3) &= 3x(2x + 3) - 4(2x + 3)\\ &= 3x(2x) + 3x(3) - 4(2x) - 4(3)\\ &= 6x^2 + 9x - 8x - 12\end{aligned}$$

Before combining like terms to find the simplest form of the answer, check the origin of each of the four terms in the sum. First, $6x^2$ is the product of the two *first* terms.

$$(3x - 4)(2x + 3) \qquad (3x)(2x) = 6x^2 \qquad \text{First terms}$$

To get $9x$, the *outside* terms are multiplied.

$$(3x - 4)(2x + 3) \qquad 3x(3) = 9x \qquad \text{Outside terms}$$

The term $-8x$ comes from the *inside* terms.

$$(3x - 4)(2x + 3) \qquad -4(2x) = -8x \qquad \text{Inside terms}$$

Finally, -12 comes from the *last* terms.

$$(3x - 4)(2x + 3) \qquad -4(3) = -12 \qquad \text{Last terms}$$

The product is found by combining these four results.

$$\begin{aligned}(3x - 4)(2x + 3) &= 6x^2 + 9x - 8x - 12\\ &= 6x^2 + x - 12\end{aligned}$$

To keep track of the order of multiplying these terms, use the initials FOIL (first, outside, inside, last). All the steps of the FOIL method can be done as follows. Try to do as many of these steps as possible in your head.

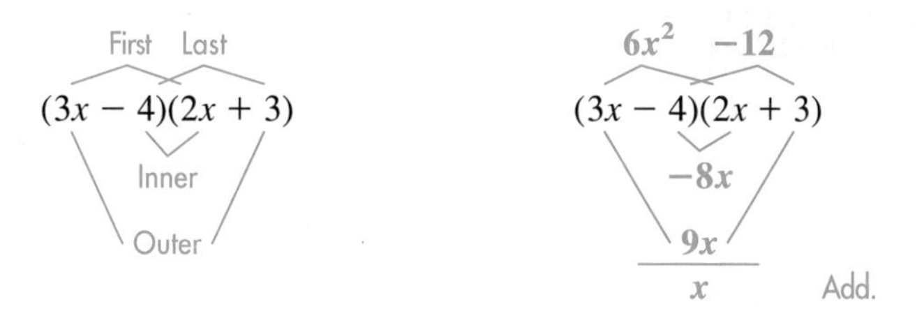

The FOIL method will be very helpful in factoring, which is discussed in the rest of this chapter.

EXAMPLE 6 USING THE FOIL METHOD

Use the FOIL method to find $(4m - 5)(3m + 1)$.

Find the product of the first terms.

$$(4m - 5)(3m + 1) \qquad (4m)(3m) = \mathbf{12m^2}$$

Multiply the outside terms.

$$(4m - 5)(3m + 1) \qquad (4m)(1) = \mathbf{4m}$$

Find the product of the inside terms.

$$(4m - 5)(3m + 1) \qquad (-5)(3m) = \mathbf{-15m}$$

Multiply the last terms.

$$(4m - 5)(3m + 1) \qquad (-5)(1) = \mathbf{-5}$$

Simplify by combining the four terms obtained above.

$$\begin{aligned}(4m - 5)(3m + 1) &= 12m^2 + 4m - 15m - 5\\ &= 12m^2 - 11m - 5\end{aligned}$$

The procedure can be written in compact form as follows.

$$\begin{array}{c} \mathbf{12m^2} \quad \mathbf{-5} \\ (4m - 5)(3m + 1) \\ \mathbf{-15m} \\ \mathbf{4m} \\ \hline \mathbf{-11m} \end{array} \quad \text{Add.}$$

Combine these four results to get $12m^2 - 11m - 5$. ■

EXAMPLE 7 USING THE FOIL METHOD

(a) $(6a - 5b)(3a + 4b) = 18a^2 + 24ab - 15ab - 20b^2$

First Outside Inside Last

$= 18a^2 + 9ab - 20b^2$

(b) $(2k + 3z)(5k - 3z) = 10k^2 - 6kz + 15kz - 9z^2$

$= 10k^2 + 9kz - 9z^2$ ■

4 A special type of binomial product that occurs frequently is the product of the sum and difference of the same two terms. By the FOIL method, the product $(x + y)(x - y)$ is

$$\begin{aligned}(x + y)(x - y) &= x^2 - xy + xy - y^2\\ &= x^2 - y^2.\end{aligned}$$

PRODUCT OF THE SUM AND DIFFERENCE OF TWO TERMS

The **product of the sum and difference of two terms** is the difference of the squares of the terms, or

$$(x + y)(x - y) = x^2 - y^2.$$

EXAMPLE 8 MULTIPLYING THE SUM AND DIFFERENCE OF TWO TERMS

Find the following products.

(a) $(p + 7)(p - 7) = p^2 - 7^2 = p^2 - 49$

(b) $(2r + 5)(2r - 5) = (2r)^2 - 5^2$
$= 2^2r^2 - 25$
$= 4r^2 - 25$

(c) $(6m + 5n)(6m - 5n) = (6m)^2 - (5n)^2$
$= 36m^2 - 25n^2$

(d) $(9y + 2z)(9y - 2z) = (9y)^2 - (2z)^2$
$= 81y^2 - 4z^2$ ■

5 Another special binomial product is the *square of a binomial*. To find the square of $x + y$, or $(x + y)^2$, multiply $x + y$ and $x + y$.

$$(x + y)^2 = (x + y)(x + y) = x^2 + xy + xy + y^2$$
$$= x^2 + 2xy + y^2$$

A similar result is true for the square of a difference, as shown next.

SQUARE OF A BINOMIAL

The **square of a binomial** is the sum of the square of the first term, twice the product of the two terms, and the square of the last term.

$$(x + y)^2 = x^2 + 2xy + y^2$$
$$(x - y)^2 = x^2 - 2xy + y^2$$

EXAMPLE 9 SQUARING A BINOMIAL

Find the following products.

(a) $(m + 7)^2 = m^2 + 2 \cdot m \cdot 7 + 7^2$
$= m^2 + 14m + 49$

(b) $(p - 5)^2 = p^2 - 2 \cdot p \cdot 5 + 5^2$
$= p^2 - 10p + 25$

(c) $(2p + 3v)^2 = (2p)^2 + 2(2p)(3v) + (3v)^2$
$= 4p^2 + 12pv + 9v^2$

(d) $(3r - 5s)^2 = (3r)^2 - 2(3r)(5s) + (5s)^2$
$= 9r^2 - 30rs + 25s^2$ ■

CAUTION As the products in the definition of the square of a binomial show,

$$(x + y)^2 \neq x^2 + y^2.$$

Also, more generally,

$$(x + y)^n \neq x^n + y^n.$$

The special products of this section are now summarized.

SPECIAL PRODUCTS

Product of the sum and difference of two terms	$(x + y)(x - y) = x^2 - y^2$
Square of a binomial	$(x + y)^2 = x^2 + 2xy + y^2$ $(x - y)^2 = x^2 - 2xy + y^2.$

The patterns for the product of the sum and difference of two terms and the square of a binomial can be used with more complicated products, as the following example shows.

EXAMPLE 10 **MULTIPLYING MORE COMPLICATED BINOMIALS**

Use the special products to multiply the following polynomials.

(a) $[(3p - 2) + 5q][(3p - 2) - 5q] = (3p - 2)^2 - (5q)^2$ Product of sum and difference of terms

$= 9p^2 - 12p + 4 - 25q^2$ Square both quantities.

(b) $[(2z + r) + 1]^2 = (2z + r)^2 + 2(2z + r)(1) + 1^2$ Square of a binomial

$= 4z^2 + 4zr + r^2 + 4z + 2r + 1$ Distributive property ■

3.4 EXERCISES

Find each of the following products. See Examples 1–7.

1. $3p(7p - 4k)$

2. $-8z(3z + 2y)$

3. $3m(5m^3 + 6m^2 - 3)$

4. $-4y^2(2y^2 + 3y - 8)$

5. $-6a^5(-2a^2 + 4a - 5)$

6. $2y^3(6y^2 - 4y^3 + 8y - 5)$

7. $4x^2y(2xy^2 + 3x^2y + 4xy)$

8. $2m^3k^2(m^2k^3 - 3mk^2 + 4k)$

9. $10rt(-3r^2t^2 + 2rt^2 - 5r^2t)$

10. $5k^3p(kp^3 - 3kp^2 + 5k^2p - k^3p)$

11. $(5m + 1)(9m - 4)$

12. $(3y - 4)(5y + 2)$

13. $(-2g + 3)(-g + 5)$

14. $(-5 + 6a)(-2 + 7a)$

15. $(-y + 1)(3y + 1)$

16. $(-4a + 11)(3a - 11)$

17. $(2a + b)(3a - b)$

18. $(5x + 2y)(3x + y)$

19. $(-8m + 5p)(3m - 5p)$

20. $(-2s + 5r)(3s + 4r)$

21. $\left(k - \frac{1}{2}m\right)\left(k + \frac{2}{3}m\right)$

22. $\left(x + \frac{5}{3}y\right)\left(x - \frac{4}{5}y\right)$

23. $\left(3w + \frac{1}{4}z\right)(w - 2z)$

24. $\left(5r - \frac{3}{5}y\right)(r + 5y)$

25. $\begin{array}{r} 3k - 2 \\ \underline{5k + 1} \end{array}$

26. $\begin{array}{r} 2m - 5 \\ \underline{m + 7} \end{array}$

27. $\begin{array}{r} 6m^2 + 2m - 1 \\ \underline{2m + 3} \end{array}$

28. $\begin{array}{r} -y^2 + 2y + 1 \\ \underline{3y - 5} \end{array}$

29. $\begin{array}{r} 2z^3 - 5z^2 + 8z - 1 \\ \underline{4z - 3} \end{array}$

30. $\begin{array}{r} 5k^3 + 2k^2 - 3k - 4 \\ \underline{2k + 5} \end{array}$

31. $\begin{array}{r} -4p^4 + 8p^3 - 9p^2 + 6p - 1 \\ \underline{3p - 2} \end{array}$

32. $\begin{array}{r} 8y^4 - 9y^3 + 3y^2 + 5y + 6 \\ \underline{2y + 5} \end{array}$

---

33. Explain in your own words how to find the product of two monomials with numerical coefficients. (See Objective 1.)

34. Explain in your own words how to find the product of two polynomials. (See Objective 2.)

35. Explain how the expressions $(x + y)^2$ and $x^2 + y^2$ differ. (See Objective 5.)

36. Does $(x + y)^4$ equal $x^4 + y^4$? Explain your answer.

Use the special products to find each product. See Examples 8–10.

37. $(4z - t)(4z + t)$

38. $(8h + k)(8h - k)$

39. $(2x + 5y)(2x - 5y)$

40. $(2c - 5d)(2c + 5d)$

41. $(5r - 3w)(5r + 3w)$

42. $(10p - 9r)(10p + 9r)$

43. $\left(x - \frac{1}{3}\right)\left(x + \frac{1}{3}\right)$

44. $\left(m - \frac{2}{5}\right)\left(m + \frac{2}{5}\right)$

45. $\left(\frac{1}{2}y - \frac{2}{3}z\right)\left(\frac{1}{2}y + \frac{2}{3}z\right)$

46. $\left(\frac{1}{4}r + \frac{1}{5}s\right)\left(\frac{1}{4}r - \frac{1}{5}s\right)$

47. $(y^3 + 5)(y^3 - 5)$

48. $(m^3 - p^3)(m^3 + p^3)$

49. $6t(t - 2s)(t + 2s)$

50. $4zw(z + 3w)(z - 3w)$

51. $(r + 6z)^2$

52. $(a - 7b)^2$

53. $(3k - 2p)^2$

54. $(3r + 2t)^2$

55. $(10a - b)^2$

56. $(5a + 9q)^2$

57. $(3ax - 5by)^2$

58. $(2mp - 3sq)^2$

59. $\left(\frac{3}{8}m - \frac{2}{3}p\right)^2$

60. $\left(\frac{5}{9}a + \frac{3}{5}b\right)^2$

61. $[(5x + 1) + 6y]^2$

62. $[(3m - 2) + p]^2$

63. $[(2a + b) - 3]^2$

64. $[(4k + h) - 4]^2$

65. $[(2a + b) - 3][(2a + b) + 3]$

66. $[(m + p) + 5][(m + p) - 5]$

67. $[(2h - k) + 8][(2h - k) - 8]$

68. $[(3m - y) + 7][(3m - y) - 7]$

69. $[5t - (3 + 2s)][5t + (3 + 2s)]$

70. $[8a - (7 - 3b)][8a + (7 - 3b)]$

Find the following products.

71. $(2a + b)(3a^2 + 2ab + b^2)$

72. $(m - 5p)(m^2 - 2mp + 3p^2)$

73. $(4z - x)(z^3 - 4z^2x + 2zx^2 - x^3)$

74. $(3r + 2s)(r^3 + 2r^2s - rs^2 + 2s^3)$

75. $(m^2 - 2mp + p^2)(m^2 + 2mp - p^2)$

76. $(3 + x + y)(-3 + x - y)$

77. $(a + b)(a + 2b)(a - 3b)$

78. $(m - p)(m - 2p)(2m + p)$

79. $(y + 2)^3$

80. $(z - 3)^3$

81. $(q - 2)^4$

82. $(r + 3)^4$

In Exercises 83–86, two expressions are given. In each case, replace x with 2 and y with 3 to show that, in general, the two expressions are not equal.

83. $(x + y)^2$; $x^2 + y^2$

84. $(x + y)^3$; $x^3 + y^3$

85. $(x + y)^4$; $x^4 + y^4$

86. $(x + y)^5$; $x^5 + y^5$

Write a mathematical expression for the area of each of the following geometric figures.

87.

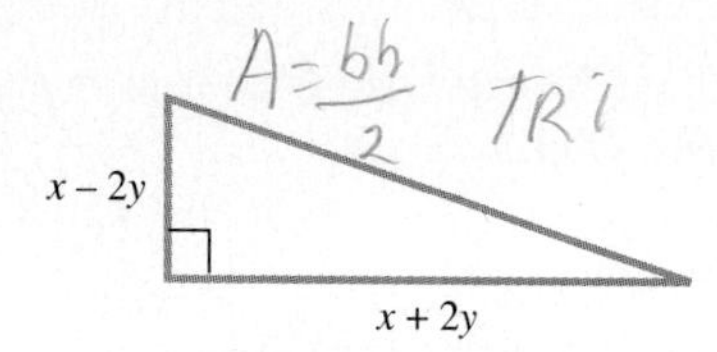

88.

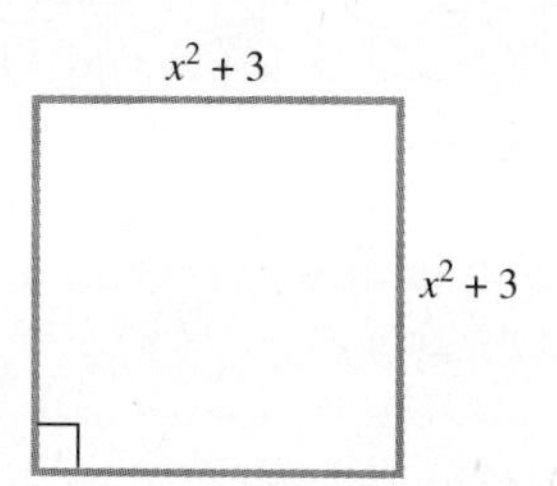

89.

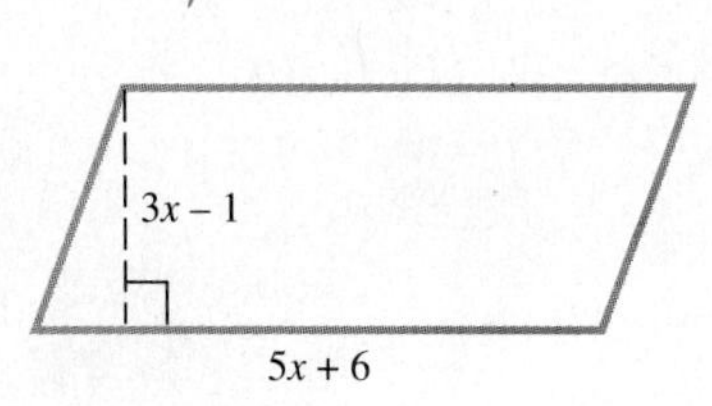

90.

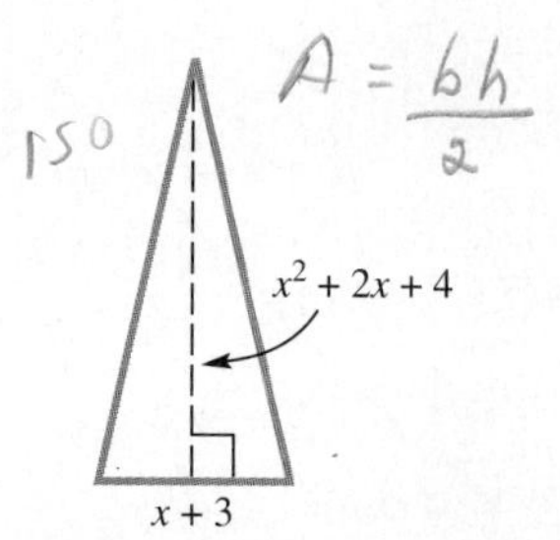

Find each product. Assume that all variables represent nonzero real numbers, with variables used as exponents restricted to integers.

91. $6b^p(2b^{p-2} + b^{2p})$

92. $-5r^a(3r^5 - 2r^{-a})$

93. $z^{m-4}(3z^{m+2} - 4z^{-m})$

94. $y^{k-5}(2y^{3-k} - 7y^{5k})$

95. $(6k^n + 1)(2k^n - 3)$

96. $(9y^z - 2)(y^z + 5)$

97. $(8k^m - 3y^{3m})(2k^{2m} + 7y^m)$

98. $(4q^{2z} - p^{5z})(3q^{5z} - p^{2z})$

99. $(y^{2n} + 1)(y^{2n} - 1)$

100. $(3z^{5p} + 2)(3z^{5p} - 2)$

PREVIEW EXERCISES

Use the distributive property to rewrite each expression. See Section 1.4.

101. $9 \cdot 6 + 9 \cdot r$

102. $8 \cdot y - 8 \cdot 5$

103. $7(2x) - 7(3z)$

104. $4(8p) + 4(9y)$

105. $x(x + 1) + 4(x + 1)$

106. $p(2p - 3) + 5(2p - 3)$

Simplify. Write answers with positive exponents. Assume all variables represent positive numbers. See Section 3.2.

107. $8z^3w(zw^2)^4$

108. $3z^4(a^2b)^5$

109. $2p^4q^{-2}(5pq)^{-1}$

110. $(5m^2n)^3(-mn^{-1})^2$

3.5 GREATEST COMMON FACTORS; FACTORING BY GROUPING

OBJECTIVES

1. WRITE INTEGERS IN PRIME FACTORED FORM.
2. FIND THE GREATEST COMMON FACTOR.
3. FACTOR OUT THE GREATEST COMMON FACTOR.
4. FACTOR BY GROUPING.

1 Integers are often written in **prime factored form,** that is, as a product of prime numbers. A **prime number** is a positive integer greater than 1 that can be divided without remainder only by itself and 1. The first few primes are 2, 3, 5, 7, 11, and 13. Each number in the factored form of an integer is a **factor** of thc integer.

EXAMPLE 1 WRITING NUMBERS IN PRIME FACTORED FORM

Write each number in prime factored form.

(a) 12

Start by dividing 12 by the first prime, 2.

$$12 = 2 \cdot 6$$

Then divide by 2 again.

$$12 = 2 \cdot 2 \cdot 3$$

All factors in this product are prime, so the prime factored form of 12 is

$$12 = 2^2 \cdot 3.$$

Notice that exponents are used for any repeated factors.

(b) $150 = 2 \cdot 75$

$= 2 \cdot 3 \cdot 25$

$= 2 \cdot 3 \cdot 5^2$ ■

2 The next example shows how to find the *greatest common factor* for a group of integers. The **greatest common factor** is the largest integer that will divide *without remainder* into every integer in the group.

EXAMPLE 2
FINDING THE GREATEST COMMON FACTOR

Find the greatest common factor for 48, 72, and 120.

Write the prime factored form of each number.

$$48 = 2^4 \cdot 3 \qquad 72 = 2^3 \cdot 3^2 \qquad 120 = 2^3 \cdot 3 \cdot 5$$

Now form the product of those primes appearing in all the factorizations, with each prime appearing the *least* number of times it appears in any of the factorizations. In this example, 2 appears four times, three times, and three times, so three is the least number of times 2 appears. The least number of times 3 appears is one, and 5 does not appear in *each* factorization. Finally,

$$\text{the greatest common factor} = 2^3 \cdot 3 = 24. \quad ■$$

The method used to find the greatest common factor in Example 2 can be extended to include terms with variables. This method is summarized below.

FINDING THE GREATEST COMMON FACTOR

Step 1 **Find prime factors.**
Write the prime factored form of the numerical coefficients of each term.

Step 2 **Select factors for the greatest common factor.**
Use each different factor (numerical or variable) that is common to all terms. Choose the *least* exponent on that factor that appears in any term.

Step 3 **Form the greatest common factor.**
Find the product of all the factors found in Step 2. This product is the greatest common factor.

Writing a polynomial as the product of two or more simpler polynomials is called **factoring** the polynomial. For example, the product of $3x$ and $5x - 2$ is $15x^2 - 6x$, and $15x^2 - 6x$ can be factored as the product $3x(5x - 2)$.

$$3x(5x - 2) = 15x^2 - 6x \qquad \text{Multiplication}$$
$$15x^2 - 6x = \mathbf{3x(5x - 2)} \qquad \text{Factoring}$$

3 The first step in factoring a polynomial is to find the greatest common factor for the terms of the polynomial. For example, the greatest common factor for $8x + 12$ is 4, since 4 is the largest number that divides into both $8x$ and 12. To factor $8x + 12$, use the distributive property.

$$8x + 12 = 4(2x) + 4(3) = 4(2x + 3)$$

As a check, multiply 4 and $2x + 3$. The result should be $8x + 12$. This process is called **factoring out** the greatest common factor.

EXAMPLE 3
FACTORING OUT THE GREATEST COMMON FACTOR

Factor out the greatest common factor.

(a) $9z - 18$
Since 9 is the greatest common factor,

$$9z - 18 = 9 \cdot z - 9 \cdot 2 = 9(z - 2).$$

(b) $56m + 35p = 7(8m + 5p)$

(c) $2y + 5$ There is no common factor other than 1.

(d) $12 + 24z$

$$\begin{aligned} 12 + 24z &= 12 \cdot 1 + 12 \cdot 2z \\ &= 12(1 + 2z) \end{aligned}$$ 12 is the greatest common factor.

(e) $r(3x + 2) - 8s(3x + 2)$
In this polynomial, the greatest common factor is the binomial $(3x + 2)$.

$$r(3x + 2) - 8s(3x + 2) = (3x + 2)(r - 8s)$$ ■

CAUTION Refer to Example 3(d). It is very common to forget to write the 1. Be careful to include it as needed. Remember to check answers by multiplication.

EXAMPLE 4
FACTORING OUT THE GREATEST COMMON FACTOR

Factor out the greatest common factor.

(a) $9x^2 + 12x^3$
The numerical part of the greatest common factor is 3, the largest number that divides into both 9 and 12. For the variable parts, x^2 and x^3, use the least exponent that appears on x; here the least exponent is 2. The greatest common factor is $3x^2$.

$$\begin{aligned} 9x^2 + 12x^3 &= 3x^2(3) + 3x^2(4x) \\ &= 3x^2(3 + 4x) \end{aligned}$$

(b) $32p^4 - 24p^3 + 40p^5$
The greatest common numerical factor is 8. Since the least exponent on p is 3, the greatest common factor is $8p^3$.

$$\begin{aligned} 32p^4 - 24p^3 + 40p^5 &= 8p^3(4p) + 8p^3(-3) + 8p^3(5p^2) \\ &= 8p^3(4p - 3 + 5p^2) \end{aligned}$$

(c) $3k^4 - 15k^7 + 24k^9 = 3k^4(1 - 5k^3 + 8k^5)$ ■

EXAMPLE 5
FACTORING OUT THE GREATEST COMMON FACTOR

Factor out the greatest common factor.

(a) $24m^3n^2 - 18m^2n + 6m^4n^3$

The numerical part of the greatest common factor is 6. Find the variable part by writing each variable with its least exponent. Here 2 is the least exponent on m that appears, while 1 is the least exponent on n. Finally, $6m^2n$ is the greatest common factor.

$$\begin{aligned} &24m^3n^2 - 18m^2n + 6m^4n^3 \\ &= (6m^2n)(4mn) + (6m^2n)(-3) + (6m^2n)(m^2n^2) \\ &= 6m^2n(4mn - 3 + m^2n^2) \end{aligned}$$

(b) $25x^2y^3 + 30xy^5 - 15x^4y^7 = 5xy^3(5x + 6y^2 - 3x^3y^4)$ ■

EXAMPLE 6
FACTORING OUT THE GREATEST COMMON FACTOR

Factor out the greatest common factor from

$$-a^3 + 3a^2 - 5a.$$

There are two ways to factor this polynomial, both of which are correct. First, a could be used as the common factor, giving

$$\begin{aligned} -a^3 + 3a^2 - 5a &= a(-a^2) + a(3a) + a(-5) \\ &= a(-a^2 + 3a - 5). \end{aligned}$$

Alternatively, $-a$ could be used as the common factor.

$$\begin{aligned} -a^3 + 3a^2 - 5a &= -a(a^2) + (-a)(-3a) + (-a)(5) \\ &= -a(a^2 - 3a + 5) \end{aligned}$$ ■

NOTE In Example 6 we showed two ways of factoring a polynomial. Sometimes, in a particular problem there will be a reason to prefer one of these forms over the other, but both are correct. The answer section in this book will usually give only one of these forms, the one where the common factor has a positive coefficient, but either is correct.

4 Sometimes a polynomial has a greatest common factor of 1, but it still may be possible to factor the polynomial by a suitable rearrangement of the terms. This process is called **factoring by grouping.**

For example, to factor the polynomial

$$ax - ay + bx - by,$$

group the terms as follows.

Terms with common factors
↓ ↓

$$(ax - ay) + (bx - by)$$

Then factor $ax - ay$ as $a(x - y)$ and factor $bx - by$ as $b(x - y)$ to get

$$ax - ay + bx - by = a(x - y) + b(x - y).$$

On the right, the common factor is $x - y$. The final factored form is

$$ax - ay + bx - by = (x - y)(a + b).$$

EXAMPLE 7
FACTORING BY GROUPING

Factor $p^2q^2 - 2q^2 + 5p^2 - 10$.

Group the terms as follows.

$$(p^2q^2 - 2q^2) + (5p^2 - 10)$$

Factor a common factor from each part to get

$$q^2(p^2 - 2) + 5(p^2 - 2),$$

which equals

$$(p^2 - 2)(q^2 + 5). \quad ■$$

CAUTION Refer to Example 7. It is a common error to stop at the step

$$q^2(p^2 - 2) + 5(p^2 - 2).$$

This expression is *not in factored form* because it is a *sum* of two terms:

$$q^2(p^2 - 2) \quad \text{and} \quad 5(p^2 - 2).$$

EXAMPLE 8
FACTORING BY GROUPING

Factor $3x - 3y - ax + ay$.

Grouping terms as above gives

$$(3x - 3y) + (-ax + ay),$$

or

$$3(x - y) + a(-x + y).$$

There is no simple common factor here; it is necessary to group the terms as follows.

$$(3x - 3y) - (ax - ay) \qquad \text{Be careful with signs.}$$

Now factor to get

$$3(x - y) - a(x - y),$$

which equals

$$(x - y)(3 - a). \quad ■$$

NOTE In Example 8, different grouping would lead to the product

$$(a - 3)(y - x).$$

As can be verified by multiplication, this is another correct factorization.

EXAMPLE 9
FACTORING BY GROUPING

Factor each of the following polynomials by grouping.

(a) $6x^2 - 4x - 15x + 10$

Work as above. Note that we must factor -5 rather than 5 from the second group in order to get a common factor of $3x - 2$.

$$\begin{aligned} 6x^2 - 4x - 15x + 10 &= 2x(3x - 2) - 5(3x - 2) && \text{Factor out } 2x \text{ and } -5. \\ &= (3x - 2)(2x - 5) && \text{Factor out } (3x - 2). \end{aligned}$$

(b) $6ax + 12bx + a + 2b$

Group the first two terms and the last two terms.

$$6ax + 12bx + a + 2b = (6ax + 12bx) + (a + 2b)$$

Now factor $6x$ from the first group, and use the identity property of multiplication to introduce the factor 1 in the second group.

$$\begin{aligned} (6ax + 12bx) + (a + 2b) &= 6x(a + 2b) + 1(a + 2b) \\ &= (a + 2b)(6x + 1) && \text{Factor out } (a + 2b). \end{aligned}$$

Again, as in Example 3(d), be careful not to forget the 1. ■

NOTE In some cases, factoring by grouping requires that the terms be rearranged before the groupings are made.

The steps used in factoring by grouping are listed below.

FACTORING BY GROUPING

Step 1 **Group terms.**
Collect the terms into groups so that each group has a common factor.

Step 2 **Factor within the groups.**
Factor out the common factor in each group.

Step 3 **Factor the entire polynomial.**
If each group now has a common factor, factor it out. If not, try a different grouping.

3.5 EXERCISES

Factor out the greatest common factor. Simplify the factors, if possible. See Examples 1–6.

1. $15k + 30$

2. $7m - 21$

3. $6p^2 + 4p$

4. $12a^3 + 9a^2$

5. $8m^2p - 4mp^2$

6. $12z^2 - 6zw^3$

7. $5xy - 8xy^2$

8. $4ab + 2a^2b$

9. $8m^4 + 6m^3 - 4m^2$

10. $2p^5 + 4p^6 - 8p^3$

11. $10t^5 - 8t^4 - 16t^3$

12. $6p^3 - 18p^2 + 9p^4$

13. $6r^2s - 3rs^2 - 9r^3s$

14. $5p^3t^2 + 25p^2t^3 + 10p^3t^4$

15. $2x^2y - 3xy^2 + 4x^2y^2$

16. $14a^3b^2 + 7a^2b - 21a^5b^3$

17. $12km^3 - 24k^3m^2 + 36k^2m^4$

18. $15m^3p^3 - 6m^3p^4 + 9mp^3$

19. $144z^{11}m^4 + 16z^3m^5 - 32z^4m^5$

20. $39a^5b^3 - 26a^7b^2 + 52a^8b^5$

21. $16z^2n^6 + 64zn^7 - 32z^3n^3$

22. $5r^3s^5 + 10r^2s^2 - 15r^4s^2$

23. $15a^3b + 12a^2c - 3ad^4$

24. $20m^4y + 30m^2y^2 + 50m^3z^2$

25. $9x^2y^2z^4 - 18x^3y^2z$

26. $30m^5p^5q^3 + 20m^6p^4q^2$

27. $12p^5q^6 + 5r^2s^2 + 24p^3r^2$

28. $2m^8k^8 + 7q^4p^5 + 14m^6p^3$

29. $14a^3b^2 + 7a^2b - 21a^5b^3 + 42ab^4$

30. $12km^3 - 24k^3m^2 + 36k^2m^4 - 60k^4m^3$

31. $-15m^3p^3 - 6m^3p^4 - 9mp^3 + 30m^2p^3$

32. $-25x^3y^2 - 20x^4y^3 + 15x^5y^4 - 50x^6y^2$

33. $(m - 9)(m + 1) + (m - 9)(m + 2)$

34. $(a + 5)(a - 6) + (a + 5)(a - 1)$

35. $(3k - 7)(k + 2) + (3k - 7)(k + 5)$

36. $(5m - 11)(2m + 5) + (5m - 11)(m + 5)$

37. $(r - 6)(2r + 1) - (r - 6)(r + 3)$

38. $(y + 2)(2y + 3) - (y + 2)(y + 1)$

39. $m^5(r + s) + m^5(t + u)$

40. $z^3(k + m) + z^3(p + q)$

41. $4(3 - x)^2 - (3 - x)^3 + 3(3 - x)$

42. $2(t - s) + 4(t - s)^2 - (t - s)^3$

43. $15(2z + 1)^3 + 10(2z + 1)^2 - 25(2z + 1)$

44. $6(a + 2b)^2 - 4(a + 2b)^3 + 12(a + 2b)^4$

45. $5(m + p)^3 - 10(m + p)^2 - 15(m + p)^4$

46. $-9a^2(p + q) - 3a^3(p + q)^2 + 6a(p + q)^3$

---

47. Write an explanation in your own words of how to find the greatest common factor of the terms of a polynomial. (See Objective 2.)

48. Which one of the following is the factorization of

$$6x^3y^4 - 12x^5y^2 + 24x^4y^8$$

that has the greatest common factor as one of the factors? (See Objective 2.)

(a) $6x^3y^2(y^2 - 2x^2 + 4xy^6)$

(b) $6xy(x^2y^3 - 2x^4y + 4x^3y^7)$

(c) $2x^3y^2(3y^2 - 6x^2 + 12xy^6)$

(d) $6x^2y^2(xy^2 - 2x^3 + 4x^2y^6)$

Factor each of the following polynomials twice. First, use a common factor with a positive coefficient, and then use a common factor with a negative coefficient. See Example 6.

49. $-2x^5 + 6x^3 + 4x^2$

50. $-5a^3 + 10a^4 - 15a^5$

51. $-32a^4m^5 - 16a^2m^3 - 64a^5m^6$

52. $-144z^{11}n^5 + 16z^3n^{11} - 32z^4n^7$

Factor by grouping. See Examples 7–9.

53. $pq + 3rq + pm + 3rm$

54. $5x + 5a + 9bx + 9ab$

55. $2b + 2c + ab + ac$

56. $3am + 3ap + 2bm + 2bp$

57. $p^2 + pq - 3py - 3yq$

58. $r^2 + 6rs - 3rt - 18st$

59. $a^2b^2 + 2b^2 - 5a^2 - 10$

60. $m^2r^2 + 8r^2 - 3m^2 - 24$

61. $x^2 - 3x + 2x - 6$

62. $y^2 - 8y + 4y - 32$

63. $3r^2 - 2r + 15r - 10$

64. $2a^2 - 6a + 7a - 21$

65. $21y^2 + 14y - 15y - 10$

66. $18q^2 + 9q - 4q - 2$

67. $16p^2 + 6pq - 8pq - 3q^2$

68. $8r^2 + 6rs - 12rs - 9s^2$

69. $14m^2 + 21mq - 2mq - 3q^2$

70. $10y^2 + 4yz - 5y - 2z$

71. $3a^3 + 3ab^2 + 2a^2b + 2b^3$

72. $16m^3 - 4m^2p^2 - 4mp + p^3$

73. $8 - 6y^3 - 12y + 9y^4$

74. $x^3y^2 + x^3 - 3y^2 - 3$

75. $1 - a + ab - b$

76. $2ab^2 - 8b^2 + a - 4$

---

77. Refer to Exercise 75 above. The factored form as given in the answer section in the back of the text is $(1 - a)(1 - b)$. As mentioned in the text, sometimes other acceptable factored forms can be given. Which one of the following is *not* a factored form of $1 - a + ab - b$? (See Objective 4.)

(a) $(a - 1)(b - 1)$ **(b)** $(-a + 1)(-b + 1)$
(c) $(-1 + a)(-1 + b)$ **(d)** $(1 - a)(b + 1)$

78. Refer to Exercise 76 above. One form of the answer is $(2b^2 + 1)(a - 4)$. Give two other acceptable factored forms of $2ab^2 - 8b^2 + a - 4$. (See Objective 4.)

Factor each of the following polynomials. Assume that all variables used as exponents represent positive integers.

79. $p^{6m} - 2p^{4m}$

80. $4r^{3z} + 8r^{5z}$

81. $q^{3k} + 2q^{2k} + 3q^k$

82. $z^{2x} - z^x + z^{3x}$

83. $y^{r+5} + y^{r+4} + y^{r+2}$

84. $8k^{2z+3} + 2k^{2z+1} + 12k^{2z}$

85. $r^pm^p + q^pm^p - r^pz^p - q^pz^p$

86. $6a^zb^z - 10b^z - 3a^zc^z + 5c^z$

Factor out the greatest common factor. Assume all variables are nonzero.

87. $3m^{-5} + m^{-3}$

88. $k^{-2} + 2k^{-4}$

89. $3p^{-3} + 2p^{-2} - 4p^{-1}$

90. $-5y^{-3} + 8y^{-2} + y^{-1}$

PREVIEW EXERCISES

Find each product by the FOIL method. See Section 3.4.

91. $(2m - 1)(3m + 2)$

92. $(4p + 5)(3p - 7)$

93. $(8r + 5s)(2r - 3s)$

94. $(7y - 2x)(3y + 5x)$

95. $(9z - 7a)(2z + 5a)$

96. $(8p + 3q)(4p - q)$

97. $(2r - 5s)(7r + 3s)$

98. $(2a + 7b)(3a + 5b)$

3.6 FACTORING TRINOMIALS

OBJECTIVES

1 FACTOR TRINOMIALS WHEN THE COEFFICIENT OF THE SQUARED TERM IS 1.

2 FACTOR TRINOMIALS WHEN THE COEFFICIENT OF THE SQUARED TERM IS NOT 1.

3 USE AN ALTERNATIVE METHOD FOR FACTORING TRINOMIALS.

4 FACTOR BY SUBSTITUTION.

The product of $x + 3$ and $x - 5$ is

$$\begin{aligned}(x + 3)(x - 5) &= x^2 - 5x + 3x - 15 \\ &= x^2 - 2x - 15.\end{aligned}$$

Also, by this result, the *factored form* of $x^2 - 2x - 15$ is $(x + 3)(x - 5)$.

Multiplication →

Factored form → $(x + 3)(x - 5) = x^2 - 2x - 15$ ← Product

← Factoring

1 Trinomials will be factored in this section, beginning with those having 1 as the coefficient of the squared term. To see how to factor these trinomials, start with an example: $x^2 - 2x - 15$. As shown below, the x^2 term came from multiplying x and x, and -15 was obtained by multiplying 3 and -5.

Product of x and x is x^2.

$$(x + 3)(x - 5) = x^2 - 2x - 15$$

Product of 3 and -5 is -15.

The $-2x$ in $x^2 - 2x - 15$ was found by multiplying the outside terms, and then the inside terms, and adding.

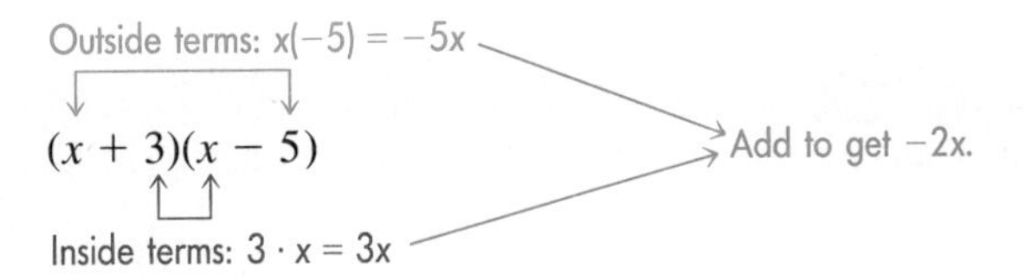

Based on this example, we factor a trinomial with 1 as the coefficient of the squared term by following these steps.

FACTORING $x^2 + bx + c$

Step 1 **Find pairs whose product is c.**
Find all pairs of numbers whose product is the third term of the trinomial (c).

Step 2 **Find pairs whose sum is b.**
Choose the pair whose sum is the coefficient of the middle term (b).

If there are no such numbers, the polynomial cannot be factored.

If a polynomial cannot be factored, it is called a **prime polynomial.**

EXAMPLE 1
FACTORING A TRINOMIAL IN $x^2 + bx + c$ FORM

Factor $r^2 + 8r + 12$.

Look for two numbers with a product of 12 and a sum of 8. Of all pairs of numbers having a product of 12, only the pair 6 and 2 has a sum of 8. Therefore,

$$r^2 + 8r + 12 = (r + 6)(r + 2).$$

Because of the commutative property, it would be equally correct to write $(r + 2)(r + 6)$. ■

EXAMPLE 2
FACTORING A TRINOMIAL IN $x^2 + bx + c$ FORM

Factor $y^2 + 2y - 35$.

Step 1 Find pairs of numbers whose product is -35.	*Step 2* Write sums of those numbers.
$-35(1)$	$-35 + 1 = -34$
$35(-1)$	$35 + (-1) = 34$
$7(-5)$	$7 + (-5) = 2 \rightarrow$ Coefficient of the middle term
$5(-7)$	$5 + (-7) = -2$

The required numbers are 7 and -5.

$$y^2 + 2y - 35 = (y + 7)(y - 5)$$

Check by finding the product of $y + 7$ and $y - 5$. ■

EXAMPLE 3
RECOGNIZING A PRIME POLYNOMIAL

Factor $m^2 + 6m + 7$.

Look for two numbers whose product is 7 and whose sum is 6. Only two pairs of integers, 7 and 1 and -7 and -1, give a product of 7. Neither of these pairs has a sum of 6, so $m^2 + 6m + 7$ cannot be factored, and is prime. ■

EXAMPLE 4
FACTORING A TRINOMIAL IN TWO VARIABLES

Factor $p^2 + 6ap - 16a^2$.

Look for two expressions whose sum is $6a$ and whose product is $-16a^2$. The quantities $8a$ and $-2a$ have the necessary sum and product, so

$$p^2 + 6ap - 16a^2 = (p + 8a)(p - 2a).$$ ■

EXAMPLE 5 FACTORING A TRINOMIAL WITH A COMMON FACTOR

Factor $16y^3 - 32y^2 - 48y$.

Start by factoring out the greatest common factor, $16y$.

$$16y^3 - 32y^2 - 48y = 16y(y^2 - 2y - 3)$$

To factor $y^2 - 2y - 3$, look for two integers whose sum is -2 and whose product is -3. The necessary integers are -3 and 1, with

$$16y^3 - 32y^2 - 48y = 16y(y - 3)(y + 1). \quad ■$$

CAUTION In factoring, always look for a common factor first. Do not forget to write the common factor as part of the answer.

2 A generalization of the method shown above can be used to factor a trinomial of the form $ax^2 + bx + c$, where the coefficient of the squared term is not equal to 1. Factor $3x^2 + 7x + 2$ to see how this method works. First, identify the values of a, b, and c.

$$\begin{array}{ccccc} ax^2 & + & bx & + & c \\ \downarrow & & \downarrow & & \downarrow \\ 3x^2 & + & 7x & + & 2 \end{array}$$

$$a = 3, \quad b = 7, \quad c = 2$$

The product ac is $3 \cdot 2 = 6$, so find integers having a product of 6 and a sum of 7 (since the middle term is 7). The necessary integers are 1 and 6, so write $7x$ as $1x + 6x$, or $x + 6x$, giving

$$3x^2 + 7x + 2 = 3x^2 + \underbrace{x + 6x}_{x + 6x = 7x} + 2.$$

Now factor by grouping.

$$= x(3x + 1) + 2(3x + 1)$$

$$3x^2 + 7x + 2 = (3x + 1)(x + 2)$$

EXAMPLE 6 FACTORING A TRINOMIAL IN $ax^2 + bx + c$ FORM

Factor $12r^2 - 5r - 2$.

Since $a = 12$, $b = -5$, and $c = -2$, the product ac is -24. Two integers whose product is -24 and whose sum is -5 are -8 and 3, with

$$12r^2 - 5r - 2 = 12r^2 + 3r - 8r - 2,$$

where $-5r$ is written as $3r - 8r$. Factor by grouping.

$$= 3r(4r + 1) - 2(4r + 1)$$

$$12r^2 - 5r - 2 = (4r + 1)(3r - 2) \quad ■$$

3 An alternative approach, the method of trying repeated combinations, is especially helpful when the product ac is large. This method is shown using the same polynomials as above.

EXAMPLE 7
FACTORING A TRINOMIAL IN $ax^2 + bx + c$ FORM

Factor each of the following.

(a) $3x^2 + 7x + 2$

To factor this polynomial, find the correct numbers to put in the blanks.

$$3x^2 + 7x + 2 = (\_\_\_x + \_\_\_)(\_\_\_x + \_\_\_)$$

Plus signs are used since all the signs in the polynomial are plus. The first two expressions have a product of $3x^2$, so they must be $3x$ and x.

$$3x^2 + 7x + 2 = (3x + \_\_\_)(x + \_\_\_)$$

The product of the two last terms must be 2, so the numbers must be 2 and 1. There is a choice. The 2 could be used with the $3x$ or with the x. Only one of these choices can give the correct middle term. Try each.

3x (outer) — 6x (outer)

$(3x + 2)(x + 1)$ — $(3x + 1)(x + 2)$

2x (inner) — x (inner)

$3x + 2x = 5x$ — $6x + x = 7x$

Wrong middle term — Correct middle term

Therefore, $3x^2 + 7x + 2 = (3x + 1)(x + 2)$.

(b) $12r^2 - 5r - 2$

The factors of 12 could be 4 and 3, 6 and 2, or 12 and 1. Try 4 and 3. If these do not work, another choice can be made.

$$12r^2 - 5r - 2 = (4r\_\_\_\_\_)(3r\_\_\_\_\_)$$

It is not possible to tell what signs to use yet. The factors of -2 are -2 and 1, or 2 and -1. Try both possibilities.

4r (outer) — 8r (outer)

$(4r - 2)(3r + 1)$ — $(4r - 1)(3r + 2)$

−6r (inner) — −3r (inner)

$4r + (-6r) = -2r$ — $8r + (-3r) = 5r$

Wrong middle term — Wrong middle term

The middle term on the right is $5r$, instead of the $-5r$ that is needed. Get $-5r$ by exchanging the middle signs.

−8r (outer)

$(4r + 1)(3r - 2)$

3r (inner)

$-8r + 3r = -5r$

Correct middle term

Thus, $12r^2 - 5r - 2 = (4r + 1)(3r - 2)$. ■

The alternative method of factoring a trinomial not having 1 as coefficient of the squared term is summarized as follows.

FACTORING $ax^2 + bx + c$

Step 1 **Find pairs whose product is *a*.**
Write all pairs of factors of the coefficient of the squared term (a).

Step 2 **Find pairs whose product is *c*.**
Write all pairs of factors of the last term (c).

Step 3 **Choose inner and outer terms.**
Use various combinations of the factors from Steps 1 and 2 until the necessary middle term is found.

If no such combinations exist, the polynomial is prime.

EXAMPLE 8 **FACTORING A TRINOMIAL IN TWO VARIABLES**

Factor $18m^2 - 19mx - 12x^2$.

There are many possible factors of both 18 and -12. As a general rule, the middle-sized factors should be tried first. Try 6 and 3 for 18, and -3 and 4 for -12.

24mx

$$(6m - 3x)(3m + 4x)$$

−9mx

$$24mx + (-9mx) = 15mx$$

Wrong middle term

Actually, this could not have been correct, since the factor $6m - 3x$ has a common factor, while the given polynomial does not.

$$(6m + 4x)(3m - 3x)$$

Common factor of 2 ↑ Common factor of 3 ↑

Both of these have a common factor, and cannot be correct.

Since 6 and 3 do not work as factors of 18, try 9 and 2 instead, with -4 and 3 as factors of -12.

$$(9m + 3x)(2m - 4x)$$

Common factors

27mx

$$(9m - 4x)(2m + 3x)$$

−8mx

$$27mx + (-8mx) = 19mx$$

The result on the right differs from the correct middle term only in the sign, so exchange the middle signs.

$$18m^2 - 19mx - 12x^2 = (9m + 4x)(2m - 3x) \quad ■$$

CAUTION As shown in Example 8, if the terms of a trinomial have no common factor (except 1), then neither of its binomial factors have a common factor. Remembering this will often help eliminate some of the trials of combinations when using this method of factoring.

EXAMPLE 9 **FACTORING A TRINOMIAL WITH A COMMON FACTOR**

Factor $16y^3 + 24y^2 - 16y$.

This polynomial has a greatest common factor of $8y$. Factor this out first.

$$16y^3 + 24y^2 - 16y = 8y(2y^2 + 3y - 2)$$

Factor $2y^2 + 3y - 2$ by either method given above.

$$16y^3 + 24y^2 - 16y = 8y(2y - 1)(y + 2)$$

Do not forget to write the common factor in front. ■

When factoring a trinomial of the form $ax^2 + bx + c$ with a negative, it is helpful to begin by factoring out -1, as shown in the next example.

EXAMPLE 10 **FACTORING $ax^2 + bx + c$, $a < 0$**

Factor $-3x^2 + 16x + 12$.

While it is possible to factor this polynomial directly, it is helpful to start by factoring out -1. Then proceed as in the earlier examples.

$$\begin{aligned} -3x^2 + 16x + 12 &= -1(3x^2 - 16x - 12) \\ &= -1(3x + 2)(x - 6) \end{aligned}$$ ■

NOTE The factored form given in Example 10 can be written in other ways. Two of them are

$$(-3x - 2)(x - 6) \qquad \text{and} \qquad (3x + 2)(-x + 6).$$

Verify that these both give the original polynomial when multiplied.

4 Sometimes a more complicated polynomial can be factored by making a substitution of one variable for another. The next examples show this **method of substitution.** The method of substitution is used when a particular polynomial appears to various powers in a more involved polynomial.

EXAMPLE 11 **FACTORING A TRINOMIAL USING SUBSTITUTION**

Factor $2(x + 3)^2 + 5(x + 3) - 12$.

Since the binomial $x + 3$ appears to the powers of 2 and 1, let the substitution variable represent $x + 3$. We may choose any letter we wish except x. Let us choose y to equal $x + 3$.

$$2(x+3)^2 + 5(x+3) - 12 = 2y^2 + 5y - 12 \qquad \text{Let } y = x + 3.$$
$$= (2y-3)(y+4) \qquad \text{Factor.}$$

Now replace y with $x + 3$ to get

$$\begin{aligned} 2(x+3)^2 + 5(x+3) - 12 &= [2(x+3) - 3][(x+3) + 4] \\ &= (2x + 6 - 3)(x + 7) \\ &= (2x+3)(x+7). \quad \blacksquare \end{aligned}$$

EXAMPLE 12
FACTORING A TRINOMIAL IN $ax^4 + bx^2 + c$ FORM

Factor $6y^4 + 7y^2 - 20$.

The variable y appears to powers in which the larger exponent is twice the smaller exponent. In a case such as this, let the substitution variable equal the smaller power. Here, let $m = y^2$. Since $y^4 = (y^2)^2 = m^2$, the given trinomial becomes

$$6m^2 + 7m - 20,$$

which is factored as follows:

$$6m^2 + 7m - 20 = (3m - 4)(2m + 5).$$

Since $m = y^2$,

$$6y^4 + 7y^2 - 20 = (3y^2 - 4)(2y^2 + 5). \quad \blacksquare$$

NOTE Some students feel comfortable enough about factoring to factor polynomials like the one in Example 12 directly, without using the substitution method.

3.6 EXERCISES

Factor each of the following trinomials completely. See Examples 1–10.

1. $c^2 + 4c - 5$
2. $d^2 + 9d + 8$
3. $p^2 + 6p + 8$
4. $m^2 + 15m + 56$
5. $a^2 - a - 12$
6. $z^2 + 2z - 35$
7. $r^2 - r - 20$
8. $y^2 - 2y - 35$
9. $x^2 - 3x - 40$
10. $a^2 - 6a - 16$
11. $k^2 - kn - 6n^2$
12. $a^2 + 3ab - 18b^2$
13. $y^2 - 3yx - 10x^2$
14. $p^2 - 2pq - 15q^2$
15. $a^2b^2 - 7ab + 12$
16. $y^2w^2 + 4yw - 21$
17. $5y^2 + y - 6$
18. $2r^2 - r - 3$
19. $3m^2 + 7m + 2$
20. $3y^2 + 14y + 8$
21. $8y^2 + 13y - 6$
22. $6x^2 + 13x + 6$
23. $18x^2 - 3x - 10$
24. $12m^2 - 8m - 15$
25. $35p^2 - 4p - 15$
26. $6m^2 - 17m - 14$
27. $12a^2 + 8ab - 15b^2$
28. $3m^2 + 7mk + 2k^2$
29. $4k^2 - 12ka + 9a^2$
30. $18a^2 - 3ab - 28b^2$

31. $35x^2 - 41xy - 24y^2$
32. $10a^2 + ab - 3b^2$
33. $8m^2 - 14mp - 39p^2$
34. $6x^2 - 5xy - 39y^2$
35. $6k^2p^2 + 13kp + 6$
36. $15z^2x^2 - 22zx - 5$
37. $12m^2 + 14m - 40$
38. $36t^2 + 30t - 50$
39. $18a^2 - 15a - 18$
40. $100r^2 - 90r + 20$
41. $6a^3 + 12a^2 - 90a$
42. $3m^4 + 6m^3 - 72m^2$
43. $13y^3 + 39y^2 - 52y$
44. $4p^3 + 24p^2 - 64p$
45. $2x^3y^3 - 48x^2y^4 + 288xy^5$
46. $6m^3n^2 - 24m^2n^3 - 30mn^4$
47. $-x^2 + 7x + 18$
48. $-p^2 - 7p - 12$
49. $-18a^2 + 17a + 15$
50. $-6x^2 + 23x + 4$
51. $-14r^3 + 19r^2 + 3r$
52. $-10h^3 + 29h^2 + 3h$

---

53. When a student was given the polynomial $4x^2 + 2x - 20$ to factor completely on a test, the student lost some credit when her answer was $(4x + 10)(x - 2)$. She complained to her teacher that when we multiply $(4x + 10)(x - 2)$, we get the original polynomial. Write a short explanation of why she lost some credit for her answer, even though the product is indeed $4x^2 + 2x - 20$. (See Objectives 2 and 3.)

54. Write an explanation as to why most people would find it more difficult to factor $36x^2 - 44x - 15$ than $37x^2 - 183x - 10$. (See Objectives 2 and 3.)

55. Find a polynomial that can be factored as $-9a(a - 5b)(2a + 7b)$.

56. Find a polynomial that can be factored as $12z^2(5z + x)(2z - x)$.

Factor each of the following completely. Assume that all variables used as exponents represent positive integers. See Examples 11 and 12.

57. $3x^4 - 14x^2 - 5$
58. $3p^4 - 8p^2 - 3$
59. $z^4 - 7z^2 - 30$
60. $k^4 + k^2 - 12$
61. $6x^4 + 5x^2 - 25$
62. $6a^4 - 11a^2 - 10$
63. $12p^4 + 28p^2r - 5r^2$
64. $2y^4 + xy^2 - 6x^2$
65. $4x^4 + 33x^2a^2 - 27a^4$
66. $2p^4 + 31p^2q^2 - 16q^4$
67. $6(p + 3)^2 + 13(p + 3) + 5$
68. $10(m - 5)^2 - 9(m - 5) - 9$
69. $6(z + k)^2 - 7(z + k) - 5$
70. $3(r + m)^2 - 10(r + m) - 25$
71. $a^2(a + b)^2 - ab(a + b)^2 - 6b^2(a + b)^2$
72. $m^2(m - p) + mp(m - p) - 2p^2(m - p)$
73. $p^2(p + q) + 4pq(p + q) + 3q^2(p + q)$
74. $2k^2(5 - y) - 7k(5 - y) + 5(5 - y)$
75. $z^2(z - x) - zx(x - z) - 2x^2(z - x)$
76. $r^2(r - s) - 5rs(s - r) - 6s^2(r - s)$
77. $p^{2n} - p^n - 6$
78. $k^{2y} + 4k^y - 5$
79. $6z^{4r} - 5z^{2r} - 4$
80. $12a^{4p} + 11a^{2p} + 2$
81. $36k^{3r} + 30k^{2r} + 4k^r$
82. $30y^{7a} - 26y^{6a} - 40y^{5a}$

PREVIEW EXERCISES

Simplify. Write answers with only positive exponents. See Sections 3.1 and 3.2.

83. -5^0

84. $\left(\frac{5}{4}\right)^{-2}$

85. $\frac{(m^2p)^{-1}}{mp^2}$

86. $\frac{(2rs)^3}{(r^{-1}s)^2}$

Find each product. See Section 3.4.

87. $(2m - 5)(2m + 5)$

88. $(3p + 2q)(3p - 2q)$

89. $(5a - 3b)^2$

90. $(2z + 5x)^2$

91. $(y - 2)(y^2 + 2y + 4)$

92. $(5a + 3)(25a^2 - 15a + 9)$

3.7 SPECIAL FACTORING

OBJECTIVES

1 FACTOR THE DIFFERENCE OF TWO SQUARES.
2 FACTOR A PERFECT SQUARE TRINOMIAL.
3 FACTOR THE DIFFERENCE OF TWO CUBES.
4 FACTOR THE SUM OF TWO CUBES.

Certain types of factoring occur so often that they deserve special study in this section.

1 As discussed earlier in this chapter, the product of the sum and difference of two terms leads to the **difference of two squares,** a pattern that is useful in factoring.

DIFFERENCE OF TWO SQUARES

$$x^2 - y^2 = (x + y)(x - y).$$

EXAMPLE 1 **FACTORING THE DIFFERENCE OF SQUARES**

Factor each difference of squares.

(a) $16m^2 - 49p^2 = (4m)^2 - (7p)^2$
$= (4m + 7p)(4m - 7p)$

(b) $81k^2 - 121a^2 = (9k)^2 - (11a)^2$
$= (9k + 11a)(9k - 11a)$

(c) $(m - 2p)^2 - 16 = (m - 2p)^2 - 4^2$
$= [(m - 2p) + 4][(m - 2p) - 4]$
$= (m - 2p + 4)(m - 2p - 4)$ ■

CAUTION Assuming no greatest common factor (except 1), it is not possible to factor (with real numbers) a *sum* of two squares such as $x^2 + 25$. In particular, $x^2 + y^2 \neq (x + y)^2$, as shown next.

2 Two other special products from Section 3.4 lead to the following rules for factoring.

PERFECT SQUARE TRINOMIAL

$$x^2 + 2xy + y^2 = (x + y)^2$$
$$x^2 - 2xy + y^2 = (x - y)^2$$

The trinomial $x^2 + 2xy + y^2$ is the square of $x + y$. For this reason, the trinomial $x^2 + 2xy + y^2$ is called a **perfect square trinomial.** In this pattern, both the first and the last terms of the trinomial must be perfect squares. In the factored form, twice the product of the first and the last terms must give the middle term of the trinomial.

EXAMPLE 2

FACTORING PERFECT SQUARE TRINOMIALS

Factor each of the following perfect square trinomials.

(a) $144p^2 - 120p + 25$

Here $144p^2 = (12p)^2$ and $25 = 5^2$. The sign on the middle term is $-$, so if $144p^2 - 120p + 25$ is a perfect square trinomial, it will have to be

$$(12p - 5)^2.$$

Take twice the product of the two terms to see if this is correct:

$$2(12p)(-5) = -120p,$$

which is the middle term of the given trinomial. Thus,

$$144p^2 - 120p + 25 = (12p - 5)^2.$$

(b) $4m^2 + 28mn + 49n^2 = (2m + 7n)^2$

Check the middle term: $2(2m)(7n) = 28mn$, which is correct.

(c) $(r + 5)^2 + 6(r + 5) + 9 = [(r + 5) + 3]^2$
$= (r + 8)^2,$

since $2(r + 5)(3) = 6(r + 5)$, the middle term.

(d) $m^2 - 8m + 16 - p^2$

The first three terms here are the square of a binomial; group them together, and factor as follows.

$$(m^2 - 8m + 16) - p^2 = (m - 4)^2 - p^2$$

The result is the difference of two squares. Factor again to get

$$(m - 4)^2 - p^2 = (m - 4 + p)(m - 4 - p). \blacksquare$$

3 The **difference of two cubes,** $x^3 - y^3$, can be factored as follows.

DIFFERENCE OF TWO CUBES

$$x^3 - y^3 = (x - y)(x^2 + xy + y^2)$$

Check this by finding the product of $x - y$ and $x^2 + xy + y^2$.

$$\begin{array}{r} x^2 + xy + y^2 \\ x - y \\ \hline -x^2y - xy^2 - y^3 \\ x^3 + x^2y + xy^2 \qquad \\ \hline x^3 \qquad\qquad\qquad - y^3 \end{array}$$

This result shows that

$$x^3 - y^3 = (x - y)(x^2 + xy + y^2).$$

EXAMPLE 3 **FACTORING THE DIFFERENCE OF CUBES**

Factor each difference of cubes.

(a) $m^3 - 8 = m^3 - 2^3 = (m - 2)(m^2 + 2m + 2^2)$
$= (m - 2)(m^2 + 2m + 4)$

(b) $27x^3 - 8y^3 = (3x)^3 - (2y)^3$
$= (3x - 2y)[(3x)^2 + (3x)(2y) + (2y)^2]$
$= (3x - 2y)(9x^2 + 6xy + 4y^2)$

(c) $1000k^3 - 27n^3 = (10k)^3 - (3n)^3$
$= (10k - 3n)[(10k)^2 + (10k)(3n) + (3n)^2]$
$= (10k - 3n)(100k^2 + 30kn + 9n^2)$ ■

4 While an expression of the form $x^2 + y^2$ (a sum of two squares) cannot be factored with real numbers, the **sum of two cubes** is factored as follows.

SUM OF TWO CUBES

$$x^3 + y^3 = (x + y)(x^2 - xy + y^2)$$

To verify this result, find the product of $x + y$ and $x^2 - xy + y^2$. Compare this pattern with the pattern for the difference of two cubes.

EXAMPLE 4 **FACTORING THE SUM OF CUBES**

Factor each sum of cubes.

(a) $r^3 + 27 = r^3 + 3^3 = (r + 3)(r^2 - 3r + 3^2)$
$= (r + 3)(r^2 - 3r + 9)$

(b) $27z^3 + 125 = (3z)^3 + 5^3 = (3z + 5)[(3z)^2 - (3z)(5) + 5^2]$
$= (3z + 5)(9z^2 - 15z + 25)$

(c) $125t^3 + 216s^6 = (5t)^3 + (6s^2)^3$
$= (5t + 6s^2)[(5t)^2 - (5t)(6s^2) + (6s^2)^2]$
$= (5t + 6s^2)(25t^2 - 30ts^2 + 36s^4)$ ■

CAUTION A common error is to think that there is a 2 as coefficient of xy when factoring the sum or difference of two cubes. There is no 2 there, so in general, expressions of the form $x^2 + xy + y^2$ and $x^2 - xy + y^2$ cannot be factored further.

The special types of factoring in this section are summarized here. *These should be memorized.*

SPECIAL TYPES OF FACTORING

Difference of two squares	$x^2 - y^2 = (x + y)(x - y)$
Perfect square trinomial	$x^2 + 2xy + y^2 = (x + y)^2$
	$x^2 - 2xy + y^2 = (x - y)^2$
Difference of two cubes	$x^3 - y^3 = (x - y)(x^2 + xy + y^2)$
Sum of two cubes	$x^3 + y^3 = (x + y)(x^2 - xy + y^2)$

3.7 EXERCISES

Factor each of the following. See Examples 1–4.

1. $x^2 - 25$

2. $9 - p^2$

3. $36m^2 - 25$

4. $4x^2 - 49$

5. $16y^2 - 81q^2$

6. $9m^2 - 100r^2$

7. $16 - 25a^2b^2$

8. $49 - 64x^2z^2$

9. $a^4 - 4b^4$

10. $m^2p^2 - 49r^2s^2$

11. $x^2 + 4x + 4$

12. $y^2 + 6y + 9$

13. $a^2 - 10a + 25$

14. $b^2 - 8b + 16$

15. $9r^2 - 6rs + s^2$

16. $4a^2 - 20ab + 25b^2$

17. $25x^2y^2 - 20xy + 4$

18. $9k^2q^2 + 24kq + 16$

19. $72m^2 - 120mp + 50p^2$

20. $100y^2 - 100yz + 25z^2$

21. $8a^3 + 1$

22. $125a^3 - 1$

23. $27x^3 - 64y^3$

24. $8a^3 + 125m^3$

25. $64x^3 + 125y^3$

26. $216z^3 - x^3$

27. $125m^3 - 8p^3$

28. $64y^3 - 1331x^3$

29. $1000 + 27r^3s^3$

30. $343 + 1000a^3b^3$

31. $64y^6 + 1$

32. $m^6 - 8$

33. The binomial $x^6 - 1$ can be considered as both a difference of squares and a difference of cubes. Factor it using both patterns, first as a difference of squares and then as a difference of cubes.

34. Compare the two results of Exercise 33. Notice that one of the factorizations is more complete than the other. Therefore, in a case where you have a choice as to using either of the patterns, which one do you think will produce the complete factorization using the methods of this section?

Factor each of the following.

35. $25 - (r + 3s)^2$

36. $81 - (2k + z)^2$

37. $m^2 - (3p - 5)^2$

38. $w^2 - (2z - 3)^2$

39. $(a + b)^2 - (a - b)^2$

40. $(c - d)^2 - (c + d)^2$

41. $(a + b)^2 + 2(a + b) + 1$

42. $(x + y)^2 + 6(x + y) + 9$

43. $(m - p)^2 + 4(m - p) + 4$

44. $(w - r)^2 + 8(w - r) + 16$

45. $(p^2 - 6p + 9) - r^2$

46. $(k^2 - 10k + 25) - z^2$

47. $9y^2 - 30y + 25 - 16x^2$

48. $25a^2 - 20a + 4 - 9b^2$

49. $r^2 - 16s^2 + 24s - 9$

50. $t^2 - 16u^2 + 8u - 1$

51. $64 - (a - b)^3$

52. $(r + 1)^3 - 1$

53. $(p - 5)^3 + 125$

54. $m^3 + (m + 3)^3$

55. $a^3 - (a - 4)^3$

56. $(p + q)^3 - (p - q)^3$

Find a value of b or c so that the following will be perfect squares.

57. $p^2 + 6p + c$

58. $y^2 - 14y + c$

59. $9z^2 - 30z + c$

60. $16r^2 + 24r + c$

61. $16q^2 + bq + 25$

62. $36x^2 + bx + 25$

Factor each of the following. Assume that all variables used as exponents represent positive integers.

63. $16m^{4x} - 9$

64. $100m^{2q} - 81$

65. $64r^{8z} - 1$

66. $4 - 49x^{4y}$

67. $100m^{2z} - 9p^{8z}$

68. $16k^{8b} - 25m^{4b}$

69. $9a^{4z} - 30a^{2z} + 25$

70. $121p^{8k} + 44p^{4k} + 4$

71. $x^{3n} - 8$

72. $216 + b^{3k}$

73. $27z^{12y} + 125x^{6y}$

74. $1000k^{15r} - m^{21r}$

PREVIEW EXERCISES

Factor completely. See Sections 3.5 and 3.6.

75. $16y^2 - 24y + 32y^3$

76. $9z^2 - 10z^5 + z^7$

77. $xy + 2y + 4x + 8$

78. $a^2b^2 + 3b^2 + 2a^2 + 6$

79. $y^2 + y - 2$

80. $m^2 - 4m - 21$

81. $6r^2 + 19rz - 7z^2$

82. $10w^2 + 19wx + 6x^2$

3.8 GENERAL METHODS OF FACTORING

OBJECTIVES

1. KNOW THE FIRST STEP IN TRYING TO FACTOR A POLYNOMIAL.
2. KNOW THE RULES FOR FACTORING BINOMIALS.
3. KNOW THE RULES FOR FACTORING TRINOMIALS.
4. KNOW THE RULES FOR FACTORING POLYNOMIALS OF MORE THAN THREE TERMS.

Polynomials are factored by using the methods discussed earlier in this chapter. This section shows how to decide which method to use in factoring a particular polynomial. The factoring process is complete when the polynomial is in the form described here.

DEFINITION OF FACTORED FORM

A polynomial is in **factored form** when the following are satisfied.

1. The polynomial is written as a product of prime polynomials with integer coefficients.
2. All the polynomial factors are prime, except that a monomial factor need not be factored completely.

The order of the factors does not matter.

For example, $9x^2(x + 2)$ is the factored form of $9x^3 + 18x^2$. Because of the second rule above, it is not necessary to factor $9x^2$ as $3 \cdot 3 \cdot x \cdot x$.

1 The first step in trying to factor a polynomial is always the same, regardless of the number of terms in the polynomial.

The first step in factoring any polynomial is to factor out any common factor.

EXAMPLE 1 FACTORING OUT A COMMON FACTOR

Factor each polynomial.

(a) $9p + 45 = 9(p + 5)$

(b) $5z^2 + 11z^3 + 9z^4 = z^2(5 + 11z + 9z^2)$

(c) $8m^2p^2 + 4mp = 4mp(2mp + 1)$ ■

2 If the polynomial to be factored is a binomial, it will be necessary to use one of the following rules.

For a **binomial** (two terms), check for the following.

Difference of two squares	$x^2 - y^2 = (x + y)(x - y)$
Difference of two cubes	$x^3 - y^3 = (x - y)(x^2 + xy + y^2)$
Sum of two cubes	$x^3 + y^3 = (x + y)(x^2 - xy + y^2)$

CAUTION The sum of two squares usually cannot be factored with real numbers.

EXAMPLE 2 FACTORING BINOMIALS

Factor each polynomial, if possible.

(a) $64m^2 - 9n^2 = (8m)^2 - (3n)^2$ Difference of two squares
$= (8m + 3n)(8m - 3n)$

(b) $8p^3 - 27 = (2p)^3 - 3^3$ Difference of two cubes
$= (2p - 3)[(2p)^2 + (2p)(3) + 3^2]$
$= (2p - 3)(4p^2 + 6p + 9)$

(c) $1000m^3 + 1 = (10m)^3 + 1^3$ Sum of two cubes
$= (10m + 1)[(10m)^2 - (10m)(1) + 1^2]$
$= (10m + 1)(100m^2 - 10m + 1)$

(d) $25m^2 + 121$ is prime. It is the sum of two squares. ■

3 If the polynomial to be factored is a trinomial, proceed as follows.

For a **trinomial** (three terms), first see if it is a perfect square trinomial of the form

$$x^2 + 2xy + y^2 = (x + y)^2,$$

or

$$x^2 - 2xy + y^2 = (x - y)^2.$$

If it is not, use the methods of Section 3.6.

EXAMPLE 3 FACTORING TRINOMIALS

Factor each polynomial.

(a) $p^2 + 10p + 25 = (p + 5)^2$ Perfect square trinomial

(b) $49z^2 - 42z + 9 = (7z - 3)^2$ Perfect square trinomial

(c) $y^2 - 5y - 6 = (y - 6)(y + 1)$
The numbers -6 and 1 have a product of -6 and a sum of -5.

(d) $r^2 + 18r + 72 = (r + 6)(r + 12)$

(e) $2k^2 - k - 6 = (2k + 3)(k - 2)$
Use either method from Section 3.6.

(f) $28z^2 + 6z - 10 = 2(14z^2 + 3z - 5)$ Factor out the common factor.
$= 2(7z + 5)(2z - 1)$ ■

4 If the polynomial to be factored has more than three terms, use the following guideline.

If the polynomial has more than three terms, try factoring by grouping.

EXAMPLE 4 **FACTORING POLYNOMIALS WITH MORE THAN THREE TERMS**

Factor each polynomial.

(a) $mn + 2n + 3m + 6 = n(m + 2) + 3(m + 2)$ Factor by grouping.
$= (m + 2)(n + 3)$

(b) $ac + bc + ad + bd = c(a + b) + d(a + b)$
$= (a + b)(c + d)$

(c) $xy^2 - y^3 + x^3 - x^2y = y^2(x - y) + x^2(x - y)$
$= (x - y)(y^2 + x^2)$

(d) $20k^3 + 4k^2 - 45k - 9 = (20k^3 + 4k^2) - (45k + 9)$ Be careful with signs.
$= 4k^2(5k + 1) - 9(5k + 1)$
$= (5k + 1)(4k^2 - 9)$ $5k + 1$ is a common factor.
$= (5k + 1)(2k + 3)(2k - 3)$ Difference of two squares

(e) $4a^2 + 4a + 1 - b^2 = (4a^2 + 4a + 1) - b^2$ Associative property
$= (2a + 1)^2 - b^2$ Perfect square trinomial
$= (2a + 1 + b)(2a + 1 - b)$ Difference of two squares ■

The steps used in factoring a polynomial are summarized here.

FACTORING A POLYNOMIAL

Step 1 Factor out any common factor.

Step 2a If the polynomial is a binomial, check to see if it is the difference of two squares, the difference of two cubes, or the sum of two cubes.

Step 2b If the polynomial is a trinomial, check to see if it is a perfect square trinomial. If it is not, factor as in Section 3.6.

Step 2c If the polynomial has more than three terms, try to factor by grouping.

Remember that the factored form can always be checked by multiplication.

3.8 EXERCISES

Factor each of the following polynomials. Assume all variables used as exponents are positive integers.

1. $100a^2 - 9b^2$

2. $10r^2 + 13r - 3$

3. $3p^4 - 3p^3 - 90p^2$

4. $k^4 - 16$

5. $3a^2pq + 3abpq - 90b^2pq$

6. $49z^2 - 16$

7. $225p^2 + 256$

8. $x^3 - 1000$

9. $6b^2 - 17b - 3$

10. $k^2 - 6k - 16$

11. $18m^3n + 3m^2n^2 - 6mn^3$

12. $6t^2 + 19tu - 77u^2$

13. $2p^2 + 11pq + 15q^2$

14. $40p - 32r$

15. $9m^2 - 45m + 18m^3$

16. $4k^2 + 28kr + 49r^2$

17. $54m^3 - 2000$

18. $mn - 2n + 5m - 10$

19. $2a^2 - 7a - 4$

20. $9m^2 - 30mn + 25n^2$

21. $kq - 9q + kr - 9r$

22. $56k^3 - 875$

23. $9r^2 + 100$

24. $16z^3x^2 - 32z^2x$

25. $8p^3 - 125$

26. $yx + yw + zx + zw$

27. $x^4 - 625$

28. $2m^2 - mn - 15n^2$

29. $p^3 + 64$

30. $48y^2z^3 - 28y^3z^4$

31. $64m^2 - 625$

32. $14z^2 - 3zk - 2k^2$

33. $12z^3 - 6z^2 + 18z$

34. $225k^2 - 36r^2$

35. $256b^2 - 400c^2$

36. $z^2 - zp - 20p^2$

37. $1000z^3 + 512$

38. $64m^2 - 25n^2$

39. $10r^2 + 23rs - 5s^2$

40. $12k^2 - 17kq - 5q^2$

41. $24p^3q + 52p^2q^2 + 20pq^3$

42. $32x^2 + 16x^3 - 24x^5$

43. $48k^4 - 243$

44. $14x^2 - 25xq - 25q^2$

45. $50p^2 - 162$

46. $y^2 + 3y - 10$

47. $12m^2rx + 4mnrx + 40n^2rx$

48. $18p^2 + 53pr - 35r^2$

49. $21a^2 - 5ab - 4b^2$

50. $x^2 - 2xy + y^2 - 4$

51. $x^2 - y^2 - 4$

52. $(5r + 2s)^2 - 6(5r + 2s) + 9$

53. $(p + 8q)^2 - 10(p + 8q) + 25$

54. $z^4 - 9z^2 + 20$

55. $21m^4 - 32m^2 - 5$

56. $(x - y)^3 - (27 - y)^3$

57. $(r + 2t)^3 + (r - 3t)^3$

58. $16x^3 + 32x^2 - 9x - 18$

59. $x^5 + 3x^4 - x - 3$

60. $2a^x + 4a^{2x} + 8a^{3x}$

61. $3m^{2k} - 7m^k - 6$

62. $2p^{2a} + 10p^a - 28$

63. $15c^{4y} + 3c^{2y} - 6c^y$

64. $x^{2p} - y^{2p}$

65. $z^{4x} - w^{2x}$

66. $k^2 + 2kp + p^2 - a^2 - 2ab - b^2$

67. $m^2 - 4m + 4 - n^2 + 6n - 9$

68. $x^8 - 1$

PREVIEW EXERCISES

Simplify. See Section 3.1.

69. $\dfrac{3^6 \cdot 3^{-1}}{3^3}$

70. $\dfrac{p^5p^{-2}}{p^{-1}p^4} \quad (p \neq 0)$

71. $\dfrac{2k^{-1}k}{k^2k^{-2}} \quad (k \neq 0)$

Solve each equation. See Section 2.1.

72. $3x + 1 = 13$

73. $5r - 2 = 8$

74. $-2q + 3 = 7$

75. $-a + 4 = 5$

76. $4p = 0$

77. $-9r = 0$

78. $-2z + 8 = 9$

79. $1 - 3y = 5$

80. $5 - 9k = -4$

3.9 SOLVING EQUATIONS BY FACTORING

OBJECTIVES

1 LEARN THE ZERO-FACTOR PROPERTY.

2 USE THE ZERO-FACTOR PROPERTY TO SOLVE EQUATIONS.

3 SOLVE APPLIED PROBLEMS THAT REQUIRE THE ZERO-FACTOR PROPERTY.

FOCUS ON PROBLEM SOLVING

If a toy rocket is launched vertically upward from ground level with an initial velocity of 128 feet per second, then its height h after t seconds is given by the equation $h = -16t^2 + 128t$ (if air resistance is neglected). How long will it take for the rocket to return to the ground?

The equations that we have solved so far in this book have been linear equations. Recall that in a linear equation, the largest power of the variable is 1. In order to solve equations of degree greater than 1, other methods must be developed. In this section we learn one of these methods, which involves factoring. This problem is Exercise 53 in the exercises for this section. After working through this section, you should be able to solve this problem.

1 Some equations that cannot be solved by other methods can be solved by factoring. This process depends on a special property of the number 0, called the **zero-factor property.**

ZERO-FACTOR PROPERTY

If two numbers have a product of 0, then at least one of the numbers must be 0. That is, if $ab = 0$, then either $a = 0$ or $b = 0$.

The zero-factor property is proved by first assuming $a \neq 0$. (If a does equal 0, then the property is proved already.) If $a \neq 0$, then $1/a$ exists, and both sides of $ab = 0$ can be multiplied by $1/a$ to get

$$\frac{1}{a} \cdot ab = \frac{1}{a} \cdot 0$$

$$b = 0.$$

Thus, if $a \neq 0$, then $b = 0$, and the property is proved.

2 The next examples show how to use the zero-factor property to solve equations.

EXAMPLE 1
USING THE ZERO-FACTOR PROPERTY TO SOLVE AN EQUATION

Solve the equation $(x + 6)(2x - 3) = 0$.

Here the product of $x + 6$ and $2x - 3$ is 0. By the zero-factor property, this can be true only if $x + 6$ equals 0 or if $2x - 3$ equals 0. That is,

$$x + 6 = 0 \quad \text{or} \quad 2x - 3 = 0.$$

Solve $x + 6 = 0$ by subtracting 6 from both sides of the equation. Solve $2x - 3 = 0$ by first adding 3 on both sides and then dividing both sides by 2.

$$\begin{aligned} x + 6 &= 0 &\quad \text{or} \quad 2x - 3 &= 0 \\ x + 6 - 6 &= 0 - 6 &\quad \text{or} \quad 2x - 3 + 3 &= 0 + 3 \\ x &= -6 &\quad \text{or} \quad 2x &= 3 \\ & & x &= \frac{3}{2} \end{aligned}$$

These solutions may be checked by first replacing x with -6 in the original equation. Then go back and replace x with $3/2$. This check shows that the solution set is $\{-6, 3/2\}$. ■

Since the product $(x + 6)(2x - 3)$ equals $2x^2 + 9x - 18$, the equation of Example 1 has a squared term and is an example of a *quadratic equation*.

QUADRATIC EQUATION

An equation that can be written in the form

$$ax^2 + bx + c = 0,$$

where $a \neq 0$, is a **quadratic equation.** The form given is called **standard form.**

Quadratic equations are discussed in more detail in Chapter 6.

EXAMPLE 2
SOLVING A QUADRATIC EQUATION BY FACTORING

Solve the equation $2p^2 + 3p = 2$.

The zero-factor property requires a product of two factors equal to 0. Get 0 on one side of the equals sign in this equation by subtracting 2 from both sides.

$$\begin{aligned} 2p^2 + 3p &= 2 \\ 2p^2 + 3p - 2 &= 0 \qquad \text{Standard form} \end{aligned}$$

Factor on the left.

$$(2p - 1)(p + 2) = 0$$

Set each factor equal to 0, and then solve each of the two resulting equations.

$$\begin{aligned} 2p - 1 &= 0 &\quad \text{or} \quad p + 2 &= 0 \\ 2p &= 1 &\quad \text{or} \quad p &= -2 \\ p &= \frac{1}{2} \end{aligned}$$

The solution set is $\{1/2, -2\}$. Check by substitution in the original equation. ■

EXAMPLE 3
SOLVING A QUADRATIC EQUATION WITH A MISSING TERM

Solve $5z^2 - 25z = 0$.

This quadratic equation has a missing term. Comparing it with the general form $ax^2 + bx + c = 0$ shows that $c = 0$. The zero-factor property still can be used, however, since

$$5z^2 - 25z = 5z(z - 5).$$

The equation

$$5z^2 - 25z = 0$$

becomes

$$5z(z - 5) = 0,$$

giving

$$5z = 0 \quad \text{or} \quad z - 5 = 0$$

$$z = 0 \quad \text{or} \quad z = 5.$$

The solutions are 0 and 5, as can be verified by substituting in the original equation. The solution set is $\{0, 5\}$. ■

EXAMPLE 4
SOLVING AN EQUATION REQUIRING REWRITING

Solve $(2q + 1)(q + 1) = 2(1 - q) + 6$.

Put the equation in the standard form $ax^2 + bx + c = 0$ by first multiplying on each side.

$$(2q + 1)(q + 1) = 2(1 - q) + 6$$
$$2q^2 + 3q + 1 = 2 - 2q + 6$$

Now combine terms and arrange them so that the right side is 0. Then factor.

$$2q^2 + 5q - 7 = 0$$
$$(2q + 7)(q - 1) = 0$$

$$2q + 7 = 0 \quad \text{or} \quad q - 1 = 0$$
$$2q = -7 \quad \text{or} \quad q = 1$$
$$q = -\frac{7}{2}$$

Check that the solution set is $\{-7/2, 1\}$. ■

In summary, use the following steps to solve an equation by factoring.

SOLVING AN EQUATION BY FACTORING

Step 1 **Write in standard form.**
Rewrite the equation if necessary so that one side is 0.

Step 2 **Factor.**
Factor the polynomial.

Step 3 **Use the zero-factor property.**
Place each variable factor equal to zero, using the zero-factor property.

Step 4 **Find the solution(s).**
Solve each equation formed in Step 3.

Step 5 **Check.**
Check each solution in the original equation.

CAUTION If $ab = 0$, then $a = 0$ or $b = 0$. However, if $ab = 6$, for example, it is not necessarily true that $a = 6$ or $b = 6$; it is very likely that *neither* $a = 6$ *nor* $b = 6$. It is important to remember that the zero-factor property works only for a product equal to *zero*.

The equations that we have solved so far in this section have all been quadratic equations. The zero-factor property can be extended to solve certain equations of degree 3 or higher, as shown in the next example.

EXAMPLE 5
SOLVING AN EQUATION OF DEGREE 3

Solve $-x^3 + x^2 = -6x$.

Start by adding $6x$ to both sides to get 0 on the right side.

$$-x^3 + x^2 + 6x = 0.$$

To make the factoring step easier, multiply both sides by -1.

$$x^3 - x^2 - 6x = 0$$

Factor the left side.

$$x(x^2 - x - 6) = 0$$
$$x(x - 3)(x + 2) = 0$$

Use the zero-factor property, extended to include the three variable factors.

$$x = 0 \quad \text{or} \quad x - 3 = 0 \quad \text{or} \quad x + 2 = 0$$
$$x = 3 \qquad\qquad x = -2$$

The solution set is $\{0, 3, -2\}$. ■

3 The next example shows an application that leads to a quadratic equation.

EXAMPLE 6
USING A QUADRATIC EQUATION IN AN APPLICATION

Some surveyors are surveying a lot that is in the shape of a parallelogram. They find that the longer sides of the parallelogram are each 8 meters longer than the distance between them. The area of the lot is 48 square meters. Find the length of the longer sides and the distance between them.

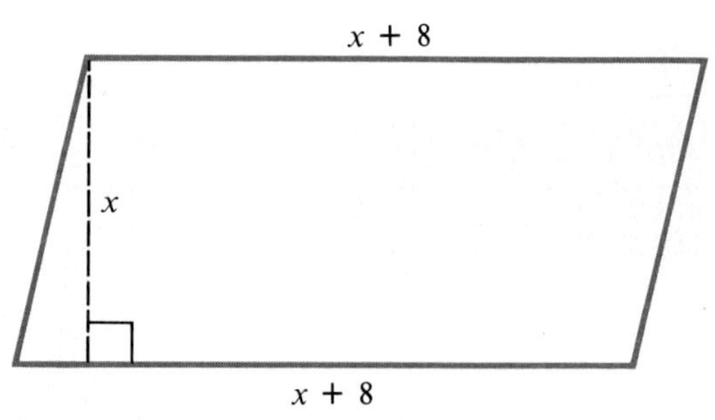

FIGURE 3.1

Let x represent the distance between the longer sides as shown in Figure 3.1. Then $x + 8$ is the length of each longer side. The area of a parallelogram is given by $A = bh$ where b is the length of the longer side and h is the distance between the longer sides. Here $b = x + 8$ and $h = x$.

$$A = bh$$

$$48 = (x + 8)x \qquad \text{Let } A = 48,\ b = x + 8,\ h = x.$$

$$48 = x^2 + 8x \qquad \text{Distributive property}$$

$$0 = x^2 + 8x - 48 \qquad \text{Subtract 48.}$$

$$0 = (x + 12)(x - 4) \qquad \text{Factor.}$$

Set each factor equal to 0.

$$x + 12 = 0 \quad \text{or} \quad x - 4 = 0$$

$$x = -12 \quad \text{or} \quad x = 4$$

A parallelogram cannot have a side with a negative length, so reject -12 as a solution. The only possible solution is 4, so the distance between the longer sides is 4 meters. The length of the longer sides is $4 + 8 = 12$ meters. ■

PROBLEM SOLVING

When applications lead to quadratic equations, it often happens that a solution of the equation does not satisfy the physical requirements of the problem, as in Example 6. In such cases, we must reject this solution of the equation as an answer to the problem. ■

3.9 EXERCISES

Find all solutions of the following equations by using the zero-factor property. See Example 1.

1. $(x - 2)(x - 3) = 0$

2. $(m + 4)(m + 5) = 0$

3. $(3m - 8)(5m + 6) = 0$

4. $(7y + 4)(3y - 5) = 0$

5. $r(r - 4)(2r + 5) = 0$

6. $3x(3x - 5)(2x + 7) = 0$

7. $(x - 5)(2x + 7)(3x - 4) = 0$

8. $(x + 9)(6x - 1)(8x + 4) = 0$

9. In trying to solve $(x + 4)(x - 1) = 1$, a student reasons that since $1 \cdot 1 = 1$, the equation is solved by solving

$$x + 4 = 1 \quad \text{or} \quad x - 1 = 1.$$

Explain the error in this reasoning. What is the correct way to solve this equation? (See Objective 2.)

10. In solving the equation $4(x - 3)(x + 7) = 0$, a student writes

$$4 = 0 \quad \text{or} \quad x - 3 = 0 \quad \text{or} \quad x + 7 = 0.$$

Then the student becomes confused, not knowing how to handle the equation $4 = 0$. Explain how to handle this, and give the solutions of the equation. (See Objective 2.)

Find all solutions of the following equations by factoring. See Examples 2–5.

11. $x^2 - x - 12 = 0$

12. $m^2 + 4m - 5 = 0$

13. $y^2 + 9y + 14 = 0$

14. $15r^2 + 7r = 2$

15. $12x^2 + 4x = 1$

16. $4a^2 + 9a + 2 = 0$

17. $3x^2 - 11x - 20 = 0$

18. $2y^2 - 12 - 4y = y^2 - 3y$

19. $3p^2 + 9p + 30 = 2p^2 - 2p$

20. $4p^2 + 16p = 0$

21. $2a^2 - 8a = 0$

22. $6m^2 - 36m = 0$

23. $-3m^2 + 27m = 0$

24. $-3m^2 + 27 = 0$

25. $-2a^2 + 8 = 0$

26. $-5p^2 + 45 = 0$

27. $-x^2 + 100 = 0$

28. $4p^2 - 16 = 0$

29. $9x^2 - 81 = 0$

30. $x(x + 3) = 4$

31. $(x + 4)(x - 6) = -16$

32. $(x + 2)(2x + 3) = 10$

33. $(w - 1)(3w + 2) = 4w$

34. $y(y + 1) - 4 = -2(y + 3)$

35. $2x(x + 2) = x(x - 3) - 12$

36. $(r - 5)(r - 3) = 3r(r - 3)$

37. $3x^3 - 5x^2 - 2x = 0$

38. $7y^3 - 22y^2 - 24y = 0$

39. $6t^3 + 5t^2 - 6t - 5 = 0$

40. $m^3 + 4m^2 - 9m - 36 = 0$

41. $-x^3 + 6x^2 + 27x = 0$

42. $-3r^3 + 5r^2 + 28r = 0$

Solve each of the following applied problems by writing a quadratic equation and then solving it. See Example 6.

43. Two numbers have a sum of 6 and a product of -16. Find the numbers.

44. The sum of two numbers is 10. The sum of their squares is 68. Find the numbers.

45. Two numbers have a sum of 12. The sum of the squares of the numbers is 90. Find the numbers.

46. If two integers are added, the result is 15. If the squares of the integers are added, the result is 125. Find the integers.

47. Two numbers have a sum of 3 and a product of -108. Find the numbers.

48. Two numbers have a sum of -6 and a product of -72. Find the numbers.

49. A table top has an area of 54 square feet. It has a length that is 3 feet more than the width. Find the width and length of the table top.

50. A building has a floor area of 140 square meters. The building has the shape of a rectangle with length 4 meters more than the width. Find the width and length of the building.

51. A sign is to have the shape of a triangle with the height 3 meters more than the length of the base. How long should the base be, if the area must be 14 square meters?

52. The height of a parallelogram is 7 feet less than its base. The area is 60 square feet. Find the height and base of the parallelogram.

53. If a toy rocket is launched vertically upward from ground level with an initial velocity of 128 feet per second, then its height h after t seconds is given by the equation $h = -16t^2 + 128t$ (if air resistance is neglected). How long will it take for the rocket to return to the ground? (*Hint:* When the rocket returns to the ground, $h = 0$.)

54. After how many seconds will the rocket in Exercise 53 be 112 feet above the ground?

55. A farmer has 300 feet of fencing and wants to enclose a rectangular area of 5000 square feet. What dimensions should she use?

56. A rectangular landfill has an area of 30,000 square feet. Its length is 200 feet more than its width. What are the dimensions of the landfill?

Use the zero-factor property to find all solutions of the following equations.

57. $2(x - 1)^2 - 7(x - 1) - 15 = 0$

58. $4(2k + 3)^2 - (2k + 3) - 3 = 0$

59. $5(3a - 1)^2 + 3 = -16(3a - 1)$

60. $2(m + 3)^2 = 5(m + 3) - 2$

61. $(x - 1)^2 - (2x - 5)^2 = 0$

62. $(3y + 1)^2 - (y - 2)^2 = 0$

63. $(2k - 3)^2 = 16k^2$

64. $9p^2 = (5p + 2)^2$

PREVIEW EXERCISES

Simplify. See Section 3.1.

65. $\dfrac{6m^5}{2m^2}$

66. $\dfrac{9y^7}{6y^3}$

67. $\dfrac{-100a^5b^2}{75a^6b^5}$

68. $\dfrac{-32r^3s^5}{48r^6s^7}$

Write each fraction with the indicated denominator.

69. $\dfrac{3}{4}, \dfrac{}{24}$

70. $\dfrac{5}{9}, \dfrac{}{63}$

71. $\dfrac{11}{15}, \dfrac{}{75}$

72. $\dfrac{7}{5}, \dfrac{}{30}$

CHAPTER 3 GLOSSARY

KEY TERMS

3.2 scientific notation A number is written in scientific notation when it is expressed in the form $a \times 10^n$, where $1 \le |a| < 10$, and n is an integer.

3.3 numerical coefficient The number in a product is called the numerical coefficient, or just the coefficient.

polynomial A polynomial is a term or a finite sum of terms in which all variables have whole number exponents and no variables appear in denominators.

descending powers A polynomial in one variable is written in descending powers of the variable if the exponents on the terms of the polynomial decrease from left to right.

trinomial A trinomial is a polynomial with exactly three terms.

binomial A binomial is a polynomial with exactly two terms.

monomial A monomial is a polynomial with exactly one term.

degree of a term The degree of a term with one variable is the exponent of the variable. The degree of a term with more than one variable is the sum of the exponents of the variables.

degree of a polynomial The degree of a polynomial is the largest degree of any of the terms of the polynomial.

negative of a polynomial The negative of a polynomial is obtained by changing the sign of each term of the polynomial.

3.5 prime factored form An integer is written in prime factored form when it is written as a product of prime numbers.

prime number A prime number is a positive integer greater than 1 that can be divided without remainder only by itself and 1.

greatest common factor The greatest common factor of a group of integers is the largest integer that will divide without remainder into every integer in the group.

factoring a polynomial Writing a polynomial as the product of two or more simpler polynomials is called factoring the polynomial.

3.6 prime polynomial If a polynomial cannot be factored, it is called a prime polynomial.

3.9 zero-factor property If two numbers have a product of 0, then at least one of the numbers must be 0.

quadratic equation in standard form An equation that can be written in the form $ax^2 + bx + c = 0$, where $a \neq 0$, is a quadratic equation. The form given is called standard form.

NEW SYMBOLS

$\boldsymbol{P(x)}$ polynomial in x (read "P of x")

CHAPTER 3 QUICK REVIEW

SECTION	CONCEPTS	EXAMPLES
3.1 INTEGER EXPONENTS	**Definitions and Rules for Exponents**	
	Product Rule: $a^m \cdot a^n = a^{m+n}$	$3^4 \cdot 3^2 = 3^6$
	Quotient Rule: $\frac{a^m}{a^n} = a^{m-n}$	$\frac{2^5}{2^3} = 2^2$
	Negative Exponent: $a^{-n} = \frac{1}{a^n}$	$5^{-2} = \frac{1}{5^2}$
	$\frac{a^{-n}}{b^{-m}} = \frac{b^m}{a^n}$	$\frac{5^{-3}}{4^{-6}} = \frac{4^6}{5^3}$
	Zero Exponent: $a^0 = 1$	$27^0 = 1, \quad (-5)^0 = 1$
3.2 FURTHER PROPERTIES OF EXPONENTS	**Power Rules**	
	$(a^m)^n = a^{mn}$	$(6^3)^4 = 6^{12}$
	$(ab)^m = a^m b^m$	$(5p)^4 = 5^4 p^4$
	$\left(\frac{a}{b}\right)^n = \frac{a^n}{b^n}$	$\left(\frac{2}{3}\right)^5 = \frac{2^5}{3^5}$
	$a^{-n} = \left(\frac{1}{a}\right)^n$	$4^{-3} = \left(\frac{1}{4}\right)^3$
	$\left(\frac{a}{b}\right)^{-n} = \left(\frac{b}{a}\right)^n$	$\left(\frac{4}{7}\right)^{-2} = \left(\frac{7}{4}\right)^2$
3.3 THE BASIC IDEAS OF POLYNOMIALS	Add or subtract polynomials by combining like terms.	Add: $(x^2 - 2x + 3) + (2x^2 - 8)$. $= 3x^2 - 2x - 5$ Subtract: $(5x^4 + 3x^2) - (7x^4 + x^2 - x)$. $= -2x^4 + 2x^2 + x$

SECTION	CONCEPTS	EXAMPLES
3.4 MULTIPLICATION OF POLYNOMIALS	Use the commutative and associative properties and the rules for exponents to multiply two polynomials.	Multiply: $(x^3 + 3x)(4x^2 - 5x + 2)$. $= 4x^5 + 12x^3 - 5x^4 - 15x^2 + 2x^3 + 6x$ $= 4x^5 - 5x^4 + 14x^3 - 15x^2 + 6x$
	Special Products $(x + y)(x - y) = x^2 - y^2$ $(x + y)^2 = x^2 + 2xy + y^2$ $(x - y)^2 = x^2 - 2xy + y^2$	$(3m + 8)(3m - 8) = 9m^2 - 64$ $(5a + 3b)^2 = 25a^2 + 30ab + 9b^2$ $(2k - 1)^2 = 4k^2 - 4k + 1$
	To multiply two binomials in general, use the FOIL method. Multiply the First terms, the Outside terms, the Inside terms, and the Last terms.	$(2x + 3)(x - 7)$ $= 2x(x) + 2x(-7) + 3x + 3(-7)$ $= 2x^2 - 14x + 3x - 21$ $= 2x^2 - 11x - 21$
3.5 GREATEST COMMON FACTORS; FACTORING BY GROUPING	**Factoring out the Greatest Common Factor** The product of the largest common numerical factor and the variable of lowest degree of every term in a polynomial is the greatest common factor of the terms of the polynomial.	Factor: $4x^2y - 50xy^2 = 2^2x^2y - 2 \cdot 5^2xy^2$. The greatest common factor is $2xy$. $4x^2y - 50xy^2 = 2xy(2x - 25y)$
	Factoring by Grouping *Step 1* Collect the terms into groups so that the terms in each group have a common factor.	$5a - 5b - ax + bx = (5a - 5b) + (-ax + bx)$
	Step 2 Factor out the common factor of the terms in each group.	$= 5(a - b) - x(a - b)$
	Step 3 If the groups now have a common factor, factor it out. If not, try a different grouping.	$= (a - b)(5 - x)$

SECTION	CONCEPTS	EXAMPLES
3.6 FACTORING TRINOMIALS	For a trinomial in the form $ax^2 + bx + c$ or $ax^2 + bxy + cy^2$:	Factor: $15x^2 + 14x - 8$.
	Step 1 Write all pairs of factors of the coefficient of the first term.	Factors of $15x^2$ are $5x$ and $3x$, $1x$ and $15x$.
	Step 2 Write all pairs of factors of the last term.	Factors of -8 are -4 and 2, 4 and -2, -1 and 8, 8 and -1.
	Step 3 Use various combinations of these factors until the necessary middle term is found. If the necessary combination does not exist, the polynomial is prime.	$15x^2 + 14x - 8 = (5x - 2)(3x + 4)$
3.7 SPECIAL FACTORING	**Difference of Two Squares**	Factor:
	$x^2 - y^2 = (x + y)(x - y)$	$4m^2 - 25n^2 = (2m)^2 - (5n)^2$ $= (2m + 5n)(2m - 5n)$
	Perfect Square Trinomials	
	$x^2 + 2xy + y^2 = (x + y)^2$ $x^2 - 2xy + y^2 = (x - y)^2$	$9y^2 + 6y + 1 = (3y + 1)^2$ $16p^2 - 56p + 49 = (4p - 7)^2$
	Difference of Two Cubes	
	$x^3 - y^3 = (x - y)(x^2 + xy + y^2)$	$8 - 27a^3 = (2 - 3a)(4 + 6a + 9a^2)$
	Sum of Two Cubes	
	$x^3 + y^3 = (x + y)(x^2 - xy + y^2)$	$64z^3 + 1 = (4z + 1)(16z^2 - 4z + 1)$
3.8 GENERAL METHODS OF FACTORING	*Step 1* Factor out any common factor. *Step 2a* If the polynomial is a binomial, check to see if it is a difference of two squares, the difference of two cubes, or the sum of two cubes.	Factor: $ak^3 + 2ak^2 - 9ak - 18a$. $= a(k^3 + 2k^2 - 9k - 18)$ $= a[(k^3 + 2k^2) - (9k + 18)]$ $= a[k^2(k + 2) - 9(k + 2)]$ $= a[(k + 2)(k^2 - 9)]$ $= a(k + 2)(k - 3)(k + 3)$

SECTION	CONCEPTS	EXAMPLES
	Step 2b If the polynomial is a trinomial, check to see if it is a perfect square trinomial. If it is not, factor as in Section 3.6. *Step 2c* If the polynomial has more than three terms, try to factor by grouping.	
3.9 SOLVING EQUATIONS BY FACTORING	*Step 1* Rewrite the equation if necessary so that one side is 0.	Solve: $2x^2 + 5x = 3.$ $2x^2 + 5x - 3 = 0$
	Step 2 Factor the polynomial.	$(2x - 1)(x + 3) = 0$
	Step 3 Set each factor equal to 0.	$2x - 1 = 0$ or $x + 3 = 0$
	Step 4 Solve each equation.	$2x = 1$ $x = -3$ $x = \frac{1}{2}$
	Step 5 Check each solution.	A check verifies that the solution set is $\{1/2, -3\}$.

CHAPTER 3 REVIEW EXERCISES

[3.1] *Identify the exponent and the base. Do not evaluate.*

1. 9^3 **2.** $(-5)^4$ **3.** -8^2 **4.** $-5z^{-2}$

Evaluate.

5. 3^5 **6.** $\left(\frac{1}{2}\right)^4$ **7.** $(-4)^3$ **8.** $\frac{3}{(-7)^{-2}}$ **9.** $\left(\frac{3}{4}\right)^{-3}$ **10.** $\left(\frac{4}{5}\right)^{-2}$

11. 6^0 **12.** $(-6)^0$ **13.** -6^0 **14.** -5^{-3} **15.** $\frac{-5}{(-2)^{-1}}$ **16.** $3^{-1} - 5^{-1}$

Simplify. Write answers with only positive exponents. Assume that all variables used as exponents represent positive integers and all other variables represent nonzero real numbers.

17. $9^{15} \cdot 9^{-7}$ **18.** $\frac{5^{-4}}{5^{-7}}$ **19.** $\frac{m^{-2}m^{-5}}{m^{-8}m^{-4}}$

20. $9(4p^3)(6p^{-7})$ **21.** $\frac{y^{2b} \cdot y^{7b}}{y^{4b-3}}$

22. Give an example to show that $(x + y)^{-1} \neq x^{-1} + y^{-1}$ by choosing specific values for x and y.

[3.2] *Simplify. Write answers with only positive exponents. Assume that all variables represent nonzero real numbers.*

23. $(3^{-4})^2$

24. $(x^{-4})^{-2}$

25. $(xy^{-3})^{-2}$

26. $(z^{-3})^3z^{-6}$

27. $(5m^{-3})^2(m^4)^{-3}$

28. $\dfrac{(3r)^2r^4}{r^{-2}r^{-3}}(9r^{-3})^{-2}$

29. $\left(\dfrac{5z^{-3}}{z^{-1}}\right)\dfrac{5}{z^2}$

30. $\left(\dfrac{6m^{-4}}{m^{-9}}\right)^{-1}\left(\dfrac{m^{-2}}{16}\right)$

31. $\left(\dfrac{3r^5}{5r^{-3}}\right)^{-2}\left(\dfrac{9r^{-1}}{2r^{-5}}\right)^3$

32. $\left(\dfrac{a^{-2}b^{-1}}{3a^2}\right)^{-2}\left(\dfrac{b^{-2}\cdot 3a^4}{2b^{-3}}\right)^{-2}\left(\dfrac{a^{-4}b^5}{a^3}\right)^{-2}$

Simplify each expression. Assume that all variables used as exponents represent integers and that all other variables represent nonzero real numbers.

33. $k^r \cdot k^{2r-3}$

34. $\dfrac{m^q \cdot m^{5q-3}}{m^{-2q}}$

35. $\dfrac{(z^{-1+a}\cdot z^{3a})^{-1}}{(z^{-a})^{-2}}$

36. Give an example to show that $(x^2 + y^2)^2 \neq x^4 + y^4$ by choosing specific values for x and y.

Write in scientific notation.

37. 3450

38. .000000076

39. .13

Write without scientific notation.

40. 2.1×10^5

41. 3.8×10^{-3}

42. 3.78×10^{-1}

Use scientific notation to compute.

43. $\dfrac{6 \times 10^4}{3 \times 10^8}$

44. $\dfrac{5 \times 10^{-2}}{1 \times 10^{-5}}$

45. $\dfrac{.000000016}{.0004}$

46. $\dfrac{.0009 \times 12{,}000}{400}$

[3.3] *Give the coefficient of each term.*

47. $14p^5$

48. $-z$

49. 29

50. $104p^3r^5$

For each polynomial, **(a)** *write in descending powers,* **(b)** *identify as* monomial, binomial, trinomial, *or* none of these, *and* **(c)** *give the degree.*

51. $9k + 11k^3 - 3k^2$

52. $14m^6 - 9m^7$

53. $-5y^4 + 3y^3 + 7y^2 - 2y$

54. $-7q^5r^3$

Find **(a)** *$P(-4)$ and* **(b)** *$P(2)$.*

55. $P(x) = -5x + 11$

56. $P(x) = -3x^2 + 7x - 1$

57. $P(x) = 2x^3 - 4x^2 + 1$

58. $P(x) = 1 - x^4 + 8x + x^2$

Combine terms where possible.

59. $8p + 11p - 6p$

60. $9y^2 + 11y^2 - 8y^3 + 4y^3$

61. $-m^5 + 2m^5 + 3m^4$

62. $-6 + 12y - 8 - y$

Add or subtract as indicated.

63. Add.

$$\begin{array}{r} -5x^2 + 7x - 3 \\ 6x^2 + 4x + 9 \\ \hline \end{array}$$

64. Subtract.

$$\begin{array}{r} -16a^3 + 8a^2 - 4a \\ 12a^3 - 4a^2 + 3a \\ \hline \end{array}$$

65. Subtract.

$$\begin{array}{r} -4y^3 \qquad + 7y - 6 \\ 2y^2 - 9y + 8 \\ \hline \end{array}$$

66. $(8y - 9) - (-4y + 2)$

67. $(12m^5 + 7m^3 + 8m^2) + (4m^3 - 9m)$

68. $(-4r^2 + 7r - 9) + (2r^2 + 5r + 6) + (r^2 + 3)$

69. $(5a^3 - 9a^2 + 11a) - (-a^3 + 6a^2 + 4a)$

70. $(2y^2 + 7y + 4) + (3y^2 - 9y + 2)$

[3.4] *Find each product.*

71. $6y(2y^2)$

72. $-9k^3(3k^2)$

73. $-2a(4a + 11)$

74. $(8y - 7)(3y + 5)$

75. $(2p^2 + 5p - 3)(p + 4)$

76. $(5r + 3s)(2r - 3s)$

77. $(2z^3 - 4z^2 + 6z - 1)(3z + 2)$

78. $(3k^2 + 5k)(4k^2 - 6)$

79. $(m - 7)(m + 3)$

80. $(2y - 3)(y + 4)$

81. $(3r - 8)(2r + 7)$

82. $(6y - 5)(6y + 5)$

83. $(3r^2 + 5)(3r^2 - 5)$

84. $\left(z + \frac{2}{3}\right)\left(z - \frac{2}{3}\right)$

85. $(y + 3z)^2$

86. $(3p - 7q)^2$

Find each product. Assume that all variables represent nonzero real numbers, with variables used as exponents restricted to integers.

87. $9y^q(2y^{q-1} - 5y^{-q})$

88. $m^{p-2}(3m^{5-p} + 4m^{-1+p})$

89. $(2z^r - 5x^{2r})(3z^r + 2x^{2r})$

90. $(k^m - 2q^m)^2$

[3.5] *Factor out the greatest common factor.*

91. $30p + 25$

92. $8y - 8z$

93. $5z + 11r$

94. $12m^2 - 6m$

95. $12a^2b + 18ab^3 - 24a^3b^2$

96. $(a + 2)(a + 1) + (a + 2)(a - 4)$

Factor by grouping.

97. $4x + 4y + mx + my$

98. $x^2 + xy + 5x + 5y$

99. $2m + 6 - am - 3a$

100. $2am - 2bm - ap + bp$

[3.6] *Factor completely.*

101. $x^2 - 11x + 24$

102. $p^2 - 2p - 35$

103. $r^2 + 15r + 54$

104. $a^2 + 16a - 36$

105. $4p^2 + 3p - 1$

106. $6m^2 + 7m - 3$

107. $-12p^2 - p + 20$

108. $-18r^2 + 3r + 10$

109. $20a^2 - 13ab + 2b^2$

110. $2r^2z - rz - 3z$

111. $r^4 + r^2 - 6$

112. $2k^4 - 5k^2 - 3$

113. $p^2(p + 2)^2 + p(p + 2)^2 - 6(p + 2)^2$

114. $10m^{4q} - m^{2q} - 3$

[3.7] *Factor completely.*

115. $9p^2 - 49$

116. $16z^2 - 121$

117. $144(r + 1)^2 - 81(s + 1)^2$

118. $p^2 + 14p + 49$

119. $9k^2 - 12k + 4$

120. $r^3 + 27$

121. $125x^3 - 1$

122. $m^6 - 1$

123. $121a^{10m} - 144b^{6m}$

124. $25k^{4b} - 30k^{2b} + 9$

125. $125r^{3k} + 1$

126. Even though $4x^2 + 36$ is the *sum* of two squares, it can be factored. Explain why, and factor it.

[3.9] *Solve each equation.*

127. $(5x - 2)(x + 1) = 0$

128. $p^2 - 5p + 6 = 0$

129. $q^2 + 2q = 8$

130. $6z^2 = 5z + 50$

131. $6r^2 + 7r = 3$

132. $8k^2 + 14k + 3 = 0$

133. $-4m^2 + 36 = 0$

134. $6y^2 - 9y = 0$

135. $(2x + 1)(x - 2) = -3$

136. $(r + 2)(r - 2) = (r - 2)(r + 3) - 2$

137. $2x^3 - x^2 - 28x = 0$

138. $-t^3 - 3t^2 + 4t + 12 = 0$

Work the following problems.

139. A triangular wall brace has the shape of a right triangle. One of the perpendicular sides is 1 foot longer than twice the other. The area enclosed by the triangle is 10.5 square feet. Find the shorter of the perpendicular sides.

140. A rectangular shed has a width that is 5 meters less than its length. The area of the floor of the shed is 50 square meters. Find the length of the shed.

141. If three times the square of a number is added to five times the number, the result is 100. Find the number.

142. If the square of a number is added to 8, the result is six times the number. Find the number.

143. Two numbers have a sum of -6 and a product of -27. Find the numbers.

144. A rectangular parking lot has a length 20 feet more than its width. Its area is 2400 square feet. What are the dimensions of the lot?

MIXED REVIEW EXERCISES

Perform the indicated operations, then simplify. Write answers with only positive exponents. Assume all variables represent nonzero real numbers.

145. $(4x + 1)(2x - 3)$

146. $\dfrac{6^{-1}y^3(y^2)^{-2}}{6y^{-4}(y^{-1})}$

147. 5^{-3}

148. $(y^6)^{-5}(2y^{-3})^{-4}$

149. $(-5 + 11w) + (6 + 5w) + (-15 - 8w^2)$

150. $7p^5(3p^4 + p^3 + 2p^2)$

151. $\dfrac{(-z^{-2})^3}{5(z^{-3})^{-1}}$

152. $-(-3)^2$

153. $\dfrac{(5z^2x^3)^2(2zx^2)^{-1}}{(-10zx^{-3})^{-2}(3z^{-1}x^{-4})^2}$

154. $(2k - 1) - (3k^2 - 2k + 6)$

Factor completely.

155. $30a + am - am^2$

156. $11k + 12k^2$

157. $8 - a^3$

158. $9x^2 + 13xy - 3y^2$

159. $15y^3 + 20y^2$

160. $25z^2 - 30zm + 9m^2$

Solve.

161. $5x^2 - 17x - 12 = 0$

162. $x^3 - x = 0$

CHAPTER 3 TEST

Evaluate.

1. $\left(\frac{3}{2}\right)^3$

2. $-(-2)^{-6}$

3. $-4^2 + (-4)^2$

Simplify. Write answers with only positive exponents.

4. $8^{-2} \cdot 8^5 \cdot 8^{-6}$

5. $\frac{y^{3r} \cdot y^{2r}}{y^{2-r}}$ $(y \neq 0,\ r \text{ an integer})$

6. $(-8)^0$

7. $\frac{(2y)^3 y^5}{y^{-2} y^{-6}}$ $(y \neq 0)$

8. $(3a^{-2}b^{-2})^{-3} \cdot (2^{-2}ab^{-3})^2$ $(a \neq 0,\ b \neq 0)$

9. $\frac{3^{-2}r^{-3}(r^{-2})^{-1}}{5(r^2)^{-3}}$ $(r \neq 0)$

10. Write 4.6×10^{-5} without scientific notation.

11. Use scientific notation to evaluate $\frac{(2{,}500{,}000)(.00003)}{(.05)(5{,}000{,}000)}$.

Perform the indicated operations in Exercises 12–16. for this week

12. $(5m^2 - 9m + 6) + (-8m^2 + 4m - 9)$

13. $(3k^3 - 5k^2 + 8k - 2) - (4k^3 + 11k + 7)$

14. $-2r - [(r - 4) - (5 - 3r)] + (9 - 4r)$

15. $(k + 4)(3k^2 + 8k - 9)$

16. $[2y + (3z - x)][2y - (3z - x)]$

17. Let $P(x) = -x^2 - 12x + 6$; find $P(-4)$.

Factor completely.

18. $ax + ay + bx + by$

19. $3x^2 + 6xy^2 + 24xy$

20. $k^2 + 10k + 24$

21. $2p^2 - 5pq + 3q^2$

22. $x^2 + 100$

23. $x^4 - 9x^2 + 20$

24. $100d^2 - 9b^6$

25. $27y^3 - 125z^6$

26. Which one of the following is not a factored form of $-x^2 - x + 12$?
(a) $(3 - x)(x + 4)$ **(b)** $-(x - 3)(x + 4)$
(c) $(-x + 3)(x + 4)$ **(d)** $(x - 3)(-x + 4)$

Solve each equation.

27. $3a^2 + 13a = 30$

28. $8r^2 = 14r - 3$

29. $4x^3 - 11x^2 - 3x = 0$

30. The length of a rectangle is 2 meters more than the width. The area of the rectangle is 63 square meters. Find the length and width of the rectangle.

CHAPTER FOUR

RATIONAL EXPRESSIONS

An oil refinery can produce gasoline, heating oil, or a combination of the two. At one refinery, the amount of heating oil produced, in hundreds of gallons per day, is given by

$$\frac{125{,}000 - 25x}{125 + 2x},$$

where x is the amount of gasoline produced in hundreds of gallons per day.

Many other useful formulas also involve the ratio of polynomials, called *rational expressions*. (Note the word "ratio" in the name.) Rational expressions play the same role in algebra as rational numbers do in arithmetic. The techniques of factoring from Chapter 3 will be very important in our work with rational expressions.

4.1 BASICS OF RATIONAL EXPRESSIONS

OBJECTIVES

1. DEFINE RATIONAL EXPRESSIONS.
2. FIND THE NUMBERS THAT MAKE A RATIONAL EXPRESSION UNDEFINED.
3. WRITE RATIONAL EXPRESSIONS IN LOWEST TERMS.
4. WRITE A RATIONAL EXPRESSION WITH A SPECIFIED DENOMINATOR.

1 In arithmetic, a rational number is the quotient of two integers, with the denominator not 0. In algebra this idea is generalized: A **rational expression** or *algebraic fraction* is the quotient of two polynomials, again with the denominator not 0. For example,

$$\frac{m+4}{m-2}, \qquad \frac{8x^2 - 2x + 5}{4x^2 + 5x}, \qquad \text{and} \qquad x^5\left(\text{or } \frac{x^5}{1}\right)$$

are all rational expressions. In other words, rational expressions are the elements of the set

$$\left\{\frac{P}{Q}\,\middle|\, P, Q \text{ polynomials, with } Q \neq 0\right\}.$$

2 Any number can be used as a replacement for the variable in a rational expression, except for values that make the denominator 0. With a zero denominator, the rational expression is undefined. For example, the number 5 cannot be used as a replacement for the x in

$$\frac{2}{x-5}$$

since 5 would make the denominator equal 0.

EXAMPLE 1
FINDING NUMBERS THAT MAKE RATIONAL EXPRESSIONS UNDEFINED

Find all numbers that make the rational expression undefined.

(a) $\dfrac{3}{7k - 14}$

The only values that cannot be used are those that make the denominator 0. Find these values by setting the denominator equal to 0 and solving the resulting equation.

$$7k - 14 = 0$$

$$7k = 14 \qquad \text{Add 14.}$$

$$k = 2 \qquad \text{Divide by 7.}$$

The number 2 cannot be used as a replacement for k; 2 makes the rational expression undefined.

(b) $\dfrac{3 + p}{p^2 - 4p + 3}$

Set the denominator equal to 0.

$$p^2 - 4p + 3 = 0$$

Factor to get

$$(p - 3)(p - 1) = 0.$$

Set each factor equal to 0.

$$p - 3 = 0 \quad \text{or} \quad p - 1 = 0$$

$$p = 3 \quad \text{or} \quad p = 1$$

Both 3 and 1 make the rational expression undefined.

(c) $\dfrac{8x + 2}{3}$

The denominator can never be 0, so any real number can replace x in the rational expression. ■

3 In arithmetic, the fraction 15/20 is written in lowest terms by dividing the numerator and denominator by 5, to get 3/4. In a similar manner, the **fundamental principle of rational numbers** is used to write rational expressions in lowest terms.

FUNDAMENTAL PRINCIPLE OF RATIONAL NUMBERS

If a/b is a rational number and if c is any nonzero real number, then

$$\frac{a}{b} = \frac{ac}{bc}.$$

(The numerator and denominator of a rational number may either be multiplied or divided by the same nonzero number without changing the value of the rational number.)

Since a rational expression is a quotient of two polynomials, and since the value of a polynomial is a real number for all values of its variables, any statement that applies to rational numbers will also apply to rational expressions. In particular, the fundamental principle may be used to write rational expressions in lowest terms. Do this as follows.

WRITING IN LOWEST TERMS

Step 1 **Factor.**
Factor both numerator and denominator to find their greatest common factor.

Step 2 **Reduce.**
Divide both numerator and denominator by their greatest common factor.

EXAMPLE 2
WRITING RATIONAL EXPRESSIONS IN LOWEST TERMS

Write each rational expression in lowest terms.

(a) $\frac{8k}{16} = \frac{k \cdot 8}{2 \cdot 8} = \frac{k}{2}$

Here we divided the numerator and denominator by 8, the greatest common factor of $8k$ and 16.

(b) $\frac{12x^3y^2}{6x^4y} = \frac{2y \cdot 6x^3y}{x \cdot 6x^3y} = \frac{2y}{x}$

Divide by the greatest common factor, $6x^3y$. (Here 3 is the least exponent on x, and 1 the least exponent on y.)

(c) $\frac{a^2 - a - 6}{a^2 + 5a + 6}$

Start by factoring the numerator and denominator.

$$\frac{a^2 - a - 6}{a^2 + 5a + 6} = \frac{(a - 3)(a + 2)}{(a + 3)(a + 2)}$$

Divide the numerator and denominator by $a + 2$ to get

$$\frac{a^2 - a - 6}{a^2 + 5a + 6} = \frac{a - 3}{a + 3}. \quad \blacksquare$$

CAUTION One of the most common errors in algebra involves incorrect use of the fundamental principle of rational numbers. Only common *factors* may be divided. For example, it would be incorrect to try to "divide" the a^2 *terms* in the numerator and denominator in Example 2(c). It is essential to *factor* before writing in lowest terms.

In the rational expression

$$\frac{a^2 - a - 6}{a^2 + 5a + 6}, \quad \text{or} \quad \frac{(a - 3)(a + 2)}{(a + 3)(a + 2)},$$

a can take any value except -3 or -2. In the rational expression

$$\frac{a-3}{a+3}$$

a cannot equal -3. Because of this,

$$\frac{a^2-a-6}{a^2+5a+6}=\frac{a-3}{a+3}$$

for all values of a except -3 or -2. From now on such statements of equality will be made with the understanding that they apply only for those real numbers that make neither denominator equal 0, and we will no longer state these restrictions.

EXAMPLE 3
WRITING RATIONAL EXPRESSIONS IN LOWEST TERMS

Write each rational expression in lowest terms.

(a) $$\frac{y^2-4}{2y+4}=\frac{(y+2)(y-2)}{2(y+2)}=\frac{y-2}{2}$$

CAUTION Remember that only common *factors* may be divided:

$$\frac{y-2}{2}\neq y \quad \text{or} \quad y-1.$$

The 2 in $y-2$ is not a *factor* of the numerator.

(b) $$\frac{m-3}{3-m}$$

In this rational expression, the numerator and denominator are exactly opposite. The given expression can be written in lowest terms by writing the denominator as $3-m=-1(m-3)$, giving

$$\frac{m-3}{3-m}=\frac{m-3}{-1(m-3)}=\frac{1}{-1}=-1.$$

(c) $$\frac{r^2-16}{4-r}=\frac{(r+4)(r-4)}{4-r}$$

$$=\frac{(r+4)(r-4)}{-1(r-4)} \qquad \text{Write } 4-r \text{ as } -1(r-4).$$

$$=\frac{r+4}{-1} \qquad \text{Lowest terms}$$

$$=-(r+4) \quad \text{or} \quad -r-4 \quad \blacksquare$$

Working as in Examples 3(b) and (c), the quotient

$$\frac{a}{-a} \quad (a\neq 0)$$

can be simplified as $$\frac{a}{-a}=\frac{a}{-1(a)}=\frac{1}{-1}=-1.$$

The following generalization applies.

The quotient of two quantities that differ only in sign is -1.

Based on this result,

$$\frac{q-7}{7-q} = -1, \qquad \frac{-5a+2b}{5a-2b} = -1,$$

but $\dfrac{r-2}{r+2}$ cannot be simplified further.

4 Just as with fractions, when rational expressions are added and subtracted it is often necessary to first rewrite them with a common denominator.

EXAMPLE 4 **WRITING RATIONAL EXPRESSIONS WITH AN INDICATED DENOMINATOR**

Write each rational expression with the indicated denominator.

(a) $\dfrac{12m}{5s}$, denominator of $20s^3m$

The original denominator of $5s$ is to be written as $20s^3m$. What can we multiply by to make the denominator $20s^3m$? That is, $5s \cdot (?) = 20s^3m$. To decide, divide $20s^3m$ by $5s$ to get $4s^2m$. Now we can use the fundamental principle and multiply the numerator and denominator by $4s^2m$.

$$\frac{12m}{5s} = \frac{12m \cdot 4s^2m}{5s \cdot 4s^2m} = \frac{48s^2m^2}{20s^3m}$$

(b) $\dfrac{m-7}{4}$, denominator of $12(m+5)$

Since $\dfrac{12(m+5)}{4} = 3(m+5)$, multiply the numerator and denominator by $3(m+5)$.

$$\frac{m-7}{4} = \frac{(m-7) \cdot 3(m+5)}{4 \cdot 3(m+5)}$$

$$= \frac{3(m-7)(m+5)}{12(m+5)} \qquad \text{Associative and commutative properties}$$

$$= \frac{3(m^2-2m-35)}{12(m+5)} \qquad \text{Multiply in numerator.}$$

$$= \frac{3m^2-6m-105}{12(m+5)} \qquad \text{Distributive property}$$

It is often useful to multiply and combine terms in the numerator, but to leave the denominator in factored form as was done here.

(c) $\dfrac{9x + y}{3x + 2y}$, denominator of $9x^2 - 4y^2$

Factor $9x^2 - 4y^2$ as $(3x + 2y)(3x - 2y)$. Then multiply the numerator and denominator of the fraction by $3x - 2y$.

$$\frac{9x + y}{3x + 2y} = \frac{(9x + y)(3x - 2y)}{(3x + 2y)(3x - 2y)}$$

$$= \frac{27x^2 - 15xy - 2y^2}{(3x + 2y)(3x - 2y)} = \frac{27x^2 - 15xy - 2y^2}{9x^2 - 4y^2} \quad \blacksquare$$

4.1 EXERCISES

Find all numbers that make each rational expression undefined. See Example 1.

1. $\dfrac{m}{m - 4}$ **2.** $\dfrac{k}{k + 9}$ **3.** $\dfrac{6 + z}{z}$

4. $\dfrac{2 - b}{b}$ **5.** $\dfrac{5p + 9}{4}$ **6.** $\dfrac{2z - 6}{5}$

7. $\dfrac{m + 5}{m^2 - m - 12}$ **8.** $\dfrac{r + 2}{r^2 + 8r + 15}$ **9.** $\dfrac{z^2 - z}{z^2 - 16}$

10. $\dfrac{y^2 + 2y - 1}{y^2 - 25}$ **11.** $\dfrac{x^2 + 5}{2x^2 + 3x - 2}$ **12.** $\dfrac{a^2 + 3a - 1}{3a^2 - 10a - 8}$

13. $\dfrac{8z - 2}{z^2 + 16}$ **14.** $\dfrac{-4p + 11}{p^2 + 25}$

Write each rational expression in lowest terms. See Examples 2 and 3.

15. $\dfrac{6m^5}{14}$ **16.** $\dfrac{25y^6}{30}$ **17.** $\dfrac{18m^5n^3}{12m^4n^5}$ **18.** $\dfrac{-8y^6z^5}{6y^7z^3}$

19. $\dfrac{2x - 2}{3x - 3}$ **20.** $\dfrac{5m + 10}{6m + 12}$ **21.** $\dfrac{4z^2 + 2z}{8z + 4}$ **22.** $\dfrac{2y - y^2}{4y - y^2}$

23. $\dfrac{a^2 - 4}{a^2 + 4a + 4}$ **24.** $\dfrac{4p^2 - 20p}{p^2 - 4p - 5}$ **25.** $\dfrac{r^2 - r - 20}{r^2 + r - 30}$

26. $\dfrac{y^2 - 2y - 15}{y^2 + 7y + 12}$ **27.** $\dfrac{2c^3 + 2c^2d - 60cd^2}{5c^3 - 30c^2d + 25cd^2}$ **28.** $\dfrac{3s^2t - 9st^2 - 54t^3}{2s^2t - 4st^2 - 48t^3}$

29. $\dfrac{a^3 - b^3}{a - b}$ **30.** $\dfrac{m^3 + n^3}{m + n}$ **31.** $\dfrac{z^3 + x^3}{z^2 - x^2}$

32. $\dfrac{p^3 - q^3}{p^2 - q^2}$ **33.** $\dfrac{xy - yw + xz - zw}{xy + yw + xz + zw}$ **34.** $\dfrac{ac + ad - bc - bd}{ab + ac - b^2 - bc}$

35. $\dfrac{6 - m}{m - 6}$

36. $\dfrac{x - 2}{2 - x}$

37. $\dfrac{x - y}{y^2 - x^2}$

38. $\dfrac{b - a}{a^2 - b^2}$

39. $\dfrac{(3 - m)(4 - n)}{(m - 3)(4 + n)}$

40. $\dfrac{(y - 4)(4 - y)}{(4 - y)(4 + y)}$

41. $\dfrac{x^2 - x}{1 - x}$

42. $\dfrac{r^2 - 16r}{16 - r}$

---

43. **(a)** Identify the two terms in both the numerator and the denominator of $\dfrac{x^2 + 4x}{x + 4}$. **(b)** Identify the two factors in both the numerator and the denominator of $\dfrac{x(x + 2)}{y(x + 2)}$.

44. Which of the following rational expressions equals -1? (See Objective 3.)

(a) $\dfrac{2x + 3}{2x - 3}$ **(b)** $\dfrac{2x - 3}{3 - 2x}$

(c) $\dfrac{2x + 3}{3 + 2x}$ **(d)** $\dfrac{2x + 3}{-2x - 3}$

Rewrite each rational expression with the indicated denominator. See Example 4.

45. $\dfrac{3}{k}$; $2k^2$

46. $\dfrac{9}{y}$; $4y^3$

47. $\dfrac{6}{z - 2}$; $5z - 10$

48. $\dfrac{4}{p + 2}$; $3p + 6$

49. $\dfrac{6a}{a + 1}$; $a^2 - 1$

50. $\dfrac{2k}{k - 4}$; $k^2 - 16$

51. $\dfrac{2x}{x - 4}$; $4 - x$

52. $\dfrac{8b}{b - 7}$; $7 - b$

53. $\dfrac{m - 1}{4}$; $16(m + 2)$

54. $\dfrac{p + 7}{3}$; $21(p - 3)$

55. $\dfrac{2z}{z - 6}$; $2z^2 - 10z - 12$

56. $\dfrac{3a}{a - 5}$; $4a^2 - 8a - 60$

57. $\dfrac{3m}{m - y}$; $y^2 - m^2$

58. $\dfrac{2k}{k - 3p}$; $9p^2 - k^2$

59. $\dfrac{4}{m + 3p}$; $m^3 + 27p^3$

60. $\dfrac{3z}{z + 2y}$; $z^3 + 8y^3$

61. $\dfrac{2r - 3s}{a + 2b}$; $3ar + as + 6br + 2bs$

62. $\dfrac{5m + 3p}{2x - 5w}$; $8mx - 20wm - 2xp + 5wp$

Show that each of the following statements is false by replacing x in each case with 2. Explain why each statement is false. (See Objective 3.)

63. $\dfrac{x + 5}{5} = x$

64. $\dfrac{x + 5}{5} = x + 1$

65. $\dfrac{2x + 3}{2x + 5} = \dfrac{3}{5}$

66. $\dfrac{x^2 + x}{x^2 - x} = -1$

67. Discuss two ways to use the fundamental principle of rational numbers. (See Objectives 3 and 4.)

Write in lowest terms.

68. $\dfrac{z^{3d} + 5z^d}{2z^{4d} + z^{3d}}$ **69.** $\dfrac{m^{3r} - m^{2r}}{3m^{2r} + 3m^r}$ **70.** $\dfrac{k^{2x} - 2k^x - 3}{k^{2x} - 4k^x + 3}$

71. $\dfrac{p^{2y} + 3p^y - 10}{p^{2y} - 4}$ **72.** $\dfrac{a^{2p} - 1}{1 - a^{2p}}$ **73.** $\dfrac{b^{2n} - 9}{9 - b^{2n}}$

74. $\dfrac{p^{-3} - 2p^{-2}}{p - 2p^2}$ **75.** $\dfrac{3m^{-1} + 2m^{-2}}{3m^3 + 2m^2}$

PREVIEW EXERCISES

Multiply or divide as indicated. See Section 1.3.

76. $\dfrac{2}{3} \cdot \dfrac{9}{7}$ **77.** $\dfrac{3}{4} \cdot \dfrac{5}{12}$ **78.** $-\dfrac{3}{8} \cdot \left(-\dfrac{16}{9}\right)$ **79.** $-\dfrac{15}{11} \cdot \left(-\dfrac{22}{5}\right)$

80. $\dfrac{5}{12} \div \dfrac{7}{18}$ **81.** $\dfrac{7}{8} \div \dfrac{3}{16}$ **82.** $-3 \div \left(-\dfrac{6}{5}\right)$ **83.** $-10 \div \left(-\dfrac{5}{4}\right)$

4.2 MULTIPLICATION AND DIVISION OF RATIONAL EXPRESSIONS

OBJECTIVES

1 MULTIPLY RATIONAL EXPRESSIONS.

2 FIND RECIPROCALS FOR RATIONAL EXPRESSIONS.

3 DIVIDE RATIONAL EXPRESSIONS.

1 Multiplication of rational expressions follows the same procedure as multiplication of rational numbers.

MULTIPLICATION OF RATIONAL NUMBERS

If a, b, c, and d are integers, with $b \neq 0$ and $d \neq 0$, then

$$\frac{a}{b} \cdot \frac{c}{d} = \frac{ac}{bd}.$$

The same idea is used to multiply two rational expressions.

MULTIPLYING RATIONAL EXPRESSIONS

Step 1 **Factor.**
Factor all numerators and denominators as completely as possible.

Step 2 **Reduce.**
Divide both numerator and denominator by any factor that both have in common.

Step 3 **Multiply.**
Multiply remaining factors in the numerator and remaining factors in the denominator.

NOTE After completing these three steps, the product should be in lowest terms. It is a good idea to check to be sure this is the case.

EXAMPLE 1 MULTIPLYING RATIONAL EXPRESSIONS

Multiply.

(a) $\frac{3x^2}{5} \cdot \frac{10}{x^3} = \frac{3x^2 \cdot 10}{5 \cdot x^3} = \frac{30x^2}{5x^3} = \frac{6 \cdot 5x^2}{x \cdot 5x^2} = \frac{6}{x}$

Divide the numerator and denominator by $5x^2$ to write the product in lowest terms. Notice that common factors in the numerator and denominator can be eliminated *before* multiplying the numerator factors and the denominator factors as follows.

$$\frac{3x^2}{5} \cdot \frac{10}{x^3} = \frac{3x^2}{5} \cdot \frac{2 \cdot 5}{x \cdot x^2} = \frac{6}{x}$$

This is the most efficient way to multiply fractions.

(b) $\frac{5p-5}{p} \cdot \frac{3p^2}{10p-10}$

Factor where possible.

$$\frac{5p-5}{p} \cdot \frac{3p^2}{10p-10} = \frac{5(p-1)}{p} \cdot \frac{3p \cdot p}{2 \cdot 5(p-1)} \qquad \text{Factor.}$$

$$= \frac{1}{1} \cdot \frac{3p}{2} \qquad \text{Divide out common factors.}$$

$$= \frac{3p}{2} \qquad \text{Multiply.}$$

(c) $\frac{k^2+2k-15}{k^2-4k+3} \cdot \frac{k^2-k}{k^2+k-20} = \frac{(k+5)(k-3)}{(k-3)(k-1)} \cdot \frac{k(k-1)}{(k+5)(k-4)}$

$$= \frac{k}{k-4}$$

(d) $(p-4) \cdot \frac{3}{5p-20}$

$$(p-4) \cdot \frac{3}{5p-20} = \frac{p-4}{1} \cdot \frac{3}{5p-20} \qquad \text{Write } p-4 \text{ as } \frac{p-4}{1}.$$

$$= \frac{p-4}{1} \cdot \frac{3}{5(p-4)} \qquad \text{Factor.}$$

$$= \frac{3}{5}$$

(e) $\dfrac{a^2b^3c^4}{(ab^2)^2c} \cdot \dfrac{(a^2b)^3c^2}{(abc^3)^2}$

Clear parentheses by using the power rules for exponents.

$$\frac{a^2b^3c^4}{(ab^2)^2c} \cdot \frac{(a^2b)^3c^2}{(abc^3)^2} = \frac{a^2b^3c^4}{a^2b^4c} \cdot \frac{a^6b^3c^2}{a^2b^2c^6}$$

Use the definition of multiplication and then the quotient rule for exponents.

$$= \frac{a^8b^6c^6}{a^4b^6c^7} \qquad \text{Product rule for exponents}$$

$$= a^{8-4}b^{6-6}c^{6-7} \qquad \text{Quotient rule for exponents}$$

$$= a^4b^0c^{-1}$$

$$= \frac{a^4}{c} \qquad b^0 = 1;\ c^{-1} = \frac{1}{c}$$

(f) $\dfrac{x^2 + 2x}{x + 1} \cdot \dfrac{x^2 - 1}{x^3 + x^2}$

$$\frac{x^2 + 2x}{x + 1} \cdot \frac{x^2 - 1}{x^3 + x^2} = \frac{x(x + 2)}{x + 1} \cdot \frac{(x + 1)(x - 1)}{x^2(x + 1)} \qquad \text{Factor where possible.}$$

$$= \frac{x(x + 2)(x + 1)(x - 1)}{x^2(x + 1)(x + 1)} \qquad \text{Multiply.}$$

Next, use the fundamental principle to write in lowest terms.

$$\frac{x^2 + 2x}{x + 1} \cdot \frac{x^2 - 1}{x^3 + x^2} = \frac{(x + 2)(x - 1)}{x(x + 1)} \quad ■$$

2 Recall that rational numbers a/b and c/d are reciprocals of each other if the numbers have a product of 1. The **reciprocal** of a rational expression can be defined in the same way: Two rational expressions are reciprocals of each other if they have a product of 1. Recall that 0 has no reciprocal.

EXAMPLE 2
FINDING THE RECIPROCAL OF A RATIONAL EXPRESSION

The following list shows several rational expressions and the reciprocals of each. In each case, check that the product of the rational expression and its reciprocal is 1.

Rational expression	*Reciprocal*
$\dfrac{5}{k}$	$\dfrac{k}{5}$
$\dfrac{m^2 - 9m}{2}$	$\dfrac{2}{m^2 - 9m}$
$\dfrac{0}{4}$	undefined ■

This example suggests the following procedure.

RECIPROCAL To find the reciprocal of a nonzero rational expression, invert the rational expression.

3 Division of rational expressions follows the rule for division of rational numbers.

DIVISION OF RATIONAL NUMBERS For rational numbers a/b and c/d, with $c/d \neq 0$,

$$\frac{a}{b} \div \frac{c}{d} = \frac{a}{b} \cdot \frac{d}{c}.$$

This result leads to the following rule.

DIVIDING RATIONAL EXPRESSIONS To divide two rational expressions, *multiply* the first by the reciprocal of the second.

EXAMPLE 3 DIVIDING RATIONAL EXPRESSIONS

Divide.

(a) $\dfrac{2z}{9} \div \dfrac{5z^2}{18} = \dfrac{2z}{9} \cdot \dfrac{18}{5z^2}$ Multiply by the reciprocal of the divisor.

$$= \frac{4}{5z}$$

(b) $\dfrac{8k-16}{3k} \div \dfrac{3k-6}{4k^2} = \dfrac{8(k-2)}{3k} \div \dfrac{3(k-2)}{4k^2}$ Factor.

$$= \frac{8(k-2)}{3k} \cdot \frac{4k^2}{3(k-2)}$$ Multiply by the reciprocal.

$$= \frac{32k}{9}$$

(c) $\dfrac{5m^2+17m-12}{3m^2+7m-20} \div \dfrac{5m^2+2m-3}{15m^2-34m+15}$

$$= \frac{(5m-3)(m+4)}{(m+4)(3m-5)} \div \frac{(5m-3)(m+1)}{(3m-5)(5m-3)}$$

$$= \frac{(5m-3)(m+4)}{(m+4)(3m-5)} \cdot \frac{(3m-5)(5m-3)}{(5m-3)(m+1)}$$

$$= \frac{5m-3}{m+1}$$ ■

EXAMPLE 4 DIVIDING RATIONAL EXPRESSIONS

Divide $\frac{m^2pq^3}{mp^4}$ by $\frac{m^5p^2q}{mpq^2}$.

Use the definitions of division and multiplication and the properties of exponents.

$$\frac{m^2pq^3}{mp^4} \div \frac{m^5p^2q}{mpq^2} = \frac{m^2pq^3}{mp^4} \cdot \frac{mpq^2}{m^5p^2q}$$

$$= \frac{m^3p^2q^5}{m^6p^6q}$$

$$= \frac{q^4}{m^3p^4} \quad \blacksquare$$

4.2 EXERCISES

Multiply or divide as indicated. Write all answers in lowest terms. See Examples 1–4.

1. $\frac{m^3}{2} \cdot \frac{4m}{m^4}$

2. $\frac{3y^4}{5y} \cdot \frac{8y^2}{9}$

3. $\frac{6a^4}{a^3} \div \frac{12a^2}{a^5}$

4. $\frac{11p^3}{p^2} \div \frac{22p^4}{p}$

5. $\frac{6y^5x^6}{y^3x^4} \cdot \frac{y^4x^2}{3y^5x^7}$

6. $\frac{6p^2q^3}{4p^3q} \cdot \frac{18p^2q^3}{12p^3q^2}$

7. $\frac{25a^2b}{60a^3b^2} \div \frac{5a^4b^2}{16a^2b}$

8. $\frac{8s^4t^2}{5t^6} \div \frac{s^3t^2}{10s^2t^4}$

9. $\frac{(-3mn)^2 \cdot (4m^2n)^3}{16m^2n^4(mn^2)^3} \div \frac{24(m^2n^2)^4}{(3m^2n^3)^2}$

10. $\frac{(-4a^2b^3)^2 \cdot (3a^2b^4)^2}{(2a^2b^3)^4 \cdot (3a^3b)^2} \div \frac{(ab)^4}{(a^2b^3)^2}$

11. $\frac{(5pq^2)^2}{60p^3q^6} \div \frac{5p^2q^2}{16p^2q^3}$

12. $\frac{(6s^2t)^2}{(5^4t^5)^3} \div \frac{s^3t^2x^2}{3x^5t}$

13. $\frac{2r}{8r+4} \cdot \frac{14r+7}{3}$

14. $\frac{6a-10}{5a} \cdot \frac{3}{9a-15}$

15. $(7k+7) \div \frac{4k+4}{5}$

16. $(8y-16) \div \frac{3y-6}{10}$

17. $(z^2-1) \cdot \frac{1}{1-z}$

18. $(y^2-4) \div \frac{2-y}{8y}$

19. $\frac{p^2-36}{p+1} \div \frac{6-p}{p}$

20. $\frac{m^2-16}{5m} \cdot \frac{2}{4-m}$

21. $\frac{6r-5s}{3r+2s} \cdot \frac{6r+4s}{5s-6r}$

22. $\frac{9y-12x}{y+x} \div \frac{4x-3y}{x+y}$

23. $\frac{m^2+6m+9}{m^2-9} \cdot \frac{m^2-6m+9}{m+3}$

24. $\frac{k^2-4k+4}{k^2-4} \div \frac{k^2+4k+4}{k+2}$

25. $\frac{2r^2+5r-3}{r^2-1} \cdot \frac{r^2-2r+1}{2r^2-7r+3}$

26. $\frac{2p^2+9p+10}{p^2+5p+6} \cdot \frac{p^2+7p+12}{2p^2+3p-5}$

27. $\dfrac{6a^2 + a - 1}{6a^2 + 5a + 1} \div \dfrac{3a^2 + 2a - 1}{3a^2 + 4a + 1}$

28. $\dfrac{2n^2 + 5mn + 2m^2}{4n^2 - m^2} \cdot \dfrac{2n^2 + mn - m^2}{n^2 + nm - 2m^2}$

29. $\dfrac{15x^2 - xy - 2y^2}{15x^2 + 11xy + 2y^2} \div \dfrac{15x^2 + 4xy - 4y^2}{15x^2 + xy - 2y^2}$

30. $\dfrac{18w^2 + 3wx - 10x^2}{6w^2 + 11wx - 10x^2} \cdot \dfrac{6w^2 + 19wx + 10x^2}{3w^2 + 11wx + 6x^2}$

31. $\dfrac{6k^2 + kr - 2r^2}{6k^2 - 5kr + r^2} \div \dfrac{3k^2 + 17kr + 10r^2}{6k^2 + 13kr - 5r^2}$

32. $\dfrac{16m^2 - 9n^2}{16m^2 - 24mn + 9n^2} \div \dfrac{16m^2 + 24mn + 9n^2}{16m^2 - 16mn + 3n^2}$

33. $\dfrac{am - an + bm - bn}{am + an - bm - bn} \div \dfrac{am - an - 3bm + 3bn}{am + an - 3bm - 3bn}$

34. $\dfrac{12 + 4y - 3x - xy}{24 + 8y - 3x - xy} \cdot \dfrac{32 + 8y - 4x - xy}{16 - 4y - 4x + xy}$

35. $\dfrac{m^3 + m + m^2 + 1}{m^3 + m^2 + mn^2 + n^2} \div \dfrac{m^3 + m + m^2n + n}{2m^2 + 2mn - mn^2 - n^3}$

36. $\dfrac{a^4 - a^3 + a^2 - a}{2a^3 + 2a^2 + a + 1} \cdot \dfrac{2a^3 - 8a^2 + a - 4}{a^3 - 4a^2 + a - 4}$

37. $\dfrac{x^2 + 2x + 1}{x - 2} \cdot \dfrac{x^2 - 5x + 6}{x + 1} \cdot \dfrac{5x - 20}{x - 3}$

38. $\dfrac{2z^2 - z - 15}{z^2 - z - 20} \cdot \dfrac{z - 5}{2z + 5} \cdot \dfrac{z^2 + 12z + 35}{z^2 + 4z - 21}$

39. $\dfrac{p^2 - q^2}{(p - q)^2} \div \dfrac{p^2 - 2pq + q^2}{p^2 - pq + q^2} \cdot \dfrac{(p - q)^4}{p^3 + q^3}$

40. $\dfrac{(g + h)^2}{g - h} \cdot \dfrac{g^3 - h^3}{g^2 - h^2} \div \dfrac{g^2 + gh + h^2}{(g - h)^2}$

41. $\dfrac{x^3 - y^3}{(x + y)(x^2 - xy + y^2)} \cdot \dfrac{x^3 + y^3}{(x - y)(x^2 + xy + y^2)}$

42. $\dfrac{(3p - q)(9p^2 + 3pq + q^2)}{27p^3 - q^3} \cdot \dfrac{(3p + q)(9p^2 - 3pq + q^2)}{27p^3 + q^3}$

43. $\dfrac{x^2 - 4x + 4 - y^2}{x^3y - 2x^2y - x^2y^2} \cdot \dfrac{x^4y}{x - 2 - y}$

44. $\dfrac{9 + 6m + m^2 - n^2}{3m^2n^2 + m^3n^2 + m^2n^3} \div \dfrac{3 + m - n}{m^4n}$

45. $\dfrac{a^2(2a + b) + 6a(2a + b) + 5(2a + b)}{3a^2(2a + b) - 2a(2a + b) - (2a + b)} \div \dfrac{a + 1}{a - 1}$

46. $\dfrac{2x^2(x - 3z) - 5x(x - 3z) + 2(x - 3z)}{4x^2(x - 3z) - 11x(x - 3z) + 6(x - 3z)} \div \dfrac{4x - 3}{4x + 1}$

---

47. In your own words, explain how to multiply rational expressions. (See Objective 1.)

48. In your own words, explain how to divide rational expressions. (See Objective 3.)

PREVIEW EXERCISES

Add or subtract as indicated. See Section 1.3.

49. $\dfrac{3}{8} + \dfrac{1}{8}$

50. $\dfrac{1}{6} + \dfrac{1}{6}$

51. $\dfrac{1}{4} + \dfrac{1}{2}$

52. $\dfrac{1}{9} + \dfrac{2}{3}$

53. $\dfrac{5}{12} - \dfrac{3}{16}$

54. $\dfrac{7}{18} - \dfrac{1}{24}$

55. $\dfrac{1}{10} - \dfrac{3}{8}$

56. $\dfrac{2}{15} - \dfrac{17}{25}$

4.3 ADDITION AND SUBTRACTION OF RATIONAL EXPRESSIONS

OBJECTIVES

1. ADD AND SUBTRACT RATIONAL EXPRESSIONS WITH THE SAME DENOMINATOR.
2. FIND A LEAST COMMON DENOMINATOR.
3. ADD AND SUBTRACT RATIONAL EXPRESSIONS WITH DIFFERENT DENOMINATORS.

1 The sum or difference of two rational expressions is found by following the definitions of addition and subtraction of rational numbers.

ADDITION AND SUBTRACTION OF RATIONAL NUMBERS

If a/b and c/b are rational numbers, then

$$\frac{a}{b} + \frac{c}{b} = \frac{a+c}{b}$$

and

$$\frac{a}{b} - \frac{c}{b} = \frac{a-c}{b}.$$

The following steps are used to add or subtract rational expressions.

ADDING OR SUBTRACTING RATIONAL EXPRESSIONS

Step 1(a) If the denominators are the same, add or subtract the numerators. Place the result over the common denominator.

Step 1(b) If the denominators are different, first find the least common denominator. Write all rational expressions with this least common denominator, and then add or subtract the numerators. Place the result over the common denominator.

Step 2 Write all answers in lowest terms.

EXAMPLE 1 ADDING AND SUBTRACTING RATIONAL EXPRESSIONS WITH THE SAME DENOMINATORS

Add or subtract as indicated.

(a) $\frac{3y}{5} + \frac{x}{5} = \frac{3y+x}{5}$

Since the denominators of these rational expressions are the same, just add the numerators.

(b) $\frac{5}{m} - \frac{7}{m} = \frac{5-7}{m} = -\frac{2}{m}$

Subtract the numerators since the denominators are the same.

(c) $\frac{7}{2r^2} - \frac{11}{2r^2} = \frac{7-11}{2r^2} = \frac{-4}{2r^2} = -\frac{2}{r^2}$ Lowest terms

(d) $\dfrac{m}{m^2 - p^2} + \dfrac{p}{m^2 - p^2} = \dfrac{m + p}{m^2 - p^2}$ Add numerators.

$= \dfrac{m + p}{(m + p)(m - p)}$ Factor.

$= \dfrac{1}{m - p}$ Lowest terms.

(e) $\dfrac{4}{x^2 + 2x - 8} + \dfrac{x}{x^2 + 2x - 8} = \dfrac{4 + x}{x^2 + 2x - 8}$

$= \dfrac{4 + x}{(x - 2)(x + 4)}$

$= \dfrac{1}{x - 2}$ ■

2 The rational expressions in each part of Example 1 had the same denominators. As mentioned above, if the rational expressions to be added or subtracted have different denominators, it is necessary to first find their **least common denominator,** an expression divisible by the denominator of each of the rational expressions. We find the least common denominator for a group of rational expressions as shown below.

FINDING THE LEAST COMMON DENOMINATOR

Step 1 Factor each denominator.

Step 2 The least common denominator is the product of all different factors from each denominator, with each factor raised to the *greatest* power that occurs in any denominator.

EXAMPLE 2 **FINDING LEAST COMMON DENOMINATORS**

Find the least common denominator for each group of denominators.

(a) $5xy^2$, $2x^3y$

The least common denominator for $5xy^2$ and $2x^3y$ is an expression that is divisible by both $5xy^2$ and $2x^3y$. The least common denominator must contain a factor of $5 \cdot 2 = 10$ to be divisible by both 5 and 2. It also must contain a factor of x^3 (which is divisible by both x and x^3), and of y^2. (Use the *greatest* power.) The least common denominator is

$$5 \cdot 2 \cdot x^3 \cdot y^2 = 10x^3y^2.$$

(b) $k - 3$, k

The least common denominator, an expression divisible by both $k - 3$ and k, is

$$k(k - 3).$$

It is often best to leave a least common denominator in factored form.

(c) $y^2 - 2y - 8, \quad y^2 + 3y + 2$

Factor the denominators to get

$$y^2 - 2y - 8 = (y - 4)(y + 2)$$
$$y^2 + 3y + 2 = (y + 2)(y + 1).$$

The least common denominator, divisible by both polynomials, is

$$(y - 4)(y + 2)(y + 1).$$

(d) $8z - 24, \quad 5z^2 - 15z$

$$8z - 24 = 8(z - 3) \quad \text{and} \quad 5z^2 - 15z = 5z(z - 3) \qquad \text{Factor.}$$

The least common denominator is

$$40z(z - 3). \quad \blacksquare$$

3 As mentioned earlier, if the rational expressions to be added or subtracted have different denominators, it is necessary to rewrite the rational expressions with their least common denominator. This is done by multiplying the numerator and the denominator of each rational expression by the factors required to get the least common denominator. This procedure is valid because each rational expression is being multiplied by a form of 1, the identity element for multiplication. The next example shows how this is done.

EXAMPLE 3 **ADDING AND SUBTRACTING RATIONAL EXPRESSIONS WITH DIFFERENT DENOMINATORS**

Add or subtract as indicated.

(a) $\dfrac{5}{2p} + \dfrac{3}{8p}$

The least common denominator for $2p$ and $8p$ is $8p$. To write the first rational expression with a denominator of $8p$, multiply by 4/4.

$$\frac{5}{2p} + \frac{3}{8p} = \frac{5 \cdot 4}{2p \cdot 4} + \frac{3}{8p} \qquad \text{Fundamental principle}$$
$$= \frac{20}{8p} + \frac{3}{8p}$$
$$= \frac{20 + 3}{8p} \qquad \text{Add numerators.}$$
$$= \frac{23}{8p}$$

(b) $\dfrac{6}{r} - \dfrac{5}{r-3}$

The least common denominator is $r(r - 3)$. Rewrite each rational expression with this denominator.

$$\frac{6}{r} - \frac{5}{r-3} = \frac{6(r-3)}{r(r-3)} - \frac{r \cdot 5}{r(r-3)} \quad \text{Fundamental principle}$$

$$= \frac{6(r-3) - 5r}{r(r-3)} \quad \text{Subtract numerators.}$$

$$= \frac{6r - 18 - 5r}{r(r-3)} \quad \text{Distributive property}$$

$$= \frac{r - 18}{r(r-3)} \quad \text{Combine terms in numerator.}$$

■

CAUTION One of the most common sign errors in algebra occurs when a rational expression with two or more terms in the numerator is being subtracted. Remember that in this situation, the subtraction sign must be distributed to *every* term in the numerator of the fraction that follows it. Read Example 4 carefully to see how this is done.

EXAMPLE 4 SUBTRACTING RATIONAL EXPRESSIONS

Subtract.

(a) $\dfrac{3}{k+1} - \dfrac{k-2}{k+3}$

The least common denominator is $(k + 1)(k + 3)$.

$$\frac{3}{k+1} - \frac{k-2}{k+3} = \frac{3(k+3)}{(k+1)(k+3)} - \frac{(k-2)(k+1)}{(k+3)(k+1)}$$

$$= \frac{3(k+3) - (k-2)(k+1)}{(k+1)(k+3)} \quad \text{Subtract.}$$

$$= \frac{3k + 9 - (k^2 - k - 2)}{(k+1)(k+3)} \quad \text{Multiply in numerator.}$$

$$= \frac{3k + 9 - k^2 + k + 2}{(k+1)(k+3)} \quad \text{Distributive property}$$

$$= \frac{-k^2 + 4k + 11}{(k+1)(k+3)} \quad \text{Combine terms.}$$

(b) $$\frac{1}{q-1} - \frac{1}{q+1} = \frac{1(q+1)}{(q-1)(q+1)} - \frac{1(q-1)}{(q+1)(q-1)} \quad \text{Get a common denominator.}$$

$$= \frac{(q+1) - (q-1)}{(q-1)(q+1)} \quad \text{Subtract.}$$

$$= \frac{q + 1 - q + 1}{(q-1)(q+1)} = \frac{2}{(q-1)(q+1)} \quad \text{Combine terms.}$$

(c) $\dfrac{m+4}{m^2-2m-3} - \dfrac{2m-3}{m^2-5m+6}$

$= \dfrac{m+4}{(m-3)(m+1)} - \dfrac{2m-3}{(m-3)(m-2)}$ Factor each denominator.

The least common denominator is $(m-3)(m+1)(m-2)$.

$= \dfrac{(m+4)(m-2)}{(m-3)(m+1)(m-2)} - \dfrac{(2m-3)(m+1)}{(m-3)(m-2)(m+1)}$

$= \dfrac{(m+4)(m-2)-(2m-3)(m+1)}{(m-3)(m+1)(m-2)}$ Subtract.

$= \dfrac{m^2+2m-8-(2m^2-m-3)}{(m-3)(m+1)(m-2)}$ Multiply.

$= \dfrac{m^2+2m-8-2m^2+m+3}{(m-3)(m+1)(m-2)}$ Distributive property

$= \dfrac{-m^2+3m-5}{(m-3)(m+1)(m-2)}$ Combine terms. ■

EXAMPLE 5 ADDING RATIONAL EXPRESSIONS Add.

$$\frac{a}{(a+1)^2} + \frac{2a}{a^2-1}$$

$= \dfrac{a}{(a+1)^2} + \dfrac{2a}{(a+1)(a-1)}$ Factor denominators.

$= \dfrac{a(a-1)}{(a+1)^2(a-1)} + \dfrac{2a(a+1)}{(a+1)(a-1)(a+1)}$ Get a common denominator.

$= \dfrac{a^2-a+2a^2+2a}{(a+1)^2(a-1)}$ Add numerators.

$= \dfrac{3a^2+a}{(a+1)^2(a-1)}$ Combine terms. ■

4.3 EXERCISES

Find the least common denominator for each of the following groups of denominators. See Example 2.

1. $k, \quad 5k^2$
2. $4r, \quad 9r^2$
3. $4x^2z^2, \quad 8x^3z^4, \quad 16xz^3$
4. $6m^6p^3, \quad 18m^4p^5, \quad 9m^2p^4$
5. $4z, \quad z-1$
6. $m-3, \quad 7m$
7. $2a+6, \quad 9a+27$
8. $3p-9, \quad 7p-21$
9. $a^2-16, \quad (a-4)^2$
10. $m^2-25, \quad (m+5)^2$
11. $8r, \quad r-1, \quad r+1$
12. $3y, \quad 2+y, \quad 2-y$
13. $x^2+7x+10, \quad x^2-25$
14. $y^2+6y+8, \quad y^2-16$
15. $p^2+6p, \quad p^2+7p+6, \quad 9p^2$
16. $5m^2-12m, \quad 5m^2+3m-36, \quad 10m^2$

17. Explain how to find the least common denominator for a group of denominators. (See Objective 2.)

Add $\dfrac{2}{(m+1)} + \dfrac{5}{m}$ *using each of the common denominators given below. Be sure all answers are given in lowest terms.*

18. $m^2(m+1)$ **19.** $m(m+1)$ **20.** $m(m+1)^2$ **21.** $m^2(m+1)^2$

22. Is it necessary to find the *least* common denominator when adding or subtracting fractions, or would any common denominator work? What advantage is there in using the *least* common denominator? See Exercises 18–21. (See Objective 3.)

Add or subtract as indicated. See Examples 1, 3, 4, and 5.

23. $\dfrac{2}{p} + \dfrac{5}{p}$ **24.** $\dfrac{8}{m} + \dfrac{4}{m}$ **25.** $\dfrac{1}{m+1} + \dfrac{m}{m+1}$ **26.** $\dfrac{k}{k-2} - \dfrac{2}{k-2}$

27. $\dfrac{6r}{r+2} + \dfrac{12}{r+2}$ **28.** $\dfrac{5m}{m-2} - \dfrac{10}{m-2}$ **29.** $\dfrac{a^2}{a+b} - \dfrac{b^2}{a+b}$ **30.** $\dfrac{y}{y^2-x^2} + \dfrac{x}{y^2-x^2}$

31. $\dfrac{5}{p^2+3p-10} + \dfrac{p}{p^2+3p-10}$ **32.** $\dfrac{2k}{2k^2+k-1} - \dfrac{1}{2k^2+k-1}$

33. $\dfrac{9}{r} + \dfrac{5}{2r}$ **34.** $\dfrac{7}{3y} + \dfrac{1}{y}$ **35.** $\dfrac{4}{3z} + \dfrac{1}{z^2}$ **36.** $\dfrac{1}{2y} + \dfrac{3}{y}$

37. $\dfrac{5}{9k} - \dfrac{3}{2k}$ **38.** $\dfrac{4}{3y} + \dfrac{1}{y^2}$ **39.** $\dfrac{1}{m-1} + \dfrac{1}{m}$ **40.** $\dfrac{2}{p-3} - \dfrac{1}{p}$

41. $\dfrac{5}{t+2} - \dfrac{3}{t-2}$ **42.** $\dfrac{3}{a-5} + \dfrac{2}{a+5}$

43. $\dfrac{5}{m-4} + \dfrac{2}{4-m}$ **44.** $\dfrac{1}{y-2} + \dfrac{3}{2-y}$

45. $\dfrac{y}{x-y} - \dfrac{y}{y-x}$ **46.** $\dfrac{p}{p-q} - \dfrac{q}{q-p}$

47. $\dfrac{7}{3a+9} - \dfrac{5}{4a+12}$ **48.** $\dfrac{-11}{5r-25} + \dfrac{4}{3r-15}$

49. $\dfrac{3}{x^2-5x+6} - \dfrac{2}{x^2-x-2}$ **50.** $\dfrac{2}{m^2-4m+4} + \dfrac{3}{m^2+m-6}$

51. $\dfrac{-4}{a^2-ab-6b^2} - \dfrac{1}{a^2+ab-2b^2}$ **52.** $\dfrac{x}{x^2+xy-2y^2} - \dfrac{3x}{x^2-3xy+2y^2}$

53. $\dfrac{r-3s}{r^2-4s^2} + \dfrac{r-s}{r^2-4rs+4s^2}$ **54.** $\dfrac{3p+2q}{3p^2+pq-2q^2} + \dfrac{3p-2q}{3p^2+5pq+2q^2}$

55. $\dfrac{3y}{y^2+yz-2z^2} + \dfrac{4y-1}{y^2-z^2}$ **56.** $\dfrac{r+s}{3r^2+2rs-s^2} - \dfrac{s-r}{6r^2-5rs+s^2}$

57. $\dfrac{3}{(p-2)^2} - \dfrac{5}{p-2} + 4$ **58.** $\dfrac{8}{(3r-1)^2} + \dfrac{2}{3r-1} - 6$

59. $\dfrac{5}{2r^2 - 3rs - 2s^2} - \dfrac{3}{3r^2 - 7rs + 2s^2} + \dfrac{4}{6r^2 + rs - s^2}$

60. $\dfrac{1}{3m^2 + mp - 2p^2} + \dfrac{4}{2m^2 + 3mp + p^2} - \dfrac{3}{6m^2 - mp - 2p^2}$

61. $\dfrac{-1}{3q + 2s} - \dfrac{4q + 3s}{27q^3 + 8s^3}$

62. $\dfrac{2}{5z - 4x} + \dfrac{2z + 3x}{125z^3 - 64x^3}$

63. $\dfrac{m + k}{m - k}\left(\dfrac{3m}{m - k} - \dfrac{2k}{k - m}\right)$

64. $\dfrac{a + 3b}{2a - b}\left(\dfrac{a}{2a - b} - \dfrac{4b}{b - 2a}\right)$

65. $\dfrac{p}{3p + 6} - \dfrac{4}{3p^2 + 6p} - \dfrac{5}{p}$

66. $\dfrac{-16z}{4z + 1} + \dfrac{1}{4z^2 + z} + \dfrac{2}{z}$

---

67. If $x = 4$ and $y = 2$, show that $\dfrac{1}{x} + \dfrac{1}{y} \neq \dfrac{1}{x + y}$. (See Objective 3.)

68. Use $x = 3$ and $y = 5$ to show that $\dfrac{1}{x} - \dfrac{1}{y} \neq \dfrac{1}{x - y}$. (See Objective 3.)

Perform the indicated operations.

69. $\left(\dfrac{5}{p + 2} - \dfrac{3}{p - 1}\right)\dfrac{p - 1}{p}$

70. $\left(\dfrac{4}{2r - 1} + \dfrac{3}{r + 3}\right)\dfrac{2r - 1}{r - 1}$

71. $\left(\dfrac{2k}{k - 3} + \dfrac{5k}{2k + 1}\right) \div \dfrac{k}{2k + 1}$

72. $\left(\dfrac{6y}{y + 2} - \dfrac{3y}{3y + 1}\right) \div \dfrac{y^2}{y + 2}$

73. $\dfrac{1}{h}\left(\dfrac{1}{(x - h)^2} - \dfrac{1}{x^2}\right)$

74. $\dfrac{1}{h}\left(\dfrac{-2}{x - h} + \dfrac{-2}{x}\right)$

75. $(a - b)^{-1} + (a + b)^{-1}$

76. $(y + z)^{-1} - (y - z)^{-1}$

77. $(2x - 1)^{-1} - (x + 2)^{-1}$

78. $(3y + 2)^{-1} + (y - 5)^{-1}$

PREVIEW EXERCISES

Simplify. See Section 1.3.

79. $\dfrac{\dfrac{2}{3} + \dfrac{1}{6}}{\dfrac{5}{9} - \dfrac{1}{3}}$

80. $\dfrac{\dfrac{1}{4} + \dfrac{3}{8}}{\dfrac{3}{2} - \dfrac{7}{8}}$

81. $\dfrac{3 + \dfrac{5}{4}}{2 - \dfrac{1}{4}}$

82. $\dfrac{2 - \dfrac{4}{3}}{\dfrac{3}{8} - 1}$

4.4 COMPLEX FRACTIONS

OBJECTIVES

1. SIMPLIFY COMPLEX FRACTIONS BY SIMPLIFYING NUMERATOR AND DENOMINATOR.
2. SIMPLIFY COMPLEX FRACTIONS BY MULTIPLYING BY A COMMON DENOMINATOR.
3. SIMPLIFY RATIONAL EXPRESSIONS WITH NEGATIVE EXPONENTS.

A **complex fraction** is an expression having a fraction in the numerator, denominator, or both. Examples of complex fractions include

$$\frac{1 + \frac{1}{x}}{2}, \qquad \frac{\frac{4}{y}}{6 - \frac{3}{y}}, \qquad \text{and} \qquad \frac{\frac{m^2 - 9}{m + 1}}{\frac{m + 3}{m^2 - 1}}.$$

1 There are two different methods for simplifying complex fractions.

SIMPLIFYING COMPLEX FRACTIONS

Method 1 Simplify the numerator and denominator separately, as much as possible. Then multiply the numerator by the reciprocal of the denominator.

In the second step, we are writing the complex fraction as a quotient of two rational expressions and dividing. Before performing this step, be sure that both the numerator and denominator are single fractions.

EXAMPLE 1 **SIMPLIFYING COMPLEX FRACTIONS BY METHOD 1**

Use Method 1 to simplify each complex fraction.

(a) $\dfrac{\frac{x + 1}{x}}{\frac{x - 1}{2x}}$

Both the numerator and the denominator are already simplified, so multiply the numerator by the reciprocal of the denominator.

$$\frac{\frac{x + 1}{x}}{\frac{x - 1}{2x}} = \frac{x + 1}{x} \div \frac{x - 1}{2x} \qquad \text{Write as a division problem.}$$

$$= \frac{x + 1}{x} \cdot \frac{2x}{x - 1} \qquad \text{Reciprocal of } \frac{x - 1}{2x}$$

$$= \frac{2(x + 1)}{x - 1}$$

(b) $$\frac{2+\frac{1}{y}}{3-\frac{2}{y}} = \frac{\frac{2y}{y}+\frac{1}{y}}{\frac{3y}{y}-\frac{2}{y}} = \frac{\frac{2y+1}{y}}{\frac{3y-2}{y}}$$ Simplify numerator and denominator.

$$= \frac{2y+1}{y} \cdot \frac{y}{3y-2}$$ Reciprocal of $\frac{3y-2}{y}$

$$= \frac{2y+1}{3y-2} \quad \blacksquare$$

2 The second method for simplifying complex fractions uses the identity property for multiplication.

SIMPLIFYING COMPLEX FRACTIONS

Method 2 Multiply the numerator and denominator of the complex fraction by the least common denominator of all the fractions appearing in the complex fraction. Then simplify the results.

EXAMPLE 2 SIMPLIFYING COMPLEX FRACTIONS BY METHOD 2

Use Method 2 to simplify each complex fraction.

(a) $$\frac{2+\frac{1}{y}}{3-\frac{2}{y}}$$

Multiply the numerator and denominator by the least common denominator of all the fractions appearing in the complex fraction. Here the least common denominator is y.

$$\frac{2+\frac{1}{y}}{3-\frac{2}{y}} = \frac{\left(2+\frac{1}{y}\right)\cdot y}{\left(3-\frac{2}{y}\right)\cdot y}$$ Multiply numerator and denominator by y.

$$= \frac{2\cdot y+\frac{1}{y}\cdot y}{3\cdot y-\frac{2}{y}\cdot y}$$ Use the distributive property.

$$= \frac{2y+1}{3y-2}$$

Compare this method of solution with that used in Example 1(b) above.

(b) $\dfrac{2p + \dfrac{5}{p-1}}{3p - \dfrac{2}{p}}$

The least common denominator is $p(p-1)$.

$$\frac{2p + \dfrac{5}{p-1}}{3p - \dfrac{2}{p}} = \frac{2p[p(p-1)] + \dfrac{5}{p-1} \cdot p(p-1)}{3p[p(p-1)] - \dfrac{2}{p} \cdot p(p-1)}$$

$$= \frac{2p[p(p-1)] + 5p}{3p[p(p-1)] - 2(p-1)}$$

$$= \frac{2p^3 - 2p^2 + 5p}{3p^3 - 3p^2 - 2p + 2} \quad ■$$

3 Rational expressions and complex fractions often involve negative exponents. To simplify such expressions, we begin by rewriting the expressions with only positive exponents.

EXAMPLE 3 SIMPLIFYING EXPRESSIONS WITH NEGATIVE EXPONENTS

Simplify each of the following.

(a) $a^{-1} + b^{-1}$

First write the expression with only positive exponents. By the definition of a negative exponent,

$$a^{-1} = \frac{1}{a} \quad \text{and} \quad b^{-1} = \frac{1}{b},$$

with

$$a^{-1} + b^{-1} = \frac{1}{a} + \frac{1}{b}.$$

Now add the two fractions. The common denominator is ab. Write each fraction with a denominator of ab.

$$a^{-1} + b^{-1} = \frac{1}{a} + \frac{1}{b} = \frac{b}{ab} + \frac{a}{ab} = \frac{b+a}{ab}$$

(b) $\dfrac{m^{-1} + p^{-2}}{2m^{-2} - p^{-1}}$

$$\frac{m^{-1} + p^{-2}}{2m^{-2} - p^{-1}} = \frac{\dfrac{1}{m} + \dfrac{1}{p^2}}{\dfrac{2}{m^2} - \dfrac{1}{p}}$$

Note that the 2 in $2m^{-2}$ is not raised to the -2 power, so $2m^{-2} = \frac{2}{m^2}$. Simplify the complex fraction by multiplying numerator and denominator by the least common denominator, m^2p^2.

$$\frac{\frac{1}{m} + \frac{1}{p^2}}{\frac{2}{m^2} - \frac{1}{p}} = \frac{m^2p^2 \cdot \frac{1}{m} + m^2p^2 \cdot \frac{1}{p^2}}{m^2p^2 \cdot \frac{2}{m^2} - m^2p^2 \cdot \frac{1}{p}}$$

$$= \frac{mp^2 + m^2}{2p^2 - m^2p}$$

(c) $$\frac{k^{-1}}{k^{-1} + 1} = \frac{\frac{1}{k}}{\frac{1}{k} + 1}$$

$$= \frac{k \cdot \frac{1}{k}}{k \cdot \frac{1}{k} + k \cdot 1} \qquad \text{Use Method 2.}$$

$$= \frac{1}{1 + k} \quad ■$$

4.4 EXERCISES

Simplify the following complex fractions by using either method. See Examples 1 and 2.

1. $\dfrac{\frac{3}{x}}{\frac{6}{x - 1}}$ **2.** $\dfrac{\frac{2}{m}}{\frac{8}{m + 4}}$ **3.** $\dfrac{\frac{4}{p + 5}}{\frac{3}{2p}}$ **4.** $\dfrac{\frac{k + 1}{2}}{\frac{3k - 1}{4}}$

5. $\dfrac{\frac{m - 1}{4m}}{\frac{m + 1}{m}}$ **6.** $\dfrac{\frac{r - s}{3}}{\frac{r + s}{6}}$ **7.** $\dfrac{\frac{4z^2x}{9}}{\frac{12xz^3}{5}}$ **8.** $\dfrac{\frac{9y^3x^4}{16}}{\frac{3y^2x^3}{8}}$

9. $\dfrac{\frac{22m^4n^5}{3m^5}}{\frac{11m^3n^2}{9m^8}}$ **10.** $\dfrac{\frac{16a^3b^4}{9a^2}}{\frac{24ab^4}{18b^2}}$ **11.** $\dfrac{\frac{k + m}{2k}}{\frac{2k - 2m}{k}}$ **12.** $\dfrac{\frac{x - 3y}{5x}}{\frac{8x - 24y}{10}}$

13. $\dfrac{1+\dfrac{1}{y}}{1-\dfrac{1}{y}}$ **14.** $\dfrac{\dfrac{2}{m}-1}{1+\dfrac{2}{m}}$ **15.** $\dfrac{\dfrac{1}{k}+\dfrac{1}{r}}{\dfrac{1}{k}-\dfrac{1}{r}}$ **16.** $\dfrac{\dfrac{2}{q}-\dfrac{3}{r}}{2+\dfrac{1}{qr}}$

17. $\dfrac{\dfrac{1}{r}-\dfrac{4}{s}}{\dfrac{s^2-16r^2}{rs}}$ **18.** $\dfrac{\dfrac{2}{x}-\dfrac{3}{y}}{\dfrac{4y^2-9x^2}{2x}}$ **19.** $\dfrac{y-\dfrac{y-3}{3}}{\dfrac{4}{9}+\dfrac{2}{3y}}$ **20.** $\dfrac{p-\dfrac{p+2}{4}}{\dfrac{3}{4}-\dfrac{5}{2p}}$

21. $\dfrac{x-\dfrac{x-1}{3}}{\dfrac{2}{3}-\dfrac{1}{6x}}$ **22.** $\dfrac{5y-\dfrac{2y+3}{2}}{\dfrac{1}{2}-\dfrac{3}{2y}}$ **23.** $\dfrac{\dfrac{a+3}{a}-\dfrac{4}{a-1}}{\dfrac{a}{a-1}+\dfrac{1}{a}}$ **24.** $\dfrac{\dfrac{k+2}{k}+\dfrac{1}{k+2}}{\dfrac{k}{k+2}+\dfrac{5}{k}}$

25. $\dfrac{\dfrac{2}{a-1}+\dfrac{3}{a}}{4-\dfrac{2}{1-a}}$ **26.** $\dfrac{\dfrac{6}{b-5}+\dfrac{5}{5-b}}{\dfrac{3}{b}+\dfrac{2}{b-5}}$

Explain why the following inequalities are true.

27. $\dfrac{\dfrac{1}{k}+3h}{\dfrac{1}{k}}\neq 3h$ **28.** $\dfrac{\dfrac{r}{t}-1}{\dfrac{r}{t}+1}\neq -1$

29. $\dfrac{m^{-1}+n^{-1}}{m^{-2}-n^{-2}}\neq\dfrac{m^2-n^2}{m+n}$ **30.** $\dfrac{a^{-1}-b^{-1}}{b^{-1}+a^{-1}}\neq\dfrac{b+a}{a-b}$

Simplify. See Example 3.

31. $\dfrac{x^{-2}}{x^{-2}-y^{-2}}$ **32.** $\dfrac{p^{-2}+q^{-2}}{q^{-2}}$ **33.** $\dfrac{m^{-1}+z^{-1}}{m^{-1}-z^{-1}}$ **34.** $\dfrac{r^{-1}+s^{-2}}{r^{-1}-s^{-1}}$

35. $(a^{-1}+b^{-1})^{-1}$ **36.** $(p^{-1}-q^{-1})^{-1}$ **37.** $(2k)^{-1}+3m^{-1}$ **38.** $5x^{-1}-(3y)^{-1}$

39. $\dfrac{\dfrac{3}{mp}-\dfrac{4}{p}+\dfrac{8}{m}}{2m^{-1}-3p^{-1}}$ **40.** $\dfrac{8k^{-1}+3q^{-1}}{\dfrac{4}{kq}-\dfrac{5}{q}+\dfrac{1}{k}}$

Simplify.

41. $2-\dfrac{1}{2+\dfrac{1}{x}}$ **42.** $1-\dfrac{3}{3-\dfrac{1}{2y}}$ **43.** $1+\dfrac{1-\dfrac{2}{p}}{1+\dfrac{2}{p}}$

44. $\dfrac{m - \dfrac{1}{1 + \dfrac{1}{m}}}{m + \dfrac{1}{1 - \dfrac{1}{m}}}$

45. $\dfrac{1}{p + \dfrac{1}{p + \dfrac{1}{1 + p}}}$

46. $1 - \dfrac{1}{1 - \dfrac{1}{1 - \dfrac{1}{x - 1}}}$

PREVIEW EXERCISES

Perform the indicated operations. See Sections 3.1 and 3.3.

47. $\dfrac{12p^7}{6p^3}$

48. $\dfrac{-9y^{11}}{3y}$

49. $\dfrac{-8a^3b^7}{6a^5b}$

50. $\dfrac{20r^3s^5}{15rs^9}$

51. Subtract.
$$\begin{array}{r} -3a^2 + 4a - 5 \\ 5a^2 + 3a - 9 \\ \hline \end{array}$$

52. Subtract.
$$\begin{array}{r} -4p^2 - 8p + 5 \\ 3p^2 + 2p + 9 \\ \hline \end{array}$$

53. Subtract.
$$\begin{array}{r} 10x^3 - 8x^2 + 4x \\ 12x^3 + 5x^2 - 7x \\ \hline \end{array}$$

54. Subtract.
$$\begin{array}{r} 2z^3 - 5z^2 + 9z \\ 8z^3 + 2z^2 - 11z \\ \hline \end{array}$$

4.5 DIVISION OF POLYNOMIALS

OBJECTIVES

1 DIVIDE A POLYNOMIAL BY A MONOMIAL.

2 DIVIDE A POLYNOMIAL BY A POLYNOMIAL OF TWO OR MORE TERMS.

1 Chapter 3 showed how to add, subtract, and multiply polynomials. Polynomial division is discussed in this section, beginning with the division of a polynomial by a monomial. (Recall that a monomial is a single term, such as $8x$, $-9m^4$, or $11y^2$.)

DIVIDING BY A MONOMIAL

To divide a polynomial by a monomial, divide each term in the polynomial by the monomial, and then write each fraction in lowest terms.

EXAMPLE 1 DIVIDING A POLYNOMIAL BY A MONOMIAL

Divide $15x^2 - 12x + 6$ by 3.

Divide each term of the polynomial by 3. Then write the result in lowest terms.

$$\frac{15x^2 - 12x + 6}{3} = \frac{15x^2}{3} - \frac{12x}{3} + \frac{6}{3}$$
$$= 5x^2 - 4x + 2$$

Check this answer by multiplying it by the divisor, 3. You should get $15x^2 - 12x + 6$ as the result.

$$\underset{\text{Divisor}}{3}\underbrace{(5x^2 - 4x + 2)}_{\text{Quotient}} = \underbrace{15x^2 - 12x + 6}_{\text{Original polynomial}}$$

■

EXAMPLE 2 **DIVIDING A POLYNOMIAL BY A MONOMIAL**

Find each quotient.

(a) $$\frac{5m^3 - 9m^2 + 10m}{5m^2} = \frac{5m^3}{5m^2} - \frac{9m^2}{5m^2} + \frac{10m}{5m^2} \qquad \text{Divide each term by } 5m^2.$$

$$= m - \frac{9}{5} + \frac{2}{m} \qquad \text{Write in lowest terms.}$$

This result is not a polynomial; the quotient of two polynomials need not be a polynomial.

(b) $$\frac{8xy^2 - 9x^2y + 6x^2y^2}{x^2y^2} = \frac{8xy^2}{x^2y^2} - \frac{9x^2y}{x^2y^2} + \frac{6x^2y^2}{x^2y^2}$$

$$= \frac{8}{x} - \frac{9}{y} + 6 \quad ■$$

2 Earlier, we saw that the quotient of two polynomials can be found by factoring and then dividing out any common factors. For instance,

$$\frac{2x^2 + 5x - 3}{4x - 2} = \frac{(2x - 1)(x + 3)}{2(2x - 1)} = \frac{x + 3}{2}.$$

When the polynomials in a quotient of two polynomials have no common factors or cannot be factored, they can be divided by a process very similar to that for dividing one whole number by another. The following examples show how this is done.

EXAMPLE 3 **DIVIDING A POLYNOMIAL BY A POLYNOMIAL**

Divide $2m^2 + m - 10$ by $m - 2$.

Write the problem, making sure that both polynomials are written in descending powers of the variables.

$$m - 2\overline{)2m^2 + m - 10}$$

Divide the first term of $2m^2 + m - 10$ by the first term of $m - 2$. Since $2m^2/m = 2m$, place this result above the division line.

$$\begin{array}{r} 2m \quad \longleftarrow \text{Result of } \frac{2m^2}{m} \\ m - 2\overline{)2m^2 + m - 10} \end{array}$$

Multiply $m - 2$ and $2m$, and write the result below $2m^2 + m - 10$.

$$\begin{array}{rl} & 2m \\ m - 2 & \overline{)2m^2 + m - 10} \\ & \underline{2m^2 - 4m} \quad \longleftarrow 2m(m - 2) = 2m^2 - 4m \end{array}$$

Now subtract $2m^2 - 4m$ from $2m^2 + m$. Do this by changing the signs on $2m^2 - 4m$ and *adding*.

$$\begin{array}{r} 2m \phantom{{}+m-10} \\ m-2\overline{)2m^2 + m - 10} \\ \underline{2m^2 - 4m} \phantom{{}-10} \\ 5m \phantom{{}-10} \end{array} \quad \leftarrow \text{Subtract.}$$

Bring down -10 and continue by dividing $5m$ by m.

$$\begin{array}{rl} 2m + 5 & \leftarrow \dfrac{5m}{m} = 5 \\ m-2\overline{)2m^2 + m - 10} & \\ \underline{2m^2 - 4m} \phantom{{}-10} & \\ 5m - 10 & \leftarrow \text{Bring down } -10. \\ \underline{5m - 10} & \leftarrow 5(m-2) = 5m - 10 \\ 0 & \leftarrow \text{Subtract.} \end{array}$$

Finally, $\dfrac{2m^2 + m - 10}{m - 2} = 2m + 5$. Check by multiplying $m - 2$ and $2m + 5$. The result should be $2m^2 + m - 10$. Since there is no remainder, this quotient could have been found by factoring and writing in lowest terms. ■

EXAMPLE 4 DIVIDING A POLYNOMIAL WITH A MISSING TERM

Divide $3x^3 - 2x + 5$ by $x - 3$.

Make sure that $3x^3 - 2x + 5$ is in descending powers of the variable. Add a term with a 0 coefficient for the missing x^2 term.

Missing term

$$x - 3\overline{)3x^3 + 0x^2 - 2x + 5}$$

Start with $3x^3/x = 3x^2$.

$$\begin{array}{rl} 3x^2 \phantom{{}+0x^2-2x+5} & \leftarrow \dfrac{3x^3}{x} = 3x^2 \\ x-3\overline{)3x^3 + 0x^2 - 2x + 5} & \\ \underline{3x^3 - 9x^2} \phantom{{}-2x+5} & \leftarrow 3x^2(x-3) \end{array}$$

Subtract by changing the signs on $3x^3 - 9x^2$ and adding.

$$\begin{array}{rl} 3x^2 \phantom{{}-2x+5} & \\ x-3\overline{)3x^3 + 0x^2 - 2x + 5} & \\ \underline{3x^3 - 9x^2} \phantom{{}-2x+5} & \\ 9x^2 \phantom{{}-2x+5} & \leftarrow \text{Subtract.} \end{array}$$

Bring down the next term.

$$\begin{array}{rl} 3x^2 \phantom{{}-2x+5} & \\ x-3\overline{)3x^3 + 0x^2 - 2x + 5} & \\ \underline{3x^3 - 9x^2} \phantom{{}-2x+5} & \\ 9x^2 - 2x \phantom{{}+5} & \leftarrow \text{Bring down } -2x. \end{array}$$

In the next step, $9x^2/x = 9x$.

$$
\begin{array}{r}
3x^2 + 9x \phantom{{}+5} \\
x-3\overline{)3x^3 + 0x^2 - 2x + 5} \\
\underline{3x^3 - 9x^2} \phantom{{}-2x+5} \\
9x^2 - 2x \phantom{{}+5} \\
\underline{9x^2 - 27x} \phantom{{}+5} \\
25x + 5
\end{array}
\quad
\begin{array}{l}
\leftarrow \dfrac{9x^2}{x} = 9x \\
\\
\\
\\
\leftarrow 9x(x-3) \\
\leftarrow \text{Subtract and bring down 5.}
\end{array}
$$

Finally, $25x/x = 25$.

$$
\begin{array}{r}
3x^2 + 9x + 25 \\
x-3\overline{)3x^3 + 0x^2 - 2x + 5} \\
\underline{3x^3 - 9x^2} \phantom{{}-2x+5} \\
9x^2 - 2x \phantom{{}+5} \\
\underline{9x^2 - 27x} \phantom{{}+5} \\
25x + 5 \\
\underline{25x - 75} \\
80
\end{array}
\quad
\begin{array}{l}
\leftarrow \dfrac{25x}{x} = 25 \\
\\
\\
\\
\\
\\
\leftarrow 25(x-3) \\
\leftarrow \text{Subtract.}
\end{array}
$$

We write the remainder of 80 as the numerator of the fraction $\dfrac{80}{x-3}$. In summary,

$$\frac{3x^3 - 2x + 5}{x - 3} = 3x^2 + 9x + 25 + \frac{80}{x - 3}.$$

Check by multiplying $x - 3$ and $3x^2 + 9x + 25$ and adding 80 to the result. You should get $3x^3 - 2x + 5$. ■

CAUTION Don't forget the + sign when adding the remainder to a quotient.

Sometimes the division process requires a fractional coefficient in the quotient.

EXAMPLE 5 **GETTING A FRACTIONAL COEFFICIENT IN THE QUOTIENT**

Divide $2p^3 + 5p^2 + p - 2$ by $2p + 2$.

$$
\begin{array}{r}
p^2 + \frac{3}{2}p - 1 \\
2p+2\overline{)2p^3 + 5p^2 + p - 2} \\
\underline{2p^3 + 2p^2} \phantom{{}+p-2} \\
3p^2 + p \phantom{{}-2} \\
\underline{3p^2 + 3p} \phantom{{}-2} \\
-2p - 2 \\
\underline{-2p - 2} \\
0
\end{array}
\quad
\leftarrow \frac{3p^2}{2p} = \frac{3}{2}p
$$

Since the remainder is 0, the quotient is $p^2 + \dfrac{3}{2}p - 1$. ■

EXAMPLE 6
DIVIDING BY A POLYNOMIAL WITH A MISSING TERM

Divide $6r^4 + 9r^3 + 2r^2 - 8r + 7$ by $3r^2 - 2$.

The polynomial $3r^2 - 2$ has a missing term. Write it as $3r^2 + 0r - 2$ and divide as usual.

$$
\begin{array}{r}
2r^2 + 3r + 2 \\
3r^2 + 0r - 2 \overline{\big)\, 6r^4 + 9r^3 + 2r^2 - 8r + 7} \\
\underline{6r^4 + 0r^3 - 4r^2} \qquad\qquad \\
9r^3 + 6r^2 - 8r \qquad \\
\underline{9r^3 + 0r^2 - 6r} \qquad \\
6r^2 - 2r + 7 \\
\underline{6r^2 + 0r - 4} \\
-2r + 11
\end{array}
$$

Since the degree of the remainder, $-2r + 11$, is less than the degree of $3r^2 - 2$, the division process is now finished. The result is written

$$2r^2 + 3r + 2 + \frac{-2r + 11}{3r^2 - 2}. \quad \blacksquare$$

4.5 EXERCISES

Divide. See Examples 1–6.

1. $\dfrac{5a^2 - 10a + 5}{5}$
2. $\dfrac{9y^2 - 12y + 27}{3}$
3. $\dfrac{8m^2 + 16m + 24}{8m}$
4. $\dfrac{22p^2 - 11p + 33}{11p}$
5. $\dfrac{12a^3 + 8a^2 - 10a}{4a^3}$
6. $\dfrac{8m^4 - 6m^3 + 4m^2}{8m^3}$
7. $\dfrac{9p^2q + 18pq^2 - 6pq}{12p^2q^2}$
8. $\dfrac{15m^2k^2 - 9mk^3 + 12m^2k}{6m^2k}$
9. $\dfrac{x^2 - 2x - 15}{x - 5}$
10. $\dfrac{p^2 - 3p - 18}{p - 6}$
11. $\dfrac{2r^2 + 13r + 21}{2r + 7}$
12. $\dfrac{10a^2 - 11a - 6}{5a + 2}$
13. $\dfrac{3m^3 + 5m^2 - 5m + 1}{3m - 1}$
14. $\dfrac{8z^3 - 6z^2 - 5z + 3}{4z + 3}$
15. $(6b^3 - 7b^2 - 3b + 1) \div (2b - 3)$
16. $(4a^3 + 5a^2 + 4a + 5) \div (4a + 5)$
17. $(a^3 - 8a^2 + 6a - 3) \div (a - 3)$
18. $(3r^3 - 22r^2 + 25r - 10) \div (3r - 1)$
19. $\dfrac{4x^3 - 18x^2 + 22x - 10}{2x^2 - 4x + 3}$
20. $\dfrac{12m^3 - 17m^2 + 30m - 10}{3m^2 - 2m + 5}$

21. $(9t^4 - 13t^3 + 23t^2 - 9t + 2) \div (t^2 - t + 2)$

22. $(2m^4 + 5m^3 - 11m^2 + 11m - 20) \div (2m^2 - m + 2)$

23. $\dfrac{8p^3 + 2p^2 + p + 18}{2p^2 + 3}$

24. $\dfrac{12z^3 + 9z^2 - 10z + 21}{3z^2 - 2}$

25. $\dfrac{4y^4 + 6y^3 + 3y - 1}{2y^2 + 1}$

26. $\dfrac{15r^4 + 3r^3 + 4r^2 + 4}{3r^2 - 1}$

27. $\dfrac{p^3 - 1}{p - 1}$

28. $\dfrac{8a^3 + 1}{2a + 1}$

29. $\dfrac{16r^4 + 1}{2r - 1}$

30. $\dfrac{81z^4 - 1}{3z + 1}$

31. $\left(2x^2 - \dfrac{7}{3}x - 1\right) \div (3x + 1)$

32. $\left(m^2 + \dfrac{7}{2}m + 3\right) \div (2m + 3)$

33. $\left(3a^2 - \dfrac{23}{4}a - 5\right) \div (4a + 3)$

34. $\left(3q^2 + \dfrac{19}{5}q - 3\right) \div (5q - 2)$

35. $(2a^5 - 3a^3 + 5a^2 - 8) \div (3a^2 + 3)$

36. $(4x^6 - 2x^3 + x^2 - 10) \div (5x^2 + 5)$

Solve each of the following problems. In Exercises 37–40, the answer is a polynomial.

37. The area of the rectangle in the figure is $6k^2 - 5k - 6$. Find L.

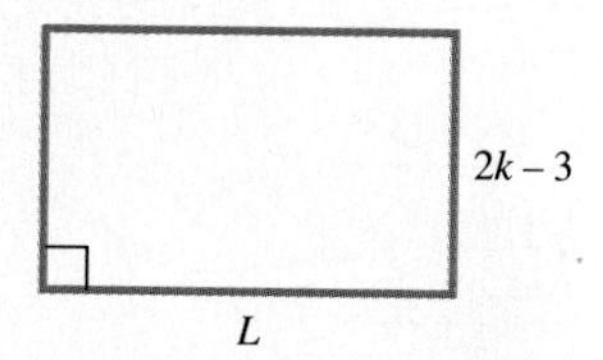

38. The volume of the box in the figure is $2p^3 + 15p^2 + 28p$. Find w.

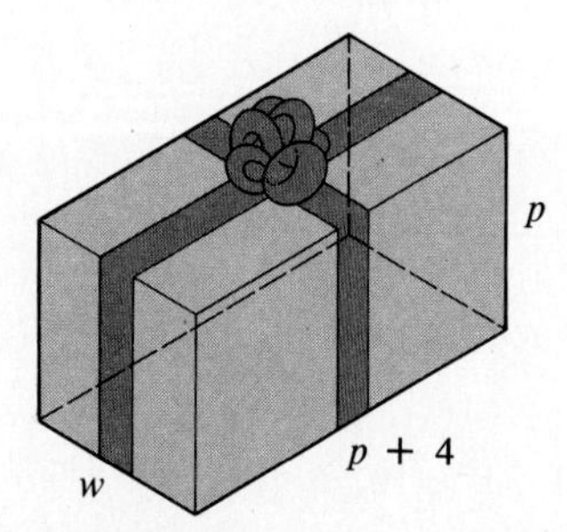

39. Suppose a car goes $2m^3 + 15m^2 + 13m - 63$ kilometers in $2m + 9$ hours. Find its rate.

40. A garden has $6z^3 - 17z^2 + 4z + 7$ flowers planted in $3z - 7$ rows, with an equal number of flowers in each row. Find the number of flowers in each row.

41. Evaluate the polynomial $p^3 - 4p^2 + 3p - 5$ when $p = -1$. Then divide $p^3 - 4p^2 + 3p - 5$ by $p + 1$. Compare the remainder to the first answer. What do these results suggest?

42. Let $P(x) = 4x^3 - 8x^2 + 13x - 2$, and $D(x) = 2x - 1$. Use division to find polynomials $Q(x)$ and $R(x)$ so that

$$P(x) = Q(x) \cdot D(x) + R(x).$$

PREVIEW EXERCISES

Evaluate each polynomial as indicated. See Section 3.3.

43. $P(x) = x^2 + 2x - 9;\quad P(1)$

44. $P(x) = 3x^2 - 5x + 7;\quad P(2)$

45. $P(x) = -x^2 + 8x + 12;\quad P(-3)$

46. $P(x) = -x^2 + 2x - 8;\quad P(-1)$

47. $P(x) = 2x^3 + x^2 - 5x - 4;\quad P(-2)$

48. $P(x) = 3x^3 - x^2 + 4x - 8;\quad P(-3)$

4.6 SYNTHETIC DIVISION

OBJECTIVES

1 USE SYNTHETIC DIVISION TO DIVIDE BY A POLYNOMIAL OF THE FORM $x - k$.

2 USE THE REMAINDER THEOREM TO EVALUATE A POLYNOMIAL.

3 DECIDE WHETHER A GIVEN NUMBER IS A SOLUTION OF AN EQUATION.

1 Many times when one polynomial is divided by a second, the second polynomial is of the form $x - k$, where the coefficient of the x term is 1. There is a shortcut way for doing these divisions. To see how this shortcut works, look first at the left below, where the division of $3x^3 - 2x + 5$ by $x - 3$ is shown. (Notice that a 0 was inserted for the missing x^2 term.)

$$\begin{array}{r}
3x^2 + 9x + 25 \\
x - 3\overline{)3x^3 + 0x^2 - 2x + 5} \\
\underline{3x^3 - 9x^2} \qquad\qquad \\
9x^2 - 2x \qquad \\
\underline{9x^2 - 27x} \qquad \\
25x + 5 \\
\underline{25x - 75} \\
80
\end{array}
\qquad\qquad
\begin{array}{rrrrr}
 & 3 & 9 & 25 \\
1 - 3\overline{)3} & 0 & -2 & 5 \\
\underline{(3)} & \underline{-9} & & \\
 & 9 & -2 & \\
 & \underline{(9)} & \underline{-27} & \\
 & & 25 & 5 \\
 & & \underline{(25)} & \underline{-75} \\
 & & & 80
\end{array}$$

On the right, exactly the same division is shown written without the variables. All the numbers in parentheses on the right are repetitions of the numbers directly above them, so they may be omitted, as shown on the left below.

$$\begin{array}{rrrr}
 & 3 & 9 & 25 \\
1 - 3\overline{)3} & 0 & -2 & 5 \\
 & \underline{-9} & & \\
 & 9 & (-2) & \\
 & & \underline{-27} & \\
 & & 25 & (5) \\
 & & & \underline{-75} \\
 & & & 80
\end{array}
\qquad\qquad
\begin{array}{rrrr}
 & 3 & 9 & 25 \\
1 - 3\overline{)3} & 0 & -2 & 5 \\
 & \underline{-9} & & \\
 & 9 & & \\
 & & \underline{-27} & \\
 & & 25 & \\
 & & & \underline{-75} \\
 & & & 80
\end{array}$$

The numbers in parentheses on the left are again repetitions of the numbers directly above them; they too may be omitted, as shown on the right above.

Now the problem can be condensed. If the 3 in the dividend is brought down to the beginning of the bottom row, the top row can be omitted, since it duplicates the bottom row.

$$\begin{array}{r|rrrr} 1-3 & 3 & 0 & -2 & 5 \\ & & -9 & -27 & -75 \\ \hline & 3 & 9 & 25 & 80 \end{array}$$

Finally, the 1 at the upper left can be omitted. Also, to simplify the arithmetic, subtraction in the second row is replaced by addition. We compensate for this by changing the -3 at upper left to its additive inverse, 3. The result of doing all this is shown below.

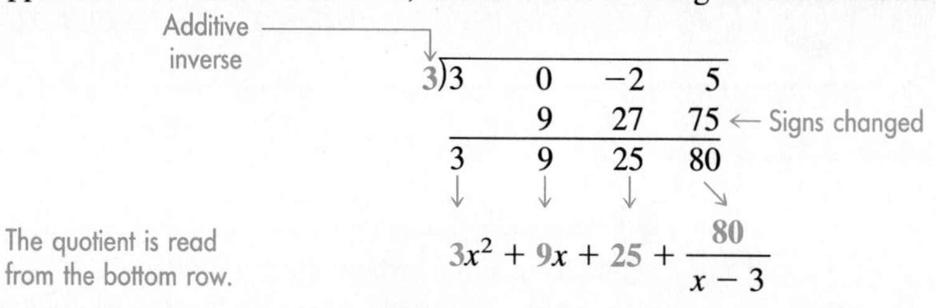

The first three numbers in the bottom row are used to obtain a polynomial with degree 1 less than the degree of the dividend. The last number gives the remainder.

SYNTHETIC DIVISION This shortcut procedure is called **synthetic division.** It is used only when dividing a polynomial by a binomial of the form $x - k$.

EXAMPLE 1 USING SYNTHETIC DIVISION

Use synthetic division to divide $5x^2 + 16x + 15$ by $x + 2$.

As mentioned above, synthetic division can be used only when dividing by a polynomial of the form $x - k$. Get $x + 2$ in this form by writing it as

$$x + 2 = x - (-2),$$

where $k = -2$. Now write the coefficients of $5x^2 + 16x + 15$, placing -2 to the left.

$x + 2$ leads to -2

$$\begin{array}{r|rrr} -2 & 5 & 16 & 15 \end{array} \leftarrow \text{Coefficients}$$

Bring down the 5, and multiply: $-2 \cdot 5 = -10$.

$$\begin{array}{r|rrr} -2 & 5 & 16 & 15 \\ & \downarrow & -10 & \\ \hline & 5 & & \end{array}$$

Add 16 and -10, getting 6. Multiply 6 and -2 to get -12.

$$\begin{array}{r|rrr} -2 & 5 & 16 & 15 \\ & & -10 & -12 \\ \hline & 5 & 6 & \end{array}$$

Add 15 and -12, getting 3.

$$\begin{array}{r|rrr} -2) & 5 & 16 & 15 \\ & & -10 & -12 \\ \hline & 5 & 6 & 3 \end{array}$$

The result is read from the bottom row.

$$\frac{5x^2 + 16x + 15}{x + 2} = 5x + 6 + \frac{3}{x + 2} \quad \blacksquare$$

EXAMPLE 2
USING SYNTHETIC DIVISION WITH MISSING TERMS

Use synthetic division to find

$$(-4x^5 + x^4 + 6x^3 + 2x^2 + 50) \div (x - 2).$$

Use the steps given above, inserting a 0 for the missing x term.

$$\begin{array}{r|rrrrrr} 2) & -4 & 1 & 6 & 2 & 0 & 50 \\ & & -8 & -14 & -16 & -28 & -56 \\ \hline & -4 & -7 & -8 & -14 & -28 & -6 \end{array}$$

Read the result from the bottom row.

$$\frac{-4x^5 + x^4 + 6x^3 + 2x^2 + 50}{x - 2} = -4x^4 - 7x^3 - 8x^2 - 14x - 28 + \frac{-6}{x - 2} \quad \blacksquare$$

2 Synthetic division can be used to evaluate polynomials. For example, in the synthetic division of Example 2, where the polynomial was divided by $x - 2$, the remainder was -6.

Replacing x in the polynomial with 2 gives

$$\begin{aligned} -4x^5 + x^4 + 6x^3 + 2x^2 + 50 &= -4 \cdot 2^5 + 2^4 + 6 \cdot 2^3 + 2 \cdot 2^2 + 50 \\ &= -4 \cdot 32 + 16 + 6 \cdot 8 + 2 \cdot 4 + 50 \\ &= -128 + 16 + 48 + 8 + 50 \\ &= -6, \end{aligned}$$

the same number as the remainder; that is, dividing by $x - 2$ produced a remainder equal to the result when x is replaced with 2. This always happens, as the following remainder theorem states.

REMAINDER THEOREM

If the polynomial $P(x)$ is divided by $x - k$, then the remainder is equal to $P(k)$.

This result is proved in more advanced courses.

EXAMPLE 3
USING THE REMAINDER THEOREM

Let $P(x) = 2x^3 - 5x^2 - 3x + 11$. Find $P(-2)$.

Use the remainder theorem; divide $P(x)$ by $x - (-2)$.

$$\begin{array}{r|rrrr} -2 & 2 & -5 & -3 & 11 \\ & & -4 & 18 & -30 \\ \hline & 2 & -9 & 15 & -19 \end{array} \leftarrow \text{Remainder}$$

By this result, $P(-2) = -19$. ■

3 The remainder theorem also can be used to show that a given number is a solution of an equation.

EXAMPLE 4
USING THE REMAINDER THEOREM

Show that -5 is a solution of the equation

$$2x^4 + 12x^3 + 6x^2 - 5x + 75 = 0.$$

One way to show that -5 is a solution is by substituting -5 for x in the equation. However, an easier way is to use synthetic division and the remainder theorem given above.

$$\text{Proposed solution} \rightarrow \begin{array}{r|rrrrr} -5 & 2 & 12 & 6 & -5 & 75 \\ & & -10 & -10 & 20 & -75 \\ \hline & 2 & 2 & -4 & 15 & 0 \end{array}$$

Since the remainder is 0, the polynomial has a value of 0 when $k = -5$, and so -5 is a solution of the given equation. ■

The synthetic division above also shows that $x - (-5)$ divides the polynomial with 0 remainder. Thus $x - (-5) = x + 5$ is a *factor* of the polynomial and

$$2x^4 + 12x^3 + 6x^2 - 5x + 75 = (x + 5)(2x^3 + 2x^2 - 4x + 15).$$

The second factor is the quotient polynomial found in the last row of the synthetic division.

4.6 EXERCISES

Use synthetic division in each of the following. See Examples 1 and 2.

1. $\dfrac{x^2 + 6x - 7}{x - 1}$
2. $\dfrac{x^2 - 2x - 15}{x + 3}$
3. $\dfrac{3a^2 + 10a + 11}{a + 2}$
4. $\dfrac{4y^2 - 5y - 20}{y - 3}$
5. $(p^2 - 3p + 5) \div (p + 1)$
6. $(z^2 + 4z - 6) \div (z - 5)$
7. $\dfrac{4a^3 - 3a^2 + 2a + 1}{a - 1}$
8. $\dfrac{5p^3 - 6p^2 + 3p + 14}{p + 1}$

9. $(6x^5 - 2x^3 - 4x^2 + 3x - 2) \div (x - 2)$

10. $(y^5 - 4y^4 + 3y^2 - 5y + 4) \div (y - 3)$

11. $(-4r^6 - 3r^5 - 3r^4 + 5r^3 - 6r^2 + 3r + 3) \div (r - 1)$

12. $(m^6 + 2m^4 - 5m + 11) \div (m - 2)$

13. $\dfrac{y^3 - 1}{y + 1}$

14. $\dfrac{m^3 - 8}{m + 2}$

15. $\dfrac{p^4 + 16}{p + 4}$

16. $\dfrac{z^4 + 81}{z + 3}$

Use the remainder theorem to find $P(k)$. See Example 3.

17. $k = 2;\quad P(x) = x^3 - 4x^2 + 8x - 1$

18. $k = -1;\quad P(y) = 2y^3 + y^2 - y + 5$

19. $k = -4;\quad P(r) = -r^3 + 8r^2 - 3r - 2$

20. $k = 3;\quad P(z) = z^3 + 5z^2 - 7z + 1$

21. $k = 3;\quad P(y) = 2y^3 - 4y^2 + 5y - 33$

22. $k = 2;\quad P(x) = x^3 - 3x^2 + 4x - 4$

23. $k = -2;\quad P(x) = x^4 - 3x^3 + 5x^2 - 2x + 5$

24. $k = -1;\quad P(t) = -t^4 + t^3 - 5t^2 + 3t - 4$

Use synthetic division to tell whether or not -2 is a solution of each of the following equations. See Example 4.

25. $x^3 - 2x^2 - 2x + 12 = 0$

26. $x^5 + 2x^4 - 3x^3 + 8x = 0$

27. $3a^3 + 2a^2 - 2a + 11 = 0$

28. $3z^3 + 10z^2 + 3z - 9 = 0$

29. $4x^5 + 3x^4 + 2x^3 + 9x^2 - 29x + 2 = 0$

30. $9k^5 + 15k^4 + 7k^2 - 2k + 6 = 0$

31. $-7w^6 + 15w^5 - w^4 + 16w^3 - 4w + 4 = 0$

32. $y^6 - 5y^5 + 2y^3 + 8y^2 + 10y + 28 = 0$

Use synthetic division to decide whether the given number is a solution for the given equation. If it is, factor the polynomial on the left of the equals sign. See Example 4.

33. $2x^3 - x^2 - 13x + 24 = 0;\quad x = -3$

34. $5p^3 + 22p^2 + p - 28 = 0;\quad p = -4$

35. $7z^3 - z^2 + 5z - 3 = 0;\quad z = 3$

36. $2r^3 + 4r^2 - r - 5 = 0;\quad r = -1$

37. $2y^4 - 5y^3 + 11y^2 - 14y + 3 = 0;\quad y = 3/2$

38. $6m^4 + m^3 - 4m^2 + 13m - 4 = 0;\quad m = 1/3$

Use synthetic division to show that the given number is a solution of the given equation. Factor the polynomial in the equation completely, then give all other rational solutions of the equation. (See Objective 3.)

39. $k = -2;\quad x^3 + x^2 - 3x - 2 = 0$

40. $k = 1;\quad 2x^3 + 2x^2 - 3x - 1 = 0$

41. $k = 4;\quad x^3 - 2x^2 - 7x - 4 = 0$

42. $k = -2;\quad x^3 - 7x - 6 = 0$

PREVIEW EXERCISES

Solve each equation. See Sections 2.1 and 3.9.

43. $7z - 2 = 5z + 4$

44. $8q + 1 = 3q - 8$

45. $4 - 3(2 + p) = 8p - 35$

46. $1 - 5(r + 3) = 7r - 2$

47. $q^2 - 4q + 3 = 0$

48. $p^2 + 5p + 4 = 0$

49. $6r^2 + r = 2$

50. $10z^2 + 17z = -3$

4.7 EQUATIONS WITH RATIONAL EXPRESSIONS

OBJECTIVES

1 SOLVE EQUATIONS WITH RATIONAL EXPRESSIONS.

2 KNOW WHEN POTENTIAL SOLUTIONS MUST BE CHECKED.

1 The easiest way to solve most equations involving rational expressions is to multiply all the terms in the equation by the least common denominator. This step will clear the equation of all denominators as the next examples show.

EXAMPLE 1 SOLVING AN EQUATION WITH RATIONAL EXPRESSIONS

Solve $\frac{2x}{5} - \frac{x}{3} = 2$.

The least common denominator for $2x/5$ and $x/3$ is 15, so multiply both sides of the equation by 15.

$$15\left(\frac{2x}{5} - \frac{x}{3}\right) = 15(2)$$

$$15\left(\frac{2x}{5}\right) - 15\left(\frac{x}{3}\right) = 15(2) \qquad \text{Distributive property}$$

$$6x - 5x = 30 \qquad \text{Multiply.}$$

$$x = 30 \qquad \text{Combine terms.}$$

By substituting 30 for x in the original equation, check that the solution set for the given equation is $\{30\}$. ■

EXAMPLE 2 SOLVING AN EQUATION WITH RATIONAL EXPRESSIONS

Solve $\frac{2}{y} - \frac{3}{2} = \frac{7}{2y}$.

Multiply both sides by the least common denominator, $2y$.

$$2y\left(\frac{2}{y} - \frac{3}{2}\right) = 2y\left(\frac{7}{2y}\right)$$

$$2y\left(\frac{2}{y}\right) - 2y\left(\frac{3}{2}\right) = 2y\left(\frac{7}{2y}\right) \qquad \text{Distributive property}$$

$$4 - 3y = 7 \qquad \text{Multiply.}$$

$$-3y = 3 \qquad \text{Subtract 4.}$$

$$y = -1 \qquad \text{Divide by } -3.$$

To see if -1 is a solution of the equation, replace y with -1 in the original equation.

$$\frac{2}{-1} - \frac{3}{2} = \frac{7}{2(-1)} \qquad ?$$

$$-\frac{4}{2} - \frac{3}{2} = -\frac{7}{2} \qquad \text{True}$$

Since -1 checks, the solution set is $\{-1\}$. ■

CAUTION When both sides of an equation are multiplied by a *variable* expression, it is possible that the resulting "solutions" are not actually solutions of the given equation.

2 Because the first step in solving a rational equation is to multiply both sides of the equation by a common denominator, in many cases it is *necessary* to check the solutions.

EXAMPLE 3
SOLVING AN EQUATION WITH NO SOLUTION

Solve $\dfrac{2}{m-3} - \dfrac{3}{m+3} = \dfrac{12}{m^2-9}$.

The least common denominator is $(m + 3)(m - 3)$, which is used to multiply both sides. Since $(m + 3)(m - 3) = 0$ when $m = 3$ or $m = -3$, the solution cannot be 3 or -3.

$$(m+3)(m-3) \cdot \frac{2}{m-3} - (m+3)(m-3) \cdot \frac{3}{m+3} = (m+3)(m-3) \cdot \frac{12}{m^2-9}$$

$$2(m+3) - 3(m-3) = 12 \qquad \text{Multiply.}$$

$$2m + 6 - 3m + 9 = 12 \qquad \text{Distributive property}$$

$$-m + 15 = 12 \qquad \text{Combine terms.}$$

$$-m = -3$$

$$m = 3$$

Since both sides were multiplied by a variable expression, we must check the potential solution, 3.

$$\frac{2}{3-3} - \frac{3}{3+3} = \frac{12}{3^2-9} \quad ?$$

$$\frac{2}{0} - \frac{3}{6} = \frac{12}{0} \quad ?$$

Division by 0 is undefined. The given equation has no solution, and the solution set is $\emptyset$. We predicted this outcome at the beginning of the solution. ■

EXAMPLE 4
SOLVING AN EQUATION WITH RATIONAL EXPRESSIONS

Solve $\dfrac{3}{p^2+p-2} - \dfrac{1}{p^2-1} = \dfrac{7}{2(p^2+3p+2)}$.

Factor to find the least common denominator $2(p - 1)(p + 2)(p + 1)$, so that $p \neq 1$, $p \neq -2$, $p \neq -1$. Multiply both sides by the least common denominator.

$$2(p-1)(p+2)(p+1)\left[\frac{3}{(p+2)(p-1)}\right] - 2(p-1)(p+2)(p+1)\left[\frac{1}{(p+1)(p-1)}\right] = 2(p-1)(p+2)(p+1)\left[\frac{7}{2(p+2)(p+1)}\right]$$

Now simplify.

$$2 \cdot 3(p+1) - 2(p+2) = 7(p-1) \quad \text{Multiply.}$$
$$6p + 6 - 2p - 4 = 7p - 7 \quad \text{Distributive property}$$
$$4p + 2 = 7p - 7 \quad \text{Combine terms.}$$
$$9 = 3p$$
$$3 = p$$

Substitute 3 for p in the original equation to check that the solution set is $\{3\}$. ■

EXAMPLE 5
SOLVING AN EQUATION THAT LEADS TO A QUADRATIC EQUATION

Solve $\dfrac{2}{3x+1} = \dfrac{1}{x} - \dfrac{6x}{3x+1}$.

Multiply both sides by $x(3x+1)$. The resulting equation is

$$2x = (3x+1) - 6x^2.$$

Now solve. Since the equation is quadratic, get 0 on the right side.

$$6x^2 - 3x + 2x - 1 = 0$$
$$6x^2 - x - 1 = 0 \quad \text{Combine terms.}$$
$$(3x+1)(2x-1) = 0 \quad \text{Factor.}$$
$$3x + 1 = 0 \quad \text{or} \quad 2x - 1 = 0 \quad \text{Zero-factor property}$$
$$x = -\frac{1}{3} \quad \text{or} \quad x = \frac{1}{2}$$

Using $-\frac{1}{3}$ in the original equation causes the denominator $3x + 1$ to equal 0, so it is not a solution. The solution set is $\left\{\frac{1}{2}\right\}$. ■

4.7 EXERCISES

Solve each of the following equations. See Examples 1–4.

1. $\dfrac{p}{4} - \dfrac{p}{8} = 2$
2. $\dfrac{m}{6} + \dfrac{m}{3} = 6$
3. $\dfrac{r+6}{3} = \dfrac{r+8}{5}$
4. $\dfrac{2k+1}{5} = \dfrac{k+1}{4}$
5. $\dfrac{3y-6}{2} = \dfrac{5y+1}{6}$
6. $\dfrac{8-p}{3} = \dfrac{6+5p}{2}$
7. $\dfrac{1}{x} = \dfrac{3}{x} - 2$
8. $\dfrac{-3}{y} = 2 + \dfrac{1}{y}$
9. $\dfrac{1}{a} + \dfrac{2}{3a} = \dfrac{1}{3}$
10. $\dfrac{2}{m} + \dfrac{3}{2m} = \dfrac{7}{6}$
11. $\dfrac{1}{r-1} + \dfrac{2}{3r-3} = \dfrac{-5}{12}$
12. $\dfrac{4}{z+2} - \dfrac{1}{3z+6} = \dfrac{11}{9}$

13. $\frac{3}{4a-8} - \frac{2}{3a-6} = \frac{1}{36}$

14. $\frac{1}{x-2} - \frac{1}{6} = \frac{2}{3x-6}$

15. $\frac{5}{6q+14} - \frac{2}{3q+7} = \frac{1}{56}$

16. $\frac{7}{2y+1} + \frac{3}{4y+2} = \frac{17}{2}$

17. $\frac{2}{x-1} + \frac{3}{x+1} = \frac{9}{x^2-1}$

18. $\frac{h-1}{h^2-4} - \frac{2}{h+2} = \frac{4}{h-2}$

19. $\frac{2}{y-5} + \frac{1}{y+5} = \frac{11}{y^2-25}$

20. $\frac{3}{m+2} - \frac{1}{m-2} = \frac{2}{m^2-4}$

Solve each of the following equations. See Example 5.

21. $y + 2 = \frac{24}{y}$

22. $m + 8 + \frac{15}{m} = 0$

23. $2 + \frac{7}{a} = \frac{4}{a^2}$

24. $3 + \frac{13}{m} = \frac{30}{m^2}$

25. $\frac{2}{p} - \frac{3}{p-2} = \frac{-3}{p^2-2p}$

26. $\frac{1}{y} - \frac{5}{y-1} = \frac{4y+2}{y^2-y}$

27. $\frac{5}{x-4} - \frac{3}{x-1} = \frac{x+11}{x^2-5x+4}$

28. $\frac{7}{z-5} - \frac{6}{z+3} = \frac{48}{z^2-2z-15}$

29. $\frac{7}{p-5} = \frac{p^2-10}{p^2-p-20} - 1$

30. $\frac{22}{2a^2-9a-5} - \frac{3}{2a+1} = \frac{2}{a-5}$

31. $\frac{1}{x+2} - \frac{5}{x^2+9x+14} = \frac{-3}{x+7}$

32. $\frac{4}{q+5} + \frac{1}{q+3} = \frac{2}{q^2+8q+15}$

33. $\frac{5m-22}{m^2-6m+9} - \frac{11}{m^2-3m} = \frac{5}{m}$

34. $\frac{r-12}{r^2-10r+25} - \frac{3}{r^2-5r} = \frac{1}{r}$

35. $\frac{2y-29}{y^2+7y-8} = \frac{5}{y+8} - \frac{2}{y-1}$

36. $\frac{3m-1}{m^2+5m-14} = \frac{1}{m-2} - \frac{2}{m+7}$

In Exercises 37–40, solve the equation in part (a), then simplify the expression in part (b).

37. **(a)** $\frac{5}{4a+1} - \frac{3}{2(4a+1)} = 1$

(b) $\frac{5}{4a+1} - \frac{3}{2(4a+1)} + 1$

38. **(a)** $\frac{7}{2y+1} + \frac{3}{4y+2} = \frac{17}{2}$

(b) $\frac{7}{2y+1} + \frac{3}{4y+2} - \frac{17}{2}$

39. **(a)** $\frac{a+3}{a} - \frac{a+4}{a+5} = \frac{15}{a^2+5a}$

(b) $\frac{a+3}{a} - \frac{a+4}{a+5} - \frac{15}{a^2+5a}$

40. **(a)** $\frac{3m-1}{m^2+5m-14} = \frac{1}{m-2} - \frac{2}{m+7}$

(b) $\frac{3m-1}{m^2+5m-14} - \frac{1}{m-2} - \frac{2}{m+7}$

---

41. How does the use of the least common denominator differ when it is used to add (or subtract) fractions compared with its use to solve an equation with fractions?

42. What values cannot be solutions in the equation

$$\frac{1}{x+2} - \frac{1}{x-3} = 4?$$

Why? (See Objective 2.)

Solve each of the following equations. In Exercises 47–50 use substitution and solve the quadratic equation first. For example, in Exercise 47 let $a = \dfrac{x}{x-1}$ *and solve the resulting quadratic equation for a. Then solve for x.*

43. $7p^{-1} + 2p^{-2} - 4 = 0$

44. $-6 + 11p^{-1} - 3p^{-2} = 0$

45. $3p^{-1} + 2(p+1)^{-1} = p^{-1}$

46. $4(x+2)^{-1} - 3x^{-1} = 10(x+2)^{-1}$

47. $\left(\dfrac{x}{x-1}\right)^2 + 2\left(\dfrac{x}{x-1}\right) - 3 = 0$

48. $2\left(\dfrac{a}{a+2}\right)^2 - \left(\dfrac{a}{a+2}\right) - 1 = 0$

49. $4\left(\dfrac{2m}{m+3}\right)^2 + 5\left(\dfrac{2m}{m+3}\right) - 6 = 0$

50. $9\left(\dfrac{y-1}{y+2}\right)^2 - 15\left(\dfrac{y-1}{y+2}\right) - 6 = 0$

PREVIEW EXERCISES

The preview exercises for this section are a combination of the work with applied problems from Section 2.3 and the work on solving equations in this section. Write an equation for each of the following, and then solve to find the unknown number.

51. Half a number is added to two thirds of the same number, giving 10.

52. Twice a number is subtracted from one third of the number, giving 5.

53. The reciprocal of a number is added to 3, giving 4.

54. When 2 is subtracted from the reciprocal of a number, the result is 1.

55. The reciprocal of 2, plus the reciprocal of 3, gives the reciprocal of what number?

56. When the reciprocals of 5 and 4 are added together, the result is the reciprocal of what number?

SUMMARY OF RATIONAL EXPRESSIONS: OPERATIONS AND EQUATIONS

A common student error is to confuse an *equation,* such as

$$\frac{x}{2} + \frac{x}{3} = -5,$$

with an *addition problem*, such as

$$\frac{x}{2} + \frac{x}{3}.$$

Equations are solved to get a numerical answer, while addition (or subtraction) problems result in simplified expressions as shown on the next page.

Solve: $\frac{x}{2} + \frac{x}{3} = -5.$

Multiply both sides by 6.

$$6\left(\frac{x}{2} + \frac{x}{3}\right) = 6(-5)$$
$$3x + 2x = -30$$
$$5x = -30$$
$$x = -6$$

Check that the solution set is $\{-6\}$.

Add: $\frac{x}{2} + \frac{x}{3}.$

The least common denominator is 6. Write both fractions with a denominator of 6.

$$\frac{x \cdot 3}{2 \cdot 3} + \frac{x \cdot 2}{3 \cdot 2}$$
$$\frac{3x}{6} + \frac{2x}{6}$$
$$\frac{3x + 2x}{6}$$
$$\frac{5x}{6}$$

EXERCISES

In the following exercises, either perform the indicated operation or solve the given equation, as appropriate.

1. $\frac{a}{2} - \frac{a}{5} = 3$

2. $\frac{12z^2x^3}{8zx^4} \cdot \frac{3x^5}{7x}$

3. $\frac{-2y}{(y-2)^2} \cdot \frac{y^2-4}{5}$

4. $\frac{3}{p} - \frac{11}{2p}$

5. $\frac{5}{m} - \frac{10}{3m}$

6. $\frac{\frac{1}{x} - \frac{1}{y}}{\frac{1}{x} + \frac{1}{y}}$

7. $\frac{3}{z} + \frac{5}{7z} = \frac{52}{7}$

8. $\frac{k}{7} + \frac{k}{3} = 10$

9. $\frac{r-5}{3} - \frac{r-2}{5} = -\frac{1}{3}$

10. $\frac{8}{5r} + \frac{1}{3r}$

11. $\frac{2}{q^2} - \frac{3}{2q}$

12. $\frac{1}{x} - \frac{1}{2x} = 1$

13. $\frac{\frac{5}{m} - \frac{2}{m+1}}{\frac{3}{m+1} + \frac{1}{m}}$

14. $\frac{2}{q+2} - \frac{3}{4q+8} = \frac{1}{4}$

15. $\frac{1}{r-2} + \frac{1}{r+1} = \frac{2}{r^2-r-2}$

16. $\frac{3p(p-2)}{p+5} \div \frac{p^2-4}{4p+20}$

17. $\frac{9}{2-x} \cdot \frac{4x-8}{5}$

18. $\frac{m-4}{3} - \frac{m+1}{2} = -\frac{11}{6}$

19. $\frac{a^2+a-2}{a^2-4} \cdot \frac{a^2-2a}{a^2+3a-4}$

20. $\frac{2q^2+5q-3}{q^2+q-6} \div \frac{10q^2-5q}{3q^3-6q^2}$

21. $\frac{5}{a^2-2a} + \frac{3}{a^2-4}$

22. $\frac{3}{k+1} + \frac{2}{5k+5} = \frac{17}{15}$

23. $\frac{m}{m-2} + 4 = \frac{2}{m-2}$

24. $\frac{r^2-1}{r^2+3r+2} \div \frac{r^2-2r+1}{r^2+2r-3}$

25. $\frac{\frac{3}{k} - \frac{5}{m}}{\frac{9m^2-25k^2}{km^2}}$

26. $\frac{4y-20}{y^2-3y-10} \cdot \frac{2y^2-3y-14}{16y^3+8y^2}$

27. $\dfrac{p^2+5p+6}{p^2-3p+2}\cdot\dfrac{p^2+3p-10}{p^2+8p+15}$

28. $\dfrac{4}{r^2-2r-8}=\dfrac{1}{2r^2+5r+2}$

29. $\dfrac{8}{3p+9}-\dfrac{2}{5p+15}=\dfrac{8}{15}$

30. $\dfrac{4y^2+11y-3}{6y^2-5y-6}\div\dfrac{4y^2-13y+3}{2y^2-9y+9}$

31. $\dfrac{2}{m^2+m-2}=\dfrac{3}{m^2+5m+6}$

32. $\dfrac{\dfrac{2}{r}+\dfrac{5}{s}}{\dfrac{4s^2-25r^2}{3rs}}$

33. $\dfrac{2}{q+1}-\dfrac{3}{q^2-q-2}=\dfrac{3}{q-2}$

34. $\dfrac{3}{r+1}+5=\dfrac{-3r}{3r+3}$

35. $\dfrac{r}{r^2+3rp+2p^2}-\dfrac{1}{r^2+5rp+6p^2}$

36. $\dfrac{7}{2p^2-8p}+\dfrac{3}{p^2-16}$

37. $\dfrac{2x^2-11xy+15y^2}{2x^2+5xy-3y^2}\cdot\dfrac{8x^2-2xy-y^2}{8x^2-18xy-5y^2}$

38. $\dfrac{-2}{a^2+2a-3}+\dfrac{5}{3a-3}=\dfrac{4}{3a+9}$

39. $\dfrac{8m^2-23mp-3p^2}{16m^2-8mp-3p^2}\div\dfrac{2m^2-3mp-9p^2}{8m^2+14mp+3p^2}$

40. $\dfrac{4y}{y^2+yz-2z^2}-\dfrac{2y}{y^2+5yz+6z^2}$

4.8 APPLICATIONS

OBJECTIVES

1. FIND THE VALUE OF AN UNKNOWN VARIABLE IN A FORMULA.
2. SOLVE A FORMULA FOR A SPECIFIED VARIABLE.
3. SOLVE APPLICATIONS ABOUT NUMBERS.
4. SOLVE APPLICATIONS ABOUT WORK.
5. SOLVE APPLICATIONS ABOUT DISTANCE.

FOCUS ON PROBLEM SOLVING

Suppose that Hortense and Mort can clean their entire house in 7 hours, while their toddler, Mimi, just by being around, can completely mess it up in only 2 hours. If Hortense and Mort clean the house while Mimi is at her grandma's, and then start cleaning up after her the minute she gets home, how long does it take until the whole place is in a shambles?

This is a problem relating the rates at which different people work at a job and the time required to do the job. As we shall see, the equation required to solve the problem involves rational expressions, so we need to use the techniques presented in the previous section. This problem is Exercise 47 in the exercises for this section. After working through this section you should be able to solve this problem.

Many common formulas involve rational expressions. Methods of working with these formulas are given in this section.

1 The first example shows how to find the value of an unknown variable in a formula.

EXAMPLE 1
FINDING THE VALUE OF A VARIABLE IN A FORMULA

In physics, the focal length, f, of a lens is given by the formula

$$\frac{1}{f} = \frac{1}{p} + \frac{1}{q},$$

where p is the distance from the object to the lens and q is the distance from the lens to the image. Find q if $p = 20$ centimeters and $f = 10$ centimeters.

Replace f with 10 and p with 20.

$$\frac{1}{f} = \frac{1}{p} + \frac{1}{q}$$

$$\frac{1}{10} = \frac{1}{20} + \frac{1}{q} \qquad \text{Let } f = 10,\ p = 20$$

Multiply both sides by the least common denominator, $20q$.

$$20q \cdot \frac{1}{10} = 20q \cdot \frac{1}{20} + 20q \cdot \frac{1}{q}$$

$$2q = q + 20$$

$$q = 20$$

The distance from the lens to thc image is 20 centimeters. ■

2 The next examples show how to solve a formula for a specified variable.

EXAMPLE 2
SOLVING A FORMULA FOR A SPECIFIED VARIABLE

Solve $\frac{1}{f} = \frac{1}{p} + \frac{1}{q}$ for p.

Begin by multiplying both sides by the common denominator fpq.

$$fpq \cdot \frac{1}{f} = fpq\left(\frac{1}{p} + \frac{1}{q}\right)$$

$$pq = fq + fp \qquad \text{Distributive property}$$

Get the terms with p (the specified variable) on the same side of the equation. To do this, subtract fp on both sides.

$$pq - fp = fq \qquad \text{Subtract } fp.$$

$$p(q - f) = fq \qquad \text{Distributive property; factor out } p.$$

$$p = \frac{fq}{q - f} \qquad \text{Divide by } q - f. \ ■$$

EXAMPLE 3 SOLVING A FORMULA FOR A SPECIFIED VARIABLE

Solve $I = \frac{nE}{R + nr}$ for n.

First, multiply both sides by $R + nr$.

$$(R + nr)I = (R + nr)\frac{nE}{R + nr}$$

$$RI + nrI = nE$$

$$RI = nE - nrI \qquad \text{Subtract } nrI.$$

$$RI = n(E - rI) \qquad \text{Distributive property; factor out } n.$$

$$\frac{RI}{E - rI} = n \qquad \text{Divide by } E - rI.$$ ■

CAUTION Refer to the steps in Examples 2 and 3 that use the distributive property. This is a step that often gives students difficulty. Remember that the variable for which you are solving *must* be a factor on one side of the equation so that in the last step, both sides are divided by the remaining factor there. The *distributive property* allows us to perform this factorization.

PROBLEM SOLVING

We are now able to solve problems that translate into equations with rational expressions. The strategy for solving these problems is the same as we have used in earlier chapters. Notice how we continue to use the six steps for problem solving from Chapter 2. ■

3 The next example shows a problem about numbers. These problems are included mainly to give practice in setting up and solving equations with rational expressions.

EXAMPLE 4 SETTING UP AND SOLVING A PROBLEM ABOUT NUMBERS

If the numerator of the fraction 6/7 is multiplied by a number, and the same number is added to the denominator, the resulting fraction equals 9/5. What is the number?

Let x represent the unknown number. From the statement of the problem, the numerator of 6/7, which is 6, is to be multiplied by x. Also, x is to be added to the denominator, 7. The result should equal 9/5. This gives the equation

the numerator of $\frac{6}{7}$ times a number → $6x$; equals → $\frac{9}{5}$; the same number added to the denominator → $7 + x$

$$\frac{6x}{7 + x} = \frac{9}{5}$$

The least common denominator is $5(7 + x)$. Multiply both sides of the equation by $5(7 + x)$.

$$5(7 + x) \cdot \frac{6x}{7 + x} = 5(7 + x) \cdot \frac{9}{5}$$
$$30x = 9(7 + x)$$
$$30x = 63 + 9x$$
$$21x = 63$$
$$x = 3$$

The required number is 3. Check this result in the words of the original problem. ■

4 Problems about work are closely related to the distance problems we discussed in Section 2.4.

PROBLEM SOLVING

People work at different rates. Let the letters r, t, and A represent the rate at which the work is done, the time required, and the amount of work accomplished, respectively. Then $A = rt$. Notice the similarity to the distance formula, $d = rt$. The amount of work is often measured in terms of jobs accomplished. Thus, if 1 job is completed, $A = 1$, and the formula gives

$$1 = rt$$
$$r = \frac{1}{t}$$

as the rate.

RATE OF WORK If a job can be accomplished in t units of time, then the rate of work is

$$\frac{1}{t} \text{ job per unit of time.}$$

In solving a work problem, we begin by using this fact to express all rates of work. ■

EXAMPLE 5
SOLVING A PROBLEM ABOUT WORK

Lindsay and Michael are working on a neighborhood cleanup. Michael can clean up all the trash in the area in 7 hours, while Lindsay can do the same job in 5 hours. How long will it take them if they work together?

Let x = the number of hours it will take the two people working together. Just as we made a chart for the distance formula, $d = rt$, we can make a chart here for $A = rt$, with $A = 1$. Since $A = 1$, the rate for each person will be $1/t$, where t is the time it takes each person to complete the job alone. For example, since Michael can clean up all the trash in 7 hours, his rate is 1/7 of the job per hour. Similarly, Lindsay's rate is 1/5 of the job per hour. Fill in the chart as shown.

WORKER	RATE	TIME WORKING TOGETHER	FRACTIONAL PART OF THE JOB DONE
Michael	$\frac{1}{7}$	x	$\frac{1}{7}x$
Lindsay	$\frac{1}{5}$	x	$\frac{1}{5}x$

Since together they complete 1 job, the sum of the fractional parts accomplished by each of them should equal 1.

$$\begin{array}{ccccc} \text{part done by Michael} & + & \text{part done by Lindsay} & = & \text{1 whole job} \\ \frac{1}{7}x & + & \frac{1}{5}x & = & 1 \end{array}$$

Solve this equation. The least common denominator is 35.

$$35\left(\frac{1}{7}x + \frac{1}{5}x\right) = 35 \cdot 1$$

$$5x + 7x = 35$$

$$12x = 35$$

$$x = \frac{35}{12}$$

Working together, Michael and Lindsay can do the entire job in 35/12 hours, or 2 hours and 55 minutes. Check this result in the original problem. ■

5 The final examples use the distance formula, $d = rt$. A method for solving applications of the distance formula was discussed in Section 2.4. The same approach applies here.

EXAMPLE 6
SOLVING A PROBLEM ABOUT DISTANCE, RATE, AND TIME

At the airport, Cheryl and Bill are walking to the gate (at the same speed) to catch their flight to Akron, Ohio. Since Bill wants a window seat, he steps onto the moving sidewalk and continues to walk while Cheryl uses the stationary sidewalk. If the sidewalk moves at 1 meter per second and Bill saves 50 seconds covering the 300-meter distance, what is their walking speed?

Let x represent their walking speed. Then Cheryl travels at x meters per second and Bill travels at $x + 1$ meters per second. Since Bill's time is 50 seconds less than Cheryl's time, express their times in terms of the known distances and the variable rates. Start with $d = rt$ and divide both sides by r to get

$$t = \frac{d}{r}.$$

For Cheryl, distance is 300 meters and the rate is x. Cheryl's time is

$$t = \frac{d}{r} = \frac{300}{x}.$$

Bill goes 300 meters at a rate of $x + 1$, so his time is

$$t = \frac{d}{r} = \frac{300}{x + 1}.$$

This information is summarized in the following chart.

Solve each formula for the specified variable. See Examples 2 and 3.

7. $I = \frac{E}{R}$; for R (electricity)

8. $r = \frac{I}{Pt}$; for t (business)

9. $F = \frac{Gm_1m_2}{d^2}$; for m_1 (physics)

10. $F = \frac{Gm_1m_2}{d^2}$; for G (physics)

11. $\frac{1}{a} = \frac{1}{b} + \frac{1}{c}$; for c (electricity)

12. $\frac{1}{f} = \frac{1}{p} + \frac{1}{q}$; for q (optics)

13. $\frac{PV}{T} = \frac{pv}{t}$; for T (chemistry)

14. $\frac{PV}{T} = \frac{pv}{t}$; for v (chemistry)

15. $a = \frac{V - v}{t}$; for V (physics)

16. $S = \frac{a(r^n - 1)}{r - 1}$; for a (mathematics)

17. $S = \frac{n}{2}(a + A)$; for a (mathematics)

18. $A = \frac{1}{2}h(b + B)$; for B (mathematics)

19. $t = a + (n - 1)d$; for n (mathematics)

20. $S = \frac{n}{2}[2a + (n - 1)d]$; for d (mathematics)

---

21. To solve the equation $m = \frac{ab}{a - b}$ for a, what is the first step? (See Objective 2.)

22. Suppose you get the equation

$$rp - rq = p + q$$

to be solved for r. What is the next step? (See Objective 2.)

In Exercises 23–26, select a variable, state what it represents, and write an equation to solve the problem. (Do not solve these problems.) See Example 4.

23. What number must be added to both the numerator and denominator of 15/17 to make the result equal 5/6?

24. What number must be added to the denominator of 15/11 to make the result equal 5/3?

25. Two integers differ by 1. The sum of their reciprocals equals 7 divided by their product. Find the integers.

26. When 7 is added to the reciprocal of twice a number, the result is the reciprocal of four times the number. Find the number.

Solve the following problems. See Examples 4–7.

27. When the reciprocal of five times a number is subtracted from 1, the result is the reciprocal of the number. Find the number.

28. Half the reciprocal of a number is added to 1/4, giving the reciprocal of seven times the number. Find the number.

29. Jim and Ali want to paint Jim's motorboat. Jim can do it in 5 hours, and Ali can do it in 6 hours. How long will it take them working together?

30. Guadalupe and Roberto are refinishing an antique chest. Working alone, Guadalupe can do the job in 12 hours. When Roberto helps, the total job takes only 8 hours. How long would it take Roberto if he worked alone?

31. Chun Lung Lau can canvass his neighborhood for a political campaign in 3 hours. His daughter needs 5 hours to canvass the same neighborhood. How long will it take them if they work together?

32. Chuck can paint a room in 8 hours when working alone. If Walt helps him, the total job takes 6 hours. How long would it take Walt if he worked alone?

33. Zubeir averages 12 miles per hour riding her bike to school. Averaging 36 miles per hour by car takes her one-half hour less time. How far does she travel to school?

	d	r	t
Bike	x	12	
Car	x	36	

34. A bus can travel 80 miles in the same time that a train goes 96 miles. The speed of the train is 10 miles per hour faster than the speed of the bus. Find both speeds.

	d	r	t
Bus	80	x	
Train	96	$x + 10$	

35. In a 500 meter race at a track meet, Sue finished 25 seconds ahead of Mei Lin. Sue's speed was 1 meter per second faster than Mei Lin's. How fast did each woman run?

36. A canal has a current of 2 miles per hour. Byron Hopkins travels in a houseboat 22 miles downstream in the same time it takes to go 16 miles upstream. Find the speed of the boat in still water.

Solve each formula for the specified variable.

37. $y = \dfrac{x + z}{a - x}$; for x

38. $F = f\left(\dfrac{v + v_o}{v - v_s}\right)$; for v

39. $A = \dfrac{Rr}{R + r}$; for R

40. $I = \dfrac{nE}{R + nr}$; for r

41. $\dfrac{E}{e} = \dfrac{R + r}{r}$; for r

42. $\dfrac{E}{e} = \dfrac{R + r}{r}$; for e

Solve each of the following problems.

43. Carroll and Elaine want to clean up an office that they share. Carroll can do the job alone in 5 hours, while Elaine can do it alone in 2 hours. How long will it take them if they work together?

44. The hot-water tap can fill a tub in 20 minutes. The cold-water tap takes 15 minutes to fill the tub. How long would it take to fill the tub with both taps open?

45. Machine A can complete a certain job in 2 hours. To speed up the work, Machine B, which could complete the job alone in 3 hours, is brought in to help. How long will it take the two machines to complete the job working together?

46. A cannery has a vat for tomato sauce. An inlet pipe can fill the vat in 12 hours, and an outlet pipe can empty the vat in 18 hours. How long will it take to fill the vat if both the outlet and inlet are open?

47. Suppose that Hortense and Mort can clean their entire house in 7 hours, while their toddler Mimi, just by being around, can completely mess it up in only 2 hours. If Hortense and Mort clean the house while Mimi is at her grandma's, and then start cleaning up after her the minute she gets home, how long does it take until the whole place is in a shambles?

48. An inlet pipe can fill a barrel of wine in 6 hours, and an outlet pipe can empty it in 8 hours. Through an error both pipes are left on. How long will it take for the barrel to fill?

49. A truck driver delivered a load to Richmond from Washington, D.C., averaging 50 miles per hour for the trip. He picked up another load in Richmond and delivered it to New York City, averaging 55 miles per hour on this part of the trip. The distance from Richmond to New York is 120 miles more than the distance from Washington to Richmond. If the entire trip required 6 hours driving time, find the distance from Washington to Richmond.

50. Janet and Russ agree to meet at the Reno airport. Janet travels 250 kilometers, and Russ travels 300 kilometers. If Russ's speed is 20 kilometers per hour greater and they both spend the same amount of time traveling, at what speed does each travel?

51. On the first part of a trip to Carmel, traveling on the freeway, Marge averaged 60 miles per hour. On the rest of the trip, which was 10 miles longer than the first part, she averaged 50 miles per hour. Find the distance to Carmel, if the first part of the trip took one-half hour less time than the second part.

52. John averages 30 miles per hour when he drives to his favorite fishing hole on the old highway, but he averages 50 miles per hour when most of his route is on the interstate. If he saves 2 hours by traveling on the interstate, how far away is the fishing hole? (Assume equal distances on each route.)

In geometry, it is shown that two triangles with corresponding angles equal, called similar triangles, *have corresponding sides proportional. For example, in the figure, angle A = angle D, angle B = angle E, and angle C = angle F, so the triangles are similar. Then the following ratios of corresponding sides are equal.*

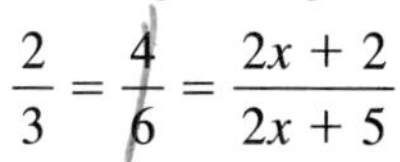

$$\frac{2}{3} = \frac{4}{6} = \frac{2x + 2}{2x + 5}$$

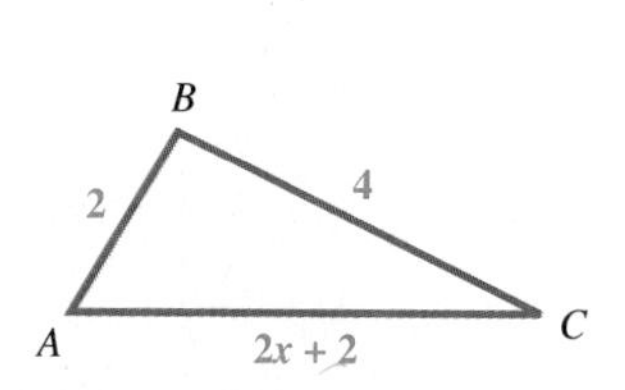

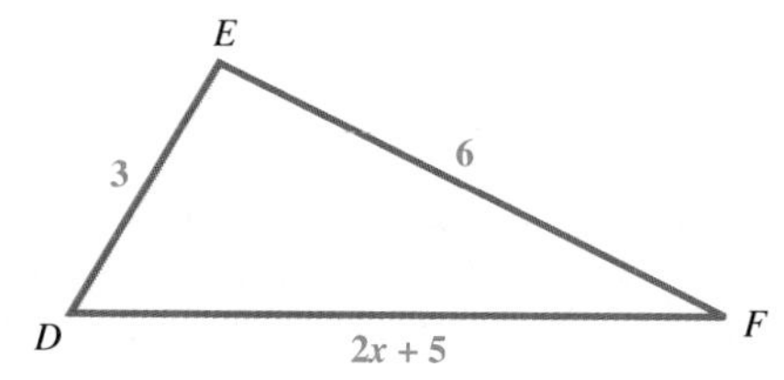

53. Solve for x using the proportion given above to find the lengths of the third sides of the triangles.

54. Suppose the triangles shown in the figure are similar. Find y and the lengths of the two longer sides of each triangle.

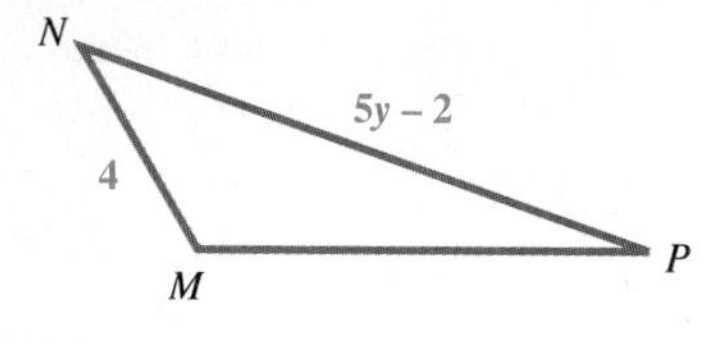

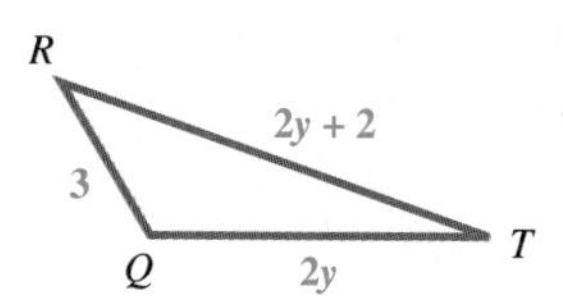

PREVIEW EXERCISES

Evaluate each expression. See Section 1.3.

55. $\sqrt{4}$ **56.** $\sqrt{9}$ **57.** $\sqrt[3]{8}$ **58.** $\sqrt[3]{27}$ **59.** $\sqrt[5]{32}$ **60.** $\sqrt[4]{16}$

CHAPTER 4 GLOSSARY

KEY TERMS

4.1 rational expression A rational expression (or algebraic fraction) is the quotient of two polynomials with denominator not 0.

4.2 reciprocals Two rational expressions are reciprocals if their product is 1.

4.3 common denominator A common denominator of several denominators is an expression that is divisible by each of these denominators.

4.4 complex fraction A fraction that has a fraction in the numerator, the denominator, or both, is called a complex fraction.

4.6 synthetic division Synthetic division is a shortcut division process used when dividing a polynomial by a binomial of the form $x - k$.

CHAPTER 4 QUICK REVIEW

SECTION	CONCEPTS	EXAMPLES
4.1 BASICS OF RATIONAL EXPRESSIONS	**Writing in Lowest Terms** *Step 1* Factor both numerator and denominator to find their greatest common factor. *Step 2* Divide both numerator and denominator by their greatest common factor.	Write in lowest terms. $\frac{2x+8}{x^2-16} = \frac{2(x+4)}{(x-4)(x+4)}$ $= \frac{2}{x-4}$
4.2 MULTIPLICATION AND DIVISION OF RATIONAL EXPRESSIONS	**Multiplying Rational Expressions** *Step 1* Factor all numerators and denominators as completely as possible. *Step 2* Divide both numerator and denominator by any factors that both have in common. *Step 3* Multiply remaining factors in the numerator, and multiply remaining factors in the denominator. *Step 4* Be certain that the product is in lowest terms.	Multiply: $\frac{x^2+2x+1}{x^2-1} \cdot \frac{5}{3x+3}$ $= \frac{(x+1)^2}{(x-1)(x+1)} \cdot \frac{5}{3(x+1)}$ $= \frac{5}{3(x-1)}$

SECTION	CONCEPTS	EXAMPLES
	Dividing Rational Expressions To divide two rational expressions, *multiply* the first by the reciprocal of the second.	Divide: $\frac{2x+5}{x-3} \div \frac{2x^2+3x-5}{x^2-9}$. $= \frac{2x+5}{x-3} \cdot \frac{(x+3)(x-3)}{(2x+5)(x-1)}$ $= \frac{x+3}{x-1}$
4.3 ADDITION AND SUBTRACTION OF RATIONAL EXPRESSIONS	**Adding or Subtracting Rational Expressions** *Step 1(a)* If the denominators are the same, add or subtract the numerators. Place the result over the common denominator. *Step 1(b)* If the denominators are different, first find the least common denominator. Write all rational expressions with this least common denominator, and then add or subtract the numerators. Place the result over the common denominator. *Step 2* Write all answers in lowest terms.	Subtract: $\frac{1}{x+6} - \frac{3}{x+2}$. $= \frac{x+2}{(x+6)(x+2)} - \frac{3(x+6)}{(x+6)(x+2)}$. $= \frac{x+2-3(x+6)}{(x+6)(x+2)}$ $= \frac{x+2-3x-18}{(x+6)(x+2)}$ $= \frac{-2x-16}{(x+6)(x+2)}$
4.4 COMPLEX FRACTIONS	**Simplifying Complex Fractions** *Method 1* Simplify the numerator and denominator separately, as much as possible. Then multiply the numerator by the reciprocal of the denominator. Write the answer in lowest terms.	Simplify the complex fraction. *Method 1* $\frac{\frac{1}{2}+\frac{1}{3}}{\frac{1}{4}-\frac{1}{2}} = \frac{\frac{3}{6}+\frac{2}{6}}{\frac{1}{4}-\frac{2}{4}}$ $= \frac{\frac{5}{6}}{\frac{-1}{4}} = \frac{5}{6} \cdot \frac{4}{-1}$ $= \frac{20}{-6} = -\frac{10}{3}$

SECTION	CONCEPTS	EXAMPLES
	Method 2 Multiply the numerator and denominator of the complex fraction by the least common denominator of all fractions appearing in the complex fraction. Then simplify the results.	*Method 2* $$\frac{\frac{1}{2}+\frac{1}{3}}{\frac{1}{4}-\frac{1}{2}} = \frac{12\left(\frac{1}{2}\right)+12\left(\frac{1}{3}\right)}{12\left(\frac{1}{4}\right)-12\left(\frac{1}{2}\right)}$$ $$= \frac{6+4}{3-6} = \frac{10}{-3} = -\frac{10}{3}$$
4.5 DIVISION OF POLYNOMIALS	**Dividing by a Monomial** To divide a polynomial by a monomial, divide each term in the polynomial by the monomial, and then write each fraction in lowest terms.	Divide: $\frac{2x^3 - 4x^2 + 6x - 8}{2x}$. $$= \frac{2x^3}{2x} - \frac{4x^2}{2x} + \frac{6x}{2x} - \frac{8}{2x}$$ $$= x^2 - 2x + 3 - \frac{4}{x}$$
	Dividing by a Polynomial Use the "long division" process.	Divide: $\frac{m^3 - m^2 + 2m + 5}{m + 1}$. $$\begin{array}{r} m^2 - 2m + 4 \\ m + 1\overline{)m^3 - m^2 + 2m + 5} \\ \underline{m^3 + m^2 } \\ -2m^2 + 2m \\ \underline{-2m^2 - 2m } \\ 4m + 5 \\ \underline{4m + 4} \\ 1 \end{array}$$ The quotient is $$m^2 - 2m + 4 + \frac{1}{m+1}.$$
4.7 EQUATIONS WITH RATIONAL EXPRESSIONS	To solve an equation involving rational expressions, multiply all the terms in the equation by the least common denominator. Then solve the resulting equation. Each potential solution must be checked to make sure that no denominator in the original equation is 0.	Solve for x. $$\frac{1}{x} + x = \frac{26}{5}$$ $$5 + 5x^2 = 26x \quad \text{Multiply by } 5x.$$ $$5x^2 - 26x + 5 = 0$$ $$(5x - 1)(x - 5) = 0$$ $$x = \frac{1}{5} \quad \text{or} \quad x = 5$$ Both check. The solution set is $\left\{\frac{1}{5}, 5\right\}$.

SECTION	CONCEPTS	EXAMPLES
4.8 APPLICATIONS	To solve a formula for a particular variable, get that variable alone on one side by following the method described in Section 4.7.	Solve for L. $c = \frac{100b}{L}$ $cL = 100b$ Multiply by L. $L = \frac{100b}{c}$ Divide by c.
	If an applied problem translates into an equation with rational expressions, solve the equation using the method described in Section 4.7.	If the same number is added to the numerator and subtracted from the denominator of $\frac{5}{7}$, the result is 1. Find the number. Let x represent the number. The equation is $\frac{5+x}{7-x} = 1.$ Multiply both sides by $7 - x$. $5 + x = 7 - x$ $2x = 2$ $x = 1$ The number is 1. To check, add 1 to 5 to get 6, and subtract 1 from 7 to get 6. Since $\frac{6}{6} = 1$, the answer is correct.

CHAPTER 4 REVIEW EXERCISES

[4.1] *Find all numbers that make the following rational expressions undefined.*

1. $\dfrac{-7}{9k + 18}$

2. $\dfrac{5r - 1}{r^2 - 7r + 10}$

3. $\dfrac{-1}{m^2 + 100}$

4. Are $\dfrac{x^2 - 16}{x + 4}$ and $x - 4$ equivalent for any value of x? Why?

Write each rational expression in lowest terms.

5. $\dfrac{25m^4p^3}{10m^5p}$

6. $\dfrac{12x^2 + x}{24x + 2}$

7. $\dfrac{y^2 + 3y - 10}{y^2 - 5y + 6}$

8. $\dfrac{25m^2 - k^2}{25m^2 - 10mk + k^2}$

9. $\dfrac{r^3 + s^3}{r + s}$

10. Does $\dfrac{2x - 2}{x - 2} = 2$? Explain.

Rewrite each rational expression with the indicated denominator.

11. $\frac{2}{r-3}$; $r^2 - 9$

12. $\frac{-4}{7m}$; $7m + 14m^2$

13. $\frac{y}{2y+1}$; $2y^2 - 5y - 3$

14. $\frac{4}{p-5}$; $5 - p$

[4.2] *Multiply or divide. Write all answers in lowest terms.*

15. $\frac{y^5}{6} \cdot \frac{9}{y^4}$

16. $\frac{25p^3q^2}{8p^4q} \div \frac{15pq^2}{16p^5}$

17. $\frac{(2y+3)^2}{5y} \cdot \frac{15y^3}{4y^2-9}$

18. $\frac{w^2-16}{w} \cdot \frac{3}{4-w}$

19. $\frac{y^2+2y}{y^2+y-2} \div \frac{y-5}{y^2+4y-5}$

20. $\frac{z^2-z-6}{z-6} \cdot \frac{z^2-6z}{z^2+2z-15}$

21. $\frac{9y^2+46y+5}{3y^2-2y-1} \cdot \frac{y^3+5y^2-6y}{y^2+11y+30}$

22. $\frac{a^3-b^3}{a^2-b^2} \div \frac{a^2+ab+b^2}{a^2+ab}$

[4.3] *Find the least common denominator for each of the following groups of denominators.*

23. p, $p + 6$

24. $32b^3$, $24b^5$

25. $9r^2$, $3r + 1$

26. $6k + 3$, $10k + 5$, $18k + 9$

27. $p^2 + 7p + 10$, $p^2 - 6p - 16$

Add or subtract as indicated.

28. $\frac{9}{m} + \frac{6}{m}$

29. $\frac{7}{3k} - \frac{16}{3k}$

30. $\frac{2}{y} - \frac{3}{8y}$

31. $\frac{3}{t-2} - \frac{5}{2-t}$

32. $\frac{2}{y+1} + \frac{6}{y-1}$

33. $\frac{2m}{m^2-25} - \frac{3}{m+5}$

34. $\frac{2z}{2z^2-zy-y^2} - \frac{3z}{3z^2-5zy+2y^2}$

35. $\frac{3r}{10r^2-3rs-s^2} + \frac{2r}{2r^2+rs-s^2}$

36. Does $\frac{1}{m} + \frac{1}{n} = \frac{1}{m+n}$? Explain why.

[4.4] *Simplify each complex fraction.*

37. $\dfrac{\frac{6}{p}}{\frac{8}{p+5}}$

38. $\dfrac{\frac{4m^5n^6}{mn}}{\frac{8m^7n^3}{m^4n^2}}$

39. $\dfrac{\frac{3}{x} - 5}{6 + \frac{1}{x}}$

40. $\dfrac{\frac{3}{p} - \frac{2}{q}}{\frac{9q^2-4p^2}{qp}}$

41. $\dfrac{\frac{y}{2} - \frac{3}{y}}{1 + \frac{y+1}{y}}$

42. $\dfrac{\frac{2}{m-3n}}{\frac{1}{3n-m}}$

43. $2g^{-1} + 3h^{-2}$

44. $\dfrac{b^{-1} + d^{-1}}{b^{-1}}$

45. $\dfrac{3x^{-1} + 2q^{-1}}{x^{-1} - 4q^{-1}}$

[4.5] *Divide.*

46. $\dfrac{18r - 32}{12}$

47. $\dfrac{5m^3 - 11m^2 + 10m}{5m}$

48. $\dfrac{20y^3x^3 + 15y^4x + 25yx^4}{10yx^2}$

49. $\dfrac{m^2 - 8m + 15}{m - 5}$

50. $\dfrac{8x^2 - 23x + 2}{x - 3}$

51. $\dfrac{5m^4 + 15m^3 - 33m^2 - 10m + 32}{5m^2 - 3}$

52. $2p - 3\overline{)2p^3 + 9p^2 - 34p + 27}$

53. $3y^2 + 1\overline{)12y^4 - 6y^3 + 7y^2 - 2y + 1}$

[4.6] *Use synthetic division for each of the following.*

54. $\dfrac{3p^2 - p - 2}{p - 1}$

55. $\dfrac{10k^2 - 3k - 15}{k + 2}$

56. $(5y^3 + 8y^2 - 6y - 4) \div (y + 2)$

57. $(2k^3 - 5k^2 + 12) \div (k - 3)$

58. $(3z^4 + 14z^3 - 22z^2 + 14z + 18) \div (z + 6)$

59. $(9a^5 + 35a^4 + 19a^2 + 18a + 15) \div (a + 4)$

60. $\dfrac{y^3 - 1}{y - 1}$

61. $\dfrac{x^6 + 1}{x + 1}$

Use synthetic division to evaluate P(k) for the given value of k.

62. $k = -1;\quad P(x) = 3x^3 - 5x^2 + 4x - 1$

63. $k = 3;\quad P(z) = z^4 - 2z^3 - 9z - 5$

Use synthetic division to decide whether or not −5 is a solution of the following equations.

64. $2w^3 + 8w^2 - 14w - 20 = 0$

65. $-3q^4 + 2q^3 + 5q^2 - 9q + 10 = 0$

[4.7] *Solve each equation.*

66. $\dfrac{y}{3} - \dfrac{y}{7} = 1$

67. $\dfrac{3z + 5}{2} = \dfrac{2z - 1}{3}$

68. $\dfrac{2}{k} + \dfrac{1}{5k} = 1$

69. $\dfrac{5y}{y + 1} - \dfrac{y}{3y + 3} = \dfrac{-56}{6y + 6}$

70. $\dfrac{2}{m - 1} + \dfrac{1}{m + 1} = \dfrac{4m + 1}{m^2 - 1}$

71. $1 - \dfrac{5}{x} + \dfrac{4}{x^2} = 0$

72. $2 + \dfrac{7}{p} = \dfrac{4}{p^2}$

73. $\dfrac{3 + p}{p^2 - 5p + 4} - \dfrac{2}{p^2 - 4p} = \dfrac{1}{p}$

[4.8] *Solve the following problems.*

74. According to a law about electrical circuits, $\frac{1}{R} = \frac{1}{S} + \frac{1}{T}$. Find R if $S = 30$ and $T = 10$.

75. Another law from physics says that $F = \frac{GmM}{d^2}$. Find G if $F = 6.67 \times 10^{-10}$, $m = 10$, $M = 25$, and $d = 5$.

76. In the law given in Exercise 75, find m if $F = 78$, $G = 6.67 \times 10^{-11}$, $M = 6 \times 10^{22}$, and $d = 1.6 \times 10^6$.

Solve each formula for the specified variable.

77. $F = \frac{GmM}{d^2}$; for M

78. $\frac{VP}{T} = \frac{vp}{t}$; for v

79. $a = \frac{V - v}{t}$; for v

80. $S = \frac{n}{2}(a + A)$; for A

81. $A = a + (n - 1)d$; for d

82. $S = \frac{n}{2}[2a + (n - 1)d]$; for a

Solve each problem.

83. What number must be added to the numerator and the denominator of 9/4 to make the result equal 3?

84. When twice the reciprocal of three times a number is subtracted from 3, the result is seven times the reciprocal of three times the number. Find the number.

85. At an Olympic swim meet, Azikiwe Ayo took first place in the 100 meter freestyle. His time was 2 seconds less than another swimmer. Find the rate for each man if their rates differed by 1/13 meters per second.

86. A sink can be filled by a cold-water tap in 8 minutes, and it can be filled by the hot-water tap in 12 minutes. How long would it take to fill the sink with both taps open?

87. A bathtub can be filled with water in 20 minutes. The drain can empty it in 30 minutes. How long will it take to fill the bathtub if the water is on and the drain is open?

88. Don Reynolds can get to the mountains using either of two roads. On the old road, he can average 30 miles per hour, and he can average 50 miles per hour on the new road. If both roads are the same length and if he gets to the mountains 1 1/5 hours sooner using the new road, how far does he drive to get to the mountains?

89. A river has a current of 4 kilometers per hour. Find the speed of Jane Gunton's boat in still water if it goes 40 kilometers downstream in the same time as it goes 24 kilometers upstream.

MIXED REVIEW EXERCISES

Perform the indicated operations.

90. $(15p^2 + 11p - 17) \div (3p - 2)$

91. $\dfrac{8}{z} - \dfrac{3}{2z^2}$

92. $\dfrac{\dfrac{r + 2s}{10}}{\dfrac{8r + 16s}{5}}$

93. $\dfrac{2p^2 - 5p - 12}{5p^2 - 18p - 8} \cdot \dfrac{25p^2 - 4}{30p - 12}$

94. $\dfrac{3m + 12}{8} \div \dfrac{5m + 20}{4}$

95. $(k^3 - 4k^2 - 15k + 18) \div (k + 3)$

96. $\dfrac{6}{5a + 10} + \dfrac{7}{6a + 12}$

97. $\dfrac{\dfrac{2}{r} - \dfrac{1}{r - 1}}{\dfrac{3}{r - 1}}$

Solve.

98. $A = \dfrac{Rr}{R + r}$; for r

99. $\dfrac{1}{m + 4} - \dfrac{3}{2m + 8} = -\dfrac{1}{2}$

100. In the law given in Exercise 74, find S, if $R = 10$ and $T = 15$.

---

101. Bob Ayo and Faustino Vallejo need to sort a pile of bottles at the recycling center. Working alone, Bob could do the entire job in 9 hours, and Faustino could do the entire job in 6 hours. How long will it take them to complete the job if they work together?

---

102. $-\dfrac{1}{3h} - \dfrac{1}{h} = -\dfrac{4}{3}$

103. Three fourths of a number is subtracted from seven sixths of the number, giving 10. Find the number.

104. $\dfrac{2z - 2}{z^2 + 3z + 2} + \dfrac{3}{z + 1} = \dfrac{4}{z + 2}$

105. $F = f\left(\dfrac{v + w}{v - z}\right)$; for z

CHAPTER 4 TEST

1. Find all values that make $\dfrac{m + 6}{2m^2 + 5m - 3}$ undefined.

2. Write $\dfrac{3p^2 + 5p}{3p^2 - 7p - 20}$ in lowest terms.

Multiply or divide.

3. $\dfrac{5k^5}{9k^2} \div \dfrac{10k^6}{8k}$

4. $\dfrac{x^2 - 7x + 12}{x^2 + 2x - 15} \cdot \dfrac{x^2 - 25}{x^2 - 16}$

5. $\dfrac{y^2 + y - 12}{y^3 + 9y^2 + 20y} \div \dfrac{y^2 - 9}{y^3 + 3y^2}$

Find the least common denominator for each group of denominators.

6. $9z$, $12z^2$, $6z^3$

7. $m^2 + m - 6$, $m^2 + 3m$

Add or subtract.

8. $\dfrac{4}{k} + \dfrac{1}{5k}$

9. $\dfrac{1}{a - b} + \dfrac{5}{b - a}$

10. $\dfrac{4}{t^2 - 16} - \dfrac{5}{2t + 8}$

Simplify each complex fraction.

11. $\dfrac{\dfrac{6}{m - 2}}{\dfrac{8}{3m - 6}}$

12. $\dfrac{x^{-2} - y^{-2}}{x^{-1} - y^{-1}}$

Divide.

13. $\dfrac{8z^3 - 16z^2 + 24z}{8z^2}$

14. $\dfrac{2q^2 - 13q - 16}{2q + 3}$

15. $\dfrac{6y^4 - 3y^3 + 5y^2 + 6y - 9}{2y + 1}$

16. Use synthetic division to divide $9x^5 + 40x^4 - 23x^3 + 8x^2 - 6x + 22$ by $x + 5$.

17. Decide whether 4 is a solution of $x^4 - 8x^3 + 21x^2 - 14x - 24 = 0$.

Solve each equation.

18. $\dfrac{1}{a} - \dfrac{1}{a - 2} = \dfrac{4}{3a}$

19. $\dfrac{2}{m - 3} - \dfrac{3}{m + 3} = \dfrac{12}{m^2 - 9}$

20. Explain the difference between $\dfrac{1}{x - 4} + \dfrac{5}{x - 1} - 2$ and $\dfrac{1}{x - 4} + \dfrac{5}{x - 1} = 2$.

Solve for the specified variable.

21. $\dfrac{2}{z} - \dfrac{5}{m} = \dfrac{1}{y}$; for m

22. $\dfrac{k}{y} = \dfrac{r + 2}{r - 2}$; for r

Solve each problem.

23. If twice the reciprocal of the larger of two consecutive integers is subtracted from the reciprocal of the smaller, the result is -5 times the reciprocal of the product of the two integers. Find the integers.

24. Wayne can do a job in 9 hours that takes Sandra 5 hours. How long would it take them if they worked together?

25. The current of a river is 3 miles per hour. Vera can go 36 miles downstream in the same time that it takes to go 24 miles upstream. Find the speed of her boat in still water.

CHAPTER FIVE

ROOTS AND RADICALS

In the last chapter we were introduced to rational expressions, which are the algebraic extensions of the rational numbers. Recall that a rational number is a number that can be written as the quotient of two integers, with denominator not zero. However, there are many numbers in the real number system that are *irrational*; that is, they cannot be written as ratios of integers. Some of the most common irrational numbers are square roots of numbers that are not perfect squares: for example, $\sqrt{3}$, $\sqrt{10}$, and $\sqrt{15}$.

Probably the first irrational number ever discovered was $\sqrt{2}$. It was discovered by the Pythagoreans, a secret Greek society of mathematicians and musicians. This discovery was a great setback to their philosophy that everything was based on the whole numbers. Pythagoreans kept their findings secret, and legend has it that members of the group who divulged this finding were sent out to sea and were never seen again.

In this chapter we begin our study of roots and radicals and their properties.

5.1 RATIONAL EXPONENTS AND RADICALS

OBJECTIVES

1 DEFINE $a^{1/n}$.

2 USE RADICAL NOTATION FOR nTH ROOTS.

3 DEFINE $a^{m/n}$.

4 WRITE RATIONAL EXPONENTIAL EXPRESSIONS AS RADICALS.

FOCUS ON PROBLEM SOLVING

The threshold weight T for a person is the weight above which the risk of death increases greatly. One researcher found that the threshold weight in pounds for men aged 40–49 is related to height in inches by the equation

$$h = 12.3T^{1/3}.$$

What height corresponds to a threshold of 216 pounds for a man in this age group?

So far in this book, our work with exponents has been limited to integers used as exponents. In the formula above, the rational number 1/3 is used as an exponent. In this section we introduce rational exponents and lead immediately into a discussion of radical notation. These ideas are very closely related, and it is difficult to talk about one without discussing the other at the same time. This problem is Exercise 95 in the exercises for this section. After working through this section, you should be able to solve this problem.

1 Recall from the first chapter that 5 is a square root of 25 because $5^2 = 25$. Also, 6 is a cube root of 216, while 2 is a seventh root of 128, and so on. The definition of nth root is now given.

nTH ROOT If n is a natural number, then b is an **nth root** of a if

$$b^n = a.$$

EXAMPLE 1
DETERMINING nTH ROOTS

(a) Since $3^2 = 9$, the number 3 is a square root of 9.

(b) 4 is a cube root of 64, since $4^3 = 64$.

(c) 2 is a fifth root of 32, since $2^5 = 32$.

(d) There is no real number whose square is -16, so there is no real number square root of -16.

(e) -5 is a cube root of -125, since $(-5)^3 = -125$. ■

In Chapter 1, nth roots were written with radicals; however, nth roots can be written using rational number exponents, with the nth root of a written $a^{1/n}$. Any definition of $a^{1/n}$ should be consistent with past properties of exponents. By one of the power rules of exponents,

$$(a^{1/n})^n = a^{(1/n)n} = a^1 = a,$$

suggesting that $a^{1/n}$ must be a number whose nth power is a.

nTH ROOT, n EVEN

If n is an *even* positive integer and if $a > 0$, then $a^{1/n}$ is the positive real number whose nth power is a:

$$(a^{1/n})^n = a.$$

The number $a^{1/n}$ is the **principal nth root** of a.

EXAMPLE 2
EVALUATING EXPONENTIALS OF THE FORM $a^{1/n}$, n EVEN

Evaluate each exponential expression.

(a) $64^{1/2} = 8$, since $64 > 0$ and $8^2 = 64$.

(b) $10{,}000^{1/4} = 10$, since $10{,}000 > 0$ and $10^4 = 10{,}000$.

(c) $-144^{1/2} = -(144^{1/2}) = -12$

(d) $(-144)^{1/2}$ is not a real number, since there is no real number whose square is -144. ■

Example 2(d) shows why the restriction $a > 0$ is necessary in the definition of $a^{1/n}$ for even integers n. With *odd* values of n, it does not matter whether the base is positive or negative.

nTH ROOT, n ODD

If a is a nonzero real number and n is an *odd* positive integer, then $a^{1/n}$ is the positive or negative real number whose nth power is a:

$$(a^{1/n})^n = a.$$

EXAMPLE 3
EVALUATING EXPONENTIALS OF THE FORM $a^{1/n}$, n ODD

Evaluate each exponential expression.

(a) $64^{1/3} = 4$, since $4^3 = 64$.

(b) $(-64)^{1/3} = -4$, since $(-4)^3 = -64$.

(c) $(-32)^{1/5} = -2$, since $(-2)^5 = -32$.

(d) $\left(\frac{1}{8}\right)^{1/3} = \frac{1}{2}$, since $\left(\frac{1}{2}\right)^3 = \frac{1}{8}$. ■

The next definition completes the definition of $a^{1/n}$.

nTH ROOT OF 0 For all positive integers n, $0^{1/n} = 0$.

2 The discussion of nth roots leads naturally to radical notation. It is common for nth roots to be written with a radical sign, first introduced in Chapter 1.

RADICAL NOTATION If n is a positive integer greater than 1, and if $a^{1/n}$ exists, then

$$a^{1/n} = \sqrt[n]{a}.$$

The number a is the **radicand,** n is the **index** or **order,** and the expression $\sqrt[n]{a}$ is a **radical.** The second root of a number, $\sqrt[2]{a}$, is its **square root.** It is customary to omit the index 2 and write the square root of a as $\sqrt{a}$.

EXAMPLE 4 **SIMPLIFYING ROOTS USING RADICAL NOTATION**

Find each root that is a real number.

(a) $\sqrt{25} = 5$, since $5^2 = 25$.

(b) $-\sqrt{25} = -5$

(c) $\sqrt{\dfrac{9}{16}} = \dfrac{3}{4}$

(d) $\sqrt[3]{125} = 5$, since $5^3 = 125$.

(e) $\sqrt[3]{-216} = -6$, since $(-6)^3 = -216$.

(f) $\sqrt[4]{-16}$ is not a real number.

(g) $\sqrt[5]{32} = 2$ ■

From our discussion so far, we have seen that exponentials of the form $a^{1/n}$ and radicals of the form $\sqrt[n]{a}$ are not real numbers if a is negative and n is even. The chart that follows summarizes some basic facts about $a^{1/n}$ and $\sqrt[n]{a}$.

BASIC FACTS ABOUT $a^{1/n}$ AND $\sqrt[n]{a}$

Let a be a real number and let n be a positive integer.

	n is even	n is odd
a is positive	$\sqrt[n]{a}$ and $a^{1/n}$ are positive	$\sqrt[n]{a}$ and $a^{1/n}$ are positive
a is negative	$\sqrt[n]{a}$ and $a^{1/n}$ are not real numbers	$\sqrt[n]{a}$ and $a^{1/n}$ are negative
a is 0	$\sqrt[n]{a}$ and $a^{1/n}$ are 0	$\sqrt[n]{a}$ and $a^{1/n}$ are 0

A given positive number has two square roots, one positive and one negative. The symbol $\sqrt{a}$, however, is used only for the *nonnegative* square root of a, the principal square root of a. The negative square root of a is written $-\sqrt{a}$. (As an abbreviation, the two square roots of the positive number a are sometimes written $\pm\sqrt{a}$, with the $\pm$ read as "positive or negative.")

A square root of a^2 (where $a \neq 0$) is a number that can be squared to give a^2. This number is either a or $-a$. Since one of these numbers is negative and one is positive, and since the symbol $\sqrt{a^2}$ represents the *nonnegative* square root, the non-negative square root must be written with absolute value bars, as $|a|$.

$\sqrt{a^2}$

For any real number a, $\sqrt{a^2} = |a|$.

This result can be generalized to any nth root.

$\sqrt[n]{a^n}$

If n is an **even** positive integer, $\sqrt[n]{a^n} = |a|$,

and if n is an **odd** positive integer, $\sqrt[n]{a^n} = a$.

EXAMPLE 5
SIMPLIFYING ROOTS WITH ABSOLUTE VALUE

Use absolute value as necessary to simplify each radical expression.

(a) $\sqrt{5^2} = |5| = 5$

(b) $\sqrt{(-5)^2} = |-5| = 5$

(c) $\sqrt[6]{(-3)^6} = |-3| = 3$

(d) $\sqrt[5]{(-4)^5} = -4$ 5 is odd; no absolute value is necessary.

(e) $-\sqrt[4]{(-9)^4} = -|-9| = -9$ ■

3 The more general expression $a^{m/n}$, where m/n is any rational number, can now be defined, starting with a rational number m/n where both m and n are positive integers, with m/n written in lowest terms. If the earlier rules of exponents are to be valid for rational exponents, then we should have

$$a^{m/n} = a^{(1/n)m} = (a^{1/n})^m,$$

provided that $a^{1/n}$ exists. This result leads to the definition of $a^{m/n}$.

$a^{m/n}$

If m and n are positive integers with m/n in lowest terms, then

$$a^{m/n} = (a^{1/n})^m,$$

provided that $a^{1/n}$ exists. If $a^{1/n}$ does not exist, then $a^{m/n}$ does not exist.

The expression $a^{-m/n}$ is defined as follows.

$a^{-m/n}$

$a^{-m/n} = \dfrac{1}{a^{m/n}}$ $(a \neq 0)$, provided that $a^{m/n}$ exists

EXAMPLE 6
EVALUATING EXPONENTIALS OF THE FORM $a^{m/n}$

Evaluate each exponential expression that is a real number.

(a) $36^{3/2} = (36^{1/2})^3 = 6^3 = 216$

(b) $125^{2/3} = (125^{1/3})^2 = 5^2 = 25$

(c) $-4^{5/2} = -(4^{5/2}) = -(4^{1/2})^5 = -(2)^5 = -32$

(d) $(-27)^{2/3} = [(-27)^{1/3}]^2 = (-3)^2 = 9$

(e) $(-100)^{3/2}$ is not a real number, since $(-100)^{1/2}$ is not a real number.

(f) $16^{-3/4} = \dfrac{1}{16^{3/4}} = \dfrac{1}{(16^{1/4})^3} = \dfrac{1}{2^3} = \dfrac{1}{8}$

(g) $25^{-3/2} = \dfrac{1}{25^{3/2}} = \dfrac{1}{(25^{1/2})^3} = \dfrac{1}{125}$ ■

An alternative definition of m/n can be obtained by using the power rule for exponents a little differently than in the definition above. If all indicated roots exist,

$$a^{m/n} = a^{m(1/n)}$$
$$= (a^m)^{1/n},$$

so that
$$\mathbf{a^{m/n} = (a^m)^{1/n}}.$$

With this result, $a^{m/n}$ can be defined in either of two ways.

$a^{m/n}$

If all indicated roots exist, then $a^{m/n} = (a^{1/n})^m = (a^m)^{1/n}$.

An expression such as $27^{2/3}$ can now be evaluated in two ways:

$$27^{2/3} = (27^{1/3})^2 = 3^2 = 9$$
$$27^{2/3} = (27^2)^{1/3} = 729^{1/3} = 9.$$

In most cases, it is easier to use $(a^{1/n})^m$.

4 Until now, expressions of the form $a^{m/n}$ have been evaluated without introducing radical notation. However, we have seen that for appropriate values of a and n, $a^{1/n} = \sqrt[n]{a}$, and so it is not difficult to extend our discussion to using radical notation for $a^{m/n}$.

RADICAL FORM OF $a^{m/n}$

If all indicated roots exist, $a^{m/n} = (\sqrt[n]{a})^m$ or $a^{m/n} = \sqrt[n]{a^m}$.

EXAMPLE 7
CONVERTING BETWEEN RATIONAL EXPONENTS AND RADICALS

Write each of the following with radicals. Assume that all variables represent positive real numbers. Use the first definition in the box above.

(a) $13^{1/2} = \sqrt{13}$

(b) $6^{3/4} = (\sqrt[4]{6})^3$

(c) $9m^{5/8} = 9(\sqrt[8]{m})^5$

(d) $6x^{2/3} - (4x)^{3/5} = 6(\sqrt[3]{x})^2 - (\sqrt[5]{4x})^3$

(e) $r^{-2/3} = \dfrac{1}{r^{2/3}} = \dfrac{1}{(\sqrt[3]{r})^2}$

(f) $(a^2 + b^2)^{1/2} = \sqrt{a^2 + b^2}$ ■

CAUTION Refer to Example 7(f). The expression $\sqrt{a^2 + b^2}$ is *not equal to* $a + b$. In general,

$$\sqrt[n]{a^n + b^n} \neq a + b.$$

EXAMPLE 8
CONVERTING BETWEEN RADICALS AND RATIONAL EXPONENTS

Replace all radicals with rational exponents. Simplify. Assume that all variables represent positive real numbers.

reduce to lowest terms

(a) $\sqrt{10} = 10^{1/2}$

(b) $\sqrt[4]{3^8} = 3^{8/4} = 3^2 = 9$

(c) $\sqrt[6]{z^6} = z$ ■

NOTE Refer to Example 8(c). It was not necessary to use absolute value bars, since the directions specifically stated that the variable represents a positive real number. Since the absolute value of the positive real number z is z itself, the answer is simply z. When working exercises involving radicals, we will often make the assumption that variables represent positive real numbers, which will eliminate the need for absolute value.

Scientific calculators have keys for powers and roots, and they are helpful when working problems with nth roots. These operations are discussed in more detail in the next section.

5.1 EXERCISES

Evaluate each of the following. Use a calculator if necessary. See Examples 1, 2, 3, and 5.

1. $121^{1/2}$ **2.** $169^{1/2}$ **3.** $441^{1/2}$ **4.** $2401^{1/2}$ **5.** $512^{1/3}$

6. $729^{1/3}$ **7.** $256^{1/4}$ **8.** $-4096^{1/4}$ **9.** $-6561^{1/4}$ **10.** $\left(\dfrac{1}{256}\right)^{1/4}$

11. $\left(\dfrac{1}{32}\right)^{1/5}$ **12.** $\left(\dfrac{4}{9}\right)^{1/2}$ **13.** $\left(\dfrac{16}{25}\right)^{1/2}$ **14.** $\left(\dfrac{64}{27}\right)^{1/3}$ **15.** $\left(\dfrac{8}{125}\right)^{1/3}$

16. $8^{2/3}$ **17.** $100^{3/2}$ **18.** $32^{2/5}$ **19.** $32^{6/5}$ **20.** $(-125)^{2/3}$

21. $-144^{3/2}$ **22.** $-49^{3/2}$ **23.** $1728^{2/3}$ **24.** $\left(\frac{1}{9}\right)^{3/2}$ **25.** $\left(\frac{16}{81}\right)^{3/4}$

26. $-\left(\frac{81}{625}\right)^{3/4}$ **27.** $-\left(\frac{9}{4}\right)^{5/2}$ **28.** $\left(\frac{27}{64}\right)^{-1/3}$ **29.** $\left(\frac{16}{625}\right)^{-1/4}$ **30.** $\left(\frac{25}{36}\right)^{-3/2}$

31. $\left(\frac{121}{36}\right)^{-3/2}$ **32.** $\left(\frac{27}{1000}\right)^{-2/3}$

---

33. Explain why $a^{1/n}$ is not a real number when a is negative and n is even. (See Objective 1.)

34. Suppose that a is positive and n is odd. Then is it true that

$$-a^{1/n} = (-a)^{1/n}$$

for all choices of a and n? (See Objective 1.)

Find each root if it is real. Use a calculator if necessary. Assume that all variables represent positive real numbers. See Example 4.

35. $\sqrt{81}$ **36.** $-\sqrt{100}$ **37.** $\sqrt[3]{64}$ **38.** $\sqrt[3]{125}$

39. $\sqrt[3]{-216}$ **40.** $\sqrt[3]{-343}$ **41.** $\sqrt[4]{81}$ **42.** $\sqrt[4]{256}$

43. $-\sqrt[4]{625}$ **44.** $-\sqrt[4]{1296}$ **45.** $\sqrt[4]{-2401}$ **46.** $\sqrt[4]{-4096}$

47. $-\sqrt[4]{6561}$ **48.** $-\sqrt[4]{10{,}000}$ **49.** $\sqrt[5]{32}$ **50.** $\sqrt[5]{243}$

51. $-\sqrt[5]{-1024}$ **52.** $-\sqrt[5]{3125}$ **53.** $-\sqrt[6]{64}$ **54.** $\sqrt[6]{-729}$

Write each of the following with radicals. Assume that all variables represent positive real numbers. See Example 7.

55. $12^{1/2}$ **56.** $3^{1/2}$ **57.** $9^{1/5}$ **58.** $8^{3/4}$

59. $(7y)^{2/3}$ **60.** $(9q)^{5/8}$ **61.** $(3py^2)^{1/4}$ **62.** $(6a^2b)^{2/3}$

63. $(2m)^{-3/2}$ **64.** $(5y)^{-3/5}$ **65.** $(2y + x)^{2/3}$ **66.** $(r + 2z)^{1/2}$

67. $(p^2 + q^2)^{1/2}$ **68.** $(b^3 + a^3)^{1/3}$ **69.** $(3m^4 + 2k^2)^{-2/3}$ **70.** $(5x^2 + 3z^3)^{-5/6}$

---

71. Show that in general, $\sqrt{a^2 + b^2} \neq a + b$ by replacing a with 3 and b with 4.

72. Suppose that someone claims that $\sqrt[n]{a^n + b^n}$ must equal $a + b$, since when we let $a = 1$ and $b = 0$, a true statement results:

$$\sqrt[n]{a^n + b^n} = \sqrt[n]{1^n + 0^n} = \sqrt[n]{1^n} = 1 = 1 + 0 = a + b.$$

Write an explanation of why this person has given faulty reasoning.

Write with rational exponents. Assume that all variables represent positive real numbers. See Example 8.

73. $\sqrt{13}$ **74.** $\sqrt{29}$ **75.** $\sqrt[3]{7}$ **76.** $\sqrt[3]{24}$ **77.** $\sqrt{2^3}$

78. $\sqrt{5^5}$ **79.** $\sqrt[3]{6^4}$ **80.** $\sqrt[4]{7^3}$ **81.** $-\sqrt[3]{8^5}$ **82.** $-\sqrt[5]{9^2}$

83. $\sqrt[4]{x^3}$ **84.** $\sqrt[5]{y^3}$ **85.** $\sqrt[7]{a^2}$ **86.** $\sqrt[4]{p^9}$

Use absolute value to simplify each of the following. *See Example 5.*

87. $\sqrt{r^6}$

88. $\sqrt{x^{10}}$

89. $\sqrt{(r-2q)^2}$

90. $\sqrt{(3a-2b)^2}$

91. $\sqrt{m^2-2mq+q^2}$

92. $\sqrt{9z^2-24zx+16x^2}$

93. $\dfrac{-5}{\sqrt{p^4-2p^2q^2+q^4}}$

94. $\dfrac{3}{\sqrt{a^4-8a^3b+16a^2b^2}}$

Work each of the following problems.

95. The threshold weight T for a person is the weight above which the risk of death increases greatly. One researcher found that the threshold weight in pounds for men aged 40–49 is related to height in inches by the equation

$$h = 12.3T^{1/3}.$$

What height corresponds to a threshold of 216 pounds for a man in this age group?

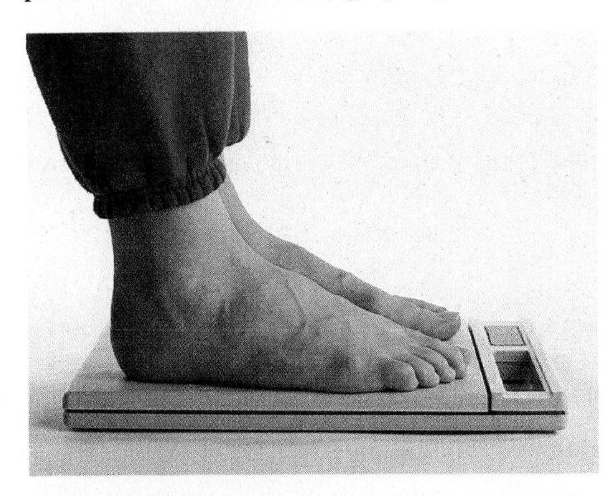

96. The length L of an animal is related to its surface area S by the equation

$$L = \left(\frac{S}{a}\right)^{1/2},$$

where a is a constant that depends on the type of animal. Find the length of an animal with a surface area of 1000 square centimeters if $a = 2/5$.

97. *Heron's Formula* states that if a triangle has sides of lengths a, b, and c, and if the semiperimeter is

$$s = \frac{1}{2}(a+b+c),$$

then the area of the triangle is

$$K = \sqrt{s(s-a)(s-b)(s-c)}.$$

Find the area of a triangle with sides measuring 4, 3, and 5 meters.

98. In accident reconstruction, the formula

$$S = 5.5\sqrt{DF}$$

is used to determine the speed S in miles per hour of a vehicle that leaves skid marks of D feet. F is a constant that represents the friction of the road surface. Let $F = .8$. Find the speed of a vehicle that left skid marks of **(a)** 50 feet; **(b)** 10 feet; and **(c)** 20 feet. Round answers to the nearest whole number.

99. The expression $\dfrac{b}{12}\sqrt{\dfrac{F}{33{,}000}}$

is used in the design of bridges. Evaluate this expression for $b = 120$ and $F = 297{,}000$.

100. The formula $C = \left(\dfrac{12}{d}\right)^{1/9}$

is used in timber connector construction. Evaluate C for $d = 3/128$.

PREVIEW EXERCISES

Use the rules for exponents to simplify each of the following. *See Sections 3.1 and 3.2.*

101. $(3x^2y^7)(-2xy^6)$

102. $(5p^3q^2r)(4pqr^5)$

103. $(3m^3n^2)^3$

104. $(6h^2j^6)^4$

105. $\dfrac{27x^4y^6}{9x^3y^2}$

106. $\dfrac{100p^9q^5s^2}{25p^8qs^2}$

5.2 MORE ABOUT RATIONAL EXPONENTS AND RADICALS

OBJECTIVES

1 USE THE RULES OF EXPONENTS WITH RATIONAL EXPONENTS.

2 USE A CALCULATOR TO FIND ROOTS.

1 In Sections 3.1 and 3.2 we introduced rules of exponents for integer exponents. In the previous section, rational exponents were defined in such a way that the power rule for exponents is still valid. With this definition of rational exponents, it can be shown that all the familiar properties of exponents are also valid for rational exponents.

RULES FOR RATIONAL EXPONENTS

Let r and s be rational numbers. For all real numbers a and b for which the indicated expressions exist:

$$a^r \cdot a^s = a^{r+s} \qquad a^{-r} = \frac{1}{a^r} \qquad \frac{a^r}{a^s} = a^{r-s} \qquad \left(\frac{a}{b}\right)^{-r} = \frac{b^r}{a^r}$$

$$(a^r)^s = a^{rs} \qquad (ab)^r = a^r b^r \qquad \left(\frac{a}{b}\right)^r = \frac{a^r}{b^r} \qquad a^{-r} = \left(\frac{1}{a}\right)^r.$$

EXAMPLE 1 APPLYING RULES FOR RATIONAL EXPONENTS

Write with only positive exponents. Assume that all variables represent positive real numbers.

(a) $2^{1/2} \cdot 2^{1/4} = 2^{1/2+1/4} = 2^{3/4}$

(b) $\dfrac{5^{2/3}}{5^{7/3}} = 5^{2/3-7/3} = 5^{-5/3} = \dfrac{1}{5^{5/3}}$

(c) $\dfrac{(x^{1/2}y^{2/3})^4}{y} = \dfrac{(x^{1/2})^4(y^{2/3})^4}{y}$ Power rule

$= \dfrac{x^2y^{8/3}}{y^1}$ Power rule

$= x^2y^{8/3-1}$ Quotient rule

$= x^2y^{5/3}$

(d) $m^{3/4}(m^{5/4} - m^{1/4}) = m^{3/4} \cdot m^{5/4} - m^{3/4} \cdot m^{1/4}$ Distributive property

$= m^{3/4+5/4} - m^{3/4+1/4}$ Product rule

$= m^{8/4} - m^{4/4}$

$= m^2 - m$

Do not make the common mistake of multiplying exponents in the first step.

(e) $\dfrac{(p^q)^{5/2}}{p^{q-1}} = \dfrac{p^{5q/2}}{p^{q-1}}$

$= p^{(5q/2)-(q-1)} = p^{(5q/2)-q+1} = p^{(3q/2)+1}$ Assume q is a rational number. ■

CAUTION Errors often occur in exercises like the ones in Example 1 because students try to convert the expressions to radical form. Remember that the rules of exponents apply here.

EXAMPLE 2 APPLYING RULES FOR RATIONAL EXPONENTS

Replace all radicals with rational exponents, and then apply the rules for rational exponents. Leave answers in exponential form. Assume that all variables represent positive real numbers.

(a) $\sqrt[3]{x^2} \cdot \sqrt[4]{x} = x^{2/3} \cdot x^{1/4}$ — Convert to rational exponents.

$= x^{2/3+1/4}$ — Use product rule for exponents.

$= x^{8/12+3/12}$ — Get a common denominator.

$= x^{11/12}$

(b) $\dfrac{\sqrt{x^3}}{\sqrt[3]{x^2}} = \dfrac{x^{3/2}}{x^{2/3}} = x^{3/2-2/3} = x^{5/6}$

(c) $\sqrt{\sqrt[4]{z}} = \sqrt{z^{1/4}} = (z^{1/4})^{1/2} = z^{1/8}$ ■

2 So far in this chapter we have discussed definitions, properties, and rules for rational exponents and radicals. An important application of radicals involves finding approximations of roots that are not rational numbers. Many expressions involving radicals and rational exponents represent *rational* numbers. For example,

$$9^{3/2} = (9^{1/2})^3 = (\sqrt{9})^3 = 3^3 = 27$$

is a rational number. However, there is no rational number equal to $\sqrt{15}$ or $\sqrt[3]{10}$, for example. These numbers are irrational. Decimal approximations of such irrational numbers can be found with scientific calculators.

To use a calculator to approximate $\sqrt{15}$, enter the number 15 and press the key marked $\boxed{\sqrt{\ }}$. The display will then read 3.8729833. (There may be fewer or more decimal places, depending upon the model of calculator used.) In this book we will show approximations rounded to three decimal places. Therefore,

$$\sqrt{15} \approx 3.873$$

where $\approx$ means "is approximately equal to."

EXAMPLE 3 FINDING APPROXIMATIONS FOR SQUARE ROOTS

Find a decimal approximation for each of the following, using a calculator.

(a) $\sqrt{39} \approx 6.245$

(b) $\sqrt{83} \approx 9.110$

(c) $-\sqrt{72} \approx -8.485$

To find the negative square root, first find the positive square root and then take its negative, using the sign changing key $\boxed{\pm}$.

(d) $\sqrt{770} \approx 27.749$

(e) $-\sqrt{420} \approx -20.494$ ■

Scientific calculators have the capability to find roots other than square roots. The key marked $\boxed{y^x}$ is the exponential key. For many scientific calculators, using the inverse key ($\boxed{\text{INV}}$) in conjunction with the exponential key will allow us to take roots. For example, to find an approximation for $\sqrt[5]{467}$ (the fifth root of 467), the typical sequence of keystrokes is as follows.

$$\boxed{4}\ \boxed{6}\ \boxed{7}\ \boxed{\text{INV}}\ \boxed{y^x}\ \boxed{5}\ \boxed{=}$$

The display should read 3.4187188. (More or fewer decimal places may show up depending upon the model of calculator.) This decimal is an approximation of $\sqrt[5]{467}$, and rounding to the nearest thousandth, we get

$$\sqrt[5]{467} \approx 3.419.$$

If your calculator does not follow the same sequence of keystrokes just described, consult the owner's manual.

EXAMPLE 4
FINDING APPROXIMATIONS FOR HIGHER ROOTS

Find a decimal approximation for each of the following, using a calculator.

(a) $\sqrt[3]{943} \approx 9.806$ **(b)** $\sqrt[4]{992} \approx 5.612$ **(c)** $\sqrt[5]{10{,}847} \approx 6.413$

(d) $629^{1/7}$

This can be found either by evaluating 1/7 in the calculator and using the $\boxed{y^x}$ key directly, with $y = 629$ and $x = 1/7$, or by writing it as $\sqrt[7]{629}$ and then using the method described above. In either case, the approximation is 2.511.

(e) $1.97^{5/2}$

Again, we can find the approximation by raising 1.97 to the 5/2 power, or by taking the square root of 1.97 and raising this result to the fifth power, since

$$1.97^{5/2} = (\sqrt{1.97})^5.$$

The approximation is 5.447. ■

5.2 EXERCISES

Use the rules of exponents to simplify each of the following. Assume that all variables represent positive real numbers. Write all answers with only positive exponents. See Example 1.

1. $9^{3/5} \cdot 9^{7/5}$ **2.** $7^{3/4} \cdot 7^{9/4}$ **3.** $\dfrac{81^{5/4}}{81^{3/4}}$ **4.** $\dfrac{125^{2/3}}{125^{1/3}}$

5. $x^{5/3} \cdot x^{-2/3}$ **6.** $m^{-8/9} \cdot m^{17/9}$ **7.** $\dfrac{k^{2/3}k^{-1}}{k^{1/3}}$ **8.** $\dfrac{z^{5/4}z^{-2}}{z^{3/4}}$

9. $(8p^9q^6)^{2/3}$ **10.** $(25m^8r^{10})^{3/2}$ **11.** $\dfrac{(r^{2/3})^2}{(r^2)^{5/3}}$ **12.** $\dfrac{(k^{5/4})^2}{(k^3)^{4/3}}$

13. $\left(\dfrac{z^{10}}{x^{12}}\right)^{1/4}$ **14.** $\left(\dfrac{r^5}{s^7}\right)^{2/5}$ **15.** $\dfrac{(m^2h)^{1/2}}{m^{3/4}h^{-1/4}}$ **16.** $\dfrac{(a^{-3}b^2)^{1/6}}{(a^2b^5)^{-1/4}}$

17. $\dfrac{p^{1/5}p^{7/10}p^{1/2}}{(p^3)^{-1/5}}$ **18.** $\dfrac{z^{1/3}z^{-2/3}z^{1/6}}{(z^{-1/6})^3}$ **19.** $\left(\dfrac{m^{-2/3}}{a^{-3/4}}\right)^4(m^{-3/8}a^{1/4})^{-2}$

20. $\left(\dfrac{b^{-3/2}}{c^{-5/3}}\right)^2(b^{-1/4}c^{-1/3})^{-1}$ **21.** $\left(\dfrac{p^{-1/4}q^{-3/2}}{3^{-1}p^{-2}q^{-2/3}}\right)^{-2}$ **22.** $\left(\dfrac{2^{-2}w^{-3/4}x^{-5/8}}{w^{3/4}x^{-1/2}}\right)^{-3}$

Multiply. Assume that all variables represent positive real numbers. See Example 1(d).

23. $p^{2/3}(p^{1/3}+2p^{4/3})$ **24.** $z^{5/8}(3z^{5/8}+5z^{11/8})$ **25.** $k^{1/4}(k^{3/2}-k^{1/2})$

26. $r^{3/5}(r^{1/2}+r^{3/4})$ **27.** $6a^{7/4}(a^{-7/4}+3a^{-3/4})$ **28.** $4m^{5/3}(m^{-2/3}-4m^{-5/3})$

29. $x^{-3/5}(x^{11/5}-4x)$ **30.** $y^{-11/7}(y^2-3y^3)$ **31.** $-z^{-5/6}(5z^3-2z^2)$

Apply the rules for exponents to simplify. Assume that x and y are positive real numbers and that a and b are rational numbers, with all indicated roots existing. See Example 1(e).

32. $y^{a/2}y^{a/3}$ **33.** $(x^{3/4})^b(x^5)^b$ **34.** $\dfrac{x^{5a/3}}{x^{a/6}}$ **35.** $\left(\dfrac{x^b}{x^{3b}}\right)^{1/2}\left(\dfrac{x^{-2b}}{x^{-4b}}\right)^{-1}$

36. $\left(\dfrac{x^{a/b}y^a}{x}\right)^{1/a}$ **37.** $\dfrac{x^a x^{a/2}}{x^{2a}}$ **38.** $\dfrac{x^{-1/b}y^{-1/a}}{x^{2/b}y^{-2/b}}$

Recall that we factor an expression like $x^4 - x^5$ by factoring out the greatest common factor, x^4, so that $x^4 - x^5 = x^4(1-x)$.

Use this idea to factor out the given common factor in the expressions in Exercises 39–46. Assume that all variables represent positive real numbers.

39. $3x^{-1/2}-4x^{1/2};\ x^{-1/2}$ **40.** $m^3-3m^{5/2};\ m^{5/2}$

41. $4t^{-1/2}+7t^{3/2};\ t^{-1/2}$ **42.** $8x^{2/3}-5x^{-1/3};\ x^{-1/3}$

43. $4p-p^{3/4};\ p^{3/4}$ **44.** $2m^{1/8}-m^{5/8};\ m^{1/8}$

45. $9k^{-3/4}-2k^{-1/4};\ k^{-3/4}$ **46.** $7z^{-5/8}-z^{-3/4};\ z^{-3/4}$

47. Find the error in the following. $-1=(-1)^1=(-1)^{2/2}=[(-1)^2]^{1/2}=1^{1/2}=1$

Choose values of x and y to show the following.

48. $(x^{1/2}+y^{1/2})^2\neq x+y$ **49.** $(x^2+y^2)^{1/2}\neq x+y$

---

50. Suppose someone claims that the following answer obtained when simplifying $a^{1/2}(a^{1/4}-a^{1/2})$ is $a^{1/8}-a^{1/4}$. Explain to the person how the error occurred, and then give the correct answer. (See Objective 1.)

Write with rational exponents, and then apply the properties of exponents. Assume that all radicands represent positive real numbers. Give answers in exponential form. See Example 2.

51. $\sqrt[5]{x^3}\cdot\sqrt[4]{x}$ **52.** $\sqrt[6]{y^5}\cdot\sqrt[3]{y^2}$ **53.** $\dfrac{\sqrt{x^5}}{x^4}$ **54.** $\dfrac{\sqrt[3]{k^5}}{\sqrt[3]{k^7}}$

55. $\sqrt{y}\cdot\sqrt[3]{y}$ **56.** $\sqrt[3]{xz}\sqrt{z}$ **57.** $\sqrt{\sqrt[3]{m}}$ **58.** $\sqrt[3]{\sqrt{k}}$

59. $\sqrt{\sqrt[3]{\sqrt[4]{x}}}$ **60.** $\sqrt[3]{\sqrt[5]{\sqrt{y}}}$

Use a calculator to find a decimal approximation for each of the following. Give answers to the nearest thousandth. See Example 3.

61. $\sqrt{7}$ **62.** $\sqrt{11}$ **63.** $-\sqrt{19}$ **64.** $-\sqrt{56}$

65. $-\sqrt{82}$ **66.** $-\sqrt{91}$ **67.** $\sqrt{150}$ **68.** $\sqrt{280}$

69. $-\sqrt{510}$ **70.** $-\sqrt{740}$ **71.** $\sqrt{890}$ **72.** $\sqrt{1000}$

Use a calculator to find a decimal approximation for each of the following. Give answers to the nearest thousandth. (Hint: In Exercise 83, find $38^{1/3}$ and then write its negative.) See Example 4.

73. $\sqrt[3]{423}$ **74.** $\sqrt[3]{555}$ **75.** $\sqrt[4]{100}$ **76.** $\sqrt[4]{250}$

77. $\sqrt[5]{23.8}$ **78.** $\sqrt[5]{98.4}$ **79.** $59^{2/3}$ **80.** $86^{7/8}$

81. $26^{-2/5}$ **82.** $104^{-5/4}$ **83.** $-38^{1/3}$ **84.** $-471^{1/4}$

---

85. Many scientific calculators are designed so that negative bases are not allowed when using the exponential key. For example, attempting to evaluate $(-4)^2$ on a popular calculator gives an error message. Write an explanation of how one can use the key to evaluate powers of negative bases by considering the sign of the final answer first.

86. Write an explanation of why the fourth root of a number can be found on a calculator by touching the square root key twice.

87. Use a calculator to find an approximation of $\sqrt[4]{\frac{2143}{22}}$, giving as many decimal places as the calculator shows. This number is the same as the decimal approximation of the irrational number π, correct to eight places.

88. Use your calculator to evaluate 5^0. Is your answer the same as you would expect to find when using the rules of exponents?

PREVIEW EXERCISES

Simplify each of the following. See Sections 3.1 and 3.2.

89. $p^8 \cdot p^{-2} \cdot p^5$ **90.** $y^3 \cdot y^{-7} \cdot y^{-9}$ **91.** $(6x^2)^{-1}(2x^3)^{-2}$

92. $(5m^{-1})^2(3m^{-2})^{-1}$ **93.** $(8r^2s)^{-1}(2r^{-1}s^{-2})^{-1}$ **94.** $(3p^{-1}q^{-2})^{-1}(2pq^{-1})^{-1}$

5.3 SIMPLIFYING RADICALS

OBJECTIVES

1. USE THE PRODUCT RULE FOR RADICALS.
2. USE THE QUOTIENT RULE FOR RADICALS.
3. SIMPLIFY RADICALS.
4. SIMPLIFY RADICALS BY USING DIFFERENT INDEXES.
5. USE THE PYTHAGOREAN FORMULA TO FIND THE LENGTH OF A SIDE OF A RIGHT TRIANGLE.

1 The product of $\sqrt{36}$ and $\sqrt{4}$ is

$$\sqrt{36} \cdot \sqrt{4} = 6 \cdot 2 = 12.$$

Multiplying 36 and 4 and then taking the square root gives

$$\sqrt{36} \cdot \sqrt{4} = \sqrt{36 \cdot 4} = \sqrt{144} = 12,$$

the same answer. This result is an example of the **product rule for radicals.**

PRODUCT RULE FOR RADICALS

If $\sqrt[n]{a}$ and $\sqrt[n]{b}$ are real numbers and n is a natural number, then

$$\sqrt[n]{a} \cdot \sqrt[n]{b} = \sqrt[n]{ab}.$$

(The product of two radicals is the radical of the product.)

The product rule can be justified using rational exponents. Since $\sqrt[n]{a} = a^{1/n}$ and $\sqrt[n]{b} = b^{1/n}$,

$$\sqrt[n]{a} \cdot \sqrt[n]{b} = a^{1/n} \cdot b^{1/n} = (ab)^{1/n} = \sqrt[n]{ab}.$$

CAUTION The product rule is applied only when the radicals have the same indexes.

EXAMPLE 1 USING THE PRODUCT RULE

Multiply. Assume that all variables represent positive real numbers.

(a) $\sqrt{5} \cdot \sqrt{7} = \sqrt{5 \cdot 7} = \sqrt{35}$

(b) $\sqrt{2} \cdot \sqrt{19} = \sqrt{2 \cdot 19} = \sqrt{38}$

(c) $\sqrt{11} \cdot \sqrt{p} = \sqrt{11p}$

(d) $\sqrt{7} \cdot \sqrt{11xyz} = \sqrt{77xyz}$

(e) $\sqrt{\frac{7}{y}} \cdot \sqrt{\frac{3}{p}} = \sqrt{\frac{21}{yp}}$ ■

EXAMPLE 2 USING THE PRODUCT RULE

Multiply. Assume that all variables represent positive real numbers.

(a) $\sqrt[3]{3} \cdot \sqrt[3]{12} = \sqrt[3]{3 \cdot 12} = \sqrt[3]{36}$

(b) $\sqrt[5]{9} \cdot \sqrt[5]{7} = \sqrt[5]{63}$

(c) $\sqrt[4]{8y} \cdot \sqrt[4]{3r^2} = \sqrt[4]{24yr^2}$

(d) $\sqrt[6]{10m^4} \cdot \sqrt[6]{5m} = \sqrt[6]{50m^5}$

(e) $\sqrt[4]{2} \cdot \sqrt[5]{2}$ cannot be simplified by the product rule, because the two indexes (4 and 5) are different. ■

2 The quotient rule for radicals is similar to the product rule.

QUOTIENT RULE FOR RADICALS

If $\sqrt[n]{a}$ and $\sqrt[n]{b}$ are real numbers, $b \neq 0$, and n is a natural number, then

$$\sqrt[n]{\frac{a}{b}} = \frac{\sqrt[n]{a}}{\sqrt[n]{b}}.$$

(The radical of a quotient is the quotient of the radicals.)

The quotient rule can be justified, like the product rule, using the properties of exponents. It, too, is used only when the radicals have the same indexes.

EXAMPLE 3
USING THE QUOTIENT RULE

Simplify. Assume that all variables represent positive real numbers.

(a) $\sqrt{\frac{16}{25}} = \frac{\sqrt{16}}{\sqrt{25}} = \frac{4}{5}$

(b) $\sqrt{\frac{7}{36}} = \frac{\sqrt{7}}{\sqrt{36}} = \frac{\sqrt{7}}{6}$

(c) $\sqrt[3]{-\frac{8}{125}} = \sqrt[3]{\frac{-8}{125}} = \frac{\sqrt[3]{-8}}{\sqrt[3]{125}} = \frac{-2}{5} = -\frac{2}{5}$

(d) $\sqrt[3]{\frac{7}{216}} = \frac{\sqrt[3]{7}}{\sqrt[3]{216}} = \frac{\sqrt[3]{7}}{6}$

(e) $\sqrt{\frac{x}{9}} = \frac{\sqrt{x}}{\sqrt{9}} = \frac{\sqrt{x}}{3}$

(f) $\sqrt{\frac{m^4}{25}} = \frac{\sqrt{m^4}}{\sqrt{25}} = \frac{m^2}{5}$ ■

3 One of the main uses of the product and quotient rules is in simplifying radicals. A radical is **simplified** if the following four conditions are met.

SIMPLIFIED RADICAL

1. The radicand has no factor raised to a power greater than or equal to the index.
2. The radicand has no fractions.
3. No denominator contains a radical.
4. Exponents in the radicand and the index of the radical have no common factor (except 1).

EXAMPLE 4
SIMPLIFYING ROOTS OF NUMBERS

Simplify.

(a) $\sqrt{24}$

Check to see if 24 is divisible by a perfect square (square of a natural number), such as 4, 9, ... Choose the largest perfect square that divides into 24. The largest such number is 4. Write 24 as the product of 4 and 6, and then use the product rule.

$$\sqrt{24} = \sqrt{4 \cdot 6} = \sqrt{4} \cdot \sqrt{6} = 2\sqrt{6}$$

(b) $\sqrt{108}$

The number 108 is divisible by the perfect square 36. If this is not obvious, try factoring 108 into prime factors.

$$\begin{aligned}\sqrt{108} &= \sqrt{2^2 \cdot 3^3} \\ &= \sqrt{2^2 \cdot 3^2 \cdot 3} \\ &= 2 \cdot 3 \cdot \sqrt{3} \qquad \text{Product rule} \\ &= 6\sqrt{3}\end{aligned}$$

(c) $\sqrt{500} = \sqrt{100 \cdot 5} = \sqrt{100} \cdot \sqrt{5} = 10\sqrt{5}$

(d) $\sqrt{10}$

No perfect square (other than 1) divides into 10, so $\sqrt{10}$ cannot be simplified further.

(e) $\sqrt[3]{16}$

Look for the largest perfect *cube* that divides into 16. The number 8 satisfies this condition, so write 16 as $8 \cdot 2$.

$$\sqrt[3]{16} = \sqrt[3]{8 \cdot 2} = \sqrt[3]{8} \cdot \sqrt[3]{2} = 2\sqrt[3]{2}$$

(f) $\sqrt[4]{162} = \sqrt[4]{81 \cdot 2}$ 81 is a perfect 4th power.

$= \sqrt[4]{81} \cdot \sqrt[4]{2}$

$= 3\sqrt[4]{2}$ ■

EXAMPLE 5 SIMPLIFYING ROOTS OF VARIABLE EXPRESSIONS

Simplify. Assume that all variables represent positive real numbers.

(a) $\sqrt{16m^3} = \sqrt{16m^2 \cdot m}$ $16m^2$ is the largest perfect square that divides $16m^3$.

$= \sqrt{16m^2} \cdot \sqrt{m}$

$= 4m\sqrt{m}$

No absolute value bars are needed around m because of the assumption that all variables represent *positive* real numbers.

(b) $\sqrt{200k^7q^8} = \sqrt{10^2 \cdot 2 \cdot (k^3)^2 \cdot k \cdot (q^4)^2}$

$= \sqrt{10^2 \cdot (k^3)^2 \cdot (q^4)^2 \cdot 2 \cdot k}$

$= 10k^3q^4\sqrt{2k}$

(c) $\sqrt{75p^6q^9} = \sqrt{(25p^6q^8)(3q)}$

$= \sqrt{25p^6q^8} \cdot \sqrt{3q}$

$= 5p^3q^4\sqrt{3q}$

(d) $\sqrt[3]{8x^4y^5} = \sqrt[3]{(8x^3y^3)(xy^2)}$ $8x^3y^3$ is the largest perfect cube that divides $8x^4y^5$.

$= \sqrt[3]{8x^3y^3} \cdot \sqrt[3]{xy^2}$

$= 2xy\sqrt[3]{xy^2}$

(e) $\sqrt[4]{32y^9} = \sqrt[4]{(16y^8)(2y)}$ $16y^8$ is the largest 4th power that divides $32y^9$.

$= \sqrt[4]{16y^8} \cdot \sqrt[4]{2y}$

$= 2y^2\sqrt[4]{2y}$ ■

NOTE From Example 5 we see that if a variable is raised to a power with an exponent divisible by 2, it is a perfect square. If it is raised to a power with an exponent divisible by 3, it is a perfect cube. In general, if it is raised to a power with an exponent divisible by n, it is a perfect nth power.

4 The conditions for a simplified radical given above state that an exponent in the radicand and the index of the radical should have no common factor. The next example shows how to simplify radicals with such common factors.

EXAMPLE 6
SIMPLIFY RADICALS BY USING SMALLER INDEXES

Simplify. Assume that all variables represent positive real numbers.

(a) $\sqrt[9]{5^6}$

Write this radical using rational exponents and then write the exponent in lowest terms. Express the answer as a radical.

$$\sqrt[9]{5^6} = 5^{6/9} = 5^{2/3} = \sqrt[3]{5^2} \quad \text{or} \quad \sqrt[3]{25}$$

(b) $\sqrt[4]{p^2} = p^{2/4} = p^{1/2} = \sqrt{p}$
Recall the assumption $p > 0$. ■

These examples suggest the following rule.

If m is an integer, n and k are natural numbers, and all indicated roots exist,

$$\sqrt[kn]{a^{km}} = \sqrt[n]{a^m}.$$

The next example shows how to simplify the product of two radicals having different indexes.

EXAMPLE 7
MULTIPLYING RADICALS WITH DIFFERENT INDEXES

Simplify $\sqrt{7} \cdot \sqrt[3]{2}$.

Since the indexes, 2 and 3, have a least common index of 6, use rational exponents to write each radical as a sixth root.

$$\sqrt{7} = 7^{1/2} = 7^{3/6} = \sqrt[6]{7^3} = \sqrt[6]{343}$$
$$\sqrt[3]{2} = 2^{1/3} = 2^{2/6} = \sqrt[6]{2^2} = \sqrt[6]{4}$$

Therefore, $\sqrt{7} \cdot \sqrt[3]{2} = \sqrt[6]{343} \cdot \sqrt[6]{4} = \sqrt[6]{1372}.$ Product rule ■

5 One useful application of radicals occurs when using the **Pythagorean formula** from geometry to find the length of one side of a right triangle when the lengths of the other sides are known.

PYTHAGOREAN FORMULA

If c is the length of the longest side of a right triangle and a and b are the lengths of the shorter sides, then

$$c^2 = a^2 + b^2.$$

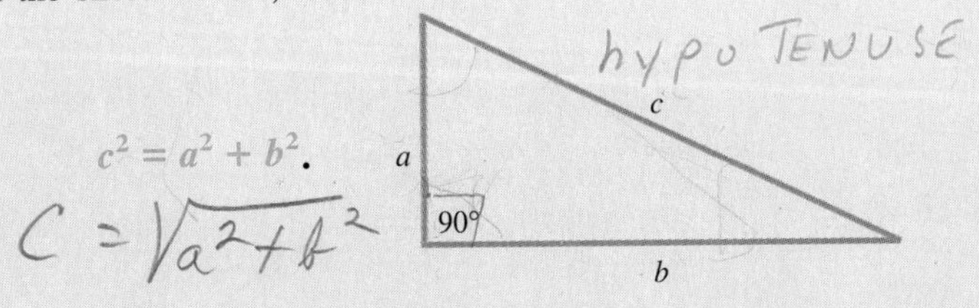

The longest side is the **hypotenuse** and the two shorter sides are the **legs** of the triangle. The hypotenuse is the side opposite the right angle.

From the formula $c^2 = a^2 + b^2$, the length of the hypotenuse is given by $c = \sqrt{a^2 + b^2}$.

EXAMPLE 8
USING THE PYTHAGOREAN FORMULA

Use the Pythagorean formula to find the length of the hypotenuse in the triangle in Figure 5.1.

By the formula, the length of the hypotenuse is

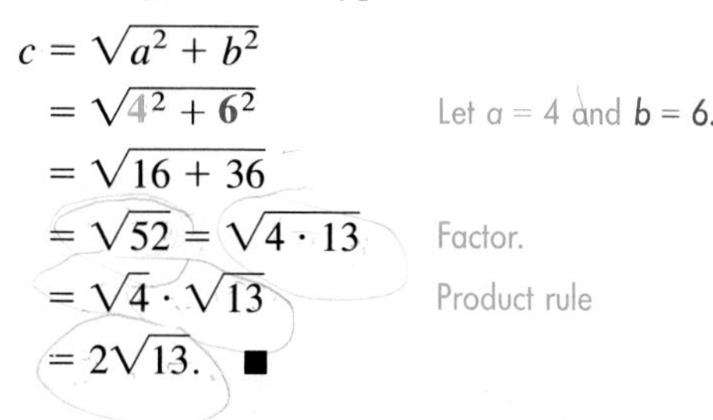

$$
\begin{aligned}
c &= \sqrt{a^2 + b^2} \\
&= \sqrt{4^2 + 6^2} && \text{Let } a = 4 \text{ and } b = 6. \\
&= \sqrt{16 + 36} \\
&= \sqrt{52} = \sqrt{4 \cdot 13} && \text{Factor.} \\
&= \sqrt{4} \cdot \sqrt{13} && \text{Product rule} \\
&= 2\sqrt{13}. \quad \blacksquare
\end{aligned}
$$

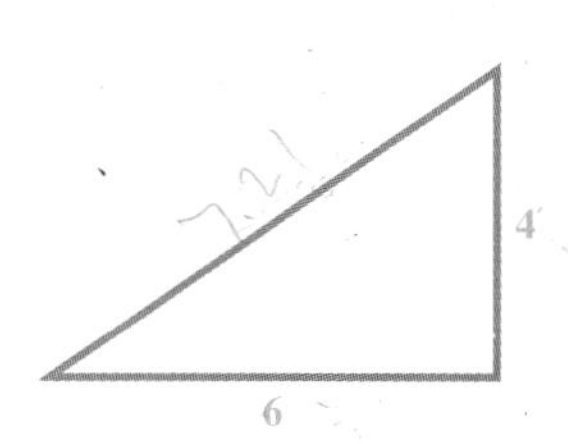

FIGURE 5.1

CAUTION When using the equation $c^2 = a^2 + b^2$, be sure that the length of the hypotenuse is substituted for c, and that the lengths of the legs are substituted for a and b. Errors often occur because values are substituted incorrectly.

5.3 EXERCISES

Multiply. See Examples 1 and 2.

1. $\sqrt{7} \cdot \sqrt{11}$ **2.** $\sqrt{2} \cdot \sqrt{5}$ **3.** $\sqrt{14} \cdot \sqrt{3}$ **4.** $\sqrt{15} \cdot \sqrt{7}$

5. $\sqrt[3]{2} \cdot \sqrt[3]{7}$ **6.** $\sqrt[3]{5} \cdot \sqrt[3]{9}$ **7.** $\sqrt[4]{8} \cdot \sqrt[4]{3}$ **8.** $\sqrt[4]{9} \cdot \sqrt[4]{5}$

Simplify each radical. Assume that all variables represent positive real numbers. See Example 3.

9. $\sqrt{\frac{16}{25}}$ **10.** $\sqrt{\frac{49}{64}}$ **11.** $\sqrt{\frac{7}{25}}$ **12.** $\sqrt{\frac{11}{49}}$ **13.** $\sqrt{\frac{r}{100}}$ **14.** $\sqrt{\frac{p}{4}}$

15. $\sqrt{\frac{m^8}{25}}$ **16.** $\sqrt{\frac{z^4}{81}}$ **17.** $\sqrt[3]{-\frac{27}{64}}$ **18.** $\sqrt[3]{\frac{125}{216}}$ **19.** $\sqrt[3]{\frac{k^2}{125}}$ **20.** $\sqrt[3]{\frac{p}{216}}$

21. Can the product rule be used to simplify $\sqrt{2} \cdot \sqrt[3]{2}$? Explain. (See Objective 1.)

22. Can the quotient rule be used to simplify $\frac{\sqrt[3]{6}}{\sqrt{2}}$? Explain. (See Objective 2.)

Simplify each of the following. Assume that all variables represent positive real numbers. See Examples 4–5.

23. $\sqrt{18}$ **24.** $-\sqrt{48}$ **25.** $\sqrt{76}$ **26.** $\sqrt{52}$

27. $\sqrt[3]{16}$ **28.** $\sqrt[3]{250}$ **29.** $\sqrt[3]{128}$ **30.** $\sqrt[3]{375}$

31. $\sqrt[4]{32}$ **32.** $\sqrt[4]{243}$ **33.** $-\sqrt[4]{1250}$ **34.** $-\sqrt[4]{512}$

35. $\sqrt[5]{128}$ **36.** $\sqrt[5]{486}$ **37.** $\sqrt{\frac{72}{25}}$ **38.** $\sqrt{\frac{80}{9}}$

39. $\sqrt[3]{\frac{32}{125}}$ **40.** $\sqrt[3]{\frac{81}{1000}}$ **41.** $\sqrt{100y^{10}}$ **42.** $\sqrt{256z^6}$

43. $-\sqrt[3]{8k^9}$ **44.** $-\sqrt[3]{27y^{15}}$ **45.** $-\sqrt{144m^{10}z^2}$ **46.** $-\sqrt{4k^2z^{18}}$

47. $-\sqrt[3]{-125m^9b^{18}c^{24}}$ **48.** $\sqrt[3]{-216y^{12}x^3z^{18}}$ **49.** $\sqrt{75y^3}$ **50.** $\sqrt{200z^3}$

51. $\sqrt{7x^5y^6}$ **52.** $\sqrt{12k^9p^{12}}$ **53.** $\sqrt[3]{8z^9r^{12}}$ **54.** $\sqrt[3]{125k^{15}n^9}$

55. $\sqrt[3]{24z^5x^9}$ **56.** $\sqrt[3]{81w^7y^8}$ **57.** $\sqrt[4]{16a^8b^{12}}$ **58.** $\sqrt[4]{81z^{16}y^{20}}$

59. $\sqrt[4]{32k^5m^{10}}$ **60.** $\sqrt[4]{162r^{15}s^{10}}$ **61.** $\sqrt{\frac{m^9}{16}}$ **62.** $\sqrt{\frac{y^{15}}{100}}$

63. $\sqrt[3]{\frac{y^{10}}{27}}$ **64.** $\sqrt[3]{\frac{r^{26}}{125}}$ **65.** $\sqrt[4]{\frac{t^{23}}{16}}$ **66.** $-\sqrt[4]{\frac{8a^9}{81}}$

67. Which one of the following expressions is *not* simplified? Tell why it is not. (See Objective 3.)

(a) $\sqrt[3]{4x}$ (b) $\frac{\sqrt{3}}{2}$

(c) $\sqrt[4]{\frac{2}{5}}$ (d) $\sqrt[5]{a^3}$

68. Which of the following expressions *is* simplified? Explain your decision. (See Objective 3.)

(a) $\sqrt[4]{81}$ (b) $\frac{\sqrt[3]{6}}{3}$

(c) $\sqrt{\frac{7}{36}}$ (d) $\sqrt[6]{a^2}$

Simplify each of the following. Assume that all variables represent positive real numbers. See Example 6.

69. $\sqrt[4]{12^2}$ **70.** $\sqrt[6]{21^3}$ **71.** $\sqrt[10]{m^{15}}$ **72.** $\sqrt[8]{y^6}$

73. $\sqrt[4]{25}$ **74.** $\sqrt[6]{8}$ **75.** $\sqrt[6]{256}$ **76.** $\sqrt[8]{64}$

Write as radicals with the same index and simplify. Assume all variables represent positive real numbers. See Example 7.

77. $\sqrt[3]{5} \cdot \sqrt{3}$ **78.** $\sqrt[3]{3} \cdot \sqrt[4]{6}$ **79.** $\sqrt[3]{2} \cdot \sqrt[5]{3}$

80. $\sqrt[4]{8} \cdot \sqrt[5]{3}$ **81.** $\sqrt{x} \cdot \sqrt[3]{x}$ **82.** $\sqrt[3]{y} \cdot \sqrt[4]{y}$

Find the missing length in each right triangle. Simplify answers if necessary. (Hint: If one leg is unknown, write the formula as $a = \sqrt{c^2 - b^2}$ or $b = \sqrt{c^2 - a^2}$.) See Example 8.

83.

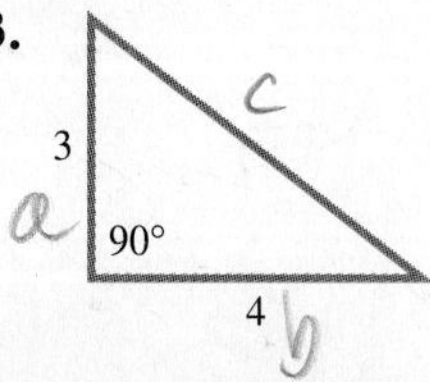

84.

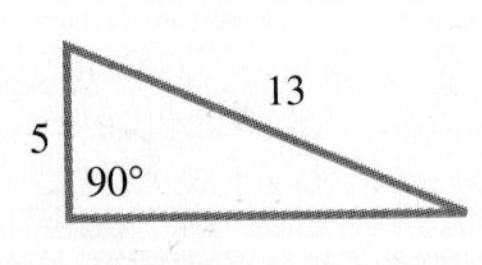

85.

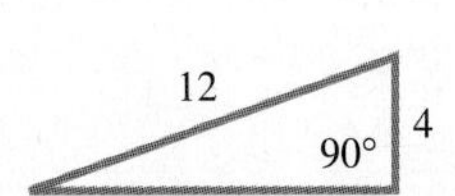

86.

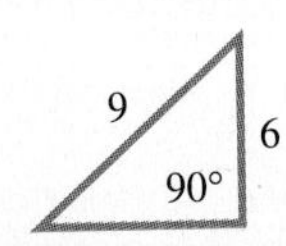

Work each problem.

87. The illumination I in footcandles produced by a light source is related to the distance d in feet from the light source by the equation $d = \sqrt{\frac{k}{I}}$, where k is a constant. If $k = 700$, how far from the light source will the illumination be 16 footcandles? Give the answer to the nearest thousandth.

88. The length of the diagonal of a box is $d = \sqrt{L^2 + W^2 + H^2}$, where L, W, and H are the length, width, and height of the box. Find the length of the diagonal of a box that is 9 1/5 inches long, 8 2/5 inches wide, and 2 inches deep. Give the answer as a simplified radical.

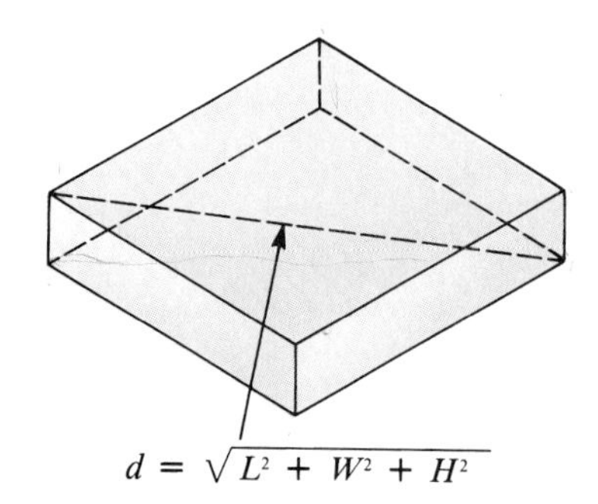

$d = \sqrt{L^2 + W^2 + H^2}$

PREVIEW EXERCISES

Simplify. See Section 3.3.

89. $8x^2 - 4x^2 + 2x^2$

90. $3p^3 - 5p^3 - 8p^2$

91. $9q + 2q - 5q^2 - q^2$

92. $7z^5 - 2z^3 + 8z^5 - z^3$

93. $16a^5 - 9a^2 + 4a$

94. $12x^4 - 5x^3 + 2x^2$

5.4 ADDING AND SUBTRACTING RADICAL EXPRESSIONS

OBJECTIVES

1 DEFINE A RADICAL EXPRESSION.

2 SIMPLIFY RADICAL EXPRESSIONS INVOLVING ADDITION AND SUBTRACTION.

1 A **radical expression** is an algebraic expression that contains radicals. For example,

$$\sqrt[4]{3} + \sqrt{6}, \qquad \sqrt{x + 2y} - 1, \qquad \text{and} \qquad \sqrt{8} - \sqrt{2r}$$

are radical expressions. The examples in the previous section discussed simplifying radical expressions that involve multiplication and division. Now we will simplify radical expressions that involve addition and subtraction.

2 An expression such as $4\sqrt{2} + 3\sqrt{2}$ can be simplified by using the distributive property.

$$4\sqrt{2} + 3\sqrt{2} = (4 + 3)\sqrt{2} = 7\sqrt{2}$$

As another example, $2\sqrt{3} - 5\sqrt{3} = (2 - 5)\sqrt{3} = -3\sqrt{3}$. This is very similar to simplifying $2x + 3x$ to $5x$ or $5y - 8y$ to $-3y$.

CAUTION Only radical expressions with the same index and the same radicand may be combined. Expressions such as $5\sqrt{3} + 2\sqrt{2}$ and $3\sqrt{3} + 2\sqrt[3]{3}$ cannot be simplified.

EXAMPLE 1 ADDING AND SUBTRACTING RADICAL EXPRESSIONS

Add or subtract the following radical expressions.

(a) $3\sqrt{24} + \sqrt{54}$

Begin by simplifying each radical. Then use the distributive property.

$$\begin{aligned} 3\sqrt{24} + \sqrt{54} &= 3\sqrt{4} \cdot \sqrt{6} + \sqrt{9} \cdot \sqrt{6} && \text{Product rule} \\ &= 3 \cdot 2\sqrt{6} + 3\sqrt{6} \\ &= 6\sqrt{6} + 3\sqrt{6} \\ &= 9\sqrt{6} && \text{Combine terms.} \end{aligned}$$

(b) $$\begin{aligned} -3\sqrt{8} + 4\sqrt{18} &= -3\sqrt{4}\sqrt{2} + 4\sqrt{9}\sqrt{2} && \text{Product rule} \\ &= -3 \cdot 2\sqrt{2} + 4 \cdot 3\sqrt{2} \\ &= -6\sqrt{2} + 12\sqrt{2} \\ &= 6\sqrt{2} \end{aligned}$$

(c) $$\begin{aligned} 2\sqrt{20x} - \sqrt{45x} &= 2\sqrt{4}\sqrt{5x} - \sqrt{9}\sqrt{5x} && \text{Product rule} \\ &= 2 \cdot 2\sqrt{5x} - 3\sqrt{5x} \\ &= 4\sqrt{5x} - 3\sqrt{5x} \\ &= \sqrt{5x} \qquad (\text{if } x \geq 0) \end{aligned}$$

(d) $2\sqrt{3} - 4\sqrt{5}$

Here the radicals differ, and are already simplified, so that $2\sqrt{3} - 4\sqrt{5}$ cannot be simplified further. ■

EXAMPLE 2 ADDING AND SUBTRACTING RADICALS WITH HIGHER INDEXES

Add or subtract the following radical expressions. Assume that all variables represent positive real numbers.

(a) $$\begin{aligned} 2\sqrt[3]{16} - 5\sqrt[3]{54} &= 2\sqrt[3]{8 \cdot 2} - 5\sqrt[3]{27 \cdot 2} \\ &= 2\sqrt[3]{8} \cdot \sqrt[3]{2} - 5\sqrt[3]{27} \cdot \sqrt[3]{2} \\ &= 2 \cdot 2 \cdot \sqrt[3]{2} - 5 \cdot 3 \cdot \sqrt[3]{2} \\ &= 4\sqrt[3]{2} - 15\sqrt[3]{2} \\ &= -11\sqrt[3]{2} \end{aligned}$$

(b) $$\begin{aligned} 2\sqrt[3]{x^2y} + \sqrt[3]{8x^5y^4} &= 2\sqrt[3]{x^2y} + \sqrt[3]{(8x^3y^3)x^2y} \\ &= 2\sqrt[3]{x^2y} + 2xy\sqrt[3]{x^2y} \\ &= (2 + 2xy)\sqrt[3]{x^2y} && \text{Distributive property} \end{aligned}$$ ■

CAUTION It is a common error to forget to write the index when working with cube roots, fourth roots, and so on. Be careful not to forget to write these indexes.

The next example shows how to simplify radical expressions involving fractions.

EXAMPLE 3 **ADDING AND SUBTRACTING RADICAL EXPRESSIONS WITH FRACTIONS**

Perform the indicated operations.

(a)
$$2\sqrt{\frac{75}{16}} + 4\frac{\sqrt{8}}{\sqrt{32}} = 2\sqrt{\frac{25 \cdot 3}{16}} + 4\frac{\sqrt{4 \cdot 2}}{\sqrt{16 \cdot 2}}$$
$$= 2\left(\frac{5\sqrt{3}}{4}\right) + 4\left(\frac{2\sqrt{2}}{4\sqrt{2}}\right)$$
$$= \frac{5\sqrt{3}}{2} + 2 \qquad \text{Multiply.}$$
$$= \frac{5\sqrt{3}}{2} + \frac{4}{2} \qquad \text{Find the common denominator.}$$
$$= \frac{5\sqrt{3} + 4}{2}$$

(b)
$$10\sqrt[3]{\frac{5}{x^6}} - 3\sqrt[3]{\frac{4}{x^9}} = 10\frac{\sqrt[3]{5}}{\sqrt[3]{x^6}} - 3\frac{\sqrt[3]{4}}{\sqrt[3]{x^9}}$$
$$= \frac{10\sqrt[3]{5}}{x^2} - \frac{3\sqrt[3]{4}}{x^3}$$
$$= \frac{10x\sqrt[3]{5}}{x^3} - \frac{3\sqrt[3]{4}}{x^3} \qquad \text{Find the common denominator.}$$
$$= \frac{10x\sqrt[3]{5} - 3\sqrt[3]{4}}{x^3} \qquad (\text{if } x \neq 0)$$ ■

5.4 EXERCISES

Simplify. Assume that all variables represent positive real numbers. See Examples 1 and 2.

1. $\sqrt{36} + \sqrt{100}$
2. $\sqrt{25} + \sqrt{81}$
3. $4\sqrt{12} - 7\sqrt{27}$
4. $3\sqrt{32} - 2\sqrt{8}$
5. $6\sqrt{18} + \sqrt{32} - 2\sqrt{50}$
6. $5\sqrt{8} - 3\sqrt{72} + 3\sqrt{50}$
7. $2\sqrt{63} - 2\sqrt{28} + 3\sqrt{7}$
8. $6\sqrt{27} - 2\sqrt{48} + \sqrt{75}$
9. $2\sqrt{5} - 3\sqrt{20} - 4\sqrt{45}$
10. $5\sqrt{54} + 2\sqrt{24} - 2\sqrt{96}$
11. $2\sqrt{40} + 6\sqrt{90} - 3\sqrt{160}$
12. $5\sqrt{28} - 3\sqrt{63} + 2\sqrt{112}$
13. $3\sqrt{2x} - \sqrt{8x} - \sqrt{72x}$
14. $4\sqrt{18k} - \sqrt{72k} + 4\sqrt{50k}$
15. $9\sqrt{3r} - 2\sqrt{12r} + 5\sqrt{27r}$
16. $-\sqrt{20z} + 2\sqrt{125z} - 3\sqrt{45z}$
17. $7q\sqrt{10} - 2q\sqrt{40} + 8q\sqrt{90}$
18. $3a\sqrt{7} + 2a\sqrt{28} - 5a\sqrt{63}$
19. $3\sqrt{72m^2} + 2\sqrt{32m^2} - 3\sqrt{18m^2}$
20. $9\sqrt{27p^2} - 4\sqrt{108p^2} - 2\sqrt{48p^2}$
21. $\sqrt[3]{54} - 2\sqrt[3]{16}$
22. $5\sqrt[3]{81} - 4\sqrt[3]{24}$
23. $2\sqrt[3]{27x} + 2\sqrt[3]{8x}$
24. $6\sqrt[3]{128m} - 3\sqrt[3]{16m}$
25. $\sqrt[3]{x^2y} - \sqrt[3]{8x^2y}$
26. $3\sqrt[3]{x^2y^2} - 2\sqrt[3]{64x^2y^2}$
27. $3x\sqrt[3]{xy^2} - 2\sqrt[3]{8x^4y^2}$
28. $6q^2\sqrt[3]{5q} - 2q\sqrt[3]{40q^4}$
29. $5\sqrt[4]{32} + 3\sqrt[4]{162}$
30. $2\sqrt[4]{512} - 4\sqrt[4]{32}$

31. $2\sqrt[4]{32a^3} + 5\sqrt[4]{2a^3}$

32. $-\sqrt[4]{16r} + 5\sqrt[4]{r}$

33. $3\sqrt[4]{x^5y} - 2x\sqrt[4]{xy}$

34. $2\sqrt[4]{m^9p^6} - 3m^2p\sqrt[4]{mp^2}$

Simplify the following radical expressions. Assume all variables represent positive real numbers. See Example 3.

35. $\sqrt{8} - \dfrac{\sqrt{64}}{\sqrt{16}}$

36. $\sqrt{48} - \dfrac{\sqrt{81}}{\sqrt{9}}$

37. $\dfrac{2\sqrt{5}}{3} + \dfrac{\sqrt{5}}{6}$

38. $\dfrac{4\sqrt{3}}{3} + \dfrac{2\sqrt{3}}{9}$

39. $\sqrt{\dfrac{8}{9}} + \sqrt{\dfrac{18}{36}}$

40. $\sqrt{\dfrac{12}{16}} + \sqrt{\dfrac{48}{64}}$

41. $\dfrac{\sqrt{32}}{3} + \dfrac{2\sqrt{2}}{3} - \dfrac{\sqrt{2}}{\sqrt{9}}$

42. $\dfrac{\sqrt{27}}{2} - \dfrac{3\sqrt{3}}{2} + \dfrac{\sqrt{3}}{\sqrt{4}}$

43. $3\sqrt{\dfrac{50}{9}} + 8\dfrac{\sqrt{2}}{\sqrt{8}}$

44. $9\sqrt{\dfrac{48}{25}} - 2\dfrac{\sqrt{2}}{\sqrt{98}}$

45. $\sqrt{\dfrac{25}{x^8}} - \sqrt{\dfrac{9}{x^6}}$

46. $\sqrt{\dfrac{100}{y^4}} + \sqrt{\dfrac{81}{y^{10}}}$

47. $3\sqrt[3]{\dfrac{m^5}{27}} - 2m\sqrt[3]{\dfrac{m^2}{64}}$

48. $2a\sqrt[4]{\dfrac{a}{16}} - 5a\sqrt[4]{\dfrac{a}{81}}$

---

49. Choose the incorrect addition and explain why it is incorrect. (See Objective 2.)
(a) $3\sqrt{5} + 2\sqrt{5} = 5\sqrt{5}$
(b) $5\sqrt{2} + 6\sqrt{3} = 11\sqrt{5}$
(c) $2\sqrt{2} + 8\sqrt{2} = 10\sqrt{2}$
(d) $6\sqrt{5} + 2\sqrt{5} = 8\sqrt{5}$

50. Choose the incorrect subtraction and explain why it is incorrect. (See Objective 2.)
(a) $6\sqrt{xy} - 2\sqrt{xy} = 4\sqrt{xy}$
(b) $3\sqrt{m} - 9\sqrt{m} = -6\sqrt{m}$
(c) $4\sqrt[3]{k} - \sqrt[3]{k} = 3\sqrt[3]{k}$
(d) $3\sqrt[3]{r} - 5\sqrt[3]{r} = -2\sqrt[3]{r}$

Work the following problems. (See Objective 2.)

51. If the lengths of the sides of a triangle are $2\sqrt{45}$, $\sqrt{75}$, and $3\sqrt{20}$ centimeters, find the perimeter.

52. A rectangular yard has a length of $\sqrt{192}$ meters and a width of $\sqrt{48}$ meters. What is the perimeter?

53. Find the perimeter of a lot with sides measuring $3\sqrt{18}$, $2\sqrt{32}$, $4\sqrt{50}$, and $5\sqrt{12}$ yards.

54. What is the perimeter of a triangle with sides measuring $3\sqrt{54}$, $4\sqrt{24}$, and $\sqrt{80}$ meters?

55. Find decimal approximations for $\sqrt{3}$ and $\sqrt{12}$. Do the approximations suggest that $\sqrt{12} = 2\sqrt{3}$?

56. Find decimal approximations for $2\sqrt{7}$ and $\dfrac{14}{\sqrt{7}}$. Do the approximations suggest that $\dfrac{14}{\sqrt{7}} = 2\sqrt{7}$?

PREVIEW EXERCISES

Simplify. See Sections 3.4 and 4.5.

57. $-6p(2p^2 - 1)$

58. $8r(6r^2 - 5)$

59. $(3a - 2b)(5a + 7b)$

60. $(4p + 7q)(8p - 9q)$

61. $(5x - y)^2$

62. $(2z - 3x)^2$

63. $(7r - 2s)(7r + 2s)$

64. $(9y - 5z)(9y + 5z)$

65. $\dfrac{8x^2 - 10x}{2x}$

66. $\dfrac{15y^3 - 9y^2}{6y}$

5.5 MULTIPLYING AND DIVIDING RADICAL EXPRESSIONS

OBJECTIVES

1. MULTIPLY RADICAL EXPRESSIONS.
2. RATIONALIZE DENOMINATORS WITH ONE RADICAL TERM.
3. RATIONALIZE DENOMINATORS WITH BINOMIALS INVOLVING RADICALS.
4. WRITE RADICAL QUOTIENTS IN LOWEST TERMS.

1 We can multiply binomial expressions involving radicals by using the FOIL (first, outside, inside, last) method. For example, the product of the binomials $(\sqrt{5} + 3)(\sqrt{6} + 1)$ is found as follows.

$$\begin{aligned}(\sqrt{5} + 3)(\sqrt{6} + 1) &= \underbrace{\sqrt{5} \cdot \sqrt{6}}_{\text{First}} + \underbrace{\sqrt{5} \cdot 1}_{\text{Outside}} + \underbrace{3 \cdot \sqrt{6}}_{\text{Inside}} + \underbrace{3 \cdot 1}_{\text{Last}} \\ &= \sqrt{30} + \sqrt{5} + 3\sqrt{6} + 3\end{aligned}$$

This result cannot be simplified further.

EXAMPLE 1 MULTIPLYING BINOMIALS INVOLVING RADICAL EXPRESSIONS

Multiply.

(a) $$\begin{aligned}(7 - \sqrt{3})(\sqrt{5} + \sqrt{2}) &= 7\sqrt{5} + 7\sqrt{2} - \sqrt{3} \cdot \sqrt{5} - \sqrt{3} \cdot \sqrt{2} \\ &= 7\sqrt{5} + 7\sqrt{2} - \sqrt{15} - \sqrt{6}\end{aligned}$$

(b) $$\begin{aligned}(\sqrt{10} + \sqrt{3})(\sqrt{10} - \sqrt{3}) &= \sqrt{10}\sqrt{10} - \sqrt{10}\sqrt{3} + \sqrt{10}\sqrt{3} - \sqrt{3}\sqrt{3} \\ &= 10 - 3 \\ &= 7\end{aligned}$$

Notice that this is an example of the kind of product that results in the difference of two squares:

$$(a + b)(a - b) = a^2 - b^2.$$

Here, $a = \sqrt{10}$ and $b = \sqrt{3}$.

(c) $$\begin{aligned}(\sqrt{7} - 3)^2 &= (\sqrt{7} - 3)(\sqrt{7} - 3) \\ &= \sqrt{7} \cdot \sqrt{7} - \sqrt{7} \cdot 3 - 3\sqrt{7} + 3 \cdot 3 \\ &= 7 - 6\sqrt{7} + 9 \\ &= 16 - 6\sqrt{7}\end{aligned}$$

(d) $$\begin{aligned}(5 - \sqrt[3]{3})(5 + \sqrt[3]{3}) &= 5 \cdot 5 + 5\sqrt[3]{3} - 5\sqrt[3]{3} - \sqrt[3]{3} \cdot \sqrt[3]{3} \\ &= 25 - \sqrt[3]{3^2} \\ &= 25 - \sqrt[3]{9}\end{aligned}$$

(e) $$\begin{aligned}(\sqrt{k} + \sqrt{y})(\sqrt{k} - \sqrt{y}) &= (\sqrt{k})^2 - (\sqrt{y})^2 \\ &= k - y \qquad (\text{if } k \geq 0 \text{ and } y \geq 0)\end{aligned}$$ ■

2 As defined earlier, a simplified radical expression should have no radical in the denominator. For example,

$$\frac{3}{\sqrt{7}}$$

is not simplified; it will not be simplified until the radical is eliminated from the denominator. Before hand-held calculators were easily available, in order to find a decimal approximation for a quotient such as $3/\sqrt{7}$, it was advantageous to rewrite the expression so that it was not necessary to divide by a decimal approximation. Although this is no longer necessary because of the availability of calculators, there are other reasons to learn how to *rationalize* a radical denominator in a quotient. The process of removing radicals from the denominator so that the denominator contains only rational numbers is called **rationalizing the denominator.** This process is explained in the following examples.

EXAMPLE 2 RATIONALIZING DENOMINATORS WITH SQUARE ROOTS

Rationalize each denominator.

(a) $\frac{3}{\sqrt{7}}$

Multiply numerator and denominator by $\sqrt{7}$. This is, in effect, multiplying by 1.

$$\frac{3}{\sqrt{7}} = \frac{3 \cdot \sqrt{7}}{\sqrt{7} \cdot \sqrt{7}}$$

Since $\sqrt{7} \cdot \sqrt{7} = \sqrt{7 \cdot 7} = \sqrt{49} = 7$,

$$\frac{3}{\sqrt{7}} = \frac{3\sqrt{7}}{7}.$$

The denominator is now a rational number.

(b) $\frac{5\sqrt{2}}{\sqrt{5}}$

Multiply the numerator and denominator by $\sqrt{5}$.

$$\frac{5\sqrt{2}}{\sqrt{5}} = \frac{5\sqrt{2} \cdot \sqrt{5}}{\sqrt{5} \cdot \sqrt{5}} = \frac{5\sqrt{10}}{5} = \sqrt{10}$$

(c) $\frac{6}{\sqrt{12}}$

Less work is involved if the radical in the denominator is simplified first.

$$\frac{6}{\sqrt{12}} = \frac{6}{\sqrt{4 \cdot 3}} = \frac{6}{2\sqrt{3}} = \frac{3}{\sqrt{3}}$$

Now rationalize the denominator by multiplying the numerator and the denominator by $\sqrt{3}$.

$$\frac{3 \cdot \sqrt{3}}{\sqrt{3} \cdot \sqrt{3}} = \frac{3\sqrt{3}}{3} = \sqrt{3} \quad ■$$

The next example shows how to rationalize denominators in expressions involving quotients under a radical sign.

EXAMPLE 3 RATIONALIZING DENOMINATORS IN ROOTS OF FRACTIONS

Simplify each of the following.

(a) $\sqrt{\frac{18}{125}}$

$$\begin{aligned}
\sqrt{\frac{18}{125}} &= \frac{\sqrt{18}}{\sqrt{125}} && \text{Quotient rule} \\
&= \frac{\sqrt{9 \cdot 2}}{\sqrt{25 \cdot 5}} && \text{Factor.} \\
&= \frac{3\sqrt{2}}{5\sqrt{5}} && \text{Product rule} \\
&= \frac{3\sqrt{2} \cdot \sqrt{5}}{5\sqrt{5} \cdot \sqrt{5}} && \text{Multiply by } \sqrt{5} \text{ in numerator and denominator.} \\
&= \frac{3\sqrt{10}}{5 \cdot 5} && \text{Product rule} \\
&= \frac{3\sqrt{10}}{25}
\end{aligned}$$

(b) $\sqrt{\frac{50m^4}{p^5}}, \quad p > 0$

$$\begin{aligned}
\sqrt{\frac{50m^4}{p^5}} &= \frac{\sqrt{50m^4}}{\sqrt{p^5}} && \text{Quotient rule} \\
&= \frac{\sqrt{25m^4 \cdot 2}}{\sqrt{p^4 \cdot p}} \\
&= \frac{5m^2\sqrt{2}}{p^2\sqrt{p}} && \text{Product rule} \\
&= \frac{5m^2\sqrt{2} \cdot \sqrt{p}}{p^2\sqrt{p} \cdot \sqrt{p}} && \text{Multiply by } \sqrt{p} \text{ in numerator and denominator.} \\
&= \frac{5m^2\sqrt{2p}}{p^2 \cdot p} && \text{Product rule} \\
&= \frac{5m^2\sqrt{2p}}{p^3} \quad \blacksquare
\end{aligned}$$

EXAMPLE 4 **RATIONALIZING A DENOMINATOR WITH A CUBE ROOT**

Simplify $\sqrt[3]{\frac{27}{16}}$.

Use the quotient rule and simplify the numerator and denominator.

$$\sqrt[3]{\frac{27}{16}} = \frac{\sqrt[3]{27}}{\sqrt[3]{16}} = \frac{3}{\sqrt[3]{8} \cdot \sqrt[3]{2}} = \frac{3}{2\sqrt[3]{2}}$$

To get a rational denominator multiply the numerator and denominator by a number that will result in a perfect cube in the radicand in the denominator. Since $2 \cdot 4 = 8$, a perfect cube, multiply the numerator and denominator by $\sqrt[3]{4}$.

$$\frac{3}{2\sqrt[3]{2}} = \frac{3 \cdot \sqrt[3]{4}}{2 \cdot \sqrt[3]{2} \cdot \sqrt[3]{4}} = \frac{3\sqrt[3]{4}}{2\sqrt[3]{8}} = \frac{3\sqrt[3]{4}}{2 \cdot 2} = \frac{3\sqrt[3]{4}}{4} \quad \blacksquare$$

CAUTION It is easy to make mistakes in problems like the one in Example 4. A typical error is to multiply numerator and denominator by $\sqrt[3]{2}$, forgetting that

$$\sqrt[3]{2} \cdot \sqrt[3]{2} \neq 2.$$

You must think of a number that, when multiplied by 2, gives a perfect *cube* under the radical. As shown in the example, 4 satisfies this condition.

3 Recall the special product

$$(a + b)(a - b) = a^2 - b^2.$$

In order to rationalize a denominator that contains a binomial expression (one that contains exactly two terms) involving radicals, such as

$$\frac{3}{1 + \sqrt{2}},$$

we must use conjugates. The **conjugate** of $1 + \sqrt{2}$ is $1 - \sqrt{2}$. In general, $a + b$ and $a - b$ are conjugates.

RATIONALIZING BINOMIAL DENOMINATORS

Whenever a radical expression has a binomial with square root radicals in the denominator, rationalize by multiplying both the numerator and the denominator by the conjugate of the denominator.

For the expression $\frac{3}{1 + \sqrt{2}}$, rationalize the denominator by multiplying both the numerator and denominator by the conjugate of the denominator, $1 - \sqrt{2}$.

$$\frac{3}{1 + \sqrt{2}} = \frac{3(1 - \sqrt{2})}{(1 + \sqrt{2})(1 - \sqrt{2})}$$

Then $(1 + \sqrt{2})(1 - \sqrt{2}) = 1^2 - (\sqrt{2})^2 = 1 - 2 = -1$. Placing -1 in the denominator gives

$$\frac{3}{1 + \sqrt{2}} = \frac{3(1 - \sqrt{2})}{-1} = -3(1 - \sqrt{2}) \quad \text{or} \quad -3 + 3\sqrt{2}.$$

EXAMPLE 5
RATIONALIZING A BINOMIAL DENOMINATOR

Rationalize the denominator in the following expressions. Assume that all variables represent positive real numbers.

(a) $\dfrac{5}{4-\sqrt{3}}$

Multiply numerator and denominator by the conjugate of the denominator, $4+\sqrt{3}$.

$$\frac{5}{4-\sqrt{3}} = \frac{5(4+\sqrt{3})}{(4-\sqrt{3})(4+\sqrt{3})}$$
$$= \frac{5(4+\sqrt{3})}{16-3}$$
$$= \frac{5(4+\sqrt{3})}{13}$$

Notice that the numerator is left in factored form. Doing this makes it easier to determine whether the expression can be reduced to lowest terms.

(b) $\dfrac{\sqrt{2}-\sqrt{3}}{\sqrt{5}+\sqrt{3}}$

Multiplication of both numerator and denominator by $\sqrt{5}-\sqrt{3}$ will rationalize the denominator.

$$\frac{\sqrt{2}-\sqrt{3}}{\sqrt{5}+\sqrt{3}} = \frac{(\sqrt{2}-\sqrt{3})(\sqrt{5}-\sqrt{3})}{(\sqrt{5}+\sqrt{3})(\sqrt{5}-\sqrt{3})}$$
$$= \frac{\sqrt{10}-\sqrt{6}-\sqrt{15}+3}{5-3}$$
$$= \frac{\sqrt{10}-\sqrt{6}-\sqrt{15}+3}{2}$$

(c) $$\frac{3}{\sqrt{5m}-\sqrt{p}} = \frac{3(\sqrt{5m}+\sqrt{p})}{(\sqrt{5m}-\sqrt{p})(\sqrt{5m}+\sqrt{p})}$$
$$= \frac{3(\sqrt{5m}+\sqrt{p})}{5m-p} \qquad (5m \neq p)$$ ■

4 The next example shows how to write radical quotients in lowest terms.

EXAMPLE 6
WRITING A RADICAL QUOTIENT IN LOWEST TERMS

Write each expression in lowest terms.

(a) $\dfrac{6+2\sqrt{5}}{4}$

Factor the numerator, and then simplify.

$$\frac{6+2\sqrt{5}}{4} = \frac{2(3+\sqrt{5})}{2\cdot 2} = \frac{3+\sqrt{5}}{2}$$

Here is an alternative method for writing this expression in lowest terms.

$$\frac{6 + 2\sqrt{5}}{4} = \frac{6}{4} + \frac{2\sqrt{5}}{4} = \frac{3}{2} + \frac{\sqrt{5}}{2} = \frac{3 + \sqrt{5}}{2}$$

(b) $\dfrac{5y - \sqrt{8y^2}}{6y} = \dfrac{5y - 2y\sqrt{2}}{6y} = \dfrac{y(5 - 2\sqrt{2})}{6y} = \dfrac{5 - 2\sqrt{2}}{6}$ (if $y > 0$) ■

CAUTION Refer to Example 6(a). A common error occurs when students try to write in lowest terms *before* factoring. Be careful to factor before writing a quotient in lowest terms.

5.5 EXERCISES

Multiply, then simplify the products. Assume that all variables represent positive real numbers. See Example 1.

1. $\sqrt{6}(3 + \sqrt{2})$

2. $\sqrt{2}(\sqrt{32} - \sqrt{9})$

3. $5(\sqrt{72} - \sqrt{8})$

4. $\sqrt{3}(\sqrt{12} + 2)$

5. $(\sqrt{7} + 3)(\sqrt{7} - 3)$

6. $(\sqrt{3} - 5)(\sqrt{3} + 5)$

7. $(\sqrt{2} - \sqrt{3})(\sqrt{2} + \sqrt{3})$

8. $(\sqrt{7} + \sqrt{3})(\sqrt{7} - \sqrt{3})$

9. $(\sqrt{8} - \sqrt{2})(\sqrt{8} + \sqrt{2})$

10. $(\sqrt{20} - \sqrt{5})(\sqrt{20} + \sqrt{5})$

11. $(\sqrt{2} + 1)(\sqrt{3} - 1)$

12. $(\sqrt{3} + 3)(\sqrt{5} - 2)$

13. $(\sqrt{11} - \sqrt{7})(\sqrt{2} + \sqrt{5})$

14. $(\sqrt{6} + \sqrt{2})(\sqrt{3} + \sqrt{2})$

15. $(2\sqrt{3} + \sqrt{5})(3\sqrt{3} - 2\sqrt{5})$

16. $(\sqrt{7} - \sqrt{11})(2\sqrt{7} + 3\sqrt{11})$

17. $(\sqrt{5} + 2)^2$

18. $(\sqrt{11} - 1)^2$

19. $(\sqrt{21} - \sqrt{5})^2$

20. $(\sqrt{6} - \sqrt{2})^2$

21. $(2 + \sqrt[3]{6})(2 - \sqrt[3]{6})$

22. $(\sqrt[3]{3} + 6)(\sqrt[3]{3} - 6)$

23. $(2 + \sqrt[3]{2})(4 - 2\sqrt[3]{2} + \sqrt[3]{4})$

24. $(\sqrt[3]{3} - 1)(\sqrt[3]{9} + \sqrt[3]{3} + 1)$

25. $(3\sqrt{x} - \sqrt{5})(2\sqrt{x} + 1)$

26. $(4\sqrt{p} + \sqrt{7})(\sqrt{p} - 9)$

27. $(3\sqrt{r} - \sqrt{s})(3\sqrt{r} + \sqrt{s})$

28. $(\sqrt{k} + 4\sqrt{m})(\sqrt{k} - 4\sqrt{m})$

29. $(\sqrt[3]{2y} - 5)(4\sqrt[3]{2y} + 1)$

30. $(\sqrt[3]{9z} - 2)(5\sqrt[3]{9z} + 7)$

Rationalize the denominators. Assume that all variables represent positive real numbers. See Examples 2–4.

31. $\dfrac{15}{\sqrt{5}}$

32. $\dfrac{20}{\sqrt{10}}$

33. $\dfrac{\sqrt{7}}{\sqrt{5}}$

34. $\dfrac{\sqrt{5}}{\sqrt{11}}$

35. $\dfrac{20\sqrt{3}}{\sqrt{5}}$

36. $\dfrac{12\sqrt{7}}{\sqrt{2}}$

37. $\dfrac{-6\sqrt{5}}{\sqrt{12}}$

38. $\dfrac{-4\sqrt{3}}{\sqrt{32}}$

39. $\sqrt{\dfrac{5}{18}}$

40. $\sqrt{\dfrac{21}{125}}$

41. $\dfrac{8\sqrt{3}}{\sqrt{k}}$

42. $\dfrac{6\sqrt{7}}{\sqrt{r}}$

43. $\frac{5\sqrt{2m}}{\sqrt{y^3}}$
44. $\frac{2\sqrt{5r}}{\sqrt{m^3}}$
45. $-\sqrt{\frac{48k^2}{z}}$
46. $-\sqrt{\frac{75m^3}{p}}$

47. $\sqrt[3]{\frac{9}{32}}$
48. $\sqrt[3]{\frac{10}{9}}$
49. $\sqrt[3]{\frac{x^6}{y}}$
50. $\sqrt[3]{\frac{m^9}{q}}$

Rationalize the denominators in each of the following. Assume that all variables represent positive real numbers and no denominators are zero. See Example 5.

51. $\frac{3}{4 + \sqrt{5}}$
52. $\frac{4}{3 - \sqrt{7}}$
53. $\frac{\sqrt{8}}{3 - \sqrt{2}}$

54. $\frac{\sqrt{27}}{2 + \sqrt{3}}$
55. $\frac{2}{3\sqrt{5} + 2\sqrt{3}}$
56. $\frac{-1}{3\sqrt{2} - 2\sqrt{7}}$

57. $\frac{2 - \sqrt{3}}{\sqrt{6} - \sqrt{5}}$
58. $\frac{5 + \sqrt{6}}{\sqrt{3} - \sqrt{2}}$
59. $\frac{m - 4}{\sqrt{m} + 2}$

60. $\frac{r - 9}{\sqrt{r} - 3}$
61. $\frac{3\sqrt{x}}{\sqrt{x} - 2\sqrt{y}}$
62. $\frac{5\sqrt{k}}{2\sqrt{k} + \sqrt{q}}$

Write each expression in lowest terms. Assume that all variables represent positive real numbers. See Example 6.

63. $\frac{30 - 20\sqrt{6}}{10}$
64. $\frac{15 - 6\sqrt{5}}{12}$
65. $\frac{3 - 3\sqrt{5}}{3}$
66. $\frac{-5 + 5\sqrt{2}}{5}$

67. $\frac{16 - 4\sqrt{8}}{12}$
68. $\frac{12 - 9\sqrt{72}}{18}$
69. $\frac{6p - \sqrt{24p^3}}{3p}$
70. $\frac{11y - \sqrt{242y^5}}{22y}$

Rationalize each denominator. Assume that all radicals represent real numbers, and no denominators are zero.

71. $\frac{1}{\sqrt{x + y}}$
72. $\frac{5}{\sqrt{m - n}}$
73. $\frac{p}{\sqrt{p + 2}}$
74. $\frac{3q}{\sqrt{5 + q}}$

---

75. What does it mean to rationalize a denominator? (See Objectives 2 and 3.)

76. What is wrong with the statement $\sqrt[3]{5} \cdot \sqrt[3]{5} = 5$? (See Objective 3.)

77. Which of the following is true?
$\frac{2 + 4\sqrt{3}}{2} = 1 + 4\sqrt{3}$ or $\frac{2 + 4\sqrt{3}}{2} = 1 + 2\sqrt{3}$
Explain. (See Objective 4.)

In calculus, it is sometimes necessary to rationalize the numerator *of a radical expression. Rationalize the numerator in each of the following. Assume that all variables represent positive real numbers.*

78. $\frac{6 - \sqrt{2}}{4}$
79. $\frac{8\sqrt{5} - 1}{6}$
80. $\frac{3\sqrt{a} + \sqrt{b}}{b}$
81. $\frac{\sqrt{p} - 3\sqrt{q}}{4q}$

Show that the following are true.

82. $(\sqrt[3]{5} - \sqrt[3]{3})(\sqrt[3]{25} + \sqrt[3]{15} + \sqrt[3]{9}) = 2$

83. $(\sqrt[3]{y} + \sqrt[3]{2})(\sqrt[3]{y^2} - \sqrt[3]{2y} + \sqrt[3]{4}) = y + 2$

Use the results of Exercises 82 and 83 to rationalize the denominators of the following.

84. $\dfrac{2}{\sqrt[3]{5} - \sqrt[3]{3}}$

85. $\dfrac{-3}{\sqrt[3]{y} + \sqrt[3]{2}}$

PREVIEW EXERCISES

Solve each equation. See Sections 2.1 and 3.9.

86. $8y - 7 = 9$

87. $-2r + 3 = 5$

88. $p^2 - 5p + 4 = 0$

89. $y^2 + 3y + 2 = 0$

90. $6m^2 = 7m + 3$

91. $15a^2 + 2 = 11a$

5.6 EQUATIONS WITH RADICALS

OBJECTIVES

1. LEARN THE POWER RULE.
2. SOLVE RADICAL EQUATIONS, SUCH AS $\sqrt{3x + 4} = 8$.
3. SOLVE RADICAL EQUATIONS, SUCH AS $\sqrt{m^2 - 4m + 9} = m - 1$, REQUIRING THE SQUARE OF A BINOMIAL.
4. SOLVE RADICAL EQUATIONS, SUCH AS $\sqrt{5m + 6} + \sqrt{3m + 4} = 2$, THAT REQUIRE SQUARING TWICE.
5. SOLVE RADICAL EQUATIONS WITH INDEXES GREATER THAN 2.

The equation $x = 1$ has only one solution. Its solution set is $\{1\}$. If we square both sides of this equation, another equation is obtained: $x^2 = 1$. This new equation has two solutions: -1 and 1. Notice that the solution of the original equation is also a solution of the squared equation. However, the squared equation has another solution, -1, that is *not* a solution of the original equation.

1 When solving equations with radicals, we will use this idea of raising both sides to a power. It is an application of the *power rule*.

POWER RULE If both sides of an equation are raised to the same power, all solutions of the original equation are also solutions of the new equation.

Read the power rule carefully; it does *not* say that all solutions to the new equation are solutions to the original equation. They may or may not be.

When the power rule is used to solve an equation, **every solution to the new equation *must* be checked in the original equation.**

Solutions that do not satisfy the original equation are called **extraneous;** they must be discarded.

2 The first example shows how to use the power rule in solving an equation.

EXAMPLE 1 USING THE POWER RULE

Solve $\sqrt{3x + 4} = 8$.

Use the power rule and square both sides of the equation to get

$$(\sqrt{3x + 4})^2 = 8^2$$
$$3x + 4 = 64$$
$$3x = 60$$
$$x = 20.$$

Check this proposed solution in the *original* equation.

$$\sqrt{3x + 4} = 8$$
$$\sqrt{3 \cdot 20 + 4} = 8 \quad ? \qquad \text{Let } x = 20.$$
$$\sqrt{64} = 8 \quad ?$$
$$8 = 8 \qquad \text{True}$$

Since 20 satisfies the *original* equation, the solution set is $\{20\}$. ■

The solution of the equation in Example 1 can be generalized to give a method for solving equations with radicals.

SOLVING AN EQUATION WITH RADICALS

Step 1 **Isolate the radical.** Make sure that one radical term is alone on one side of the equation.

Step 2 **Apply the power rule.** Raise each side of the equation to a power that is the same as the index of the radical.

Step 3 **Solve.** Solve the resulting equation; if it still contains a radical, repeat Steps 1 and 2.

Step 4 **Check.** It is essential that all potential solutions be checked in the original equation.

CAUTION Be careful not to skip Step 4 or you may get an incorrect solution set.

EXAMPLE 2 USING THE POWER RULE

Solve $\sqrt{5q - 1} + 3 = 0$.

To get the radical alone on one side, subtract 3 from both sides.

$$\sqrt{5q - 1} = -3$$

Now square both sides.

$$(\sqrt{5q - 1})^2 = (-3)^2$$
$$5q - 1 = 9$$
$$5q = 10$$
$$q = 2$$

The potential solution, 2, must be checked by substituting it in the original equation.

$$\sqrt{5q-1}+3=0$$
$$\sqrt{5\cdot 2-1}+3=0 \quad ? \qquad \text{Let } q=2.$$
$$3+3=0 \qquad \text{False}$$

This false result shows that 2 is not a solution of the original equation; it is extraneous. The solution set is $\emptyset$. ■

NOTE We could have determined after the first step that the equation in Example 2 had no solution. The equation $\sqrt{5q-1}=-3$ has no solution because the expression on the left cannot be negative.

3 The next examples involve finding the square of a binomial. Recall from Chapter 3 that $(x+y)^2=x^2+2xy+y^2$.

EXAMPLE 3 USING THE POWER RULE; SQUARING A BINOMIAL

Solve $\sqrt{4-x}=x+2$.

Square both sides; the square of $x+2$ is $(x+2)^2=x^2+4x+4$.

$$(\sqrt{4-x})^2=(x+2)^2$$
$$4-x=x^2+4x+4 \qquad \text{Twice the product of 2 and } x$$
$$0=x^2+5x \qquad \text{Subtract 4 and add } x.$$
$$0=x(x+5) \qquad \text{Factor.}$$
$$x=0 \quad \text{or} \quad x+5=0 \qquad \text{Zero-factor property}$$
$$x=-5.$$

Check each potential solution in the original equation.

If $x=0$,
$$\sqrt{4-x}=x+2$$
$$\sqrt{4-0}=0+2 \quad ?$$
$$\sqrt{4}=2 \quad ?$$
$$2=2. \qquad \text{True}$$

If $x=-5$,
$$\sqrt{4-x}=x+2$$
$$\sqrt{4-(-5)}=-5+2 \quad ?$$
$$\sqrt{9}=-3 \quad ?$$
$$3=-3. \qquad \text{False}$$

The solution set is $\{0\}$. The other potential solution, -5, is extraneous. ■

CAUTION Errors are often made when a radical equation involves squaring a binomial, as in Example 3. The middle term is often omitted. Remember that $(x+2)^2\neq x^2+4$ because $(x+2)^2=x^2+4x+4$.

EXAMPLE 4
USING THE POWER RULE; SQUARING A BINOMIAL

Solve $\sqrt{m^2 - 4m + 9} = m - 1$.

Square both sides. The square of the binomial $m - 1$ is $(m - 1)^2 = m^2 - 2(m)(1) + 1^2$.

$$(\sqrt{m^2 - 4m + 9})^2 = (m - 1)^2$$
$$m^2 - 4m + 9 = m^2 - 2m + 1$$

Twice the product of m and 1

Subtract m^2 and 1 from both sides. Then add $4m$ to both sides, to get

$$8 = 2m$$
$$4 = m.$$

Check this potential solution in the original equation.

$$\sqrt{m^2 - 4m + 9} = m - 1$$
$$\sqrt{4^2 - 4 \cdot 4 + 9} = 4 - 1 \quad ? \qquad \text{Let } m = 4.$$
$$3 = 3 \qquad \text{True}$$

The solution set of the original equation is $\{4\}$. ■

4 The next example shows an equation where both sides must be squared twice.

EXAMPLE 5
USING THE POWER RULE; SQUARING TWICE

Solve $\sqrt{5m + 6} + \sqrt{3m + 4} = 2$.

Start by getting one radical alone on one side of the equation. Do this by subtracting $\sqrt{3m + 4}$ from both sides.

$$\sqrt{5m + 6} = 2 - \sqrt{3m + 4}$$

Now square both sides.

$$(\sqrt{5m + 6})^2 = (2 - \sqrt{3m + 4})^2$$
$$5m + 6 = 4 - 4\sqrt{3m + 4} + (3m + 4)$$

Twice the product of 2 and $\sqrt{3m + 4}$

This equation still contains a radical, so it will be necessary to square both sides again. Before doing this, isolate the radical term on the right.

$$5m + 6 = 8 + 3m - 4\sqrt{3m + 4}$$
$$2m - 2 = -4\sqrt{3m + 4} \qquad \text{Subtract 8 and } 3m.$$
$$m - 1 = -2\sqrt{3m + 4} \qquad \text{Divide by 2.}$$

Now square both sides again.

$$(m - 1)^2 = (-2\sqrt{3m + 4})^2$$
$$(m - 1)^2 = (-2)^2(\sqrt{3m + 4})^2 \qquad (ab)^2 = a^2b^2$$
$$m^2 - 2m + 1 = 4(3m + 4)$$
$$m^2 - 2m + 1 = 12m + 16 \qquad \text{Distributive property}$$

This equation is quadratic and may be solved with the zero-factor property. Start by getting 0 on one side of the equation; then factor.

$$m^2 - 14m - 15 = 0$$
$$(m - 15)(m + 1) = 0$$

By the zero-factor property,

$$m - 15 = 0 \quad \text{or} \quad m + 1 = 0$$
$$m = 15 \quad \text{or} \quad m = -1.$$

Check each of these potential solutions in the original equation. Only -1 works, so the solution set, $\{-1\}$, has only one element. ■

5 The power rule also works for powers greater than 2, as the next example shows.

EXAMPLE 6 USING THE POWER RULE FOR POWERS GREATER THAN 2

Solve $\sqrt[4]{z + 5} = \sqrt[4]{2z - 6}$.

Raise both sides to the fourth power.

$$(\sqrt[4]{z + 5})^4 = (\sqrt[4]{2z - 6})^4$$
$$z + 5 = 2z - 6$$
$$11 = z$$

Check this result in the original equation.

$$\sqrt[4]{z + 5} = \sqrt[4]{2z - 6}$$
$$\sqrt[4]{11 + 5} = \sqrt[4]{2 \cdot 11 - 6} \quad ? \quad \text{Let } z = 11.$$
$$\sqrt[4]{16} = \sqrt[4]{16} \quad \text{True}$$

The solution set is $\{11\}$. ■

5.6 EXERCISES

Solve each of the following equations. For those involving exponents, change to radical form first. See Examples 1–4.

1. $\sqrt{r - 2} = 3$
2. $\sqrt{q + 1} = 7$
3. $\sqrt{6k - 1} = 1$
4. $\sqrt{7m - 3} = 5$
5. $a^{1/2} + 5 = 0$
6. $m^{1/2} - 8 = 0$
7. $\sqrt{3k + 1} - 4 = 0$
8. $\sqrt{5z + 1} - 11 = 0$
9. $4 - \sqrt{x - 2} = 0$
10. $9 - \sqrt{4k + 1} = 0$
11. $\sqrt{9a - 4} = \sqrt{8a + 1}$
12. $\sqrt{4p - 2} = \sqrt{3p + 5}$
13. $2\sqrt{x} = \sqrt{3x + 4}$
14. $2\sqrt{m} = \sqrt{5m - 16}$
15. $3\sqrt{z - 1} = 2\sqrt{2z + 2}$
16. $5\sqrt{4a + 1} = 3\sqrt{10a + 25}$
17. $k = \sqrt{k^2 + 4k - 20}$
18. $p = \sqrt{p^2 - 3p + 18}$
19. $a = (a^2 + 3a + 9)^{1/2}$
20. $z = (z^2 - 4z - 8)^{1/2}$
21. $(x^2 + 3x - 3)^{1/2} = x + 1$
22. $(y^2 - 2y + 6)^{1/2} = y + 2$
23. $\sqrt{k^2 + 2k + 9} = k + 3$
24. $\sqrt{a^2 - 3a + 3} = a - 1$

25. $\sqrt{r^2 + 9r + 3} = -r$

26. $\sqrt{p^2 - 15p + 15} = p - 5$

27. $\sqrt{z^2 + 12z - 4} + 4 - z = 0$

28. $\sqrt{m^2 + 3m + 12} - m - 2 = 0$

Solve each of the following equations. See Examples 5 and 6.

29. $\sqrt[3]{2x + 5} = \sqrt[3]{6x + 1}$

30. $\sqrt[4]{p - 1} = 2$

31. $\sqrt[3]{a^2 + 5a + 1} = \sqrt[3]{a^2 + 4a}$

32. $\sqrt[3]{r^2 + 2r + 8} = \sqrt[3]{r^2}$

33. $\sqrt[3]{2m - 1} = \sqrt[3]{m + 13}$

34. $\sqrt[3]{2k - 11} - \sqrt[3]{5k + 1} = 0$

35. $\sqrt[4]{a + 8} = \sqrt[4]{2a}$

36. $\sqrt[4]{z + 11} = \sqrt[4]{2z + 6}$

37. $(x - 8)^{1/3} + 2 = 0$

38. $(r + 1)^{1/3} + 1 = 0$

39. $(2k - 5)^{1/4} + 4 = 0$

40. $(8z - 3)^{1/4} + 2 = 0$

41. $\sqrt{k + 2} - \sqrt{k - 3} = 1$

42. $\sqrt{r + 6} - \sqrt{r - 2} = 2$

43. $\sqrt{x + 2} + \sqrt{x - 1} = 3$

44. $\sqrt{y + 4} + \sqrt{y - 4} = 4$

45. $(3p + 4)^{1/2} - (2p - 4)^{1/2} = 2$

46. $(5y + 6)^{1/2} - (y + 3)^{1/2} = 3$

47. $(2p - 1)^{2/3} = p^{1/3}$

48. $(m^2 - 2m)^{1/3} - m^{1/3} = 0$

49. $\sqrt{2\sqrt{x + 11}} = \sqrt{4x + 2}$

50. $\sqrt{1 + \sqrt{24 - 10x}} = \sqrt{3x + 5}$

---

51. What is wrong with the following "solution"? (See Objective 3.)

$$\sqrt{a + 6} = a + 2$$
$$a + 6 = a^2 + 4$$
$$0 = a^2 - a - 2$$
$$0 = (a - 2)(a + 1)$$
$$a - 2 = 0 \quad \text{or} \quad a + 1 = 0$$
$$a = 2 \quad \text{or} \quad a = -1$$

52. Without working it out, you should be able to tell that the equation $\sqrt{x - 1} + 3 = 0$ has no solution. Why? (See Objective 2.)

Work each problem. Give answers rounded to the nearest hundredth.

53. Carpenters stabilize wall frames with a diagonal brace as shown in the figure. The length of the brace is given by $L = \sqrt{H^2 + W^2}$. If the bottom of the brace is attached 9 feet from the corner and the brace is 12 feet long, how far up the corner post should it be nailed?

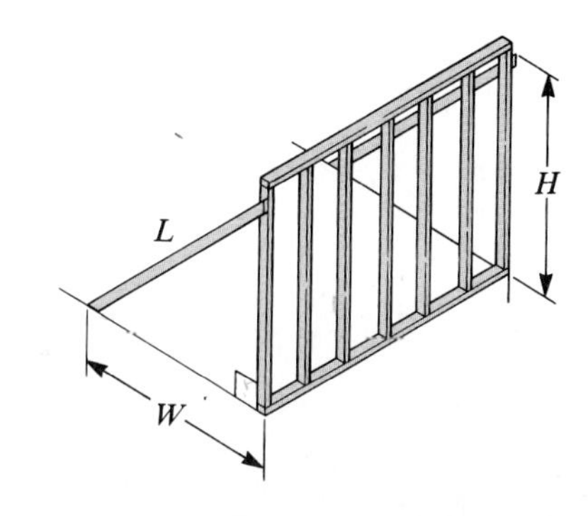

54. The period of a pendulum in seconds depends on its length L in feet, and is given by $P = 2\pi\sqrt{\frac{L}{32}}$. How long is a pendulum with a period of 2 seconds? Use 3.14 for π.

PREVIEW EXERCISES

Perform the indicated operations. See Sections 3.3 and 3.4.

55. $(5 + 9x) + (-4 - 8x)$

56. $(12 + 7y) - (-3 + 2y)$

57. $(x + 3)(2x - 5)$

Simplify each radical expression. Rationalize the denominator. See Section 5.5.

58. $\dfrac{2}{4 + \sqrt{3}}$

59. $\dfrac{-7}{5 - \sqrt{2}}$

60. $\dfrac{\sqrt{2} + \sqrt{7}}{\sqrt{5} + \sqrt{3}}$

5.7 COMPLEX NUMBERS

OBJECTIVES

1. SIMPLIFY NUMBERS OF THE FORM $\sqrt{-b}$, WHERE $b > 0$.
2. RECOGNIZE A COMPLEX NUMBER.
3. ADD AND SUBTRACT COMPLEX NUMBERS.
4. MULTIPLY COMPLEX NUMBERS.
5. DIVIDE COMPLEX NUMBERS.
6. FIND POWERS OF i.

As discussed in Chapter 1, the set of real numbers includes as subsets many other number sets (the rational numbers, integers, and natural numbers, for example). In this section a new set of numbers is introduced that includes the set of real numbers as a subset.

1 The equation $x^2 + 1 = 0$ has no real number solutions, since any solution must be a number whose square is -1. In the set of real numbers all squares are nonnegative numbers, because the product of either two positive numbers or two negative numbers is positive. To provide a solution for the equation $x^2 + 1 = 0$, a new number i is defined so that

$$i^2 = -1.$$

That is, i is a number whose square is -1. This definition of i makes it possible to define any square root of a negative number as follows.

For any positive real number b, $\quad \sqrt{-b} = i\sqrt{b}$.

EXAMPLE 1 SIMPLIFYING SQUARE ROOTS OF NEGATIVE NUMBERS

Write each number as a product of a real number and i.

(a) $\sqrt{-100} = i\sqrt{100} = 10i$

(b) $\sqrt{-2} = \sqrt{2}i = i\sqrt{2}$ ■

CAUTION It is easy to mistake $\sqrt{2}i$ for $\sqrt{2i}$, with the i under the radical. For this reason, it is common to write $\sqrt{2}i$ as $i\sqrt{2}$.

When finding a product such as $\sqrt{-4} \cdot \sqrt{-9}$, the product rule for radicals cannot be used, since that rule applies only when both radicals represent real numbers. For this reason, always change $\sqrt{-b}$ $(b > 0)$ to the form $i\sqrt{b}$ before performing any multiplications or divisions. For example,

$$\sqrt{-4} \cdot \sqrt{-9} = i\sqrt{4} \cdot i\sqrt{9} = i \cdot 2 \cdot i \cdot 3 = 6i^2.$$

Since $i^2 = -1$,

$$6i^2 = 6(-1) = -6.$$

An *incorrect* use of the product rule for radicals would give a wrong answer.

$$\sqrt{-4} \cdot \sqrt{-9} = \sqrt{(-4)(-9)} = \sqrt{36} = 6 \qquad \text{Incorrect}$$

EXAMPLE 2 **MULTIPLYING SQUARE ROOTS OF NEGATIVE NUMBERS**

Multiply.

(a) $\sqrt{-3} \cdot \sqrt{-7} = i\sqrt{3} \cdot i\sqrt{7} = i^2\sqrt{3 \cdot 7} = (-1)\sqrt{21} = -\sqrt{21}$

(b) $\sqrt{-2} \cdot \sqrt{-8} = i\sqrt{2} \cdot i\sqrt{8} = i^2\sqrt{2 \cdot 8} = (-1)\sqrt{16} = (-1)4 = -4$

(c) $\sqrt{-5} \cdot \sqrt{6} = i\sqrt{5} \cdot \sqrt{6} = i\sqrt{30}$ ■

The methods used to find products also apply to quotients, as the next example shows.

EXAMPLE 3 **DIVIDING SQUARE ROOTS OF NEGATIVE NUMBERS**

Divide.

(a) $\dfrac{\sqrt{-75}}{\sqrt{-3}} = \dfrac{i\sqrt{75}}{i\sqrt{3}} = \sqrt{\dfrac{75}{3}} = \sqrt{25} = 5$

(b) $\dfrac{\sqrt{-32}}{\sqrt{8}} = \dfrac{i\sqrt{32}}{\sqrt{8}} = i\sqrt{\dfrac{32}{8}} = i\sqrt{4} = 2i$ ■

2 With the new number i and the real numbers, a new set of numbers can be formed that includes the real numbers as a subset. The *complex numbers* are defined as follows.

COMPLEX NUMBER

If a and b are real numbers, then any number of the form $a + bi$ is called a **complex number.**

In the complex number $a + bi$, the number a is called the **real part** and b is called the **imaginary part.** When $b = 0$, $a + bi$ is a real number, so the real numbers are a subset of the complex numbers. Complex numbers with $b \neq 0$ are called **imaginary numbers.** In spite of their name, imaginary numbers are very useful in applications, particularly in work with electricity.

The relationships among the various sets of numbers discussed in this book are shown in Figure 5.2.

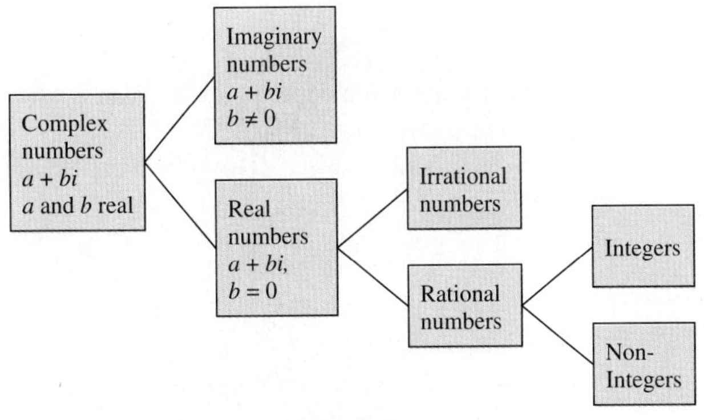

FIGURE 5.2

3 The commutative, associative, and distributive properties for real numbers are also valid for complex numbers. To add complex numbers, add their real parts and add their imaginary parts.

ADDING COMPLEX NUMBERS

The **sum** of the complex numbers $a + bi$ and $c + di$ is

$$(a + bi) + (c + di) = (a + c) + (b + d)i.$$

EXAMPLE 4 ADDING COMPLEX NUMBERS

Add.

(a) $(2 + 3i) + (6 + 4i) = (2 + 6) + (3 + 4)i$ — Commutative, associative, and distributive properties

$= 8 + 7i$

(b) $5 + (9 - 3i) = (5 + 9) + (0 - 3)i$

$= 14 - 3i$ ■

Since the sum of the complex numbers $8 - 2i$ and $-8 + 2i$ is $0 + 0i$, or just 0, these two numbers are additive inverses of each other. Additive inverses are used to subtract complex numbers.

SUBTRACTING COMPLEX NUMBERS

To **subtract** two complex numbers, *add* the first complex number and the additive inverse of the second.

EXAMPLE 5 SUBTRACTING COMPLEX NUMBERS

Subtract.

(a) $(6 + 5i) - (3 + 2i) = (6 + 5i) + (-3 - 2i)$ — Additive inverse

$= 3 + 3i$

(b) $(7 - 3i) - (8 - 6i) = (7 - 3i) + (-8 + 6i)$

$= -1 + 3i$

(c) $(-9 + 4i) - (-9 + 8i) = (-9 + 4i) + (9 - 8i)$

$= 0 - 4i$ ■

In Example 5(c), the answer was written as $0 - 4i$ rather than just $-4i$. A complex number written in the form $a + bi$, like $0 - 4i$, is in **standard form.** In this section, most answers will be given in standard form.

4 Complex numbers of the form $a + bi$ have the same form as a binomial, so the product of two complex numbers can be found by using the FOIL method for multiplying binomials. (Recall that FOIL stands for *First-Outside-Inside-Last*.) The next example shows how to apply this method.

EXAMPLE 6 MULTIPLYING COMPLEX NUMBERS

(a) Multiply $3 + 5i$ and $4 - 2i$.

Use the FOIL method.

$$(3 + 5i)(4 - 2i) = \underset{\text{First}}{3(4)} + \underset{\text{Outside}}{3(-2i)} + \underset{\text{Inside}}{5i(4)} + \underset{\text{Last}}{5i(-2i)}$$

Now simplify. (Remember that $i^2 = -1$.)

$$\begin{aligned}(3 + 5i)(4 - 2i) &= 12 - 6i + 20i - 10i^2 \\ &= 12 + 14i - 10(-1) \qquad \text{Let } i^2 = -1. \\ &= 12 + 14i + 10 \\ &= 22 + 14i\end{aligned}$$

(b) $$\begin{aligned}(2 + 3i)(1 - 5i) &= 2(1) + 2(-5i) + 3i(1) + 3i(-5i) \\ &= 2 - 10i + 3i - 15i^2 \\ &= 2 - 7i - 15(-1) \\ &= 2 - 7i + 15 \\ &= 17 - 7i \quad ■\end{aligned}$$

The two complex numbers $a + bi$ and $a - bi$ are called **conjugates** of each other. The product of a complex number and its conjugate is always a real number, as shown here.

$$\begin{aligned}(a + bi)(a - bi) &= a \cdot a - abi + abi - b^2i^2 \\ &= a^2 - b^2(-1) \\ &= a^2 + b^2\end{aligned}$$

$$(a + bi)(a - bi) = a^2 + b^2$$

For example, $(3 + 7i)(3 - 7i) = 3^2 + 7^2 = 9 + 49 = 58$.

5 Conjugates are used to find the quotient of two complex numbers. Divisions of complex numbers are usually expressed as fractions of the form

$$\frac{a + bi}{c + di}.$$

To divide complex numbers, *multiply the numerator and the denominator by the conjugate of the denominator.*

EXAMPLE 7 DIVIDING COMPLEX NUMBERS

Find the quotients.

(a) $\dfrac{4 - 3i}{5 + 2i}$

Multiply the numerator and the denominator by the conjugate of the denominator. The conjugate of $5 + 2i$ is $5 - 2i$.

$$\begin{aligned}\frac{4 - 3i}{5 + 2i} &= \frac{(4 - 3i)(5 - 2i)}{(5 + 2i)(5 - 2i)} \\ &= \frac{20 - 8i - 15i + 6i^2}{5^2 + 2^2} \\ &= \frac{14 - 23i}{29} \quad \text{or} \quad \frac{14}{29} - \frac{23}{29}i\end{aligned}$$

The final result is given in standard form.

(b) $\dfrac{1+i}{i}$

The conjugate of i is $-i$. Multiply the numerator and the denominator by $-i$.

$$\begin{aligned}\frac{1+i}{i} &= \frac{(1+i)(-i)}{i(-i)}\\ &= \frac{-i-i^2}{-i^2}\\ &= \frac{-i-(-1)}{-(-1)}\\ &= \frac{-i+1}{1}\\ &= 1-i \quad \blacksquare\end{aligned}$$

6 The fact that $i^2 = -1$ can be used to find higher powers of i, as shown in the following examples.

$$i^3 = i \cdot i^2 = i(-1) = -i \qquad i^6 = i^2 \cdot i^4 = (-1) \cdot 1 = -1$$
$$i^4 = i^2 \cdot i^2 = (-1)(-1) = 1 \qquad i^7 = i^3 \cdot i^4 = (-i) \cdot 1 = -i$$
$$i^5 = i \cdot i^4 = i \cdot 1 = i \qquad i^8 = i^4 \cdot i^4 = 1 \cdot 1 = 1$$

A few powers of i are listed here.

POWERS OF i

$i^1 = i$	$i^5 = i$	$i^9 = i$	$i^{13} = i$
$i^2 = -1$	$i^6 = -1$	$i^{10} = -1$	$i^{14} = -1$
$i^3 = -i$	$i^7 = -i$	$i^{11} = -i$	$i^{15} = -i$
$i^4 = 1$	$i^8 = 1$	$i^{12} = 1$	$i^{16} = 1$

As these examples suggest, the powers of i rotate through the four numbers i, -1, $-i$, and 1. Larger powers of i can be simplified by using the fact that $i^4 = 1$. For example, $i^{75} = (i^4)^{18} \cdot i^3 = 1^{18} \cdot i^3 = 1 \cdot i^3 = i^3 = -i$. This example suggests a quick method for simplifying large powers of i.

SIMPLIFYING LARGE POWERS OF i

Step 1 Divide the exponent by 4.

Step 2 Observe the remainder obtained in Step 1. The large power of i is the same as i raised to the power determined by this remainder. Refer to the chart above to complete the simplification. (If the remainder is 0, the power simplifies to $i^0 = 1$.)

EXAMPLE 8 **SIMPLIFYING POWERS OF i**

Find each power of i.

(a) $i^{12} = (i^4)^3 = 1^3 = 1$

(b) i^{39}

We can use the guidelines shown above. Start by dividing 39 by 4 (Step 1).

$$\begin{array}{r} 9 \\ 4\overline{)39} \\ \underline{36} \\ 3 \leftarrow \text{Remainder} \end{array}$$

The remainder is 3. So $i^{39} = i^3 = -i$ (Step 2).

Another way of simplifying i^{39} is as follows.

$$i^{39} = i^{36} \cdot i^3 = (i^4)^9 \cdot i^3 = 1^9 \cdot (-i) = -i$$

(c) $i^{-2} = \dfrac{1}{i^2} = \dfrac{1}{-1} = -1$

(d) $i^{-1} = \dfrac{1}{i}$

To simplify this quotient, multiply the numerator and the denominator by the conjugate of i, which is $-i$.

$$i^{-1} = \frac{1}{i} = \frac{1(-i)}{i(-i)} = \frac{-i}{-i^2} = \frac{-i}{-(-1)} = \frac{-i}{1} = -i \quad ■$$

5.7 EXERCISES

Simplify. See Examples 1–3.

1. $\sqrt{-144}$
2. $\sqrt{-196}$
3. $-\sqrt{-225}$
4. $-\sqrt{-400}$
5. $\sqrt{-3}$
6. $\sqrt{-19}$
7. $\sqrt{-75}$
8. $\sqrt{-125}$
9. $4\sqrt{-12} - 3\sqrt{-3}$
10. $6\sqrt{-18} + 2\sqrt{-32}$
11. $2\sqrt{-45} + 4\sqrt{-80}$
12. $-3\sqrt{-28} + 2\sqrt{-63}$
13. $\sqrt{-5} \cdot \sqrt{-5}$
14. $\sqrt{-3} \cdot \sqrt{-3}$
15. $\sqrt{-9} \cdot \sqrt{-36}$
16. $\sqrt{-4} \cdot \sqrt{-81}$
17. $\sqrt{-16} \cdot \sqrt{-100}$
18. $\sqrt{-81} \cdot \sqrt{-121}$
19. $\dfrac{\sqrt{-200}}{\sqrt{-100}}$
20. $\dfrac{\sqrt{-50}}{\sqrt{-2}}$
21. $\dfrac{\sqrt{-54}}{\sqrt{6}}$
22. $\dfrac{\sqrt{-90}}{\sqrt{10}}$
23. $\dfrac{\sqrt{-288}}{\sqrt{-8}}$
24. $\dfrac{\sqrt{-48} \cdot \sqrt{-3}}{\sqrt{-2}}$

---

25. Why is it incorrect to use the product rule for radicals to multiply $\sqrt{-3} \cdot \sqrt{-12}$? (See Objective 1.)
26. In your own words describe the relationship between complex numbers and real numbers. (See Objective 2.)

Add or subtract as indicated. Give answers in standard form. See Examples 4 and 5.

27. $(6 + 2i) + (4 + 3i)$ **28.** $(1 + i) + (2 - i)$ **29.** $(3 - 2i) + (-4 + 5i)$

30. $(7 + 15i) + (-11 - 14i)$ **31.** $(5 - i) + (-5 + i)$ **32.** $(-2 + 6i) + (2 - 6i)$

33. $(6 - 3i) - (-2 - 2i)$ **34.** $(5 - 8i) - (1 + 7i)$ **35.** $(5 + 5i) - (-2 - 3i)$

36. $(-1 - i) - (-3 - 4i)$ **37.** $(-3 - 4i) - (-1 - i)$ **38.** $(-2 - 3i) - (5 + 5i)$

39. $(-4 + 11i) + (-2 - 4i) + (-3i)$ **40.** $(4 + i) + (1 + 5i) + (3 + 2i)$

41. $[(7 + 3i) - (4 - 2i)] + (3 - i)$ **42.** $[(3 + 2i) + (-4 - i)] - (2 + 5i)$

Multiply. Give answers in standard form. See Example 6.

43. $(8i)(2i)$ **44.** $(2i)(5i)$ **45.** $(4i)(7i)$ **46.** $(-3i)(-2i)$

47. $(-6i)(-5i)$ **48.** $2i(4 - 3i)$ **49.** $6i(1 - i)$ **50.** $(3 + 2i)(-1 + 2i)$

51. $(7 + 3i)(-5 + i)$ **52.** $(4 + 3i)(6 + 5i)$ **53.** $(-1 + 3i)(1 + 2i)$ **54.** $(2 + 2i)^2$

55. $(-1 + 3i)^2$ **56.** $(5 + 6i)(5 - 6i)$ **57.** $(2 + 3i)(2 - 3i)$ **58.** $(4 + 2i)(4 - 2i)$

59. $(1 + i)(1 - i)$ **60.** $(5 - 2i)(5 + 2i)$ **61.** $(8 - 9i)(8 + 9i)$ **62.** $(3 + 7i)(3 - 7i)$

Find the quotients. Give answers in standard form. See Example 7.

63. $\frac{2}{1 + i}$ **64.** $\frac{-5}{2 - i}$ **65.** $\frac{i}{2 + i}$ **66.** $\frac{-i}{3 - 2i}$ **67.** $\frac{2 - 2i}{1 + i}$ **68.** $\frac{2 - 3i}{2 + 3i}$

69. $\frac{1 - i}{1 + i}$ **70.** $\frac{-1 + 5i}{-3 + 2i}$ **71.** $\frac{2 + 6i}{-2 + i}$ **72.** $\frac{4 - 3i}{1 + 2i}$ **73.** $\frac{-9}{2i}$ **74.** $\frac{-6}{5i}$

Find each of the following powers of i. See Example 8.

75. i^6 **76.** i^9 **77.** i^{17} **78.** i^{98} **79.** i^{-3} **80.** i^{-5} **81.** i^{-10} **82.** i^{-40}

Perform the indicated operations. Give answers in standard form.

83. $\frac{3}{2 - i} + \frac{5}{1 + i}$ **84.** $\frac{2}{3 + 4i} + \frac{4}{1 - i}$ **85.** $\left(\frac{2 + i}{2 - i} + \frac{i}{1 + i}\right)i$ **86.** $\left(\frac{4 - i}{1 + i} - \frac{2i}{2 + i}\right)4i$

Work each problem.

Ohm's law for the current I in a circuit with voltage E, resistance R, capacitive reactance X_c, and inductive reactance X_L is

$$I = \frac{E}{R + (X_L - X_c)i}.$$

87. Find I in standard form if $E = 2 + 3i$, $R = 5$, $X_L = 4$, and $X_c = 3$.

88. Find E if $I = 1 - i$, $R = 2$, $X_L = 3$, and $X_c = 1$.

89. Show: $1 - 5i$ is a solution of $x^2 - 2x + 26 = 0$.

90. Show: $3 - 2i$ is a solution of $p^2 - 6p + 13 = 0$.

PREVIEW EXERCISES

Give all square roots of each number. See Section 5.1.

91. 25 **92.** 81 **93.** 45

Solve each equation. See Section 3.9.

94. $x^2 - 2x - 24 = 0$ **95.** $x^2 + 3x - 40 = 0$ **96.** $2x^2 - 5x = 7$

CHAPTER 5 GLOSSARY

KEY TERMS

5.1 ***n*th root** An nth root of a number a is a number b such that $b^n = a$.

principal *n*th root For a positive number a and even value of n, the principal nth root of a is the positive nth root of a.

radicand, index In the expression $\sqrt[n]{a}$, a is the radicand and n is the index (or *order*).

radical The expression $\sqrt[n]{a}$ is a radical.

5.3 Pythagorean formula The Pythagorean formula states that in a right triangle the square of the length of the hypotenuse equals the sum of the squares of the lengths of the legs.

hypotenuse The hypotenuse of a right triangle is the longest side, opposite the right angle.

legs The legs of a right triangle are the two shorter sides.

5.4 radical expression A radical expression is an algebraic expression that contains radicals.

5.5 rationalizing the denominator The process of removing radicals from the denominator so that the denominator contains only rational quantities is called rationalizing the denominator.

conjugate For real numbers a and b, $a + b$ and $a - b$ are conjugates.

5.6 extraneous solution An extraneous solution of a radical equation is a solution of $x = a^2$ that is not a solution of $\sqrt{x} = a$.

5.7 complex number A complex number is a number that can be written in the form $a + bi$, where a and b are real numbers.

real part The real part of $a + bi$ is a.

imaginary part The imaginary part of $a + bi$ is b.

imaginary number A complex number $a + bi$ with $b \neq 0$ is called an imaginary number.

standard form A complex number written as $a + bi$, where a and b are real numbers, is in standard form.

NEW SYMBOLS

$a^{1/n}$ a to the power $\frac{1}{n}$

$\sqrt[n]{a}$ principal nth root of a

$\pm$ positive or negative

$a^{m/n}$ a to the power $\frac{m}{n}$

$\approx$ is approximately equal to

i a number whose square is -1

CHAPTER 5 QUICK REVIEW

SECTION	CONCEPTS	EXAMPLES
5.1 RATIONAL EXPONENTS AND RADICALS	$a^{1/n} = b$ means $b^n = a$. $a^{1/n}$ is the principal nth root of a. $a^{1/n} = \sqrt[n]{a}$ whenever $\sqrt[n]{a}$ exists.	The two square roots of 64 are $\sqrt{64} = 8$ and $-\sqrt{64} = -8$. Of these, 8 is the principal square root of 64. $25^{1/2} = \sqrt{25} = 5$ $(-64)^{1/3} = \sqrt[3]{-64} = -4$

SECTION	CONCEPTS	EXAMPLES
	$\sqrt[n]{a^n} = \|a\|$ if n is even. $\sqrt[n]{a^n} = a$ if n is odd. If m and n are positive integers with m/n in lowest terms, then $$a^{m/n} = (a^{1/n})^m$$ provided that $a^{1/n}$ is a real number.	$\sqrt[3]{-27} = -3 \quad \sqrt[4]{(-2)^4} = \|-2\| = 2$ $8^{5/3} = (8^{1/3})^5 = 2^5 = 32$
5.2 MORE ABOUT RATIONAL EXPONENTS AND RADICALS	All of the usual rules for exponents are valid for rational exponents.	$5^{-1/2} \cdot 5^{1/4} = 5^{-1/2+1/4} = 5^{-1/4} = \dfrac{1}{5^{1/4}}$ $(y^{2/5})^{10} = y^4$ $\dfrac{x^{-1/3}}{x^{-1/2}} = x^{-1/3-(-1/2)}$ $= x^{-1/3+1/2} = x^{1/6} \quad (x > 0)$
5.3 SIMPLIFYING RADICALS	**Product and Quotient Rules for Radicals** If $\sqrt[n]{a}$ and $\sqrt[n]{b}$ are real numbers, and if n is a natural number, $$\sqrt[n]{a} \cdot \sqrt[n]{b} = \sqrt[n]{ab}$$ and $$\sqrt[n]{\frac{a}{b}} = \frac{\sqrt[n]{a}}{\sqrt[n]{b}} \quad (b \neq 0).$$	Apply the product and quotient rules. $\sqrt{3} \cdot \sqrt{7} = \sqrt{21}$ $\sqrt[5]{x^3y} \cdot \sqrt[5]{xy^2} = \sqrt[5]{x^4y^3}$ $\dfrac{\sqrt{x^5}}{\sqrt{x^4}} = \sqrt{\dfrac{x^5}{x^4}} = \sqrt{x} \quad (x > 0)$
	Simplified Radical **1.** The radicand has no factor raised to a power greater than or equal to the index. **2.** The radicand has no fractions. **3.** No denominator contains a radical. **4.** Exponents in the radicand and the index of the radical have no common factor (except 1). That is, $$\sqrt[kn]{a^{km}} = \sqrt[n]{a^m}$$ if all roots exist.	Simplify each radical. $\sqrt{18} = \sqrt{9 \cdot 2} = 3\sqrt{2}$ $\sqrt[3]{54x^5y^3} = \sqrt[3]{27x^3y^3 \cdot 2x^2} = 3xy\sqrt[3]{2x^2}$ $\sqrt{\dfrac{7}{4}} = \dfrac{\sqrt{7}}{\sqrt{4}} = \dfrac{\sqrt{7}}{2}$ $\sqrt[9]{x^3} = x^{3/9} = x^{1/3}$ or $\sqrt[3]{x}$ $\sqrt[8]{64} = \sqrt[8]{2^6} = \sqrt[4]{2^3} = \sqrt[4]{8}$

SECTION	CONCEPTS	EXAMPLES
5.4 ADDING AND SUBTRACTING RADICAL EXPRESSIONS	Only radical expressions with the same index and the same radicand may be combined.	Perform the indicated operations. $3\sqrt{17} + 2\sqrt{17} - 8\sqrt{17} = (3 + 2 - 8)\sqrt{17} = -3\sqrt{17}$ $\sqrt[3]{2} - \sqrt[3]{250} = \sqrt[3]{2} - 5\sqrt[3]{2} = -4\sqrt[3]{2}$ $\left.\begin{matrix}\sqrt{15} + \sqrt{30}\\ \sqrt{3} + \sqrt[3]{3}\end{matrix}\right\}$ cannot be further simplified.
5.5 MULTIPLYING AND DIVIDING RADICAL EXPRESSIONS	Radical expressions may often be multiplied by using the FOIL method. Special products from Section 3.4 may apply.	Multiply. $(\sqrt{2} + \sqrt{7})(\sqrt{3} - \sqrt{6})$ $= \sqrt{6} - \sqrt{12} + \sqrt{21} - \sqrt{42}$ $= \sqrt{6} - 2\sqrt{3} + \sqrt{21} - \sqrt{42}$ $(\sqrt{5} - \sqrt{10})(\sqrt{5} + \sqrt{10}) = 5 - 10 = -5$ $(\sqrt{3} - \sqrt{2})^2 = 3 - 2\sqrt{3} \cdot \sqrt{2} + 2 = 5 - 2\sqrt{6}$
	Rationalize the denominator by multiplying both the numerator and denominator by the same expression.	Rationalize the denominator. $\frac{\sqrt{7}}{\sqrt{5}} = \frac{\sqrt{7}}{\sqrt{5}} \cdot \frac{\sqrt{5}}{\sqrt{5}} = \frac{\sqrt{35}}{5}$ $\frac{\sqrt[3]{2}}{\sqrt[3]{4}} = \frac{\sqrt[3]{2}}{\sqrt[3]{4}} \cdot \frac{\sqrt[3]{2}}{\sqrt[3]{2}} = \frac{\sqrt[3]{4}}{\sqrt[3]{8}} = \frac{\sqrt[3]{4}}{2}$ $\frac{4}{\sqrt{5} - \sqrt{2}} = \frac{4}{\sqrt{5} - \sqrt{2}} \cdot \frac{\sqrt{5} + \sqrt{2}}{\sqrt{5} + \sqrt{2}} = \frac{4(\sqrt{5} + \sqrt{2})}{5 - 2} = \frac{4(\sqrt{5} + \sqrt{2})}{3}$
	To write a quotient involving radicals, such as $\frac{5 + 15\sqrt{6}}{10}$ in lowest terms, factor the numerator and denominator, and then divide both by the greatest common factor.	$\frac{5 + 15\sqrt{6}}{10} = \frac{5(1 + 3\sqrt{6})}{5 \cdot 2} = \frac{1 + 3\sqrt{6}}{2}$

SECTION	CONCEPTS	EXAMPLES
5.6 EQUATIONS WITH RADICALS	**Solving Equations with Radicals** *Step 1* Make sure that one radical is alone on one side of the equals sign. *Step 2* Raise each side of the equation to a power that is the same as the index of the radical. *Step 3* Solve the resulting equation; if it still contains a radical, repeat Steps 1 and 2. *Step 4* It is essential that all potential solutions be checked in the *original* equation. Potential solutions that do not check are *extraneous;* they are not part of the solution set.	Solve $\sqrt{2x+3} - x = 0$. $\sqrt{2x+3} = x$ $2x + 3 = x^2$ $x^2 - 2x - 3 = 0$ $(x-3)(x+1) = 0$ $x - 3 = 0$ or $x + 1 = 0$ $x = 3$ $\quad$ $x = -1$ A check shows that 3 is a solution, but -1 is extraneous. The solution set is $\{3\}$.
5.7 COMPLEX NUMBERS	$i^2 = -1$ For any positive number b, $\sqrt{-b} = i\sqrt{b}$. To perform a multiplication such as $\sqrt{-3} \cdot \sqrt{-27}$, first change each factor to the form $i\sqrt{b}$, then multiply. The same procedure applies to quotients such as $\dfrac{\sqrt{-18}}{\sqrt{-2}}$. **Adding and Subtracting Complex Numbers** $(a + bi) + (c + di)$ $= (a + c) + (b + d)i$ $(a + bi) - (c + di)$ $= (a - c) + (b - d)i$	Simplify. $\sqrt{-3} \cdot \sqrt{-27} = i\sqrt{3} \cdot i\sqrt{27}$ $= i^2\sqrt{81}$ $= -1 \cdot 9 = -9$ $\dfrac{\sqrt{-18}}{\sqrt{-2}} = \dfrac{i\sqrt{18}}{i\sqrt{2}} = \dfrac{\sqrt{18}}{\sqrt{2}} = \sqrt{9} = 3$ Perform the indicated operations. $(5 + 3i) + (8 - 7i) = 13 - 4i$ $(5 + 3i) - (8 - 7i) = (5 + 3i) + (-8 + 7i)$ $= -3 + 10i$

SECTION	CONCEPTS	EXAMPLES
	Multiplying and Dividing Complex Numbers Multiply complex numbers by using the FOIL method. Divide complex numbers by multiplying the numerator and the denominator by the conjugate of the denominator.	$(2+i)(5-3i) = 10-6i+5i-3i^2$ $= 10-i-3(-1)$ $= 10-i+3$ $= 13-i$ $\frac{2}{3+i} = \frac{2}{3+i}\cdot\frac{3-i}{3-i}$ $= \frac{2(3-i)}{9-i^2}$ $= \frac{2(3-i)}{10}$ $= \frac{3-i}{5} = \frac{3}{5}-\frac{1}{5}i$

CHAPTER 5 REVIEW EXERCISES

Assume all variables represent positive numbers throughout these review exercises.

[5.1] *Evaluate each of the following.*

1. $100^{3/2}$ **2.** $16^{5/4}$ **3.** $1000^{-2/3}$ **4.** $-\left(\frac{36}{25}\right)^{3/2}$

Find each of the following roots.

5. $\sqrt{1764}$ **6.** $-\sqrt{289}$ **7.** $\sqrt[3]{-125}$ **8.** $-\sqrt[4]{256}$ **9.** $\sqrt[5]{-32}$ **10.** $\sqrt[6]{729}$

Write as radicals.

11. $5^{1/3}$ **12.** $8^{1/4}$ **13.** $(2k)^{2/5}$ **14.** $(m+3n)^{1/2}$ **15.** $(3a+b)^{-5/3}$

Write with rational exponents.

16. $\sqrt{15}$ **17.** $\sqrt[3]{9}$ **18.** $\sqrt[4]{5^3}$ **19.** $\sqrt[5]{p^4}$

[5.2] *Use the rules for exponents to simplify each of the following. Write answers with only positive exponents.*

20. $5^{1/4}\cdot 5^{7/4}$ **21.** $\frac{96^{2/3}}{96^{-1/3}}$ **22.** $\frac{(a^{1/3})^4}{a^{2/3}}$ **23.** $\left(\frac{m^3}{m^5}\right)^{1/2}$

24. $\frac{y^{-1/3}\cdot y^{5/6}}{y}$ **25.** $\left(\frac{z^{-1}x^{-3/5}}{2^{-2}z^{-1/2}x}\right)^{-1}$ **26.** $r^{-1/2}(r+r^{3/2})$ **27.** $2y^{-2/3}(y^{4/3}-5y^{-1})$

Simplify each of the following by first writing each radical in exponential form. Give answers as radicals.

28. $\sqrt[8]{s^4}$ **29.** $\sqrt[6]{r^9}$ **30.** $\frac{\sqrt{p^5}}{p^2}$ **31.** $\sqrt[4]{k^3}\cdot\sqrt{k^3}$

32. $\sqrt{3^{17}}$ **33.** $\sqrt[3]{m^5}\cdot\sqrt[3]{m^8}$ **34.** $-\sqrt[3]{y^{11}}$ **35.** $\sqrt[4]{\sqrt[3]{z}}$

Use a calculator to find a decimal approximation for each of the following. Give answers to the nearest thousandth.

36. $\sqrt{42}$ **37.** $-\sqrt{756}$ **38.** $\sqrt[3]{888}$ **39.** $\sqrt[4]{60}$

40. $\sqrt[3]{-21}$ **41.** $36^{2/3}$ **42.** $500^{-3/4}$ **43.** $-\sqrt{7090}$

44. $-500^{4/3}$ **45.** $-28^{-1/2}$

[5.3] *Simplify each of the following.*

46. $\sqrt{6} \cdot \sqrt{13}$ **47.** $\sqrt{5} \cdot \sqrt{r}$ **48.** $\sqrt[3]{6} \cdot \sqrt[3]{5}$ **49.** $\sqrt{125}$

50. $\sqrt[3]{108}$ **51.** $\sqrt{100y^7}$ **52.** $\sqrt{18m^4n^5}$ **53.** $\sqrt[3]{64p^8q^5}$

54. $\sqrt{\frac{49}{81}}$ **55.** $\sqrt{\frac{y^3}{144}}$ **56.** $\sqrt[3]{\frac{m^{15}}{27}}$ **57.** $\sqrt[3]{\frac{r^2}{8}}$

58. $\sqrt[6]{15^3}$ **59.** $\sqrt[4]{p^6}$ **60.** $\sqrt[3]{2} \cdot \sqrt[4]{5}$ **61.** $\sqrt{x} \cdot \sqrt[5]{x}$

Find the missing length in each right triangle. Simplify answers if possible.

62.

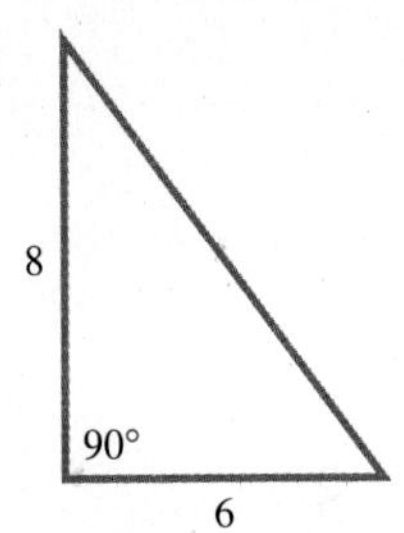

63.

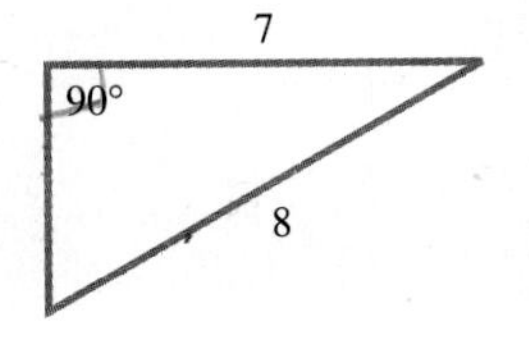

[5.4] *Simplify each of the following.*

64. $2\sqrt{8} - 3\sqrt{50}$ **65.** $-5\sqrt{18} + 12\sqrt{72}$ **66.** $8\sqrt{80} - 3\sqrt{45}$

67. $3\sqrt[3]{54} + 5\sqrt[3]{16}$ **68.** $-6\sqrt[4]{32} + \sqrt[4]{512}$ **69.** $8\sqrt[3]{x^3y^2} - 2x\sqrt[3]{y^2}$

70. $\frac{3}{\sqrt{16}} - \frac{\sqrt{5}}{2}$ **71.** $\frac{4}{\sqrt{25}} + \frac{\sqrt{5}}{4}$ **72.** $\frac{\sqrt{45}}{3} + \frac{2}{\sqrt{9}} - \frac{1}{\sqrt{25}}$

[5.5] *Multiply.*

73. $(\sqrt{3} + 1)(\sqrt{3} - 2)$ **74.** $(\sqrt{7} + \sqrt{5})(\sqrt{7} - \sqrt{5})$ **75.** $(3\sqrt{2} + 1)(2\sqrt{2} - 3)$

76. $(\sqrt{13} - \sqrt{2})^2$ **77.** $(\sqrt[3]{2} + 3)(\sqrt[3]{4} - 3\sqrt[3]{2} + 9)$ **78.** $(\sqrt[3]{4y} - 1)(\sqrt[3]{4y} + 3)$

Rationalize the denominator and simplify. Assume that no denominator is zero.

79. $\frac{2}{\sqrt{7}}$ **80.** $\frac{-5}{\sqrt{11}}$ **81.** $\frac{\sqrt{6}}{\sqrt{8}}$ **82.** $\frac{-6\sqrt{3}}{\sqrt{2}}$

83. $\frac{8\sqrt{5z}}{\sqrt{2k}}$ **84.** $-\sqrt[3]{\frac{9}{25}}$ **85.** $\sqrt[3]{\frac{32}{m}}$ **86.** $\sqrt[3]{\frac{108m^3}{p^5}}$

87. $\frac{1}{2 - \sqrt{7}}$ **88.** $\frac{-5}{\sqrt{6} - \sqrt{3}}$ **89.** $\frac{2\sqrt{z}}{\sqrt{z} - 2\sqrt{x}}$ **90.** $\frac{5\sqrt{3r} + \sqrt{5s}}{\sqrt{3r} - 2\sqrt{5s}}$

Write in lowest terms.

91. $\dfrac{2 - 2\sqrt{5}}{8}$ **92.** $\dfrac{4 - 8\sqrt{8}}{12}$ **93.** $\dfrac{6k + \sqrt{27k^3}}{3k}$

94. A friend wants to rationalize the denominator of the fraction $\dfrac{5 + \sqrt{7}}{2 + \sqrt{3}}$, and he decides to multiply the numerator and denominator by $\sqrt{3}$. Why will his plan not work?

95. Another friend wants to rationalize the denominator of the fraction $\dfrac{5}{\sqrt[3]{6}}$, and she decides to multiply the numerator and denominator by $\sqrt[3]{6}$. Why will her plan not work?

[5.6] *Solve each of the following equations.*

96. $\sqrt{8y + 9} = 5$ **97.** $\sqrt{2z - 3} - 3 = 0$ **98.** $3\sqrt{m} = \sqrt{10m - 9}$

99. $2\sqrt{k} = \sqrt{3k + 25}$ **100.** $\sqrt{r^2 - 6r + 9} = r + 7$ **101.** $\sqrt{p^2 + 3p + 7} = p + 2$

102. $\sqrt[3]{5m - 1} = \sqrt[3]{3m - 2}$ **103.** $(b + 6)^{1/4} = (2b)^{1/4}$ **104.** $(r + 2)^{1/2} - (r - 3)^{1/2} = 1$

[5.7] *Simplify each of the following.*

105. $\sqrt{-25}$ **106.** $\sqrt{-49}$ **107.** $\sqrt{-200}$ **108.** $2\sqrt{-50} - 6\sqrt{-18}$

Perform the indicated operations. Write each answer in standard form.

109. $(-2 + 5i) + (-8 - 7i)$ **110.** $(4 - 9i) + (-1 + 2i)$ **111.** $(-6 - i) - (2 - 5i)$

112. $\sqrt{-25} \cdot \sqrt{-81}$ **113.** $\dfrac{\sqrt{-72}}{\sqrt{-8}}$ **114.** $(2 + 3i)(1 - i)$

115. $(5 + 4i)(2 - 8i)$ **116.** $\dfrac{6 - 5i}{2 + 3i}$

Find each power of i.

117. i^{11} **118.** i^{-13}

MIXED REVIEW EXERCISES

Simplify.

119. $-\sqrt{27y} + 2\sqrt{75y}$ **120.** $-\sqrt{-841}$ **121.** $-8^{2/3}$

122. $\sqrt[4]{7} \cdot \sqrt[4]{3}$ **123.** $2\sqrt{54m^3} + 5\sqrt{96m^3}$ **124.** $\dfrac{(p^{5/4})^2}{p^{-3/2}}$

125. $\sqrt{169a^2b^4}$ **126.** $\sqrt[3]{\sqrt{k}}$ **127.** $\sqrt[3]{216}$

128. $\dfrac{k^{2/3}k^{-1/2}k^{3/4}}{2(k^2)^{-1/4}}$ **129.** $\dfrac{3\sqrt{7p}}{\sqrt{y}}$ **130.** $\dfrac{-1}{\sqrt{12}}$

131. $\left(\dfrac{1}{32}\right)^{-7/5}$ **132.** $-\sqrt[3]{27z^{12}}$ **133.** $\sqrt[4]{7^9}$

134. $\sqrt{20}$

Perform the indicated operations.

135. $(\sqrt{11} + 3\sqrt{5})(\sqrt{11} + 5\sqrt{5})$

136. $\sqrt{-5} \cdot \sqrt{-7}$

137. $\dfrac{3 - i}{2 + i}$

138. $(5 + 4i) - (-9 - 3i)$

139. $\dfrac{\sqrt{50}}{\sqrt{-2}}$

140. $(\sqrt{13} - \sqrt{2})^2$

CHAPTER 5 TEST

Simplify each of the following. Assume that all variables represent positive real numbers.

1. $27^{4/3}$

2. $\left(\dfrac{16}{25}\right)^{-3/2}$

3. $\dfrac{9^{-3/4}}{9^{-1/4}}$

4. $\left(\dfrac{a^{-2} \cdot a^{-1/2}}{a^{-3/4} \cdot a}\right)^{-1}$

5. Multiply: $r^{-2/3}(r^{5/3} + r^{-1/3})$.

Find each of the following roots. Assume that all variables represent nonnegative real numbers. In Exercise 9, use a calculator and give the answer to the nearest thousandth.

6. $\sqrt[3]{-1000}$

7. $\sqrt{100y^{14}}$

8. $-\sqrt[4]{16m^4p^{12}}$

9. $\sqrt[3]{494}$

10. Simplify: $\sqrt[3]{a^4} \cdot \sqrt[3]{a^7}$.

Simplify each of the following. Assume that all variables represent nonnegative real numbers.

11. $\sqrt{108x^5}$

12. $\sqrt[4]{16a^5b^{11}}$

13. $\sqrt[3]{3} \cdot \sqrt[3]{18}$

14. $8\sqrt{20} + 3\sqrt{80} - 2\sqrt{500}$

15. $(\sqrt{5} - \sqrt{3})(3\sqrt{5} + \sqrt{3})$

16. $(2\sqrt{5} - 3)(\sqrt{5} + 6)$

Rationalize each denominator.

17. $-\sqrt{\dfrac{81}{40}}$

18. $\dfrac{3}{\sqrt[3]{4}}$

19. $\dfrac{-8}{\sqrt{7} - \sqrt{5}}$

Solve each of the following equations.

20. $4\sqrt{p} = \sqrt{13p + 27}$

21. $\sqrt{y^2 - 5y + 3} = 4 + y$

Perform the indicated operations. Give answers in standard form.

22. $(-6 + 4i) - (8 - 7i) + 10i$

23. $(1 + 5i)(3 + i)$

24. $\dfrac{1 + 2i}{3 - i}$

25. Is $\sqrt[3]{a^3 + b^3} = a + b$? Explain.

CHAPTER SIX

QUADRATIC EQUATIONS AND INEQUALITIES

The height in feet at t seconds of an object thrown upward with an initial velocity of 32 feet per second is

$$h = 32t - 16t^2.$$

Second-degree (quadratic) equations are often used to express the relationship between distance and time as in this example. In this chapter we continue the study of equations begun in Chapter 2 and touched on again in each succeeding chapter. Earlier chapters dealt primarily with linear equations. Now in this chapter, several methods are introduced for solving second-degree equations.

6.1 COMPLETING THE SQUARE

OBJECTIVES

1. EXTEND THE ZERO-FACTOR PROPERTY TO COMPLEX NUMBERS.
2. LEARN THE SQUARE ROOT PROPERTY.
3. SOLVE QUADRATIC EQUATIONS OF THE FORM $(ax - b)^2 = c$ BY USING THE SQUARE ROOT PROPERTY.
4. SOLVE QUADRATIC EQUATIONS BY COMPLETING THE SQUARE.

In Section 3.9, we solved quadratic equations by using the zero-factor property. In this section, we introduce additional methods for solving quadratic equations.

Recall that a *quadratic equation* is defined as follows.

QUADRATIC EQUATION An equation that can be written in the form $ax^2 + bx + c = 0$, where a, b, and c are real numbers, with $a \neq 0$, is a **quadratic equation in standard form.**

For example, $m^2 - 4m + 5 = 0$ and $3q^2 = 4q - 8$ are quadratic equations, with the first of these in standard form.

1 The zero-factor property stated in Chapter 3 was restricted to real number factors. Now that property can be extended to complex number factors. - A + b i

ZERO-FACTOR PROPERTY If a and b are complex numbers and if $ab = 0$, then either $a = 0$ or $b = 0$, or both.

For example, the equation $x^2 + 4 = 0$ can now be solved with the zero-factor property as follows.

$$x^2 + 4 = 0$$

$$x^2 - (-4) = 0 \qquad \text{Write as a difference.}$$

$$(x + \sqrt{-4})(x - \sqrt{-4}) = 0 \qquad \text{Factor.}$$

$$x + \sqrt{-4} = 0 \quad \text{or} \quad x - \sqrt{-4} = 0 \qquad \text{Zero-factor property}$$

$$x = -\sqrt{-4} \qquad x = \sqrt{-4}$$

$$x = -2i \qquad x = 2i \qquad \sqrt{-4} = 2i$$

The solution set is $\{-2i, 2i\}$.

2 Although factoring is the simplest way to solve quadratic equations, not every quadratic equation can be solved easily by factoring. In this section and the next, other methods of solving quadratic equations are developed based on the following property.

SQUARE ROOT PROPERTY

If a and b are complex numbers and if $a^2 = b$, then $a = \sqrt{b}$ or $a = -\sqrt{b}$.

To see why this property works, write $a^2 = b$ as $a^2 - b = 0$. Then factor to get $(a - \sqrt{b})(a + \sqrt{b}) = 0$. Place each factor equal to zero to get $a = \sqrt{b}$ or $a = -\sqrt{b}$. For example, by the square root property, $x^2 = 16$ implies that $x = 4$ or $x = -4$.

CAUTION Do not forget that if $b \neq 0$, using the square root property always produces *two* square roots, one positive and one negative.

EXAMPLE 1 USING THE SQUARE ROOT PROPERTY

Solve $r^2 = 5$.

From the square root property, since $(\sqrt{5})^2 = 5$,

$$r = \sqrt{5} \qquad \text{or} \qquad r = -\sqrt{5}$$

and the solution set is $\{\sqrt{5}, -\sqrt{5}\}$. ■

Recall from Chapter 5 that roots such as those in Example 1 are sometimes abbreviated with the symbol $\pm$ (read "positive or negative"); with this symbol the solutions in Example 1 would be written $\pm\sqrt{5}$.

3 The square root property can be used to solve more complicated equations, such as

$$(x - 5)^2 = 36,$$

by substituting $(x - 5)^2$ for a^2 and 36 for b in the square root property to get

$$x - 5 = 6 \qquad \text{or} \qquad x - 5 = -6$$

$$x = 11 \qquad\qquad x = -1.$$

Check that both 11 and -1 satisfy the given equation so that the solution set is $\{11, -1\}$.

EXAMPLE 2 USING THE SQUARE ROOT PROPERTY

Solve $(2a - 3)^2 = 18$.

By the square root property,

$$2a - 3 = \sqrt{18} \qquad \text{or} \qquad 2a - 3 = -\sqrt{18},$$

from which

$$2a = 3 + \sqrt{18} \qquad \text{or} \qquad 2a = 3 - \sqrt{18}$$

$$a = \frac{3 + \sqrt{18}}{2} \qquad \text{or} \qquad a = \frac{3 - \sqrt{18}}{2}.$$

Since $\sqrt{18} = \sqrt{9 \cdot 2} = 3\sqrt{2}$, the solution set can be written

$$\left\{\frac{3 + 3\sqrt{2}}{2}, \frac{3 - 3\sqrt{2}}{2}\right\}. \quad ■$$

The next example shows an equation that has imaginary solutions.

EXAMPLE 3
USING THE SQUARE ROOT PROPERTY

Solve $(b + 2)^2 = -16$.

Begin with

$$b + 2 = \sqrt{-16} \quad \text{or} \quad b + 2 = -\sqrt{-16}.$$

Since $\sqrt{-16} = 4i$,

$$b + 2 = 4i \quad \text{or} \quad b + 2 = -4i$$
$$b = -2 + 4i \quad \text{or} \quad b = -2 - 4i.$$

The solution set is $\{-2 + 4i, -2 - 4i\}$. ■

4 The square root property can be used to solve any quadratic equation by writing it in the form $(x + k)^2 = n$. That is, the left side of the equation must be rewritten as a perfect square trinomial (that can be factored as $(x + k)^2$) and the right side must be a constant. Rewriting the equation in this form is called **completing the square.**

For example,

$$m^2 + 8m + 10 = 0$$

is a quadratic equation that cannot be solved easily by factoring. To get a perfect square trinomial on the left side of the equation $m^2 + 8m + 10 = 0$, first subtract 10 from both sides:

$$m^2 + 8m = -10.$$

The left side should be a perfect square, say $(m + k)^2$. Since $(m + k)^2 = m^2 + 2mk + k^2$, comparing $m^2 + 8m$ with $m^2 + 2mk$ shows that

$$2mk = 8m$$
$$k = 4.$$

If $k = 4$, then $(m + k)^2$ becomes $(m + 4)^2$, or $m^2 + 8m + 16$. Get the necessary $+16$ by adding 16 on both sides so that

$$m^2 + 8m = -10$$

becomes

$$m^2 + 8m + 16 = -10 + 16.$$

Factor the left side. It should be a perfect square. Since $m^2 + 8m + 16$ factors as $(m + 4)^2$, the equation becomes

$$(m + 4)^2 = 6.$$

This equation can be solved with the square root property:

$$m + 4 = \sqrt{6} \quad \text{or} \quad m + 4 = -\sqrt{6},$$

leading to the solution set $\{-4 + \sqrt{6}, -4 - \sqrt{6}\}$.

Based on the work of this example, an equation of the form $x^2 + px = q$ can be converted into an equation of the form $(x + k)^2 = n$ by adding the square of half the coefficient of x to both sides of the equation.

In summary, to find the solutions of $ax^2 + bx + c = 0$ (with $a \neq 0$) by completing the square, proceed as follows.

COMPLETING THE SQUARE

Step 1 If $a \neq 1$, divide both sides by a.
Step 2 Rewrite the equation so that both terms containing variables are on one side of the equals sign and the constant is on the other side.
Step 3 Take half the coefficient of x and square it.
Step 4 Add the square to both sides of the equation.
Step 5 One side should now be a perfect square trinomial. Factor that side and write it as the square of a binomial. Simplify the other side.
Step 6 Use the square root property to complete the solution.

memorize

NOTE Steps 1 and 2 can be done in either order. With some problems, it is more convenient to do Step 2 first.

EXAMPLE 4 SOLVING A QUADRATIC EQUATION BY COMPLETING THE SQUARE

Solve $k^2 + 5k - 1 = 0$ by completing the square.

Follow the steps listed above. Since $a = 1$, Step 1 is not needed here. Begin by adding 1 to both sides.

$$k^2 + 5k = 1 \qquad \text{Step 2}$$

Take half of 5 (the coefficient of the first-degree term) and square the result.

$$\frac{1}{2} \cdot 5 = \frac{5}{2} \quad \text{and} \quad \left(\frac{5}{2}\right)^2 = \frac{25}{4} \qquad \text{Step 3}$$

Add the square to each side of the equation to get

$$k^2 + 5k + \frac{25}{4} = 1 + \frac{25}{4}. \qquad \text{Step 4}$$

Write the left side as a perfect square and add on the right. Then use the square root property.

$$\left(k + \frac{5}{2}\right)^2 = \frac{29}{4} \qquad \text{Step 5}$$

$$k + \frac{5}{2} = \sqrt{\frac{29}{4}} \quad \text{or} \quad k + \frac{5}{2} = -\sqrt{\frac{29}{4}} \qquad \text{Step 6}$$

Simplify $\sqrt{\frac{29}{4}}$: $\sqrt{\frac{29}{4}} = \frac{\sqrt{29}}{\sqrt{4}} = \frac{\sqrt{29}}{2}$.

$$k + \frac{5}{2} = \frac{\sqrt{29}}{2} \quad \text{or} \quad k + \frac{5}{2} = \frac{-\sqrt{29}}{2}$$

$$k = -\frac{5}{2} + \frac{\sqrt{29}}{2} \quad \text{or} \quad k = -\frac{5}{2} - \frac{\sqrt{29}}{2}$$

Combine terms on the right to get the two solutions,

$$k = \frac{-5 + \sqrt{29}}{2} \quad \text{or} \quad k = \frac{-5 - \sqrt{29}}{2}.$$

The solution set is $\left\{\frac{-5 + \sqrt{29}}{2}, \frac{-5 - \sqrt{29}}{2}\right\}$. ■

EXAMPLE 5
SOLVING A QUADRATIC EQUATION BY COMPLETING THE SQUARE

Solve $2a^2 - 4a - 5 = 0$.

Go through the steps for completing the square. First divide both sides of the equation by 2 to make the coefficient of the second-degree term equal to 1, getting

$$a^2 - 2a - \frac{5}{2} = 0. \qquad \text{Step 1}$$

$$a^2 - 2a = \frac{5}{2} \qquad \text{Step 2}$$

$$\frac{1}{2}(-2) = \mathbf{-1} \text{ and } (-1)^2 = 1. \qquad \text{Step 3}$$

$$a^2 - 2a + \mathbf{1} = \frac{5}{2} + \mathbf{1} \qquad \text{Step 4}$$

$$(a - 1)^2 = \frac{7}{2} \qquad \text{Step 5}$$

$$a - 1 = \sqrt{\frac{7}{2}} \quad \text{or} \quad a - 1 = -\sqrt{\frac{7}{2}} \qquad \text{Step 6}$$

$$a = 1 + \sqrt{\frac{7}{2}} \qquad a = 1 - \sqrt{\frac{7}{2}}.$$

Since $\sqrt{\frac{7}{2}} = \frac{\sqrt{14}}{2}$,

$$a = 1 + \frac{\sqrt{14}}{2} \quad \text{or} \quad a = 1 - \frac{\sqrt{14}}{2}.$$

Add the two terms in each solution as follows:

$$1 + \frac{\sqrt{14}}{2} = \frac{2}{2} + \frac{\sqrt{14}}{2} = \frac{2 + \sqrt{14}}{2}$$

$$1 - \frac{\sqrt{14}}{2} = \frac{2}{2} - \frac{\sqrt{14}}{2} = \frac{2 - \sqrt{14}}{2}.$$

The solution set is $\left\{\frac{2 + \sqrt{14}}{2}, \frac{2 - \sqrt{14}}{2}\right\}$. ■

6.1 EXERCISES

1. Which of the following are quadratic equations?
(a) $2x - 5 = 7$ **(b)** $2x^2 - 5x = 7$ **(c)** $m^2 = 3m$ **(d)** $2p^3 + 5p^2 + 1 = 0$

Use the square root property to solve each equation. (All the solutions for these equations are real numbers.) See Examples 1 and 2.

2. $b^2 = 49$ **3.** $n^2 = 100$ **4.** $t^2 = 13$

5. $y^2 = 11$ **6.** $k^2 = 12$ **7.** $w^2 = 18$

8. $(p + 6)^2 = 9$ **9.** $(q - 2)^2 = 25$ **10.** $(3a - 1)^2 = 3$

11. $(2x + 4)^2 = 5$ **12.** $(4p + 1)^2 = 12$ **13.** $(5k - 2)^2 = 24$

Find the imaginary number solutions of the following equations. See Example 3.

14. $m^2 = -9$ **15.** $r^2 = -144$ **16.** $a^2 = -72$ **17.** $p^2 = -18$

18. $(x - 5)^2 = -1$ **19.** $(z + 2)^2 = -2$ **20.** $(2m - 1)^2 = -3$ **21.** $(3k + 2)^2 = -5$

---

22. What is wrong with the following "solution" using the zero-factor property? (See Objective 1.)

$$(x - i)(x - 2i) = 1$$

$$x - i = 1 \quad \text{or} \quad x - 2i = 1$$

$$x = 1 + i \quad \text{or} \quad x = 1 + 2i$$

23. What is wrong with the following "solution" using the square root property? (See Objective 2.)

$$a^2 = 9$$

$$a = 3$$

Decide what number must be added to each expression to make it a perfect square trinomial.

24. $p^2 + 6p$ **25.** $z^2 + 12z$ **26.** $y^2 - 10y$ **27.** $q^2 - 8q$

28. $r^2 - 5r$ **29.** $y^2 - 9y$ **30.** $x^2 + 11x$ **31.** $z^2 + 3z$

Solve each equation by completing the square. (All the solutions for these equations are real numbers.) See Examples 4 and 5.

32. $x^2 - 2x - 15 = 0$ **33.** $m^2 - 4m = 32$

34. $2y^2 + y = 15$ **35.** $2z^2 - 7z = 15$

36. $2m^2 + 5m - 3 = 0$ **37.** $3a^2 + 2a = 1$

38. $x^2 - 2x - 1 = 0$ **39.** $p^2 - 4p + 1 = 0$

40. $m^2 = -6m - 7$ **41.** $r^2 + 8r + 14 = 0$

42. $q^2 + \frac{7}{2}q = 2$ **43.** $r^2 + \frac{4}{3}r = \frac{4}{3}$

Solve each equation by completing the square. (Some of these equations have imaginary number solutions.) See Examples 4 and 5.

44. $s^2 + 13 = 6s$ **45.** $m^2 + 2m + 2 = 0$ **46.** $9a^2 - 24a = -13$

47. $x^2 - x - 1 = 0$ **48.** $m^2 + 6m + 10 = 0$ **49.** $x^2 + 4x + 13 = 0$

50. $25m^2 - 20m - 1 = 0$ **51.** $3p^2 + 3 = 8p$ **52.** $25y^2 + 46 = 70y$

53. $2y^2 - 4y = 5$ **54.** $4m^2 - 12m + 13 = 0$ **55.** $25p^2 - 50p + 29 = 0$

56. Can any quadratic equation be solved by completing the square? Explain. (See Objective 4.)

Solve for x. Assume that a and b represent positive real numbers.

57. $x^2 = 4b$ **58.** $x^2 - b = 0$

59. $9x^2 - 25a = 0$ **60.** $4x^2 = b^2 + 16$

61. $x^2 - a^2 - 36 = 0$ **62.** $(5x - 2b)^2 = 3a$

PREVIEW EXERCISES

Evaluate $\sqrt{b^2 - 4ac}$ for the following values of a, b, and c. See Section 1.3.

63. $a = 3, b = 1, c = -1$ **64.** $a = 4, b = 11, c = -3$ **65.** $a = 18, b = -25, c = -3$

66. $a = 6, b = 7, c = 2$ **67.** $a = 7, b = 5, c = -5$ **68.** $a = 3, b = 9, c = -4$

6.2 THE QUADRATIC FORMULA

OBJECTIVES

1. SOLVE QUADRATIC EQUATIONS BY USING THE QUADRATIC FORMULA.
2. SOLVE QUADRATIC EQUATIONS WITH IMAGINARY NUMBER SOLUTIONS USING THE QUADRATIC FORMULA.
3. SOLVE APPLICATIONS BY USING THE QUADRATIC FORMULA.

FOCUS ON PROBLEM SOLVING

A janitorial service provides two people to clean an office building. Working together, the two can clean the building in 5 hours. One person is new to the job and takes 2 hours longer than the other person to clean the building working alone. How long would it take the experienced worker to clean the building working alone?

This problem about work rates is similar to the work problems discussed in Chapter 4, except that it results in a quadratic equation that cannot be solved by factoring. This problem is Exercise 41 in the exercises for this section. After working through this section, you should be able to solve this problem.

The examples in the previous section showed that any quadratic equation can be solved by completing the square. However, completing the square can be tedious and time consuming. In this section the method of completing the square is used to solve the general quadratic equation $ax^2 + bx + c = 0$, where a, b, and c are complex numbers and $a \neq 0$. The solution of this general equation can then be used as a formula to find the solution of any specific quadratic equation.

To solve $ax^2 + bx + c = 0$ by completing the square (assuming $a > 0$ for now), we follow the steps given in Section 6.1.

$$ax^2 + bx + c = 0$$

$$x^2 + \frac{b}{a}x + \frac{c}{a} = 0 \qquad \text{Divide each side by } a. \text{ (Step 1)}$$

$$x^2 + \frac{b}{a}x = -\frac{c}{a} \qquad \text{Subtract } \frac{c}{a} \text{ from each side. (Step 2)}$$

$$\frac{1}{2}\left(\frac{b}{a}\right) = \frac{b}{2a}; \quad \left(\frac{b}{2a}\right)^2 = \frac{b^2}{4a^2} \qquad \text{(Step 3)}$$

$$x^2 + \frac{b}{a}x + \frac{b^2}{4a^2} = -\frac{c}{a} + \frac{b^2}{4a^2} \qquad \text{Add } \frac{b^2}{4a^2} \text{ to each side. (Step 4)} \qquad (1)$$

$$x^2 + \frac{b}{a}x + \frac{b^2}{4a^2} = \left(x + \frac{b}{2a}\right)^2 \qquad \text{Write the left side as a square.}$$

Using this result, and rearranging the right side, equation (1) can be written as

$$\left(x + \frac{b}{2a}\right)^2 = \frac{b^2}{4a^2} + \frac{-c}{a}. \qquad \text{(Step 5)} \qquad \textbf{(2)}$$

$$\frac{b^2}{4a^2} + \frac{-c}{a} = \frac{b^2}{4a^2} + \frac{-4ac}{4a^2} = \frac{b^2 - 4ac}{4a^2}$$

Combine terms on the right side.

Finally, equation (2) becomes

$$\left(x + \frac{b}{2a}\right)^2 = \frac{b^2 - 4ac}{4a^2}.$$

$$x + \frac{b}{2a} = \sqrt{\frac{b^2 - 4ac}{4a^2}} \quad \text{or} \quad x + \frac{b}{2a} = -\sqrt{\frac{b^2 - 4ac}{4a^2}}.$$

Square root property (Step 6)

Since

$$\sqrt{\frac{b^2 - 4ac}{4a^2}} = \frac{\sqrt{b^2 - 4ac}}{\sqrt{4a^2}} = \frac{\sqrt{b^2 - 4ac}}{2a},$$

the result can be expressed as

$$x + \frac{b}{2a} = \frac{\sqrt{b^2 - 4ac}}{2a} \quad \text{or} \quad x + \frac{b}{2a} = \frac{-\sqrt{b^2 - 4ac}}{2a}$$

$$x = \frac{-b}{2a} + \frac{\sqrt{b^2 - 4ac}}{2a} \quad \text{or} \quad x = \frac{-b}{2a} - \frac{\sqrt{b^2 - 4ac}}{2a}$$

$$x = \frac{-b + \sqrt{b^2 - 4ac}}{2a} \quad \text{or} \quad x = \frac{-b - \sqrt{b^2 - 4ac}}{2a}.$$

If $a < 0$, it can be shown that the same two solutions are obtained. The result is the **quadratic formula,** often abbreviated as follows.

QUADRATIC FORMULA

The solutions of $ax^2 + bx + c = 0$ $(a \neq 0)$ are

$$x = \frac{-b \pm \sqrt{b^2 - 4ac}}{2a}.$$

CAUTION Notice in the quadratic formula that the square root is added to or subtracted from the value of $-b$ *before* dividing by $2a$.

1 To use the quadratic formula, first write the given equation in the form $ax^2 + bx + c = 0$; then identify the values of a, b, and c and substitute them into the quadratic formula, as shown in the next examples.

EXAMPLE 1
USING THE QUADRATIC FORMULA

Solve $6x^2 - 5x - 4 = 0$.

First identify the letters a, b, and c of the general quadratic equation. Here a, the coefficient of the second-degree term, is 6, while b, the coefficient of the first-degree term, is -5, and the constant, c, is -4. Substitute these values into the quadratic formula.

$$x = \frac{-b \pm \sqrt{b^2 - 4ac}}{2a}$$

$$x = \frac{-(-5) \pm \sqrt{(-5)^2 - 4(6)(-4)}}{2(6)} \qquad a = 6, b = -5, c = -4$$

$$= \frac{5 \pm \sqrt{25 + 96}}{12}$$

$$= \frac{5 \pm \sqrt{121}}{12} = \frac{5 \pm 11}{12}$$

This last statement leads to two solutions, one from + and one from −. Using + and − in turn gives

$$x = \frac{5 + 11}{12} = \frac{16}{12} = \frac{4}{3} \quad \text{or} \quad x = \frac{5 - 11}{12} = \frac{-6}{12} = -\frac{1}{2}.$$

Check each of these solutions by substitution in the original equation. The solution set is $\{4/3, -1/2\}$. ■

EXAMPLE 2
USING THE QUADRATIC FORMULA

Solve $4r^2 = 8r - 1$.

Rewrite the equation in standard form as $4r^2 - 8r + 1 = 0$ and identify $a = 4$, $b = -8$, and $c = 1$. Now use the quadratic formula.

$$r = \frac{-b \pm \sqrt{b^2 - 4ac}}{2a}$$

$$r = \frac{-(-8) \pm \sqrt{(-8)^2 - 4(4)(1)}}{2(4)} \qquad a = 4, b = -8, c = 1$$

$$= \frac{8 \pm \sqrt{64 - 16}}{8} = \frac{8 \pm \sqrt{48}}{8}$$

$$= \frac{8 \pm 4\sqrt{3}}{8} = \frac{4(2 \pm \sqrt{3})}{8} = \frac{2 \pm \sqrt{3}}{2}$$

The solution set for $4r^2 = 8r - 1$ is $\left\{\frac{2 + \sqrt{3}}{2}, \frac{2 - \sqrt{3}}{2}\right\}$. ■

2 The solutions of the equations in the next two examples are imaginary numbers.

EXAMPLE 3 USING THE QUADRATIC FORMULA

Solve $9q^2 + 5 = 6q$.

After writing the equation in the standard form, as $9q^2 - 6q + 5 = 0$, identify $a = 9$, $b = -6$, and $c = 5$. Then use the quadratic formula.

$$q = \frac{-(-6) \pm \sqrt{(-6)^2 - 4(9)(5)}}{2(9)}$$

$$= \frac{6 \pm \sqrt{-144}}{18} = \frac{6 \pm 12i}{18} \qquad \sqrt{-144} = 12i$$

$$= \frac{6(1 \pm 2i)}{18} = \frac{1 \pm 2i}{3}$$

The solution set is $\left\{\frac{1}{3} + \frac{2}{3}i, \frac{1}{3} - \frac{2}{3}i\right\}$. ■

EXAMPLE 4 USING THE QUADRATIC FORMULA

Solve $ix^2 - 5x + 2i = 0$.

Here $a = i$, $b = -5$, $c = 2i$. Substituting these values into the quadratic formula gives

$$x = \frac{5 \pm \sqrt{25 - (4)(i)(2i)}}{2i} \qquad a = i,\ b = -5,\ c = 2i$$

$$= \frac{5 \pm \sqrt{25 - 8i^2}}{2i}$$

$$= \frac{5 \pm \sqrt{25 + 8}}{2i} \qquad i^2 = \ 1$$

$$= \frac{5 \pm \sqrt{33}}{2i}.$$

Write the solutions without i in the denominator by multiplying the numerator and the denominator by $-i$ as follows.

$$\frac{5 \pm \sqrt{33}}{2i} \cdot \frac{-i}{-i} = \frac{-5i \pm i\sqrt{33}}{-2i^2} = \frac{-5i \pm i\sqrt{33}}{2}$$

The solution set is $\left\{\frac{-5i + i\sqrt{33}}{2}, \frac{-5i - i\sqrt{33}}{2}\right\}$. ■

3 We can now use the quadratic formula to solve applied problems that produce quadratic equations that cannot be solved by factoring.

PROBLEM SOLVING

In Chapter 4 we solved problems about work rates. Recall, a person's work rate is $\frac{1}{t}$ part of the job per hour, where t is the time in hours required to do the complete job. Thus, the fractional part of the job the person will do in x hours is $\frac{1}{t}x$. ■

EXAMPLE 5 SOLVING A WORK PROBLEM

Two mechanics take 4 hours to repair a car. If each worked alone, one of them could do the job in 1 hour less time than the other. How long would it take the slower one to complete the job alone?

Let x represent the number of hours for the slower mechanic to complete the job alone. Then the faster mechanic could do the entire job in $x - 1$ hours. Together, they do the job in 4 hours. This information is shown in the following chart.

WORKER	RATE	TIME WORKING TOGETHER (IN HOURS)	FRACTIONAL PART OF THE JOB DONE
Faster mechanic	$\frac{1}{x-1}$	4	$\frac{1}{x-1} \cdot 4$
Slower mechanic	$\frac{1}{x}$	4	$\frac{1}{x} \cdot 4$

The sum of the fractional parts done by each should equal 1 (the whole job).

$$\underset{\text{slower mechanic}}{\text{part done by}} + \underset{\text{faster mechanic}}{\text{part done by}} = \text{1 whole job}$$

$$\frac{4}{x} + \frac{4}{x-1} = 1$$

Multiply both sides by the common denominator, $x(x - 1)$, to get

$$4(x - 1) + 4x = x(x - 1)$$
$$4x - 4 + 4x = x^2 - x.$$

Simplify to get

$$0 = x^2 - 9x + 4.$$

Now use the quadratic formula.

$$x = \frac{9 \pm \sqrt{81 - 16}}{2} = \frac{9 \pm \sqrt{65}}{2} \qquad a = 1, b = -9, c = 4$$

From a calculator $\sqrt{65} \approx 8.062$, so $x \approx \frac{9 \pm 8.062}{2}$.

Using the + sign gives $x \approx 8.5$, while the − sign leads to $x \approx .5$. (The numbers were rounded to the nearest tenth.) Only the solution 8.5 makes sense in the original equation. (Why?) Thus, the slower mechanic can do the job in about 8.5 hours and the faster in about $8.5 - 1 = 7.5$ hours. ■

6.2 EXERCISES

Use the quadratic formula to find the real number solutions of each of the following equations. See Examples 1 and 2.

1. $4x^2 - 8x + 1 = 0$ **2.** $m^2 + 2m - 5 = 0$ **3.** $2x^2 + 4x + 1 = 0$ **4.** $2z^2 + 3z - 1 = 0$

5. $m^2 + 18 = 10m$ **6.** $p^2 + 6p + 4 = 0$ **7.** $2y^2 = 2y + 1$ **8.** $9r^2 + 6r = 1$

9. $q^2 - 1 = q$ **10.** $2p^2 - 4p = 5$ **11.** $5m^2 + 8m + 2 = 0$ **12.** $3r^2 - r - 1 = 0$

13. $5r^2 - 2r - 1 = 0$ **14.** $5q^2 - 2q - 2 = 0$ **15.** $p^2 + \frac{p}{3} = \frac{2}{3}$ **16.** $\frac{x^2}{4} - \frac{x}{2} = 1$

17. $4k(k + 1) = 1$ **18.** $(r - 1)(4r) = 19$ **19.** $(g + 2)(g - 3) = 1$ **20.** $(y - 5)(y + 2) = 6$

21. $3x^2 + 2x = 2$ **22.** $26r - 2 = 3r^2$ **23.** $y = \frac{5(5 - y)}{3(y + 1)}$ **24.** $k = \frac{k + 15}{3(k - 1)}$

---

25. What is wrong with the following solution of $5x^2 - 5x + 1 = 0$?

$$x = \frac{5 \pm \sqrt{25 - 4(5)(1)}}{2(5)} \qquad a = 5, b = -5, c = 1$$
$$= \frac{5 \pm \sqrt{5}}{10}$$
$$= \frac{1}{2} \pm \sqrt{5}$$

26. Is $\frac{-b \pm \sqrt{b^2 - 4ac}}{2a} = -b \pm \frac{\sqrt{b^2 - 4ac}}{2a}$? Explain.

27. Can the quadratic formula be used to solve $2y^2 - 5 = 0$? Explain. (See Objective 1.)

28. Can $4m^2 + 3m = 0$ be solved by the quadratic formula? Explain. (See Objective 1.)

Use the quadratic formula to find the imaginary number solutions of the following equations. See Examples 3 and 4.

29. $3x^2 + 4x + 2 = 0$ **30.** $2k^2 + 3k = -2$

31. $m^2 - 6m + 14 = 0$ **32.** $p^2 + 4p + 11 = 0$

33. $4z^2 = 4z - 7$ **34.** $9a^2 + 7 = 6a$

35. $m^2 + 1 = -m$ **36.** $y^2 = 2y - 2$

37. $4ix - 3x^2 = 0$ **38.** $5m^2 - 8im = 0$

39. $2iz^2 - 3z + 2i = 0$ **40.** $r^2 - ir + 12 = 0$

Solve each problem. If necessary, use a calculator and round your answer to the nearest tenth. See Example 5.

41. A janitorial service provides two people to clean an office building. Working together, the two can clean the building in 5 hours. One person is new to the job and would take 2 hours longer than the other person to clean the building working alone. How long would it take the experienced worker to clean the building working alone?

42. Working together, two people can cut a large lawn in 2 hours. One person can do the job alone in 1 hour more than the other. How long would it take the faster person to do the job?

Photo for Exercise 41

43. Joann Johnson can clean all the snow from a strip of driveway in 4 hours less time than it takes Wing Tom. Working together, the job takes 8/3 hours. How long would it take each person working alone?

44. Ben Whitney can work through a stack of invoices in 1 hour less time than Arnold Parker can. Working together they take 3/2 hours. How long would it take each person working alone?

45. Tom and Dick are planting flats of spring flowers. Working alone, Tom would take 2 hours longer than Dick to plant the flowers. Working together, they do the job in 12 hours. How long would it have taken each person working alone?

46. Computer A can run a certain job in 1 hour less time than computer B. Together, they can complete the job in 10 hours. How long would it take each computer working alone?

Use the quadratic formula to solve each of the following equations. Exercises 51 and 52 have imaginary number solutions.

47. $p^2 + \sqrt{2}p = 4$

48. $\sqrt{3}p^2 = p + 2\sqrt{3}$

49. $2\sqrt{5}m^2 + 3m - 4\sqrt{5} = 0$

50. $6p^2 - 14 = 7\sqrt{7}p$

51. $ix^2 + x + 2i = 0$

52. $6ix^2 - 11x = 3i$

PREVIEW EXERCISES

In Exercises 53–56, evaluate $b^2 - 4ac$, $-b/a$, and c/a. See Sections 1.3 and 5.3.

53. $a = 4$, $b = -2$, $c = -3$

54. $a = 5$, $b = -8$, $c = -1$

55. $a = \sqrt{2}$, $b = -5$, $c = \sqrt{2}$

56. $a = 3\sqrt{3}$, $b = 2$, $c = -\sqrt{3}$

Complete each of the following. See Section 2.1.

57. If $x = 3$, then $x - ? = 0$.

58. If $x = -2$, then $x + ? = 0$.

59. If $x = -\frac{1}{4}$, then $x + ? = 0$.

60. If $x = \frac{2}{5}$, then $x - ? = 0$.

6.3 THE DISCRIMINANT; SUM AND PRODUCT OF THE SOLUTIONS

OBJECTIVES

1. FIND THE DISCRIMINANT OF A QUADRATIC EQUATION.
2. USE THE DISCRIMINANT TO DETERMINE THE NUMBER AND TYPE OF SOLUTIONS.
3. USE THE DISCRIMINANT TO DECIDE WHETHER A QUADRATIC TRINOMIAL CAN BE FACTORED.
4. FIND THE SUM AND PRODUCT OF THE SOLUTIONS OF A QUADRATIC EQUATION.
5. WRITE A QUADRATIC EQUATION GIVEN ITS SOLUTIONS.

1 The solutions of the quadratic equation $ax^2 + bx + c = 0$ are

$$x = \frac{-b \pm \sqrt{b^2 - 4ac}}{2a}.$$

If a, b, and c are integers, the type of solutions of a quadratic equation (that is, rational, irrational, or imaginary) is determined by the quantity under the square root sign, $b^2 - 4ac$. Because it distinguishes among the three types of solutions, the quantity $b^2 - 4ac$ is called the **discriminant.** By calculating the discriminant before solving a quadratic equation, we can predict whether the solutions will be rational numbers, irrational numbers, or imaginary numbers. This can be useful in an applied problem, for example, where irrational or imaginary number solutions are not acceptable.

DISCRIMINANT

The discriminant of $ax^2 + bx + c = 0$ is given by $b^2 - 4ac$. If a, b, and c are integers, then the type of solution is determined as follows.

DISCRIMINANT	SOLUTIONS
Positive, and the square of an integer	Two different rational solutions
Positive, but not the square of an integer	Two different irrational solutions
Zero	One rational solution
Negative	Two different imaginary solutions

2 The next examples show how to use the discriminant to decide what type of solution a quadratic equation has.

EXAMPLE 1 USING THE DISCRIMINANT

Find the discriminant for the following equations and decide whether the solutions are rational, irrational, or imaginary numbers.

(a) $6x^2 - x - 15 = 0$

The discriminant is found by evaluating $b^2 - 4ac$.

$$b^2 - 4ac = (-1)^2 - 4(6)(-15) \qquad a = 6,\ b = -1,\ c = -15$$
$$= 1 + 360 = 361$$

A calculator shows that $361 = 19^2$, a perfect square, and since a, b, and c are integers, the solutions will be two different rational numbers. Verify this by using the quadratic formula to find the solutions.

(b) $3m^2 - 4m = 5$

First write the equation as $3m^2 - 4m - 5 = 0$. Then $a = 3$, $b = -4$, $c = -5$, and the discriminant is

$$b^2 - 4ac = (-4)^2 - 4(3)(-5) = 16 + 60 = 76.$$

Because 76 is not the square of an integer, $\sqrt{76}$ is irrational. From this and from the fact that a, b, and c are integers, the equation will have two different irrational solutions, one using $\sqrt{76}$ and one using $-\sqrt{76}$.

(c) $4x^2 + x + 1 = 0$

Since $a = 4$, $b = 1$, and $c = 1$, the discriminant is

$$(1)^2 - 4(4)(1) = -15.$$

Since the discriminant is negative and a, b, and c are integers, the quadratic equation $4x^2 + x + 1 = 0$ will have two imaginary number solutions. ■

EXAMPLE 2
USING THE DISCRIMINANT

Find k so that $9x^2 + kx + 4 = 0$ will have only one rational number solution.

The equation will have only one rational number solution if the discriminant is 0. Here, since $a = 9$, $b = k$, and $c = 4$, the discriminant is

$$b^2 - 4ac = k^2 - 4(9)(4) = k^2 - 144.$$

Set this result equal to 0 and solve for k.

$$k^2 - 144 = 0$$
$$k^2 = 144$$
$$k = 12 \quad \text{or} \quad k = -12$$

The equation will have only one rational number solution if $k = 12$ or $k = -12$. ■

3 It can be shown that a quadratic trinomial can be factored with rational coefficients only if the corresponding quadratic equation has rational solutions. Thus, the discriminant can be used to decide whether or not a given trinomial is factorable.

EXAMPLE 3
DECIDING WHETHER A TRINOMIAL IS FACTORABLE

Decide whether or not the following trinomials can be factored.

(a) $24x^2 + 7x - 5$

To decide whether the solutions of $24x^2 + 7x - 5 = 0$ are rational numbers, evaluate the discriminant.

$$b^2 - 4ac = 7^2 - 4(24)(-5) = 49 + 480 = 529 = 23^2$$

Since 529 is a perfect square, the solutions are rational numbers and the trinomial can be factored. Verify that

$$24x^2 + 7x - 5 = (3x - 1)(8x + 5).$$

(b) $11m^2 - 9m + 12$

The discriminant is $b^2 - 4ac = (-9)^2 - 4(11)(12) = -447$. This number is negative, so the corresponding quadratic equation has imaginary number solutions and therefore the trinomial cannot be factored. ■

4 We can develop two interesting properties of the solutions of a quadratic equation $ax^2 + bx + c = 0$, $a \neq 0$. If

$$x_1 = \frac{-b + \sqrt{b^2 - 4ac}}{2a} \quad \text{and} \quad x_2 = \frac{-b - \sqrt{b^2 - 4ac}}{2a},$$

then the sum of the solutions is

$$x_1 + x_2 = \frac{-b + \sqrt{b^2 - 4ac}}{2a} + \frac{-b - \sqrt{b^2 - 4ac}}{2a}$$
$$= \frac{-b + \sqrt{b^2 - 4ac} - b - \sqrt{b^2 - 4ac}}{2a} = \frac{-2b}{2a} = -\frac{b}{a}.$$

The sum of the solutions is given by $-b/a$.

The product of the solutions is

$$x_1x_2 = \left(\frac{-b + \sqrt{b^2 - 4ac}}{2a}\right)\left(\frac{-b - \sqrt{b^2 - 4ac}}{2a}\right).$$

Using the rule $(x + y)(x - y) = x^2 - y^2$ with $x = -b$ and $y = \sqrt{b^2 - 4ac}$ gives the product of the numerators.

$$\begin{aligned} x_1x_2 &= \frac{(-b)^2 - (\sqrt{b^2 - 4ac})^2}{(2a)^2} \\ &= \frac{b^2 - (b^2 - 4ac)}{4a^2} \\ &= \frac{b^2 - b^2 + 4ac}{4a^2} = \frac{4ac}{4a^2} = \frac{c}{a} \end{aligned}$$

The product of the solutions is c/a. These results are summarized as follows.

SUM AND PRODUCT OF THE SOLUTIONS

If x_1 and x_2 are the solutions of the quadratic equation $ax^2 + bx + c = 0$ $(a \neq 0)$, then

$$x_1 + x_2 = -\frac{b}{a} \quad \text{and} \quad x_1x_2 = \frac{c}{a}.$$

EXAMPLE 4 **FINDING THE SUM AND PRODUCT OF THE SOLUTIONS OF A QUADRATIC EQUATION**

Find the sum and product of the solutions of $2x^2 = 5x - 3$.

First write the equation as $2x^2 - 5x + 3 = 0$. Then $a = 2$, $b = -5$, and $c = 3$. The sum of the solutions is

$$-\frac{b}{a} = -\frac{-5}{2} = \frac{5}{2},$$

and the product of the solutions is

$$\frac{c}{a} = \frac{3}{2}. \quad ■$$

A good way to use these properties is in checking the solutions of a quadratic equation.

EXAMPLE 5 **VERIFYING THE SOLUTIONS OF A QUADRATIC EQUATION**

Use the sum and product properties to verify that the proposed solutions of each equation are correct.

(a) $5x^2 - 2x - 8 = 0$; solutions are $\dfrac{1 + \sqrt{41}}{5}$ and $\dfrac{1 - \sqrt{41}}{5}$.

The sum of the solutions should equal $-b/a$ and their product should equal c/a. Here $a = 5$, $b = -2$, and $c = -8$.

$$-\frac{b}{a} = -\frac{-2}{5} = \frac{2}{5} \quad \text{and} \quad \frac{c}{a} = \frac{-8}{5} = -\frac{8}{5}$$

The sum of the solutions is

$$\frac{1+\sqrt{41}}{5}+\frac{1-\sqrt{41}}{5}=\frac{1+\sqrt{41}+1-\sqrt{41}}{5}=\frac{2}{5}.$$

Their product is

$$\left(\frac{1+\sqrt{41}}{5}\right)\left(\frac{1-\sqrt{41}}{5}\right)=\frac{(1)^2-(\sqrt{41})^2}{25}=\frac{1-41}{25}=\frac{-40}{25}=-\frac{8}{5}.$$

Since these results agree with $-b/a$ and c/a found above, the proposed solutions are correct. Less work was involved using this method of checking solutions than by substituting the solutions in the original equation.

(b) $-2x^2-5ix=12$; solutions are $3i/2$ and $-4i$.

Write the equation as $2x^2+5ix+12=0$ to find $a=2$, $b=5i$, and $c=12$. Then the sum of the solutions is

$$\frac{3i}{2}-4i=\frac{3i}{2}-\frac{8i}{2}=-\frac{5i}{2}=-\frac{b}{a}.$$

The product is

$$\frac{3i}{2}(-4i)=-6i^2=6=\frac{12}{2}=\frac{c}{a}.$$

The solutions $3i/2$ and $-4i$ are correct. ■

5 The next example shows how the solutions of a quadratic equation can be used to write the equation.

EXAMPLE 6
WRITING AN EQUATION FROM ITS SOLUTION SET

Write a quadratic equation with the given solution set.

(a) $\left\{\frac{2}{3}, -\frac{4}{5}\right\}$

From the given solutions,

$$x=\frac{2}{3} \quad \text{or} \quad x=-\frac{4}{5}$$

$$x-\frac{2}{3}=0 \quad \text{or} \quad x+\frac{4}{5}=0.$$

By the zero-factor property, working backwards,

$$\left(x-\frac{2}{3}\right)\left(x+\frac{4}{5}\right)=0$$

$$x^2+\frac{4}{5}x-\frac{2}{3}x-\frac{8}{15}=0. \qquad \text{Use FOIL.}$$

Multiply by the common denominator to get

$$15x^2 + 12x - 10x - 8 = 0$$

or

$$15x^2 + 2x - 8 = 0.$$

Use the sum and product of the solutions to verify that the solutions of this quadratic equation are 2/3 and $-4/5$.

(b) $\left\{\frac{2\sqrt{5}}{5}, -\sqrt{5}\right\}$

$$x = \frac{2\sqrt{5}}{5} \quad \text{or} \quad x = -\sqrt{5}$$

$$x - \frac{2\sqrt{5}}{5} = 0 \quad \text{or} \quad x + \sqrt{5} = 0$$

$$\left(x - \frac{2\sqrt{5}}{5}\right)(x + \sqrt{5}) = 0 \qquad \text{Zero-factor property, working backwards}$$

$$x^2 + \sqrt{5}x - \frac{2\sqrt{5}}{5}x - 2 = 0 \qquad \text{Use FOIL.}$$

$$5x^2 + 5\sqrt{5}x - 2\sqrt{5}x - 10 = 0 \qquad \text{Multiply by the common denominator.}$$

$$5x^2 + 3\sqrt{5}x - 10 = 0 \qquad \text{Combine terms.}$$

Use the sum and product properties to show that the solutions of $5x^2 + 3\sqrt{5}x - 10 = 0$ are $2\sqrt{5}/5$ and $-\sqrt{5}$. ■

6.3 EXERCISES

*Use the discriminant to determine whether the following equations have solutions that are **(a)** two different rational numbers, **(b)** exactly one rational number, **(c)** two different irrational numbers, or **(d)** two different imaginary numbers. In Exercises 7–9, first multiply both sides by the common denominator so that the coefficients will be integers. See Example 1.*

1. $x^2 + 4x + 4 = 0$

2. $2y^2 - y + 1 = 0$

3. $4x^2 - 4x = -3$

4. $25a^2 + 20a = 2$

5. $9r^2 - 30r + 15 = 0$

6. $25m^2 - 70m + 49 = 0$

7. $\frac{x^2}{4} + \frac{3x}{2} = 5$

8. $\frac{1}{5}y^2 - \frac{2}{5}y + \frac{3}{5} = 0$

9. $\frac{2}{3}m^2 + \frac{7}{9}m - \frac{1}{3} = 0$

Find the value of a, b, or c so that each equation will have exactly one rational solution. See Example 2.

10. $p^2 + bp + 25 = 0$

11. $r^2 - br + 49 = 0$

12. $am^2 + 8m + 1 = 0$

13. $ay^2 + 24y + 16 = 0$

14. $9y^2 - 30y + c = 0$

15. $4m^2 + bm + 9 = 0$

16. $ax^2 + 42x + 49 = 0$

17. $9y^2 + by + 4 = 0$

18. $16m^2 - 40m + c = 0$

19. Is it possible for the solution of a quadratic equation with integer coefficients to include just one irrational number? Why? (See Objective 1.) yes

20. Can the solution of a quadratic equation with integer coefficients include one real and one imaginary number? Why? (See Objective 1.) yes

Use the discriminant to tell which polynomials can be factored. If a polynomial can be factored, factor it. See Example 3.

21. $6k^2 + 7k - 5$

22. $8r^2 - 2r - 21$

23. $24z^2 - 14z - 5$

24. $8k^2 + 38k - 33$

25. $9r^2 - 11r + 8$

26. $15m^2 - 17m - 12$

27. $20n^2 + 64n - 21$

28. $30w^2 + 41w - 6$

29. $12p^2 - 15p - 11$

Without solving, find the sum and product of the solutions to the following equations. See Example 4.

30. $x^2 - 6x - 7 = 0$

31. $x^2 + 6x - 8 = 0$

32. $5x^2 = 2x + 1$

33. $4x^2 - 3x = 0$

34. $9x^2 - x = 0$

35. $4x^2 = 25$

36. $16x^2 = 49$

37. $\frac{1}{4}x^2 + \frac{5}{2}x - 3 = 0$

38. $\frac{3}{5}x^2 - \frac{2}{5}x - 1 = 0$

Use the sum and product properties to decide whether x_1 and x_2 are solutions of the given equation. See Example 5.

39. $6x^2 - 11x + 4 = 0;\quad x_1 = \frac{1}{2},\ x_2 = \frac{4}{3}$

40. $9x^2 + 21x + 10 = 0;\quad x_1 = -\frac{2}{3},\ x_2 = -\frac{5}{3}$

41. $3x^2 + 5x = 1;\quad x_1 = \frac{-5 + \sqrt{37}}{6},\ x_2 = \frac{-5 - \sqrt{37}}{6}$

42. $x^2 = 6x + 3;\quad x_1 = 3 + 2\sqrt{3},\ x_2 = 2 + 3\sqrt{3}$

43. $2x^2 = 3x - 4;\quad x_1 = \frac{3 + \sqrt{23}}{4},\ x_2 = \frac{3 - \sqrt{23}}{4}$

44. $4x^2 = 2x - 1;\quad x_1 = \frac{1}{4} + \frac{\sqrt{3}}{2}i,\ x_2 = \frac{1}{4} - \frac{\sqrt{3}}{2}i$

Find quadratic equations for Exercises 45–54 whose solutions satisfy the given conditions. See Example 6.

45. Solution set $\left\{-\frac{1}{2}, \frac{3}{5}\right\}$

46. Solution set $\left\{\frac{1}{2}, \frac{2}{3}\right\}$

47. Solution set $\{2\sqrt{5}, -2\sqrt{5}\}$

48. Solution set $\left\{\frac{2 + \sqrt{3}}{2}, \frac{2 - \sqrt{3}}{2}\right\}$

49. Solution set $\{4i, -4i\}$

50. Solution set $\{1 + 3i, 1 - 3i\}$

51. $x_1 + x_2 = \frac{3}{5},\ x_1x_2 = -\frac{4}{3}$

52. $x_1 + x_2 = -\frac{2}{9},\ x_1x_2 = \frac{5}{4}$

53. $x_1 + x_2 = \frac{\sqrt{3}}{4}, x_1x_2 = 2$

54. $x_1 + x_2 = -3\sqrt{2}, x_1x_2 = \frac{-\sqrt{2}}{2}$

55. One solution of $4x^2 + bx - 3 = 0$ is $-5/2$. Find b and the other solution.

56. One solution of $3x^2 - 7x + c = 0$ is $1/3$. Find c and the other solution.

Find p in each of the following equations using the given condition.

57. $3px^2 - 5x + p = 0$; sum of the roots is 5

58. $2py^2 + 4y + 6p = 0$; sum of the roots is 4/3

59. $pm^2 - 4m + p + 2 = 0$; product of the roots is 3/5

60. $2pa^2 + a - 2p + 3 = 0$; product of the roots is $-3/2$

PREVIEW EXERCISES

Use the substitution property to make the indicated substitutions. See Section 1.2.

61. Let $y^2 = a$ in $y^4 + 3y^2 - 5$.

62. Let $x^2 = b$ in $4x^4 - 5x^2 + 2$.

63. Let $2x - 1 = m$ in $3(2x - 1)^2 + 4(2x - 1) - 1$.

64. Let $p^3 = k$ in $3p^6 + 2p^3 - 4$.

Solve each equation. See Section 5.6.

65. $\sqrt{x - 1} = 6$

66. $\sqrt{2x + 3} = 9$

67. $\sqrt{4x - 3} + 2 = 7$

68. $\sqrt{2x + 1} + \sqrt{x + 3} = 0$

6.4 EQUATIONS QUADRATIC IN FORM

OBJECTIVES

1 SOLVE AN EQUATION WITH FRACTIONS BY WRITING IT IN QUADRATIC FORM.

2 USE QUADRATIC EQUATIONS TO SOLVE APPLIED PROBLEMS.

3 SOLVE AN EQUATION WITH RADICALS BY WRITING IT IN QUADRATIC FORM.

4 SOLVE AN EQUATION THAT IS QUADRATIC IN FORM BY SUBSTITUTION.

FOCUS ON PROBLEM SOLVING

The distance from Jackson to Lodi is about 40 miles, as is the distance from Lodi to Hilmar. Rico drove from Jackson to Lodi during the rush hour, stopped in Lodi for a root beer, and then drove on to Hilmar at a rate 10 miles per hour faster. Driving time for the entire trip was 88 minutes. Find his speed from Jackson to Lodi.

This is a typical distance-rate-time problem and is Exercise 41 in the exercises for this section. Combining the work done in earlier chapters with the methods of solving equations introduced in this section will enable you to solve the problem.

Four methods have now been introduced for solving quadratic equations written in the form $ax^2 + bx + c = 0$. The following chart gives some advantages and disadvantages of each method.

METHODS FOR SOLVING QUADRATIC EQUATIONS

METHOD	ADVANTAGES	DISADVANTAGES
Factoring	Usually the fastest method	Not all polynomials are factorable; some factorable polynomials are hard to factor.
Square root property	Simplest method for solving equations of the form $(x + a)^2 = b$	Few equations are given in this form.
Completing the square	None for solving equations (the procedure is useful in other areas of mathematics)	It requires more steps than other methods.
Quadratic formula	Can always be used	It is more difficult than factoring because of the square root.

1 A variety of nonquadratic equations can be written in the form of a quadratic equation and solved by using these methods. For example, some equations with fractions lead to quadratic equations. As you solve the equations in this section try to decide which is the best method for each equation.

EXAMPLE 1 **WRITING AN EQUATION WITH FRACTIONS IN QUADRATIC FORM**

Solve $\dfrac{1}{x} + \dfrac{1}{x-1} = \dfrac{7}{12}$.

Clear the equation of fractions by multiplying each term by the common denominator, $12x(x-1)$.

$$12x(x-1)\frac{1}{x} + 12x(x-1)\frac{1}{x-1} = 12x(x-1)\frac{7}{12}$$

$$12(x-1) + 12x = 7x(x-1)$$

$$12x - 12 + 12x = 7x^2 - 7x \quad \text{Distributive property}$$

$$24x - 12 = 7x^2 - 7x \quad \text{Combine terms.}$$

A quadratic equation must be in the standard form $ax^2 + bx + c = 0$ before it can be solved. Combine terms and arrange them so that one side of the equation is zero. This equation can be solved by factoring.

$$0 = 7x^2 - 31x + 12 \quad \text{Subtract 24x, add 12.}$$

$$0 = (7x-3)(x-4) \quad \text{Factor.}$$

Setting each factor equal to 0 and solving the two linear equations gives the solutions 3/7 and 4. Check by substituting these solutions in the original equation. The solution set is $\{3/7, 4\}$. ■

2 Earlier we solved distance-rate-time (or motion) problems that led to linear equations or rational equations. Now we can extend that work to motion problems that lead to quadratic equations. Writing an equation to solve the problem will be done just as it was earlier. Distance-rate-time applications often lead to equations with fractions, as in the next example.

EXAMPLE 2
SOLVING A MOTION PROBLEM

A riverboat for tourists averages 12 miles per hour in still water. It takes the boat 1 hour, 4 minutes to go 6 miles upstream and return. Find the speed of the current. See Figure 6.1.

FIGURE 6.1

For a problem about rate (or speed), use the distance formula, $d = rt$.

Let $\qquad x =$ the speed of the current;

$12 - x =$ the rate upstream;

$12 + x =$ the rate downstream.

The rate upstream is the difference of the speed of the boat in still water and the speed of the current, or $12 - x$. The speed downstream is, in the same way, $12 + x$. To find the time, rewrite the formula $d = rt$ as

$$t = \frac{d}{r}.$$

This information was used to complete the following chart.

	d	r	t
Upstream	6	$12 - x$	$\frac{6}{12 - x}$
Downstream	6	$12 + x$	$\frac{6}{12 + x}$

Times in hours

The total time, 1 hour and 4 minutes, can be written as

$$1 + \frac{4}{60} = 1 + \frac{1}{15} = \frac{16}{15} \text{ hours.}$$

Since the time upstream plus the time downstream equals $\frac{16}{15}$ hours,

$$\frac{6}{12 - x} + \frac{6}{12 + x} = \frac{16}{15}.$$

Now multiply both sides of the equation by the common denominator $15(12 - x)(12 + x)$ and solve the resulting quadratic equation.

$$15(12 + x)6 + 15(12 - x)6 = 16(12 - x)(12 + x)$$
$$90(12 + x) + 90(12 - x) = 16(144 - x^2)$$
$$1080 + 90x + 1080 - 90x = 2304 - 16x^2 \quad \text{Distributive property}$$
$$2160 = 2304 - 16x^2 \quad \text{Combine terms.}$$
$$16x^2 = 144$$
$$x^2 = 9$$

Solve $x^2 = 9$ by using the square root property to get the two solutions

$$x = 3 \quad \text{or} \quad x = -3.$$

The speed of the current cannot be -3, so the solution is $x = 3$ miles per hour. ■

3 In Section 5.6 we saw that some equations with radicals lead to quadratic equations.

EXAMPLE 3
SOLVING RADICAL EQUATIONS THAT LEAD TO QUADRATIC EQUATIONS

Solve each equation.

(a) $k = \sqrt{6k - 8}$

This equation is not quadratic. However, squaring both sides of the equation gives $k^2 = 6k - 8$, which is a quadratic equation that can be solved by factoring.

$$k^2 = 6k - 8$$
$$k^2 - 6k + 8 = 0$$
$$(k - 4)(k - 2) = 0$$
$$k = 4 \quad \text{or} \quad k = 2 \quad \text{Proposed solutions}$$

Check both of these numbers in the original (and *not* the squared) equation to be sure they are solutions.

If $k = 4$,
$$4 = \sqrt{6(4) - 8} \quad ?$$
$$4 = \sqrt{16} \quad ?$$
$$4 = 4. \quad \text{True}$$

If $k = 2$,
$$2 = \sqrt{6(2) - 8} \quad ?$$
$$2 = \sqrt{4} \quad ?$$
$$2 = 2. \quad \text{True}$$

Both numbers check, so the solution set is $\{2, 4\}$.

(b) $x + \sqrt{x} = 6$

$$\sqrt{x} = 6 - x \quad \text{Isolate the radical on one side.}$$
$$x = 36 - 12x + x^2 \quad \text{Square both sides.}$$
$$0 = x^2 - 13x + 36 \quad \text{Get 0 on one side.}$$
$$0 = (x - 4)(x - 9) \quad \text{Factor.}$$
$$x - 4 = 0 \quad \text{or} \quad x - 9 = 0 \quad \text{Set each factor equal to 0.}$$
$$x = 4 \qquad\qquad x = 9$$

Check both potential solutions.

If $x = 4$,
$$4 + \sqrt{4} = 6 \quad ?$$
$$6 = 6. \quad \text{True}$$

If $x = 9$,
$$9 + \sqrt{9} = 6 \quad ?$$
$$12 = 6. \quad \text{False}$$

The solution set is $\{4\}$. ■

4 An equation that can be written in the form $au^2 + bu + c = 0$, for $a \neq 0$ and u an algebraic expression, is called **quadratic in form.**

EXAMPLE 4 SOLVING EQUATIONS THAT ARE QUADRATIC IN FORM

Solve each equation.

(a) $2(4m - 3)^2 + 7(4m - 3) + 5 = 0$

Because of the repeated quantity $4m - 3$, this equation is quadratic in form with $u = 4m - 3$. (Any letter, except m, could be used instead of u.) Write

$$2(4m - 3)^2 + 7(4m - 3) + 5 = 0$$

as

$$2u^2 + 7u + 5 = 0. \quad \text{Let } 4m - 3 = u.$$
$$(2u + 5)(u + 1) = 0 \quad \text{Factor.}$$

By the zero-factor property, the solutions to $2u^2 + 7u + 5 = 0$ are

$$u = -\frac{5}{2} \quad \text{or} \quad u = -1.$$

To find m, substitute $4m - 3$ for u.

$$4m - 3 = -\frac{5}{2} \quad \text{or} \quad 4m - 3 = -1$$
$$4m = \frac{1}{2} \quad \text{or} \quad 4m = 2$$
$$m = \frac{1}{8} \quad \text{or} \quad m = \frac{1}{2}$$

The solution set of the original equation is $\left\{\frac{1}{8}, \frac{1}{2}\right\}$.

(b) $y^4 = 6y^2 - 3$.

First write the equation as

$$y^4 - 6y^2 + 3 = 0.$$

Then substitute u for y^2 and u^2 for y^4 to get $u^2 - 6u + 3 = 0$, with $a = 1$, $b = -6$, and $c = 3$. By the quadratic formula,

$$\begin{aligned} u = y^2 &= \frac{6 \pm \sqrt{36 - 12}}{2} \qquad a = 1,\ b = -6,\ c = 3 \\ &= \frac{6 \pm \sqrt{24}}{2} \\ &= \frac{6 \pm 2\sqrt{6}}{2} \\ &= 3 \pm \sqrt{6}. \end{aligned}$$

Find y by using the square root property as follows.

$$\begin{aligned} y^2 &= 3 + \sqrt{6} &\quad \text{or} \quad y^2 &= 3 - \sqrt{6} \\ y &= \pm\sqrt{3 + \sqrt{6}} &\quad \text{or} \quad y &= \pm\sqrt{3 - \sqrt{6}} \end{aligned}$$

The solution set contains four numbers:

$$\{\sqrt{3 + \sqrt{6}}, -\sqrt{3 + \sqrt{6}}, \sqrt{3 - \sqrt{6}}, -\sqrt{3 - \sqrt{6}}\}. \quad \blacksquare$$

EXAMPLE 5

SOLVING EQUATIONS THAT ARE QUADRATIC IN FORM

Solve each equation.

(a) Solve $4x^6 + 1 = 5x^3$.

Let $x^3 = u$. Then $x^6 = u^2$. Substitute into the given equation.

$$\begin{aligned} 4x^6 + 1 &= 5x^3 \\ 4u^2 + 1 &= 5u \qquad \text{Let } x^6 = u^2,\ x^3 = u. \\ 4u^2 - 5u + 1 &= 0 \qquad \text{Get 0 on one side.} \\ (4u - 1)(u - 1) &= 0 \qquad \text{Factor.} \\ u = \frac{1}{4} \quad &\text{or} \quad u = 1 \end{aligned}$$

Replace u with x^3 to get $\quad x^3 = \frac{1}{4} \quad$ or $\quad x^3 = 1$.

From these equations,

$$x = \sqrt[3]{\frac{1}{4}} = \frac{\sqrt[3]{1}}{\sqrt[3]{4}} = \frac{1}{\sqrt[3]{4}} \cdot \frac{\sqrt[3]{2}}{\sqrt[3]{2}} = \frac{\sqrt[3]{2}}{2} \quad \text{or} \quad x = \sqrt[3]{1} = 1.$$

This method of substitution will give only the real number solutions for equations with polynomials of degree $2n$ where n is odd. However, it gives all complex number solutions for equations with polynomials of degree $2n$ where n is even. The real number solution set of $4x^6 + 1 = 5x^3$ is $\{\sqrt[3]{2}/2, 1\}$. $\blacksquare$

(b) $2a^{2/3} - 11a^{1/3} + 12 = 0$

Let $a^{1/3} = u$; then $a^{2/3} = u^2$. Substitute into the given equation.

$$2u^2 - 11u + 12 = 0 \qquad \text{Let } a^{1/3} = u,\ a^{2/3} = u^2.$$

$$(2u - 3)(u - 4) = 0 \qquad \text{Factor.}$$

$$2u - 3 = 0 \quad \text{or} \quad u - 4 = 0$$

$$u = \frac{3}{2} \quad \text{or} \quad u = 4$$

$$a^{1/3} = \frac{3}{2} \quad \text{or} \quad a^{1/3} = 4 \qquad u = a^{1/3}$$

$$a = \left(\frac{3}{2}\right)^3 = \frac{27}{8} \quad \text{or} \quad a = 4^3 = 64 \qquad \text{Cube both sides.}$$

(Recall that equations with fractional exponents were solved in Section 5.6 by raising both sides to the same power.) Check that the solution set is $\{27/8, 64\}$. ■

6.4 EXERCISES

Decide whether factoring, the quadratic formula, or the square root property is most appropriate for solving each of the following quadratic equations. DO NOT SOLVE THE EQUATIONS.

1. $(2x + 3)^2 = 4$

2. $4x^2 - 3x = 1$

3. $y^2 + 5y - 8 = 0$

4. $2k^2 + 3k = 1$

5. $3m^2 = 2 - 5m$

6. $p^2 = 5$

Solve each equation. See Example 1.

7. $1 + \frac{3}{x} - \frac{28}{x^2} = 0$

8. $4 + \frac{7}{k} = \frac{2}{k^2}$

9. $3(p + 1) - 5 = \frac{2}{p + 1}$

10. $1 + \frac{5}{b - 3} = -2b$

11. $3 = \frac{1}{y + 2} + \frac{2}{(y + 2)^2}$

12. $1 + \frac{2}{3y + 2} = \frac{15}{(3y + 2)^2}$

13. $\frac{6}{p} = 2 + \frac{p}{p + 1}$

14. $\frac{k}{2 - k} + \frac{2}{k} = 5$

15. $\frac{2r}{r - 3} + \frac{4}{r} - 6 = 0$

Find all solutions by first squaring. Be sure to check all answers. See Example 3.

16. $x\sqrt{2} = \sqrt{5x - 2}$

17. $y\sqrt{3} = \sqrt{2 - y}$

18. $t\sqrt{3} = \sqrt{t + 1}$

19. $p = \sqrt{p + 3}$

20. $\sqrt{2x + 3} = 2 + \sqrt{x - 2}$

21. $\sqrt{m + 1} = -1 + \sqrt{2m}$

22. $(3q - 2)^{1/2} + (6q + 3)^{1/2} = 4$

23. $(2m)^{1/2} + (m - 1)^{1/2} = 1$

24. $(-z)^{1/2} - (5 + z)^{1/2} = 1$

25. $(2 - p)^{1/2} - (8 + p)^{1/2} = 2$

Solve each equation by using substitution. See Example 4.

26. $2(3k - 1)^2 + 5(3k - 1) + 2 = 0$

27. $3(2p + 2)^2 - 7(2p + 2) - 6 = 0$

28. $3(2x + 5)^2 + 2(2x + 5) = 2$

29. $\left(x - \frac{1}{2}\right)^2 + 5\left(x - \frac{1}{2}\right) - 4 = 0$

30. $(4x + 5)^2 + 3(4x + 5) + 2 = 0$

31. $2(5x + 3)^2 = (5x + 3) + 28$

32. $z^4 - 37z^2 + 36 = 0$

33. $9m^4 - 25m^2 + 16 = 0$

34. $4q^4 - 13q^2 + 9 = 0$

35. $4z^4 = 41z^2 - 100$

36. $2x^4 + x^2 - 3 = 0$

37. $4k^4 + 5k^2 + 1 = 0$

---

38. What is wrong with the following "solution"? (See Objective 4.)

$$2(m - 1)^2 - 3(m - 1) + 1 = 0$$

$$2u^2 - 3u + 1 = 0 \quad \text{Let } u = m - 1.$$

$$(2u - 1)(u - 1) = 0$$

$$2u - 1 = 0 \quad \text{or} \quad u - 1 = 0$$

$$u = \frac{1}{2} \qquad u = 1$$

Solution set: $\left\{\frac{1}{2}, 1\right\}$

Solve each applied problem. See Example 2.

39. Lupe Valverde flew her plane at a constant speed for 6 hours. She traveled 810 miles with the wind, then turned around and traveled 720 miles against the wind. The wind speed was a constant 15 miles per hour. Find the speed of the plane.

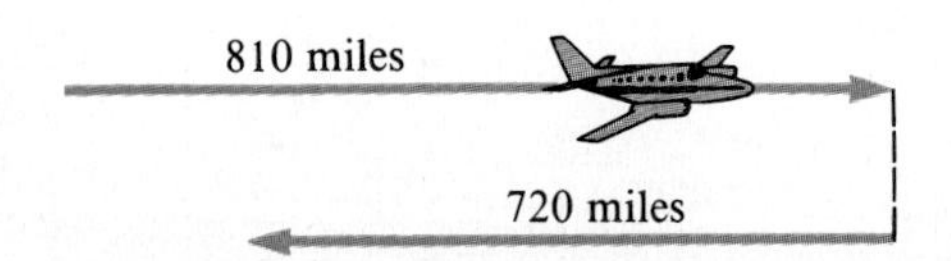

40. In the river, Yoshiaki found that he could go 16 miles downstream and then 4 miles back upstream at top speed in a total of 48 minutes. What was the top speed of Yoshiaki's boat if the current was 15 miles per hour?

41. The distance from Jackson to Lodi is about 40 miles, as is the distance from Lodi to Hilmar. Rico drove from Jackson to Lodi during the rush hour, stopped in Lodi for a root beer, and then drove on to Hilmar at a rate 10 miles per hour faster. Driving time for the entire trip was 88 minutes. Find his speed from Jackson to Lodi.

42. In Canada, Medicine Hat and Cranbrook are 300 kilometers apart. Harry rides his Honda 20 kilometers per hour faster than Karen rides her Yamaha. Find Harry's average speed if he travels from Cranbrook to Medicine Hat in 1 1/4 hours less time than Karen.

43. Two pipes together can fill a large tank in 2 hours. One of the pipes, used alone, takes 3 hours longer than the other to fill the tank. How long would each pipe take to fill the tank alone?

44. A washing machine can be filled in 6 minutes if both the hot and cold water taps are fully opened. To fill the washer with hot water alone takes 9 minutes longer than filling with cold water alone. How long does it take to fill the tank with cold water?

Solve each equation for any real number solutions. See Example 5.

45. $x^6 - 9x^3 + 8 = 0$

46. $x^6 - 28x^3 + 27 = 0$

47. $2m^6 + 11m^3 + 5 = 0$

48. $8x^6 + 513x^3 + 64 = 0$

49. $x^3 - 8 = 0$

50. $x^3 - 27 = 0$

51. $2 - (y - 1)^{-1} = 6(y - 1)^{-2}$

52. $3 = 2(p - 1)^{-1} + (p - 1)^{-2}$

53. $(x^2 + x)^2 = 8(x^2 + x) - 12$

54. $2(1 + \sqrt{x})^2 = 11(1 + \sqrt{x}) - 12$

55. Explain how substitution is used to solve an equation. (See Objective 4.)

PREVIEW EXERCISES

Solve each equation for the specified variable. See Section 2.2.

56. $P = 2L + 2W$; for L

57. $A = \frac{1}{2}bh$; for h

58. $F = \frac{9}{5}C + 32$; for C

59. $S = 2\pi rh + 2\pi r^2$; for h

60. $\frac{m^2}{2} + \frac{p^2}{4} = 1$; for m^2

61. $\frac{r^2 - 6}{3} = q^2$; for r^2

6.5 FORMULAS AND APPLICATIONS

OBJECTIVES

1. SOLVE FORMULAS FOR VARIABLES INVOLVING SQUARES AND SQUARE ROOTS.
2. SOLVE APPLIED PROBLEMS ABOUT MOTION ALONG A STRAIGHT LINE.
3. SOLVE APPLIED PROBLEMS USING THE PYTHAGOREAN FORMULA.
4. SOLVE APPLIED PROBLEMS USING FORMULAS FOR AREA.

FOCUS ON PROBLEM SOLVING

At a point 30 meters from the base of a tower, the distance to the top of the tower is 2 meters more than twice the height of the tower. Find the height of the tower.

It is sometimes necessary to find the height of an object that cannot be measured directly. In this section we shall see how the Pythagorean formula, which was introduced in Chapter 5, can be used for this purpose. This problem is Exercise 19 in the exercises at the end of this section. After working through this section, you should be able to solve this problem.

1 Many useful formulas have a variable that is squared or under a radical. The methods presented earlier in this chapter can be used to solve for such variables.

EXAMPLE 1
SOLVING FOR A VARIABLE INVOLVING A SQUARE OR A SQUARE ROOT

(a) Solve $w = \dfrac{kFr}{v^2}$ for v.

Begin by multiplying each side by v^2 to clear the equation of fractions. Then solve first for v^2.

$$w = \frac{kFr}{v^2}$$

$$v^2w = kFr \qquad \text{Multiply each side by } v^2.$$

$$v^2 = \frac{kFr}{w} \qquad \text{Divide each side by } w.$$

$$v = \pm\sqrt{\frac{kFr}{w}} \qquad \text{Square root property}$$

$$v = \frac{\pm\sqrt{kFr}}{\sqrt{w}} \cdot \frac{\sqrt{w}}{\sqrt{w}} = \frac{\pm\sqrt{kFrw}}{w} \qquad \text{Rationalize the denominator.}$$

(b) Solve $d = \sqrt{\dfrac{4A}{\pi}}$ for A.

First square both sides to eliminate the radical.

$$d = \sqrt{\frac{4A}{\pi}}$$

$$d^2 = \frac{4A}{\pi} \qquad \text{Square both sides.}$$

$$\pi d^2 = 4A \qquad \text{Multiply both sides by } \pi.$$

$$\frac{\pi d^2}{4} = A \qquad \text{Divide both sides by 4.}$$

Check the solution in the original equation. ■

EXAMPLE 2
SOLVING FOR A SQUARED VARIABLE

Solve $s = 2t^2 + kt$ for t.

Since the equation has terms with t^2 and t, write it in the standard quadratic form $ax^2 + bx + c = 0$, with t as the variable instead of x.

$$s = 2t^2 + kt$$
$$0 = 2t^2 + kt - s$$

Now use the quadratic formula with $a = 2$, $b = k$, and $c = -s$.

$$t = \frac{-k \pm \sqrt{k^2 - 4(2)(-s)}}{2(2)} \qquad a = 2,\ b = k,\ c = -s$$

$$= \frac{-k \pm \sqrt{k^2 + 8s}}{4}$$

The solutions are $t = \dfrac{-k + \sqrt{k^2 + 8s}}{4}$ and $t = \dfrac{-k - \sqrt{k^2 + 8s}}{4}$. ■

PROBLEM SOLVING

The following examples show that it is important to check all proposed solutions of applied problems against the information of the original problem. Numbers that are valid solutions of the equation may not satisfy the physical conditions of the problem. ■

2 The next example shows how quadratic equations can be used to solve problems about motion in a straight line.

EXAMPLE 3 SOLVING A STRAIGHT LINE MOTION PROBLEM

The position of a projectile moving in a straight line is given by the formula

$$s = 2t^2 - 5t,$$

where s is the position from a starting point in feet and t is the time in seconds the projectile has been in motion. How many seconds will it take for the projectile to move 12 feet?

We must find t when $s = 12$. First substitute 12 for s in the original equation, and then solve for t.

$$12 = 2t^2 - 5t \qquad \text{Let } s = 12.$$

$$0 = 2t^2 - 5t - 12 \qquad \text{Get 0 on one side.}$$

$$0 = (2t + 3)(t - 4) \qquad \text{Factor.}$$

$$t = -\frac{3}{2} \quad \text{or} \quad t = 4$$

Here, the negative answer is not of interest. (In some problems of this type it might be.) It will take 4 seconds for the projectile to move 12 feet. ■

3 The Pythagorean formula is used in the solution of the next example.

EXAMPLE 4 USING THE PYTHAGOREAN FORMULA

Two cars left an intersection at the same time, one heading due north, the other due west. Some time later, they were exactly 100 miles apart. The car headed north had gone 20 miles farther than the car headed west. How far had each car traveled?

Let x be the distance traveled by the car headed west. Then $x + 20$ is the distance traveled by the car headed north. These distances are shown in Figure 6.2. The cars are 100 miles apart, so the hypotenuse of the right triangle equals 100 and the two legs are equal to x and $x + 20$. By the Pythagorean formula,

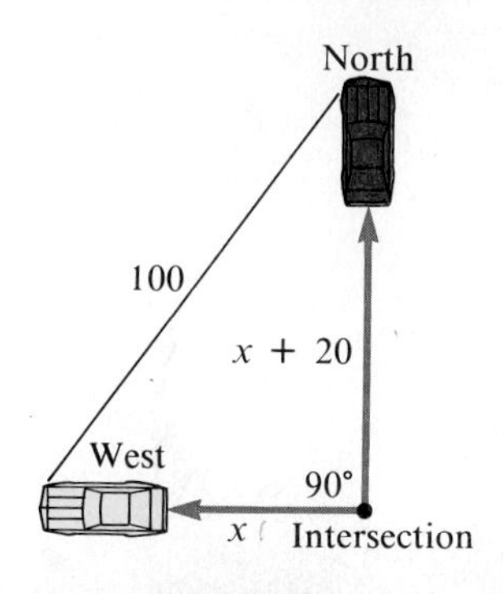

FIGURE 6.2

$$c^2 = a^2 + b^2.$$

$$100^2 = x^2 + (x + 20)^2$$

$$10{,}000 = x^2 + x^2 + 40x + 400 \qquad \text{Square the binomial.}$$

$$0 = 2x^2 + 40x - 9600 \qquad \text{Get 0 on one side.}$$

$$0 = 2(x^2 + 20x - 4800) \qquad \text{Factor out the common factor.}$$

$$0 = x^2 + 20x - 4800 \qquad \text{Divide both sides by 2.}$$

Use the quadratic formula to find x.

$$x = \frac{-20 \pm \sqrt{400 - 4(1)(-4800)}}{2} \qquad a = 1,\ b = 20,\ c = -4800$$

$$= \frac{-20 \pm \sqrt{19{,}600}}{2}$$

From a calculator, $\sqrt{19{,}600} = 140$, so

$$x = \frac{-20 \pm 140}{2}.$$

The solutions are $x = 60$ or $x = -80$. Discard the negative solution. The required distances are 60 miles and $60 + 20 = 80$ miles. ■

4 Formulas for area often lead to quadratic equations, as shown next.

EXAMPLE 5 SOLVING AN AREA PROBLEM

A reflecting pool in a park is 20 feet wide and 30 feet long. The park gardener wants to plant a strip of grass of uniform width around the edge of the pool. She has enough seed to cover 336 square feet. How wide will the strip be?

The pool is shown in Figure 6.3. If x represents the unknown width of the grass strip, the width of the large rectangle is given by $20 + 2x$ (the width of the pool plus two grass strips), and the length is given by $30 + 2x$. The area of the large rectangle is given by the product of its length and width, $(20 + 2x)(30 + 2x)$. The area of the pool is $20 \cdot 30 = 600$ square feet. The area of the large rectangle, minus the area of the pool, should equal the area of the grass strip. Since the area of the grass strip is to be 336 square feet, the equation is

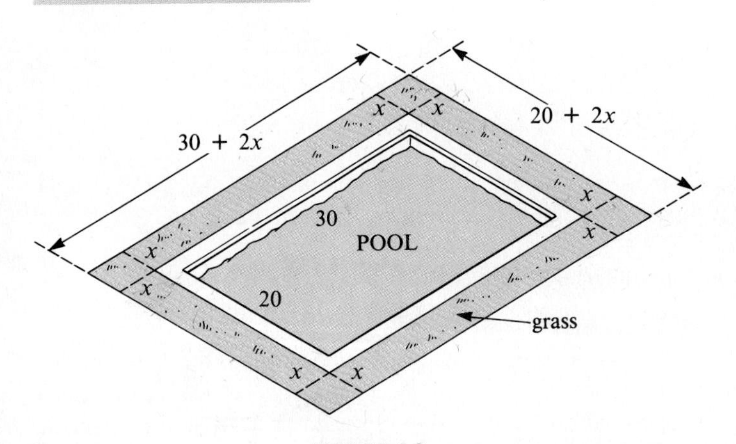

FIGURE 6.3

Area of rectangle ↓ ; Area of pool ↓ ; Area of grass ↓

$$(20 + 2x)(30 + 2x) - 600 = 336$$

$$600 + 100x + 4x^2 - 600 = 336 \qquad \text{Multiply.}$$

$$4x^2 + 100x - 336 = 0 \qquad \text{Combine terms.}$$

$$x^2 + 25x - 84 = 0 \qquad \text{Divide by 4.}$$

$$(x + 28)(x - 3) = 0 \qquad \text{Factor.}$$

$$x = -28 \quad \text{or} \quad x = 3.$$

The width cannot be -28 feet, so $x = 3$. The grass strip should be 3 feet wide. ■

6.5 EXERCISES

Solve each equation for the indicated variable. (Leave ± in the answers where appropriate). See Examples 1 and 2.

1. $R = \dfrac{k}{d^2}$; for d

2. $I = \dfrac{ks}{d^2}$; for d

3. $L = \dfrac{kd^4}{h^2}$; for h

4. $F = \dfrac{kA}{v^2}$; for v

5. $V = \dfrac{1}{3}\pi r^2 h$; for r

6. $F = \dfrac{kwv^2}{r}$; for v

7. $F = \dfrac{k}{\sqrt{d}}$; for d

8. $D = \sqrt{kh}$; for h

9. $M = \sqrt{\dfrac{2L}{p}}$; for p

10. $p = \sqrt{\dfrac{kL}{g}}$; for L

11. $S = 2\pi rh + \pi r^2$; for r

12. $V = \pi(r^2 + R^2)h$; for r

Solve the following applied problems using the methods for solving quadratic equations. Where applicable, round answers to the nearest tenth. See Examples 3–5.

13. A search light moves horizontally, back and forth along a wall with the distance of the light from a starting point at t minutes given by $s = 100t^2 - 300t$. How long will it take before the light returns to the starting point?

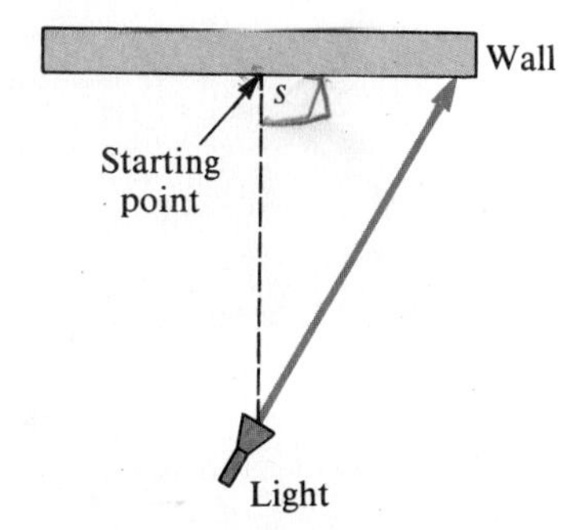

14. An object is projected directly upward from the ground. After t seconds its distance in feet above the ground is $s = 144t - 16t^2$.

(a) After how many seconds will the object be 128 feet above the ground? (*Hint:* Look for a common factor before solving the equation.)

(b) When does the object strike the ground?

15. The formula $D = 100t - 13t^2$ gives the distance in feet a car going approximately 68 miles per hour will skid in t seconds. Find the time it would take for the car to skid 190 feet. (*Hint:* Your answer must be less than the time it takes the car to stop, which is 3.8 seconds.)

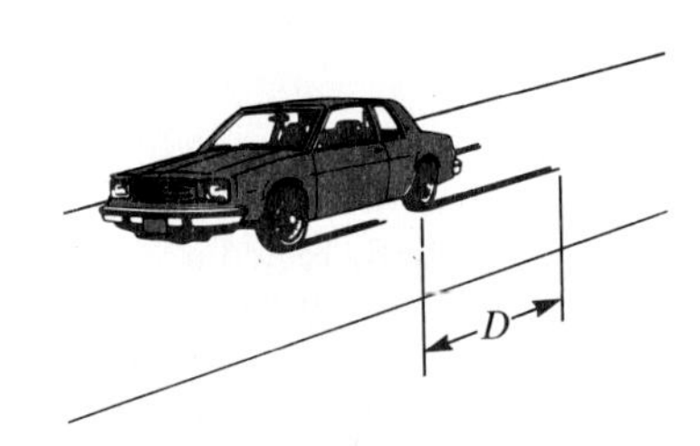

16. The formula in Exercise 15 becomes $D = 73t - 13t^2$ for a car going 50 miles per hour. Find the time for such a car to skid 100 feet. (*Hint:* The car will stop at about 2.8 seconds.)

17. A toy manufacturer needs a piece of plastic in the shape of a right triangle with the longer leg 2 centimeters more than twice as long as the shorter leg, and the hypotenuse 1 centimeter more than the longer leg. How long should the three sides of the triangular piece be?

18. A developer owns a piece of land enclosed on three sides by streets, giving it the shape of a right triangle. The hypotenuse is 8 meters longer than the longer leg, and the shorter leg is 9 meters shorter than the hypotenuse. Find the lengths of the three sides of the property.

19. At a point 30 meters from the base of a tower, the distance to the top of the tower is 2 meters more than twice the height of the tower. Find the height of the tower.

20. Two pieces of a large wooden puzzle fit together to form a rectangle with a length 1 centimeter less than twice the width. The diagonal, where the two pieces meet, is 2.5 centimeters in length. Find the length and width of the rectangle.

21. A 13-foot ladder is leaning against a house. The distance from the bottom of the ladder to the house is 7 feet less than the distance from the top of the ladder to the ground. How far is the bottom of the ladder from the house?

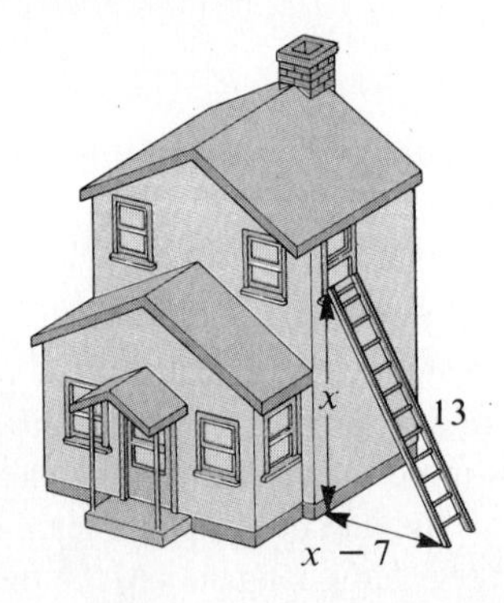

22. A rectangular office building is 2 meters longer than twice its width. The area of the ground floor is 60 square meters. Find the length and width of the building.

23. A rectangular piece of sheet metal has a length that is 4 inches less than twice the width. A square piece 2 inches on a side is cut from each corner. The sides are then turned up to form an uncovered box of volume 256 cubic inches. Find the length and width of the original piece of metal.

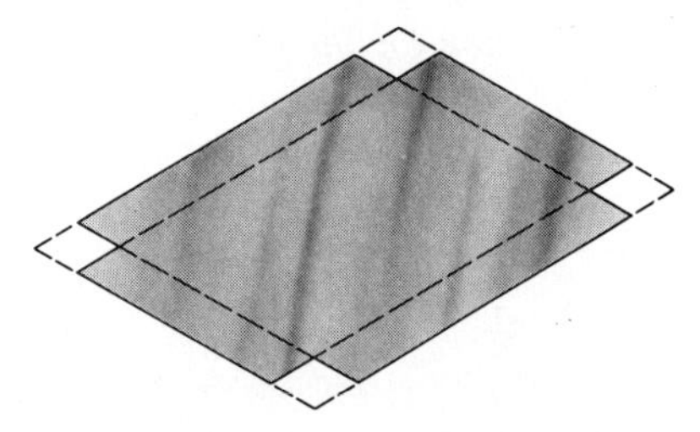

24. Jan and Russ want to buy a rug for a room that is 15 by 20 feet. They want to leave an even strip of flooring uncovered around the edges of the room. How wide a strip will they have if they buy a rug with an area of 234 square feet?

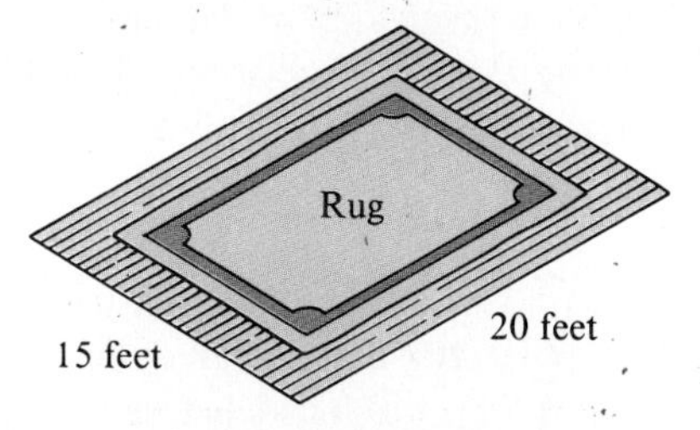

25. Minh has a workshop 30 by 40 feet. She wants to spread redwood bark in a strip of uniform width around the workshop. She has enough bark for 296 square feet. How wide can the strip of bark be?

26. Arif's backyard is 20 by 30 meters. He wants to put a flower garden in the middle of the backyard, leaving a strip of grass of uniform width around the flower garden. To be happy, Arif must have 184 square meters of grass. Under these conditions, what will the length and width of the garden be?

27. In one area the demand for videotapes is $1000/p$ per day, where p is the rental price in dollars per tape. The supply is $500p - 100$ per day. At what price does supply equal demand?

28. A certain bakery has found that the daily demand for croissants is $3200/p$, where p is the price of a croissant in cents. The daily supply is $3p - 200$. Find the price at which supply and demand are equal.

Use a calculator with a square root key to find the unknown lengths of the sides in each right triangle. Round to the nearest thousandth.

29.

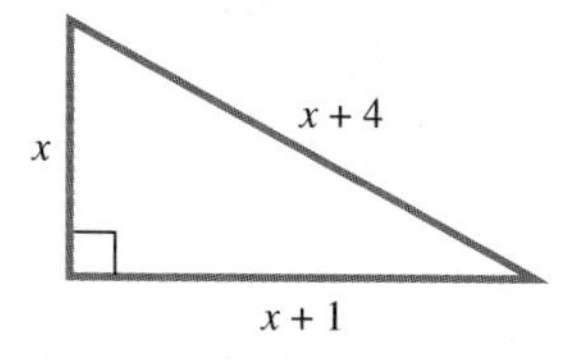

30.

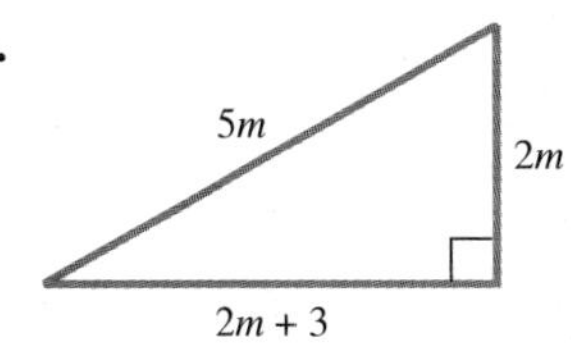

Solve each equation for the indicated variable.

31. $p = \dfrac{E^2R}{(r+R)^2}$; for R

32. $S(6S - t) = t^2$; for S

33. $10p^2c^2 + 7pcr = 12r^2$; for r

34. $S = vt + \dfrac{1}{2}gt^2$; for t

35. $LI^2 + RI + \dfrac{1}{c} = 0$; for I

36. $P = EI - RI^2$; for I

---

37. Describe the Pythagorean formula in your own words.

38. Describe the steps used to solve an equation for a variable that is squared. (See Objective 1.)

39. Describe the steps used to solve an equation for a variable that is under a radical sign. (See Objective 1.)

Recall that the corresponding sides of similar triangles are proportional. Use this fact to find the lengths of the indicated sides of the following pairs of similar triangles. Check all possible solutions in both triangles. Sides of a triangle cannot be negative.

40. side AC = __________

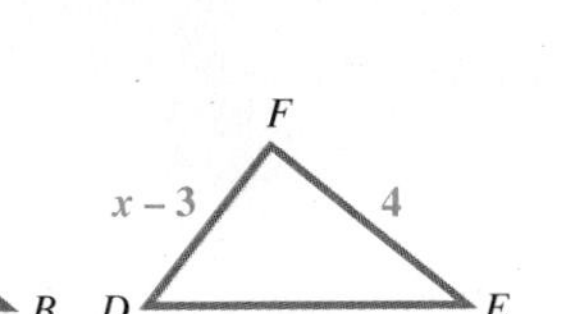

41. side RQ = __________

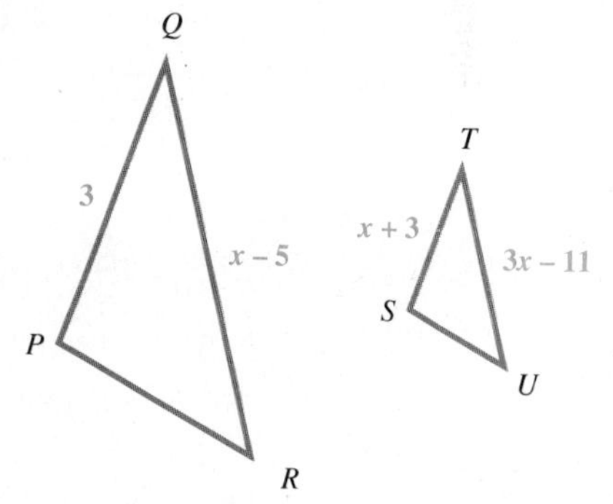

PREVIEW EXERCISES

Graph the following intervals on the number line. See Sections 1.2 and 2.5.

42. $[1, 4]$

43. $(-3, 0)$

44. $(-1, 5]$

45. $(-\infty, 1] \cup [4, \infty)$

46. $(-\infty, -2) \cup (1, \infty)$

47. $(-\infty, 0) \cup \left[\dfrac{3}{2}, \infty\right)$

6.6 NONLINEAR INEQUALITIES

OBJECTIVES

1 SOLVE QUADRATIC INEQUALITIES.

2 SOLVE POLYNOMIAL INEQUALITIES OF DEGREE 3 OR MORE.

3 SOLVE RATIONAL INEQUALITIES.

We have discussed methods of solving linear inequalities (earlier) and methods of solving quadratic equations (in this chapter). Now this work can be extended to include solving *quadratic inequalities*.

QUADRATIC INEQUALITY

A **quadratic inequality** can be written in the form

$$ax^2 + bx + c < 0 \quad \text{or} \quad ax^2 + bx + c > 0,$$

where a, b, and c are real numbers, with $a \neq 0$.

As before, $<$ and $>$ may be replaced with $\leq$ and $\geq$ as necessary.

1 A method for solving quadratic inequalities is shown in the next example.

EXAMPLE 1 SOLVING A QUADRATIC INEQUALITY

Solve $x^2 - x - 12 > 0$.

First, solve the quadratic *equation*

$$x^2 - x - 12 = 0$$

by factoring.

$$(x - 4)(x + 3) = 0$$

$$x - 4 = 0 \quad \text{or} \quad x + 3 = 0$$

$$x = 4 \quad \text{or} \quad x = -3$$

The numbers 4 and -3 divide the number line into three regions, as shown in Figure 6.4. (Be careful to put the smaller number on the left.)

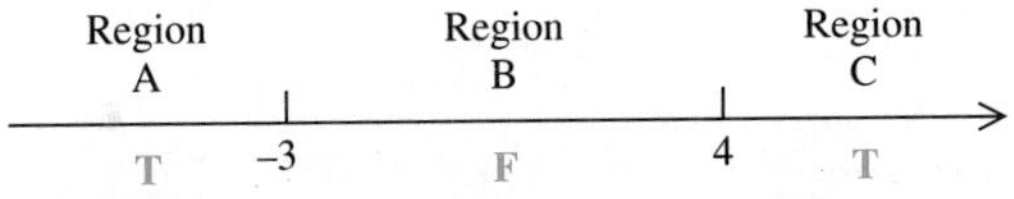

Region A | Region B | Region C

T -3 F 4 T

FIGURE 6.4

If one number in a region satisfies the inequality, then all the region's numbers will satisfy the inequality. Choose any number from Region A in Figure 6.4 (any number less than -3). Substitute this number for x in the inequality $x^2 - x - 12 > 0$. If the result is true, then all numbers in Region A satisfy the original inequality.

Let us choose -5 from Region A. Substitute -5 into $x^2 - x - 12 > 0$, getting

$$(-5)^2 - (-5) - 12 > 0 \quad ?$$

$$25 + 5 - 12 > 0 \quad ?$$

$$18 > 0. \quad \text{True}$$

Since -5 from Region A satisfies the inequality, all numbers from Region A are solutions.

Try 0 from Region B. If $x = 0$, then

$$0^2 - 0 - 12 > 0 \quad ?$$
$$-12 > 0. \qquad \text{False}$$

The numbers in Region B are *not* solutions. Verify that the number 5 satisfies the inequality, so that the numbers in Region C are also solutions to the inequality.

Based on these results (shown by the colored letters in Figure 6.4), the solution set includes the numbers in Regions A and C, as shown on the graph in Figure 6.5. The solution set is written

$$(-\infty, -3) \cup (4, \infty). \quad ■$$

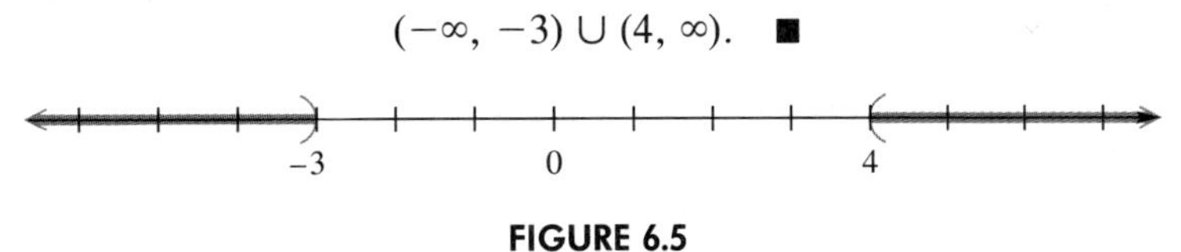

FIGURE 6.5

In summary, a quadratic inequality is solved by following these steps.

SOLVING A QUADRATIC INEQUALITY

Step 1 Write the inequality as an equation and solve the equation.

Step 2 Place the numbers found in Step 1 on a number line. These numbers divide the number line into regions.

Step 3 Substitute a number from each region into the inequality to determine the intervals that make the inequality true. All numbers in those intervals that make the inequality true are in the solution set. A graph of the solution set will usually look like one of these.

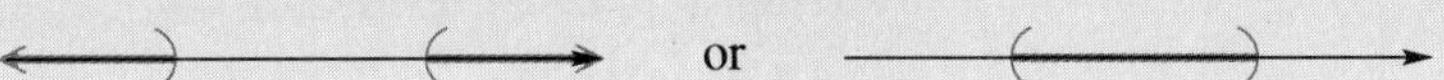

EXAMPLE 2
SOLVING A QUADRATIC INEQUALITY

Solve $q^2 - 4q - 5 \le 0$.

The corresponding quadratic equation can be solved by factoring.

$$q^2 - 4q - 5 = 0$$
$$(q + 1)(q - 5) = 0$$
$$q + 1 = 0 \quad \text{or} \quad q - 5 = 0$$
$$q = -1 \quad \text{or} \quad q = 5$$

These numbers, -1 and 5, divide the number line into three regions, as shown in Figure 6.6.

Region A, Region B, Region C

F, -1, T, 5, F

FIGURE 6.6

Check a number from each region as follows.

If $q = -2$, then $q^2 - 4q - 5 = 4 + 8 - 5 = 7 \leq 0$. False
If $q = 0$, then $q^2 - 4q - 5 = 0 - 0 - 5 = -5 \leq 0$. True
If $q = 6$, then $q^2 - 4q - 5 = 36 - 24 - 5 = 7 \leq 0$. False

Only the numbers in Region B (together with the endpoints -1 and 5) satisfy the inequality $q^2 - 4q - 5 \leq 0$, giving the solution $[-1, 5]$. A graph of the solution set is shown in Figure 6.7. ■

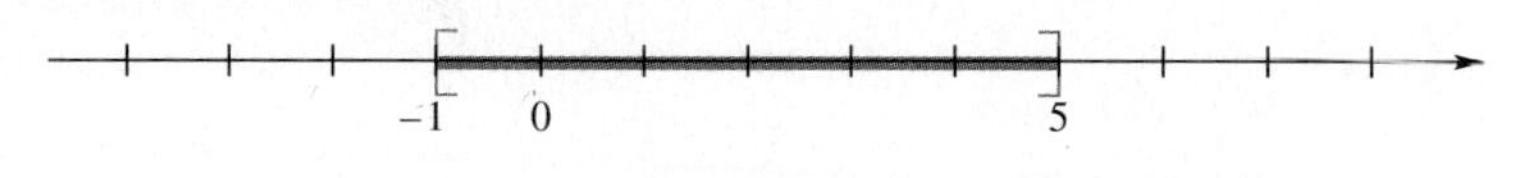

FIGURE 6.7

2 Higher-degree polynomial inequalities that are factorable can be solved in the same way as quadratic inequalities.

EXAMPLE 3
SOLVING A THIRD-DEGREE POLYNOMIAL INEQUALITY

Solve $(x - 1)(x + 2)(x - 4) < 0$.

This is a *cubic* (third-degree) inequality rather than a quadratic inequality, but it can be solved by the method shown above and by extending the zero-factor property to more than two factors. Begin by setting the factored polynomial *equal* to 0 and solving the equation.

$$(x - 1)(x + 2)(x - 4) = 0$$

$$x - 1 = 0 \quad \text{or} \quad x + 2 = 0 \quad \text{or} \quad x - 4 = 0$$

$$x = 1 \quad \text{or} \quad x = -2 \quad \text{or} \quad x = 4$$

Locate the numbers -2, 1, and 4 on a number line as in Figure 6.8 to determine the regions A, B, C, and D.

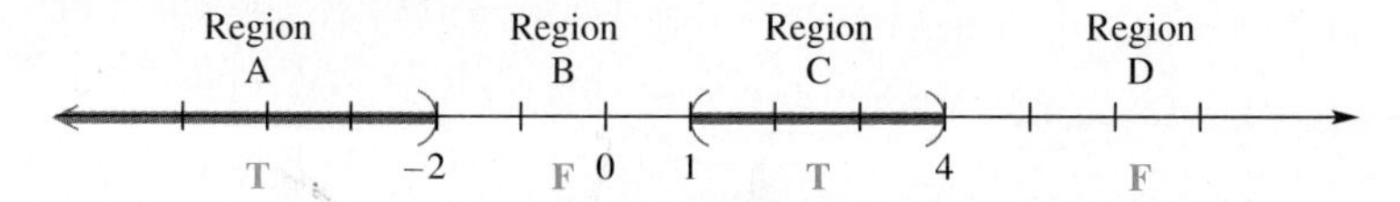

FIGURE 6.8

Substitute a number from each region into the original inequality to determine which regions satisfy the inequality. These results are shown below the number line in Figure 6.8. For example, in Region A, using $x = -3$ gives

$$(-3 - 1)(-3 + 2)(-3 - 4) < 0 \quad ?$$

$$(-4)(-1)(-7) < 0 \quad ?$$

$$-28 < 0. \quad \text{True}$$

The numbers in Region A are in the solution set. Verify that the numbers in Region C are also in the solution set, which is written

$$(-\infty, -2) \cup (1, 4).$$

The solution set is graphed in Figure 6.8. ■

3 Inequalities involving fractions, called **rational inequalities,** are solved in a similar manner, by going through the following steps.

SOLVING RATIONAL INEQUALITIES

Step 1 Write the inequality as an equation and solve the equation.
Step 2 Set the denominator equal to zero and solve that equation.
Step 3 Use the solutions from Steps 1 and 2 to divide a number line into regions.
Step 4 Test a number from each region by substitution in the inequality to determine the intervals that satisfy the inequality.
Step 5 Be sure to exclude any values that make the denominator equal to zero.

CAUTION Don't forget Step 2. Any number that makes the denominator zero *must* separate two regions on the number line.

EXAMPLE 4 SOLVING A RATIONAL INEQUALITY

Solve the inequality $\dfrac{-1}{p-3} > 1$.

Write the corresponding equation and solve it. (Step 1)

$$\frac{-1}{p-3} = 1$$

$$-1 = p - 3 \qquad \text{Multiply by the common denominator.}$$

$$2 = p$$

Find the number that makes the denominator 0. (Step 2)

$$p - 3 = 0$$

$$p = 3$$

These two numbers, 2 and 3, divide a number line into three regions. (Step 3) (See Figure 6.9.)

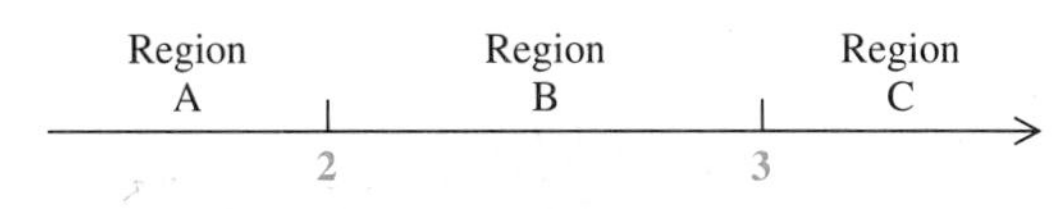

FIGURE 6.9

Testing one number from each region in the given inequality shows that the solution set is the interval (2, 3). (Step 4) This interval does not include any value that might make the denominator of the original inequality equal to zero. (Step 5)

A graph of the solution set is given in Figure 6.10. ■

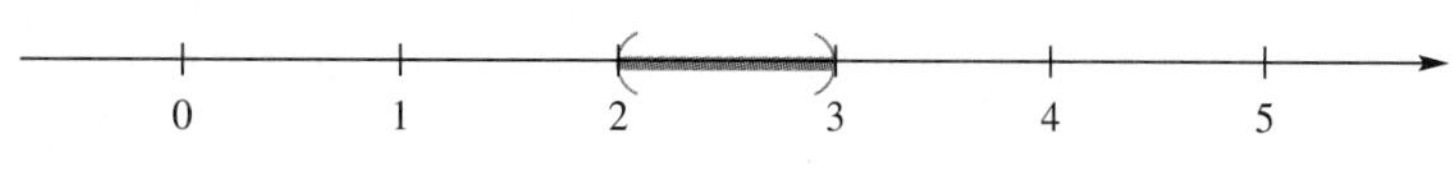

FIGURE 6.10

EXAMPLE 5
SOLVING A RATIONAL INEQUALITY

Solve $\dfrac{m-2}{m+2} \le 2.$

Write the corresponding equation and solve it. (Step 1)

$$\frac{m-2}{m+2} = 2$$

$$m - 2 = 2(m + 2) \qquad \text{Multiply by the common denominator.}$$

$$m - 2 = 2m + 4 \qquad \text{Distributive property}$$

$$-6 = m$$

Set the denominator equal to zero and solve the equation. (Step 2)

$$m + 2 = 0$$

$$m = -2$$

The numbers -6 and -2 determine three regions. (Step 3) Test one number from each region to see that the solution set is

$$(-\infty, -6] \cup (-2, \infty). \qquad \text{Step 4}$$

The number -6 satisfies the equality in $\le$, but -2 cannot be used as a solution since it makes the denominator equal to 0. (Step 5)

The graph of the solution set is shown in Figure 6.11. ■

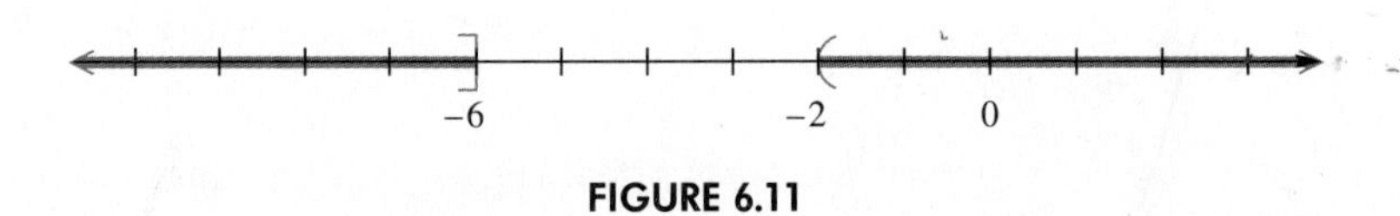

FIGURE 6.11

Special cases of quadratic inequalities may occur, such as those discussed in the next example.

EXAMPLE 6
SOLVING SPECIAL CASES

Solve $(2y - 3)^2 > -1$.

Since $(2y - 3)^2$ is never negative, it is always greater than -1. Thus, the solution is the set of all real numbers, $(-\infty, \infty)$. In the same way, there is no solution for $(2y - 3)^2 < -1$ and the solution set is $\emptyset$. ■

6.6 EXERCISES

Solve each of the following inequalities, then graph each solution. See Examples 1 and 2.

1. $(2x + 1)(x - 5) > 0$
2. $(m + 6)(m - 2) \ge 0$
3. $(r + 4)(r - 6) \le 0$
4. $m^2 - 3m - 10 > 0$
5. $3r^2 + 10r \le 8$
6. $9p^2 + 3p \ge 2$
7. $2y^2 + y < 15$
8. $6x^2 + x > 1$
9. $4y^2 + 7y + 3 < 0$
10. $3r^2 + 16r + 5 < 0$
11. $y^2 - 6y + 6 \ge 0$
12. $3k^2 - 6k + 2 \le 0$
13. $2x^2 + 4x + 1 > 0$
14. $3n^2 - 5n + 1 > 0$
15. $y^2 - 4y \ge 0$
16. $a^2 + 2a < 0$
17. $3k^2 - 5k \le 0$
18. $2z^2 + 3z > 0$

Solve each of the following inequalities, then graph each solution. See Example 3.

19. $(p-1)(p-2)(p-3)<0$ **20.** $(2r+1)(3r-2)(4r+5)<0$ **21.** $(a-4)(2a+3)(3a-1)\le 0$

22. $(z+2)(4z-3)(2z+5)\ge 0$ **23.** $(2k+1)(3k-1)(k+4)>0$ **24.** $(3m+5)(2m+1)(m-3)<0$

Solve each of the following inequalities, then graph each solution. See Examples 4–6.

25. $\frac{8}{x-2}\ge 2$ **26.** $\frac{20}{y-1}\le 1$ **27.** $\frac{-6}{p-5}\ge 10$ **28.** $\frac{3}{m-1}\le 1$

29. $\frac{1-p}{4-p}>0$ **30.** $\frac{x+1}{x-5}<0$ **31.** $\frac{a}{a+2}\ge 2$ **32.** $\frac{m}{m+5}\le 1$

33. $\frac{t}{2t-3}<-1$ **34.** $\frac{y}{3y-2}>-2$ **35.** $\frac{x^2-4}{3x}\ge 1$ **36.** $\frac{m^2-6}{5m}<1$

37. $\frac{2x-3}{x^2+1}\ge 0$ **38.** $\frac{9x-8}{4x^2+25}<0$ **39.** $\frac{(3x-5)^2}{x+2}>0$ **40.** $\frac{(5x-3)^2}{2x+1}\le 0$

Work each problem.

41. A company's profits in hundreds of thousands of dollars are given by

$$P = 2x^2 - 25x + 75,$$

where x is the number of units sold. For what values of x does the company make a profit?

42. The velocity of a particle in feet per second is given by

$$v = 4t^2 - 38t + 84,$$

where t is time in seconds. Find the times when velocity is negative.

43. Explain why a rational inequality like those in the examples can have at most one endpoint included in the solution set. (See Objective 3.)

PREVIEW EXERCISES

Suppose that $3x - 2y = 12$. *Find* y *for the following values of* x. *See Section 1.3.*

44. 0 **45.** 4 **46.** -2 **47.** -3

CHAPTER 6 GLOSSARY

KEY TERMS

6.2 quadratic formula The quadratic formula is a formula for solving quadratic equations.

6.3 discriminant The discriminant is the quantity under the radical in the quadratic formula.

6.4 quadratic in form An equation that can be written as a quadratic equation is called quadratic in form.

6.6 quadratic inequality A quadratic inequality is an inequality that can be written in the form $ax^2 + bx + c < 0$ or $ax^2 + bx + c > 0$, or with $\le$ or $\ge$.

rational inequality A rational inequality has a fraction with a variable denominator.

CHAPTER 6 QUICK REVIEW

SECTION	CONCEPTS	EXAMPLES
6.1 COMPLETING THE SQUARE	**Square Root Property** If a and b are complex numbers and if $a^2 = b$, then $a = \sqrt{b}$ or $a = -\sqrt{b}$.	Solve $(x-1)^2 = 8$. $x - 1 = \pm\sqrt{8} = \pm 2\sqrt{2}$ $x = 1 \pm 2\sqrt{2}$ Solution set: $\{1 \pm 2\sqrt{2}\}$
	Completing the Square *Step 1* If $a \neq 1$, divide both sides by a.	Solve $2x^2 - 4x - 18 = 0$. $x^2 - 2x - 9 = 0$
	Step 2 Rewrite the equation so that both terms containing variables are on one side of the equals sign and the constant is on the other side.	$x^2 - 2x = 9$
	Step 3 Take half the coefficient of x and square it.	$\left[\frac{1}{2}(-2)\right]^2 = (-1)^2 = 1$
	Step 4 Add the square to both sides of the equation.	$x^2 - 2x + 1 = 9 + 1$
	Step 5 One side should now be a perfect square trinomial. Factor and write it as the square of a binomial. Simplify the other side.	$(x-1)^2 = 10$
	Step 6 Use the square root property to complete the solution.	$x - 1 = \pm\sqrt{10}$ $x = 1 \pm \sqrt{10}$ Solution set: $\{1 \pm \sqrt{10}\}$
6.2 THE QUADRATIC FORMULA	**Quadratic Formula** The solutions of $ax^2 + bx + c = 0$ $(a \neq 0)$ are $x = \frac{-b \pm \sqrt{b^2 - 4ac}}{2a}$.	Solve $3x^2 + 5x + 2 = 0$. $x = \frac{-5 \pm \sqrt{5^2 - 4(3)(2)}}{2(3)}$ $x = -1$ or $x = -\frac{2}{3}$ Solution set: $\left\{-1, -\frac{2}{3}\right\}$

SECTION	CONCEPTS	EXAMPLES
6.3 THE DISCRIMINANT; SUM AND PRODUCT OF THE SOLUTIONS	If a, b, and c are integers, then the discriminant, $b^2 - 4ac$, of $ax^2 + bx + c = 0$ determines the type of solutions as follows.	
	Discriminant / **Solutions**	
	Positive square of an integer — 2 rational solutions	For $x^2 + 3x - 10 = 0$, the discriminant is $3^2 - 4(1)(-10) = 49$. There are 2 rational solutions.
	Positive, not square of an integer — 2 irrational solutions	For $2x^2 + 5x + 1 = 0$, the discriminant is $5^2 - 4(2)(1) = 17$. There are 2 irrational solutions.
	Zero — 1 rational solution	For $9x^2 - 6x + 1 = 0$, the discriminant is $(-6)^2 - 4(9)(1) = 0$. There is 1 rational solution.
	Negative — 2 imaginary solutions	For $4x^2 + x + 1 = 0$, the discriminant is $1^2 - 4(4)(1) = -15$. There are 2 imaginary solutions.
	If the solutions of $ax^2 + bx + c = 0$ are x_1 and x_2, then $x_1 + x_2 = -\frac{b}{a}$ and $x_1x_2 = \frac{c}{a}$.	The solutions of $2x^2 + 5x - 3$ are 1/2 and -3. Sum: $\frac{1}{2} + (-3) = -\frac{5}{2} = -\frac{b}{a}$ Product: $\frac{1}{2}(-3) = -\frac{3}{2} = \frac{-3}{2} = \frac{c}{a}$
6.5 FORMULAS AND APPLICATIONS	To solve a formula for a squared variable, proceed as follows.	
	(a) The variable appears only to the second degree. Isolate the squared variable on one side of the equation, then use the square root property.	Solve $A = \frac{2mp}{r^2}$ for r. $r^2A = 2mp$ Multiply by r^2. $r^2 = \frac{2mp}{A}$ Divide by A. $r = \pm\sqrt{\frac{2mp}{A}}$ Take square roots. $r = \pm\frac{\sqrt{2mpA}}{A}$ Rationalize.

SECTION	CONCEPTS	EXAMPLES
	(b) The variable appears to the first and second degree. Write the equation in standard quadratic form, then use the quadratic formula to solve.	Solve $m^2 + rm = t$ for m. $m^2 + rm - t = 0$ Standard form $m = \frac{-r \pm \sqrt{r^2 - 4(1)(-t)}}{2(1)}$ $a = 1, b = r, c = -t$ $m = \frac{-r \pm \sqrt{r^2 + 4t}}{2}$
6.6 NONLINEAR INEQUALITIES	**Solving a Quadratic Inequality** *Step 1* Write the inequality as an equation and solve. *Step 2* Place the numbers found in Step 1 on a number line. These numbers divide the line into regions. *Step 3* Substitute a number from each region into the inequality to determine the intervals that belong in the solution set—those intervals containing numbers that make the inequality true.	Solve $2x^2 + 5x + 2 < 0$. $2x^2 + 5x + 2 = 0$ $x = -\frac{1}{2}, x = -2$  $x = -3$ makes it false; $x = -1$ makes it true; $x = 0$ makes it false. Solution: $\left(-2, -\frac{1}{2}\right)$
	Solving a Rational Inequality *Step 1* Write the inequality as an equation and solve the equation. *Step 2* Set the denominator equal to zero and solve that equation. *Step 3* Use the solutions from Steps 1 and 2 to divide a number line into regions. *Step 4* Test a number from each region in the inequality to determine the regions that satisfy the inequality. *Step 5* Exclude any values that make the denominator zero.	Solve $\frac{x}{x+2} > 4$. $\frac{x}{x+2} = 4$; $x = -\frac{8}{3}$ $x + 2 = 0$ $x = -2$  -4 makes it false; $-\frac{7}{3}$ makes it true; 0 makes it false. The solution is $\left(-\frac{8}{3}, -2\right)$, since -2 makes the denominator 0 and $-\frac{8}{3}$ gives a false sentence.

CHAPTER 6 REVIEW EXERCISES

[6.1] *Solve each of the following equations by the square root property. Complete the square first, if necessary.*

1. $t^2 = 16$ **2.** $z^2 - 50 = 0$ **3.** $(2x + 5)^2 = 49$ **4.** $(5r + 3)^2 = 72$

***5.** $(3r - 2)^2 = -16$ ***6.** $(2y - 3)^2 = -2$ **7.** $x^2 - 4x = 15$ **8.** $2r^2 - 3r + 1 = 0$

[6.2] *Solve each of the following equations by the quadratic formula.*

9. $2y^2 - y = 21$ **10.** $k^2 + 5k = 7$ **11.** $2s^2 + 3s = 10$

12. $9p^2 + 49 = 42p$ **13.** $m(2m - 7) = 3(m^2 + 1)$ **14.** $a(5a - 3) = 2(a^2 + 5)$

***15.** $ix^2 - 4x + i = 0$ ***16.** $2y^2 - 4iy + 3 = 0$

Solve each applied problem.

17. A paint-mixing machine has two inlet pipes. One takes 1 hour longer than the other to fill the tank. Together they fill the tank in 3 hours. How long would it take each of them alone to fill the tank? Give your answer to the nearest tenth of an hour.

18. A new machine processes a batch of checks in one hour less time than an old one. How long would it take the old machine to process a batch of checks that the two machines together could process in 2 hours? Round your answer to the nearest tenth.

19. Judy Lewis can write a report for her boss in 2 hours less time than Jim Katz can. Working together, they can do the report in 24/7 hours. How long would it take each person working alone to write the report?

[6.3] *Use the discriminant to predict whether the solutions to the following equations are* **(a)** *two different rational numbers,* **(b)** *exactly one rational number,* **(c)** *two different irrational numbers, or* **(d)** *two different imaginary numbers.*

20. $a^2 + 5a + 1 = 0$ **21.** $4c^2 + 4c = 3$ **22.** $4x^2 = 6x - 9$

23. $y^2 + 2y + 5 = 0$ **24.** $9z^2 - 30z + 25 = 0$ **25.** $11m^2 - 7m - 9 = 0$

Use the discriminant to tell which polynomials can be factored. If a polynomial can be factored, factor it.

26. $24x^2 + 74x + 45$ **27.** $35x^2 + 69x - 35$

Without solving, give the sum and product of the solutions of each of the following equations.

28. $x^2 - 4x + 7 = 0$ **29.** $2x^2 - 5x = 8$

Write quadratic equations that have the following solution sets.

30. $\{2, -5\}$ **31.** $\left\{-2, -\frac{1}{3}\right\}$

[6.4] *Solve each equation. In Exercise 40, give only real number solutions.*

32. $\frac{16}{r + 1} + r = 9$ **33.** $4 = \frac{1}{t + 2} + \frac{3}{(t + 2)^2}$ **34.** $2 - \frac{1}{3 - k} = \frac{15}{(3 - k)^2}$

---

*Exercises identified with asterisks have imaginary number solutions.

35. $2r = -\sqrt{2 - 7r}$

36. $z = \sqrt{\frac{5z + 3}{2}}$

37. $y = -\sqrt{\frac{2 - 3y}{5}}$

***38.** $3p^4 - p^2 - 2 = 0$

***39.** $5m^4 - 2m^2 - 3 = 0$

40. $x^6 - 10x^3 + 16 = 0$

41. $8(2a + 3)^2 + 2(2a + 3) - 1 = 0$

Solve each problem.

42. John paddled his canoe 10 miles upstream, then paddled back. If the speed of the current was 3 miles per hour and the total trip took 3 1/2 hours, what was John's speed?

43. Ann Bezzone drove 8 kilometers to pick up her friend Lenore, and then drove 11 kilometers to a shopping mall at a speed 15 kilometers per hour faster. If Ann's total travel time was 24 minutes, what was her speed on the trip to pick up Lenore?

[6.5] *Solve each formula for the indicated variable. (Give answers with ± where appropriate.)*

44. $S = \frac{Id^2}{k}$; for d

45. $k = \frac{rF}{wv^2}$; for v

46. $p = \sqrt{\frac{kL}{g}}$; for k

47. $mt^2 = 3mt + 5$; for t

Solve each problem.

48. The expression $s = 16t^2 + 15t$ gives the distance in feet an object thrown off a building has fallen in t seconds. Find the time t when the object has fallen 25 feet. Round the answer to the nearest hundredth.

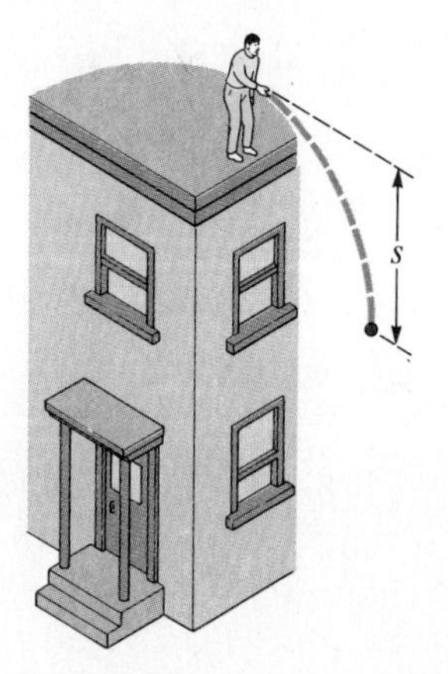

49. A small plastic rectangle that is part of a puzzle must be 3 centimeters shorter than 5 times its width. Its area must be 2 square centimeters. Find the length and width of the rectangle.

50. A large machine requires a part in the shape of a right triangle with a hypotenuse 9 feet less than twice the length of the longer leg. The shorter leg must be 3/4 the length of the longer leg. Find the lengths of the three sides of the part.

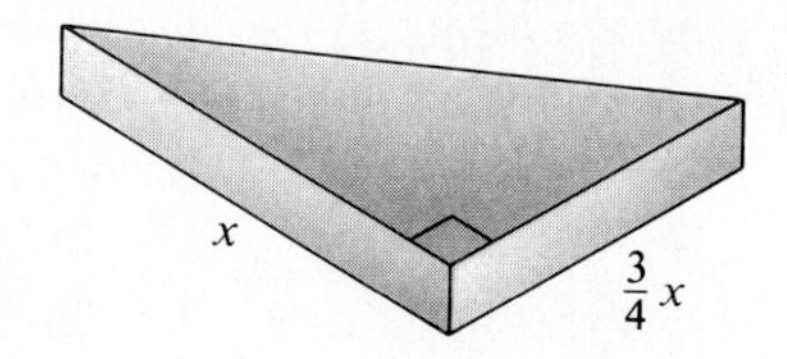

51. Masami wants to buy a mat for a photograph that measures 14 inches by 20 inches. She wants to have an even border around the picture when it is mounted on the mat. If the area of the mat she chooses is 352 square inches, how wide will the border be?

52. An excursion boat traveled 20 miles upriver and then traveled back. If the current was 10 miles per hour and the entire trip took 2 hours, find the speed of the boat in still water.

53. Two cars traveled from Elmhurst to Oakville, a distance of 100 miles. One car traveled 10 miles per hour faster than the other and arrived 5/6 of an hour sooner. Find the speed of the faster car.

54. The manager of a fast food outlet has determined that the demand for frozen yogurt is $25/p$ units per day, where p is the price (in dollars) per unit. The supply is $70p + 15$ units per day. Find the price at which supply and demand are equal.

55. Use the formula $A = P(1 + r)^2$ to find the interest rate r at which a deposit of $P = \$10,000$ will increase to an amount A of $11,664 in 2 years.

[6.6] *Solve each of the following inequalities, then graph each solution.*

56. $(2p + 3)(p - 1) > 0$

57. $k^2 + k < 12$

58. $2p^2 + 5p > 3$

59. $12y^2 + 5y - 3 \geq 0$

60. $3m^2 - 5m < 0$

61. $(2p - 1)(p + 2)(5p - 4) < 0$

62. $\dfrac{2}{x + 1} \geq 5$

63. $\dfrac{y + 1}{2y - 1} < 3$

64. $\dfrac{r^2 - 15}{r} \leq -2$

MIXED REVIEW EXERCISES

Solve by any method.

65. $m = \sqrt{3m + 10}$

66. $V = (r^2 + R^2)h;$ for R

***67.** $3p^2 - 6p + 4 = 0$

68. $x^4 + 14 = 9x^2$

69. $5y^2 - 8y + 3 = 0$

70. $(r - 1)(2r + 3)(r + 7) \geq 0$

71. Two pipes together can fill a large vat in 3 hours. One of the pipes, used alone, takes 2 hours longer than the other to fill the vat. How long would each pipe take to fill the vat alone? Give your answers to the nearest tenth.

72. Phong paddled his canoe 10 miles upstream, then paddled back. If the speed of the current was 3 miles per hour and the total trip took 3 1/2 hours, what was Phong's speed?

73. $(t - 1)(t + 2) = 5$

74. $D = \sqrt{\dfrac{xy}{3}};$ for y

75. $6 + \dfrac{19}{m} = -\dfrac{15}{m^2}$

76. $4q^2 + 12q + 9 > 0$

***77.** $3z^2 = 4z - 2$

78. $(b^2 - 2b)^2 = 11(b^2 - 2b) - 24$

79. A student gave the following solution to the equation $b^2 = 12$.

$$b^2 = 12$$
$$b = \sqrt{12}$$
$$b = 2\sqrt{3}$$

What is wrong with this solution?

80. Explain in your own words how to use the quadratic formula.

CHAPTER 6 TEST

Solve by the square root property. Complete the square first, if necessary.

1. $r^2 = 8$

2. $(3m + 2)^2 = 4$

3. $z^2 - 2z = 1$

4. $2x = \sqrt{\dfrac{5x + 2}{3}}$

Solve by the quadratic formula in Exercises 5–6.

5. $2m^2 - 3m + 1 = 0$

***6.** $3r^2 + 4r + 5 = 0$

7. Use the discriminant to predict the number and type of solutions for $2p^2 + 5 = 4p$.

8. Without solving, give the sum and product of the solutions of the equation in Exercise 7.

Solve by any method.

9. $\dfrac{12}{(p - 1)^2} = 1 + \dfrac{1}{p - 1}$

***10.** $y^4 = 6y^2 + 27$

11. $12 = (2p - 1)^2 + (2p - 1)$

---

12. Mario and Luis are cutting firewood. Luis can cut a cord of firewood in 2 hours less time than Mario. Together they can do the job in 5 hours. How long will it take each of them if they work alone? Give your answers to the nearest tenth.

Solve each formula for the indicated variable. (Give answers with $\pm$.)

13. $S = 4\pi r^2$; for r

14. $A = \pi r\sqrt{r^2 + h^2}$; for h

15. $r = kp(1 - p)$; for p

Work each problem.

16. Ken has a pool 20 by 24 feet. He wants to put grass in a strip of uniform width around the pool. He has enough grass seed for 192 square feet. How wide will the strip be?

17. A lot is in the shape of a right triangle. The shortest side measures 50 meters. The longest side is 110 meters less than twice the middle side. How long is the middle side?

Solve. Graph each solution set.

18. $2x^2 + 7x < 15$

19. $\dfrac{4}{p - 1} \geq 1$

20. What is wrong with the following solution for $x^4 - 12x^2 + 32 = 0$?

Let $u = x^2$. $\quad u^2 - 12u + 32 = 0$

$(u - 4)(u - 8) = 0$

$u - 4 = 0 \quad$ or $\quad u - 8 = 0$

$u = 4 \qquad\qquad u = 8$

Solution set: $\{4, 8\}$

---

*Exercises identified with asterisks have imaginary number solutions.

CHAPTER SEVEN

THE STRAIGHT LINE

René Descartes, a French mathematician of the seventeenth century, is credited with giving us an indispensible method of locating a point in a plane. The rectangular, or Cartesian, coordinate system of locating a point using two number lines intersecting at right angles at their origins is introduced in this chapter. We will also study methods of graphing equations in two variables.

Howard Eves, in *An Introduction to the History of Mathematics* (6th edition), relates the tale of how the idea of rectangular coordinates (analytic geometry) supposedly came to Descartes:

> Another story, perhaps on a par with the story of Isaac Newton and the falling apple, says that the initial flash of analytic geometry came to Descartes when watching a fly crawling about on the ceiling near a corner of his room. It struck him that the path of the fly on the ceiling could be described if only one knew the relation connecting the fly's distances from two adjacent walls. Even though this second story may be apocryphal, it has good pedagogic value.

7.1 THE RECTANGULAR COORDINATE SYSTEM

OBJECTIVES

1 PLOT ORDERED PAIRS.

2 FIND ORDERED PAIRS THAT SATISFY A GIVEN EQUATION.

3 GRAPH LINES.

4 FIND x- AND y-INTERCEPTS.

5 RECOGNIZE EQUATIONS OF VERTICAL OR HORIZONTAL LINES.

6 USE THE DISTANCE FORMULA.

1 Each of the pairs of numbers (1, 2), (−1, 5), and (3, 7) is an example of an **ordered pair**; that is, a pair of numbers written within parentheses in which the order of the numbers is important. The two numbers are the **components** of the ordered pair. An ordered pair is graphed using two number lines that intersect at right angles at the zero points, as shown in Figure 7.1 on the next page. The common zero point is called the **origin.** The horizontal line, the ***x*-axis,** represents the first number in an ordered pair, and the vertical line, the ***y*-axis,** represents the second. The x-axis and the y-axis make up a **rectangular** (or Cartesian) **coordinate system.**

Locate, or **plot,** the point on the graph that corresponds to the ordered pair (3, 1) by going three units from zero to the right along the x-axis, and then one unit up parallel to the y-axis. The point corresponding to the ordered pair (3, 1) is labeled A in Figure 7.2. The point (4, −1) is labeled B, (−5, 6) is labeled C, and (−4, −5) is labeled D. Point E corresponds to (−5, 0). The phrase "the point corresponding to the ordered pair (3, 1)" is often abbreviated "the point (3, 1)." The numbers in an ordered pair are called the **coordinates** of the corresponding point.

CAUTION The parentheses used to represent an ordered pair are also used to represent an open interval (introduced in Chapter 1.) In general, there is no confusion between these symbols because the context of the discussion tells us whether we are discussing ordered pairs or open intervals.

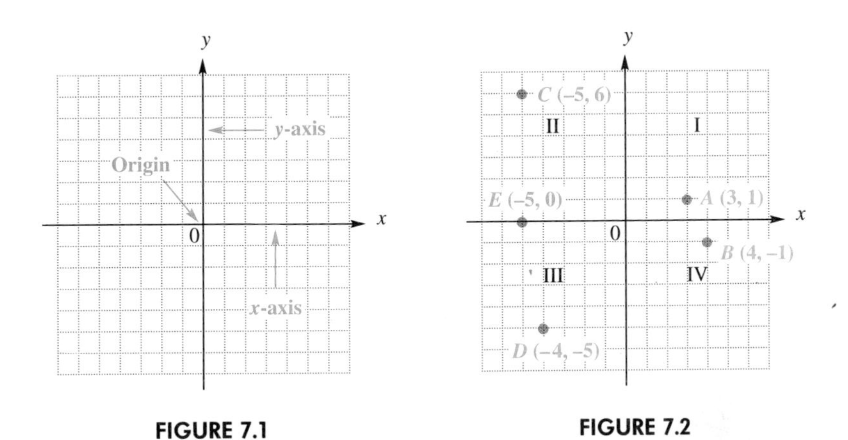

FIGURE 7.1

FIGURE 7.2

The four regions of the graph shown in Figure 7.2 are called **quadrants I, II, III,** and **IV,** reading counterclockwise from the upper right quadrant. The points of the x-axis and y-axis themselves do not belong to any quadrant. For example, point E in Figure 7.2 belongs to no quadrant.

2 An equation with two variables will have solutions written as ordered pairs. Unlike linear equations in a single variable as introduced in Chapter 2, equations with two variables will, in general, have an infinite number of solutions.

To find the ordered pairs that satisfy the equation, select any number for one of the variables, substitute it into the equation for that variable, and then solve for the other variable. For example, suppose $x = 0$ in the equation $2x + 3y = 6$. Then, by substitution,

$$2x + 3y = 6$$

becomes

$$2(0) + 3y = 6 \qquad \text{Let } x = 0.$$
$$0 + 3y = 6$$
$$3y = 6$$
$$y = 2,$$

giving the ordered pair (0, 2). Other ordered pairs satisfying $2x + 3y = 6$ include $(6, -2)$ and $(3, 0)$, for example.

EXAMPLE 1
COMPLETING ORDERED PAIRS

Complete the following ordered pairs for $2x + 3y = 6$.

(a) $(-3, \quad)$

Let $x = -3$. Substitute into the equation.

$$2(-3) + 3y = 6 \qquad \text{Let } x = -3.$$
$$-6 + 3y = 6$$
$$3y = 12$$
$$y = 4$$

The ordered pair is $(-3, 4)$.

(b) $(\quad, -4)$

Replace y with -4.

$$2x + 3y = 6$$
$$2x + 3(-4) = 6 \qquad \text{Let } y = -4.$$
$$2x - 12 = 6$$
$$2x = 18$$
$$x = 9$$

The ordered pair is $(9, -4)$. ■

3 The equation $2x + 3y = 6$ is graphed by first plotting all the ordered pairs mentioned above. These are shown in Figure 7.3(a). The resulting points appear to lie on a straight line. If all the ordered pairs that satisfy the equation $2x + 3y = 6$ were graphed, they would form a straight line. In fact, the graph of any first-degree equation in two variables is a straight line. The graph of $2x + 3y = 6$ is the line shown in Figure 7.3(b).

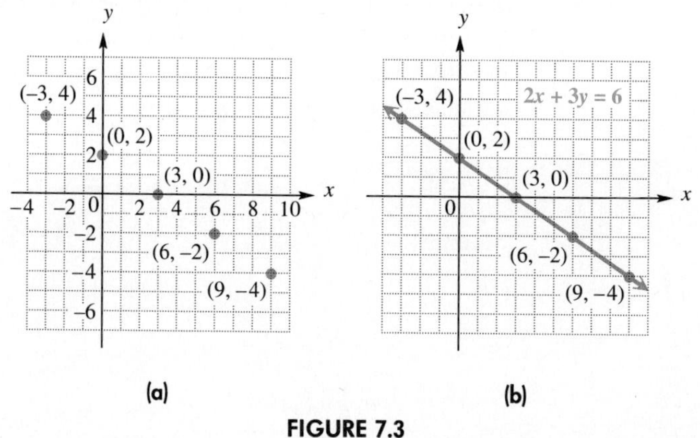

FIGURE 7.3

All first-degree equations with two variables have straight-line graphs. Since a straight line is determined if any two different points on the line are known, finding two different points is enough to graph the line.

STANDARD FORM OF A LINEAR EQUATION IN TWO VARIABLES

An equation that can be written in the form

$$Ax + By = C \qquad (A \text{ and } B \text{ not both } 0)$$

is a linear equation. This form is called the **standard form.**

4 Two points that are useful for graphing lines are the x- and y-intercepts. The ***x*-intercept** is the point (if any) where the line crosses the x-axis, and the ***y*-intercept** is the point (if any) where the line crosses the y-axis. (*Note:* In many texts, the intercepts are defined as numbers, and not points. However, in this book we will refer to intercepts as points.)

Intercepts can be found as follows.

INTERCEPTS Let $y = 0$ to find the x-intercept; let $x = 0$ to find the y-intercept.

EXAMPLE 2
FINDING INTERCEPTS

Find the x- and y-intercepts of $4x - y = -3$, and graph the equation.

Find the x-intercept by letting $y = 0$.

$$4x - 0 = -3 \qquad \text{Let } y = 0.$$
$$4x = -3$$
$$x = -\frac{3}{4} \qquad x\text{-intercept is } \left(-\frac{3}{4}, 0\right).$$

For the y-intercept, let $x = 0$.

$$4(0) - y = -3 \qquad \text{Let } x = 0.$$
$$-y = -3$$
$$y = 3 \qquad y\text{-intercept is } (0, 3).$$

The intercepts are the two points $(-3/4, 0)$ and $(0, 3)$. Use these two points to draw the graph, as shown in Figure 7.4. ■

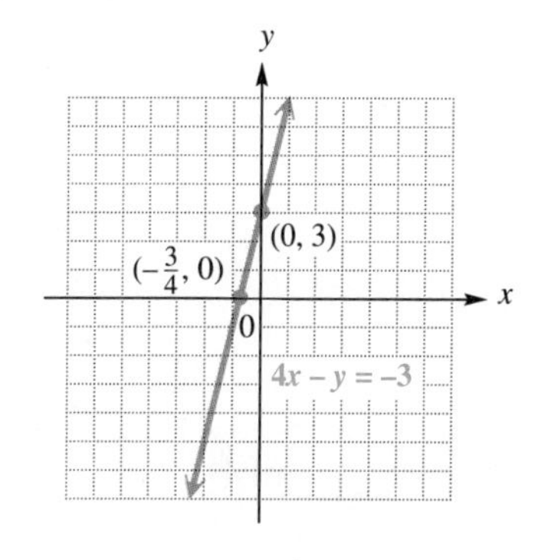

FIGURE 7.4

NOTE While two points, such as the two intercepts in Figure 7.4, are sufficient to graph a straight line, it is a good idea to use a third point to guard against errors. Verify that $(-1, -1)$ also lies on the graph of $4x - y = -3$.

5 The next example shows that a graph can fail to have an x-intercept, which is why the phrase "if any" was added when discussing intercepts.

EXAMPLE 3
GRAPHING A HORIZONTAL LINE

Graph $y = 2$.

Writing $y = 2$ as $0x + 1y = 2$ shows that any value of x, including $x = 0$, gives $y = 2$, making the y-intercept $(0, 2)$. Since y is always 2, there is no value of x corresponding to $y = 0$, and so the graph has no x-intercept. The graph, shown in Figure 7.5, is a horizontal line. ■

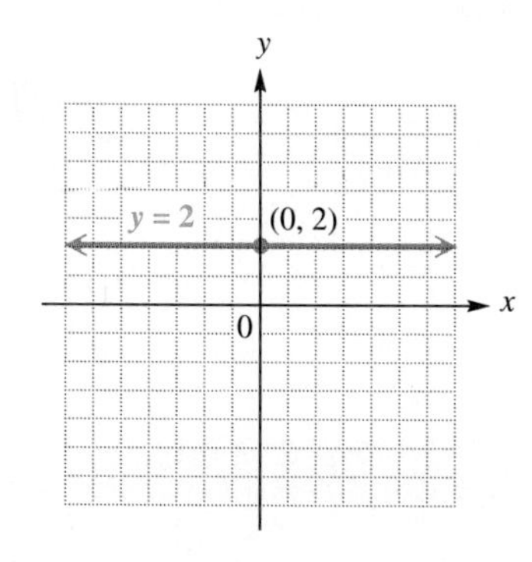

FIGURE 7.5

It is also possible for a graph to have no y-intercept, as in the next example.

EXAMPLE 4
GRAPHING A VERTICAL LINE

Graph $x = -1$.

The form $1x + 0y = -1$ shows that every value of y leads to $x = -1$, and so no value of y makes $x = 0$. The graph, therefore, has no y-intercept. The only way a straight line can have no y-intercept is to be vertical, as shown in Figure 7.6. ■

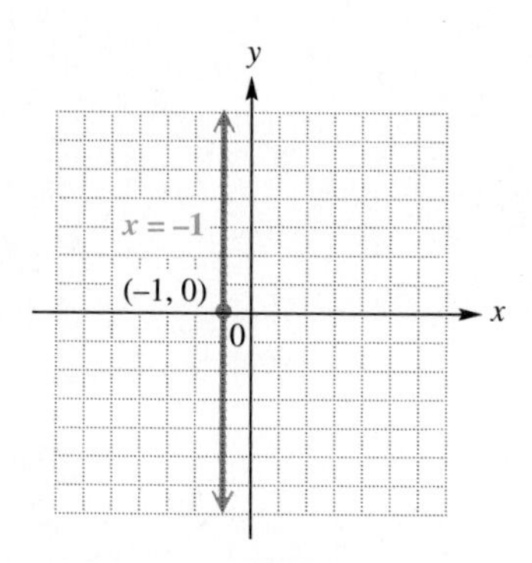

FIGURE 7.6

The line graphed in the next example has both x-intercept and y-intercept at the origin.

EXAMPLE 5
GRAPHING A LINE THAT PASSES THROUGH THE ORIGIN

Graph $x + 2y = 0$.

Find the x-intercept by letting $y = 0$.

$$x + 2y = 0$$
$$x + 2(0) = 0 \qquad \text{Let } y = 0.$$
$$x + 0 = 0$$
$$x = 0 \qquad x\text{-intercept is } (0, 0).$$

To find the y-intercept, let $x = 0$.

$$x + 2y = 0$$
$$0 + 2y = 0 \qquad \text{Let } x = 0.$$
$$y = 0 \qquad y\text{-intercept is } (0, 0).$$

Both intercepts are the same ordered pair, (0, 0). Another point is needed to graph the line. Choose any number for x (or for y), say $x = 4$, and solve for y.

$$x + 2y = 0$$
$$4 + 2y = 0 \qquad \text{Let } x = 4.$$
$$2y = -4$$
$$y = -2$$

This gives the ordered pair (4, −2). These two points lead to the graph shown in Figure 7.7. ■

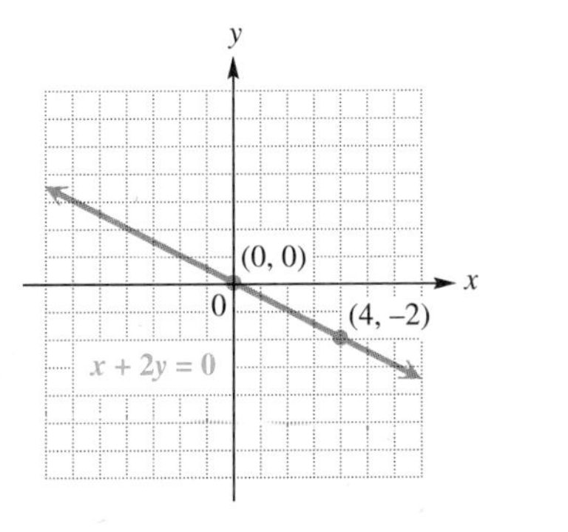

FIGURE 7.7

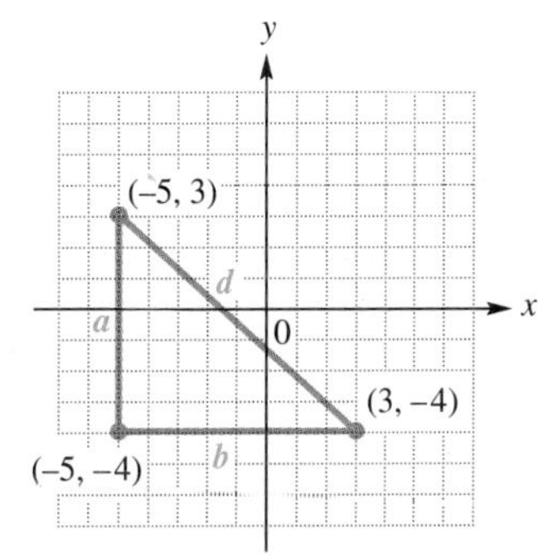

FIGURE 7.8

6 Figure 7.8 shows the points (3, −4) and (−5, 3). To find the distance between these points, use the Pythagorean formula from geometry, first mentioned in Chapter 5.

The vertical line through (−5, 3) and the horizontal line through (3, −4) intersect at the point (−5, −4). Thus, the point (−5, −4) becomes the vertex of the right angle in a right triangle. By the Pythagorean formula, the square of the length of the hypotenuse, d, of the right triangle in Figure 7.8, is equal to the sum of the squares of the lengths of the two legs a and b:

$$d^2 = a^2 + b^2.$$

The length a is the difference between the coordinates of the endpoints. Since the x-coordinate of both points is −5, the side is vertical, and we can find a by finding the difference between the y-coordinates. Subtract −4 from 3 to get a positive value for a.

$$a = 3 - (-4) = 7$$

Similarly, find b by subtracting -5 from 3.

$$b = 3 - (-5) = 8.$$

Substituting these values into the formula, we have

$$d^2 = a^2 + b^2$$
$$d^2 = 7^2 + 8^2 \qquad \text{Let } a = 7 \text{ and } b = 8.$$
$$d^2 = 49 + 64$$
$$d^2 = 113$$
$$d = \sqrt{113}. \qquad \text{Use the square root property.}$$

Therefore, the distance between $(-5, 3)$ and $(3, -4)$ is $\sqrt{113}$.

NOTE It is customary to leave the distance in radical form. Do not use a calculator to get an approximation unless you are specifically directed to do so.

This result can be generalized. Figure 7.9 shows the two different points (x_1, y_1) and (x_2, y_2). To find a formula for the distance d between these two points, notice that the distance between (x_2, y_2) and (x_2, y_1) is given by $a = y_2 - y_1$, and the distance between (x_1, y_1) and (x_2, y_1) is given by $b = x_2 - x_1$. From the Pythagorean formula,

$$d^2 = (x_2 - x_1)^2 + (y_2 - y_1)^2,$$

and by using the square root property, the distance formula is obtained.

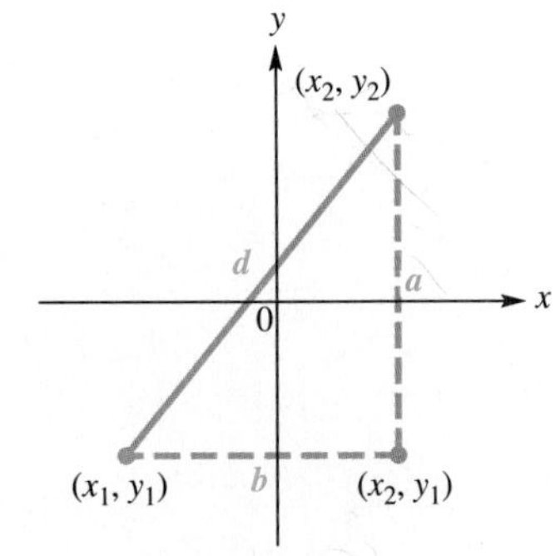

FIGURE 7.9

DISTANCE FORMULA

The distance between the points (x_1, y_1) and (x_2, y_2) is

$$d = \sqrt{(x_2 - x_1)^2 + (y_2 - y_1)^2}.$$

This result is called the **distance formula.**

The small numbers 1 and 2 in the ordered pairs (x_1, y_1) and (x_2, y_2) are called *subscripts*. We read x_1 as "x sub 1." Subscripts are used to distinguish between different values of a variable that have a common property. For example, in the ordered pairs $(-3, 5)$ and $(6, 4)$, -3 can be designated as x_1 and 6 as x_2. Their common property is that they are both x components of ordered pairs. This idea is used in the next example.

EXAMPLE 6 USING THE DISTANCE FORMULA

Find the distance between $(-3, 5)$ and $(6, 4)$.

When using the distance formula to find the distance between two points, designating the points as (x_1, y_1) and (x_2, y_2) is arbitrary. Let us choose $(x_1, y_1) = (-3, 5)$ and $(x_2, y_2) = (6, 4)$.

$$\begin{aligned} d &= \sqrt{(x_2 - x_1)^2 + (y_2 - y_1)^2} \\ &= \sqrt{(6 - (-3))^2 + (4 - 5)^2} \qquad x_2 = 6,\ y_2 = 4,\ x_1 = -3,\ y_1 = 5 \\ &= \sqrt{9^2 + (-1)^2} \\ &= \sqrt{82} \end{aligned}$$ ■

In geometry, points that lie on the same straight line are said to be **collinear.** The distance formula can be used to determine if three points are collinear, as shown in Example 7.

EXAMPLE 7 DETERMINING WHETHER THREE POINTS ARE COLLINEAR

Are the points $A(3, 1)$, $B(6, 2)$, and $C(9, 3)$ collinear?

If these three points are collinear, the sum of the lengths of the two shorter line segments should equal the length of the longest line segment. The distance formula can be used to find the lengths of the three line segments.

$$\begin{aligned} AB &= \sqrt{(6 - 3)^2 + (2 - 1)^2} = \sqrt{10} \\ BC &= \sqrt{(9 - 6)^2 + (3 - 2)^2} = \sqrt{10} \\ AC &= \sqrt{(9 - 3)^2 + (3 - 1)^2} = \sqrt{40} = 2\sqrt{10} \end{aligned}$$

Since $AB + BC = AC$, the three points are collinear. ■

7.1 EXERCISES

Name the quadrant in which each point is located.

1. $(1, 5)$ **2.** $(-2, 4)$ **3.** $(-3, -2)$ **4.** $(5, -1)$

5. $(2, -3)$ **6.** $(-7, -4)$ **7.** $(-1, 4)$ **8.** $(0, 4)$

Locate the following points on a rectangular coordinate system.

9. $(2, 3)$ **10.** $(-1, 2)$ **11.** $(-3, -2)$ **12.** $(1, -4)$ **13.** $(0, 5)$

14. $(-2, -4)$ **15.** $(-2, 4)$ **16.** $(3, 0)$ **17.** $(-2, 0)$ **18.** $(3, -3)$

In each exercise, complete the given ordered pairs for the equation. Then graph the equation. See Example 1.

19. $2x + y = 5$; $(0,\ \ )$, $(\ \ , 0)$, $(1,\ \ )$, $(\ \ , 1)$

20. $3x - 4y = 24$; $(0,\ \ )$, $(\ \ , 0)$, $(6,\ \ )$, $(\ \ , -3)$

21. $x - y = 4$; $(0,\ \ )$, $(\ \ , 0)$, $(2,\ \ )$, $(\ \ , -1)$

22. $x + 3y = 12$; $(0,\ \ )$, $(\ \ , 0)$, $(3,\ \ )$, $(\ \ , 6)$

23. $4x + 5y = 20$; $(0,\ \ )$, $(\ \ , 0)$, $(3,\ \ )$, $(\ \ , 2)$

24. $2x - 5y = 12$; $(0,\ \ )$, $(\ \ , 0)$, $(\ \ , -2)$, $(-2,\ \ )$

25. $3x + 2y = 8$; $(0,\ \ )$, $(\ \ , 0)$, $(2,\ \ )$, $(\ \ , -2)$

26. $5x + y = 12$; $(0,\ \ )$, $(\ \ , 0)$, $(\ \ , -3)$, $(2,\ \ )$

---

27. Explain how to find the x-intercept of a linear equation in two variables. (See Objective 4.)

28. Explain how to find the y-intercept of a linear equation in two variables. (See Objective 4.)

29. Which one of the following has as its graph a horizontal line? (See Objective 5.)
(a) $2y = 6$ **(b)** $2x = 6$
(c) $x - 4 = 0$ **(d)** $x + y = 0$

30. What is the minimum number of points that must be determined in order to graph a linear equation in two variables? (See Objective 4.)

For each equation, find the x-intercept and the y-intercept. Then graph the equation. See Examples 2–5.

31. $3x + 2y = 12$ **32.** $2x + 5y = 10$ **33.** $5x + 6y = 10$ **34.** $3y + x = 6$

35. $2x - y = 5$ **36.** $3x - 2y = 4$ **37.** $x - 3y = 2$ **38.** $y - 4x = 3$

39. $y + x = 0$ **40.** $2x - y = 0$ **41.** $3x = y$ **42.** $x = -4y$

43. $x = 2$ **44.** $y = -3$ **45.** $y = 4$ **46.** $x = -2$

Find the distance between each of the following pairs of points. See Example 6.

47. $(3, 4)$ and $(-2, 1)$

48. $(-2, 1)$ and $(3, -2)$

49. $(-2, 4)$ and $(3, -2)$

50. $(1, -5)$ and $(6, 3)$

51. $(-3, 7)$ and $(2, -4)$

52. $(0, 5)$ and $(-3, 12)$

53. $(\sqrt{2}, \sqrt{6})$ and $(-2\sqrt{2}, 4\sqrt{6})$

54. $(\sqrt{7}, 9\sqrt{3})$ and $(-\sqrt{7}, 4\sqrt{3})$

55. $(a, a - b)$ and $(b, a + b)$

56. $(m + n, n)$ and $(m - n, m)$

---

57. As given in the text, the distance formula is expressed with a radical. Write the distance formula using rational exponents.

58. An alternate form of the distance formula is

$$d = \sqrt{(x_1 - x_2)^2 + (y_1 - y_2)^2}.$$

Compare this to the form given in this section, and explain why the two forms are equivalent.

Use the distance formula to determine whether or not the following points are collinear. See Example 7.

59. $(0, -4)$, $(2, 2)$, $(5, 11)$ **60.** $(1, -2)$, $(3, -1)$, $(5, 0)$ **61.** $(1, 1)$, $(5, 7)$, $(7, 10)$

62. $(0, 2)$, $(3, 0)$, $(-3, 4)$ **63.** $(0, 5)$, $(5, -5)$, $(10, -14)$ **64.** $(0, 6)$, $(4, -5)$, $(-2, 12)$

Find the perimeters of the triangles determined by the following sets of points.

65. $(4, 4)$, $(-4, -2)$, $(4, -2)$

66. $(2, 6)$, $(-3, -3)$, $(6, 2)$

67. $(1, 3)$, $(-2, -4)$, $(1, -2)$

68. $(-3, -2)$, $(7, 4)$, $(1, 14)$

Determine whether or not each of the following groups of ordered pairs represents a right triangle. (Hint: If a and b are the shorter sides of a triangle and d is the longest side, and if $d^2 = a^2 + b^2$, then the triangle is a right triangle.)

69. (0, 6), (9, −6), (−3, 0)

70. (3, 2), (12, −10), (0, −4)

71. (−2, −2), (8, 4), (2, 14)

72. (5, 6), (0, −3), (9, 2)

*The **midpoint** of the line segment joining the two points (x_1, y_1) and (x_2, y_2) is the point $(\bar{x}, \bar{y})$ where*

$$\bar{x} = \frac{x_1 + x_2}{2} \quad \text{and} \quad \bar{y} = \frac{y_1 + y_2}{2}.$$

Find the midpoint of the line segment PQ.

73. $P(3, 5)$, $Q(-4, 7)$

74. $P(8, -6)$, $Q(2, 3)$

75. $P(0, 4)$, $Q(-3, -1)$

76. $P(1, -7)$, $Q(-8, 4)$

77. $P(7, 2)$, $Q(-3, -8)$

78. $P(10, -4)$, $Q(6, 5)$

---

79. Recently the U.S. population has been growing according to the equation

$$y = 1.7x + 230$$

where y gives the population (in millions) in year x, measured from year 1980. For example, in 1980 $x = 0$ and $y = 1.7(0) + 230 = 230$. This means that the population was about 230 million in 1980. To find the population in 1985, let $x = 5$, and so on. Find the population in each of the following years.

(a) 1982 **(b)** 1985 **(c)** 1990

(d) In what year will the population reach 315 million?

(e) Graph the equation.

80. It is estimated that y, the number of items of a particular commodity (in millions) sold in the United States in year x, where x represents the number of years since 1990, is given by

$$y = 1.71x + 2.98.$$

That is, $x = 0$ represents 1990, $x = 1$ represents 1991, and so on. Find the number of items sold in the following years.

(a) 1990 **(b)** 1991 **(c)** 1992

(d) In what year will about 10 million items be sold?

(e) Graph the equation.

In Exercises 81–82, translate each statement into an equation. Then graph the equation.

81. The x-value is 7 more than the y-value.

82. The y-value is twice the x-value.

Use the Pythagorean formula to solve each problem.

83. A 20-foot ladder is leaning against a house. The bottom of the ladder is 12 feet from the house. How far will the top of the ladder be above the ground?

84. A surveyor wishes to determine the length of the sides of a lot that has the shape of a right triangle. He has found the lengths of the two longer sides to be 15 meters and 17 meters. What is the length of the third side?

PREVIEW EXERCISES

Find each quotient. See Section 1.3.

85. $\dfrac{5-3}{6-2}$ **86.** $\dfrac{-4-2}{5-7}$ **87.** $\dfrac{-3-(-5)}{4-(-1)}$

88. $\dfrac{6-(-2)}{-3-(-1)}$ **89.** $\dfrac{-2-(-3)}{-1-4}$ **90.** $\dfrac{-8-2}{6-(-3)}$

7.2 THE SLOPE OF A LINE

OBJECTIVES

1. FIND THE SLOPE OF A LINE GIVEN TWO POINTS ON THE LINE.
2. FIND THE SLOPE OF A LINE GIVEN THE EQUATION OF THE LINE.
3. GRAPH A LINE GIVEN ITS SLOPE AND A POINT ON THE LINE.
4. USE SLOPE TO DECIDE WHETHER TWO LINES ARE PARALLEL OR PERPENDICULAR.
5. SOLVE PROBLEMS INVOLVING AVERAGE RATE OF CHANGE.

FOCUS ON PROBLEM SOLVING

In one state the fine for driving 10 miles per hour over the speed limit is $35 and the fine for driving 15 miles per hour over the speed limit is $42. What is the average rate of change in the fine for each mile per hour over the speed limit?

Associated with every nonvertical line is a number called the slope of the line, which measures the ratio of its vertical change to its horizontal change. The slope formula gives the average rate of change of y per unit change of x. By studying the concept of slope, we can determine how a line is oriented in the plane. The problem above is an application of the slope concept, and is Exercise 51 in the exercises for this section. After working through this section, you should be able to solve this problem.

1 Two different points determine a line. A line also can be determined by a point on the line and some measure of the "steepness" of the line. The most useful measure of the steepness of a line is called the *slope* of the line. One way to get a measure of the steepness of a line is to compare the vertical change in the line (the *rise*) to the horizontal change (the *run*) while moving along the line from one fixed point to another.

Suppose that (x_1, y_1) and (x_2, y_2) are two different points on a line. Then, going along the line from (x_1, y_1) to (x_2, y_2), the y-value changes from y_1 to y_2, an amount equal to $y_2 - y_1$. As y changes from y_1 to y_2, the value of x changes from x_1 to x_2 by the amount $x_2 - x_1$. See Figure 7.10. The ratio of the change in y to the change in x is called the **slope** of the line, with the letter m used for the slope.

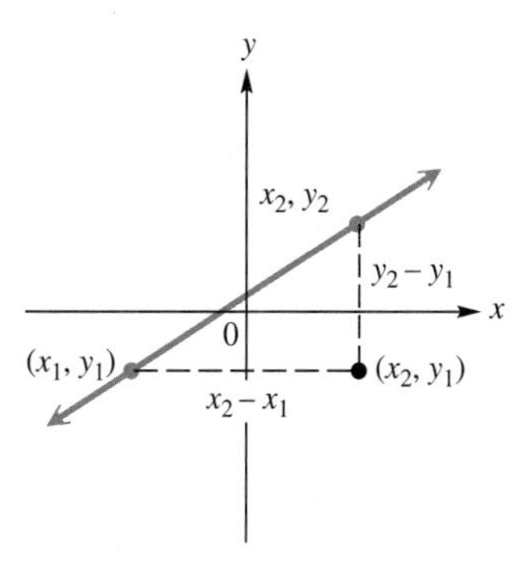

FIGURE 7.10

SLOPE

If $x_1 \neq x_2$, the slope of the line through the distinct points (x_1, y_1) and (x_2, y_2) is

$$m = \frac{\text{change in } y}{\text{change in } x} = \frac{y_2 - y_1}{x_2 - x_1}.$$

EXAMPLE 1
USING THE DEFINITION OF SLOPE

Find the slope of the line through the points $(2, -1)$ and $(-5, 3)$.

If $(2, -1) = (x_1, y_1)$ and $(-5, 3) = (x_2, y_2)$, then

$$m = \frac{y_2 - y_1}{x_2 - x_1}$$

$$= \frac{3 - (-1)}{-5 - (2)} = \frac{4}{-7} = -\frac{4}{7}.$$

See Figure 7.11. On the other hand, if $(2, -1) = (x_2, y_2)$ and $(-5, 3) = (x_1, y_1)$, the slope would be

$$m = \frac{-1 - 3}{2 - (-5)} = \frac{-4}{7} = -\frac{4}{7},$$

the same answer. This example suggests that the slope is the same no matter which point is considered first. Also, using similar triangles from geometry, it can be shown that the slope is the same no matter which two different points on the line are chosen. ■

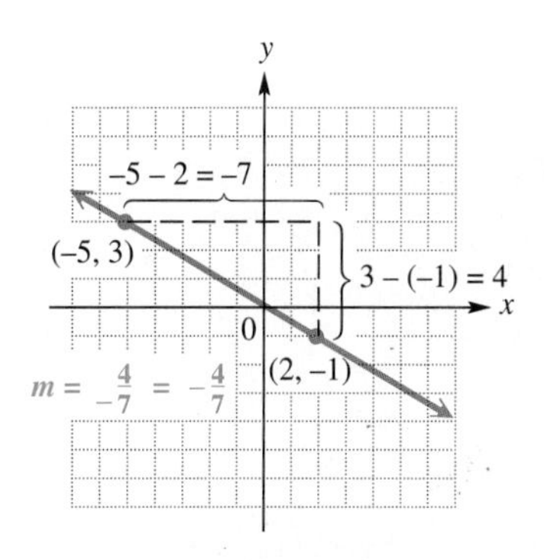

FIGURE 7.11

CAUTION In calculating the slope, be careful to subtract the y-values and the x-values in the *same* order.

Correct		Incorrect	
$\dfrac{y_2 - y_1}{x_2 - x_1}$ or	$\dfrac{y_1 - y_2}{x_1 - x_2}$	$\dfrac{y_2 - y_1}{x_1 - x_2}$ or	$\dfrac{y_1 - y_2}{x_2 - x_1}$

2 When the equation of a line is given, its slope can be found from the definition of slope by first finding two different points on the line.

EXAMPLE 2 FINDING THE SLOPE OF A LINE

Find the slope of the line $4x - y = 8$.

The intercepts can be used as the two different points needed to find the slope. Let $y = 0$ to find that the x-intercept is $(2, 0)$. Then let $x = 0$ to find that the y-intercept is $(0, -8)$. Use these two points in the slope formula. The slope is

$$m = \frac{-8 - 0}{0 - 2} = \frac{-8}{-2} = 4. \quad \blacksquare$$

EXAMPLE 3 FINDING THE SLOPE OF A LINE

Find the slope of each of the following lines.

(a) $x = -3$

By inspection, $(-3, 5)$ and $(-3, -4)$ are two points that satisfy the equation $x = -3$. Use these two points to find the slope.

$$m = \frac{-4 - 5}{-3 - (-3)} = \frac{-9}{0}$$

Since division by zero is undefined, the slope is undefined. This is why the definition of slope includes the restriction that $x_1 \neq x_2$.

(b) $y = 5$

Find the slope by selecting two different points on the line, such as $(3, 5)$ and $(-1, 5)$, and by using the definition of slope.

$$m = \frac{5 - 5}{3 - (-1)} = \frac{0}{4} = 0 \quad \blacksquare$$

As shown in Section 7.1, $x = -3$ has a graph that is a vertical line, and $y = 5$ has a graph that is a horizontal line. Generalizing from those results and the results of Example 3 above, we can make the following statements about vertical and horizontal lines.

VERTICAL AND HORIZONTAL LINES

A vertical line has an equation of the form $x = k$, where k is a real number, and its slope is undefined. A horizontal line has an equation of the form $y = k$, and its slope is 0.

3 Examples 4 and 5 show how to graph a straight line by using the slope and one point that the line contains.

EXAMPLE 4
USING THE SLOPE AND A POINT TO GRAPH A LINE

Graph the line that has slope 2/3 and goes through the point $(-1, 4)$.

First locate the point $(-1, 4)$ on a graph as shown in Figure 7.12. Then, from the definition of slope,

$$m = \frac{\textbf{change in } y}{\textbf{change in } x} = \frac{2}{3}.$$

Move *up* 2 units in the y-direction and then 3 units to the *right* in the x-direction to locate another point on the graph (labeled P). The line through $(-1, 4)$ and P is the required graph. ■

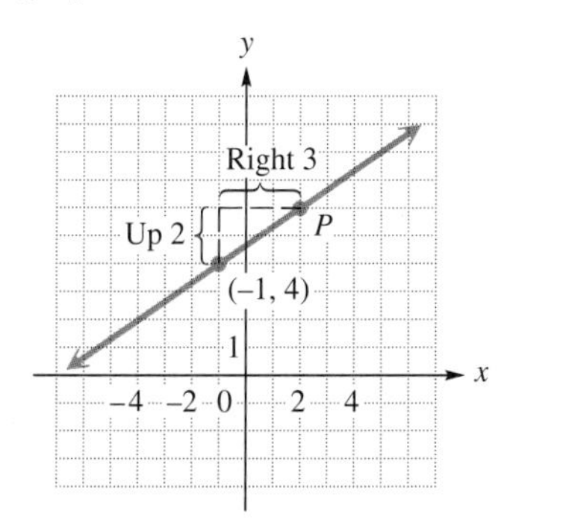

FIGURE 7.12

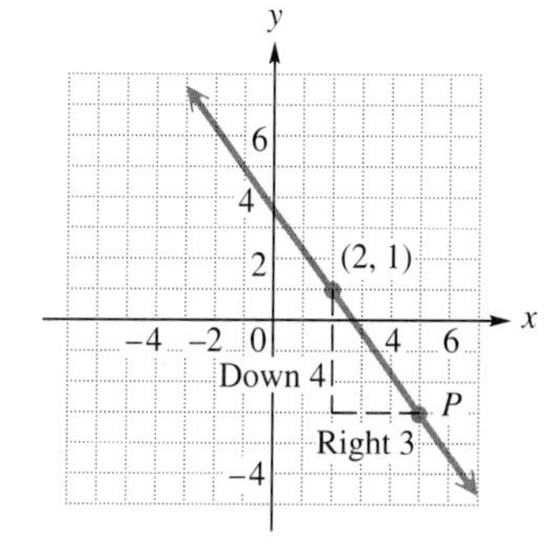

FIGURE 7.13

EXAMPLE 5
USING THE SLOPE AND A POINT TO GRAPH A LINE

Graph the line through $(2, 1)$ that has slope $-4/3$.

Start by locating the point $(2, 1)$ on the graph. Find a second point on the line by using the definition of slope.

$$\text{slope} = \frac{\text{change in } y}{\text{change in } x} = \frac{-4}{3}$$

Move *down* 4 units from $(2, 1)$ and then 3 units to the *right*. Draw a line through this second point and $(2, 1)$, as shown in Figure 7.13. The slope also could be written as

$$\frac{\text{change in } y}{\text{change in } x} = \frac{4}{-3}.$$

In that case the second point is located *up* 4 units and 3 units to the *left*. Verify that this approach produces the same line. ■

In Example 4, the slope of the line is the *positive* number 2/3. The graph of the line in Figure 7.12 goes up from left to right. The line in Example 5 has a *negative* slope, $-4/3$. As Figure 7.13 shows, its graph goes down from left to right. These facts suggest the following generalization.

A positive slope indicates that the line goes up from left to right; a negative slope indicates that the line goes down from left to right.

Figure 7.14 shows lines of positive, zero, negative, and undefined slopes.

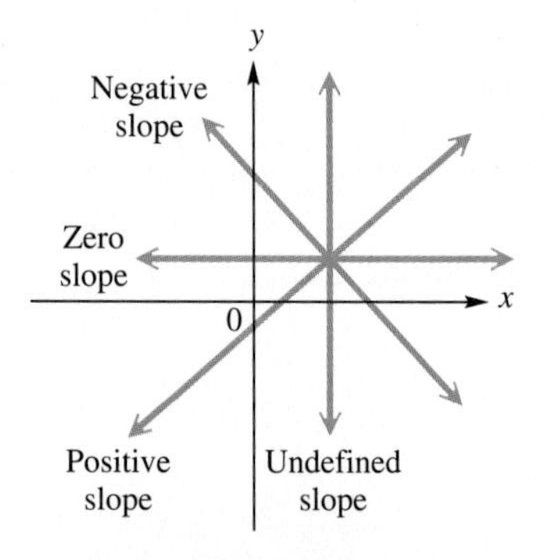

FIGURE 7.14

4 The slopes of a pair of parallel or perpendicular lines are related in a special way. The slope of a line measures the steepness of a line. Since parallel lines have equal steepness, their slopes also must be equal. Also, lines with the same slope are parallel.

SLOPES OF PARALLEL LINES

Two nonvertical lines with the same slope are parallel; two nonvertical parallel lines have the same slope.

EXAMPLE 6

DETERMINING WHETHER TWO LINES ARE PARALLEL

Are the lines L_1, through $(-2, 1)$ and $(4, 5)$, and L_2, through $(3, 0)$ and $(0, -2)$, parallel?

The slope of L_1 is

$$m_1 = \frac{5 - 1}{4 - (-2)} = \frac{4}{6} = \frac{2}{3}.$$

The slope of L_2 is

$$m_2 = \frac{-2 - 0}{0 - 3} = \frac{-2}{-3} = \frac{2}{3}.$$

Since the slopes are equal, the lines are parallel. ■

Recall that perpendicular lines are lines which meet at right angles. It can be shown that the slopes of perpendicular lines have a product of -1. For example, if the slope of a line is $3/4$, then any line perpendicular to it has slope $-4/3$, because $(3/4)(-4/3) = -1$.

SLOPES OF PERPENDICULAR LINES

If neither is vertical, perpendicular lines have slopes that are negative reciprocals; that is, their product is -1. Also, lines with slopes that are negative reciprocals are perpendicular.

EXAMPLE 7
DETERMINING WHETHER TWO LINES ARE PERPENDICULAR

Are the lines with equations $2y = 3x - 6$ and $2x + 3y = -6$ perpendicular?

Find the slope of each line by first finding two points on the line. The points $(0, -3)$ and $(2, 0)$ are on the first line. The slope is

$$m_1 = \frac{0 - (-3)}{2 - 0} = \frac{3}{2}.$$

The second line goes through $(-3, 0)$ and $(0, -2)$ and has slope

$$m_2 = \frac{-2 - 0}{0 - (-3)} = -\frac{2}{3}.$$

Since the product of the slopes of the two lines is $(3/2)(-2/3) = -1$, the lines are perpendicular. ■

5 We have seen how the slope of a line is the ratio of the change in y (vertical change) to the change in x (horizontal change). This idea can be extended to real-life situations as follows: the slope gives the average rate of change of y per unit of change in x, where the value of y is dependent upon the value of x. The next example illustrates this idea of average rate of change. We assume a linear relationship between x and y.

EXAMPLE 8
INTERPRETING SLOPE AS AVERAGE RATE OF CHANGE

An environmental researcher finds that when a certain chemical pollutant is introduced into a large lake, the reproduction of redfish declines. In a given period of time, dumping five tons of the chemical results in a redfish population of 24,000. Also, dumping fifteen tons of the chemical results in a redfish population of 10,000. Let y be the redfish population when x tons of pollutant are introduced into the river, and find the average rate of change of y per unit change in x.

Since 5 tons of chemical yields a population of 24,000, $(x_1, y_1) = (5, 24{,}000)$. Similarly, since 15 tons yields a population of 10,000, $(x_2, y_2) = (15, 10{,}000)$. The average rate of change of y per unit change in x is found by using the slope formula.

$$\begin{array}{l}\text{Average rate of change}\\ \text{of } y \text{ per unit change in } x\end{array} = \frac{y_2 - y_1}{x_2 - x_1} = \frac{10{,}000 - 24{,}000}{15 - 5} = \frac{-14{,}000}{10} = -1400$$

The result, -1400, indicates that there is a decrease of 1400 fish for every ton of pollutant introduced into the river. Geometrically, this would indicate that the line joining the points $(5, 24{,}000)$ and $(15, 10{,}000)$ has slope -1400. ■

7.2 EXERCISES

Graph the line through each pair of points. Find the slope in each case. See Example 1.

1. (2, −3) and (1, 5) **2.** (4, −1) and (−2, −6) **3.** (6, 3) and (5, 4) **4.** (6, −2) and (5, 1)

5. (3, 3) and (−5, −6) **6.** (1, 2) and (−4, 6) **7.** (2, 5) and (−4, 5) **8.** (−4, 2) and (−4, 8)

In Exercises 9–16, tell whether the slope of the line is positive, negative, zero, or undefined. (See Objective 3.)

9.

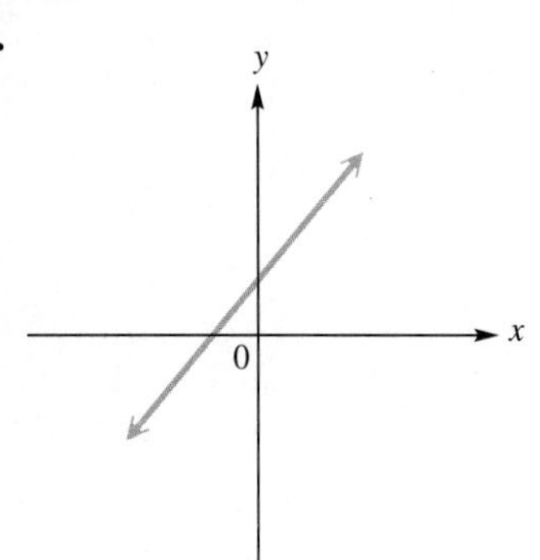

10.

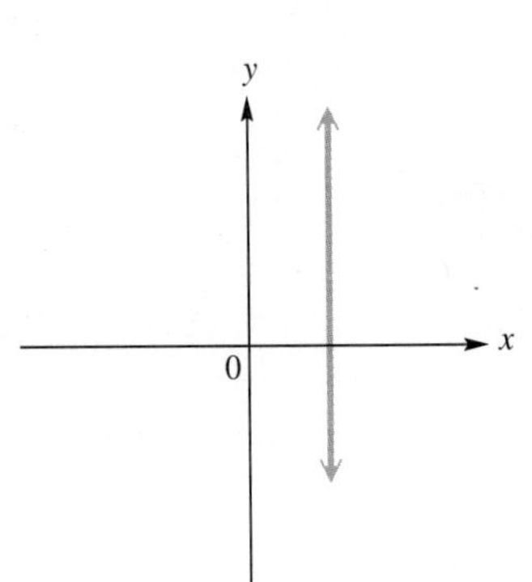

11.

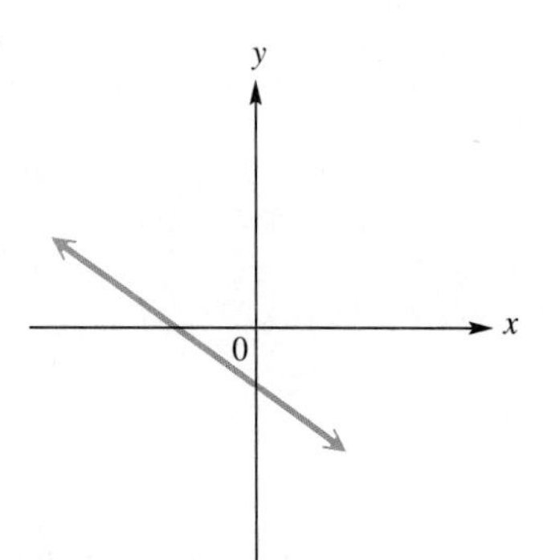

12.

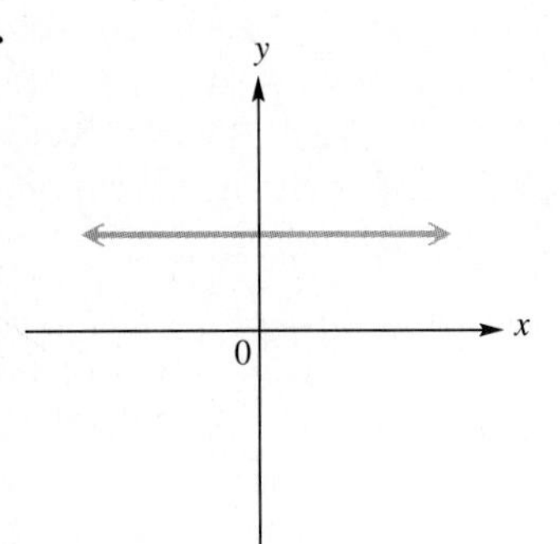

13.

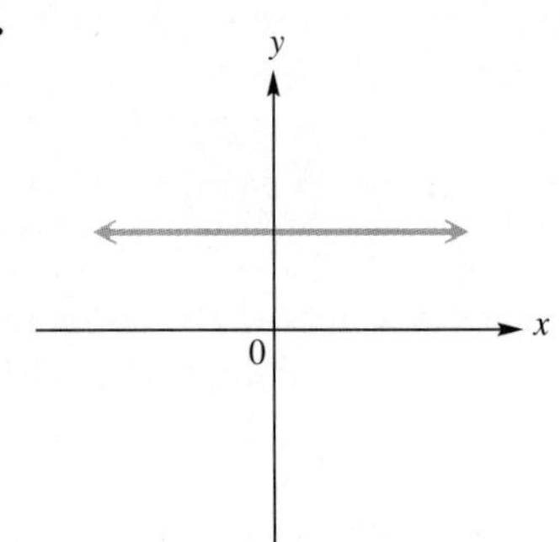

14.

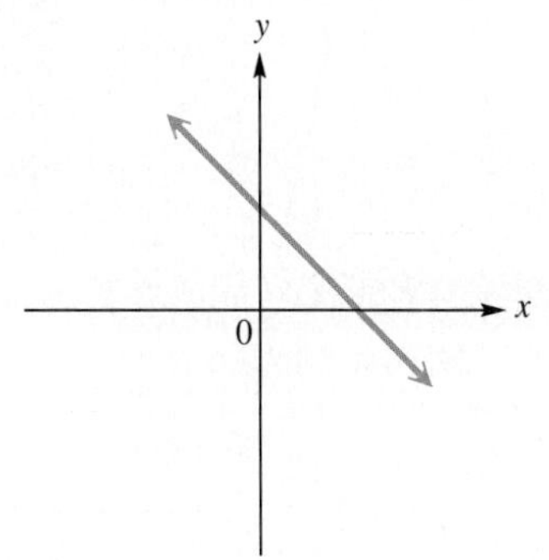

15.

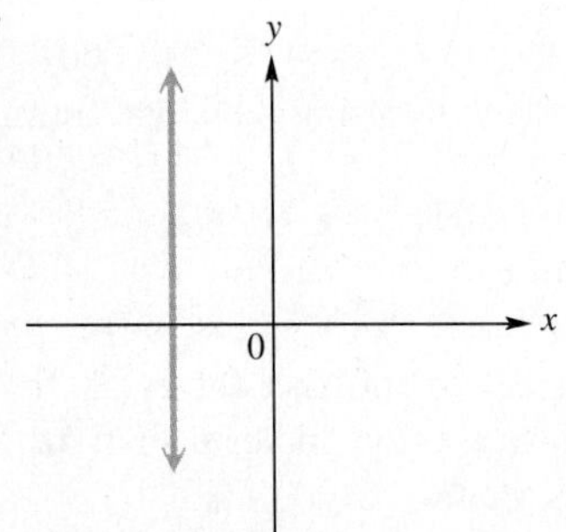

16.

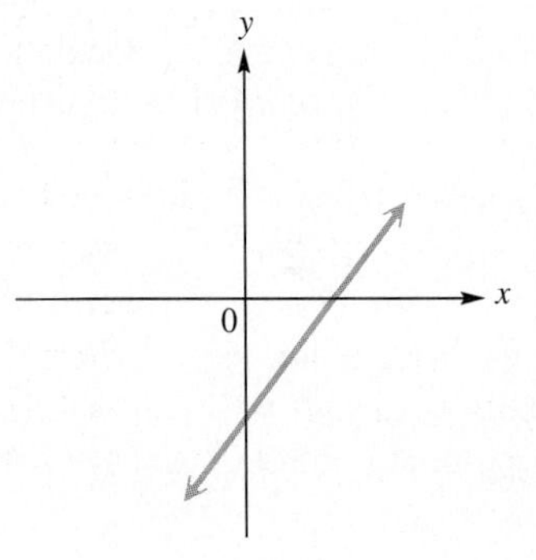

Find the slope of each of the following lines, and sketch the graph. See Examples 1 and 2.

17. $2x + 4y = 5$
18. $3x - 6y = 12$
19. $-x + y = 5$
20. $x + y = 1$
21. $6x - 5y = 30$
22. $4x + 3y = 12$

---

23. Explain the procedure for graphing a straight line using its slope and a point on the line. (See Objective 3.)

24. A vertical line has equation _____ $= k$, for some constant k; a horizontal line has equation _____ $= k$, for some constant k. (See Objective 2.)

Use the method of Examples 4 and 5 in the text to graph each of the following lines.

25. $m = \frac{1}{2}$, through $(-3, 2)$
26. $m = \frac{2}{3}$, through $(0, 1)$
27. $m = -\frac{5}{4}$, through $(-2, -1)$
28. $m = -\frac{3}{2}$, through $(-1, -2)$
29. $m = -2$, through $(-1, -4)$
30. $m = 3$, through $(1, 2)$
31. $m = 0$, through $(2, -5)$
32. undefined slope, through $(-3, 1)$
33. undefined slope, through $(-4, 1)$
34. $m = 0$, through $(5, 3)$

---

35. If a line has slope $-4/9$, then any line parallel to it has slope _____, and any line perpendicular to it has slope _____. (See Objective 4.)

36. What is the slope of any line perpendicular to a line with undefined slope? (See Objective 4.)

Decide which pairs of lines are parallel. See Example 6.

37. $3x = y$ and $2y - 6x = 5$
38. $2x + 5y = -8$ and $6 + 2x = 5y$
39. $4x + y = 0$ and $5x - 8 = 2y$
40. $x = 6$ and $6 - x = 8$
41. The line through $(4, 6)$ and $(-8, 7)$ and the line through $(7, 4)$ and $(-5, 5)$
42. The line through $(9, 15)$ and $(-7, 12)$ and the line through $(-4, 8)$ and $(-20, 5)$

Decide which pairs of lines are perpendicular. See Example 7.

43. $4x - 3y = 8$ and $4y + 3x = 12$
44. $2x = y + 3$ and $2y + x = 3$
45. $4x - 3y = 5$ and $3x - 4y = 2$
46. $5x - y = 7$ and $5x = 3 + y$
47. $2x + y = 1$ and $x - y = 2$
48. $2y - x = 3$ and $y + 2x = 1$

In each of the following problems, use the idea of average rate of change. Assume a linear relationship in each case. See Example 8.

49. At 4:00 A.M. the temperature was 10° C. By 4:00 P.M. it had risen to 21° C. Find the average rate of change of the temperature per hour.

50. Family income in the United States has steadily increased for many years (primarily due to inflation). In 1970 the median family income was about $10,000 a year. In 1985 it was about $25,000 a year. Find the average rate of change of family income over that period.

51. In one state the fine for driving 10 miles per hour over the speed limit is \$35 and the fine for driving 15 miles per hour over the speed limit is \$42. What is the average rate of change in the fine for each mile per hour over the limit?

52. On the third day of the rotation diet, Lynn Elliott weighed 92.5 kilograms. By the eleventh day, she weighed 90.9 kilograms. What was her average rate of weight loss per day?

Solve the following problems using your knowledge of the slopes of parallel and perpendicular lines.

53. Show that $(-13, -9)$, $(-11, -1)$, $(2, -2)$, and $(4, 6)$ are the vertices of a parallelogram. (*Hint:* A parallelogram is a four-sided figure with opposite sides parallel.)

54. Is the figure with vertices at $(-11, -5)$, $(-2, -19)$, $(12, -10)$, and $(3, 4)$ a parallelogram? Is it a rectangle? (A rectangle is a parallelogram with a right angle.)

Slopes can be used to determine if three points A, B, and C are collinear. If the slopes of AB and AC are the same, then the points are collinear. Use this idea to decide whether each group of three points lie on the same line.

55. $(1, 3)$, $(-2, 9)$, $(4, -2)$

56. $(6, -1)$, $(-2, -5)$, $(4, -2)$

57. $(3, 4)$, $(-2, -1)$, $(2, 3)$

58. $(-1, 2)$, $(-3, -1)$, $(5, 2)$

PREVIEW EXERCISES

Solve each equation for y. See Section 2.2.

59. $2x + 3y = 8$

60. $5x - y = 12$

61. $-3x = 4y - 7$

62. $8 - 3y = 6x$

Write each of the following equations without fractions in the form $Ax + By = C$. Combine terms if possible. See Section 2.2.

63. $y - 7 = -\frac{2}{3}(x - 3)$

64. $y - (-2) = \frac{3}{2}(x - 5)$

65. $y - (-1) = \frac{5}{3}[x - (-4)]$

66. $y - 7 = -\frac{1}{4}[x - (-3)]$

7.3 LINEAR EQUATIONS

OBJECTIVES

1 WRITE THE EQUATION OF A LINE GIVEN ITS SLOPE AND A POINT ON THE LINE.

2 WRITE THE EQUATION OF A LINE GIVEN TWO POINTS ON THE LINE.

3 WRITE THE EQUATION OF A LINE GIVEN ITS SLOPE AND y-INTERCEPT.

4 FIND THE SLOPE AND y-INTERCEPT OF A LINE GIVEN ITS EQUATION.

5 WRITE THE EQUATION OF A LINE PARALLEL OR PERPENDICULAR TO A GIVEN LINE.

6 FIND AN EQUATION RELATING TWO UNKNOWNS IN A LINEAR RELATIONSHIP.

FOCUS ON PROBLEM SOLVING

Erin is a biology student. She has heard that the number of times a cricket chirps in one minute can be used to find the temperature. In an experiment, she finds that a cricket chirps 40 times per minute when the temperature is 50° F, and 80 times per minute when the temperature is 60° F. What is the temperature when the cricket chirps 20 times per minute?

Many real-world situations can be described by straight-line graphs. By finding an equation that relates chirps to temperature, we can solve the problem stated above. It is Exercise 75 in the exercises for this section. After working through this section, you should be able to solve this problem.

1 If the slope of a line and a particular point on the line are known, it is possible to find an equation of the line. Suppose that the slope of a line is m, (x_1, y_1) is a particular point on the line, and (x, y) is any other point on the line. Then by the definition of slope,

$$m = \frac{y - y_1}{x - x_1}.$$

Multiplying both sides by $x - x_1$ gives the following result, called the *point-slope form* of the equation of the line.

POINT-SLOPE FORM

The equation of the line through (x_1, y_1) with slope m is written in **point-slope form** as

$$y - y_1 = m(x - x_1).$$

EXAMPLE 1 USING THE POINT-SLOPE FORM

Find the equation of the line with slope 1/3, going through the point $(-2, 5)$.

Use the point-slope form of the equation of a line, with $(x_1, y_1) = (-2, 5)$ and $m = 1/3$.

$$y - y_1 = m(x - x_1)$$

$$y - 5 = \frac{1}{3}[(x - (-2)] \qquad \text{Let } y_1 = 5,\ m = \frac{1}{3},\ x_1 = -2.$$

$$y - 5 = \frac{1}{3}(x + 2)$$

$$3y - 15 = x + 2 \qquad \text{Multiply by 3.}$$

or $$x - 3y = -17 \quad \blacksquare$$

It is convenient to have an agreement on the form in which a linear equation should be written. In Section 7.1, we defined *standard form* for a linear equation as

$$Ax + By = C.$$

In addition, from now on, let us agree that A, B and C will be integers, with $A \geq 0$. For example, the final equation found in Example 1, $x - 3y = -17$, is written in standard form.

CAUTION The definition of "standard form" is not standard from one text to another. Any linear equation can be written in many different (all equally correct) forms. For example, the equation $2x + 3y = 8$ can be written as $2x = 8 - 3y$, $3y = 8 - 2x$, $x + (3/2)y = 4$, $4x + 6y = 16$, and so on. In addition to writing it in the form $Ax + By = C$ (with $A \geq 0$), let us agree that the form $2x + 3y = 8$ is preferred over any multiples of both sides, such as $4x + 6y = 16$.

2 If two points on a line are known, it is possible to find an equation of the line. First, find the slope using the slope formula, and then use the slope with one of the given points in the point-slope form. This is illustrated in the next example.

EXAMPLE 2 **FINDING AN EQUATION OF A LINE WHEN TWO POINTS ARE KNOWN**

Find an equation of the line through the points $(-4, 3)$ and $(5, -7)$.

First find the slope, using the definition.

$$m = \frac{-7 - 3}{5 - (-4)} = -\frac{10}{9}$$

Use either $(-4, 3)$ or $(5, -7)$ as (x_1, y_1) in the point-slope form of the equation of a line. If $(-4, 3)$ is used, then $-4 = x_1$ and $3 = y_1$.

$$y - y_1 = m(x - x_1)$$

$$y - 3 = -\frac{10}{9}[x - (-4)] \qquad \text{Let } y_1 = 3,\ m = -\frac{10}{9},\ x_1 = -4.$$

$$y - 3 = -\frac{10}{9}(x + 4)$$

$$9(y - 3) = -10(x + 4) \qquad \text{Multiply by 9.}$$

$$9y - 27 = -10x - 40 \qquad \text{Distributive property}$$

$$10x + 9y = -13$$

On the other hand, if $(5, -7)$ were used, the equation becomes

$$y - (-7) = -\frac{10}{9}(x - 5)$$

$$9y + 63 = -10x + 50$$

$$10x + 9y = -13,$$

the same equation. In summary, the line through $(-4, 3)$ and $(5, -7)$ has equation $10x + 9y = -13$. ■

In the previous section, we saw that vertical and horizontal lines have special equations. We can analyze this further. Notice that the point-slope form does not apply to a vertical line, since the slope of a vertical line is undefined. A vertical line through the point (k, y), where k is a constant and y represents any real number, has equation $x = k$.

A horizontal line has slope 0. From the point-slope form, the equation of a horizontal line through the point (x, k), where x is any real number and k is a constant, is

$$\begin{aligned} y - y_1 &= m(x - x_1) \\ y - k &= 0(x - x) && y_1 = k,\ x_1 = x \\ y - k &= 0 \\ y &= k. \end{aligned}$$

In summary, horizontal and vertical lines have the following special equations.

EQUATIONS OF VERTICAL AND HORIZONTAL LINES

If x and y are any real numbers and k is a constant, the vertical line through (k, y) has equation $x = k$, and the horizontal line through (x, k) has equation $y = k$.

3 Suppose that the slope m of a line is known, and the y-intercept of the line is $(0, b)$. Then substituting into the point-slope form gives

$$\begin{aligned} y - y_1 &= m(x - x_1) \\ y - b &= m(x - 0) && x_1 = 0,\ y_1 = b \\ y - b &= mx \\ y &= mx + b. && \text{Add } b \text{ to both sides.} \end{aligned}$$

This last result is known as the *slope-intercept form* of the equation of the line.

SLOPE-INTERCEPT FORM

The equation of a line with slope m and y-intercept $(0, b)$ is written in **slope-intercept form** as

$$y = \underset{\substack{\uparrow \\ \text{Slope}}}{m}x + \underset{\substack{\uparrow \\ y\text{-intercept is } (0,\, b)}}{b}$$

EXAMPLE 3 USING THE SLOPE-INTERCEPT FORM

Find the equation of the line with slope $-4/5$ and y-intercept $(0, -2)$.

Here $m = -4/5$ and $b = -2$. Substitute these values into the slope-intercept form.

$$y = mx + b$$

$$y = -\frac{4}{5}x - 2$$

Clear of fractions by multiplying both sides by 5:

$$5y = -4x - 10,$$

or

$$4x + 5y = -10. \quad ■$$

4 In the previous section, the slope of the line $4x - y = 8$ was found to be 4 by first getting two points on the line and then using the definition of slope. For an alternative method of finding the slope of this line, transform $4x - y = 8$ into the form $y = mx + b$.

$$4x - y = 8$$
$$-y = -4x + 8 \quad \text{Subtract } 4x.$$
$$y = 4x - 8 \quad \text{Multiply by } -1.$$

When the equation is solved for y, the coefficient of x is the slope. Writing the equation in this form also tells us the y-intercept. In this case, it is $(0, -8)$.

EXAMPLE 4
FINDING THE SLOPE AND y-INTERCEPT FROM THE EQUATION

Write $3y + 2x = 9$ in slope-intercept form, and find the slope and y-intercept.

Put the equation in slope-intercept form by solving for y.

$$3y + 2x = 9$$
$$3y = -2x + 9$$
$$y = -\frac{2}{3}x + 3$$

From the slope-intercept form, the slope is $-2/3$ and the y-intercept is $(0, 3)$. ■

NOTE The importance of the slope-intercept form of a linear equation cannot be overemphasized. First, every linear equation (of a non-vertical line) has a *unique* (one and only one) slope-intercept form. Second, in Section 7.6 we will study linear *functions,* where the slope-intercept form is necessary in specifying such functions.

5 As mentioned in the last section, parallel lines have the same slope and perpendicular lines have slopes with a product of -1. These results are used in the next example.

EXAMPLE 5
FINDING EQUATIONS OF PARALLEL OR PERPENDICULAR LINES

Find an equation of the line passing through the point $(-4, 5)$, and **(a)** parallel to the line $2x + 3y = 6$; **(b)** perpendicular to the line $2x + 3y = 6$.

(a) The slope of the graph of $2x + 3y = 6$ can be found by solving for y.

$$2x + 3y = 6$$
$$3y = -2x + 6 \quad \text{Subtract } 2x \text{ on both sides.}$$
$$y = -\frac{2}{3}x + 2 \quad \text{Divide both sides by 3.}$$

The slope is given by the coefficient of x, so $m = -2/3$.

Slope ↓

$$y = -\frac{2}{3}x + 2$$

This means that the required equation of the line through $(-4, 5)$ and parallel to $2x + 3y = 6$ has slope $-2/3$. Now use the point-slope form, with $(x_1, y_1) = (-4, 5)$ and $m = -2/3$.

$$y - 5 = -\frac{2}{3}[x - (-4)] \qquad y_1 = 5,\ m = -\frac{2}{3},\ x_1 = -4$$

$$y - 5 = -\frac{2}{3}(x + 4)$$

$$3(y - 5) = -2(x + 4) \qquad \text{Multiply by 3.}$$

$$3y - 15 = -2x - 8 \qquad \text{Distributive property}$$

$$2x + 3y = 7 \qquad \text{Standard form}$$

(b) To be perpendicular to the line $2x + 3y = 6$, a line must have a slope that is the negative reciprocal of $-2/3$, which is $3/2$. Use the point $(-4, 5)$ and slope $3/2$ in the point-slope form of the equation.

$$y - 5 = \frac{3}{2}[x - (-4)] \qquad y_1 = 5,\ m = \frac{3}{2},\ x_1 = -4$$

$$2y - 10 = 3[x + 4] \qquad \text{Multiply by 2.}$$

$$2y - 10 = 3x + 12 \qquad \text{Distributive property}$$

$$-3x + 2y = 22$$

$$3x - 2y = -22 \qquad \text{Standard form}$$ ■

6 The final example shows how the familiar formula relating Celsius and Fahrenheit temperatures is obtained.

EXAMPLE 6

FINDING THE FORMULA RELATING CELSIUS AND FAHRENHEIT TEMPERATURES

There is a linear relationship between Celsius and Fahrenheit temperatures. It is common knowledge that when $C = 0°$, $F = 32°$, and when $C = 100°$, $F = 212°$.

(a) Use this information to solve for F in terms of C.

Think of ordered pairs of temperatures (C, F), where C and F represent corresponding Celsius and Fahrenheit temperatures. The equation that relates the two scales has a straight-line graph that contains the points $(0, 32)$ and $(100, 212)$. The slope of this line can be found by using the slope formula.

$$m = \frac{212 - 32}{100 - 0} = \frac{180}{100} = \frac{9}{5}$$

Now, think of the point-slope form of the equation in terms of C and F, where C replaces x and F replaces y. Use $m = 9/5$, and $(C_1, F_1) = (0, 32)$.

$$F - F_1 = m(C - C_1)$$

$$F - 32 = \frac{9}{5}(C - 0) \qquad F_1 = 32,\ m = \frac{9}{5},\ C_1 = 0$$

$$F - 32 = \frac{9}{5}C$$

$$F = \frac{9}{5}C + 32 \qquad \text{Solve for } F.$$

This last result gives F in terms of C.

(b) Find the Fahrenheit temperature when the Celsius temperature is 50°.

Let $C = 50$ in the formula found in part (a) to find the corresponding Fahrenheit temperature.

$$F = \frac{9}{5}C + 32$$

$$F = \frac{9}{5}(50) + 32 \qquad \text{Let } C = 50.$$

$$F = 90 + 32$$

$$F = 112$$

When the Celsius temperature is 50°, the Fahrenheit temperature is 112°. ■

NOTE The formula found in part (a) of Example 6 can be solved for C in terms of F to get the alternate form, $C = \frac{5}{9}(F - 32)$.

A summary of the various forms of linear equations follows.

SUMMARY OF FORMS OF LINEAR EQUATIONS

$Ax + By = C$	**Standard form** (Neither A nor B is 0.) Slope is $-\frac{A}{B}$. x-intercept is $\left(\frac{C}{A}, 0\right)$. y-intercept is $\left(0, \frac{C}{B}\right)$.
$x = k$	**Vertical line** Undefined slope x-intercept is $(k, 0)$.
$y = k$	**Horizontal line** Slope is 0. y-intercept is $(0, k)$.
$y = mx + b$	**Slope-intercept form** Slope is m. y-intercept is $(0, b)$.
$y - y_1 = m(x - x_1)$	**Point-slope form** Slope is m. Line passes through (x_1, y_1).

7.3 EXERCISES

Find equations of the lines satisfying the following conditions. Write the equations in standard form. See Example 1.

1. $m = -\frac{3}{4}$, through $(-2, 5)$

2. $m = -\frac{5}{6}$, through $(4, -3)$

3. $m = -2$, through $(1, 5)$

4. $m = 1$, through $(-2, 3)$

5. $m = \frac{1}{2}$, through $(7, 4)$

6. $m = \frac{1}{4}$, through $(1, -2)$

7. $m = 0$, through $(-3, 2)$

8. $m = 0$, through $(1, 5)$

9. $m = 4$, x-intercept $(3, 0)$

10. $m = -5$, x-intercept $(-2, 0)$

---

11. Explain why the point-slope form of an equation cannot be used to find the equation of a vertical line. (See Objective 2.)

12. Which one of the following equations is in standard form, according to the definition of standard form given in this text? (See Objective 1.)
(a) $3x + 2y - 6 = 0$
(b) $y = 5x - 12$
(c) $2y = 3x + 4$
(d) $6x - 5y = 12$

Find equations for the following lines. (Hint: What kind of line has undefined slope?)

13. Undefined slope, through $(2, 8)$

14. Undefined slope, through $(-4, 1)$

15. Vertical, through $(-7, 1)$

16. Vertical, through $(3, -9)$

Find equations of the lines passing through the following pairs of points. Write the equations in standard form. See Example 2.

17. $(3, 4)$ and $(2, 6)$

18. $(5, -2)$ and $(-3, 1)$

19. $(6, 1)$ and $(-2, 5)$

20. $(4, -2)$ and $(1, 3)$

21. $(1, 1)$ and $(0, -4)$

22. $(3, -4)$ and $(-2, 2)$

23. $\left(-\frac{2}{5}, \frac{2}{5}\right)$ and $\left(\frac{4}{3}, \frac{2}{3}\right)$

24. $\left(\frac{3}{4}, \frac{8}{3}\right)$ and $\left(\frac{2}{5}, \frac{2}{3}\right)$

25. $(2, 5)$ and $(1, 5)$

26. $(-2, 2)$ and $(4, 2)$

27. $(1, \sqrt{5})$ and $(3, 2\sqrt{5})$

28. $(-4, \sqrt{2})$ and $(5, -\sqrt{2})$

Find equations of the lines satisfying the following conditions. Write the equations in slope-intercept form. See Example 3.

29. $m = 5$, $b = 4$

30. $m = -2$, $b = 1$

31. $m = -\frac{2}{3}$, $b = \frac{1}{2}$

32. $m = -\frac{5}{8}$, $b = \frac{1}{4}$

33. Slope $\frac{2}{5}$, y-intercept $(0, -1)$

34. Slope $-\frac{3}{4}$, y-intercept $(0, 2)$

35. Slope 0, y-intercept $(0, 4)$

36. Slope 0, y-intercept $(0, -3)$

Write the following equations in slope-intercept form, and then find the slope of the line and the y-intercept. See Example 4.

37. $x + y = 8$ **38.** $x - y = 2$ **39.** $5x + 2y = 10$ **40.** $6x - 5y = 18$

41. $2x - 3y = 5$ **42.** $4x + 3y = 10$ **43.** $-5x - 3y = 4$ **44.** $-2x - 7y = 15$

45. $8x + 11y = 9$ **46.** $4x + 13y = 19$

In Exercises 47–54, choose the one of the four graphs given here which most closely resembles the graph of the given equation. Each equation is given in the form $y = mx + b$, so consider the signs of m and b in making your choice.

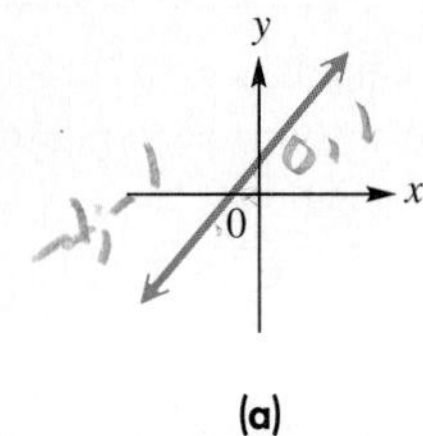

(a) (b)

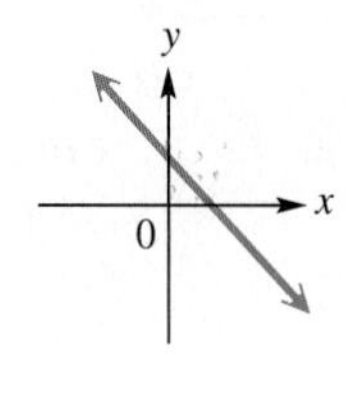

(c)

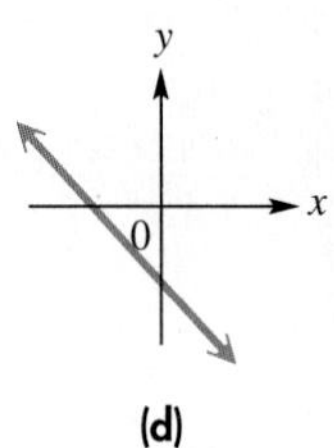

(d)

47. $y = 3x + 6$ **48.** $y = 4x + 5$ **49.** $y = -3x + 6$ **50.** $y = -4x + 5$

51. $y = 3x - 6$ **52.** $y = 4x - 5$ **53.** $y = -3x - 6$ **54.** $y = -4x - 5$

In Exercises 55–62, choose the one of the four graphs given here which most closely resembles the graph of the given equation. (See Objective 2.)

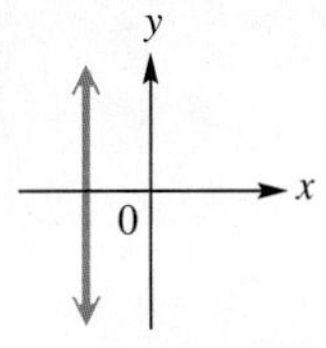

(a)

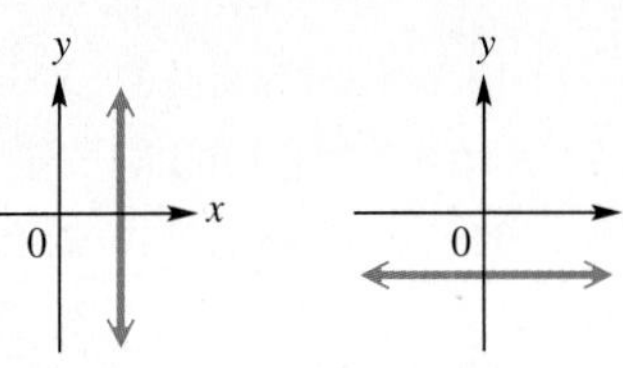

(b)

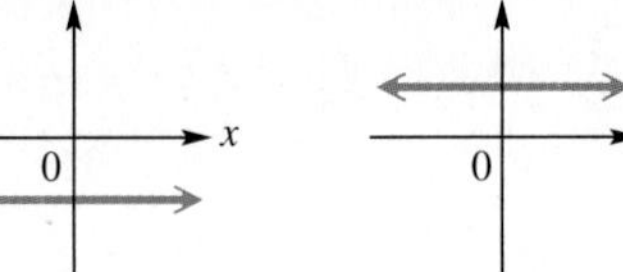

(c) (d)

55. $y + 2 = 0$ **56.** $y + 4 = 0$ **57.** $x + 3 = 0$ **58.** $x + 7 = 0$

59. $y - 2 = 0$ **60.** $y - 4 = 0$ **61.** $x - 3 = 0$ **62.** $x - 7 = 0$

Find equations of the lines satisfying the following conditions. Write the equations in standard form. See Example 5.

63. Parallel to $3x - y = 8$ and through $(-7, 3)$

64. Parallel to $2x + 5y = 10$ and through $(4, 7)$

65. Parallel to $-x + 2y = 3$ and through $(-2, -2)$

66. Through $(-1, 3)$ and perpendicular to $3x + 2y = 6$

67. Through (8, 5) and perpendicular to $2x - y = 4$

68. Through (2, -7) and perpendicular to $5x + 2y = 7$

69. Parallel to $y = 4$ and through (-2, 7)

70. Parallel to $x - 2 = 0$ and through (8, 4)

Many real-world situations can be described approximately by a straight-line graph. One way to find the equation of such a line is to use two typical data points from the graph and the point-slope form of the equation of a line. Assume that these problems have straight-line graphs. See Example 6.

71. A company finds that it can make a total of 20 generators for \$13,900, and that 10 generators cost \$7500.
- **(a)** Write an equation that gives the total cost y to produce x generators.
- **(b)** Predict the cost to produce 12 generators.

72. The sales of a small company were \$27,000 in its second year of business and \$63,000 in its fifth year.
- **(a)** Write an equation giving the sales y in year x.
- **(b)** Estimate the sales in the fourth year.

73. A weekly magazine had 28 pages of advertising one week that produced revenue of \$9700. Another week, \$18,500 was produced by 34 pages of advertising.
- **(a)** Write an equation for revenue y from x pages of advertising.
- **(b)** What amount of revenue would be produced by 30 pages of advertising?

74. The owner of a variety store found that in 1985, year 0, his profits were \$28,000, while in 1992, year 7, profits had increased to \$42,000.
- **(a)** Write an equation expressing the profit y in terms of the year x.
- **(b)** Estimate profits in 1993.

75. Erin is a biology student. She has heard that the number of times a cricket chirps in one minute can be used to find the temperature. In an experiment, she finds that a cricket chirps 40 times per minute when the temperature is 50° F, and 80 times per minute when the temperature is 60° F. What is the temperature when the cricket chirps 20 times per minute?

76. In Exercise 75, when will the cricket stop chirping?

PREVIEW EXERCISES

Solve each inequality. See Section 2.5.

77. $2x + 5 < 6$ **78.** $-x + 10 > 8$ **79.** $6 - 3x \leq 12$

80. $4x - 2 \geq 8$ **81.** $-5x - 5 > 3$ **82.** $12 - 8x < 12$

7.4 LINEAR INEQUALITIES

OBJECTIVES

1 GRAPH LINEAR INEQUALITIES.

2 GRAPH THE INTERSECTION OF TWO LINEAR INEQUALITIES.

3 GRAPH THE UNION OF TWO LINEAR INEQUALITIES.

4 GRAPH FIRST-DEGREE ABSOLUTE VALUE INEQUALITIES.

1 Linear inequalities with one variable were graphed on the number line in Chapter 2. In this section linear inequalities in two variables are graphed on a rectangular coordinate system.

LINEAR INEQUALITY

An inequality that can be written as

$$Ax + By < C \qquad \text{or} \qquad Ax + By > C,$$

where A, B, and C are real numbers and A and B are not both 0, is a **linear inequality in two variables.**

Also, $\leq$ and $\geq$ may replace $<$ and $>$ in the definition.

A line divides the plane into three regions: the line itself and the two half-planes on either side of the line. Recall that the graphs of linear inequalities in one variable are intervals on the number line that sometimes include an end-point. The graphs of linear inequalities in two variables are *regions* in the real number plane and may include a *boundary line*. The **boundary line** for the inequality $Ax + By < C$ or $Ax + By > C$ is the graph of the *equation* $Ax + By = C$. To graph a linear inequality, we go through the following steps.

GRAPHING A LINEAR INEQUALITY

Step 1 **Draw the boundary.**
Draw the graph of the straight line that is the boundary. Make the line solid if the inequality involves $\leq$ or $\geq$; make the line dashed if the inequality involves $<$ or $>$.

Step 2 **Choose a test point.**
Choose any point not on the line as a test point.

Step 3 **Shade the appropriate region.**
Shade the region that includes the test point if it satisfies the original inequality; otherwise, shade the region on the other side of the boundary line.

EXAMPLE 1 GRAPHING A LINEAR INEQUALITY

Graph $3x + 2y \geq 6$.

First graph the straight line $3x + 2y = 6$. The graph of this line, the boundary of the graph of the inequality, is shown in Figure 7.15. The graph of the inequality $3x + 2y \geq 6$ includes the points of the line $3x + 2y = 6$, and either the points *above* the line $3x + 2y = 6$ or the points *below* that line. To decide which, select any point not on the line $3x + 2y = 6$ as a test point. The origin, (0, 0), is often a good choice. Substitute the values from the test point (0, 0) for x and y in the inequality $3x + 2y > 6$.

$$3(0) + 2(0) > 6 \quad ?$$
$$0 > 6 \quad \text{False}$$

Since the result is false, (0, 0) does not satisfy the inequality, and so the solution includes all points on the other side of the line. This region is shaded in Figure 7.15. ■

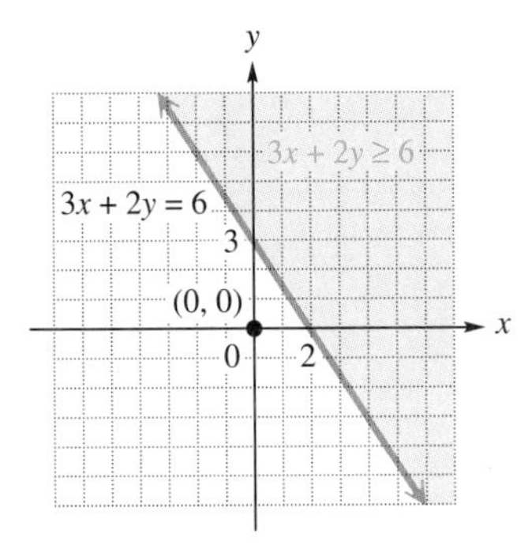

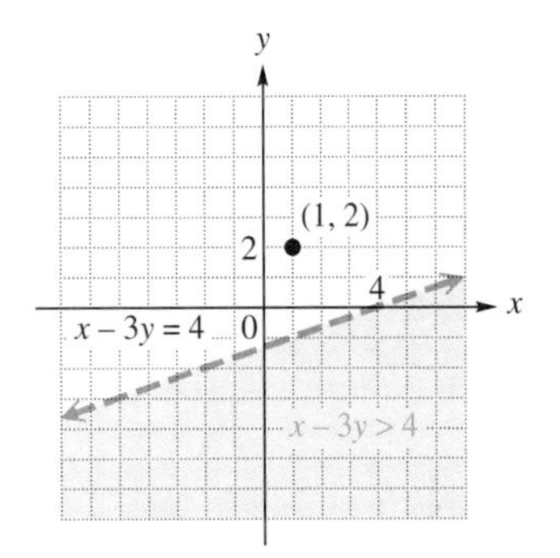

FIGURE 7.16

EXAMPLE 2 GRAPHING A LINEAR INEQUALITY

Graph $x - 3y > 4$.

First graph the boundary line, $x - 3y = 4$. The graph is shown in Figure 7.16. The points of the boundary line do not belong to the inequality $x - 3y > 4$ (since the inequality symbol is $>$ and not $\geq$). For this reason, the line is dashed. To decide which side of the line is the graph of the solution, choose any point that is not on the line, say (1, 2). Substitute 1 for x and 2 for y in the original inequality.

$$1 - 3(2) > 4 \quad ?$$
$$-5 > 4 \quad \text{False}$$

Because of this false result, the solution lies on the side of the boundary line that does *not* contain the test point (1, 2). The solution, graphed in Figure 7.16, includes only those points in the shaded region (not those on the line). ■

2 In Section 2.6 we discussed how the words "and" and "or" are used with compound inequalities. In that section, the inequalities were in a single variable. Those ideas can be extended to include inequalities in two variables. If a pair of inequalities is joined with the word "and," it is interpreted as the intersection of the solutions of the inequalities. The graph of the intersection of two or more inequalities is the region of the plane where all points satisfy all of the inequalities at the same time.

EXAMPLE 3
GRAPHING THE INTERSECTION OF TWO INEQUALITIES

Graph:

$$2x + 4y \geq 5 \qquad \text{and} \qquad x \geq 1.$$

To begin, we graph each of the two inequalities $2x + 4y \geq 5$ and $x \geq 1$ separately. The graph of $2x + 4y \geq 5$ is shown in Figure 7.17(a), and the graph of $x \geq 1$ is shown in Figure 7.17(b).

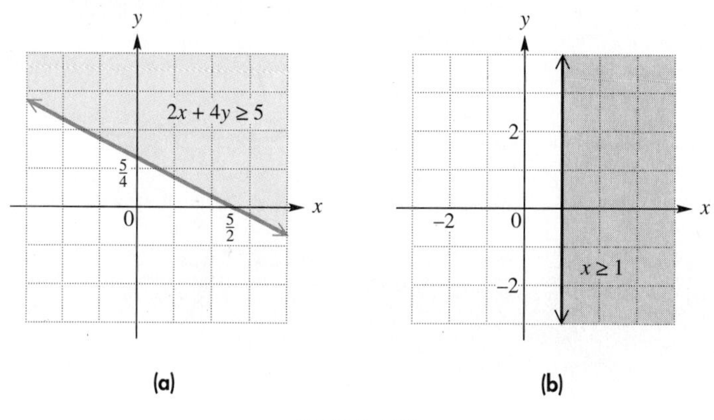

FIGURE 7.17

In practice, the two graphs in Figure 7.17 are graphed on the same axes. Then use heavy shading to identify the intersection of the graphs, as shown in Figure 7.18. ■

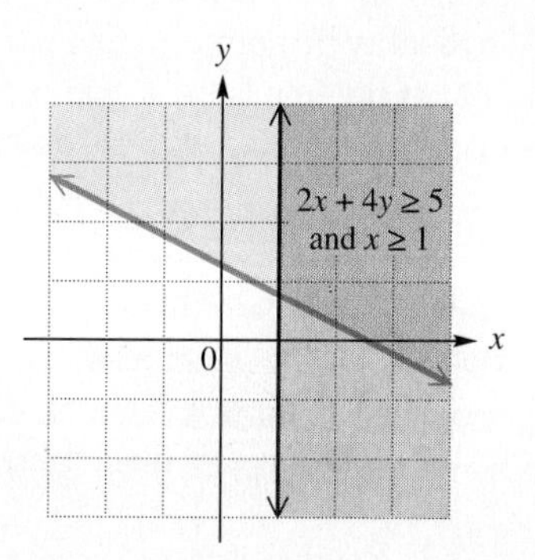

FIGURE 7.18

3 When two inequalities are joined by the word "or," we must find the union of the graphs of the inequalities. The graph of the union of two inequalities includes all of the points that satisfy either inequality.

EXAMPLE 4 GRAPHING THE UNION OF TWO INEQUALITIES

Graph:

$$2x + 4y \geq 5 \qquad \text{or} \qquad x \geq 1.$$

The graphs of $2x + 4y \geq 5$ and $x \geq 1$ are shown in Figures 7.17(a) and 7.17(b). The graph of the union includes all points that belong to either inequality. The graph of the union is the shaded region in Figure 7.19. ■

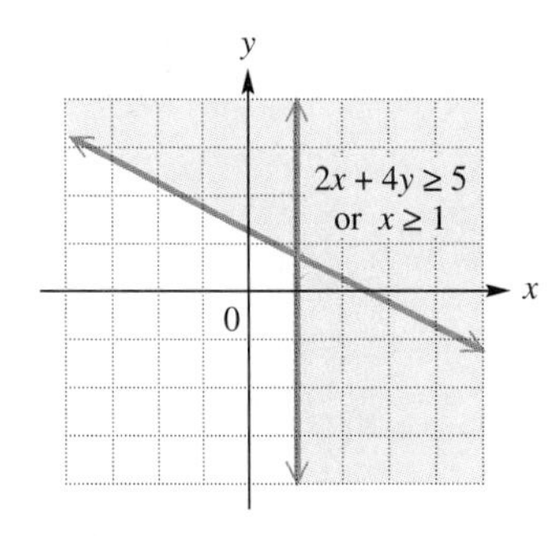

FIGURE 7.19

4 An absolute value inequality should first be written as an intersection or a union as shown in Chapter 2. Then the methods given above can be used to graph the inequality.

EXAMPLE 5 GRAPHING AN ABSOLUTE VALUE INEQUALITY

Graph $|x| \leq 4$.

Recall from Section 2.7 that $|x| \leq 4$ is equivalent to $-4 \leq x \leq 4$. The boundary lines are the vertical lines $x = -4$ and $x = 4$. The graph of the solution is the shaded region between these lines as shown in Figure 7.20. ■

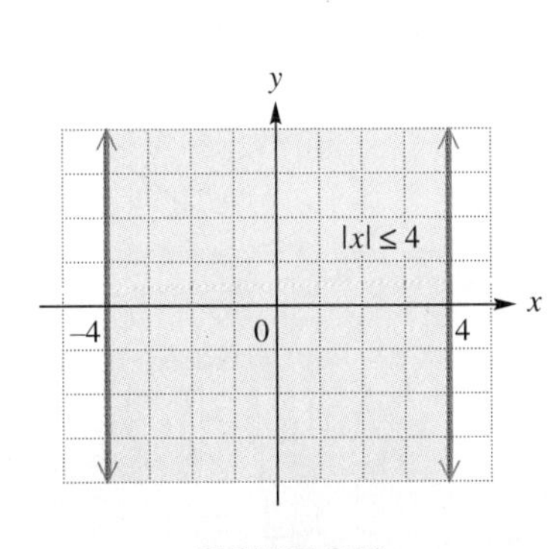

FIGURE 7.20

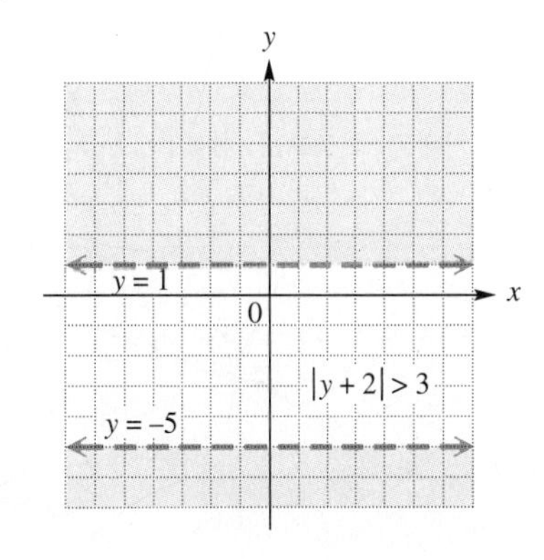

FIGURE 7.21

EXAMPLE 6 GRAPHING AN ABSOLUTE VALUE INEQUALITY

Graph $|y + 2| > 3$.

The equation of the boundary, $|y + 2| = 3$, can be rewritten as

$$y + 2 = 3 \qquad \text{or} \qquad y + 2 = -3$$
$$y = 1 \qquad \text{or} \qquad y = -5.$$

From this, $y = 1$ and $y = -5$ are boundary lines. Checking points from each of the three regions determined by the horizontal boundary lines gives the graph shown in Figure 7.21. ■

7.4 EXERCISES

Graph each linear inequality. See Examples 1 and 2.

1. $x + y \le 2$
2. $-x \le 3 - y$
3. $4x - y \le 5$
4. $3x + y \ge 6$
5. $x + 3y \ge -2$
6. $4x + 6y \le -3$
7. $x + 2y \le -5$
8. $2x - 4y \le 3$
9. $4x - 3y < 12$
10. $5x + 3y > 15$
11. $x + y > 0$
12. $y < x$
13. $x < 1$
14. $y \ge -3$

---

15. Explain how to determine whether a dashed line or a solid line is used for the boundary when graphing a linear inequality. (See Objective 1.)

16. Explain the different uses of the words "and" and "or" as they pertain to graphing inequalities. (See Objectives 2 and 3.)

Graph each of the following. See Example 3.

17. $x + y \le 1$ and $x \ge 0$
18. $3x - 4y \le 6$ and $y \ge 1$
19. $2x - y \ge 1$ and $3x + 2y \ge 6$
20. $x + 3y \ge 6$ and $3x - 4y \le 12$
21. $-x - y < 5$ and $x - y \le 3$
22. $6x - 4y < 8$ and $x + 2y \ge 4$

Graph each of the following. See Example 4.

23. $2x - y \le 4$ or $x + 2 \ge 3$
24. $x + y \ge -2$ or $y \le 5$
25. $5x - 2y + 10 \le 0$ or $x \le 2 - y$
26. $x > -5 + y$ or $3x + 4y > 12$
27. $2x + 3y > 6$ or $y < 8 + x$
28. $4x - y < 8$ or $2y > -4 + x$

---

29. Write $|x| > 1$ as an equivalent "and" or "or" statement. (See Objective 4.)

30. Write $|x| < 1$ as an equivalent "and" or "or" statement. (See Objective 4.)

Graph each of the following. See Examples 5 and 6.

31. $|x| > 1$
32. $|x| < 3$
33. $|y| < 5$
34. $|y| > 4$
35. $|x| < 4$
36. $|y| > 1$
37. $|x + 1| < 2$
38. $|y - 2| > 3$
39. $|y - 3| < 2$
40. $|y + 3| > 5$

Graph each of the following.

41. $|y| \ge x$ and $x \ge 0$
42. $|x| \ge y$ and $y \ge 0$
43. $|x + 2| > y$ and $x + y \ge 4$
44. $|y - 1| < x$ and $2x - y < 6$

Graph each of the following.

45. $|y| \le x$ or $|y| \le 2$
46. $|x| \ge y$ or $|x| \ge 2$
47. $|x - 3| \le 4$ or $|y - 2| \le 2$
48. $|x + 1| \ge 3$ or $|y - 5| \le 1$

PREVIEW EXERCISES

Find k, given that $y = 1$ and $x = 3$. See Section 2.1.

49. $y = kx$
50. $y = kx^2$
51. $y = \frac{k}{x}$
52. $y = \frac{k}{x^2}$

7.5 VARIATION

OBJECTIVES

1 WRITE AN EQUATION EXPRESSING DIRECT VARIATION.

2 FIND THE CONSTANT OF VARIATION AND SOLVE DIRECT VARIATION PROBLEMS.

3 SOLVE INVERSE VARIATION PROBLEMS.

4 SOLVE JOINT VARIATION PROBLEMS.

5 SOLVE COMBINED VARIATION PROBLEMS.

FOCUS ON PROBLEM SOLVING

The period of a pendulum varies directly as the square root of the length of the pendulum and inversely as the square root of the acceleration due to gravity. Find the period when the length is 4 feet and the acceleration due to gravity is 32 feet per second per second, if the period is 1.06π seconds when the length is 9 feet and the acceleration due to gravity is 32 feet per second per second.

The concept of variation is introduced in this section. Terms such as *varies directly* and *varies inversely* must be understood before variation problems can be solved. The problem above is Exercise 35 in the exercises for this section. After working through this section, you should be able to solve this problem.

1 The circumference of a circle is given by the formula $C = 2\pi r$, where r is the radius of the circle. As the formula shows, the circumference is always a constant multiple of the radius (C is always found by multiplying r by the constant 2π). Because of this, the circumference is said to *vary directly* as the radius.

DIRECT VARIATION

y **varies directly as** x if there exists some constant k such that

$$y = kx.$$

Also, y is said to be **proportional** to x. The number k is called the **constant of variation.** In direct variation, for $k > 0$, as the value of x increases, the value of y also increases. Similarly, as x decreases, y also decreases.

Consider the following example of direct variation. If Tom earns $8 per hour, his wages vary directly as the number of hours he works. If y represents his total wages and x the number of hours he has worked, then

$$y = 8x.$$

Here k, the constant of variation, is 8.

2 The following example shows how to find the value of the constant k.

EXAMPLE 1
FINDING THE CONSTANT OF VARIATION

Suppose that y varies directly as z, and $y = 50$ when $z = 100$. Find k and the equation connecting y and z.

Since y varies directly as z,

$$y = kz$$

for some constant k. Use the fact that $y = 50$ when $z = 100$ to find k. Substitute these values into the equation $y = kz$.

$$50 = k \cdot 100 \qquad \text{Let } y = 50 \text{ and } z = 100.$$

Now solve for k.

$$k = \frac{50}{100} = \frac{1}{2}$$

The variables y and z are related by the equation

$$y = \frac{1}{2}z. \quad ■$$

EXAMPLE 2
SOLVING A DIRECT VARIATION PROBLEM

Hooke's law for an elastic spring states that the distance a spring stretches is proportional to the force applied. If a force of 150 newtons* stretches a certain spring 8 centimeters, how much will a force of 400 newtons stretch the spring?

If d is the distance the spring stretches and f is the force applied, then $d = kf$ for some constant k. Since a force of 150 newtons stretches the spring 8 centimeters,

$$d = kf \qquad \text{Formula}$$
$$8 = k \cdot 150 \qquad d = 8, f = 150$$
$$k = \frac{8}{150} = \frac{4}{75}, \qquad \text{Find } k.$$

and $d = \frac{4}{75}f$.

For a force of 400 newtons,

$$d = \frac{4}{75}(400) \qquad \text{Let } f = 400.$$
$$= \frac{64}{3}.$$

The spring will stretch 64/3 centimeters if a force of 400 newtons is applied. ■

The direct variation equation $y = kx$ is a linear equation. However, other kinds of variation involve polynomial equations of higher degree. That is, one variable can be directly proportional to a power of another variable.

*A newton is a unit of measure of force used in physics.

DIRECT VARIATION AS A POWER

y* varies directly as the *n*th power of *x if there exists a real number k such that

$$y = kx^n.$$

An example of direct variation as a power involves the area of a circle. The formula for the area of a circle is

$$A = \pi r^2.$$

Here, π is the constant of variation, and the area varies directly as the square of the radius.

EXAMPLE 3
SOLVING A PROBLEM INVOLVING DIRECT VARIATION AS A POWER

The distance a body falls from rest varies directly as the square of the time it falls (here we disregard air resistance). If an object falls 64 feet in 2 seconds, how far will it fall in 8 seconds?

If d represents the distance the object falls and t the time it takes to fall,

$$d = kt^2$$

for some constant k. To find the value of k, use the fact that the object falls 64 feet in 2 seconds.

$$d = kt^2 \quad \text{Formula}$$
$$64 = k(2)^2 \quad \text{Let } d = 64 \text{ and } t = 2.$$
$$k = 16 \quad \text{Find } k.$$

With this result, the variation equation becomes

$$d = 16t^2.$$

Now let $t = 8$ to find the number of feet the object will fall in 8 seconds.

$$d = 16(8)^2 \quad \text{Let } t = 8.$$
$$= 1024$$

The object will fall 1024 feet in 8 seconds. ■

3 In direct variation where $k > 0$, as x increases, y increases, and similarly as x decreases, y decreases. Another type of variation is *inverse variation.*

INVERSE VARIATION

y* varies inversely as *x if there exists a real number k such that

$$y = \frac{k}{x}.$$

Also, ***y* varies inversely as the *n*th power of *x*** if there exists a real number k such that

$$y = \frac{k}{x^n}.$$

For inverse variation, for $k > 0$, as x increases, y decreases, and as x decreases, y increases.

An example of inverse variation can be found by looking at the formula for the area of a parallelogram. In its usual form, the formula is

$$A = bh.$$

Dividing both sides by b gives

$$h = \frac{A}{b}.$$

Here, h (height) varies inversely as b (base), with A (the area) serving as the constant of variation. For example, if a parallelogram has an area of 72 square inches, the values of b and h might be any of the following:

$$\begin{array}{ll} b = 2,\ h = 36 & b = 12,\ h = 6 \\ b = 3,\ h = 24 & b = 9,\ \ h = 8 \\ b = 4,\ h = 18 & b = 8,\ \ h = 9. \end{array}$$

Notice that in the first group, as b increases, h decreases. In the second group, as b decreases, h increases.

EXAMPLE 4
SOLVING AN INVERSE VARIATION PROBLEM

The weight of an object above the earth varies inversely as the square of its distance from the center of the earth. A space vehicle in an elliptical orbit has a maximum distance from the center of the earth (apogee) of 6700 miles. Its minimum distance from the center of the earth (perigee) is 4090 miles. See Figure 7.22. If an astronaut in the vehicle weighs 57 pounds at its apogee, what does the astronaut weigh at its perigee?

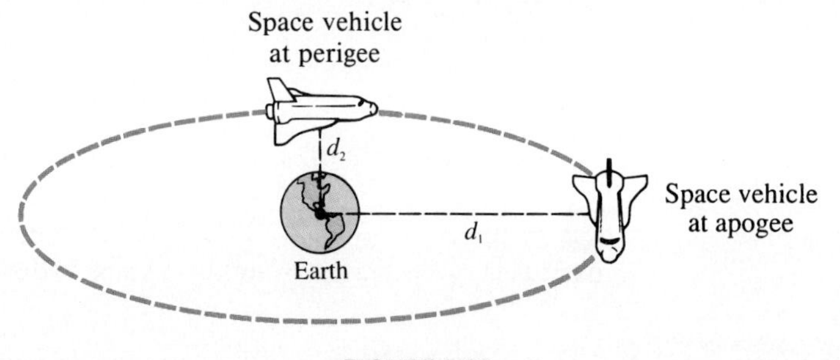

FIGURE 7.22

If w is the weight and d is the distance from the center of the earth, then

$$w = \frac{k}{d^2}$$

for some constant k. At the apogee the astronaut weighs 57 pounds and the distance from the center of the earth is 6700 miles. Use these values to find k.

$$57 = \frac{k}{(6700)^2} \qquad \text{Let } w = 57 \text{ and } d = 6700.$$
$$k = 57(6700)^2$$

Then the weight at the perigee with $d = 4090$ miles is

$$w = \frac{57(6700)^2}{(4090)^2} \approx 153 \text{ pounds.} \quad ■$$

NOTE The approximate answer in Example 4, 153 pounds, was obtained by using a calculator. A calculator will often be helpful in performing operations required in variation problems.

4 It is common for one variable to depend on several others. For example, if one variable varies as the product of several other variables (perhaps raised to powers), the first variable is said to **vary jointly** as the others.

EXAMPLE 5
SOLVING A JOINT VARIATION PROBLEM

The strength of a rectangular beam varies jointly as its width and the square of its depth. If the strength of a beam 2 inches wide by 10 inches deep is 1000 pounds per square inch, what is the strength of a beam 4 inches wide and 8 inches deep?

If S represents the strength, w the width, and d the depth, then

$$S = kwd^2$$

for some constant, k. Since $S = 1000$ if $w = 2$ and $d = 10$,

$$1000 = k(2)(10)^2. \qquad \text{Let } S = 1000,\ w = 2, \text{ and } d = 10.$$

Solving this equation for k gives

$$1000 = k \cdot 2 \cdot 100$$
$$1000 = 200k$$
$$k = 5,$$

so

$$S = 5wd^2.$$

Find S when $w = 4$ and $d = 8$ by substitution in $S = 5wd^2$.

$$S = 5(4)(8)^2 \qquad \text{Let } w = 4 \text{ and } d = 8.$$
$$= 1280$$

The strength of the beam is 1280 pounds per square inch. ■

5 There are many combinations of direct and inverse variation. The final example shows a typical **combined variation** problem.

EXAMPLE 6
SOLVING A COMBINED VARIATION PROBLEM

The maximum load that a cylindrical column with a circular cross section can hold varies directly as the fourth power of the diameter of the cross section and inversely as the square of the height. A 9-meter column 1 meter in diameter will support 8 metric tons. How many metric tons can be supported by a column 12 meters high and 2/3 meter in diameter?

Let L represent the load, d the diameter, and h the height. Then

$$L = \frac{kd^4}{h^2}.$$

← Load varies directly as the 4th power of the diameter.
← Load varies inversely as the square of the height.

Now find k. Let $h = 9$, $d = 1$, and $L = 8$.

$$8 = \frac{k(1)^4}{9^2} \qquad h = 9,\ d = 1,\ L = 8$$

Solve for k.

$$8 = \frac{k}{81}$$

$$k = 648$$

Substitute 648 for k in the first equation.

$$L = \frac{648d^4}{h^2}$$

Now find L when $h = 12$ and $d = 2/3$ by substituting the values into the last equation.

$$L = \frac{648\left(\frac{2}{3}\right)^4}{12^2} \qquad h = 12,\ d = \frac{2}{3}$$

$$L = \frac{648\left(\frac{16}{81}\right)}{144}$$

$$= 648 \cdot \frac{16}{81} \cdot \frac{1}{144} = \frac{8}{9}$$

The maximum load is about 8/9 metric ton. ■

7.5 EXERCISES

Solve each of the following. See Examples 1–5.

1. If x varies directly as r, and $x = 9$ when $r = 4$, find x when r is 10.
2. If p varies directly as y, and $p = 4$ when $y = 3$, find p when y is 7.
3. If a varies directly as the square of b, and $a = 4$ when $b = 3$, find a when b is 2.
4. If h varies directly as the square of m, and $h = 15$ when $m = 5$, find h when $m = 7$.
5. If z varies inversely as w, and $z = 5$ when $w = 1/2$, find z when $w = 8$.

6. If t varies inversely as s, and $t = 3$ when $s = 5$, find t when $s = 3$.

7. If p varies jointly as q and r^2, and $p = 100$ when $q = 2$ and $r = 3$, find p when $q = 5$ and $r = 2$.

8. If f varies jointly as g^2 and h, and $f = 25$ when $g = 4$ and $h = 2$, find f when $g = 3$ and $h = 6$.

9. Suppose z varies jointly as x and y^2 and inversely as w. Also, $z = 3/8$ when $x = 2$, $y = 3$, and $w = 12$. Find z, if $x = 4$, $y = 1$, and $w = 6$.

10. Repeat Exercise 9, for $x = 7$, $y = 2$, and $w = 10$.

11. For $k > 0$, if y varies directly as x, when x increases, y ________, and when x decreases, y ________. (See Objective 1.)

12. For $k > 0$, if y varies inversely as x, when x increases, y ________, and when x decreases, y ________. (See Objective 3.)

In Exercises 13–20, tell whether the equation represents direct, inverse, joint, or combined variation. (See Objectives 1, 3, 4, and 5.)

13. $y = \frac{4}{x}$

14. $y = \frac{7}{x}$

15. $y = 6x$

16. $y = 3x$

17. $y = xz^2$

18. $y = x^2z$

19. $y = \frac{x}{zw}$

20. $y = \frac{x^2}{zw^3}$

21. Explain the difference between inverse variation and direct variation. (See Objectives 1 and 3.)

22. What is meant by the constant of variation in a direct variation problem? If you were to graph the linear equation $y = kx$ for some nonnegative constant k, what role does the value of k play in the graph? (*Hint:* See Section 7.2.)

Work each of the following problems. Use a calculator if necessary. See Examples 2–6.

23. For a body falling freely from rest (disregarding air resistance), the distance the body falls varies directly as the square of the time. If an object is dropped from the top of a tower 576 feet high and hits the ground in 6 seconds, how far did it fall in the first 4 seconds?

24. The current in a simple electrical circuit is inversely proportional to the resistance. If the current is 20 amps (an *amp* is a unit for measuring current) when the resistance is 5 ohms, find the current when the resistance is 8 ohms.

25. The pressure exerted by a certain liquid at a given point varies directly as the depth of the point beneath the surface of the liquid. The pressure at 30 meters is 80 newtons per square centimeter. What pressure is exerted at 50 meters?

26. The force required to compress a spring is proportional to the change in length of the spring. If a force of 20 newtons is required to compress a certain spring 2 centimeters, how much force is required to compress the spring from 20 centimeters to 10 centimeters?

27. The force with which the earth attracts an object above the earth's surface varies inversely with the square of the distance of the object from the center of the earth. If an object 4000 miles from the center of the earth is attracted with a force of 160 pounds, find the force of attraction if the object were 4500 miles from the center of the earth.

28. The illumination produced by a light source varies inversely as the square of the distance from the source. If the illumination produced 4 meters from a certain light source is 50 footcandles, find the illumination produced 9 meters from the same source.

29. The distance that a person can see to the horizon from a point above the surface of the earth varies directly as the square root of the height of that point (disregarding mountains, smog, and haze). If a person 144 meters above the surface of the earth can see for 18 kilometers to the horizon, how far can a person see to the horizon from a point 1600 meters high?

30. Two electrons repel each other with a force inversely proportional to the square of the distance between them. When the electrons are 5×10^{-10} meters apart, they repel one another with a force of 100 units. How far apart are they when the force of repulsion is 196 units?

31. A meteorite approaching the earth has a velocity inversely proportional to the square root of its distance from the center of the earth. If the velocity is 5 kilometers per second when the distance is 8100 kilometers from the center of the earth, find the velocity at a distance of 6400 kilometers.

32. The volume of a gas varies inversely as the pressure and directly as the temperature. (Temperature must be measured in *degrees Kelvin* (K), a unit of measurement used in physics.) If a certain gas occupies a volume of 1.3 liters at 300 K and a pressure of 18 newtons per square centimeter, find the volume at 340 K and a pressure of 24 newtons per square centimeter.

33. The force of the wind blowing on a vertical surface varies jointly as the area of the surface and the square of the velocity. If a wind of 40 miles per hour exerts a force of 50 pounds on a surface of 1/2 square foot, how much force will a wind of 80 miles per hour place on a surface of 2 square feet?

34. It is shown in engineering that the maximum load a cylindrical column of circular cross section can hold varies directly as the fourth power of the diameter and inversely as the square of the height. If a column 9 feet high and 3 feet in diameter will support a load of 8 tons, how great a load will be supported by a column 12 feet high and 2 feet in diameter?

35. The period of a pendulum varies directly as the square root of the length of the pendulum and inversely as the square root of the acceleration due to gravity. Find the period when the length is 4 feet and the acceleration due to gravity is 32 feet per second per second, if the period is 1.06π seconds when the length is 9 feet and the acceleration due to gravity is 32 feet per second per second.

36. The force needed to keep a car from skidding on a curve varies inversely as the radius of the curve and jointly as the weight of the car and the square of the speed. If 242 pounds of force keep a 2000-pound car from skidding on a curve of radius 500 feet at 30 miles per hour, what force would keep the same car from skidding on a curve of radius 750 feet at 50 miles per hour?

37. The maximum load of a horizontal beam that is supported at both ends varies directly as the width and the square of the height and inversely as the length between the supports. A beam 6 meters long, .1 meter wide, and .06 meter high supports a load of 360 kilograms. What is the maximum load supported by a beam 16 meters long, .2 meter wide, and .08 meter high?

38. The number of long distance phone calls between two cities in a certain time period varies directly as the populations p_1 and p_2 of the cities, and inversely as the distance between them. If 80,000 calls are made between two cities 400 miles apart, with populations of 70,000 and 100,000, how many calls are made between cities with populations of 50,000 and 75,000 that are 250 miles apart?

39. The collision impact of an automobile varies jointly as its mass and the square of its speed. Suppose a 2000-pound car traveling at 55 miles per hour has a collision impact of 6.1. What is the collision impact of the same car at 65 miles per hour?

40. The absolute temperature T of an ideal gas varies jointly as its pressure P and its volume V. If $T = 250$ when $P = 25$ pounds per square centimeter and $V = 50$ cubic centimeters, find T when $P = 50$ pounds per square centimeter and $V = 25$ cubic centimeters.

Photo for Exercise 35

PREVIEW EXERCISES

For each of the following, evaluate y when x = 3. See Section 7.1.

41. $y = -7x + 12$ **42.** $y = -5x - 4$ **43.** $y = 3x - 8$ **44.** $y = 2x - 9$

For each of the following, solve for y. See Section 2.2.

45. $3x - 7y = 8$ **46.** $2x - 4y = 7$ **47.** $\frac{1}{2}x - 4y = 5$ **48.** $\frac{3}{4}x + 2y = 9$

7.6 INTRODUCTION TO FUNCTIONS; LINEAR FUNCTIONS

OBJECTIVES

1 DEFINE AND IDENTIFY RELATIONS AND FUNCTIONS.
2 FIND THE DOMAIN AND RANGE OF A RELATION.
3 USE THE VERTICAL LINE TEST.
4 USE $f(x)$ NOTATION.
5 IDENTIFY AND GRAPH LINEAR FUNCTIONS.

FOCUS ON PROBLEM SOLVING

Forensic scientists use the lengths of certain bones to calculate the height of a person. Two bones often used are the tibia (t), the bone from the ankle to the knee, and the femur (f), the bone from the knee to the hip socket. A person's height (h) is determined from the lengths of these bones using functions defined by the following formulas. All measurements are in centimeters.

For men: $h = 69.09 + 2.24f$

$h = 81.69 + 2.39t$

For women: $h = 61.41 + 2.32f$

$h = 72.57 + 2.53t$

Find the height of a man with a femur measuring 56 centimeters.

The function concept is one of the most important in mathematics. Earlier in this chapter we learned that for a linear equation $y = mx + b$, substituting a value for x gives us a value for y. Since there is only one y-value obtained for every x-value, we have a situation that describes a function.

The problem above is Exercise 65 in the exercises for this section. After working through this section, you should be able to solve this problem.

It is often useful to describe one quantity in terms of another. For example, the growth of a plant is related to the amount of light it receives, the demand for a product is related to the price of the product, the cost of a trip is related to the distance traveled, and so on. To represent these corresponding quantities, it is helpful to use ordered pairs.

For example, we can indicate the relationship between the demand for a product and its price by writing ordered pairs in which the first number represents the price and the second number represents the demand. The ordered pair (5, 1000) then could indicate a demand for 1000 items when the price of the item is \$5. Since the demand depends on the price charged, we place the price first and the demand second. The ordered pair is an abbreviation for the sentence "If the price is 5 (dollars), then the demand is for 1000 (items)." Similarly, the ordered pairs (3, 5000) and (10, 250) show that a price of \$3 produces a demand for 5000 items, and a price of \$10 produces a demand for 250 items.

In this example, the demand depends on the price of the item. For this reason, demand is called the *dependent variable,* and price the *independent variable*. Generalizing, if the value of the variable y depends on the variable x, then y is the **dependent variable** and x the **independent variable.**

1 Since related quantities can be written using ordered pairs, the concept of *relation* can be defined as follows.

RELATION A **relation** is a set of ordered pairs.

A special kind of relation, called a *function,* is very important in mathematics and its applications.

FUNCTION A **function** is a relation in which, for each value of the first component of the ordered pairs, there is exactly one value of the second component.

EXAMPLE 1 **DETERMINING WHETHER A RELATION IS A FUNCTION**

Determine whether each of the following relations is a function.

$$F = \{(1, 2), (-2, 5), (3, -1)\}$$
$$G = \{(-2, 1), (-1, 0), (0, 1), (1, 2), (2, 2)\}$$
$$H = \{(-4, 1), (-2, 1), (-2, 0)\}$$

Relations F and G are functions, because for each x-value, there is only one y-value. Notice that in G, the last two ordered pairs have the same y-value. This does not violate the definition of function, since each first component (x-value) has only one second component (y-value).

Relation H is not a function, because the last two ordered pairs have the same x-value, but different y-values. ■

EXAMPLE 2 **RECOGNIZING A FUNCTION EXPRESSED AS A MAPPING**

A function can also be expressed as a correspondence or *mapping* from one set to another. The mapping in Figure 7.23 is a function that assigns to a state its population (in millions) expected by the year 2000. ■

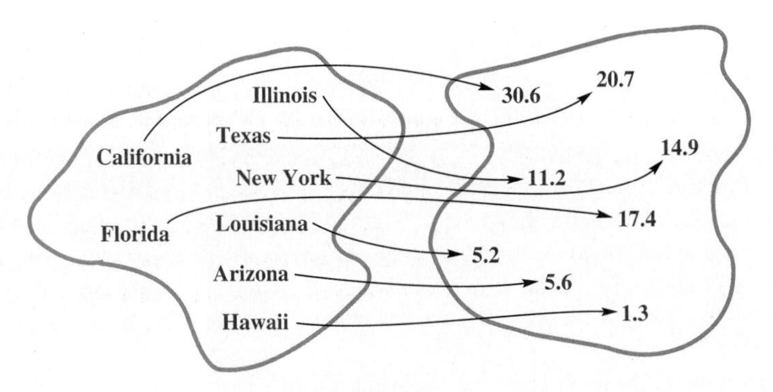

FIGURE 7.23

2 The set of all first components (x-values) of the ordered pairs of a relation is called the **domain** of the relation, and the set of all second components (y-values) is called the **range.** For example, the domain of function F in Example 1 is $\{1, -2, 3\}$; the range is $\{2, 5, -1\}$. Also, the domain of function G is $\{-2, -1, 0, 1, 2\}$ and the range is $\{0, 1, 2\}$. Domains and ranges can also be defined in terms of independent and dependent variables.

DOMAIN AND RANGE

In a relation, the set of all values of the independent variable (x) is the **domain;** the set of all values of the dependent variable (y) is the **range.**

EXAMPLE 3 **FINDING DOMAINS AND RANGES**

Give the domain and range of each function.

(a) $\{(3, -1), (4, 2), (0, 5)\}$

The domain, the set of x-values, is $\{3, 4, 0\}$; the range is the set of y-values, $\{-1, 2, 5\}$.

(b) The function in Figure 7.23

The domain is {Illinois, Texas, California, New York, Florida, Louisiana, Arizona, Hawaii} and the range is $\{1.3, 5.2, 5.6, 11.2, 14.9, 17.4, 20.7, 30.6\}$. ■

The **graph** of a relation is the graph of its ordered pairs. The graph gives a picture of the relation.

EXAMPLE 4 **FINDING DOMAINS AND RANGES FROM GRAPHS**

Three relations are graphed in Figure 7.24. Give the domain and range of each.

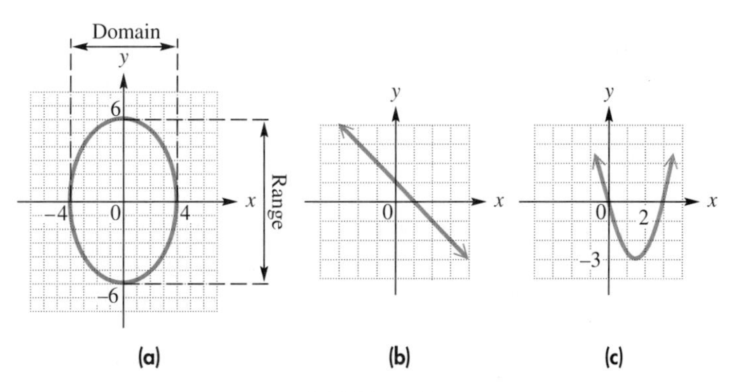

FIGURE 7.24

(a) In Figure 7.24(a), the x-values of the points on the graph include all numbers between -4 and 4, inclusive. The y-values include all numbers between -6 and 6, inclusive. Using interval notation,

the domain is $[-4, 4]$;

the range is $[-6, 6]$.

(b) In Figure 7.24(b), the arrowheads indicate that the line extends indefinitely left and right, as well as up and down. Therefore, both the domain and the range are the set of all real numbers, written $(-\infty, \infty)$.

(c) In Figure 7.24(c), the arrowheads indicate that the graph extends indefinitely left and right, as well as upward. The domain is $(-\infty, \infty)$. Because there is a least y-value, -3, the range includes all numbers greater than or equal to -3, written $[-3, \infty)$. ■

Relations are often defined by equations such as $y = 2x + 3$ and $y^2 = x$. It is sometimes necessary to determinc the domain of a relation when given the equation that defines the relation. In this book, the following agreement on the domain of a relation is assumed.

AGREEMENT ON DOMAIN The domain of a relation is assumed to be all real numbers that produce real numbers when substituted for the independent variable.

To illustrate this agreement, suppose that we consider the function $y = 2x + 3$. Since any real number can be used as a replacement for x in $y = 2x + 3$, the domain of this function is the set of real numbers. As another example, the function defined by $y = 1/x$ has all real numbers except 0 as a domain, since the denominator cannot be 0.

EXAMPLE 5 IDENTIFYING FUNCTIONS AND DETERMINING DOMAINS

For each of the following, decide whether it defines a function and give the domain.

(a) $y = \sqrt{2x - 1}$

Here, for any choice of x in the domain, there is exactly one corresponding value for y (the radical is a nonnegative number), so this equation defines a function. Since the radicand cannot be negative,

$$\begin{aligned} 2x - 1 &\geq 0 \\ 2x &\geq 1 \\ x &\geq \frac{1}{2}. \end{aligned}$$

The domain is $\left[\frac{1}{2}, \infty\right)$.

(b) $y^2 = x$

The ordered pairs (16, 4) and (16, -4) both satisfy this equation. Since one value of x, 16, corresponds to two values of y, 4 and -4, this equation does not define a function. Solving $y^2 = x$ for y gives $y = \sqrt{x}$ or $y = -\sqrt{x}$, which shows that two values of y correspond to each positive value of x. Because x is equal to the square of y, the values of x must always be nonnegative. The domain is $[0, \infty)$.

(c) $y \leq x - 1$

By definition, y is a function of x if a value of x leads to exactly one value of y. In this example, a particular value of x, say 1, corresponds to many values of y. The ordered pairs (1, 0), (1, -1), (1, -2), (1, -3), and so on, all satisfy the inequality.

For this reason, this inequality does not define a function. Any number can be used for x, so the domain is the set of real numbers, $(-\infty, \infty)$.

(d) $y = \dfrac{5}{x^2 - 1}$

Given any value of x in the domain, we find y by squaring x, subtracting 1, then dividing the result into 5. This process produces exactly one value of y for each x-value, so this equation defines a function. The domain includes all real numbers except those that make the denominator zero. We find these numbers by setting the denominator equal to zero and solving for x.

$$x^2 - 1 = 0$$
$$x^2 = 1$$
$$x = 1 \quad \text{or} \quad x = -1 \qquad \text{Square root property}$$

Thus, the domain includes all real numbers except 1 and -1. In interval notation this is written as

$$(-\infty, -1) \cup (-1, 1) \cup (1, \infty). \quad \blacksquare$$

3 In a function each value of x leads to only one value of y, so that any vertical line drawn through the graph of a function would have to cut the graph in at most one point. This is the **vertical line test for a function.**

VERTICAL LINE TEST If a vertical line cuts the graph of a relation in more than one point, then the relation does not represent a function.

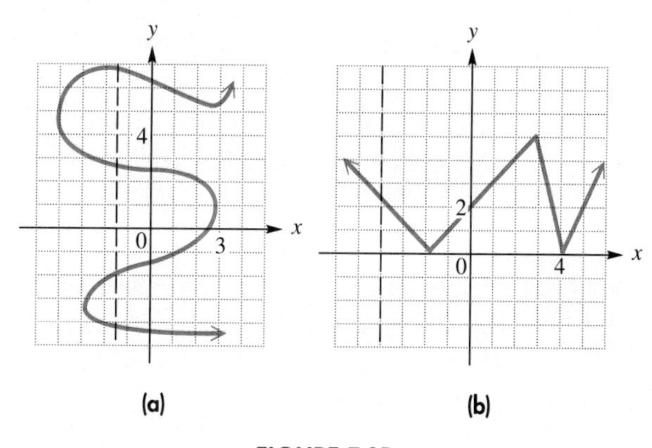

FIGURE 7.25

For example, the graph shown in Figure 7.25(a) is not the graph of a function, since a vertical line can cut the graph in more than one point, while the graph in Figure 7.25(b) does represent a function.

4 To say that y is a function of x means that for each value of x from the domain of the function, there is exactly one value of y. To emphasize that *y is a function of x,* or that y depends on x, it is common to write

$$y = f(x),$$

with $f(x)$ read "f of x." (In this special notation, the parentheses do not indicate multiplication.) The letter f stands for *function*. For example, if $y = 9x - 5$, we emphasize that y is a function of x by writing $y = 9x - 5$ as

$$f(x) = 9x - 5.$$

This **functional notation** can be used to simplify certain statements. For example, if $y = 9x - 5$, then replacing x with 2 gives

$$\begin{aligned} y &= 9 \cdot 2 - 5 \\ &= 18 - 5 \\ &= 13. \end{aligned}$$

The statement "if $x = 2$, then $y = 13$" is abbreviated with functional notation as

$$f(2) = 13.$$

Also, $f(0) = 9 \cdot 0 - 5 = -5$, and $f(-3) = -32$.

These ideas and the symbols used to represent them can be explained as follows.

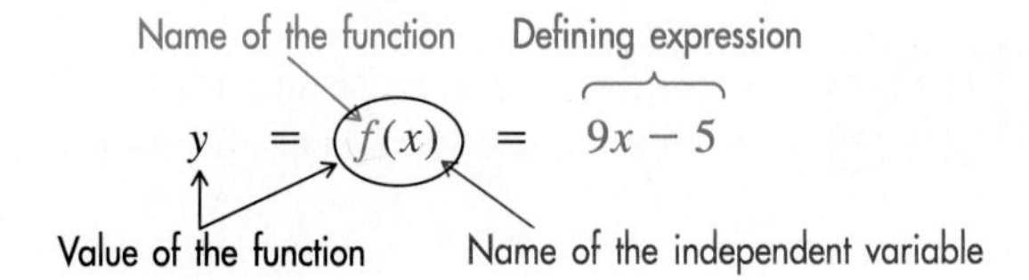

CAUTION The symbol $f(x)$ *does not* indicate "f times x," but represents the y-value for the indicated x-value. As shown above, $f(2)$ is the y-value that corresponds to the x-value 2.

EXAMPLE 6 USING FUNCTIONAL NOTATION

Let $f(x) = -x^2 + 5x - 3$. Find the following.

(a) $f(2)$

Replace x with 2.

$$f(2) = -2^2 + 5 \cdot 2 - 3 = -4 + 10 - 3 = 3$$

(b) $f(-1)$

$$f(-1) = -(-1)^2 + 5(-1) - 3 = -1 - 5 - 3 = -9$$

(c) $f(q)$

Replace x with q.

$$f(q) = -q^2 + 5q - 3$$

The replacement of one variable with another is important in later courses. ■

If a function is defined by an equation with x and y, not with functional notation, use the following to find $f(x)$.

FINDING AN EXPRESSION FOR $f(x)$

If an equation that defines a function is given with x and y, to find $f(x)$:

1. solve the equation for y;
2. replace y with $f(x)$.

EXAMPLE 7
WRITING EQUATIONS USING FUNCTIONAL NOTATION

Rewrite each equation using functional notation; then find $f(-2)$ and $f(a)$.

(a) $y = x^2 + 1$
This equation is already solved for y. Since $y = f(x)$,

$$f(x) = x^2 + 1.$$

To find $f(-2)$, let $x = -2$:

$$\begin{aligned} f(-2) &= (-2)^2 + 1 \\ &= 4 + 1 \\ &= 5. \end{aligned}$$

Find $f(a)$ by letting $x = a$: $f(a) = a^2 + 1$.

(b) $x - 4y = 5$
First solve $x - 4y = 5$ for y.

$$\begin{aligned} x - 4y &= 5 \\ x - 5 &= 4y \\ y &= \frac{x-5}{4} \qquad \text{so} \qquad f(x) = \frac{1}{4}x - \frac{5}{4} \end{aligned}$$

Now find $f(-2)$ and $f(a)$.

$$f(-2) = \frac{1}{4}(-2) - \frac{5}{4} = -\frac{7}{4}$$

and

$$f(a) = \frac{1}{4}a - \frac{5}{4} \quad ■$$

5 The first two-dimensional graphing was of straight lines. Straight-line graphs (except for vertical lines) are graphs of *linear functions.*

LINEAR FUNCTION

A function that can be written in the form

$$f(x) = mx + b$$

for real numbers m and b is a **linear function.**

Recall from Section 7.3 that m is the slope of the line and $(0, b)$ is the y-intercept.

EXAMPLE 8
GRAPHING LINEAR FUNCTIONS

Write as a linear function and graph.

(a) $y = 3 - 2x$
Replacing y with $f(x)$ gives the linear function

$$f(x) = -2x + 3.$$

Begin the graph of the function by locating the y-intercept $(0, 3)$. From the y-intercept use the slope $-2 = -2/1$ to go down 2 and right 1. This second point is used to get the graph in Figure 7.26. The vertical line test confirms that this is the graph of a function.

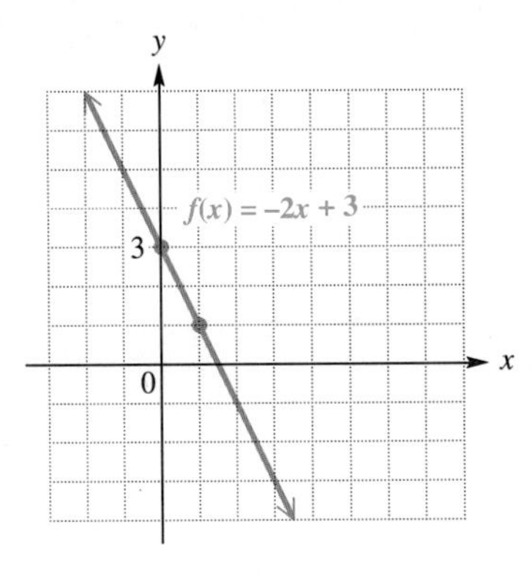

FIGURE 7.26

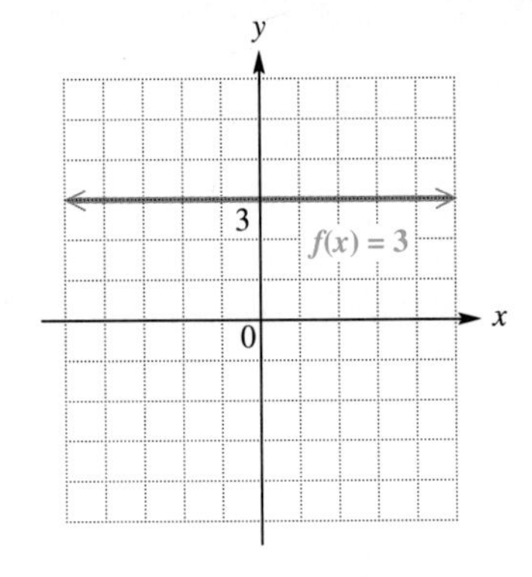

FIGURE 7.27

(b) $y - 3 = 0$
Rewrite this equation as $\quad y = 3 \quad$ or $\quad f(x) = 3.$

The graph, shown in Figure 7.27, is a horizontal straight line. ■

A linear function of the form $f(x) = k$ is sometimes called a **constant function.** The domain of any linear function is $(-\infty, \infty)$. The range of a nonconstant linear function is $(-\infty, \infty)$, while the range of the constant function $f(x) = k$ is $\{k\}$.

NOTE In Section 7.5 we saw that y varies directly as x if there is some constant k such that $y = kx$. Since this equation defines a linear function, direct variation is an important application of linear functions.

In summary, three variations of the definition of function are given here.

VARIATIONS OF THE DEFINITION OF FUNCTION

1. A **function** is a relation in which, for each value of the first component of the ordered pairs, there is exactly one value of the second component.
2. A **function** is a set of ordered pairs in which no first component is repeated.
3. A **function** is a rule or correspondence that assigns exactly one range value to each domain value.

7.6 EXERCISES

Give the domain and range of each relation. Identify any functions. See Examples 1–3.

1. $\{(5, 1), (3, 2), (4, 9), (7, 6)\}$

2. $\{(8, 0), (5, 4), (9, 3), (3, 8)\}$

3. $\{(2, 4), (0, 2), (2, 5)\}$

4. $\{(9, -2), (-3, 5), (9, 2)\}$

5. $\{(-3, 1), (4, 1), (-2, 7)\}$

6. $\{(-12, 5), (-10, 3), (8, 3)\}$

7. $\{(1, 3), (4, 7), (0, 6), (7, 2)\}$

8. $\{(8, 5), (3, 9), (-2, 11), (5, 3)\}$

9.

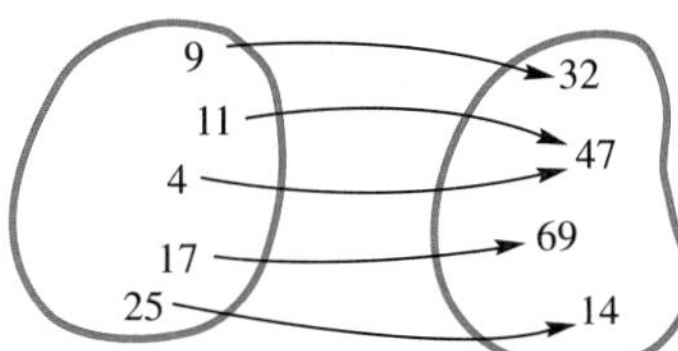

10.

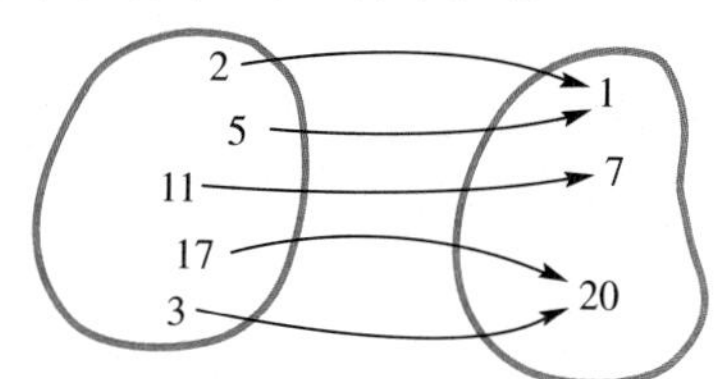

11.

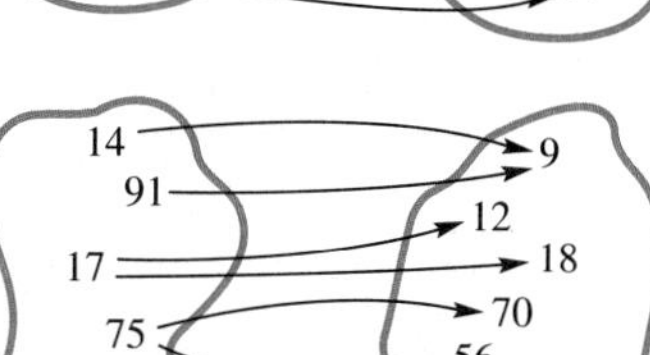

12.

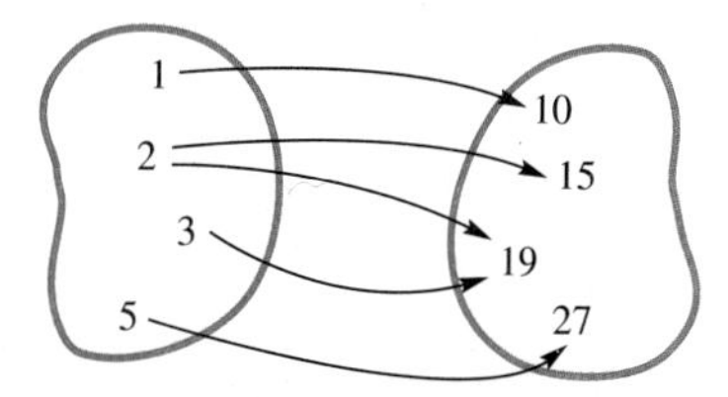

---

13. In your own words, write a definition of *function*. (See Objective 1.)

14. What is the *domain* of a function? What is the *range* of a function? (See Objective 2.)

Identify any of the following graphs that represent functions. (Hint: Use the vertical line test.) Give the domain and range of each relation. See Example 4.

15.

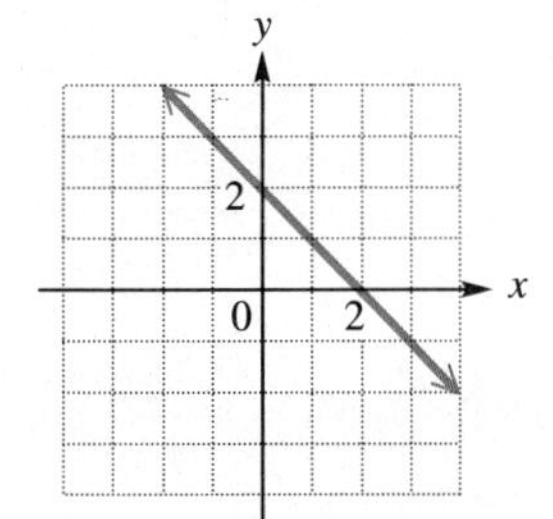

16.

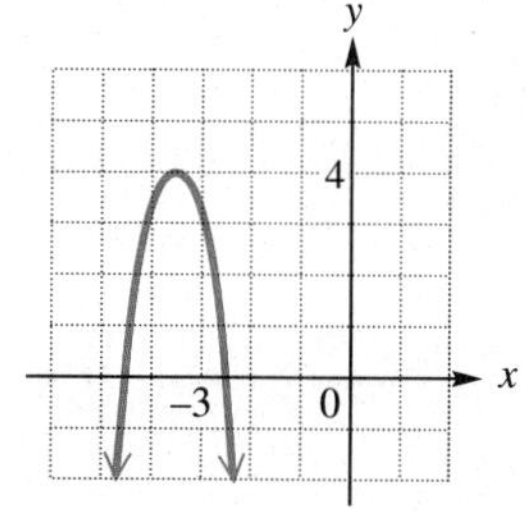

17.

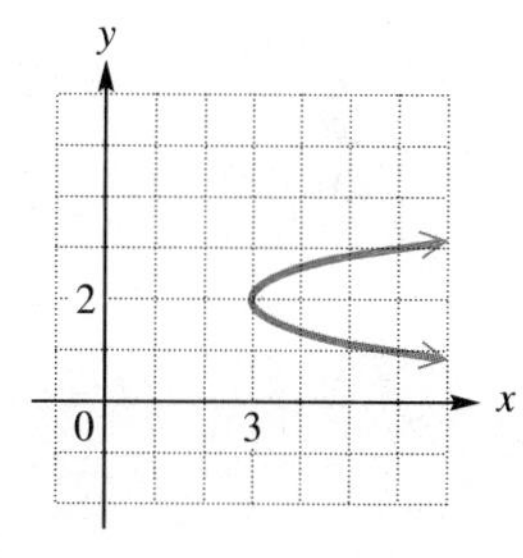

18.

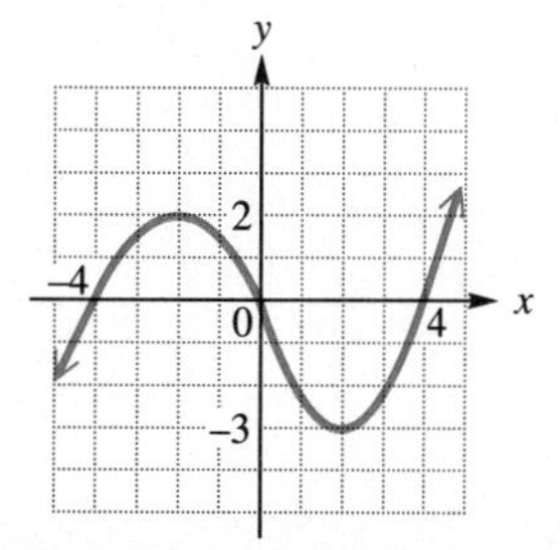

19.

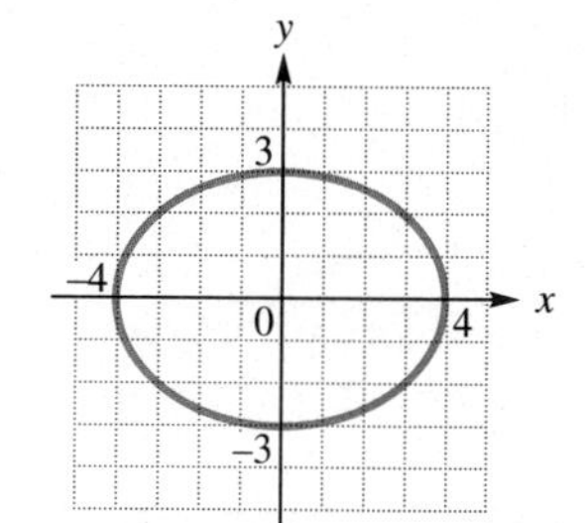

20.

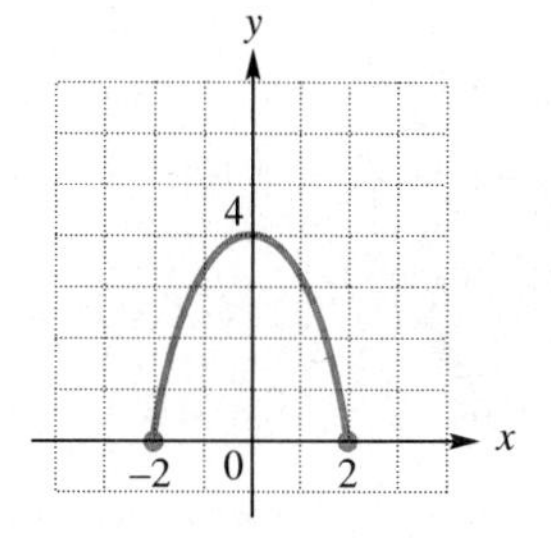

Give the domain of each relation defined as follows. Identify any that are functions. See Example 5.

21. $y = 9x + 2$

22. $y = 3x - 6$

23. $y = x^2$

24. $y = x^3$

25. $x = \sqrt{y}$

26. $x = y^4$

27. $y < x$

28. $y > x$

29. $y = \sqrt{x - 3}$

30. $y = \sqrt{x + 5}$

31. $y = |x|$

32. $y = \sqrt[3]{x}$

33. $y = \dfrac{x + 1}{x - 7}$

34. $y = \dfrac{2x + 3}{x + 5}$

35. $y = \dfrac{3}{x^2 - 16}$

36. $y = \dfrac{-4}{x^2 - 25}$

37. $y = \dfrac{-4}{x^2 + 1}$

38. $y = \dfrac{2}{x^2 + 9}$

39. $y = \dfrac{x + 2}{x^2 - 5x + 4}$

40. $y = \dfrac{x - 3}{x^2 + 6x + 8}$

Let $f(x) = 3 + 2x$ and $g(x) = x^2 - 2$. Find the following. See Example 6.

41. $f(1)$

42. $f(4)$

43. $g(2)$

44. $g(0)$

45. $g(-1)$

46. $g(-3)$

47. $f(-8)$

48. $f(-5)$

Sketch the graph of each linear function. Give the domain and range. See Example 8.

49. $f(x) = -2x + 5$

50. $g(x) = 4x - 1$

51. $h(x) = \dfrac{1}{2}x + 2$

52. $F(x) = -\dfrac{1}{4}x + 1$

53. $G(x) = 2x$

54. $H(x) = -3x$

55. $f(x) = 5$

56. $g(x) = -4$

Write an expression for $f(x)$ for each of the following. Give the domain and range. See Example 7.

57. $x + y = 4$

58. $x - y = 3$

59. $2x + 3y = 6$

60. $3x - 2y = -6$

61. Which of the following defines a linear function? (See Objective 5.)

(a) $y = \dfrac{x - 5}{4}$ **(b)** $y = \sqrt[3]{x}$ **(c)** $y = x^2$ **(d)** $y = x^{1/2}$

62. Which of the functions in Exercise 61 has domain $[0, \infty)$?

63. The graph shown here depicts spot prices in dollars per barrel for West Texas intermediate crude oil over a 10-day period during October, 1990.

(a) Is this the graph of a function?

(b) What is the domain?

(c) What is the range? (Round to nearest half dollar.)

(d) Estimate the price on October 24.

(e) On what day was oil at its highest price? Its lowest price?

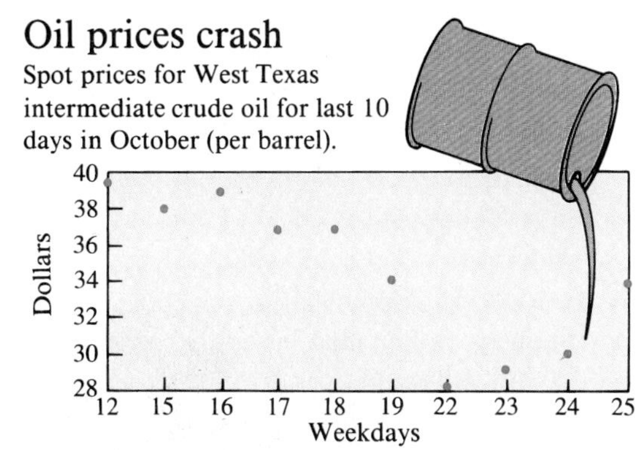

Source: Oil Buyer's Guide International of Lakewood, N.J.

64. The graph shown here depicts gasoline prices (per gallon) in the New Orleans metropolitan area for one month in late 1990.

(a) Is this the graph of a function?

(b) What is the domain?

(c) What is the range?

(d) Estimate the price on October 12.

(e) By how much had gasoline prices risen from the invasion of Kuwait on August 2 to October 25? (*Hint:* Look at the inserts.)

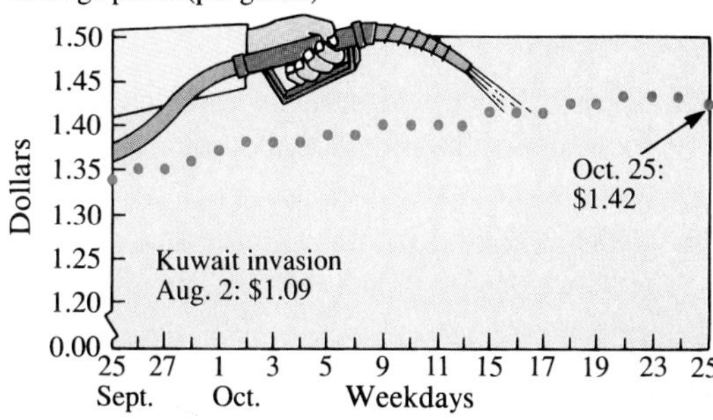

*Regular unleaded, average from 14 stations in the metro area.

Forensic scientists use the lengths of certain bones to calculate the height of a person. Two bones often used are the tibia (t), the bone from the ankle to the knee, and the femur (f), the bone from the knee to the hip socket. A person's height (h) is determined from the lengths of these bones using functions defined by the following formulas. All measurements are in centimeters.

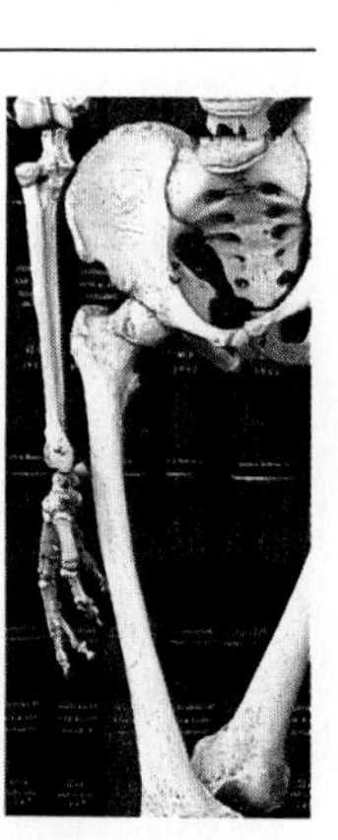

For men: $h = 69.09 + 2.24f$ or $h = 81.69 + 2.39t$

For women: $h = 61.41 + 2.32f$ or $h = 72.57 + 2.53t$

65. Find the height of a man with a femur measuring 56 centimeters.

66. Find the height of a man with a tibia measuring 40 centimeters.

67. Find the height of a woman with a femur measuring 50 centimeters.

68. Find the height of a woman with a tibia measuring 36 centimeters.

Federal regulations set standards for the size of the quarters of marine mammals. A pool to house sea otters must have a volume of "the square of the sea otter's average adult length (in meters) multiplied by 3.14 and by .91 meters." If x represents the sea otter's average adult length and f(x) represents the volume of the corresponding pool size, this formula can be written as $f(x) = (.91)(3.14)x^2$. *Find the volume of the pool for each of the following adult lengths (in meters). Round answers to the nearest hundredth.*

69. .8 **70.** 1.0 **71.** 1.2 **72.** 1.5

PREVIEW EXERCISES

Solve each equation. See Section 2.1.

73. $3(2x - 1) - x = 4$

74. $-2(y - 1) + 3y = 8$

75. $-5(2m + 3) - m = 10$

76. $4(3k + 1) - 8k = -7$

77. $2\left(\frac{3x - 1}{2}\right) - 2x = -4$

78. $3\left(\frac{5p + 4}{3}\right) + p = 6$

79. $-4\left(\frac{t - 3}{3}\right) + 2t = 8$

80. $-2\left(\frac{3n + 1}{5}\right) + n = -4$

CHAPTER 7 GLOSSARY

KEY TERMS

7.1 origin When two number lines intersect at a right angle, the origin is the common zero point.

x-axis The horizontal number line in a rectangular coordinate system is called the x-axis.

y-axis The vertical number line in a rectangular coordinate system is called the y-axis.

rectangular (Cartesian) coordinate system Two number lines that intersect at a right angle at their zero points form a rectangular coordinate system.

components The components of an ordered pair are the two numbers in the ordered pair.

plot To plot an ordered pair is to locate it on a rectangular coordinate system.

coordinates The numbers in an ordered pair are called the coordinates of the corresponding point.

quadrant A quadrant is one of the four regions in the plane determined by a rectangular coordinate system.

linear equation in two variables A linear equation in two variables is a first-degree equation with two variables.

standard form A linear equation is in standard form when written as $Ax + By = C$, with $A \geq 0$, and A, B, and C integers.

x-intercept The point where a line crosses the x-axis is the x-intercept.

y-intercept The point where a line crosses the y-axis is the y-intercept.

collinear Points that lie on the same straight line are collinear.

7.2 slope The ratio of the change in y compared to the change in x along a line is the slope of the line.

7.4 boundary line In the graph of a linear inequality, the boundary line separates the region that satisfies the inequality from the region that does not satisfy the inequality.

7.6 dependent variable If the quantity y depends on x, then y is called the dependent variable in an equation relating x and y.

independent variable If y depends on x, then x is the independent variable in an equation relating x and y.

relation A relation is a set of ordered pairs.

function A function is a relation in which each value of the first component, x, corresponds to exactly one value of the second component, y.

domain The domain of a relation is the set of first components (x-values) of the ordered pairs of a relation.

range The range of a relation is the set of second components (y-values) of the ordered pairs of a relation.

graph of a relation The graph of a relation is the graph of the ordered pairs of the relation.

vertical line test The vertical line test says that if a vertical line cuts the graph of a relation in more than one point, then the relation is not a function.

linear function A function that can be written in the form $f(x) = mx + b$ for real numbers m and b is a linear function.

NEW SYMBOLS

(a, b) ordered pair

x_1 a specific value of the variable x (read "x sub one")

m slope

$f(x)$ function of x (read "f of x")

CHAPTER 7 QUICK REVIEW

SECTION	CONCEPTS	EXAMPLES
7.1 THE RECTANGULAR COORDINATE SYSTEM	**Finding Intercepts** To find the x-intercept, let $y = 0$. To find the y-intercept, let $x = 0$.	The graph of $2x + 3y = 12$ has x-intercept $(6, 0)$ and y-intercept $(0, 4)$.
	Distance Formula The distance between (x_1, y_1) and (x_2, y_2) is $d = \sqrt{(x_2 - x_1)^2 + (y_2 - y_1)^2}$.	The distance between $(3, -2)$ and $(-1, 1)$ is $\sqrt{(-1 - 3)^2 + [1 - (-2)]^2}$ $= \sqrt{(-4)^2 + 3^2} = \sqrt{16 + 9}$ $= \sqrt{25} = 5$.
7.2 THE SLOPE OF A LINE	If $x_2 \neq x_1$, then $m = \frac{\text{change in } y}{\text{change in } x} = \frac{y_2 - y_1}{x_2 - x_1}$.	For $2x + 3y = 12$, $m = \frac{4 - 0}{0 - 6} = -\frac{2}{3}$.
	A vertical line has undefined slope.	$x = 3$ has undefined slope.
	A horizontal line has 0 slope.	$y = -5$ has $m = 0$.
	Parallel lines have equal slopes.	$y = 2x + 3$ $4x - 2y = 6$ $m = 2$ $m = 2$ Lines are parallel.
	The slopes of perpendicular lines are negative reciprocals with a product of -1.	$y = 3x - 1$ $x + 3y = 4$ $m = 3$ $m = -\frac{1}{3}$ Lines are **perpendicular**.

SECTION	CONCEPTS	EXAMPLES
7.3 LINEAR EQUATIONS	**Standard Form** $Ax + By = C$ **Vertical Line** $x = k$ **Horizontal Line** $y = k$ **Slope-Intercept Form** $y = mx + b$ **Point-Slope Form** $y - y_1 = m(x - x_1)$	$2x - 5y = 8$ $x = -1$ $y = 4$ $y = 2x + 3$ $m = 2$, y-intercept is $(0, 3)$. $y - 3 = 4(x - 5)$ $(5, 3)$ is on the line, $m = 4$.
7.4 LINEAR INEQUALITIES	**Graphing a Linear Inequality** *Step 1* Draw the graph of the line that is the boundary. Make the line solid if the inequality involves $\leq$ or $\geq$; make the line dashed if the inequality involves $<$ or $>$. *Step 2* Choose any point not on the line as a test point. *Step 3* Shade the region that includes the test point if the test point satisfies the original inequality; otherwise, shade the region on the other side of the boundary line.	Graph $2x - 3y \leq 6$. Draw the graph of $2x - 3y = 6$. Use a solid line because of $\leq$. Choose $(1, 2)$. $2(1) - 3(2) = 2 - 6 \leq 6$ True Shade the side of the line that includes $(1, 2)$. 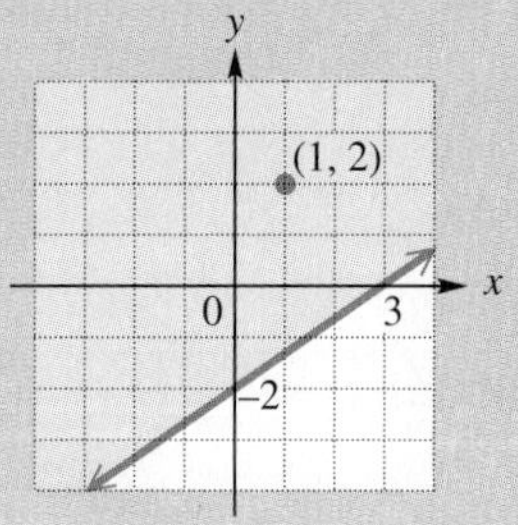
7.5 VARIATION	If there is some constant k such that: $y = kx^n$, then y varies directly as x^n; $y = \frac{k}{x^n}$, then y varies inversely as x^n.	The area of a circle **varies directly as** the square of the radius. $A = kr^2$ Pressure **varies inversely as** volume. $p = \frac{k}{V}$

SECTION	CONCEPTS	EXAMPLES
7.6 INTRODUCTION TO FUNCTIONS; LINEAR FUNCTIONS	To evaluate a function using functional notation (that is, $f(x)$ notation) for a given value of x, substitute the value wherever x appears.	If $f(x) = x^2 - 7x + 12$, then $f(1) = 1^2 - 7(1) + 12 = 6.$
	To write the equation that defines a function in functional notation, solve the equation for y. Then replace y with $f(x)$.	Given: $2x + 3y = 12.$ $3y = -2x + 12$ $y = -\frac{2}{3}x + 4$ $f(x) = -\frac{2}{3}x + 4$
	To graph the linear function $f(x) = mx + b$, replace $f(x)$ with y and graph $y = mx + b$, as in Section 7.2.	

CHAPTER 7 REVIEW EXERCISES

[7.1] *Complete the given ordered pairs for each equation. Then graph the equation.*

1. $3x + 2y = 10$; $(0, \quad)$, $(\quad , 0)$, $(2, \quad)$, $(\quad , -2)$

2. $x - y = 8$; $(2, \quad)$, $(\quad , -3)$, $(3, \quad)$, $(\quad , -2)$

Find the x- and y-intercepts and then graph each of the following equations.

3. $4x - 3y = 12$

4. $5x + 7y = 28$

5. $2x + 5y = 20$

6. $x - 4y = 8$

Find the distance between each of the pairs of points in Exercises 7–10.

7. $(1, 5)$ and $(-2, 3)$

8. $(-4, -2)$ and $(3, 0)$

9. (m, n) and $(m + 1, n - 3)$

10. $(a, b - 2)$ and $(a + 4, b - 5)$

11. Find the perimeter of the triangle determined by the points $(-1, 3)$, $(7, 2)$, and $(9, -6)$.

12. Use the distance formula to decide whether the points $(1, 5)$, $(-3, -3)$, and $(4, 6)$ lie on the same line.

13. Explain how the signs of the x- and y-coordinates of a point determine the quadrant in which the point lies.

[7.2] *Find the slope of each line in Exercises 14–21.*

14. Through $(-1, 2)$ and $(4, -5)$

15. Through $(0, 3)$ and $(-2, 4)$

16. $y = 2x + 3$

17. $3x - 4y = 5$

18. $x = 5$

19. Parallel to $3y = 2x + 5$

20. Perpendicular to $3x - y = 4$

21. Through $(-1, 5)$ and $(-1, -4)$

22. Use the slope formula to show that the figure with vertices (corners) at $(-2, -1)$, $(-2, 3)$, $(4, 3)$ and $(4, -1)$ is a rectangle. (*Hint:* A rectangle has opposite sides parallel and adjacent sides perpendicular.)

Tell whether the line has positive, negative, zero, or undefined slope in Exercises 23–25.

23.

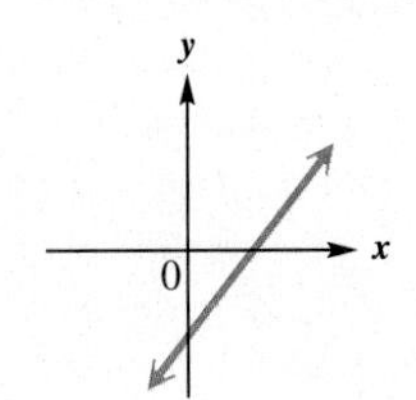

24.

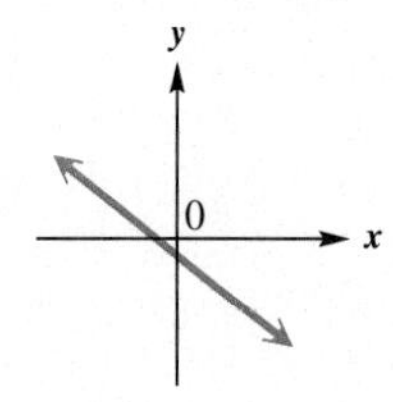

25.

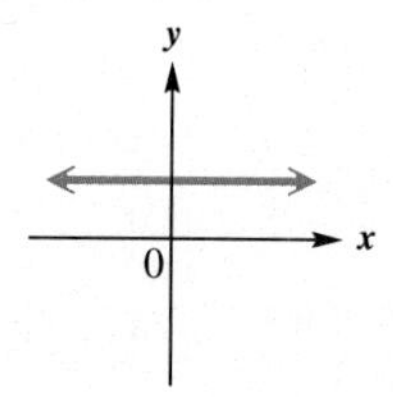

[7.3] *Write an equation for each line in standard form.*

26. Slope $-\frac{1}{3}$, y-intercept $(0, -1)$

27. Slope 0, y-intercept $(0, -2)$

28. Slope $-\frac{4}{3}$, through $(2, 7)$

29. Slope 3, through $(-1, 4)$

30. Vertical, through $(2, 5)$

31. Through $(2, -5)$ and $(1, 4)$

32. Through $(-3, -1)$ and $(2, 6)$

33. Parallel to $4x - y = 3$ and through $(7, -1)$

34. Perpendicular to $2x - 5y = 7$ and through $(4, 3)$

35. The graph of $x - 6 = 0$ is which one of the following?

(a) a vertical line with x-intercept $(-6, 0)$
(b) a vertical line with x-intercept $(6, 0)$
(c) a horizontal line with y-intercept $(0, -6)$
(d) a horizontal line with y-intercept $(0, 6)$

[7.4] *Use the ideas of Section 7.4 to answer Exercise 36.*

36. Explain the procedure used to graph a linear inequality in two variables.

Graph each inequality in Exercises 37–42.

37. $3x - 2y \leq 5$ **38.** $5x + y > 7$ **39.** $y \leq 2$ **40.** $-3 \leq x < 5$ **41.** $|x| > 2$ **42.** $|y - 3| < 4$

43. Graph: $2x + y \leq 1$ and $x > 2y$

44. Graph: $x - y \geq 2$ or $x \geq 2$

[7.5] *Solve the following problems.*

45. If m varies directly as p^2 and inversely as q, and $m = 32$ when $p = 8$ and $q = 10$, find m when $p = 12$ and $q = 15$.

46. If x varies jointly as y and z and inversely as $\sqrt{w}$, and $x = 12$ when $y = 3$, $z = 8$, and $w = 36$, find x when $y = 5$, $z = 4$, and $w = 25$.

47. The cost of a dam is proportional to the square of the height and the length of the crest (distance across the top). If a dam with a height of 100 meters and a crest of 600 meters costs 1.5 million dollars, how much would it cost to build a dam 80 meters high with a crest of 500 meters?

48. The volume of a sphere varies directly as the cube of the radius. If the volume of a sphere with a radius of 3 meters is 36π cubic meters, find the volume of a sphere with a radius of 5 meters.

[7.6] *Give the domain and range of each relation. Identify any functions.*

49. $\{(1, 5), (2, 7), (3, 9), (4, 11)\}$

50. $\{(-4, 2), (-4, -2), (1, 5), (1, -5)\}$

51.

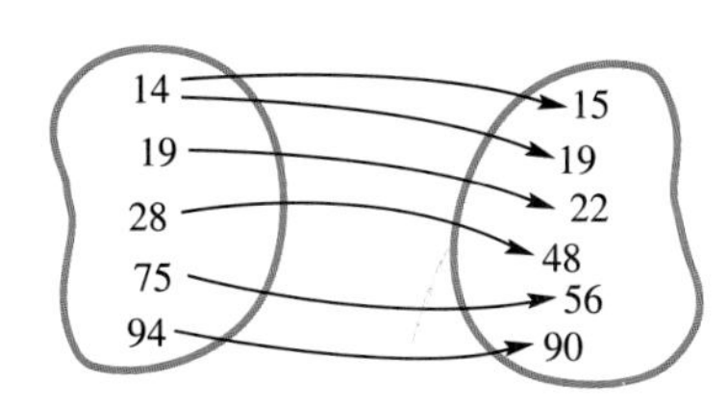

52.

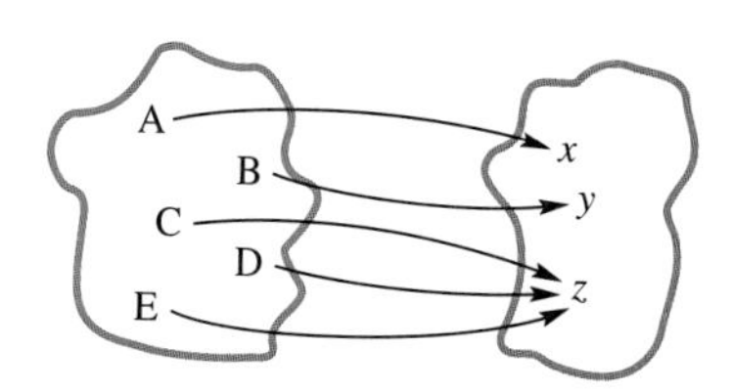

53.

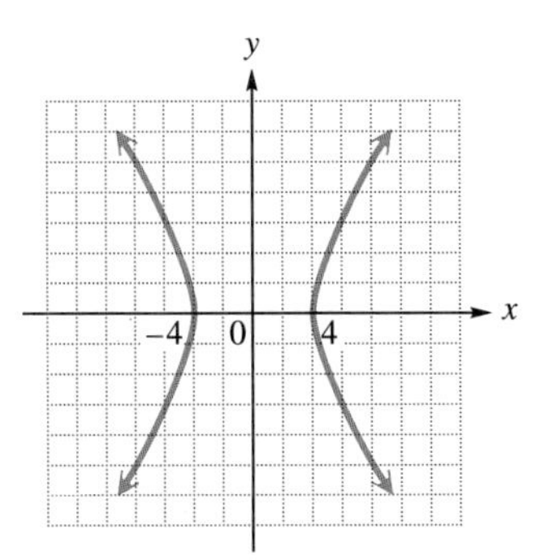

54.

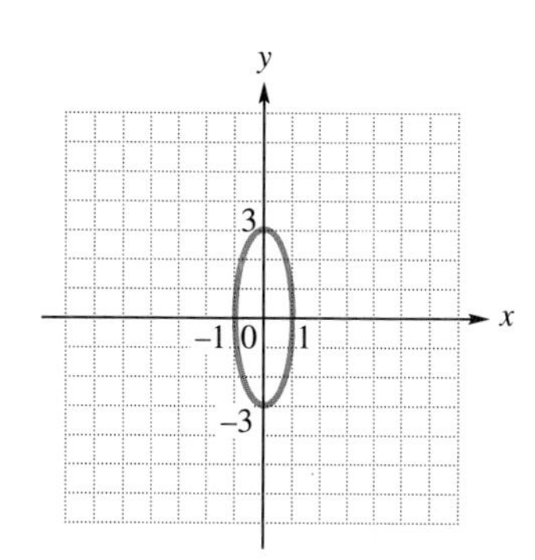

Give the domain for each of the following and identify any functions.

55. $3y = 2x - 1$

56. $|y| = x$

57. $y \leq 2x + 3$

58. $y = \dfrac{2}{x - 1}$

Given $f(x) = x^2 + 2x - 1$, *find the following.*

59. $f(-1)$

60. $f(0)$

61. $f(3)$

62. $f\left(-\dfrac{1}{2}\right)$

63. $f(k)$

64. $f[f(0)]$

Identify any graphs of functions. Use the vertical line test.

65.

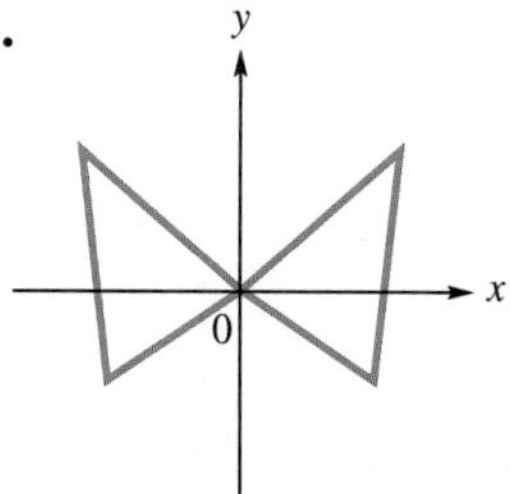

66.

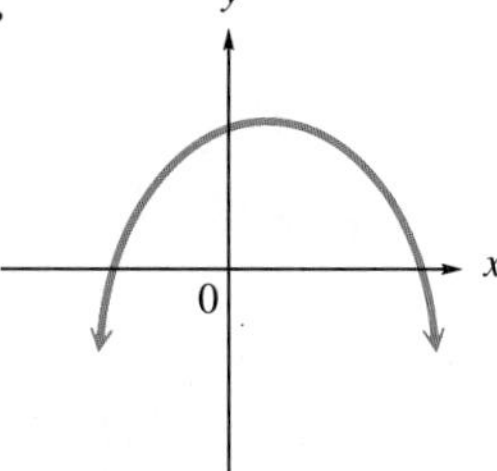

67.

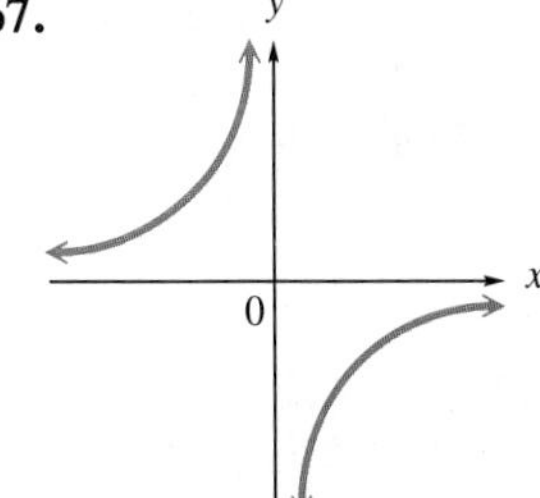

---

68. Graph the linear function $f(x) = -3x - 6$.

69. Suppose that $2x - 5y = 7$ defines a function. If $y = f(x)$, which one of the following defines the same function?

(a) $f(x) = \dfrac{7 - 2x}{5}$

(b) $f(x) = \dfrac{-7 - 2x}{5}$

(c) $f(x) = \dfrac{-7 + 2x}{5}$

(d) $f(x) = \dfrac{7 + 2x}{5}$

70. Can the graph of a linear function have undefined slope? Explain.

CHAPTER 7 TEST

1. Find the slope of the line through the points (6, 4) and (−1, 2).

For each line in Exercises 2–5 find the x- and y-intercepts, and graph the equation.

2. $3x - 2y = 20$ **3.** $x - 5y = 8$ **4.** $y = 3$ **5.** $x = -2$

6. Find the distance between the points (−3, 5) and (4, −2).

Use the slopes to decide whether the following pairs of lines are parallel, perpendicular, *or* neither.

7. $5x - y = 7$ and $5y = -x + 3$

8. $2y = 3x + 7$ and $3y = 2x + 9$

Find the equation of each line in Exercises 9–13, and write it in standard form.

9. $m = -3$, through (4, −1)

10. $m = -\frac{5}{8}$, through (−3, 7)

11. Through (2, 4) and (−2, 6)

12. Horizontal, through (−3, 4)

13. Perpendicular to $3x + 5y = 12$, through (−7, 2)

14. Which one of the following has positive slope, and negative y-coordinate for its y-intercept?

(a)

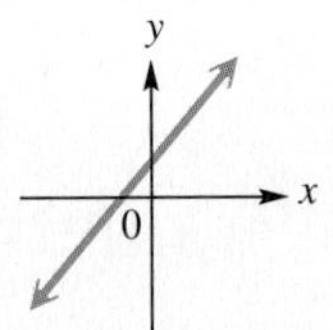

(b)

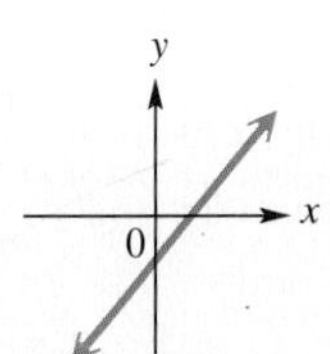

(c)

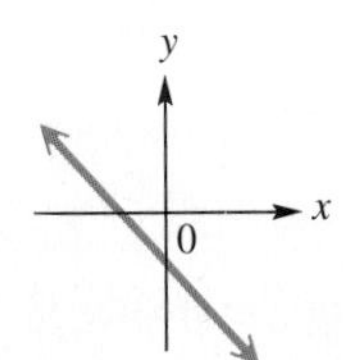

(d)

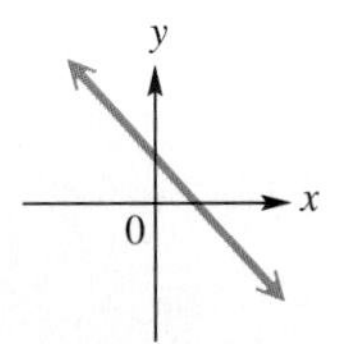

Graph each line described in Exercises 15–16.

15. The line through (−1, 3), with slope $-\frac{2}{5}$

16. The line with undefined slope, through (2, −4)

Graph the inequalities in Exercises 17–18.

17. $3x + 4y \leq 12$

18. $2x > y - 4$

19. Graph: $y < 2x - 1$ and $x - y < 3$

20. Graph: $|x| \leq 2$ or $|y| \leq 2$

Solve each problem.

21. Suppose that y varies directly as x and inversely as z. If $y = 3/2$ when $x = 3$ and $z = 10$, find y when $x = 4$ and $z = 12$.

22. The current in a simple electrical circuit is inversely proportional to the resistance. If the current is 80 amps when the resistance is 30 ohms, find the current when the resistance is 12 ohms.

23. Which of the following does not define a function?

(a) {(0, 1), (−2, 3), (4, 8)} **(b)** $y = 2x - 6$

(c) $y = \sqrt{x + 2}$

(d)

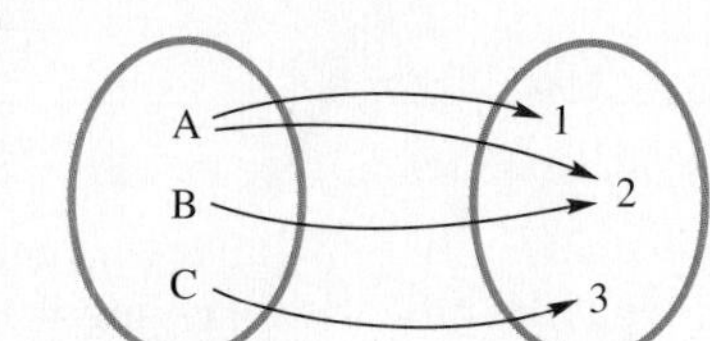

24. If $f(x) = -x^2 + 2x - 1$, find $f(1)$.

25. Graph the linear function $f(x) = \frac{2}{3}x - 1$. What is its domain? What is its range?

CHAPTER EIGHT

GRAPHING RELATIONS AND FUNCTIONS

According to a familiar saying, "A picture is worth a thousand words." Adding another way to convey information to a verbal description enhances it. For this reason, it is helpful in understanding a relation or function to see a "picture" of the set of ordered pairs that make up the relation or function.

A great deal can be learned about a function or relation by studying its graph. In Chapter 7 we discussed graphs of first-degree (linear) functions. In this chapter we show how to graph many second-degree functions and relations, as well as additional nonlinear first-degree functions and relations. With the availability of graphing calculators and computers, it is less important to be able to hand draw graphs of relations. However, a knowledge of the general characteristics of certain basic types of functions and relations is essential even for graphing with a calculator or computer.

Before continuing our study of graphs of relations, we consider ways in which simple basic functions can be combined to produce more complicated functions.

8.1 THE ALGEBRA OF FUNCTIONS

OBJECTIVES

1 FORM A NEW FUNCTION BY ADDING, SUBTRACTING, MULTIPLYING, OR DIVIDING FUNCTIONS.

2 DETERMINE THE DOMAINS OF FUNCTIONS FORMED AS IN OBJECTIVE 1.

3 FORM COMPOSITE FUNCTIONS.

FOCUS ON PROBLEM SOLVING

When a thermal inversion layer is over a city (as happens often in Los Angeles), pollutants cannot rise vertically but are trapped below the layer and must disperse horizontally. Assume that a factory smokestack begins emitting a pollutant at 8 A.M. Assume that the pollutant disperses horizontally over a circular area. If t represents the time, in hours, since the factory began emitting pollutants ($t = 0$ represents 8 A.M.), assume that the radius of the circle of pollution is $r(t) = 2t$ miles. Let $A(r) = \pi r^2$ represent the area of a circle of radius r. Find and interpret $(A \circ r)(t)$.

In this problem, the area of a circle is expressed as a function of the radius and the radius is expressed as a function of time. These two functions can be combined in a certain way to form a new function, the *composite* of the area and radius functions, that expresses area as a function of time. This method of combining functions is useful in many applications. This problem is Exercise 53 in the exercises for this section. After working through the section, you should be able to solve this problem.

When a company accountant sits down to estimate the firm's overhead, the first step might be to find functions representing the cost of materials, labor charges, equipment maintenance, and so on. The sum of these various functions could then be used to find the total overhead for the company. This example illustrates a use for a function that is the sum of two or more simpler functions.

1 Given two functions f and g, their *sum*, written $f + g$, is defined as

$$(f + g)(x) = f(x) + g(x),$$

for all x such that both $f(x)$ and $g(x)$ exist. Similar definitions can be given for the difference, $f - g$, product, fg, and quotient, f/g, of functions; however, the quotient,

$$\left(\frac{f}{g}\right)(x) = \frac{f(x)}{g(x)},$$

is defined only for those values of x where both $f(x)$ and $g(x)$ exist, with the additional condition that $g(x) \neq 0$. The various operations on functions are defined as follows.

OPERATIONS ON FUNCTIONS

If f and g are functions, then for all values of x for which both $f(x)$ and $g(x)$ exist,
the sum of f and g is

$$(f + g)(x) = f(x) + g(x),$$

the difference of f and g is

$$(f - g)(x) = f(x) - g(x),$$

the product of f and g is

$$(fg)(x) = f(x) \cdot g(x),$$

and the quotient of f and g is

$$\left(\frac{f}{g}\right)(x) = \frac{f(x)}{g(x)}, \quad \text{where } g(x) \neq 0.$$

NOTE The condition $g(x) \neq 0$ in the definition of the quotient means that the domain of $(f/g)(x)$ consists of all values of x for which $f(x)$ is defined and $g(x)$ is not zero.

EXAMPLE 1 USING THE OPERATIONS ON FUNCTIONS

Let $f(x) = x^2 + 1$, and $g(x) = 3x + 5$. Find each of the following.

(a) $(f + g)(1)$
Since $f(1) = 1^2 + 1 = 2$ and $g(1) = 3(1) + 5 = 8$, use the definition above to get

$$(f + g)(1) = f(1) + g(1) = 2 + 8 = 10.$$

(b) $(f - g)(-3) = [(-3)^2 + 1] - [3(-3) + 5]$
$= 10 - (-4) = 14$

(c) $(fg)(5) = f(5) \cdot g(5) = [5^2 + 1] \cdot [3(5) + 5]$
$= 26 \cdot 20 = 520$

(d) $\left(\frac{f}{g}\right)(0) = \frac{f(0)}{g(0)} = \frac{0^2 + 1}{3 \cdot 0 + 5} = \frac{1}{5}$ ■

EXAMPLE 2 USING THE OPERATIONS ON FUNCTIONS

Let $f(x) = 8x - 9$ and $g(x) = \sqrt{2x - 1}$.

(a) $(f + g)(x) = f(x) + g(x) = 8x - 9 + \sqrt{2x - 1}$

(b) $(f - g)(x) = f(x) - g(x) = 8x - 9 - \sqrt{2x - 1}$

(c) $(fg)(x) = f(x) \cdot g(x) = (8x - 9)\sqrt{2x - 1}$

(d) $\left(\frac{f}{g}\right)(x) = \frac{f(x)}{g(x)} = \frac{8x - 9}{\sqrt{2x - 1}}$ ■

2 In Example 2 the domain of f is the set of all real numbers, while the domain of $g(x) = \sqrt{2x - 1}$ includes just those real numbers that make $2x - 1 \geq 0$; the domain of g is the interval $[1/2, \infty)$. The domain of $f + g$, $f - g$, and fg is thus $[1/2, \infty)$. With f/g, the denominator cannot be zero, so the value 1/2 is excluded from the domain. The domain of f/g is $(1/2, \infty)$.

The domains of $f + g$, $f - g$, fg, and f/g are summarized below. (Recall that the intersection of two sets is the set of all elements belonging to *both* sets.)

DOMAINS OF $f + g$, $f - g$, fg, f/g

For functions f and g, the domains of $f + g$, $f - g$, and fg include all real numbers in the intersection of the domains of f and g, while the domain of f/g includes those real numbers in the intersection of the domains of f and g for which $g(x) \neq 0$.

3 The sketch in Figure 8.1 shows a function f that assigns to each element x of set X some element y of set Y. Suppose also that a function g takes each element of set Y and assigns a value z of set Z. Using both f and g, then, an element x in X is assigned to an element z in Z. The result of this process is a new function h, which takes an element x in X and assigns an element z in Z. This function h is called the *composition* of functions g and f, written $g \circ f$, and defined as follows.

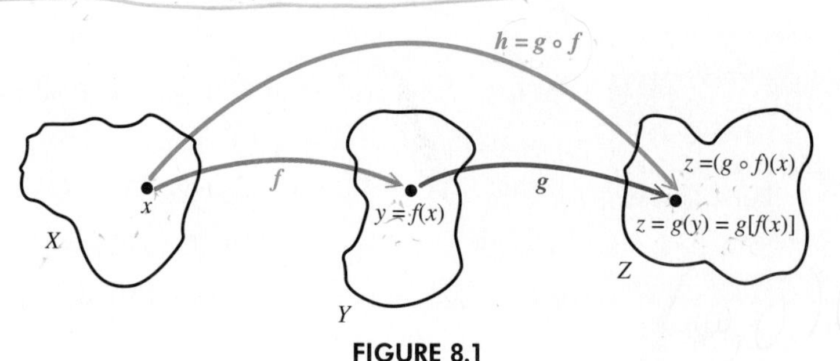

FIGURE 8.1

DEFINITION OF COMPOSITION OF FUNCTIONS

If f and g are functions, then the **composite function,** or **composition,** of g and f is defined by

$$(g \circ f)(x) = g[f(x)]$$

for all x in the domain of f such that $f(x)$ is in the domain of g.

Read $f \circ g$ as "f of g".

Suppose an oil well off the California coast is leaking, with the leak spreading oil in a circular layer over the surface. At any time t, in minutes, after the beginning of the leak, the radius of the circular oil slick is $r(t) = 5t$ feet. Since $A(r) = \pi r^2$ gives the area of a circle of radius r, the area can be expressed as a function of time by substituting $5t$ for r in $A(r) = \pi r^2$ to get

$$A(r) = \pi r^2$$
$$A[r(t)] = \pi(5t)^2 = 25\pi t^2$$

The function $A[r(t)]$ is a composite function of the functions A and r.

EXAMPLE 3
EVALUATING COMPOSITE FUNCTIONS

Given $f(x) = 2x - 1$ and $g(x) = \dfrac{4}{x - 1}$, find each of the following.

(a) $f[g(2)]$

First find $g(2)$. Since

$$g(x) = \frac{4}{x - 1},$$

$$g(2) = \frac{4}{2 - 1} = \frac{4}{1} = 4.$$

Now find $f[g(2)] = f(4)$.

$$f(x) = 2x - 1$$

$$f[g(2)] = f(4) = 2(4) - 1 = 7$$

(b) $g[f(-3)]$

$$f(-3) = 2(-3) - 1 = -7$$

$$g[f(-3)] = g(-7) = \frac{4}{-7 - 1} = \frac{4}{-8} = -\frac{1}{2} \quad \blacksquare$$

EXAMPLE 4
FINDING COMPOSITE FUNCTIONS

Let $f(x) = 4x + 1$ and $g(x) = 2x^2 + 5x$. Find each of the following.

(a) $(g \circ f)(x)$

By definition, $(g \circ f)(x) = g[f(x)]$. Using the given functions,

$$\begin{aligned}
(g \circ f)(x) = g[f(x)] &= g(4x + 1) && f(x) = 4x + 1 \\
&= 2(4x + 1)^2 + 5(4x + 1) && g(x) = 2x^2 + 5x \\
&= 2(16x^2 + 8x + 1) + 20x + 5 && \text{Multiply.} \\
&= 32x^2 + 16x + 2 + 20x + 5 && \text{Distributive property} \\
&= 32x^2 + 36x + 7. && \text{Combine terms.}
\end{aligned}$$

(b) $(f \circ g)(x)$

If we use the definition above with f and g interchanged, $(f \circ g)(x)$ becomes $f[g(x)]$, with

$$\begin{aligned}
(f \circ g)(x) &= f[g(x)] \\
&= f(2x^2 + 5x) && g(x) = 2x^2 + 5x \\
&= 4(2x^2 + 5x) + 1 && f(x) = 4x + 1 \\
&= 8x^2 + 20x + 1. && \text{Distributive property} \quad \blacksquare
\end{aligned}$$

As Example 4 shows, it is not always true that $f \circ g = g \circ f$. In fact, two composite functions are equal only for a special class of functions, discussed in Chapter 10. In Example 4, the domain of both composite functions is the set of all real numbers.

CAUTION In general, the composite function $f \circ g$ is not the same as the product fg. For example, with f and g defined as in Example 4,

$$f \circ g = 8x^2 + 20x + 1$$

but

$$fg = (4x + 1)(2x^2 + 5x) = 8x^3 + 22x^2 + 5x.$$

EXAMPLE 5
FINDING COMPOSITE FUNCTIONS

Let $f(x) = 1/x$ and $g(x) = \sqrt{3 - x}$. Find $f \circ g$ and $g \circ f$. Give the domain of each.

First find $f \circ g$.

$$\begin{aligned}(f \circ g)(x) &= f[g(x)] \\ &= f(\sqrt{3 - x}) \\ &= \frac{1}{\sqrt{3 - x}}\end{aligned}$$

The radical $\sqrt{3 - x}$ is a nonzero real number only when $3 - x > 0$ or $x < 3$, so that the domain of $f \circ g$ is the interval $(-\infty, 3)$.

Use the same functions to find $g \circ f$, as follows.

$$\begin{aligned}(g \circ f)(x) &= g[f(x)] \\ &= g\left(\frac{1}{x}\right) && f(x) = \frac{1}{x} \\ &= \sqrt{3 - \frac{1}{x}} && g(x) = \sqrt{3 - x} \\ &= \sqrt{\frac{3x - 1}{x}} && \text{Write as a single fraction.}\end{aligned}$$

The domain of $g \circ f$ is the set of all real numbers x such that $x \neq 0$ and $3 - f(x) \geq 0$. By the methods of Section 6.6, the domain of $g \circ f$ is the set $(-\infty, 0) \cup [1/3, \infty)$. ■

EXAMPLE 6
FINDING THE FUNCTIONS THAT FORM A GIVEN COMPOSITE

Find functions f and g such that

$$(f \circ g)(x) = (x^2 - 5)^3 - 4(x^2 - 5) + 3.$$

One pair of functions that will work is

$$f(x) = x^3 - 4x + 3 \quad \text{and} \quad g(x) = x^2 - 5.$$

Then

$$\begin{aligned}(f \circ g)(x) &= f[g(x)] \\ &= f(x^2 - 5) \\ &= (x^2 - 5)^3 - 4(x^2 - 5) + 3.\end{aligned}$$

There are other pairs of functions f and g that also work. For instance,

$$f(x) = (x - 5)^3 - 4(x - 5) + 3$$

and

$$g(x) = x^2. \quad ■$$

8.1 EXERCISES

For each pair of functions defined as follows, find **(a)** $f + g$, **(b)** $f - g$, **(c)** fg, *and* **(d)** f/g. *Give the domain of each. See Example 2.*

1. $f(x) = 4x - 1, \quad g(x) = 6x + 3$

2. $f(x) = 9 - 2x, \quad g(x) = -5x + 2$

3. $f(x) = 3x^2 - 2x, \quad g(x) = x^2 - 2x + 1$

4. $f(x) = 6x^2 - 11x, \quad g(x) = x^2 - 4x - 5$

5. $f(x) = \sqrt{2x + 5}, \quad g(x) = \sqrt{4x - 9}$

6. $f(x) = \sqrt{11x - 3}, \quad g(x) = \sqrt{2x - 15}$

Let $f(x) = 4x^2 - 2x$ *and let* $g(x) = 8x + 1$. *Find each of the following. See Examples 1–4.*

7. $(f + g)(3)$ **8.** $(f + g)(-5)$ **9.** $(fg)(4)$ **10.** $(fg)(-3)$ **11.** $\left(\frac{f}{g}\right)(-1)$

12. $\left(\frac{f}{g}\right)(4)$ **13.** $(f + g)(m)$ **14.** $(f - g)(2k)$ **15.** $(f \circ g)(2)$ **16.** $(f \circ g)(-5)$

17. $(g \circ f)(2)$ **18.** $(g \circ f)(-5)$ **19.** $(f \circ g)(k)$ **20.** $(g \circ f)(5z)$

Find $f \circ g$ *and* $g \circ f$ *and their domains for each pair of functions defined as follows. See Examples 4 and 5.*

21. $f(x) = 8x + 12, \quad g(x) = 3x - 1$

22. $f(x) = -6x + 9, \quad g(x) = 5x + 7$

23. $f(x) = 5x + 3, \quad g(x) = -x^2 + 4x + 3$

24. $f(x) = 4x^2 + 2x + 8, \quad g(x) = x + 5$

25. $f(x) = \frac{1}{x}, \quad g(x) = x^2$

26. $f(x) = \frac{2}{x^4}, \quad g(x) = 2 - x$

27. $f(x) = \sqrt{x + 2}, \quad g(x) = 8x - 6$

28. $f(x) = 9x - 11, \quad g(x) = 2\sqrt{x + 2}$

29. $f(x) = \frac{1}{x - 5}, \quad g(x) = \frac{2}{x}$

30. $f(x) = \frac{8}{x - 6}, \quad g(x) = \frac{4}{3x}$

The graphs of functions f and g are shown. Use these graphs to find the values in Exercises 31–38.

31. $f(1) + g(1)$ **32.** $f(4) - g(3)$

33. $f(-2) \cdot g(4)$ **34.** $\frac{f(4)}{g(2)}$

35. $(f \circ g)(2)$ **36.** $(g \circ f)(2)$

37. $(g \circ f)(-4)$ **38.** $(f \circ g)(-2)$

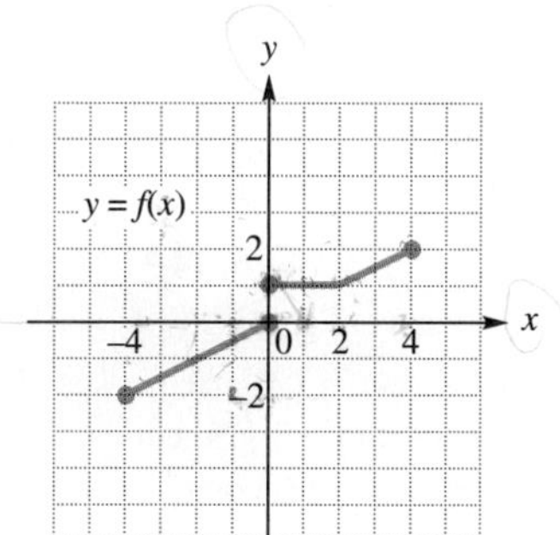

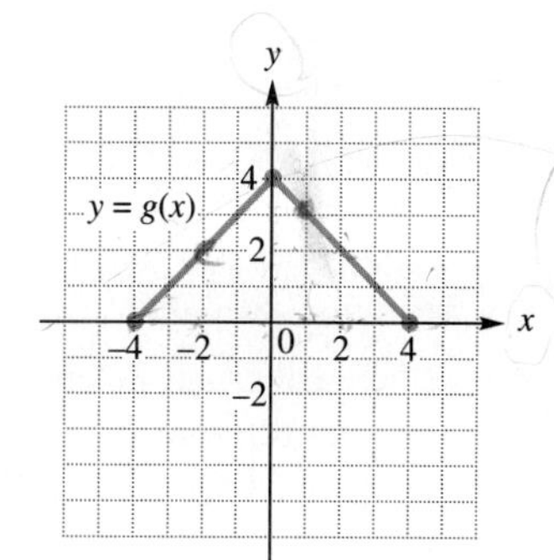

For each pair of functions defined as follows, show that $(f \circ g)(x) = x$ *and* $(g \circ f)(x) = x$.

39. $f(x) = 8x, \quad g(x) = \frac{1}{8}x$

40. $f(x) = \frac{3}{4}x, \quad g(x) = \frac{4}{3}x$

41. $f(x) = 8x - 11, \quad g(x) = \dfrac{x + 11}{8}$

42. $f(x) = \dfrac{x - 3}{4}, \quad g(x) = 4x + 3$

43. $f(x) = x^3 + 6, \quad g(x) = \sqrt[3]{x - 6}$

44. $f(x) = \sqrt[5]{x - 9}, \quad g(x) = x^5 + 9$

In each of the following exercises, a function h is given. Find functions f and g such that $h(x) = (f \circ g)(x)$. Many such pairs of functions exist. See Example 6.

45. $h(x) = (6x - 2)^2$

46. $h(x) = (11x^2 + 12x)^2$

47. $h(x) = \sqrt{x^2 - 1}$

48. $h(x) = \dfrac{1}{x^2 + 2}$

49. $h(x) = \dfrac{(x - 2)^2 + 1}{5 - (x - 2)^2}$

50. $h(x) = (x + 2)^3 - 3(x + 2)^2$

---

51. Suppose the demand for a certain brand of vacuum cleaner is given by

$$D(p) = \frac{-p^2}{100} + 500,$$

where p is the price in dollars. If the price, in terms of the cost, c, is expressed as

$$p(c) = 2c - 10,$$

find $D(c)$, the demand in terms of the cost.

52. Suppose the population P of a certain species of fish depends on the number x (in hundreds) of a smaller kind of fish which serves as its food supply, so that

$$P(x) = 2x^2 + 1.$$

Suppose, also, that the number x (in hundreds) of the smaller species of fish depends upon the amount a (in appropriate units) of its food supply, a kind of plankton. Suppose

$$x = f(a) = 3a + 2.$$

Find $(P \circ f)(a)$, the relationship between the population P of the large fish and the amount a of plankton available.

53. When a thermal inversion layer is over a city (as happens often in Los Angeles), pollutants cannot rise vertically but are trapped below the layer and must disperse horizontally. Assume that a factory smokestack begins emitting a pollutant at 8 A.M. Assume that the pollutant disperses horizontally over a circular area. If t represents the time, in hours, since the factory began emitting pollutants ($t = 0$ represents 8 A.M.), assume that the radius of the circle of pollution is $r(t) = 2t$ miles. Let

$A(r) = \pi r^2$ represent the area of a circle of radius r. Find and interpret $(A \circ r)(t)$.

54. An oil well off the Gulf Coast is leaking, with the leak spreading oil over the surface as a circle. At any time t, in minutes, after the beginning of the leak, the radius of the circular oil slick on the surface is $r(t) = 4t$ feet. Let $A(r) = \pi r^2$ represent the area of a circle of radius r. Find and interpret $(A \circ r)(t)$.

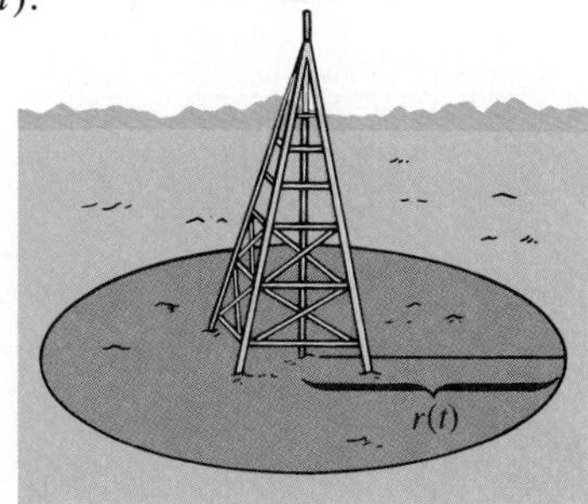

55. In your own words, explain how to find the function values for $f + g$. (See Objective 1.)

56. In your own words, explain how to find the function values for fg. (See Objective 1.)

PREVIEW EXERCISES

Complete each table of values and graph the function. See Section 7.1.

57. $y = x$

x	-3	0	3
y			

58. $y = -x$

x	-3	0	3
y			

59. $y = x - 3$

x	-2	0	2
y			

60. $y = -x + 3$

x	-2	0	2
y			

61. $y = -\frac{1}{2}x + 5$

x	-4	0	4
y			

62. $y = \frac{2}{3}x - 1$

x	-3	0	6
y			

8.2 QUADRATIC FUNCTIONS; PARABOLAS

OBJECTIVES

1. GRAPH $f(x) = x^2$.
2. GRAPH $f(x) = ax^2$.
3. GRAPH TRANSLATIONS OF $f(x) = ax^2$.
4. GRAPH $f(x) = a(x - h)^2 + k$.
5. USE THE GEOMETRIC DEFINITION OF A PARABOLA.

In Chapter 7 we discussed first-degree (linear) functions, those where the highest power of the variable is 1. In this section, we will look at *quadratic functions,* which are second-degree functions.

DEFINITION OF QUADRATIC FUNCTION

A function f is a **quadratic function** if

$$f(x) = ax^2 + bx + c,$$

where a, b, and c are real numbers, with $a \neq 0$.

x	$f(x)$
-2	4
-1	1
0	0
1	1
2	4

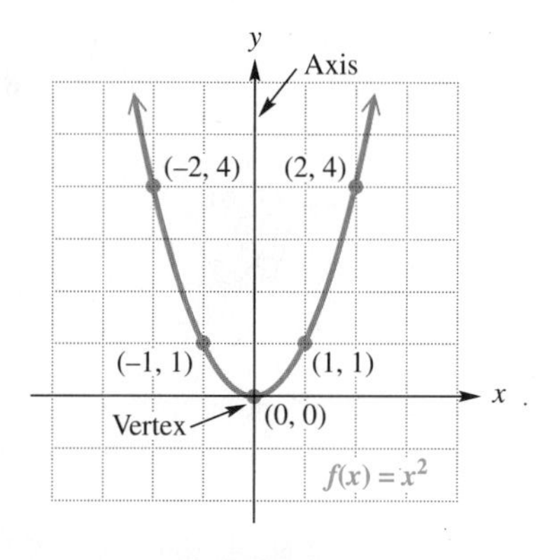

FIGURE 8.2

1 The simplest second-degree function is defined by $f(x) = x^2$. This function can be graphed by finding several ordered pairs that satisfy the equation: for example, (0, 0), (1, 1), (−1, 1), (2, 4), (−2, 4), (1/2, 1/4), (−1/2, 1/4), (3/2, 9/4), and (−3/2, 9/4). Plotting these points and drawing a smooth curve through them gives the graph shown in Figure 8.2. This graph is called a **parabola.** (The formal definition of a parabola is given at the end of this section.) Every quadratic function has a graph that is a parabola.

Parabolas are symmetric about a line (the y-axis in Figure 8.2). Intuitively, this means that if the graph were folded along the line of symmetry, the two sides would coincide. We discuss symmetry in more detail later in this chapter. The line of symmetry for a parabola is called the **axis** of the parabola. The point where the axis intersects the parabola is the **vertex** of the parabola. The vertex is the lowest (or highest) point of a vertical parabola.

Parabolas have many practical applications. For example, the reflectors of solar ovens and flashlights are made by revolving a parabola about its axis. The **focus** of a parabola is a point on its axis that determines the curvature. See Figure 8.3. (The focus is discussed in more detail later in this section.) When the parabolic reflector of a solar oven is aimed at the sun, the light rays bounce off the reflector and collect at the focus, creating an intense temperature at that point. On the other hand, when a lightbulb is placed at the focus of a parabolic reflector, light rays reflect out parallel to the axis.

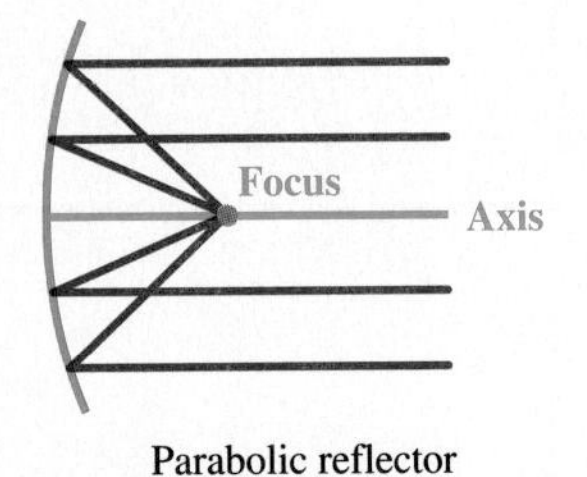

Parabolic reflector

FIGURE 8.3

There are several ways to use the algebra of functions to get variations of $f(x) = x^2$.

1. $f(x) = ax^2$ Multiply $g(x) = a$ and $h(x) = x^2$ to get $f(x)$.
2. $f(x) = x^2 + k$ Add $h(x) = x^2$ and $g(x) = k$ to get $f(x)$.
3. $f(x) = (x - h)^2$ Form the composite function $H[g(x)]$ with $H(x) = x^2$ and $g(x) = x - h$.
4. $f(x) = a(x - h)^2 + k$ Form $H[g(x)]$ with $H(x) = ax^2 + k$ and $g(x) = x - h$.

Each of these functions has a graph that is a parabola, but in each case the graph is modified from that of $f(x) = x^2$ as shown in the next examples.

2 The first example shows the result of changing $f(x) = x^2$ to $f(x) = ax^2$, for a nonzero constant a.

EXAMPLE 1
GRAPHING PARABOLAS OF THE FORM $f(x) = ax^2$

Graph the functions defined as follows.

(a) $g(x) = -x^2$

For a given value of x, the corresponding value of $g(x)$ will be the negative of what it was for $f(x) = x^2$. (See the table of values with Figure 8.4.) Because of this, the graph of $g(x) = -x^2$ is the same shape as that of $f(x)$, but opens downward. See Figure 8.4. This is generally true; the graph of $f(x) = ax^2 + bx + c$ opens downward whenever $a < 0$.

x	y
−2	−4
−1	−1
0	0
1	−1
2	−4

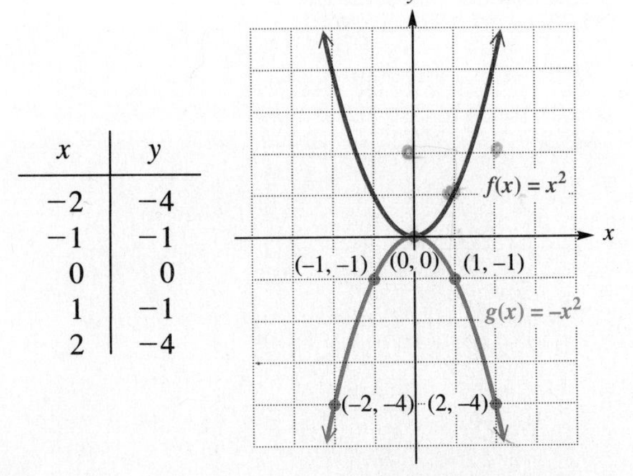

FIGURE 8.4

(b) $g(x) = \dfrac{1}{2}x^2$

Choose a value of x, and then find $g(x)$. The coefficient 1/2 will cause the resulting value of $g(x)$ to be smaller than for $f(x) = x^2$, making the parabola wider than $f(x) = x^2$. See Figure 8.5. In both parabolas of this example, the axis is the vertical line $x = 0$ and the vertex is the origin (0, 0). ■

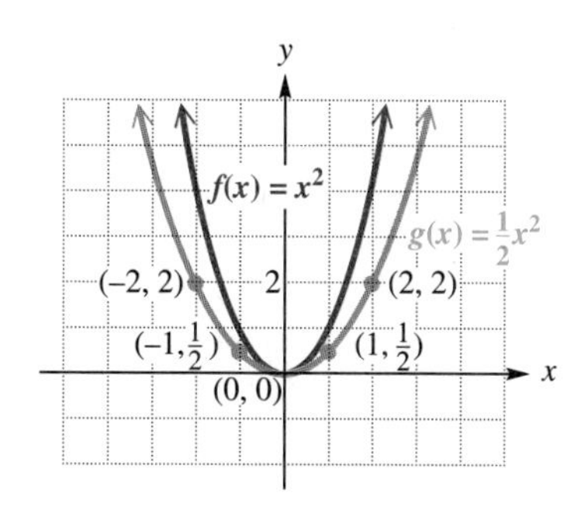

FIGURE 8.5

3 The next few examples show the results of horizontal and vertical shifts, called **translations,** of the graph of $f(x) = x^2$.

EXAMPLE 2
GRAPHING A PARABOLA WITH A VERTICAL SHIFT

Graph $g(x) = x^2 - 4$.

By comparing the tables of values for $g(x) = x^2 - 4$ and $f(x) = x^2$ shown with Figure 8.6, we can see that for corresponding x-values, the y-values of g are each 4 less than those for f. Thus, the graph of $f(x) = x^2 - 4$ is the same as that of $f(x) = x^2$, but translated 4 units down. See Figure 8.6. The vertex of this parabola (here the lowest point) is at $(0, -4)$. The axis of the parabola is the vertical line $x = 0$. ■

$g(x) = x^2 - 4$

x	y
−2	0
−1	−3
0	−4
1	−3
2	0

$f(x) = x^2$

x	y
−2	4
−1	1
0	0
1	1
2	4

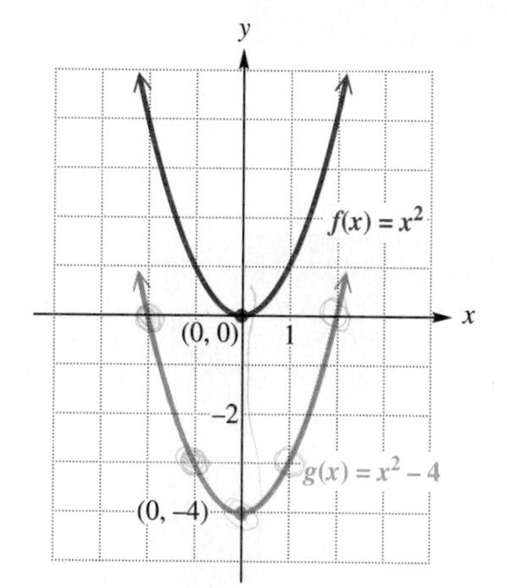

FIGURE 8.6

EXAMPLE 3
GRAPHING A PARABOLA WITH A HORIZONTAL SHIFT

Graph $g(x) = (x - 4)^2$.

Comparing the tables of values shown with Figure 8.7 shows that the graph of $g(x) = (x - 4)^2$ is the same as that of $f(x) = x^2$, but translated 4 units right. The vertex is at (4, 0). As shown in Figure 8.7, the axis of this parabola is the vertical line $x = 4$. ■

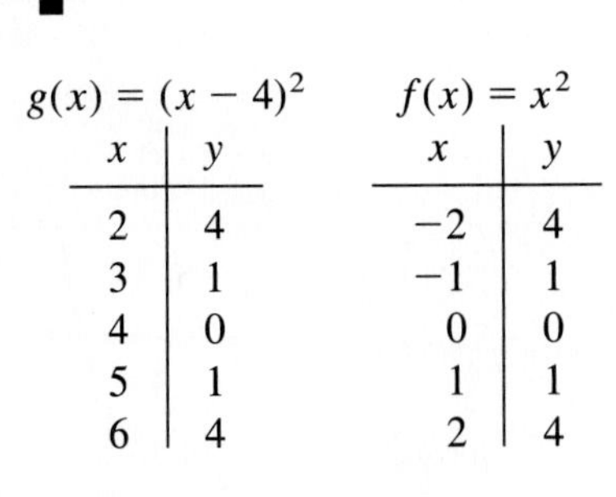

$g(x) = (x - 4)^2$

x	y
2	4
3	1
4	0
5	1
6	4

$f(x) = x^2$

x	y
−2	4
−1	1
0	0
1	1
2	4

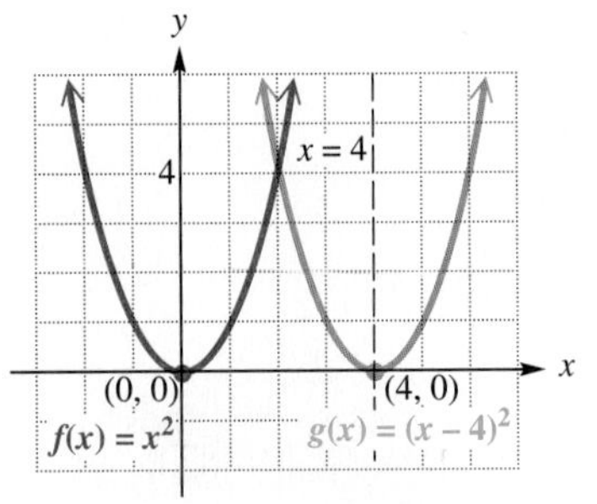

FIGURE 8.7

CAUTION Errors frequently occur when horizontal shifts are involved. In order to determine the direction and magnitude of horizontal shifts, find the value that would cause the expression $x - h$ to equal 0. For example, the graph of $f(x) = (x - 5)^2$ would be shifted 5 units to the *right,* because +5 would cause $x - 5$ to equal 0. On the other hand, the graph of $f(x) = (x + 4)^2$ would be shifted 4 units to the *left,* because −4 would cause $x + 4$ to equal 0.

4 Examples 1–3 suggest the following general principles for graphing $f(x) = a(x - h)^2 + k$.

GENERAL PRINCIPLES

1. The graph of $f(x) = a(x - h)^2 + k$, where $a \neq 0$, is a parabola with vertex (h, k), and the vertical line $x = h$ as axis.
2. The graph opens upward if a is positive and downward if a is negative.
3. The graph is wider than $f(x) = x^2$ if $0 < |a| < 1$. The graph is narrower than $f(x) = x^2$ if $|a| > 1$.

EXAMPLE 4
USING THE GENERAL PRINCIPLES TO GRAPH A PARABOLA

Graph $f(x) = -2(x + 3)^2 + 4$.

The parabola opens downward (because $a < 0$), and is narrower than the graph of $f(x) = x^2$, since $a = -2$, and $|-2| > 1$. This parabola has vertex at (−3, 4), as shown in Figure 8.8. To complete the graph, we plotted the ordered pairs (−4, 2) and (−2, 2). ■

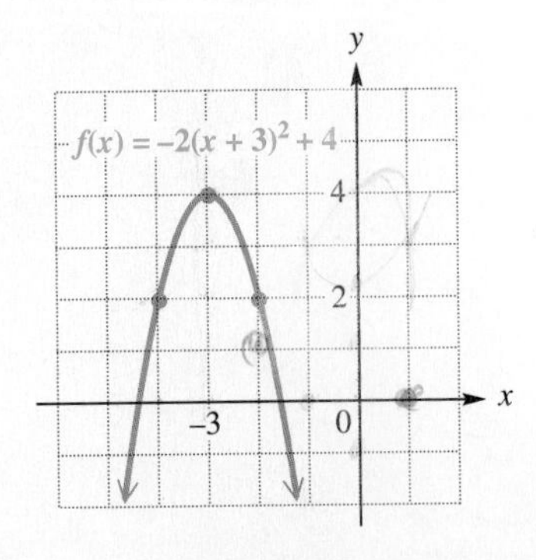

FIGURE 8.8

5 Geometrically, a parabola is defined as the set of all points in a plane that are equally distant from a fixed point and a fixed line not containing the point. The point is called the **focus** and the line is the **directrix.** The line through the focus and perpendicular to the directrix is the axis of the parabola. The point on the axis that is equally distant from the focus and the directrix is the vertex of the parabola. See Figure 8.9.

The parabola in Figure 8.9 has the point $(0, p)$ as focus and the line $y = -p$ as directrix. The vertex is $(0, 0)$. Let (x, y) be any point on the parabola. The distance from (x, y) to the directrix is $|y - (-p)|$, while the distance from (x, y) to $(0, p)$ is $\sqrt{(x - 0)^2 + (y - p)^2}$. Since (x, y) is equally distant from the directrix and the focus,

$$|y - (-p)| = \sqrt{(x - 0)^2 + (y - p)^2}.$$

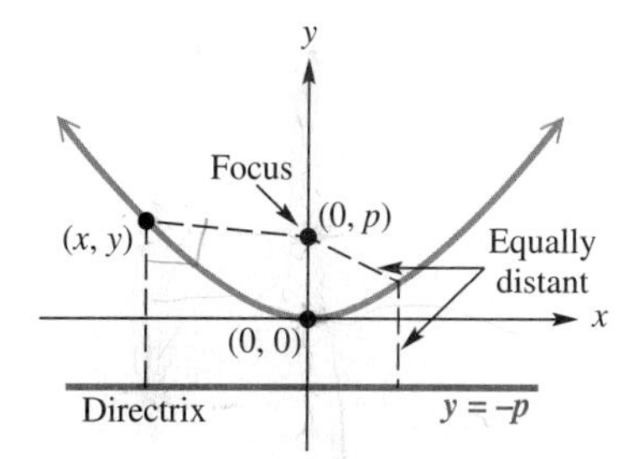

FIGURE 8.9

Square both sides, getting

$$(y + p)^2 = x^2 + (y - p)^2$$

$$y^2 + 2py + p^2 = x^2 + y^2 - 2py + p^2 \quad \text{Square of a binomial}$$

$$4py = x^2, \quad \text{Combine terms.}$$

the equation of the parabola with focus $(0, p)$ and directrix $y = -p$. Solving $4py = x^2$ for y gives

$$y = \frac{1}{4p}x^2,$$

so that $1/(4p) = a$ when the equation is written as $y = ax^2 + bx + c$.

This result could be extended to a parabola with vertex at (h, k), focus p units above (h, k), and directrix p units below (h, k), or to a parabola with vertex at (h, k), focus p units below (h, k), and directrix p units above (h, k).

EXAMPLE 5 **USING THE GEOMETRIC DEFINITION OF A PARABOLA**

Use the geometric definition to find the equation of the parabola with focus at $(0, 3)$ and directrix $y = -3$.

Since the vertex is halfway between the focus and the directrix, the vertex is $(0, 0)$, as shown in Figure 8.10. The distance between the focus and the vertex is $p = 3$. Thus, the equation is

$$y = \frac{1}{4p}x^2$$

$$y = \frac{1}{4(3)}x^2$$

$$y = \frac{1}{12}x^2$$

or $\quad 12y = x^2.$ ■

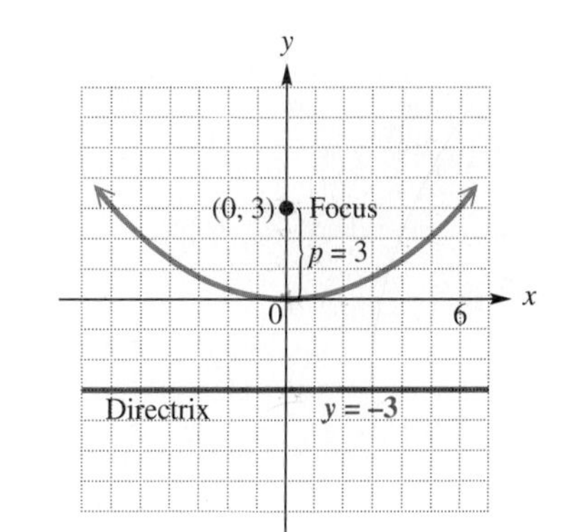

FIGURE 8.10

8.2 EXERCISES

Identify the vertex of each parabola. See Examples 1–4.

1. $f(x) = x^2 - 5$ **2.** $f(x) = x^2 + 7$ **3.** $f(x) = 5x^2$ **4.** $f(x) = -8x^2$

5. $f(x) = (x + 4)^2$ **6.** $f(x) = (x + 2)^2$ **7.** $f(x) = (x - 9)^2 + 12$ **8.** $f(x) = (x - 7)^2 - 3$

9. $f(x) = (x - t)^2 + r$ **10.** $f(x) = (x + w)^2 - v$

For each quadratic function, tell whether the graph opens upward or downward, and whether the graph is wider, narrower, or the same as $f(x) = x^2$. See Examples 1 and 4.

11. $f(x) = x^2 - 3$ **12.** $f(x) = x^2 + 6$ **13.** $f(x) = -5x^2$

14. $f(x) = \frac{1}{3}x^2 - 1$ **15.** $f(x) = \frac{3}{4}x^2 + 4$ **16.** $f(x) = -12x^2$

---

17. Explain what causes the graph of a quadratic function of the form $f(x) = ax^2$ to be wider than the graph of $f(x) = x^2$ or narrower than the graph of $f(x) = x^2$. (See Objective 2.)

18. For $f(x) = a(x - h)^2 + k$, in what quadrant is the vertex if:
(a) $h > 0, k > 0$; **(b)** $h > 0, k < 0$;
(c) $h < 0, k > 0$; **(d)** $h < 0, k < 0$?
(See Objective 4.)

In Exercises 19–26, match the graph of each equation with the graph at the right that it most closely resembles. (See Objectives 3 and 4.)

19. $f(x) = x^2 + 2$

20. $g(x) = x^2 - 5$

21. $h(x) = -x^2 + 4$

22. $k(x) = -x^2 - 4$

23. $F(x) = (x - 1)^2$

24. $G(x) = (x + 1)^2$

25. $H(x) = (x - 1)^2 + 1$

26. $K(x) = (x + 1)^2 + 1$

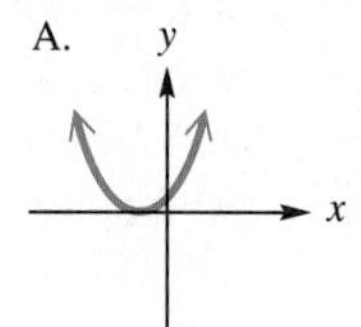

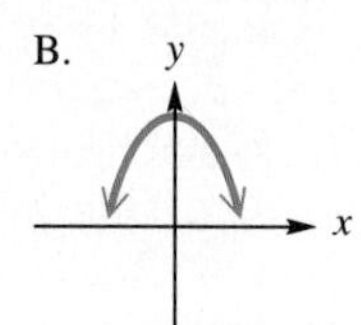

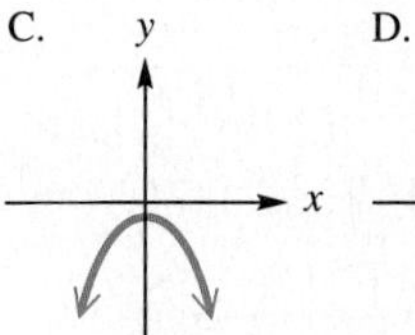

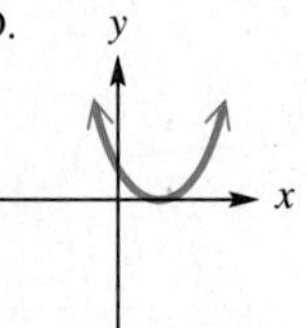

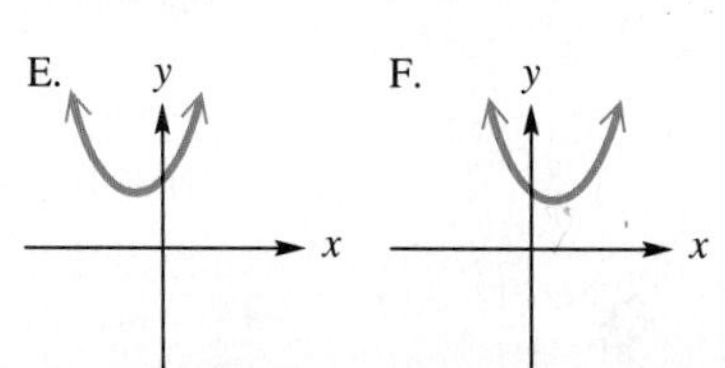

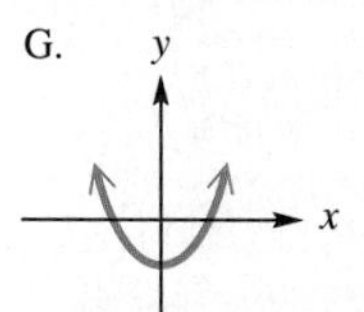

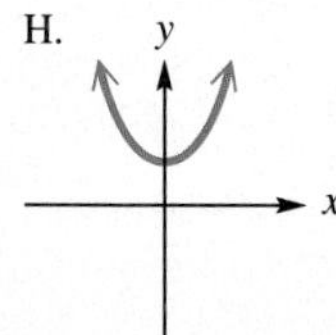

Sketch the graph of each parabola. Plot at least two points in addition to the vertex. See Examples 1–4.

27. $f(x) = 3x^2$ **28.** $f(x) = -2x^2$ **29.** $f(x) = -\frac{1}{4}x^2$ **30.** $f(x) = \frac{1}{3}x^2$

31. $f(x) = x^2 - 1$ **32.** $f(x) = x^2 + 3$ **33.** $f(x) = -x^2 + 2$ **34.** $f(x) = -x^2 - 4$

35. $f(x) = 2x^2 - 2$ **36.** $f(x) = -3x^2 + 1$ **37.** $f(x) = (x - 4)^2$ **38.** $f(x) = (x - 3)^2$

39. $f(x) = 3(x + 1)^2$ **40.** $f(x) = -2(x + 1)^2$ **41.** $f(x) = (x + 1)^2 - 2$ **42.** $f(x) = (x - 2)^2 + 3$

43. $f(x) = 2(x - 1)^2 - 3$ **44.** $f(x) = -3(x + 4)^2 + 5$ **45.** $f(x) = -3(x + 2)^2 + 2$

46. $f(x) = -2(x + 1)^2 - 3$ **47.** $f(x) = \frac{2}{3}(x - 1)^2 - 2$ **48.** $f(x) = \frac{5}{4}(x - 2)^2 - 3$

Use the geometric definition of a parabola to find the equation of each of the following parabolas. See Example 5.

49. focus $(0, 2)$, directrix $y = -2$ **50.** focus $(0, -1/2)$, directrix $y = 1/2$

51. focus $(0, -1)$, directrix $y = 1$ **52.** focus $(0, 1/4)$, directrix $y = -1/4$

53. focus $(0, 4)$, directrix $y = 2$ **54.** focus $(0, -1)$, directrix $y = -2$

PREVIEW EXERCISES

Solve each quadratic equation by completing the square. Give only real number answers. See Section 6.1.

55. $x^2 + 6x - 3 = 0$ **56.** $x^2 + 8x - 4 = 0$ **57.** $2x^2 - 12x = 5$

58. $3x^2 - 12x = 10$ **59.** $-x^2 - 3x + 2 = 0$ **60.** $-2x^2 + 5x - 4 = 0$

8.3 MORE ABOUT PARABOLAS AND THEIR APPLICATIONS

OBJECTIVES

1. FIND THE VERTEX OF A VERTICAL PARABOLA.
2. GRAPH A QUADRATIC FUNCTION.
3. USE THE DISCRIMINANT TO FIND THE NUMBER OF x-INTERCEPTS OF A VERTICAL PARABOLA.
4. USE QUADRATIC FUNCTIONS TO SOLVE PROBLEMS INVOLVING MAXIMUM OR MINIMUM VALUE.
5. GRAPH HORIZONTAL PARABOLAS.

FOCUS ON PROBLEM SOLVING

Morgan's Department Store wants to construct a rectangular parking lot on land bordered on one side by a highway. It has 280 feet of fencing that is to be used to fence off the other three sides. What should be the dimensions of the lot if the enclosed area is to be a maximum?

At first glance it would seem that this problem has nothing to do with parabolas. However, in order to solve it, a quadratic function is needed. This problem is an example of an *optimization* problem, many of which are found in calculus courses. However, calculus is not needed to solve this one. This problem is Exercise 37 in the exercises for this section. After working through this section, you should be able to solve this problem.

1 When the equation of a parabola is given in the form $f(x) = ax^2 + bx + c$, it is necessary to locate the vertex in order to sketch an accurate graph. This can be done in two ways. The first is by completing the square, as shown in Examples 1 and 2. The second is by using a formula which can be derived by completing the square.

EXAMPLE 1

COMPLETING THE SQUARE TO FIND THE VERTEX

Find the vertex of the graph of $f(x) = x^2 - 4x + 5$.

To find the vertex, we need to express $x^2 - 4x + 5$ in the form $(x - h)^2 + k$. This is done by completing the square. (Recall that this process was introduced in Section 6.1.) To simplify the notation, replace $f(x)$ by y.

$$y = x^2 - 4x + 5$$

$$y - 5 = x^2 - 4x \qquad \text{Get the constant term on the left.}$$

$$y - 5 + 4 = x^2 - 4x + 4 \qquad \text{Half of } -4 \text{ is } -2;\ (-2)^2 = 4. \text{ Add 4 to both sides.}$$

$$y - 1 = (x - 2)^2 \qquad \text{Combine terms on the left and factor on the right.}$$

$$y = (x - 2)^2 + 1 \qquad \text{Add 1 to both sides.}$$

Now write the original equation as $f(x) = (x - 2)^2 + 1$. As shown in Section 8.2, the vertex of this parabola is (2, 1). ■

EXAMPLE 2

COMPLETING THE SQUARE TO FIND THE VERTEX

Find the vertex of the graph of $y = -3x^2 + 6x - 1$.

We must complete the square on $-3x^2 + 6x$. Because the x^2 term has a coefficient other than 1, divide both sides by this coefficient, and then proceed as in Example 1.

$$y = -3x^2 + 6x - 1$$

$$\frac{y}{-3} = x^2 - 2x + \frac{1}{3} \qquad \text{Divide both sides by } -3.$$

$$\frac{y}{-3} - \frac{1}{3} = x^2 - 2x \qquad \text{Move the constant term to the left.}$$

$$\frac{y}{-3} - \frac{1}{3} + 1 = x^2 - 2x + 1 \qquad \text{Half of } -2 \text{ is } -1;\ (-1)^2 = 1. \text{ Add 1 to both sides.}$$

$$\frac{y}{-3} + \frac{2}{3} = (x - 1)^2 \qquad \text{Combine constants on the left and factor on the right.}$$

$$\frac{y}{-3} = (x - 1)^2 - \frac{2}{3} \qquad \text{Subtract } \frac{2}{3} \text{ from both sides.}$$

$$y = -3(x - 1)^2 + 2 \qquad \text{Multiply by } -3 \text{ to get into desired form.}$$

The vertex is (1, 2). ■

A formula for the vertex of the graph of the quadratic function $y = ax^2 + bx + c$ can be found by completing the square for the general form of the equation.

$$y = ax^2 + bx + c \quad (a \neq 0)$$

$$\frac{y}{a} = x^2 + \frac{b}{a}x + \frac{c}{a} \qquad \text{Divide by } a.$$

$$\frac{y}{a} - \frac{c}{a} = x^2 + \frac{b}{a}x \qquad \text{Subtract } \frac{c}{a}.$$

$$\frac{y}{a} - \frac{c}{a} + \frac{b^2}{4a^2} = x^2 + \frac{b}{a}x + \frac{b^2}{4a^2} \qquad \text{Add } \frac{b^2}{4a^2}.$$

$$\frac{y}{a} + \frac{b^2 - 4ac}{4a^2} = \left(x + \frac{b}{2a}\right)^2 \qquad \text{Combine terms on left and factor on right.}$$

$$\frac{y}{a} = \left(x + \frac{b}{2a}\right)^2 - \frac{b^2 - 4ac}{4a^2} \qquad \text{Get } y \text{ term alone on the left.}$$

$$y = a\left(x + \frac{b}{2a}\right)^2 + \frac{4ac - b^2}{4a} \qquad \text{Multiply by } a.$$

$$y = a\left[x - \underbrace{\left(\frac{-b}{2a}\right)}_{h}\right]^2 + \underbrace{\frac{4ac - b^2}{4a}}_{k}$$

The final equation shows that the vertex (h, k) can be expressed in terms of a, b, and c. However, it is not necessary to memorize the expression for k, since it can be obtained by replacing x by $\frac{-b}{2a}$. Using function notation, the y-value of the vertex is $f\left(\frac{-b}{2a}\right)$.

VERTEX FORMULA

The graph of the quadratic function $f(x) = ax^2 + bx + c$ has its vertex at

$$\left(\frac{-b}{2a}, f\left(\frac{-b}{2a}\right)\right),$$

and the axis of the parabola is the line $x = \frac{-b}{2a}$.

EXAMPLE 3 USING THE FORMULA TO FIND THE VERTEX

Use the vertex formula to find the vertex of the graph of the function

$$f(x) = x^2 - x - 6.$$

For this function, $a = 1$, $b = -1$, and $c = -6$. The x-coordinate of the vertex of the parabola is given by

$$\frac{-b}{2a} = \frac{-(-1)}{2(1)} = \frac{1}{2}.$$

The y-coordinate is $f\left(\frac{-b}{2a}\right) = f\left(\frac{1}{2}\right)$.

$$f\left(\frac{1}{2}\right) = \left(\frac{1}{2}\right)^2 - \frac{1}{2} - 6 = \frac{1}{4} - \frac{1}{2} - 6 = -\frac{25}{4}$$

Finally, the vertex is $\left(\frac{1}{2}, -\frac{25}{4}\right)$. ■

2 Graphs of quadratic functions were introduced in Section 8.2. A more general approach involving intercepts and the vertex is given here.

GRAPHING A QUADRATIC FUNCTION $f(x) = ax^2 + bx + c$

Step 1 **Find the y-intercept.**
Find the y-intercept by evaluating $f(0)$.

Step 2 **Find the x-intercepts.**
Find the x-intercepts, if any, by solving $f(x) = 0$.

Step 3 **Find the vertex.**
Find the vertex either by using the formula or by completing the square.

Step 4 **Complete the graph.**
Find and plot additional points as needed, using the symmetry about the axis.

Verify that the graph opens upward (if $a > 0$) or opens downward (if $a < 0$).

EXAMPLE 4 USING THE STEPS FOR GRAPHING A QUADRATIC FUNCTION

Graph the quadratic function $f(x) = x^2 - x - 6$.

Begin by finding the y-intercept.

$$f(x) = x^2 - x - 6$$
$$f(0) = 0^2 - 0 - 6 \quad \text{Find } f(0).$$
$$f(0) = -6$$

The y-intercept is $(0, -6)$. Now find any x-intercepts.

$$f(x) = x^2 - x - 6$$
$$0 = x^2 - x - 6 \quad \text{Let } f(x) = 0.$$
$$0 = (x - 3)(x + 2) \quad \text{Factor.}$$
$$x - 3 = 0 \quad \text{or} \quad x + 2 = 0 \quad \text{Set each factor equal to 0 and solve.}$$
$$x = 3 \quad \text{or} \quad x = -2$$

The x-intercepts are $(3, 0)$ and $(-2, 0)$. The vertex, found in Example 3, is $\left(\frac{1}{2}, -\frac{25}{4}\right)$. Plot the points found so far, and plot any additional points as needed. The symmetry of the graph is helpful here. The graph is shown in Figure 8.11. ■

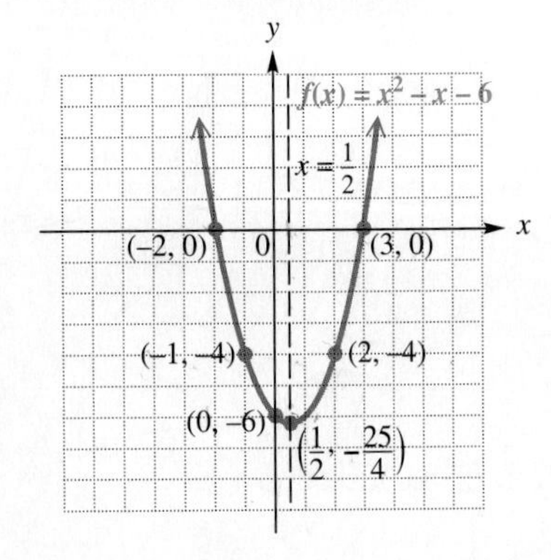

FIGURE 8.11

3 The graph of a quadratic function may have one x-intercept, two x-intercepts, or no x-intercepts, as shown in Figure 8.12.

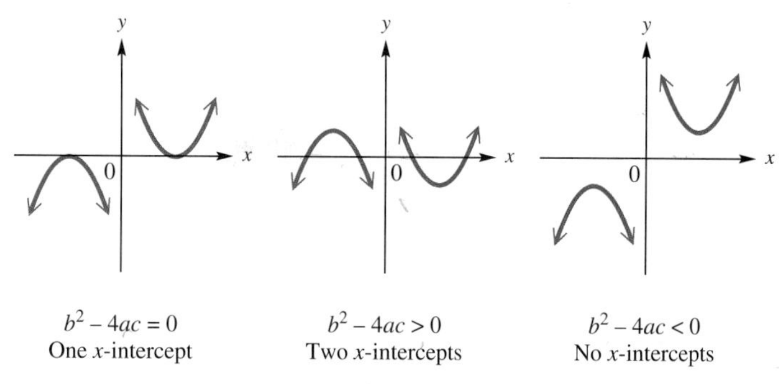

FIGURE 8.12

Recall from Section 6.3 that the value of $b^2 - 4ac$ is called the *discriminant* of the quadratic equation $ax^2 + bx + c = 0$. It can be used to determine the number of real solutions of a quadratic equation. In a similar way, the discriminant of a quadratic *function* can be used to determine the number of x-intercepts of its graph. If the discriminant is positive, the parabola will have two x-intercepts. If the discriminant is 0, there will be only one x-intercept, and it will be the vertex of the parabola. If the discriminant is negative, the graph will have no x-intercepts.

EXAMPLE 5
USING THE DISCRIMINANT TO DETERMINE THE NUMBER OF x-INTERCEPTS

Determine the number of x-intercepts of the graph of each quadratic function. Use the discriminant.

(a) $f(x) = 2x^2 + 3x - 5$

The discriminant is $b^2 - 4ac$. Here $a = 2$, $b = 3$, and $c = -5$, so

$$b^2 - 4ac = 9 - 4(2)(-5) = 49.$$

Since the discriminant is positive, the parabola has two x-intercepts.

(b) $f(x) = -3x^2 - 1$

In this equation, $a = -3$, $b = 0$, and $c = -1$. The discriminant is

$$b^2 - 4ac = 0 - 4(-3)(-1) = -12.$$

The discriminant is negative and so the graph has no x-intercepts.

(c) $f(x) = 9x^2 + 6x + 1$

Here, $a = 9$, $b = 6$, and $c = 1$. The discriminant is

$$b^2 - 4ac = 36 - 4(9)(1) = 0.$$

The parabola has only one x-intercept (its vertex) since the value of the discriminant is 0. ■

4 As we have seen, the vertex of a vertical parabola is either the highest or the lowest point on the parabola. The y-value of the vertex gives the maximum or minimum value of y, while the x-value tells where that maximum or minimum occurs.

PROBLEM SOLVING

In many practical problems we want to know the largest or smallest value of some quantity. When that quantity can be expressed as a quadratic function $f(x) = ax^2 + bx + c$, as in the next example, the vertex can be used to find the desired value. ■

EXAMPLE 6 FINDING THE MAXIMUM AREA OF A RECTANGULAR REGION

A farmer has 120 feet of fencing. He wants to put a fence around three sides of a rectangular plot of land next to a river. Find the maximum area he can enclose.

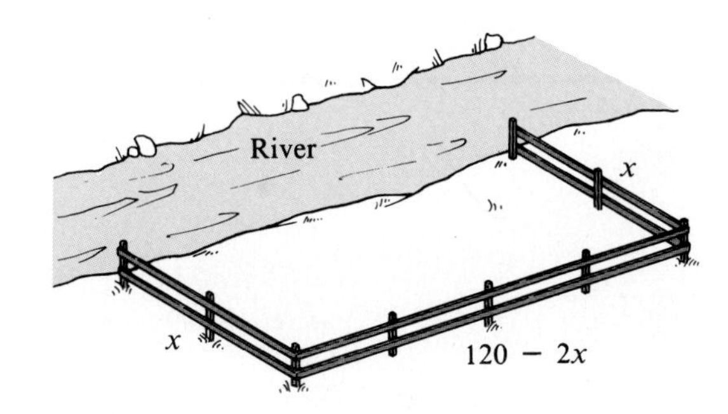

FIGURE 8.13

Figure 8.13 shows the plot. Let x represent its width. Then, since there are 120 feet of fencing,

$$x + x + \text{length} = 120 \qquad \text{Sum of the sides is 120 feet.}$$
$$2x + \text{length} = 120 \qquad \text{Combine terms.}$$
$$\text{length} = 120 - 2x. \qquad \text{Subtract } 2x.$$

The area is given by the product of the length and width, or

$$A = (120 - 2x)x = 120x - 2x^2.$$

To make the area (and thus $120x - 2x^2$) as large as possible, first find the vertex of the parabola $A = 120x - 2x^2$. Do this by completing the square on $120x - 2x^2$.

$$A = 120x - 2x^2$$
$$= -2x^2 + 120x$$
$$\frac{A}{-2} = x^2 - 60x \qquad \text{Divide by } -2.$$
$$\frac{A}{-2} + 900 = x^2 - 60x + 900 \qquad \text{Add 900 on both sides.}$$
$$\frac{A}{-2} + 900 = (x - 30)^2 \qquad \text{Factor.}$$
$$\frac{A}{-2} = (x - 30)^2 - 900 \qquad \text{Subtract 900.}$$
$$A = -2(x - 30)^2 + 1800 \qquad \text{Multiply by } -2.$$

The graph is a parabola that opens downward, and its vertex is (30, 1800). The vertex of the graph shows that the maximum area will be 1800 square feet. This area will occur if x, the width of the plot, is 30 feet. ■

CAUTION Be careful when interpreting the meanings of the coordinates of the vertex in problems involving maximum or minimum values. The first coordinate, x, gives the value for which the *function value* is a maximum or a minimum. It is always necessary to read the problem carefully to determine whether you are asked to find the value of the independent variable, the function value, or both.

5 If x and y are exchanged in the equation $y = ax^2 + bx + c$, the equation becomes $x = ay^2 + by + c$. Because of the interchange of the roles of x and y, these parabolas are horizontal (with horizontal lines as axes), compared with the vertical ones graphed previously.

EXAMPLE 7 **GRAPHING A HORIZONTAL PARABOLA**

Graph $x = (y - 2)^2 - 3$.

This graph has its vertex at $(-3, 2)$, since the roles of x and y are reversed. It opens to the right, the positive x-direction, and has the same shape as $y = x^2$. Plotting a few additional points gives the graph shown in Figure 8.14. ■

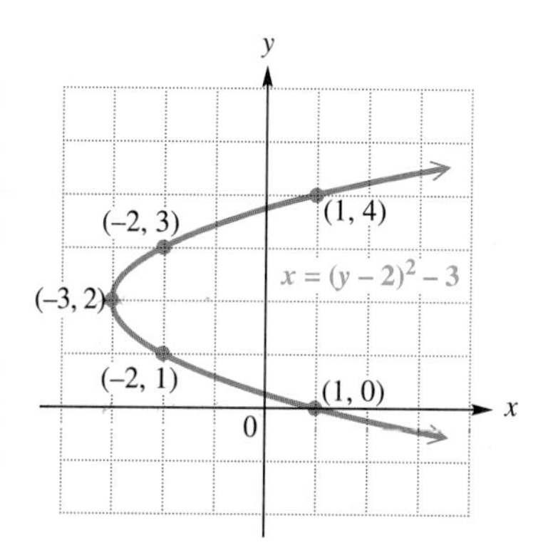

FIGURE 8.14

When a quadratic equation is given in the form $x = ay^2 + by + c$, completing the square on y will put the equation into a form in which the vertex can be identified.

EXAMPLE 8 **COMPLETING THE SQUARE TO GRAPH A HORIZONTAL PARABOLA**

Graph $x = -2y^2 + 4y - 3$.

Complete the square on the right to express the equation in the form $x = a(y - k)^2 + h$.

$$\frac{x}{-2} = y^2 - 2y + \frac{3}{2} \qquad \text{Divide by } -2.$$

$$\frac{x}{-2} - \frac{3}{2} = y^2 - 2y \qquad \text{Subtract } \frac{3}{2}.$$

$$\frac{x}{-2} - \frac{3}{2} + 1 = y^2 - 2y + 1 \qquad \text{Add 1.}$$

$$\frac{x}{-2} - \frac{1}{2} = (y - 1)^2 \qquad \text{Factor on the right; add on the left.}$$

$$\frac{x}{-2} = (y - 1)^2 + \frac{1}{2} \qquad \text{Add } \frac{1}{2}.$$

$$x = -2(y - 1)^2 - 1 \qquad \text{Multiply by } -2.$$

Because of the negative coefficient (-2), the graph opens to the left (the negative x direction). Because $|-2| > 1$, the graph is narrower than $y = x^2$. As shown in Figure 8.15, the vertex is $(-1, 1)$. ■

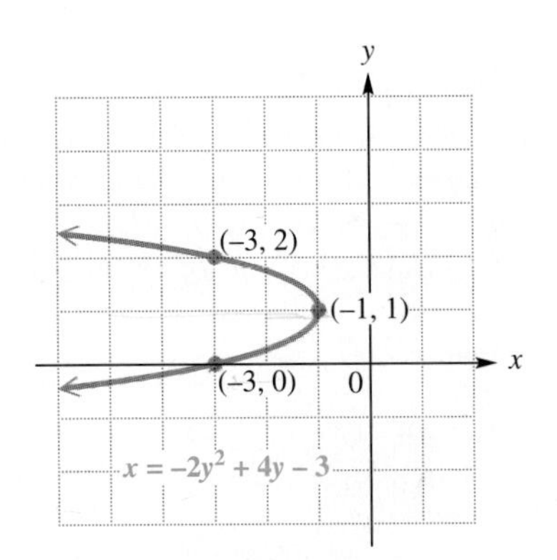

FIGURE 8.15

CAUTION Only quadratic equations that are solved for y are examples of functions. The graphs of the equations in Examples 7 and 8 are not graphs of functions. They do not satisfy the conditions of the vertical line test. Furthermore, the vertex formula given earlier in the section does not apply to parabolas with horizontal axes.

In summary, the graphs of parabolas studied in this section and the previous one fall into the following categories.

GRAPHS OF PARABOLAS

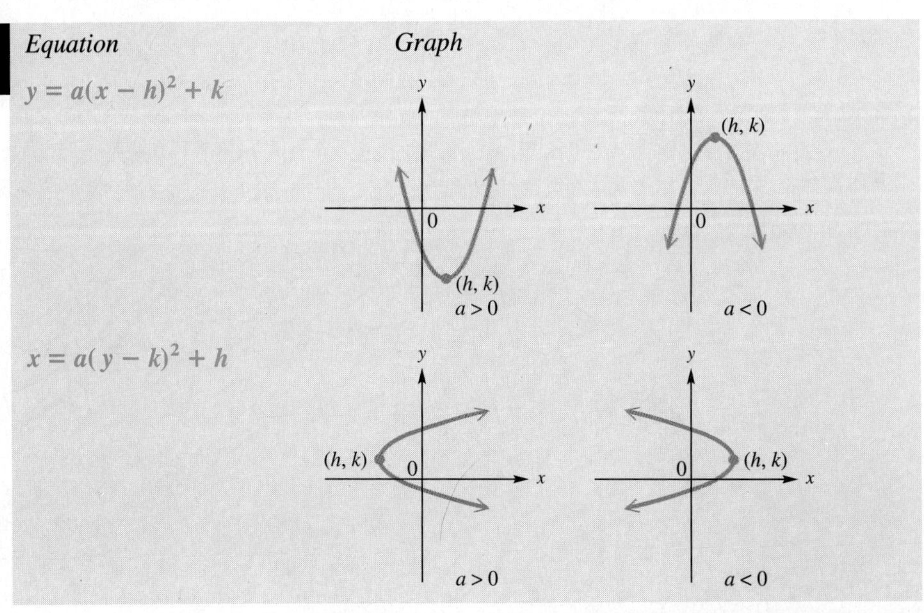

8.3 EXERCISES

Find the vertex of each parabola. For each equation, decide whether the graph opens upward, downward, to the left, or to the right, and state whether it is wider, narrower, or the same shape as the graph of $y = x^2$. If it is a vertical parabola, use the discriminant to determine the number of x-intercepts. See Examples 1–3, 5, 7, and 8.

1. $y = x^2 + 8x - 5$
2. $y = x^2 - 2x + 4$
3. $y = -x^2 + 3x + 3$
4. $y = -x^2 + 5x - 1$
5. $y = 2x^2 - 6x + 3$
6. $y = 3x^2 + 12x - 10$
7. $x = 4y^2 + 8y + 2$
8. $x = -3y^2 + 12y - 9$
9. $x = 2y^2 - 8y + 9$
10. $x = 3y^2 + 6y + 1$

---

11. How can you determine just by looking at the equation of a parabola whether it has a vertical or a horizontal axis? (See Objectives 2 and 5.)

12. Why can't the graph of a quadratic function be a parabola that opens to the left or right? (See Objective 5.)

13. How can you tell the number of x-intercepts of the graph of a quadratic function without graphing the function? (See Objective 3.)

14. If the vertex of the graph of a quadratic function is $(1, -3)$, and the graph opens downward, how many x-intercepts does the graph have? (See Objective 3.)

Graph each parabola by first completing the square. See Examples 4, 7, and 8.

15. $f(x) = x^2 + 8x + 14$
16. $f(x) = x^2 + 10x + 23$
17. $f(x) = x^2 + 4x - 3$
18. $f(x) = x^2 + 2x - 4$
19. $y = 3x^2 - 9x + 8$
20. $y = 2x^2 + 6x - 1$
21. $y = -2x^2 + 4x + 5$
22. $y = -5x^2 - 10x + 2$
23. $x = y^2 - 6y + 4$
24. $x = 2y^2 + 4y - 5$
25. $x = 4y^2 + 8y + 2$
26. $x = -3y^2 + 12y - 9$

Solve each of the following problems. See Example 6.

27. Christina Santiago runs a taco stand. Her past records indicate that the cost of operating the stand is given by

$$c = 2x^2 - 28x + 160,$$

where x is the units of tacos sold daily and c is in dollars. Find the number of units of tacos she must sell to produce the lowest cost, and find this lowest cost.

28. Hilary Langlois owns a video store. He has found that the profits of the store are approximately given by

$$p = -x^2 + 16x + 34,$$

where p represents profit in dollars, and x is the units of videos rented daily. Find the number of units of videos that he should rent daily to produce the maximum profit. Also find the maximum profit.

29. Suppose the price $p(x)$ of a product in dollars is related to the demand x for the product by the equation

$$p(x) = 980x - 5x^2,$$

where x is measured in hundreds. Find the value of x for which $p(x)$ is a maximum. What is this maximum value?

30. The number of mosquitoes $m(x)$, in millions, in a certain area of Mississippi depends on the July rainfall, x, in inches, approximately as follows.

$$m(x) = 10x - x^2$$

Find the rainfall that will produce the maximum number of mosquitoes. How many mosquitoes are produced by that amount of rainfall?

31. After experimentation, two Pacific Institute physics students find that when a bottle of California wine is shaken several times, held upright, and uncorked, its cork travels according to the equation

$$s(t) = -16t^2 + 64t + 3,$$

where s is its height in feet above the ground t seconds after being released. After how many seconds will it reach its maximum height? What is the maximum height?

32. Professor Blakemore has found that the number of students attending his intermediate algebra class is approximated by

$$S(x) = -x^2 + 20x + 80,$$

where x is the number of hours that the University Center is open daily. Find the number of hours that the center should be open so that the number of students attending class is a maximum. What is this maximum number of students?

33. If an object is thrown upward with an initial velocity of 32 feet per second, then its height after t seconds is given by

$$h = 32t - 16t^2.$$

Find the maximum height attained by the object. Find the number of seconds it takes the object to hit the ground.

34. Of all pairs of numbers whose sum is 80, find the pair with the maximum product. (*Hint:* Let x and $80 - x$ represent the two numbers. Write a quadratic equation for the product.)

35. The length and width of a rectangle have a sum of 52 meters. What width will lead to the maximum area? (*Hint:* Let x represent the width and $52 - x$ the length. Write a quadratic equation for the area.)

36. For a trip to a resort, a charter bus company charges a fare of \$48 per person, plus \$2 per person for each unsold seat on the bus. If the bus has 42 seats and x represents the number of unsold seats, find the following:

(a) an expression for the total revenue R from the trip (*Hint:* Multiply the total number riding, $42 - x$, by the price per ticket, $48 + 2x$);

(b) the graph for the expression from part (a);

(c) the number of unsold seats that produces the maximum revenue;

(d) the maximum revenue.

37. Morgan's Department Store wants to construct a rectangular parking lot on land bordered on one side by a highway. It has 280 feet of fencing that is to be used to fence off the other three sides. What should be the dimensions of the lot if the enclosed area is to be a maximum?

38. If air resistance is neglected, a projectile shot straight upward with an initial velocity of 40 meters per second will be at a height s in meters given by the function

$$s(t) = -4.9t^2 + 40t,$$

where t is the number of seconds elapsed after projection. After how many seconds will it reach its maximum height, and what is this maximum height? Round off your answers to the nearest tenth.

PREVIEW EXERCISES

For each of the following, find **(a)** $f(-x)$ *and* **(b)** $-f(x)$. *See Section 7.6.*

39. $f(x) = 2x + 3$

40. $f(x) = -x + 10$

41. $f(x) = 4x^2 + 3$

42. $f(x) = 2(x - 1)^2$

43. $f(x) = -x^2 + 2x$

44. $f(x) = x^2 - x$

8.4 SYMMETRY; INCREASING/DECREASING FUNCTIONS

OBJECTIVES

1 TEST FOR SYMMETRY WITH RESPECT TO AN AXIS.

2 TEST FOR SYMMETRY WITH RESPECT TO THE ORIGIN.

3 DECIDE IF A FUNCTION IS INCREASING OR DECREASING ON AN INTERVAL.

The parabolas graphed in the previous two sections were symmetric with respect to the axis of the parabola, a vertical or horizontal line through the vertex. The graphs of many other relations also are symmetric with respect to a line or a point. As we saw when graphing parabolas, the presence of symmetry is helpful in drawing graphs.

1 Figure 8.16(a) shows a graph that is symmetric with respect to the y-axis. As suggested by Figure 8.16(a), for a graph to be symmetric with respect to the y-axis, the point $(-x, y)$ must be on the graph whenever (x, y) is on the graph.

Similarly, if the graph in Figure 8.16(b) were folded in half along the x-axis, the portion from the top would exactly match the portion from the bottom. Such a graph is symmetric with respect to the x-axis. As the graph suggests, symmetry with respect to the x-axis means that the point $(x, -y)$ must be on the graph whenever the point (x, y) is on the graph.

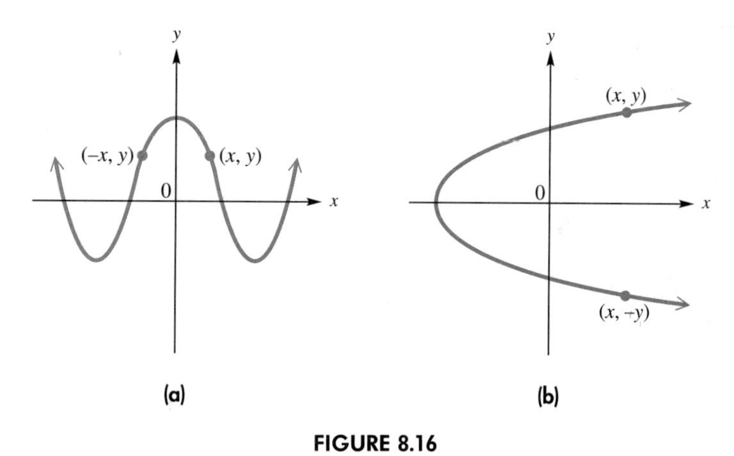

FIGURE 8.16

The following tests tell when a graph is symmetric with respect to the x-axis or y-axis.

SYMMETRY WITH RESPECT TO AN AXIS

The graph of a relation is **symmetric with respect to the y-axis** if the replacement of x with $-x$ results in an equivalent equation.

The graph of a relation is **symmetric with respect to the x-axis** if the replacement of y with $-y$ results in an equivalent equation.

EXAMPLE 1
TESTING FOR SYMMETRY WITH RESPECT TO AN AXIS

Test for symmetry with respect to the x-axis and to the y-axis.

(a) $y = x^2 + 4$

Replace x with $-x$:

$$y = x^2 + 4 \qquad \text{becomes} \qquad y = (-x)^2 + 4 = x^2 + 4.$$

The result is the same as the original equation, so the graph (shown in Figure 8.17) is symmetric with respect to the y-axis. The graph is *not* symmetric with respect to the x-axis, since replacing y with $-y$ gives

$$-y = x^2 + 4 \qquad \text{or} \qquad y = -x^2 - 4,$$

which is not equivalent to the original equation.

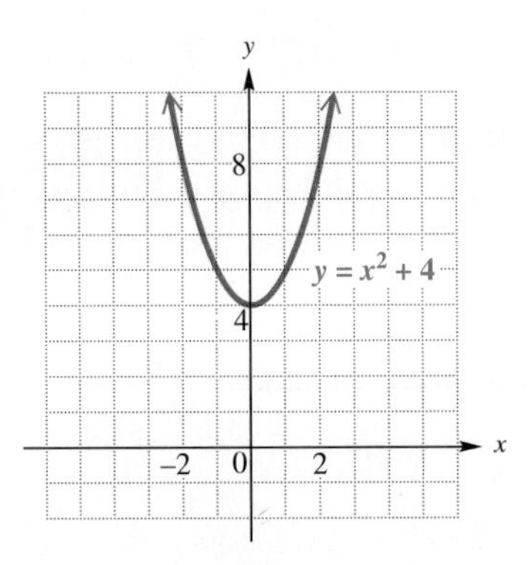

FIGURE 8.17

FIGURE 8.18

(b) $x = y^2 - 3$

Replace y with $-y$ to get $x = (-y)^2 - 3 = y^2 - 3$, the same as the original equation. The graph is symmetric with respect to the x-axis, as shown in Figure 8.18. The graph is not symmetric with respect to the y-axis.

(c) $2x + y = 4$

Replace x with $-x$ and then replace y with $-y$; in neither case does an equivalent equation result. This graph is neither symmetric with respect to the x-axis nor to the y-axis.

(d) $x = |y|$

Replacing x with $-x$ gives $-x = |y|$, which is not equivalent to the original equation. The graph is not symmetric to the y-axis. Replacing y with $-y$ gives $x = |-y| = |y|$. Thus, the graph is symmetric to the x-axis. ■

2 Another kind of symmetry occurs when it is possible to rotate a graph 180° about the origin and have the result coincide exactly with the original graph. Symmetry of this type is called *symmetry with respect to the origin*. It can be shown that rotating a graph 180° is equivalent to saying that the point $(-x, -y)$ is on the graph whenever (x, y) is on the graph. Figure 8.19 shows two graphs that are symmetric with respect to the origin. A test for this type of symmetry follows.

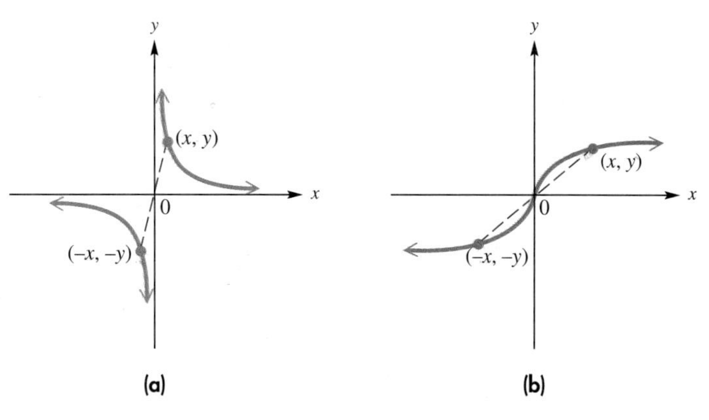

FIGURE 8.19

SYMMETRY WITH RESPECT TO THE ORIGIN

The graph of a relation is **symmetric with respect to the origin** if the replacement of both x with $-x$ and y with $-y$ results in an equivalent equation.

EXAMPLE 2 TESTING FOR SYMMETRY WITH RESPECT TO THE ORIGIN

For each of the following equations, decide if the graph is symmetric with respect to the origin.

(a) $3x = 5y$

Replace x with $-x$ and y with $-y$ in the equation.

$$3(-x) = 5(-y) \quad \text{Substitute } -x \text{ for } x, -y \text{ for } y.$$
$$-3x = -5y \quad \text{Multiply.}$$
$$3x = 5y \quad \text{Multiply both sides by } -1.$$

Since the equation is equivalent to $3x = 5y$, the graph has symmetry with respect to the origin. See Figure 8.20.

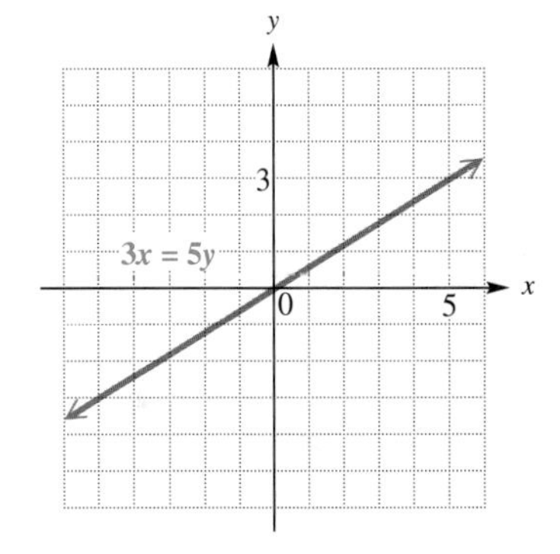

FIGURE 8.20

(b) $2y = 3x^2$

Substituting $-x$ for x and $-y$ for y gives

$$2(-y) = 3(-x)^2 \quad \text{Replace } y \text{ with } -y, x \text{ with } -x.$$
$$-2y = 3x^2, \quad \text{Multiply.}$$

which is not equivalent to the original equation. As Figure 8.21 shows, the graph is not symmetric with respect to the origin. ■

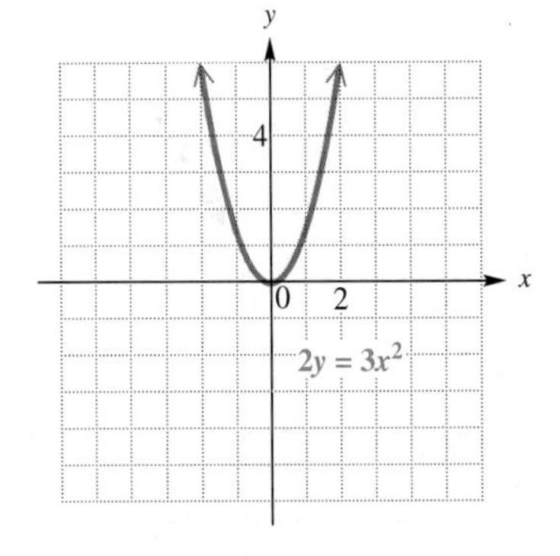

FIGURE 8.21

A summary of the tests for symmetry is given below.

TESTS FOR SYMMETRY

	Symmetric with respect to		
	x-axis	*y-axis*	*origin*
Test	Replace y with $-y$.	Replace x with $-x$.	Replace x with $-x$ and replace y with $-y$.
Example	y, x, 0	y, x, 0	y, x, 0

3 Intuitively a function is said to be *increasing* if its graph goes up from left to right. The function graphed in Figure 8.22(a) is an increasing function. On the other hand, a function is *decreasing* if its graph goes down from left to right, like the function in Figure 8.22(b). The function graphed in Figure 8.22(c) is neither an increasing function nor a decreasing function. However it is increasing on the interval $(-\infty, -1)$ and decreasing on the interval $(-1, \infty)$.

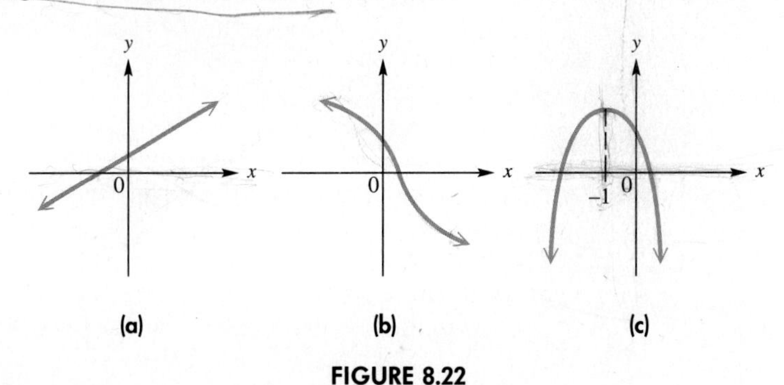

FIGURE 8.22

In the following definition of a function increasing or decreasing on an interval, I represents any interval of real numbers.

INCREASING AND DECREASING FUNCTIONS

Let f be a function, with x_1 and x_2 in an interval I in the domain of f. Then

f is **increasing** on I if $f(x_1) < f(x_2)$ whenever $x_1 < x_2$;

f is **decreasing** on I if $f(x_1) > f(x_2)$ whenever $x_1 < x_2$.

EXAMPLE 3 DETERMINING INCREASING OR DECREASING INTERVALS

Give the intervals where the following functions are increasing or decreasing.

(a) The function graphed in Figure 8.23(a) is decreasing on $(-\infty, -2)$ and $(1, \infty)$. The function is increasing on $(-2, 1)$.

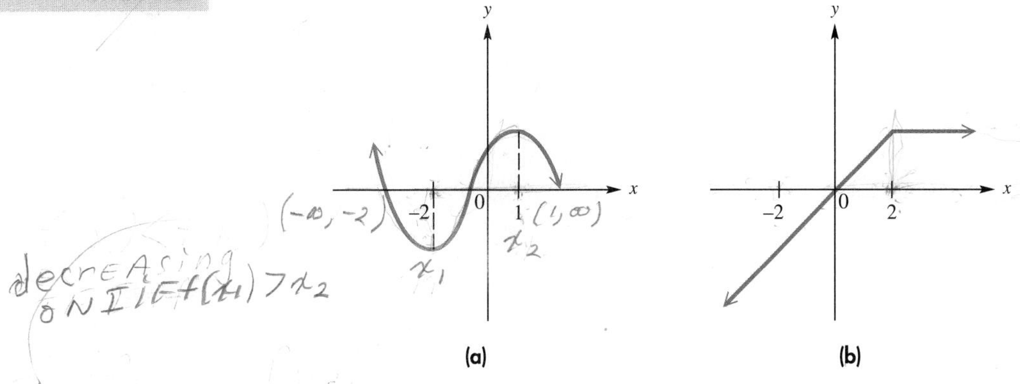

FIGURE 8.23

(b) The function in Figure 8.23(b) is increasing on $(-\infty, 2)$. On the interval $(2, \infty)$ the function is neither increasing nor decreasing. ■

A function that is neither increasing nor decreasing on an interval is said to be **constant** on that interval.

CAUTION When identifying intervals over which a function is increasing, decreasing, or constant, remember that we are interested in identifying *domain* intervals. Range values do not appear in these stated intervals.

8.4 EXERCISES

Plot the following points, and then use the same axes to plot the points that are symmetric to the given point with respect to the following elements: ***(a)*** *x-axis,* ***(b)*** *y-axis,* ***(c)*** *origin.*

1. $(5, -3)$ **2.** $(-6, 1)$ **3.** $(-4, -2)$ **4.** $(-8, 3)$ **5.** $(-8, 0)$ **6.** $(0, -3)$

In Exercises 7–18, use the tests for symmetry to decide whether the graph of each relation is symmetric with respect to the x-axis, the y-axis, or the origin. See Examples 1–2. Do not graph.

7. $x^2 + y^2 = 5$ **8.** $y^2 = 4 - x^2$ **9.** $y = x^2 - 8x$ **10.** $y = 4x - x^2$

11. $y = |x|$ **12.** $y = |x| + 1$ **13.** $y = x^3$ **14.** $y = -x^3$

15. $y = \dfrac{1}{1 + x^2}$ **16.** $y = \dfrac{-1}{x^2 + 9}$ **17.** $xy = 2$ **18.** $xy = -6$

Sketch examples of graphs that satisfy the following conditions.

19. Symmetric with respect to the x-axis but not to the y-axis

20. Symmetric with respect to the y-axis but not to the x-axis

21. Symmetric with respect to the origin but to neither the x-axis nor the y-axis

For each of the following give the intervals where f is increasing or decreasing. In Exercises 26 and 27, first graph the function. See Example 3.

22.

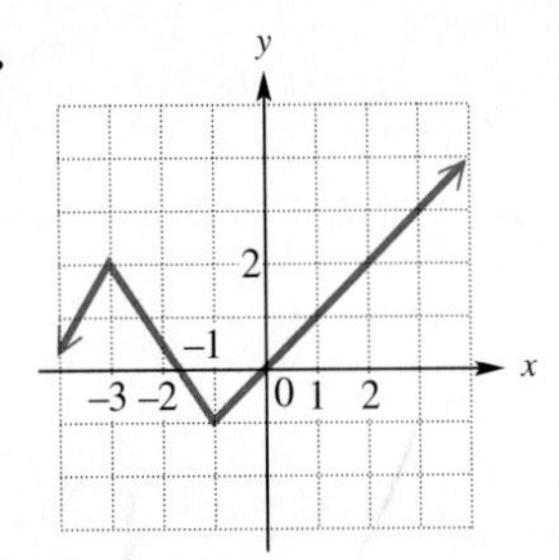

23.

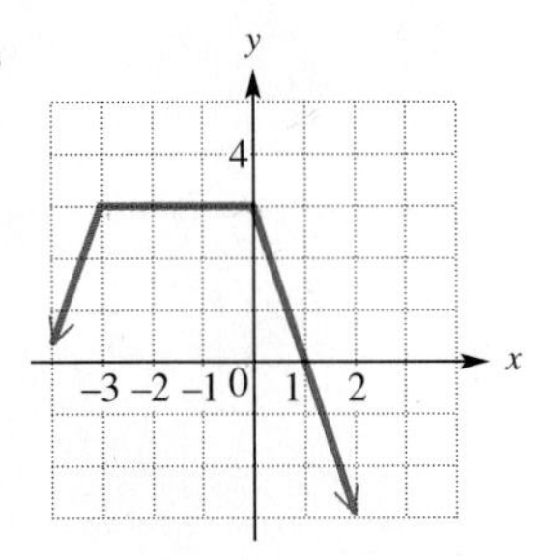

24.

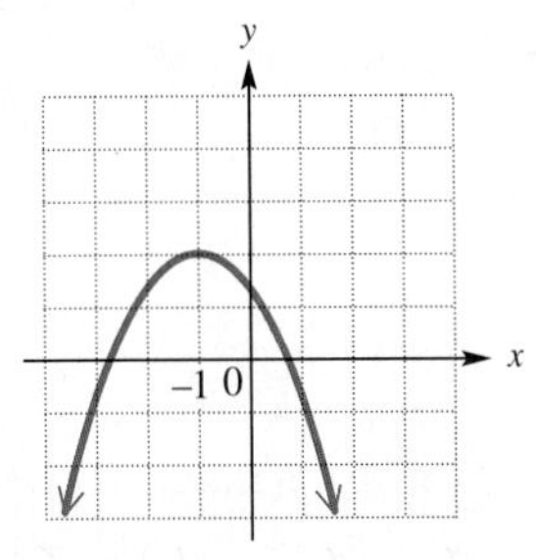

25.

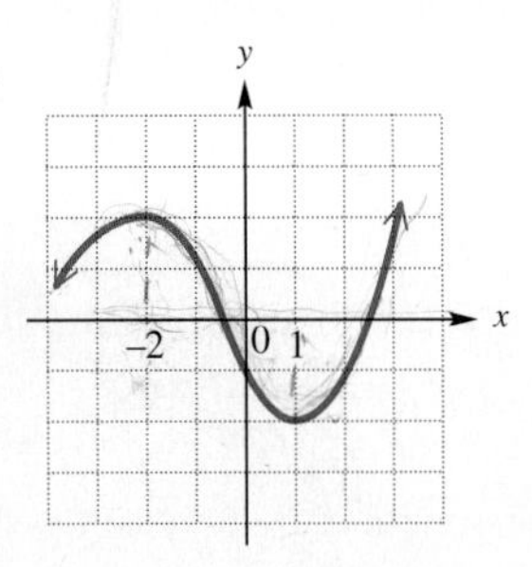

26. $f(x) = 3 - x^2$

27. $f(x) = 2x^2 + 1$

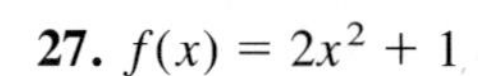

In Exercises 28–33 decide whether each figure is symmetric to **(a)** *the given line;* **(b)** *the given point. (See Objectives 1 and 2.)*

28.

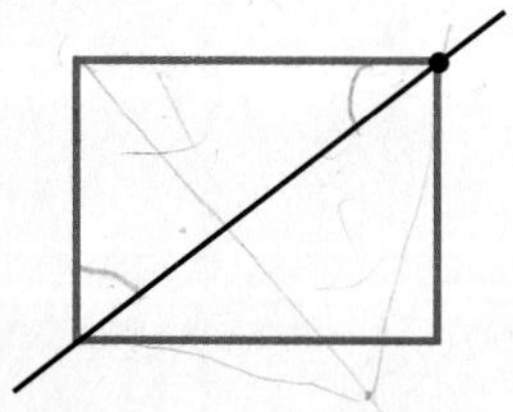

29.

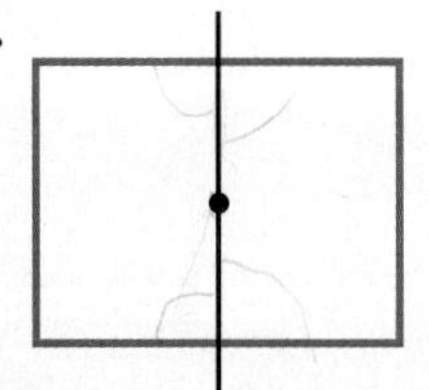

30.

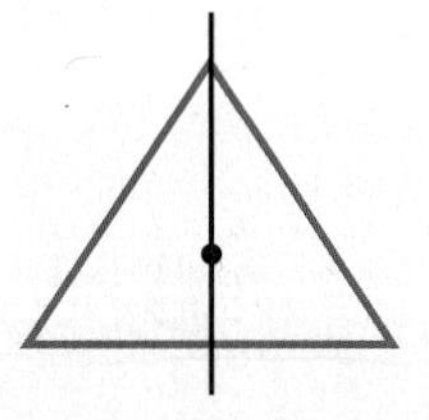

31.

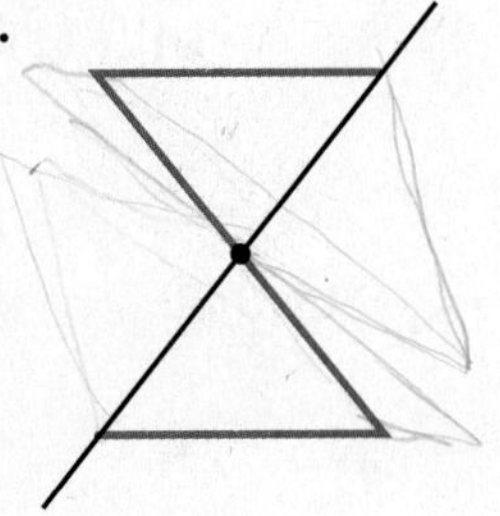

32.

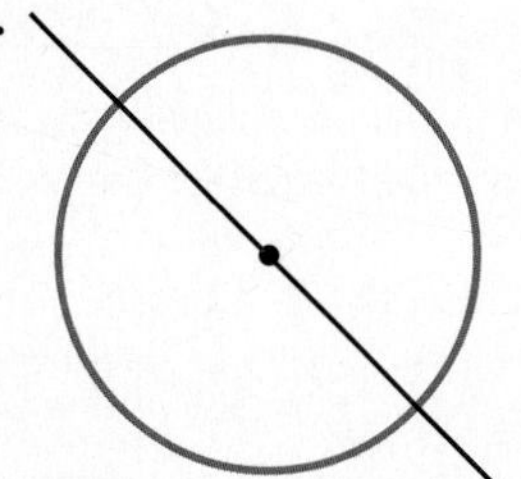

33.

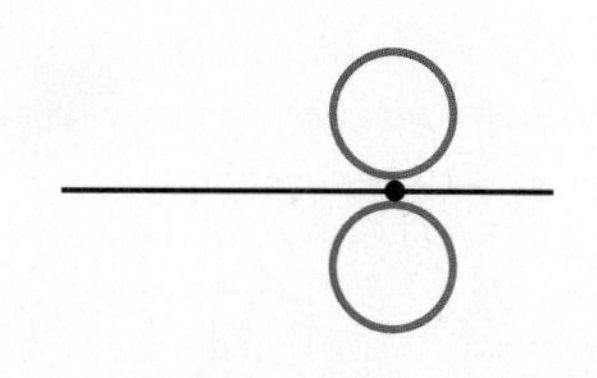

Suppose that $f(2) = 3$. For each assumption in Exercises 34–36, find another value of the function. (See Objectives 1 and 2.)

34. $f(x)$ is symmetric with respect to the origin.

35. $f(x)$ is symmetric with respect to the line $x = 3$.

36. $f(x)$ is symmetric with respect to the y-axis.

Complete the left half of the graph of $f(x)$ in the figure for each of the following conditions. (See Objectives 1 and 2.)

37. $f(-x) = f(x)$

38. $f(-x) = -f(x)$

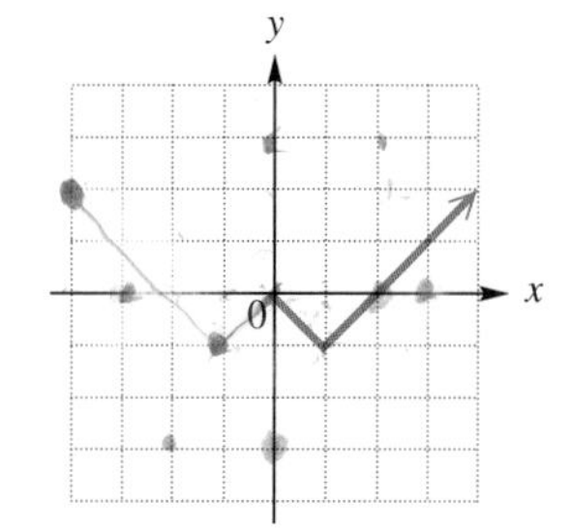

39. A graph that is symmetric with respect to both the x-axis and the y-axis is also symmetric to the origin. Explain why.

40. Is it possible for a *function* (excluding $f(x) = 0$) to be symmetric with respect to the x-axis? Why? (See Objective 1.)

PREVIEW EXERCISES

Find the distance between each pair of points. See Section 7.1.

41. $(2, -1)$ and $(4, 3)$

42. $(5, 6)$ and $(-2, -3)$

43. $(-4, 7)$ and $(-1, -8)$

44. (x, y) and $(2, -1)$

45. (x, y) and $(-4, -3)$

46. (x, y) and (h, k)

8.5 THE CIRCLE AND THE ELLIPSE

OBJECTIVES

1. FIND THE EQUATION OF A CIRCLE GIVEN THE CENTER AND RADIUS.
2. DETERMINE THE CENTER AND RADIUS OF A CIRCLE GIVEN ITS EQUATION.
3. FIND THE EQUATION OF A CIRCLE GIVEN INFORMATION OTHER THAN THE CENTER AND RADIUS.
4. RECOGNIZE THE EQUATION OF AN ELLIPSE.
5. GRAPH ELLIPSES.

Earlier the second-degree relations $y = ax^2 + bx + c$ and $x = ay^2 + by + c$ $(a \neq 0)$ were discussed. These relations have just one second-degree term. This section begins a discussion of second-degree relations that have both x^2 and y^2 terms, starting with those second-degree relations that have a circle for a graph.

1 A **circle** is the set of all points in a plane that lie a fixed distance from a fixed point. The fixed point is called the **center** and the fixed distance is called the **radius.** The distance formula can be used to find an equation of a circle.

EXAMPLE 1
FINDING AN EQUATION OF A CIRCLE AND GRAPHING IT

Find an equation of the circle with radius 3 and center at (0, 0), and graph the circle.

If the point (x, y) is on the circle, the distance from (x, y) to the center (0, 0) is 3, as shown in Figure 8.24. By the distance formula,

$$\sqrt{(x_2 - x_1)^2 + (y_2 - y_1)^2} = d$$

$$\sqrt{(x - 0)^2 + (y - 0)^2} = 3$$

$$x^2 + y^2 = 9. \quad \text{Square both sides.}$$

An equation of this circle is $x^2 + y^2 = 9$. ■

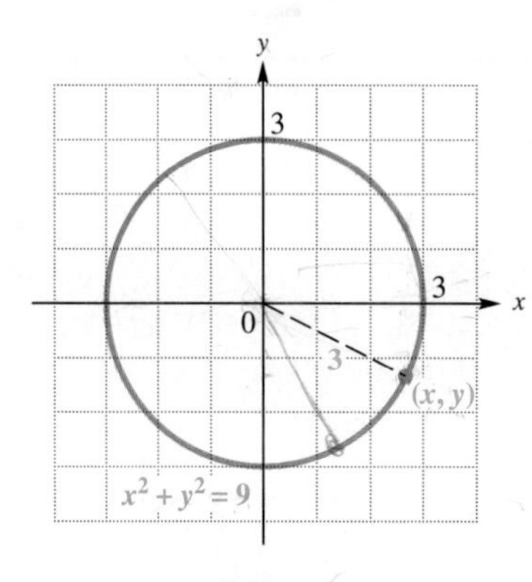

FIGURE 8.24

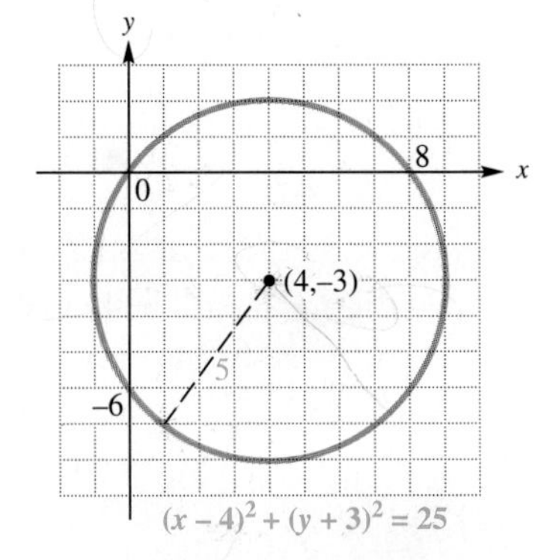

FIGURE 8.25

EXAMPLE 2
FINDING AN EQUATION OF A CIRCLE AND GRAPHING IT

Find an equation for the circle that has its center at (4, −3) and radius 5, and graph the circle.

Again use the distance formula.

$$\sqrt{(x - 4)^2 + (y + 3)^2} = 5$$

$$(x - 4)^2 + (y + 3)^2 = 25 \quad \text{Square both sides.}$$

The graph of this circle is shown in Figure 8.25. ■

Examples 1 and 2 can be generalized to get an equation of a circle with radius r and center at (h, k). If (x, y) is a point on the circle, the distance from the center (h, k) to the point (x, y) is r. Then by the distance formula,

$$\sqrt{(x - h)^2 + (y - k)^2} = r.$$

Squaring both sides gives the following equation of a circle.

EQUATION OF A CIRCLE

$$(x - h)^2 + (y - k)^2 = r^2$$

is an equation of a circle of radius r with center at (h, k).

EXAMPLE 3
USING THE FORM OF THE EQUATION OF A CIRCLE

Find an equation of the circle with center at $(-1, 2)$ and radius 4.

Use the last result, with $h = -1$, $k = 2$, and $r = 4$ to get

$$(x - h)^2 + (y - k)^2 = r^2$$
$$[x - (-1)]^2 + (y - 2)^2 = (4)^2$$
$$(x + 1)^2 + (y - 2)^2 = 16. \quad ■$$

2 In the equation found in Example 2, multiplying out $(x - 4)^2$ and $(y + 3)^2$ and then combining like terms gives

$$(x - 4)^2 + (y + 3)^2 = 25$$
$$x^2 - 8x + 16 + y^2 + 6y + 9 = 25$$
$$x^2 + y^2 - 8x + 6y = 0.$$

This result suggests that an equation that has both x^2 and y^2 terms may represent a circle. The next example shows how to tell, using the method of completing the square.

EXAMPLE 4
COMPLETING THE SQUARE TO FIND THE CENTER AND RADIUS

Graph $x^2 + y^2 + 2x + 6y - 15 = 0$.

Since the equation has x^2 and y^2 terms with equal coefficients, its graph might be that of a circle. To find the center and radius, complete the square on x and y as follows.

$$x^2 + y^2 + 2x + 6y = 15 \qquad \text{Move the constant to the right.}$$
$$(x^2 + 2x \quad) + (y^2 + 6y \quad) = 15 \qquad \text{Rewrite in anticipation of completing the square.}$$
$$(x^2 + 2x + 1) + (y^2 + 6y + 9) = 15 + 1 + 9 \qquad \text{Complete the square in both x and y.}$$
$$(x + 1)^2 + (y + 3)^3 = 25 \qquad \text{Factor on the left and add on the right.}$$

The last equation shows that the graph is a circle with center at $(-1, -3)$ and radius 5. The graph is shown in Figure 8.26. ■

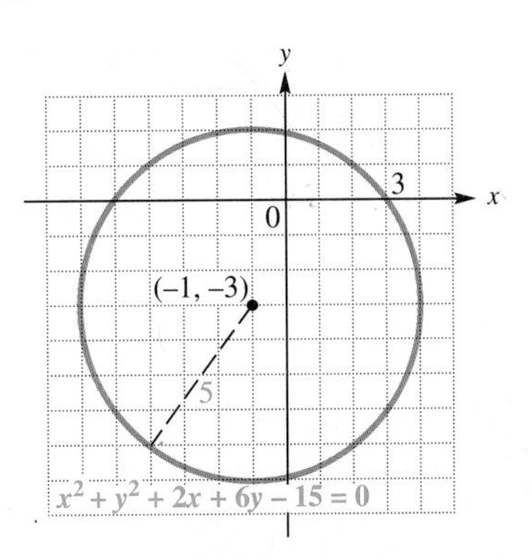

FIGURE 8.26

NOTE If the procedure of Example 4 leads to an equation of the form $(x - h)^2 + (y - k)^2 = 0$, the graph is the single point (h, k). If the constant on the right side is negative, the equation has no graph.

3 If the center and one point on a circle are known, the equation can be found using the form given in the box above. This form of the equation also can be used to find the equation of a circle when two points on the circle are known.

EXAMPLE 5 **FINDING THE EQUATION OF A CIRCLE**

Find the equation of a circle that satisfies the given conditions.

(a) Center at $(-1, 2)$, through $(2, 2)$

Use the equation $(x - h)^2 + (y - k)^2 = r^2$. Here, $h = -1$, $k = 2$, and r must be found. Since the point $(2, 2)$ is on the circle, its coordinates must satisfy the equation of the circle, so let $x = 2$ and $y = 2$. This gives

$$(x - h)^2 + (y - k)^2 = r^2$$

$$(2 - (-1))^2 + (2 - 2)^2 = r^2 \qquad x = 2,\ y = 2,\ h = -1,\ k = 2$$

$$3^2 + 0^2 = r^2$$

$$r = 3, \qquad \text{Square root property, } r \geq 0$$

and so the equation of the circle is

$$(x - (-1))^2 + (y - 2)^2 = 3^2 \qquad h = -1,\ k = 2,\ r = 3$$

$$(x + 1)^2 + (y - 2)^2 = 9.$$

(b) The endpoints of a diameter are $(-2, 5)$ and $(4, 1)$

As shown in Figure 8.27 the center of the circle is the midpoint of the diameter. In the exercises for Section 7.1, the midpoint of the line segment with endpoints at (x_1, y_1) and (x_2, y_2) was given as $\left(\frac{x_1 + x_2}{2}, \frac{y_1 + y_2}{2}\right)$. That is, the x-value of the midpoint is the average of the x-values of the endpoints and the y-value is the average of the y-values of the endpoints. Here, the center will have coordinates

$$\left(\frac{-2 + 4}{2}, \frac{5 + 1}{2}\right) = (1, 3).$$ CENTER

The distance from either endpoint to the center gives the radius. Using $(-2, 5)$,

$$r = \sqrt{(-2 - 1)^2 + (5 - 3)^2} = \sqrt{9 + 4} = \sqrt{13}.$$

Thus, the equation of the circle is

$$(x - h)^2 + (y - k)^2 = r^2$$

$$(x - 1)^2 + (y - 3)^2 = (\sqrt{13})^2$$

$$(x - 1)^2 + (y - 3)^2 = 13. \quad ■$$

2/2

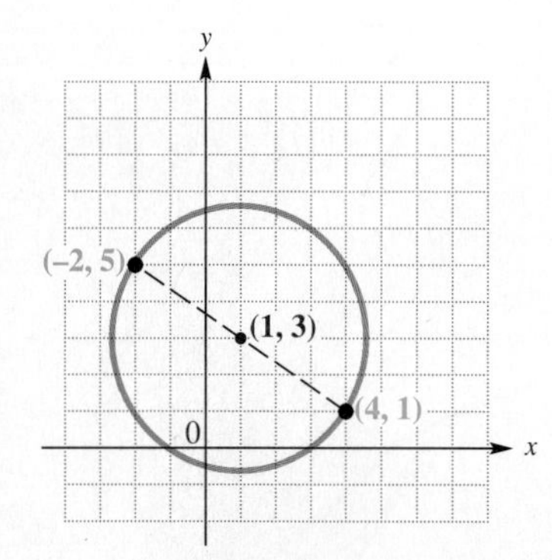

FIGURE 8.27

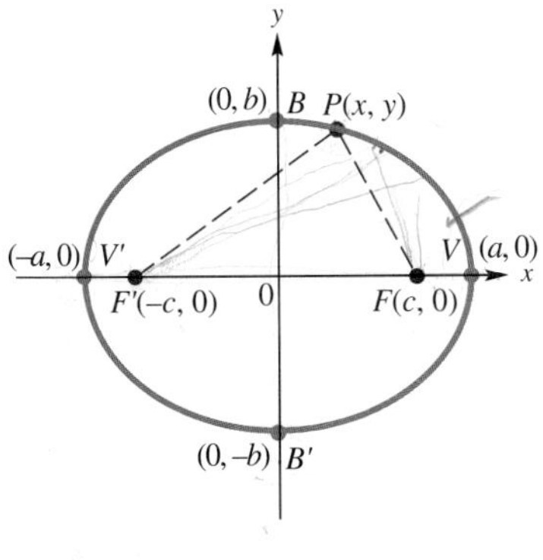

FIGURE 8.28

4 As the earth travels around the sun over a year's time, it traces out a curve called an ellipse. An **ellipse** is the set of all points in a plane the sum of whose distances from two fixed points is constant. These fixed points are called **foci** (singular: *focus*). Figure 8.28 shows an ellipse whose foci are $(c, 0)$ and $(-c, 0)$, with x-intercepts $(a, 0)$ and $(-a, 0)$ and y-intercepts $(0, b)$ and $(0, -b)$. The origin is the **center** of the ellipse. Points V and V' are the **vertices** of the ellipse and the line segment connecting V and V' is the **major axis.** The foci always lie on the major axis. The line segment from B to B' is the **minor axis.**

From the definition above, it can be shown by the distance formula that an ellipse has the following equation.

EQUATION OF AN ELLIPSE

The ellipse whose x-intercepts are $(a, 0)$ and $(-a, 0)$ and whose y-intercepts are $(0, b)$ and $(0, -b)$ has an equation of the form

$$\frac{x^2}{a^2} + \frac{y^2}{b^2} = 1.$$

The proof of this is included in the exercises. (See Exercise 51.) Note that a circle is a special case of an ellipse, where $a^2 = b^2$.

The paths of the earth and other planets around the sun are approximately ellipses; the sun is at one focus and a point in space is at the other. The orbits of communication satellites and other space vehicles are elliptical.

An ellipse is the graph of a relation. As suggested by the graph in Figure 8.28, if the ellipse has equation $(x^2/a^2) + (y^2/b^2) = 1$, the domain is $[-a, a]$ and the range is $[-b, b]$. Notice that the ellipse in Figure 8.28 is symmetric with respect to the x-axis, the y-axis, and the origin. More generally, every ellipse is symmetric with respect to its major axis, its minor axis, and its center.

5 To graph an ellipse, plot the four intercepts and sketch the ellipse through the intercepts.

EXAMPLE 6
GRAPHING AN ELLIPSE

Graph $\frac{x^2}{49} + \frac{y^2}{36} = 1.$

The x-intercepts for this ellipse are $(7, 0)$ and $(-7, 0)$. The y-intercepts are $(0, 6)$ and $(0, -6)$. Plotting the intercepts and sketching the ellipse through them gives the graph in Figure 8.29. ■

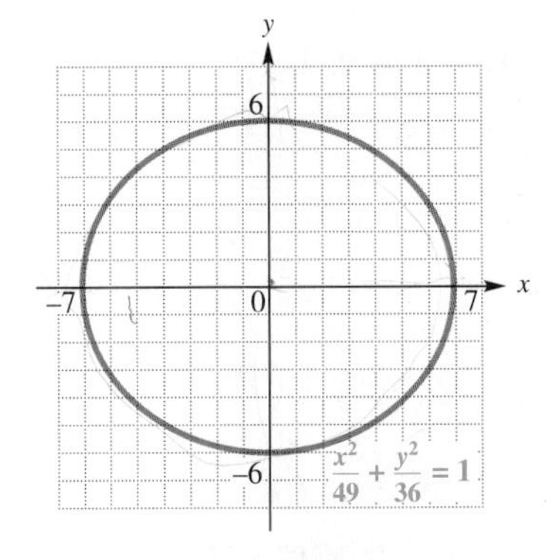

FIGURE 8.29

EXAMPLE 7
GRAPHING AN ELLIPSE

Graph $\dfrac{x^2}{36} + \dfrac{y^2}{121} = 1$.

The x-intercepts for this ellipse are (6, 0) and (−6, 0), and the y-intercepts are (0, 11) and (0, −11). The graph has been sketched in Figure 8.30. ■

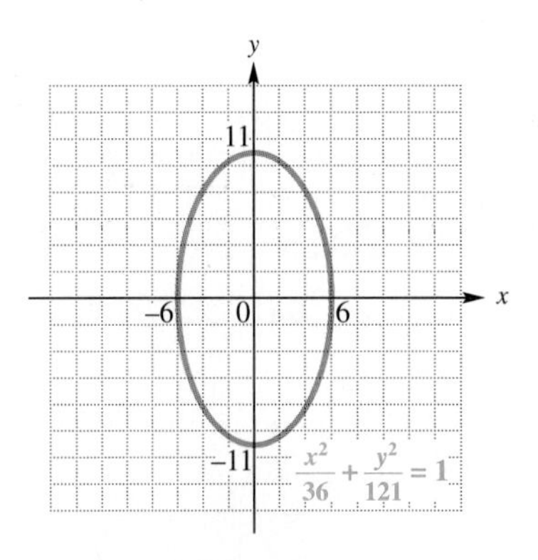

FIGURE 8.30

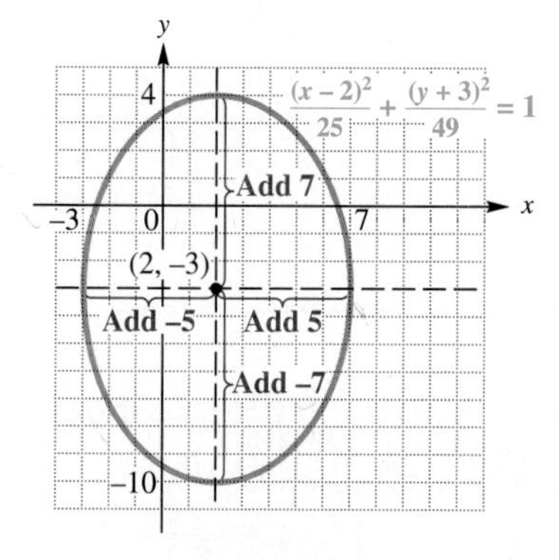

FIGURE 8.31

As with the graphs of parabolas and circles, the graph of an ellipse may be shifted horizontally and vertically, as in the next example.

EXAMPLE 8
GRAPHING AN ELLIPSE SHIFTED HORIZONTALLY AND VERTICALLY

Graph $\dfrac{(x-2)^2}{25} + \dfrac{(y+3)^2}{49} = 1$.

Just as the circle with equation $(x-2)^2 + (y+3)^2 = 1$ has its center at (2, −3), so does this ellipse. Figure 8.31 shows that the graph goes through the four points (2, 4), (7, −3), (2, −10), and (−3, −3). The x-values of these points are found by adding $\pm a = \pm 5$ to 2, and the y-values come from adding $\pm b = \pm 7$ to −3. ■

8.5 EXERCISES

Write the equation of each circle. See Examples 1–3.

1. Center at (2, 4), radius 5
2. Center at (−1, 5), radius 3
3. Center at (0, 3), radius $\sqrt{2}$
4. Center at (1, 0), radius $\sqrt{3}$
5. Center at (0, −1), radius 4
6. Center at (−2, −1), radius 1

---

7. What is the center of a circle that has equation $x^2 + y^2 = r^2$ $(r \neq 0)$? (See Objective 1.)
8. How many points are there on the graph of $(x-4)^2 + (y+1)^2 = 0$? (See Objective 2.)
9. How many points are there on the graph of $(x-4)^2 + (y+1)^2 = -1$? (See Objective 2.)
10. Explain why a circle is a special case of an ellipse. (See Objective 4.)

Find the center and radius of each circle. See Example 4.

11. $x^2 + y^2 + 6x - 2y + 6 = 0$
12. $x^2 + y^2 - 10x - 8y - 23 = 0$
13. $x^2 + y^2 + 8x + 4y - 29 = 0$
14. $x^2 + y^2 - 6x - 4y + 9 = 0$
15. $x^2 + y^2 - 2x + 6y + 1 = 0$
16. $x^2 + y^2 - 6x + 10y - 2 = 0$

Graph each of the following. Label the center if not centered at the origin. See Examples 1, 2, and 4.

17. $x^2 + y^2 = 16$ **18.** $x^2 + y^2 = 9$ **19.** $2x^2 + 2y^2 = 8$ **20.** $3x^2 + 3y^2 = 15$

21. $y^2 = 144 - x^2$ **22.** $4x^2 = 16 - 4y^2$ **23.** $(x - 1)^2 + (y + 3)^2 = 4$

24. $(x + 2)^2 + (y - 4)^2 = 9$ **25.** $(x + 4)^2 + (y + 5)^2 = 36$ **26.** $(x - 3)^2 + (y + 5)^2 = 25$

Graph each ellipse. Label the intercepts. See Examples 6 and 7.

27. $\frac{x^2}{4} + \frac{y^2}{9} = 1$ **28.** $\frac{x^2}{16} + \frac{y^2}{25} = 1$ **29.** $\frac{x^2}{9} + \frac{y^2}{16} = 1$ **30.** $\frac{x^2}{36} + \frac{y^2}{9} = 1$

31. $\frac{x^2}{16} + \frac{y^2}{4} = 1$ **32.** $\frac{x^2}{49} + \frac{y^2}{81} = 1$ **33.** $\frac{x^2}{9/4} + \frac{y^2}{25/16} = 1$ **34.** $\frac{x^2}{81/4} + \frac{y^2}{25/4} = 1$

Graph each ellipse. Label the center. These ellipses have centers shifted from the origin. See Example 8.

35. $\frac{(x + 1)^2}{64} + \frac{(y - 2)^2}{49} = 1$ **36.** $\frac{(x - 4)^2}{9} + \frac{(y + 2)^2}{4} = 1$

37. $\frac{(x - 2)^2}{16} + \frac{(y - 1)^2}{9} = 1$ **38.** $\frac{(x + 3)^2}{25} + \frac{(y + 2)^2}{36} = 1$

In Exercises 39–46, find an equation of a circle that satisfies the given conditions. See Example 5.

39. Center (0, 0), through $(1, \sqrt{7})$ **40.** Center (0, 0), through (3, 1)

41. Center (4, −2), through (8, 2) **42.** Center (−1, −3), through (2, 1)

43. Endpoints of a diameter at (6, 3) and (−2, 5)

44. Endpoints of a diameter at (1, −4) and (−5, 2)

45. Center at (4, 3) and tangent to (just touching) the y-axis (*Hint:* Make a sketch.)

46. Center at (−3, −2) and tangent to the x-axis

---

47. It is possible to sketch an ellipse on a piece of poster board by fastening two ends of a length of string, pulling the string taut with a pencil, and tracing a curve, as shown in the figure. Explain why this method works.

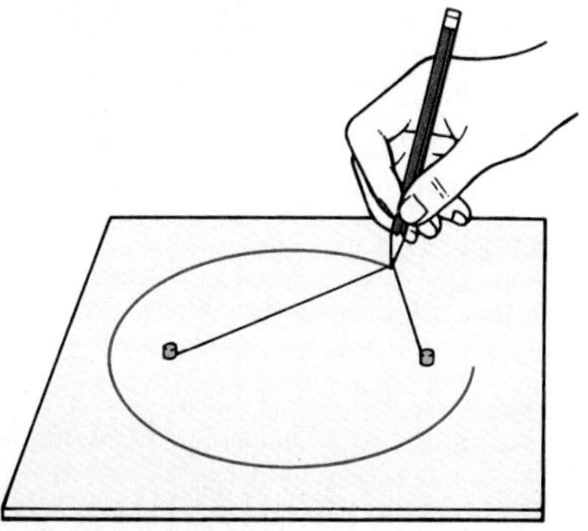

48. How can the method of Exercise 47 be modified to draw a circle?

Work each problem.

49. The orbit of Venus around the sun (one of the foci) is an ellipse with equation

$$\frac{x^2}{5013} + \frac{y^2}{4970} = 1,$$

where x and y are measured in millions of miles.

(a) Find the farthest distance between Venus and the sun.

(b) Find the smallest distance between Venus and the sun.

(*Hint:* See Figure 8.28 and use the fact that $c^2 = a^2 - b^2$.)

50. A one-way road passes under an overpass in the form of half of an ellipse, 15 feet high at the center and 20 feet wide. Assuming a truck is 12 feet wide, what is the tallest truck that can pass under the overpass?

51. **(a)** Suppose that $(c, 0)$ and $(-c, 0)$ are the foci of an ellipse and that the sum of the distances from any point (x, y) of the ellipse to the two foci is $2a$. See the figure. Show that the equation of the resulting ellipse is

$$\frac{x^2}{a^2} + \frac{y^2}{a^2 - c^2} = 1.$$

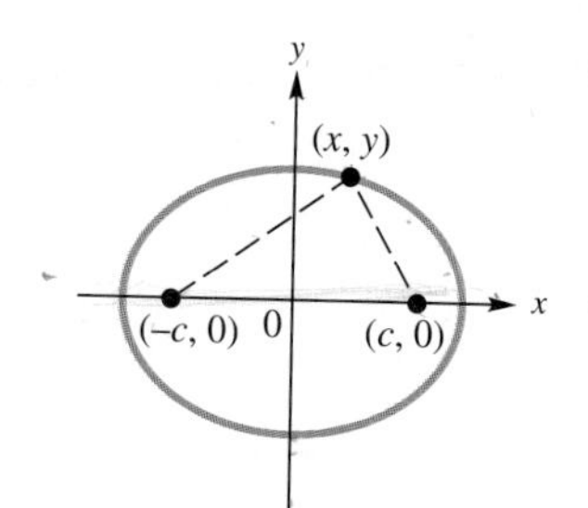

(b) Show that in the equation in part (a), the x-intercepts are $(a, 0)$ and $(-a, 0)$.

(c) Let $b^2 = a^2 - c^2$, and show that $(0, b)$ and $(0, -b)$ are the y-intercepts in the equation in part (a).

52. Use the result of Exercise 51(a) to find an equation of an ellipse with foci $(3, 0)$ and $(-3, 0)$, where the sum of the distances from any point of the ellipse to the two foci is 10.

PREVIEW EXERCISES

Find the intercepts of the graph of each equation. See Section 7.1. Assume that a, b, m, and n represent nonzero real numbers.

53. $2x + 3y = 6$ **54.** $x - 3y = 9$ **55.** $4x - 3y = 5$ **56.** $-x + 2y = 7$

57. $ax + by = c$ **58.** $mx + ny = p$ **59.** $y = 4$ **60.** $x + 1 = 0$

8.6 THE HYPERBOLA; SQUARE ROOT FUNCTIONS

OBJECTIVES

1. RECOGNIZE THE EQUATION OF A HYPERBOLA.
2. GRAPH HYPERBOLAS BY USING THE ASYMPTOTES.
3. IDENTIFY CONIC SECTIONS BY THEIR EQUATIONS.
4. GRAPH SQUARE ROOT FUNCTIONS.

This section begins by introducing another second-degree relation, the *hyperbola*.

1 A **hyperbola** is the set of all points in a plane such that the absolute value of the *difference* of the distances from two fixed points (called **foci**) is constant. Figure 8.32 shows a hyperbola; it can be shown, using the distance formula and the definition above, that this hyperbola is given by the equation

$$\frac{x^2}{16} - \frac{y^2}{12} = 1.$$

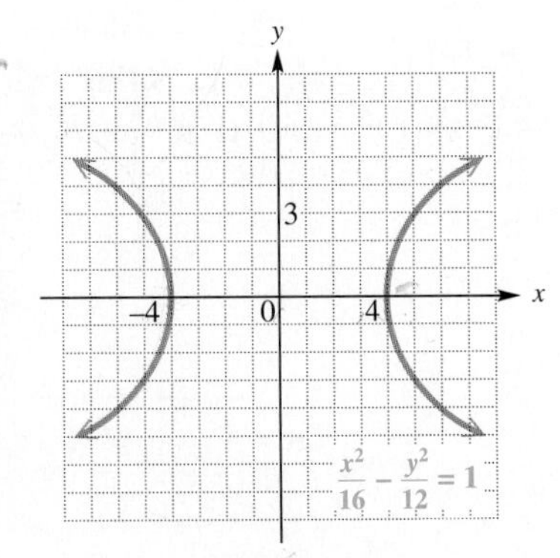

FIGURE 8.32

To find the x-intercepts, let $y = 0$.

$$\frac{x^2}{16} - \frac{0^2}{12} = 1$$
$$\frac{x^2}{16} = 1$$
$$x^2 = 16$$
$$x = \pm 4$$

The x-intercepts are $(4, 0)$ and $(-4, 0)$. To find any y-intercepts, let $x = 0$.

$$\frac{0^2}{16} - \frac{y^2}{12} = 1$$
$$\frac{-y^2}{12} = 1$$
$$y^2 = -12$$

Because there are no *real* solutions to $y^2 = -12$, the graph has no y-intercepts.

Figure 8.33 gives the graph of

$$\frac{y^2}{25} - \frac{x^2}{9} = 1.$$

Here the y-intercepts are $(0, 5)$ and $(0, -5)$, and there are no x-intercepts.

These examples suggest the following summary.

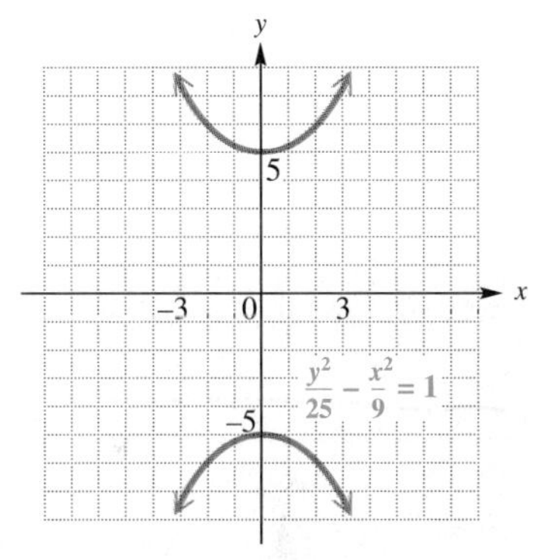

FIGURE 8.33

EQUATIONS OF HYPERBOLAS

A hyperbola with x-intercepts $(a, 0)$ and $(-a, 0)$ has an equation of the form

$$\frac{x^2}{a^2} - \frac{y^2}{b^2} = 1,$$

and a hyperbola with y-intercepts $(0, b)$ and $(0, -b)$ has an equation of the form

$$\frac{y^2}{b^2} - \frac{x^2}{a^2} = 1.$$

2 Starting with

$$\frac{x^2}{a^2} - \frac{y^2}{b^2} = 1$$

and solving for y gives

$$\frac{x^2}{a^2} - 1 = \frac{y^2}{b^2} \qquad \text{Rearrange terms.}$$

$$\frac{x^2 - a^2}{a^2} = \frac{y^2}{b^2} \qquad \text{Combine terms on the left.}$$

$$y = \pm\frac{b}{a}\sqrt{x^2 - a^2}. \; (*) \qquad \text{Square root; multiply by } b.$$

If x^2 is very large in comparison to a^2, the difference $x^2 - a^2$ would be very close to x^2. If this happens, then the points satisfying equation (*) above would be very close to one of the lines

$$y = \pm\frac{b}{a}x.$$

Thus, as $|x|$ gets larger and larger, the points of the hyperbola $x^2/a^2 - y^2/b^2 = 1$ come closer and closer to the lines $y = (\pm b/a)x$. These lines, called the *asymptotes* of the hyperbola, are very helpful when graphing the hyperbola. The asymptotes can be found as follows.

ASYMPTOTES OF HYPERBOLAS

The extended diagonals of the rectangle with corners at the points (a, b), $(-a, b)$, $(-a, -b)$ and $(a, -b)$ are the **asymptotes** of either of the hyperbolas

$$\frac{x^2}{a^2} - \frac{y^2}{b^2} = 1 \quad \text{or} \quad \frac{y^2}{b^2} - \frac{x^2}{a^2} = 1.$$

This rectangle is called the **fundamental rectangle.** It can be shown using the methods of Chapter 7 that the equations of these asymptotes are

$$y = \frac{b}{a}x \quad \text{and} \quad y = -\frac{b}{a}x,$$

as mentioned above.

EXAMPLE 1 GRAPHING A HYPERBOLA

Graph $\dfrac{x^2}{16} - \dfrac{y^2}{25} = 1.$

Here $a = 4$ and $b = 5$. The x-intercepts are (4, 0) and (−4, 0). The four points (4, 5), (−4, 5), (−4, −5), and (4, −5) are the corners of the rectangle that determine the asymptotes, as shown in Figure 8.34. The equations of the asymptotes are $y = \pm\dfrac{5}{4}x$, and the hyperbola approaches these lines as x and y get larger and larger in absolute value. ■

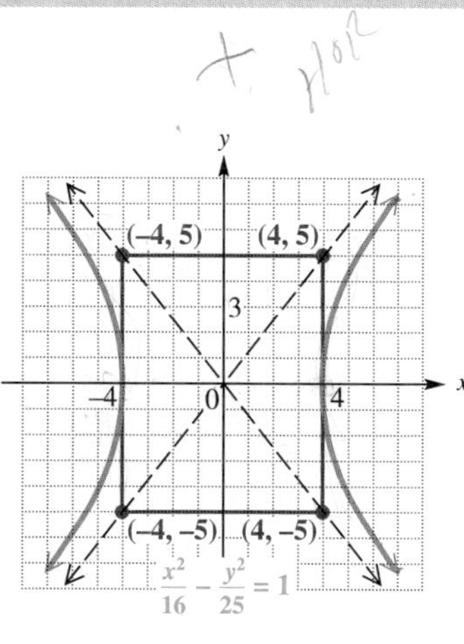

FIGURE 8.34

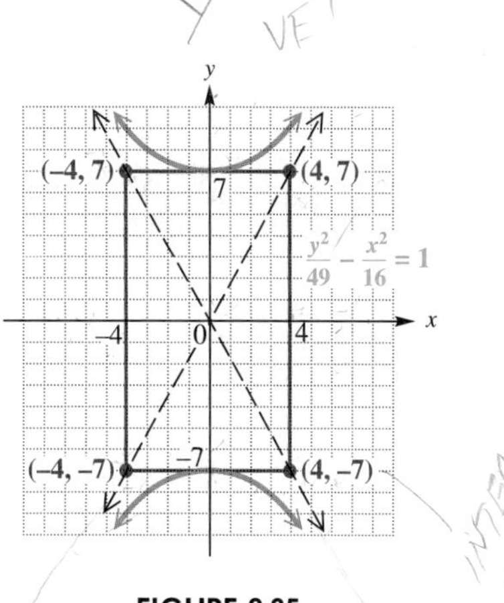

FIGURE 8.35

EXAMPLE 2 GRAPHING A HYPERBOLA

Graph $\frac{y^2}{49} - \frac{x^2}{16} = 1$.

This hyperbola has y-intercepts $(0, 7)$ and $(0, -7)$. The asymptotes are the extended diagonals of the rectangle with corners at $(4, 7)$, $(-4, 7)$, $(-4, -7)$, and $(4, -7)$, and have equations $y = \pm\frac{7}{4}x$. See Figure 8.35. ■

In summary, to graph either of the two forms of hyperbolas, $\frac{x^2}{a^2} - \frac{y^2}{b^2} = 1$ or $\frac{y^2}{b^2} - \frac{x^2}{a^2} = 1$, follow these steps.

GRAPHING A HYPERBOLA

Step 1 **Find the intercepts.**
Locate the intercepts: at $(a, 0)$ and $(-a, 0)$ if the x^2 term has a positive coefficient, or at $(0, b)$ and $(0, -b)$ if the y^2 term has a positive coefficient.

Step 2 **Find the fundamental rectangle.**
Locate the corners of the fundamental rectangle at (a, b), $(a, -b)$, $(-a, -b)$ and $(-a, b)$.

Step 3 **Sketch the asymptotes.**
The extended diagonals of the rectangle are the asymptotes of the hyperbola, and have equations $y = \pm\frac{b}{a}x$. Sketch these asymptotes.

Step 4 **Draw the graph.**
Sketch each branch of the hyperbola through an intercept and approaching (but not touching) the asymptotes.

CAUTION When sketching the graph of a hyperbola, be sure that the branches do not touch the asymptotes.

In Section 8.5 we showed that the center of a circle or an ellipse may be shifted away from the origin. The same is true of hyperbolas, as shown in the next example.

EXAMPLE 3
GRAPHING A HYPERBOLA SHIFTED AWAY FROM THE ORIGIN

Graph $\dfrac{(y+2)^2}{9} - \dfrac{(x+3)^2}{4} = 1.$

This hyperbola has the same graph as

$$\frac{y^2}{9} - \frac{x^2}{4} = 1,$$

but the graph is centered at $(-3, -2)$. The vertices are located 3 units above and below the center, with y-values of $3 - 2 = 1$ and $-3 - 2 = -5$. Both have an x-value of -3. See Figure 8.36. ■

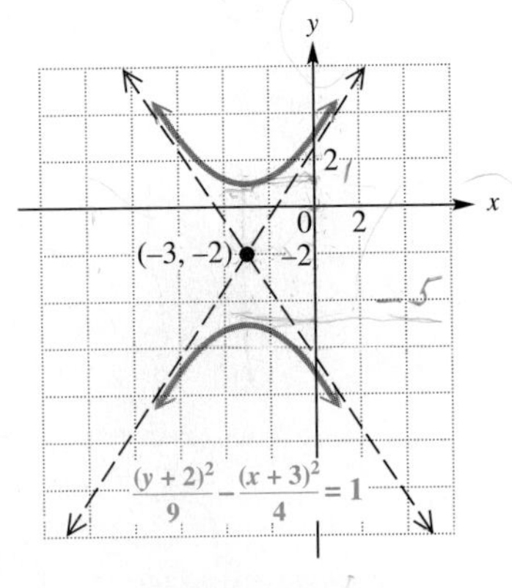

FIGURE 8.36

3 The graphs of the second-degree relations studied in this chapter, parabolas, circles, ellipses, and hyperbolas, are called **conic sections** since each is the intersection of a cone and a plane, as shown in Figure 8.37.

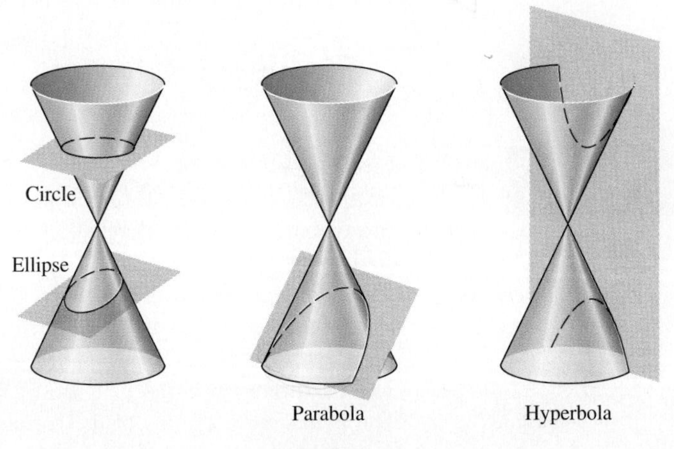

FIGURE 8.37

Rewriting a second-degree equation in one of the forms given for ellipses, hyperbolas, circles, or parabolas makes it possible to determine when the graph is one of these figures. A summary of the equations and graphs of the conic sections follows.

Equation	*Graph*	*Description*	*Identification*
$y = a(x - h)^2 + k$	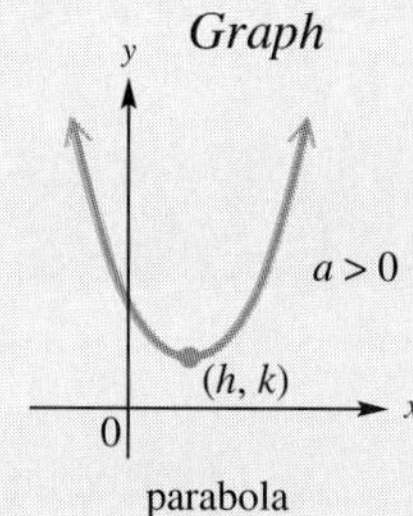 parabola	Opens upward if $a > 0$, downward if $a < 0$. Vertex is at (h, k).	x^2 term y is not squared.
$x = a(y - k)^2 + h$	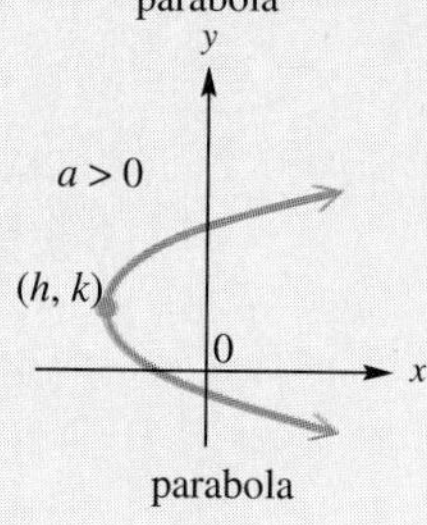 parabola	Opens to right if $a > 0$, to left if $a < 0$. Vertex is at (h, k).	y^2 term x is not squared.
$(x - h)^2 + (y - k)^2 = r^2$	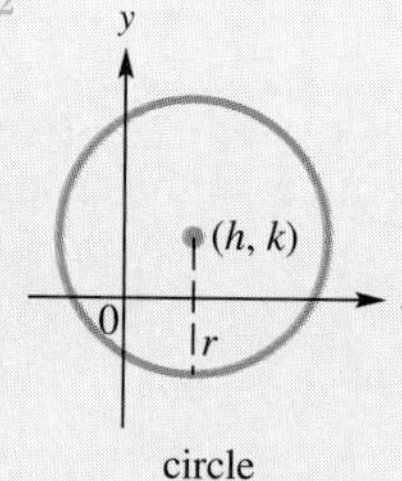 circle	Center at (h, k), radius r	x^2 and y^2 terms have the same positive coefficient.
$\frac{x^2}{a^2} + \frac{y^2}{b^2} = 1$	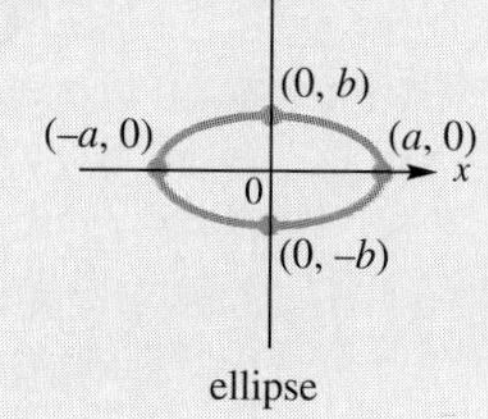 ellipse	x-intercepts are $(a, 0)$ and $(-a, 0)$. y-intercepts are $(0, b)$ and $(0, -b)$.	x^2 and y^2 terms have different positive coefficients.
$\frac{x^2}{a^2} - \frac{y^2}{b^2} = 1$	y $(-a, 0)$ $(a, 0)$ x 0 hyperbola	x-intercepts are $(a, 0)$ and $(-a, 0)$. Asymptotes are found from (a, b), $(a, -b)$, $(-a, -b)$, and $(-a, b)$.	x^2 has a positive coefficient. y^2 has a negative coefficient.
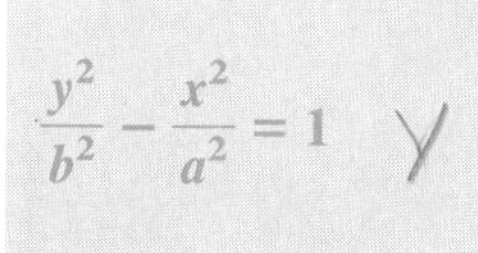 $\frac{y^2}{b^2} - \frac{x^2}{a^2} = 1$	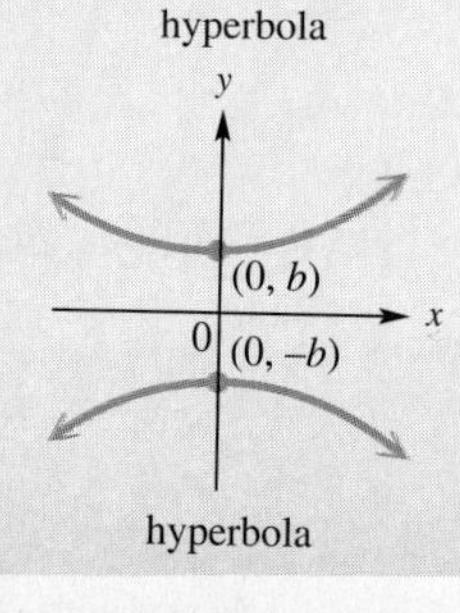 hyperbola	y-intercepts are $(0, b)$ and $(0, -b)$. Asymptotes are found from (a, b), $(a, -b)$, $(-a, -b)$, and $(-a, b)$.	y^2 has a positive coefficient. x^2 has a negative coefficient.

It can be shown that all conic sections of the types presented in this chapter have equations of the form

$$Ax^2 + Bx + Cy^2 + Dy + E = 0,$$

where either A or C must be nonzero. The special characteristics of each of the equations of the conic sections are summarized below.

EQUATIONS OF CONIC SECTIONS

Conic section	*Characteristic*	*Example*
Parabola	Either $A = 0$ or $C = 0$, but not both	$x^2 = y + 4$ $x = (y - 2)^2 + 1$
Circle	$A = C \neq 0$	$x^2 + y^2 = 16$
Ellipse	$A \neq C$, $AC > 0$	$\frac{x^2}{16} + \frac{y^2}{25} = 1$
Hyperbola	$AC < 0$	$x^2 - y^2 = 1$

EXAMPLE 4

DETERMINING THE TYPE OF A CONIC SECTION FROM ITS EQUATION

What kind of graph is given by the equation $9x^2 = 108 + 12y^2$? Sketch the graph.

Dividing both sides by 108 and rewriting so that both variables are on the same side of the equation gives

$$9x^2 = 108 + 12y^2$$

$$\frac{x^2}{12} = 1 + \frac{y^2}{9} \qquad \text{Divide by 108.}$$

$$\frac{x^2}{12} - \frac{y^2}{9} = 1. \qquad \text{Subtract } \frac{y^2}{9}.$$

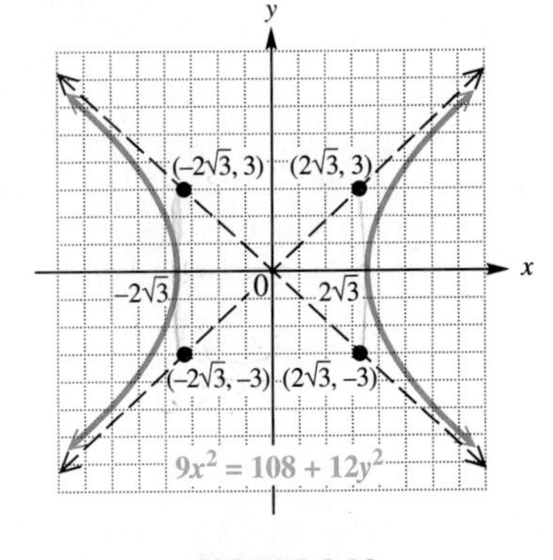

FIGURE 8.38

Because of the minus sign, the graph is a hyperbola, with x-intercepts $(2\sqrt{3}, 0)$ and $(-2\sqrt{3}, 0)$, since $\sqrt{12} = 2\sqrt{3}$. The asymptotes go through the points $(2\sqrt{3}, 3)$, $(2\sqrt{3}, -3)$, $(-2\sqrt{3}, -3)$, and $(-2\sqrt{3}, 3)$. See Figure 8.38. ■

EXAMPLE 5

IDENTIFYING CONIC SECTIONS FROM THEIR EQUATIONS

Identify the graph of each equation.

(a) $x^2 = y - 3$

Only one of the two variables is squared (x), so this is the vertical parabola $y = x^2 + 3$.

(b) $x^2 = y^2 + 9$

Since both variables are squared, collect the variable terms on the same side of the equation and divide each term by 9.

$$x^2 = y^2 + 9$$

$$x^2 - y^2 = 9 \qquad \text{Subtract } y^2.$$

$$\frac{x^2}{9} - \frac{y^2}{9} = 1 \qquad \text{Divide by 9.}$$

The minus sign indicates that this is a hyperbola with x-intercepts $(3, 0)$ and $(-3, 0)$.

(c) $x^2 = 9 - y^2$

Get the variable terms on the same side of the equation.

$$x^2 + y^2 = 9$$

Since the squared variable terms have the same coefficient and are added, this equation represents the circle with center at the origin and radius 3. ■

4 Horizontal parabolas, circles, ellipses, and hyperbolas are examples of relations that do not satisfy the conditions of a function. Recall that no vertical line will intersect the graph of a function in more than one point. However, by considering only a part of the graph of each of these figures, we can have the graph of a function, as seen in Figure 8.39.

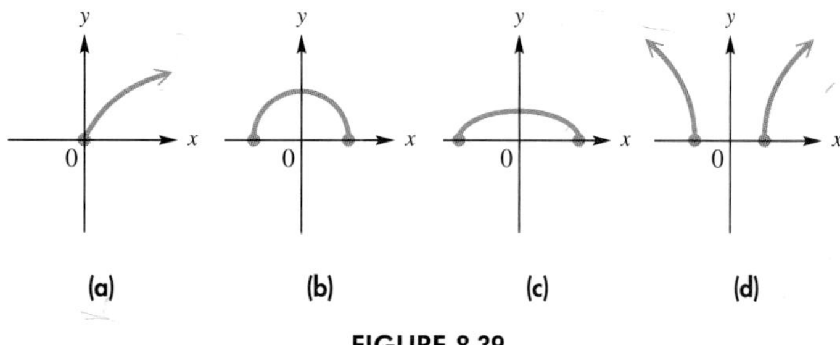

FIGURE 8.39

In parts (a), (b), and (c) of Figure 8.39, the top portion of a conic section is shown (parabola, circle, and ellipse, respectively). In part (d), the top two portions of a hyperbola are shown. In each case, the graph is the graph of a function, since the graph satisfies the conditions of the vertical line test.

In order to obtain equations for the graphs shown in Figure 8.39, the idea of a square root function is introduced.

SQUARE ROOT FUNCTION

A function of the form

$$f(x) = \sqrt{u}$$

for an algebraic expression u, with $u \geq 0$, is called a **square root function.**

NOTE The square root function $f(x) = \sqrt{u}$ is a composite function with $h(x) = \sqrt{x}$, $g(x) = u$ and $f(x) = h[g(x)] = h(u) = \sqrt{u}$. For example, if $h(x) = \sqrt{x}$ and $g(x) = u = 25 - x^2$, then $f(x) = h[g(x)] = \sqrt{g(x)} = \sqrt{25 - x^2}$.

EXAMPLE 6
GRAPHING A SQUARE ROOT FUNCTION

Graph $f(x) = \sqrt{25 - x^2}$.

Replace $f(x)$ with y and square both sides to get the equation

$$y^2 = 25 - x^2, \qquad \text{or} \qquad x^2 + y^2 = 25.$$

This is the graph of a circle with center at (0, 0) and radius 5. Since $f(x)$, or y, represents a principal square root in the original equation, $f(x)$ must be nonnegative. This restricts the graph to the upper half of the circle with endpoints at $(-5, 0)$ and $(5, 0)$, as shown in Figure 8.40. Use the graph and the vertical line test to verify that it is indeed a function. ■

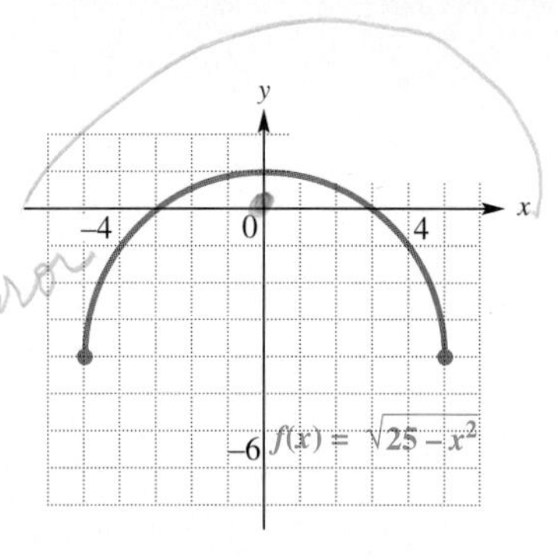

FIGURE 8.40

NOTE Refer to Figure 8.40. Had we wanted to graph the function

$$g(x) = -\sqrt{25 - x^2},$$

the graph of $g(x)$ would be obtained simply by reflecting the graph of $f(x)$ across the x-axis. This would ensure that the y-values were nonpositive, as indicated by the $-$ sign in the rule for $g(x)$.

EXAMPLE 7
GRAPHING A SQUARE ROOT FUNCTION

Graph $\dfrac{y}{6} = -\sqrt{1 - \dfrac{x^2}{16}}$.

Square both sides to get an equation whose form is known.

$$\frac{y^2}{36} = 1 - \frac{x^2}{16} \qquad \text{Square both sides.}$$

$$\frac{x^2}{16} + \frac{y^2}{36} = 1 \qquad \text{Add } \frac{x^2}{16}.$$

This is the equation of an ellipse with x-intercepts (4, 0) and $(-4, 0)$ and y-intercepts (0, 6) and $(0, -6)$. Since $\frac{y}{6}$ equals a negative square root in the original equation, y must be nonpositive, restricting the graph to the lower half of the ellipse, as shown in Figure 8.41. Verify that this is the graph of a function, using the vertical line test. ■

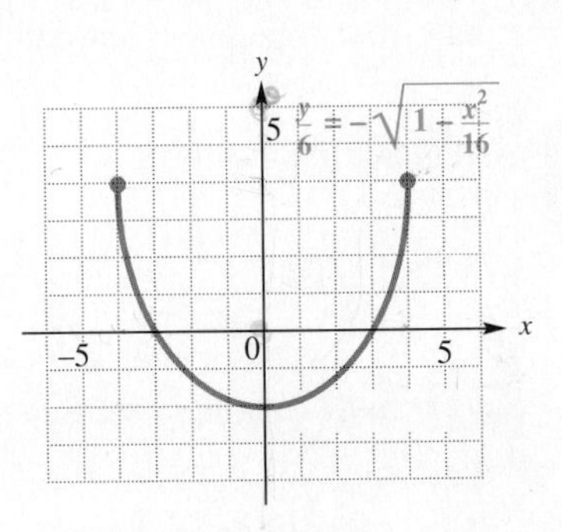

FIGURE 8.41

8.6 EXERCISES

Graph each hyperbola. See Examples 1–3.

1. $\dfrac{x^2}{25} - \dfrac{y^2}{9} = 1$ **2.** $\dfrac{y^2}{16} - \dfrac{x^2}{16} = 1$ **3.** $\dfrac{y^2}{49} - \dfrac{x^2}{36} = 1$ **4.** $\dfrac{x^2}{144} - \dfrac{y^2}{49} = 1$

5. $\dfrac{x^2}{64} - \dfrac{y^2}{100} = 1$ **6.** $\dfrac{x^2}{4} - \dfrac{y^2}{25} = 1$ **7.** $\dfrac{x^2}{36} - \dfrac{y^2}{36} = 1$ **8.** $\dfrac{y^2}{9} - \dfrac{x^2}{4} = 1$

9. $\dfrac{y^2}{4/9} - \dfrac{x^2}{9/4} = 1$ **10.** $\dfrac{x^2}{81/16} - \dfrac{y^2}{49/4} = 1$ **11.** $\dfrac{64x^2}{9} - \dfrac{25y^2}{36} = 1$ **12.** $\dfrac{121x^2}{25} - \dfrac{16y^2}{9} = 1$

13. $\dfrac{(x-1)^2}{9} - \dfrac{(y+3)^2}{25} = 1$ **14.** $\dfrac{(x+3)^2}{16} - \dfrac{(y-2)^2}{36} = 1$ **15.** $\dfrac{(x-3)^2}{16} - \dfrac{(y+2)^2}{49} = 1$

16. $\dfrac{(y-5)^2}{4} - \dfrac{(x+1)^2}{9} = 1$ **17.** $\dfrac{(y+1)^2}{25} - \dfrac{(x-3)^2}{36} = 1$ **18.** $\dfrac{(x+2)^2}{16} - \dfrac{(y+2)^2}{25} = 1$

In Exercises 19–30, identify the graph of the equation as one of the four conic sections. (It may be necessary to transform the equation into a more recognizable form.) See the chart in this section if necessary. Then sketch the graph of the equation. See Examples 4 and 5.

19. $x^2 + y^2 = 25$ **20.** $9x^2 = 36 + 4y^2$ **21.** $x^2 = 16 - 4y^2$ **22.** $4x^2 + 9y^2 = 36$

23. $2x^2 - y = 0$ **24.** $x^2 + 2y^2 = 8$ **25.** $y^2 = 144 - x^2$ **26.** $x^2 - y^2 = 36$

27. $x^2 + 9y^2 = 9$ **28.** $25x^2 + 9y^2 = 225$ **29.** $25x^2 = 225 + 9y^2$ **30.** $4x^2 - 25y^2 = 100$

---

31. Write an explanation of how you can tell whether the branches of a hyperbola open up and down or right and left. (See Objective 1.)

32. Explain why the graph of a hyperbola as defined in this section does not satisfy the conditions for the graph of a function. (See Objective 4.)

Graph each of the following square root functions. See Examples 6 and 7.

33. $f(x) = \sqrt{x+5}$ **34.** $f(x) = \sqrt{x-1}$ **35.** $f(x) = -\sqrt{9-x}$ **36.** $f(x) = -\sqrt{x-4}$

37. $f(x) = \sqrt{\dfrac{x+6}{3}}$ **38.** $f(x) = \sqrt{\dfrac{5-x}{2}}$ **39.** $y = \sqrt{16-x^2}$ **40.** $y = -\sqrt{4-4x^2}$

41. $\dfrac{y}{4} = -\sqrt{1 - \dfrac{x^2}{9}}$ **42.** $y = -\sqrt{36-x^2}$ **43.** $\dfrac{y}{6} = \sqrt{1 - \dfrac{x^2}{36}}$ **44.** $y = \sqrt{36-4x^2}$

Work each of the following problems.

45. An arch has the shape of half an ellipse. If the equation of the ellipse is $100x^2 + 324y^2 = 32{,}400$, where x and y are in meters, **(a)** how high is the center of the arch? **(b)** how wide is the arch across the bottom? (See the figure.)

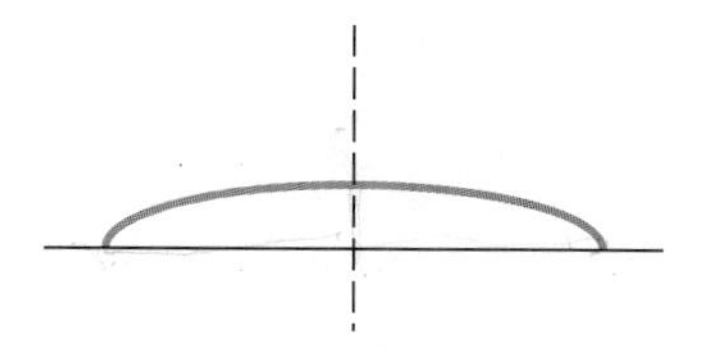

46. A pair of buildings in a sports complex is shaped and positioned like a portion of the branches of the hyperbola $400x^2 - 625y^2 = 250{,}000$, where x and y are in meters. **(a)** How far apart are the buildings at their closest point? **(b)** Find the distance d in the figure.

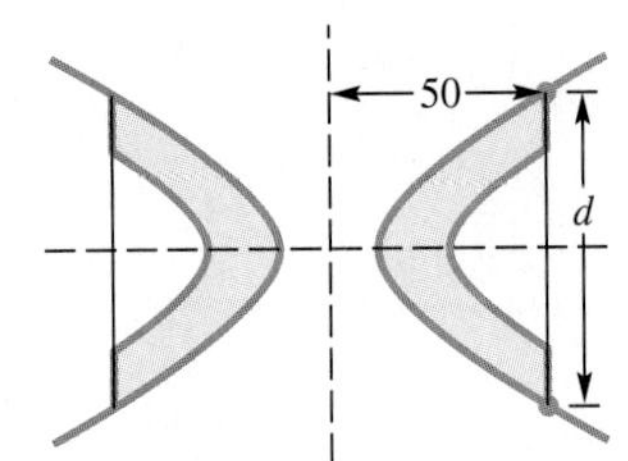

Solve in the manner indicated. See Figure 8.37.

47. (a) Show by a sketch how a cone can be cut by a plane in exactly one point. (For this reason, a point is sometimes called a *degenerate circle* or *degenerate ellipse*.)

(b) Show by a sketch how a cone can be cut by a plane to produce exactly one straight line. (For this reason, a straight line is sometimes called a *degenerate parabola*.)

48. Following the definitions in Exercise 47, what is a *degenerate hyperbola?*

49. Suppose that a hyperbola has center at the origin, foci at $(-c, 0)$ and $(c, 0)$, and the absolute value of the difference between the distances from any point (x, y) of the hyperbola to the two foci is $2a$. See the figure. Let $b^2 = c^2 - a^2$, and show that an equation of the hyperbola is

$$\frac{x^2}{a^2} - \frac{y^2}{b^2} = 1.$$

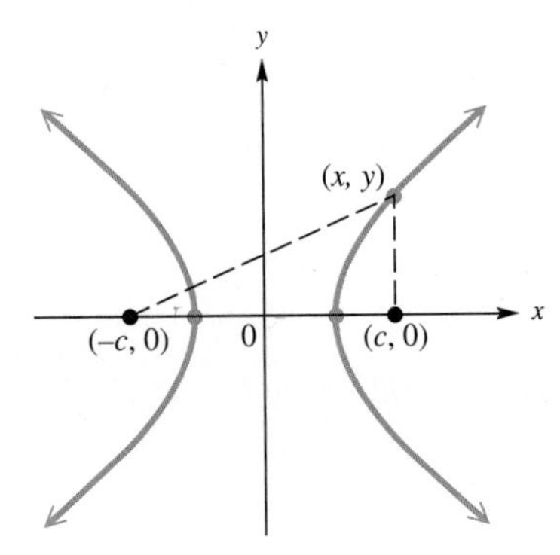

50. Use the result of Exercise 49 to find an equation of a hyperbola with center at the origin, foci at $(-2, 0)$ and $(2, 0)$, and absolute value of the difference between the distances from any point of the hyperbola to the two foci equal to 2.

PREVIEW EXERCISES

Evaluate each of the following if $x = 3$; if $x = 5$. See Section 1.1.

51. $|2x|$

52. $|-x|$

53. $|5 - x|$

54. $|x + 4|$

55. $|3x + 1|$

56. $|-2x - 4|$

8.7 OTHER USEFUL FUNCTIONS

OBJECTIVES

1 GRAPH ABSOLUTE VALUE FUNCTIONS.

2 GRAPH FUNCTIONS DEFINED PIECEWISE.

3 GRAPH GREATEST INTEGER FUNCTIONS.

4 GRAPH STEP FUNCTIONS.

FOCUS ON PROBLEM SOLVING

When a diabetic takes long-acting insulin, the insulin reaches its peak effect on the blood sugar level in about 3 hours. This effect remains fairly constant for 5 hours, then declines, and is very low until the next injection. In a typical patient, the level of insulin might be given by the following function.

$$i(t) = \begin{cases} 40t + 100 & \text{if } 0 \leq t \leq 3 \\ 220 & \text{if } 3 < t \leq 8 \\ -80t + 860 & \text{if } 8 < t \leq 10 \\ 60 & \text{if } 10 < t \leq 24 \end{cases}$$

Here $i(t)$ is the blood sugar level, in appropriate units, at time t measured in hours from the time of the injection. Chuck takes his insulin at 6 A.M. Find the blood sugar level at each of the following times. **(a)** 7 A.M. **(b)** 9 A.M. **(c)** 10 A.M. **(d)** noon **(e)** 2 P.M. **(f)** 5 P.M. **(g)** midnight **(h)** Graph $y = i(t)$.

Relationships like the insulin level described above cannot be expressed with a single equation. Such relationships are often functions, however, and we can apply the information about functions studied up to this point in working with these functions. In this section several functions of this type will be examined. This problem is Exercise 37 in the exercises for this section. After working through the section, you should be able to solve this problem.

In this section we discuss several useful functions that are neither linear nor quadratic. The graphs of these functions are made of *portions* of different straight lines or of straight lines and curves.

1 One common function that is not linear and not quadratic is the **absolute value function,** defined by $f(x) = |x|$, or $y = |x|$. Since $|x|$ can be found for any real number x, the domain is $(-\infty, \infty)$. Also, $|x| \geq 0$ for any real number x, so the range is $[0, \infty)$. When $x \geq 0$, then $y = |x| = x$, so for nonnegative values of x we graph $y = x$. On the other hand, if $x < 0$, then $y = |x| = -x$, and we graph $y = -x$ for negative values of x. The final graph is shown in Figure 8.42. By the vertical line test, the graph is that of a function. Notice that the graph of $f(x) = |x|$ is symmetric with respect to the y-axis. Like the parabola, the graphs of absolute value functions always have an axis of symmetry.

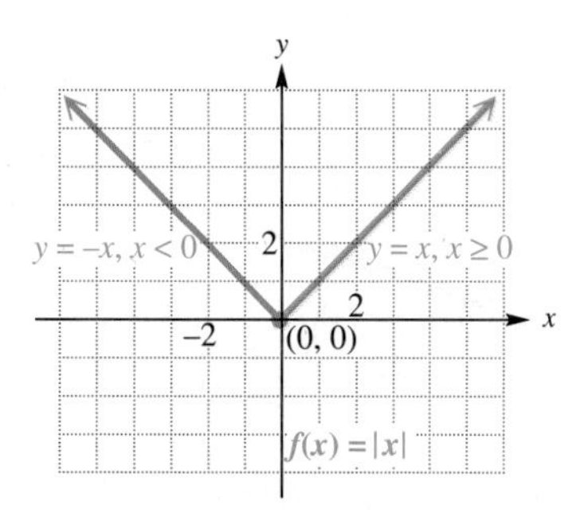

FIGURE 8.42

EXAMPLE 1 GRAPHING AN ABSOLUTE VALUE FUNCTION

Graph $f(x) = |x - 2|$.

Again, the domain is $(-\infty, \infty)$, and the range is $[0, \infty)$. Since $y \geq 0$, the lowest point on the graph (the "vertex") is found by setting $y = 0$ to get $(2, 0)$. Graphing the point $(2, 0)$ and a few other selected points leads to the graph in Figure 8.43. This graph has the same shape as that of $y = |x|$, but the "vertex" point is translated 2 units to the right, from $(0, 0)$ to $(2, 0)$. The axis of symmetry is the vertical line $x = 2$, through $(2, 0)$. ∎

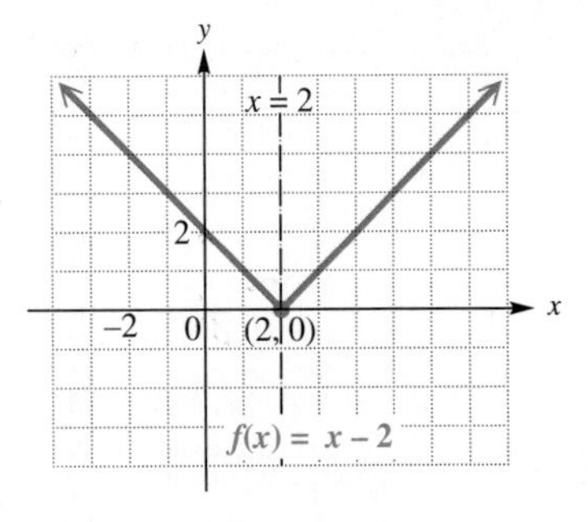

FIGURE 8.43

EXAMPLE 2 GRAPHING AN ABSOLUTE VALUE FUNCTION

Graph $f(x) = |3x + 4| + 1$.

The value of $f(x) = y$ is always greater than or equal to 1, since $|3x + 4| \geq 0$. Thus, the y-value of the "vertex" is 1. The x-value of the "vertex" can be found by substituting 1 for y in the equation.

$$y = |3x + 4| + 1$$

$$1 = |3x + 4| + 1 \qquad \text{Let } y = 1.$$

$$0 = |3x + 4| \qquad \text{Subtract 1.}$$

$$x = -\frac{4}{3} \qquad \text{Solve } 3x + 4 = 0.$$

Plotting a few other ordered pairs leads to the graph in Figure 8.44. The graph shows the "vertex" translated to $(-4/3, 1)$. The axis of symmetry is $x = -4/3$. The coefficient of x, 3, determines the slopes of the two rays that form the graph. One has slope 3, and the other has slope -3. Because the absolute value of the slopes is greater than 1, the lines are steeper than the lines that form the graph of $y = |x|$. ∎

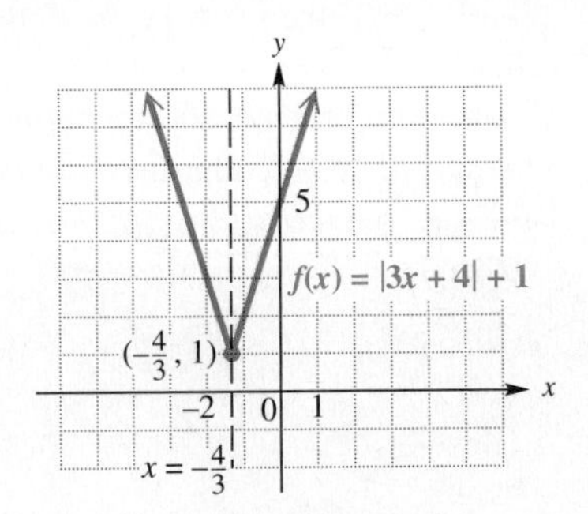

FIGURE 8.44

2 The graphs of the absolute value functions of Examples 1 and 2 are made up of portions of two different straight lines. Such functions, called **functions defined piecewise,** are often defined with different equations for different parts of the domain.

EXAMPLE 3 GRAPHING A FUNCTION DEFINED PIECEWISE

Graph the function

$$f(x) = \begin{cases} x + 1 & \text{if } x \leq 2 \\ -2x + 7 & \text{if } x > 2. \end{cases}$$

We must graph each part of the domain separately. If $x \leq 2$, this portion of the graph has an endpoint at $x = 2$. Find the y-value by substituting 2 for x in $y = x + 1$ to get $y = 3$. Another point is needed to graph this part of the graph. Choose an x-value less than 2. Choosing $x = -1$ gives $y = -1 + 1 = 0$. Draw the graph through (2, 3) and $(-1, 0)$ as a ray with an endpoint at (2, 3). Graph the ray for $x > 2$ similarly. This ray will have an open endpoint when $x = 2$ and $y = -2(2) + 7 = 3$. Choosing $x = 4$ gives $y = -2(4) + 7 = -1$. The ray through (2, 3) and $(4, -1)$ completes the graph. In this example the two rays meet at (2, 3), although this is not always the case. The graph is shown in Figure 8.45. ■

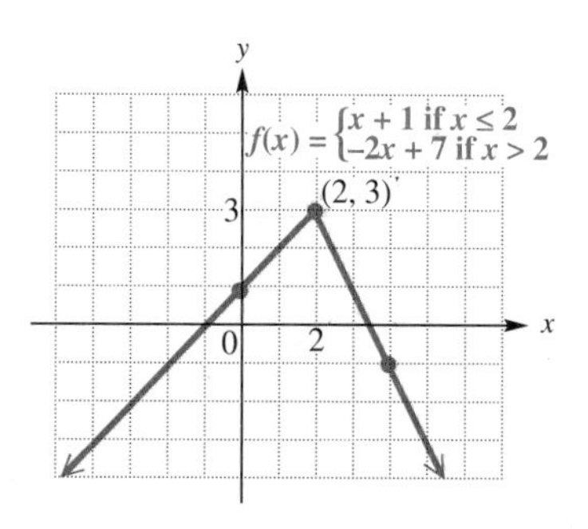

FIGURE 8.45

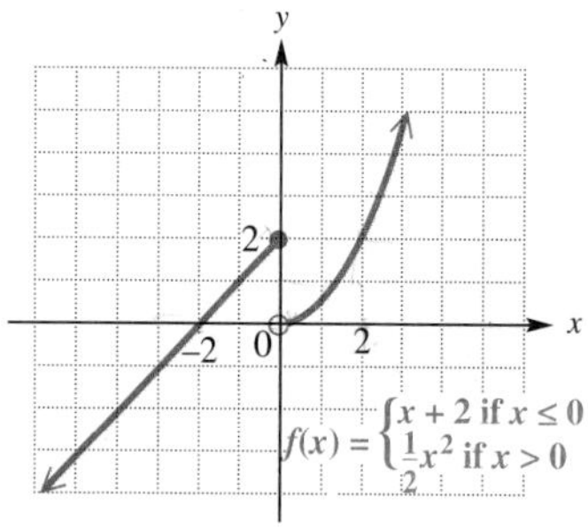

FIGURE 8.46

EXAMPLE 4 GRAPHING A FUNCTION DEFINED PIECEWISE

Graph

$$f(x) = \begin{cases} x + 2 & \text{if } x \leq 0 \\ \dfrac{1}{2}x^2 & \text{if } x > 0. \end{cases}$$

Graph the line $y = x + 2$, choosing x so that $x \leq 0$, with a solid endpoint at (0, 2). The line has slope 1 and y-intercept (0, 2). Then graph $y = \frac{1}{2}x^2$ for $x > 0$. This graph will be half of a parabola with an open endpoint at (0, 0). See Figure 8.46. Notice that in this example, the two portions do not meet. ■

3 Another type of function with a graph composed of line segments is the **greatest integer function,** written $f(x) = [\![x]\!]$, and defined as follows:

$[\![x]\!]$ is the greatest integer less than or equal to x.

For example, $[\![8]\!] = 8$, $[\![-5]\!] = -5$, $[\![\pi]\!] = 3$, $[\![12\ 1/9]\!] = 12$, $[\![-2.001]\!] = -3$, and so on.

EXAMPLE 5

GRAPHING THE GREATEST INTEGER FUNCTION

Graph $f(x) = [\![x]\!]$.

For any value of x in the interval $[0, 1)$, $[\![x]\!] = 0$. Also, for x in $[1, 2)$, $[\![x]\!] = 1$. This process continues; for x in $[2, 3)$, the value of $[\![x]\!]$ is 2. The values of y are constant between integers, but they jump at integer values of x. This makes the graph, shown in Figure 8.47, a series of line segments. In each case, the left endpoint of the segment is included, and the right endpoint is excluded. The domain of the function is $(-\infty, \infty)$, while the range is the set of integers, $\{\ldots, -2, -1, 0, 1, 2, \ldots\}$. ■

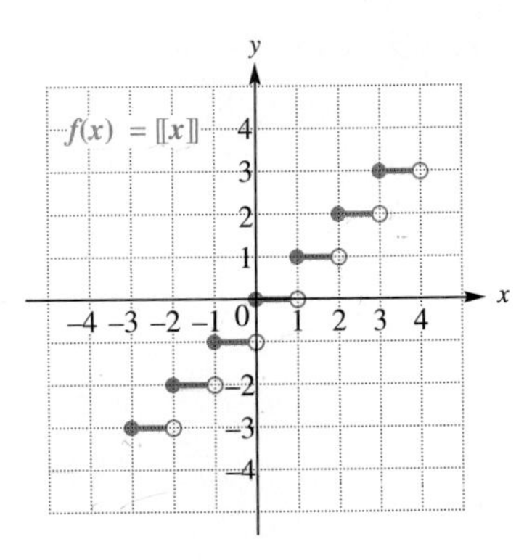

FIGURE 8.47

4 Each of the graphs in Examples 5, 6, and 7 is made up of a series of horizontal line segments. (See Figures 8.47, 8.48, and 8.49.) The functions producing these graphs are called **step functions.**

EXAMPLE 6

GRAPHING A GREATEST INTEGER FUNCTION

Graph $f(x) = \left[\!\!\left[\frac{1}{2}x + 1\right]\!\!\right]$.

If x is in the interval $[0, 2)$, then $y = 1$. For x in $[2, 4)$, $y = 2$, and so on. The graph is shown in Figure 8.48. Again, the domain is $(-\infty, \infty)$. The range is $\{\ldots, -1, 0, 1, 2, \ldots\}$. ■

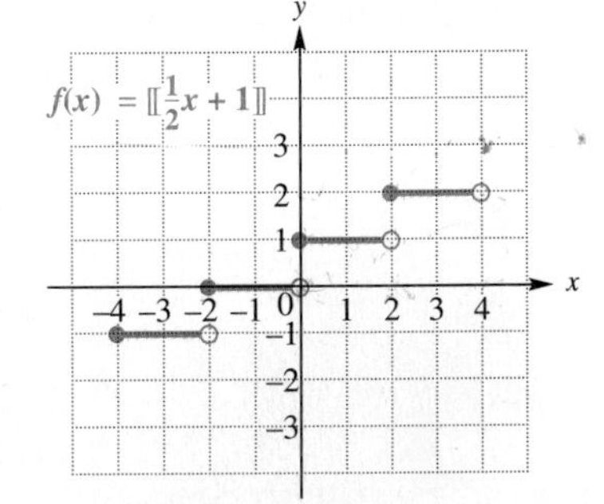

FIGURE 8.48

The greatest integer function can be used to describe many common pricing practices encountered in everyday life, as shown in the next example.

EXAMPLE 7

APPLYING THE GREATEST INTEGER FUNCTION

An express mail company charges \$10 for a package weighing 1 pound or less. Each additional pound or part of a pound costs \$3 more. Find the cost to send a package weighing 2 pounds; 2.5 pounds; 5.8 pounds. Graph the ordered pairs (pounds, cost). Is this the graph of a function?

The cost for a package weighing 2 pounds is \$10 for the first pound and \$3 for the second pound, for a total of \$13. For a 2.5 pound package, the cost will be the same as for 3 pounds: $10 + 2(3) = 16$, or \$16. A 5.8 pound package will cost the same as a 6 pound package: $10 + 5(3) = 25$, or \$25. The graph of this step function is shown in Figure 8.49. Notice that the right endpoints are included in this case, instead of the left endpoints. ■

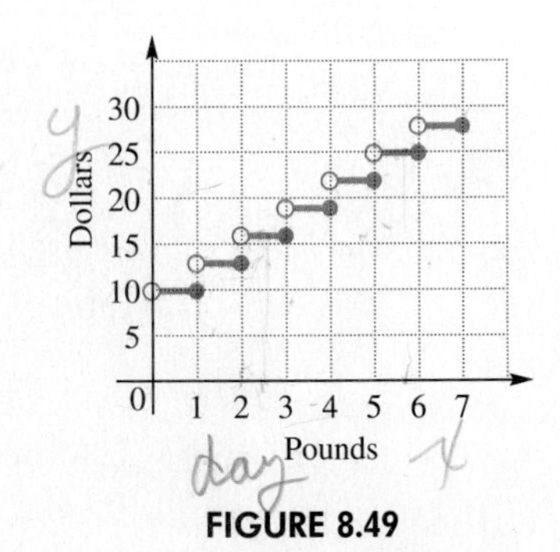

FIGURE 8.49

8.7 EXERCISES

Graph each of the functions defined as follows. See Examples 1 and 2.

1. $f(x) = |x + 1|$ **2.** $f(x) = |x - 1|$ **3.** $f(x) = |2 - x|$ **4.** $f(x) = |-3 - x|$

5. $y = |x| + 4$ **6.** $y = 2|x| - 1$ **7.** $y = 3|x - 2| - 1$ **8.** $y = \frac{1}{2}|x + 3| + 1$

For each of the following, find **(a)** $f(-5)$, **(b)** $f(-1)$, **(c)** $f(0)$, **(d)** $f(3)$, *and* **(e)** $f(5)$. *See Example 3.*

9. $f(x) = \begin{cases} 2x & \text{if } x \le -1 \\ x - 1 & \text{if } x > -1 \end{cases}$

10. $f(x) = \begin{cases} x - 2 & \text{if } x < 3 \\ 4 - x & \text{if } x \ge 3 \end{cases}$

11. $f(x) = \begin{cases} 3x + 5 & \text{if } x \le 0 \\ 4 - 2x & \text{if } 0 < x < 2 \\ x & \text{if } x \ge 2 \end{cases}$

12. $f(x) = \begin{cases} 4x + 1 & \text{if } x < 2 \\ 3x & \text{if } 2 \le x \le 5 \\ 3 - 2x & \text{if } x > 5 \end{cases}$

Graph each of the functions defined as follows. See Examples 3 and 4.

13. $f(x) = \begin{cases} x - 1 & \text{if } x \le 3 \\ 2 & \text{if } x > 3 \end{cases}$

14. $f(x) = \begin{cases} 6 - x & \text{if } x \le 3 \\ 3x - 6 & \text{if } x > 3 \end{cases}$

15. $f(x) = \begin{cases} 4 - x & \text{if } x < 2 \\ 1 + 2x & \text{if } x \ge 2 \end{cases}$

16. $f(x) = \begin{cases} -2 & \text{if } x \ge 1 \\ 2 & \text{if } x < 1 \end{cases}$

17. $f(x) = \begin{cases} 2x + 1 & \text{if } x \ge 0 \\ x & \text{if } x < 0 \end{cases}$

18. $f(x) = \begin{cases} 5x - 4 & \text{if } x \ge 1 \\ x & \text{if } x < 1 \end{cases}$

19. $f(x) = \begin{cases} 2 + x & \text{if } x < -4 \\ -x^2 & \text{if } x \ge -4 \end{cases}$

20. $f(x) = \begin{cases} -2x & \text{if } x \le 2 \\ -x^2 & \text{if } x > 2 \end{cases}$

21. $f(x) = \begin{cases} |x| & \text{if } x > -2 \\ x^2 - 2 & \text{if } x \le -2 \end{cases}$

22. $f(x) = \begin{cases} |x| - 1 & \text{if } x > -1 \\ x^2 - 1 & \text{if } x \le -1 \end{cases}$

Graph each of the following. See Examples 5 and 6.

23. $f(x) = [\![-x]\!]$ **24.** $f(x) = [\![2x]\!]$ **25.** $f(x) = [\![2x - 1]\!]$ **26.** $f(x) = [\![3x + 1]\!]$

27. $f(x) = [\![3x]\!]$ **28.** $f(x) = [\![3x]\!] + 1$ **29.** $f(x) = [\![3x]\!] - 1$ **30.** $f(x) = x - [\![x]\!]$

Work the following problems. See Example 7.

31. Suppose a chain-saw rental firm charges a fixed \$4 sharpening fee plus \$7 per day or fraction of a day. Let $S(x)$ represent the cost of renting a saw for x days. Find the value of
(a) $S(1)$, **(b)** $S(1.25)$, and **(c)** $S(3.5)$.
(d) Graph $y = S(x)$.
(e) Give the domain and range of S.

32. A mail order firm charges postage of 30¢ to mail one ounce or less, and then 27¢ for each additional ounce or fraction of an ounce. Let $M(x)$ be the cost of mailing a package weighing x ounces. Find
(a) $M(.75)$, **(b)** $M(1.6)$, and **(c)** $M(4)$.
(d) Graph $y = M(x)$.
(e) Give the domain and range of M.

33. Use the greatest integer function and write an expression for the number of ounces for which postage will be charged on a letter weighing x ounces. (See Exercise 32.)

34. Montreal taxi rates in a recent year were 90¢ for the first 1/9 mile and 10¢ for each additional 1/9 mile or fraction of 1/9. Let $C(x)$ be the cost for a taxi ride of x 1/9 miles. Find
(a) $C(1)$, **(b)** $C(2.3)$, and **(c)** $C(8)$.
(d) Graph $y = C(x)$.
(e) Give the domain and range of C.

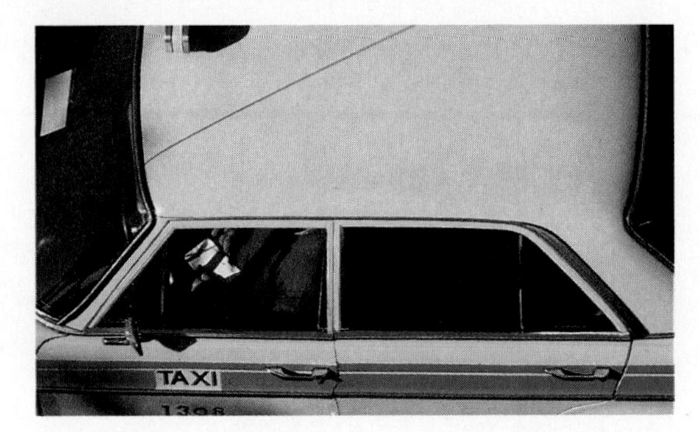

35. For a lift truck rental of no more than three days, the charge is $300. An additional charge of $75 is made for each day or portion of a day after three. Graph the ordered pairs (number of days, cost).

36. A car rental costs $37 for one day, which includes 50 free miles. Each additional 25 miles or portion (of 25 miles) costs $10. Graph the ordered pairs (miles, cost).

37. When a diabetic takes long-acting insulin, the insulin reaches its peak effect on the blood sugar level in about 3 hours. This effect remains fairly constant for 5 hours, then declines, and is very low until the next injection. In a typical patient, the level of insulin might be given by the following function.

$$i(t) = \begin{cases} 40t + 100 & \text{if } 0 \le t \le 3 \\ 220 & \text{if } 3 < t \le 8 \\ -80t + 860 & \text{if } 8 < t \le 10 \\ 60 & \text{if } 10 < t \le 24 \end{cases}$$

Here $i(t)$ is the blood sugar level, in appropriate units, at time t measured in hours from the time of the injection. Chuck takes his insulin at 6 A.M. Find the blood sugar level at each of the following times.
(a) 7 A.M. **(b)** 9 A.M.
(c) 10 A.M. **(d)** noon
(e) 2 P.M. **(f)** 5 P.M.
(g) midnight **(h)** Graph $y = i(t)$.

38. To rent a midsized car from Avis costs $30 per day or fraction of a day. If you pick up the car in Lansing and drop it in West Lafayette, there is a fixed $50 dropoff charge. Let $C(x)$ represent the cost of renting the car for x days, taking it from Lansing to West Lafayette. Find each of the following.
(a) $C(3/4)$ **(b)** $C(9/10)$
(c) $C(1)$ **(d)** $C(1\ 5/8)$
(e) $C(2.4)$ **(f)** Graph $y = C(x)$.
(g) Is C a function? **(h)** Is C a linear function?

39. Explain why the graph of an absolute value function typically has the shape of a "V". (See Objective 1.)

40. Explain how to determine the y-values of the greatest integer function for negative x-values. (See Objective 3.)

41. When functions are defined piecewise, what two points *must* be plotted? Explain. (See Objective 2.)

PREVIEW EXERCISES

Divide. See Section 4.5.

42. $\dfrac{2x^2 + x - 15}{2x - 5}$ **43.** $\dfrac{3m^3 - 2m^2 - 7m - 2}{3m + 1}$ **44.** $\dfrac{4y^3 + 15y^2 + y - 10}{y + 3}$

Use synthetic division in each of the following. See Section 4.6.

45. $\dfrac{2p^2 - p - 6}{p - 2}$ **46.** $\dfrac{3k^2 + 7k + 5}{k + 1}$ **47.** $\dfrac{2z^3 - 8z^2 - 14z + 16}{z - 3}$

CHAPTER 8 GLOSSARY

KEY TERMS

8.2 quadratic function A function f is quadratic if $f(x) = ax^2 + bx + c$, where a, b, and c are real numbers with $a \neq 0$.

parabola The graph of a quadratic function is a parabola. A parabola is defined as the set of all points in a plane that are equidistant from a fixed point and a fixed line not containing the point.

axis of a parabola The line of symmetry for a parabola is called the axis of the parabola.

vertex of a parabola The intersection of a parabola with its axis is called the vertex of the parabola.

focus of a parabola The focus of a parabola is a point on the axis that determines the curvature of the figure.

translation A translation is a horizontal or vertical shift of a graph.

directrix The directrix of a parabola is a fixed line that, together with the focus, determines the parabola.

8.5 circle A circle is the set of all points in a plane that lie a fixed distance from a fixed point.

center The fixed point mentioned in the definition of a circle is the center of the circle.

radius The radius of a circle is the fixed distance between the center and any point on the circle.

ellipse An ellipse is the set of all points in a plane the sum of whose distances from two fixed points is constant.

foci of an ellipse The foci of an ellipse are two fixed points that determine an ellipse.

center of an ellipse The center of an ellipse is the point halfway between the foci of the ellipse.

vertices of an ellipse The x- or y-intercepts of the graph of an ellipse with center at the origin are called the vertices.

major axis The longer line segment joining the vertices on an ellipse is its major axis.

minor axis The shorter line segment joining the other intercepts of an ellipse is its minor axis.

8.6 hyperbola A hyperbola is the set of all points in a plane such that the absolute value of the difference of the distances from two fixed points is constant.

asymptotes of a hyperbola The two intersecting lines that the branches of a hyperbola approach are called asymptotes of the hyperbola.

fundamental rectangle The fundamental rectangle of a hyperbola is the rectangle with vertices at (a, b), $(a, -b)$, $(-a, b)$, $(-a, -b)$.

conic sections Graphs that result from cutting an infinite cone with a plane are called conic sections.

square root function A function of the form $f(x) = \sqrt{u}$ for an algebraic expression u, with $u \geq 0$, is called a square root function.

8.7 absolute value function The absolute value function is defined by $f(x) = |x|$, for all real numbers x.

functions defined piecewise Functions that have different equations defining different parts of their domains are called functions defined piecewise.

greatest integer function The greatest integer function defines y as the greatest integer less than or equal to x.

step function A step function has a graph that is a series of horizontal line segments.

NEW SYMBOLS

$f \circ g$ composite function

$[\![x]\!]$ greatest integer function

CHAPTER 8 QUICK REVIEW

SECTION	CONCEPTS	EXAMPLES
8.1 THE ALGEBRA OF FUNCTIONS	**Operations on Functions** $(f+g)(x) = f(x) + g(x)$ $(f-g)(x) = f(x) - g(x)$ $(fg)(x) = f(x) \cdot g(x)$ $\left(\frac{f}{g}\right)(x) = \frac{f(x)}{g(x)}, \quad g(x) \neq 0$ **Composition of Functions** If f and g are functions, then the composite function of g and f is $(g \circ f)(x) = g[f(x)]$.	If $f(x) = 3x^2 + 2$ and $g(x) = \sqrt{x}$, then $(f+g)(x) = 3x^2 + 2 + \sqrt{x}$ $(f-g)(x) = 3x^2 + 2 - \sqrt{x}$ $(fg)(x) = (3x^2 + 2)(\sqrt{x})$ $\left(\frac{f}{g}\right)(x) = \frac{3x^2 + 2}{\sqrt{x}}, \; x > 0.$ If $g(x) = \sqrt{x}$ and $f(x) = x^2 - 1$, then the composite function of g and f is $(g \circ f)(x) = \sqrt{x^2 - 1}$.
8.2 QUADRATIC FUNCTIONS; PARABOLAS	**1.** The graph of the quadratic function $f(x) = a(x - h)^2 + k, \; a \neq 0$ is a parabola with vertex at (h, k) and the vertical line $x = h$ as axis. **2.** The graph opens upward if a is positive and downward if a is negative. **3.** The graph is wider than $f(x) = x^2$ if $0 < \|a\| < 1$ and narrower if $\|a\| > 1$.	Graph: $f(x) = -(x + 3)^2 + 1$. 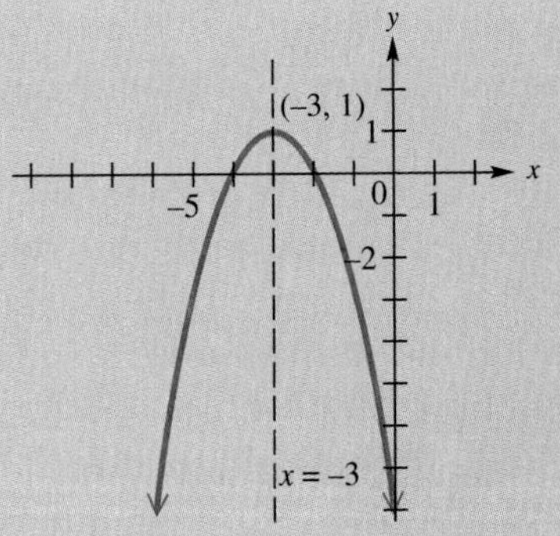
8.3 MORE ABOUT PARABOLAS AND THEIR APPLICATIONS	The vertex of the graph of $f(x) = ax^2 + bx + c, \; a \neq 0$, may be found by completing the square. The vertex has coordinates $\left(-\frac{b}{2a}, f\left(-\frac{b}{2a}\right)\right)$. Steps in graphing a quadratic function: *Step 1* Find the y-intercept by evaluating $f(0)$. *Step 2* Find any x-intercepts by solving $f(x) = 0$. *Step 3* Find the vertex either by using the formula or by completing the square.	Graph: $f(x) = x^2 + 4x + 3$. The vertex is $(-2, -1)$. Since $f(0) = 3$, the y-intercept is $(0, 3)$. The solutions of $x^2 + 4x + 3 = 0$ are -1 and -3, so the x-intercepts are $(-1, 0)$ and $(-3, 0)$. 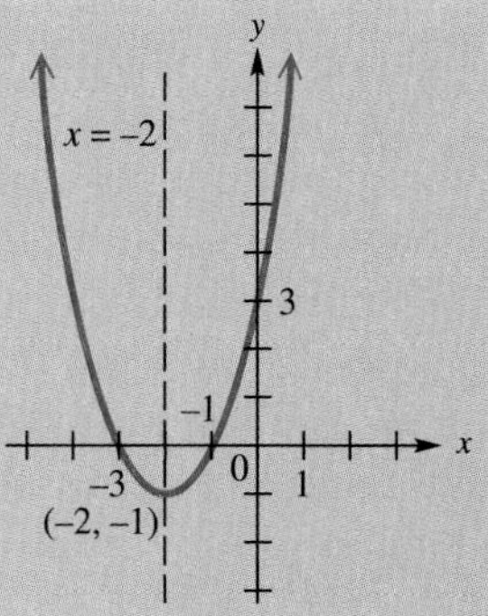

SECTION	CONCEPTS	EXAMPLES
	Step 4 Find and plot any additional points as needed, using the symmetry about the axis. Verify that the graph opens upward (if $a > 0$) or opens downward (if $a < 0$). If the discriminant, $b^2 - 4ac$, is positive, the graph of $f(x) = ax^2 + bx + c$ has two x-intercepts; if zero, one x-intercept; if negative, no x-intercepts. The graph of $x = ay^2 + by + c$ is a horizontal parabola, opening to the right if $a > 0$, or to the left if $a < 0$.	Graph: $x = 2y^2 + 6y + 5$. 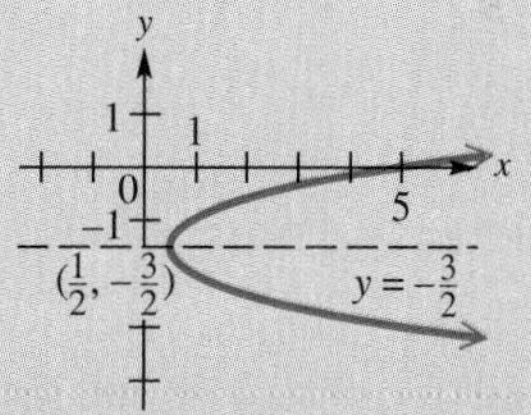
8.4 SYMMETRY; INCREASING/ DECREASING FUNCTIONS	To decide whether the graph of a function is symmetric with respect to the following, perform the indicated test. **(a)** The x-axis Replace y with $-y$. **(b)** The y-axis Replace x with $-x$. **(c)** The origin Replace x with $-x$ and y with $-y$.	Test each function for symmetry. **(a)** $x = y^2 - 5$ $x = (-y)^2 - 5 = y^2 - 5$ The graph is symmetric to the x-axis. **(b)** $y = -2x^2 + 1$ $y = -2(-x)^2 + 1 = -2x^2 + 1$ The graph is symmetric to the y-axis. **(c)** $x^2 + y^2 = 4$ $(-x)^2 + (-y)^2 = 4$ $x^2 + y^2 = 4$ The graph is symmetric to the origin (and to the x-axis and y-axis).

SECTION	CONCEPTS	EXAMPLES
	A function f is increasing on an interval if $f(x_1) < f(x_2)$ whenever $x_1 < x_2$. A function f is decreasing on an interval if $f(x_1) > f(x_2)$ whenever $x_1 < x_2$. If f is neither increasing nor decreasing on an interval, it is constant there.	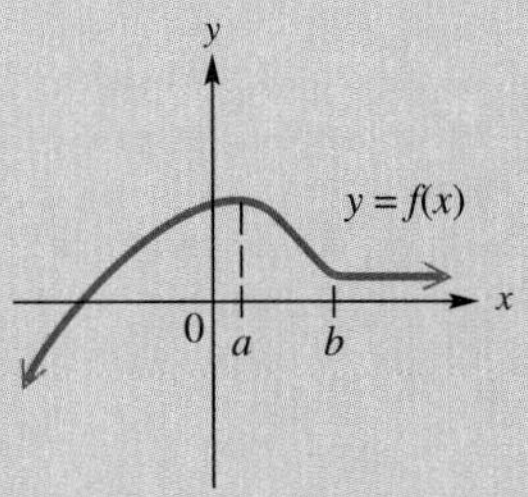 f is increasing on $(-\infty, a)$. f is decreasing on (a, b). f is constant on (b, ∞).
8.5 THE CIRCLE AND THE ELLIPSE	The circle with radius r and center at (h, k) has an equation of the form $(x-h)^2 + (y-k)^2 = r^2$. The ellipse whose x-intercepts are $(a, 0)$ and $(-a, 0)$ and whose y-intercepts are $(0, b)$ and $(0, -b)$ has an equation of the form $\frac{x^2}{a^2} + \frac{y^2}{b^2} = 1$.	The circle $(x+2)^2 + (y-3)^2 = 25$ has center $(-2, 3)$ and radius 5. 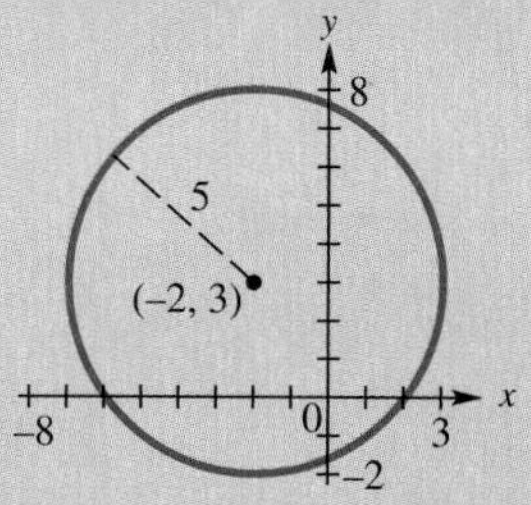 Graph: $\frac{x^2}{9} + \frac{y^2}{4} = 1$. 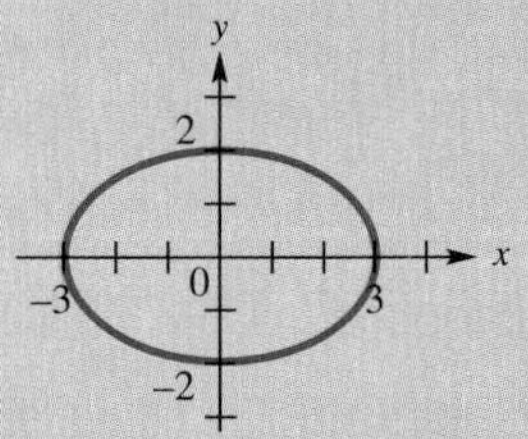
8.6 THE HYPERBOLA; SQUARE ROOT FUNCTIONS	A hyperbola with x-intercepts $(a, 0)$ and $(-a, 0)$ has an equation of the form $\frac{x^2}{a^2} - \frac{y^2}{b^2} = 1$ and a hyperbola with y-intercepts $(0, b)$ and $(0, -b)$ has an equation of the form $\frac{y^2}{b^2} - \frac{x^2}{a^2} = 1$	Graph: $\frac{x^2}{4} - \frac{y^2}{4} = 1$. 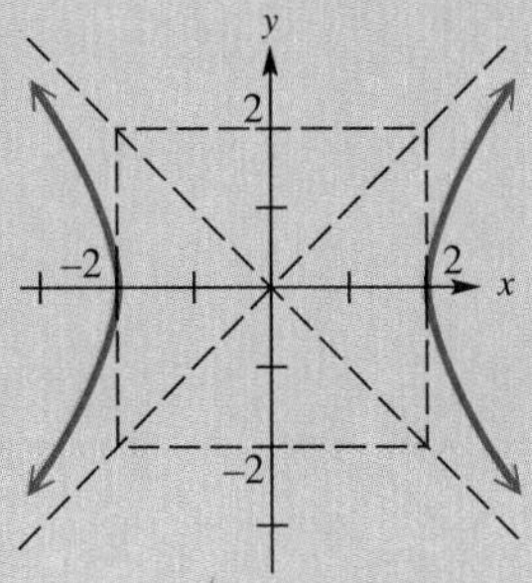

SECTION	CONCEPTS	EXAMPLES
	The extended diagonals of the fundamental rectangle with corners at the points (a, b), $(-a, b)$, $(-a, -b)$, and $(a, -b)$ are the asymptotes of these hyperbolas.	The fundamental rectangle has corners at $(2, 2)$, $(-2, 2)$, $(-2, -2)$, and $(2, -2)$.
	To graph a square root function, square both sides of the equation so that it can be recognized as a conic section. Then graph only the part indicated by the sign of the expression.	Graph: $y = -\sqrt{4 - x^2}$. Square both sides and rearrange terms to get $x^2 + y^2 = 4.$ This equation has a circle as its graph. However, graph only the lower half of the circle, since the original equation indicates that y cannot be positive. 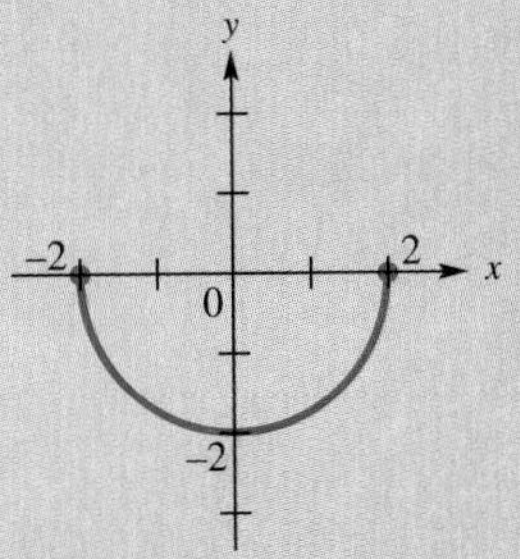
8.7 OTHER USEFUL FUNCTIONS	**Absolute Value Function** $f(x) = \lvert x \rvert$ It is symmetric to a vertical axis through the "vertex."	Graph: $f(x) = 2\lvert x - 1 \rvert + 3$. 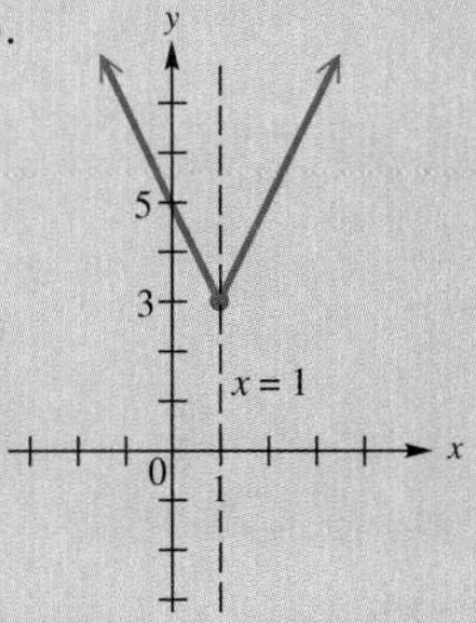
	Functions Defined Piecewise Graph each portion with an open or solid endpoint as appropriate.	Graph: $f(x) = \begin{cases} x - 2 & \text{if } x \geq 1 \\ 3x & \text{if } x < 1. \end{cases}$ 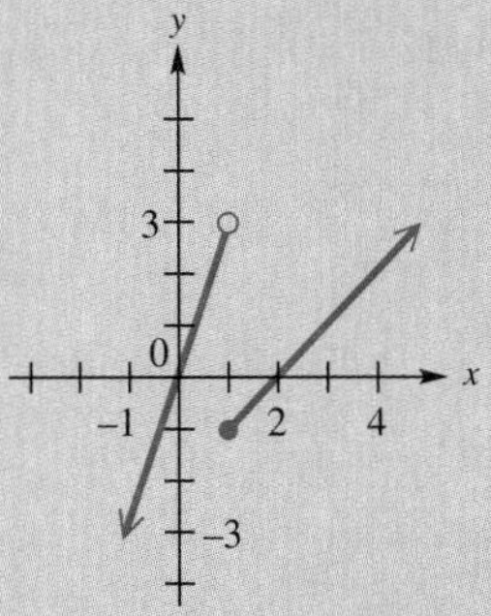

SECTION	CONCEPTS	EXAMPLES
	Greatest Integer Function $f(x) = [\![x]\!]$ y is the greatest integer less than or equal to x.	Graph: $f(x) = [\![2x - 1]\!]$.

CHAPTER 8 REVIEW EXERCISES

[8.1] *Given $f(x) = x^2 - 2x$ and $g(x) = 5x + 3$, find the following. Give the domain of each.*

1. $(f + g)(x)$

2. $(f - g)(x)$

3. $(fg)(x)$

4. $\left(\frac{f}{g}\right)(x)$

5. $(g \circ f)(x)$

6. $(f \circ g)(x)$

For $f(x) = 2x - 3$ and $g(x) = \sqrt{x}$, find the following.

7. $(f - g)(4)$

8. $\left(\frac{f}{g}\right)(9)$

9. $(fg)(5)$

10. $(f + g)(h)$

11. $(fg)(2b)$

12. $(g \circ f)(2)$

13. $(f \circ g)(-1)$

14. $(f \circ g)(k)$

[8.2–8.3] *Identify the vertex of each parabola. Graph.*

15. $f(x) = -5x^2$

16. $f(x) = 3x^2 - 2$

17. $f(x) = 6 - 2x^2$

18. $y = (x + 2)^2$

19. $y = (x - 3)^2 - 7$

20. $y = \frac{4}{3}(x - 2)^2 + 1$

21. $y = x^2 + 3x + 2$

22. $y = -3x^2 + 4x - 2$

23. $x = 2y^2 + 3$

24. $x = (y - 1)^2 + 2$

25. $x = -y^2 + 3y - 5$

26. $x = 2y^2 + 5y + 4$

27. Explain how the discriminant can be used to determine the number of x-intercepts of the graph of a quadratic function.

28. Which one of the following would most closely resemble the graph of $f(x) = a(x - h)^2 + k$, if $a < 0$, $h > 0$, and $k < 0$?

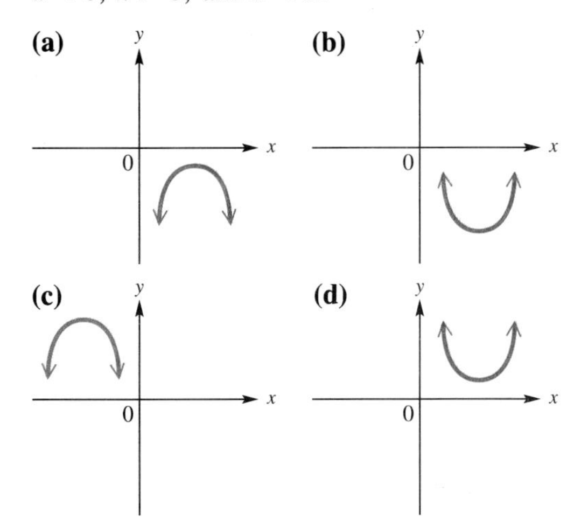

A projectile is fired into the air. Its height (in feet) above the ground t seconds after firing is

$$s = -16t^2 + 160t.$$

29. Find the number of seconds required for the projectile to reach maximum height.

30. Find the maximum height above the ground that the projectile reaches.

Solve the following problems.

31. Find the length and width of a rectangle having a perimeter of 100 meters if the area is to be a maximum.

32. Find the two numbers whose sum is 10 and whose product is a maximum.

[8.4] *Use the tests for symmetry to determine any symmetries of the graph of each relation. Do not graph.*

33. $2x^2 - y^2 = 4$

34. $3x^2 + 4y^2 = 12$

35. $2x - y^2 = 8$

36. $y = 2x^2 + 3$

37. $y = 2\sqrt{x} - 4$

38. $x = -3\sqrt{y}$

39. $y = \dfrac{1}{x^2}$

40. $y = \dfrac{4}{x^2 - 1}$

41. $xy = 4$

Give the intervals where each of the following functions is increasing, decreasing, or constant.

42.

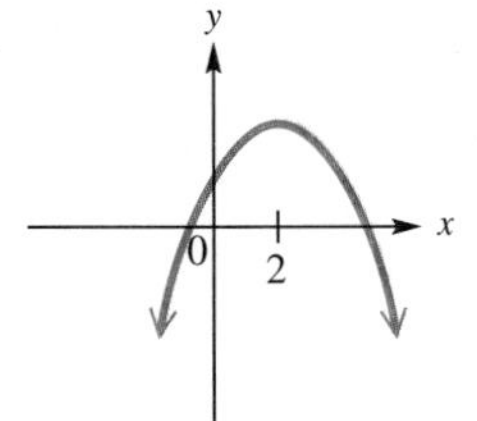

43.

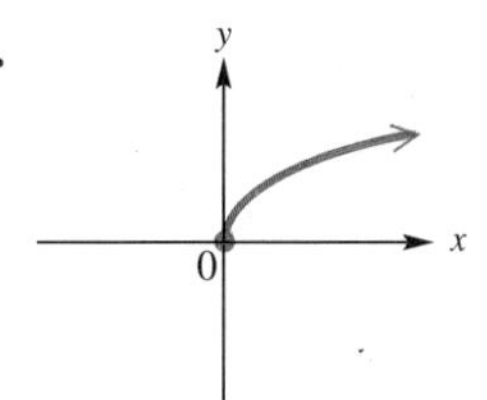

44.

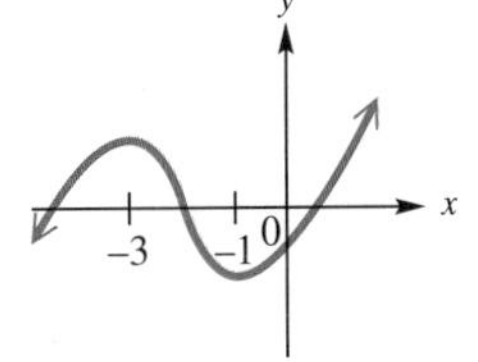

45.

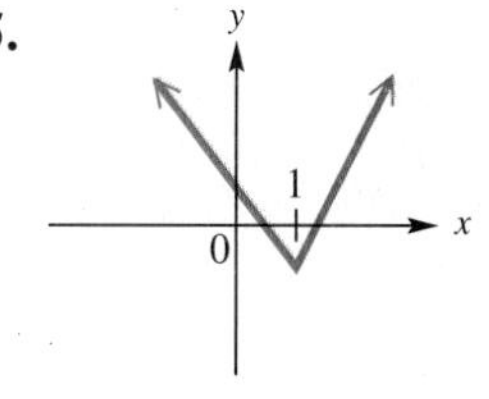

[8.5] *Write the equations of the following circles.*

46. Center at $(-3, 4)$, radius 5

47. Center at $(6, -1)$, radius 3

48. Center at the origin, through $(4, 4\sqrt{3})$

49. Center at $\left(-\dfrac{1}{2}, \dfrac{2}{3}\right)$, through $\left(-\dfrac{5}{2}, \dfrac{4}{3}\right)$

Find the center and radius of each circle.

50. $x^2 + y^2 + 8x - 8y + 23 = 0$

51. $x^2 + y^2 - 10x - 4y - 7 = 0$

52. $3x^2 + 3y^2 - 6x - 12y - 21 = 0$

53. $5x^2 + 5y^2 - 5x + 10y = 8$

Graph each of the following.

54. $x^2 + y^2 = 9$

55. $(x - 2)^2 + (y + 3)^2 = 16$

56. $x^2 - 4x + y^2 + 8y + 4 = 0$

57. $\dfrac{x^2}{9} + \dfrac{y^2}{25} = 1$

58. $\dfrac{x^2}{64} + \dfrac{y^2}{9} = 1$

59. $\dfrac{(x + 1)^2}{9} + \dfrac{(y - 2)^2}{16} = 1$

[8.6] *Graph each of the following.*

60. $\dfrac{x^2}{16} - \dfrac{y^2}{36} = 1$

61. $\dfrac{x^2}{9} - \dfrac{y^2}{4} = 1$

62. $x^2 - y^2 = 16$

63. $\dfrac{(x - 2)^2}{36} - \dfrac{(y + 3)^2}{16} = 1$

64. $f(x) = \sqrt{16 - x^2}$

65. $y = \sqrt{4 + x^2}$

Identify each of the following relations as a parabola, circle, ellipse, or hyperbola. Do not graph.

66. $y = x^2 - 2$

67. $x^2 + y^2 = 16$

68. $x^2 + 9y^2 = 18$

69. $\dfrac{x^2}{25} + \dfrac{y^2}{36} = 1$

70. $\dfrac{x^2}{16} - \dfrac{y^2}{9} = 1$

71. $8x^2 + 8y^2 = 128$

72. $x + y^2 = 2$

73. $y^2 - x^2 = 1$

74. $y = x^2 - 4x + 2$

75. $x^2 + y^2 - 6x + 6y = 2$

76. $\dfrac{(x - 3)^2}{25} + \dfrac{(y - 4)^2}{16} = 1$

77. $\dfrac{(x - 5)^2}{4} - \dfrac{(y - 2)^2}{9} = 1$

[8.7] *Graph each function.*

78. $f(x) = 2|x| + 3$

79. $f(x) = |x - 2|$

80. $f(x) = -|x - 1|$

81. $f(x) = \begin{cases} -x & \text{if } x \le 0 \\ x^2 & \text{if } x > 0 \end{cases}$

82. $f(x) = \begin{cases} -x^2 + 1 & \text{if } x \le 1 \\ -x + 1 & \text{if } x > 1 \end{cases}$

83. $f(x) = \begin{cases} 2x + 1 & \text{if } x \le -1 \\ x + 3 & \text{if } x > -1 \end{cases}$

84. $f(x) = [\![-x]\!]$

85. $f(x) = [\![x + 1]\!]$

86. $f(x) = [\![x]\!] + 1$

MIXED REVIEW EXERCISES

Graph each relation.

87. $3x^2 + y^2 = 36$

88. $f(x) = -\sqrt{2 - x}$

89. $f(x) = \begin{cases} x^2 - 2 & \text{if } x < 1 \\ 2x - 3 & \text{if } x \ge 1 \end{cases}$

90. $f(x) = |2x + 1|$

91. $\dfrac{x^2}{25} + \dfrac{y^2}{81} = 1$

92. $\dfrac{x^2}{4} - y^2 = 1$

93. $(x + 5)^2 + y^2 = 9$

94. $2x^2 - y^2 = 16$

95. $y^2 = x - 1$

CHAPTER 8 TEST

For $f(x) = 3x - 1$ and $g(x) = -x^2 - 1$, find the following.

1. $g(1)$

2. $(f + g)(-2)$

3. $\left(\frac{f}{g}\right)(3)$

4. $(f \circ g)(2)$

5. Graph the quadratic function $f(x) = -x^2 + 4$. Identify the vertex.

6. The position in feet of a moving object at time t in seconds is given by

$$d = -2t^2 + 12t + 32.$$

Find the maximum position it reaches and the time it takes to reach that position.

Use the tests for symmetry to determine all symmetries for each of the following relations.

7. $y = 5 - 2x$

8. $y = -x^2 + 1$

9. $y = x^3 - x$

10. $x^2 + 4y^2 = 4$

11. Give all intervals where the graph is increasing, decreasing, or constant.

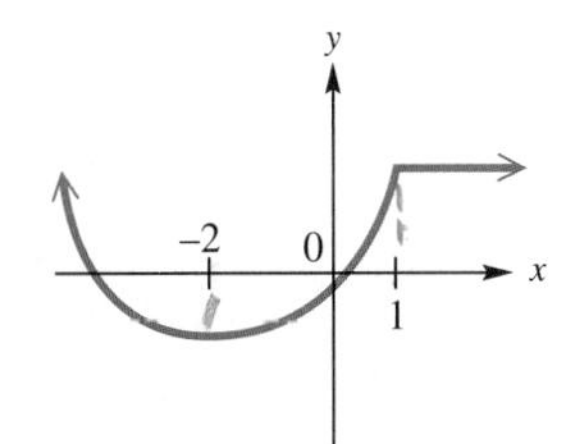

12. Find the center and radius of the circle whose equation is $(x + 1)^2 + (y - 2)^2 = 9$. Sketch its graph.

13. Find the center and the radius of the circle whose equation is

$$x^2 - 2x + y^2 - 6y = 6.$$

Graph each relation.

14. $f(x) = \sqrt{16 - x^2}$

15. $x = 3 - y^2$

16. $9x^2 + 16y^2 = 144$

17. $4x^2 - 25y^2 = 100$

Identify each equation as a parabola, hyperbola, ellipse, or circle.

18. $2x^2 + 2y^2 = 36$

19. $6x^2 - 4y^2 = 12$

20. $3y^2 + 3x = 8$

21. $9x^2 + 25y^2 = 225$

22. In the equation of a parabola that is a function, which variable is squared? In which direction may the parabola open?

Graph each function.

23. $f(x) = \frac{1}{3}|x + 2| - 1$

24. $f(x) = [\![2x]\!]$

25. $f(x) = \begin{cases} -x & \text{if } x \leq 2 \\ x - 4 & \text{if } x > 2 \end{cases}$

THE GRAPHING CALCULATOR

With a graphing calculator in graphing mode, it is possible to get a wide variety of graphs of *functions* quickly and easily. Most graphing calculators have some built-in graphs for basic functions. One of these is the graph of x^2. By touching the x^2 key and the "execute", =, or "graph" key, the graph appears automatically. For user defined graphs, it is necessary to first use the range key to set the domain and range that will be shown on the screen. The user may choose the minimum and maximum x-values and the scale (the number of units between tick marks) for the x-axis, and the minimum and maximum y-values and the scale for the y-axis. The equations (formulas) are input using the variable x (see your instruction booklet for how this is done), the operation keys, and the numerals. It is necessary to clear the screen with the appropriate key before entering a new graph, unless you want to superimpose graphs.

Relations that are not functions can be graphed by superimposing the graphs of two functions. For example, to graph $x = y^2$, solve for y to get $y = \pm\sqrt{x}$, which represents the two functions $y = \sqrt{x}$ and $y = -\sqrt{x}$. Graph $y = \sqrt{x}$ using the $\sqrt{\ }$ key, then graph $y = -\sqrt{x}$. The two graphs will both show and, combined, they give the graph of $x = y^2$, a horizontal parabola.

You may want to use the following relations to experiment with your calculator. These relations have been selected from examples in the text, so you can compare your calculator graphs with those given in the example figures. The figure number of the graph in the text is given with each relation. There are a few with no corresponding figure.

1. $y = -x^2$ (Figure 8.4)

2. $y = x^2 - 4$ (Figure 8.6)

3. $y = (x - 4)^2$ (Figure 8.7)

4. $y = -2(x + 3)^2 + 4$ (Figure 8.8)

5. $y = x^2 - x - 6$ (Figure 8.11)

6. $x = -2y^2$

7. $x = y^2 - 1$

8. $f(x) = \sqrt{25 - x^2}$ (Figure 8.40)

9. $\dfrac{y}{6} = -\sqrt{1 + \dfrac{x^2}{16}}$

10. $\dfrac{x^2}{49} + \dfrac{y^2}{36} = 1$ (Figure 8.29)

11. $\dfrac{x^2}{36} + \dfrac{y^2}{121} = 1$ (Figure 8.30)

12. $\dfrac{x^2}{16} - \dfrac{y^2}{25} = 1$ (Figure 8.34)

13. $\dfrac{y^2}{49} - \dfrac{x^2}{16} = 1$ (Figure 8.35)

14. $9x^2 = 108 + 12y^2$ (Figure 8.38)

15. $f(x) = |x - 2|$ (Figure 8.43)

16. $f(x) = |3x + 4| + 1$ (Figure 8.44)

17. $f(x) = \begin{cases} x + 1 & \text{if } x \le 2 \\ -2x + 7 & \text{if } x > 2 \end{cases}$ (Figure 8.45)

18. $f(x) = \begin{cases} x + 2 & \text{if } x \le 0 \\ \dfrac{1}{2}x^2 & \text{if } x > 0 \end{cases}$ (Figure 8.46)

CHAPTER NINE

POLYNOMIAL AND RATIONAL FUNCTIONS AND THEIR ZEROS

In Chapter 7 we studied linear functions, and in Chapter 8 we studied quadratic functions. These are functions of the form $f(x) = mx + b$ and $f(x) = ax^2 + bx + c$, and are the simplest examples of a larger group of functions known as *polynomial functions*. In this chapter we will study polynomial functions, although a more complete analysis of polynomial functions requires the concepts of calculus.

Rational expressions were introduced in Chapter 4, and in this chapter we will also look at functions that are defined by such expressions. They are called *rational functions*. They, too, are studied in more detail in calculus.

Polynomial and rational functions can be used to predict outcomes of various activities, as illustrated in the problems featured in the Focus on Problem Solving entries in this chapter.

9.1 POLYNOMIAL FUNCTIONS AND THEIR GRAPHS

OBJECTIVES

1. GRAPH POLYNOMIAL FUNCTIONS OF THE FORM $P(x) = ax^n$.
2. GRAPH POLYNOMIAL FUNCTIONS OF THE FORM $P(x) = ax^n + k$, $P(x) = a(x - h)^n$, AND $P(x) = a(x - h)^n + k$.
3. GRAPH POLYNOMIAL FUNCTIONS THAT ARE FACTORED OR ARE EASILY FACTORABLE INTO LINEAR FACTORS.
4. LEARN THE RELATIONSHIPS AMONG x-INTERCEPT OF A GRAPH, ZERO OF A FUNCTION, AND SOLUTION OF AN EQUATION.

FOCUS ON PROBLEM SOLVING

During the early part of the twentieth century, the deer population of the Kaibab Plateau in Arizona experienced a rapid increase, because hunters had reduced the number of natural predators. The increase in population depleted the food resources and eventually caused the population to decline. For the period from 1905 to 1930, the deer population was approximated by

$$D(x) = -.125x^5 + 3.125x^4 + 4000,$$

where x is time in years from 1905.

(a) Use a calculator to find enough points to graph $D(x)$.

(b) From the graph, over what period of time (from 1905 to 1930) was the population increasing? relatively stable? decreasing?

Functions defined by polynomials can be used to predict information about naturally occurring phenomena. Knowing how to graph polynomial functions helps in interpreting valuable information about the functions. In this section we look at some simple polynomial functions, and develop some methods for graphing them. This problem is Exercise 25 in the exercises for this section. After working through this section, you should be able to solve this problem.

Polynomial functions are defined as follows.

DEFINITION OF POLYNOMIAL FUNCTION

A **polynomial function of degree *n*** is a function defined by

$$P(x) = a_n x^n + a_{n-1}x^{n-1} + \ldots + a_1 x + a_0,$$

for complex numbers $a_n, a_{n-1}, \ldots, a_1$, and a_0, where $a_n \neq 0$.

$P(x)$ is used instead of $f(x)$ in this chapter to emphasize that the functions are polynomial functions. The polynomials discussed in Chapter 3 had only real-number coefficients. In this chapter the domain of the coefficients is extended to include all complex numbers.* The number a_n is the **leading coefficient** of $P(x)$. If all the coefficients of a polynomial are 0, the polynomial equals 0 and is called the **zero polynomial.** The zero polynomial has no degree. A polynomial of the form $P(x) = a_0$ for a nonzero complex number a_0 has degree 0, however.

1 We begin the discussion of graphing polynomial functions with the graphs of polynomial functions defined by equations of the form $P(x) = ax^n$.

EXAMPLE 1 GRAPHING FUNCTIONS OF THE FORM $P(x) = ax^n$

Graph each function.

(a) $P(x) = x^3$

Choose several values for x, and find the corresponding values of $P(x)$, or y, as shown in the upper table in Figure 9.1. Plot the resulting ordered pairs and connect the points with a smooth curve. The graph of $P(x) = x^3$ is shown in blue in Figure 9.1.

(b) $P(x) = x^5$

Work as in part (a) of this example to get the graph shown in red in Figure 9.1. Both the graph of $P(x) = x^3$ and the graph of $P(x) = x^5$ are symmetric with respect to the origin.

$P(x) = x^3$

x	$P(x)$
-2	-8
-1	-1
0	0
1	1
2	8

$P(x) = x^5$

x	$P(x)$
-1.5	-7.6
-1	-1
0	0
1	1
1.5	7.6

FIGURE 9.1

*Recall that the set of complex numbers includes both real numbers and imaginary numbers such as $5i$ or $6 + i\sqrt{7}$.

(c) $P(x) = x^4$, $P(x) = x^6$

Some typical ordered pairs for the graphs of $P(x) = x^4$ and $P(x) = x^6$ are given in the tables in Figure 9.2. These graphs are symmetric with respect to the y-axis, as is the graph of $P(x) = ax^2$ for a nonzero real number a. ■

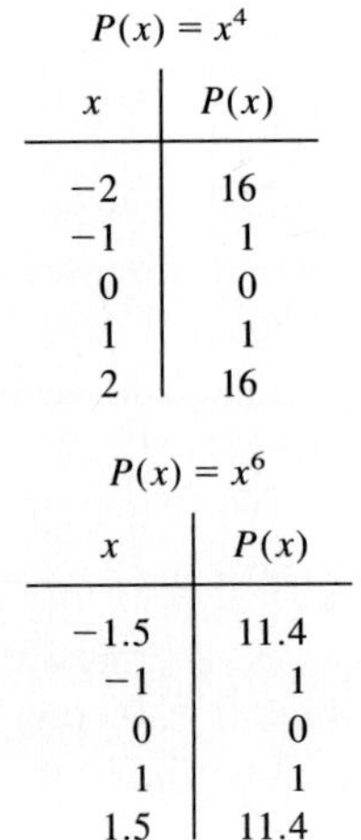

$P(x) = x^4$

x	$P(x)$
-2	16
-1	1
0	0
1	1
2	16

$P(x) = x^6$

x	$P(x)$
-1.5	11.4
-1	1
0	0
1	1
1.5	11.4

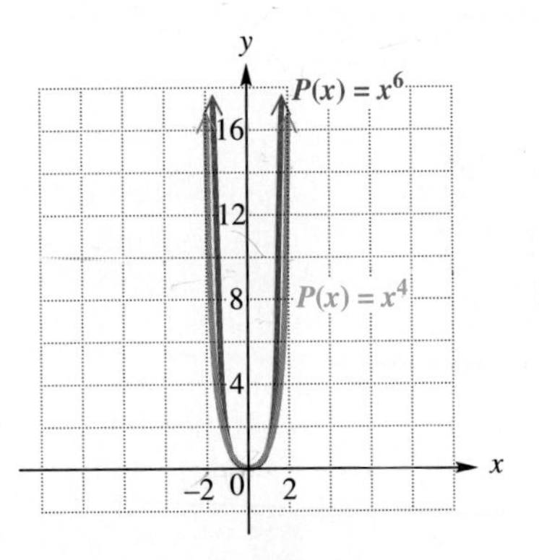

FIGURE 9.2

As with the graph of $y = ax^2$ in Section 8.2, the value of a in $P(x) = ax^n$ affects the width of the graph. When $|a| > 1$, the graph is narrower than the graph of $P(x) = x^n$; when $0 < |a| < 1$, the graph is broader. The graph of $P(x) = -ax^n$ is reflected about the x-axis as compared to the graph of $P(x) = ax^n$.

EXAMPLE 2
EXAMINING THE EFFECT OF a FOR $P(x) = ax^n$

Graph each of the following functions.

(a) $P(x) = \frac{1}{2}x^3$

The graph is broader than that of $P(x) = x^3$ but has the same general shape. It includes the points $(-2, -4)$, $(-1, -1/2)$, $(0, 0)$, $(1, 1/2)$, and $(2, 4)$. See Figure 9.3.

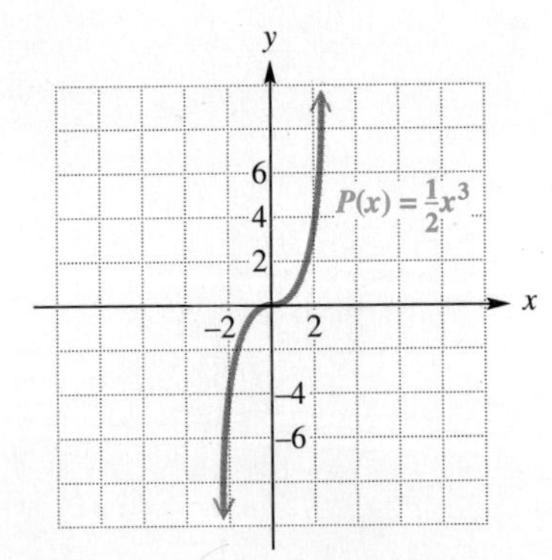

FIGURE 9.3

(b) $P(x) = -\frac{3}{2}x^4$

The following table gives some ordered pairs for this function.

x	-2	-1	0	1	2
$P(x)$	-24	$-\frac{3}{2}$	0	$-\frac{3}{2}$	-24

The graph is shown in Figure 9.4. This graph is narrower than that of $P(x) = x^4$, since $|-3/2| > 1$, and opens downward instead of upward, since $-3/2 < 0$. ■

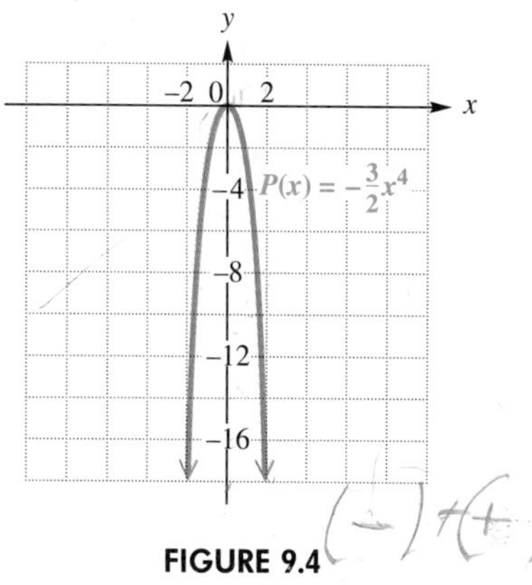

FIGURE 9.4

2 Compared with the graph of $P(x) = ax^n$, the graph of $P(x) = ax^n + k$ is translated (shifted) k units up if $k > 0$ and $|k|$ units down if $k < 0$. Also, the graph of $P(x) = a(x - h)^n$ is translated h units to the right if $h > 0$ and $|h|$ units to the left if $h < 0$, when compared with the graph of $P(x) = ax^n$.

The graph of $P(x) = a(x - h)^n + k$ shows a combination of these translations. The effects here are the same as those we saw with quadratic functions in Section 8.2.

EXAMPLE 3
EXAMINING VERTICAL AND HORIZONTAL SHIFTS

Graph each of the following.

(a) $P(x) = x^5 - 2$

The graph will be the same as that of $P(x) = x^5$, but translated down 2 units. See Figure 9.5.

(b) $P(x) = (x + 1)^6$

This function P has a graph like that of $P(x) = x^6$, but since $x + 1 = x - (-1)$, it is translated one unit to the left as shown in Figure 9.6.

(c) $P(x) = -(x - 1)^3 + 3$

The negative sign causes the graph to be reflected about the x-axis when compared with the graph of $P(x) = x^3$. As shown in Figure 9.7, the graph is also translated 1 unit to the right and 3 units up. ■

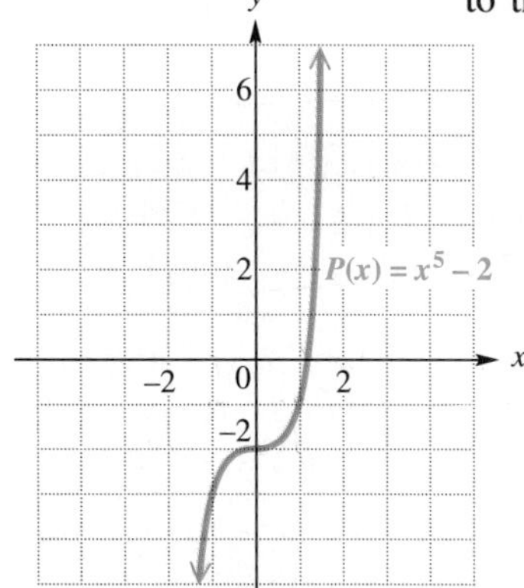

FIGURE 9.5

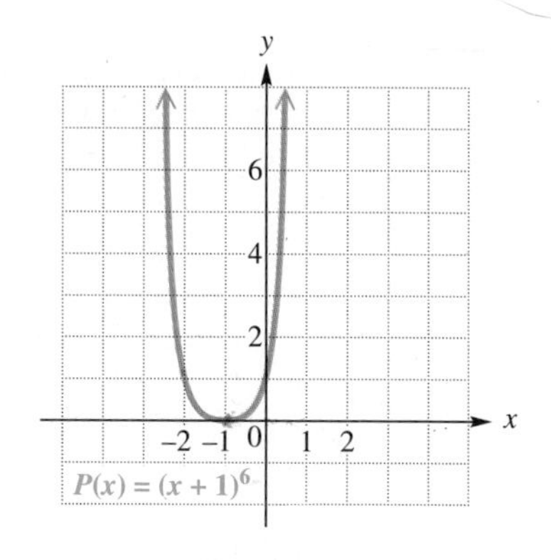

FIGURE 9.6

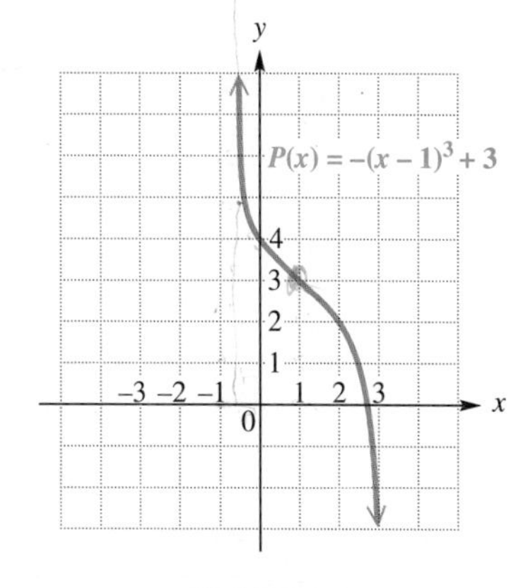

FIGURE 9.7

Generalizing from the graphs in Examples 1–3, the domain of a polynomial function is the set of all real numbers. The range of a polynomial function of odd degree is also the set of all real numbers. Some typical graphs of polynomial functions of odd degree are shown in Figure 9.8.

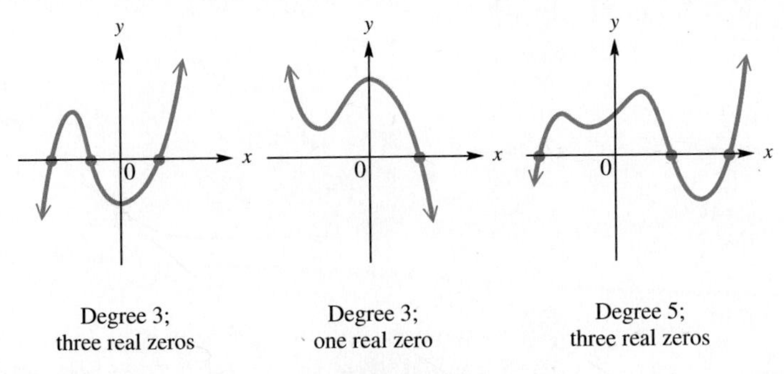

FIGURE 9.8

A polynomial function of even degree will have a range that takes the form $(-\infty, k]$ or $[k, \infty)$ for some real number k. Figure 9.9 shows two typical graphs of polynomial functions of even degree.

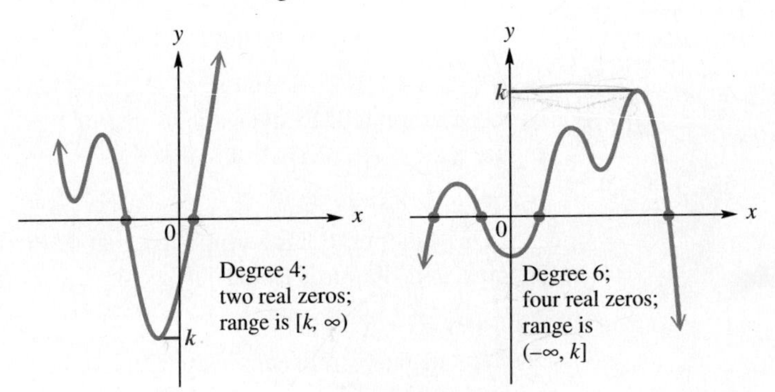

FIGURE 9.9

The values of x that satisfy $P(x) = 0$ are called the **zeros** of $P(x)$. Methods for finding zeros of first- and second-degree polynomials were given in earlier chapters. In later sections of this chapter, several theorems are presented that help to find, or at least approximate, the zeros of polynomials of higher degree. The graphs in Figure 9.8 suggest that every polynomial function P of odd degree has at least one real zero.

3 It is difficult to accurately graph most polynomial functions without using calculus. A large number of points must be plotted to get a reasonably accurate graph. If a polynomial function P can be factored into linear factors, however, its graph can be approximated by plotting only a few points.

EXAMPLE 4 GRAPHING A POLYNOMIAL FUNCTION GIVEN IN FACTORED FORM

Graph $P(x) = (2x + 3)(x - 1)(x + 2)$.

Multiplying out the expression on the right would show that P is a third-degree polynomial, also called a *cubic* polynomial. Sketch the graph of P by first setting each of the three factors equal to 0 and solving the resulting equations to find the zeros of the function.

$$2x + 3 = 0 \quad \text{or} \quad x - 1 = 0 \quad \text{or} \quad x + 2 = 0$$

$$x = -\frac{3}{2} \qquad x = 1 \qquad x = -2$$

The three zeros, $-3/2$, 1, and -2, divide the x-axis into four intervals:

$$(-\infty, -2), \left(-2, -\frac{3}{2}\right), \left(-\frac{3}{2}, 1\right), (1, \infty).$$

These intervals are shown in Figure 9.10.

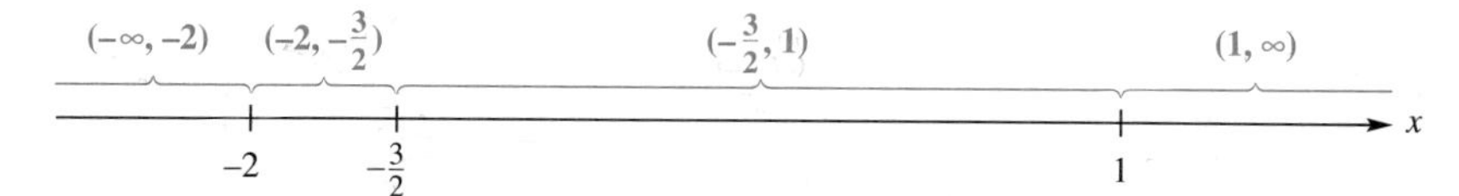

FIGURE 9.10

Since the three zeros give the only x-intercepts, the values of $P(x)$ in the intervals shown in Figure 9.10 are either always positive or always negative. To find the sign of $P(x)$ in each interval, select a value of x in the interval and determine by substitution whether the function values are positive or negative in that interval. A typical selection of test points and the results of the tests are shown below. (When 0 lies in an interval, it is a good idea to use it as a test point. This will give the y-intercept.)

Interval	*Test point*	*Value of P(x)*	*Sign of P(x)*
$(-\infty, -2)$	-3	-12	Negative
$(-2, -3/2)$	$-7/4$	$11/32$	Positive
$(-3/2, 1)$	0	-6	Negative
$(1, \infty)$	2	28	Positive

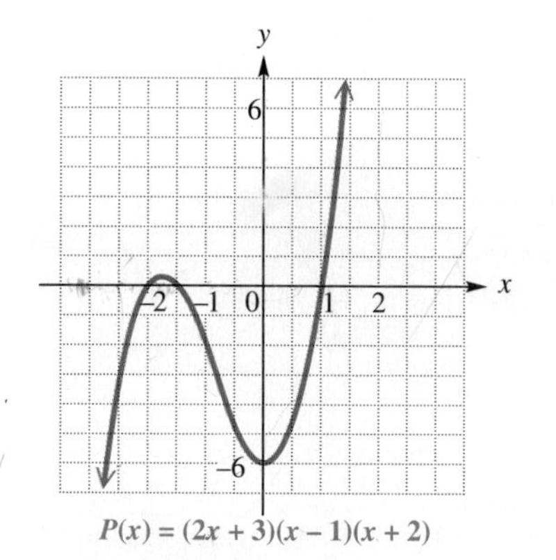

$P(x) = (2x + 3)(x - 1)(x + 2)$

FIGURE 9.11

When the values of $P(x)$ are negative, the graph is below the x-axis, and when $P(x)$ takes on positive values, the graph is above the x-axis. The y-intercept is $(0, -6)$. These results suggest that the graph looks like the sketch in Figure 9.11. The sketch could be improved by plotting additional points in each region. ■

EXAMPLE 5 **GRAPHING A POLYNOMIAL FUNCTION REQUIRING FACTORING**

Sketch the graph of $P(x) = 3x^4 + x^3 - 2x^2$.

The polynomial can be factored as follows:

$$3x^4 + x^3 - 2x^2 = x^2(3x^2 + x - 2)$$
$$= x^2(3x - 2)(x + 1).$$

The zeros, 0, 2/3, and -1, divide the x-axis into four intervals:

$$(-\infty, -1), \quad (-1, 0), \quad (0, 2/3), \quad \text{and} \quad (2/3, \infty).$$

Determine the sign of $P(x)$ (and thus the location of the graph relative to the x-axis) in each region by substituting a value for x from each region to get the following information. Since $P(0) = 0$, the origin is the y-intercept.

Interval	*Test point*	*Value of P(x)*	*Location relative to x-axis*
$(-\infty, -1)$	-2	32	Above
$(-1, 0)$	$-1/2$	$-7/16$	Below
$(0, 2/3)$	$1/3$	$-4/27$	Below
$(2/3, \infty)$	1	2	Above

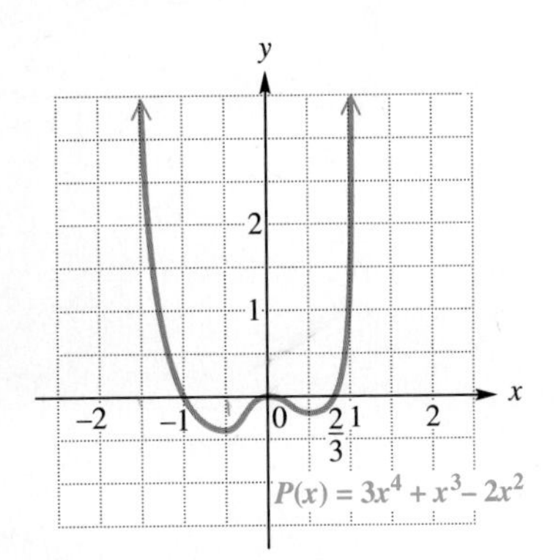

FIGURE 9.12

With the values of x used for the test points and the corresponding values of y, sketch the graph as shown in Figure 9.12. ■

4 There are important relationships among the following ideas:

1. the **x-intercepts of the graph** of $y = P(x)$;
2. the **zeros of the function** $P(x)$; and,
3. the **solutions of the equation** $P(x) = 0$.

Consider the function from Example 4, $P(x) = (2x + 3)(x - 1)(x + 2)$. When the factors are multiplied out, the function is $P(x) = 2x^3 + 5x^2 - x - 6$. The graph of this function, shown in Figure 9.11, has x-intercepts $(-2, 0)$, $(-3/2, 0)$, and $(1, 0)$. Since -2, $-3/2$, and 1 are the x-values for which the function is 0, they are the zeros of $P(x)$. Furthermore, since $P(-2) = P(-3/2) = P(1) = 0$, they are also solutions of the equation $P(x) = 2x^3 + 5x^2 - x - 6 = 0$. The discussion above can be summarized as follows.

x-INTERCEPTS, ZEROS, AND SOLUTIONS

If the point $(a, 0)$ is an x-intercept of the graph of the function $y = P(x)$, then a is a zero of $P(x)$ and a is a solution of the equation $P(x) = 0$.

EXAMPLE 6 **DESCRIBING RELATIONSHIPS AMONG x-INTERCEPTS, ZEROS, AND SOLUTIONS**

Describe the relationships among the x-intercepts of the graph of $P(x) = 3x^4 + x^3 - 2x^2$, the zeros of $P(x)$, and the solutions of the equation $3x^4 + x^3 - 2x^2 = 0$.

This function was graphed in Example 5. See Figure 9.12. From the graph, we see that the x-intercepts are $(-1, 0)$, $(0, 0)$, and $(2/3, 0)$. Therefore, -1, 0, and 2/3 are zeros of $P(x)$, and -1, 0, and 2/3 are solutions of the equation $3x^4 + x^3 - 2x^2 = 0$. ■

9.1 EXERCISES

*Each of the polynomial functions defined below is symmetric with respect to a line or a point. For each function, **(a)** sketch the graph, and **(b)** give the line or point of symmetry. See Examples 1–3.*

1. $P(x) = \frac{1}{4}x^6$ **2.** $P(x) = -\frac{2}{3}x^5$ **3.** $P(x) = -\frac{5}{4}x^5$ **4.** $P(x) = 2x^4$

5. $P(x) = \frac{1}{2}x^3 + 1$ **6.** $P(x) = -x^4 + 2$ **7.** $P(x) = -(x + 1)^3$ **8.** $P(x) = \frac{1}{3}(x + 3)^4$

9. $P(x) = (x - 1)^4 + 2$ **10.** $P(x) = (x + 2)^3 - 1$

Graph each of the following. See Examples 4 and 5.

11. $P(x) = 2x(x - 3)(x + 2)$ **12.** $P(x) = x^2(x + 1)(x - 1)$

13. $P(x) = x^2(x - 2)(x + 3)^2$ **14.** $P(x) = x^2(x - 5)(x + 3)(x - 1)$

15. $P(x) = x^3 - x^2 - 2x$ **16.** $P(x) = -x^3 - 4x^2 - 3x$

17. $P(x) = (x + 2)(x - 1)(x + 1)$ **18.** $P(x) = (x - 4)(x + 2)(x - 1)$

19. $P(x) = (3x - 1)(x + 2)^2$ **20.** $P(x) = (4x + 3)(x + 2)^2$

21. $P(x) = x^3 + 5x^2 - x - 5$ (*Hint:* Factor by grouping.)

22. $P(x) = x^3 + x^2 - 36x - 36$

---

23. Which one of the following is not an example of a polynomial function? (See Objectives 1 and 2.)
(a) $P(x) = x^2$ **(b)** $P(x) = (x + 1)^3$
(c) $P(x) = \frac{1}{x}$ **(d)** $P(x) = 2x^5$

24. Write a short explanation of how the values of a, h, and k affect the graph of $P(x) = a(x - h)^n + k$ in comparison to the graph of $P(x) = x^n$. (See Objective 2.)

25. During the early part of the twentieth century, the deer population of the Kaibab Plateau in Arizona experienced a rapid increase, because hunters had reduced the number of natural predators. The increase in population depleted the food resources and eventually caused the population to decline. For the period from 1905 to 1930, the deer population was approximated by

$$D(x) = -.125x^5 + 3.125x^4 + 4000,$$

where x is time in years from 1905.
(a) Use a calculator to find enough points to graph $D(x)$.
(b) From the graph, over what period of time (from 1905 to 1930) was the population increasing? relatively stable? decreasing?

26. The pressure of the oil in a reservoir tends to drop with time. By taking sample pressure readings for a particular oil reservoir, petroleum engineers have found that the change in pressure is given by

$$P(t) = t^3 - 25t^2 + 200t,$$

where t is time in years from the date of the first reading.
(a) Graph $P(t)$.
(b) For what time period is the change in pressure (drop) increasing? decreasing?

For each of the graphs in Exercises 27 and 28, state the relationships among the x-intercepts of the graph of $P(x)$, the zeros of the function $P(x)$, and the solutions of the equation $P(x) = 0$. See Example 6.

27.

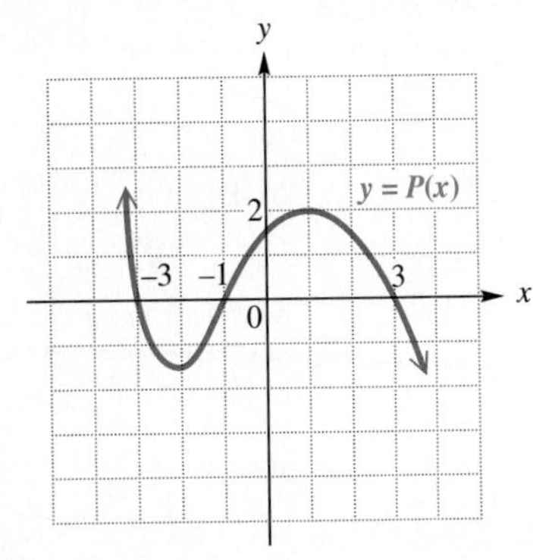

28.

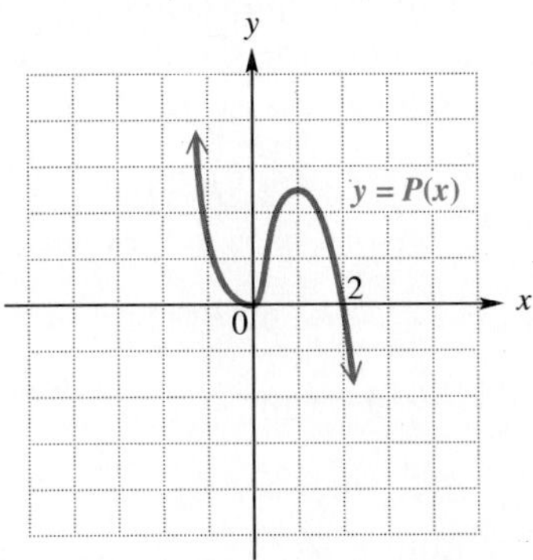

Some polynomial functions are examples of even *or* odd *functions, according to the following definition.*

A function $y = f(x)$ is an **even** function if $f(-x) = f(x)$ for all x in the domain of f. A function $y = f(x)$ is an **odd** function if $f(-x) = -f(x)$ for all x in the domain of f.

For example, $P(x) = x^2$ is an even function, because $P(-x) = (-x)^2 = x^2 = P(x)$. Also, $P(x) = x^3$ is an odd function, because $P(-x) = (-x)^3 = -x^3 = -P(x)$. A function may be neither even nor odd.

Decide whether the polynomial functions defined as follows are even, odd, or neither.

29. $P(x) = 2x^3$

30. $P(x) = -4x^5$

31. $P(x) = .2x^4$

32. $P(x) = -x^6$

33. $P(x) = -x^5$

34. $P(x) = 3x^2 + 2x$

35. $P(x) = 2x^3 + 3x^2$

36. $P(x) = 4x^3 - x$

37. $P(x) = x^4 + 3x^2 + 5$

---

38. Refer to the discussion of symmetry in Section 8.4 and fill in the blanks with the correct responses. By the definition of an even function, if (a, b) lies on the graph of an even function, so does $(-a, b)$. Therefore, the graph of an even function is symmetric with respect to the __________. If (a, b) lies on the graph of an odd function, then by definition, so does $(-a, -b)$. Therefore, the graph of an odd function is symmetric with respect to the __________.

39. Give an example (if possible) of a function whose graph is symmetric with respect to the y-axis.

40. Give an example (if possible) of a function whose graph is symmetric with respect to both the x-axis and the origin. (Do not consider $f(x) = 0$.)

Approximate maximum or minimum values of polynomial functions can be found for given intervals by first evaluating the function at the left endpoint of the interval, then adding .1 to the value of x and reevaluating the polynomial. Keep doing this until the right endpoint of the interval is reached. Then identify the approximate maximum and minimum value for the polynomial on the interval.

41. $P(x) = x^3 + 4x^2 - 8x - 8$, $[-3.8, -3]$

42. $P(x) = x^3 + 4x^2 - 8x - 8$, $[.3, 1]$

43. $P(x) = 2x^3 - 5x^2 - x + 1$, $[-1, 0]$

44. $P(x) = 2x^3 - 5x^2 - x + 1$, $[1.4, 2]$

45. $P(x) = x^4 - 7x^3 + 13x^2 + 6x - 28$, $[-2, -1]$

46. $P(x) = x^4 - 7x^3 + 13x^2 + 6x - 28$, $[2, 3]$

---

47. The graphs of $P(x) = x$, $P(x) = x^3$, and $P(x) = x^5$ all pass through the points $(-1, -1)$, $(0, 0)$, and $(1, 1)$. Draw a graph showing these three functions for the domain $[-1, 1]$. Use your result to decide how the graph of $P(x) = x^{11}$ would look for the same interval. (See Objective 1.)

48. The graphs of $P(x) = x^2$, $P(x) = x^4$, and $P(x) = x^6$ all pass through the points $(-1, 1)$, $(0, 0)$, and $(1, 1)$. Draw a graph showing these three functions for the domain $[-1, 1]$. Use your result to decide how the graph of $P(x) = x^{12}$ would look for the same interval. (See Objective 1.)

PREVIEW EXERCISES

Divide using synthetic division. See Section 4.6.

49. $\dfrac{x^3 + 2x^2 - 17x - 10}{x + 5}$

50. $\dfrac{a^4 + 4a^3 + 2a^2 + 9a + 4}{a + 4}$

51. $\dfrac{4m^3 - 3m - 2}{m + 1}$

52. $\dfrac{3q^3 - 4q + 2}{q - 1}$

Use synthetic division to decide whether or not the given number is a zero of the given polynomial. See Section 4.6.

53. 2; $P(x) = x^2 - 2x - 8$

54. 4; $P(r) = 2r^3 - 6r^2 - 9r + 4$

9.2 ZEROS OF POLYNOMIALS

OBJECTIVES

1. DECIDE IF $x - k$ IS A FACTOR OF THE POLYNOMIAL $P(x)$.
2. DETERMINE A POLYNOMIAL GIVEN ITS ZEROS AND A PARTICULAR FUNCTION VALUE.
3. FIND ALL THE ZEROS OF A POLYNOMIAL IF SOME OF ITS ZEROS ARE GIVEN.
4. FACTOR A POLYNOMIAL INTO LINEAR FACTORS GIVEN SOME OF ITS ZEROS.

Recall the remainder theorem given in Section 4.6:

> If the polynomial $P(x)$ is divided by $x - k$, then the remainder is equal to $P(k)$.

For example, for the polynomial

$$P(x) = 5x^3 - 6x^2 - 28x - 2,$$

$P(-2)$ can be found by dividing $P(x)$ by $x - (-2) = x + 2$, using synthetic division, as follows.

$$\begin{array}{r|rrrr} -2 & 5 & -6 & -28 & -2 \\ & & -10 & 32 & -8 \\ \hline & 5 & -16 & 4 & -10 \end{array} \leftarrow P(-2)$$

By the remainder theorem, the remainder in the synthetic division, -10, equals $P(-2)$. Also, $P(1)$ is found similarly.

$$\begin{array}{r|rrrr} 1 & 5 & -6 & -28 & -2 \\ & & 5 & -1 & -29 \\ \hline & 5 & -1 & -29 & -31 \end{array}$$

Since the remainder is -31, $P(1) = -31$.

1 By the remainder theorem, if $P(k) = 0$, then the remainder when $P(x)$ is divided by $x - k$ is zero. This means that $x - k$ is a factor of $P(x)$. Conversely, if $x - k$ is a factor of $P(x)$, then $P(k)$ must equal 0. This is summarized in the following theorem.

FACTOR THEOREM The polynomial $x - k$ is a factor of the polynomial $P(x)$ if and only if $P(k) = 0$.

EXAMPLE 1 DECIDING WHETHER $x - k$ IS A FACTOR OF $P(x)$ (REAL COEFFICIENTS)

Is $x - 1$ a factor of $P(x) = 2x^4 + 3x^2 - 5x + 7$?

By the factor theorem, $x - 1$ will be a factor of $P(x)$ only if $P(1) = 0$. Use synthetic division and the remainder theorem to decide.

$$\begin{array}{r|rrrrr} 1 & 2 & 0 & 3 & -5 & 7 \\ & & 2 & 2 & 5 & 0 \\ \hline & 2 & 2 & 5 & 0 & 7 \end{array}$$

Since the remainder is 7, $P(1) = 7$, not 0, so $x - 1$ is not a factor of $P(x)$. ■

EXAMPLE 2 DECIDING WHETHER $x - k$ IS A FACTOR OF $P(x)$ (COMPLEX COEFFICIENTS)

Is $x - i$ a factor of $P(x) = 3x^3 + (-4 - 3i)x^2 + (5 + 4i)x - 5i$?

The only way $x - i$ can be a factor of $P(x)$ is for $P(i)$ to be 0. Decide if this is the case by using synthetic division.

$$\begin{array}{r|rrrr} i & 3 & -4-3i & 5+4i & -5i \\ & & 3i & -4i & 5i \\ \hline & 3 & -4 & 5 & 0 \end{array}$$

Since the remainder is 0, $P(i) = 0$, and $x - i$ is a factor of $P(x)$. The other factor is the quotient $3x^2 - 4x + 5$, so $P(x)$ can be factored as

$$P(x) = (x - i)(3x^2 - 4x + 5). \quad ■$$

The next theorem says that every polynomial of degree 1 or more has a zero, so that every such polynomial can be factored. This theorem was first proved by the mathematician Carl F. Gauss in his doctoral thesis in 1799 when he was 22 years old. Although many proofs of this result have been given, all of them involve mathematics beyond the algebra in this book, so no proof is included here.

FUNDAMENTAL THEOREM OF ALGEBRA

Every polynomial of degree 1 or more has at least one complex zero.

From the fundamental theorem, if $P(x)$ is of degree 1 or more then there is some number k such that $P(k) = 0$. By the factor theorem, then

$$P(x) = (x - k) \cdot Q(x)$$

for some polynomial $Q(x)$. The fundamental theorem and the factor theorem can be used to factor $Q(x)$ in the same way. Assuming that $P(x)$ has degree n and repeating this process n times gives

$$P(x) = a(x - k_1)(x - k_2) \ldots (x - k_n),$$

where a is the leading coefficient of $P(x)$. Each of these factors leads to a zero of $P(x)$, so $P(x)$ has the n zeros $k_1, k_2, k_3, \ldots, k_n$. This result can be used to prove the next theorem. The proof is left for the exercises. See Exercise 69.

ZEROS OF A POLYNOMIAL

A polynomial of degree n has at most n distinct zeros.

The theorem says that there exist *at most n* distinct zeros. For example, the polynomial function $P(x) = x^3 + 3x^2 + 3x + 1 = (x + 1)^3$ is of degree 3 but has only one zero, -1. Actually, the zero -1 occurs three times, since there are three factors of $x + 1$; this zero is called a *zero of* **multiplicity** *3*.

EXAMPLE 3 **DETERMINING THE MULTIPLICITY OF ZEROS**

Determine the zeros of the polynomial function

$$P(x) = (x - 5)^4(x + 1)^2x^5$$

and give their multiplicities.

Since $(x - 5)^4$ is the largest power of $(x - 5)$ that appears in the factored form of $P(x)$, 5 is a zero of multiplicity 4. Because $(x + 1)^2$ can be written $[x - (-1)]^2$, -1 is a zero of multiplicity 2. Finally, because $x^5 = (x - 0)^5$, 0 is a zero of multiplicity 5. ■

2 The Fundamental Theorem of Algebra and the results concerning zeros and their multiplicities can be used to find a polynomial, given its zeros and a particular function value.

EXAMPLE 4 DETERMINING POLYNOMIALS THAT SATISFY GIVEN CONDITIONS (REAL ZEROS)

Find a polynomial $P(x)$ of degree 3 that satisfies the following conditions.

(a) Zeros of -1, 2, and 4; $P(1) = 3$

These three zeros give $x - (-1) = x + 1$, $x - 2$, and $x - 4$ as factors of $P(x)$. Since $P(x)$ is to be of degree 3, these are the only possible factors by the theorem just above. Therefore, $P(x)$ has the form

$$P(x) = a(x + 1)(x - 2)(x - 4)$$

for some nonzero real number a. To find a, use the fact that $P(1) = 3$.

$$\begin{aligned} P(1) = a(1 + 1)(1 - 2)(1 - 4) &= 3 \\ a(2)(-1)(-3) &= 3 \\ 6a &= 3 \\ a &= \frac{1}{2} \end{aligned}$$

Thus, $$P(x) = \frac{1}{2}(x + 1)(x - 2)(x - 4),$$

or $$P(x) = \frac{1}{2}x^3 - \frac{5}{2}x^2 + x + 4.$$

(b) -2 is a zero of multiplicity 3; $P(-1) = 4$

The polynomial $P(x)$ has the form

$$\begin{aligned} P(x) &= a(x + 2)(x + 2)(x + 2) \\ &= a(x + 2)^3. \end{aligned}$$

Since $P(-1) = 4$,

$$\begin{aligned} P(-1) = a(-1 + 2)^3 &= 4 \\ a(1)^3 &= 4 \\ a &= 4, \end{aligned}$$

and $P(x) = 4(x + 2)^3 = 4x^3 + 24x^2 + 48x + 32$. ■

CAUTION In Example 4(a), it would be *wrong* to clear of fractions by multiplying through by 2. The result would then be $2 \cdot P(x)$ and not $P(x)$.

The remainder theorem can be used to show that both $2 + i$ and $2 - i$ are zeros of $P(x) = x^3 - x^2 - 7x + 15$. It is not a coincidence that both $2 + i$ and its conjugate $2 - i$ are zeros of this polynomial. If $a + bi$ is a zero of a polynomial function with *real* coefficients, then so is $a - bi$. This is given as the next theorem. The proof is left for the exercises. See Exercise 70.

CONJUGATE ZEROS THEOREM If $P(x)$ is a polynomial having only real coefficients and if $a + bi$ is a zero of $P(x)$, then the conjugate $a - bi$ is also a zero of $P(x)$.

CAUTION The requirement that the polynomial have only real coefficients is very important. For example, $P(x) = x - (1 + i)$ has $1 + i$ as a zero, but the conjugate $1 - i$ is not a zero.

EXAMPLE 5
DETERMINING A POLYNOMIAL SATISFYING CERTAIN CONDITIONS (COMPLEX ZEROS)

Find a polynomial of lowest degree having real coefficients and zeros 3 and $2 + i$.

The complex number $2 - i$ also must be a zero, so there are at least three zeros, 3, $2 + i$, and $2 - i$. For the polynomial to be of lowest degree these must be the only zeros. Then by the factor theorem there must be three factors, $x - 3$, $x - (2 + i)$, and $x - (2 - i)$. A polynomial of lowest degree is

$$\begin{aligned} P(x) &= (x - 3)[x - (2 + i)][x - (2 - i)] \\ &= (x - 3)(x - 2 - i)(x - 2 + i) \\ &= x^3 - 7x^2 + 17x - 15. \end{aligned}$$

Other polynomials, such as $2(x^3 - 7x^2 + 17x - 15)$ or $\sqrt{5}(x^3 - 7x^2 + 17x - 15)$, for example, also satisfy the given conditions on zeros. The information on zeros given in the problem is not enough to give a specific value for the leading coefficient. ■

3 The conjugate zeros theorem helps predict the number of real zeros of polynomials with real coefficients. A polynomial of odd degree n, where $n \geq 1$, with real coefficients must have at least one real zero (since zeros of the form $a + bi$, where $b \neq 0$, occur in conjugate pairs.) On the other hand, a polynomial of even degree n with real coefficients need have no real zeros but may have up to n real zeros.

EXAMPLE 6
FINDING ALL ZEROS OF A POLYNOMIAL GIVEN ONE ZERO (COMPLEX ZEROS)

Find all zeros of $P(x) = x^4 - 7x^3 + 18x^2 - 22x + 12$, given that $1 - i$ is a zero.

Since $1 - i$ is a zero and the coefficients are real numbers, by the conjugate zeros theorem $1 + i$ is also a zero. The remaining zeros are found by first dividing the original polynomial by $x - (1 - i)$.

$$\begin{array}{r|rrrrr} 1 - i & 1 & -7 & 18 & -22 & 12 \\ & & 1 - i & -7 + 5i & 16 - 6i & -12 \\ \hline & 1 & -6 - i & 11 + 5i & -6 - 6i & 0 \end{array}$$

Rather than go back to the original equation, divide the quotient from the first division by $x - (1 + i)$ as follows.

$$\begin{array}{r|rrrr} 1+i & 1 & -6-i & 11+5i & -6-6i \\ & & 1+i & -5-5i & 6+6i \\ \hline & 1 & -5 & 6 & 0 \end{array}$$

Find the zeros of the quadratic polynomial $x^2 - 5x + 6$ by solving the equation $x^2 - 5x + 6 = 0$. Factoring the polynomial shows that the zeros are 2 and 3, so the four zeros of $P(x)$ are $1 - i$, $1 + i$, 2, and 3. ■

4 The theorems given in this section can be used to factor a polynomial into linear factors (factors of the form $ax - b$) when one or more of its zeros are known.

EXAMPLE 7 **FACTORING A POLYNOMIAL GIVEN INFORMATION ABOUT ITS ZEROS**

Factor $P(x)$ into linear factors, given that k is a zero of $P(x)$.

(a) $P(x) = 6x^3 + 19x^2 + 2x - 3$; $k = -3$

Since $k = -3$ is a zero of $P(x)$, $x - (-3) = x + 3$ is a factor. Use synthetic division to divide $P(x)$ by $x + 3$.

$$\begin{array}{r|rrrr} -3 & 6 & 19 & 2 & -3 \\ & & -18 & -3 & 3 \\ \hline & 6 & 1 & -1 & 0 \end{array}$$

The quotient is $6x^2 + x - 1$, so

$$P(x) = (x + 3)(6x^2 + x - 1).$$

Factor $6x^2 + x - 1$ as $(2x + 1)(3x - 1)$ to get

$$P(x) = (x + 3)(2x + 1)(3x - 1),$$

where all factors are linear.

(b) $P(x) = 3x^3 + (-1 + 3i)x^2 + (-12 + 5i)x + 4 - 2i$; $k = 2 - i$

One factor is $x - (2 - i)$ or $x - 2 + i$. Divide $P(x)$ by $x - (2 - i)$.

$$\begin{array}{r|rrrr} 2-i & 3 & -1+3i & -12+5i & 4-2i \\ & & 6-3i & 10-5i & -4+2i \\ \hline & 3 & 5 & -2 & 0 \end{array}$$

By the division algorithm,

$$P(x) = (x - 2 + i)(3x^2 + 5x - 2).$$

Factor $3x^2 + 5x - 2$ as $(3x - 1)(x + 2)$; then a linear factored form of $P(x)$ is

$$P(x) = (x - 2 + i)(3x - 1)(x + 2). \quad ■$$

NOTE In Example 7(b), the conjugate $2 + i$ is not also a zero, because $P(x)$ has some imaginary coefficients.

9.2 EXERCISES

Use the factor theorem to decide whether the second polynomial is a factor of the first. See Examples 1 and 2.

1. $4x^2 + 2x + 42;\ x - 3$

2. $-3x^2 - 4x + 2;\ x + 2$

3. $x^3 + 2x^2 - 3;\ x - 1$

4. $2x^3 + x + 2;\ x + 1$

5. $3x^3 - 12x^2 - 11x - 20;\ x - 5$

6. $4x^3 + 6x^2 - 5x - 2;\ x + 2$

7. $2x^4 + 5x^3 - 2x^2 + 5x + 3;\ x + 3$

8. $5x^4 + 16x^3 - 15x^2 + 8x + 16;\ x + 4$

For each of the following polynomial functions, give all zeros and their multiplicities. See Example 3.

9. $P(x) = x^4(x + 3)^5(x - 8)^2$

10. $P(x) = x^9(x - 4)^6(x + 7)^4$

11. $P(x) = (4x - 7)^3(x - 5)$

12. $P(x) = (5x + 1)^3(x + 1)$

13. $P(x) = (x - i)^4(x + i)^4$

14. $P(x) = (x + 3i)^9(x - 3i)^9$

For each of the following, find a polynomial of degree 3 with real coefficients that satisfies the given conditions. See Example 4.

15. Zeros of -3, -1, and 4; $P(2) = 5$

16. Zeros of 1, -1, and 0; $P(2) = -3$

17. Zeros of -2, 1, and 0; $P(-1) = -1$

18. Zeros of 2, 5, and -3; $P(1) = -4$

19. Zeros of 3, i, and $-i$; $P(2) = 50$

20. Zeros of -2, i, and $-i$; $P(-3) = 30$

For each of the following, find a polynomial of lowest degree with real coefficients having the given zeros. See Example 5.

21. $5 + i$ and $5 - i$

22. $3 - 2i$ and $3 + 2i$

23. 2, $1 - i$, and $1 + i$

24. -3, $2 - i$, and $2 + i$

25. $1 + \sqrt{2}$, $1 - \sqrt{2}$, and 1

26. $1 - \sqrt{3}$, $1 + \sqrt{3}$, and -2

27. $2 + i$, $2 - i$, 3, and -1

28. $3 + 2i$, $3 - 2i$, -1, and 2

29. 2 and $3 + i$

30. -1 and $4 - 2i$

31. $1 - \sqrt{2}$, $1 + \sqrt{2}$, and $1 - i$

32. $2 + \sqrt{3}$, $2 - \sqrt{3}$, and $2 + 3i$

33. $2 - i$, $3 + 2i$

34. $5 + i$, $4 - i$

35. 4, $1 - 2i$, $3 + 4i$

36. -1, $1 + \sqrt{2}$, $1 - \sqrt{2}$, $1 + 4i$

37. $1 + 2i$, 2 (multiplicity 2)

38. $2 + i$, -3 (multiplicity 2)

For each of the following polynomials, one zero is given. Find all others. See Example 6.

39. $P(x) = x^3 - x^2 - 4x - 6;\quad 3$

40. $P(x) = x^3 - 5x^2 + 17x - 13;\quad 1$

41. $P(x) = x^3 + x^2 - 4x - 24;\quad -2 + 2i$

42. $P(x) = x^3 + x^2 - 20x - 50;\quad -3 + i$

43. $P(x) = 2x^3 - 2x^2 - x - 6;\quad 2$

44. $P(x) = 2x^3 - 5x^2 + 6x - 2;\quad 1 + i$

45. $P(x) = x^4 + 5x^2 + 4;\quad -i$

46. $P(x) = x^4 + 10x^3 + 27x^2 + 10x + 26;\quad i$

47. $P(x) = x^4 - 3x^3 + 6x^2 + 2x - 60;\quad 1 + 3i$

48. $P(x) = x^4 - 6x^3 - x^2 + 86x + 170;\quad 5 + 3i$

Factor $P(x)$ into linear factors given that k is a zero of $P(x)$. See Example 7.

49. $P(x) = 2x^3 - 3x^2 - 17x + 30;\quad k = 2$

50. $P(x) = 2x^3 - 3x^2 - 5x + 6;\quad k = 1$

51. $P(x) = 6x^3 + 25x^2 + 3x - 4;\quad k = -4$

52. $P(x) = 8x^3 + 50x^2 + 47x - 15;\quad k = -5$

53. $P(x) = x^3 + (7 - 3i)x^2 + (12 - 21i)x - 36i;\quad k = 3i$

54. $P(x) = 2x^3 + (3 + 2i)x^2 + (1 + 3i)x + i;\quad k = -i$

55. $P(x) = 2x^3 + (3 - 2i)x^2 + (-8 - 5i)x + 3 + 3i; \quad k = 1 + i$

56. $P(x) = 6x^3 + (19 - 6i)x^2 + (16 - 7i)x + 4 - 2i; \quad k = -2 + i$

---

57. Show that -2 is a zero of multiplicity 2 of $P(x) = x^4 + 2x^3 - 7x^2 - 20x - 12$, and find all other complex zeros. Then write $P(x)$ in factored form.

58. Show that -1 is a zero of multiplicity 3 of $P(x) = x^5 + 9x^4 + 33x^3 + 55x^2 + 42x + 12$, and find all other complex zeros. Then write $P(x)$ in factored form.

59. What are the possible numbers of real zeros (counting multiplicities) for a polynomial function with real coefficients of degree five? (See Objective 3.)

60. Explain why a polynomial of degree four with real coefficients has either 0, 2, or 4 real zeros. (See Objective 3.)

61. Explain why it is not possible for a polynomial of degree 3 with real coefficients to have zeros of 1, 2, and $1 + i$. (See Objective 2.)

62. Show that the zeros of the polynomial $P(x) = x^3 + ix^2 - (7 - i)x + (6 - 6i)$ are $1 - i$, 2, and -3. Does the conjugate zeros theorem apply? Why or why not? (See Objective 2.)

63. The displacement at time t of a particle moving along a straight line is given by

$$s(t) = t^3 - 2t^2 - 5t + 6,$$

where t is in seconds and s is measured in centimeters. The displacement is 0 after 1 second has elapsed. At what other times (positive) is the displacement 0?

64. For headphone radios, the cost function (in thousands of dollars) is given by

$$C(x) = 2x^3 - 9x^2 + 17x - 4,$$

and the revenue function (in thousands of dollars) is given by $R(x) = 5x$, where x is the number of items (in hundred thousands) produced. Cost equals revenue if 200,000 items are produced ($x = 2$). Find all other break-even points.

If c and d are complex numbers, prove each of the following statements. These statements are used in Exercise 71. (Hint: Let $c = a + bi$ and $d = m + ni$ and form the conjugates, the sums, and the products. The notation $\bar{c}$ represents the conjugate of c.)

65. $\overline{c + d} = \bar{c} + \bar{d}$

66. $\overline{cd} = \bar{c} \cdot \bar{d}$

67. $\bar{a} = a$ for any real number a

68. $\overline{c^n} = (\bar{c})^n$, n is a positive integer

Use the theorems presented in this section to prove each of the following statements.

69. A polynomial of degree n has at most n distinct zeros. (*Hint:* Use an indirect proof, where you assume the opposite of the statement is true and show that it leads to a contradiction.)

70. If $a + bi$ is a zero of a polynomial $P(x)$ having only real coefficients, then $a - bi$ is also a zero of $P(x)$.

71. Complete the proof of the conjugate zeros theorem, outlined below. Assume that

$$P(x) = a_n x^n + a_{n-1}x^{n-1} + \ldots + a_1 x + a_0,$$

where all coefficients are real numbers.

(a) If the complex number z is a zero of $P(x)$, find $P(z)$.

(b) Take the conjugate of both sides of the result from part (a).

(c) Use generalizations of the properties given in Exercises 65–68 on the result of part (b) to show that $a_n(\bar{z})^n + a_{n-1}(\bar{z})^{n-1} + \ldots + a_1(\bar{z}) + a_0 = 0$.

(d) Why does the result in part (c) mean that $\bar{z}$ is a zero of $P(x)$?

PREVIEW EXERCISES

Solve each of the following equations. See Sections 2.1 and 6.4.

72. $\frac{3x-2}{7}=\frac{x+2}{5}$

73. $\frac{2p+5}{5}=\frac{p+2}{3}$

74. $4-\frac{11}{x}-\frac{3}{x^2}=0$

9.3 RATIONAL ZEROS OF POLYNOMIAL FUNCTIONS

OBJECTIVES

1 STATE AND UNDERSTAND THE RATIONAL ZEROS THEOREM.

2 FIND ALL RATIONAL ZEROS OF A POLYNOMIAL WITH INTEGER COEFFICIENTS.

3 FIND ALL RATIONAL ZEROS OF A POLYNOMIAL WITH RATIONAL NUMBER COEFFICIENTS.

By the fundamental theorem of algebra, every polynomial of degree 1 or more has a zero. However, the fundamental theorem merely says that such a zero exists. It gives no help at all in identifying zeros. Other theorems can be used to find any rational zeros of polynomials with rational coefficients or to find decimal approximations of any irrational zeros.

1 The next theorem gives a useful method for finding a set of possible zeros of a polynomial with integer coefficients.

RATIONAL ZEROS THEOREM

Let $P(x) = a_nx^n + a_{n-1}x^{n-1} + \cdots + a_1x + a_0$, where $a_n \neq 0$, be a polynomial with integer coefficients. If p/q is a rational number written in lowest terms and if p/q is a zero of $P(x)$, then p is a factor of the constant term a_0 and q is a factor of the leading coefficient a_n.

Proof

$P(p/q) = 0$ since p/q is a zero of $P(x)$, so

$$a_n(p/q)^n + a_{n-1}(p/q)^{n-1} + \cdots + a_1(p/q) + a_0 = 0.$$

This also can be written as

$$a_n(p^n/q^n) + a_{n-1}(p^{n-1}/q^{n-1}) + \cdots + a_1(p/q) + a_0 = 0.$$

Multiply both sides of this last result by q^n and add $-a_0q^n$ to both sides.

$$a_np^n + a_{n-1}p^{n-1}q + \cdots + a_1pq^{n-1} = -a_0q^n$$

Factoring out p gives

$$p(a_np^{n-1} + a_{n-1}p^{n-2}q + \cdots + a_1q^{n-1}) = -a_0q^n.$$

This result shows that $-a_0q^n$ equals the product of the two factors, p and $(a_np^{n-1} + \cdots + a_1q^{n-1})$. For this reason, p must be a factor of $-a_0q^n$. Since it was assumed that p/q is written in lowest terms, p and q have no common factor other than 1, so p is not a factor of q^n. Thus p must be a factor of a_0. In a similar way it can be shown that q is a factor of a_n.

2 In the next example we use the rational zeros theorem to find all rational zeros of a polynomial with integer coefficients.

EXAMPLE 1
FINDING THE RATIONAL ZEROS OF A POLYNOMIAL WITH INTEGER COEFFICIENTS

Find all rational zeros of $P(x) = 2x^4 - 11x^3 + 14x^2 - 11x + 12$.

If p/q is to be a rational zero of $P(x)$, by the rational zeros theorem p must be a factor of $a_0 = 12$ and q must be a factor of $a_4 = 2$. The possible values of p are ± 1, ± 2, ± 3, ± 4, ± 6, or ± 12, while q must be ± 1 or ± 2. The possible rational zeros are found by forming all possible quotients of the form p/q; any rational zero of $P(x)$ will come from the list

$$\pm 1,\ \pm 1/2,\ \pm 2,\ \pm 3,\ \pm 3/2,\ \pm 4,\ \pm 6,\ \text{or}\ \pm 12.$$

Though none of these numbers may be zeros, if $P(x)$ has any rational zeros, they will be in the list above. These proposed zeros can be checked by synthetic division. Doing so shows that 4 is a zero.

$$\begin{array}{r|rrrrr} 4 & 2 & -11 & 14 & -11 & 12 \\ & & 8 & -12 & 8 & -12 \\ \hline & 2 & -3 & 2 & -3 & 0 \end{array} \quad \leftarrow P(4) = 0$$

As a fringe benefit of this calculation, the simpler polynomial $Q(x) = 2x^3 - 3x^2 + 2x - 3$ can be used to find the remaining zeros. Any rational zero of $Q(x)$ will have a numerator of ± 3 or ± 1 and a denominator of ± 1 or ± 2 and so will come from the list

$$\pm 3,\ \pm 3/2,\ \pm 1,\ \pm 1/2.$$

Again use synthetic division and trial and error to find that 3/2 is a zero.

$$\begin{array}{r|rrrr} 3/2 & 2 & -3 & 2 & -3 \\ & & 3 & 0 & 3 \\ \hline & 2 & 0 & 2 & 0 \end{array} \quad \leftarrow Q(3/2) = 0$$

The quotient is $2x^2 + 2$, which, by the quadratic formula, has i and $-i$ as zeros. They are imaginary zeros, however. The rational zeros of the polynomial function $P(x) = 2x^4 - 11x^3 + 14x^2 - 11x + 12$ are 4 and 3/2. ■

EXAMPLE 2
FINDING RATIONAL ZEROS AND FACTORING A POLYNOMIAL

Find all rational zeros of $P(x) = 6x^4 + 7x^3 - 12x^2 - 3x + 2$, and factor the polynomial.

For a rational number p/q to be a zero of $P(x)$, p must be a factor of $a_0 = 2$ and q must be a factor of $a_4 = 6$. Thus, p can be ± 1 or ± 2 and q can be ± 1, ± 2, ± 3, or ± 6. The rational zeros, p/q, must come from the following list.

$$\pm 1,\ \pm 2,\ \pm 1/2,\ \pm 1/3,\ \pm 1/6,\ \pm 2/3$$

Check 1 first because it is easy.

$$\begin{array}{r|rrrrr} 1 & 6 & 7 & -12 & -3 & 2 \\ & & 6 & 13 & 1 & -2 \\ \hline & 6 & 13 & 1 & -2 & 0 \end{array}$$

The 0 remainder shows that 1 is a zero. Now use the quotient polynomial $6x^3 + 13x^2 + x - 2$ and synthetic division to find that -2 is also a zero.

$$\begin{array}{r|rrrr} -2 & 6 & 13 & 1 & -2 \\ & & -12 & -2 & 2 \\ \hline & 6 & 1 & -1 & 0 \end{array}$$

The new quotient polynomial is $6x^2 + x - 1$. Use the quadratic formula or factor to solve the equation $6x^2 + x - 1 = 0$. The remaining two zeros are $1/3$ and $-1/2$.

Factor the polynomial $P(x)$ in the following way. Since the four zeros of $P(x) = 6x^4 + 7x^3 - 12x^2 - 3x + 2$ are 1, -2, $1/3$, and $-1/2$, the corresponding factors are $x - 1$, $x + 2$, $x - 1/3$, and $x + 1/2$, and

$$P(x) = a_4(x - 1)(x + 2)\left(x - \frac{1}{3}\right)\left(x + \frac{1}{2}\right).$$

Since $a_4 = 6 = 2 \cdot 3$, fractions can be cleared in the last two factors by writing the product as

$$\begin{aligned} P(x) &= 6(x - 1)(x + 2)\left(x - \frac{1}{3}\right)\left(x + \frac{1}{2}\right) \\ &= (x - 1)(x + 2)\left[(3)\left(x - \frac{1}{3}\right)\right]\left[(2)\left(x + \frac{1}{2}\right)\right] \\ &= (x - 1)(x + 2)(3x - 1)(2x + 1). \quad \blacksquare \end{aligned}$$

NOTE We found by using synthetic division that 1 is a zero of $P(x) = 6x^4 + 7x^3 - 12x^2 - 3x + 2$ in Example 2. An easy way to determine whether 1 is a zero of a polynomial is as follows: *If the sum of the coefficients of a polynomial is 0, then 1 is a zero of the polynomial.* Since any power of 1 is 1, when each term of the polynomial is evaluated, the answer is the coefficient of the term; the result above follows. Verify this in Example 2. (This does not help in determining other zeros, however, as synthetic division does.)

3 Rational zeros of a polynomial with rational coefficients can be found by first multiplying the polynomial by a number that will clear it of all fractions, then using the rational zeros theorem.

EXAMPLE 3
FINDING RATIONAL ZEROS OF A POLYNOMIAL WITH FRACTIONS AS COEFFICIENTS

Find all rational zeros of $\quad P(x) = x^4 - \frac{1}{6}x^3 + \frac{2}{3}x^2 - \frac{1}{6}x - \frac{1}{3}.$

Find the values of x that make $P(x) = 0$,

or
$$x^4 - \frac{1}{6}x^3 + \frac{2}{3}x^2 - \frac{1}{6}x - \frac{1}{3} = 0.$$

Multiply both sides by 6 to eliminate all fractions.

$$6x^4 - x^3 + 4x^2 - x - 2 = 0$$

This polynomial will have the same zeros as $P(x)$. The possible rational zeros are of the form p/q where p is ± 1, or ± 2, and q is ± 1, ± 2, ± 3, or ± 6. Then p/q may be

$$\pm 1,\ \pm 2,\ \pm 1/2,\ \pm 1/3,\ \pm 1/6, \text{ or } \pm 2/3.$$

Use synthetic division to find that $-1/2$ and $2/3$ are zeros.

$$\begin{array}{r|rrrrr} -\frac{1}{2} & 6 & -1 & 4 & -1 & -2 \\ & & -3 & 2 & -3 & 2 \\ \hline & 6 & -4 & 6 & -4 & 0 \end{array}$$

$$\begin{array}{r|rrrr} \frac{2}{3} & 6 & -4 & 6 & -4 \\ & & 4 & 0 & 4 \\ \hline & 6 & 0 & 6 & 0 \end{array}$$

The final quotient is $Q(x) = 6x^2 + 6 = 6(x^2 + 1)$. This polynomial can be solved by the quadratic formula; its zeros are i and $-i$. Since these zeros are imaginary numbers, there are just two rational zeros: $-1/2$ and $2/3$. ■

CAUTION Remember, the rational zeros theorem can be used only if the coefficients of the polynomial are integers. Polynomial functions with rational coefficients can be rewritten with integer coefficients in order to use the theorem, but the theorem cannot be used with polynomial functions with irrational or imaginary coefficients.

9.3 EXERCISES

Give all possible rational zeros for the following polynomials.

1. $P(x) = 6x^3 + 17x^2 - 31x - 1$

2. $P(x) = 15x^3 + 61x^2 + 2x - 1$

3. $P(x) = 12x^3 + 20x^2 - x - 2$

4. $P(x) = 12x^3 + 40x^2 + 41x + 3$

5. $P(x) = 2x^3 + 7x^2 + 12x - 8$

6. $P(x) = 2x^3 + 20x^2 + 68x - 40$

7. $P(x) = x^4 + 4x^3 + 3x^2 - 10x + 50$

8. $P(x) = x^4 - 2x^3 + x^2 + 18$

9. Discuss a major drawback of the rational zeros theorem.

10. Can the rational zeros theorem be used for a polynomial with irrational coefficients? (See Objective 3.)

Find all rational zeros of the following polynomials. See Examples 1 and 2.

11. $P(x) = x^3 - 2x^2 - 13x - 10$

12. $P(x) = x^3 + 5x^2 + 2x - 8$

13. $P(x) = x^3 + 6x^2 - x - 30$

14. $P(x) = x^3 - x^2 - 10x - 8$

15. $P(x) = x^3 + 9x^2 - 14x - 24$

16. $P(x) = x^3 + 3x^2 - 4x - 12$

17. $P(x) = x^4 + 9x^3 + 21x^2 - x - 30$

18. $P(x) = x^4 + 4x^3 - 7x^2 - 34x - 24$

Find the rational zeros of the following polynomials; then write each in factored form. See Example 2.

19. $P(x) = 6x^3 + 17x^2 - 31x - 12$

20. $P(x) = 15x^3 + 61x^2 + 2x - 8$

21. $P(x) = 12x^3 + 20x^2 - x - 6$

22. $P(x) = 12x^3 + 40x^2 + 41x + 12$

23. $P(x) = 2x^3 + 7x^2 + 12x - 8$

24. $P(x) = 2x^3 + 20x^2 + 68x - 40$

25. $P(x) = 2x^4 + 3x^3 - 4x^2 - 3x + 2$

26. $P(x) = x^4 - 2x^3 + x^2 + 18$

27. $P(x) = 3x^4 + 5x^3 - 10x^2 - 20x - 8$

28. $P(x) = 6x^4 + x^3 - 7x^2 - x + 1$

Find all rational zeros of the following polynomials. See Example 3.

29. $P(x) = x^3 - \frac{4}{3}x^2 - \frac{13}{3}x - 2$

30. $P(x) = x^3 + x^2 - \frac{16}{9}x + \frac{4}{9}$

31. $P(x) = x^4 + \frac{1}{4}x^3 + \frac{11}{4}x^2 + x - 5$

32. $P(x) = \frac{10}{7}x^4 - x^3 - 7x^2 + 5x - \frac{5}{7}$

33. $P(x) = \frac{1}{3}x^5 + x^4 - \frac{5}{3}x^3 - \frac{11}{3}x^2 + 4$

34. $P(x) = x^5 + x^4 - \frac{37}{4}x^2 + \frac{9}{4}x + \frac{9}{4}$

For each of the following polynomial functions, find all rational zeros and factor the polynomial. Then graph the function using the method described in Section 9.1.

35. $P(x) = 2x^3 - 5x^2 - x + 6$

36. $P(x) = 3x^3 + x^2 - 10x - 8$

37. $P(x) = x^3 + x^2 - 8x - 12$

38. $P(x) = x^3 + 6x^2 - 32$

39. Show that $P(x) = x^2 - 2$ has no rational zeros, so $\sqrt{2}$ must be irrational.

40. Show that $P(x) = x^2 - 5$ has no rational zeros, so $\sqrt{5}$ must be irrational.

41. Show that $P(x) = x^4 + 5x^2 + 4$ has no rational zeros.

42. Show that $P(x) = x^5 - 3x^3 + 5$ has no rational zeros.

43. Show that any integer zeros of a polynomial function must be factors of the constant term a_0.

44. If k is a zero of the polynomial function $P(x)$, then $P(k) =$ __________, a factor of $P(x)$ is __________, and an x-intercept of the graph of P is __________.

PREVIEW EXERCISES

*Use synthetic division to find **(a)** $P(2)$ and **(b)** $P(-3)$ for each of the following polynomials. See Section 4.6.*

45. $P(x) = 2x^3 - 3x^2 + 9x - 4$

46. $P(x) = -3x^3 - 2x^2 + 5x + 10$

47. $P(x) = x^4 - x^2 + 2$

48. $P(x) = x^3 - 7x + 3$

9.4 REAL ZEROS OF POLYNOMIAL FUNCTIONS

OBJECTIVES

1 DETERMINE THE POSSIBLE NUMBERS OF POSITIVE AND NEGATIVE REAL ZEROS OF A POLYNOMIAL FUNCTION USING DESCARTES' RULE OF SIGNS.

2 LOCATE A REAL ZERO OF A POLYNOMIAL FUNCTION BETWEEN TWO NUMBERS USING THE INTERMEDIATE VALUE THEOREM.

3 DECIDE WHICH REAL NUMBERS ARE GREATER THAN OR LESS THAN ALL REAL ZEROS OF A POLYNOMIAL FUNCTION USING THE BOUNDEDNESS THEOREM.

4 USE APPROXIMATE REAL ZEROS TO GRAPH POLYNOMIAL FUNCTIONS THAT CANNOT BE FACTORED WITH RATIONAL COEFFICIENTS.

FOCUS ON PROBLEM SOLVING

A technique for measuring cardiac output depends on the concentration of a dye in the bloodstream after a known amount is injected into a vein near the heart. For a normal heart, the concentration of dye in the bloodstream at time x (in seconds) is given by the function

$$g(x) = -.006x^4 + .140x^3 - .053x^2 + 1.79x.$$

(a) Find $g(20)$.
(b) Graph $g(x)$.

In Section 9.1 we saw an application of a polynomial function to the prediction of deer population. Here we see another biological application: predicting the concentration of a dye in the bloodstream. This problem is Exercise 45 in the exercises for this section. After working through this section, you should be able to graph the function and solve this problem.

Every polynomial function of degree 1 or more has a zero. However, the fundamental theorem of algebra does not say whether a polynomial has *real* zeros. Even if it does have real zeros, often their exact values cannot be determined. This section discusses methods of determining how many real zeros a polynomial may have, and explains methods for approximating any real zeros.

The recent strides made in computer technology and graphing calculators have made some of the material in this section of limited value. However, the concepts presented here allow students to understand the ideas of finding real zeros of polynomial functions. Once these ideas are mastered, students may then wish to investigate the use of computers and graphing calculators to find real zeros of polynomial functions. Learning the methods of this section first will help students realize the power of today's technology.

1 *Descartes' rule of signs,* stated below, gives a test for finding the number of positive or negative real zeros of a given polynomial.

DESCARTES' RULE OF SIGNS

Let $P(x)$ be a polynomial with real coefficients and terms in descending powers of x.

(a) The number of positive real zeros of $P(x)$ either equals the number of variations in sign occurring in the coefficients of $P(x)$, or is less than the number of variations by a positive even integer.

(b) The number of negative real zeros of $P(x)$ either equals the number of variations in sign of $P(-x)$, or is less than the number of variations by a positive even integer.

In the theorem, variation in sign is a change from positive to negative or negative to positive in successive terms of the polynomial. Missing terms (those with 0 coefficients) are counted as no change in sign and can be ignored.

For the purposes of this theorem, zeros of multiplicity k count as k zeros. For example,

$$P(x) = (x - 1)^4 = +x^4 \underset{1}{-} 4x^3 \underset{2}{+} 6x^2 \underset{3}{-} 4x \underset{4}{+} 1$$

has 4 changes of sign. By Descartes' rule of signs, $P(x)$ has either 4, 2, or 0 positive real zeros. In this case there are 4, and each of the 4 positive real zeros is 1.

EXAMPLE 1
USING DESCARTES' RULE OF SIGNS

Find the number of positive and negative real zeros possible for the polynomial

$$P(x) = x^4 - 6x^3 + 8x^2 + 2x - 1.$$

$P(x)$ has 3 variations in sign:

$$+x^4 \underset{1}{-} 6x^3 \underset{2}{+} 8x^2 + 2x \underset{3}{-} 1.$$

Thus, by Descartes' rule of signs, $P(x)$ has either 3 or $3 - 2 = 1$ positive real zeros. Since

$$\begin{aligned} P(-x) &= (-x)^4 - 6(-x)^3 + 8(-x)^2 + 2(-x) - 1 \\ &= x^4 + 6x^3 + 8x^2 - 2x - 1 \end{aligned}$$

has only one variation in sign, $P(x)$ has only one negative real zero. ■

EXAMPLE 2
USING DESCARTES' RULE OF SIGNS

Find the number of positive and negative real zeros possible for

$$Q(x) = x^5 + 5x^4 + 3x^2 + 2x + 1.$$

The polynomial $Q(x)$ has no variations in sign and so has no positive real zeros. Here

$$Q(-x) = -x^5 + 5x^4 + 3x^2 - 2x + 1$$

has three variations in sign, so $Q(x)$ has either 3 or 1 negative real zeros. The other zeros are imaginary numbers. ■

NOTE Every polynomial function of odd degree having real coefficients must have at least one real zero. This is convenient to remember when using Descartes' rule of signs.

2 Much of our work in locating real zeros uses the following result, which is related to the fact that graphs of polynomial functions are unbroken curves, with no gaps or sudden jumps. The proof requires advanced methods, so it is not given here.

INTERMEDIATE VALUE THEOREM FOR POLYNOMIALS

If $P(x)$ defines a polynomial function with *only real coefficients,* and if for real numbers a and b, $P(a)$ and $P(b)$ are opposite in sign, then there exists at least one real zero between a and b.

This theorem helps to identify intervals where zeros of polynomials are located. For example, in Figure 9.13, $P(a)$ and $P(b)$ are opposite in sign, so 0 is between $P(a)$ and $P(b)$. Then, by the intermediate value theorem, there must be a number c between a and b such that $P(c) = 0$.

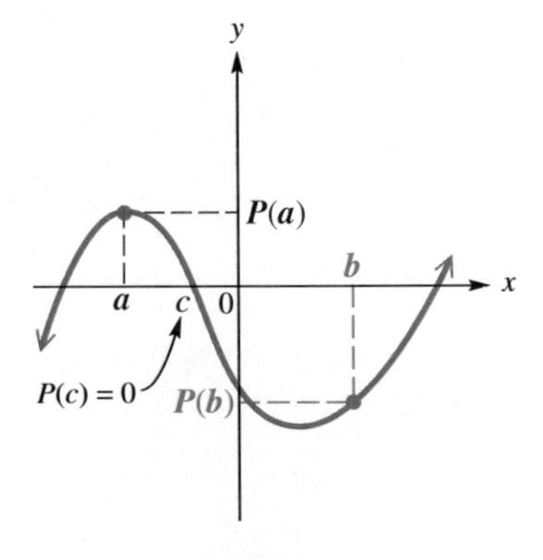

FIGURE 9.13

EXAMPLE 3 USING THE INTERMEDIATE VALUE THEOREM

Does $P(x) = x^3 - 2x^2 - x + 1$ have any real zeros between 2 and 3?

Use synthetic division to find $P(2)$ and $P(3)$.

$$\begin{array}{r|rrrr} 2 & 1 & -2 & -1 & 1 \\ & & 2 & 0 & -2 \\ \hline & 1 & 0 & -1 & -1 \end{array} \qquad \begin{array}{r|rrrr} 3 & 1 & -2 & -1 & 1 \\ & & 3 & 3 & 6 \\ \hline & 1 & 1 & 2 & 7 \end{array}$$

The results show that $P(2) = -1$ and $P(3) = 7$. Since $P(2)$ is negative but $P(3)$ is positive, by the intermediate value theorem, there must be a real zero between 2 and 3. ■

CAUTION Be careful how you interpret the intermediate value theorem. If $P(a)$ and $P(b)$ are *not* opposite in sign, it does not necessarily mean that there is no zero between a and b. For example, in Figure 9.14, $P(a)$ and $P(b)$ are both negative, but -3 and -1, which are between a and b, are zeros of $P(x)$.

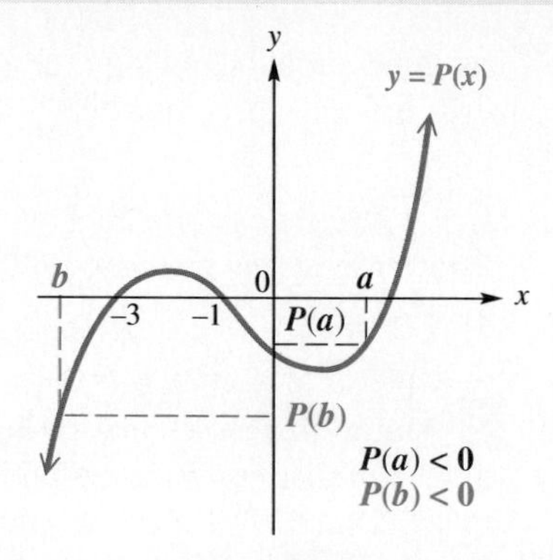

FIGURE 9.14

The intermediate value theorem for polynomials is helpful in limiting the search for real zeros to a smaller and smaller interval. In Example 3 the theorem was used to show that the polynomial $P(x) = x^3 - 2x^2 - x + 1$ has a real zero between 2 and 3. The theorem then could be used repeatedly to express the zero more accurately.

3 As suggested by the graphs of Figure 9.15, if the values of $|x|$ in a polynomial get larger and larger, then so will the values of $|y|$. This is used in the next theorem, the boundedness theorem, which shows how the bottom row of a synthetic division can be used to place upper and lower bounds on the possible real zeros of a polynomial.

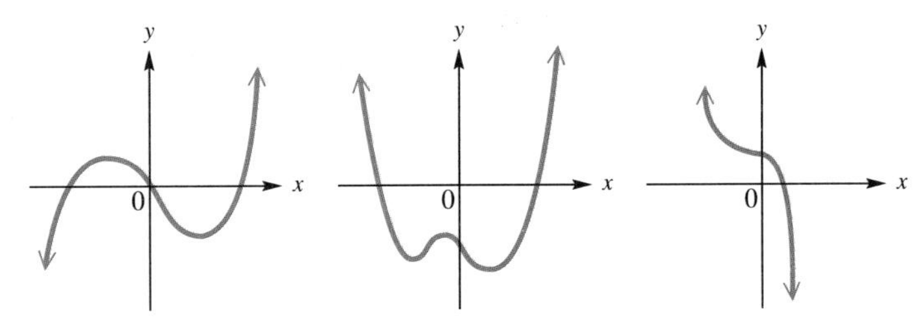

FIGURE 9.15

BOUNDEDNESS THEOREM

Let $P(x)$ be a polynomial with *real* coefficients and with a *positive* leading coefficient. If $P(x)$ is divided synthetically by $x - c$, and

(a) if $c > 0$ and all numbers in the bottom row of the synthetic division are nonnegative, then $P(x)$ has no zero greater than c;

(b) if $c < 0$ and the numbers in the bottom row of the synthetic division alternate in sign (with 0 considered positive or negative, as needed), then $P(x)$ has no zero less than c.

EXAMPLE 4
USING THE BOUNDEDNESS THEOREM

Approximate the real zeros of $P(x) = x^4 - 6x^3 + 8x^2 + 2x - 1$.

Use Descartes' rule of signs first. From Example 1, $P(x)$ has either three or one positive real zeros and one negative real zero. Next, check for rational zeros. The only possible rational zeros are ± 1.

$$\begin{array}{r|rrrrr} 1 & 1 & -6 & 8 & 2 & -1 \\ & & 1 & -5 & 3 & 5 \\ \hline & 1 & -5 & 3 & 5 & 4 \end{array} \quad \leftarrow P(1) = 4$$

$$\begin{array}{r|rrrrr} -1 & 1 & -6 & 8 & 2 & -1 \\ & & -1 & 7 & -15 & 13 \\ \hline & 1 & -7 & 15 & -13 & 12 \end{array} \quad \leftarrow P(-1) = 12$$

Neither 1 nor -1 is a zero.

We now use the two theorems in this section to search in some consistent way for the location of irrational real zeros. The leading coefficient of $P(x)$ is positive and the numbers in the last row of the second synthetic division above alternate in sign. Since

$-1 < 0$, by the boundedness theorem -1 is less than or equal to any zero of $P(x)$. By the synthetic division above, -1 is not a zero of $P(x)$. Also, $P(-1) = 12 > 0$. By substitution, or synthetic division, $P(0) = -1 < 0$. Thus the one negative real zero is between -1 and 0.

Try $c = -.5$. Divide $P(x)$ by $x + .5$.

$$\begin{array}{r|rrrrr} -.5 & 1 & -6 & 8 & 2 & -1 \\ & & -.5 & 3.25 & -5.625 & 1.8125 \\ \hline & 1 & -6.5 & 11.25 & -3.625 & .8125 \end{array}$$

Since $P(-.5) = .8125 > 0$ and $P(0) = -1 < 0$, there is a real zero between $-.5$ and 0.

Now try $c = -.4$.

$$\begin{array}{r|rrrrr} -.4 & 1 & -6 & 8 & 2 & -1 \\ & & -.4 & 2.56 & -4.224 & .8896 \\ \hline & 1 & -6.4 & 10.56 & -2.224 & -.1104 \end{array}$$

Since $P(-.5)$ is positive, but $P(-.4)$ is negative, there is a zero between $-.5$ and $-.4$. The value of $P(-.4)$ is closer to zero than $P(-.5)$, so it is probably safe to say that, to one decimal place of accuracy, $-.4$ is a real zero of $P(x)$. A more accurate result can be found, if desired, by continuing this process.

Find the remaining real zeros of $P(x)$ by using synthetic division to find $P(1)$, $P(2)$, $P(3)$, and so on, until a change in sign is noted. It is helpful to use the shortened form of synthetic division shown below. Only the last row of the synthetic division is shown for each division. The first row of the chart is used for each division and the work in the second row of the division is done mentally.

x					$P(x)$	
	1	−6	8	2	−1	
−1	1	−7	15	−13	12	
						← Zero between −1 and 0
0	1	−6	8	2	−1	
						← Zero between 0 and 1
1	1	−5	3	5	4	
2	1	−4	0	2	3	
						← Zero between 2 and 3
3	1	−3	−1	−1	−4	
						← Zero between 3 and 4
4	1	−2	0	2	7	

Since the polynomial is degree 4, there are no more than four zeros. Expand the table to find the real zeros to the nearest tenth. For example, for the zero between 0 and 1, work as follows. Start halfway between 0 and 1 with $x = .5$. Since $P(.5) > 0$ and $P(0) < 0$, try $x = .4$ next, and so on.

x					$P(x)$	
	1	−6	8	2	−1	
.5	1	−5.5	5.25	4.63	1.31	
.4	1	−5.6	5.76	4.30	.72	
.3	1	−5.7	6.29	3.89	.17	← Zero between .3 and .2
.2	1	−5.8	6.84	3.37	−.33	

The value $P(.3) = .17$ is closer to 0 than $P(.2) = -.33$, so to the nearest tenth, the zero is .3. Use synthetic division to verify that the remaining two zeros are approximately 2.4 and 3.7. ■

Many of today's scientific calculators have programming capabilities. It would be fairly simple to program the function of Example 4 into one of these calculators, and evaluate the polynomial for successive integer values. When it is found, for example, that a zero lies between −1 and 0, then this interval could be subdivided and a closer approximation then obtained. By repeating this process, zeros can be found to whatever accuracy is required.

4 If a polynomial of degree 3 or more cannot be factored, in order to graph the function without the use of a graphing calculator you will need to plot a large number of points. The theorems studied in this chapter are helpful in deciding which points to plot and in finding the ordered pairs.

EXAMPLE 5
GRAPHING A FUNCTION DEFINED BY A NON-FACTORABLE POLYNOMIAL

Graph the function defined by $P(x) = 8x^3 - 12x^2 + 2x + 1$.

By Descartes' rule of signs, $P(x)$ has two or zero positive real zeros and one negative real zero. Locate the real zeros using synthetic division. Start with $x = 0$ and then find $P(1)$, $P(2)$, $P(3)$, and so on until a row of the synthetic division is all positive, indicating there are no zeros greater than the x-value for that row. Then do the same thing in the negative direction; find $P(-1)$, $P(-2)$, and so on until alternating signs in the last row of the synthetic division indicate the number that is less than or equal to all the real zeros. As shown in the shortened form of synthetic division below, a sign change in the value of $P(x)$ indicates a zero.

x				$P(x)$	*Ordered Pair*	
	8	−12	2	1		
2	8	4	10	21	(2, 21)	
						← Zero
1	8	−4	−2	−1	(1, −1)	
						← Zero
0	8	−12	2	1	(0, 1)	
						← Zero
−1	8	−20	22	−21	(−1, −21)	

All three real zeros have been located. As expected, two are positive and one is negative. Since the numbers in the row for $x = 2$ are all positive and $2 > 0$, 2 is greater than any real zero of $P(x)$. Also, -1 is less than any real zero of $P(x)$.

By the intermediate value theorem, there is a zero between 0 and 1 and between -1 and 0, as well as between 1 and 2. The polynomial function is graphed by plotting the points from the chart and then drawing a continuous curve through them, as shown in Figure 9.16. ■

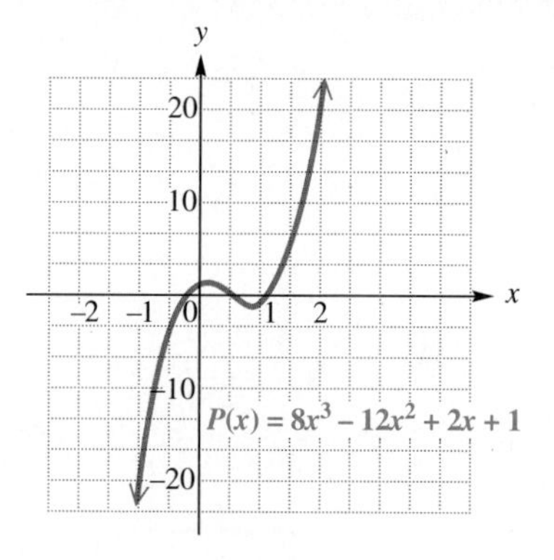

FIGURE 9.16

EXAMPLE 6
GRAPHING A FUNCTION DEFINED BY A NON-FACTORABLE POLYNOMIAL

Graph $P(x) = 3x^4 - 14x^3 + 54x - 3$.

Use Descartes' rule of signs to see that there are three positive real zeros or one positive real zero and one negative real zero. The points to plot are found using synthetic division to make a table like the one shown below. Start with $x = 0$ and work up through the positive integers until a row with all nonnegative numbers is found. Then work down through the negative integers until a row with alternating signs is found.

x					$P(x)$	*Ordered Pair*
	3	−14	0	54	−3	
5	3	1	5	79	392	(5, 392)
4	3	−2	−8	22	85	(4, 85)
3	3	−5	−15	9	24	(3, 24)
2	3	−8	−16	22	41	(2, 41)
1	3	−11	−11	43	40	(1, 40)
0	3	−14	0	54	−3	(0, −3)
−1	3	−17	17	37	−40	(−1, −40)
−2	3	−20	40	−26	49	(−2, 49)

Since the row in the chart for $x = 5$ contains all positive numbers, the polynomial has no zero greater than 5. Also, since the row for $x = -2$ has numbers that alternate in sign, there is no zero less than -2. By the changes in sign of $P(x)$, the polynomial has zeros between 0 and 1 and between -2 and -1. Plotting the points found above and drawing a continuous curve through them gives the graph in Figure 9.17. ■

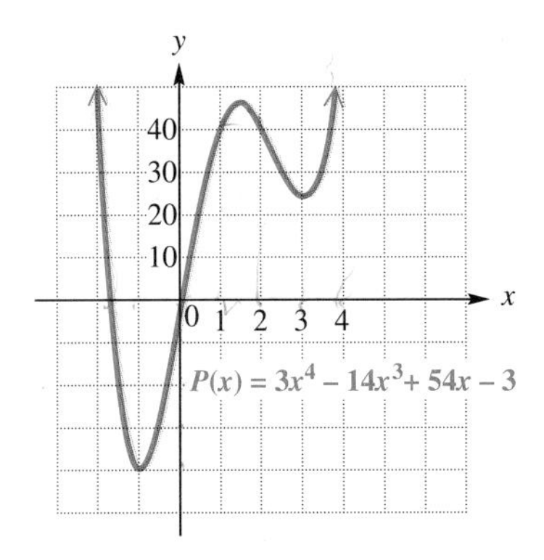

FIGURE 9.17

9.4 EXERCISES

A calculator will be helpful in working many of the exercises in this section.

Use Descartes' rule of signs to find the number of positive and negative real zeros possible for each of the following polynomials. See Examples 1 and 2.

1. $P(x) = 2x^3 - 4x^2 + 2x + 7$

2. $P(x) = x^3 + 2x^2 + x - 10$

3. $P(x) = 5x^4 + 3x^2 + 2x - 9$

4. $P(x) = 3x^4 + 2x^3 - 8x^2 - 10x - 1$

5. $P(x) = x^5 + 3x^4 - x^3 + 2x + 3$

6. $P(x) = 2x^5 - x^4 + x^3 - x^2 + x + 5$

Use the intermediate value theorem for polynomials to show that each of the following polynomials has a real zero between the numbers given. See Example 3.

7. $P(x) = 3x^2 - 2x - 6$; 1 and 2

8. $P(x) = x^3 + x^2 - 5x - 5$; 2 and 3

9. $P(x) = 2x^3 - 8x^2 + x + 16$; 2 and 2.5

10. $P(x) = 3x^3 + 7x^2 - 4$; 1/2 and 1

11. $P(x) = 2x^4 - 4x^2 + 3x - 6$; 2 and 1.5

12. $P(x) = x^4 - 4x^3 - x + 1$; 1 and .3

---

13. Suppose that a polynomial function $P(x)$ is defined in such a way that $P(2) = -4$ and $P(2.5) = 2$. What conclusion does the intermediate value theorem allow you to make? (See Objective 2.)

14. Suppose that a polynomial function $P(x)$ is defined in such a way that $P(3) = -4$ and $P(4) = -10$. Can we be certain that there is no zero between 3 and 4? Explain. (See Objective 2.)

Show that the real zeros of each of the following polynomial functions satisfy the given conditions.

15. $P(x) = x^4 - x^3 + 3x^2 - 8x + 8$; no real zero greater than 2

16. $P(x) = 2x^5 - x^4 + 2x^3 - 2x^2 + 4x - 4$; no real zero greater than 1

17. $P(x) = x^4 + x^3 - x^2 + 3$; no real zero less than -2

18. $P(x) = x^5 + 2x^3 - 2x^2 + 5x + 5$; no real zero less than -1

19. $P(x) = 3x^4 + 2x^3 - 4x^2 + x - 1$; no real zero greater than 1

20. $P(x) = 3x^4 + 2x^3 - 4x^2 + x - 1$; no real zero less than -2

21. $P(x) = x^5 - 3x^3 + x + 2$; no real zero greater than 2

22. $P(x) = x^5 - 3x^3 + x + 2$; no real zero less than -3

*For each of the following polynomials, **(a)** Find the number of positive and negative real zeros that are possible. See Examples 1 and 2. **(b)** Approximate each zero as a decimal to the nearest tenth. See Example 4.*

23. $P(x) = x^3 + 3x^2 - 2x - 6$

24. $P(x) = x^3 + x^2 - 5x - 5$

25. $P(x) = x^3 - 4x^2 - 5x + 14$

26. $P(x) = x^3 + 9x^2 + 34x + 13$

27. $P(x) = x^3 + 6x - 13$

28. $P(x) = 4x^3 - 3x^2 + 4x - 5$

29. $P(x) = 4x^4 - 8x^3 + 17x^2 - 2x - 14$

30. $P(x) = 3x^4 - 4x^3 - x^2 + 8x - 2$

31. $P(x) = -x^4 + 2x^3 + 3x^2 + 6$

32. $P(x) = -2x^4 - x^2 + x - 5$

Graph each of the following polynomial functions. See Examples 5 and 6.

33. $P(x) = x^3 - 7x - 6$

34. $P(x) = x^3 + x^2 - 4x - 4$

35. $P(x) = x^4 - 5x^2 + 6$

36. $P(x) = x^3 - 3x^2 - x + 3$

37. $P(x) = 6x^3 + 11x^2 - x - 6$

38. $P(x) = x^4 - 2x^2 - 8$

39. $P(x) = -x^3 + 6x^2 - x - 14$

40. $P(x) = 6x^4 - x^3 - 23x^2 - 4x + 12$

The following polynomials have zeros in the given intervals. Approximate these zeros to the nearest hundredth.

41. $P(x) = x^4 + x^3 - 6x^2 - 20x - 16$; $[3.2, 3.3]$ and $[-1.4, -1.1]$

42. $P(x) = x^4 - 2x^3 - 2x^2 - 18x + 5$; $[.2, .4]$ and $[3.7, 3.8]$

43. $P(x) = x^4 - 4x^3 - 20x^2 + 32x + 12$; $[-4, -3]$, $[-1, 0]$, $[1, 2]$, and $[6, 7]$

44. $P(x) = x^4 - 4x^3 - 44x^2 + 160x - 80$; $[-7, -6]$, $[0, 1]$, $[2, 3]$, and $[7, 8]$

---

45. A technique for measuring cardiac output depends on the concentration of a dye in the bloodstream after a known amount is injected into a vein near the heart. For a normal heart, the concentration of dye in the bloodstream at time x (in seconds) is given by the function

$$g(x) = -.006x^4 + .140x^3 - .053x^2 + 1.79x.$$

(a) Find $g(20)$.
(b) Graph $g(x)$.

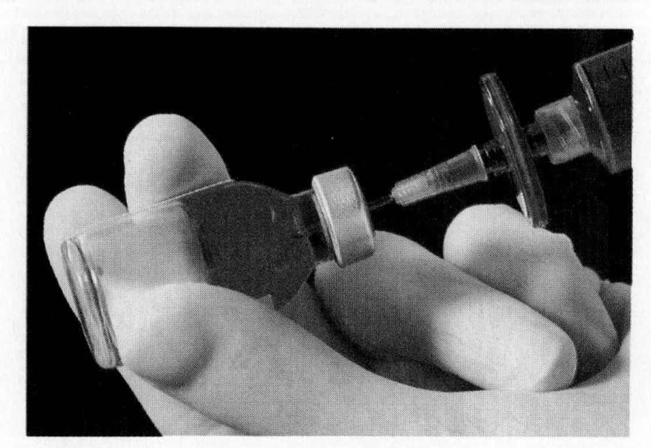

46. The polynomial function

$$A(x) = -.015x^3 + 1.058x$$

gives the approximate alcohol concentration (in tenths of a percent) in an average person's bloodstream x hours after drinking about 8 ounces of

100-proof whiskey. The function is approximately valid for x in the interval $[0, 8]$.

(a) Graph $A(x)$.

(b) Using the graph you drew for part (a), estimate the time of maximum alcohol concentration.

(c) In one state, a person is legally drunk if the blood alcohol concentration exceeds .08%. Use the graph from part (a) to estimate the period in which the average person is legally drunk.

PREVIEW EXERCISES

*For each of the following rational expressions **(a)** find the numbers that make the expression equal to zero, and **(b)** find the numbers that make the expression undefined. See Section 4.1.*

47. $\dfrac{p-3}{p+4}$

48. $\dfrac{t+5}{t-8}$

49. $\dfrac{x^2-9}{2x+5}$

50. $\dfrac{x^2-25}{3x-7}$

51. $\dfrac{2r^2+r-21}{3r^2-17r-6}$

52. $\dfrac{5x^2-14x-3}{2x^2-15x-8}$

9.5 RATIONAL FUNCTIONS

OBJECTIVES

1. DEFINE *RATIONAL FUNCTION.*
2. GRAPH RATIONAL FUNCTIONS USING REFLECTION AND TRANSLATION.
3. FIND ASYMPTOTES OF A RATIONAL FUNCTION.
4. GRAPH RATIONAL FUNCTIONS WHERE THE DEGREE OF THE NUMERATOR IS LESS THAN THE DEGREE OF THE DENOMINATOR.
5. GRAPH RATIONAL FUNCTIONS WHERE THE DEGREES OF THE NUMERATOR AND THE DENOMINATOR ARE EQUAL.
6. GRAPH RATIONAL FUNCTIONS WHERE THE DEGREE OF THE NUMERATOR IS GREATER THAN THE DEGREE OF THE DENOMINATOR.
7. GRAPH A RATIONAL FUNCTION THAT IS NOT IN LOWEST TERMS.

FOCUS ON PROBLEM SOLVING

Antique-car owners often enter their cars in a *concours d'elegance* in which a maximum of 100 points can be awarded to a particular car. Points are awarded for the general attractiveness of the car. The function

$$C(x) = \frac{10x}{49(101-x)}$$

expresses the cost, in thousands of dollars, of restoring a car so that it will win x points. Graph the function.

A function defined by a rational expression is called a rational function. Just as polynomial functions can be used to predict information, rational functions can be used in a similar manner. The problem above is Exercise 51 in the exercises for this section. After working through this section, you should be able to graph this function.

1 We begin with a definition of *rational function.*

DEFINITION OF RATIONAL FUNCTION

A function of the form

$$f(x) = \frac{p(x)}{q(x)},$$

where $p(x)$ and $q(x)$ are polynomial functions, $q(x) \neq 0$, is called a **rational function.**

Since any values of x such that $q(x) = 0$ are excluded from the domain of a rational function, this type of function often has a graph that has one or more breaks in it.

2 The simplest rational function with a variable denominator is

$$f(x) = \frac{1}{x}.$$

The domain of this function is the set of all real numbers except 0. The number 0 cannot be used as a value of x, but for graphing it is helpful to find the values of $f(x)$ for some values of x close to 0. The following table shows what happens to $f(x)$ as x gets closer and closer to 0 from either side.

x approaches 0.

x	-1	$-.1$	$-.01$	$-.001$	$.001$	$.01$	$.1$	1
$f(x)$	-1	-10	-100	-1000	1000	100	10	1

$|f(x)|$ gets larger and larger.

The table suggests that $|f(x)|$ gets larger and larger as x gets closer and closer to 0, which is written in symbols as

$$|f(x)| \to \infty \text{ as } x \to 0.$$

(The symbol $x \to 0$ means that x approaches as close as desired to 0, without necessarily ever being equal to 0.) Since x cannot equal 0, the graph of $f(x) = 1/x$ will never intersect the vertical line $x = 0$. This line is called a *vertical asymptote.*

On the other hand, as $|x|$ gets larger and larger, the values of $f(x) = 1/x$ get closer and closer to 0, as shown in the table below.

x	$-10{,}000$	-1000	-100	-10	10	100	1000	$10{,}000$
$f(x)$	$-.0001$	$-.001$	$-.01$	$-.1$	$.1$	$.01$	$.001$	$.0001$

SEE p433

Letting $|x|$ get larger and larger without bound (written $|x| \to \infty$) causes the graph of $y = f(x) = 1/x$ to move closer and closer to the horizontal line $y = 0$. This line is called a *horizontal asymptote.*

If the point (a, b) lies on the graph of $f(x) = 1/x$, then so does the point $(-a, -b)$. Therefore, the graph of f is symmetric with respect to the origin. Choosing some positive values of x and finding the corresponding values of $f(x)$ gives the first-quadrant part of the graph shown in Figure 9.18. The other part of the graph (in the third quadrant) can be found by symmetry.

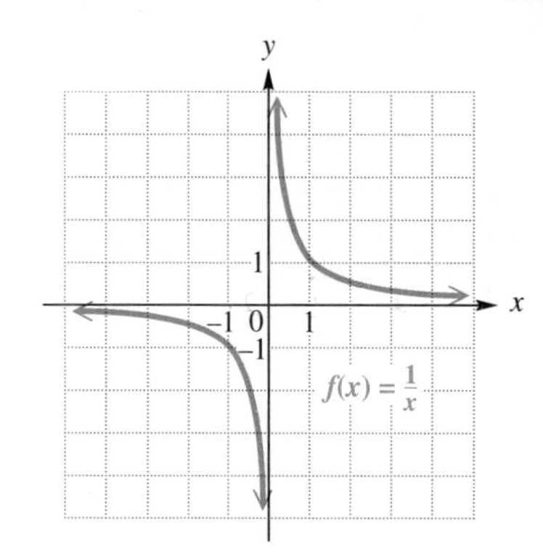

FIGURE 9.18

EXAMPLE 1 **GRAPHING A RATIONAL FUNCTION USING REFLECTION**

Graph $f(x) = -\dfrac{2}{x}$.

The expression on the right side of the equation can be rewritten so that

$$f(x) = -2 \cdot \frac{1}{x}.$$

Compared to $f(x) = 1/x$, the graph will be reflected about the x-axis (because of the negative sign), and each point will be twice as far from the x-axis. See Figure 9.19. ■

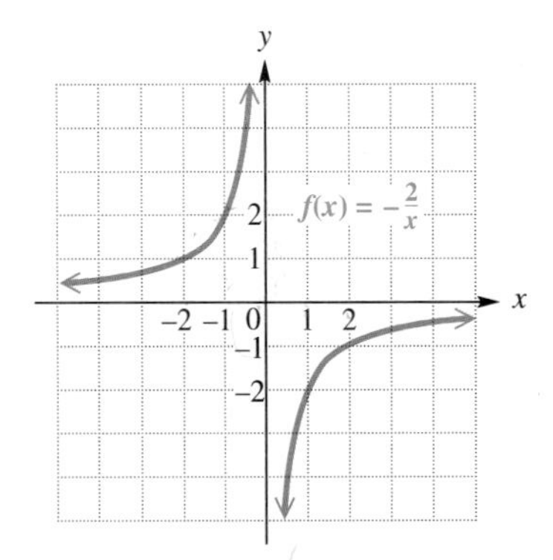

FIGURE 9.19

EXAMPLE 2 **GRAPHING A RATIONAL FUNCTION USING TRANSLATION**

Graph $f(x) = \dfrac{2}{1 + x}$.

The domain of this function is the set of all real numbers except -1. As shown in Figure 9.20, the graph is that of $f(x) = 1/x$, translated 1 unit to the left, with each y-value doubled. This can be seen by writing the expression as

$$f(x) = 2 \cdot \frac{1}{x - (-1)}. \quad ■$$

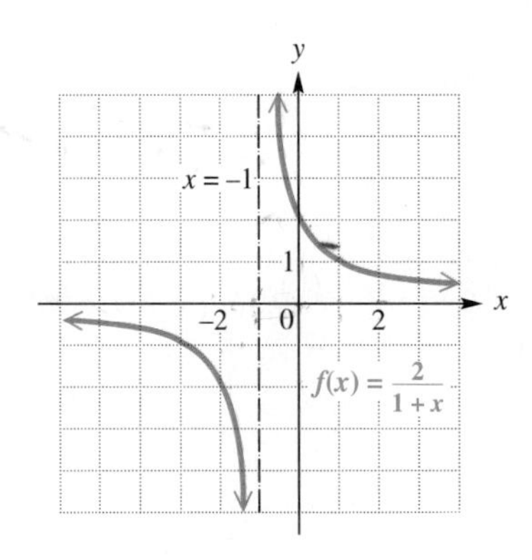

FIGURE 9.20

3 The examples above suggest the following definitions of vertical and horizontal asymptotes.

DEFINITIONS OF VERTICAL AND HORIZONTAL ASYMPTOTES

For the rational function with $f(x) = \dfrac{p(x)}{q(x)}$, written in lowest terms,

if $|f(x)| \to \infty$ as $x \to a$, then the line $x = a$ is a **vertical asymptote**; and if $f(x) \to a$ as $|x| \to \infty$, then the line $y = a$ is a **horizontal asymptote**.

Locating asymptotes is an important part of sketching the graphs of rational functions. Vertical asymptotes are found by setting the denominator of a rational function equal to zero and then solving. Horizontal asymptotes (and, in some cases, oblique asymptotes) are found by considering what happens to $f(x)$ as $|x| \to \infty$. The next example shows how to find asymptotes.

EXAMPLE 3 FINDING ASYMPTOTES OF RATIONAL FUNCTIONS

For each rational function, find all asymptotes.

(a) $f(x) = \dfrac{x + 1}{(2x - 1)(x + 3)}$

To find the vertical asymptotes, set the denominator equal to zero and solve.

$$(2x - 1)(x + 3) = 0$$

$$2x - 1 = 0 \quad \text{or} \quad x + 3 = 0 \qquad \text{Zero-factor property}$$

$$x = \frac{1}{2} \quad \text{or} \quad x = -3$$

The equations of the vertical asymptotes are $x = 1/2$ and $x = -3$.

To find the equation of the horizontal asymptote, we divide each term by the largest power of x in the expression. Begin by multiplying the factors in the denominator to get

$$f(x) = \frac{x + 1}{(2x - 1)(x + 3)} = \frac{x + 1}{2x^2 + 5x - 3}.$$

Now divide each term in the numerator and denominator by x^2, since 2 is the largest exponent on x. This gives

$$f(x) = \frac{\dfrac{x}{x^2} + \dfrac{1}{x^2}}{\dfrac{2x^2}{x^2} + \dfrac{5x}{x^2} - \dfrac{3}{x^2}} = \frac{\dfrac{1}{x} + \dfrac{1}{x^2}}{2 + \dfrac{5}{x} - \dfrac{3}{x^2}}.$$

As $|x|$ gets larger and larger, the quotients $1/x$, $1/x^2$, $5/x$, and $3/x^2$ all approach 0, and the value of $f(x)$ approaches

$$\frac{0 + 0}{2 + 0 - 0} = \frac{0}{2} = 0.$$

The line $y = 0$ (that is, the x-axis) is therefore the horizontal asymptote.

(b) $f(x) = \dfrac{2x + 1}{x - 3}$

Set the denominator equal to zero to find that the vertical asymptote has the equation $x = 3$. To find the horizontal asymptote, divide each term in the rational expression by x, since the greatest power of x in the expression is 1.

$$f(x) = \frac{2x + 1}{x - 3} = \frac{\frac{2x}{x} + \frac{1}{x}}{\frac{x}{x} - \frac{3}{x}} = \frac{2 + \frac{1}{x}}{1 - \frac{3}{x}}$$

As $|x|$ gets larger and larger, both $1/x$ and $3/x$ approach 0, and $f(x)$ approaches

$$\frac{2 + 0}{1 - 0} = \frac{2}{1} = 2,$$

so the line $y = 2$ is the horizontal asymptote.

(c) $f(x) = \dfrac{x^2 + 1}{x - 2}$

Setting the denominator equal to zero shows that the vertical asymptote has the equation $x = 2$. If we divide by the largest power of x as before (x^2 in this case), we see that there is no horizontal asymptote because

$$f(x) = \frac{\frac{x^2}{x^2} + \frac{1}{x^2}}{\frac{x}{x^2} - \frac{2}{x^2}} = \frac{1 + \frac{1}{x^2}}{\frac{1}{x} - \frac{2}{x^2}}$$

does not approach any real number as $|x| \to \infty$, since $1/0$ is undefined. This will happen whenever the degree of the numerator is greater than the degree of the denominator. In such cases divide the denominator into the numerator to write the expression in another form. Using synthetic division gives

$$\begin{array}{r|rrr} 2 & 1 & 0 & 1 \\ & & 2 & 4 \\ \hline & 1 & 2 & 5. \end{array}$$

The function can now be written as

$$f(x) = \frac{x^2 + 1}{x - 2} = x + 2 + \frac{5}{x - 2}.$$

For very large values of $|x|$, $5/(x - 2)$ is close to 0, and the graph approaches the line $y = x + 2$. This line is an **oblique asymptote** (neither vertical nor horizontal) for the function.

In general, if the degree of its numerator is exactly one more than the degree of its denominator, a rational function may have an oblique asymptote. The equation of this asymptote is found by dividing the numerator by the denominator and disregarding the remainder. ■

The results of Example 3 can be summarized as shown below.

DETERMINING ASYMPTOTES

In order to find asymptotes of a rational function written in lowest terms, use the following procedures.

1. **Vertical Asymptotes**

   Find any vertical asymptotes by setting the denominator equal to 0 and solving for x. If a is a zero of the denominator, then the line $x = a$ is a vertical asymptote.

2. **Other Asymptotes**

   Determine any other asymptotes. There are three possibilities:

   (a) If the numerator has lower degree than the denominator, there is a horizontal asymptote, $y = 0$ (the x-axis).

   (b) If the numerator and denominator have the same degree, and the function is of the form

$$f(x) = \frac{a_n x^n + \ldots + a_0}{b_n x^n + \ldots + b_0}, \quad \text{where } b_n \neq 0,$$

   dividing by x^n in the numerator and denominator produces the horizontal asymptote

$$y = \frac{a_n}{b_n}.$$

   (c) If the numerator is of degree exactly one more than the denominator, there may be an oblique asymptote. To find it, divide the numerator by the denominator and disregard any remainder. Set the rest of the quotient equal to y to get the equation of the asymptote.

NOTE The graph of a rational function may have more than one vertical asymptote, or it may have none at all. The graph cannot intersect any vertical asymptote. There can be only one other (non-vertical) asymptote, and the graph *may* intersect that asymptote. This will be seen in Example 6.

The following procedure can be used to graph rational functions reduced to lowest terms.

GRAPHING RATIONAL FUNCTIONS

Let $f(x) = \dfrac{p(x)}{q(x)}$ be a rational function written in lowest terms. To sketch its graph, follow the steps below.

1. Find any vertical asymptotes.
2. Find any horizontal or oblique asymptote.
3. Find the y-intercept by evaluating $f(0)$.
4. Find the x-intercepts, if any, by solving $f(x) = 0$. (These will be the zeros of the numerator, $p(x)$.)
5. Determine whether the graph will intersect its non-vertical asymptote by solving $f(x) = k$, where k is the y-value of the non-vertical asymptote.
6. Plot a few selected points, as necessary. Choose an x-value in each interval of the domain as determined by the vertical asymptotes and x-intercepts.
7. Complete the sketch.

4 The next example shows how the above guidelines can be used to graph a rational function.

EXAMPLE 4
GRAPHING A RATIONAL FUNCTION; DEGREE OF NUMERATOR < DEGREE OF DENOMINATOR

Graph $f(x) = \dfrac{x+1}{(2x-1)(x+3)}$.

Step 1 As shown in Example 3(a), the vertical asymptotes have equations $x = 1/2$ and $x = -3$.

Step 2 Again, as shown in Example 3(a), the horizontal asymptote is the x-axis.

Step 3 Since $f(0) = \dfrac{0+1}{(2(0)-1)((0)+3)} = -\dfrac{1}{3}$, the y-intercept is $(0, -1/3)$.

Step 4 The x-intercept is found by solving $f(x) = 0$.

$$\frac{x+1}{(2x-1)(x+3)} = 0$$

$$x + 1 = 0 \qquad \text{Multiply by } (2x-1)(x+3).$$

$$x = -1$$

The x-intercept is $(-1, 0)$.

Step 5 To determine whether the graph intersects its horizontal asymptote, solve

$$f(x) = 0.$$

↑ y-value of horizontal asymptote

Since the horizontal asymptote is the x-axis, the solution of this equation was found in Step 4. The graph intersects its horizontal asymptote at $(-1, 0)$.

Step 6 Plot a point in each of the intervals determined by the x-intercepts and vertical asymptotes, $(-\infty, -3)$, $(-3, -1)$, $(-1, 1/2)$, and $(1/2, \infty)$ to get an idea of how the graph behaves in each region.

Step 7 Complete the sketch. Keep in mind that the graph approaches its asymptotes as the points on the graph become farther away from the origin. The graph is shown in Figure 9.21. ■

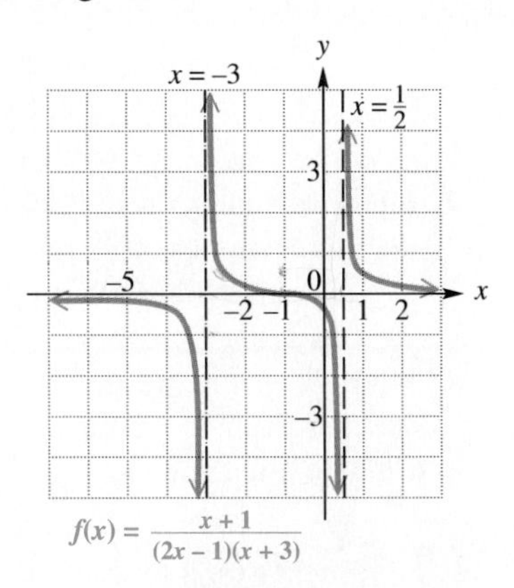

FIGURE 9.21

5 In the remaining examples, we will not specifically number the steps.

EXAMPLE 5
GRAPHING A RATIONAL FUNCTION; DEGREE OF NUMERATOR = DEGREE OF DENOMINATOR

Graph $f(x) = \dfrac{2x + 1}{x - 3}$.

As shown in Example 3(b), the equation of the vertical asymptote is $x = 3$ and the equation of the horizontal asymptote is $y = 2$. Since $f(0) = -1/3$, the y-intercept is $(0, -1/3)$. The solution of $f(x) = 0$ is $-1/2$, so the only x-intercept is $(-1/2, 0)$. The graph does not intersect its horizontal asymptote, since $f(x) = 2$ has no solution. (Verify this.) The points $(-4, 1)$ and $(6, 13/3)$ are on the graph and can be used to complete the sketch, as shown in Figure 9.22. ■

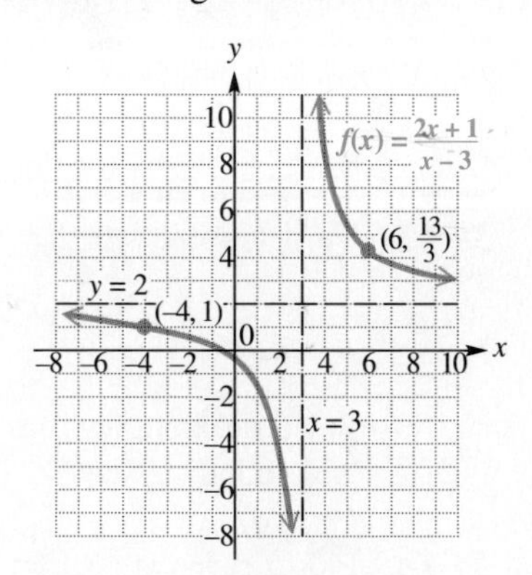

FIGURE 9.22

EXAMPLE 6
GRAPHING A RATIONAL FUNCTION; DEGREE OF NUMERATOR = DEGREE OF DENOMINATOR

Graph $f(x) = \dfrac{3(x+1)(x-2)}{(x+4)^2}$.

The only vertical asymptote is the line $x = -4$. To find any horizontal asymptotes, multiply the factors in the numerator and denominator.

$$f(x) = \frac{3x^2 - 3x - 6}{x^2 + 8x + 16}$$

As explained in the guidelines above, the equation of the horizontal asymptote can be shown to be

$$y = \frac{3}{1} \quad \begin{matrix} \leftarrow \text{Leading coefficient of numerator} \\ \leftarrow \text{Leading coefficient of denominator} \end{matrix}$$

or $y = 3$. The y-intercept is $(0, -3/8)$ and the x-intercepts are $(-1, 0)$ and $(2, 0)$. By setting $f(x) = 3$ and solving, we can find the point where the graph intersects the horizontal asymptote.

$$\begin{aligned} f(x) &= \frac{3x^2 - 3x - 6}{x^2 + 8x + 16} \\ 3 &= \frac{3x^2 - 3x - 6}{x^2 + 8x + 16} \\ 3x^2 - 3x - 6 &= 3x^2 + 24x + 48 && \text{Multiply by } x^2 + 8x + 16. \\ -3x - 6 &= 24x + 48 && \text{Subtract } 3x^2. \\ -27x &= 54 \\ x &= -2 \end{aligned}$$

The graph intersects its horizontal asymptote at $(-2, 3)$.

Some other points that lie on the graph are $(-10, 9)$, $(-3, 30)$, and $(5, 2/3)$. These can be used to complete the graph, shown in Figure 9.23. ■

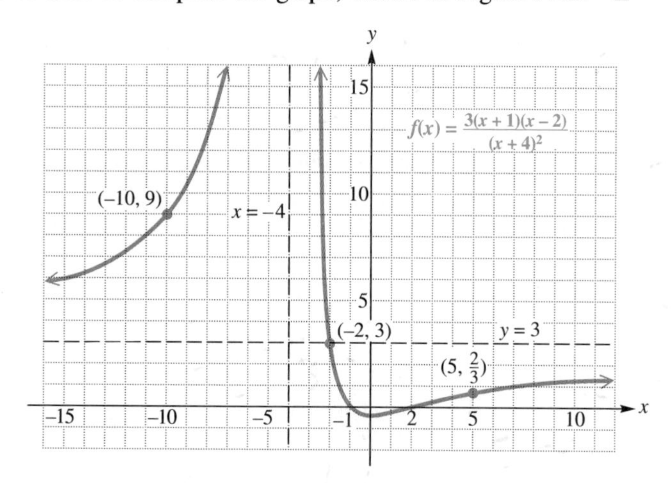

FIGURE 9.23

6 The next example discusses a rational function with the degree of its numerator greater than the degree of its denominator.

EXAMPLE 7
GRAPHING A RATIONAL FUNCTION; DEGREE OF NUMERATOR > DEGREE OF DENOMINATOR

Graph $f(x) = \dfrac{x^2 + 1}{x - 2}$.

As shown in Example 3(c), the vertical asymptote has the equation $x = 2$, and the graph has an oblique asymptote with the equation $y = x + 2$. The y-intercept is $(0, -1/2)$, and the graph has no x-intercepts, since the numerator, $x^2 + 1$, has no real zeros. It can be shown that the graph does not intersect its oblique asymptote. Using the intercepts, asymptotes, the points $(4, 17/2)$ and $(-1, -2/3)$, and the general behavior of the graph near its asymptotes, we obtain the graph shown in Figure 9.24. ■

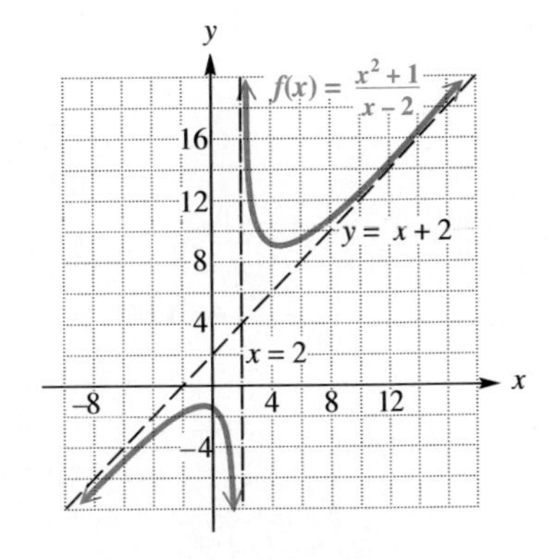

FIGURE 9.24

7 As mentioned earlier, a rational function must be in lowest terms before we can use the methods discussed in this section to determine the graph. The next example gives a typical rational function that is not in lowest terms.

EXAMPLE 8
GRAPHING A RATIONAL FUNCTION THAT IS NOT IN LOWEST TERMS

Graph $f(x) = \dfrac{x^2 - 4}{x - 2}$.

Start by noticing that the domain of this function cannot contain 2. The rational expression $(x^2 - 4)/(x - 2)$ can be reduced to lowest terms by factoring the numerator, and using the fundamental principle.

$$\frac{x^2 - 4}{x - 2} = \frac{(x + 2)(x - 2)}{x - 2} = x + 2 \quad (x \neq 2)$$

Therefore, the graph of this function will be the same as the graph of $y = x + 2$ (a straight line), with the exception of the point with x-value 2. A "hole" appears in the graph at (2, 4). See Figure 9.25. ■

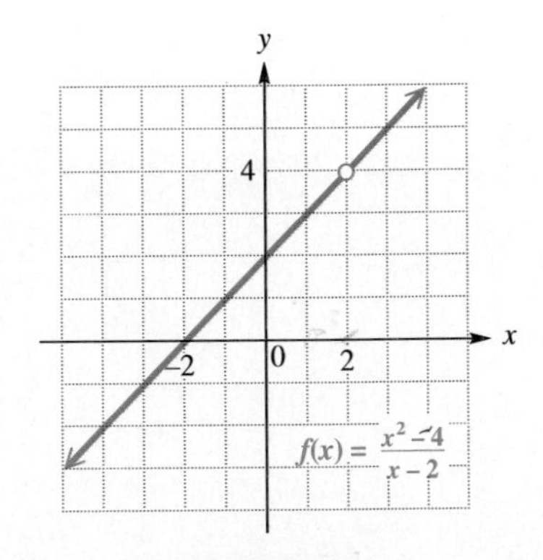

FIGURE 9.25

9.5 EXERCISES

Use reflections and translations to graph each of the following rational functions. See Examples 1 and 2.

1. $f(x) = \dfrac{2}{x}$

2. $f(x) = -\dfrac{3}{x}$

3. $f(x) = \dfrac{1}{x+2}$

4. $f(x) = \dfrac{1}{x-3}$

5. $f(x) = \dfrac{1}{x} + 1$

6. $f(x) = \dfrac{1}{x} - 2$

---

7. Sketch the following graphs and compare them with the graph of $f(x) = 1/x^2$.

(a) $f(x) = \dfrac{1}{(x-3)^2}$ **(b)** $f(x) = -\dfrac{2}{x^2}$

(c) $f(x) = \dfrac{-2}{(x-3)^2}$

8. Describe in your own words what is meant by an *asymptote* of a rational function. (See Objectives 2 and 3.)

Give the equations of any vertical, horizontal, or oblique asymptotes for the rational functions in Exercises 9–22. See Example 3.

9. $f(x) = \dfrac{2}{x-5}$

10. $f(x) = \dfrac{-1}{x+2}$

11. $f(x) = \dfrac{-8}{3x-7}$

12. $f(x) = \dfrac{5}{4x-9}$

13. $f(x) = \dfrac{2-x}{x+2}$

14. $f(x) = \dfrac{x-4}{5-x}$

15. $f(x) = \dfrac{3x-5}{2x+9}$

16. $f(x) = \dfrac{4x+3}{3x-7}$

17. $f(x) = \dfrac{2}{x^2-4x+3}$

18. $f(x) = \dfrac{-5}{x^2-3x-10}$

19. $f(x) = \dfrac{x^2-1}{x+3}$

20. $f(x) = \dfrac{x^2+4}{x-1}$

21. $f(x) = \dfrac{(x-3)(x+1)}{(x+2)(2x-5)}$

22. $f(x) = \dfrac{3(x+2)(x-4)}{(5x-1)(x-5)}$

---

23. Which one of the following rational functions does not have a vertical asymptote? (See Objective 3.)

(a) $f(x) = \dfrac{1}{x^2+2}$ **(b)** $f(x) = \dfrac{1}{x^2-2}$

(c) $f(x) = \dfrac{3}{x^2}$ **(d)** $f(x) = \dfrac{2x+1}{x-8}$

24. Which one of the following rational functions does not have a horizontal asymptote? (See Objectives 3 and 7.)

(a) $f(x) = \dfrac{2x-7}{x+3}$ **(b)** $f(x) = \dfrac{3x}{x^2-9}$

(c) $f(x) = \dfrac{x^2-9}{x+3}$ **(d)** $f(x) = \dfrac{x+5}{(x+2)(x-3)}$

Graph each of the following. See Examples 4–8.

25. $f(x) = \dfrac{4}{5+3x}$

26. $f(x) = \dfrac{1}{(x-2)(x+4)}$

27. $f(x) = \dfrac{3}{(x+4)^2}$

28. $f(x) = \dfrac{3x}{(x+1)(x-2)}$

29. $f(x) = \dfrac{2x+1}{(x+2)(x+4)}$

30. $f(x) = \dfrac{5x}{x^2-1}$

31. $f(x) = \dfrac{-x}{x^2 - 4}$ **32.** $f(x) = \dfrac{3x}{x - 1}$ **33.** $f(x) = \dfrac{4x}{1 - 3x}$

34. $f(x) = \dfrac{x + 1}{x - 4}$ **35.** $f(x) = \dfrac{x - 5}{x + 3}$ **36.** $f(x) = \dfrac{x}{x^2 - 9}$

37. $f(x) = \dfrac{3x}{x^2 - 16}$ **38.** $f(x) = \dfrac{x^2 - 9}{x^2 - 4}$ **39.** $f(x) = \dfrac{x^2 - 1}{x^2 - 4}$

40. $f(x) = \dfrac{2x^2 + 3}{x - 4}$ **41.** $f(x) = \dfrac{x^2 + 1}{x + 3}$ **42.** $f(x) = \dfrac{x^2 + 2x}{2x - 1}$

43. $f(x) = \dfrac{x^2 - x}{x + 2}$ **44.** $f(x) = \dfrac{(x - 3)(x + 1)}{(x - 1)^2}$ **45.** $f(x) = \dfrac{x(x - 2)}{(x + 3)^2}$

46. $f(x) = \dfrac{(x - 5)(x - 2)}{x^2 + 9}$ **47.** $f(x) = \dfrac{1}{x^2 + 1}$ **48.** $f(x) = \dfrac{-9}{x^2 + 9}$ **49.** $f(x) = \dfrac{x^2 - 16}{x + 4}$

---

50. The figures below show the four ways that a rational function can approach the vertical line $x = 2$ as an asymptote. Identify the graph of each of the following rational functions.

(a) $f(x) = \dfrac{1}{(x - 2)^2}$ **(b)** $f(x) = \dfrac{1}{x - 2}$

(c) $f(x) = \dfrac{-1}{x - 2}$ **(d)** $f(x) = \dfrac{-1}{(x - 2)^2}$

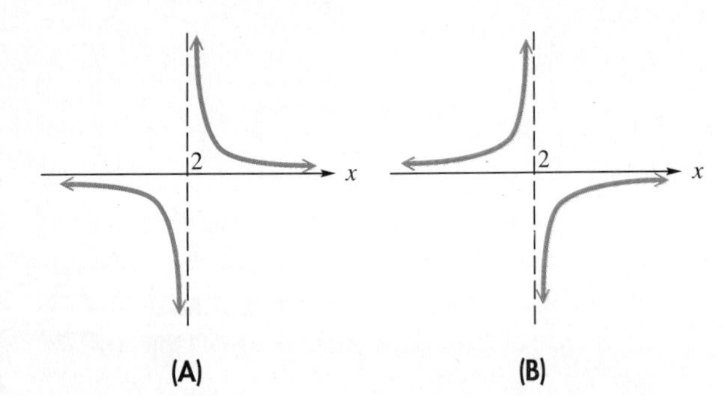

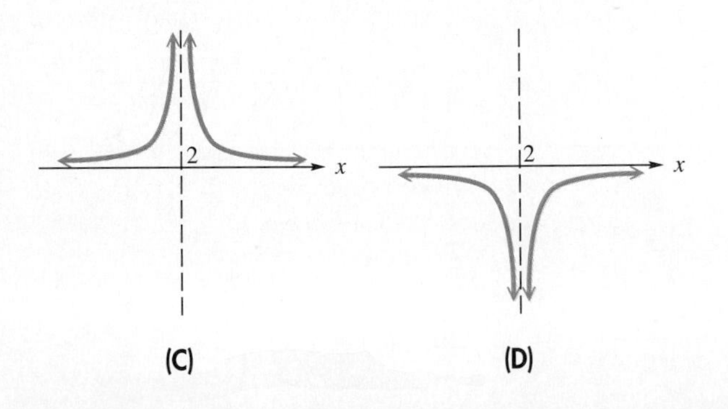

Work the following problems.

51. Antique-car owners often enter their cars in a *concours d'elegance* in which a maximum of 100 points can be awarded to a particular car. Points are awarded for the general attractiveness of the car. The function

$$C(x) = \frac{10x}{49(101 - x)}$$

expresses the cost, in thousands of dollars, of restoring a car so that it will win x points. Graph the function.

52. Suppose the average cost per unit in thousands of dollars, $C(x)$, to produce x units of margarine is given by

$$C(x) = \frac{500}{x + 30}.$$

(a) Find $C(10)$, $C(20)$, $C(50)$, $C(75)$, and $C(100)$.

(b) Would a more reasonable domain for C be $(0, \infty)$ or $[0, \infty)$? Why?
(c) Graph $C(x)$.

53. In situations involving environmental pollution, a cost-benefit model expresses cost as a function of the percentage of pollutant removed from the environment. Suppose a cost-benefit model is expressed as

$$y = \frac{6.7x}{100 - x},$$

where y is the cost in thousands of dollars of removing x percent of a certain pollutant.
(a) Graph the function.
(b) Is it possible, according to this function, to remove all of the pollutant?

54. In a recent year, the cost per ton, y, to build an oil tanker of x thousand deadweight tons was approximated by

$$y = \frac{110{,}000}{x + 225}.$$

(a) Find y for $x = 25$, $x = 50$, $x = 100$, $x = 200$, $x = 300$, and $x = 400$.
(b) Graph the function.

In recent years the economist Arthur Laffer has been a center of controversy because of his ***Laffer curve,*** *an idealized version of which is shown here.*

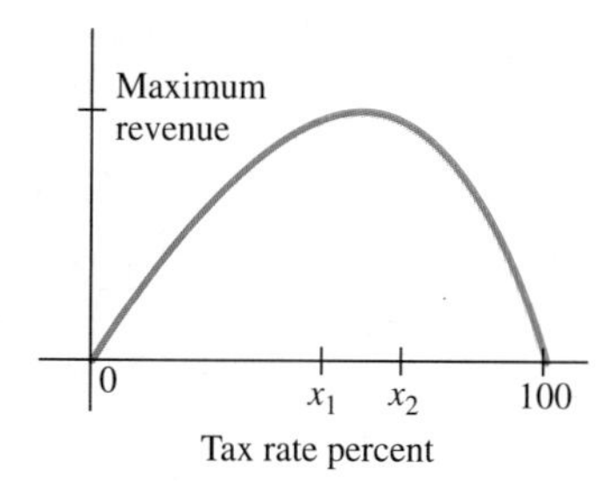

According to this curve, increasing a tax rate, say from x_1 percent to x_2 percent on the graph above, can actually lead to a decrease in government revenue. All economists agree on the endpoints, 0 revenue at tax rates of both 0% and 100%, but there is much disagreement on the location of the rate x_1 that produces maximum revenue.

55. Suppose an economist studying the Laffer curve produces the rational function defined by

$$y = \frac{80x - 8000}{x - 110},$$

with y giving government revenue in tens of millions of dollars for a tax rate of x percent, with the function valid for $55 \le x \le 100$. Find the revenue for the following tax rates.
(a) 55% **(b)** 60% **(c)** 70%
(d) 90% **(e)** 100% **(f)** Graph the function.

56. Suppose an economist studies a different tax, this time producing

$$y = \frac{60x - 6000}{x - 120}$$

where y is government revenue in millions of dollars from a tax rate of x percent, with the function valid for $50 \le x \le 100$. Find the revenue from the following tax rates.
(a) 50%
(b) 60%
(c) 80%
(d) 100%
(e) Graph the function.

57. Let $f(x) = p(x)/q(x)$ be a rational function reduced to lowest terms. Suppose that the degree of $p(x)$ is m and the degree of $q(x)$ is n. Write an explanation of how you would determine the nonvertical asymptote in each of the following situations. (See Objective 3.)
(a) $m < n$ **(b)** $m = n$ **(c)** $m > n$

58. Suppose that a friend tells you that the graph of

$$f(x) = \frac{x^2 - 25}{x + 5}$$

has a vertical asymptote with equation $x = -5$. Is this correct? If not, describe the behavior of the graph at $x = -5$. (See Objective 7.)

PREVIEW EXERCISES

Evaluate each of the following. See Sections 3.1 and 5.1.

59. $25^{1/2}$ **60.** 32^0 **61.** $\left(\frac{3}{4}\right)^3$ **62.** 5^{-2}

63. 7^0 **64.** $64^{5/6}$ **65.** $\left(\frac{8}{125}\right)^{4/3}$ **66.** $\left(\frac{27}{64}\right)^{-2/3}$

CHAPTER 9 GLOSSARY

KEY TERMS

9.1 polynomial function of degree n A polynomial function of degree n is a function defined by

$$P(x) = a_n x^n + a_{n-1}x^{n-1} + \ldots + a_1 x + a_0,$$

for complex numbers $a_n, a_{n-1}, \ldots, a_1, a_0$, where $a_n \neq 0$.

leading coefficient The number a_n in the definition of polynomial function is the leading coefficient of the polynomial.

zero polynomial If all the coefficients of a polynomial are 0, the polynomial equals 0 and is called the zero polynomial.

zeros of a polynomial The values of x that satisfy $P(x) = 0$ are called the zeros of $P(x)$.

even function A function $y = f(x)$ is an even function if $f(-x) = f(x)$ for all x in the domain of f.

odd function A function $y = f(x)$ is an odd function if $f(-x) = -f(x)$ for all x in the domain of f.

9.2 multiplicity The polynomial $P(x) = (x - 3)^2(x + 4)^5$, for example, has a zero of 3 with multiplicity 2 and a zero of -4 with multiplicity 5.

9.5 rational function A function of the form $f(x) = p(x)/q(x)$, where $p(x)$ and $q(x)$ are polynomial functions and $q(x) \neq 0$, is called a rational function.

vertical asymptote If $f(x)$ is a rational function and if $|f(x)| \to \infty$ as $x \to a$, then the line $x = a$ is a vertical asymptote.

horizontal asymptote If $f(x)$ is a rational function and if $f(x) \to a$ as $|x| \to \infty$, then the line $y = a$ is a horizontal asymptote.

oblique asymptote If $f(x)$ is a rational function and if $f(x) \to mx + b$ as $|x| \to \infty$, then the line $y = mx + b$ is an oblique asymptote.

NEW SYMBOLS

$\bar{c}$ the conjugate of the complex number c

$x \to a$ x approaches a

CHAPTER 9 QUICK REVIEW

SECTION	CONCEPTS	EXAMPLES
9.1 POLYNOMIAL FUNCTIONS AND THEIR GRAPHS	The graph of the function $$P(x) = a(x - h)^n + k$$ can be found by considering the effects of the constants a, h, and k on the graph of $$y = x^n.$$	Sketch the graph of $$P(x) = -(x + 2)^4 + 1.$$ 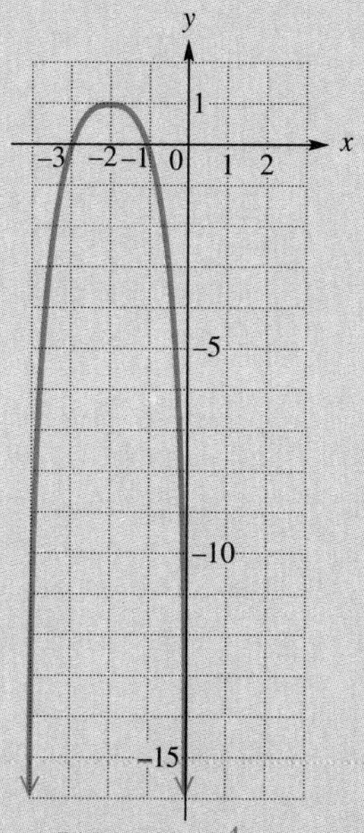 $P(x) = -(x + 2)^4 + 1$
	To graph a polynomial function in general, find x-intercepts and y-intercepts. Choose a value in each region determined by the x-intercepts to determine whether the graph is above or below the x-axis.	Sketch the graph of $$P(x) = (x + 2)(x - 1)(x + 3).$$ 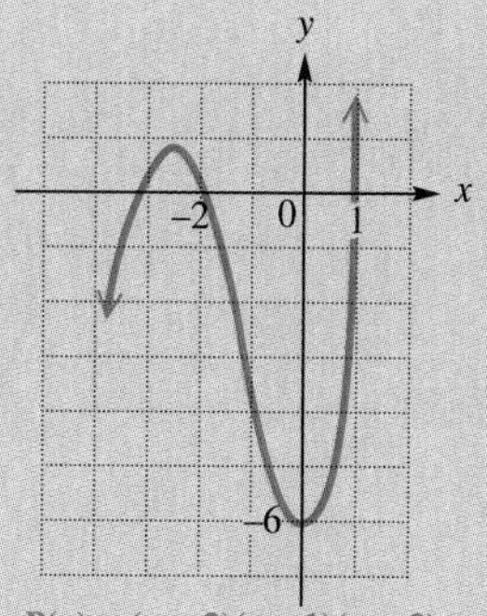 $P(x) = (x + 2)(x - 1)(x + 3)$

SECTION	CONCEPTS	EXAMPLES
9.2 ZEROS OF POLYNOMIALS	**Factor Theorem** The polynomial $x - k$ is a factor of the polynomial $P(x)$ if and only if $P(k) = 0$.	For the polynomial function $$P(x) = x^3 + x + 2,$$ $P(-1) = 0$. Therefore, $x - (-1)$, or $x + 1$, is a factor of $P(x)$. Also, since $x - 1$ is a factor of $$Q(x) = x^3 - 1,$$ $$Q(1) = 0.$$
	Fundamental Theorem of Algebra Every polynomial of degree 1 or more has at least one complex zero. A polynomial of degree n has at most n distinct zeros.	$P(x) = x^3 + x + 2$ has at least 1 and at most 3 zeros.
	Conjugate Zeros Theorem If $P(x)$ is a polynomial having only real coefficients and if $a + bi$ is a zero of $P(x)$, then the conjugate $a - bi$ is also a zero of $P(x)$.	Since $1 + 2i$ is a zero of $$P(x) = x^3 - 5x^2 + 11x - 15,$$ its conjugate $1 - 2i$ is a zero as well.
9.3 RATIONAL ZEROS OF POLYNOMIAL FUNCTIONS	**Rational Zeros Theorem** Let $P(x) = a_n x^n + a_{n-1}x^{n-1} + \ldots + a_1 x + a_0$, where $a_n \neq 0$, be a polynomial with integer coefficients. If p/q is a rational number written in lowest terms and if p/q is a zero of $P(x)$, then p is a factor of the constant term a_0 and q is a factor of the leading coefficient a_n.	The only rational numbers that can possibly be zeros of $$P(x) = 2x^3 - 9x^2 - 4x - 5$$ are ± 1, $\pm \frac{1}{2}$, ± 5, and $\pm \frac{5}{2}$. By synthetic division, it can be shown that the only rational zero of $P(x)$ is 5. $5\overline{)2 \quad -9 \quad -4 \quad -5}$ $\quad\quad 10 \quad 5 \quad 5$ $\quad 2 \quad 1 \quad 1 \quad 0$ ← P(5)

SECTION	CONCEPTS	EXAMPLES
9.4 REAL ZEROS OF POLYNOMIAL FUNCTIONS	**Descartes' Rule of Signs** Let $P(x)$ be a polynomial with *real coefficients* and terms in descending powers of x. **(a)** The number of positive real zeros of $P(x)$ either equals the number of variations in sign occurring in the coefficients of $P(x)$, or is less than the number of variations by a positive even integer. **(b)** The number of negative real zeros of $P(x)$ either equals the number of variations in sign of $P(-x)$, or is less than the number of variations by a positive even integer.	Given that $P(x) = x^3 - 8x^2 + 4,$ since there are two sign changes, there are either 2 or 0 positive real zeros. Since $P(-x) = -x^3 - 8x^2 + 4$ has one sign change, there is 1 negative real zero.
	Intermediate Value Theorem for Polynomials If $P(x)$ defines a polynomial function with *only real coefficients,* and if for real numbers a and b, $P(a)$ and $P(b)$ are opposite in sign, then there exists at least one real zero between a and b.	For the polynomial function $P(x) = -x^4 + 2x^3 + 3x^2 + 6,$ $P(3.1) = 2.0599$ and $P(3.2) = -2.6016.$ Since $P(3.1) > 0$ and $P(3.2) < 0$, there exists at least one real zero between 3.1 and 3.2.
	Boundedness Theorem Let $P(x)$ be a polynomial with *real* coefficients and with a *positive* leading coefficient. If $P(x)$ is divided synthetically by $x - c$, and **(a)** if $c > 0$ and all numbers in the bottom row of the synthetic division are nonnegative, then $P(x)$ has no zero greater than c; **(b)** if $c < 0$ and the numbers in the bottom row of the synthetic division alternate in sign (with 0 considered positive or negative, as needed), then $P(x)$ has no zero less than c.	$P(x) = x^3 - x^2 - 8x + 12$ has no zero greater than 4 and no zero less than -4. $\begin{array}{r\|rrrr} 4 & 1 & -1 & -8 & 12 \\ & & 4 & 12 & 16 \\ \hline & 1 & 3 & 4 & 28 \end{array}$ ← All positive $\begin{array}{r\|rrrr} -4 & 1 & -1 & -8 & 12 \\ & & -4 & 20 & -48 \\ \hline & 1 & -5 & 12 & -36 \end{array}$ ← Alternating signs

SECTION	CONCEPTS	EXAMPLES
9.5 RATIONAL FUNCTIONS	To graph a rational function in lowest terms, find asymptotes and intercepts. Determine whether the graph intersects the nonvertical asymptote. Plot a few points, as necessary, to complete the sketch.	Graph the rational function $$f(x) = \frac{x^2 - 1}{(x + 3)(x - 2)}.$$ 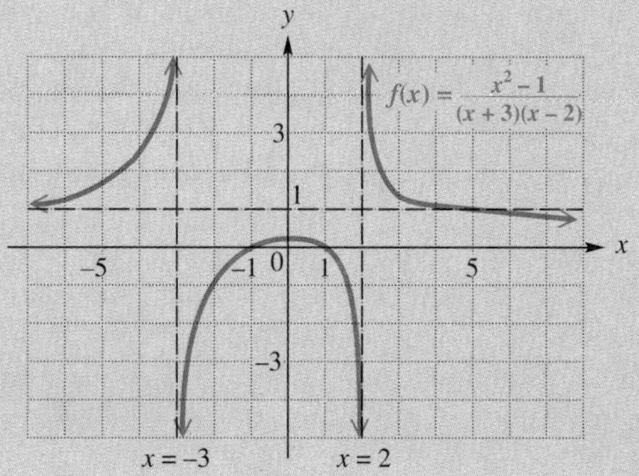
	If a rational function is not reduced to lowest terms, there may be a ''hole'' in the graph.	Graph the rational function $$f(x) = \frac{x^2 - 1}{x + 1}.$$ 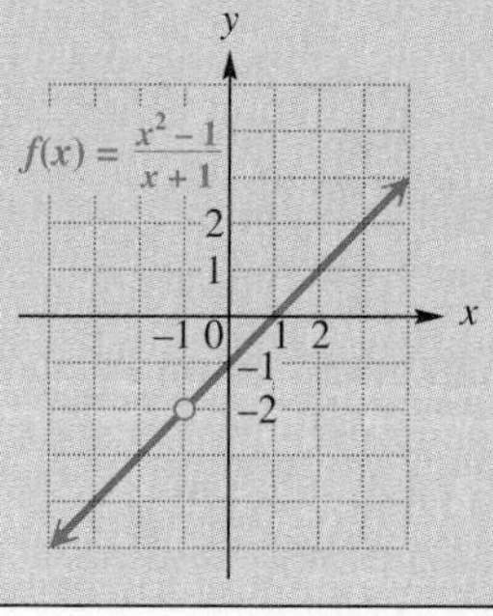

CHAPTER 9 REVIEW EXERCISES

[9.1] *Graph each of the following polynomial functions.*

1. $P(x) = x^3 + 5$

2. $P(x) = 1 - x^4$

3. $P(x) = x^2(2x + 1)(x - 2)$

4. $P(x) = (4x - 3)(3x + 2)(x - 1)$

5. $P(x) = 2x^3 + 13x^2 + 15x$

6. $P(x) = x(x - 1)(x + 2)(x - 3)$

7. $P(x) = x^3 + 2x^2 - 16x - 32$

8. $P(x) = -x^3 - 3x^2 + 9x + 27$

9. For the polynomial function $P(x) = x^3 - 3x^2 - 7x + 12$, $P(4) = 0$. Therefore, we can say that 4 is a ________ of the function, 4 is a ________ of the equation $x^3 - 3x^2 - 7x + 12 = 0$, and that $(4, 0)$ is a(n) ________ of the graph of the function.

10. Which one of the following is not a polynomial function?
(a) $f(x) = x^2$ **(b)** $f(x) = 2x + 5$
(c) $f(x) = 1/x$ **(d)** $f(x) = x^{100}$

[9.2] *In Exercises 11–14, find a polynomial of lowest degree having the following zeros.*

11. $-1, 4, 7$

12. $8, 2, 3$

13. $-\sqrt{7}, \sqrt{7}, 2, -1$

14. $1 + \sqrt{5}, 1 - \sqrt{5}, -4, 1$

---

15. Is -1 a zero of $P(x) = 2x^4 + x^3 - 4x^2 + 3x + 1$?

16. Is -2 a zero of $P(x) = 2x^4 + x^3 - 4x^2 + 3x + 1$?

17. Is $x + 1$ a factor of $P(x) = x^3 + 2x^2 + 3x - 1$?

18. Is $x + 1$ a factor of $P(x) = 2x^3 - x^2 + x + 4$?

19. Find a polynomial of degree 3 with -2, 1, and 4 as zeros, and $P(2) = 16$.

20. Find a polynomial of degree 4 with real coefficients having 1, -1, and $3i$ as zeros, and $P(2) = 39$.

21. Find a lowest-degree polynomial with real coefficients having zeros 2, -2, and $-i$.

22. Find a lowest-degree polynomial with real coefficients having zeros 2, -3, and $5i$.

23. Find a polynomial of lowest degree with real coefficients having -3 and $1 - i$ as zeros.

24. Find all zeros of $P(x) = x^4 - 3x^3 - 8x^2 + 22x - 24$, given that $1 - i$ is a zero, and factor $P(x)$.

25. Find all zeros of $P(x) = x^4 - 6x^3 + 14x^2 - 24x + 40$, given that $3 + i$ is a zero, and factor $P(x)$.

26. Find all zeros of $P(x) = x^4 + x^3 - x^2 + x - 2$, given that 1 and -2 are zeros, and factor $P(x)$.

[9.3] *Find all rational zeros of each polynomial in Exercises 27–30.*

27. $P(x) = 2x^3 - 9x^2 - 6x + 5$

28. $P(x) = 3x^3 - 10x^2 - 27x + 10$

29. $P(x) = x^3 - \dfrac{17}{6}x^2 - \dfrac{13}{3}x - \dfrac{4}{3}$

30. $P(x) = 8x^4 - 14x^3 - 29x^2 - 4x + 3$

For the polynomial functions in Exercises 31 and 32, find all rational zeros and factor the polynomial. Then graph the function.

31. $P(x) = x^3 - 5x^2 - 13x - 7$

32. $P(x) = 3x^3 - 2x^2 - 19x - 6$

33. Use a polynomial to show that $\sqrt{11}$ is irrational.

34. Show that $P(x) = x^3 - 9x^2 + 2x - 5$ has no rational zeros.

[9.4] *Show that the following polynomials have real zeros satisfying the given conditions.*

35. $P(x) = 3x^3 - 8x^2 + x + 2$, zero in $[-1, 0]$ and $[2, 3]$

36. $P(x) = 4x^3 - 37x^2 + 50x + 60$, zero in $[2, 3]$ and $[7, 8]$

37. $P(x) = x^3 + 2x^2 - 22x - 8$, zero in $[-1, 0]$ and $[-6, -5]$

38. $P(x) = 2x^4 - x^3 - 21x^2 + 51x - 36$, has no real zero greater than 4

39. $P(x) = 6x^4 + 13x^3 - 11x^2 - 3x + 5$, has no zero greater than 1 or less than -3

In Exercises 40 and 41, approximate the real zeros of each polynomial as decimals to the nearest tenth.

40. $P(x) = 2x^3 - 11x^2 - 2x + 2$

41. $P(x) = x^4 - 4x^3 - 5x^2 + 14x - 15$

42. **(a)** Find the number of positive and negative zeros of $P(x) = x^3 + 3x^2 - 4x - 2$.
(b) Show that $P(x)$ has a zero between -4 and -3. Approximate this zero to the nearest tenth.

Graph each of the following polynomial functions.

43. $P(x) = 2x^3 - 11x^2 - 2x + 2$ (See Exercise 40.)

44. $P(x) = x^4 - 4x^3 - 5x^2 + 14x - 15$ (See Exercise 41.)

45. $P(x) = x^3 + 3x^2 - 4x - 2$ (See Exercise 42.)

46. $P(x) = 2x^4 - 3x^3 + 4x^2 + 5x - 1$

[9.5] *Graph each of the following rational functions.*

47. $f(x) = \frac{8}{x}$

48. $f(x) = \frac{2}{3x - 1}$

49. $f(x) = \frac{4x - 2}{3x + 1}$

50. $f(x) = \frac{6x}{(x - 1)(x + 2)}$

51. $f(x) = \frac{2x}{x^2 - 1}$

52. $f(x) = \frac{x^2 + 4}{x + 2}$

53. $f(x) = \frac{x^2 - 1}{x}$

54. $f(x) = \frac{x^2 + 6x + 5}{x - 3}$

55. $f(x) = \frac{4x^2 - 9}{2x + 3}$

56. Under what conditions will the graph of a rational function, reduced to lowest terms, have an oblique asymptote?

MIXED REVIEW EXERCISES

Graph each of the following functions.

57. $f(x) = -\frac{1}{x^3}$

58. $f(x) = (x - 2)^4$

59. $f(x) = \frac{3 - 4x}{2x + 1}$

60. $f(x) = (x + 3)^2(x - 1)$

61. $f(x) = 2x^5 - 3x^4 + x^2 - 2$

62. $f(x) = \frac{(x + 4)(2x + 5)}{x - 1}$

63. $f(x) = \frac{x^3 + 1}{x + 1}$

64. $f(x) = 3x^3 + 2x^2 - 27x - 18$

CHAPTER 9 TEST

Graph the polynomial functions in Exercises 1 and 2.

1. $P(x) = (1 - x)^4$ **2.** $P(x) = (x + 1)(x - 2)x$

3. Find a lowest degree polynomial with -1, -2, and 4 as zeros.

4. Is 5 a zero of $P(x) = x^4 - 3x^3 - 2x^2 + x - 5$?

5. Is $x + 2$ a factor of $P(x) = 2x^3 + x^2 - x + 10$?

6. Find a polynomial of degree 4 with real coefficients and 2, -1, and $-i$ as zeros, and with $P(3) = 80$.

7. Find all zeros of $P(x) = x^4 - 5x^3 - 3x^2 + 43x - 60$, given that $2 - i$ is a zero.

8. Factor $P(x) = 2x^3 - x^2 - 13x - 6$, given that -2 is a zero.

9. For the polynomial function $P(x) = 6x^3 - 25x^2 + 12x + 7$

(a) list all rational numbers that can possibly be zeros of $P(x)$;

(b) find all rational zeros of $P(x)$.

10. Explain why the polynomial function $P(x) = x^4 + x^2 + 1$ cannot have any negative zeros. (*Hint:* Look at the exponents.)

11. Show that $P(x) = 6x^3 - 25x^2 + 12x + 7$ has a real zero in $[-1, 0]$.

12. Show that $P(x) = 2x^4 - 3x^3 + 4x^2 - 5x - 1$ has no real zeros greater than 2 or less than -1.

13. Find the possible numbers of positive and negative real zeros of $P(x) = x^4 + 3x^3 - x^2 - 2x + 1$.

14. Show that $P(x) = 2x^4 - 3x^3 + 4x^2 - 5x - 1$ has a real zero in $[1, 2]$ and approximate it to the nearest tenth.

15. Graph the polynomial function $P(x) = x^3 - 5x^2 + 8x - 4$.

Graph each of the following rational functions.

16. $f(x) = \dfrac{-2}{x + 3}$

17. $f(x) = \dfrac{3x - 1}{x - 2}$

18. $f(x) = \dfrac{x^2 - 1}{x^2 - 9}$

19. $f(x) = \dfrac{9 - x^2}{3 - x}$

20. Which one of the following functions has no x-intercepts?

(a) $f(x) = (x - 2)(x + 3)^2$ **(b)** $f(x) = \dfrac{x + 7}{x - 2}$

(c) $f(x) = x^3 - x$ **(d)** $f(x) = \dfrac{1}{x^2 + 4}$

THE GRAPHING CALCULATOR

Graphing calculators can easily be used to graph polynomial and rational functions. In Example 4 in Section 9.1, we graphed the polynomial function

$$P(x) = (2x + 3)(x - 1)(x + 2)$$

by using intercepts and test points in each region formed by the x-intercepts. The graph is shown in Figure 9.11. Use a graphing calculator to graph this function, with the minimum and maximum x-values of -4 and 2, and minimum and maximum y-values of -6 and 6. Compare the result to the figure in the text.

The rational function

$$f(x) = \frac{3(x + 1)(x - 2)}{(x + 4)^2}$$

was analyzed in Example 6 in Section 9.5. Program your graphing calculator to graph this function, using minimum and maximum x-values of -15 and 12, and minimum and maximum y-values of -1 and 15. Compare the display to Figure 9.23.

Some graphing calculators have the capability of "zooming in" on a particular portion of the graph of a function. If your calculator has this capability, zoom in on the portion of the graph of the rational function above in the vicinity of the origin to see how the function changes from decreasing to increasing. You might also have the capability to "trace" the graph, while the calculator displays the coordinates of the points on the graph.

In calculus we are often required to find the highest and lowest points on a graph in a particular region of the domain. By experimenting with your graphing calculator, you can begin to appreciate how these marvels of technology can help us to find excellent approximations of the coordinates of these points.

You may want to use the following functions to experiment with your calculator. These functions have been selected from examples in the text, so you can compare your calculator graphs with those given in the example figures. The figure number of the graph in the text is given with each function.

1. $P(x) = x^3$ (Figure 9.1)

2. $P(x) = x^6$ (Figure 9.2)

3. $P(x) = \frac{1}{2}x^3$ (Figure 9.3)

4. $P(x) = -\frac{3}{2}x^4$ (Figure 9.4)

5. $P(x) = x^5 - 2$ (Figure 9.5)

6. $P(x) = (x + 1)^6$ (Figure 9.6)

7. $P(x) = -(x - 1)^3 + 3$ (Figure 9.7)

8. $P(x) = 3x^4 + x^3 - 2x^2$ (Figure 9.12)

9. $P(x) = 8x^3 - 12x^2 + 2x + 1$ (Figure 9.16)

10. $P(x) = 3x^4 - 14x^3 + 54x - 3$ (Figure 9.17)

11. $f(x) = \frac{1}{x}$ (Figure 9.18)

12. $f(x) = -\frac{2}{x}$ (Figure 9.19)

13. $f(x) = \frac{2}{1 + x}$ (Figure 9.20)

14. $f(x) = \frac{x + 1}{(2x - 1)(x+3)}$ (Figure 9.21)

15. $f(x) = \frac{2x + 1}{x - 3}$ (Figure 9.22)

16. $f(x) = \frac{x^2 + 1}{x - 2}$ (Figure 9.24)

17. Graph $f(x) = \frac{x^2 - 4}{x - 2}$ on your graphing calculator and compare your result with Figure 9.25.

CHAPTER TEN

EXPONENTIAL AND LOGARITHMIC FUNCTIONS

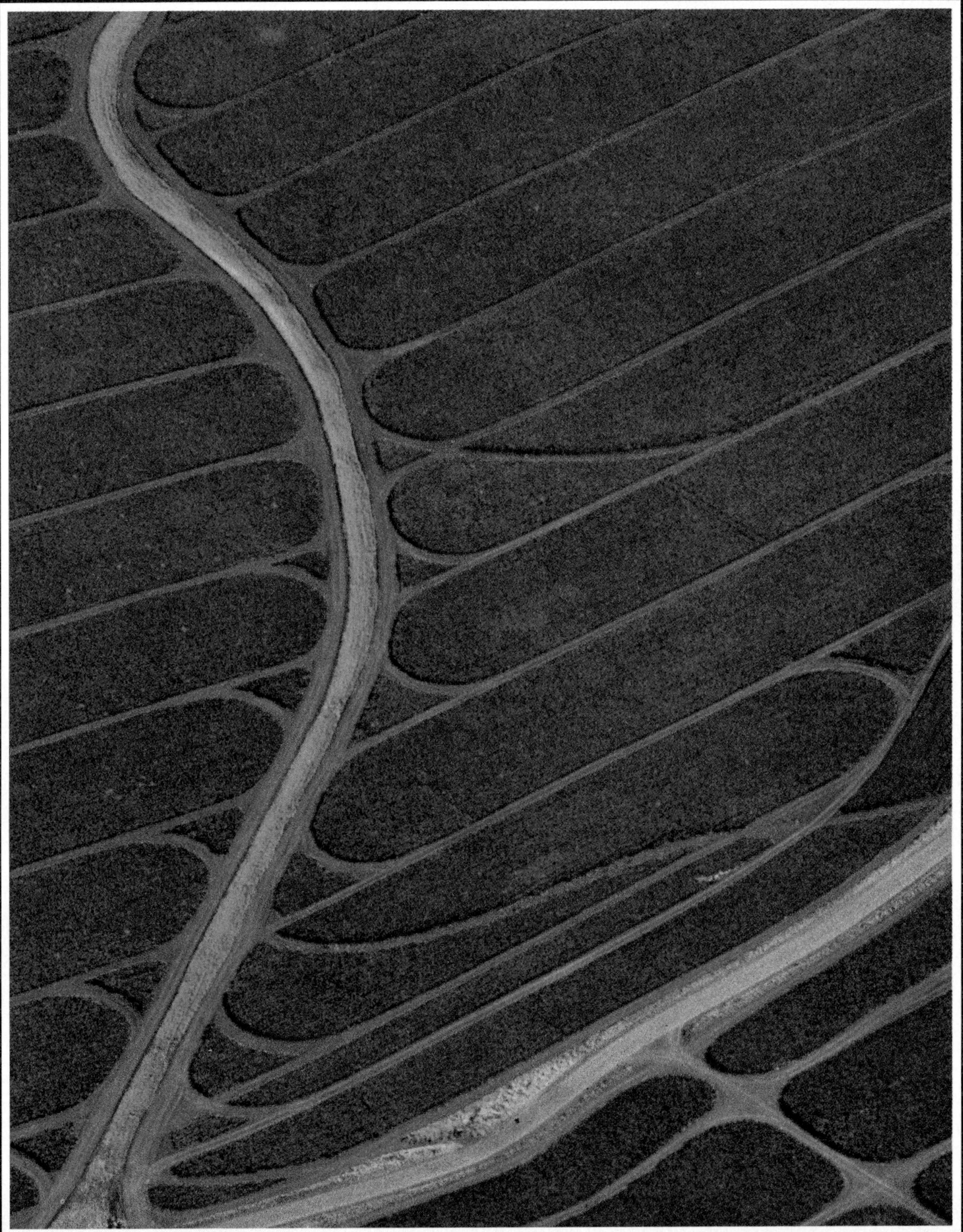

Long before the days of calculators and computers, the search for making calculations easier was an on-going process. Machines built by Charles Babbage and Blaise Pascal, a system of "rods" used by John Napier, and slide rules were the forerunners of today's mechanical marvels. The invention of logarithms by John Napier in the 16th century was a great breakthrough in the search for easier calculations.

While logarithms no longer are used for computations, they play an important role in higher mathematics. They are closely related to functions that have the independent variable in the exponent (exponential functions), since their relationship is that of inverse functions. Exponential and logarithmic functions are introduced in this chapter. Until now, we have worked with *algebraic functions* that involve only the basic operations of addition, subtraction, multiplication, division, and taking roots. The exponential and logarithmic functions in this chapter go beyond (transcend) these basic operations, so they are called *transcendental functions*.

10.1 INVERSE FUNCTIONS

OBJECTIVES

1. DECIDE WHETHER A FUNCTION IS ONE-TO-ONE.
2. SHOW THAT TWO FUNCTIONS ARE INVERSES.
3. FIND THE EQUATION OF THE INVERSE OF A FUNCTION.
4. GRAPH THE INVERSE OF f FROM THE GRAPH OF f.

A calculator with the following keys will be very helpful in this chapter.

$$\boxed{y^x}, \quad \boxed{10^x} \text{ or } \boxed{\log x}, \quad \boxed{e^x} \text{ or } \boxed{\ln x}$$

We will explain how these keys are used at appropriate places in the chapter.

Addition and subtraction are inverse operations: starting with a number x, adding 5, and then subtracting 5 gives x back as a result. Similarly, some functions are inverses of each other. For example, the functions

$$f(x) = 8x \qquad \text{and} \qquad g(x) = \frac{1}{8}x$$

are inverses since each "undoes" the other. For example, if a value of x such as $x = 12$ is chosen, so that

$$f(12) = 8 \cdot 12 = 96,$$

calculating $g(96)$ gives

$$g(96) = \frac{1}{8} \cdot 96 = 12.$$

Thus, $$g[f(12)] = 12.$$

Also, $f[g(12)] = 12$. For these functions f and g, it can be shown that

$$f[g(x)] = x \quad \text{and} \quad g[f(x)] = x$$

for any value of x.

This section will show how to start with a function such as $f(x) = 8x$ and obtain the inverse function $g(x) = (1/8)x$. Not all functions have inverse functions. The only functions that do have inverse functions are *one-to-one functions*.

1 For the function $y = 5x - 8$, any two different values of x produce two different values of y. On the other hand, for the function $y = x^2$, two *different* values of x can lead to the *same* value of y; for example, both $x = 4$ and $x = -4$ give $y = 4^2 = (-4)^2 = 16$. A function such as $y = 5x - 8$, where different elements from the domain always lead to different elements from the range, is called a *one-to-one function*.

DEFINITION OF ONE-TO-ONE FUNCTION

A function f is a **one-to-one function** if, for elements a and b from the domain of f,

$$a \neq b \quad \text{implies} \quad f(a) \neq f(b).$$

EXAMPLE 1

DECIDING WHETHER A FUNCTION IS ONE-TO-ONE

Decide whether each of the following functions is one-to-one.

(a) $f(x) = -4x + 12$

Suppose that $a \neq b$. Then $-4a \neq -4b$, and $-4a + 12 \neq -4b + 12$. Thus, the fact that $a \neq b$ implies that $f(a) \neq f(b)$, so f is one-to-one.

(b) $f(x) = \sqrt{25 - x^2}$

If $x = 3$, then

$$f(3) = \sqrt{25 - 3^2} = \sqrt{16} = 4.$$

For $x = -3$,

$$f(-3) = \sqrt{25 - (-3)^2} = \sqrt{16} = 4.$$

Both $x = 3$ and $x = -3$ lead to $y = 4$, so f is not one-to-one. ■

As shown in Example 1(b), a way to show that a function is *not* one-to-one is to produce a pair of unequal numbers that lead to the same function value.

In addition, there is a useful graphical test that tells whether or not a function is one-to-one. This *horizontal line test* for one-to-one functions can be summarized as follows.

HORIZONTAL LINE TEST

If each horizontal line cuts the graph of a function in no more than one point, then the function is one-to-one.

EXAMPLE 2
USING THE HORIZONTAL LINE TEST

Use the horizontal line test to determine whether the graphs in Figures 10.1 and 10.2 are graphs of one-to-one functions.

(a)

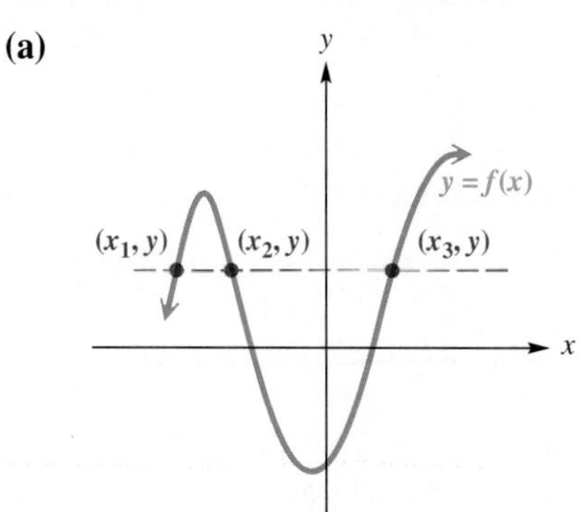

FIGURE 10.1

Note that each point where the horizontal line cuts the graph has the same value of y but a different value of x. Since more than one (here three) different values of x lead to the same value of y, the function is not one-to-one.

(b)

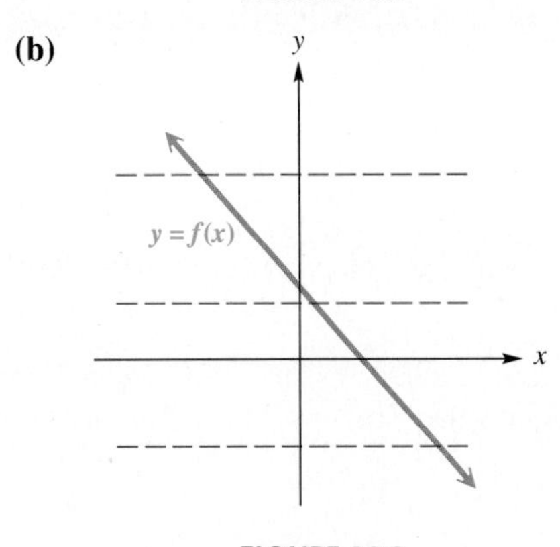

FIGURE 10.2

Every horizontal line will intersect the graph in Figure 10.2 in exactly one point. This function is one-to-one. ■

The function graphed in Figure 10.2 is decreasing on its entire domain. Any function that is increasing or decreasing on its domain is one-to-one and has an inverse.

2 As mentioned earlier, certain pairs of one-to-one functions ''undo'' one another. For example, if

$$f(x) = 8x + 5 \quad \text{and} \quad g(x) = \frac{x - 5}{8}$$

$$f(10) = 8 \cdot 10 + 5 = 85 \quad \text{and} \quad g(85) = \frac{85 - 5}{8} = 10.$$

Starting with 10, we ''applied'' function f and then ''applied'' function g to the result, which gave back the number 10. See Figure 10.3.

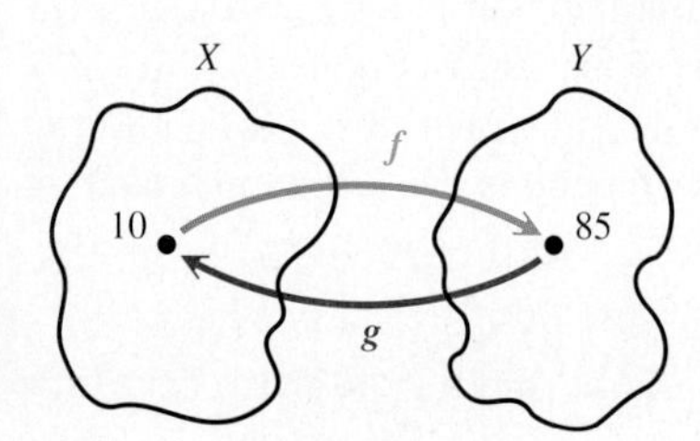

FIGURE 10.3

Similarly, for these same functions, check that

$$f(3) = \mathbf{29} \quad \text{and} \quad g(\mathbf{29}) = 3,$$
$$f(-5) = -35 \quad \text{and} \quad g(-35) = -5,$$
$$g(2) = -\frac{3}{8} \quad \text{and} \quad f\left(-\frac{3}{8}\right) = 2.$$

In particular, for these functions,

$$f[g(13)] = 13 \quad \text{and} \quad g[f(13)] = 13.$$

In fact, for *any* value of x,

$$f[g(x)] = x \quad \text{and} \quad g[f(x)] = x,$$

or

$$(f \circ g)(x) = x \quad \text{and} \quad (g \circ f)(x) = x.$$

Because of this property, g is called the *inverse* of f.

DEFINITION OF INVERSE FUNCTION

Let f be a one-to-one function. Then g is the **inverse function** of f if

$$(f \circ g)(x) = x \quad \text{for every } x \text{ in the domain of } g,$$

and

$$(g \circ f)(x) = x \quad \text{for every } x \text{ in the domain of } f.$$

EXAMPLE 3 DECIDING WHETHER TWO FUNCTIONS ARE INVERSES

Let $f(x) = x^3 - 1$, and let $g(x) = \sqrt[3]{x + 1}$. Is g the inverse function of f?

Use the definition to get

$$(f \circ g)(x) = f[g(x)] = (\sqrt[3]{x + 1})^3 - 1$$
$$= x + 1 - 1 = x$$
$$(g \circ f)(x) = g[f(x)] = \sqrt[3]{(x^3 - 1) + 1} = \sqrt[3]{x^3} = x.$$

Since both $f[g(x)] = x$ and $g[f(x)] = x$, function g is the inverse function of f. Also, f is the inverse function of g. ■

A special notation is often used for inverse functions: if g is the inverse function of f, then g can be written as f^{-1} (read "f-inverse"). In Example 3,

$$f^{-1}(x) = \sqrt[3]{x + 1}.$$

CAUTION Do not confuse the -1 in f^{-1} with a negative exponent. The symbol $f^{-1}(x)$ does not represent $1/f(x)$; it represents the inverse function of f. Keep in mind that a function f can have an inverse function f^{-1} if and only if f is one-to-one.

The definition of inverse function can be used to show that the domain of f equals the range of f^{-1}, and the range of f equals the domain of f^{-1}. See Figure 10.4.

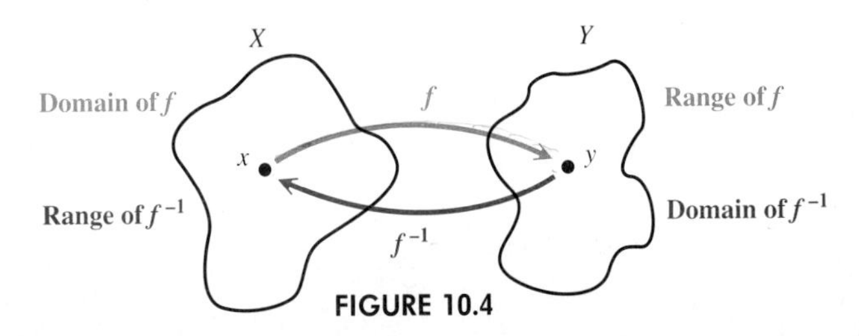

FIGURE 10.4

3 In the example at the beginning of this section, for the inverse functions f and g, $f(12) = 96$ and $g(96) = 12$; that is $(12, 96)$ belonged to f and $(96, 12)$ belonged to g. The inverse of any one-to-one function f can be found by exchanging the components of the ordered pairs of f. The equation of the inverse of a function defined by $y = f(x)$ also is found by exchanging x and y. For example, if $f(x) = 7x - 2$, then $y = 7x - 2$. The function f is one-to-one, so that f^{-1} exists. The ordered pairs in f^{-1} have the form (y, x), where $x = f^{-1}(y)$. Solve the given equation for x, as follows.

$$y = 7x - 2$$
$$7x = y + 2$$
$$x = \frac{y + 2}{7},$$

or, since $x = f^{-1}(y)$,

$$f^{-1}(y) = \frac{y + 2}{7}.$$

It is common to use x to represent an element from the domain of a function. Replacing y with x gives

$$f^{-1}(x) = \frac{x + 2}{7}.$$

In summary, finding the equation of an inverse function involves the following steps.

FINDING AN EQUATION FOR f^{-1}

1. Check that the function f defined by $y = f(x)$ is a one-to-one function.
2. Solve for x. Let $x = f^{-1}(y)$.
3. Exchange x and y to get $y = f^{-1}(x)$.
4. Check that $(f \circ f^{-1})(x) = x$ and $(f^{-1} \circ f)(x) = x$.

EXAMPLE 4

FINDING THE INVERSE OF A FUNCTION

For each of the functions defined as follows, find any inverse function.

(a) $f(x) = \dfrac{4x + 6}{5}$

This function is one-to-one and thus has an inverse. Let $y = f(x)$, and solve for x, getting

$$y = \frac{4x + 6}{5}$$
$$5y = 4x + 6 \qquad \text{Multiply by 5.}$$
$$5y - 6 = 4x \qquad \text{Subtract 6.}$$
$$\frac{5y - 6}{4} = x. \qquad \text{Divide by 4.}$$

Finally, exchange x and y, and let $y = f^{-1}(x)$, to get

$$\frac{5x - 6}{4} = y,$$

or

$$f^{-1}(x) = \frac{5x - 6}{4}.$$

The domain and range of both f and f^{-1} are the set of real numbers. In function f, the value of y is found by multiplying x by 4, adding 6 to the product, then dividing that sum by 5. In the equation for the inverse, x is *multiplied* by 5, then 6 is *subtracted*, and the result is *divided* by 4. This shows how an inverse function is used to "undo" what a function does to the variable x.

(b) $f(x) = x^3 - 1$

Two different values of x will produce two different values of $x^3 - 1$, so the function is one-to-one and has an inverse. To find the inverse, first solve $y = x^3 - 1$ for x, as follows.

$$y = x^3 - 1$$

$$y + 1 = x^3 \qquad \text{Add 1.}$$

$$\sqrt[3]{y + 1} = x \qquad \text{Take cube roots.}$$

Exchange x and y, giving

$$\sqrt[3]{x + 1} = y,$$

or

$$f^{-1}(x) = \sqrt[3]{x + 1}.$$

In Example 3, we verified that $\sqrt[3]{x + 1}$ is the inverse of $x^3 - 1$.

(c) $f(x) = x^2$

We can find two different values of x that give the same value of y. For example, both $x = 4$ and $x = -4$ give $y = 16$. Therefore, the function is not one-to-one and thus has no inverse function. ■

4 Suppose f and f^{-1} are inverse functions, and $f(a) = b$ for real numbers a and b. Then, by the definition of inverse, $f^{-1}(b) = a$. This shows that if a point (a, b) is on the graph of f, then (b, a) will belong to the graph of f^{-1}. As shown in Figure 10.5, the points (a, b) and (b, a) are symmetric with respect to the line $y = x$. Thus, the graph of f^{-1} can be obtained from the graph of f by reflecting the graph of f about the line $y = x$.

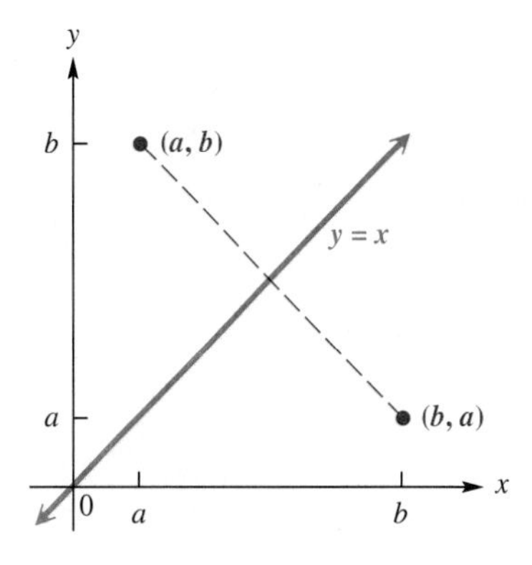

FIGURE 10.5

For example, Figure 10.6 shows the graph of $f(x) = x^3 - 1$ in blue and the graph of $f^{-1}(x) = \sqrt[3]{x + 1}$ in red. These graphs are symmetric with respect to the line $y = x$.

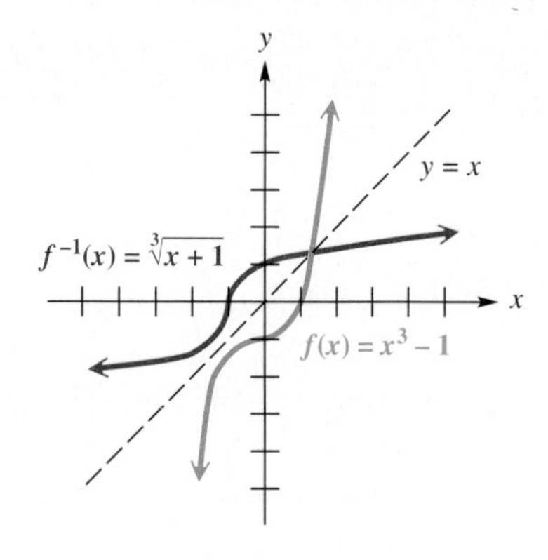

FIGURE 10.6

EXAMPLE 5
FINDING THE INVERSE OF A FUNCTION WITH A RESTRICTED DOMAIN

Let $f(x) = \sqrt{x + 5}$ with domain $[-5, \infty)$. Find $f^{-1}(x)$.

The function f is one-to-one and has an inverse function. To find this inverse function, start with

$$y = \sqrt{x + 5}$$

and solve for x, to get

$$y^2 = x + 5 \qquad \text{Square both sides.}$$
$$y^2 - 5 = x. \qquad \text{Subtract 5.}$$

Exchanging x and y gives

$$x^2 - 5 = y.$$

However, $x^2 - 5$ is not $f^{-1}(x)$. In the definition of f above, the domain was given as $[-5, \infty)$. The range of f is $[0, \infty)$. As mentioned above, the range of f equals the domain of f^{-1}, so f^{-1} must be given as

$$f^{-1}(x) = x^2 - 5, \qquad \text{domain } [0, \infty).$$

As a check, the range of f^{-1}, $[-5, \infty)$, equals the domain of f. Graphs of f and f^{-1} are shown in Figure 10.7. The line $y = x$ is included on the graph to show that the graphs of f and f^{-1} are mirror images with respect to this line. ■

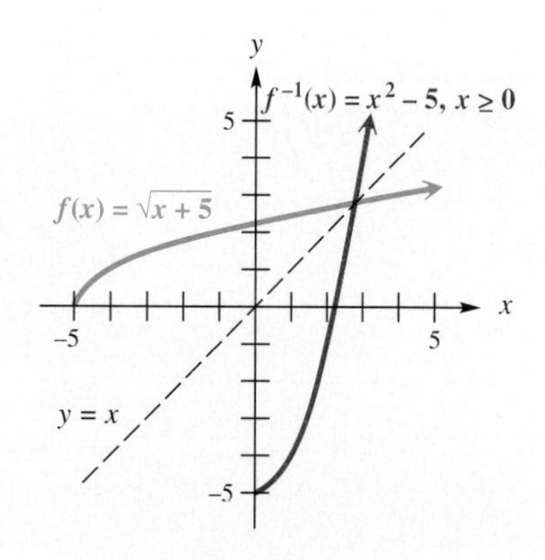

FIGURE 10.7

10.1 EXERCISES

Decide whether each of the following functions is one-to-one. See Examples 1 and 2.

1.

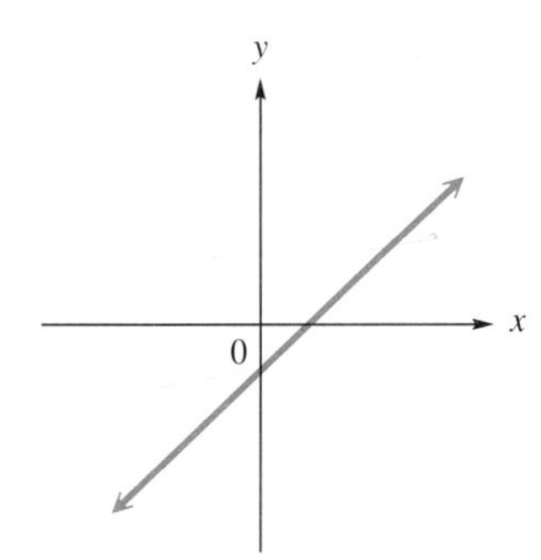

2.

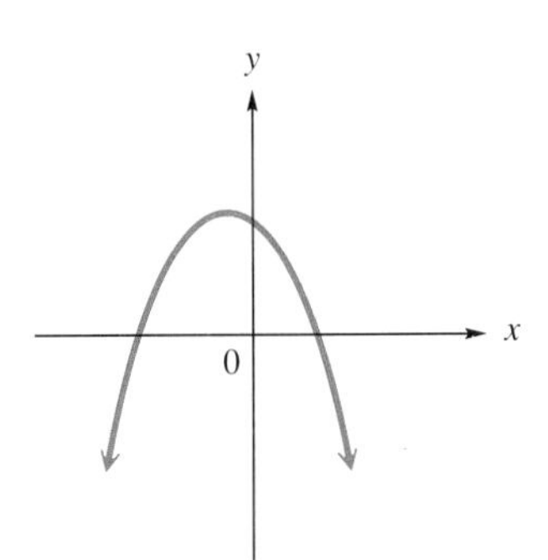

3.

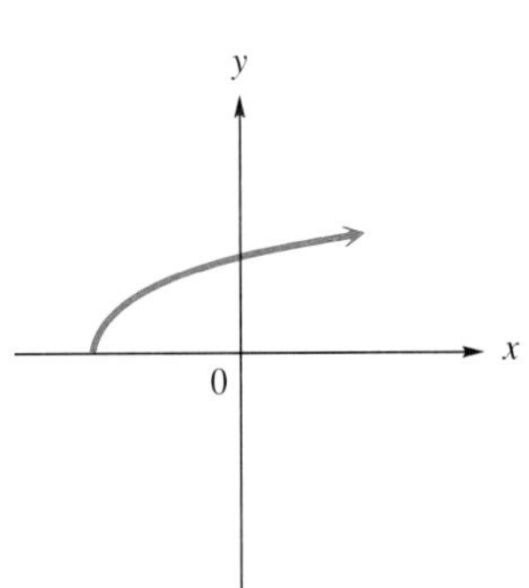

4.

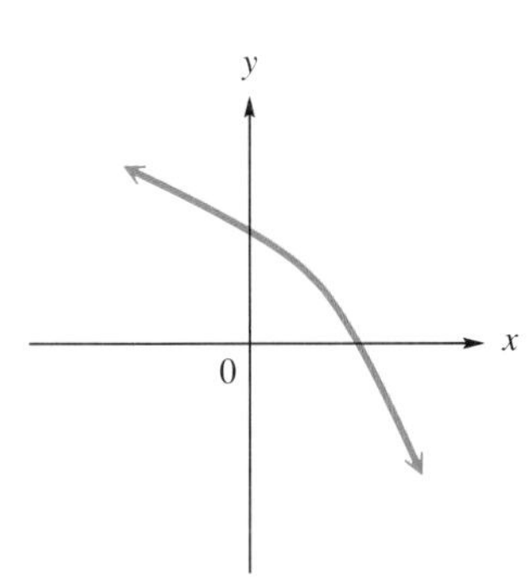

5.

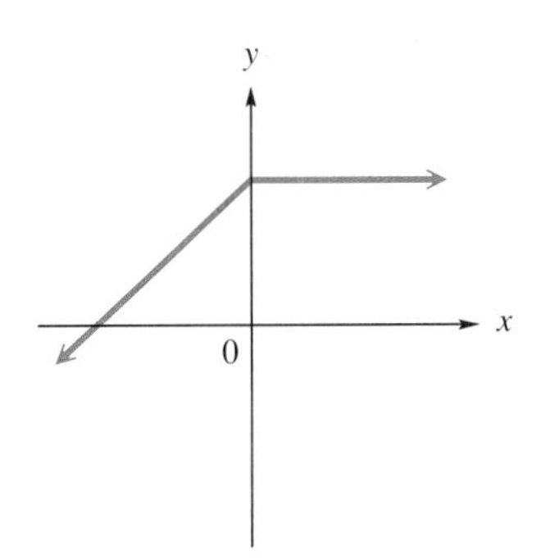

6.

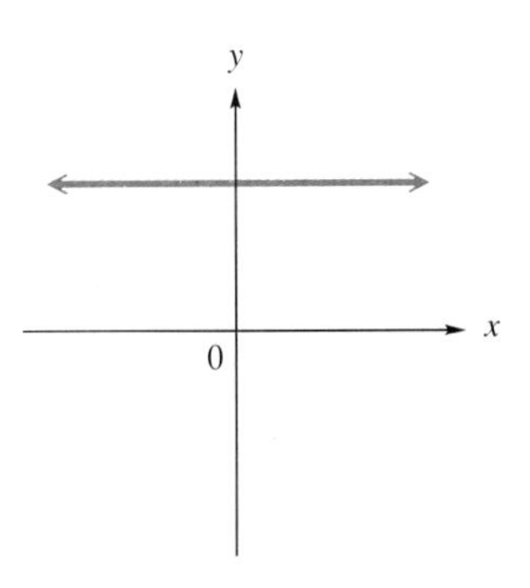

7. $y = 4x - 5$

8. $x + y = 0$

9. $y = 6 - x$

10. $y = -x^2$

11. $y = (x - 2)^2$

12. $y = -(x + 3)^2 - 8$

13. $y = \sqrt{36 - x^2}$

14. $y = -\sqrt{100 - x^2}$

15. $y = |25 - x^2|$

16. $y = -|16 - x^2|$

17. $y = x^3 - 1$

18. $y = -\sqrt[3]{x + 5}$

19. $y = (\sqrt{x} + 1)^2$

20. $y = (3 - 2\sqrt{x})^2$

21. $y = \dfrac{1}{x + 2}$

22. $y = \dfrac{-4}{x - 8}$

23. $y = 9$

24. $y = -4$

In Exercises 25–30, an everyday activity is described. Keeping in mind that an inverse operation "undoes" what the operation does, describe the inverse activity.

25. putting on your shoes

26. getting into a car

27. going to sleep

28. turning the television on

29. earning money

30. digging a hole in the ground

31. Try to visualize the graph of each of the following functions, and tell which one does not have an inverse. (See Objective 1.)

(a) $f(x) = -4x + 6$

(b) $f(x) = x^2$

(c) $f(x) = x^2, \quad x \geq 0$

(d) $f(x) = \sqrt{4 - x^2}, \quad x \geq 0$

32. Explain why the function $f(x) = x^4$ does not have an inverse, and give examples of ordered pairs to illustrate your explanation. (See Objective 1.)

Decide whether the functions in each pair are inverses of each other. See Examples 3 and 5.

33.

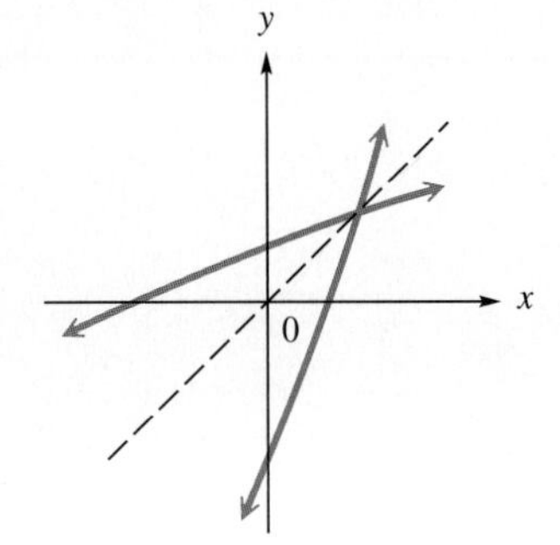

34.

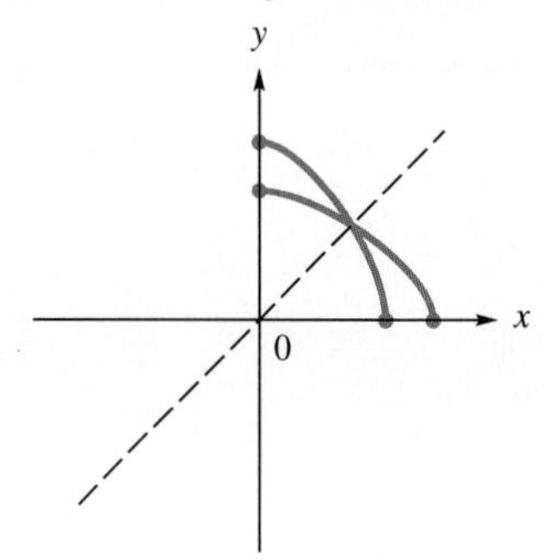

35.

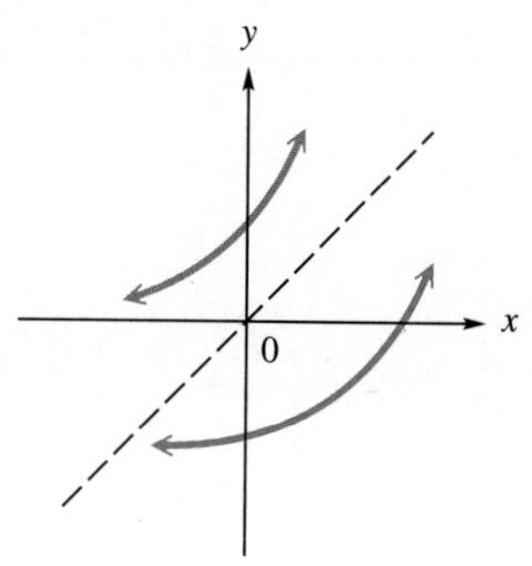

36.

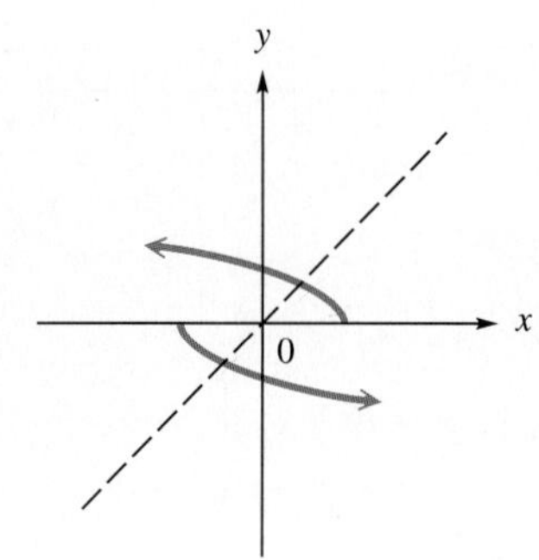

37.

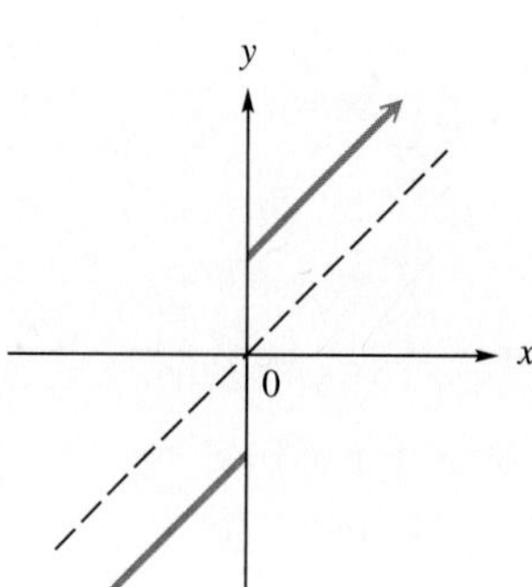

38.

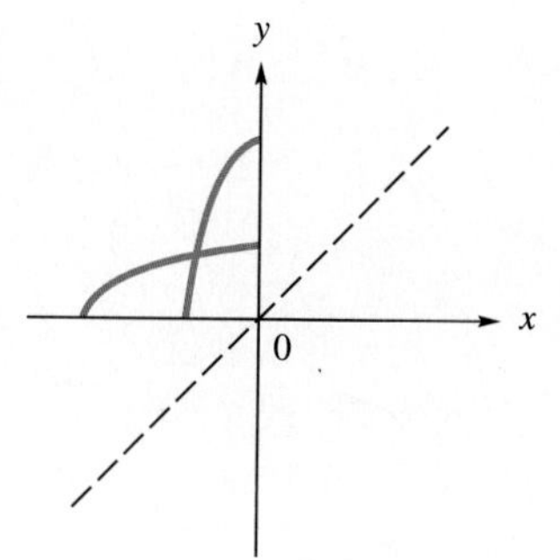

39. $f(x) = -\frac{3}{11}x, \quad g(x) = -\frac{11}{3}x$

40. $f(x) = 2x + 4, \quad g(x) = \frac{1}{2}x - 2$

41. $f(x) = 5x - 5, \quad g(x) = \frac{1}{5}x + 1$

42. $f(x) = 8x - 7, \quad g(x) = \frac{x + 8}{7}$

43. $f(x) = \frac{1}{x}, \quad g(x) = \frac{1}{x}$

44. $f(x) = 4x, \quad g(x) = -\frac{1}{4}x$

45. $f(x) = \sqrt{x + 8}$, domain $[-8, \infty)$, and $g(x) = x^2 - 8$, domain $[0, \infty)$

46. $f(x) = x^2 + 3$, domain $[0, \infty)$, and $g(x) = \sqrt{x - 3}$, domain $[3, \infty)$

47. $f(x) = |x - 1|$, domain $[-1, \infty)$, and $g(x) = |x + 1|$, domain $[1, \infty)$

48. $f(x) = -|x + 5|$, domain $[-5, \infty)$, and $g(x) = |x - 5|$, domain $[5, \infty)$

Graph the inverse of each one-to-one function.

49.

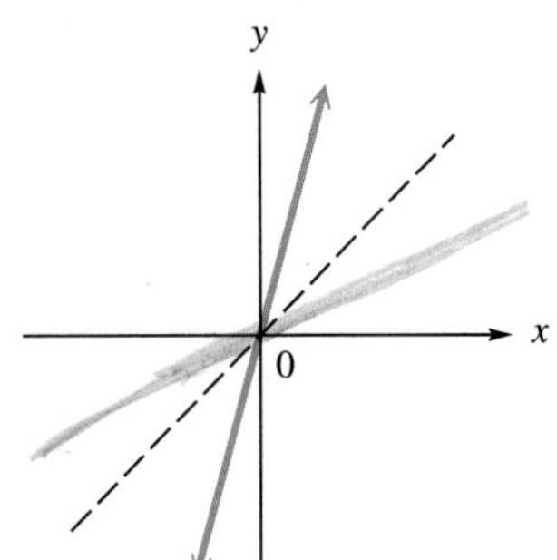

50.

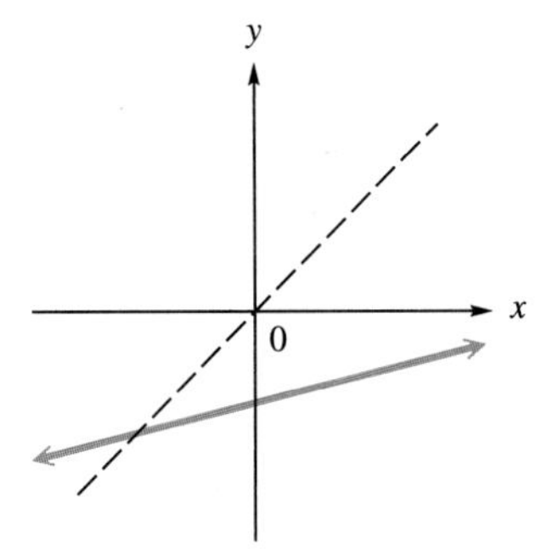

51.

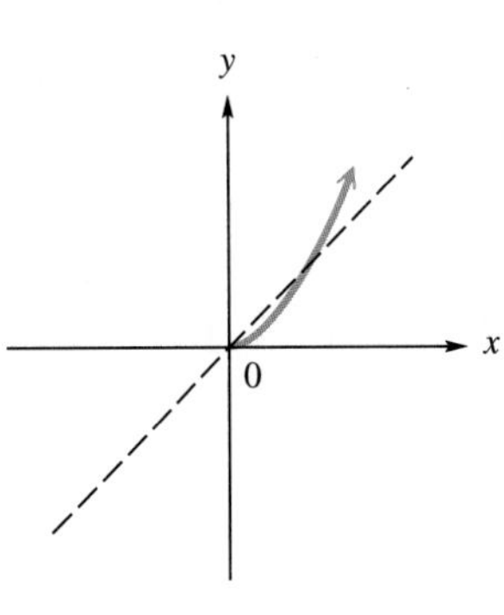

52.

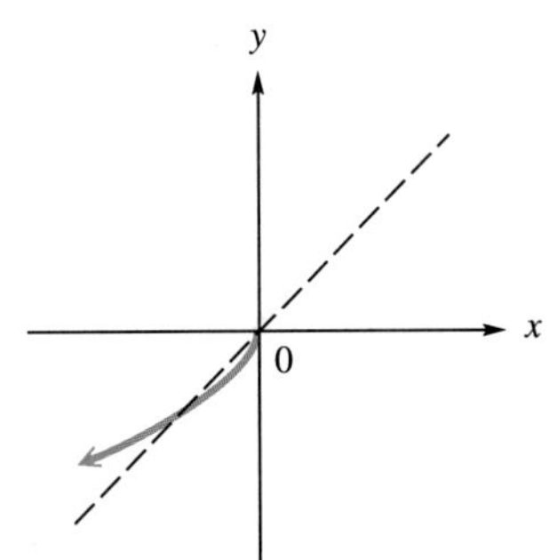

53.

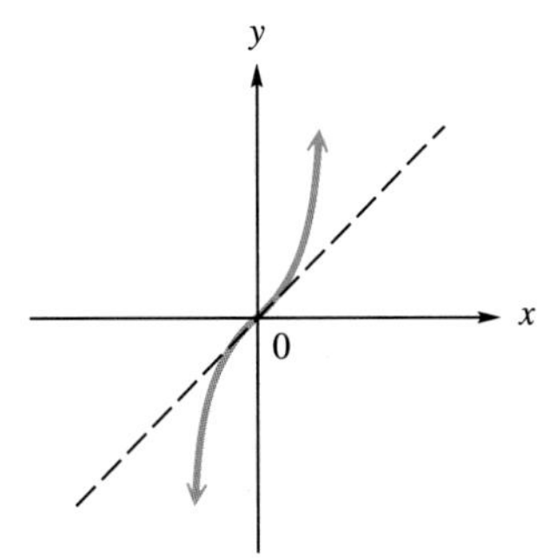

54.

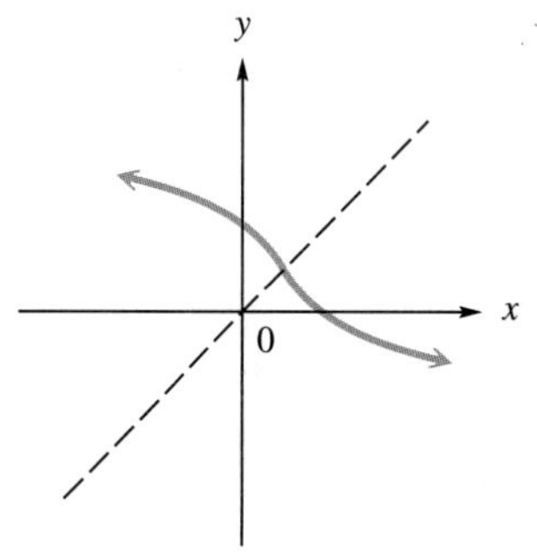

For each function defined as follows that is one-to-one, write an equation for the inverse function in the form of $y = f^{-1}(x)$, and then graph f and f^{-1} on the same axes. See Examples 4 and 5.

55. $y = 3x - 4$ **56.** $y = 4x - 5$ **57.** $y = \frac{1}{3}x$ **58.** $y = -\frac{2}{5}x$ **59.** $y = x^3 + 1$

60. $y = -x^3 - 2$ **61.** $x + 4y = 12$ **62.** $3x + y = 9$ **63.** $y = x^2$ **64.** $y = -x^2 + 2$

65. $y = \frac{1}{x}$ **66.** $xy = 4$ **67.** $xy = 3$ **68.** $y = \frac{2}{x}$

69. $f(x) = 4 - x^2$, domain $(-\infty, 0]$ **70.** $f(x) = \sqrt{6 + x}$, domain $[-6, \infty)$

71. $f(x) = \sqrt{9 - x}$, domain $(-\infty, 9]$ **72.** $f(x) = \sqrt{x^2 - 9}$, domain $(-\infty, -3]$

73. $f(x) = -\sqrt{x^2 - 16}$, domain $[4, \infty)$ **74.** $f(x) = (x - 1)^2$, domain $(-\infty, 1]$

PREVIEW EXERCISES

Evaluate each of the following. See Sections 3.1 and 5.1.

75. 3^{-2} **76.** $4^{-1/2}$ **77.** $8^{2/3}$ **78.** $1000^{-2/3}$ **79.** $81^{-5/4}$ **80.** $625^{3/4}$

10.2 EXPONENTIAL FUNCTIONS

OBJECTIVES

1 LEARN THE ADDITIONAL PROPERTIES OF EXPONENTIAL EXPRESSIONS.
2 SOLVE EXPONENTIAL EQUATIONS.
3 GRAPH EXPONENTIAL FUNCTIONS.
4 SOLVE COMPOUND INTEREST PROBLEMS.
5 USE EXPONENTIAL FUNCTIONS IN GROWTH AND DECAY APPLICATIONS.

FOCUS ON PROBLEM SOLVING

A small business estimates that the value y of a copy machine is decreasing according to the function

$$y = 5000(2)^{-.15t},$$

where t is the number of years that have elapsed since the machine was purchased, and y is in dollars.

(a) What was the original value of the machine? (*Hint:* Let $t = 0$.)
(b) What is the value of the machine 5 years after purchase? Give your answer to the nearest dollar.
(c) What is the value of the machine 10 years after purchase? Give your answer to the nearest dollar.
(d) Graph the function.

The equation in the problem above is different from those studied earlier in this book because it has a variable in an exponent. This is an example of an exponential function. We develop a method of solving such equations in this section. This problem is Exercise 45 in the exercises for this section. After working through this section, you should be able to solve this problem.

Recall from Section 5.1 the definition of a^r, where r is a rational number: if $r = m/n$, then for appropriate values of m and n,

$$a^{m/n} = (\sqrt[n]{a})^m.$$

For example,

$$16^{3/4} = (\sqrt[4]{16})^3 = 2^3 = 8,$$

$$27^{-1/3} = \frac{1}{27^{1/3}} = \frac{1}{\sqrt[3]{27}} = \frac{1}{3},$$

and

$$64^{-1/2} = \frac{1}{64^{1/2}} = \frac{1}{\sqrt{64}} = \frac{1}{8}.$$

In this section the definition of a^r is extended to include all real (not just rational) values of the exponent r. For example, the new symbol $2^{\sqrt{3}}$ might be evaluated by approximating the exponent $\sqrt{3}$ by the numbers 1.7, 1.73, 1.732, and so on. Since these decimals approach the value of $\sqrt{3}$ more and more closely, it seems reasonable that $2^{\sqrt{3}}$ should be approximated more and more closely by the numbers $2^{1.7}$, $2^{1.73}$, $2^{1.732}$, and so on. (Recall, for example, that $2^{1.7} = 2^{17/10} = \sqrt[10]{2^{17}}$.) In fact, this is exactly how $2^{\sqrt{3}}$ is defined (in a more advanced course).

1 With this interpretation of real exponents, all rules and theorems for exponents are valid for real-number exponents as well as rational ones. In addition to the properties of exponents presented earlier, several new properties are used in this chapter. For example, if $y = 2^x$, then each real value of x leads to exactly one value of y, and therefore, $y = 2^x$ defines a function. Furthermore,

$$\text{if } 3^x = 3^4, \text{ then } x = 4,$$

and for $p > 0$,

$$\text{if } p^2 = 3^2, \text{ then } p = 3.$$

Also, $$4^2 < 4^3 \quad \text{but} \quad \left(\frac{1}{2}\right)^2 > \left(\frac{1}{2}\right)^3,$$

so that when $a > 1$, increasing the exponent on a leads to a *larger* number, but if $0 < a < 1$, increasing the exponent on a leads to a *smaller* number.

These properties are generalized below. Proofs of the properties are not given here, as they require more advanced mathematics.

ADDITIONAL PROPERTIES OF EXPONENTS

(a) If $a > 0$ and $a \neq 1$, then a^x is a distinct real number for all real numbers x.
(b) If $a > 0$ and $a \neq 1$, then $a^b = a^c$ if and only if $b = c$.
(c) If $a > 1$ and $m < n$, then $a^m < a^n$.
(d) If $0 < a < 1$ and $m < n$, then $a^m > a^n$.

Properties (a) and (b) require $a > 0$ so that a^x is always defined. For example, $(-6)^x$ is not a real number if $x = 1/2$. This means that a^x will always be positive, since a is positive. In part (a), $a \neq 1$ because $1^x = 1$ for every real-number value of x, so that each value of x does not lead to a distinct real number. For part (b) to hold, a must not equal 1 since, for example, $1^4 = 1^5$, even though $4 \neq 5$. In part (b), the expression "if and only if" means both

$$\text{if } a^b = a^c \text{ then } b = c,$$

and

$$\text{if } b = c \text{ then } a^b = a^c.$$

2 Up to this point, all equations that we have solved have had the variable as a base; all exponents have been constants. An **exponential equation** is an equation with a variable exponent such as

$$\left(\frac{1}{3}\right)^x = 81.$$

Property (b) given earlier is used to solve some exponential equations as in the following example.

EXAMPLE 1 **SOLVING AN EXPONENTIAL EQUATION**

Solve $\left(\frac{1}{3}\right)^x = 81$.

If the bases were the same on each side of the equation, we could solve for x. First, write $1/3$ as 3^{-1}, so that $(1/3)^x = (3^{-1})^x = 3^{-x}$. Since $81 = 3^4$,

$$\left(\frac{1}{3}\right)^x = 81$$

becomes

$$3^{-x} = 3^4$$

By property (b),

$$-x = 4, \quad \text{or} \quad x = -4.$$

Check that the solution set of the given equation is $\{-4\}$. ■

EXAMPLE 2
SOLVING AN EQUATION WITH EXPONENTS

Find b if $b^{4/3} = 81$.

Raise both sides of the equation to the 3/4 power, since $(b^{4/3})^{3/4} = b^1 = b$.

$$\begin{aligned} 81 &= b^{4/3} \\ \pm 81^{3/4} &= (b^{4/3})^{3/4} \\ (\pm\sqrt[4]{81})^3 &= b^1 \\ (\pm 3)^3 &= b \\ \pm 27 &= b \end{aligned}$$

Find both positive and negative 4th roots.

Check both proposed solutions in the original equation.

$$\begin{aligned} 27^{4/3} &= (\sqrt[3]{27})^4 \\ &= 3^4 \\ &= 81 \end{aligned} \qquad \begin{aligned} (-27)^{4/3} &= (\sqrt[3]{-27})^4 \\ &= (-3)^4 \\ &= 81 \end{aligned}$$

Both solutions check, so the solution set is $\{-27, 27\}$. ■

CAUTION Remember that raising both sides of an equation to the same power may produce false "solutions." Thus, it was *necessary* to check both answers in Example 2.

The methods used to solve the equations in Examples 1 and 2 would not work with an equation like $3^x = 12$. A more general method for solving exponential equations is given in Section 10.6.

We can now define a one-to-one exponential function f with domain the set of real numbers (not just the rationals).

EXPONENTIAL FUNCTION

If $a > 0$ and $a \neq 1$, then

$$f(x) = a^x$$

defines the **exponential function** with base a.

NOTE The two restrictions on a in the definition of exponential function are important. The restriction that a must be positive is necessary so that the function can be defined for all real numbers x. For example, letting a be negative ($a = -2$, for instance) and letting $x = 1/2$ would give the expression $(-2)^{1/2}$, which is not real. The other restriction, $a \neq 1$, is necessary because 1 raised to any power is equal to 1, and the function would then be the linear function $f(x) = 1$, which is not one-to-one.

3 Exponential functions can be graphed by finding several ordered pairs that belong to the function. Plotting these points and connecting them with a smooth curve gives the graph.

EXAMPLE 3
GRAPHING AN EXPONENTIAL FUNCTION

Graph $f(x) = 2^x$.

Choose some values of x and find the corresponding values of $f(x)$.

x	-3	-2	-1	0	1	2	3
$f(x)$	1/8	1/4	1/2	1	2	4	8

Plotting these points and drawing a smooth curve through them gives the graph in Figure 10.8. The domain of the function is the set of real numbers and the range is the set of all positive numbers. The x-axis is a horizontal asymptote. ■

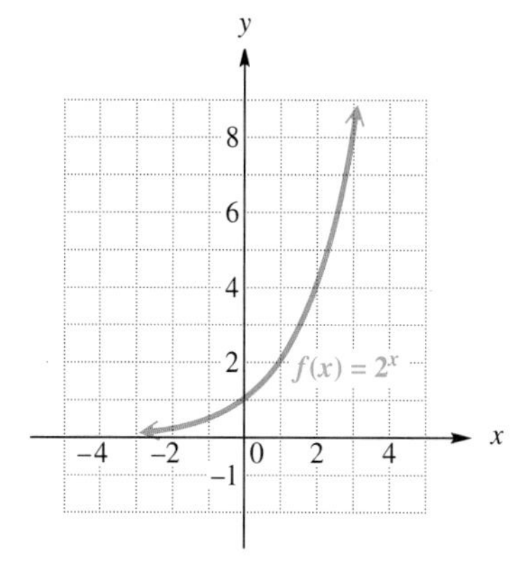

FIGURE 10.8

EXAMPLE 4
GRAPHING AN EXPONENTIAL FUNCTION

Graph $g(x) = (1/2)^x$.

Again, find some sample points.

x	-3	-2	-1	0	1	2	3
$g(x)$	8	4	2	1	1/2	1/4	1/8

The graph, shown in Figure 10.9, is very similar to that of $f(x) = 2^x$ in Figure 10.8, with the same domain and range, except that here as x gets larger, y *decreases*.

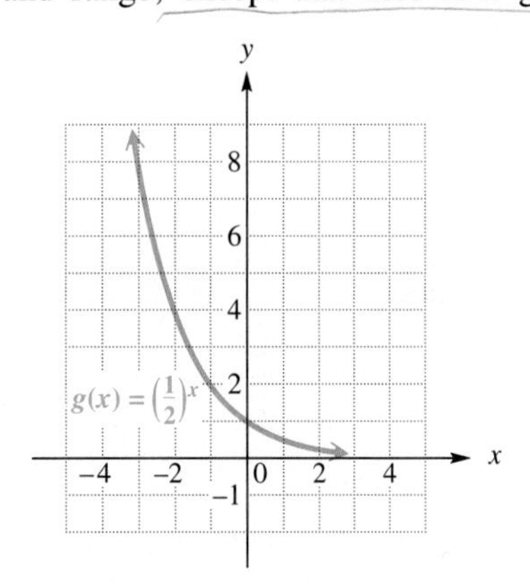

FIGURE 10.9

Using the horizontal line test with the graphs in Figures 10.8 and 10.9 confirms that they are the graphs of one-to-one functions. ■

The graph of $f(x) = 2^x$ is typical of graphs of $f(x) = a^x$ where $a > 1$. For larger values of a, the graphs rise more steeply, but the general shape is similar to the graph in Figure 10.8. When $0 < a < 1$ the graph decreases like the graph of $g(x) = (1/2)^x$ in Figure 10.9. In Figure 10.10 the graphs of several typical exponential functions illustrate these facts. (Notice that the decreasing functions are labeled with a^{-x}, where $a > 1$.)

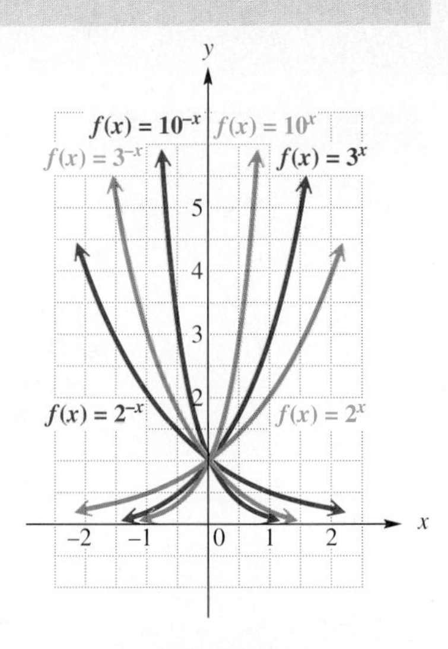

FIGURE 10.10

Based on Examples 3 and 4, the following generalizations can be made about the graphs of exponential functions defined by $f(x) = a^x$.

GRAPH OF $f(x) = a^x$

1. The point (0, 1) is on the graph.
2. If $a > 1$, f is an increasing function; if $0 < a < 1$, f is a decreasing function.
3. The x-axis is a horizontal asymptote.
4. The domain is $(-\infty, \infty)$ and the range is $(0, \infty)$.

We can use composition of functions to produce more general exponential functions. Suppose $h(u) = ka^u$, where k is a constant and $u = g(x)$. Let $f(x) = h(g(x))$. Then

$$f(x) = h(g(x)) = ka^{g(x)}.$$

For example, if $a = 5$, $g(x) = 2x + 3$, and $k = 4$, then

$$f(x) = 4 \cdot 5^{2x+3}.$$

EXAMPLE 5 **GRAPHING A COMPOSITE EXPONENTIAL FUNCTION**

Graph $f(x) = 2^{-x+2}$.

A table of values is given below.

x	−1	0	1	2	3
$f(x)$	8	4	2	1	1/2

Plotting these points and then connecting them with a smooth curve gives the graph in Figure 10.11. Notice that this graph goes down as x increases, while the graph of $f(x) = 2^x$ goes up. This happens because the negative sign in front of the x makes $f(x)$ get smaller as x gets larger. The graph is also translated 2 units to the right when compared to the graph of $f(x) = 2^{-x}$. ■

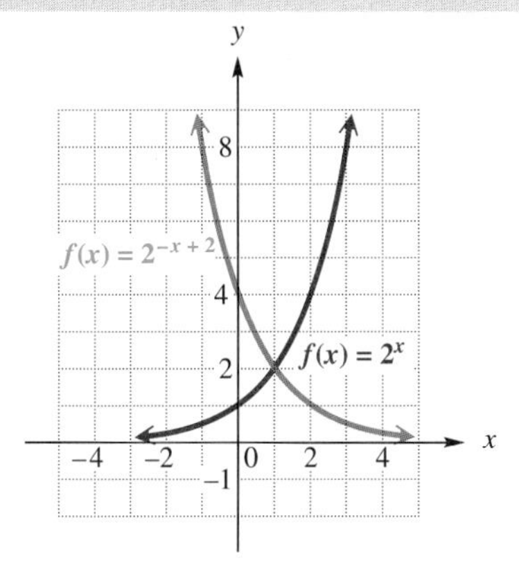

FIGURE 10.11

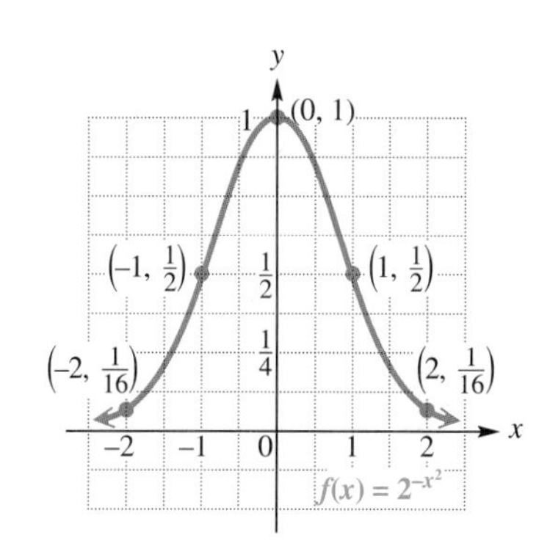

FIGURE 10.12

EXAMPLE 6 **GRAPHING A COMPOSITE EXPONENTIAL FUNCTION**

Graph $f(x) = 2^{-x^2}$.

Write $f(x) = 2^{-x^2}$ as $f(x) = 1/(2^{x^2})$ to find ordered pairs that belong to the function. Some ordered pairs are shown in the following table of values.

x	−2	−1	0	1	2
y	1/16	1/2	1	1/2	1/16

As the table suggests, $0 \leq y \leq 1$ for all values of x. Plotting these points and drawing a smooth curve through them gives the graph in Figure 10.12. This graph is symmetric with respect to the y-axis and has the x-axis as a horizontal asymptote. ■

4 The formula for compound interest (interest paid on both principal and interest) is an important application of exponential functions.

COMPOUND INTEREST

If P dollars is deposited in an account paying a rate of interest i compounded (paid) m times per year, then after n years the account will contain A dollars, where

$$A = P\left(1 + \frac{i}{m}\right)^{mn}.$$

EXAMPLE 7 **SOLVING A COMPOUND INTEREST PROBLEM**

Suppose \$1000 is deposited in an account paying 8% per year compounded quarterly (four times a year). Find the total amount in the account after 10 years if no withdrawals are made. Find the amount of interest earned.

Use the compound interest formula from above with $P = 1000$, $i = .08$, $m = 4$, and $n = 10$.

$$A = P\left(1 + \frac{i}{m}\right)^{mn}$$

$$A = 1000\left(1 + \frac{.08}{4}\right)^{4(10)}$$

$$= 1000(1 + .02)^{40} = 1000(1.02)^{40}$$

The number $(1.02)^{40}$ can be found in financial tables or by using a calculator with a y^x key. With a calculator, enter 1.02, press the key labeled y^x (or x^y), and then enter

40 and press the "=" key. To five decimal places, $(1.02)^{40} = 2.20804$. The amount on deposit after 10 years is

$$A = 1000(1.02)^{40} = 1000(2.20804) = 2208.04,$$

or \$2208.04. The amount of interest earned is

$$\$2208.04 - \$1000.00 = \$1208.04. \quad \blacksquare$$

In the formula for compound interest, A is sometimes called the **future value** and P the **present value.**

EXAMPLE 8
FINDING THE PRESENT VALUE

An accountant wants to buy a new computer in three years that will cost \$20,000. **(a)** How much should be deposited now, at 6% interest compounded annually, to give the required \$20,000 in three years?

Since the money deposited should amount to \$20,000 in three years, \$20,000 is the future value of the money. To find the present value P of \$20,000 (the amount to deposit now), use the compound interest formula with $A = 20{,}000$, $i = .06$, $m = 1$, and $n = 3$.

$$A = P\left(1 + \frac{i}{m}\right)^{mn}$$

$$20{,}000 = P\left(1 + \frac{.06}{1}\right)^{1(3)} \qquad \text{Substitute into the formula.}$$

$$= P(1.06)^3$$

$$\frac{20{,}000}{(1.06)^3} = P \qquad \text{Divide each side by } (1.06)^3.$$

$$P = 16{,}792.39 \qquad \text{Use a calculator.}$$

The accountant must deposit \$16,792.39.

(b) If only \$15,000 is available to deposit now, what annual interest rate is required for it to increase to \$20,000 in three years?

Here $P = 15{,}000$, $A = 20{,}000$, $m = 1$, $n = 3$, and i is unknown. Substitute the known values into the compound interest formula and solve for i.

$$A = P\left(1 + \frac{i}{m}\right)^{mn}$$

$$20{,}000 = 15{,}000\left(1 + \frac{i}{1}\right)^{1(3)}$$

$$\frac{4}{3} = (1 + i)^3 \qquad \text{Divide both sides by 15,000.}$$

$$\left(\frac{4}{3}\right)^{1/3} = 1 + i \qquad \text{Take the cube root on both sides.}$$

$$\left(\frac{4}{3}\right)^{1/3} - 1 = i \qquad \text{Subtract 1 on both sides.}$$

$$i \approx .10 \qquad \text{Use a calculator.}$$

An interest rate of 10% will produce enough interest to increase the $15,000 deposit to the $20,000 needed at the end of three years. ■

Perhaps the single most useful base for an exponential function is the irrational number e. Base e exponential functions provide a good model for many natural, as well as economic, phenomena. The letter e was chosen to represent this number in honor of the Swiss mathematician Leonard Euler (pronounced "oiler") (1707–83). Applications of the exponential function with base e are given later in this chapter.

To see one way the number e is used, begin with the formula for compound interest. Suppose that a lucky investment produces annual interest of 100%, so that $i = 1.00$, or $i = 1$. Suppose also that only $1 can be deposited at this rate, and for only one year. Then $P = 1$ and $n = 1$. Substituting into the formula for compound interest gives

$$P\left(1 + \frac{i}{m}\right)^{mn} = 1\left(1 + \frac{1}{m}\right)^{m(1)}$$
$$= \left(1 + \frac{1}{m}\right)^{m}.$$

If interest is compounded annually, making $m = 1$, the total amount on deposit is

$$\left(1 + \frac{1}{m}\right)^{m} = \left(1 + \frac{1}{1}\right)^{1}$$
$$= 2^{1} = 2,$$

so an investment of $1 becomes $2 in one year. As interest is compounded more and more often, the value of this expression will increase.

A calculator with a y^x key was used to get the following results. These results have been rounded when necessary to five decimal places.

Frequency of Compounding	m	$\left(1 + \frac{1}{m}\right)^{m}$
annually	1	2
semiannually	2	2.25
quarterly	4	2.44141
monthly	12	2.61304
daily	365	2.71457
hourly	8760	2.71813
every minute	525,600	2.71828

The table suggests that as m increases, the value of $(1 + 1/m)^m$ gets closer and closer to some fixed number. It turns out that this is indeed the case. This fixed number is called e.

VALUE OF e

To nine decimal places,

$$e \approx 2.718281828.$$

Scientific and business calculators give values of e^x. To obtain a specific value, key in the number x and then press the key labeled e^x. (If your calculator does not have an e^x key, press the INV key and then the ln x key. The reason that this method works will be apparent in Section 3 of this chapter.) Also, Table 2 in this book gives various values of e^x.

In Figure 10.13 the functions defined by $f(x) = 2^x$, $f(x) = e^x$, and $f(x) = 3^x$ are graphed for comparison.

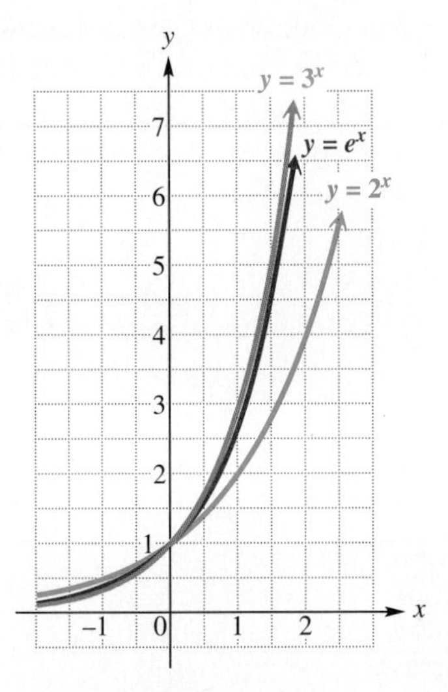

FIGURE 10.13

5 It can be shown that in many situations involving growth or decay of a population, the amount or number present at time t can be closely approximated by an exponential function with base e.

EXAMPLE 9 SOLVING AN APPLICATION OF AN EXPONENTIAL FUNCTION

Suppose the population of a southwestern city is

$$P(t) = 10{,}000e^{.04t},$$

where t represents time measured in years. The population at time $t = 0$ is

$$\begin{aligned} P(0) &= 10{,}000e^{(.04)0} \\ &= 10{,}000e^{0} \\ &= 10{,}000(1) \\ &= 10{,}000. \end{aligned}$$

The population of the city is 10,000 at time $t = 0$, written $P_0 = 10{,}000$. In year $t = 5$ the population is

$$\begin{aligned} P(5) &= 10{,}000e^{(.04)5} \\ &= 10{,}000e^{.2}. \end{aligned}$$

The number $e^{.2}$ can be found in Table 2 or by using a suitable calculator. By either of these methods, $e^{.2} = 1.22140$ (to five decimal places), so

$$P(5) = 10{,}000(1.22140) = 12{,}214.$$

Thus, in five years the population of the city will be about 12,200. A graph of $P(t) = 10{,}000e^{.04t}$ is shown in Figure 10.14. ■

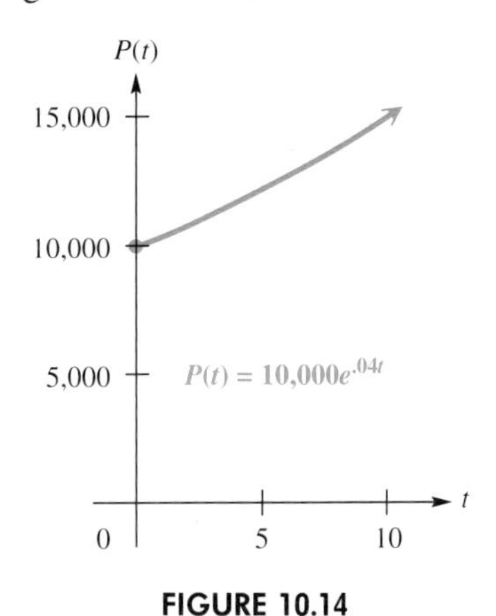

FIGURE 10.14

10.2 EXERCISES

Solve each of the following equations. See Examples 1 and 2.

1. $4^x = 2$

2. $125^r = 5$

3. $\left(\frac{1}{2}\right)^k = 4$

4. $\left(\frac{2}{3}\right)^x = \frac{9}{4}$

5. $2^{3-y} = 8$

6. $5^{2p+1} = 25$

7. $\frac{1}{27} = b^{-3}$

8. $\frac{1}{81} = k^{-4}$

9. $4 = r^{2/3}$

10. $z^{5/2} = 32$

11. $125^{-x} = 25^{3x}$

12. $216^{3-a} = 36^a$

13. $\left(\frac{1}{8}\right)^{-2p} = 2^{p+3}$

14. $3^{-h} = \left(\frac{1}{27}\right)^{1-2h}$

15. $\left(\frac{1}{2}\right)^{-x} = \left(\frac{1}{4}\right)^{x+1}$

16. $\left(\frac{2}{3}\right)^{k-1} = \left(\frac{81}{16}\right)^{k+1}$

Use a calculator to evaluate each of the following. Express your answers to the nearest thousandth.

17. $4.2^{1.458}$

18. $3.4^{2.422}$

19. $8^{-.965}$

20. $6^{-.113}$

---

21. Try to evaluate $(-2)^4$ on a calculator. You will probably get an error message, since the exponential function key on most calculators does not allow negative bases. Discuss the concept introduced in this section that is closely related to this "peculiarity" of scientific calculators. (See Objective 1.)

22. Explain why the exponential equation $4^x = 6$ cannot be solved using the method explained in this section. (See Objective 2.)

23. Graph f for each of the following. Compare the graphs to that of $f(x) = 2^x$. See Examples 3–5.
(a) $f(x) = 2^x + 1$ **(b)** $f(x) = 2^x - 4$
(c) $f(x) = 2^{x+1}$ **(d)** $f(x) = 2^{x-4}$

24. Graph f for each of the following. See Examples 3–5.
(a) $f(x) = 3^{-x} - 2$ **(b)** $f(x) = 3^{-x} + 4$
(c) $f(x) = 3^{-x-2}$ **(d)** $f(x) = 3^{-x+4}$

Graph f for each of the following. See Examples 3–6.

25. $f(x) = 3^x$ **26.** $f(x) = 4^x$ **27.** $f(x) = 3^{-x}$ **28.** $f(x) = (3/2)^x$
29. $f(x) = 2.718^{-x}$ **30.** $f(x) = 2.718^x$ **31.** $f(x) = 2^{|x|}$ **32.** $f(x) = 2^{-|x|}$
33. $f(x) = 2^x + 2^{-x}$ **34.** $f(x) = (1/2)^x + (1/2)^{-x}$

Suppose f is an exponential function of the form $f(x) = a^x$ and $f(3) = 2$. Determine the function values in Exercises 35–37. (See Objective 3.)

35. $f(-3)$ **36.** $f(1)$ **37.** $f(9)$

38. What two points on the graph of $f(x) = a^x$ can be found without any computation? (See Objective 3.)

39. For $a > 1$, how does the graph of $f(x) = a^x$ change as a increases? What if $0 < a < 1$? (See Objective 3.)

40. For $a > 0$ and $a \neq 1$, if $a^x = a^y$, then $x = y$. Give an example to show why the restriction $a \neq 1$ is necessary for this statement to be true. (See Objective 1.)

41. What is the domain of the exponential function $f(x) = a^x$, $a > 0$, $a \neq 1$? (See Objective 3.)

Use a calculator to help graph f in Exercises 42–44.

42. $f(x) = \dfrac{e^x - e^{-x}}{2}$ **43.** $f(x) = \dfrac{e^x + e^{-x}}{2}$ **44.** $f(x) = x \cdot 2^x$

Work each problem. See Examples 7–9.

45. A small business estimates that the value y of a copy machine is decreasing according to the function

$$y = 5000(2)^{-.15t},$$

where t is the number of years that have elapsed since the machine was purchased, and y is in dollars.

(a) What was the original value of the machine? (*Hint:* Let $t = 0$.)
(b) What is the value of the machine 5 years after purchase? Give your answer to the nearest dollar.
(c) What is the value of the machine 10 years after purchase? Give your answer to the nearest dollar.
(d) Graph the function.

46. Refer to the function in Exercise 45. When will the value of the machine be \$2500? (*Hint:* Let $y =$ 2500, divide both sides by 5000, and use the method of Example 1.)

47. Suppose that the population of a city is given by $P(t)$, where

$$P(t) = 1{,}000{,}000e^{.02t},$$

and t represents time measured in years from some initial year. Find each of the following.

(a) P_0 **(b)** $P(2)$ **(c)** $P(4)$ **(d)** $P(10)$
(e) Graph $y = P(t)$.

48. Suppose the quantity in grams of a radioactive substance present at time t is

$$Q(t) = 500e^{-.05t}.$$

Let t be time measured in days from some initial day. Find the quantity present at each of the following times.

(a) $t = 0$ **(b)** $t = 4$ **(c)** $t = 8$ **(d)** $t = 20$
(e) Graph $y = Q(t)$.

49. Experiments have shown that the sales of a product, under relatively stable market conditions, but in the absence of promotional activities such as advertising, tend to decline at a constant yearly rate. This rate of sales decline varies considerably from product to product, but seems to remain the same for any particular product. The sales decline can be expressed by a function of the form

$$S(t) = S_0e^{-at},$$

where $S(t)$ is the rate of sales at time t measured in years, S_0 is the rate of sales at time $t = 0$, and a is the sales decay constant.

(a) Suppose the sales decay constant for a particular product is $a = .10$. Let $S_0 = 50{,}000$ and find $S(1)$ and $S(3)$.

(b) Find $S(2)$ and $S(10)$ if $S_0 = 80{,}000$ and $a = .05$.

50. Suppose \$10,000 is deposited at 12% interest for 3 years. Find the interest earned if the interest is compounded

(a) annually; **(b)** quarterly; **(c)** daily (365 days).

Use the formula for compound interest and a calculator to find the future value of each amount. See Example 7.

51. \$4292 at 6% compounded annually for 10 years

52. \$10,765 at 11% compounded semiannually for 7 years

53. \$8906.54 at 5% compounded semiannually for 9 years

54. \$56,780 at 5.3% compounded quarterly for 23 quarters

Find the present value for the following future values. See Example 8(a).

55. \$10,000, if interest is 12% compounded semiannually for 5 years

56. \$25,000, if interest is 6% compounded quarterly for 11 quarters

57. \$45,678.93, if interest is 9.6% compounded monthly for 11 months

58. \$123,788, if interest is 8.7% compounded daily for 195 days

Find the required annual interest rate in percent to the nearest tenth in each of the following. See Example 8(b).

59. \$25,000 compounded annually for 2 years to yield \$31,360

60. \$65,000 compounded monthly for 6 months to yield \$65,325

61. \$1200 compounded quarterly for 5 years to yield \$1780

62. \$15,000 compounded semiannually for 4 semiannual periods to yield \$19,000

The graph shown here accompanied the article "Is Our World Warming?" which appeared in the October, 1990, issue of National Geographic. *It shows projected temperature increases using two graphs: one an exponential-type curve, and the other linear. From the graph approximate the increase* **(a)** *for the exponential curve, and* **(b)** *for the linear graph for each of the following years.*

63. 2000

64. 2010

65. 2020

66. 2040

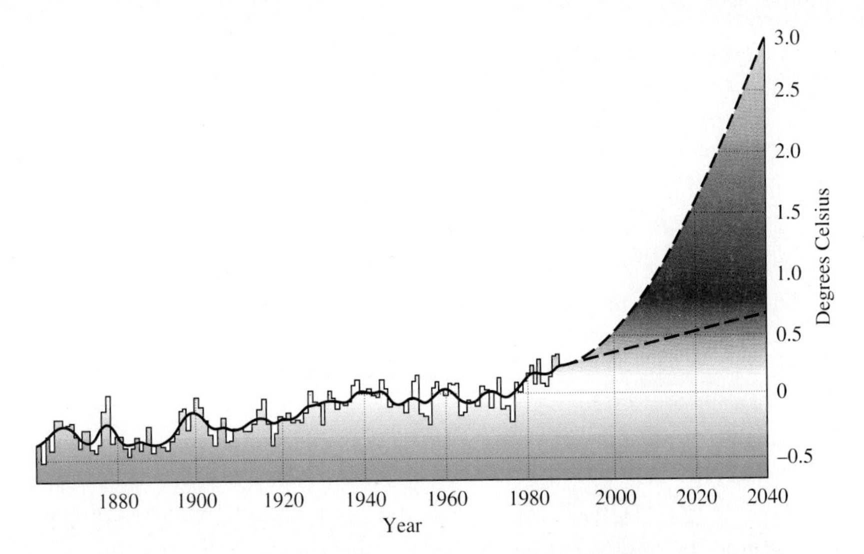

Note: Zero equals average global temperature for the period 1950–1979.
Graph by Dale D. Glasgow. © National Geographic Society.

Solve each of the following problems. See Example 9.

67. *Escherichia coli* is a strain of bacteria that occurs naturally in many different organisms. Under certain conditions, the number of bacteria present in a colony is

$$E(t) = E_0 \cdot 2^{t/30},$$

where $E(t)$ is the number of bacteria present t minutes after the beginning of an experiment, and E_0 is the number present when $t = 0$. Let $E_0 =$ 2,400,000 and find the number of bacteria at the following times.
(a) $t = 5$ **(b)** $t = 10$ **(c)** $t = 60$ **(d)** $t = 120$

68. The higher a student's grade-point average, the fewer applications that the student need send to medical schools (other things being equal). Using information given in a guidebook for prospective medical students, we constructed the function $y = 540e^{-1.3x}$ for the number of applications a student should send out. Here y is the number of applications for a student whose grade-point average is x. The domain of x is the interval [2.0, 4.0]. Find the number of applications that should be sent out by students having the following grade-point averages.
(a) 2.0 **(b)** 3.0 **(c)** 3.5 **(d)** 3.9

In calculus, it is shown that

$$e^x = 1 + x + \frac{x^2}{2 \cdot 1} + \frac{x^3}{3 \cdot 2 \cdot 1} + \frac{x^4}{4 \cdot 3 \cdot 2 \cdot 1} + \frac{x^5}{5 \cdot 4 \cdot 3 \cdot 2 \cdot 1} + \cdots.$$

69. Use the terms shown here and replace x with 1 to approximate $e^1 = e$ to three decimal places. Then check your results in Table 2 or with a calculator.

70. Use the terms shown here and replace x with $-.05$ to approximate $e^{-.05}$ to four decimal places. Check your results in Table 2 or with a calculator.

PREVIEW EXERCISES

Evaluate each expression. Write the answers with radicals in Exercises 75–78. See Sections 3.1 and 5.1.

71. 2^{-3} **72.** $\left(\frac{1}{2}\right)^{-4}$ **73.** $36^{1/2}$ **74.** $\left(\frac{27}{8}\right)^{1/3}$ **75.** $3^{5/2}$ **76.** $5^{3/2}$ **77.** $2^{-4/3}$ **78.** $6^{-2/5}$

10.3 LOGARITHMIC FUNCTIONS

OBJECTIVES

1 DEFINE A LOGARITHM.

2 CONVERT BETWEEN EXPONENTIAL AND LOGARITHMIC STATEMENTS.

3 SOLVE LOGARITHMIC EQUATIONS OF THE FORM $\log_a b = k$ FOR a, b, OR k.

4 GRAPH LOGARITHMIC FUNCTIONS.

5 USE LOGARITHMIC FUNCTIONS IN APPLICATIONS.

FOCUS ON PROBLEM SOLVING

The number of fish in an aquarium is given by

$$f(t) = 8 \log_5 (2t + 5),$$

where t is time in months. Find the number of fish present at the following times.

(a) $t = 0$ **(b)** $t = 10$ **(c)** $t = 60$ **(d)** Graph f.

Just as exponential functions can describe growth and decay, logarithmic functions can be applied similarly. The important relationship between inverse functions is seen in this section, because a logarithmic function is the inverse of an exponential function. This problem is Exercise 61 in the exercises for this section. After working through this section, you should be able to solve this problem.

1 The previous section dealt with exponential functions of the form $y = a^x$ for all positive values of a, where $a \neq 1$. As mentioned there, the horizontal line test shows that exponential functions are one-to-one, and thus have inverse functions. In this section we discuss the inverses of exponential functions. The equation defining the inverse of a function is found by exchanging x and y in the equation that defines the function. Doing so with $y = a^x$ gives

$$x = a^y$$

as the equation of the inverse function of the exponential function defined by $y = a^x$. This equation can be solved for y by using the following definition.

DEFINITION OF LOGARITHM

For all real numbers y, and all positive numbers a and x, where $a \neq 1$,

$$y = \log_a x \quad \text{if and only if} \quad x = a^y.$$

The abbreviation **log** is used for **logarithm.** Read $\log_a x$ as "the logarithm of x to the base a." This key definition should be memorized. It is helpful to remember the location of the base and exponent in each form.

Logarithmic form: $y = \log_a x$ (Exponent: y; Base: a)

Exponential form: $x = a^y$ (Exponent: y; Base: a)

In working with logarithmic form and exponential form it is always helpful to remember the following.

A **logarithm** is an exponent; $\log_a x$ is the exponent on the base a that yields the number x.

2 The definition of logarithm can be used to write exponential statements in logarithmic form and logarithmic statements in exponential form.

EXAMPLE 1
CONVERTING BETWEEN EXPONENTIAL AND LOGARITHMIC FORM

The list below shows several pairs of equivalent statements. The same statement is written in both exponential and logarithmic form.

Exponential form	*Logarithmic form*
$2^3 = 8$	$\log_2 8 = 3$
$(1/2)^{-4} = 16$	$\log_{1/2} 16 = -4$
$10^5 = 100{,}000$	$\log_{10} 100{,}000 = 5$
$3^{-4} = 1/81$	$\log_3 (1/81) = -4$
$5^1 = 5$	$\log_5 5 = 1$
$(3/4)^0 = 1$	$\log_{3/4} 1 = 0$ ■

For any positive real number b, we know that $b^1 = b$ and $b^0 = 1$. Writing these two statements in logarithmic form gives the following two properties of logarithms.

For any positive real number b, $b \neq 1$,

$$\log_b b = 1 \quad \text{and} \quad \log_b 1 = 0.$$

EXAMPLE 2
USING PROPERTIES OF LOGARITHMS

(a) $\log_7 7 = 1$

(b) $\log_{\sqrt{2}} \sqrt{2} = 1$

(c) $\log_9 1 = 0$

(d) $\log_{.2} 1 = 0$ ■

3 A **logarithmic equation** is an equation with a logarithm in at least one term. Logarithmic equations of the form $\log_a b = k$ can be solved for any of the three variables by first writing the equation in exponential form.

EXAMPLE 3
SOLVING LOGARITHMIC EQUATIONS

Solve the following equations.

(a) $\log_4 x = -2$

By the definition of logarithm, $\log_4 x = -2$ is equivalent to $4^{-2} = x$. Then

$$x = 4^{-2} = \frac{1}{4^2} = \frac{1}{16}.$$

The solution set is $\{1/16\}$.

(b) $\log_{1/2} 16 = y$

Write the statement in exponential form.

$$\log_{1/2} 16 = y \quad \text{is equivalent to} \quad \left(\frac{1}{2}\right)^y = 16.$$

Now write both sides of the equation with the same base. Since $1/2 = 2^{-1}$ and $16 = 2^4$,

$$\left(\frac{1}{2}\right)^y = 16$$

becomes

$$(2^{-1})^y = 2^4$$

$$2^{-y} = 2^4 \qquad \text{Power rule for exponents}$$

$$-y = 4 \qquad \text{Property of exponentials}$$

$$y = -4. \qquad \text{Multiply both sides by } -1.$$

The solution set is $\{-4\}$.

(c) $\log_x \dfrac{8}{27} = 3$

First, write the expression in exponential form.

$$x^3 = \frac{8}{27}$$

$$x^3 = \left(\frac{2}{3}\right)^3 \qquad \frac{8}{27} = \left(\frac{2}{3}\right)^3$$

$$x = \frac{2}{3} \qquad \text{Take the cube root on each side.}$$

The solution set is $\{2/3\}$. ■

We can now define the logarithmic function with base a as follows.

LOGARITHMIC FUNCTION

If $a > 0$, $a \neq 1$, and $x > 0$, then

$$f(x) = \log_a x$$

defines the **logarithmic function with base *a*.**

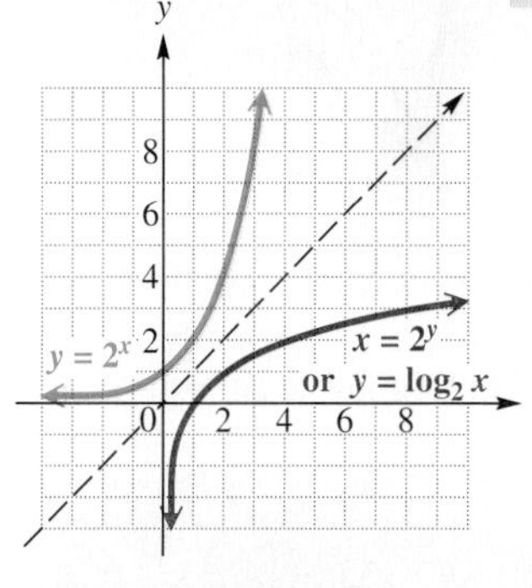

FIGURE 10.15

Exponential and logarithmic functions are inverses of each other. Since the domain of an exponential function is the set of all real numbers, the range of a logarithmic function also will be the set of all real numbers. In the same way, both the range of an exponential function and the domain of a logarithmic function are the set of all positive real numbers, so logarithms can be found for positive numbers only.

The graph of $y = 2^x$ is shown in Figure 10.15. The graph of its inverse is found by reflecting the graph of $y = 2^x$ about the line $y = x$. The graph of the inverse function, defined by $y = \log_2 x$, shown in red, has the y-axis as a vertical asymptote.

4 To graph a logarithmic function, it is helpful to write it in exponential form first. Then plot selected ordered pairs to determine the graph.

EXAMPLE 4 GRAPHING A LOGARITHMIC FUNCTION

Graph $y = \log_{1/2} x$.

Writing $y = \log_{1/2} x$ in its exponential form as $x = (1/2)^y$ helps to identify ordered pairs that satisfy the equation. Here it is easier to choose values for y and find the corresponding values of x. Doing this gives the following pairs.

x	1/4	1/2	1	2	4	8
y	2	1	0	−1	−2	−3

Plotting these points (be careful to get them in the right order) and connecting them with a smooth curve gives the graph in Figure 10.16. Note that this graph also has the y-axis as a vertical asymptote. ■

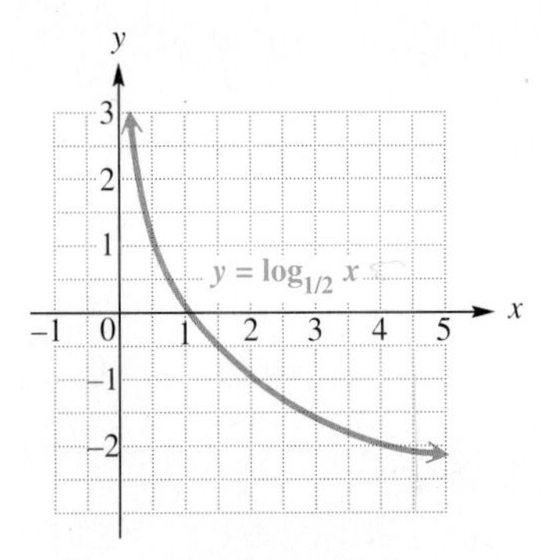

FIGURE 10.16

The graph in Figure 10.16 is typical of the graphs of logarithmic functions with base $0 < a < 1$. The graph in Figure 10.15 of $x = 2^y$, which is equivalent to $y = \log_2 x$, is typical of graphs of logarithmic functions with base $a > 1$.

Based on the graphs of the functions $y = \log_2 x$ in Figure 10.15 and $y = \log_{1/2} x$ in Figure 10.16, the following generalizations can be made about the graphs of logarithmic functions of the form $f(x) = \log_a x$.

GRAPH OF $f(x) = \log_a x$

1. The point (1, 0) is always on the graph.
2. If $a > 1$, f is an increasing function; if $0 < a < 1$, f is a decreasing function.
3. The y-axis is a vertical asymptote.
4. The domain is $(0, \infty)$ and the range is $(-\infty, \infty)$.

Compare these generalizations to the similar ones for the exponential functions in Section 10.2.

More general logarithmic functions can be obtained by forming the composition of $h(x) = \log_a x$ with $g(x)$ to get

$$f(x) = h[g(x)] = \log_a [g(x)].$$

The next examples illustrate such functions.

EXAMPLE 5 GRAPHING A COMPOSITE LOGARITHMIC FUNCTION

Graph $f(x) = \log_2 (x - 1)$.

The domain of this function is $(1, \infty)$, since logarithms can be found only for positive numbers. The graph of $f(x) = \log_2 (x - 1)$ is the same as the graph of $f(x) = \log_2 x$, translated one unit to the right. See Figure 10.17. ■

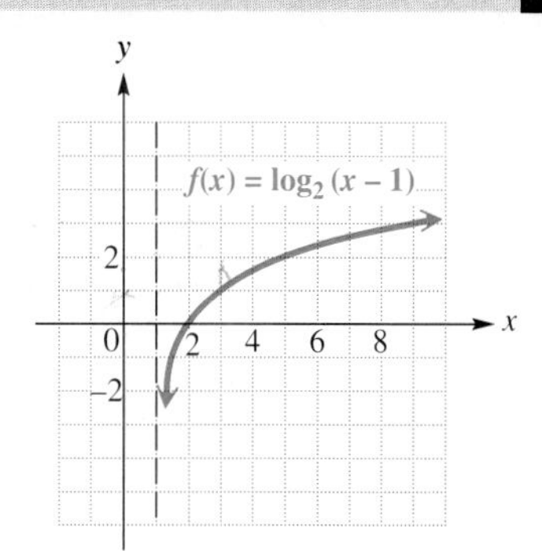

FIGURE 10.17

EXAMPLE 6 GRAPHING A COMPOSITE LOGARITHMIC FUNCTION

Graph $y = \log_3 |x|$.

Write $y = \log_3 |x|$ in exponential form as $3^y = |x|$ to help identify some ordered pairs that satisfy the equation. Here, it is easier to choose y-values and find the corresponding x-values. Doing so gives the following ordered pairs.

x	-3	-1	$-1/3$	$1/3$	1	3
y	1	0	-1	-1	0	1

Plotting these points (be careful to get them in the right order) and connecting them with two smooth curves gives the graph in Figure 10.18. Because of $|x|$, the graph is symmetric with respect to the y-axis. ■

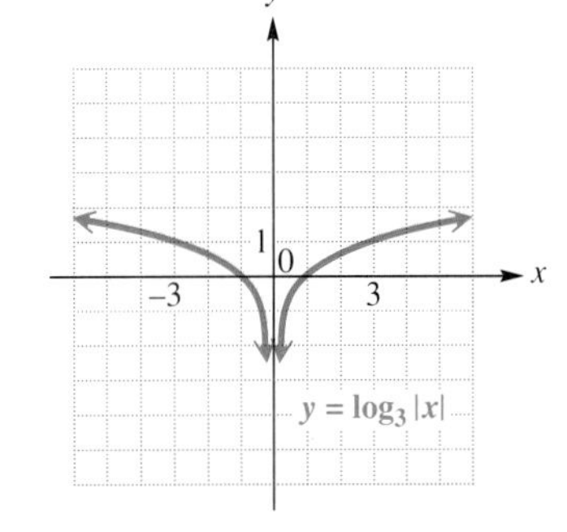

FIGURE 10.18

5 Logarithmic functions, like exponential functions, are used in applications to describe growth and decay. As shown in the previous section, a quantity that grows exponentially grows faster and faster. On the other hand, logarithmic functions describe growth that is taking place at a slower and slower rate. A similar statement applies to exponential and logarithmic decay.

EXAMPLE 7 SOLVING AN APPLICATION OF A LOGARITHMIC FUNCTION

Sales of a new product are approximated by $S = 100 + 30 \log_3 (2t + 1)$, where S is sales in thousands of units and t is the number of years after the product is introduced.

(a) What were the sales after 1 year?

Let $t = 1$ and find S.

$$\begin{aligned} S &= 100 + 30 \log_3 (2t + 1) \\ &= 100 + 30 \log_3 (2 \cdot 1 + 1) && \text{Let } t = 1. \\ &= 100 + 30 \log_3 3 \\ &= 100 + 30(1) && \log_3 3 = 1 \\ S &= 130 \end{aligned}$$

Sales were 130 thousand units after 1 year.

(b) Find the sales after 13 years.
Substitute $t = 13$ into the expression for S.

$$\begin{aligned} S &= 100 + 30 \log_3 (2t + 1) \\ &= 100 + 30 \log_3 (2 \cdot 13 + 1) && \text{Let } t = 13. \\ &= 100 + 30 \log_3 27 \\ &= 100 + 30(3) && \text{Log}_3\ 27 = 3 \\ &= 190 \end{aligned}$$

After 13 years, sales had increased to 190 thousand units.

(c) Graph S.
Use the two ordered pairs (1, 130) and (13, 190) found above. Check that (0, 100) and (40, 220) also satisfy the equation. Use these ordered pairs and a knowledge of the general shape of the graph of a logarithmic function to get the graph in Figure 10.19. ■

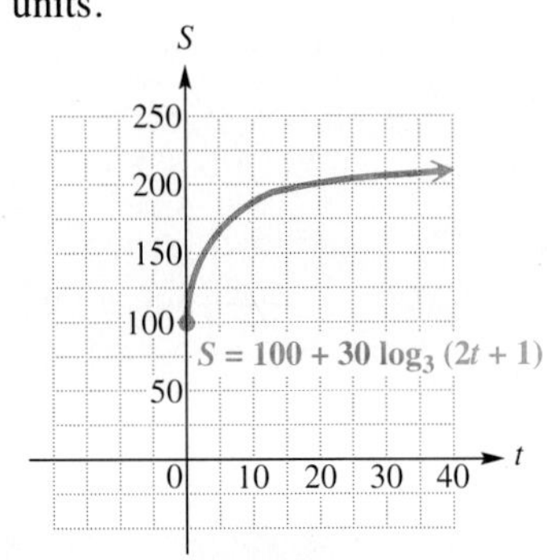

FIGURE 10.19

10.3 EXERCISES

Write each exponential statement in logarithmic form. See Example 1.

1. $3^4 = 81$ **2.** $2^5 = 32$ **3.** $10^4 = 10{,}000$ **4.** $8^2 = 64$

5. $\left(\frac{1}{2}\right)^{-4} = 16$ **6.** $\left(\frac{2}{3}\right)^{-3} = 27/8$ **7.** $10^{-4} = .0001$ **8.** $\left(\frac{1}{100}\right)^{-2} = 10{,}000$

Write each logarithmic statement in exponential form. See Example 1.

9. $\log_6 36 = 2$ **10.** $\log_5 5 = 1$ **11.** $\log_{\sqrt{3}} 81 = 8$ **12.** $\log_4 \left(\frac{1}{64}\right) = -3$

13. $\log_{10} .0001 = -4$ **14.** $\log_3 \sqrt[3]{9} = 2/3$ **15.** $\log_m k = n$ **16.** $\log_2 r = y$

Find the value of each of the following expressions. (Hint: Begin by setting the expression equal to y.) See Examples 2 and 3(b).

17. $\log_5 25$ **18.** $\log_3 81$ **19.** $\log_8 8$ **20.** $\log_7 1$

21. $\log_{10} .001$ **22.** $\log_6 \frac{1}{216}$ **23.** $\log_4 \frac{\sqrt[3]{4}}{2}$ **24.** $\log_9 \frac{\sqrt[4]{27}}{3}$

25. $\log_6 36^4$ **26.** $\log_5 125^2$ **27.** $2^{\log_2 9}$ **28.** $8^{\log_8 11}$

Suppose f is a logarithmic function defined by $y = \log_a x$ and $f(3) = 2$. Determine the function values in Exercises 29–31. (See Objective 3.)

29. $f(1/3)$ **30.** $f(1)$ **31.** $f(9)$

32. Does the fact that (5, 4) is on the graph of the logarithmic function of base a mean that $5 = \log_a 4$ or that $4 = \log_a 5$? (See Objective 4.)

---

33. Explain why 1 is not allowed as a base for a logarithm. (See Objective 1.)

34. Explain why logarithms of negative numbers are not defined. (See Objective 1.)

35. Compare the summary of facts about the graph of $f(x) = a^x$ in Section 10.2 with the similar summary of facts about the graph of $f(x) = \log_a x$ in this section. Make a list of the facts that reinforce the idea that these are inverse functions. (See Objective 3 in Section 10.2 and Objective 4 in this section.)

36. Explain why, for any valid base a, the logarithm to the base a of 1 is always 0. (See Objective 1.)

Solve each of the equations in Exercises 37–50. See Examples 2 and 3.

37. $y = \log_6 216$

38. $\log_5 x = -3$

39. $\log_x 9 = 1/2$

40. $\log_m 125 = -3$

41. $\log_x 64 = -6$

42. $\log_4 x = 5/2$

43. $\log_x 1/8 = -3$

44. $\log_p .1 = 1$

45. $\log_m 3 = 1/2$

46. $\log_{1/2} x = -3$

47. $\log_2 x = 0$

48. $\log_k 4 = 1$

49. $\log_{12} P = 1$

50. $\log_b 1 = 0$

---

51. Graph f for each of the following. Compare the graphs to that of $f(x) = \log_2 x$. See Examples 4–6.

(a) $f(x) = (\log_2 x) + 3$
(b) $f(x) = \log_2 (x + 3)$
(c) $f(x) = |\log_2 (x + 3)|$

52. Graph f for each of the following. Compare the graphs to that of $f(x) = \log_{1/2} x$. See Examples 4–6.

(a) $f(x) = (\log_{1/2} x) - 2$
(b) $f(x) = \log_{1/2} (x - 2)$
(c) $f(x) = |\log_{1/2} (x - 2)|$

Graph f for each of the following. See Examples 4–6.

53. $f(x) = \log_3 x$

54. $f(x) = \log_{10} x$

55. $f(x) = \log_{1/2} (1 - x)$

56. $f(x) = \log_{1/3} (3 - x)$

57. $f(x) = \log_2 x^2$

58. $f(x) = \log_3 (x - 1)$

59. $f(x) = x \log_{10} x$

60. $f(x) = x^2 \log_{10} x$

Work the following problems. See Example 7.

61. The number of fish in an aquarium is given by

$$f(t) = 8 \log_5 (2t + 5),$$

where t is time in months. Find the number of fish present when

(a) $t = 0$;
(b) $t = 10$;
(c) $t = 60$.
(d) Graph $y = f(t)$.

62. A company analyst has found that total sales, in thousands of dollars, after a major advertising campaign, are

$$S = 100 \log_2 (x + 2),$$

where x is time in weeks after the campaign was introduced. Find the sales for

(a) $x = 0$; **(b)** $x = 2$; **(c)** $x = 6$.
(d) Graph S.

63. The population of an animal species that is introduced into a certain area may grow rapidly at first but then grow more slowly as time goes on. A logarithmic function can provide an excellent description of such growth. Suppose that the population of foxes, $F(t)$, in an area t months after the foxes were introduced there is

$$F(t) = 500 \log_{10} (2t + 3).$$

Use a calculator with a log key to find the population of foxes at the following times. (For example, to find $\log_{10} 3$, touch 3, then log x. Some calculators may require a different sequence of strokes.)
(a) when they are first released into the area (that is, when $t = 0$);
(b) after 3 months;
(c) after 15 months.
(d) Graph $y = F(t)$.

64. A population of hares in a specific area is growing according to the function

$$H = 500 \log_3 (2t + 3),$$

where t is in years after the population was introduced into the area. Find the number of hares for the following times.
(a) $t = 0$ **(b)** $t = 3$ **(c)** $t = 12$
(d) Graph H.

The loudness of sounds is measured in a unit called a ***decibel.*** *To measure with this unit, an intensity of* I_0 *is assigned to a very faint sound, called the* ***threshold sound.*** *If a particular sound has intensity* I, *then the decibel rating of this louder sound is*

$$10 \cdot \log_{10} \frac{I}{I_0}.$$

65. Find the decibel ratings of sounds having the following intensities.
(a) $100I_0$ **(b)** $1000I_0$
(c) $100{,}000I_0$ **(d)** $1{,}000{,}000I_0$

66. Find the decibel ratings of the following sounds, having intensities as given. (You will need a calculator with a log key.) Round answers to the nearest whole number.
(a) whisper, $115I_0$
(b) busy street, $9{,}500{,}000I_0$
(c) a heavy truck that is 20 meters away, $1{,}200{,}000{,}000I_0$
(d) rock music, $895{,}000{,}000{,}000I_0$
(e) jetliner at takeoff, $109{,}000{,}000{,}000{,}000I_0$

67. The intensity of an earthquake, measured on the *Richter scale,* is $\log_{10} (I/I_0)$, where I_0 is the intensity of an earthquake of a certain (small) size. Find the Richter scale ratings of earthquakes having the following intensities.
(a) $1000I_0$ **(b)** $1{,}000{,}000I_0$ **(c)** $100{,}000{,}000I_0$

68. (a) The San Francisco earthquake of 1906 had a Richter scale rating of 8.3. Use a calculator with a y^x key to express the intensity of this earthquake as a multiple of I_0. (See Exercise 67.)
(b) In 1989, the San Francisco region experienced an earthquake with a Richter scale rating of 7.1. Express the intensity of this earthquake as a multiple of I_0.
(c) Compare the intensity of the two San Francisco earthquakes discussed above.

PREVIEW EXERCISES

Use the properties of exponents to simplify the following expressions. See Sections 3.1 and 3.2.

69. $2^4 \cdot 2^5$

70. $3^{-2} \cdot 3^8$

71. $\frac{4^{-1}}{4^5}$

72. $\frac{5^8}{5^{-2}}$

73. $(2^4)^3$

74. $(5x^2)^4$

75. $\frac{p^5p^{-3}}{(p^3)^2} \quad (p \neq 0)$

76. $\frac{2^3m^4(m^2)^3}{m^5} \quad (m \neq 0)$

10.4 PROPERTIES OF LOGARITHMS

OBJECTIVES

1. USE THE PROPERTY FOR THE LOGARITHM OF A PRODUCT.
2. USE THE PROPERTY FOR THE LOGARITHM OF A QUOTIENT.
3. USE THE PROPERTY FOR THE LOGARITHM OF A POWER.
4. USE THE PROPERTIES OF LOGARITHMS TO WRITE LOGARITHMIC EXPRESSIONS IN ALTERNATIVE FORMS.

Logarithms have been used as an aid to numerical calculation for several hundred years. Today the widespread use of calculators has made the use of logarithms for calculation obsolete. However, logarithms are very important in applications and in further work in mathematics. The properties that make logarithms so useful are given in this section.

1 One way in which logarithms simplify problems is by changing a problem of multiplication into one of addition. This is done with the following property.

LOGARITHM OF A PRODUCT

If x, y, and b are positive numbers, where $b \neq 1$, then

$$\log_b xy = \log_b x + \log_b y.$$

In words, the logarithm of a product is equal to the sum of the logarithms of the factors.

Proof

Recall that

$$\log_b x = m \quad \text{means} \quad b^m = x$$
$$\log_b y = n \quad \text{means} \quad b^n = y.$$

Then, by substitution,

$$xy = b^m \cdot b^n = b^{m+n},$$

using the product rule for exponents. Now go back to logarithmic form.

$$xy = b^{m+n} \quad \text{means} \quad \log_b xy = m + n$$

Substituting for $m + n$ gives

$$\log_b xy = \log_b x + \log_b y,$$

which is the desired result. ■

NOTE The word statement of the property for the logarithm of a product can also be stated by replacing "logarithm" with "exponent," and the rule then becomes the familiar rule for multiplying exponential expressions: The *exponent* of a product is equal to the sum of the *exponents* of the factors.

EXAMPLE 1
USING THE PROPERTY FOR THE LOGARITHM OF A PRODUCT

Use the property for the logarithm of a product to rewrite the following (assume $x > 0$).

(a) $\log_5 (6 \cdot 9)$
By the property for the logarithm of a product,

$$\log_5 (6 \cdot 9) = \log_5 6 + \log_5 9.$$

(b) $\log_7 8 + \log_7 12$

$$\log_7 8 + \log_7 12 = \log_7 (8 \cdot 12) = \log_7 96$$

(c) $\log_3 3x$

$$\log_3 3x = \log_3 3 + \log_3 x$$
$$\log_3 3x = 1 + \log_3 x \qquad \text{Log}_3\ 3 = 1$$

(d) $\log_4 x^3$
Since $x^3 = x \cdot x \cdot x$,

$$\begin{aligned}\log_4 x^3 &= \log_4 (x \cdot x \cdot x)\\ &= \log_4 x + \log_4 x + \log_4 x\\ &= 3 \log_4 x. \quad \blacksquare\end{aligned}$$

2 The rule for division is similar to the rule for multiplication.

LOGARITHM OF A QUOTIENT

If x, y, and b are positive numbers, where $b \neq 1$, then

$$\log_b \frac{x}{y} = \log_b x - \log_b y.$$

In words, the logarithm of a quotient is equal to the logarithm of the numerator minus the logarithm of the denominator.

The proof of this rule is similar to the proof of the product rule. (See Exercise 49.)

EXAMPLE 2
USING THE PROPERTY FOR THE LOGARITHM OF A QUOTIENT

Use the property for the logarithm of a quotient to rewrite the following.

(a) $\log_4 \frac{7}{9} = \log_4 7 - \log_4 9$

(b) If $x > 0$, then $\log_5 6 - \log_5 x = \log_5 \frac{6}{x}$.

(c) $\log_3 \frac{27}{5} = \log_3 27 - \log_3 5$

$$= 3 - \log_3 5 \qquad \text{Log}_3\ 27 = 3 \quad \blacksquare$$

3 The next property gives a method for evaluating powers and roots of numbers such as

$$2^{\sqrt{2}}, \qquad (\sqrt{2})^{3/4}, \qquad (.032)^{5/8}, \qquad \text{and} \qquad \sqrt[5]{12}.$$

This property makes it possible to find approximations for numbers that could not be evaluated before. Example 1(d) showed that $\log_4 x^3 = 3 \log_4 x$, which suggests the following generalization.

LOGARITHM OF A POWER

If x and b are positive real numbers, where $b \neq 1$, and if r is any real number, then

$$\log_b x^r = r(\log_b x).$$

In words, the logarithm of a number to a power is equal to the exponent times the logarithm of the number.

Proof

Let $\log_b x = m$. Then, changing to exponential form,

$$b^m = x.$$

Now raise both sides to the power r.

$$(b^m)^r = x^r$$
$$b^{mr} = x^r$$

Change back to logarithmic form.

$$\log_b x^r = mr = rm$$

Substitute $\log_b x$ for m, to get the desired result.

$$\log_b x^r = r \log_b x \quad \blacksquare$$

As examples of this result,

$$\log_b m^5 = 5 \log_b m \qquad \text{and} \qquad \log_3 5^{3/4} = \frac{3}{4} \log_3 5.$$

As a special case of the rule above, let $r = \dfrac{1}{p}$, so that $\log_b \sqrt[p]{x} = \log_b x^{1/p} = \dfrac{1}{p} \log_b x$. For example, using this result, with $x > 0$,

$$\log_b \sqrt[5]{x} = \frac{1}{5} \log_b x \qquad \text{and} \qquad \log_b \sqrt[3]{x^4} = \frac{4}{3} \log_b x.$$

EXAMPLE 3
USING THE PROPERTY FOR THE LOGARITHM OF A POWER

Use the property for the logarithm of a power to rewrite each of the following. Assume $a > 0$, $b > 0$, $x > 0$, $a \neq 1$, and $b \neq 1$.

(a) $\log_3 5^2 = 2 \log_3 5$ **(b)** $\log_a x^4 = 4 \log_a x$

(c) $\log_b \sqrt{7}$

When using the property for the logarithm of a power with logarithms of expressions involving radicals, begin by rewriting the radical expression with a rational exponent, as shown in Section 5.1.

$$\log_b \sqrt{7} = \log_b 7^{1/2} \qquad \sqrt{x} = x^{1/2}$$
$$= \frac{1}{2} \log_b 7 \qquad \text{Logarithm of a power}$$

(d) $\log_2 \sqrt[5]{x^2} = \log_2 x^{2/5} = \frac{2}{5} \log_2 x$ ■

Two special properties involving both exponential and logarithmic expressions come directly from the fact that logarithmic and exponential functions are inverses of each other.

THEOREM ON INVERSES

If $b > 0$ and $b \neq 1$, then

$$b^{\log_b x} = x \quad (x > 0) \quad \text{and} \quad \log_b b^x = x.$$

Proof

To prove the first statement, let $y = \log_b x$, which can be written in exponential form as $b^y = x$. Now replace y with $\log_b x$, to get $b^{\log_b x} = x$. The proof of the second statement is similar. (See Exercise 50.) ■

EXAMPLE 4 USING THE THEOREM ON INVERSES

Find the value of the following logarithmic expressions.

(a) $\log_5 5^4$

Since $\log_b b^x = x$, $\log_5 5^4 = \mathbf{4}$.

(b) $\log_3 9$

Since $9 = 3^2$, $\log_3 9 = \log_3 3^2 = \mathbf{2}$. The property $\log_b b^x = x$ was used in the last step.

(c) $4^{\log_4 10} = \mathbf{10}$ ■

4 The properties of logarithms are useful for writing expressions in an alternative form. This use of logarithms is important in calculus.

EXAMPLE 5 WRITING LOGARITHMS IN ALTERNATIVE FORMS

Use the properties of logarithms to rewrite each expression. Assume all variables represent positive real numbers.

(a) $\log_4 4x^3 = \log_4 4 + \log_4 x^3$ Logarithm of a product

$= 1 + 3 \log_4 x$ Logarithm of a power; $\log_4 4 = 1$

(b) $\log_7 \sqrt{\dfrac{p}{q}} = \log_7 \left(\dfrac{p}{q}\right)^{1/2}$

$= \dfrac{1}{2} \log_7 \dfrac{p}{q}$ Logarithm of a power

$= \dfrac{1}{2}(\log_7 p - \log_7 q)$ Logarithm of a quotient

(c) $\log_5 \dfrac{a}{bc} = \log_5 a - \log_5 (bc)$ Logarithm of a quotient

$= \log_5 a - (\log_5 b + \log_5 c)$ Logarithm of a product

$= \log_5 a - \log_5 b - \log_5 c$ Distributive property

Notice the careful use of parentheses in the second step. Since we are subtracting the logarithm of a product, and it is being rewritten as a sum of two terms, parentheses *must* be placed around the sum.

(d) $\log_8 (2p + 3r)$ cannot be rewritten by the properties of logarithms. ■

CAUTION Remember that there is no property of logarithms to rewrite logarithms of a *sum* or *difference*. For example, we *cannot* write $\log_b (x + y)$ in terms of $\log_b x$ and $\log_b y$.

In the next example, we use numerical values for $\log_2 5$ and $\log_2 3$. While we use the equals sign to give these values, they are actually just approximations, since most logarithms of this type are irrational numbers. While it would be more correct to use the symbol $\approx$, we will simply use = with the understanding that the values are correct to four decimal places.

EXAMPLE 6 USING THE PROPERTIES OF LOGARITHMS WITH NUMERICAL VALUES

Given that $\log_2 5 = 2.3219$ and $\log_2 3 = 1.5850$, evaluate the following.

(a) $\log_2 15$

Since 15 is $3 \cdot 5$, use the property for the logarithm of a product.

$$\log_2 15 = \log_2 3 \cdot 5$$
$$= \log_2 3 + \log_2 5$$
$$= 1.5850 + 2.3219$$
$$= 3.9069$$

(b) $\log_2 .6$

Since $.6 = 6/10 = 3/5$, the property for the logarithm of a quotient gives

$$\log_2 .6 = \log_2 \frac{3}{5}$$
$$= \log_2 3 - \log_2 5$$
$$= 1.5850 - 2.3219$$
$$= -.7369.$$

(c) $\log_2 27$

Use the property for the logarithm of a power and the fact that $27 = 3^3$.

$$\begin{aligned}\log_2 27 &= \log_2 3^3\\ &= 3 \log_2 3\\ &= 3(1.5850)\\ &= 4.7550 \quad \blacksquare\end{aligned}$$

EXAMPLE 7
DECIDING WHETHER STATEMENTS ABOUT LOGARITHMS ARE TRUE

Decide whether each of the following statements is true or false.

(a) $\log_2 8 - \log_2 4 = \log_2 4$

Since $\log_2 8 = 3$ and $\log_2 4 = 2$, by substitution into the given statement we get

$$3 - 2 = 2,$$

which is false.

(b) $\log_3 (\log_2 8) = \dfrac{\log_7 49}{\log_8 64}$

Evaluate both sides. Since $\log_2 8 = 3$,

$$\log_3 (\log_2 8) = \log_3 (3) = 1.$$

Also,

$$\frac{\log_7 49}{\log_8 64} = \frac{2}{2} = 1.$$

The statement is true. ■

10.4 EXERCISES

Express each of the following as a sum, difference, or product of logarithms, or as a single number if possible. Assume that all variables represent positive numbers. See Examples 1–5.

1. $\log_6 3/4$

2. $\log_6 5/8$

3. $\log_2 8^{1/2}$

4. $\log_3 9^{3/2}$

5. $\log_3 3^2$

6. $2^{\log_2 5}$

7. $\log_5 \dfrac{5x}{y}$

8. $\log_4 \dfrac{16p}{3q}$

9. $\log_2 \dfrac{5\sqrt{7}}{3}$

10. $\log_2 \sqrt{mn}$

11. $\log_3 \sqrt{\dfrac{ma}{b}}$

12. $\log_5 \dfrac{\sqrt[3]{x} \cdot \sqrt[4]{y}}{z}$

13. $\log_5 (9x + 4y)$

14. $\log_3 (2x + 3y)$

---

15. Refer to the "note" following the word statement of the logarithm of a product rule in this section. Now, state the rule for the logarithm of a quotient in words, replacing "logarithm" with "exponent." (See Objective 2.)

16. Explain why the statement for the logarithm of a power requires that x be a positive real number. (See Objective 3.)

17. Refer to Example 7(a). Change the right side of the equation so that the statement becomes true. Use a logarithmic expression. (See Objective 2.)

18. What is wrong with the following "proof" that $\log_2 16$ does not exist?

$$\log_2 16 = \log_2 (-4)(-4)$$
$$= \log_2 (-4) + \log_2 (-4)$$

Since the logarithm of a negative number is not defined, the final step cannot be evaluated, and so $\log_2 16$ does not exist. (See Objective 1.)

Write each of the following expressions as a single logarithm. Assume that all variables represent positive real numbers. See Examples 1–5.

19. $\log_a x + \log_a y - \log_a m$

20. $(\log_b k - \log_b m) - \log_b a$

21. $2 \log_m a - 3 \log_m b^2$

22. $\frac{1}{2} \log_y p^3q^4 - \frac{2}{3} \log_y p^4q^3$

23. $-\frac{3}{4} \log_x a^6b^8 + \frac{2}{3} \log_x a^9b^3$

24. $\log_a (pq^2) + 2 \log_a \frac{p}{q}$

25. $\log_b (x + 2) + \log_b 7x - \log_b 8$

26. $\log_h (2m + 3) + \log_h 2m - \log_h 3$

27. $2 \log_a (z + 1) + \log_a (3z + 2)$

28. $\log_b (2y + 5) - \frac{1}{2} \log_b (y + 3)$

29. $-\frac{2}{3} \log_5 5m^2 + \frac{1}{2} \log_5 25m^2$

30. $-\frac{3}{4} \log_3 16p^4 - \frac{2}{3} \log_3 8p^3$

Given that $\log_{10} 2 = .3010$ *and* $\log_{10} 3 = .4771$ *(correct to four decimal places), evaluate each of the following. See Example 6.*

31. $\log_{10} 6$

32. $\log_{10} 12$

33. $\log_{10} 16$

34. $\log_{10} 36$

35. $\log_{10} 180$

36. $\log_{10} 40$

37. Which is larger, $\log_7 2$ or $\log_6 2$? Explain.

38. Which is larger, $\log_{1/3} 2$ or $\log_{1/4} 2$? Explain.

Let $\log_b A = 2$ *and* $\log_b B = -4$. *Find the values of each pair of expressions. Are they the same or different?*

39. $\log_b AB$ and $(\log_b A) \cdot (\log_b B)$

40. $\log_b \frac{A}{B}$ and $\frac{\log_b A}{\log_b B}$

41. $\log_b B^2$ and $(\log_b B)^2$

42. $\log_b \sqrt[3]{AB}$ and $\sqrt[3]{\log_b AB}$

Answer true *or* false *for Exercises 43–48. See Example 7.*

43. $\log_{10} 2 + \log_{10} 3 = \log_{10} (2 + 3)$

44. $\log_6 24 - \log_6 4 = 1$

45. $\log_8 (\log_5 60 - \log_5 12) = 0$

46. $\log_4 (\log_{10} 100) = \frac{\log_5 5}{\log_5 25}$

47. $\frac{\log_{10} 8}{\log_{10} 16} = \frac{1}{2}$

48. $\frac{\log_{10} 10}{\log_{100} 10} = 2$

49. Prove the quotient rule for logarithms.

50. Prove: $\log_b b^x = x \quad (b > 0,\ b \neq 1)$

PREVIEW EXERCISES

Evaluate each expression. See Section 10.3.

51. $\dfrac{\log_2 4}{\log_2 8}$

52. $\dfrac{\log_3 27}{\log_3 1/3}$

53. $\dfrac{\log_{10} .1}{\log_{10} 100}$

54. $\dfrac{\log_{10} 10{,}000}{\log_{10} .001}$

10.5 EVALUATING LOGARITHMS; NATURAL LOGARITHMS

OBJECTIVES

1. EVALUATE COMMON LOGARITHMS BY USING A CALCULATOR.
2. FIND THE ANTILOGARITHM OF A COMMON LOGARITHM.
3. USE COMMON LOGARITHMS IN AN APPLICATION.
4. EVALUATE NATURAL LOGARITHMS USING A CALCULATOR.
5. FIND THE ANTILOGARITHM OF A NATURAL LOGARITHM.
6. USE NATURAL LOGARITHMS IN AN APPLICATION.
7. USE THE CHANGE-OF-BASE RULE.

FOCUS ON PROBLEM SOLVING

The number of species in a sample is given by

$$S(n) = .36 \ln\left(1 + \frac{n}{.36}\right).$$

Here n is the number of individuals in the sample and the constant .36 indicates the diversity of species in the community. Find $S(n)$ for the following values of n.

(a) 100 **(b)** 200 **(c)** 150 **(d)** 10

This problem involves the natural logarithm function defined by $f(x) = \ln x$, which has base e ($e \approx 2.718$). The number e is one of the most important numbers in mathematics. Base e logarithms occur in many applications from the natural and social sciences. This problem is Exercise 65 in the exercises for this section. After working through this section, you should be able to solve this problem.

As mentioned earlier, logarithms are important in many applications of mathematics to everyday problems, particularly in biology, engineering, economics, and social science. In this section we show how to find numerical approximations for logarithms. Traditionally base 10 logarithms have been used most extensively, since our number system is base 10. Logarithms to base 10 are called **common logarithms** and $\log_{10} x$ is abbreviated as simply $\log x$, where the base is understood to be 10.

While extensive tables of common logarithms have been available for many years, today common logarithms are nearly always evaluated using calculators. Nevertheless, a table of common logarithms is included in this text, and instructions for using the table are given in Appendix C.

1 The next example gives the results of evaluating some common logarithms using a calculator with a log key. (This may be a second function key on some calculators.) Just enter the number, then touch the log key. (For some calculators, this order may be different.) We will give all logarithms to four decimal places.

EXAMPLE 1
EVALUATING COMMON LOGARITHMS

Evaluate each logarithm.

(a) $\log 327.1 = 2.5147$

(b) $\log 437{,}000 = 5.6405$

(c) $\log .0615 = -1.2111$

In Example 1(c), $\log .0615 = -1.2111$, a negative result. The common logarithm of a number between 0 and 1 is always negative because the logarithm is the exponent on 10 that produces the number. For example,

$$10^{-1.2111} = .0615.$$

If the exponent (the logarithm) were positive, the result would be greater than 1, since $10^0 = 1$. ■

2 In Example 1(a), the number 327.1, whose common logarithm is 2.5147, is called the common **antilogarithm** (abbreviated *antilog*) of 2.5147. In the same way 437,000, in Example 1(b), is the common antilogarithm of 5.6405. In general, in the expression $y = \log_{10} x$, x is the antilog.

NOTE From Example 1,

$$\log \overbrace{327.1}^{\text{antilogarithm}} = \overbrace{2.5147}^{\text{logarithm}}.$$

In exponential form, this is written as

$$10^{\overbrace{2.5147}^{\text{logarithm}}} = \overbrace{327.1}^{\text{antilogarithm}},$$

which shows that the logarithm 2.5147 is just an exponent. A common logarithm is the exponent on 10 that produces the antilogarithm.

EXAMPLE 2
FINDING COMMON ANTILOGARITHMS

Find the common antilogarithm of each of the following.

(a) 2.6454

Use the fact that for an antilogarithm N, $10^{2.6454} = N$. Some calculators have a 10^x key. With such a calculator, enter 2.6454, then touch the 10^x key to get the antilogarithm:

$$N \approx 442.$$

On other calculators, antilogarithms are found using the INV and log keys: enter 2.6454, touch the INV key, then touch the log key to get 442. INV indicates "inverse," so the INV key followed by the log key gives the inverse of the logarithm, which is the exponential expression 10^x.

(b) -2.3686
The antilogarithm is N, where

$$10^{-2.3686} = N.$$

Using the 10^x key or the INV and log keys gives

$$N \approx .00428.$$

(c) 1.5203
The antilogarithm is 33.14 to four significant figures. ■

Most calculators will work by one of the methods we have described. There are a few that may not. If yours is one of those, see your instruction booklet. Your teacher also may be able to help you.

3 In chemistry, the **pH** of a solution is defined as follows.

$$\text{pH} = -\log\,[H_3O^+],$$

where $[H_3O^+]$ is the hydronium ion concentration in moles per liter.

The pH is a measure of the acidity or alkalinity of a solution, with water, for example, having a pH of 7. In general, acids have pH numbers less than 7, and alkaline solutions have pH values greater than 7.

EXAMPLE 3
FINDING pH

Find the pH of grapefruit with a hydronium ion concentration of 6.3×10^{-4}.
Use the definition of pH.

$$\begin{aligned} \text{pH} &= -\log\,(6.3 \times 10^{-4}) \\ &= -(\log 6.3 + \log 10^{-4}) \qquad \text{Logarithm of a product} \\ &= -[.7993 - 4] \\ &= -.7993 + 4 \approx 3.2 \end{aligned}$$

It is customary to round pH values to the nearest tenth. ■

EXAMPLE 4
FINDING HYDRONIUM ION CONCENTRATION

Find the hydronium ion concentration of drinking water with a pH of 6.5.

$$\begin{aligned} \text{pH} = 6.5 &= -\log\,[H_3O^+] \\ \log\,[H_3O^+] &= -6.5 \end{aligned}$$

Since the antilogarithm of -6.5 is 3.2×10^{-7},

$$[H_3O^+] = 3.2 \times 10^{-7}.$$ ■

4 The most important logarithms used in applications are **natural logarithms,** which have as base the number e. The number e is irrational, like π: recall, $e \approx 2.7182818$. Logarithms to base e are called natural logarithms because they occur in biology and the social sciences in natural situations that involve growth or decay. The base e logarithm of x is written $\ln x$ (read "el en x"). A graph of $y = \ln x$, the natural logarithm function, is given in Figure 10.20.

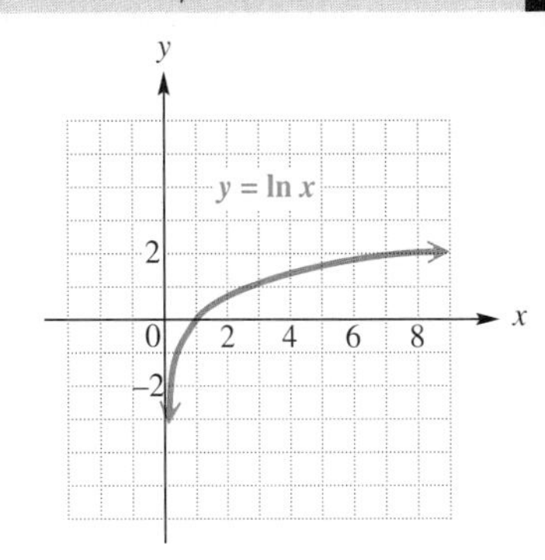

FIGURE 10.20

Natural logarithms can be found with a calculator that has an ln key or with a table of natural logarithms. A table of natural logarithms is given in Table 2. Reading directly from this table,

$$\ln 55 = 4.0073,$$

$$\ln 1.9 = 0.6419,$$

and

$$\ln .4 = -.9163.$$

The next example shows how to find natural logarithms with a calculator that has a key labeled $\ln x$.

EXAMPLE 5 **FINDING NATURAL LOGARITHMS**

Find each of the following logarithms to four significant digits.

(a) $\ln .5841$

Enter .5841 and touch the ln key to get

$$\ln .5841 \approx -.5377.$$

(Again, with some calculators, the steps are reversed.) As with common logarithms, a number between 0 and 1 has a negative natural logarithm.

(b) $\ln 192.7 = 5.261$

(c) $\ln 10.84 = 2.383$ ■

If your calculator has an e^x key, but not a key labeled $\ln x$, find natural logarithms by entering the number, touching the INV key, and then touching the e^x key. This works because $y = e^x$ is the inverse function of $y = \ln x$ (or $y = \log_e x$).

5 Antilogarithms of natural logarithms are found as for common logarithms.

$$\text{If} \quad \underset{\text{logarithm}}{\ln x = y}, \quad \text{then} \quad \underset{\text{antilogarithm}}{e^y = x},$$

since the base of natural logarithms is e. Some calculators have a key labeled e^x. With others, use the INV key and the $\ln x$ key to evaluate a natural antilogarithm.

EXAMPLE 6
FINDING NATURAL ANTILOGARITHMS

Find the natural antilogarithm of each of the following.

(a) 2.5017
Let N be the antilogarithm. Then

$$\ln N = 2.5017.$$

Find N by entering 2.5017, then touching the INV and $\ln x$ keys, getting

$$N = 12.20.$$

Alternatively, to use the e^x key, enter 2.5017 and touch the e^x key to get 12.20.

(b) -1.429
The antilogarithm is .2395.

(c) .0053
The antilogarithm is 1.0053. ■

Since the antilog of $\ln x$ is e^x, Table 2 also gives values of e^x. Use the antilogs found in Example 6 to practice using the table values.

6 The next example shows a social science application of natural logarithms.

EXAMPLE 7
APPLYING NATURAL LOGARITHMS

The number of years, $N(r)$, since two independently evolving languages split off from a common ancestral language is approximated by

$$N(r) = -5000 \ln r$$

where r is the percent of words from the ancestral language common to both languages now. Find $N(r)$ if $r = 70\%$.

Write 70% as .7 and find $N(.7)$.

$$N(.7) = -5000 \ln .7$$
$$\approx -5000(-.3567) \approx 1783$$

Approximately 1800 years have passed since the two languages separated. ■

7 A calculator (or a table) can be used to approximate the values of common logarithms (base 10) or natural logarithms (base e). However, in some applications it is convenient to use logarithms to other bases. The following rule is used to convert logarithms from one base to another.

CHANGE-OF-BASE RULE

If $a > 0$, $a \neq 1$, $b > 0$, $b \neq 1$, and $x > 0$, then

$$\log_a x = \frac{\log_b x}{\log_b a}.$$

NOTE As an aid in remembering the change-of-base rule, notice that x is "above" a on both sides of the equation.

Proof

Let $\log_a x = m$.

$$\begin{aligned} \log_a x &= m \\ a^m &= x && \text{Change to exponential form.} \\ \log_b (a^m) &= \log_b x && \text{Take logarithms on both sides.} \\ m \log_b a &= \log_b x && \text{Use the power property.} \\ (\log_a x)(\log_b a) &= \log_b x && \text{Substitute for } m. \\ \log_a x &= \frac{\log_b x}{\log_b a} && \text{Divide both sides by } \log_b a. \end{aligned}$$ ■

Any positive number other than 1 can be used for base b in the change of base rule, but usually the only practical bases are e and 10, since calculators (or tables) give logarithms only for these two bases.

The next example shows how the change-of-base rule is used to find logarithms to bases other than 10 or e with a calculator.

EXAMPLE 8
USING THE CHANGE-OF-BASE RULE

Find each logarithm using a calculator.

(a) $\log_5 12$

Use common logarithms and the rule for change-of-base.

$$\log_5 12 = \frac{\log 12}{\log 5} \approx \frac{1.0792}{.6990}$$

Use a calculator to evaluate this quotient.

$$\log_5 12 \approx \frac{1.0792}{.6990} \approx 1.5440$$

(b) $\log_2 134$

Use natural logarithms and the change-of-base rule.

$$\begin{aligned} \log_2 134 &= \frac{\ln 134}{\ln 2} \\ &\approx \frac{4.8978}{.6931} \\ &\approx 7.0661 \end{aligned}$$ ■

NOTE In Example 8, logarithms that were evaluated in intermediate steps, such as log 12 and ln 2, were shown to four decimal places. However, the final answers were obtained *without* rounding off these intermediate values, using all the digits obtained with the calculator. In general, it is best to wait until the final step to round off the answer; otherwise, a build-up of round-off error may cause the final answer to have an incorrect final decimal place digit.

The change-of-base rule can be used to find natural logarithms with a calculator that has the log x key, but not the ln x key. For example,

$$\ln 5 = \log_e 5 = \frac{\log 5}{\log e} \approx \frac{\log 5}{\log 2.718},$$

since $e \approx 2.718$. Similarly, if a calculator has only the ln x key, then the change-of-base rule could be used to find common logarithms.

An actual application of base 2 logarithms is given next.

EXAMPLE 9 FINDING THE DIVERSITY OF A SPECIES

One measure of the diversity of the species in an ecological community is given by the formula

$$H = -[P_1 \log_2 P_1 + P_2 \log_2 P_2 + \ldots + P_n \log_2 P_n],$$

where $P_1, P_2 \ldots, P_n$ are the proportions of a sample belonging to each of n species found in the sample. For example, in a community with two species, where there are 90 of one species and 10 of the other, $P_1 = 90/100 = .9$ and $P_2 = 10/100 = .1$. Thus,

$$H = -[.9 \log_2 .9 + .1 \log_2 .1].$$

Verify that $\log_2 .1 \approx -3.32$ and $\log_2 .9 \approx -.152$, so that

$$H \approx -[(.9)(-.152) + (.1)(-3.32)] \approx .469. \quad \blacksquare$$

10.5 EXERCISES

You will need to use a calculator for most of the problems in this exercise set.

Find each logarithm to four decimal places. See Examples 1 and 5. (Hint: Use the properties of logarithms first in Exercises 20–24.)

1. log 278
2. log 3460
3. log 3.28
4. log 57.3
5. log 9.83
6. log 589
7. log .327
8. log .0763
9. log .000672
10. log .00382
11. log 675,000
12. log 371,000,000
13. ln 5
14. ln 6
15. ln 697
16. ln 4.83
17. ln 1.72
18. ln 38.5
19. ln 97.6
20. $\ln e^2$
21. $\ln e^{3.1}$
22. $\ln 5e^3$
23. $\ln 4e^2$
24. $\ln 3e^{-1}$

25. Enter -1 into a scientific calculator and touch the log key. What happens? Explain why this happens.

26. Repeat Exercise 25, but use the ln key instead.

27. Which one of the following is not equal to 1? Do not use a calculator.
(a) log 1 **(b)** log 10 **(c)** ln e **(d)** $(\ln 10)^0$

28. Which one of the following is a negative number? Do not use a calculator.
(a) log 125 **(b)** ln 5 **(c)** log 1.348 **(d)** ln .322

Find the common antilogarithm of each of the following numbers. Give answers to three significant digits. See Example 2.

29. .5340 **30.** .9309 **31.** 2.6571

32. $-.0691$ **33.** -1.118 **34.** -2.3893

Find the natural antilogarithm of each of the following numbers. Give answers to three significant digits. See Example 6.

35. 3.8494 **36.** 1.7938 **37.** 1.3962

38. $-.0259$ **39.** $-.4168$ **40.** -1.2190

Use common logarithms or natural logarithms to find each of the following to four decimal places. See Example 8.

41. $\log_5 11$ **42.** $\log_6 15$ **43.** $\log_3 1.89$

44. $\log_4 7.21$ **45.** $\log_8 9.63$ **46.** $\log_2 3.42$

47. $\log_{11} 47.3$ **48.** $\log_{18} 51.2$ **49.** $\log_{50} 31.3$

50. Refer to Example 8 in this section. Work part (a) using natural logarithms, and work part (b) using common logarithms. Verify that the same final answers are obtained using these other bases.

Use the formula $\text{pH} = -\log [H_3O^+]$ *to find the pH of substances with the given hydronium ion concentrations. See Example 3.*

51. milk, 4×10^{-7} **52.** sodium hydroxide (lye), 3.2×10^{-14}

53. limes, 1.6×10^{-2} **54.** crackers, 3.9×10^{-9}

Use the formula for pH to find the hydronium ion concentrations of substances with the given pH values. See Example 4.

55. shampoo, 5.5 **56.** beer, 4.8 **57.** soda pop, 2.7 **58.** wine, 3.4

Refer to Example 7 for Exercises 59–62. How many years have elapsed since the split if the following percents of the words of the ancestral language are common to both languages today?

59. 90% **60.** 50%

Find r if two languages split from a common ancestral language the following number of years ago.

61. about 1000 **62.** about 500

Work the following problems. For Exercises 63 and 64, see Example 9.

63. Suppose a sample of a small community shows two species with 50 individuals each. Find the index of diversity H.

64. A virgin forest in northwestern Pennsylvania has 4 species of large trees with the following proportions of each: hemlock, .521; beech, .324; birch, .081; maple, .074. Find the index of diversity H.

65. The number of species in a sample is given by

$$S(n) = .36 \ln\left(1 + \frac{n}{.36}\right).$$

Here n is the number of individuals in the sample and the constant .36 indicates the diversity of species in the community.

Find $S(n)$ for the following values of n.
(a) 100 **(b)** 200 **(c)** 150 **(d)** 10

66. In Exercise 65, find $S(n)$ if .36 is changed to .88. Use the following values of n.
(a) 50
(b) 100
(c) 250

In the central Sierra Nevada mountains of California, the percent of moisture that falls as snow rather than rain is approximated reasonably well by

$$p = 86.3 \ln h - 680,$$

where p is the percent of snow moisture at an altitude h in feet. (Assume $h \geq 3000$.)

67. Find the percent of moisture that falls as snow at the following altitudes.
(a) 3000 feet
(b) 4000 feet
(c) 7000 feet

68. Graph p.

PREVIEW EXERCISES

Solve the following equations. See Sections 2.1, 4.7, 5.6, and 6.1.

69. $(3m + 2)(4) = \left(\frac{4m - 1}{5}\right)(2)$

70. $3(y - 4) = -2(2y + 5)$

71. $\frac{x + 2}{x} = 5$

72. $\frac{3z}{2z - 1} = 4$

73. $\sqrt{k + 2} = 7$

74. $(2x + 3)^3 = 27$

75. $x^2 = 48$

76. $3p^2 = 15$

10.6 EXPONENTIAL AND LOGARITHMIC EQUATIONS

OBJECTIVES

1 SOLVE EXPONENTIAL EQUATIONS.
2 SOLVE LOGARITHMIC EQUATIONS.
3 SOLVE CONTINUOUS COMPOUNDING PROBLEMS.
4 SOLVE EXPONENTIAL GROWTH AND DECAY PROBLEMS.

Carbon 14 is a radioactive form of carbon that is found in all living plants and animals. After a plant or animal dies, the radioactive Carbon 14 disintegrates according to the function

$$y = y_0 e^{-.0001212t},$$

where y is the amount of Carbon 14 after t years.

(a) If an initial sample contains $y_0 = 10$ grams of Carbon 14, how many grams will be present after 3000 years?

(b) How many grams of the initial 10 gram sample will be present after 6000 years?

(c) What is the half-life of Carbon 14?

You have probably heard of the Carbon 14 dating process used to determine the age of fossils. The method is based on the exponential decay function in the problem above. This problem is Exercise 55 in the exercises for this section. After working through this section, you should be able to solve this problem.

FOCUS ON PROBLEM SOLVING

As mentioned at the beginning of this chapter, exponential and logarithmic functions are important in many useful applications of mathematics. Using these functions in applications often requires solving exponential and logarithmic equations. Some simple equations were solved in earlier sections of this chapter. More general methods for solving these equations depend on the properties below. These properties follow from the fact that exponential and logarithmic functions are one-to-one. Property 1 was given and used to solve exponential equations in Section 10.2.

PROPERTIES OF LOGARITHMIC AND EXPONENTIAL FUNCTIONS

For $b > 0$ and $b \neq 1$:

1. $x = y$ if and only if $b^x = b^y$.
2. If $x > 0$ and $y > 0$, $x = y$ if and only if $\log_b x = \log_b y$.

1 The first examples illustrate a general method, using Property 2, for solving exponential equations.

EXAMPLE 1 SOLVING EXPONENTIAL EQUATIONS

Solve each equation.

(a) $3^x = 12$

We have seen that Property 1 cannot be used to solve this equation, so we apply Property 2. While any appropriate base can be used to apply Property 2, the best practical base to use is either base 10 or base e. Taking base e (natural) logarithms of both sides of the equation gives

$$\ln 3^x = \ln 12$$

$$x \ln 3 = \ln 12 \qquad \text{Logarithm of a power}$$

$$x = \frac{\ln 12}{\ln 3}. \qquad \text{Divide each side by } \ln 3.$$

This is the exact solution. A decimal approximation for x can be found with a calculator.

$$x = \frac{\ln 12}{\ln 3} = \frac{2.4849}{.4771} = 2.262,$$

correct to three decimal places. The solution set is $\{2.262\}$.

A calculator with a y^x key can be used to check this answer. Evaluate $3^{2.262}$; the result should be approximately 12. This step verifies that, to the nearest thousandth, the solution set is $\{2.262\}$.

(b) Solve $3^{2x-1} = 4^{x+2}$.

Taking natural logarithms on both sides gives

$$\ln 3^{2x-1} = \ln 4^{x+2}.$$

Now use properties of logarithms.

$$(2x - 1) \ln 3 = (x + 2) \ln 4 \qquad \text{Logarithm of a power}$$

$$2x \ln 3 - \ln 3 = x \ln 4 + 2 \ln 4 \qquad \text{Distributive property}$$

$$2x \ln 3 - x \ln 4 = 2 \ln 4 + \ln 3 \qquad \text{Add } \ln 3 \text{ and subtract } x \ln 4.$$

Factor out x on the left to get

$$x(2 \ln 3 - \ln 4) = 2 \ln 4 + \ln 3$$

or

$$x = \frac{2 \ln 4 + \ln 3}{2 \ln 3 - \ln 4}.$$

Using the properties of logarithms, this can be expressed as

$$x = \frac{\ln 16 + \ln 3}{\ln 9 - \ln 4}$$

or finally,

$$x = \frac{\ln 48}{\ln \frac{9}{4}} = \frac{\ln 48}{\ln 2.25}.$$

This quotient could be approximated by a decimal if desired:

$$x = \frac{\ln 48}{\ln 2.25} \approx \frac{3.8712}{.8109} \approx 4.774.$$

To the nearest thousandth, the solution set is $\{4.774\}$. ■

CAUTION Be careful when evaluating a quotient like $\frac{\ln 12}{\ln 3}$ in Example 1(a): $\frac{\ln 12}{\ln 3}$ is **not** equal to ln 4, since $\ln 4 = 1.3863$, but $\frac{\ln 12}{\ln 3} = 2.262$.

EXAMPLE 2 SOLVING AN EXPONENTIAL EQUATION

Solve $e^{-2 \ln x} = \frac{1}{16}$.

First use the property for the logarithm of a power to rewrite the exponent on the left side of the equation.

$$e^{\ln x^{-2}} = \frac{1}{16} \quad \text{Logarithm of a power}$$

$$x^{-2} = \frac{1}{16} \quad \text{Theorem on inverses}$$

$$x^{-2} = 4^{-2} \quad \frac{1}{16} = \frac{1}{4^2} = 4^{-2}$$

$$x = 4 \quad \text{Property 1}$$

Check this answer by substituting in the original equation to see that the solution set is $\{4\}$. ■

In summary, exponential equations can be solved by one of the following methods, depending upon the form of the equation.

SOLVING EXPONENTIAL EQUATIONS

1. Using Property 2, take logarithms to the same base on each side; then use the property for the logarithm of a power.
2. Write both sides as exponential expressions with the same base or the same exponent; then use Property 1.

2 The next examples illustrate ways to solve logarithmic equations. The properties of logarithms given in Section 10.4 are useful here, as is Property 2.

EXAMPLE 3 SOLVING A LOGARITHMIC EQUATION

Solve $\log_3 (x + 6) - \log_3 (x + 2) = \log_3 x$.

Using a property of logarithms, rewrite the equation as

$$\log_3 \frac{x + 6}{x + 2} = \log_3 x. \quad \text{Logarithm of a quotient}$$

Then

$$\frac{x + 6}{x + 2} = x \quad \text{Property 2}$$

$$x + 6 = x(x + 2) \quad \text{Multiply each side by } x + 2.$$

$$x + 6 = x^2 + 2x$$

$$x^2 + x - 6 = 0 \quad \text{Get 0 on one side.}$$

$$(x + 3)(x - 2) = 0 \quad \text{Factor.}$$

$$x = -3 \quad \text{or} \quad x = 2. \quad \text{Zero-factor property}$$

The negative solution ($x = -3$) cannot be used since it is not in the domain of $\log_3 x$. For this reason, the only valid solution is the positive number 2, giving the solution set $\{2\}$. ■

CAUTION Recall that the domain of $y = \log_b x$ is $(0, \infty)$. For this reason, it is always necessary to check that the solution of a logarithmic equation yields only logarithms of *positive numbers* in the original equation.

EXAMPLE 4
SOLVING A LOGARITHMIC EQUATION

Solve $\log(3x + 2) + \log(x - 1) = 1$.

Since $\log x$ is an abbreviation for $\log_{10} x$, and $1 = \log_{10} 10$ or $\log 10$, replace 1 with $\log 10$ on the right side.

$$
\begin{aligned}
\log(3x + 2) + \log(x - 1) &= \log 10 \\
\log(3x + 2)(x - 1) &= \log 10 && \text{Logarithm of a product} \\
(3x + 2)(x - 1) &= 10 && \text{Property 2} \\
3x^2 - x - 2 &= 10 \\
3x^2 - x - 12 &= 0
\end{aligned}
$$

Now use the quadratic formula to get

$$x = \frac{1 \pm \sqrt{1 + 144}}{6}.$$

If $x = (1 - \sqrt{145})/6$, then $x - 1 < 0$; therefore, $(1 - \sqrt{145})/6$ is not in the domain and this proposed solution must be discarded, giving the solution set

$$\left\{\frac{1 + \sqrt{145}}{6}\right\}. \quad ■$$

The definition of logarithm could have been used in Example 4 by first writing

$$
\begin{aligned}
\log(3x + 2) + \log(x - 1) &= 1 \\
\log_{10}(3x + 2)(x - 1) &= 1 && \text{Logarithm of a product} \\
(3x + 2)(x - 1) &= 10^1, && \text{Definition of logarithm}
\end{aligned}
$$

then continuing as shown above.

EXAMPLE 5
SOLVING A LOGARITHMIC EQUATION

Solve $\log_2 (x + 5)^3 = 4$.

Use the definition of logarithm to write the exponential expression

$$
\begin{aligned}
(x + 5)^3 &= 2^4. \\
x + 5 &= \sqrt[3]{2^4} && \text{Take the cube root on each side.} \\
x &= -5 + \sqrt[3]{2^4} && \text{Subtract 5 on each side.} \\
x &= -5 + \sqrt[3]{16}
\end{aligned}
$$

Verify that the solution satisfies the original equation, so the solution set is $\{-5 + \sqrt[3]{16}\}$. ■

EXAMPLE 6 SOLVING A LOGARITHMIC EQUATION

Solve $\ln e^{\ln x} - \ln (x - 3) = \ln 2$.

On the left, since $e^{\ln x} = x$, by the theorem on inverses,

$$\ln x - \ln (x - 3) = \ln 2 \quad \text{Substitution}$$

$$\ln \frac{x}{x - 3} = \ln 2 \quad \text{Logarithm of a quotient}$$

$$\frac{x}{x - 3} = 2 \quad \text{Property 2}$$

$$x = 2x - 6$$

$$6 = x.$$

Verify that the solution set is $\{6\}$. ■

In summary, a logarithmic equation may be solved using one or both of the following methods, where u and v are variable expressions and a and b are appropriate constants.

SOLVING LOGARITHMIC EQUATIONS

1. If $\log_a u = b$, solve by using the definition of logarithm to write the expression in exponential form as $u = a^b$.
2. If $\log_a u = \log_a v$, use Property 2 to get $u = v$. It may be necessary to use the properties of logarithms to first get the equation in the form $\log_a u = \log_a v$.

3 The compound interest formula

$$A = P\left(1 + \frac{i}{m}\right)^{mn}$$

was discussed in Section 2 of this chapter. The table presented there shows that increasing the frequency of compounding makes smaller and smaller differences in the amount of interest earned. In fact, it can be shown that even if interest is compounded at intervals of time as small as one chooses (such as each hour, each minute, or each second), the total amount of interest earned will be only slightly more than for daily compounding. This is true even for a process called **continuous compounding,** which can be described loosely as compounding every instant. As suggested in Section 2, the value of the expression $(1 + 1/m)^m$ approaches e as m gets larger. Because of this, the formula for continuous compounding involves the number e.

CONTINUOUS COMPOUNDING

If P dollars is deposited at a rate of interest r compounded continuously for t years, the final amount on deposit is

$$A = Pe^{rt}$$

dollars.

EXAMPLE 7 **SOLVING A CONTINUOUS COMPOUNDING PROBLEM**

Suppose \$5000 is deposited in an account paying 8% compounded continuously for five years. Find the total amount on deposit at the end of five years.

Let $P = 5000$, $t = 5$, and $r = .08$. Then

$$A = 5000e^{.08(5)} = 5000e^{.4}.$$

From Table 2 or a calculator, $e^{.4} \approx 1.49182$, and

$$A = 5000(1.49182) = 7459.10,$$

or \$7459.10. Check that daily compounding would have produced a compound amount about 30¢ less. ■

EXAMPLE 8 **SOLVING A CONTINUOUS COMPOUNDING PROBLEM**

How long will it take for the money in an account that is compounded continuously at 8% interest to double?

Use the formula for continuous compounding, $A = Pe^{rt}$, to find the time t that makes $A = 2P$. Substitute $2P$ for A and .08 for r; then solve for t.

$$A = Pe^{rt}$$

$$2P = Pe^{.08t} \quad \text{Substitute.}$$

$$2 = e^{.08t} \quad \text{Divide by } P.$$

Taking natural logarithms on both sides gives

$$\ln 2 = \ln e^{.08t}.$$

Use the property $\ln e^x = x$ to get $\ln e^{.08t} = .08t$.

$$\ln 2 = .08t \quad \text{Substitute.}$$

$$\frac{\ln 2}{.08} = t \quad \text{Divide by .08.}$$

$$8.664 = t$$

It will take about 8 2/3 years for the amount to double. ■

4 Applications involving exponential growth and decay were studied in Sections 10.2 and 10.3. In many cases, quantities grow or decay according to a function defined by an exponential expression with base e. The next example illustrates this.

EXAMPLE 9 **SOLVING AN EXPONENTIAL DECAY PROBLEM**

Nuclear energy derived from radioactive isotopes can be used to supply power to space vehicles. The output of the radioactive power supply for a certain satellite is given by the function

$$y = 40e^{-.004t},$$

where y is in watts and t is the time in days.

(a) How much power will be available at the end of 180 days?

Let $t = 180$ in the formula.

$$y = 40e^{-.004(180)}$$

$$y \approx 19.5 \qquad \text{Use a calculator.}$$

About 19.5 watts will be left.

(b) How long will it take for the amount of power to be half of its original strength?

The original amount of power is 40 watts. (Why?) Since half of 40 is 20, replace y with 20 in the formula, and solve for t.

$$20 = 40e^{-.004t}$$

$$.5 = e^{-.004t} \qquad \text{Divide by 40.}$$

$$\ln .5 = \ln e^{-.004t}$$

$$\ln .5 = -.004t \qquad \ln e^k = k$$

$$t = \frac{\ln .5}{-.004}$$

$$t \approx 173 \qquad \text{Use a calculator.}$$

After about 173 days, the amount of available power will be half of its original amount. ■

In Examples 8 and 9(b), we found the amount of time that it would take for an amount to double and to become half of its original amount. These are examples of *doubling time* and *half-life*. The **doubling time** of a quantity that grows exponentially is the amount of time that it takes for any initial amount to grow to twice its value. Similarly, the **half-life** of a quantity that decays exponentially is the amount of time that it takes for any initial amount to decay to half its value.

10.6 EXERCISES

A calculator with log and ln keys will be needed to work many of the exercises in this section.

Solve the following equations. When necessary, give answers as decimals rounded to the nearest thousandth. See Examples 1 and 2.

1. $3^x = 6$

2. $4^x = 12$

3. $3^{a+2} = 5$

4. $5^{2-x} = 12$

5. $6^{1-2k} = 8$

6. $3^{2m-5} = 13$

7. $e^{k-1} = 4$

8. $e^{2-y} = 12$

9. $2e^{5a+2} = 8$

10. $10e^{3z-7} = 5$

11. $2^x = -3$

12. $(1/4)^p = -4$

13. $e^{2x} \cdot e^{5x} = e^{14}$

14. $e^{\ln x} = 3$

15. $e^{-\ln x} = 2$

16. $e^{3 \ln x} = \frac{1}{8}$

17. $e^{\ln x + \ln (x-2)} = 8$

18. $e^{2 \ln x - \ln (x+2)} = \frac{8}{3}$

19. $6^{1-2k} = 8^{k+2}$

20. $3^{2m-5} = 13^{1-m}$

21. $e^{k-1} = 4^{2k}$

22. $e^{2-y} = 2^{y+2}$

23. $100(1 + .02)^{3+n} = 150$

24. $500(1 + .05)^{p/4} = 200$

Solve the following equations. When necessary give answers as decimals rounded to the nearest thousandth. See Examples 3–6.

25. $\log (t - 1) = 1$

26. $\log q^2 = 1$

27. $\log (x - 3) = 1 - \log x$

28. $\log (z - 6) = 2 - \log (z + 15)$

29. $\ln (y + 2) = \ln (y - 7) + \ln 4$

30. $\ln p - \ln (p + 1) = \ln 5$

31. $\ln (5 + 4y) - \ln (3 + y) = \ln 3$

32. $\ln m + \ln (2m + 5) = \ln 7$

33. $\ln x + 1 = \ln (x - 4)$

34. $\ln (4x - 2) = \ln 4 - \ln (x - 2)$

35. $\ln (3 - y) - \ln (2y + 1) = \ln (4y)$

36. $\ln (4k + 5) - \ln (k + 2) = \ln (3k)$

37. $\log_5 (r + 2) + \log_5 (r - 2) = 1$

38. $\log_4 (z + 3) + \log_4 (z - 3) = 1$

39. $\log_3 (a - 3) = 1 + \log_3 (a + 1)$

40. $\log w + \log (3w - 13) = 1$

41. $\ln e^x - \ln e^3 = \ln e^5$

42. $\ln e^x - 2 \ln e = \ln e^4$

43. $\log_2 \sqrt{2y^2} - 1 = 1/2$

44. $\log_2 (\log_2 x) = 1$

45. $\log z = \sqrt{\log z}$

46. $\log x^2 = (\log x)^2$

---

47. Suppose that you overhear the following statement: "I must reject any negative answer that I obtain when I solve an equation involving logarithms." Is this correct? Write an explanation of why it is or is not correct. (See Objective 2.)

48. What values of x could not possibly be solutions of the following equation?

$$\log (4x - 7) + \log (x^2 + 4) = 0$$

(See Objective 2.)

Work the following problems. In Exercises 49 and 50, refer to the formula for compound interest in Section 10.2. See Examples 7 and 8.

49. Suppose \$56,890 is deposited at 5.1% for 3 years. Find the interest earned by
(a) daily, **(b)** hourly, and
(c) continuous compounding.

50. If \$34,678.79 is deposited at 5.6% for 5 years, find the interest earned by
(a) daily, **(b)** hourly, and
(c) continuous compounding.

51. How much money must be deposited at 5% compounded continuously to amount to \$8000 in 4 years?

52. What amount will grow to \$12,000 in 5 years if interest is compounded continuously at 6%?

53. How long will it take for \$1000 to double if the money is compounded continuously at the following rates?
(a) 5% **(b)** 6% **(c)** 8%

54. At what interest rate will \$5000 grow to \$8000 if compounded continuously for 3 years?

Solve the following problems. See Example 9.

55. Carbon 14 is a radioactive form of carbon that is found in all living plants and animals. After a plant or animal dies, the radioactive Carbon 14 disintegrates according to the function

$$y = y_0 e^{-.0001212t},$$

where y is the amount of Carbon 14 after t years.

(a) If an initial sample contains $y_0 = 10$ grams of Carbon 14, how many grams will be present after 3000 years?

(b) How many grams of the initial 10 gram sample will be present after 6000 years?
(c) What is the half-life of Carbon 14?

56. Suppose an Egyptian mummy is discovered in which the amount of Carbon 14 is about half the original amount, $(1/2)y_0$. About how long ago did the Egyptian die? See Exercise 55.

57. Paint from the Lascaux caves of France contains 15% of the normal amount of Carbon 14 $(.15y_0)$. Estimate the age of the caves.

58. A round table hanging in Winchester Castle (in England) was alleged to belong to King Arthur, who lived in the 5th century. A chemical analysis recently showed that the table had 91% of the amount of radiocarbon present in living wood. How old is the table?

59. The amount of a radioactive substance present at time t measured in seconds is $A(t) = 5000e^{-.02t}$, where $A(t)$ is measured in grams. Find the half-life of the substance.

60. Find the half-life of a radioactive specimen if the amount y in grams present at time t in days is $y = 1000e^{-.4t}$.

61. The growth of bacteria in food products makes it necessary to time-date some products (such as milk) so that they will be sold and consumed before the bacteria count is too high. Suppose for a certain product that the number of bacteria present is given by

$$f(t) = 500e^{.1t},$$

under certain storage conditions, where t is time in days after packing of the product and the value of $f(t)$ is in millions. Find the number of bacteria present at each of the following times.

(a) 2 days **(b)** 1 week
(c) Suppose the product cannot be safely eaten after the bacteria count reaches 3,000,000,000. How long will this take?
(d) If $t = 0$ corresponds to January 1, what date should be placed on the product?

62. Suppose the number of rabbits in a colony is $y = y_0e^{.4t}$, where t represents time in months and y_0 is the rabbit population when $t = 0$.
(a) If $y_0 = 100$, find the number of rabbits present at time $t = 4$.
(b) How long will it take for the number of rabbits to triple?

63. A midwestern city finds its residents moving to the suburbs. Its population is declining according to the relationship $P = P_0e^{-.04t}$, where t is time measured in years and P_0 is the population at time $t = 0$. Assume that $P_0 = 1{,}000{,}000$.
(a) Find the population at time $t = 1$.
(b) Estimate the time it will take for the population to be reduced to 750,000.
(c) How long will it take for the population to be cut in half?

64. A large cloud of radioactive debris from a nuclear explosion has floated over the Pacific Northwest, contaminating much of the hay supply. Consequently, farmers in the area are concerned that the cows who eat this hay will give contaminated milk. (The tolerance level for radioactive iodine in milk is 0.) The percent of the initial amount of radioactive iodine still present in the hay after t days is approximated by $P(t) = 100e^{-.1t}$. Some scientists feel that the hay is safe after the percent of radioactive iodine has declined to 10% of the original amount. Find the number of days before the hay can be used.

PREVIEW EXERCISES

Solve each equation. See Section 2.1.

65. $3(2x - 1) - x = 4$

66. $-2(y - 1) + 3y = 8$

67. $-5(2m + 3) - m = 10$

68. $4(3k + 1) - 8k = -7$

69. $2\left(\frac{3x - 1}{2}\right) - 2x = -4$

70. $3\left(\frac{5p + 4}{3}\right) + p = 6$

71. $-4\left(\frac{t - 3}{3}\right) + 2t = 8$

72. $-2\left(\frac{3n + 1}{5}\right) + n = -4$

CHAPTER 10 GLOSSARY

KEY TERMS

10.1 one-to-one function A one-to-one function is a function in which each x-value corresponds to just one y-value and each y-value corresponds to just one x-value.

inverse of a function f If f is a one-to-one function, the inverse of f is the set of all ordered pairs of the form (y, x), where (x, y) belongs to f.

10.2 exponential equation An equation involving an exponential, where the variable is in the exponent, is an exponential equation.

10.3 logarithm A logarithm is an exponent; $\log_a x$ is the exponent on the base a that gives the number x.

logarithmic equation A logarithmic equation is an equation with a logarithm in at least one term.

10.5 common logarithm A common logarithm is a logarithm to the base 10.

antilogarithm An antilogarithm is the number that corresponds to a given logarithm.

natural logarithm A natural logarithm is a logarithm to the base e.

NEW SYMBOLS

f^{-1} the inverse of f

e a constant, approximately 2.7182818

$\log_a x$ the logarithm of x to the base a

$\log x$ common (base 10) logarithm of x

$\ln x$ natural (base e) logarithm of x

CHAPTER 10 QUICK REVIEW

SECTION	CONCEPTS	EXAMPLES
10.1 INVERSE FUNCTIONS	**Horizontal Line Test** If a horizontal line intersects the graph of a function in no more than one point, then the function is one-to-one.	Find f^{-1} if $f(x) = 2x - 3$. The graph of f is a straight line, so f is one-to-one by the horizontal line test.
	Inverse Functions For a one-to-one function f defined by an equation $y = f(x)$, solve for x. Let $x = f^{-1}(y)$. Exchange x and y to get $y = f^{-1}(x)$. Check that $(f \circ f^{-1})(x) = x$ and $(f^{-1} \circ f)(x) = x$.	Solve for x in the equation $y = 2x - 3$. $y = 2x - 3$ $y + 3 = 2x$ $\frac{y+3}{2} = x$ $\frac{y+3}{2} = f^{-1}(y)$ Let $x = f^{-1}(y)$. Replace y with x. $\frac{x+3}{2} = f^{-1}(x)$

SECTION	CONCEPTS	EXAMPLES
	The graph of f^{-1} is a mirror image of the graph of f with respect to the line $y = x$.	The graphs of a function f and its inverse f^{-1} are given below. 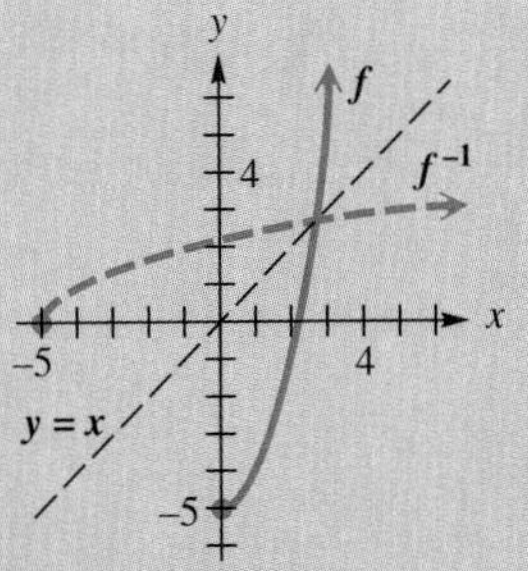
10.2 EXPONENTIAL FUNCTIONS	For $a > 0$, $a \neq 1$, $f(x) = a^x$ is an exponential function with base a.	$f(x) = 3^x$ is an exponential function with base 3. Its graph is shown here. 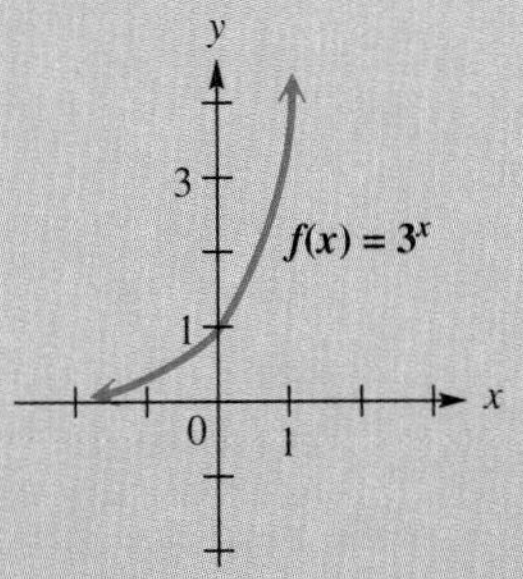
10.3 LOGARITHMIC FUNCTIONS	For $a > 0$, $x > 0$, $a \neq 1$, $y = \log_a x$ has the same meaning as $a^y = x$. For $b > 0$, $b \neq 1$, $\log_b b = 1$ and $\log_b 1 = 0$. For $a > 0$, $a \neq 1$, $x > 0$, $g(x) = \log_a x$ is the logarithmic function with base a.	$y = \log_2 x$ means $x = 2^y$. $\log_3 3 = 1$ $\log_5 1 = 0$ $g(x) = \log_3 x$ is the logarithmic function with base 3. Its graph is shown here. 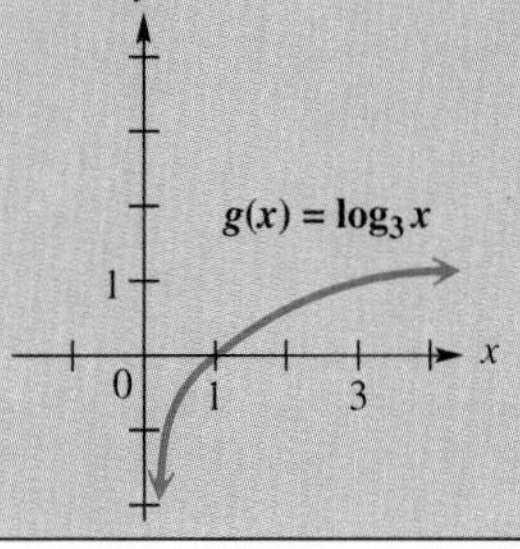

SECTION	CONCEPTS	EXAMPLES
10.4 PROPERTIES OF LOGARITHMS	**Logarithm of a Product** $\log_a xy = \log_a x + \log_a y$ **Logarithm of a Quotient** $\log_a \frac{x}{y} = \log_a x - \log_a y$ **Logarithm of a Power** $\log_a x^r = r \log_a x$ **Theorem on Inverses** $b^{\log_b x} = x$ and $\log_b b^x = x$	$\log_2 3m = \log_2 3 + \log_2 m \quad (m > 0)$ $\log_5 \frac{9}{4} = \log_5 9 - \log_5 4$ $\log_{10} 2^3 = 3 \log_{10} 2$ $6^{\log_6 10} = 10 \qquad \log_3 3^4 = 4$
10.5 EVALUATING LOGARITHMS; NATURAL LOGARITHMS	**Change-of-Base Rule** If $a > 0$, $a \neq 1$, $b > 0$, $b \neq 1$, $x > 0$, then $\log_a x = \frac{\log_b x}{\log_b a}$.	$\log_3 17 = \frac{\ln 17}{\ln 3} = \frac{\log 17}{\log 3}$ ≈ 2.579
10.6 EXPONENTIAL AND LOGARITHMIC EQUATIONS	To solve exponential equations, use these properties ($b > 0$, $b \neq 1$). **1.** If $b^x = b^y$, then $x = y$. **2.** If $x = y$, $(x > 0, y > 0)$, then $\log_b x = \log_b y$.	Solve: $2 = 2$. $3x = 5$ $x = \frac{5}{3}$ The solution set is $\{5/3\}$. Solve: $5^m = 8$. $\log 5^m = \log 8$ $m \log 5 = \log 8$ $m = \frac{\log 8}{\log 5}$ $\frac{\log 8}{\log 5} \approx 1.29$. The solution set is $\{1.29\}$.

SECTION	CONCEPTS	EXAMPLES
	To solve logarithmic equations, use these properties, where $b > 0$, $b \neq 1$, $x > 0$, $y > 0$. First use the properties of 10.4, if necessary, to get the equation in the proper form.	
	1. If $\log_b x = \log_b y$, then $x = y$.	Solve: $\log_3 2x = \log_3 (x + 1)$. $2x = x + 1$ $x = 1$ The solution set is $\{1\}$.
	2. If $\log_b x = y$, then $b^y = x$.	Solve: $\log_2 (3a - 1) = 4$. $3a - 1 = 2^4 = 16$ $3a = 17$ $a = \frac{17}{3}$ The solution set is $\left\{\frac{17}{3}\right\}$.

CHAPTER 10 REVIEW EXERCISES

[10.1] *For each equation of a one-to-one function, find the equation of the inverse.*

1. $3y = x - 1$

2. $6x - 5y = 11$

3. $f(x) = x^2 - 4$

4. $f(x) = 2 - 3x^2$

5. $y + 1 = (x + 2)^3$

6. $x - 3 = (y + 2)^2$

7. $f(x) = \sqrt{2x + 5}$

8. $y = \sqrt{3 - x}$

Graph the inverse of each function on the same axes as the function.

9.

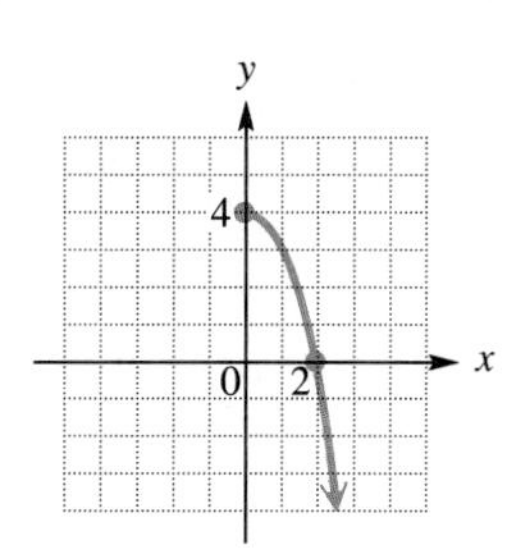

10.

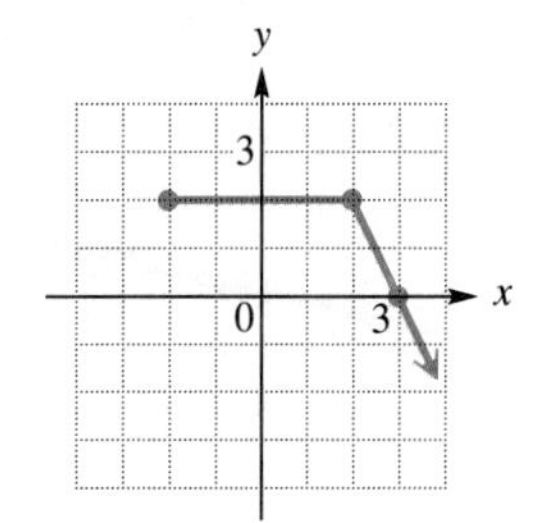

11. $3y = x - 1$

12. $f(x) = x^3 - 4$

13. True or false: Any linear function of the form $f(x) = mx + b$, $m \neq 0$, has an inverse.

14. How can you tell from the graph of a function whether it is one-to-one?

[10.2] *Graph each of the following.*

15. $y = 4^x$

16. $y = \left(\frac{1}{2}\right)^x$

17. $y = 3^{2x}$

18. $y = 2^{-x+1}$

---

19. If \$4000 is left at 6% interest compounded semi-annually, how much will be in the account at the end of 5 years?

20. A population of laboratory mice is growing according to the equation

$$y = 10e^{.5t},$$

where y is the population at time t in years.

(a) How many mice were present at time $t = 0$?
(b) How many mice will there be in 3 years?
(c) How many years will it take for the population to double?

[10.3] *Graph each of the following.*

21. $y = \log_4 x$

22. $y = \log_{1/2} x$

Convert to exponential form.

23. $\log_3 81 = 4$

24. $\log_5 \frac{1}{5} = -1$

25. $\log_8 2 = \frac{1}{3}$

Convert to logarithmic form.

26. $4^3 = 64$

27. $9^{1/2} = 3$

28. $2^{-3} = \frac{1}{8}$

[10.4] *Express each of the following as a sum, difference, or product of logarithms. Assume that all variables represent positive numbers.*

29. $\log_5 \frac{3}{10}x$

30. $\log_3 m^2 p\sqrt{g}$

31. $\log_2 \frac{5k}{3r^3}$

32. $\log_8 \frac{7m}{64}$

Write each of the following as a single logarithm. Assume that all variables represent positive numbers.

33. $\log_a 2x + \log_a 3$

34. $3 \log_b p - 2 \log_b q$

35. $\log_5 (x - 1) - \log_5 (x^2 - 1) \quad (x > 1)$

36. $\log_{10} (2x + 3) + \log_{10} 5x$

37. $4 \log_2 x - 3 \log_2 x$

38. $\log_{12} 2p - \log_{12} 5r$

[10.5] *Evaluate each logarithm correct to four decimal places. A calculator will be needed for many of the exercises in this set.*

39. $\log 2.95$

40. $\log 432$

41. $\log .0714$

42. $\log .16$

43. $\log 10^{.2}$

44. $\log 1$

45. $\ln 8$

46. $\ln 25$

47. $\ln e$

48. $\ln e^{.1}$

49. $\ln e^4$

50. $\ln 1$

Find each of the following correct to four decimal places.

51. $\log_6 10$ **52.** $\log_3 2.51$ **53.** $\log_\pi 12$ **54.** $\log_9 \sqrt{2}$

55. $\log 4e^2$ **56.** $\log 2e^{1.3}$ **57.** $\ln 4e^{1.2}$ **58.** $\ln 3e^{5.2}$

[10.6] *Solve each of the following equations.*

59. $4^{m+1} = 5$ **60.** $8^{2y-1} = 3^y$ **61.** $\log (3x - 1) = \log 10$

62. $\log_3 (p + 2) - \log_3 p = \log_3 2$ **63.** $\log (2x + 3) = \log x + 1$ **64.** $\ln (\ln e^{-x}) = \ln 3$

---

65. How much will \$10,000 compounded continuously at 6% annual interest amount to in three years?

66. Historically, the consumption of electricity has increased at a continuous rate of 6% per year. If it continued to increase at this rate, find the number of years before twice as much electricity would be needed.

67. The population of wolves in a certain area of the Arctic tundra is

$$P = 1000e^{.05t},$$

where t is time in years. Find the population at the following times.

(a) after 2 years **(b)** after 5 years

(c) after 10 years

68. How long will it take for the population in Exercise 67 to double?

MIXED REVIEW EXERCISES

Solve.

69. Use the formula $S = C(1 - r)^n$, where C is the original cost of the item, r is the annual depreciation, and n is the useful life of the item in years, to find the scrap value of equipment costing \$50,000, with a useful life of 10 years, if the annual depreciation is 10%.

70. Use the formula $S = C(1 - r)^n$ to find the constant annual depreciation rate of a machine that has lost 1/4 of its value in 2 years.

71. $27^{x+2} = 81$

72. $\log_3 k = -2$

73. If \$20,000 is deposited at 7% annual interest compounded quarterly, how much will be in the account after 5 years, assuming no withdrawals are made?

74. The height in meters of the members of a certain tribe is approximated by

$$h = .5 + \log t,$$

where t is the tribe member's age in years, and $1 \leq t \leq 20$. Estimate the height of a tribe member who is 5 years old.

75. $\log_m 125 = 3$

76. $2^{y-3} = 8$

CHAPTER 10 TEST

1. True or false: $y = x^2 + 4$ defines a one-to-one function.

2. Find the inverse of the one-to-one function $f(x) = \sqrt[3]{4 - x}$.

3. Graph the inverse of f, given the graph of f below.

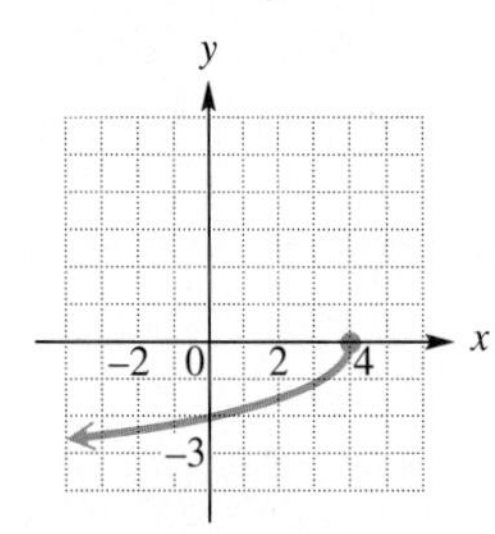

Graph the following functions.

4. $y = 2^x$ **5.** $y = \left(\frac{1}{3}\right)^x$ **6.** $y = \log_3 x$ **7.** $y = \log_{1/2} x$

Given that $\log M = 2.3156$ and $\log N = .1827$, evaluate each of the following.

8. $\log MN$ **9.** $\log \frac{M}{N}$ **10.** $\log \sqrt{M}$

Find the following.

11. $\log_6 36$ **12.** $\log_2 (1/64)$ **13.** $\log_{81} (1/27)$ **14.** $\log_5 5$

15. $\log 246$ **16.** $\log .000317$ **17.** $\log_2 10$ **18.** $\ln 89.1$

Solve the following equations.

19. $\log_9 x = \frac{5}{2}$ **20.** $3^{2x} = 3^{1-x}$ **21.** $2^m = 14$

22. $\log_2 (x + 3) + \log_2 (x - 1) = \log_2 5$ **23.** $\log_4 (2x + 4) - \log_4 (x - 2) = 1$

Solve the following problems.

24. Suppose that a culture of bacteria grows according to

$$C(t) = 8000e^{.4t},$$

where $C(t)$ is the number of bacteria present at time t (measured in hours).

(a) How many bacteria will be present at time $t = 0$?

(b) How many bacteria will be present at time $t = 5$?

(c) When will the culture contain 16,000 bacteria?

25. Suppose that \$3500 is deposited at 8% annual interest and no withdrawals are made. How much will be in the account after 4 years if interest is compounded

(a) quarterly **(b)** continuously?

THE GRAPHING CALCULATOR

One way to decide whether two functions are inverses is to graph them both on the same grid, using the same domain and range with the same x-scale and y-scale. Recall, if they are inverse functions, their graphs will be symmetric to the line $y = x$. Remember that *both* expressions must define *functions*, if they are to be inverse functions.

The functions $f(x) = 10^x$, $f(x) = e^x$, $f(x) = \log_{10} x$, and $f(x) = \log_e x = \ln x$ are built in to graphing calculators and can be graphed by touching the appropriate key after touching the graph key. Variations of these functions can be graphed as follows. To graph an exponential function with any base other than 10 or e, first use the range key to set the domain, range, and scales. Then touch the graph key and the required keys to input the function. For example, to graph $f(x) = 3^{x^2}$ with the TI-81 calculator, touch the following sequence of keys.

$y =$ 3 ^ x x^2 graph

(There may be slight differences with some calculators.) To graph $f(x) = 3^{-x^2}$, we can use the inverse function key x^{-1}, along with parentheses, as follows.

$y =$ (3 ^ x x^2) x^{-1} graph

Logarithmic functions to bases other than 10 or e can be graphed by using the change of base theorem. For example, to graph

$$f(x) = \log_3 (x - 1) = \frac{\ln (x - 1)}{\ln 3}$$

with base e logarithms, use the following sequence.

$y =$ ln (x − 1) ÷ ln 3 graph

You may want to use the following functions to experiment with your calculator. Most of these functions have been selected from examples in the text, so you can compare your calculator graphs with those given in the example figures. The figure number of the graph in the text is given with each function.

1. $f(x) = x^3 - 1$ and $g(x) = \sqrt[3]{x + 1}$ (Figure 10.6)

2. $f(x) = \sqrt{x + 5}$ and $g(x) = x^2 - 5,\ x > 0$ (Figure 10.7)

3. $f(x) = 8x - 7$ and $g(x) = \dfrac{x + 8}{7}$ (10.1 Exercise 42)

4. $f(x) = 4x - 3$ and $g(x) = \dfrac{x + 3}{4}$

5. $f(x) = 2^x$ (Figure 10.8)

6. $g(x) = \left(\dfrac{1}{2}\right)^x$ (Figure 10.9)

7. $h(x) = 2^{-x+2}$ (Figure 10.11)

8. $F(x) = 2^{-x^2}$ (Figure 10.12)

9. $f(x) = \log_2 x$ (Figure 10.15)

10. $g(x) = \log_{1/2} x$ (Figure 10.16)

11. $k(x) = \log_2 (x - 1)$ (Figure 10.17)

12. $h(x) = \ln (2x + 1)$

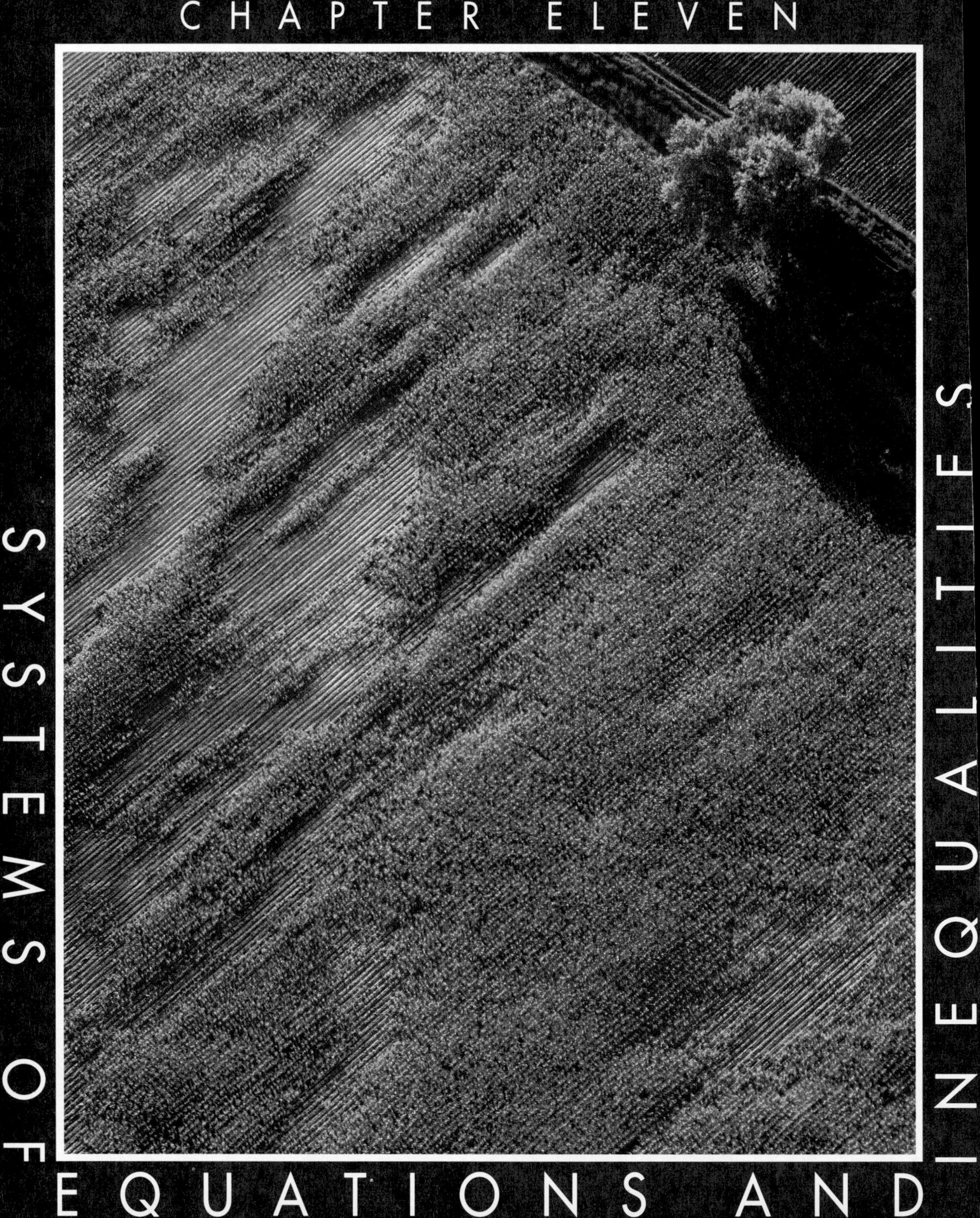
CHAPTER ELEVEN
SYSTEMS OF
EQUATIONS AND
INEQUALITIES

Many of the applied problems studied earlier are more easily solved by writing two or more equations and finding a solution common to this set of equations. For example, suppose that a health food store stocks two brands of wheat germ, which cost the store \$3 and \$2 per pound. The shipping costs for the two brands are \$1 and \$1.50 per pound, respectively. The manager can spend a total of \$250 to buy the wheat germ and \$100 to ship it. How many pounds of each brand should be ordered? Since two quantities are to be found here, it is useful to use two variables and write two equations involving those variables. In this chapter we show how to solve such problems.

11.1 LINEAR SYSTEMS OF EQUATIONS IN TWO VARIABLES

OBJECTIVES

1. SOLVE LINEAR SYSTEMS BY THE ELIMINATION METHOD.
2. SOLVE LINEAR SYSTEMS BY THE SUBSTITUTION METHOD.
3. IDENTIFY LINEAR SYSTEMS THAT ARE INCONSISTENT OR THAT HAVE DEPENDENT EQUATIONS.
4. SOLVE LINEAR SYSTEMS THAT HAVE FRACTIONS.

Many applications of mathematics require the simultaneous solution of a large number of equations having many variables. Such a set of equations is called a **system of equations.** The definition of a linear equation given earlier can be extended to more variables: Any equation of the form

$$a_1x_1 + a_2x_2 + \ldots + a_nx_n = b$$

for real numbers $a_1, a_2, \ldots, a_n$ (not all of which are 0), and b, is a **linear equation.** If all the equations in a system are linear, the system is a **system of linear equations,** or a **linear system.**

In Figure 11.1, the two linear equations $x + y = 5$ and $2x - y = 4$ are graphed on the same axes. Notice that they intersect at the point (3, 2). Because (3, 2) is the only ordered pair that satisfies both equations at the same time, we say that {(3, 2)} is the solution set of the system

$$x + y = 5$$
$$2x - y = 4.$$

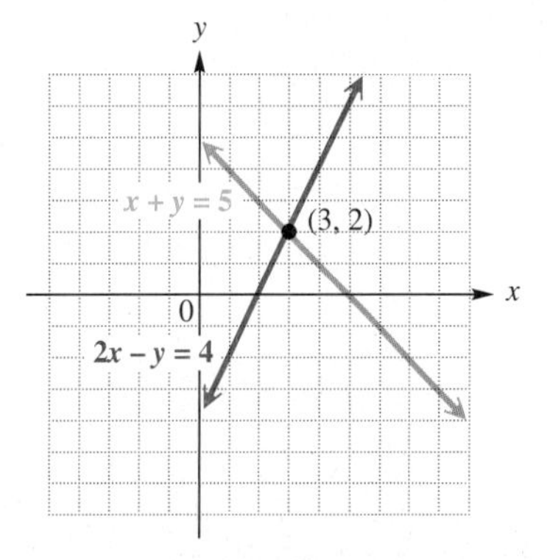

FIGURE 11.1

Since the graph of a linear equation is a straight line, there are three possibilities for the solution of a system of two linear equations, as shown in Figure 11.2.

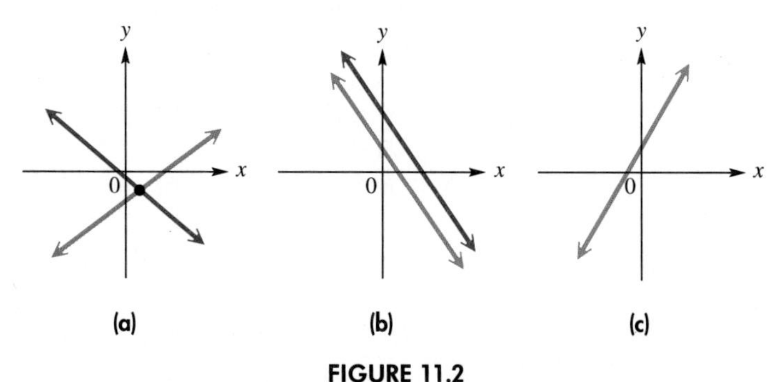

FIGURE 11.2

GRAPHS OF A LINEAR SYSTEM

1. The two graphs intersect in a single point. The coordinates of this point give the solution of the system. This is the most common case. See Figure 11.2(a).
2. The graphs are parallel lines. In this case the system is **inconsistent;** that is, there is no solution common to both equations of the system, and the solution set is $\emptyset$. See Figure 11.2(b).
3. The graphs are the same line. In this case the equations are **dependent,** since any solution of one equation of the system is also a solution of the other. The solution set is an infinite set of ordered pairs representing the points on the line. See Figure 11.2(c).

1 It is possible to find the solution of a system of equations by graphing. However, since it can be hard to read exact coordinates from a graph, an algebraic method is usually used to solve a system. One such algebraic method, called the **elimination method** (or the **addition method**), is explained in the following examples.

EXAMPLE 1
SOLVING A SYSTEM BY ELIMINATION

Solve the system

$$2x + 3y = 2 \qquad (1)$$
$$x - 3y = 10. \qquad (2)$$

The elimination method involves combining the two equations so that one variable is eliminated. This is done using the following fact.

If $a = b$ and $c = d$, then $a + c = b + d$.

Adding corresponding sides of equations (1) and (2) gives

$$\begin{array}{r} 2x + 3y = 2 \\ x - 3y = 10 \\ \hline 3x \qquad = 12. \end{array}$$

Dividing both sides of the equation $3x = 12$ by 3, we get

$$x = 4.$$

To find y, replace x with 4 in either equation (1) or equation (2). Using (1) gives

$$\begin{aligned} 2x + 3y &= 2 \\ 2(4) + 3y &= 2 && \text{Let } x = 4. \\ 8 + 3y &= 2 && \text{Multiply.} \\ 3y &= -6 && \text{Subtract 8.} \\ y &= -2. && \text{Divide by 3.} \end{aligned}$$

The solution of the system is $x = 4$ and $y = -2$, written as the ordered pair $(4, -2)$. The solution set is $\{(4, -2)\}$. Check this solution by substituting 4 for x and -2 for y in both equations of the original system. ■

By adding the equations in Example 1, the variable y was eliminated because the coefficients of y were opposites. In many cases the coefficients will *not* be opposites. In these cases it is necessary to transform one or both equations so that the coefficients of one of the variables are opposites. The general method of solving a system by the elimination method is summarized as follows.

SOLVING LINEAR SYSTEMS BY ELIMINATION

Step 1 Write both equations in the form $Ax + By = C$.

Step 2 Multiply one or both equations by appropriate numbers so that the sum of the coefficients of either x or y is zero.

Step 3 Add the new equations. The sum should be an equation with just one variable.

Step 4 Solve the equation from Step 3.

Step 5 Substitute the result of Step 4 into either of the given equations and solve for the other variable.

Step 6 Check the solution in both of the given equations.

EXAMPLE 2
SOLVING A SYSTEM BY ELIMINATION

Solve the system

$$5x - 2y = 4 \qquad \textbf{(3)}$$
$$2x + 3y = 13. \qquad \textbf{(4)}$$

Both equations are in the form $Ax + By = C$. Suppose that we wish to eliminate the variable x. In order to do this, multiply equation (3) by 2 and equation (4) by -5.

$$\begin{aligned} 10x - 4y &= 8 && \text{2 times each side of equation (3)} \\ -10x - 15y &= -65 && \text{−5 times each side of equation (4)} \end{aligned}$$

Now add.

$$\begin{array}{r} 10x - 4y = 8 \\ \underline{-10x - 15y = -65} \\ -19y = -57 \\ y = 3 \end{array}$$

To find x, substitute 3 for y in either equation (3) or equation (4). Substituting in equation (4) gives

$$\begin{aligned} 2x + 3y &= 13 \\ 2x + 3(3) &= 13 && \text{Let } y = 3. \\ 2x + 9 &= 13 \\ 2x &= 4 && \text{Subtract 9.} \\ x &= 2. && \text{Divide by 2.} \end{aligned}$$

The solution set of the system is $\{(2, 3)\}$. Check this solution in both equations of the given system. ■

2 Linear systems can also be solved by the **substitution method.** The substitution method is most useful in solving linear systems in which one variable has a coefficient of 1. However, as shown in Section 11.4, the substitution method is the best choice for solving many *nonlinear* systems.

The method of solving a system by substitution is summarized as follows.

SOLVING LINEAR SYSTEMS BY SUBSTITUTION

Step 1 Solve one of the equations for either variable. (If one of the variables has coefficient 1 or -1, choose it, since the substitution method is usually easier this way.)

Step 2 Substitute for that variable in the other equation. The result should be an equation with just one variable.

Step 3 Solve the equation from Step 2.

Step 4 Substitute the result from Step 3 into the equation from Step 1 to find the value of the other variable.

Step 5 Check the solution in both of the given equations.

The next two examples illustrate this method.

EXAMPLE 3
SOLVING A SYSTEM BY SUBSTITUTION

Solve the system

$$\frac{x}{2} + \frac{y}{3} = \frac{13}{6} \qquad \textbf{(5)}$$

$$4x - y = -1. \qquad \textbf{(6)}$$

Write equation (5) without fractions by multiplying both sides by the common denominator, 6.

$$6 \cdot \frac{x}{2} + 6 \cdot \frac{y}{3} = 6 \cdot \frac{13}{6} \qquad \text{Multiply by 6.}$$

$$3x + 2y = 13 \qquad \text{Distributive property} \qquad \textbf{(7)}$$

The new system is

$$\begin{aligned} 3x + 2y &= 13 && (7) \\ 4x - y &= -1. && (6) \end{aligned}$$

To use the substitution method, first solve one of the equations for either x or y. Since the coefficient of y in equation (6) is -1, it is easiest to solve for y in equation (6).

$$-y = -1 - 4x$$
$$y = 1 + 4x.$$

Substitute $1 + 4x$ for y in equation (7), and solve for x.

$$3x + 2(1 + 4x) = 13 \quad \text{Let } y = 1 + 4x.$$
$$3x + 2 + 8x = 13 \quad \text{Distributive property}$$
$$11x = 11 \quad \text{Combine terms; subtract 2.}$$
$$x = 1 \quad \text{Divide by 11.}$$

Since $y = 1 + 4x$, let $x = 1$ to get

$$y = 1 + 4(1) = 5.$$

Check that the solution set is $\{(1, 5)\}$. ■

EXAMPLE 4 SOLVING A SYSTEM BY SUBSTITUTION

Solve the system

$$4x - 3y = 7 \qquad \textbf{(8)}$$
$$3x - 2y = 6. \qquad \textbf{(9)}$$

If the substitution method is to be used, one equation must be solved for one of the two variables. Let us solve equation (9) for x.

$$3x = 2y + 6$$
$$x = \frac{2y + 6}{3}$$

Now substitute $\frac{2y + 6}{3}$ for x in equation (8).

$$4x - 3y = 7$$
$$4\left(\frac{2y + 6}{3}\right) - 3y = 7$$

Multiply both sides of the equation by the common denominator 3 to eliminate the fraction.

$$4(2y + 6) - 9y = 21 \quad \text{Multiply by 3.}$$
$$8y + 24 - 9y = 21 \quad \text{Distributive property}$$
$$24 - y = 21 \quad \text{Combine terms.}$$
$$-y = -3 \quad \text{Add } -24.$$
$$y = 3 \quad \text{Divide by } -1.$$

Since $x = \frac{2y + 6}{3}$ and $y = 3$,

$$x = \frac{2(3) + 6}{3} = \frac{6 + 6}{3} = 4,$$

and the solution set is $\{(4, 3)\}$. ■

The substitution method is not usually the best choice for a system like the one in Example 4. However, it is sometimes necessary when solving a system of *nonlinear* equations to proceed as shown in this example.

3 The next examples illustrate special cases that may result when systems are solved. (We will use the elimination method, but the same conclusions will be true when the substitution method is used.)

EXAMPLE 5 RECOGNIZING AN INCONSISTENT SYSTEM

Solve the system

$$3x - 2y = 4 \qquad \textbf{(10)}$$
$$-6x + 4y = 7. \qquad \textbf{(11)}$$

The variable x can be eliminated by multiplying both sides of equation (10) by 2 and then adding.

$$\begin{array}{rl} 6x - 4y = 8 & \text{2 times equation (10)} \\ \underline{-6x + 4y = 7} & \\ 0 = 15 & \text{False} \end{array}$$

Both variables were eliminated here, leaving the false statement $0 = 15$, a signal that these two equations have no solutions in common. The system is inconsistent, with an empty solution set. In slope-intercept form, the equations would show that the graphs are parallel lines because they have the same slope but different y-intercepts. ■

EXAMPLE 6 SOLVING A SYSTEM OF DEPENDENT EQUATIONS

Solve the system

$$-4x + y = 2 \qquad \textbf{(12)}$$
$$8x - 2y = -4. \qquad \textbf{(13)}$$

Eliminate x by multiplying both sides of equation (12) by 2 and then adding it to equation (13).

$$\begin{array}{rl} -8x + 2y = 4 & \text{2 times equation (12)} \\ \underline{8x - 2y = -4} & \\ 0 = 0 & \text{True} \end{array}$$

This true statement, $0 = 0$, indicates that a solution of one equation is also a solution of the other, so the solution set is an infinite set of ordered pairs. The graphs of the equations are the same line, since the slopes and y-intercepts are equal. The two equations are dependent.

We will write the solution of a system of dependent equations as an ordered pair by expressing x in terms of y as follows. Choose either equation and solve for x. Choosing equation (12) gives

$$-4x + y = 2$$
$$x = \frac{2 - y}{-4} = \frac{y - 2}{4}.$$

The solution set is written as

$$\left\{\left(\frac{y-2}{4}, y\right)\right\}.$$

By selecting values for y and calculating the corresponding values for x, individual ordered pairs of the solution set can be found. For example, if $y = -2$, $x = (-2 - 2)/4 = -1$ and the ordered pair $(-1, -2)$ is a solution. ■

Examples 5 and 6 suggest the following summary.

SPECIAL CASES FOR SYSTEMS

Consider the system of equations

$$a_1x + b_1y = c_1$$
$$a_2x + b_2y = c_2,$$

with $a_2 \neq 0$, $b_2 \neq 0$, and $c_2 \neq 0$. The graphs of the equations are parallel lines and the system is inconsistent if

$$\frac{a_1}{a_2} = \frac{b_1}{b_2} \neq \frac{c_1}{c_2}.$$

(See Figure 11.2(b).) The graphs of the equations are the same line and the equations are dependent if

$$\frac{a_1}{a_2} = \frac{b_1}{b_2} = \frac{c_1}{c_2}.$$

(See Figure 11.2(c).)

4 If one or more of the equations in a system involve fractions, multiply by the least common denominator before solving.

EXAMPLE 7

SOLVING A SYSTEM WITH FRACTIONS AS COEFFICIENTS

Solve the system

$$\frac{x}{2} + \frac{y}{3} = \frac{13}{6} \qquad \textbf{(14)}$$

$$\frac{x}{5} - \frac{y}{3} = \frac{-22}{15}. \qquad \textbf{(15)}$$

Write the equations without fractions by multiplying both sides of equation (14) by 6 and both sides of equation (15) by 15.

$$6 \cdot \frac{x}{2} + 6 \cdot \frac{y}{3} = 6 \cdot \frac{13}{6} \qquad \text{6 times equation (14)}$$

$$15 \cdot \frac{x}{5} - 15 \cdot \frac{y}{3} = 15 \cdot \frac{-22}{15} \qquad \text{15 times equation (15)}$$

Multiplying gives the equivalent system

$$3x + 2y = 13 \qquad \textbf{(16)}$$

$$3x - 5y = -22. \qquad \textbf{(17)}$$

This system can be solved by elimination. Multiply equation (16) by -1 and add to equation (17).

$$\begin{aligned} -3x - 2y &= -13 && \text{−1 times equation (16)} \\ 3x - 5y &= -22 && \\ \hline -7y &= -35 && \text{Add.} \\ y &= 5 && \text{Divide by −7.} \end{aligned}$$

To solve for x, substitute into either of the original equations to find that $x = 1$. The solution set of the system is $\{(1, 5)\}$. ■

11.1 EXERCISES

Decide whether the systems in Exercises 1–10 have a single solution, no solution, or infinitely many solutions.

1. One line has positive slope and one line has negative slope.

2. One line has slope 0 and one line has undefined slope.

3. Both lines have slope 0 and have the same y-intercept.

4. Both lines have undefined slope and have the same x-intercept.

5. Both lines have undefined slope and have different x-intercepts.

6. Both lines have slope 0 and have different y-intercepts.

7. $y = 3x + 6$
$y = 3x + 9$

8. $y = 4x + 8$
$y = 4x + 12$

9. $x + y = 4$
$kx + ky = 4k \ (k \neq 0)$

10. $x + y = 10$
$kx + ky = 9k \ (k \neq 0)$

Use the elimination method to solve the following systems of linear equations. State which pairs of equations are inconsistent or dependent. See Examples 1, 2, and 5–7.

11. $x + y = 10$
$2x - y = 5$

12. $3x - 2y = 4$
$3x + y = -2$

13. $2x - 3y = 3$
$2x + 2y = 8$

14. $6x + y = 5$
$5x + y = 3$

15. $2x - 5y = 10$
$3x + y = 15$

16. $4x - y = 3$
$-2x + 3y = 1$

17. $7x + 2y = 3$
$-14x - 4y = -6$

18. $4x + 2y = 3$
$-3x - 3y = 0$

19. $-3x + 5y = 2$
$2x - 3y = 1$

20. $4x - 16y = 4$
$x - 4y = 1$

21. $5x + 3y = 1$
$-3x - 4y = 6$

22. $6x - 6y = 5$
$x - y = 8$

23. $2x = 5y + 10$
$10y = 15 + 4x$

24. $4x = 2y + 3$
$y = 2x + 1$

25. $\frac{x}{2} + \frac{y}{3} = -\frac{1}{3}$
$\frac{x}{2} + 2y = -7$

26. $x + 5y = 3$
$\frac{2}{3}x + \frac{10}{3}y = 2$

27. $\frac{1}{4}x + \frac{5}{4}y = -\frac{1}{2}$
$-x - 5y = 2$

28. $\frac{x}{3} + \frac{y}{5} = 2$
$\frac{x}{3} - \frac{y}{2} = -\frac{1}{3}$

29. Refer to Example 2. What other numbers might equations (3) and (4) be multiplied by to eliminate y by adding the two equations? (See Objective 1.)

30. What makes a system of equations inconsistent? (See Objective 3.)

Solve these systems of linear equations by the substitution method. See Examples 3 and 4.

31. $4x + y = 9$
$x - 2y = 0$

32. $3x - 4y = -22$
$-3x + y = 0$

33. $2x = y + 6$
$y = 5x$

34. $-3x = -y - 5$
$x = -2y$

35. $x - 4y = 2$
$2x = 8y + 1$

36. $3x + 5y = 17$
$4x - y = -8$

37. $5x - 4y = 9$
$3 + x = 2y$

38. $6x - y = -9$
$4 + y = -7x$

39. $x = 3y + 5$
$y = \frac{2}{3}x$

40. $3x = \frac{3}{4}y - 2$
$x = \frac{1}{4}y$

41. $\frac{x}{2} + \frac{y}{3} = 3$
$2y = 3x$

42. $\frac{x}{4} - \frac{y}{5} = 9$
$y = 5x$

43. $-2x + 3y = -1$
$4y = x + 2$

44. $6x = 2y + 6$
$3x - 2y = 3$

---

45. What must be true for a system of two linear equations (in two variables) to have no solution? (See Objective 3.)

46. What must be true for a system of two linear equations (in two variables) to have an infinite number of solutions? (See Objective 3.)

In the following systems, let $p = 1/x$ and $q = 1/y$. Substitute, solve for p and q, and then find x and y. (Hint: $3/x = 3 \cdot 1/x = 3p$.)

47. $\frac{3}{x} + \frac{4}{y} = \frac{5}{2}$
$\frac{5}{x} - \frac{3}{y} = \frac{7}{4}$

48. $\frac{4}{x} - \frac{9}{y} = -1$
$-\frac{7}{x} + \frac{6}{y} = -\frac{3}{2}$

49. $\frac{2}{x} - \frac{5}{y} = \frac{3}{2}$
$\frac{4}{x} + \frac{1}{y} = \frac{4}{5}$

50. $\frac{2}{x} + \frac{3}{y} = \frac{11}{2}$
$-\frac{1}{x} + \frac{2}{y} = -1$

Solve by any method. Here a and b represent nonzero constants.

51. $x = -y$
$y + x = 6$

52. $x = -6y$
$x - y = 14$

53. $x + 2y = 4$
$x - \frac{1}{2}y = 4$

54. $y - \frac{1}{3}x = 0$
$2x - 3y = 6$

55. $ax + by = 2$
$-ax + 2by = 1$

56. $2ax - y = 3$
$y = 5ax$

57. $3ax + 2y = 1$
$-ax + y = 2$

58. $ax + by = c$
$bx + ay = c$

Work the following problems that involve graphs of linear systems.

59. The following graph shows the expected annual return on \$10,000 for two different investment opportunities.

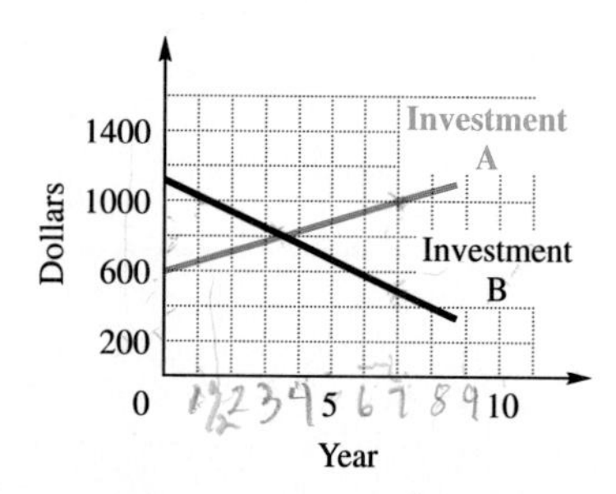

(a) What is the annual return for each investment after seven years?

(b) In what year is the return from the two investments the same? What is the return at that time?

(c) How many years does it take in each case to get an annual return of \$1000?

60. Julio Nicolai compared the monthly interest costs for two types of home mortgages: a fixed-rate mortgage and a graduated-payment mortgage. His results are shown in the figure.

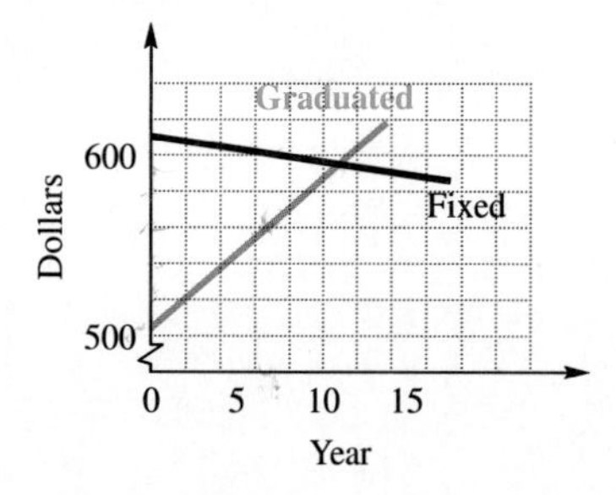

(a) In how many years will the monthly interest costs be equal for the two plans?

(b) What is the monthly interest in the seventh year for each plan?

(c) In what year will the monthly interest cost be \$590 for each plan?

61. The following graph shows a company's costs to produce computer parts and the revenue from the sale of computer parts.

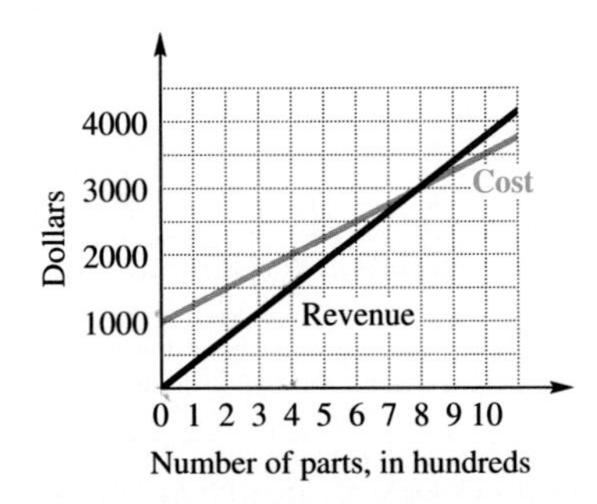

(a) At what production level does the cost equal the revenue? What is the revenue at that point?

(b) Profit is revenue less cost. Estimate the profit on the sale of 400 parts.

(c) The cost to produce 0 items is called the *fixed cost*. Find the fixed cost.

62. The following figure shows graphs that represent the supply and demand for one brand of frozen yogurt at various prices per half-gallon.

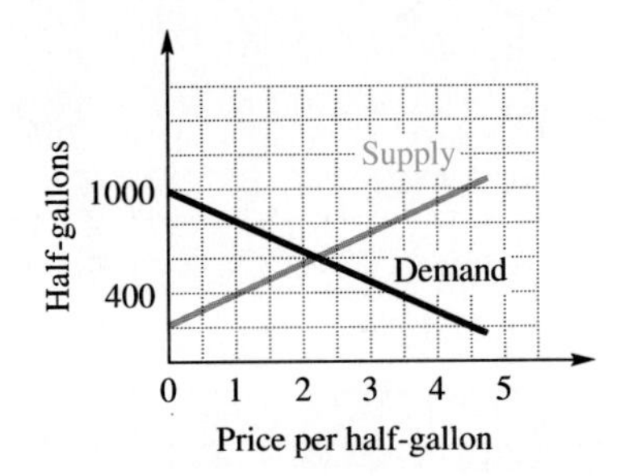

(a) At what price does supply equal demand?

(b) For how many half-gallons does supply equal demand?

(c) What are the supply and demand at a price of \$1 per half-gallon?

PREVIEW EXERCISES

Multiply both sides of each equation by the given number. See Section 2.1.

63. $2x - 4y + z = 8;$ by -3

64. $3x + 6y - 2z = 12;$ by 4

65. $-x + 3y - 2z = 18;$ by -2

Add. See Section 3.3.

66.
$$\begin{array}{r} 3x + 2y - z = 7 \\ \underline{5x - 2y + 2z = 10} \end{array}$$

67.
$$\begin{array}{r} x + 5y - 3z = 12 \\ \underline{2x + 6y + 3z = -15} \end{array}$$

68.
$$\begin{array}{r} -2x - 4y + 5z = 20 \\ \underline{2x + 3y - 4z = 15} \end{array}$$

11.2 LINEAR SYSTEMS OF EQUATIONS IN THREE VARIABLES

OBJECTIVES

1. SOLVE LINEAR SYSTEMS WITH THREE EQUATIONS AND THREE VARIABLES BY THE ELIMINATION METHOD.
2. SOLVE LINEAR SYSTEMS WITH THREE EQUATIONS AND THREE VARIABLES WHERE SOME OF THE EQUATIONS HAVE MISSING TERMS.
3. SOLVE LINEAR SYSTEMS WITH THREE EQUATIONS AND THREE VARIABLES THAT ARE INCONSISTENT OR THAT INCLUDE DEPENDENT EQUATIONS.

A solution of an equation in three variables, such as $2x + 3y - z = 4$, is called an **ordered triple** and is written (x, y, z). For example, the ordered triples $(1, 1, 1)$ and $(10, -3, 7)$ are each solutions of $2x + 3y - z = 4$, since the numbers in these ordered triples satisfy the equation when used as replacements for x, y, and z, respectively.

In the rest of this chapter, the term *linear equation* is extended to equations of the form $Ax + By + Cz + \ldots + Dw = K$. For example, $2x + 3y - 5z = 7$ and $x - 2y - z + 3w - 2v = 8$ are linear equations, the first having three variables, and the second having five variables.

In this section we discuss the solution of a system of linear equations in three variables such as

$$\begin{aligned} 4x + 8y + z &= 2 \\ x + 7y - 3z &= -14 \\ 2x - 3y + 2z &= 3. \end{aligned}$$

Theoretically, a system of this type can be solved by graphing. However, the graph of a linear equation with three variables is a *plane* and not a line. Since the graph of each equation of the system is a plane, which requires three-dimensional graphing, this method is not practical. However, it does serve to illustrate the number of solutions possible for such systems, as Figure 11.3 shows. (Although Figure 11.3 shows one example of each possible type of solution, other examples could be given.)

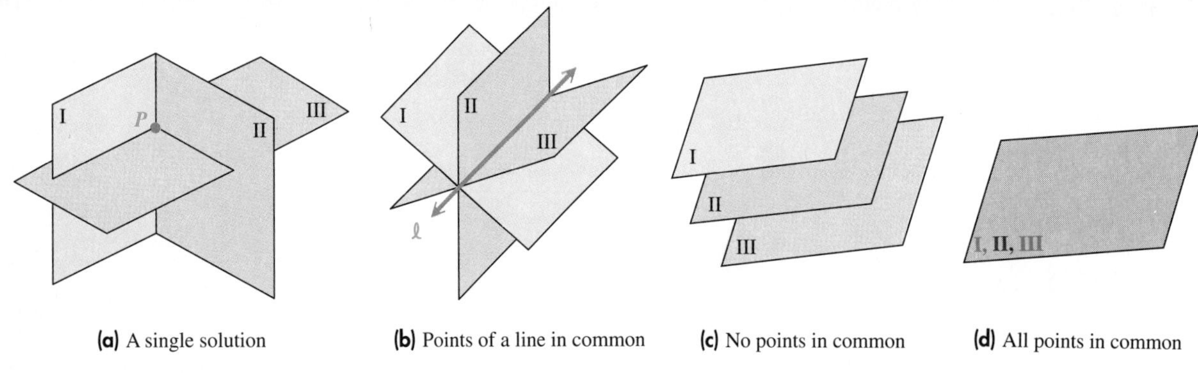

(a) A single solution **(b)** Points of a line in common **(c)** No points in common **(d)** All points in common

FIGURE 11.3

GRAPHS OF LINEAR SYSTEMS IN THREE VARIABLES

1. The three planes may meet at a single, common point that is the solution of the system. See Figure 11.3(a).
2. The three planes may have the points of a line in common so that the set of points that satisfy the equation of the line is the solution of the system. See Figure 11.3(b).
3. The planes may have no points common to all three so that there is no solution for the system. See Figure 11.3(c).
4. The three planes may coincide so that the solution of the system is the set of all points on a plane. See Figure 11.3(d).

1 Since a graphic solution of a system of equations of three variables is impractical, these systems are solved with an extension of the elimination method, summarized as follows.

SOLVING LINEAR SYSTEMS IN THREE VARIABLES BY ELIMINATION

Step 1 Use the elimination method to eliminate any variable from any two of the given equations. The result is an equation in two variables.

Step 2 Eliminate the *same* variable from any *other* two equations. The result is an equation in the same two variables as in Step 1.

Step 3 Use the elimination method to eliminate a second variable from the two equations in two variables that result from Steps 1 and 2. The result is an equation in one variable that gives the value of that variable.

Step 4 Substitute the value of the variable found in Step 3 into either of the equations in two variables to find the value of the second variable.

Step 5 Use the values of the two variables from Steps 3 and 4 to find the value of the third variable by substituting into any of the original equations.

EXAMPLE 1

SOLVING A SYSTEM IN THREE VARIABLES

Solve the system

$$4x + 8y + z = 2 \quad \textbf{(1)}$$

$$x + 7y - 3z = -14 \quad \textbf{(2)}$$

$$2x - 3y + 2z = 3. \quad \textbf{(3)}$$

To begin we must choose a pair of equations and eliminate a single variable from them. Let us eliminate x by multiplying equation (2) by -4 and adding it to equation (1).

$$\begin{array}{rl} 4x + 8y + z = 2 & (1) \\ -4x - 28y + 12z = 56 & -4 \text{ times equation (2)} \\ \hline -20y + 13z = 58 & \end{array} \qquad \textbf{(4)}$$

Next, choose a different pair of equations and eliminate x from them. Multiply equation (2) by -2 and add to equation (3).

$$\begin{array}{rl} -2x - 14y + 6z = 28 & -2 \text{ times equation (2)} \\ 2x - 3y + 2z = 3 & (3) \\ \hline -17y + 8z = 31 & \end{array} \qquad \textbf{(5)}$$

Now eliminate y from equations (4) and (5). Multiply equation (4) by 17 and equation (5) by -20 and add.

$$\begin{array}{rl} -340y + 221z = 986 & 17 \text{ times equation (4)} \\ 340y - 160z = -620 & -20 \text{ times equation (5)} \\ \hline 61z = 366 & \\ z = 6 & \end{array} \qquad \textbf{(6)}$$

From equation (6) we see that the value of z in the solution is 6. To find the value of y, we may go back to either equation (4) or equation (5). Let us go back to equation (4).

$$\begin{array}{rl} -20y + 13z = 58 & (4) \\ -20y + 13(6) = 58 & \text{Let } z = 6. \\ -20y + 78 = 58 & \\ -20y = -20 & \\ y = 1 & \end{array}$$

Now go back to any of the original equations to find x. Choosing equation (1) gives

$$\begin{array}{rl} 4x + 8(1) + 6 = 2 & \text{Let } y = 1 \text{ and } z = 6 \text{ in equation (1).} \\ 4x + 14 = 2 & \\ 4x = -12 & \\ x = -3. & \end{array}$$

The ordered triple $(-3, 1, 6)$ is the only solution of the system. Check that the solution satisfies all three equations in the system, so that the solution set is $\{(-3, 1, 6)\}$. ■

NOTE In Example 1, we used a process that yielded a system of equations (1), (4), and (6) that is equivalent to the original system.

$$\begin{array}{rl} 4x + 8y + z = 2 & (1) \\ -20y + 13z = 58 & (4) \\ z = 6 & (6) \end{array}$$

Notice the "triangular" form here. While it is not absolutely necessary to eliminate x first and then y to find z, it is a good preparation for our later work with matrices in Chapter 12.

2 Systems that have some equations with missing terms can be solved using the method described in the next example.

EXAMPLE 2 **SOLVING A SYSTEM OF EQUATIONS WITH MISSING TERMS**

Solve the system

$$6x - 12y = -5 \qquad (7)$$
$$8y + z = 0 \qquad (8)$$
$$9x - z = 12. \qquad (9)$$

Since equation (9) is missing the variable y, one way to begin the solution is to eliminate y again with equations (7) and (8). Multiply both sides of equation (7) by 2 and both sides of equation (8) by 3, and then add.

$$\begin{array}{rl} 12x - 24y = -10 & \text{Multiply both sides of (7) by 2.} \\ 24y + 3z = 0 & \text{Multiply both sides of (8) by 3.} \\ \hline 12x + 3z = -10 & \end{array}$$

Use this result, together with equation (9), to eliminate z. Multiply both sides of equation (9) by 3. This gives

$$\begin{array}{rl} 27x - 3z = 36 & \text{Multiply both sides of (9) by 3.} \\ 12x + 3z = -10 & \\ \hline 39x = 26. & \end{array}$$

$$x = \frac{26}{39} = \frac{2}{3}.$$

Substitution into equation (9) gives

$$9x - z = 12 \qquad (9)$$
$$9\left(\frac{2}{3}\right) - z = 12 \qquad \text{Let } x = \frac{2}{3}.$$
$$6 - z = 12$$
$$z = -6.$$

Substitution of -6 for z in equation (8) gives

$$8y + z = 0 \qquad (8)$$
$$8y + (-6) = 0 \qquad \text{Let } z = -6.$$
$$8y = 6$$
$$y = \frac{3}{4}.$$

Check in each of the original equations of the system to verify that the solution set of the system is $\{(2/3, 3/4, -6)\}$. ■

3 Linear systems with three variables may be inconsistent or may include dependent equations. The next examples illustrate these cases.

EXAMPLE 3
SOLVING AN INCONSISTENT SYSTEM WITH THREE VARIABLES

Solve the system

$$2x - 4y + 6z = 5 \quad \textbf{(10)}$$
$$-x + 3y - 2z = -1 \quad \textbf{(11)}$$
$$x - 2y + 3z = 1. \quad \textbf{(12)}$$

Eliminate x by adding equations (11) and (12) to get the equation

$$y + z = 0.$$

Now, to eliminate x again, multiply both sides of equation (12) by -2 and add the result to equation (10).

$$\begin{array}{rl} -2x + 4y - 6z = -2 & \text{−2 times equation (12)} \\ 2x - 4y + 6z = 5 & \text{(10)} \\ \hline 0 = 3 & \text{False} \end{array}$$

The resulting false statement indicates that the system is inconsistent and has no solution. In this particular system the graphs of equations (10) and (12) are parallel to each other. The solution set is $\emptyset$. ■

EXAMPLE 4
SOLVING A SYSTEM OF DEPENDENT EQUATIONS

Solve the system

$$2x + 4y + 2z = 8 \quad \textbf{(13)}$$
$$x + 2y + z = 4 \quad \textbf{(14)}$$
$$3x - y - 4z = -9. \quad \textbf{(15)}$$

Notice that equation (13) is obtained by multiplying equation (14) on both sides by 2. Therefore, the graphs of these two equations are the same plane. Equation (15) is neither parallel to nor a multiple of equation (14), and therefore the plane representing equation (15) will intersect the plane representing equations (13) and (14) in a line. (From geometry, if two distinct planes intersect, they will intersect in a line. Visualize, for example, the intersection of a wall and a ceiling in a room.)

To find a way to represent the set of ordered triples that form the solution of this system, eliminate x by multiplying both sides of equation (14) by -3 and adding to equation (15). (Either y or z could have been eliminated instead.)

$$\begin{array}{rl} -3x - 6y - 3z = -12 & \text{−3 times equation (14)} \\ 3x - y - 4z = -9 & \text{(15)} \\ \hline -7y - 7z = -21 & \textbf{(16)} \end{array}$$

Now solve equation (16) for z.

$$-7y - 7z = -21$$
$$-7z = 7y - 21$$
$$z = -y + 3$$

This gives z in terms of y. Now, express x also in terms of y by solving equation (14) for x and substituting $-y + 3$ for z in the result.

$$\begin{aligned} x + 2y + z &= 4 \\ x &= -2y - z + 4 \\ x &= -2y - (-y + 3) + 4 \\ x &= -y + 1 \end{aligned}$$

The system has an infinite number of solutions. For any value of y, the value of z is given by $-y + 3$ and x equals $-y + 1$. For example, if $y = 1$, then $x = -1 + 1 = 0$ and $z = -1 + 3 = 2$, giving the solution $(0, 1, 2)$.

Had equation (16) been solved for y instead of z, the solution would have had a different form but would have led to the same set of solutions. The form of the solution set found above, $\{(-y + 1, y, -y + 3)\}$, is said to have **y arbitrary.** Had we solved the system in such a way as to have z arbitrary, then the solution set would be of the form $\{(-2 + z, 3 - z, z)\}$. By choosing $z = 2$, one solution would be $(0, 1, 2)$, which was verified earlier. ■

11.2 EXERCISES

Solve each system of equations. See Examples 1 and 2.

1. $\begin{aligned} 2x + y + z &= 3 \\ 3x - y + z &= -2 \\ 4x - y + 2z &= 0 \end{aligned}$

2. $\begin{aligned} x + y + z &= 6 \\ 2x + 3y - z &= 7 \\ 3x - y - z &= 6 \end{aligned}$

3. $\begin{aligned} 3x + 2y + z &= 4 \\ 2x - 3y + 2z &= -7 \\ x + 4y - z &= 10 \end{aligned}$

4. $\begin{aligned} -3x + y - z &= 8 \\ -4x + 2y + 3z &= -3 \\ 2x + 3y - 2z &= -1 \end{aligned}$

5. $\begin{aligned} 2x + 5y + 2z &= 9 \\ 4x - 7y - 3z &= 7 \\ 3x - 8y - 2z &= 9 \end{aligned}$

6. $\begin{aligned} 5x - 2y + 3z &= 13 \\ 4x + 3y + 5z &= -10 \\ 2x + 4y - 2z &= -22 \end{aligned}$

7. $\begin{aligned} x + y - z &= -2 \\ 2x - y + z &= -5 \\ x - 2y + 3z &= 4 \end{aligned}$

8. $\begin{aligned} x + 3y - 6z &= 7 \\ 2x - y + z &= 1 \\ x + 2y + 2z &= -1 \end{aligned}$

9. $\begin{aligned} 2x + 3y - z &= 1 \\ x + 2y + 2z &= 5 \\ x - y + z &= 6 \end{aligned}$

10. $\begin{aligned} 2x - 3y + 2z &= -10 \\ 5x + 2y + 3z &= -2 \\ 4x + 6y + 5z &= 2 \end{aligned}$

11. $\begin{aligned} 2x - 3y + 2z &= -1 \\ x + 2y &= 14 \\ x - 3z &= -5 \end{aligned}$

12. $\begin{aligned} 2x - y + 3z &= 0 \\ x + 2y - z &= 5 \\ 2y + z &= 1 \end{aligned}$

13. $\begin{aligned} 4x + 2y - 3z &= 6 \\ x - 4y + z &= -4 \\ -x + 2z &= 2 \end{aligned}$

14. $\begin{aligned} -x + y &= 1 \\ y - z &= 14 \\ x - 3z &= -2 \end{aligned}$

15. $\begin{aligned} x + y &= 1 \\ 2x - z &= 0 \\ y + 2z &= -2 \end{aligned}$

16. $\begin{aligned} 2x + y &= 6 \\ 3y - 2z &= -4 \\ 3x - 5z &= -7 \end{aligned}$

Solve each system of equations. Give solutions of dependent equations with z arbitrary. See Examples 1–4.

17. $\begin{aligned} x + 3y + 2z &= 4 \\ x - y - z &= 3 \\ 3x + 9y + 6z &= 5 \end{aligned}$

18. $\begin{aligned} 2x + y - z &= 6 \\ x - y - z &= 6 \\ -x - \frac{1}{2}y + \frac{1}{2}z &= -3 \end{aligned}$

19. $\begin{aligned} x - 2y &= 0 \\ 3y + z &= -1 \\ 4x - z &= 11 \end{aligned}$

20. $\begin{aligned} 2x - 5y - z &= 3 \\ 5x + 14y - z &= -11 \\ 7x + 9y - 2z &= -5 \end{aligned}$

21. $\begin{aligned} x + 4y - z &= 3 \\ x + 3y + 2z &= -5 \\ 3x + 12y - 3z &= 9 \end{aligned}$

22. $\begin{aligned} 2x + 2y - 6z &= 5 \\ 3x - y + z &= 2 \\ -x - y + 3z &= 4 \end{aligned}$

23. $\begin{aligned} x - 2y + 4z &= -10 \\ -3x + 6y - 12z &= 20 \\ 2x + 5y + z &= 12 \end{aligned}$

24. $\begin{aligned} 4x - 2y + z &= 0 \\ x + 3y - 2z &= 0 \\ 2x + y - 5z &= 0 \end{aligned}$

25. $\begin{aligned} 2x - 4y + z &= 0 \\ x - 3y - z &= 0 \\ 3x - y + 2z &= 0 \end{aligned}$

26. $\begin{aligned} -5x + 5y - 20z &= -40 \\ x - y + 4z &= 8 \\ 3x + y + 6z &= 12 \end{aligned}$

27. $\begin{aligned} 2x - 8y + 2z &= -10 \\ -x + 4y - z &= 5 \\ x + 2y - 3z &= -5 \end{aligned}$

28. $\begin{aligned} -5x - y + z &= 3 \\ 8x + 2y - z &= -5 \\ 3x + y + 5z &= 3 \end{aligned}$

Extend the methods of this section to solve the systems in Exercises 29 and 30.

29. $\begin{aligned} x + y + z - w &= 5 \\ 2x + y - z + w &= 3 \\ x - 2y + 3z + w &= 18 \\ -x - y + z + 2w &= 8 \end{aligned}$

30. $\begin{aligned} 3x + y - z + 2w &= 9 \\ x + y + 2z - w &= 10 \\ x - y - z + 3w &= -2 \\ -x + y - z + w &= -6 \end{aligned}$

Solve each problem.

31. Find the values of m and b if the points $(2, -1)$ and $(4, 5)$ lie on the line with equation $y = mx + b$. Then give the equation of the line.

32. Find the equation, in the form $y = mx + b$, of the line through $(3, -2)$ and $(-1, 2)$.

33. Find the values of a, b, and c so that the points $(1, 2)$, $(-1, 0)$, and $(-2, 8)$ lie on the graph of the equation $y = ax^2 + bx + c$.

34. Find the values of a, b, and c so that the points $(0, 3)$, $(-1, 0)$, and $(3, 24)$ lie on the graph of the equation $y = ax^2 + bx + c$.

35. Discuss why it is necessary to eliminate the same variable in the first two steps of the elimination method with three equations and three variables. (See Objective 1.)

36. In Step 3 of the elimination method for solving systems in three variables, does it matter which variable is eliminated? Explain. (See Objective 1.)

PREVIEW EXERCISES

Solve each of the following problems. See Sections 2.3 and 2.4.

37. To make a 10% acid solution for chemistry class, Juanita wants to mix some 5% solution with 10 centiliters of 20% solution. How much 5% solution should she use?

38. The width of a rectangle is 4 units less than the length. The perimeter is 64 units. Find the length.

39. In a 100-mile bicycle race, the winner finished one hour before the person who came in last. The person in last place averaged 20 miles per hour. Find the winner's average speed.

40. The sum of three numbers is 16. The largest number is -3 times the smallest, while the middle number is 4 less than the largest. Find the three numbers.

41. The sum of the three angle measures of a triangle is 180°. The largest angle is twice the measure of the smallest, and the third angle measures 20° less than the largest. Find the measures of the three angles.

42. Margaret Maggio has a collection of pennies, dimes, and quarters. The number of dimes is one less than twice the number of pennies. If there are 27 coins in all worth a total of \$4.20, how many of each kind of coin is in the collection?

11.3 APPLICATIONS OF LINEAR SYSTEMS OF EQUATIONS

OBJECTIVES

1. SOLVE GEOMETRY PROBLEMS USING TWO VARIABLES.
2. SOLVE MONEY PROBLEMS USING TWO VARIABLES.
3. SOLVE MIXTURE PROBLEMS USING TWO VARIABLES.
4. SOLVE DISTANCE-RATE-TIME PROBLEMS USING TWO VARIABLES.
5. SOLVE PROBLEMS WITH THREE UNKNOWNS USING A SYSTEM OF THREE EQUATIONS.

FOCUS ON PROBLEM SOLVING

An investor plans to invest a total of $15,000 in two accounts, one paying 8% annual simple interest, and the other 7%. If he wants to earn $1100 annual interest, how much should he invest at each rate?

This is a typical investment problem that can be solved using a system of equations. As we saw earlier, investment problems fall into a broad class of problems including mixture problems and problems about money. The strategies used earlier, together with the methods of solving a linear system are used to solve these problems. This problem is Exercise 15 in the exercises for this section. After working through this section, you should be able to solve this problem.

PROBLEM SOLVING

Many problems involve more than one unknown quantity. Although some problems with two unknowns can be solved using just one variable, many times it is easier to use two variables. To solve a problem with two unknowns, we write two equations that relate the unknown quantities. The system formed by the pair of equations then can be solved using the methods of Section 11.1.

The following steps give a strategy for solving problems using more than one variable.

SOLVING AN APPLIED PROBLEM BY WRITING A SYSTEM OF EQUATIONS

Step 1 **Determine what you are to find.**
Assign a variable for each unknown and *write down* what it represents.

Step 2 **Write down other information.**
If appropriate, draw a figure or a diagram and label it using the variables from Step 1. Use a chart or a box diagram to summarize the information.

Step 3 **Write a system of equations.**
Write as many equations as there are unknowns.

Step 4 **Solve the system.**

Step 5 **Answer the question(s).**
Be sure you have answered all questions posed.

Step 6 **Check.**
Check your solution(s) in the original problem. Be sure your answer makes sense. ■

1 Problems about the perimeter of a geometric figure often involve two unknowns. The next example shows how to write a system of equations to solve such a problem.

EXAMPLE 1
SOLVING A GEOMETRY PROBLEM

The length of a rectangular house is to be 6 meters more than its width. Find the length and width of the house if the perimeter must be 48 meters.

Begin by sketching a rectangle to represent the foundation of the house. Let x = the length and y = the width. See Figure 11.4.

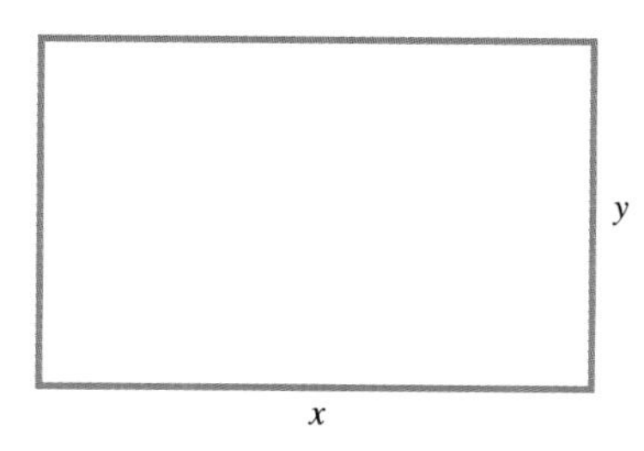

FIGURE 11.4

The length, x, is 6 meters more than the width, y. Therefore,

$$x = 6 + y.$$

The formula for the perimeter of a rectangle is $P = 2L + 2W$. Here $P = 48$, $L = x$, and $W = y$, so

$$48 = 2x + 2y.$$

The length and width can now be found by solving the system

$$\begin{aligned} x &= 6 + y \\ 48 &= 2x + 2y. \end{aligned}$$

Solve this system to find that the width is 9 meters and the length 15 meters. Be sure to check the solution in the words of the original problem. ■

2 Another type of problem that often leads to a system of equations is one about different amounts of money.

EXAMPLE 2
SOLVING A PROBLEM ABOUT MONEY

For an art project Kay bought 8 sheets of colored paper and 3 marker pens for \$6.50. She later needed 2 more sheets of paper and 2 different colored pens. These items cost \$3.00. Find the cost of 1 marker pen and 1 sheet of colored paper.

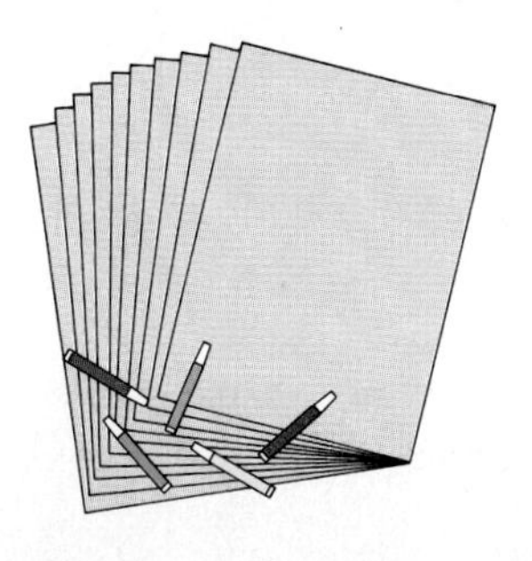

Let x represent the cost of a sheet of paper and y represent the cost of a pen. For the first purchase $8x$ represents the cost of the paper and $3y$ the cost of the pens. The total cost was \$6.50, so

$$8x + 3y = 6.50.$$

For the second purchase,

$$2x + 2y = 3.00.$$

We can solve the system by multiplying both sides of the second equation by -4 and adding the result to the first equation.

$$\begin{array}{r} 8x + 3y = 6.50 \\ -8x - 8y = -12.00 \\ \hline -5y = -5.50 \\ y = 1.10 \end{array}$$

By substituting 1.10 for y in either of the equations, verify that $x = .40$. Kay paid \$.40 for a sheet of colored paper and \$1.10 for a pen. ■

3 We solved mixture problems earlier using one variable. Many mixture problems can be solved more easily by writing and solving a system of equations.

EXAMPLE 3
SOLVING A MIXTURE PROBLEM

How many ounces of 5% hydrochloric acid and of 20% hydrochloric acid must be combined to get 10 ounces of solution that is 12.5% hydrochloric acid?

Let x represent the number of ounces of 5% solution and y represent the number of ounces of 20% solution. A table summarizes the given information.

KIND OF SOLUTION	OUNCES OF SOLUTION	OUNCES OF ACID
5%	x	$.05x$
20%	y	$.20y$
12.5%	10	$(.125)10$

When the x ounces of 5% solution and the y ounces of 20% solution are combined, the total number of ounces is 10, so that

$$x + y = 10. \quad \textbf{(1)}$$

The ounces of acid in the 5% solution, $.05x$, plus the ounces of acid in the 20% solution, $.20y$, should equal the total number of ounces of acid in the mixture, which is $(.125)10$. That is,

$$.05x + .20y = (.125)10. \quad \textbf{(2)}$$

Eliminate x by first multiplying both sides of equation (2) by 100 to clear it of decimals, and then multiplying both sides of equation (1) by -5. Then add the results.

$$\begin{array}{rl} 5x + 20y = 125 & \text{Multiply both sides of (2) by 100.} \\ -5x - 5y = -50 & \text{Multiply both sides of (1) by } -5. \\ \hline 15y = 75 & \\ y = 5 & \end{array}$$

Since $y = 5$ and $x + y = 10$, x is also 5. Therefore, 5 ounces each of the 5% and the 20% solutions are required. ■

4 Constant rate applications require the distance formula, $d = rt$, where d is distance, r is rate (or speed), and t is time. These applications often lead to a system of equations, as in the next example.

EXAMPLE 4
SOLVING A MOTION PROBLEM

A car travels 250 kilometers in the same time that a truck travels 225 kilometers. If the speed of the car is 8 kilometers per hour faster than the speed of the truck, find both speeds.

A table is useful for organizing the information in problems about distance, rate, and time. Fill in the given information for each vehicle (in this case, distance) and use variables for the unknown speeds (rates) as follows.

	d	r	t
Car	250	x	
Truck	225	y	

The problem states that the car travels 8 kilometers per hour faster than the truck. Since the two speeds are x and y,

$$x = y + 8.$$

The table shows nothing about time. To get an expression for time, solve the distance formula, $d = rt$, for t to get

$$\frac{d}{r} = t.$$

The two times can be written as $\frac{250}{x}$ and $\frac{225}{y}$. Since both vehicles travel for the same time,

$$\frac{250}{x} = \frac{225}{y}.$$

This is not a linear equation. However, multiplying both sides by xy gives

$$250y = 225x,$$

which is linear. Now solve the system

$$x = y + 8$$
$$250y = 225x.$$

The substitution method can be used. Replace x with $y + 8$ in the second equation of this system.

$$\begin{aligned} 250y &= 225(y + 8) && \text{Let } x = y + 8. \\ 250y &= 225y + 1800 && \text{Distributive property} \\ 25y &= 1800 \\ y &= 72 \end{aligned}$$

Since $x = y + 8$, the value of x is $72 + 8 = 80$. It is important to check the solution in the original problem since one of the equations had variable denominators. Checking verifies that the speeds are 80 kilometers per hour for the car and 72 kilometers per hour for the truck. ■

5 Some applications involve three unknowns. When three variables are used, three equations are necessary to solve the problem. We can then use the methods of Section 11.2 to solve the system. The next two examples illustrate this.

EXAMPLE 5

SOLVING A PROBLEM ABOUT FOOD PRICES

Joe Schwartz bought apples, hamburger, and milk at the grocery store. Apples cost \$.70 a pound, hamburger was \$1.50 a pound, and milk was \$.80 a quart. He bought twice as many pounds of apples as hamburger. The number of quarts of milk was one more than the number of pounds of hamburger. If his total bill was \$8.20, how much of each item did he buy?

First choose variables to represent the three unknowns.

Let x = the number of pounds of apples;

y = the number of pounds of hamburger;

z = the number of quarts of milk.

Next, use the information in the problem to write three equations. Since Joe bought twice as many pounds of apples as hamburger,

$$x = 2y$$

or

$$x - 2y = 0.$$

The number of quarts of milk amounted to one more than the number of pounds of hamburger, so

$$z = 1 + y$$

or

$$-y + z = 1.$$

Multiplying the cost of each item by the amount of that item and adding gives the total bill.

$$.70x + 1.50y + .80z = 8.20$$

Multiply both sides of this equation by 10 to clear it of decimals.

$$7x + 15y + 8z = 82$$

Use the method shown in the last section to solve the system

$$\begin{aligned} x - 2y &= 0 \\ -y + z &= 1 \\ 7x + 15y + 8z &= 82. \end{aligned}$$

Verify that the solution is (4, 2, 3). Now go back to the statements defining the variables to decide what the numbers of the solution represent. Doing this shows that Joe bought 4 pounds of apples, 2 pounds of hamburger, and 3 quarts of milk. ■

Business problems involving production sometimes require the solution of a system of equations. The final example shows how to set up such a system.

EXAMPLE 6
SOLVING A BUSINESS PRODUCTION PROBLEM

A company produces three color television sets, models X, Y, and Z. Each model X set requires 2 hours of electronics work, 2 hours of assembly time, and 1 hour of finishing time. Each model Y requires 1, 3, and 1 hours of electronics, assembly, and finishing time, respectively. Each model Z requires 3, 2, and 2 hours of the same work, respectively. There are 100 hours available for electronics, 100 hours available for assembly, and 65 hours available for finishing per week. How many of each model should be produced each week if all available time must be used?

Let x = the number of model X produced per week;
y = the number of model Y produced per week;
z = the number of model Z produced per week.

A chart is useful for organizing the information in a problem of this type.

	MODEL X	MODEL Y	MODEL Z	TOTALS
Hours of electronics work	2	1	3	100
Hours of assembly time	2	3	2	100
Hours of finishing time	1	1	2	65

The x model X sets require $2x$ hours of electronics, the y model Y sets require $1y$ (or y) hours of electronics, and the z model Z sets require $3z$ hours of electronics. Since 100 hours are available for electronics,

$$2x + y + 3z = 100.$$

Similarly, from the fact that 100 hours are available for assembly,

$$2x + 3y + 2z = 100,$$

and the fact that 65 hours are available for finishing leads to the equation

$$x + y + 2z = 65.$$

Solve the system

$$\begin{aligned} 2x + y + 3z &= 100 \\ 2x + 3y + 2z &= 100 \\ x + y + 2z &= 65 \end{aligned}$$

to find $x = 15$, $y = 10$, and $z = 20$. The company should produce 15 model X, 10 model Y, and 20 model Z sets per week. ■

Notice the advantage of setting up the chart in Example 6. By reading across, we can easily determine the coefficients and the constants in the system.

11.3 EXERCISES

*For each problem, (**a**) select variables to represent the two unknowns, (**b**) write two equations using the two variables, and (**c**) solve the resulting system.*

1. The perimeter of a triangle is 39 centimeters. Two sides have the same length. The remaining side is 9 centimeters less than twice the length of either of the other sides. Find the lengths of the sides of the triangle.

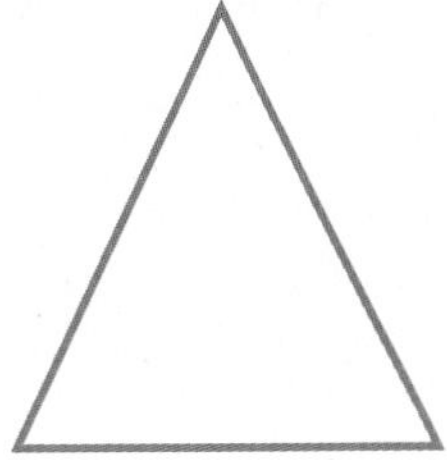

2. The perimeter of a rectangle is 80 inches. The width is 1 inch more than one-half the length. Find the length and width of the rectangle.

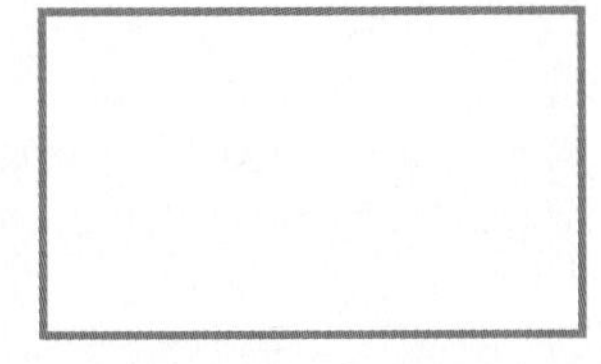

3. The side of a square is 4 centimeters longer than the side of an equilateral triangle. The perimeter of the square is 22 inches more than the perimeter of the triangle. Find the lengths of a side of the square and a side of the triangle.

4. The length of a rectangle is 7 feet more than the width. If the length were decreased by 3 feet and the width were increased by 2 feet, the perimeter would be 44 feet. Find the length and width of the original rectangle.

5. Bobby and Melody are planning to move and need to buy some cardboard boxes. At Smith's they can buy either 10 small and 12 medium boxes for $42, or 5 small and 10 medium boxes for $29. Find the cost of each size box.

6. At the Eastside Nut Ranch, 5 pounds of peanuts and 6 pounds of cashews cost $70, and 3 pounds of peanuts and 7 pounds of cashews cost $76. Find the prices charged by the ranch for a pound of peanuts and a pound of cashews.

7. A factory makes use of two basic machines, A and B, to turn out two different products, yarn and thread. Each unit of yarn requires 1 hour on machine A and 2 hours on machine B, and each unit of thread requires 1 hour on A and 1 hour on B. Machine A runs 8 hours per day, and machine B runs 14 hours per day. How many units each of yarn and thread should the factory make to keep its machines running at capacity?

8. A company that makes personal computers has found that each standard model requires 4 hours to manufacture the electronics and 2 hours to manufacture the case. The top-of-the-line model requires 5 hours for the electronics and 1.5 hours for the case. On a particular production run, the company has available 200 hours in the electronics department and 80 hours in the cabinet department. How many of each model can be made?

Solve each problem. See Examples 1–4.

9. Harry bought 2 kilograms of dark clay and 3 kilograms of light clay, paying $2.40 for the clay. He later needed 1 kilogram more of dark clay and 2 kilograms more of light clay, costing $1.40 altogether. How much did he pay for each type of clay?

10. A biologist wants to grow two types of algae, green and brown. She has 15 kilograms of nutrient X and 26 kilograms of nutrient Y. A vat of green algae needs 2 kilograms of nutrient X and 3 kilograms of nutrient Y, while a vat of brown algae needs 1 kilogram of nutrient X and 2 kilograms of nutrient Y. How many vats of each type of algae should the biologist grow in order to use all the nutrients?

11. Mabel Johnston bought apples and oranges at DeVille's Grocery. She bought 6 pounds of fruit. Oranges cost $.90 per pound, while apples cost $.70 per pound. If she spent a total of $5.20, how many pounds of each kind of fruit did she buy?

12. Joshua Rogers has been saving dimes and quarters. He has 50 coins in all. If the total value is $10.70, how many dimes and how many quarters does he have?

13. A teller at South Savings and Loan received a checking account deposit in twenty-dollar bills and fifty-dollar bills. She received a total of 35 bills, and the amount of the deposit was $1600. How many of each denomination were deposited?

14. Tickets to the senior play at Slidell High School cost $2.50 for general admission or $2.00 with a student identification. If 92 people paid to see a performance and $203 was collected, how many of each type of admission were sold?

15. An investor plans to invest a total of $15,000 in two acounts, one paying 8% annual simple interest, and the other 7%. If he wants to earn $1100 annual interest, how much should he invest at each rate?

RATE	AMOUNT	INTEREST
8%	x	
7%	y	
Total	$15,000	$1100

16. A total of $6000 is invested, part at 8% simple interest and part at 4%. If the annual return from the two investments is the same, how much is invested at each rate?

RATE	AMOUNT	INTEREST
8%	x	
4%	y	
Total	$6000	

17. Teresita de Zayas invested $56,000, part at 14% annual simple interest and the rest at 7%. Her annual interest from the two investments was the same as if she had invested the entire amount at 10%. How much did she invest at each rate?

18. Stu Hobbs has money invested in two accounts. He has $3000 more invested at 7% annual simple interest than he has at 5%. If he receives a total of $450 annual interest from the two investments, how much does he have invested at each rate?

19. Pure acid is to be added to a 10% acid solution to obtain 9 liters of a 20% acid solution. What amounts of each should be used?

KIND	AMOUNT	PURE ACID
100%	x	
10%	y	
20%	9	

20. How many gallons each of 25% alcohol and 35% alcohol should be mixed to get 10 gallons of 32% alcohol?

KIND	AMOUNT	PURE ALCOHOL
25%	x	
35%	y	
32%	10	

21. A truck radiator holds 24 liters of fluid. How much pure antifreeze must be added to a mixture that is 4% antifreeze in order to fill the radiator with a mixture that is 20% antifreeze?

22. A grocer plans to mix candy that sells for $1.20 a pound with candy that sells for $2.40 a pound to get a mixture that he plans to sell for $1.65 a pound. How much of the $1.20 and $2.40 candy should he use if he wants 20 pounds of the mix?

23. A party mix is made by adding nuts that sell for $2.50 a kilogram to a cereal mixture that sells for $1 a kilogram. How much of each should be added to get 60 kilograms of a mix that will sell for $1.70 a kilogram?

24. A popular fruit drink is made by mixing fruit juice and soda. Such a mixture with 50% juice is to be mixed with another mixture that is 30% juice to get 100 liters of a mixture which is 45% juice. How much of each should be used?

25. A freight train and an express train leave towns 390 kilometers apart, traveling toward one another. The express train travels 30 kilometers per hour faster than the freight train. They pass one another 3 hours later. What are their speeds?

26. Two cars start out from the same spot and travel in opposite directions. At the end of 6 hours, they are 690 kilometers apart. If one car travels 15 kilometers per hour faster than the other, what are their speeds?

27. A train travels 150 kilometers in the same time that a plane covers 400 kilometers. If the speed of the plane is 20 kilometers per hour less than 3 times the speed of the train, find both speeds.

28. In his motorboat, Tri travels upstream at top speed to his favorite fishing spot, a distance of 18 miles, in 1 hour. Returning, he finds that the trip downstream, still at top speed, takes only 3/4 hour. What is the speed of the current? What is the speed of Tri's boat in still water? (*Hint:* Let b represent the speed of the boat in still water and let c represent the speed of the current.)

Solve each problem. See Examples 5 and 6.

29. Find three numbers whose sum is 32, if the second is 1 more than the first, and the first is 4 less than the third.

30. Find three numbers whose sum is 8, if the first number is 26 less than the sum of the second and third, and the second number is 11 more than the first.

31. The sum of the measures of the angles of a triangle is 180°. The measure of the largest angle is 12° less than the sum of the measures of the other two. The smallest angle measures 58° less than the largest. Find the measures of the angles.

32. The sum of the measures of the angles of a triangle is 180°. In a certain triangle, the measure of the second angle is 10° more than three times the first. The third angle measure is equal to the sum of the measures of the other two. Find the measures of the three angles.

33. The perimeter of a triangle is 56 inches. The longest side measures 12 inches less than the sum of the other two sides. Three times the shortest side is 26 inches more than the longest side. Find the lengths of the three sides.

34. The perimeter of a triangle is 70 centimeters. The longest side is 6 centimeters less than the sum of the other two sides. Twice the shortest side is 4 centimeters more than the longest side. Find the length of each side of the triangle.

35. Kasey has a collection of nickels, dimes, and quarters. She has 10 more dimes than nickels, and there are 4 fewer quarters than dimes and nickels together. The total value of the coins is $7.05. How many of each coin does she have?

36. Barbara has some fives, tens, and twenties. Altogether she has 27 bills worth $255. There are 6 fewer fives than tens. Find the number of each type of bill she has.

37. To meet his sales quota, Jesus Barreto must sell 20 new cars. He must sell 4 more sub-compact cars

than compact cars, and one more full-size car than compact cars. How many of each type must he sell?

38. A manufacturer supplies three wholesalers, A, B, and C. The output from a day's production is 160 cases of goods. She must send wholesaler A three times as many cases as she sends B, and she must send wholesaler C 80 cases less than she provides A and B together. How many cases should she send to each wholesaler to distribute the entire day's production to them?

39. Three kinds of tickets are available for a Rhonda Rock concert: "up close," "in the middle," and "far out." "Up close" tickets cost \$2 more than "in the middle" tickets, while "in the middle" tickets cost \$1 more than "far out" tickets. Twice the cost of an "up close" ticket is \$1 less than 3 times the cost of a "far out" seat. Find the price of each kind of ticket.

40. A shop manufactures three kinds of bolts, types A, B, and C. Production restrictions require them to make 5 units more type C bolts than the total of the other types and twice as many type B bolts as type A. The shop must produce a total of 245 units of bolts per day. How many units of each type can be made per day?

41. Nancy Dunn has inherited \$80,000 from her uncle. She invests part of the money in a video rental firm which produces a return of 7% per year, and divides the rest equally between a tax-free bond at 6% a year and a money market fund at 12% a year. Her annual return on these investments is \$6800. How much is invested in each?

42. Sonny's Flooring, Inc., borrowed in three loans for major renovations. The company borrowed a total of \$75,000. Some of the money was borrowed at 8% interest, and \$30,000 more than that amount was borrowed at 10%. The rest was borrowed at 7%. How much was borrowed at each rate if the total annual simple interest was \$6950?

43. The owner of a tea shop wants to mix three kinds of tea to make 100 ounces of a mixture that will sell for \$.83 an ounce. He uses Orange Pekoe, which sells for \$.80 an ounce, Irish Breakfast, for \$.85 an ounce, and Earl Grey, for \$.95 an ounce. If he wants to use twice as much Orange Pekoe as Irish Breakfast, how much of each kind of tea should he use?

44. The manager of a candy store wants to feature a special Easter candy mixture of jelly beans, small chocolate eggs, and marshmallow chicks. She plans to make 15 pounds of mix to sell at \$1 a pound. Jelly beans sell for \$.80 a pound, chocolate eggs for \$2 a pound, and marshmallow chicks for \$1 a pound. She will use twice as many pounds of jelly beans as eggs and chicks combined and five times as many pounds of jelly beans as chocolate eggs. How many pounds of each candy should she use?

45. Write an applied problem similar to Exercise 5. Give the solution.

46. Write an applied problem similar to Exercise 33. Give the solution.

PREVIEW EXERCISES

Solve each equation. See Section 6.4.

47. $2x^4 - 5x^2 - 3 = 0$

48. $3x^4 + 26x^2 + 35 = 0$

49. $x^4 - 16 = 0$

50. $x^4 - 81 = 0$

51. $r^4 - 7r^2 + 12 = 0$

52. $p^4 - 14p^2 + 45 = 0$

11.4 NONLINEAR SYSTEMS OF EQUATIONS

OBJECTIVES

1. SOLVE A NONLINEAR SYSTEM BY SUBSTITUTION.
2. USE THE ELIMINATION METHOD TO SOLVE A SYSTEM WITH TWO SECOND-DEGREE EQUATIONS.
3. SOLVE A SYSTEM THAT REQUIRES A COMBINATION OF METHODS.

FOCUS ON PROBLEM SOLVING

A company has found that the price p (in dollars) of its scientific calculator is related to the supply x (in thousands) by the equation $px = 16$. Also, the price is related to the demand x (in thousands) for the calculator by the equation $p = 10x + 12$. The *equilibrium price* is the value of p where demand equals supply. Find the equilibrium price and the supply/demand at that price by solving a system of equations.

The concept of supply and demand plays an important part in the study of economics. Typically, as the price of an item increases, the demand for the item decreases, while the supply increases. It is helpful for a company to know the equilibrium price; that is, the price where demand equals supply. In order to find the equilibrium price in this problem, it is necessary to know how to solve a nonlinear system of equations. This problem is Exercise 55 in the exercises for this section. After working through this section, you should be able to solve this problem.

An equation that cannot be put into the form $Ax + By = C$ is called a **nonlinear equation.** A **nonlinear system of equations** includes at least one nonlinear equation. Graphs of nonlinear equations were studied in Chapter 8.

When solving nonlinear systems, it will be helpful to visualize the types of graphs of the equations of the system in order to determine the possible number of points of intersection. For example, if a system contains two equations where the graph of one is a parabola and the graph of the other is a line, then there may be 0, 1, or 2 points of intersection. This is illustrated in Figure 11.5.

1 Nonlinear systems can be solved by the elimination method, the substitution method, or a combination of the two. The following examples show the use of these methods for solving nonlinear systems. The substitution method is usually the most useful when one of the equations is linear. The first two examples illustrate this kind of system.

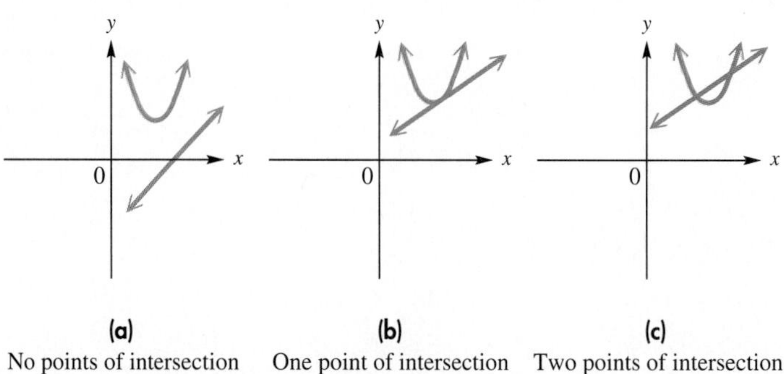

(a) No points of intersection (b) One point of intersection (c) Two points of intersection

FIGURE 11.5

EXAMPLE 1
USING SUBSTITUTION WHEN ONE EQUATION IS LINEAR

Solve the system

$$x^2 + y^2 = 9 \quad (1)$$
$$2x - y = 3. \quad (2)$$

The graph of equation (1) is a circle and the graph of equation (2) is a line. Visualizing the possibilities indicates that there may be 0, 1, or 2 points of intersection. See Figure 11.6. When solving a system of this type, it is best to solve the linear equation for one of the two variables; then substitute the resulting expression into the nonlinear equation to obtain an equation in one variable. Solving equation (2) for y gives

$$2x - y = 3 \quad (2)$$
$$y = 2x - 3. \quad (3)$$

Substitute $2x - 3$ for y in equation (1) to get

$$x^2 + (2x - 3)^2 = 9$$
$$x^2 + 4x^2 - 12x + 9 = 9 \qquad \text{Square } 2x - 3.$$
$$5x^2 - 12x = 0. \qquad \text{Combine terms.}$$

Solve this quadratic equation by factoring.

$$x(5x - 12) = 0 \qquad \text{Common factor is } x.$$
$$x = 0 \quad \text{or} \quad x = \frac{12}{5}$$

Let $x = 0$ in the equation $y = 2x - 3$ to get $y = -3$. If $x = 12/5$, then $y = 9/5$. The solution set of the system is $\{(0, -3), (12/5, 9/5)\}$. The graph of the system, shown in Figure 11.6 confirms the two points of intersection. ■

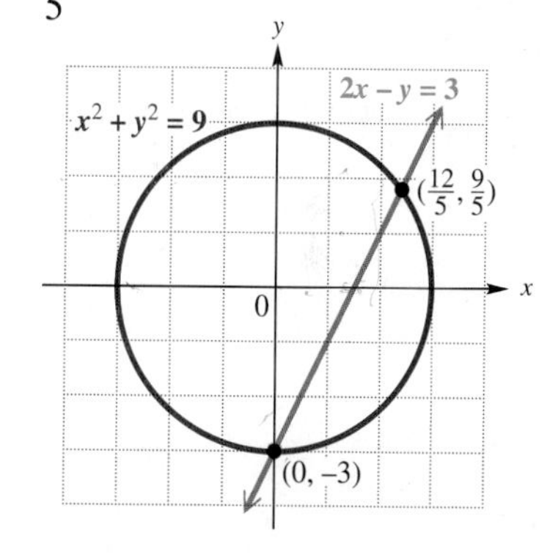

FIGURE 11.6

EXAMPLE 2
USING SUBSTITUTION WHEN ONE EQUATION IS LINEAR

Solve the system

$$6x - y = 5 \quad (4)$$
$$xy = 4. \quad (5)$$

The graph of equation (4) is a line, and although we have not specifically mentioned equations like equation (5), it can be shown by plotting points that its graph is a hyperbola. Once again, visualizing a line and a hyperbola indicates that there may be 0, 1, or 2 points of intersection.

Since neither equation has a squared term here, solve either equation for one of the variables and then substitute the result into the other equation. Solving $xy = 4$ for x gives $x = 4/y$. Substituting $4/y$ for x in equation (4) gives

$$6\left(\frac{4}{y}\right) - y = 5.$$

Clear fractions by multiplying both sides by y, noting the restriction that y cannot be 0. Then solve for y.

$$\frac{24}{y} - y = 5$$

$$24 - y^2 = 5y \qquad \text{Multiply by } y.$$

$$0 = y^2 + 5y - 24 \qquad \text{Standard form}$$

Solve this quadratic equation by factoring.

$$0 = (y - 3)(y + 8)$$

$$y = 3 \quad \text{or} \quad y = -8$$

Substitute these results into $x = 4/y$ to obtain the corresponding values of x.

If $y = 3$, then $x = \frac{4}{3}$.

If $y = -8$, then $x = -\frac{1}{2}$.

The solution set has two ordered pairs: $\{(4/3, 3), (-1/2, -8)\}$. The graph in Figure 11.7 shows that there are two points of intersection. ■

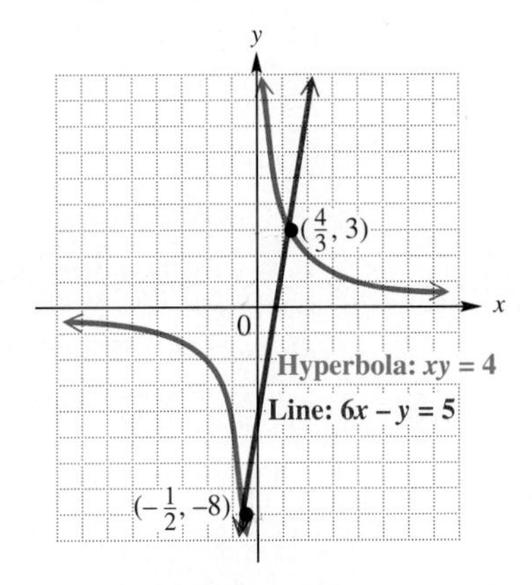

FIGURE 11.7

2 The elimination method is often useful when both equations are second-degree equations. This method is shown in the following example.

EXAMPLE 3 **SOLVING A NONLINEAR SYSTEM BY ELIMINATION**

Solve the system

$$x^2 + y^2 = 9 \qquad (6)$$

$$2x^2 - y^2 = -6. \qquad (7)$$

The graph of equation (6) is a circle, while the graph of equation (7) is a hyperbola. Analyzing the possibilities will lead to the conclusion that there may be 0, 1, 2, 3, or 4 points of intersection.

Adding the two equations will eliminate y, leaving an equation that can be solved for x.

$$\begin{array}{rl} x^2 + y^2 = 9 & (6) \\ 2x^2 - y^2 = -6 & (7) \\ \hline 3x^2 \quad = 3 & \\ x^2 = 1 & \\ x = 1 \quad \text{or} \quad x = -1 & \end{array}$$

Each value of x gives corresponding values for y when substituted into one of the original equations. Using equation (6) gives the following.

If $x = 1$,

$$(1)^2 + y^2 = 9$$
$$y^2 = 8$$
$$y = \sqrt{8} \quad \text{or} \quad -\sqrt{8}$$
$$y = 2\sqrt{2} \quad \text{or} \quad -2\sqrt{2}.$$

If $x = -1$,

$$(-1)^2 + y^2 = 9$$
$$y^2 = 8$$
$$y = 2\sqrt{2} \quad \text{or} \quad -2\sqrt{2}.$$

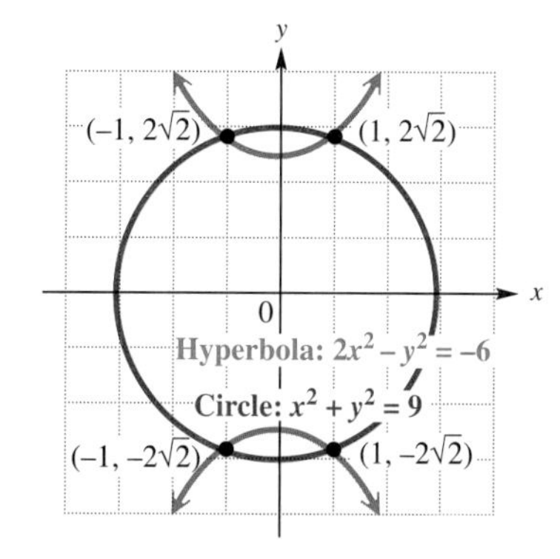

FIGURE 11.8

The solution set has four ordered pairs: $\{(1, 2\sqrt{2}), (1, -2\sqrt{2}), (-1, 2\sqrt{2}), (-1, -2\sqrt{2})\}$. Figure 11.8 shows the four points of intersection. ■

3 The next example shows a system of second-degree equations that requires a combination of methods to solve.

EXAMPLE 4 SOLVING A NONLINEAR SYSTEM BY A COMBINATION OF METHODS

Solve the system

$$x^2 + 2xy - y^2 = 7 \qquad \textbf{(8)}$$
$$x^2 - y^2 = 3. \qquad \textbf{(9)}$$

While we have not graphed equations like equation (8), its graph is a hyperbola. The graph of equation (9) is also a hyperbola. Two hyperbolas may have 0, 1, 2, 3, or 4 points of intersection.

The elimination method can be used here in combination with the substitution method. Begin by eliminating the squared terms by multiplying both sides of equation (9) by -1 and then adding the result to equation (8).

$$\begin{array}{rl} x^2 + 2xy - y^2 = 7 & (8) \\ -x^2 \qquad + y^2 = -3 & -1 \text{ times equation (9)} \\ \hline 2xy \qquad = 4 & \end{array}$$

Next, solve $2xy = 4$ for y. (Either variable would do.)

$$2xy = 4$$

$$y = \frac{2}{x} \qquad \textbf{(10)}$$

Now substitute $y = 2/x$ into one of the original equations. It is easier to do this with equation (9).

$$x^2 - y^2 = 3 \qquad (9)$$

$$x^2 - \left(\frac{2}{x}\right)^2 = 3$$

$$x^2 - \frac{4}{x^2} = 3$$

Clear the equation of fractions by multiplying both sides by x^2 and solve. Notice that the equation is then quadratic in form.

$$x^4 - 4 = 3x^2$$

$$x^4 - 3x^2 - 4 = 0 \qquad \text{Subtract } 3x^2.$$

$$(x^2 - 4)(x^2 + 1) = 0 \qquad \text{Factor.}$$

$$x^2 - 4 = 0 \quad \text{or} \quad x^2 + 1 = 0 \qquad \text{Zero-factor property}$$

$$x^2 = 4 \quad \text{or} \quad x^2 = -1$$

$$x = 2 \quad \text{or} \quad x = -2 \qquad x = i \quad \text{or} \quad x = -i$$

Substituting the four values of x from above into equation (10) gives the corresponding values for y.

If $x = 2$, then $y = 1$. If $x = i$, then $y = -2i$.

If $x = -2$, then $y = -1$. If $x = -i$, then $y = 2i$.

Note that if you substitute the x-values found above into equations (8) or (9) instead of into equation (10), you get extraneous solutions. It is always wise to check all solutions in both of the given equations. There are four ordered pairs in the solution set, two real and two imaginary:

$$\{(2, 1), (-2, -1), (i, -2i), (-i, 2i)\}.$$

The graph of the system, shown in Figure 11.9, shows only the two real intersection points because the graph is in the real number plane. The two ordered pairs with imaginary coordinates are solutions of the system, but do not show up on the graph. ■

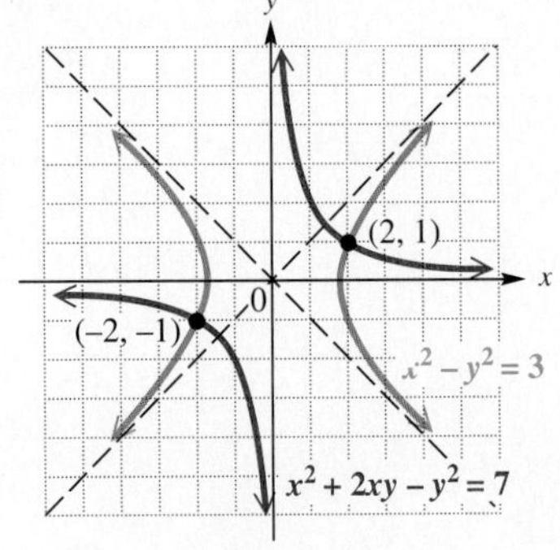

FIGURE 11.9

NOTE In several examples of this section, we analyzed the possible number of points of intersection of the graphs in each system. However, in some examples we worked with equations whose graphs had not been studied. Keep in mind that it is not absolutely essential to visualize the number of points of intersection in order to solve the system. Without having studied equations like equations (5) and (8), it would be unrealistic to expect you to know that their graphs are hyperbolas. Visualizing the geometry of the graphs is only an aid to solving these systems, and not an essential part of the solution process.

EXAMPLE 5

SOLVING A SYSTEM WITH AN ABSOLUTE VALUE EQUATION

Solve the system

$$x^2 + y^2 = 16 \qquad \textbf{(11)}$$
$$|x| + y = 4. \qquad \textbf{(12)}$$

The substitution method is required here. Equation (12) can be rewritten as $|x| = 4 - y$. Then, using the definition of absolute value gives

$$x = 4 - y \quad \text{or} \quad x = -(4 - y) = y - 4. \qquad \textbf{(13)}$$

(Since $|x| \geq 0$ for all real x, $4 - y \geq 0$, or $4 \geq y$.) Substituting from either part of (13) into equation (11) gives the same result.

$$(4 - y)^2 + y^2 = 16 \quad \text{or} \quad (y - 4)^2 + y^2 = 16$$

Since $(4 - y)^2 = (y - 4)^2 = 16 - 8y + y^2$, either equation becomes

$$(16 - 8y + y^2) + y^2 = 16$$
$$2y^2 - 8y = 0$$
$$2y(y - 4) = 0$$
$$y = 0 \quad \text{or} \quad y = 4.$$

From equation (13),

if $y = 0$, then $x = 4 - 0$ or $x = 0 - 4$.
$x = 4$ $\qquad$ $x = -4$

if $y = 4$, then $x = 4 - 4$ or $x = 4 - 4$.
$x = 0$ $\qquad$ $x = 0$

The solution set, $\{(4, 0), (-4, 0), (0, 4)\}$, includes the points of intersection shown in Figure 11.10. ■

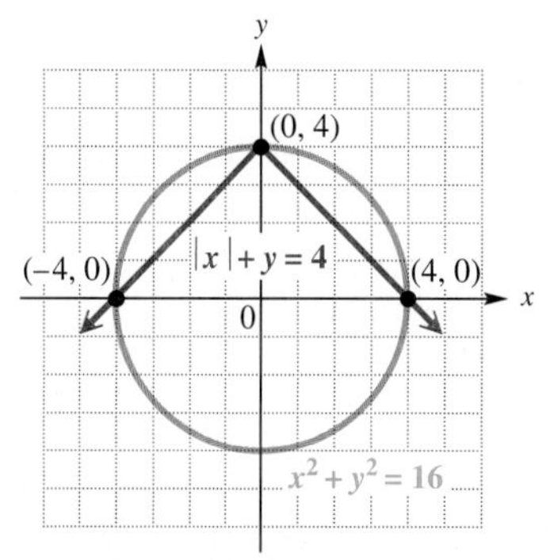

FIGURE 11.10

11.4 EXERCISES

For each of Exercises 1–16, draw a sketch of the two graphs described, with the indicated number of points of intersection. (There may be more than one way to do this in some cases.) See Figure 11.5.

1. a line and a circle; no points

2. a line and a circle; one point

3. a line and a circle; two points

4. a line and an ellipse; no points

5. a line and an ellipse; one point

6. a line and an ellipse; two points

7. a line and a hyperbola; no points

8. a line and a hyperbola; one point

9. a line and a hyperbola; two points

10. a circle and an ellipse; one point

11. a circle and an ellipse; four points

12. a parabola and an ellipse; one point

13. a parabola and an ellipse; four points

14. a parabola and a hyperbola; two points

15. a parabola and a hyperbola; four points

16. a circle and a hyperbola; two points

Solve the following systems by the substitution method. See Examples 1 and 2.

17. $y = x^2 + 2x$
$y = x$

18. $y = 2x^2 - x$
$2y = 4x$

19. $y = x^2 - 4x + 4$
$x + y = 2$

20. $y = x^2 + 2x + 1$
$x - y = -7$

21. $x^2 + y^2 = 1$
$x + 2y = 1$

22. $x^2 + 3y^2 = 3$
$x = 3y$

23. $xy = 2$
$x + 3y = -5$

24. $xy = -12$
$x + 3y = 5$

25. $xy = -6$
$x + y = -1$

26. $xy = 4$
$x + y = 5$

27. $y = 2x^2 + 4x$
$y = x^2 - 3x - 10$

28. $y = 2x^2$
$y = 8x^2 - 2x - 4$

29. $4x^2 - y^2 = 7$
$y = 2x^2 - 3$

30. $y = x^2 - 3$
$x^2 + y^2 = 9$

31. $x^2 - xy + y^2 = 0$
$x - 2y = 1$

32. $x^2 - 3x + y^2 = 4$
$2x - y = 3$

---

33. Write an explanation of the steps you would use to solve the system

$$x^2 + y^2 = 25$$
$$y = x - 1$$

by the substitution method. (See Objective 1.)

34. Write an explanation of the steps you would use to solve the system

$$x^2 + y^2 = 12$$
$$x^2 - y^2 = 13$$

by the elimination method. (See Objective 2.)

Solve the following systems either by the elimination method or by a combination of the elimination and substitution methods. See Examples 3, 4, and 5.

35. $x^2 + y^2 = 4$
$x^2 - y^2 = 4$

36. $x^2 + y^2 = 26$
$x^2 - y^2 = 2$

37. $2x^2 + 3y^2 = 6$
$x^2 + 3y^2 = 3$

38. $3x^2 + y^2 = 13$
$4x^2 - 3y^2 = 13$

39. $5x^2 - 2y^2 = -13$
$3x^2 + 4y^2 = 39$

40. $6x^2 + y^2 = 9$
$3x^2 + 4y^2 = 36$

41. $xy + y^2 = 3$
$2x^2 - xy - y^2 = 5$

42. $xy - y^2 = 2$
$3x^2 - xy + y^2 = 25$

43. $x^2 + 2xy - y^2 = 7$
$x^2 - y^2 = 3$

44. $2x^2 + 3xy + 2y^2 = 21$
$x^2 + y^2 = 6$

45. $3x^2 + 2xy - 3y^2 = 5$
$-x^2 - 3xy + y^2 = 3$

46. $-2x^2 + 7xy - 3y^2 = 4$
$2x^2 - 3xy + 3y^2 = 4$

47. $x = |y|$
$x^2 + y^2 = 18$

48. $2x + |y| = 4$
$x^2 + y^2 = 5$

49. $y = |x - 1|$
$y = x^2 - 4$

50. $2x^2 - y^2 = 4$
$|x| = |y|$

Write a nonlinear system of equations and solve in Exercises 51–56.

51. The sum of the squares of two numbers is 8. The product of the two numbers is 4. Find the numbers.

52. The sum of the squares of two numbers is 26. The difference of the squares of the same two numbers is 24. Find the numbers.

53. The area of a rectangular rug is 84 square feet and its perimeter is 38 feet. Find the length and width of the rug.

54. Find the length and width of a rectangular room whose perimeter is 50 meters and whose area is 100 square meters.

55. A company has found that the price p (in dollars) of its scientific calculator is related to the supply x (in thousands) by the equation $px = 16$. Also, the price is related to the demand x (in thousands) for the calculator by the equation $p = 10x + 12$. The *equilibrium price* is the value of p where demand equals supply. Find the equilibrium price and the supply/demand at that price by solving a system of equations.

56. The calculator company in Exercise 55 has also determined that the cost y to make x (thousand) calculators is $y = 4x^2 + 36x + 20$, while the revenue y from the sale of x (thousand) calculators is $36x^2 - 3y = 0$. Find the *break-even point,* where cost just equals revenue, by solving a system of equations.

PREVIEW EXERCISES

Graph each inequality. See Section 7.4.

57. $2x - y \leq 4$

58. $-x + 3y > 9$

59. $4x + 2y > 8$

60. $-5x + 3y \leq 15$

61. $2x \leq y$

62. $y \geq -4$

11.5 SECOND-DEGREE INEQUALITIES, SYSTEMS, AND LINEAR PROGRAMMING

OBJECTIVES

1 GRAPH SECOND-DEGREE INEQUALITIES.

2 GRAPH THE SOLUTION SET OF A SYSTEM OF INEQUALITIES.

3 SOLVE LINEAR PROGRAMMING PROBLEMS BY GRAPHING.

FOCUS ON PROBLEM SOLVING

The manufacturing process requires that oil refineries manufacture at least 2 gallons of gasoline for each gallon of fuel oil. To meet the winter demand for fuel oil, at least 3 million gallons a day must be produced. The demand for gasoline is no more than 6.4 million gallons per day. If the price of gasoline is $1.90 and the price of fuel oil is $1.50 per gallon, how much of each should be produced to maximize revenue?

This is an example of an *optimization* problem. Optimization problems involve maximizing or minimizing quantities. Here, we are asked to maximize revenue; in another situation, we might be required to minimize cost. In order to solve this problem, we use an application of systems of linear inequalities known as *linear programming*. This problem is Exercise 57 in the exercises for this section. After working through this section, you should be able to solve this problem.

1 The linear inequality $3x + 2y \leq 5$ is graphed by first graphing the boundary line $3x + 2y = 5$. **Second-degree inequalities** such as $x^2 + y^2 \leq 36$ are graphed in much the same way. The boundary of $x^2 + y^2 \leq 36$ is the graph of the equation $x^2 + y^2 = 36$, a circle with radius 6 and center at the origin, as shown in Figure 11.11. As with linear inequalities, the inequality $x^2 + y^2 \leq 36$ will include either the points outside the circle or the points inside the circle. Decide which region to shade by substituting any point not on the circle, such as (0, 0), into the inequality. Since $0^2 + 0^2 < 36$ is a true statement, the inequality includes the points inside the circle, the shaded region in Figure 11.11.

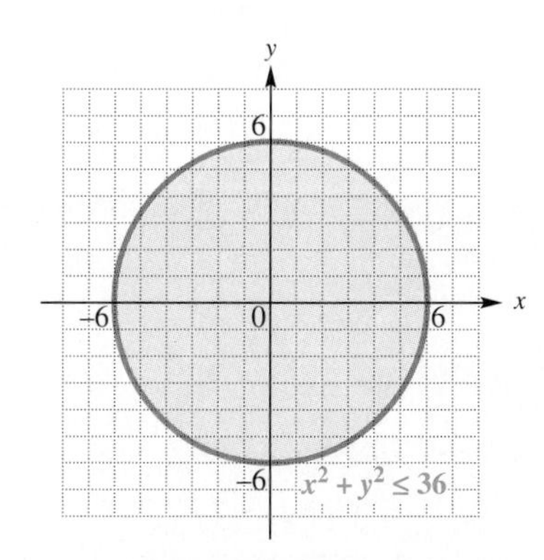

FIGURE 11.11

EXAMPLE 1
GRAPHING A SECOND-DEGREE INEQUALITY

Graph $y < -2(x - 4)^2 - 3$.

The boundary, $y = -2(x - 4)^2 - 3$, is a parabola opening downward with vertex at $(4, -3)$. Using the point $(0, 0)$ as a test point gives

$$0 < -2(0 - 4)^2 - 3 \quad ?$$
$$0 < -32 - 3 \quad ?$$
$$0 < -35. \quad \text{False}$$

Because the final inequality is a false statement, the points in the region containing (0, 0) do not satisfy the inequality. So the solution includes all points in the region that does not include (0, 0). Figure 11.12 shows the final graph; the parabola is drawn with a dashed line since the points of the parabola itself do not satisfy the inequality. ■

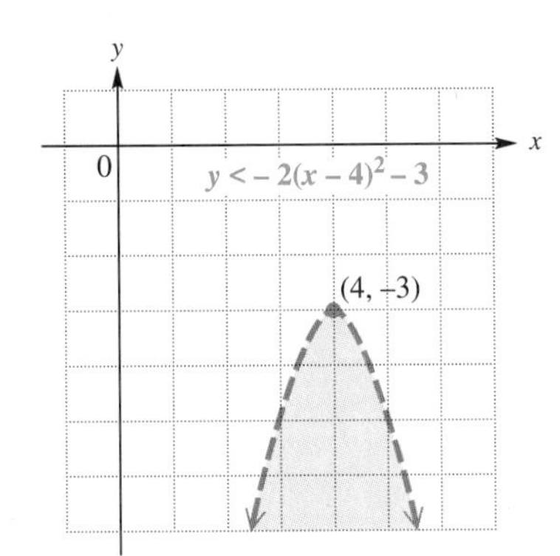

FIGURE 11.12

EXAMPLE 2
GRAPHING A SECOND-DEGREE INEQUALITY

Graph $16y^2 \leq 144 + 9x^2$.

First rewrite the inequality as follows.

$$16y^2 - 9x^2 \leq 144$$
$$\frac{y^2}{9} - \frac{x^2}{16} \leq 1 \quad \text{Divide by 144.}$$

This form of the inequality shows that the boundary is the hyperbola

$$\frac{y^2}{9} - \frac{x^2}{16} = 1.$$

The desired region will either be the region between the branches of the hyperbola or the regions above the top branch and below the bottom branch. The boundary itself will also be included in the graph, because of the symbol $\leq$.

Test the region between the branches by choosing (0, 0) as a test point. Substitute into the original inequality.

$$16y^2 \leq 144 + 9x^2$$
$$16(0)^2 \leq 144 + 9(0)^2 \quad ?$$
$$0 \leq 144 + 0 \quad ?$$
$$0 \leq 144 \quad \text{True}$$

Since the test point (0, 0) satisfies the inequality $16y^2 \leq 144 + 9x^2$, the region between the branches containing (0, 0) is shaded, as shown in Figure 11.13. ■

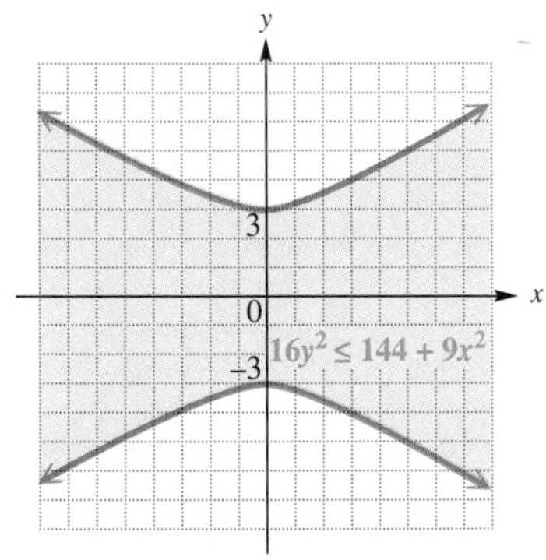

FIGURE 11.13

2 If two or more inequalities are considered at the same time, at least one of which is nonlinear, we have a **nonlinear system of inequalities.** In order to find the solution set of a nonlinear system, we find the intersection of the graphs of the inequalities in the system. This is just an extension of the method used in Section 7.4 to graph the pairs of linear inequalities joined by the word *and*.

EXAMPLE 3
GRAPHING A NONLINEAR SYSTEM OF INEQUALITIES

Graph the solution set of

$$2x + 3y > 6$$
$$x^2 + y^2 < 16.$$

Begin by graphing the solution of $2x + 3y > 6$. Using the methods of Section 7.4, we find that the boundary line is the graph of $2x + 3y = 6$, and is a dashed line because of the symbol $>$. The test point (0, 0) leads to a false statement in the inequality $2x + 3y > 6$, so shade the region above the line, as shown in Figure 11.14.

Using the methods of Examples 1 and 2, we find that the graph $x^2 + y^2 < 16$ is the interior of a dashed circle centered at the origin with radius 4. This is shown in Figure 11.15.

Finally, to get the graph of the solution set of the system, find the intersection of the graphs of the two inequalities. The overlapping region in Figure 11.16 is the solution set. ■

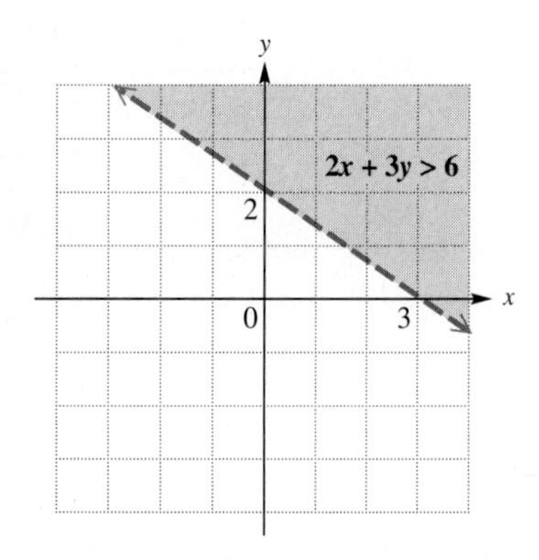

FIGURE 11.14

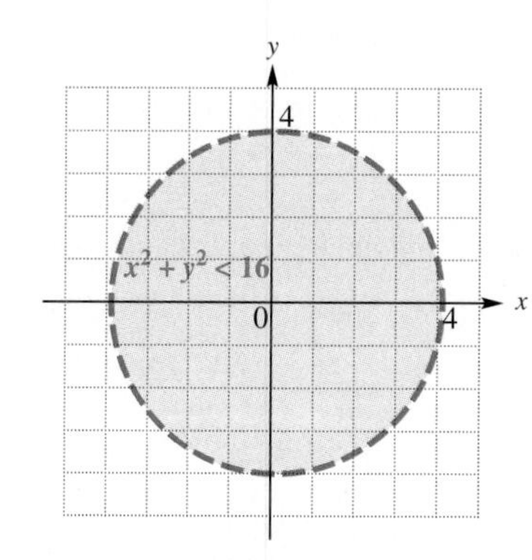

FIGURE 11.15

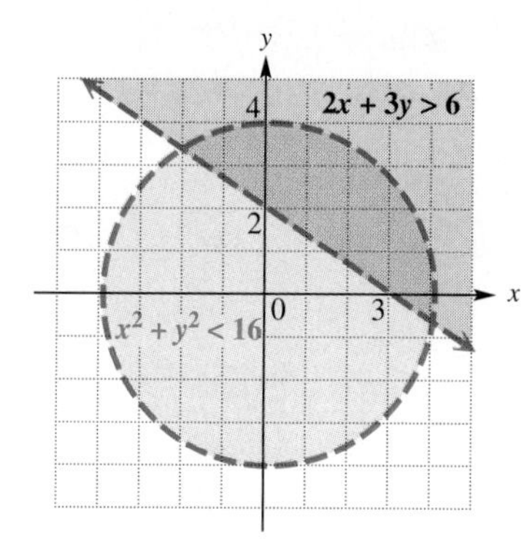

FIGURE 11.16

EXAMPLE 4
GRAPHING A SYSTEM WITH THREE INEQUALITIES

Graph the solution of the system

$$y \geq x^2 - 2x + 1$$
$$2x^2 + y^2 > 4$$
$$y < 4.$$

The graph of $y = x^2 - 2x + 1$ is a parabola with vertex at (1, 0). Those points in the interior of the parabola satisfy the condition $y > x^2 - 2x + 1$. Thus points on the parabola or in the interior are in the solution of $y \geq x^2 - 2x + 1$. The graph of $2x^2 + y^2 = 4$ is an ellipse. To satisfy the inequality $2x^2 + y^2 > 4$, a point must lie outside the ellipse. The graph of $y < 4$ includes all points below the line $y = 4$. Finally, the graph of the system is the shaded region in Figure 11.17 that lies outside the ellipse, inside or on the boundary of the parabola, and below the line $y = 4$. ■

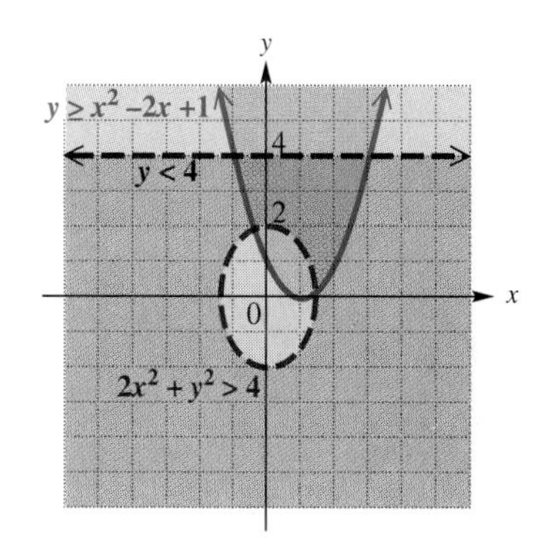

FIGURE 11.17

While the system in the next example does not contain a nonlinear inequality, it is different from those that we have solved in both this section and Section 7.4. It contains four linear inequalities, and will prepare us to solve linear programming problems, discussed at the end of this section.

EXAMPLE 5
SOLVING A SYSTEM OF SEVERAL LINEAR INEQUALITIES

Graph the solution set of the system

$$2x + 3y \geq 12$$
$$7x + 4y \geq 28$$
$$y \leq 6$$
$$x \leq 5.$$

The graph is obtained by graphing the four inequalities on the same axes and shading the region common to all four as shown in Figure 11.18. As the graph shows, the boundary lines are all solid. ■

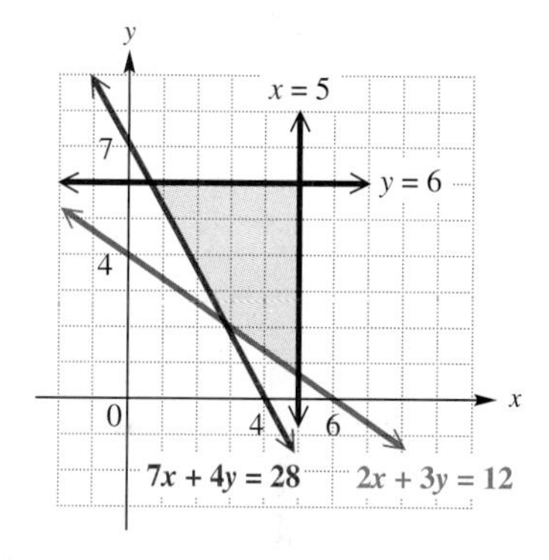

FIGURE 11.18

3 We now look at an application of systems of inequalities.

PROBLEM SOLVING

An important application of mathematics to business and social science is called linear programming. **Linear programming** is used to find such things as minimum cost and maximum profit. The basic ideas of this technique are shown in the following example. ■

The Smith Company makes two products, tape decks and amplifiers. Each tape deck gives a profit of \$3, while each amplifier produces \$7 profit. The company must manufacture at least one tape deck per day to satisfy one of its customers, but no more than five because of production problems. Also, the number of amplifiers produced cannot exceed six per day. As a further requirement, the number of tape decks cannot exceed the number of amplifiers. How many of each should the company manufacture in order to obtain the maximum profit?

To begin, translate the statement of the problem into symbols by assuming

$$x = \text{number of tape decks to be produced daily}$$
$$y = \text{number of amplifiers to be produced daily.}$$

According to the statement of the problem given above, the company must produce at least one tape deck (one or more), so

$$x \geq 1.$$

Since no more than 5 tape decks may be produced,

$$x \leq 5.$$

Since no more than 6 amplifiers may be made in one day,

$$y \leq 6.$$

The requirement that the number of tape decks may not exceed the number of amplifiers translates as

$$x \leq y.$$

The number of tape decks and of amplifiers cannot be negative, so

$$x \geq 0 \quad \text{and} \quad y \geq 0.$$

These restrictions, or **constraints,** that are placed on production form the system of inequalities

$$x \geq 1, \quad x \leq 5, \quad y \leq 6, \quad x \leq y, \quad x \geq 0, \quad y \geq 0.$$

The maximum possible profit that the company can make, subject to these constraints, is found by sketching the graph of the solution of the system. See Figure 11.19. The only feasible values of x and y are those that satisfy all constraints. These values correspond to points that lie on the boundary or in the shaded region, called the **region of feasible solutions.**

Since each tape deck gives a profit of \$3, the daily profit from the production of x tape decks is $3x$ dollars. Also, the profit from the production of y amplifiers will be $7y$ dollars per day. The total daily profit is thus

$$\text{profit} = 3x + 7y.$$

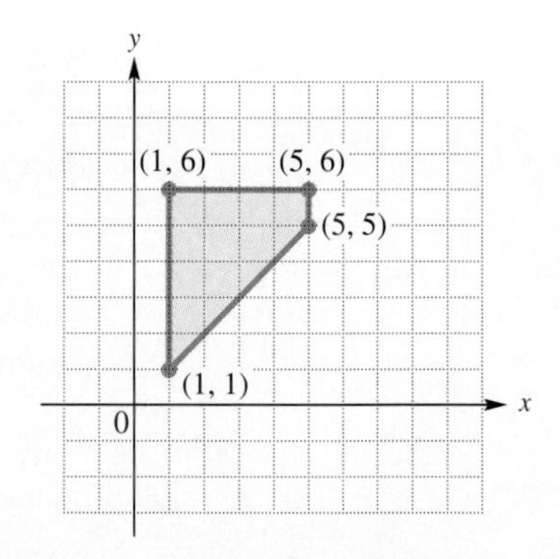

FIGURE 11.19

The problem of the Smith Company may now be stated as follows: find values of x and y in the shaded region of Figure 11.19 that will produce the maximum possible value of $3x + 7y$.

It can be shown that any optimum value (maximum or minimum) will always occur at a **vertex** (or **corner**) **point** of the region of feasible solutions. Locate the point (x, y) that gives the maximum profit by checking the coordinates of the vertex points, shown in Figure 11.19 and listed below. Find the profit that corresponds to each coordinate pair and choose the one that gives the maximum profit.

Point	*Profit* $= 3x + 7y$
(1, 1)	$3(1) + 7(1) = 10$
(1, 6)	$3(1) + 7(6) = 45$
(5, 6)	$3(\mathbf{5}) + 7(\mathbf{6}) = \mathbf{57}$ ← Maximum
(5, 5)	$3(5) + 7(5) = 50$

The maximum profit of \$57 is obtained when 5 tape decks and 6 amplifiers are produced each day.

EXAMPLE 6
SOLVING A PROBLEM USING LINEAR PROGRAMMING

Robin, who is ill, takes vitamin pills. Each day, she must have at least 16 units of vitamin A, at least 5 units of vitamin B_1, and at least 20 units of vitamin C. She can choose between red pills, costing 10¢ each, which contain 8 units of A, 1 of B_1, and 2 of C; and blue pills, costing 20¢ each, which contain 2 units of A, 1 of B_1, and 7 of C. How many of each pill should she buy in order to minimize her cost and yet fulfill her daily requirements?

Let x represent the number of red pills to buy, and let y represent the number of blue pills to buy. Then the cost in pennies per day is given by

$$\text{cost} = 10x + 20y.$$

Since Robin buys x of the 10¢ pills and y of the 20¢ pills, she gets vitamin A as follows: 8 units from each red pill and 2 units from each blue pill. Altogether, she gets $8x + 2y$ units of A per day. Since she must get at least 16 units,

$$8x + 2y \geq 16.$$

Each red pill or each blue pill supplies 1 unit of vitamin B_1. Robin needs at least 5 units per day, so

$$x + y \geq 5.$$

For vitamin C, the inequality is

$$2x + 7y \geq 20.$$

Also, $x \geq 0$ and $y \geq 0$, since Robin cannot buy negative numbers of the pills.

The total cost of the pills can be minimized by finding the solution of the system of inequalities formed by the constraints. (See Figure 11.20.) The solution to this minimizing problem will also occur at a vertex point. Check the coordinates of the vertex points in the cost function to find the lowest cost.

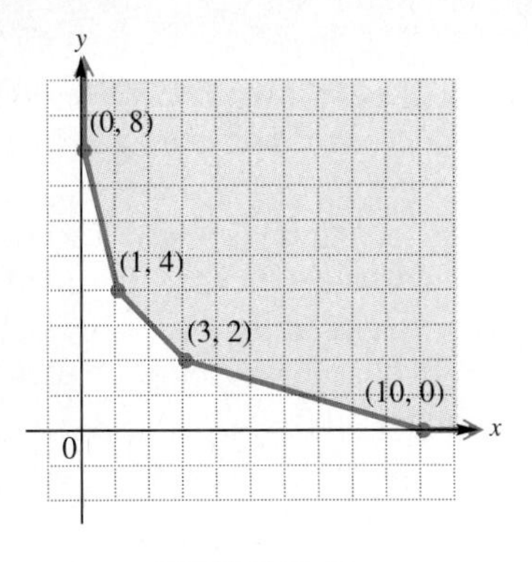

FIGURE 11.20

Point	*Cost* $= 10x + 20y$
(10, 0)	$10(10) + 20(0) = 100$
(3, 2)	$10(3) + 20(2) = 70$ ← Minimum
(1, 4)	$10(1) + 20(4) = 90$
(0, 8)	$10(0) + 20(8) = 160$

Robin's best bet is to buy 3 red pills and 2 blue ones, for a total cost of 70¢ per day. She receives minimum amounts of vitamins B_1 and C but an excess of vitamin A. Even with an excess of A, this is still the best buy. ■

PROBLEM SOLVING

To solve a linear programming problem in general, use the following steps.

SOLVING A LINEAR PROGRAMMING PROBLEM

1. Write the objective function and all necessary constraints.
2. Graph the feasible region.
3. Identify all vertex points.
4. Find the value of the objective function at each vertex point.
5. The solution is given by the vertex producing the optimum value of the objective function. ■

11.5 EXERCISES

Graph each of the following inequalities. See Examples 1 and 2.

1. $y < x^2$

2. $y^2 < 36 - x^2$

3. $x^2 \geq 4 - 2y^2$

4. $y \geq x^2 - 4$

5. $y^2 \leq 16 + x^2$

6. $x^2 \leq 36 + 4y^2$

7. $y \leq x^2 + 4x + 6$

8. $y > 2x^2 - 4x + 3$

9. $9x^2 > 4y^2 - 36$

10. $9x^2 \leq 4y^2 + 36$

11. $y^2 \leq 4 - 2x^2$

12. $4x^2 + 9y^2 < 36$

13. $2x^2 + 2y^2 \geq 98$

14. $4y^2 \leq 64 - 16x^2$

15. $x \leq y^2 - 4y + 1$

16. $x \geq 2y^2 + 6y - 1$

17. $y^2 - 9x^2 \geq 9$

18. $16x^2 < 9y^2 + 144$

19. Write an explanation of how to graph the solution set of a nonlinear inequality. (See Objective 1.)

20. Write an explanation of how to graph the solution set of a system of nonlinear inequalities. (See Objective 2.)

Graph each of the following systems of inequalities. See Examples 3–5.

21. $3x + 2y < 12$
$x - 3y < 6$

22. $x - 2y > -4$
$2x + 3y < 8$

23. $2x - 3y \leq 6$
$4x + 5y \geq 10$

24. $3x - y \leq 0$
$x - 2y \leq 8$

25. $x \leq 4$
$y \leq 6$

26. $x \geq 2$
$y \leq -4$

27. $2x - y > 0$
$y > x^2$

28. $y > x^2 - 2$
$y < -x^2 + 2$

29. $x^2 - y^2 \geq 1$
$\frac{x^2}{9} + \frac{y^2}{4} \leq 1$

30. $(x-1)^2 + (y-2)^2 < 4$
$(x-2)^2 + y^2 < 4$

31. $y \leq (x-2)^2 + 1$
$\frac{x^2}{4} + \frac{y^2}{9} \leq 1$

32. $-x^2 + \frac{y^2}{4} \geq 1$
$-4 \leq y \leq 4$

33. $x \geq -3$
$y \leq 2$
$y > x - 3$

34. $2x - 5y \geq 10$
$x + 3y \geq 6$
$y \leq 2$

35. $2y + x \geq -5$
$y \leq 3 + x$
$x \leq 0$
$y \leq 0$

36. $2x + 3y \leq 12$
$2x + 3y > -6$
$3x + y < 4$
$x \geq 0$
$y \geq 0$

37. $y \geq 3^x$
$y \geq 2$

38. $y \leq \left(\frac{1}{2}\right)^x$
$y \geq 4$

39. $y \leq |x + 2|$
$x \geq 0$

40. $y \leq \log x$
$y \geq |x - 2|$

41. $e^{-x} - y \leq 1$
$x - 2y \geq 4$

42. $\ln x - y \geq 1$
$x^2 - 2x - y \leq 1$

---

43. Which one of the following is a description of the graph of the solution set of the system below? (See Objective 2.)

$$x^2 + y^2 < 25$$
$$y > -2$$

(a) all points outside the circle $x^2 + y^2 = 25$ and above the line $y = -2$
(b) all points outside the circle $x^2 + y^2 = 25$ and below the line $y = -2$
(c) all points inside the circle $x^2 + y^2 = 25$ and above the line $y = -2$
(d) all points inside the circle $x^2 + y^2 = 25$ and below the line $y = -2$

44. Fill in the blank with the appropriate response. (See Objective 2.) The graph of the system

$$y > x^2 + 1$$
$$\frac{x^2}{9} + \frac{y^2}{4} > 1$$
$$y < 5$$

consists of all points $\underset{\text{(inside/outside)}}{\underline{\quad ? \quad}}$ the parabola $y = x^2 + 1$, $\underset{\text{(inside/outside)}}{\underline{\quad ? \quad}}$ the ellipse $\frac{x^2}{9} + \frac{y^2}{4} = 1$, and $\underset{\text{(above/below)}}{\underline{\quad ? \quad}}$ the line $y = 5$.

The graphs in Exercises 45–48 show regions of feasible solutions. Find the maximum and minimum values of the given expressions. See Example 6.

45. $3x + 5y$

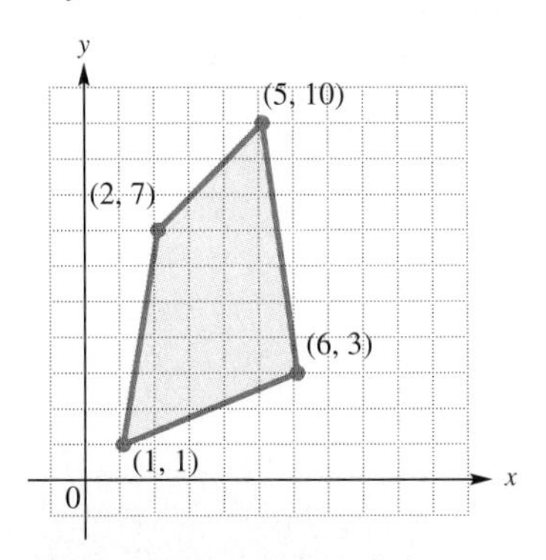

46. $6x + y$

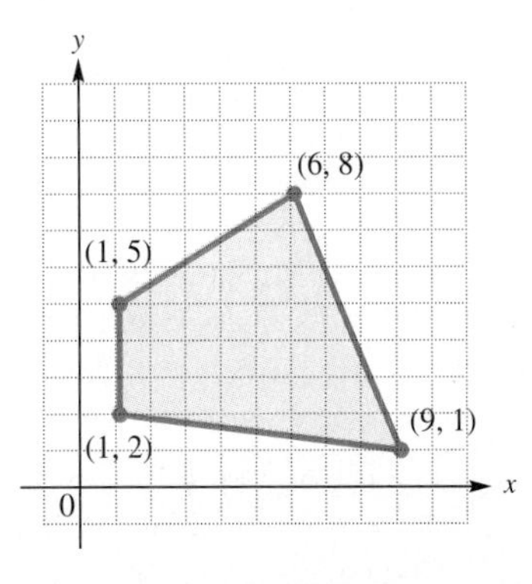

47. $40x + 75y$

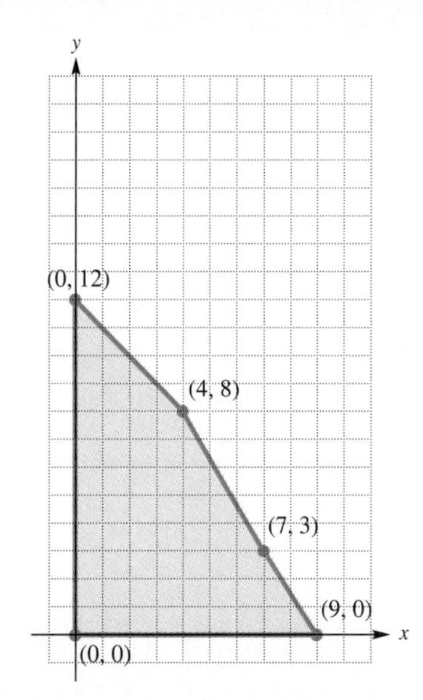

48. $35x + 125y$

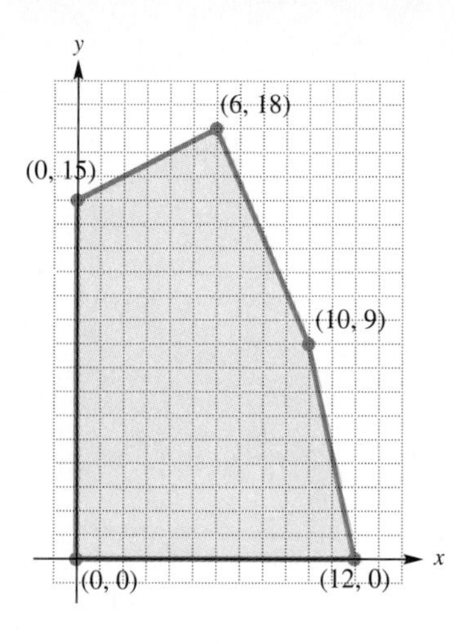

In Exercises 49–52, use graphical methods to solve each problem. See Example 6.

49. Find $x \geq 0$ and $y \geq 0$ such that
$$2x + 3y \leq 6$$
$$4x + y \leq 6$$
and $5x + 2y$ is maximized.

50. Find $x \geq 0$ and $y \geq 0$ such that
$$x + y \leq 10$$
$$5x + 2y \geq 20$$
$$2y \geq x$$
and $x + 3y$ is minimized.

51. Find $x \geq 2$ and $y \geq 5$ such that
$$3x - y \geq 12$$
$$x + y \leq 15$$
and $2x + y$ is minimized.

52. Find $x \geq 10$ and $y \geq 20$ such that
$$2x + 3y \leq 100$$
$$5x + 4y \leq 200$$
and $x + 3y$ is maximized.

Solve each of the following problems. See Example 6.

53. Farmer Jones raises only pigs and geese. He wants to raise no more than 16 animals with no more than 12 geese. He spends \$50 to raise a pig and \$20 to raise a goose. He has \$500 available for this purpose. Find the maximum profit he can make if he makes a profit of \$80 per goose and \$40 per pig.

54. A wholesaler of party goods wishes to display her products at a convention of social secretaries in such a way that she gets the maximum number of inquiries about her whistles and hats. Her booth at the convention has 12 square meters of floor space to be used for display purposes. A display unit for hats requires 2 square meters; and for whistles, 4 square meters. Experience tells the wholesaler that she should never have more than a total of 5 units of whistles and hats on display at one time. If she receives three inquiries for each unit of hats and two inquiries for each unit of whistles on display, how many of each should she display in order to get the maximum number of inquiries?

55. An office manager wants to buy some filing cabinets. She knows that cabinet #1 costs \$20 each, requires 6 square feet of floor space, and holds 8 cubic feet of files. Cabinet #2 costs \$40 each, requires 8 square feet of floor space, and holds 12 cubic feet. She can spend no more than \$280 due to budget limitations, while her office has room for no more than 72 square feet of cabinets. She wants the maximum storage capacity within the limits imposed by funds and space. How many of each type of cabinet should she buy?

56. In a small town in South Carolina, zoning rules require that the window space (in square feet) in a house be at least one-sixth of the space used up by solid walls. The cost to heat the house is 10¢ for each square foot of solid walls and 40¢ for each square foot of windows. Find the maximum total area (windows plus walls) if $80 is available to pay for heat.

57. The manufacturing process requires that oil refineries manufacture at least 2 gallons of gasoline for each gallon of fuel oil. To meet the winter demand for fuel oil, at least 3 million gallons a day must be produced. The demand for gasoline is no more than 6.4 million gallons per day. If the price of gasoline is $1.90 and the price of fuel oil is $1.50 per gallon, how much of each should be produced to maximize revenue?

58. Seall Manufacturing Company makes color television sets. It produces a bargain set that sells for $100 profit and a deluxe set that sells for $150 profit. On the assembly line the bargain set requires 3 hours, while the deluxe set takes 5 hours. The cabinet shop spends 1 hour on the cabinet for the bargain set and 3 hours on the cabinet for the deluxe set. Both sets require 2 hours of time for testing and packing. On a particular production run the Seall Company has available 3900 work hours on the assembly line, 2100 work hours in the cabinet shop, and 2200 work hours in the testing and packing department. How many sets of each type should it produce to make maximum profit? What is the maximum profit?

Photo for Exercise 57

PREVIEW EXERCISES

Find the value of each expression. See Section 1.3.

59. $(-3)(1) - 2(4) - (-1)(-5)$

60. $(-4)(-2) + (2)(-3) - (-1)(-6)$

61. $3[(-2)(2) - (-3)(1)]$

62. $2[(3)(2) - (-2)(1)]$

63. $[(-1)(4) + (-4)(-2)] - [(-2)(3) - (-3)(1)]$

64. $[(5)(-4) + (-1)(-6)] + [(-2)(-4) - (-1)(3)]$

CHAPTER 11 GLOSSARY

KEY TERMS

11.1 **system of equations** Two or more equations that are to be solved at the same time form a system of equations.

linear system A linear system is a system of equations that contains only linear equations.

inconsistent system A system is inconsistent if it has no solution.

dependent equations Dependent equations are equations whose graphs are the same line.

elimination (or addition) method The elimination (or addition) method of solving a system of equations involves the elimination of a variable by adding the two equations.

substitution method The substitution method of solving a system involves substituting an expression for one variable into an equation with two variables, in order to get an equation with just one variable.

11.2 ordered triple A solution of an equation in three variables is called an ordered triple.

11.4 nonlinear equation An equation that cannot be put into the form $Ax + By = C$ is called a nonlinear equation.

nonlinear system of equations A nonlinear system of equations is a system with at least one nonlinear equation.

11.5 second-degree inequality A second-degree inequality is an inequality with at least one variable of degree two and no variable of degree greater than two.

nonlinear system of inequalities A nonlinear system of inequalities consists of two or more inequalities considered at the same time, where at least one of them is nonlinear.

linear programming Linear programming is an application of linear inequalities, used to find such things as minimum cost and maximum profit.

constraints Restrictions that are placed on the variables in a linear programming application are called constraints.

region of feasible solutions The shaded region in a linear programming application is called the region of feasible solutions.

vertex point The optimum value in a linear programming application will always occur at a vertex point (or corner point) of the region of feasible solutions.

CHAPTER 11 QUICK REVIEW

SECTION	CONCEPTS	EXAMPLES
11.1 LINEAR SYSTEMS OF EQUATIONS IN TWO VARIABLES	**Solving Linear Systems of Two Equations by Elimination**	Solve by elimination.
	Step 1 Write both equations in the form $Ax + By = C$.	$5x + y = 2$ $2x - 3y = 11$
	Step 2 Multiply one or both equations by appropriate numbers so that the sum of the coefficients of either x or y is zero.	To eliminate y, multiply the top equation by 3, and add.
	Step 3 Add the new equations. The sum should be an equation with just one variable.	$15x + 3y = 6$ $2x - 3y = 11$ $17x = 17$ $x = 1$
	Step 4 Solve the equation from Step 3.	Let $x = 1$ in the top equation, and solve for y.

SECTION	CONCEPTS	EXAMPLES
	Step 5 Substitute the result of Step 4 into either of the given equations and solve for the other variable.	$5(1) + y = 2$ $y = -3$
	Step 6 Check the solution in both of the given equations.	Check to verify that $\{(1, -3)\}$ is the solution set.
	Solving Linear Systems of Two Equations by Substitution	Solve by substitution. $4x - y = 7$ $3x + 2y = 30$
	Step 1 Solve one of the equations for either variable.	Solve for y in the top equation. $y = 4x - 7$
	Step 2 Substitute for that variable in the other equation. The result should be an equation with just one variable. *Step 3* Solve the equation from Step 2.	Substitute $4x - 7$ for y in the bottom equation, and solve for x. $3x + 2(4x - 7) = 30$ $3x + 8x - 14 = 30$ $11x - 14 = 30$ $11x = 44$ $x = 4$
	Step 4 Substitute the result from Step 3 into the equation from Step 1 to find the value of the other variable.	Substitute 4 for x in the equation $y = 4x - 7$ to find that $y = 9$.
	Step 5 Check the solution in both of the given equations.	Check to see that $\{(4, 9)\}$ is the solution set.
11.2 LINEAR SYSTEMS OF EQUATIONS IN THREE VARIABLES	**Solving Linear Systems in Three Variables** *Step 1* Use the elimination method to eliminate any variable from any two of the given equations. The result is an equation in two variables.	Solve the system $x + 2y - z = 6$ $x + y + z = 6$ $2x + y - z = 7.$ Add the first and second equations; z is eliminated and the result is $2x + 3y = 12$.
	Step 2 Eliminate the *same* variable from any *other* two equations. The result is an equation in the same two variables as in Step 1.	Eliminate z again by adding the second and third equations to get $3x + 2y = 13$. Now solve the system $2x + 3y = 12 \quad (*)$ $3x + 2y = 13.$

SECTION	CONCEPTS	EXAMPLES
	Step 3 Use the elimination method to eliminate a second variable from the two equations in two variables that result from Steps 1 and 2. The result is an equation in one variable that gives the value of that variable.	To eliminate x, multiply the top equation by -3 and the bottom equation by 2. $-6x - 9y = -36$ $6x + 4y = 26$ $-5y = -10$ $y = 2$
	Step 4 Substitute the value of the variable found in Step 3 into either of the equations in two variables to find the value of the second variable.	Let $y = 2$ in equation (*). $2x + 3(2) = 12$ $2x + 6 = 12$ $2x = 6$ $x = 3$
	Step 5 Use the values of the two variables from Steps 3 and 4 to find the value of the third variable by substituting into any of the original equations.	Let $y = 2$ and $x = 3$ in any of the original equations to find $z = 1$. The solution set is $\{(3, 2, 1)\}$.
11.3 APPLICATIONS OF LINEAR SYSTEMS OF EQUATIONS	To solve an applied problem with two (three) unknowns, write two (three) equations that relate the unknowns. Then solve the system.	The perimeter of a rectangle is 18 feet. The length is 3 feet more than twice the width. Find the dimensions of the rectangle. Let x represent the length and y represent the width. From the perimeter formula, one equation is $2x + 2y = 18$. From the problem, another equation is $x = 3 + 2y$. Now solve the system $2x + 2y = 18$ $x = 3 + 2y$. The solution of the system is (7, 2). Therefore, the length is 7 feet and the width is 2 feet.

SECTION	CONCEPTS	EXAMPLES
11.4 NONLINEAR SYSTEMS OF EQUATIONS	Nonlinear systems can be solved by the substitution method, the elimination method, or a combination of the two.	Solve the system $x^2 + 2xy - y^2 = 14$ $x^2 - y^2 = -16. \quad (*)$ Multiply equation (*) by -1 and use elimination. $x^2 + 2xy - y^2 = 14$ $-x^2 \quad + y^2 = 16$ $2xy = 30$ $xy = 15$ Solve for y to obtain $y = 15/x$, and substitute into equation (*). $x^2 - \left(\frac{15}{x}\right)^2 = -16$ This simplifies to $x^2 - \frac{225}{x^2} = -16.$ Multiply by x^2 and get one side equal to 0. $x^4 + 16x^2 - 225 = 0$ Factor and solve. $(x^2 - 9)(x^2 + 25) = 0$ $x = \pm 3 \qquad x = \pm 5i$ Find corresponding y values to get the solution set $\{(3, 5), (-3, -5), (5i, -3i), (-5i, 3i)\}$.
11.5 SECOND-DEGREE INEQUALITIES, SYSTEMS, AND LINEAR PROGRAMMING	To graph a second-degree inequality, graph the corresponding equation as a boundary and use test points to determine which region(s) form the solution. Shade the appropriate region(s).	Graph $y \geq x^2 - 2x + 3$.

SECTION	CONCEPTS	EXAMPLES
	The solution set of a system of inequalities is the intersection of the solution sets of the individual inequalities.	Graph the solution set of the system $3x - 5y > -15$ $x^2 + y^2 \leq 25.$ 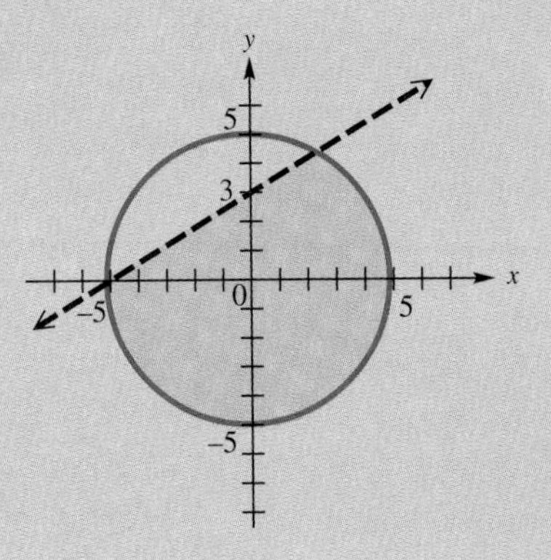
	To solve a linear programming problem, write the objective function and all constraints, graph the feasible region, identify all vertex points (corner points), and find the value of the objective function at each vertex point. Choose the required maximum or minimum value accordingly.	The feasible region for $x + 2y \leq 14$ $3x + 4y \leq 36$ $x \geq 0$ $y \geq 0$ is given here. Maximize the objective function $8x + 12y$. 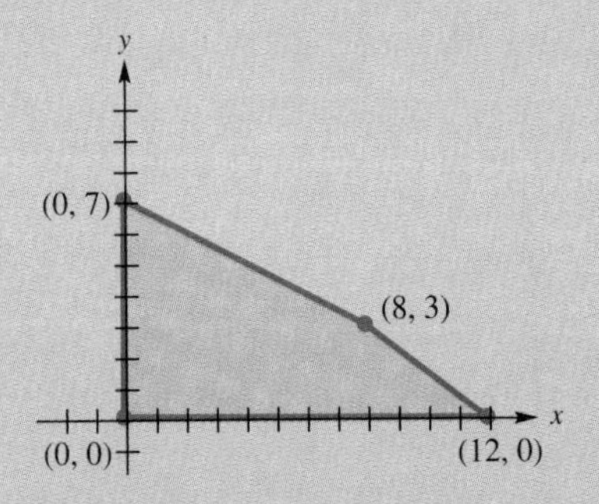 *Vertex point* — *Value of* $8x + 12y$ (0, 0) — 0 (0, 7) — 84 (12, 0) — 96 **(8, 3)** — **100 (maximum)** The objective function is maximized for $x = 8$ and $y = 3$.

CHAPTER 11 REVIEW EXERCISES

[11.1] *Solve the following systems of equations by the elimination method.*

1. $5x - 3y = 19$
   $4x + y = 5$
2. $-x + 4y = 15$
   $2x + y = 6$
3. $6x + 5y = 4$
   $4x - 2y = -8$
4. $2x + 5y = 10$
   $4x + 10y = 3$

Solve the following systems of equations by the substitution method.

5. $-3x - y = 4$
   $x = \frac{2}{3}y$
6. $4x = 2y + 3$
   $x = \frac{1}{2}y$
7. $5x - 2y = 2$
   $x + 6y = 26$
8. $3x - 2y = 11$
   $4x + 3y = 26$
9. $\frac{2}{x} + \frac{4}{y} = 0$
   $\frac{5}{x} - \frac{6}{y} = -8$

10. Can a system of two linear equations have exactly two solutions? Explain.

[11.2] *Solve the following systems of equations.*

11. $2x + 3y - z = -16$
    $x + 2y + 2z = -3$
    $3x - y - z = 5$
12. $4x - y = 2$
    $3y + z = 9$
    $x + 2z = 7$
13. $-x + 2y + 3z = 5$
    $2x + 3y + z = 12$
    $3x - 6y - 9z = -10$
14. $-3x + y + z = 8$
    $4x + 2y + 3z = 15$
    $-6x + 2y + 2z = 10$

[11.3] *Translate each problem into a system of equations and then solve.*

15. On a 10-day trip Gale rented a car for \$31 a day at weekend rates and \$40 a day at weekday rates. If his rental bill was \$364, how many days did he rent at each rate?

16. Sweet's Candy Store is offering a special mix for Valentine's Day. Ms. Sweet will mix some \$2 a pound candy with some \$1 a pound candy to get 50 pounds of mix that she will sell at \$1.30 a pound. How many pounds of each should be used?

17. A plane flies 560 miles in 1.75 hours traveling with the wind. The return trip later against the same wind takes the plane 2 hours. Find the speed of the plane and the speed of the wind.

18. The length of a rectangle is 3 meters more than the width. The perimeter is 42 meters. Find the length and width of the rectangle.

19. The sum of the measures of the angles of a triangle is 180°. One angle measures 10° less than the sum of the other two. The measure of the middle-sized angle is the average of the other two. Find the measures of the three angles.

20. How many liters each of 8%, 10%, and 20% hydrogen peroxide should be mixed together to get 8 liters of 12.5% solution, if the amount of 20% solution used must be 2 liters less than the amount of 8% solution used?

21. Find the values of a, b, and c so that the points $(1, 1)$, $(0, -4)$, and $(-2, 4)$ lie on the graph of the equation $y = ax^2 + bx + c$.

22. Sue sells real estate. On three recent sales, she made a 10% commission, a 6% commission, and a 5% commission. Her total commissions on these sales were \$8500, and she sold property worth \$140,000. If the 5% sale amounted to the sum of the other two, what were the three sales prices?

23. A farmer wishes to satisfy her fertilizer needs with three brands, A, B, and C. She needs to apply a total of 26.4 pounds of nitrogen, 28 pounds of potash, and 26.8 pounds of sulfate of ammonia. Brand A contains 8% nitrogen, 5% potash, and 10% sulfate of ammonia. Brand B contains 6% nitrogen, 10% potash, and 6% sulfate of ammonia. Brand C contains 10% nitrogen, 8% potash, and 5% sulfate of ammonia. How much of each fertilizer should be used?

[11.4] *Solve each system.*

24. $y = 2x - x^2$
$x + 2y = -3$

25. $x^2 + y^2 = 13$
$y = 3x - 3$

26. $x^2 - 4y^2 = -12$
$3x + y = 4$

27. $xy = 12$
$3x + 2y = -18$

28. $4x^2 + 3y^2 = 24$
$2x^2 - y^2 = 12$

29. $2y - 2 = x^2 + 3x$
$x - y = -3$

30. $3x^2 + y^2 = 19$
$2x - y = -2$

31. $xy = 15$
$2x - y = 1$

32. $x^2 + y^2 = 13$
$2x^2 + y^2 = 9$

33. $xy = 3$
$y = 2x - 1$

Give all possible numbers of points of intersection of the pair of graphs given in Exercises 34–37.

34. a circle and a line

35. a circle and a hyperbola

36. a parabola and a line

37. a parabola and a hyperbola

[11.5] *Graph each inequality.*

38. $y \le (x - 2)^2$

39. $x^2 + y^2 \le 25$

40. $4y^2 \ge 100 + 25x^2$

Graph the solution of each of the systems of inequalities.

41. $3x + 2y \ge 10$
$x - 2y \ge 5$

42. $\frac{y^2}{4} - \frac{x^2}{9} \ge 1$
$x^2 + y^2 \le 36$

43. $|x| \le 2$
$|y| > 1$
$4x^2 + 9y^2 \le 36$

44. Find $x \ge 0$ and $y \ge 0$ such that

$$3x + 2y \le 6$$
$$-2x + 4y \le 8$$

and $2x + 5y$ is maximized.

Set up a system of inequalities for Exercises 45 and 46; then graph the solution of the system.

45. A bakery makes both cakes and cookies. Each batch of cakes requires two hours in the oven and three hours in the decorating room. Each batch of cookies needs one and a half hours in the oven and two thirds of an hour in the decorating room. The oven is available no more than 15 hours a day, while the decorating room can be used no more than 13 hours a day.

46. A company makes two kinds of pizza, basic and plain. Basic contains cheese and beef, while plain contains onions and beef. The company sells at least three units a day of basic, and at least two units of plain. The beef costs \$5 per unit for basic, and \$4 per unit for plain. They can spend no more than \$50 per day on beef. Dough for basic is \$2 per unit, while dough for plain is \$1 per unit. The company can spend no more than \$16 per day on dough.

47. How many batches of cakes and cookies should the bakery of Exercise 45 make in order to maximize profits if cookies produce a profit of \$20 per batch and cakes produce a profit of \$30 per batch?

48. How many units of each kind of pizza should the company of Exercise 46 make in order to maximize profits if basic sells for \$20 per unit and plain for \$15 per unit?

MIXED REVIEW EXERCISES

Solve.

49. $5x - 3y + 2z = -5$
$2x + 2y - z = 4$
$4x - y + z = -1$

50. $x > -2$
$y < 5$
$\frac{x^2}{9} + \frac{y^2}{36} \le 1$

51. $4x + 5y = 5$
$3x + 7y = -6$

52. $x^2 + y^2 = 8$
$2x^2 - y^2 = 4$

53. Three kinds of tea worth \$4.60, \$5.75, and \$6.50 per pound are to be mixed to get 20 pounds of tea worth \$5.25 per pound. The amount of \$4.60 tea used is to be equal to the total amount of the other two kinds together. How many pounds of each tea should be used?

54. The sum of three numbers is 23. The second number is 3 more than the first. The sum of the first and twice the third is 4. Find the three numbers.

55. $7x - 10y = 11$
$\frac{3}{2}x - 5y = 8$

56. $9x^2 + 16y^2 \ge 144$
$x^2 - y^2 \ge 16$

57. A gold merchant has some 12 carat gold (12/24 pure gold), and some 22 carat gold (22/24 pure). How many grams of each should be mixed to get 25 grams of 15 carat gold?

58. $xy = -24$
$2x^2 - y^2 = -4$

CHAPTER 11 TEST

Solve each system.

1. $2x + 3y = 10$
$-3x + 2y = 11$

2. $3x + 4y = -7$
$2x - y = 10$

3. $3x + 4y = 8$
$6x = 7 - 8y$

4. $2x - 3y = -8$
$x = y - 3$

5. $\frac{2}{3}x + \frac{1}{3}y = \frac{5}{3}$
$\frac{1}{4}x - \frac{5}{2}y = \frac{17}{2}$

6. $\frac{1}{x} + \frac{3}{y} = \frac{9}{8}$
$-\frac{2}{x} + \frac{1}{y} = \frac{1}{12}$

7. $2x - y + z = 9$
$3x + y - 2z = 4$
$x + y - 4z = -6$

8. $4x - 2y = -8$
$3y - 5z = 14$
$2x + z = -10$

9. What is an inconsistent system? What are dependent equations?

Translate each problem into a system of equations and then solve.

10. A pharmacist needs 100 gallons of 50% alcohol solution. She has 30% alcohol solution and 80% alcohol solution that she can mix. How many gallons of each will be required to make 100 gallons of 50% alcohol solution?

11. Two cars start from points 400 miles apart and travel toward each other. They meet after 4 hours. Find the average speed of each car if one travels 20 miles per hour faster than the other.

12. The perimeter of a triangle is 23 centimeters. The difference between the largest side and the smallest side is 4 centimeters. The sum of the two smaller sides is 3 centimeters more than the largest side. Find the lengths of the three sides.

13. If a nonlinear system of equations contains two equations whose graphs are both parabolas, how many possible points of intersection of the graphs are there?

Solve each nonlinear system.

14. $x + 4y = 10$
$xy = 4$

15. $y + 2 = 3x$
$x^2 + y^2 = 2$

16. $x^2 + y^2 = 16$
$2x^2 - y^2 = 8$

17. $y = x^2 + 2$
$y = 2x - 5$

Graph the solution set in Exercises 18 and 19.

18. $y \leq x^2 + 2x - 1$

19. $3x^2 + 2y^2 \leq 24$
$x^2 + y^2 \geq 4$

---

20. The J. J. Gravois Company designs and sells two types of rings: the VIP and the SST. They can produce up to 24 rings each day using up to 60 total hours of labor. It takes 3 hours to make one VIP ring, and 2 hours to make one SST ring. How many of each type of ring should be made daily in order to maximize the company's profit, if the profit on a VIP ring is $30 and the profit on an SST ring is $40?

THE GRAPHING CALCULATOR

Graphing calculators have the capability of graphing more than one function on the same set of axes. Thus, we can use them to graph systems of equations. Figure 11.1 in Section 11.1 shows that the solution of the system

$$x + y = 5$$
$$2x - y = 4$$

is (3, 2). This can easily be verified on a graphing calculator by graphing both lines on the same set of axes, and then zooming in on the point of intersection. By using the tracing capability of the calculator, we can find the coordinates of the point.

However, a small problem arises when we begin this procedure. Graphing calculators require that we enter the equation in function form. Therefore, we must rewrite each equation in the form $y = mx + b$ (since they are both linear), and then enter the equation into the calculator. The system above is equivalent to

$$y = -x + 5$$
$$y = 2x - 4.$$

Here we see an important use of the function concept.

Suppose that we wish to solve the system

$$x^2 + y^2 = 9$$
$$2x - y = 3$$

as shown in Example 1 in Section 11.4. We notice that the graph of $x^2 + y^2 = 9$ is a circle, and the equation does not define a function. As explained at the end of Chapter 8, this can be handled by solving for y to get

$$y = \pm\sqrt{9 - x^2}.$$

Now the system can be solved by entering three functions:

$$y = \sqrt{9 - x^2}$$
$$y = -\sqrt{9 - x^2}$$
$$y = 2x - 3 \qquad \text{(equivalent to } 2x - y = 3\text{).}$$

As shown in Figure 11.6, the solutions are (12/5, 9/5) and (10, 3). By zooming and tracing, the calculator should give you the x-value 2.4 and the y-value 1.8 for the first of these.

You may want to use the following systems to experiment with your calculator. Most have been selected from examples and figures in the text, so you can compare your calculator results with those given there. The example or figure number is given with each system.

1. $2x + 3y = 2$
$x - 3y = 10$ (Section 11.1, Example 1)

2. $5x - 2y = 4$
$2x + 3y = 13$ (Section 11.1, Example 2)

3. $3x - 2y = 4$
$-6x + 4y = 7$ (Section 11.1, Example 5)

4. $-4x + y = 2$
$8x - 2y = -4$ (Section 11.1, Example 6)

5. $\frac{x}{2} + \frac{y}{3} = \frac{13}{6}$
$4x - y = -1$ (Section 11.1, Example 3)

6. $4x - 3y = 7$
$3x - 2y = 6$ (Section 11.1, Example 4)

7. $6x - y = 5$
$xy = 4$ (Figure 11.7)

8. $x^2 + y^2 = 9$
$2x^2 - y^2 = -6$ (Figure 11.8)

9. Use a graphing calculator to show that the system

$$x^2 + y^2 = 4$$
$$x^2 + y^2 = 25$$

consists of two concentric (having the same center) circles, and thus has no solutions.

10. Use a graphing calculator to graph these four functions on the same set of axes.

$$y = x^2$$
$$y = x - 4$$
$$y = x - \frac{1}{4}$$
$$y = x + 2$$

This will show how a line may intersect a parabola in either 0, 1, or 2 points.

CHAPTER TWELVE

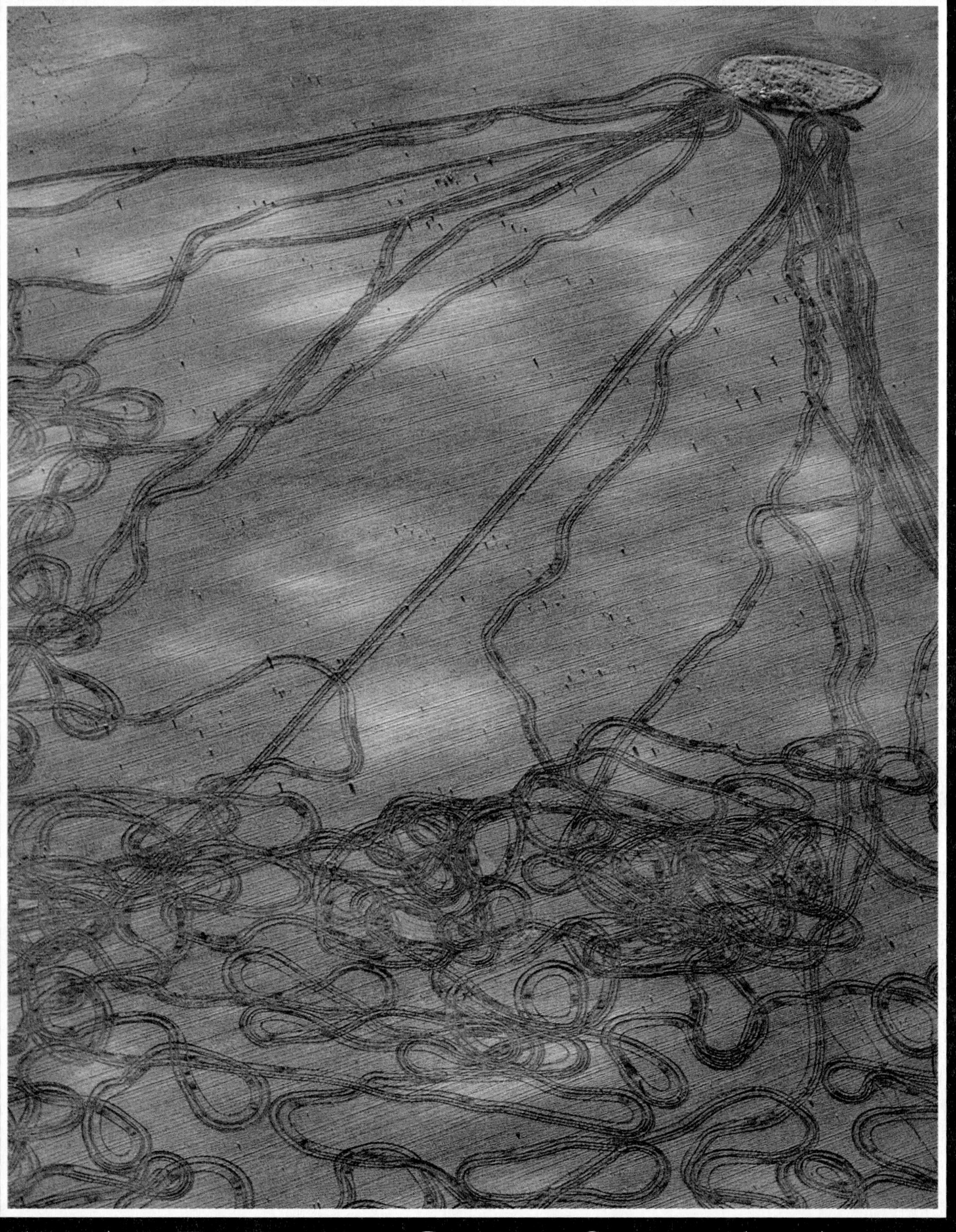

MATRICES AND DETERMINANTS

Often, a great deal of information must be presented in compact form. One way to do this is with a table. When the information is to be manipulated mathematically, a *matrix,* which can be thought of as a simplified table, is used. For example, airlines use matrices to allocate and utilize reserve crews at minimum cost. Matrices have been of interest to mathematicians for some time, but are now also important in the fields of life science, management, and the social sciences.

This chapter presents some of the basic ideas of matrices and *determinants,* numbers associated with matrices. Then we show how matrices and determinants are used to solve linear systems of equations.

12.1 MATRICES AND DETERMINANTS

OBJECTIVES

1. UNDERSTAND THE TERMINOLOGY OF MATRICES.
2. EVALUATE 2×2 DETERMINANTS.
3. USE EXPANSION BY MINORS ABOUT THE FIRST COLUMN TO EVALUATE DETERMINANTS.
4. USE EXPANSION BY MINORS ABOUT ANY ROW OR COLUMN TO EVALUATE DETERMINANTS.
5. EVALUATE 4×4 OR LARGER DETERMINANTS.

A **matrix** (plural: matrices) is a rectangular array of numbers enclosed by brackets or parentheses in which the position of each number is meaningful. This is similar to the idea of an ordered pair, and, in fact, is an extension of that idea. Each number in the array is an **element** of the matrix.

1 Matrices are classified by their size, that is, by the number of rows and columns that they contain. For example, the matrix

$$\begin{bmatrix} 2 & 7 & -5 \\ 3 & -6 & 0 \end{bmatrix}$$

Rows

Columns

has two rows (horizontal) and three columns (vertical), and is called a 2×3 (read "2 by 3") matrix. A matrix with m rows and n columns is an $m \times n$ matrix. The number of rows is always given first.

EXAMPLE 1 CLASSIFYING MATRICES BY SIZE

(a) The matrix $\begin{bmatrix} 6 & 5 \\ 3 & 4 \\ 5 & -1 \end{bmatrix}$ is a 3×2 matrix.

(b) $\begin{bmatrix} 5 & 8 & 9 \\ 0 & 5 & -3 \\ -4 & 0 & 5 \end{bmatrix}$ is a 3×3 matrix.

(c) $[1 \quad 6 \quad 5 \quad -2 \quad 5]$ is a 1×5 matrix.

(d) $\begin{bmatrix} 3 \\ -5 \\ 0 \\ 2 \end{bmatrix}$ is a 4×1 matrix. ■

A matrix having the same number of rows as columns is called a **square matrix.** The matrix given in Example 1(b) above is a square matrix, as are

$$\begin{bmatrix} -5 & 6 \\ 8 & 3 \end{bmatrix} \quad \text{and} \quad \begin{bmatrix} 0 & 0 & 0 & 0 \\ -2 & 4 & 1 & 3 \\ 0 & 0 & 0 & 0 \\ -5 & -4 & 1 & 8 \end{bmatrix}.$$

A matrix containing only one row is called a **row matrix.** The matrix in Example 1(c) is a row matrix, as are

$$[5 \quad 8], \qquad [6 \quad -9 \quad 2], \qquad \text{and} \qquad [-4 \quad 0 \quad 0 \quad 0].$$

Finally, a matrix of only one column, as in Example 1(d), is a **column matrix.**

2 Every square matrix A is associated with a real number called the **determinant** of A, written $\delta(A)$.

DEFINITION OF DETERMINANT OF A 2 × 2 MATRIX

The **determinant of a 2 × 2 matrix** A,

$$A = \begin{bmatrix} a_{11} & a_{12} \\ a_{21} & a_{22} \end{bmatrix},$$

is defined as

$$\delta(A) = |A| = \begin{vmatrix} a_{11} & a_{12} \\ a_{21} & a_{22} \end{vmatrix} = a_{11}a_{22} - a_{21}a_{12}.$$

NOTE Notice that matrices are enclosed with square brackets, while determinants are denoted with vertical bars. Also, the matrix is an *array* of numbers, but its determinant is a *single number*.

EXAMPLE 2 EVALUATING THE DETERMINANT OF A 2 × 2 MATRIX

If $P = \begin{bmatrix} -3 & 4 \\ 6 & 8 \end{bmatrix}$, then

$$\delta(P) = \begin{vmatrix} -3 & 4 \\ 6 & 8 \end{vmatrix} = -3(8) - 6(4) = -48. \quad ■$$

The definition of a determinant can be extended to a 3×3 matrix as follows.

DEFINITION OF DETERMINANT OF A 3 × 3 MATRIX

The **determinant of a 3 × 3 matrix** A,

$$A = \begin{bmatrix} a_{11} & a_{12} & a_{13} \\ a_{21} & a_{22} & a_{23} \\ a_{31} & a_{32} & a_{33} \end{bmatrix},$$

is defined as

$$\delta(A) = \begin{vmatrix} a_{11} & a_{12} & a_{13} \\ a_{21} & a_{22} & a_{23} \\ a_{31} & a_{32} & a_{33} \end{vmatrix} = (a_{11}a_{22}a_{33} + a_{12}a_{23}a_{31} + a_{13}a_{21}a_{32}) - (a_{31}a_{22}a_{13} + a_{32}a_{23}a_{11} + a_{33}a_{21}a_{12}).$$

3 An easy method for calculating 3 × 3 determinants is found by rearranging and factoring the terms given above to get

$$\begin{vmatrix} a_{11} & a_{12} & a_{13} \\ a_{21} & a_{22} & a_{23} \\ a_{31} & a_{32} & a_{33} \end{vmatrix} = a_{11}(a_{22}a_{33} - a_{32}a_{23}) - a_{21}(a_{12}a_{33} - a_{32}a_{13}) + a_{31}(a_{12}a_{23} - a_{22}a_{13}). \quad (*)$$

Each of the quantities in parentheses represents the determinant of a 2 × 2 matrix that is the part of the 3 × 3 matrix remaining when the row and column of the multiplier are eliminated, as shown below.

$$a_{11}(a_{22}a_{33} - a_{32}a_{23}) \qquad \begin{bmatrix} a_{11} & a_{12} & a_{13} \\ a_{21} & a_{22} & a_{23} \\ a_{31} & a_{32} & a_{33} \end{bmatrix}$$

$$a_{21}(a_{12}a_{33} - a_{32}a_{13}) \qquad \begin{bmatrix} a_{11} & a_{12} & a_{13} \\ a_{21} & a_{22} & a_{23} \\ a_{31} & a_{32} & a_{33} \end{bmatrix}$$

$$a_{31}(a_{12}a_{23} - a_{22}a_{13}) \qquad \begin{bmatrix} a_{11} & a_{12} & a_{13} \\ a_{21} & a_{22} & a_{23} \\ a_{31} & a_{32} & a_{33} \end{bmatrix}$$

The determinant of each 2 × 2 matrix is called a **minor** of the associated element in the 3 × 3 matrix. Thus, the minors of a_{11}, a_{21}, and a_{31} are as shown below.

Element	a_{11}	a_{21}	a_{31}
Minor	$\begin{vmatrix} a_{22} & a_{23} \\ a_{32} & a_{33} \end{vmatrix}$	$\begin{vmatrix} a_{12} & a_{13} \\ a_{32} & a_{33} \end{vmatrix}$	$\begin{vmatrix} a_{12} & a_{13} \\ a_{22} & a_{23} \end{vmatrix}$

EXAMPLE 3

FINDING THE MINORS OF THE ELEMENTS OF A MATRIX

For the matrix

$$\begin{bmatrix} 6 & 2 & 4 \\ 8 & 9 & 3 \\ 1 & 2 & 0 \end{bmatrix},$$

find the minor of each of the following elements.

(a) 6

Since 6 is in the first row and first column of the matrix, mentally remove that row and column. The minor is the determinant of the remaining 2×2 matrix.

$$\text{minor of } 6 = \begin{vmatrix} 9 & 3 \\ 2 & 0 \end{vmatrix} = 9 \cdot 0 - 2 \cdot 3 = -6$$

(b) 9

Eliminate the second row and second column, which include the element 9. The determinant of the remaining 2×2 matrix is the minor of 9.

$$\text{minor of } 9 = \begin{vmatrix} 6 & 4 \\ 1 & 0 \end{vmatrix} = 6 \cdot 0 - 1 \cdot 4 = -4$$

(c) 1

The minor of this element is

$$\text{minor of } 1 = \begin{vmatrix} 2 & 4 \\ 9 & 3 \end{vmatrix} = 2 \cdot 3 - 9 \cdot 4 = -30. \quad \blacksquare$$

The determinant of the 3×3 matrix can be evaluated by multiplying each element in the first column by its minor and combining the products as indicated in equation (*) given above. This is called the **expansion of the determinant by minors about the first column.**

EXAMPLE 4 EVALUATING THE DETERMINANT OF A 3 x 3 MATRIX

Evaluate the determinant of

$$\begin{bmatrix} 1 & 3 & -2 \\ -1 & -2 & -3 \\ 1 & 1 & 2 \end{bmatrix}$$

by expanding about the first column.

Using the procedure above,

$$\begin{vmatrix} 1 & 3 & -2 \\ -1 & -2 & -3 \\ 1 & 1 & 2 \end{vmatrix} = 1\begin{vmatrix} -2 & -3 \\ 1 & 2 \end{vmatrix} - (-1)\begin{vmatrix} 3 & -2 \\ 1 & 2 \end{vmatrix} + 1\begin{vmatrix} 3 & -2 \\ -2 & -3 \end{vmatrix}$$

$$= 1[(-2)(2) - (1)(-3)] + 1[(3)(2) - (1)(-2)]$$
$$+ 1[(3)(-3) - (-2)(-2)]$$
$$= 1(-1) + 1(8) + 1(-13)$$
$$= -1 + 8 - 13 = -6. \quad \blacksquare$$

4 To get equation (*) the terms could have been rearranged and factored differently by factoring out the three elements of the second or third columns or of any of the three rows of the array. Therefore, expanding by minors about any row or any column results in the same value for the determinant. The following array of signs is helpful in determining the correct signs for the terms of other expansions.

ARRAY OF SIGNS FOR A 3 × 3 DETERMINANT

$$\begin{matrix} + & - & + \\ - & + & - \\ + & - & + \end{matrix}$$

The signs alternate in each row and each column, beginning with + in the first-row, first-column position, so the array of signs can be reproduced when needed. If the determinant is expanded about the second column, for example, the first term would have a minus sign, the second a plus sign, and the third a minus sign. These signs are independent of the sign of the corresponding element of the matrix. The sign array can be extended for larger matrices.

EXAMPLE 5 **EVALUATING A DETERMINANT USING EXPANSION BY MINORS**

Evaluate the determinant of Example 4 using expansion by minors about the second column.

Using the methods described above,

$$\begin{vmatrix} 1 & 3 & -2 \\ -1 & -2 & -3 \\ 1 & 1 & 2 \end{vmatrix} = -3\begin{vmatrix} -1 & -3 \\ 1 & 2 \end{vmatrix} + (-2)\begin{vmatrix} 1 & -2 \\ 1 & 2 \end{vmatrix} - 1\begin{vmatrix} 1 & -2 \\ -1 & -3 \end{vmatrix}$$

$$= -3(1) - 2(4) - 1(-5)$$

$$= -3 - 8 + 5$$

$$= -6. \quad \blacksquare$$

CAUTION Be very careful to keep track of all negative signs when evaluating determinants. Work carefully, writing down each step as in the examples. Skipping steps frequently leads to errors in these computations.

5 The method of expansion by minors can be extended to find the determinant of any $n \times n$ matrix. For a larger matrix, the sign array is also extended. For example, the signs for a 4×4 matrix are arranged as follows.

ARRAY OF SIGNS FOR A 4 × 4 DETERMINANT

$$\begin{matrix} + & - & + & - \\ - & + & - & + \\ + & - & + & - \\ - & + & - & + \end{matrix}$$

EXAMPLE 6 **EVALUATING THE DETERMINANT OF A 4 × 4 MATRIX**

Evaluate

$$\begin{vmatrix} -1 & -2 & 3 & 2 \\ 0 & 1 & 4 & -2 \\ 3 & -1 & 4 & 0 \\ 2 & 1 & 0 & 3 \end{vmatrix}.$$

Expanding by minors about the fourth row gives

$$-2\begin{vmatrix}-2 & 3 & 2\\ 1 & 4 & -2\\ -1 & 4 & 0\end{vmatrix} + 1\begin{vmatrix}-1 & 3 & 2\\ 0 & 4 & -2\\ 3 & 4 & 0\end{vmatrix} - 0\begin{vmatrix}-1 & -2 & 2\\ 0 & 1 & -2\\ 3 & -1 & 0\end{vmatrix} + 3\begin{vmatrix}-1 & -2 & 3\\ 0 & 1 & 4\\ 3 & -1 & 4\end{vmatrix}$$

$$= -2(6) + 1(-50) - 0 + 3(-41)$$

$$= -185. \quad \blacksquare$$

Each of the four determinants in Example 6 must be evaluated by expansion of three minors, with much work required to get the final value. Determinants of large matrices can be evaluated quickly and easily, however, with the aid of a computer. Many calculators have a key, or a program, for finding determinants. Even without such a key, a calculator can help to avoid computational errors when calculating determinants.

12.1 EXERCISES

Find the size of each of the following matrices. Identify any square, column, or row matrices. See Example 1.

1. $\begin{bmatrix}-4 & 8\\ 2 & 3\end{bmatrix}$

2. $\begin{bmatrix}-9 & 6 & 2\\ 4 & 1 & 8\end{bmatrix}$

3. $\begin{bmatrix}-6 & 8 & 0 & 0\\ 4 & 1 & 9 & 2\\ 3 & -5 & 7 & 1\end{bmatrix}$

4. $[8 \quad -2 \quad 4 \quad 6 \quad 3]$

5. $\begin{bmatrix}2\\ 4\end{bmatrix}$

6. $\begin{bmatrix}-9 & 6 & 5 & 1 & 2\\ -4 & 0 & 8 & 7 & 3\end{bmatrix}$

7. $\begin{bmatrix}-5\\ -8\\ -4\\ -6\end{bmatrix}$

8. $[-9]$

9. $\begin{bmatrix}-4 & 2 & 3\\ -8 & 2 & 1\\ 4 & 6 & 8\end{bmatrix}$

10. $\begin{bmatrix}-4 & 2\\ 3 & 5\end{bmatrix}$

Evaluate each of the following determinants. See Example 2.

11. $\begin{vmatrix}5 & 8\\ 2 & -4\end{vmatrix}$

12. $\begin{vmatrix}-3 & 0\\ 0 & 9\end{vmatrix}$

13. $\begin{vmatrix}-1 & -2\\ 5 & 3\end{vmatrix}$

14. $\begin{vmatrix}6 & -4\\ 0 & -1\end{vmatrix}$

15. $\begin{vmatrix}9 & 3\\ -3 & -1\end{vmatrix}$

16. $\begin{vmatrix}0 & 2\\ 1 & 5\end{vmatrix}$

17. $\begin{vmatrix}3 & 4\\ 5 & -2\end{vmatrix}$

18. $\begin{vmatrix}-9 & 7\\ 2 & 6\end{vmatrix}$

19. $\begin{vmatrix}0 & 4\\ 4 & 0\end{vmatrix}$

20. $\begin{vmatrix}1 & 0\\ 0 & 2\end{vmatrix}$

21. $\begin{vmatrix}8 & 3\\ 8 & 3\end{vmatrix}$

22. $\begin{vmatrix}9 & -4\\ -4 & 9\end{vmatrix}$

23. $\begin{vmatrix}x & 4\\ 8 & 2\end{vmatrix}$

24. $\begin{vmatrix}k & 3\\ 0 & 4\end{vmatrix}$

25. $\begin{vmatrix}y & 2\\ 8 & y\end{vmatrix}$

26. $\begin{vmatrix}3 & 8\\ m & n\end{vmatrix}$

27. $\begin{vmatrix}x & y\\ y & x\end{vmatrix}$

28. $\begin{vmatrix}2m & 8n\\ 8n & 2m\end{vmatrix}$

29. $\begin{vmatrix}p & 0\\ 3 & p\end{vmatrix}$

30. $\begin{vmatrix}0 & 0\\ 5k & -8k\end{vmatrix}$

31. Explain how to evaluate a 2×2 determinant. (See Objective 2.)

32. In your own words, define the minor of an element of a matrix. (See Objective 3.)

Evaluate each of the following determinants. See Examples 3–5.

33. $\begin{vmatrix} 1 & 0 & 0 \\ 0 & -1 & 0 \\ 1 & 0 & 1 \end{vmatrix}$

34. $\begin{vmatrix} -2 & 0 & 1 \\ 0 & 1 & 0 \\ 0 & 0 & -1 \end{vmatrix}$

35. $\begin{vmatrix} -2 & 0 & 0 \\ 4 & 0 & 1 \\ 3 & 4 & 2 \end{vmatrix}$

36. $\begin{vmatrix} 3 & -2 & 0 \\ 0 & -1 & 1 \\ 4 & 0 & 2 \end{vmatrix}$

37. $\begin{vmatrix} 1 & 2 & 0 \\ -1 & 2 & -1 \\ 0 & 1 & 4 \end{vmatrix}$

38. $\begin{vmatrix} 2 & 1 & -1 \\ 4 & 7 & -2 \\ 2 & 4 & 0 \end{vmatrix}$

39. $\begin{vmatrix} 10 & 2 & 1 \\ -1 & 4 & 3 \\ -3 & 8 & 10 \end{vmatrix}$

40. $\begin{vmatrix} 7 & -1 & 1 \\ 1 & -7 & 2 \\ -2 & 1 & 1 \end{vmatrix}$

41. $\begin{vmatrix} 1 & -2 & 3 \\ 0 & 0 & 0 \\ 1 & 10 & -12 \end{vmatrix}$

42. $\begin{vmatrix} 2 & 3 & 0 \\ 1 & 9 & 0 \\ -1 & -2 & 0 \end{vmatrix}$

43. $\begin{vmatrix} 3 & 3 & -1 \\ 2 & 6 & 0 \\ -6 & -6 & 2 \end{vmatrix}$

44. $\begin{vmatrix} 5 & -3 & 2 \\ -5 & 3 & -2 \\ 1 & 0 & 1 \end{vmatrix}$

45. $\begin{vmatrix} 3 & 2 & 0 \\ 0 & 1 & x \\ 2 & 0 & 0 \end{vmatrix}$

46. $\begin{vmatrix} 0 & 3 & y \\ 0 & 4 & 2 \\ 1 & 0 & 1 \end{vmatrix}$

47. $\begin{vmatrix} 5 & 0 & 1 \\ 0 & 3 & m \\ 1 & 2 & 0 \end{vmatrix}$

48. $\begin{vmatrix} a & 2 & 0 \\ 1 & 0 & 6 \\ 0 & a & 0 \end{vmatrix}$

Solve each of the following equations for x.

49. $\begin{vmatrix} -2 & 0 & 1 \\ -1 & 3 & x \\ 5 & -2 & 0 \end{vmatrix} = 3$

50. $\begin{vmatrix} 4 & 3 & 0 \\ 2 & 0 & 1 \\ -3 & x & -1 \end{vmatrix} = 5$

51. $\begin{vmatrix} 5 & 3x & -3 \\ 0 & 2 & -1 \\ 4 & -1 & x \end{vmatrix} = -7$

52. $\begin{vmatrix} 2x & 1 & -1 \\ 0 & 4 & x \\ 3 & 0 & 2 \end{vmatrix} = x$

---

53. Explain why a determinant with a row or column of zeros has a value of zero.

54. Does a determinant with two identical rows have a value of zero? (*Hint:* Make up several such determinants and evaluate them to draw a conclusion.)

Find the value of each of the following determinants. See Example 6.

55. $\begin{vmatrix} 4 & 0 & 0 & 2 \\ -1 & 0 & 3 & 0 \\ 2 & 4 & 0 & 1 \\ 0 & 0 & 1 & 2 \end{vmatrix}$

56. $\begin{vmatrix} -2 & 0 & 4 & 2 \\ 3 & 6 & 0 & 4 \\ 0 & 0 & 0 & 3 \\ 9 & 0 & 2 & -1 \end{vmatrix}$

57. $\begin{vmatrix} 1 & 1 & 0 & 1 \\ 2 & 1 & 0 & 2 \\ 0 & 1 & -1 & 1 \\ 1 & -1 & 1 & 1 \end{vmatrix}$

58. $\begin{vmatrix} 2 & 7 & 0 & -1 \\ 1 & 0 & 1 & 3 \\ 2 & 4 & -1 & -1 \\ -1 & 1 & 0 & 8 \end{vmatrix}$

Determinants can be used to find the area of a triangle given the coordinates of its vertices. Given a triangle PQR with vertices (x_1, y_1), (x_3, y_3), and (x_2, y_2), *as shown in the following figure, we can introduce segments PM, RN, and QS perpendicular to the x-axis, forming trapezoids PMNR, RNSQ, and PMSQ. Recall that the area of a trapezoid is given by half the sum of the parallel bases times the altitude. For example, the area of trapezoid PMSQ equals* $(1/2)(y_1 + y_3)(x_3 - x_1)$. *The area of triangle PQR can be found by subtracting the area of PMSQ from the sum of the areas of PMNR and RNSQ. Thus, for the area A of triangle PQR,*

$$A = (1/2)(x_3y_1 - x_1y_3 + x_2y_3 - x_3y_2 + x_1y_2 - x_2y_1).$$

By evaluating the determinant

$$\begin{vmatrix} x_1 & y_1 & 1 \\ x_2 & y_2 & 1 \\ x_3 & y_3 & 1 \end{vmatrix},$$

it can be shown that the area of the triangle is given by the absolute value *of A, where*

$$A = \frac{1}{2}\begin{vmatrix} x_1 & y_1 & 1 \\ x_2 & y_2 & 1 \\ x_3 & y_3 & 1 \end{vmatrix}.$$

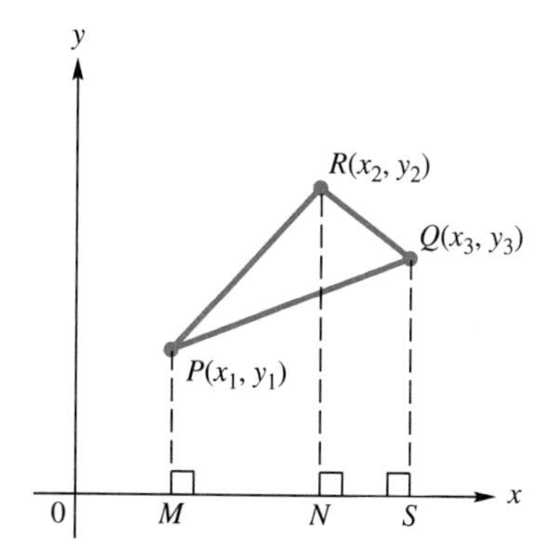

Use the given formula to find the area of the triangles with vertices at P, Q, and R.

59. $P(0, 1)$, $Q(2, 0)$, $R(1, 3)$

60. $P(2, 5)$, $Q(-1, 3)$, $R(4, 0)$

61. $P(2, -2)$, $Q(0, 0)$, $R(-3, -4)$

62. $P(4, 7)$, $Q(5, -2)$, $R(1, 1)$

63. $P(3, 8)$, $Q(-1, 4)$, $R(0, 1)$

64. $P(-3, -1)$, $Q(4, 2)$, $R(3, -3)$

Determinants can be used to find the equation of a line passing through two given points. The next two exercises show how to do this.

65. Expand the determinant below and show that the result is the equation of the line through (2, 3), and (−1, 4).

$$\begin{vmatrix} x & y & 1 \\ 2 & 3 & 1 \\ -1 & 4 & 1 \end{vmatrix} = 0$$

66. (a) Write the equation of the line through the points (x_1, y_1) and (x_2, y_2) using the point-slope formula.

(b) Expanding the determinant in the equation below, show that the equation is equivalent to the equation in part (a).

$$\begin{vmatrix} x & y & 1 \\ x_1 & y_1 & 1 \\ x_2 & y_2 & 1 \end{vmatrix} = 0$$

PREVIEW EXERCISES

Give the property that justifies each statement. A statement may involve more than one property. See Section 1.4.

67. $3 \cdot 5 = 5 \cdot 3$

68. $0 \cdot a + 0 \cdot b + 0 \cdot c = 0$

69. $-5[2 \cdot 1 - 4(-3)] = -5(2 \cdot 1) - (-5)[(4)(-3)]$

70. $1 \cdot 3 + 5 \cdot 6 = 5 \cdot 6 + 1 \cdot 3$

12.2 PROPERTIES OF DETERMINANTS

OBJECTIVE

1 USE THE PROPERTIES OF DETERMINANTS TO SIMPLIFY THE PROCESS OF EVALUATING DETERMINANTS.

Some determinants can be evaluated only after much tedious calculation. The more calculation is involved, the greater the chance for error. To evaluate the determinant of a 3×3 matrix, it is necessary to evaluate three different determinants and then combine them correctly. For a determinant of a 4×4 matrix, twelve determinants must be found. This section gives several properties of determinants that make them easier to calculate. These properties of determinants are summarized at the end of this section.

1 The value of

$$\begin{vmatrix} 3 & 0 \\ 5 & 0 \end{vmatrix}$$

is $3 \cdot 0 - 5 \cdot 0 = 0$. Also, expanded about the second row,

$$\begin{vmatrix} 1 & 2 & 3 \\ 0 & 0 & 0 \\ 4 & 6 & 8 \end{vmatrix} = -0\begin{vmatrix} 2 & 3 \\ 6 & 8 \end{vmatrix} + 0\begin{vmatrix} 1 & 3 \\ 4 & 8 \end{vmatrix} - 0\begin{vmatrix} 1 & 2 \\ 4 & 6 \end{vmatrix} = 0.$$

The fact that the matrix for each of these determinants has a row or column of zeros leads to a determinant of 0. These examples suggest the following property.

DETERMINANT PROPERTY 1

If every element in a row or column of a matrix is 0, then the determinant equals 0.

Proof

Expand the determinant about the 0 row or column. Each term of this expansion has a 0 factor, so each term is 0, and the sum therefore is 0. For example,

$$\begin{vmatrix} a & b & c \\ 0 & 0 & 0 \\ d & e & f \end{vmatrix} = -0\begin{vmatrix} b & c \\ e & f \end{vmatrix} + 0\begin{vmatrix} a & c \\ d & f \end{vmatrix} - 0\begin{vmatrix} a & b \\ d & e \end{vmatrix} = 0. \quad \blacksquare$$

EXAMPLE 1
APPLYING PROPERTY 1

$$\begin{vmatrix} 2 & 4 \\ 0 & 0 \end{vmatrix} = 0 \quad \text{and} \quad \begin{vmatrix} -3 & 7 & 0 \\ 4 & 9 & 0 \\ -6 & 8 & 0 \end{vmatrix} = 0 \quad \blacksquare$$

Suppose the matrix

$$\begin{bmatrix} 1 & 5 \\ 3 & 7 \end{bmatrix}$$

is rewritten as

$$\begin{bmatrix} 1 & 3 \\ 5 & 7 \end{bmatrix},$$

so the rows of the first matrix are the columns of the second matrix. Evaluating the determinant of each matrix gives

$$\begin{vmatrix} 1 & 5 \\ 3 & 7 \end{vmatrix} = 1 \cdot 7 - 3 \cdot 5 = -8$$

$$\begin{vmatrix} 1 & 3 \\ 5 & 7 \end{vmatrix} = 1 \cdot 7 - 5 \cdot 3 = -8.$$

The two determinants are the same, suggesting the next property of determinants.

DETERMINANT PROPERTY 2 If corresponding rows and columns of a matrix are interchanged, the determinant is not changed.

A proof of Property 2 is requested in Exercise 37.

EXAMPLE 2 APPLYING PROPERTY 2 Given

$$A = \begin{bmatrix} 2 & 1 & 6 \\ 3 & 0 & 5 \\ -4 & 6 & 9 \end{bmatrix} \quad \text{and} \quad B = \begin{bmatrix} 2 & 3 & -4 \\ 1 & 0 & 6 \\ 6 & 5 & 9 \end{bmatrix},$$

by Property 2, since the rows of the first matrix are the columns of the second,

$$\begin{vmatrix} 2 & 1 & 6 \\ 3 & 0 & 5 \\ -4 & 6 & 9 \end{vmatrix} = \begin{vmatrix} 2 & 3 & -4 \\ 1 & 0 & 6 \\ 6 & 5 & 9 \end{vmatrix}. \quad ■$$

Exchanging the rows of the matrix

$$\begin{bmatrix} 1 & 6 \\ 5 & 3 \end{bmatrix} \quad \text{gives} \quad \begin{bmatrix} 5 & 3 \\ 1 & 6 \end{bmatrix}.$$

While the first determinant is

$$\begin{vmatrix} 1 & 6 \\ 5 & 3 \end{vmatrix} = 1 \cdot 3 - 5 \cdot 6 = -27,$$

the second determinant is

$$\begin{vmatrix} 5 & 3 \\ 1 & 6 \end{vmatrix} = 5 \cdot 6 - 1 \cdot 3 = 27.$$

The second determinant is the negative of the first. This suggests the third property of determinants.

DETERMINANT PROPERTY 3 Interchanging two rows (or columns) of a matrix changes the sign of the determinant.

Exercise 38 asks for a proof of Property 3.

EXAMPLE 3 APPLYING PROPERTY 3

(a) Interchange the two columns of

$$\begin{bmatrix} 2 & 5 \\ 3 & 4 \end{bmatrix} \quad \text{to get} \quad \begin{bmatrix} 5 & 2 \\ 4 & 3 \end{bmatrix}.$$

By Property 3, since

$$\begin{vmatrix} 2 & 5 \\ 3 & 4 \end{vmatrix} = -7,$$

then

$$\begin{vmatrix} 5 & 2 \\ 4 & 3 \end{vmatrix} = 7.$$

(b) $\begin{vmatrix} 2 & 1 & 6 \\ 3 & 0 & 5 \\ -4 & 6 & 9 \end{vmatrix} = -\begin{vmatrix} -4 & 6 & 9 \\ 3 & 0 & 5 \\ 2 & 1 & 6 \end{vmatrix}$ ■

Multiplying each element of the second row of the matrix

$$\begin{bmatrix} 2 & -3 \\ 4 & 1 \end{bmatrix}$$

by -5 gives the new matrix

$$\begin{bmatrix} 2 & -3 \\ 4(-5) & 1(-5) \end{bmatrix} = \begin{bmatrix} 2 & -3 \\ -20 & -5 \end{bmatrix}.$$

The determinants of these two matrices are

$$\begin{vmatrix} 2 & -3 \\ 4 & 1 \end{vmatrix} = 2 \cdot 1 - 4(-3) = 14$$

and

$$\begin{vmatrix} 2 & -3 \\ -20 & -5 \end{vmatrix} = 2(-5) - (-20)(-3) = -70.$$

Since $-70 = -5(14)$, the new determinant is -5 times the original determinant. The next property generalizes this idea. A proof is requested in Exercise 39.

DETERMINANT PROPERTY 4 If every element of a row (or column) of a matrix is multiplied by the real number k, then the determinant of the new matrix is k times the determinant of the original matrix.

EXAMPLE 4
APPLYING PROPERTY 4

By Property 4, if

$$A = \begin{bmatrix} -2 & 3 & 7 \\ 0 & 5 & 2 \\ -16 & 6 & 4 \end{bmatrix} \quad \text{and} \quad B = \begin{bmatrix} 1 & 3 & 7 \\ 0 & 5 & 2 \\ 8 & 6 & 4 \end{bmatrix},$$

then $\delta(A) = -2 \cdot \delta(B)$, since the first column of matrix A is -2 times the first column of matrix B. ■

The matrix

$$\begin{bmatrix} 2 & 1 & 2 \\ 5 & 10 & 5 \\ 3 & 6 & 3 \end{bmatrix}$$

has first and third columns that are identical. Expanding the determinant about the second column gives

$$\begin{aligned} \begin{vmatrix} 2 & 1 & 2 \\ 5 & 10 & 5 \\ 3 & 6 & 3 \end{vmatrix} &= -1\begin{vmatrix} 5 & 5 \\ 3 & 3 \end{vmatrix} + 10\begin{vmatrix} 2 & 2 \\ 3 & 3 \end{vmatrix} - 6\begin{vmatrix} 2 & 2 \\ 5 & 5 \end{vmatrix} \\ &= -1(15 - 15) + 10(6 - 6) - 6(10 - 10) \\ &= 0. \end{aligned}$$

This result suggests the next property.

DETERMINANT PROPERTY 5

The determinant of a matrix with two identical rows (or columns) equals 0.

EXAMPLE 5
APPLYING PROPERTY 5

Since two rows are identical, the determinant of

$$\begin{bmatrix} -4 & 2 & 3 \\ 0 & 1 & 6 \\ -4 & 2 & 3 \end{bmatrix}$$

equals 0. ■

Multiply each element of the second row of the matrix

$$\begin{bmatrix} -3 & 5 \\ \mathbf{1} & \mathbf{2} \end{bmatrix}$$

by 3 and add the result to the corresponding elements of the first row.

$$\begin{bmatrix} -3 + \mathbf{1}(3) & 5 + \mathbf{2}(3) \\ 1 & 2 \end{bmatrix} = \begin{bmatrix} -3 + 3 & 5 + 6 \\ 1 & 2 \end{bmatrix} = \begin{bmatrix} 0 & 11 \\ 1 & 2 \end{bmatrix}$$

Verify that the determinant of the new matrix is the same as the determinant of the original matrix, -11. This idea, which is generalized below, is perhaps the most useful property of determinants presented in this section.

DETERMINANT PROPERTY 6 The determinant of a matrix is unchanged if a multiple of a row (or column) of the matrix is added to the corresponding elements of another row (or column).

EXAMPLE 6 APPLYING PROPERTY 6 Multiply each element of the first column of the matrix

$$\begin{bmatrix} -2 & 4 & 1 \\ 2 & 1 & 5 \\ 3 & 0 & 2 \end{bmatrix}$$

by 3, and add the results to the third column to get the new matrix

$$\begin{bmatrix} -2 & 4 & 1+3(-2) \\ 2 & 1 & 5+3(2) \\ 3 & 0 & 2+3(3) \end{bmatrix} = \begin{bmatrix} -2 & 4 & -5 \\ 2 & 1 & 11 \\ 3 & 0 & 11 \end{bmatrix}.$$

By Property 6, the determinants of these two matrices are the same. Verify that each is equal to 37. ■

The following examples show how the properties of determinants can be used to simplify the calculation of determinants.

EXAMPLE 7 USING THE PROPERTIES TO EVALUATE A DETERMINANT Without expanding, show that the determinant of the following matrix is zero.

$$\begin{bmatrix} 2 & 5 & -1 \\ 1 & -15 & 3 \\ -2 & 10 & -2 \end{bmatrix}$$

Each element in the second column equals -5 times the corresponding element in the third column. Multiply the elements of the third column by 5 and then add the results to the corresponding elements in the second column to get a matrix with an equivalent determinant,

$$\begin{bmatrix} 2 & 0 & -1 \\ 1 & 0 & 3 \\ -2 & 0 & -2 \end{bmatrix}.$$

By Property 1, the determinant of this matrix is zero. ■

EXAMPLE 8 USING THE PROPERTIES TO EVALUATE A DETERMINANT Evaluate

$$\begin{vmatrix} 4 & 2 & 1 & 0 \\ -2 & 4 & -1 & 7 \\ -5 & 2 & 3 & 1 \\ 6 & 4 & -3 & 2 \end{vmatrix}.$$

Use Property 6 to change the first row (any row or column could be used) of the matrix to a row in which every element but one is zero. Begin by multiplying the elements of the second column by -2 and adding the results to the first column, replacing the first column with this new column. The result is a 0 for the first element in the first row.

$$\begin{bmatrix} 0 & 2 & 1 & 0 \\ -10 & 4 & -1 & 7 \\ -9 & 2 & 3 & 1 \\ -2 & 4 & -3 & 2 \end{bmatrix}$$

The first column replaced by the sum of -2 times the second column and the first column

To get a zero for the second element in the first row, multiply the elements of the third column by -2 and add to the second column.

$$\begin{bmatrix} 0 & 0 & 1 & 0 \\ -10 & 6 & -1 & 7 \\ -9 & -4 & 3 & 1 \\ -2 & 10 & -3 & 2 \end{bmatrix}$$

The second column replaced by the sum of -2 times the third column and the second column

Since the first row now has only one nonzero number, expand the determinant about the first row to get

$$1\begin{vmatrix} -10 & 6 & 7 \\ -9 & -4 & 1 \\ -2 & 10 & 2 \end{vmatrix} = \begin{vmatrix} -10 & 6 & 7 \\ -9 & -4 & 1 \\ -2 & 10 & 2 \end{vmatrix}.$$

Repeat the process with the 3×3 matrix

$$\begin{bmatrix} -10 & 6 & 7 \\ -9 & -4 & 1 \\ -2 & 10 & 2 \end{bmatrix}.$$

Change the third column to a column with two zeros, as shown below.

$$\begin{bmatrix} 53 & 34 & 0 \\ -9 & -4 & 1 \\ -2 & 10 & 2 \end{bmatrix}$$

The first row replaced by the sum of -7 times the second row and the first row

$$\begin{bmatrix} 53 & 34 & 0 \\ -9 & -4 & 1 \\ 16 & 18 & 0 \end{bmatrix}$$

The third row replaced by the sum of -2 times the second row and the third row

Expand the determinant about the third column to evaluate.

$$-1\begin{vmatrix} 53 & 34 \\ 16 & 18 \end{vmatrix} = -1(954 - 544) = -410 \quad ■$$

NOTE When applying Property 6, work with sums of *rows* to get a *column* with only one nonzero number or with sums of *columns* to get a *row* with one nonzero number.

The following summary lists the properties of determinants presented in this section.

PROPERTIES OF DETERMINANTS

1. If every element in a row or column of a matrix is 0, then the determinant equals 0.
2. If corresponding rows and columns of a matrix are interchanged, the determinant is not changed.
3. Interchanging two rows (or columns) of a matrix changes the sign of the determinant.
4. If every element of a row (or column) of a matrix is multiplied by the real number k, then the determinant of the new matrix is k times the determinant of the original matrix.
5. The determinant of a matrix with two identical rows (or columns) equals 0.
6. The determinant of a matrix is unchanged if a multiple of a row (or column) of the matrix is added to the corresponding elements of another row (or column).

12.2 EXERCISES

Give the property that tells why each of the following determinants equals 0. (More than one may be used.) See Examples 1, 4, and 5.

1. $\begin{vmatrix} 2 & 3 \\ 2 & 3 \end{vmatrix}$ **2.** $\begin{vmatrix} -5 & -5 \\ 6 & 6 \end{vmatrix}$ **3.** $\begin{vmatrix} 2 & 0 \\ 3 & 0 \end{vmatrix}$ **4.** $\begin{vmatrix} -8 & 0 \\ -6 & 0 \end{vmatrix}$

5. $\begin{vmatrix} -1 & 2 & 4 \\ 4 & -8 & -16 \\ 3 & 0 & 5 \end{vmatrix}$ **6.** $\begin{vmatrix} 2 & -8 & 3 \\ 0 & 2 & -1 \\ -6 & 24 & -9 \end{vmatrix}$ **7.** $\begin{vmatrix} 3 & 6 & 6 \\ 2 & 0 & 4 \\ 1 & 4 & 2 \end{vmatrix}$ **8.** $\begin{vmatrix} 1 & 0 & 0 \\ 1 & 0 & 1 \\ 3 & 0 & 0 \end{vmatrix}$

9. $\begin{vmatrix} m & 2 & 2m \\ 3n & 1 & 6n \\ 5p & 6 & 10p \end{vmatrix}$ **10.** $\begin{vmatrix} 7z & 8x & 2y \\ z & x & y \\ 7z & 7x & 7y \end{vmatrix}$

11. Explain why the determinant of a matrix with two identical rows (or columns) equals 0. See Examples 1–6.

Identify the appropriate properties from this section to tell why each of the following is true. Do not evaluate the determinants. See Examples 2–6.

12. $\begin{vmatrix} 2 & 1 & 6 \\ 3 & 0 & 2 \\ 4 & 1 & 8 \end{vmatrix} = \begin{vmatrix} 2 & 3 & 4 \\ 1 & 0 & 1 \\ 6 & 2 & 8 \end{vmatrix}$ **13.** $\begin{vmatrix} 4 & -2 \\ 3 & 8 \end{vmatrix} = \begin{vmatrix} 4 & 3 \\ -2 & 8 \end{vmatrix}$

14. $\begin{vmatrix} 2 & 6 \\ 3 & 5 \end{vmatrix} = -\begin{vmatrix} 3 & 5 \\ 2 & 6 \end{vmatrix}$ **15.** $\begin{vmatrix} -1 & 8 & 9 \\ 0 & 2 & 1 \\ 3 & 2 & 0 \end{vmatrix} = -\begin{vmatrix} 8 & -1 & 9 \\ 2 & 0 & 1 \\ 2 & 3 & 0 \end{vmatrix}$

16. $3\begin{vmatrix} 6 & 0 & 2 \\ 4 & 1 & 3 \\ 2 & 8 & 6 \end{vmatrix} = \begin{vmatrix} 6 & 0 & 2 \\ 4 & 3 & 3 \\ 2 & 24 & 6 \end{vmatrix}$

17. $-\frac{1}{2}\begin{vmatrix} 5 & -8 & 2 \\ 3 & -6 & 9 \\ 2 & 4 & 4 \end{vmatrix} = \begin{vmatrix} 5 & 4 & 2 \\ 3 & 3 & 9 \\ 2 & -2 & 4 \end{vmatrix}$

18. $\begin{vmatrix} 3 & -4 \\ 2 & 5 \end{vmatrix} = \begin{vmatrix} 3 & -4 \\ 5 & 1 \end{vmatrix}$

19. $\begin{vmatrix} -1 & 6 \\ 3 & -5 \end{vmatrix} = \begin{vmatrix} -1 & 6 \\ 2 & 1 \end{vmatrix}$

20. $\begin{vmatrix} 5 & 8 \\ 2 & -1 \end{vmatrix} = \begin{vmatrix} 5 & -2 \\ 2 & -5 \end{vmatrix}$

21. $\begin{vmatrix} 13 & 5 \\ 6 & 1 \end{vmatrix} = \begin{vmatrix} -2 & 5 \\ 3 & 1 \end{vmatrix}$

22. $\begin{vmatrix} 2 & 5 & 8 \\ 1 & 0 & 2 \\ 4 & 3 & 5 \end{vmatrix} = \begin{vmatrix} 2 & 5 & 8 \\ 1 & 0 & 2 \\ 7 & 3 & 11 \end{vmatrix}$

23. $2\begin{vmatrix} 4 & 2 & -1 \\ m & 2n & 3p \\ 5 & 1 & 0 \end{vmatrix} = \begin{vmatrix} 4 & 2 & -1 \\ 2m & 4n & 6p \\ 5 & 1 & 0 \end{vmatrix}$

24. $\begin{vmatrix} 3 & 5 & 0 \\ 2 & 1 & 3 \\ -5 & 1 & 6 \end{vmatrix} = \begin{vmatrix} 3 & 5 & 0 \\ 2+3k & 1+5k & 3+0k \\ -5 & 1 & 6 \end{vmatrix}$

25. $\begin{vmatrix} -4 & 2 & 1 \\ 3 & 0 & 5 \\ -1 & 4 & -2 \end{vmatrix} = \begin{vmatrix} -4 & 2 & 1+(-4)k \\ 3 & 0 & 5+3k \\ -1 & 4 & -2+(-1)k \end{vmatrix}$

Use Property 6 to find each of the following determinants. See Examples 7 and 8.

26. $\begin{vmatrix} 2 & 4 \\ 3 & 6 \end{vmatrix}$

27. $\begin{vmatrix} -5 & 10 \\ 6 & -12 \end{vmatrix}$

28. $\begin{vmatrix} 4 & 8 & 0 \\ -1 & -2 & 1 \\ 2 & 4 & 3 \end{vmatrix}$

29. $\begin{vmatrix} 6 & 8 & -12 \\ -1 & 16 & 2 \\ 4 & 0 & -8 \end{vmatrix}$

30. $\begin{vmatrix} 3 & 1 & 2 \\ 2 & 3 & 1 \\ 1 & 0 & -2 \end{vmatrix}$

31. $\begin{vmatrix} -2 & 2 & 3 \\ 0 & 2 & 1 \\ -1 & 4 & 0 \end{vmatrix}$

32. $\begin{vmatrix} -4 & 2 & 3 \\ 2 & 0 & 1 \\ 0 & 4 & 2 \end{vmatrix}$

33. $\begin{vmatrix} 6 & 3 & 2 \\ 1 & 0 & 2 \\ -1 & 4 & 1 \end{vmatrix}$

34. $\begin{vmatrix} 1 & 0 & 2 & 2 \\ 2 & 4 & 1 & -1 \\ 1 & -3 & 1 & 0 \\ 1 & 1 & 0 & 1 \end{vmatrix}$

35. $\begin{vmatrix} 2 & -1 & 1 & 0 \\ 1 & 1 & 0 & 1 \\ 0 & -1 & 1 & 1 \\ 1 & 2 & 1 & 2 \end{vmatrix}$

---

36. The note following Example 8 advises working with sums of rows to get a column with only one nonzero number. Why is this good advice? What happens if you work with columns to get a column with just one nonzero number?

Use the determinant

$$\begin{vmatrix} a & b & c \\ d & e & f \\ g & h & j \end{vmatrix}$$

to prove the following properties of determinants by finding the values of the original determinant and the new determinant and comparing them.

37. Property 2

38. Property 3

39. Property 4

PREVIEW EXERCISES

Solve each system of equations. See Sections 11.1 and 11.2.

40. $x + 5y = 21$
$7x + 2y = 48$

41. $-2x + 4y = 6$
$6x - y = 15$

42. $2x + 3y + z = 17$
$5x + 2z = 3$
$x + 7y + 3z = 46$

43. $x + 2y - z = -3$
$2x + 3y = -1$
$-x - y + 3z = 8$

12.3 SOLUTION OF LINEAR SYSTEMS OF EQUATIONS BY DETERMINANTS—CRAMER'S RULE

OBJECTIVES

1. UNDERSTAND THE DERIVATION OF CRAMER'S RULE.
2. APPLY CRAMER'S RULE TO A LINEAR SYSTEM WITH TWO EQUATIONS AND TWO VARIABLES.
3. APPLY CRAMER'S RULE TO A LINEAR SYSTEM WITH THREE EQUATIONS AND THREE VARIABLES.

1 The elimination method is used in this section to solve the general system of two equations with two variables,

$$a_1x + b_1y = c_1 \qquad (1)$$
$$a_2x + b_2y = c_2. \qquad (2)$$

The result is a formula that can be used for any system of two equations with two unknowns. To get this general solution, eliminate y and solve for x by first multiplying both sides of equation (1) by b_2 and both sides of equation (2) by $-b_1$. Then add these results and solve for x.

$$a_1b_2x + b_1b_2y = c_1b_2 \qquad b_2 \text{ times both sides of equation (1)}$$
$$-a_2b_1x - b_1b_2y = -c_2b_1 \qquad -b_1 \text{ times both sides of equation (2)}$$
$$(a_1b_2 - a_2b_1)x = c_1b_2 - c_2b_1$$

$$x = \frac{c_1b_2 - c_2b_1}{a_1b_2 - a_2b_1}$$

Solve for y by multiplying both sides of equation (1) by $-a_2$ and equation (2) by a_1 and then adding the two equations.

$$-a_1a_2x - a_2b_1y = -a_2c_1 \qquad -a_2 \text{ times both sides of (1)}$$
$$a_1a_2x + a_1b_2y = a_1c_2 \qquad a_1 \text{ times both sides of (2)}$$
$$(a_1b_2 - a_2b_1)y = a_1c_2 - a_2c_1$$

$$y = \frac{a_1c_2 - a_2c_1}{a_1b_2 - a_2b_1}$$

Both numerators and the common denominator of these values for x and y can be written as determinants, since

$$c_1b_2 - c_2b_1 = \begin{vmatrix} c_1 & b_1 \\ c_2 & b_2 \end{vmatrix}, \qquad a_1c_2 - a_2c_1 = \begin{vmatrix} a_1 & c_1 \\ a_2 & c_2 \end{vmatrix},$$

$$\text{and} \qquad a_1b_2 - a_2b_1 = \begin{vmatrix} a_1 & b_1 \\ a_2 & b_2 \end{vmatrix}.$$

Using these determinants, the solutions for x and y become

$$x = \frac{\begin{vmatrix} c_1 & b_1 \\ c_2 & b_2 \end{vmatrix}}{\begin{vmatrix} a_1 & b_1 \\ a_2 & b_2 \end{vmatrix}} \quad \text{and} \quad y = \frac{\begin{vmatrix} a_1 & c_1 \\ a_2 & c_2 \end{vmatrix}}{\begin{vmatrix} a_1 & b_1 \\ a_2 & b_2 \end{vmatrix}}, \quad \text{if } \begin{vmatrix} a_1 & b_1 \\ a_2 & b_2 \end{vmatrix} \neq 0.$$

For convenience, we will denote the three determinants in the solution as

$$\begin{vmatrix} a_1 & b_1 \\ a_2 & b_2 \end{vmatrix} = D, \qquad \begin{vmatrix} c_1 & b_1 \\ c_2 & b_2 \end{vmatrix} = D_x, \qquad \text{and} \qquad \begin{vmatrix} a_1 & c_1 \\ a_2 & c_2 \end{vmatrix} = D_y.$$

NOTE The elements of D are the four coefficients of the variables in the given system, the elements of D_x are obtained by replacing the coefficients of x in D by the respective constants, and the elements of D_y are obtained by replacing the coefficients of y in D by the respective constants.

These results are summarized as **Cramer's rule.**

CRAMER'S RULE FOR 2 × 2 SYSTEMS

Given the system

$$a_1x + b_1y = c_1$$
$$a_2x + b_2y = c_2,$$

with $a_1b_2 - a_2b_1 \neq 0$,

then
$$x = \frac{D_x}{D} \quad \text{and} \quad y = \frac{D_y}{D},$$

where

$$D_x = \begin{vmatrix} c_1 & b_1 \\ c_2 & b_2 \end{vmatrix}, \quad D_y = \begin{vmatrix} a_1 & c_1 \\ a_2 & c_2 \end{vmatrix}, \quad \text{and} \quad D = \begin{vmatrix} a_1 & b_1 \\ a_2 & b_2 \end{vmatrix} \neq 0.$$

2 Cramer's rule is used to solve a system of linear equations by evaluating the three determinants D, D_x, and D_y and then writing the appropriate quotients for x and y.

CAUTION As indicated above, Cramer's rule does not apply if $D = 0$. When $D = 0$, the system is inconsistent or has dependent equations. For this reason, it is a good idea to evaluate D first.

EXAMPLE 1
APPLYING CRAMER'S RULE TO A 2 × 2 SYSTEM

Use Cramer's rule to solve the system

$$5x + 7y = -1$$
$$6x + 8y = 1.$$

By Cramer's rule, $x = D_x/D$ and $y = D_y/D$. As mentioned in the caution above, we find D first. If $D \neq 0$, we then find D_x and D_y.

$$D = \begin{vmatrix} 5 & 7 \\ 6 & 8 \end{vmatrix} = 5(8) - 6(7) = -2$$

$$D_x = \begin{vmatrix} -1 & 7 \\ 1 & 8 \end{vmatrix} = (-1)(8) - (1)(7) = -15$$

$$D_y = \begin{vmatrix} 5 & -1 \\ 6 & 1 \end{vmatrix} = 5(1) - (6)(-1) = 11$$

From Cramer's rule, $x = \dfrac{D_x}{D} = \dfrac{-15}{-2} = \dfrac{15}{2}$

and $$y = \frac{D_y}{D} = \frac{11}{-2} = -\frac{11}{2}.$$

The solution set is $\{(15/2, -11/2)\}$, as can be verified by substituting in the given system. ■

3 Cramer's rule can be generalized to systems of three equations in three variables (or n equations in n variables).

CRAMER'S RULE FOR 3 × 3 SYSTEMS

Given the system

$$a_1x + b_1y + c_1z = d_1$$
$$a_2x + b_2y + c_2z = d_2$$
$$a_3x + b_3y + c_3z = d_3,$$

then $$x = \frac{D_x}{D}, \quad y = \frac{D_y}{D}, \quad \text{and} \quad z = \frac{D_z}{D},$$

where $$D_x = \begin{vmatrix} d_1 & b_1 & c_1 \\ d_2 & b_2 & c_2 \\ d_3 & b_3 & c_3 \end{vmatrix}, \qquad D_y = \begin{vmatrix} a_1 & d_1 & c_1 \\ a_2 & d_2 & c_2 \\ a_3 & d_3 & c_3 \end{vmatrix},$$

$$D_z = \begin{vmatrix} a_1 & b_1 & d_1 \\ a_2 & b_2 & d_2 \\ a_3 & b_3 & d_3 \end{vmatrix}, \qquad D = \begin{vmatrix} a_1 & b_1 & c_1 \\ a_2 & b_2 & c_2 \\ a_3 & b_3 & c_3 \end{vmatrix} \neq 0.$$

EXAMPLE 2
APPLYING CRAMER'S RULE TO A 3 × 3 SYSTEM

Use Cramer's rule to solve the system

$$x + y - z + 2 = 0$$
$$2x - y + z + 5 = 0$$
$$x - 2y + 3z - 4 = 0.$$

For Cramer's rule, the system must be rewritten in the form

$$\begin{aligned} x + y - z &= -2 \\ 2x - y + z &= -5 \\ x - 2y + 3z &= 4. \end{aligned}$$

Verify that the required determinants are as follows:

$$D = \begin{vmatrix} 1 & 1 & -1 \\ 2 & -1 & 1 \\ 1 & -2 & 3 \end{vmatrix} = -3, \qquad D_x = \begin{vmatrix} -2 & 1 & -1 \\ -5 & -1 & 1 \\ 4 & -2 & 3 \end{vmatrix} = 7,$$

$$D_y = \begin{vmatrix} 1 & -2 & -1 \\ 2 & -5 & 1 \\ 1 & 4 & 3 \end{vmatrix} = -22, \qquad D_z = \begin{vmatrix} 1 & 1 & -2 \\ 2 & -1 & -5 \\ 1 & -2 & 4 \end{vmatrix} = -21.$$

Thus

$$x = \frac{D_x}{D} = \frac{7}{-3} = -\frac{7}{3}, \qquad y = \frac{D_y}{D} = \frac{-22}{-3} = \frac{22}{3},$$

and

$$z = \frac{D_z}{D} = \frac{-21}{-3} = 7,$$

so the solution set is $\{(-7/3, 22/3, 7)\}$. ■

CAUTION As shown in Example 2, each equation in the system must be written in the form $ax + by + cz + \cdots = k$ before using Cramer's rule.

EXAMPLE 3
APPLYING CRAMER'S RULE WHEN $D = 0$

Use Cramer's rule to solve the system

$$\begin{aligned} 2x - 3y + 4z &= 10 \\ 6x - 9y + 12z &= 24 \\ x + 2y - 3z &= 5. \end{aligned}$$

First find D.

$$D = \begin{vmatrix} 2 & -3 & 4 \\ 6 & -9 & 12 \\ 1 & 2 & -3 \end{vmatrix} = 2\begin{vmatrix} -9 & 12 \\ 2 & -3 \end{vmatrix} - 6\begin{vmatrix} -3 & 4 \\ 2 & -3 \end{vmatrix} + 1\begin{vmatrix} -3 & 4 \\ -9 & 12 \end{vmatrix}$$
$$= 2(3) - 6(1) + 1(0)$$
$$= 0$$

As mentioned above, Cramer's rule does not apply if $D = 0$. When $D = 0$, the system either is inconsistent or contains dependent equations. Use the elimination method to tell which is the case. Verify that *this* system is inconsistent, so the solution set is $\emptyset$. ■

Cramer's rule can be extended to 4×4 or larger systems.

12.3 EXERCISES

Use Cramer's rule to solve each of the following systems of equations. If $D = 0$, use another method to determine the solution. See Examples 1 and 3.

1. $x + y = 4$
$2x - y = 2$

2. $3x + 2y = -4$
$2x - y = -5$

3. $4x + 3y = -7$
$2x + 3y = -11$

4. $4x - y = 0$
$2x + 3y = 14$

5. $5x + 4y = 10$
$3x - 7y = 6$

6. $3x + 2y = -4$
$5x - y = 2$

7. $2x - 3y = -5$
$x + 5y = 17$

8. $x + 9y = -15$
$3x + 2y = 5$

9. $3x + 2y = 4$
$6x + 4y = 8$

10. $1.5x + 3y = 5$
$2x + 4y = 3$

11. $12x + 8y = 3$
$15x + 10y = 9$

12. $15x - 10y = 5$
$9x + 6y = 3$

Use Cramer's rule to solve each of the following systems of equations. If $D = 0$, use another method to complete the solution. See Examples 2 and 3.

13. $4x - y + 3z = -3$
$3x + y + z = 0$
$2x - y + 4z = 0$

14. $5x + 2y + z = 15$
$2x - y + z = 9$
$4x + 3y + 2z = 13$

15. $2x - y + 4z = -2$
$3x + 2y - z = -3$
$x + 4y + 2z = 17$

16. $x + y + z = 4$
$2x - y + 3z = 4$
$4x + 2y - z = -15$

17. $4x - 3y + z = -1$
$5x + 7y + 2z = -2$
$3x - 5y - z = 1$

18. $2x - 3y + z = 8$
$-x - 5y + z = -4$
$3x - 5y + 2z = 12$

19. $x + 2y + 3z = 4$
$4x + 3y + 2z = 1$
$-x - 2y - 3z = 0$

20. $2x - y + 3z = 1$
$-2x + y - 3z = 2$
$5x - y + z = 2$

21. $-2x - 2y + 3z = 4$
$5x + 7y - z = 2$
$2x + 2y - 3z = -4$

22. $-3x + 2y - 2z = 4$
$4x + y + z = 5$
$3x - 2y + 2z = 1$

23. $2x + 3y = 13$
$2y - z = 5$
$x + 2z = 4$

24. $3x - z = -10$
$y + 4z = 8$
$x + 2z = -1$

25. $5x - y = -4$
$3x + 2z = 4$
$4y + 3z = 22$

26. $3x + 5y = -7$
$2x + 7z = 2$
$4y + 3z = -8$

27. $x + 2y = 10$
$3x + 4z = 7$
$-y - z = 1$

28. $5x - 2y = 3$
$4y + z = 8$
$x + 2z = 4$

---

29. In your own words, explain what it means in applying Cramer's rule if $D = 0$. (See Objective 2.)

30. Describe D_x, D_y, and D_z in terms of the coefficients and constants in the given system of equations. (See Objective 1.)

Use Cramer's rule to solve each of the following systems.

31. $x + 3y - 2z - w = 9$
$4x + y + z + 2w = 2$
$-3x - y + z - w = -5$
$x - y - 3z - 2w = 2$

32. $3x + 2y - w = 0$
$2x + z + 2w = 5$
$x + 2y - z = -2$
$2x - y + z + w = 2$

33. $x + y - z + w = 2$
$x - y + z + w = 4$
$-2x + y + 2z - w = -5$
$x + 3z + 2w = 5$

34. $x + 2y - z + w = 8$
$2x - y + 2w = 8$
$y + 3z = 5$
$x - z = 4$

Solve each system for x and y using Cramer's rule. Assume a and b are nonzero constants.

35. $bx + y = a^2$
$ax + y = b^2$

36. $ax + by = \frac{b}{a}$
$x + y = \frac{1}{b}$

37. $b^2x + a^2y = b^2$
$ax + by = a$

38. $x + \frac{1}{b}y = b$
$\frac{1}{a}x + y = a$

PREVIEW EXERCISES

Give the additive inverse and reciprocal for each of the following numbers. See Sections 1.1 and 1.4.

39. -2

40. 4

41. 27

42. $\frac{2}{3}$

43. $-\frac{4}{5}$

44. $-\frac{7}{3}$

12.4 SOLUTION OF LINEAR SYSTEMS OF EQUATIONS BY MATRICES

OBJECTIVES

1. WRITE THE AUGMENTED MATRIX FOR A SYSTEM OF EQUATIONS.
2. USE ROW TRANSFORMATIONS TO SOLVE A SYSTEM WITH TWO EQUATIONS.
3. USE ROW TRANSFORMATIONS TO SOLVE A SYSTEM WITH THREE EQUATIONS.
4. USE ROW TRANSFORMATIONS TO SOLVE INCONSISTENT SYSTEMS OR SYSTEMS WITH DEPENDENT EQUATIONS.

Several methods for solving linear systems have now been shown. This section describes how to solve these systems by matrix methods. Matrix methods are particularly suitable for computer solutions of larger systems of equations having many unknowns.

1 As an example, we start with the system

$$\begin{aligned} x + 3y + 2z &= 1 \\ 2x + y - z &= 2 \\ x + y + z &= 2, \end{aligned}$$

and write the coefficients of the variables and the constants as a matrix, called the **augmented matrix** of the system.

$$\left[\begin{array}{rrr|r} 1 & 3 & 2 & 1 \\ 2 & 1 & -1 & 2 \\ 1 & 1 & 1 & 2 \end{array}\right]$$

The vertical line, which is optional, is used only to separate the coefficients from the constants. This matrix has 3 rows (horizontal) and 4 columns (vertical). To refer to a number in the matrix, use its row and column numbers. For example, the number 3 is in the first row, second column position.

The rows of this matrix can be treated just like the equations of a system of linear equations. Since the augmented matrix is nothing more than a short form of the system, any transformation of the matrix that results in an equivalent system of equations can be performed. Operations that produce such transformations are given below.

MATRIX ROW TRANSFORMATIONS

For any augmented matrix of a system of linear equations, the following row transformations will result in the matrix of an equivalent system.

1. Any two rows may be interchanged.
2. The elements of any row may be multiplied by a nonzero real number.
3. Any row may be changed by adding to its elements a multiple of the corresponding elements of another row.

EXAMPLE 1
USING THE ROW TRANSFORMATIONS

(a) The first row transformation is used to change the matrix

$$\begin{bmatrix} 1 & 3 & 5 \\ \mathbf{0} & \mathbf{1} & \mathbf{2} \\ 1 & -1 & -2 \end{bmatrix} \quad \text{to} \quad \begin{bmatrix} \mathbf{0} & \mathbf{1} & \mathbf{2} \\ 1 & 3 & 5 \\ 1 & -1 & -2 \end{bmatrix}$$

by interchanging the first two rows.

(b) Using the second row transformation with $k = -2$ changes

$$\begin{bmatrix} 1 & 3 & 5 \\ 0 & 1 & 2 \\ 1 & -1 & -2 \end{bmatrix} \quad \text{to} \quad \begin{bmatrix} -2 & -6 & -10 \\ 0 & 1 & 2 \\ 1 & -1 & -2 \end{bmatrix},$$

where the elements of the first row of the original matrix were multiplied by -2.

(c) The third row transformation is used to change

$$\begin{bmatrix} 1 & 3 & 5 \\ 0 & 1 & 2 \\ 1 & -1 & -2 \end{bmatrix} \quad \text{to} \quad \begin{bmatrix} 0 & 4 & 7 \\ 0 & 1 & 2 \\ 1 & -1 & -2 \end{bmatrix},$$

by multiplying each element in the third row of the original matrix by -1 and adding the results to the corresponding elements in the first row of that matrix. That is, the elements in the new first row were found as follows.

$$\begin{bmatrix} 1 + \mathbf{1}(-1) & 3 + (\mathbf{-1})(-1) & 5 + (\mathbf{-2})(-1) \\ 0 & 1 & 2 \\ 1 & -1 & -2 \end{bmatrix} = \begin{bmatrix} 0 & 4 & 7 \\ 0 & 1 & 2 \\ 1 & -1 & -2 \end{bmatrix}$$

Rows two and three were left unchanged. ■

2 If the word "row" is replaced by "equation," it can be seen that the three row transformations also apply to a system of equations, so that a system of equations can be solved by transforming its corresponding matrix into the matrix of an equivalent, simpler system. The goal is a matrix in the form

$$\left[\begin{array}{cc|c} 1 & 0 & a \\ 0 & 1 & b \end{array}\right] \quad \text{or} \quad \left[\begin{array}{ccc|c} 1 & 0 & 0 & a \\ 0 & 1 & 0 & b \\ 0 & 0 & 1 & c \end{array}\right]$$

for systems with two or three equations respectively. Notice that on the left of the vertical bar there are ones down the diagonal from upper left to lower right and zeros elsewhere in the matrices. When these matrices are rewritten as systems of equations, the values of the variables are known. The **Gauss-Jordan method** is a systematic way of using the matrix row transformations to change the augmented matrix of a system into the form that shows its solution. The following examples will illustrate this method.

EXAMPLE 2
USING THE GAUSS-JORDAN METHOD

Solve the linear system

$$3x - 4y = 1$$
$$5x + 2y = 19.$$

The equations should all be in the same form, with the variable terms in the same order on the left, and the constant term on the right. Begin by writing the augmented matrix.

$$\left[\begin{array}{cc|c} 3 & -4 & 1 \\ 5 & 2 & 19 \end{array}\right]$$

The goal is to transform this augmented matrix into one in which the value of the variables will be easy to see. That is, since each column in the matrix represents the coefficients of one variable, the augmented matrix should be transformed so that it is of the form

$$\left[\begin{array}{cc|c} 1 & 0 & k \\ 0 & 1 & j \end{array}\right]$$

for real numbers k and j. Once the augmented matrix is in this form, the matrix can be rewritten as a linear system to get

$$x = k$$
$$y = j.$$

The necessary transformations are performed as follows. It is best to work in columns beginning in each column with the element that is to become 1. In the augmented matrix,

$$\left[\begin{array}{cc|c} 3 & -4 & 1 \\ 5 & 2 & 19 \end{array}\right],$$

there is a 3 in the first row, first column position. Use transformation 2, multiplying each entry in the first row by 1/3 to get a 1 in this position. (This step is abbreviated as (1/3)R1.)

$$\left[\begin{array}{rr|r} 1 & -4/3 & 1/3 \\ 5 & 2 & 19 \end{array}\right] \qquad \frac{1}{3}\text{R1}$$

Get 0 in the second row, first column by multiplying each element of the first row by −5 and adding the result to the corresponding element in the second row, using transformation 3.

$$\left[\begin{array}{rr|r} 1 & -4/3 & 1/3 \\ 0 & 26/3 & 52/3 \end{array}\right] \qquad -5\text{R1} + \text{R2}$$

Get 1 in the second row, second column by multiplying each element of the second row by 3/26, using transformation 2.

$$\left[\begin{array}{rr|r} 1 & -4/3 & 1/3 \\ 0 & 1 & 2 \end{array}\right] \qquad \frac{3}{26}\text{R2}$$

Finally, get 0 in the first row, second column by multiplying each element of the second row by 4/3 and adding the result to the corresponding element in the first row.

$$\left[\begin{array}{rr|r} 1 & 0 & 3 \\ 0 & 1 & 2 \end{array}\right] \qquad \frac{4}{3}\text{R2} + \text{R1}$$

This last matrix corresponds to the system

$$x = 3$$
$$y = 2,$$

that has the solution set {(3, 2)}. This solution could have been read directly from the third column of the final matrix. ■

3 A linear system with three equations is solved in a similar way. Row transformations are used to get 1s down the diagonal from left to right and 0s above and below each 1.

EXAMPLE 3
USING THE GAUSS-JORDAN METHOD

Use the Gauss-Jordan method to solve the system

$$\begin{aligned} x - y + 5z &= -6 \\ 3x + 3y - z &= 10 \\ x + 3y + 2z &= 5. \end{aligned}$$

Since the system is in proper form, begin by writing the augmented matrix of the linear system.

$$\left[\begin{array}{rrr|r} 1 & -1 & 5 & -6 \\ 3 & 3 & -1 & 10 \\ 1 & 3 & 2 & 5 \end{array}\right]$$

The final matrix is to be of the form

$$\left[\begin{array}{ccc|c} 1 & 0 & 0 & m \\ 0 & 1 & 0 & n \\ 0 & 0 & 1 & p \end{array}\right],$$

where m, n, and p are real numbers. This final form of the matrix gives the system $x = m$, $y = n$, and $z = p$, so the solution set is $\{(m, n, p)\}$.

There is already a 1 in the first row, first column. Get a 0 in the second row of the first column by multiplying each element in the first row by -3 and adding the result to the corresponding element in the second row, using transformation 3.

$$\left[\begin{array}{ccc|c} 1 & -1 & 5 & -6 \\ 0 & 6 & -16 & 28 \\ 1 & 3 & 2 & 5 \end{array}\right] \quad -3R1 + R2$$

Now, to change the last element in the first column to 0, use transformation 3 and multiply each element of the first row by -1, then add the results to the corresponding elements of the third row.

$$\left[\begin{array}{ccc|c} 1 & -1 & 5 & -6 \\ 0 & 6 & -16 & 28 \\ 0 & 4 & -3 & 11 \end{array}\right] \quad -1R1 + R3$$

The same procedure is used to transform the second and third columns. For both of these columns perform the additional step of getting 1 in the appropriate position of each column. Do this by multiplying the elements of the row by the reciprocal of the number in that position.

$$\left[\begin{array}{ccc|c} 1 & -1 & 5 & -6 \\ 0 & 1 & -8/3 & 14/3 \\ 0 & 4 & -3 & 11 \end{array}\right] \quad \frac{1}{6}R2$$

$$\left[\begin{array}{ccc|c} 1 & 0 & 7/3 & -4/3 \\ 0 & 1 & -8/3 & 14/3 \\ 0 & 4 & -3 & 11 \end{array}\right] \quad R2 + R1$$

$$\left[\begin{array}{ccc|c} 1 & 0 & 7/3 & -4/3 \\ 0 & 1 & -8/3 & 14/3 \\ 0 & 0 & 23/3 & -23/3 \end{array}\right] \quad -4R2 + R3$$

$$\left[\begin{array}{ccc|c} 1 & 0 & 7/3 & -4/3 \\ 0 & 1 & -8/3 & 14/3 \\ 0 & 0 & 1 & -1 \end{array}\right] \quad \frac{3}{23}R3$$

$$\left[\begin{array}{ccc|c} 1 & 0 & 0 & 1 \\ 0 & 1 & -8/3 & 14/3 \\ 0 & 0 & 1 & -1 \end{array}\right] \quad -\frac{7}{3}R3 + R1$$

$$\left[\begin{array}{ccc|c} 1 & 0 & 0 & 1 \\ 0 & 1 & 0 & 2 \\ 0 & 0 & 1 & -1 \end{array}\right] \quad \frac{8}{3}R3 + R2$$

The linear system associated with this final matrix is

$$\begin{aligned} x &= 1 \\ y &= 2 \\ z &= -1, \end{aligned}$$

and the solution set is $\{(1, 2, -1)\}$. ■

4 The next two examples show how to recognize inconsistent systems or systems with dependent equations when solving such systems using row operations.

EXAMPLE 4 RECOGNIZING AN INCONSISTENT SYSTEM

Use the Gauss-Jordan method to solve the system

$$\begin{aligned} x + y &= 2 \\ 2x + 2y &= 5. \end{aligned}$$

Write the augmented matrix.

$$\left[\begin{array}{cc|c} 1 & 1 & 2 \\ 2 & 2 & 5 \end{array}\right]$$

Multiply the elements in the first row by -2 and add the result to the corresponding elements in the second row.

$$\left[\begin{array}{cc|c} 1 & 1 & 2 \\ 0 & 0 & 1 \end{array}\right] \quad -2\text{R1} + \text{R2}$$

The next step would be to get a 1 in the second row, second column. Because of the zeros, it is impossible to go further. Since the second row corresponds to the equation

$$0x + 0y = 1$$

which has no solution, the system is inconsistent, and the solution set is $\emptyset$. ■

EXAMPLE 5 SOLVING A SYSTEM WITH DEPENDENT EQUATIONS

Use the Gauss-Jordan method to solve the system

$$\begin{aligned} 2x - 5y + 3z &= 1 \\ x - 2y - 2z &= 8. \end{aligned}$$

Recall from Section 11.2 that a system with two equations and three variables has an infinite number of solutions. The Gauss-Jordan method can be used to give the solution with just one arbitrary variable. Start with the augmented matrix

$$\left[\begin{array}{ccc|c} 2 & -5 & 3 & 1 \\ 1 & -2 & -2 & 8 \end{array}\right].$$

Exchange rows to get a 1 in the first row, first column position.

$$\left[\begin{array}{ccc|c} 1 & -2 & -2 & 8 \\ 2 & -5 & 3 & 1 \end{array}\right]$$

Now multiply each element in the first row by -2 and add to the corresponding element in the second row.

$$\left[\begin{array}{ccc|c} 1 & -2 & -2 & 8 \\ 0 & -1 & 7 & -15 \end{array}\right] \quad -2R1 + R2$$

Multiply each element in the second row by -1.

$$\left[\begin{array}{ccc|c} 1 & -2 & -2 & 8 \\ 0 & 1 & -7 & 15 \end{array}\right] \quad -1R2$$

Multiply each element in the second row by 2 and add to the corresponding element in the first row.

$$\left[\begin{array}{ccc|c} 1 & 0 & -16 & 38 \\ 0 & 1 & -7 & 15 \end{array}\right] \quad 2R2 + R1$$

It is not possible to go further with the Gauss-Jordan method. The equations that correspond to the final matrix are

$$x - 16z = 38 \qquad \text{and} \qquad y - 7z = 15.$$

Solve these equations for x and y, respectively.

$$x - 16z = 38 \qquad\qquad y - 7z = 15$$
$$x = 16z + 38 \qquad\qquad y = 7z + 15$$

The solution can now be written with z arbitrary, as

$$\{(16z + 38, 7z + 15, z)\}. \quad \blacksquare$$

The cases that might occur when matrix methods are used to solve a system of linear equations are summarized below.

1. If the number of rows with nonzero elements to the left of the vertical line is equal to the number of variables in the system, then the system has a single solution.
2. If one of the rows has the form $[0 \quad 0 \ldots 0 \mid a]$ with $a \neq 0$, then the system has no solution.
3. If there are fewer rows in the matrix containing nonzero elements than the number of variables, then there are infinitely many solutions for the system. These solutions should be given in terms of arbitrary variables. See Example 5.

Although only examples with two variables or three variables have been shown, the Gauss-Jordan method can be extended to solve linear systems with more variables. As the number of variables increases, the process quickly becomes very tedious and the opportunity for error increases. The method, however, is suitable for use by computers and a fairly large system can be solved quickly in that way.

12.4 EXERCISES

Use the third row transformation to change each of the following matrices as indicated. See Example 1.

1. $\begin{bmatrix} 2 & 4 \\ 4 & 7 \end{bmatrix}$; -2 times row 1 added to row 2

2. $\begin{bmatrix} -1 & 4 \\ 7 & 0 \end{bmatrix}$; 7 times row 1 added to row 2

3. $\begin{bmatrix} 1 & 1/5 \\ 5 & 2 \end{bmatrix}$; -5 times row 1 added to row 2

4. $\begin{bmatrix} 5 & 7 \\ 2 & -4 \end{bmatrix}$; 4/7 times row 1 added to row 2

5. $\begin{bmatrix} 1 & 5 & 6 \\ -2 & 3 & -1 \\ 4 & 7 & 0 \end{bmatrix}$; 2 times row 1 added to row 2

6. $\begin{bmatrix} 2 & 5 & 6 \\ 4 & -1 & 2 \\ 3 & 7 & 1 \end{bmatrix}$; -6 times row 3 added to row 1

7. $\begin{bmatrix} -3 & 1 & -4 \\ 2 & 1 & 3 \\ -7 & 5 & 2 \end{bmatrix}$; -5 times row 2 added to row 3

8. $\begin{bmatrix} 4 & 10 & -8 \\ 7 & 4 & 3 \\ -1 & 1 & 0 \end{bmatrix}$; -4 times row 3 added to row 2

Write the augmented matrix for each of the following systems. Do not solve the system.

9. $2x + 3y = 11$
$x + 2y = 8$

10. $3x + 5y = -13$
$2x + 3y = -9$

11. $x + 5y = 6$
$x + 2y = 8$

12. $2x + 7y = 1$
$5x = -15$

13. $2x + y + z = 3$
$3x - 4y + 2z = -7$
$x + y + z = 2$

14. $4x - 2y + 3z = 4$
$3x + 5y + z = 7$
$5x - y + 4z = 7$

15. $x + y = 2$
$2y + z = -4$
$z = 2$

16. $x = 6$
$y + 2z = 2$
$x - 3z = 6$

17. $x = 5$
$y = -2$
$z = 3$

18. $x = 8$
$y + z = 6$
$z = 2$

Write the system of equations associated with each of the following augmented matrices. Do not try to solve.

19. $\left[\begin{array}{cc|c} 1 & 0 & 2 \\ 0 & 1 & 3 \end{array}\right]$

20. $\left[\begin{array}{cc|c} 1 & 0 & 5 \\ 0 & 1 & -3 \end{array}\right]$

21. $\left[\begin{array}{cc|c} 2 & 1 & 1 \\ 3 & -2 & -9 \end{array}\right]$

22. $\left[\begin{array}{cc|c} 1 & -5 & -18 \\ 6 & 2 & 20 \end{array}\right]$

23. $\left[\begin{array}{ccc|c} 1 & 0 & 0 & 2 \\ 0 & 1 & 0 & 3 \\ 0 & 0 & 1 & -2 \end{array}\right]$

24. $\left[\begin{array}{ccc|c} 1 & 0 & 1 & 4 \\ 0 & 1 & 0 & 2 \\ 0 & 0 & 1 & 3 \end{array}\right]$

25. $\left[\begin{array}{ccc|c} 3 & 2 & 1 & 1 \\ 0 & 2 & 4 & 22 \\ -1 & -2 & 3 & 15 \end{array}\right]$

26. $\left[\begin{array}{ccc|c} 2 & 1 & 3 & 12 \\ 4 & -3 & 0 & 10 \\ 5 & 0 & -4 & -11 \end{array}\right]$

Use the Gauss-Jordan method to solve each of the following systems of equations. See Examples 2–5.

27. $x + y = 5$
$x - y = -1$

28. $x + 2y = 5$
$2x + y = -2$

29. $x + y = -3$
$2x - 5y = -6$

30. $3x - 2y = 4$
$3x + y = -2$

31. $2x - 3y = 10$
$2x + 2y = 5$

32. $4x + y = 5$
$2x + y = 3$

33. $2x - 5y = 10$
$4x - 5y = 15$

34. $4x - y = 3$
$-2x + 3y = 1$

35. $2x - 3y = 2$
$4x - 6y = 1$

36. $x + 2y = 1$
$2x + 4y = 3$

37. $6x - 3y = 1$
$-12x + 6y = -2$

38. $x - y = 1$
$-x + y = -1$

39. $x + y = -1$
$y + z = 4$
$x + z = 1$

40. $x - z = -3$
$y + z = 9$
$x + z = 7$

41. $x + y - z = 6$
$2x - y + z = -9$
$x - 2y + 3z = 1$

42. $x + 3y - 6z = 7$
$2x - y + 2z = 0$
$x + y + 2z = -1$

43. $-x + y = -1$
$y - z = 6$
$x + z = -1$

44. $x + y = 1$
$2x - z = 0$
$y + 2z = -2$

45. $2x - y + 3z = 0$
$x + 2y - z = 5$
$2y + z = 1$

46. $4x + 2y - 3z = 6$
$x - 4y + z = -4$
$-x + 2z = 2$

---

47. Compare the use of an augmented matrix as a shorthand way of writing a system of linear equations and the use of synthetic division as a shorthand way to divide polynomials.

48. Compare the use of the third row operation on a matrix and the elimination method of solving a system of linear equations.

Use the Gauss-Jordan method to solve each of the following systems.

49. $3x + 2y - w = 0$
$2x + z + 2w = 5$
$x + 2y - z = -2$
$2x - y + z + w = 2$

50. $x + 3y - 2z - w = 9$
$4x + y + z + 2w = 2$
$-3x - y + z - w = -5$
$x - y - 3z - 2w = 2$

Solve each of the following problems by first setting up a system of equations. Use the Gauss-Jordan method.

51. George Esquibel deposits some money in a bank account paying 5% per year. He uses some additional money, amounting to 1/3 the amount placed in the bank, to buy bonds paying 6% per year. With the balance of his funds he buys an 8% certificate of deposit. The first year his investments bring a return of \$690. If the total of the investments is \$10,000, how much is invested at each rate?

52. To get necessary funds for a planned expansion, a small company took out three loans totaling \$25,000. The company was able to borrow some of the money at 8%. It borrowed \$2000 more than 1/2 the amount of the 8% loan at 10%, and the rest at 9%. The total annual interest was \$2220. How much did the company borrow at each rate?

53. One day a service station sold 400 gallons of premium gasoline, 150 gallons of regular gasoline, and 130 gallons of super gasoline for a total of \$909. The next day 380 gallons of premium, 170 gallons of regular, and 150 gallons of super were sold for \$931. The difference in price per gallon of super and regular is one-half the difference in price per gallon of premium and regular. How much does this station charge for each type of gasoline?

54. A biologist has three salt solutions: some 5% solution, some 15% solution, and some 25% solution. She needs to mix some of each to get 50 liters of 20% solution. Since she has a lot of the 5% solution, she wants to use twice as much of that as the 15% solution. How much of each solution should she use?

Solve each of the systems in Exercises 55–60 by the Gauss-Jordan method. Let z be the arbitrary variable if necessary. See Example 5.

55. $\begin{aligned} x - 3y + 2z &= 10 \\ 2x - y - z &= 8 \end{aligned}$

56. $\begin{aligned} 3x + y - z &= 12 \\ x + 2y + z &= 10 \end{aligned}$

57. $\begin{aligned} x + 2y - z &= 0 \\ 3x - y + z &= 6 \\ -2x - 4y + 2z &= 0 \end{aligned}$

58. $\begin{aligned} 3x + 5y - z &= 0 \\ 4x - y + 2z &= 1 \\ -6x - 10y + 2z &= 0 \end{aligned}$

59. $\begin{aligned} x - 2y + z &= 5 \\ -2x + 4y - 2z &= 2 \\ 2x + y - z &= 2 \end{aligned}$

60. $\begin{aligned} 3x + 6y - 3z &= 12 \\ -x - 2y + z &= 16 \\ x + y - 2z &= 20 \end{aligned}$

---

61. At rush hour, substantial traffic congestion is encountered at the traffic intersections shown in the following figure.

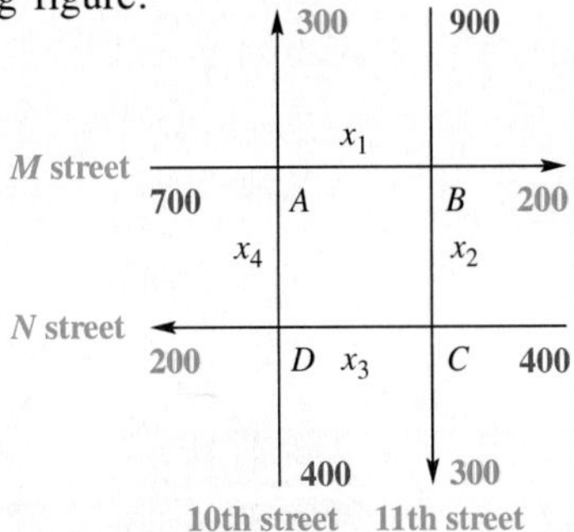

The city wishes to improve the signals at these corners so as to speed the flow of traffic. The traffic engineers first gather data. As the figure shows, 700 cars per hour come down M Street to intersection A; 300 cars per hour come to intersection A on 10th Street. A total of x_1 of these cars leave A on M Street, while x_4 cars leave A on 10th Street. The number of cars entering A must equal the number leaving, so that

$$x_1 + x_4 = 700 + 300$$

or

$$x_1 + x_4 = 1000. \qquad (1)$$

For intersection B, x_1 cars enter B on M Street, and x_2 cars enter B on 11th Street. The figure shows that 900 cars leave B on 11th while 200 leave on M. We have

$$x_1 + x_2 = 900 + 200$$

or

$$x_1 + x_2 = 1100. \qquad (2)$$

At intersection C, 400 cars enter on N Street, 300 on 11th Street, while x_2 leave on 11th Street and x_3 leave on N Street. This gives

$$x_2 + x_3 = 400 + 300$$

or

$$x_2 + x_3 = 700. \qquad (3)$$

Finally, intersection D has x_3 cars entering on N and x_4 entering on 10th. There are 400 leaving D on 10th and 200 leaving on N.

(a) Set up an equation for intersection D.

(b) Use equations (1), (2), (3), and your answer to part (a) to set up a system of equations. Solve the system by the Gauss-Jordan method.

(c) Since you got a row of zeros, the system of equations is dependent and does not have a unique solution. Solve equation (1) for x_1 and substitute the result into equation (2).

(d) Solve equation (1) for x_4. What is the largest possible value of x_1 so that x_4 is not negative?

(e) Your answer to part (c) should be $x_2 - x_4 = 100$. Solve the equation for x_4 and then find the smallest possible value of x_2 so that x_4 is not negative.

(f) Your answer to part (a) should be $x_3 + x_4 = 600$. Solve the equation for x_4 and then find the largest possible values of x_3 and x_4 so that neither is negative.

(g) From your answers to parts (d)–(f), what is the maximum value of x_4 so that all the equations are satisfied and all variables are nonnegative? Of x_3? Of x_2? Of x_1?

PREVIEW EXERCISES

Solve each system of equations. See Sections 11.1 and 11.4.

62. $12 = a + 7d$
$27 = a + 14d$

63. $-18 = a + 4d$
$16 = a + 10d$

64. $64 = ar^4$
$4 = ar^2$

65. $ar^2 = 27$
$ar^5 = 729$

CHAPTER 12 GLOSSARY

KEY TERMS

12.1 matrix A matrix is a rectangular array of numbers enclosed by brackets or parentheses in which the position of each number is meaningful.

element of a matrix Each number in a matrix is called an element of the matrix.

square matrix A square matrix has the same number of rows as columns.

row matrix A row matrix has just one row.

column matrix A column matrix has just one column.

determinant Every square matrix is associated with a real number called the determinant of the matrix.

minor The minor of an element of an $n \times n$ matrix is the determinant of an $(n - 1) \times (n - 1)$ matrix, which is a part of the $n \times n$ matrix.

expansion by minors Expansion by minors is a method of evaluating a determinant.

12.3 Cramer's rule Cramer's rule is a method of solving a system of equations using determinants.

12.4 augmented matrix A matrix with a vertical bar that separates the columns into two groups is called an augmented matrix.

Gauss-Jordan method The Gauss-Jordan method is a systematic way of using matrix row transformations to solve a system of equations.

NEW SYMBOLS

$\begin{bmatrix} a & b & c \\ d & e & f \end{bmatrix}$ matrix with 2 rows, 3 columns

$\delta(A)$ determinant of matrix A

$\begin{vmatrix} a & b \\ c & d \end{vmatrix}$ determinant of a 2×2 matrix

$\begin{vmatrix} a & b & c \\ d & e & f \\ g & h & i \end{vmatrix}$ determinant of a 3×3 matrix

CHAPTER 12 QUICK REVIEW

SECTION	CONCEPTS	EXAMPLES
12.1 MATRICES AND DETERMINANTS	**Value of a 2 × 2 Determinant** $\begin{vmatrix} a_{11} & a_{12} \\ a_{21} & a_{22} \end{vmatrix} = a_{11}a_{22} - a_{21}a_{12}.$	$\begin{vmatrix} 3 & 4 \\ -2 & 6 \end{vmatrix} = (3)(6) - (-2)(4) = 26$
	Determinants larger than 2 × 2 are evaluated by expansion by minors about a column or row.	Evaluate $\begin{vmatrix} 2 & -3 & -2 \\ -1 & -4 & -3 \\ -1 & 0 & 2 \end{vmatrix}$ by expanding about the second column.
	Array of Signs for a 3 × 3 Determinant $\begin{matrix} + & - & + \\ - & + & - \\ + & - & + \end{matrix}$	$\begin{vmatrix} 2 & -3 & -2 \\ -1 & -4 & -3 \\ -1 & 0 & 2 \end{vmatrix}$ $= -(-3)(-5) + (-4)(2) - (0)(-8)$ $= -15 - 8 + 0$ $= -23$
12.2 PROPERTIES OF DETERMINANTS	**1.** If every element in a row or column of a matrix is 0, then the determinant equals 0.	$\begin{vmatrix} 1 & 2 & 3 \\ 0 & 0 & 0 \\ 5 & -2 & 4 \end{vmatrix} = 0$
	2. If corresponding rows and columns of a matrix are interchanged, the determinant is not changed.	$\begin{vmatrix} 5 & 4 & 8 \\ 7 & 1 & 0 \\ 2 & 5 & 3 \end{vmatrix} = \begin{vmatrix} 5 & 7 & 2 \\ 4 & 1 & 5 \\ 8 & 0 & 3 \end{vmatrix}$
	3. Interchanging two rows (or columns) of a matrix changes the sign of the determinant.	$\begin{vmatrix} 4 & 8 & 2 \\ 1 & 5 & 7 \\ 0 & 6 & 1 \end{vmatrix} = -\begin{vmatrix} 8 & 4 & 2 \\ 5 & 1 & 7 \\ 6 & 0 & 1 \end{vmatrix}$

SECTION	CONCEPTS	EXAMPLES
	4. If every element of a row (or column) of a matrix is multiplied by the real number k, then the determinant of the new matrix is k times the determinant of the original matrix.	$\begin{vmatrix} 1 & 5 & 1 \\ 2k & 3k & 4k \\ 0 & 6 & 7 \end{vmatrix} = k\begin{vmatrix} 1 & 5 & 1 \\ 2 & 3 & 4 \\ 0 & 6 & 7 \end{vmatrix}$
	5. The determinant of a matrix with two identical rows (or columns) equals 0.	$\begin{vmatrix} 1 & 6 & 8 \\ 2 & 5 & 3 \\ 2 & 5 & 3 \end{vmatrix} = 0$
	6. The determinant of a matrix is unchanged if a multiple of a row (or column) of the matrix is added to the corresponding elements of another row (or column).	$\begin{vmatrix} 0 & 3 & 5 \\ 1 & 6 & 3 \\ 2 & 3 & 4 \end{vmatrix} = \begin{vmatrix} 0 & 3 & 5 \\ 1 & 6 & 3 \\ 0 & -9 & -2 \end{vmatrix}$ $-2R2 + R3$
12.3 SOLUTION OF LINEAR SYSTEMS OF EQUATIONS BY DETERMINANTS—CRAMER'S RULE	**Cramer's Rule for 2 × 2 Systems** Given the system $a_1x + b_1y = c_1$ $a_2x + b_2y = c_2$ with $a_1b_2 - a_2b_1 = D \neq 0$, then $x = \dfrac{\begin{vmatrix} c_1 & b_1 \\ c_2 & b_2 \end{vmatrix}}{\begin{vmatrix} a_1 & b_1 \\ a_2 & b_2 \end{vmatrix}} = \dfrac{D_x}{D}$ and $y = \dfrac{\begin{vmatrix} a_1 & c_1 \\ a_2 & c_2 \end{vmatrix}}{\begin{vmatrix} a_1 & b_1 \\ a_2 & b_2 \end{vmatrix}} = \dfrac{D_y}{D}.$	Solve using Cramer's rule. $x - 2y = -1$ $2x + 5y = 16$ $x = \dfrac{\begin{vmatrix} -1 & -2 \\ 16 & 5 \end{vmatrix}}{\begin{vmatrix} 1 & -2 \\ 2 & 5 \end{vmatrix}} = \dfrac{-5 + 32}{5 + 4} = \dfrac{27}{9} = 3$ $y = \dfrac{\begin{vmatrix} 1 & -1 \\ 2 & 16 \end{vmatrix}}{\begin{vmatrix} 1 & -2 \\ 2 & 5 \end{vmatrix}} = \dfrac{16 + 2}{5 + 4} = \dfrac{18}{9} = 2$ The solution set is $\{(3, 2)\}$.

<table>
<tr><th>SECTION</th><th>CONCEPTS</th><th>EXAMPLES</th></tr>
<tr><td></td><td>

Cramer's Rule for 3 × 3 Systems

Given the system

$$a_1x + b_1y + c_1z = d_1$$
$$a_2x + b_2y + c_2z = d_2$$
$$a_3x + b_3y + c_3z = d_3$$

with

$$D_x = \begin{vmatrix} d_1 & b_1 & c_1 \\ d_2 & b_2 & c_2 \\ d_3 & b_3 & c_3 \end{vmatrix}, \quad D_y = \begin{vmatrix} a_1 & d_1 & c_1 \\ a_2 & d_2 & c_2 \\ a_3 & d_3 & c_3 \end{vmatrix}, \quad D_z = \begin{vmatrix} a_1 & b_1 & d_1 \\ a_2 & b_2 & d_2 \\ a_3 & b_3 & d_3 \end{vmatrix},$$

$$D = \begin{vmatrix} a_1 & b_1 & c_1 \\ a_2 & b_2 & c_2 \\ a_3 & b_3 & c_3 \end{vmatrix} \neq 0,$$

then

$$x = \frac{D_x}{D}, \quad y = \frac{D_y}{D}, \quad z = \frac{D_z}{D}.$$

</td><td>

Solve using Cramer's rule.

$$3x + 2y + z = -5$$
$$x - y + 3z = -5$$
$$2x + 3y + z = 0$$

Using the methods of expansion by minors, it can be shown that $D_x = 45$, $D_y = -30$, $D_z = 0$, and $D = -15$. Therefore,

$$x = \frac{D_x}{D} = \frac{45}{-15} = -3,$$
$$y = \frac{D_y}{D} = \frac{-30}{-15} = 2,$$
$$z = \frac{D_z}{D} = \frac{0}{-15} = 0.$$

The solution set is $\{(-3, 2, 0)\}$.

</td></tr>
<tr><td>12.4 SOLUTION OF LINEAR SYSTEMS OF EQUATIONS BY MATRICES</td><td>

For any augmented matrix of a system of linear equations, the following row transformations will result in the matrix of an equivalent system.

1. Any two rows may be interchanged.
2. The elements of any row may be multiplied by a nonzero real number.
3. Any row may be changed by adding to its elements a multiple of the corresponding elements of another row.

</td><td>

$$\left[\begin{array}{ccc|c} 2 & 1 & 4 & 7 \\ 3 & 0 & 6 & 9 \\ 1 & 5 & 6 & 8 \end{array}\right] \rightarrow \left[\begin{array}{ccc|c} 1 & 5 & 6 & 8 \\ 3 & 0 & 6 & 9 \\ 2 & 1 & 4 & 7 \end{array}\right]$$

Interchange R1 and R3.

$$\left[\begin{array}{ccc|c} 1 & 5 & 6 & 8 \\ 3 & 0 & 6 & 9 \\ 2 & 1 & 4 & 7 \end{array}\right] \rightarrow \left[\begin{array}{ccc|c} 1 & 5 & 6 & 8 \\ 1 & 0 & 2 & 3 \\ 2 & 1 & 4 & 7 \end{array}\right] \tfrac{1}{3}\text{R2}$$

$$\left[\begin{array}{ccc|c} 1 & 5 & 6 & 8 \\ 1 & 0 & 2 & 3 \\ 2 & 1 & 4 & 7 \end{array}\right] \rightarrow \left[\begin{array}{ccc|c} 1 & 5 & 6 & 8 \\ 0 & -5 & -4 & -5 \\ 2 & 1 & 4 & 7 \end{array}\right] -1\text{R1} + \text{R2}$$

</td></tr>
</table>

CHAPTER 12 REVIEW EXERCISES

[12.1 *Evaluate each of the following determinants.*

1. $\begin{vmatrix} -1 & 8 \\ 2 & 9 \end{vmatrix}$
2. $\begin{vmatrix} -2 & 4 \\ 0 & 3 \end{vmatrix}$
3. $\begin{vmatrix} 4 & 6 \\ 6 & 9 \end{vmatrix}$
4. $\begin{vmatrix} 3 & 7 \\ -2 & 1 \end{vmatrix}$
5. $\begin{vmatrix} -2 & 4 & 1 \\ 3 & 0 & 2 \\ -1 & 0 & 3 \end{vmatrix}$
6. $\begin{vmatrix} -1 & 2 & 3 \\ 4 & 0 & 3 \\ 5 & -1 & 2 \end{vmatrix}$
7. $\begin{vmatrix} 2 & -1 & 0 \\ 0 & 3 & 1 \\ -2 & 0 & 1 \end{vmatrix}$
8. $\begin{vmatrix} -1 & 7 & 2 \\ 3 & 0 & 5 \\ -1 & 2 & 6 \end{vmatrix}$

Solve each of the following determinant equations for x.

9. $\begin{vmatrix} -3 & 2 \\ 1 & x \end{vmatrix} = 5$
10. $\begin{vmatrix} 3x & 7 \\ -x & 4 \end{vmatrix} = 8$
11. $\begin{vmatrix} 2 & 5 & 0 \\ 1 & 3x & -1 \\ 0 & 2 & 0 \end{vmatrix} = 4$
12. $\begin{vmatrix} 6x & 2 & 0 \\ 1 & 5 & 3 \\ x & 2 & -1 \end{vmatrix} = 2x$

[12.2] *Give the property that justifies each of the following statements.*

13. $\begin{vmatrix} 8 & 9 & 2 \\ 0 & 0 & 0 \\ 3 & 1 & 4 \end{vmatrix} = 0$
14. $\begin{vmatrix} 4 & 6 \\ 3 & 5 \end{vmatrix} = \begin{vmatrix} 4 & 3 \\ 6 & 5 \end{vmatrix}$
15. $\begin{vmatrix} 8 & 2 \\ 4 & 3 \end{vmatrix} = 2\begin{vmatrix} 4 & 1 \\ 4 & 3 \end{vmatrix}$
16. $\begin{vmatrix} 4 & 6 & 2 \\ -3 & 8 & -5 \\ 4 & 6 & 2 \end{vmatrix} = 0$
17. $\begin{vmatrix} 5 & -1 & 2 \\ 3 & -2 & 0 \\ -4 & 1 & 2 \end{vmatrix} = \begin{vmatrix} 5 & -1 & 2 \\ 8 & -3 & 2 \\ -4 & 1 & 2 \end{vmatrix}$
18. $\begin{vmatrix} 8 & 2 & -5 \\ -3 & 1 & 4 \\ 2 & 0 & 5 \end{vmatrix} = -\begin{vmatrix} 8 & -5 & 2 \\ -3 & 4 & 1 \\ 2 & 5 & 0 \end{vmatrix}$

Use Determinant Property 6 to evaluate the following.

19. $\begin{vmatrix} 5 & -1 & 2 \\ 3 & -2 & 0 \\ -4 & 1 & 2 \end{vmatrix}$
20. $\begin{vmatrix} 8 & 2 & -5 \\ -3 & 1 & 4 \\ 2 & 0 & 5 \end{vmatrix}$

[12.3] *Use Cramer's rule to solve the following systems of equations.*

21. $3x - 4y = 5$; $2x + y = 8$
22. $4x - 3y = 12$; $2x + 6y = 15$
23. $5x + 6y = 10$; $4x - 3y = 12$
24. $2x - 5y = 8$; $3x + 4y = 10$
25. $x + 2y + 5z = 9$; $2x - y = 3$; $3x + 2y + z = 9$
26. $4x + y + z = 11$; $x - y - z = 4$; $y + 2z = 0$
27. $-x + 3y - 4z = 2$; $2x + 4y + z = 3$; $3x - z = 9$
28. $5x + y = 10$; $3x + 2y + z = -3$; $-y - 2z = -13$

[12.4] *Use the Gauss-Jordan method to solve each of the following systems.*

29. $2x + 3y = 10$; $-3x + y = 18$
30. $5x + 2y = -10$; $3x - 5y = -6$
31. $3x + y = -7$; $x - y = -5$

32. $3x + 5y = 17$
$4x = y - 8$

33. $x - z = -3$
$y + z = 6$
$2x - 3z = -9$

34. $2x - y + 4z = -1$
$-3x + 5y - z = 5$
$2x + 3y + 2z = 3$

35. $x + 3y + 2z = 11$
$3x + 7y + 4z = 23$
$5x + 3y - 5z = -14$

36. $5x - 8y + z = 1$
$3x - 2y + 4z = 3$
$10x - 16y + 2z = 3$

CHAPTER 12 TEST

Evaluate the following determinants.

1. $\begin{vmatrix} -3 & 2 \\ -1 & 4 \end{vmatrix}$

2. $\begin{vmatrix} 1 & 2 \\ 5 & 4 \end{vmatrix}$

3. $\begin{vmatrix} 10 & 0 \\ -1 & 2 \end{vmatrix}$

4. $\begin{vmatrix} -1 & 7 & 2 \\ 3 & 0 & 5 \\ -1 & 2 & 6 \end{vmatrix}$

5. $\begin{vmatrix} 2 & 5 & 0 \\ 1 & -3 & 4 \\ 0 & 2 & -1 \end{vmatrix}$

6. $\begin{vmatrix} 0 & 5 & 3 \\ 7 & 0 & -2 \\ 0 & -1 & 6 \end{vmatrix}$

Decide if each of the following is true or false. Give a property of determinants to justify each true statement.

7. $\begin{vmatrix} 5 & 8 \\ -1 & 3 \end{vmatrix} = \begin{vmatrix} 5 & -1 \\ 8 & 3 \end{vmatrix}$

8. $-2\begin{vmatrix} 7 & 4 \\ 3 & 1 \end{vmatrix} = \begin{vmatrix} -14 & -8 \\ -6 & -2 \end{vmatrix}$

9. $\begin{vmatrix} 1 & 2 & 3 \\ 5 & 4 & 6 \\ 0 & -2 & 1 \end{vmatrix} = -\begin{vmatrix} 2 & 1 & 3 \\ 4 & 5 & 6 \\ -2 & 0 & 1 \end{vmatrix}$

10. $\begin{vmatrix} 2 & 0 & 5 \\ 1 & 4 & 6 \\ 3 & 1 & 2 \end{vmatrix} = \begin{vmatrix} 2 & 0 & 5 \\ 1 & 4 & 6 \\ 0 & -11 & -16 \end{vmatrix}$

Use determinant Property 6 to find the value of each determinant.

11. $\begin{vmatrix} 7 & 4 & 1 \\ -3 & 1 & -1 \\ 2 & 5 & 3 \end{vmatrix}$

12. $\begin{vmatrix} 5 & 2 & 1 \\ 0 & 4 & -1 \\ 3 & 1 & 0 \end{vmatrix}$

Solve by Cramer's rule.

13. $2x + 6y = 3$
$-3x + y = 8$

14. $3x + 2y = 4$
$5x + 5y = 9$

15. $x + y + z = 4$
$2x - z = -5$
$3y + z = 9$

16. $2x + y + z = 9$
$-x + 4y + 3z = 2$
$x - y + 2z = -3$

Solve by the Gauss-Jordan method.

17. $x + y = 4$
$2x + 3y = 10$

18. $2x + y = 5$
$3x - 2y = 4$

19. $x + y + z = 1$
$2x - y = -2$
$3y + z = 2$

20. $3x - 2y + 4z = 1$
$4x + y - 5z = 2$
$-6x + 4y - 8z = -2$

CHAPTER THIRTEEN

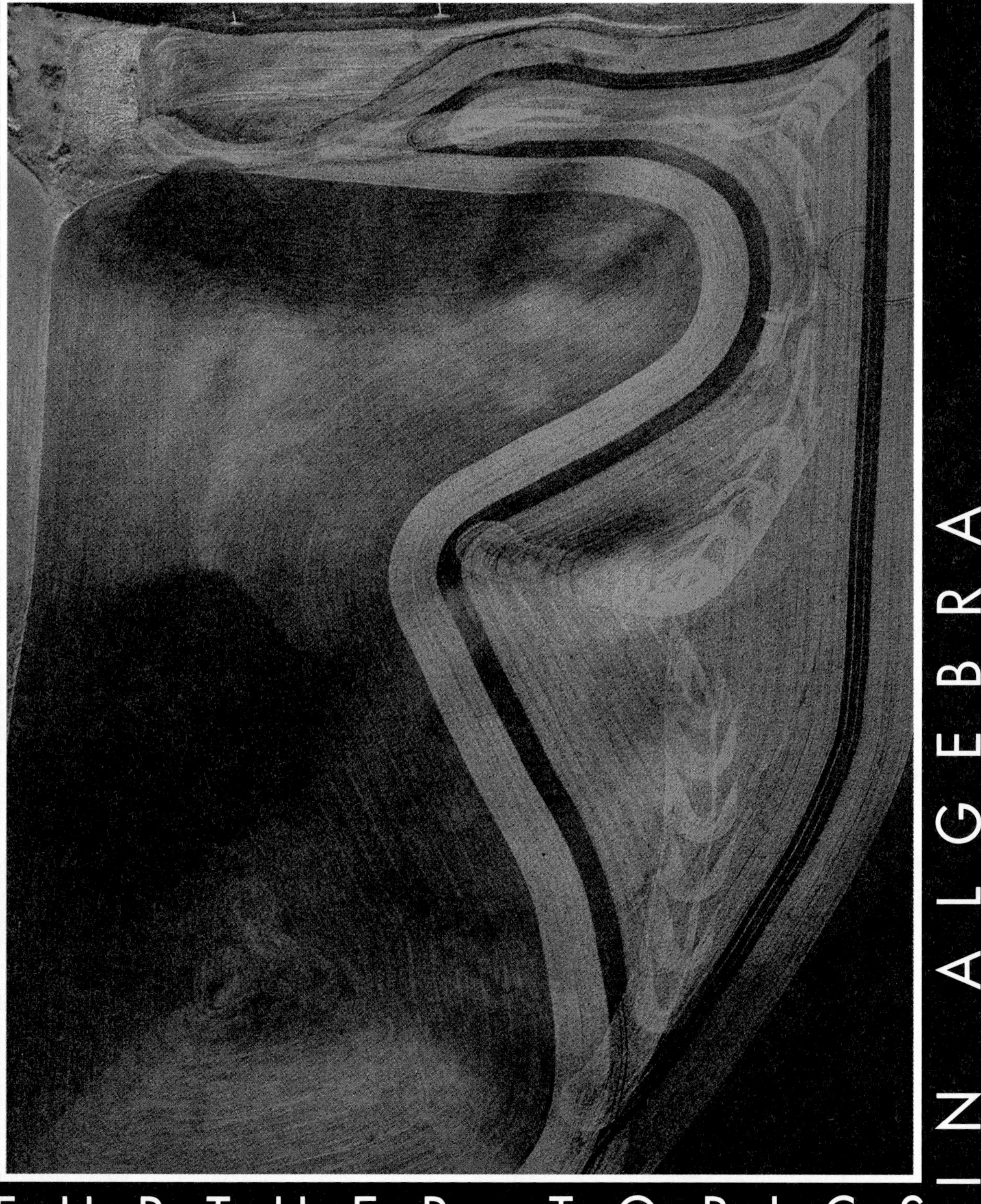

FURTHER TOPICS IN ALGEBRA

This chapter concludes our study of algebra for college students by discussing a variety of topics: sequences, series, the binomial theorem, mathematical induction, permutations, combinations, and probability. A *sequence* is a list of terms in a specific order, such as 1, 3, 5, 7, 9, ... A *series* is the sum of the terms of a sequence. The *binomial theorem* allows us to find the expansion of a binomial raised to a power, such as $(2x - y)^5$, without actually having to perform the multiplications. *Mathematical induction* is a powerful method that is used to prove statements that are true for all positive integers. Counting principles (*permutations* and *combinations*) involve finding the number of possible arrangements or groupings of a set of objects. We use them to answer questions like "How many different poker hands can be dealt with a 52-card deck?" and "How many license plates can be made using two letters followed by three numbers?" *Probability theory* gives us information of the likelihood of the occurrence of certain events.

13.1 INTRODUCTION TO SEQUENCES; ARITHMETIC SEQUENCES

OBJECTIVES

1. DEFINE *SEQUENCE*.
2. FIND TERMS OF A SEQUENCE GIVEN THE GENERAL TERM OR A RECURSION FORMULA.
3. FIND THE COMMON DIFFERENCE AND TERMS FOR AN ARITHMETIC SEQUENCE.
4. FIND SPECIFIC TERMS OF AN ARITHMETIC SEQUENCE.

1 If the domain of a function f is the set of positive integers, the range elements can be *ordered*, as $f(1)$, $f(2)$, $f(3)$, and so on. This ordered list of numbers is called a **sequence.** Since the letter x has been used to suggest real numbers, the variable n is used instead with sequences to suggest the positive integer domain. Although a sequence may be defined by $f(n) = 2n + 3$, for example, it is customary to use a_n instead of $f(n)$ and write $a_n = 2n + 3$.

2 The elements of a sequence, called the **terms** of the sequence, are written in order as $a_1, a_2, a_3, \ldots$. The **general term,** or ***n*th term,** of the sequence is a_n. The general term of a sequence is used to find any term of the sequence. For example, if $a_n = n + 1$, then $a_2 = 2 + 1 = 3$.

A sequence is a **finite sequence** if the domain is the set $\{1, 2, 3, 4, \ldots, n\}$, where n is a positive integer. An **infinite sequence** has the set of all positive integers as its domain. For example, the sequence of positive even integers,

$$2, 4, 6, 8, 10, 12, 14, \ldots,$$

is infinite, but the sequence of days in June,

$$1, 2, 3, 4, \ldots, 29, 30,$$

is finite.

EXAMPLE 1
FINDING THE TERMS OF A SEQUENCE FROM THE GENERAL TERM

Write the first five terms for each of the following sequences.

(a) $a_n = \dfrac{n+1}{n+2}$

Replacing n, in turn, with 1, 2, 3, 4, and 5 gives

$$\frac{2}{3}, \quad \frac{3}{4}, \quad \frac{4}{5}, \quad \frac{5}{6}, \quad \frac{6}{7}.$$

(b) $a_n = (-1)^n \cdot n$

Replace n, in turn, with 1, 2, 3, 4, and 5, to get

$$a_1 = (-1)^1 \cdot 1 = -1$$
$$a_2 = (-1)^2 \cdot 2 = 2$$
$$a_3 = (-1)^3 \cdot 3 = -3$$
$$a_4 = (-1)^4 \cdot 4 = 4$$
$$a_5 = (-1)^5 \cdot 5 = -5.$$

(c) $b_n = \dfrac{(-1)^n}{2^n}$

Here $b_1 = -1/2$, $b_2 = 1/4$, $b_3 = -1/8$, $b_4 = 1/16$, and $b_5 = -1/32$. ■

Sequences can be defined using recursion formulas. A **recursion formula** defines the nth term of a sequence in terms of the previous term. For example, if the first term of a sequence is $a_1 = 2$ and the nth term is $a_n = a_{n-1} + 3$, then $a_2 = a_1 + 3 = 2 + 3 = 5$, $a_3 = a_2 + 3 = 5 + 3 = 8$, and so on.

EXAMPLE 2
USING A RECURSION FORMULA

Find the first four terms for the sequences defined as follows.

(a) $a_1 = 4$; $a_{n+1} = 2a_n + 1$

We have $a_1 = 4$, so

$$a_2 = 2 \cdot a_1 + 1 = 2 \cdot 4 + 1 = 9,$$
$$a_3 = 2 \cdot a_2 + 1 = 2 \cdot 9 + 1 = 19,$$

and

$$a_4 = 2 \cdot a_3 + 1 = 2 \cdot 19 + 1 = 39.$$

(b) $a_1 = 2$; $a_{n+1} = a_n + n - 1$

$$a_1 = 2$$
$$a_2 = a_1 + 1 - 1 = 2 + 0 = 2 \qquad n = 1$$
$$a_3 = a_2 + 2 - 1 = 2 + 1 = 3 \qquad n = 2$$
$$a_4 = a_3 + 3 - 1 = 3 + 2 = 5 \qquad n = 3$$

■

3 A sequence in which each term after the first is obtained by adding a fixed number to the preceding term is called an **arithmetic sequence** (or **arithmetic progression**). The fixed number that is added is called the **common difference.** The sequence

$$5, 9, 13, 17, 21, \ldots$$

is an arithmetic sequence since each term after the first is obtained by adding 4 to the previous term. That is,

$$\begin{aligned} 9 &= 5 + 4 \\ 13 &= 9 + 4 \\ 17 &= 13 + 4 \\ 21 &= 17 + 4, \end{aligned}$$

and so on. The common difference is 4.

If the common difference of an arithmetic sequence is d, then by the definition of an arithmetic sequence,

$$a_{n+1} - a_n = d$$

for every positive integer n in its domain.

EXAMPLE 3
FINDING THE COMMON DIFFERENCE FOR AN ARITHMETIC SEQUENCE

Find the common difference, d, for the arithmetic sequence

$$-9, -7, -5, -3, -1, \ldots.$$

Since this sequence is arithmetic, d can be found by choosing any two adjacent terms and subtracting the first from the second. Choosing -7 and -5 gives

$$d = -5 - (-7) = 2.$$

Choosing -9 and -7 would give $d = -7 - (-9) = 2$, the same result. ■

If a_1 and d are known, then all the terms of an arithmetic sequence can be found.

EXAMPLE 4
FINDING TERMS OF AN ARITHMETIC SEQUENCE

Write the first five terms for each of the following arithmetic sequences.

(a) The first term is 7, and each succeeding term is found by adding -3 to the preceding term.

Here

$$\begin{aligned} a_1 &= 7, \\ a_2 &= 7 + (-3) = 4, \\ a_3 &= 4 + (-3) = 1, \\ a_4 &= 1 + (-3) = -2, \\ a_5 &= -2 + (-3) = -5. \end{aligned}$$

(b) $a_1 = -12$, $d = 5$

Starting with a_1, add d to each term to get the next term.

$$a_1 = -12$$
$$a_2 = -12 + d = -12 + 5 = -7$$
$$a_3 = -7 + d = -7 + 5 = -2$$
$$a_4 = -2 + d = -2 + 5 = 3$$
$$a_5 = 3 + d = 3 + 5 = 8 \quad \blacksquare$$

4 If a_1 is the first term of an arithmetic sequence and d is the common difference, then the terms of the sequence are given by

$$a_1 = a_1$$
$$a_2 = a_1 + d$$
$$a_3 = a_2 + d = a_1 + d + d = a_1 + 2d$$
$$a_4 = a_3 + d = a_1 + 2d + d = a_1 + 3d$$
$$a_5 = a_1 + 4d$$
$$a_6 = a_1 + 5d,$$

and, by this pattern, $a_n = a_1 + (n - 1)d$.

nTH TERM OF AN ARITHMETIC SEQUENCE

In an arithmetic sequence with first term a_1 and common difference d, the nth term, a_n, is given by

$$a_n = a_1 + (n - 1)d.$$

EXAMPLE 5 **USING THE FORMULA FOR THE nTH TERM OF AN ARITHMETIC SEQUENCE**

Find a_{13} and a_n for the arithmetic sequence

$$-3, 1, 5, 9, \ldots.$$

Here $a_1 = -3$ and $d = 1 - (-3) = 4$. To find a_{13}, substitute 13 for n in the preceding formula.

$$a_{13} = a_1 + (13 - 1)d$$
$$a_{13} = -3 + (12)4$$
$$a_{13} = -3 + 48$$
$$a_{13} = 45$$

Find a_n by substituting values for a_1 and d in the formula for a_n.

$$a_n = -3 + (n - 1) \cdot 4$$
$$a_n = -3 + 4n - 4 \qquad \text{Distributive property}$$
$$a_n = 4n - 7 \quad \blacksquare$$

EXAMPLE 6 **USING THE FORMULA FOR THE nTH TERM OF AN ARITHMETIC SEQUENCE**

Find a_{18} and a_n for the arithmetic sequence having $a_2 = 9$ and $a_3 = 15$.

Find d first: $d = a_3 - a_2 = 15 - 9 = 6$.

Since
$$a_2 = a_1 + d,$$
$$9 = a_1 + 6 \quad \text{and} \quad a_1 = 3.$$

Then $a_{18} = 3 + (18 - 1) \cdot 6$

$a_{18} = 3 + 17 \cdot 6 = 105,$

and $a_n = 3 + (n - 1) \cdot 6$

$a_n = 3 + 6n - 6$ Distributive property

$a_n = 6n - 3.$ ■

EXAMPLE 7
USING THE FORMULA FOR THE nTH TERM OF AN ARITHMETIC SEQUENCE

Suppose that an arithmetic sequence has $a_8 = -16$ and $a_{16} = -40$. Find a_1.

Since $a_8 = a_1 + (8 - 1)d$, replacing a_8 with -16 gives $-16 = a_1 + 7d$ or $a_1 = -16 - 7d$. Similarly, $-40 = a_1 + 15d$ or $a_1 = -40 - 15d$. From these two equations, using the substitution method from Chapter 11,

$$-16 - 7d = -40 - 15d,$$

so $d = -3$. To find a_1, substitute -3 for d in $a_1 = -16 - 7d$:

$a_1 = -16 - 7d$

$a_1 = -16 - 7(-3)$ Let $d = -3$.

$a_1 = 5.$

We could have used $a_1 = -40 - 15d$ instead. ■

13.1 EXERCISES

Write the first five terms of each of the following sequences. See Example 1.

1. $a_n = 6n + 4$ **2.** $a_n = 3n - 2$ **3.** $a_n = 2^n$ **4.** $a_n = 3^n$

5. $a_n = (-1)^{n+1}$ **6.** $a_n = (-1)^n(n + 2)$ **7.** $a_n = \dfrac{2n}{n + 3}$ **8.** $a_n = \dfrac{-4}{n + 5}$

9. $a_n = \dfrac{8n - 4}{2n + 1}$ **10.** $a_n = \dfrac{-3n + 6}{n + 1}$ **11.** $a_n = (-2)^n(n)$ **12.** $a_n = (-1/2)^n(n^{-1})$

13. $a_n = \dfrac{n^2 + 1}{n^2 + 2}$ **14.** $a_n = \dfrac{(-1)^{n-1}(n + 1)}{n + 2}$ **15.** $a_n = x^n$ **16.** $a_n = n \cdot x^{-n}$

Find the first ten terms for the sequences defined as follows. See Example 2.

17. $a_1 = 4, \quad a_n = a_{n-1} + 5$ **18.** $a_1 = -3, \quad a_n = a_{n-1} + 2$ **19.** $a_1 = -9, \quad a_n = a_{n-1} - 4$

20. $a_1 = -8, \quad a_n = a_{n-1} - 7$ **21.** $a_1 = 2, \quad a_n = 2 \cdot a_{n-1}$ **22.** $a_1 = -3, \quad a_n = 2 \cdot a_{n-1}$

23. $a_1 = 1, \quad a_2 = 1, \quad a_n = a_{n-1} + a_{n-2}$ (the Fibonacci sequence)

24. $a_1 = 3, \quad a_2 = 2, \quad a_n = a_{n-1} - a_{n-2}$

For each of the following arithmetic sequences, write the indicated number of terms. See Example 4.

25. $a_1 = 4, d = 2, n = 5$ **26.** $a_1 = 6, d = 8, n = 4$ **27.** $a_2 = 9, d = -2, n = 4$

28. $a_3 = 7, d = -4, n = 4$ **29.** $a_3 = -2, d = -4, n = 4$ **30.** $a_2 = -12, d = -6, n = 5$

31. $a_3 = 6, a_4 = 12, n = 6$ **32.** $a_5 = 8, a_6 = 5, n = 6$

For each of the following sequences that are arithmetic, find d and a_n. See Examples 3 and 5.

33. 12, 17, 22, 27, 32, 37, . . .

34. 8, 17, 26, 35, 44, 53, . . .

35. 18, 15, 12, 9, 6, . . .

36. 30, 24, 18, 12, . . .

37. $-19, -12, -5, 2, 9, \dots$

38. $-30, -20, -12, -6, -2, \dots$

39. $x, x + m, x + 2m, x + 3m, x + 4m, \dots$

40. $k + p, k + 2p, k + 3p, k + 4p, \dots$

41. $2z + m, 2z, 2z - m, 2z - 2m, 2z - 3m, \dots$

42. $3r - 4z, 3r - 3z, 3r - 2z, 3r - z, 3r, 3r + z, \dots$

Find a_8 and a_n for each of the following arithmetic sequences. See Example 6.

43. $a_1 = 5, d = 2$

44. $a_1 = -3, d = -4$

45. $a_3 = 2, d = 1$

46. $a_4 = 5, d = -2$

47. $a_1 = 8, a_2 = 6$

48. $a_1 = 6, a_2 = 3$

49. $a_{10} = 6, a_{12} = 15$

50. $a_{15} = 8, a_{17} = 2$

51. $a_1 = x, a_2 = x + 3$

52. $a_2 = y + 1, d = -3$

53. $a_6 = 2m, a_7 = 3m$

54. $a_5 = 4p + 1, a_7 = 6p + 7$

---

55. Give an example of an arithmetic sequence with $a_4 = 12$. (See Objective 1.)

56. Explain in your own words what is meant by *arithmetic sequence*. (See Objective 3.)

57. Which one of the following is not an arithmetic sequence? (See Objective 3.)

(a) 4, 6, 8, 10, . . . **(b)** -2, 6, 14, 22, . . .
(c) 1/2, 1, 3/2, 2, . . . **(d)** 5, 10, 20, 40, . . .

58. Refer to the sequence in Exercise 57 that is not arithmetic. Explain in your own words how each term after the first is determined by using the previous term of the sequence.

Find a_1 for each of the following arithmetic sequences. See Example 7.

59. $a_9 = 47, \quad a_{15} = 77$

60. $a_{10} = 50, \quad a_{20} = 110$

61. $a_{15} = 168, \quad a_{16} = 180$

62. $a_{10} = -54, \quad a_{17} = -89$

Explain why each of the following sequences is arithmetic.

63. log 2, log 4, log 8, log 16, log 32, . . .

64. log 12, log 36, log 108, log 324, . . .

*In statistics, the **arithmetic mean** of a list of numbers, $x_1, x_2, x_3, \dots, x_n$ is defined as*

$$\bar{x} = \frac{x_1 + x_2 + x_3 + \cdots + x_n}{n}.$$

That is, to find the arithmetic mean, add the numbers and divide this sum by the number of terms in the list. For example, to find the arithmetic mean of 9, 12, 18, 24, 20, 22, *and* 14, *add these numbers and divide by $n = 7$.*

$$\bar{x} = \frac{9 + 12 + 18 + 24 + 20 + 22 + 14}{7} = \frac{119}{7} = 17$$

Find the arithmetic mean for each of the following.

65. 10, 9, 6, 12, 8, 7, 7, 7, 8, 6

66. 124, 132, 129, 135, 130

67. 28.1, 29.2, 31.5, 26.0, 28.9, 30.4

68. .002, .005, .001, .003, .004, .002

PREVIEW EXERCISES

Evaluate ar^n for the given values of a, r, and n. See Section 2.2.

69. $a = 2, \quad r = 3, \quad n = 2$

70. $a = 3, \quad r = 2, \quad n = 4$

71. $a = 4, \quad r = \frac{1}{2}, \quad n = 3$

72. $a = 5, \quad r = \frac{1}{4}, \quad n = 2$

73. $a = .1, \quad r = .1, \quad n = 5$

74. $a = .2, \quad r = .2, \quad n = 4$

13.2 GEOMETRIC SEQUENCES

OBJECTIVES

1 DEFINE *GEOMETRIC SEQUENCE.*

2 FIND THE COMMON RATIO FOR A GEOMETRIC SEQUENCE.

3 FIND SPECIFIC TERMS OF A GEOMETRIC SEQUENCE.

1 A **geometric sequence** (or **geometric progression**) is a sequence in which each term after the first is obtained by multiplying the preceding term by a constant nonzero real number, called the **common ratio.** An example of a geometric sequence is 2, 8, 32, 128, . . . in which the first term is 2 and the common ratio is 4.

2 If the common ratio of a geometric sequence is r, then by the definition of a geometric sequence,

$$\frac{a_{n+1}}{a_n} = r$$

for every positive integer n in its domain. By this definition, the common ratio can be found by choosing any term except the first and dividing it by the preceding term.

EXAMPLE 1 FINDING THE COMMON RATIO FOR A GEOMETRIC SEQUENCE

Find the common ratio, r, for the geometric sequence

$$15, \frac{15}{2}, \frac{15}{4}, \frac{15}{8}, \ldots .$$

Find r by choosing *any* two adjacent terms and dividing the second one by the first. Choosing the second and third terms of the sequence gives

$$r = \frac{15}{4} \div \frac{15}{2} = \frac{1}{2}.$$

Additional terms of the sequence can be found by multiplying each successive term by 1/2. ■

As mentioned above, the geometric sequence, 2, 8, 32, 128, . . . has $r = 4$. Notice that

$$8 = 2 \cdot 4$$
$$32 = 8 \cdot 4 = (2 \cdot 4) \cdot 4 = 2 \cdot 4^2$$
$$128 = 32 \cdot 4 = (2 \cdot 4^2) \cdot 4 = 2 \cdot 4^3.$$

To generalize this result, assume that a geometric sequence has first term a_1 and common ratio r. The second term can be written as $a_2 = a_1 r$, the third as $a_3 = a_2 r = (a_1 r)r = a_1 r^2$, and so on. Following this pattern, the nth term is $a_n = a_1 r^{n-1}$.

nTH TERM OF A GEOMETRIC SEQUENCE

In the geometric sequence with first term a_1 and common ratio r, the nth term is

$$a_n = a_1 r^{n-1}.$$

3 The above formula is used to find a particular term of a geometric sequence in the next example.

EXAMPLE 2
USING THE FORMULA FOR THE nTH TERM OF A GEOMETRIC SEQUENCE

Find the sixth term of the geometric sequence with first term 8, and common ratio $-1/2$.

In this sequence, we have

$$a_1 = 8,$$

$$r = -\frac{1}{2},$$

and

$$n = 6.$$

Substituting into the formula, we get

$$a_n = a_1 r^{n-1}$$

$$a_6 = 8\left(-\frac{1}{2}\right)^{6-1}$$

$$= 8\left(-\frac{1}{2}\right)^5 = 8\left(-\frac{1}{32}\right) = -\frac{1}{4}. \quad \blacksquare$$

EXAMPLE 3
USING THE FORMULA FOR THE nTH TERM OF A GEOMETRIC SEQUENCE

Find a_5 and a_n for each of the following geometric sequences.

(a) 4, 12, 36, 108, . . .

The first term, a_1, is 4. Find r by choosing any term except the first and dividing it by the preceding term. For example,

$$r = 36/12 = 3.$$

Since $a_4 = 108$, $a_5 = 3 \cdot 108 = 324$. The fifth term also could be found using the formula for a_n, $a_n = a_1 r^{n-1}$, and replacing n with 5, r with 3, and a_1 with 4.

$$a_5 = 4 \cdot (3)^{5-1} = 4 \cdot 3^4 = 324$$

By the formula,

$$a_n = 4 \cdot 3^{n-1}.$$

(b) 64, 32, 16, 8, . . .
Here $r = 8/16 = 1/2$, and $a_4 = 8$, so $a_5 = (8)(1/2) = 4$. Alternatively,

$$a_5 = 64\left(\frac{1}{2}\right)^{5-1} = 64\left(\frac{1}{16}\right) = 4.$$

Also,
$$a_n = 64\left(\frac{1}{2}\right)^{n-1}.$$
■

EXAMPLE 4
USING THE FORMULA FOR THE *n*TH TERM OF A GEOMETRIC SEQUENCE

Find a_1 and r for each of the following geometric sequences.

(a) The third term is 20 and the sixth term is -160.
Use the formula for the nth term of a geometric sequence.

$$\text{For } n = 3, \quad a_3 = a_1 r^2 = 20;$$
$$\text{for } n = 6, \quad a_6 = a_1 r^5 = -160.$$

We have $a_1 r^2 = 20$, so that $a_1 = 20/r^2$. Substitute this in the second equation and solve for r.

$$a_1 r^5 = -160$$
$$\left(\frac{20}{r^2}\right) r^5 = -160$$
$$20r^3 = -160$$
$$r^3 = -8$$
$$r = -2 \qquad \sqrt[3]{-8} = -2$$

Since $a_1 r^2 = 20$ and $r = -2$, we have $a_1 = \dfrac{20}{r^2} = \dfrac{20}{(-2)^2} = 5$.

(b) $a_5 = 15$ and $a_7 = 375$
First substitute $n = 5$ and then $n = 7$ into $a_n = a_1 r^{n-1}$.

$$a_5 = a_1 r^4 = 15 \qquad \text{and} \qquad a_7 = a_1 r^6 = 375$$

Solve the first equation for a_1 to get $a_1 = 15/r^4$. Then substitute for a_1 in the second equation.

$$a_1 r^6 = 375$$
$$\frac{15}{r^4} \cdot r^6 = 375$$
$$15r^2 = 375$$
$$r^2 = 25$$
$$r = \pm 5 \qquad \text{Square root property}$$

Either 5 or -5 can be used for r. To find a_1, use

$$a_1 = \frac{15}{r^4}.$$

Replace r with ± 5.

$$a_1 = \frac{15}{(\pm 5)^4} = \frac{15}{625} = \frac{3}{125}$$

There are two sequences that satisfy the given conditions: one with $a_1 = 3/125$ and $r = 5$ and the other with $a_1 = 3/125$ and $r = -5$. ■

13.2 EXERCISES

For each of the following, write the first n terms of the geometric sequence that satisfy the given conditions.

1. $a_1 = 2, \quad r = 3, \quad n = 4$
2. $a_1 = 4, \quad r = 2, \quad n = 5$
3. $a_1 = 1/2, \quad r = 4, \quad n = 4$
4. $a_1 = 2/3, \quad r = 6, \quad n = 3$
5. $a_1 = -2, \quad r = -3, \quad n = 4$
6. $a_1 = -4, \quad r = 2, \quad n = 5$
7. $a_1 = 3125, \quad r = 1/5, \quad n = 7$
8. $a_1 = 729, \quad r = 2/3, \quad n = 5$
9. $a_1 = -1, \quad r = -1, \quad n = 6$
10. $a_1 = -1, \quad r = 1, \quad n = 5$
11. $a_3 = 6, \quad a_4 = 12, \quad n = 5$
12. $a_2 = 9, \quad a_3 = 3, \quad n = 4$

Find a_5 and a_n for each of the following geometric sequences. See Example 3.

13. $a_1 = 4, \quad r = 3$
14. $a_1 = 8, \quad r = 4$
15. $a_1 = -2, \quad r = 3$
16. $a_1 = -5, \quad r = 4$
17. $a_1 = -3, \quad r = -5$
18. $a_1 = -4, \quad r = -2$
19. $a_2 = 3, \quad r = 2$
20. $a_3 = 6, \quad r = 3$
21. $a_4 = 64, \quad r = -4$
22. $a_4 = 81, \quad r = -3$

For each of the following sequences that are geometric, find r and a_n. (In Exercises 31–34, r is given.) See Examples 1, 3, and 4.

23. 6, 12, 24, 48, . . .
24. 4, 16, 64, 256, . . .
25. 3/4, 3/2, 3, 6, 12, . . .
26. 5/6, 5/3, 10/3, 20/3, 40/3, . . .
27. −4, 2, −1, 1/2, . . .
28. 49, −7, 1, −1/7, . . .
29. 18, 20, 24, 32, 48, . . .
30. −7, −5, −3, −1, 1, 3, . . .
31. $a_3 = 9, \quad r = 2$
32. $a_5 = 6, \quad r = 1/2$
33. $a_3 = -2, \quad r = 3$
34. $a_2 = 5, \quad r = 1/2$

---

35. Give an example of a geometric sequence with $a_3 = 15$. (See Objective 1.)

36. Explain in your own words what is meant by *geometric sequence*. (See Objective 1.)

37. Which one of the following is not a geometric sequence? (See Objective 1.)
(a) 1, −2, 4, −8, . . .
(b) 1, 10, 100, 1000, . . .
(c) 3, 3/2, 3/4, 3/8, . . .
(d) 1, 1, 2, 3, 5, 8, . . .

38. Refer to the sequence in Exercise 37 that is not geometric. Explain in your own words how each term after the first two terms is determined. (This is a famous sequence known as the Fibonacci sequence.)

Find a_1 and r for each of the following geometric sequences. See Example 4.

39. $a_2 = 6, \quad a_6 = 486$
40. $a_3 = -12, \quad a_6 = 96$
41. $a_2 = 64, \quad a_8 = 1$
42. $a_2 = 100, \quad a_5 = 1/10$

Explain why the following sequences are geometric.

43. $\log 6, \log 6^2, \log 6^4, \log 6^8, \ldots$
44. $\log 2, \log 4, \log 16, \log 256, \ldots$

PREVIEW EXERCISES

Evaluate each of the following sums.

45. $2 + 4 + 6 + 8 + 10$

46. $-8 + (-2) + 4 + 10 + 16$

47. $10 - 5 + 0 - 5 + 10$

48. $1/2 + 1/4 + 1/8 + 1/16 + 1/32$

13.3 SERIES AND APPLICATIONS

OBJECTIVES

1. USE SUMMATION NOTATION.
2. FIND THE SUM OF A SPECIFIED NUMBER OF TERMS OF AN ARITHMETIC SEQUENCE.
3. FIND THE SUM OF A SPECIFIED NUMBER OF TERMS OF A GEOMETRIC SEQUENCE.
4. SOLVE APPLIED PROBLEMS INVOLVING SEQUENCES.

FOCUS ON PROBLEM SOLVING

Fruit and vegetable dealer Olive Greene paid 10¢ per pound for 10,000 pounds of onions. Each week the price she charges increases by .1¢ per pound, while her onions lose 5% of their weight. If she sells all the onions at the end of the 6th week, does she make or lose money? How much?

The first two sections of this chapter showed how to find specific terms of sequences. In this section we develop formulas to find sums of terms of sequences. The sum of the terms of a sequence is called a *series*. With these formulas we are able to solve problems that lead to sequences and series. This problem is Exercise 75 in the exercises of this section. After working through this section, you should be able to help Olive Greene determine whether she makes or loses money.

1 The sum of the terms of a sequence is called a **series.** A compact shorthand notation can be used to write a series. For example,

$$1 + 5 + 9 + 13 + 17 + 21$$

is the sum of the first six terms in the sequence with general term $a_n = 4n - 3$. This series can be written as

$$\sum_{i=1}^{6} (4i - 3)$$

(read "the sum from $i = 1$ to 6 of $4i - 3$").

The Greek letter sigma, Σ, is used to mean "sum." To evaluate this sum, replace i in $4i - 3$ first by 1, then 2, then 3, until finally i is replaced with 6.

$$\begin{aligned}\sum_{i=1}^{6} (4i - 3) &= (4 \cdot 1 - 3) + (4 \cdot 2 - 3) + (4 \cdot 3 - 3) \\ &\quad + (4 \cdot 4 - 3) + (4 \cdot 5 - 3) + (4 \cdot 6 - 3) \\ &= 1 + 5 + 9 + 13 + 17 + 21 = 66\end{aligned}$$

CAUTION In this notation, i is called the **index of summation.** Do not confuse this use of i with the use of i to represent an imaginary number. Other letters may be used for the index of summation.

EXAMPLE 1 USING SUMMATION NOTATION

Evaluate each of the following sums.

(a) $$\sum_{i=1}^{4} i^2(i + 1) = 1^2(1 + 1) + 2^2(2 + 1) + 3^2(3 + 1) + 4^2(4 + 1)$$
$$= 1 \cdot 2 + 4 \cdot 3 + 9 \cdot 4 + 16 \cdot 5$$
$$= 2 + 12 + 36 + 80$$
$$= 130$$

(b) $$\sum_{j=3}^{6} \frac{j+1}{j-2} = \frac{3+1}{3-2} + \frac{4+1}{4-2} + \frac{5+1}{5-2} + \frac{6+1}{6-2}$$
$$= \frac{4}{1} + \frac{5}{2} + \frac{6}{3} + \frac{7}{4} = \frac{41}{4}$$ ■

2 Suppose someone borrows \$3000 and agrees to pay \$100 per month plus interest of 1% per month on the unpaid balance until the loan is paid off. The first month he pays \$100 to reduce the loan, plus interest of (.01)3000 = 30 dollars. The second month he pays another \$100 toward the loan and interest of (.01)2900 = 29 dollars. Since the loan is reduced by \$100 each month, his interest payments decrease by (.01)100 = 1 dollar each month, forming the arithmetic sequence

$$30, 29, 28, \ldots, 3, 2, 1.$$

The total amount of interest paid is given by the sum of the terms of this sequence. A formula can be developed to find this sum without adding all thirty numbers directly.

If an arithmetic sequence has terms $a_1, a_2, a_3, a_4 \ldots, a_n$, and S_n is defined as the sum of the first n terms of the sequence, then

$$S_n = a_1 + a_2 + a_3 + \ldots + a_n.$$

A formula for S_n can be found by writing the sum of the first n terms as follows.

$$S_n = a_1 + [a_1 + d] + [a_1 + 2d] + \ldots + [a_1 + (n - 1)d]$$

Next, write this same sum in reversed order.

$$S_n = [a_1 + (n - 1)d] + [a_1 + (n - 2)d] + \ldots + [a_1 + d] + a_1$$

Now add the corresponding sides of these two equations.

$$S_n + S_n = (a_1 + [a_1 + (n - 1)d]) + ([a_1 + d] + [a_1 + (n - 2)d])$$
$$+ \ldots + ([a_1 + (n - 1)d] + a_1)$$

From this,

$$2S_n = [2a_1 + (n - 1)d] + [2a_1 + (n - 1)d] + \ldots + [2a_1 + (n - 1)d].$$

Since there are n of the $[2a_1 + (n - 1)d]$ terms on the right,

$$2S_n = n[2a_1 + (n - 1)d]$$

$$S_n = \frac{n}{2}[2a_1 + (n - 1)d].$$

Using the formula $a_n = a_1 + (n - 1)d$, S_n can also be written as

$$S_n = \frac{n}{2}[a_1 + a_1 + (n - 1)d],$$

or

$$S_n = \frac{n}{2}(a_1 + a_n).$$

A summary of this work with sums of arithmetic sequences follows.

SUM OF THE FIRST n TERMS OF AN ARITHMETIC SEQUENCE

If an arithmetic sequence has first term a_1 and common difference d, then the sum of the first n terms is given by

$$S_n = \frac{n}{2}(a_1 + a_n)$$

or

$$S_n = \frac{n}{2}[2a_1 + (n - 1)d].$$

The first formula is used when the first and last terms are known; otherwise the second formula is used.

Either one of these formulas can be used to find the total interest on the \$3000 loan discussed above. In the sequence of interest payments, $a_1 = 30$, $d = -1$, $n = 30$, and $a_n = 1$. Choosing the first formula,

$$S_n = \frac{n}{2}(a_1 + a_n),$$

gives

$$S_{30} = \frac{30}{2}(30 + 1) = 15(31) = 465,$$

so a total of \$465 interest will be paid over the 30 months.

EXAMPLE 2 **USING THE SUM FORMULA (ARITHMETIC SEQUENCE)**

Find S_{12} for the arithmetic sequence

$$-9, -5, -1, 3, 7, \ldots.$$

Using $a_1 = -9$, $n = 12$, and $d = 4$ in the second formula,

$$S_n = \frac{n}{2}[2a_1 + (n - 1)d],$$

gives

$$S_{12} = \frac{12}{2}[2(-9) + 11(4)] = 6(-18 + 44) = 156. \quad ■$$

EXAMPLE 3 USING THE SUM FORMULA (ARITHMETIC SEQUENCE)

Find the sum of the first 60 positive integers.

Here $n = 60$, $a_1 = 1$ and $a_{60} = 60$. Use the first formula for S_n.

$$S_n = \frac{n}{2}(a_1 + a_n)$$

$$S_{60} = \frac{60}{2}(1 + 60) = 30 \cdot 61 = 1830 \quad \blacksquare$$

EXAMPLE 4 USING THE SUM FORMULA (ARITHMETIC SEQUENCE)

The sum of the first 17 terms of an arithmetic sequence is 187. If $a_{17} = -13$, find a_1 and d.

Use the first formula for S_n, with $n = 17$, to find a_1.

$$_{17} = \frac{17}{2}(a_1 + a_{17}) \qquad \text{Let } n = 17.$$

$$187 = \frac{17}{2}(a_1 - 13) \qquad \text{Let } S_{17} = 187,\ a_{17} = -13.$$

$$22 = a_1 - 13 \qquad \text{Multiply by } \frac{2}{17}.$$

$$a_1 = 35$$

Since $a_{17} = a_1 + (17 - 1)d$,

$$-13 = 35 + 16d \qquad \text{Let } a_{17} = -13,\ a_1 = 35.$$

$$-48 = 16d$$

$$d = -3. \quad \blacksquare$$

Any sum of the form

$$\sum_{i=1}^{n} (mi + p),$$

where m and p are real numbers, represents the sum of the terms of an arithmetic sequence having first term

$$a_1 = m(1) + p = m + p$$

and common difference $d = m$. These sums can be evaluated by the formulas in this section, as shown by the next example.

EXAMPLE 5 FINDING SUMS OF ARITHMETIC SEQUENCES USING SUMMATION NOTATION

Find each of the following sums.

(a) $\sum_{i=1}^{10} (4i + 8)$

This sum represents the sum of the first *ten* terms of the arithmetic sequence having

$$a_1 = 4 \cdot 1 + 8 = 12,$$

$$n = 10,$$

and

$$a_n = a_{10} = 4 \cdot 10 + 8 = 48.$$

Thus

$$\sum_{i=1}^{10} (4i + 8) = S_{10} = \frac{10}{2}(12 + 48) = 5(60) = 300.$$

(b) $\sum_{j=1}^{15} (9 - j)$

Since $n = 15$, $a_1 = 9 - 1 = 8$, and $a_{15} = 9 - 15 = -6$,

$$\sum_{j=1}^{15} (9 - j) = S_{15} = \frac{15}{2}[8 + (-6)] = \frac{15}{2}(2) = 15. \quad \blacksquare$$

3 Just as formulas were developed to find the sum of the first n terms of an arithmetic sequence, the same can be done for a geometric sequence. We begin by first writing the sum, S_n, as

$$S_n = a_1 + a_2 + a_3 + \ldots + a_n.$$

This can also be written as

$$S_n = a_1 + a_1r + a_1r^2 + \ldots + a_1r^{n-1}. \qquad \textbf{(1)}$$

If $r = 1$, then $S_n = na_1$, which is a correct formula for this case. If $r \neq 1$, multiply both sides of equation (1) by r, obtaining

$$rS_n = a_1r + a_1r^2 + a_1r^3 + \ldots + a_1r^n. \qquad \textbf{(2)}$$

If equation (2) is subtracted from equation (1), the result is

$$\begin{aligned} S_n &= a_1 + a_1r + a_1r^2 + \ldots + a_1r^{n-1} \\ rS_n &= \qquad\; a_1r + a_1r^2 + \ldots + a_1r^{n-1} + a_1r^n \\ \hline S_n - rS_n &= a_1 \qquad\qquad\qquad\qquad\qquad\quad - a_1r^n \end{aligned}$$

Subtract.

or $S_n(1 - r) = a_1(1 - r^n)$, Factor.

which finally gives

$$S_n = \frac{a_1(1 - r^n)}{1 - r}, \quad \text{where } r \neq 1.$$

Divide by $1 - r$.

This discussion is summarized as follows.

SUM OF THE FIRST n TERMS OF A GEOMETRIC SEQUENCE

If a geometric sequence has first term a_1 and common ratio r, then the sum of the first n terms is given by

$$S_n = \frac{a_1(1 - r^n)}{1 - r}, \quad \text{where } r \neq 1.$$

The next example shows how this formula is used.

EXAMPLE 6 USING THE SUM FORMULA (GEOMETRIC SEQUENCE)

(a) Find S_5 for the geometric sequence

$$3, 6, 12, 24, 48.$$

Here $a_1 = 3$ and $r = 2$. Using the formula above,

$$S_5 = \frac{3(1 - 2^5)}{1 - 2} = \frac{3(1 - 32)}{-1} = \frac{3(-31)}{-1} = 93.$$

(b) Find S_4 for the geometric sequence

$$10, 2, \frac{2}{5}, \ldots .$$

Here $a_1 = 10$, $r = 1/5$, and $n = 4$. Substitute these values into the formula for S_n.

$$S_4 = \frac{10\left[1 - \left(\frac{1}{5}\right)^4\right]}{1 - \frac{1}{5}} \qquad a_1 = 10, r = 1/5, n = 4$$

$$= \frac{10\left(1 - \frac{1}{625}\right)}{\frac{4}{5}}$$

$$= 10\left(\frac{624}{625}\right) \cdot \frac{5}{4}$$

$$S_4 = \frac{312}{25} \quad ■$$

A sum of the form

$$\sum_{i=1}^{n} m \cdot p^i$$

represents the sum of the terms of a geometric sequence having first term

$$a_1 = m \cdot p^1 = mp$$

and common ratio $r = p$. These sums can be found by the formula for S_n given above.

EXAMPLE 7
FINDING SUMS OF GEOMETRIC SEQUENCES USING SUMMATION NOTATION

Find each of the following sums.

(a) $\sum_{i=1}^{7} 2 \cdot 3^i$

In this sum, $a_1 = 2 \cdot 3^1 = 6$ and $r = 3$. Thus,

$$\sum_{i=1}^{7} 2 \cdot 3^i = S_7 = \frac{6(1 - 3^7)}{1 - 3} = \frac{6(1 - 2187)}{-2} = \frac{6(-2186)}{-2} = 6558.$$

(b) $$\sum_{i=1}^{5} \left(\frac{3}{4}\right)^i = S_5 = \frac{\frac{3}{4}\left[1 - \left(\frac{3}{4}\right)^5\right]}{1 - \frac{3}{4}}$$

$$= \frac{\frac{3}{4}\left(1 - \frac{243}{1024}\right)}{\frac{1}{4}}$$

$$= 3\left(\frac{781}{1024}\right) = \frac{2343}{1024}$$ ■

4 The formulas developed so far in this chapter will help us to solve applied problems that lead to sequences.

PROBLEM SOLVING

In some cases, we might be asked to find a particular term of an arithmetic or geometric sequence that appears in an applied problem. We would then use the appropriate formula for finding a_n (developed in the earlier sections). If the problem requires finding the sum of a specified number of terms of an arithmetic or geometric sequence, we then use one of the formulas developed in this section. It will be important to read carefully to determine whether we want a specific term or a sum of terms. ■

EXAMPLE 8
SOLVING AN APPLIED PROBLEM INVOLVING AN ARITHMETIC SEQUENCE

A child building a tower with blocks uses 15 for the bottom row. Each row has 2 fewer blocks than the previous row. Suppose that there are 8 rows in the tower.

(a) How many blocks are used for the top row?

The number of blocks in each row forms an arithmetic sequence with $a_1 = 15$ and $d = -2$. Find a_n for $n = 8$ by using the formula $a_n = a_1 + (n - 1)d$.

$$a_8 = 15 + (8 - 1)(-2) = 1$$

There is just one block in the top row.

(b) What is the total number of blocks in the tower?

Here we must find the sum of the terms of the arithmetic sequence formed. We have $a_1 = 15$, $n = 8$, and $a_8 = 1$ (as found in part (a)). Use the formula

$$S_n = \frac{n}{2}(a_1 + a_n).$$

$$S_8 = \frac{8}{2}(15 + 1) = 4(16) = 64.$$

There are 64 blocks in the tower. ■

NOTE In Example 8(b), we could have used the other form of the formula for the sum of the first n terms of an arithmetic sequence,

$$S_n = \frac{n}{2}[2a_1 + (n - 1)d],$$

with $n = 8$, $a_1 = 15$, and $d = -2$.

EXAMPLE 9
SOLVING AN APPLIED PROBLEM INVOLVING A GEOMETRIC SEQUENCE

An insect population is growing in such a way that each generation is 1.5 times as large as the previous generation. Suppose there are 100 insects in the first generation.

(a) How many will there be in the fourth generation?

The population can be written as a geometric sequence with a_1 as the first-generation population, a_2 the second-generation population, and so on. Then the fourth generation population will be a_4. Using the formula for a_n, where $n = 4$, $r = 1.5$, and $a_1 = 100$, gives

$$a_4 = a_1 r^3 = 100(1.5)^3 = 100(3.375) = 337.5.$$

In the fourth generation, the population will number about 338 insects.

(b) What will be the total number of insects in the first four generations?

Since we must find the sum of the first four terms of this geometric sequence, we use the formula

$$S_n = \frac{a_1(1 - r^n)}{1 - r},$$

with $n = 4$, $a_1 = 100$, and $r = 1.5$. Use a calculator to get

$$S_4 = \frac{100(1 - 1.5^4)}{1 - 1.5} = \frac{100(1 - 5.0625)}{-.5} = 812.5.$$

The total population for the four generations will amount to about 813 insects. ■

CAUTION Be careful that you choose the correct term in problems like Example 9(a).

A sequence of equal payments made at equal periods of time is called an **annuity.** The sum of the payments and interest on the payments is the **future value** of the annuity.

EXAMPLE 10
FINDING THE FUTURE VALUE OF AN ANNUITY

To save money for a trip to Europe, Marge deposited $1000 at the end of each year for four years in an account paying 6% interest. What is the future value of this annuity?

The first payment will earn interest for 3 years, the second payment for 2 years, and the third payment for 1 year. The last payment earns no interest. The total amount is

$$1000(1.06)^3 + 1000(1.06)^2 + 1000(1.06) + 1000.$$

This is the sum of a geometric sequence with first term (starting at the end of the sum as written above) $a_1 = 1000$ and common ratio $r = 1.06$. Using the formula for S_4, the sum of four terms, gives

$$S_4 = \frac{1000[1 - (1.06)^4]}{1 - 1.06}$$
$$\approx \frac{1000[1 - 1.2625]}{-.06}$$
$$= 4374.62. \qquad \text{Use a calculator.}$$

The future value of the annuity is $4374.62. ■

13.3 EXERCISES

Evaluate each of the following sums. See Example 1.

1. $\sum_{i=1}^{5} (2i + 1)$ **2.** $\sum_{i=1}^{6} (3i - 2)$ **3.** $\sum_{j=1}^{9} j$ **4.** $\sum_{j=1}^{4} \frac{1}{j}$

5. $\sum_{i=1}^{5} (i + 1)^{-1}$ **6.** $\sum_{i=1}^{4} i^i$ **7.** $\sum_{i=1}^{3} i^{-i}$ **8.** $\sum_{k=1}^{4} (k + 1)^2$

9. $\sum_{k=1}^{6} (-1)^k \cdot k$ **10.** $\sum_{i=1}^{7} (-1)^{i+1} \cdot i^2$ **11.** $\sum_{i=3}^{7} \frac{i}{2}$ **12.** $\sum_{i=5}^{9} \frac{i + 1}{3}$

Find the sum of the first ten terms for each of the following arithmetic sequences. See Example 2.

13. $a_1 = 8$, $d = 3$ **14.** $a_1 = -9$, $d = 4$ **15.** $a_3 = 5$, $a_4 = 8$ **16.** $a_2 = 9$, $a_4 = 13$

17. 5, 9, 13, . . . **18.** 8, 6, 4, . . . **19.** $3\frac{1}{2}, 5, 6\frac{1}{2}, \ldots$ **20.** $2\frac{1}{2}, 3\frac{3}{4}, 5, \ldots$

21. $a_1 = 10$, $a_{10} = 5\frac{1}{2}$ **22.** $a_1 = -8$, $a_{10} = -5/4$

Find a_1 and d for each of the following arithmetic sequences. See Example 4.

23. $S_{20} = 1090, \quad a_{20} = 102$

24. $S_{31} = 5580, \quad a_{31} = 360$

25. $S_{12} = -108, \quad a_{12} = -19$

26. $S_{25} = 650, \quad a_{25} = 62$

Evaluate each of the following sums. See Example 5.

27. $\sum_{i=1}^{3} (i + 4)$

28. $\sum_{i=1}^{5} (i - 8)$

29. $\sum_{j=1}^{10} (2j + 3)$

30. $\sum_{j=1}^{15} (5j - 9)$

31. $\sum_{i=1}^{12} (7i + 2)$

32. $\sum_{j=1}^{10} (8j - 3)$

33. $\sum_{i=1}^{12} (-5 - 8i)$

34. $\sum_{k=1}^{19} (-3 - 4k)$

35. $\sum_{i=1}^{1000} i$

36. $\sum_{k=1}^{2000} k$

37. $\sum_{i=6}^{15} (4i - 2)$

Find the sum of the first five terms for each of the following geometric sequences. See Example 6.

38. 3, 6, 12, 24, . . .

39. 5, 20, 80, 320, . . .

40. 12, -6, 3, $-3/2$, . . .

41. 18, -3, 1/2, $-1/12$, . . .

42. 9, -6, 4, . . .

43. 50, -10, 2, . . .

44. $a_1 = 4, \quad r = 2$

45. $a_1 = 3, \quad r = 3$

46. $a_2 = 1/3, \quad r = 3$

47. $a_2 = -1, \quad r = 2$

Find each of the following sums. See Example 7.

48. $\sum_{i=1}^{4} 2^i$

49. $\sum_{j=1}^{6} 3^j$

50. $\sum_{i=1}^{4} (-3)^i$

51. $\sum_{i=1}^{4} (-3^i)$

52. $\sum_{i=1}^{8} 64(1/2)^i$

53. $\sum_{i=1}^{6} 81(2/3)^i$

54. $\sum_{j=1}^{4} (2/5)^j$

55. $\sum_{i=1}^{4} (-3/4)^i$

56. $\sum_{i=1}^{4} 6(3/2)^i$

57. $\sum_{k=1}^{5} 9(5/3)^k$

58. $\sum_{i=3}^{6} 2^i$

59. $\sum_{i=4}^{7} 3^i$

60. $\sum_{i=1}^{4} 5(3/5)^{i-1}$

61. $\sum_{i=1}^{5} 256(-3/4)^{i-1}$

Solve each problem. See Examples 3 and 8.

62. Find the sum of all the integers from 51 to 71.

63. Find the sum of all the integers from -8 to 30.

64. If a clock strikes the proper number of chimes each hour on the hour, how many times will it chime in a month of 30 days?

65. A stack of telephone poles has 30 in the bottom row, 29 in the next, and so on, with one pole in the top row. How many poles are in the stack?

66. A sky diver falls 10 meters during the first second, 20 meters during the second, 30 meters during the third, and so on. How many meters will the diver fall during the tenth second? During the first ten seconds?

67. Deepwell Drilling Company charges a flat $100 set-up charge, plus $5 for the first foot, $6 for the second, $7 for the third, and so on. Find the total charge for a 70-foot well.

68. An object falling under the force of gravity falls about 16 feet the first second, 48 feet during the next second, 80 feet during the third second, and so on. How far would the object fall during the eighth second? What is the total distance the object would fall in eight seconds?

69. The population of a city was 49,000 five years ago. Each year the zoning commission permits an increase of 580 in the population. What will the population be five years from now?

70. A super slide of uniform slope is to be built on a level piece of land. There are to be twenty equally spaced supports, with the longest support 15 meters long and the shortest 2 meters long. Find the total length of all the supports.

71. How much material would be needed for the rungs of a ladder of 31 rungs, if the rungs taper uniformly from 18 inches to 28 inches?

Solve each problem. See Example 9.

72. Suppose you could save $1 on January 1, $2 on January 2, $4 on January 3, and so on. What amount would you save on January 31?

73. See Exercise 72. What would be the total amount of your savings during January? (*Hint:* $2^{31} =$ 2,147,483,648.)

74. Richland Oil has a well that produced $4,000,000 of income its first year. Each year thereafter, the well produced half as much as it did the previous year. What total amount of income would be produced by the well in 6 years?

75. Fruit and vegetable dealer Olive Greene paid 10¢ per pound for 10,000 pounds of onions. Each week the price she charges increases by .1¢ per pound, while her onions lose 5% of their weight. If she sells all the onions after 6 weeks, does she make or lose money? How much?

76. The final step in processing a black and white photographic print is to immerse the print in a chemical called ''fixer.'' The print is then washed in running water. Under certain conditions, 98% of the fixer in a print will be removed with 15 minutes of washing. How much of the original fixer would then be left after one hour?

77. A scientist has a vat containing 100 liters of a pure chemical. Twenty liters are drained and replaced with water. After complete mixing, 20 liters of the mixture are drained and replaced with water. What will be the strength of the mixture after nine such drainings?

78. The half-life of a radioactive substance is the time it takes for half the substance to decay. Suppose that the half-life of a substance is 3 years and that 10^{15} molecules of the substance are present initially. How many molecules will be present after 15 years?

79. Each year a machine loses 20% of the value it had at the beginning of the year. Find the value of a machine at the end of 6 years if it cost $100,000 new.

Find the future value of each of the following annuities. See Example 10.

80. Payments of $1000 at the end of each year for 9 years at 8% interest compounded annually.

81. Payments of $800 at the end of each year for 12 years at 10% interest compounded annually.

82. Payments of $2430 at the end of each year for 10 years at 11% interest compounded annually.

83. Payments of $1500 at the end of each year for 6 years at 12% interest compounded annually.

Photo for Exercise 75

PREVIEW EXERCISES

Evaluate the expression $\dfrac{a}{1-r}$ *for the following values of a and r. See Section 2.2.*

84. $a = 4, \quad r = 1/2$

85. $a = -3, \quad r = -1/4$

86. $a = 10, \quad r = .1$

87. $a = -10, \quad r = -.1$

13.4 SUMS OF INFINITE GEOMETRIC SEQUENCES

OBJECTIVES

1 FIND THE SUM OF THE TERMS OF AN INFINITE GEOMETRIC SEQUENCE.

2 USE THE SUM FORMULA TO WRITE A REPEATING DECIMAL AS A QUOTIENT OF INTEGERS.

FOCUS ON PROBLEM SOLVING

A pendulum bob swings through an arc 40 centimeters long on its first swing. Each swing thereafter, it swings only 80% as far as on the previous swing. How far will it swing altogether before coming to a complete stop?

Under certain conditions, an infinite geometric sequence will have a sum. In this section we discuss what conditions are necessary for this, and develop a formula that allows us to determine this sum. The problem above can be solved by using this formula. It is Exercise 53 in the exercises for this section. After working through this section, you should be able to solve this problem.

1 In the preceding section the sum of the first n terms of a geometric sequence was given as

$$S_n = \sum_{i=1}^{n} a_i = \frac{a_1(1 - r^n)}{1 - r},$$

where a_1 is the first term and r ($r \neq 1$) is the common ratio. Now consider an infinite geometric sequence such as

$$2, 1, \frac{1}{2}, \frac{1}{4}, \frac{1}{8}, \frac{1}{16}, \ldots$$

with first term 2 and common ratio 1/2. Using the formula for S_n gives the following sequence.

$$S_1 = 2, \quad S_2 = 3, \quad S_3 = \frac{7}{2}, \quad S_4 = \frac{15}{4}, \quad S_5 = \frac{31}{8}, \quad S_6 = \frac{63}{16}$$

These sums seem to be getting closer and closer to the number 4. In fact, by selecting a value of n large enough, it is possible to make S_n as close as desired to 4. This idea is expressed as

$$\lim_{n \to \infty} S_n = 4.$$

(Read: "the limit of S_n as n increases without bound is 4.") For no value of n is $S_n = 4$. However, if n is large enough, then S_n is as close to 4 as desired.*

Since

$$\lim_{n \to \infty} S_n = 4,$$

the number 4 is called the *sum* of the infinite geometric sequence

$$2, 1, \frac{1}{2}, \frac{1}{4}, \ldots$$

and

$$2 + 1 + \frac{1}{2} + \frac{1}{4} + \frac{1}{8} + \ldots = 4.$$

*These phrases "large enough" and "as close as desired" are not nearly precise enough for mathematicians; much of a standard calculus course is devoted to making them more precise.

EXAMPLE 1
FINDING THE SUM OF AN INFINITE GEOMETRIC SEQUENCE

Find $1 + \frac{1}{3} + \frac{1}{9} + \frac{1}{27} + \ldots$.

Use the formula for the sum of the first n terms of a geometric sequence to get

$$S_1 = 1, \quad S_2 = \frac{4}{3}, \quad S_3 = \frac{13}{9}, \quad S_4 = \frac{40}{27},$$

and in general

$$S_n = \frac{1\left[1 - \left(\frac{1}{3}\right)^n\right]}{1 - \frac{1}{3}}.$$

The chart below shows the value of $(1/3)^n$ for larger and larger values of n.

n	1	10	100	200
$\left(\frac{1}{3}\right)^n$	$\frac{1}{3}$	1.69×10^{-5}	1.94×10^{-48}	3.76×10^{-96}

As n gets larger and larger, $(1/3)^n$ gets closer and closer to 0. That is,

$$\lim_{n\to\infty} \left(\frac{1}{3}\right)^n = 0,$$

making it reasonable that

$$\lim_{n\to\infty} S_n = \lim_{n\to\infty} \frac{1\left[1 - \left(\frac{1}{3}\right)^n\right]}{1 - \frac{1}{3}} = \frac{1(1-0)}{1 - \frac{1}{3}} = \frac{1}{\frac{2}{3}} = \frac{3}{2}.$$

Hence,

$$1 + \frac{1}{3} + \frac{1}{9} + \frac{1}{27} + \ldots = \frac{3}{2}. \quad \blacksquare$$

If a geometric sequence has a first term a_1 and a common ratio r, then

$$S_n = \frac{a_1(1 - r^n)}{1 - r} \qquad (r \neq 1)$$

for every positive integer n. If $-1 < r < 1$, then $\lim_{n\to\infty} r^n = 0$, and

$$\lim_{n\to\infty} S_n = \frac{a_1(1 - 0)}{1 - r} = \frac{a_1}{1 - r}.$$

This quotient, $a_1/(1 - r)$, is called the **sum of an infinite geometric sequence.** The limit $\lim_{n\to\infty} S_n$ is often expressed as S_∞ or $\sum_{i=1}^{\infty} a_i$. These results lead to the following definition.

SUM OF THE TERMS OF AN INFINITE GEOMETRIC SEQUENCE

The sum of an infinite geometric sequence with first term a_1 and common ratio r, where $-1 < r < 1$, is given by

$$S_\infty = \frac{a_1}{1-r}.$$

If $|r| > 1$, the terms get larger and larger, so there is no limit as $n \to \infty$. Hence the sequence will not have a sum.

EXAMPLE 2
USING THE SUM FORMULA

(a) Find the sum $-\frac{3}{4} + \frac{3}{8} - \frac{3}{16} + \frac{3}{32} - \frac{3}{64} + \cdots$.

The first term is $a_1 = -3/4$. To find r, divide any two adjacent terms. For example,

$$r = \frac{a_3}{a_2} = \frac{-\frac{3}{16}}{\frac{3}{8}} = -\frac{1}{2}.$$

Since $-1 < r < 1$, the formula in this section applies, and

$$S_\infty = \frac{a_1}{1-r} = \frac{-\frac{3}{4}}{1-\left(-\frac{1}{2}\right)} = -\frac{1}{2}.$$

(b) $\displaystyle\sum_{i=1}^{\infty}\left(\frac{3}{5}\right)^i = \frac{\frac{3}{5}}{1-\frac{3}{5}} = \frac{3}{2}$

Notice that the upper limit is ∞, indicating that this is an infinite sequence. ■

2 The formula in this section can be used to convert repeating decimals (which represent rational numbers) to fractions of the form p/q, where p and q are integers, with $q \neq 0$.

EXAMPLE 3
WRITING A REPEATING DECIMAL AS A QUOTIENT OF INTEGERS

Write each repeating decimal in the form p/q, where p and q are integers.

(a) .090909 . . .

This decimal can be written as

$$.090909\ldots = .09 + .0009 + .000009 + \cdots,$$

which is the sum of the terms of an infinite geometric sequence having $a_1 = .09$ and $r = .01$. The sum of this sequence is given by

$$S_\infty = \frac{a_1}{1-r} = \frac{.09}{1-.01} = \frac{.09}{.99} = \frac{1}{11}.$$

Thus, .090909 . . . = 1/11.

(b) 2.5121212 . . .
Write the number as

$$2.5121212\ldots = 2.5 + .012 + .00012 + .0000012 + \ldots$$
$$= 2.5 + (.012 + .00012 + .0000012 + \ldots).$$

Beginning with the second term, this is the sum of the terms of an infinite geometric sequence with $a_1 = .012$ and $r = .01$, so 2.5121212 . . . can be written as

$$2.5 + \frac{.012}{1 - .01} = 2.5 + \frac{.012}{.990} = 2.5 + \frac{2}{165} = \frac{829}{330}. \quad ■$$

13.4 EXERCISES

Find r for each of the following infinite geometric sequences. Identify any whose sums would exist. See Example 2.

1. 9, 18, 36, 72, 144, . . .
2. 3, 9, 27, 81, . . .
3. 10, 100, 1000, 10,000, . . .
4. −8, −4, −2, −1, −1/2, . . .
5. 12, 6, 3, 3/2, . . .
6. −8, −16, −32, −64, . . .
7. −1, −1/2, −1/4, −1/8, . . .
8. .9, .09, .009, .0009, . . .

---

9. Under what conditions will an infinite geometric sequence have a sum? (See Objective 1.)

10. In Exercise 8, the sum of the terms may be written as

$$.9 + .09 + .009 + .0009 + \ldots,$$

which is also written as .9999 . . . . The sum does exist, since $r = .1$. Use the formula to find this sum, and then complete this statement: The value of .9999 . . . is __________. (See Objective 2.)

Find each of the following sums by using the formula of this section where it applies. See Example 2.

11. $16 + 4 + 1 + \ldots$
12. $81 + 27 + 9 + 3 + 1 + \ldots$
13. $100 + 10 + 1 + \ldots$
14. $128 + 64 + 32 + \ldots$
15. $90 + 30 + 10 + \ldots$
16. $25 + 5 + 1 + \ldots$
17. $256 - 128 + 64 - 32 + 16 - \ldots$
18. $120 - 60 + 30 - 15 + \ldots$
19. $108 - 36 + 12 - 4 + \ldots$
20. $10{,}000 - 1000 + 100 - 10 + 1 - \ldots$
21. $\frac{3}{4} + \frac{3}{8} + \frac{3}{16} + \ldots$
22. $\frac{4}{5} + \frac{2}{5} + \frac{1}{5} + \ldots$
23. $3 - \frac{3}{2} + \frac{3}{4} - \ldots$
24. $9 - 3 + 1 - \ldots$
25. $\frac{1}{3} - \frac{2}{9} + \frac{4}{27} - \frac{8}{81} + \ldots$
26. $1 + \frac{1}{1.01} + \frac{1}{(1.01)^2} + \ldots$
27. $\frac{1}{36} + \frac{1}{30} + \frac{1}{25} + \ldots$
28. $1 + \frac{1}{2^2} + \frac{1}{2^4} + \ldots$

29. $\sum_{i=1}^{\infty} (1/4)^i$

30. $\sum_{i=1}^{\infty} (9/10)^i$

31. $\sum_{i=1}^{\infty} (1.2)^i$

32. $\sum_{i=1}^{\infty} (1.001)^i$

33. $\sum_{i=1}^{\infty} (-1/4)^i$

34. $\sum_{i=1}^{\infty} (.3)^i$

35. $\sum_{i=1}^{\infty} 5^{-i}$

36. $\sum_{i=1}^{\infty} (-1/2)^i$

37. $\sum_{i=1}^{\infty} 10^{-i}$

38. $\sum_{i=1}^{\infty} 4^{-i}$

39. $\sum_{i=1}^{\infty} (1/2)^{-i}$

40. $\sum_{i=1}^{\infty} (3/4)^{-i}$

---

41. Which one of the following does not have a sum? (See Objective 1.)

(a) $\sum_{i=1}^{10} \left(\frac{1}{2}\right)^i$ **(b)** $\sum_{i=1}^{\infty} \left(\frac{3}{2}\right)^i$

(c) $\sum_{i=1}^{\infty} \left(\frac{1}{2}\right)^i$ **(d)** $\sum_{i=1}^{1000} 3 \cdot 4^i$

42. Refer to Exercise 10. The same result can be obtained a different way. Use the fact that $1/3 = .3333\ldots$ to develop an argument that leads to the same conclusion.

Express each repeating decimal in the form p/q, where p and q are integers. See Example 3.

43. .55555 . . .

44. .77777 . . .

45. .121212 . . .

46. .858585 . . .

47. .313131 . . .

48. .909090 . . .

49. .508508508 . . .

50. .613613613 . . .

Use the formula of this section to solve each problem.

51. Mitzi drops a ball from a height of 10 meters and notices that on each bounce the ball returns to about 3/4 of its previous height. About how far will the ball travel before it comes to rest? (*Hint:* Consider the sum of two sequences.)

52. A sugar factory receives an order for 1000 units of sugar. The production manager thus orders production of 1000 units of sugar. He forgets, however, that the production of sugar requires some sugar (to prime the machines, for example), and so he ends up with only 900 units of sugar. He then orders an additional 100 units, and receives only 90 units. A further order for 10 units produces 9 units. Finally seeing he is wrong, the manager decides to try mathematics. He views the production process as an infinite geometric progression with $a_1 = 1000$ and $r = .1$. Using this, find the number of units of sugar that he should have ordered originally.

53. A pendulum bob swings through an arc 40 centimeters long on its first swing. Each swing thereafter, it swings only 80% as far as on the previous swing. How far will it swing altogether before coming to a complete stop?

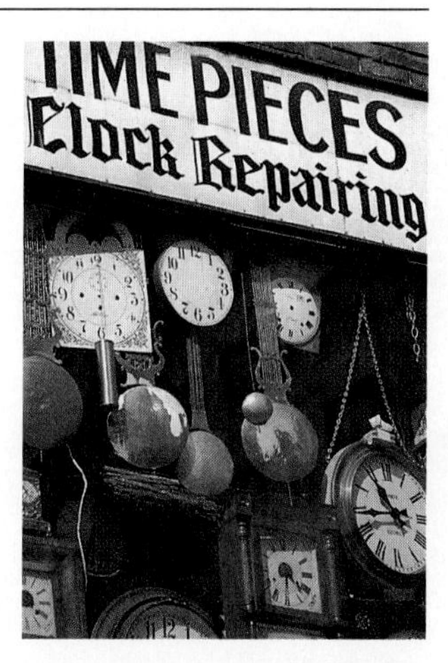

54. After a person pedaling a bicycle removes his or her feet from the pedals, the wheel rotates 400 times the first minute. As it continues to slow down, each minute it rotates only 3/4 as many times as in the previous minute. How many times will the wheel rotate before coming to a complete stop?

55. A sequence of equilateral triangles is constructed. The first triangle has sides 2 meters in length. To get the second triangle, midpoints of the sides of the original triangle are connected. If this process could be continued indefinitely, what would be the total perimeter of all the triangles?

56. What would be the total area of all the triangles in Exercise 55, disregarding the overlapping?

PREVIEW EXERCISES

Simplify. See Section 3.4.

57. $(3x + 2y)^2$

58. $(4x - 3y)^2$

59. $(a + b)^3$

60. $(x - y)^4$

13.5 THE BINOMIAL THEOREM

OBJECTIVES

1. LEARN HOW TO CONSTRUCT PASCAL'S TRIANGLE.
2. EVALUATE FACTORIALS.
3. EVALUATE THE BINOMIAL COEFFICIENT $\binom{n}{r}$ FOR SPECIFIC VALUES OF n AND r.
4. EXPAND $(x + y)^n$ USING THE BINOMIAL THEOREM.
5. FIND THE rTH TERM IN THE BINOMIAL EXPANSION OF $(x + y)^n$.

1 In this section we introduce a method of evaluating binomial expressions of the form $(x + y)^n$, where n is a whole number. Some expansions of $(x + y)^n$, for various values of n, are given below:

$$(x + y)^0 = 1$$
$$(x + y)^1 = x + y$$
$$(x + y)^2 = x^2 + 2xy + y^2$$
$$(x + y)^3 = x^3 + 3x^2y + 3xy^2 + y^3$$
$$(x + y)^4 = x^4 + 4x^3y + 6x^2y^2 + 4xy^3 + y^4$$
$$(x + y)^5 = x^5 + 5x^4y + 10x^3y^2 + 10x^2y^3 + 5xy^4 + y^5$$

and so on. Studying these results reveals a pattern that can be used to write a general expression for $(x + y)^n$.

Notice that after the special case $(x + y)^0 = 1$, each expression begins with x raised to the same power as the binomial itself. That is, the expansion of $(x + y)^1$ has a first term of x^1, $(x + y)^2$ has a first term of x^2, $(x + y)^3$ has a first term of x^3, and so on. Also, the last term in each expansion is y to the same power as the binomial. Thus the expansion of $(x + y)^n$ should begin with the term x^n and end with the term y^n.

Also, the exponents on x decrease by one in each term after the first, while the exponents on y, beginning with y in the second term, increase by one in each succeeding term. That is, the *variables* in the terms of the expansion of $(x + y)^n$ have the following pattern.

$$x^n, \quad x^{n-1}y, \quad x^{n-2}y^2, \quad x^{n-3}y^3, \; \ldots, \; xy^{n-1}, \quad y^n$$

This pattern suggests that the sum of the exponents on x and y in each term is n. For example, in the third term in the list above, the variable is $x^{n-2}y^2$, and the sum of the exponents is $n - 2 + 2 = n$.

Now examine the *coefficients* in the terms of the expansions shown above. Writing the coefficients alone gives the following pattern.

PASCAL'S TRIANGLE

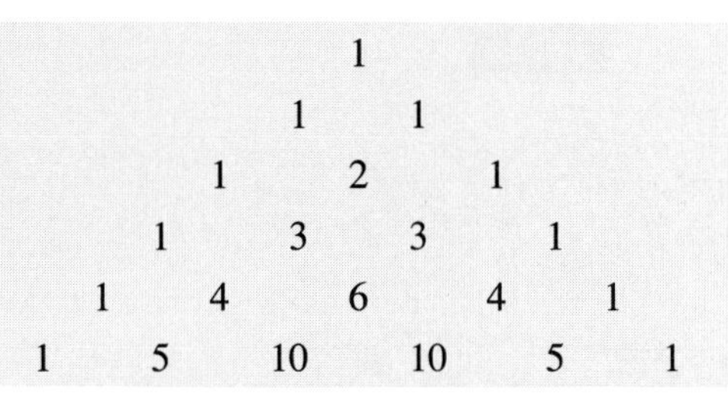

With the coefficients arranged in this way, it can be seen that each number in the triangle is the sum of the two numbers directly above it (one to the right and one to the left.) For example, in the fifth row from the top, 1 is the sum of 1, the only number above it, 4 is the sum of 1 and 3, 6 is the sum of 3 and 3, and so on. This triangular array of numbers is called **Pascal's triangle,** in honor of the seventeenth-century mathematician Blaise Pascal (1623–1662), one of the first to use it extensively.

To get the coefficients for $(x + y)^6$, include a seventh row in the array of numbers given above. Adding adjacent numbers, we find that the seventh row is

$$1 \quad 6 \quad 15 \quad 20 \quad 15 \quad 6 \quad 1.$$

Using these coefficients, the expansion of $(x + y)^6$ is

$$(x + y)^6 = x^6 + 6x^5y + 15x^4y^2 + 20x^3y^3 + 15x^2y^4 + 6xy^5 + y^6.$$

2 Although it is possible to use Pascal's triangle to find the coefficients of $(x + y)^n$ for any positive integer value of n, this becomes impractical for large values of n because of the need to write out all the preceding rows. A more efficient way of finding these coefficients uses factorial notation. The number $n!$ (read "n-factorial"), is defined as follows.

n-FACTORIAL ($n!$)

For any positive integer n:

$$n! = n(n - 1)(n - 2) \ldots (3)(2)(1),$$

and

$$0! = 1.$$

NOTE $0!$ is defined to equal 1 for convenience in later work.

EXAMPLE 1 EVALUATING FACTORIALS

Evaluate each factorial.

(a) $5! = 5 \cdot 4 \cdot 3 \cdot 2 \cdot 1 = 120$

(b) $7! = 7 \cdot 6 \cdot 5 \cdot 4 \cdot 3 \cdot 2 \cdot 1 = 5040$

(c) $2! = 2 \cdot 1 = 2$

(d) $1! = 1$

(e) $0! = 1$ (by definition) ■

NOTE Many scientific calculators have a key for factorials.

3 Now look at the coefficients of the expression

$$(x + y)^5 = x^5 + 5x^4y + 10x^3y^2 + 10x^2y^3 + 5xy^4 + y^5.$$

The coefficient on the second term, $5x^4y$, is 5, and the exponents on the variables are 4 and 1. Note that

$$5 = \frac{5!}{4!1!}.$$

The coefficient on the third term is 10, with exponents of 3 and 2, and

$$10 = \frac{5!}{3!2!}.$$

The last term (the sixth term) can be written as $y^5 = 1x^0y^5$, with coefficient 1, and exponents of 0 and 5. Since $0! = 1$, check that

$$1 = \frac{5!}{0!5!}.$$

Generalizing from these examples, we find that the coefficient for the term of the expansion of $(x + y)^n$ in which the variable part is x^ry^{n-r} (where $r \leq n$) will be

$$\frac{n!}{r!(n-r)!}.$$

This number, called a **binomial coefficient,** is often written as $\binom{n}{r}$ (read "n above r" or "n choose r.")

BINOMIAL COEFFICIENT

For nonnegative integers n and r, with $r \leq n$, the symbol $\binom{n}{r}$ is defined as

$$\binom{n}{r} = \frac{n!}{r!(n-r)!}.$$

These binomial coefficients are just numbers from Pascal's triangle. For example, $\binom{3}{0}$ is the first number in the fourth row, and $\binom{7}{4}$ is the fifth number in the eighth row.

EXAMPLE 2 EVALUATING BINOMIAL COEFFICIENTS

Evaluate each binomial coefficient.

(a) $\binom{6}{2} = \dfrac{6!}{2!(6-2)!} = \dfrac{6!}{2!4!} = \dfrac{6 \cdot 5 \cdot 4 \cdot 3 \cdot 2 \cdot 1}{2 \cdot 1 \cdot 4 \cdot 3 \cdot 2 \cdot 1} = 15$

(b) $\binom{8}{0} = \dfrac{8!}{0!(8-0)!} = \dfrac{8!}{0!8!} = \dfrac{8!}{1 \cdot 8!} = 1$

(c) $\binom{10}{10} = \dfrac{10!}{10!(10-10)!} = \dfrac{10!}{10!0!} = 1$ ■

NOTE Many scientific calculators have a key for the binomial coefficient $\binom{n}{r}$. However, the symbol $\binom{n}{r}$ may not be used to identify the key; it may be labeled C_r^n, ${}_nC_r$, or $C(n, r)$. More about this notation will be discussed in Section 13.8. Consult the owner's manual of your calculator to learn how to use this key.

Our conjectures about the expansion of $(x + y)^n$ are summarized below.

EXPANSION OF $(x + y)^n$

1. There are $(n + 1)$ terms in the expansion.
2. The first term is x^n, and the last term is y^n.
3. The exponent on x decreases by 1 and the exponent on y increases by 1 in each succeeding term.
4. The sum of the exponents on x and y in any term is n.
5. The coefficient of $x^r y^{n-r}$ is $\binom{n}{r}$.

4 These observations about the expansion of $(x + y)^n$ for any positive integer value of n suggest the **binomial theorem.**

BINOMIAL THEOREM

For any positive integer n:

$$(x + y)^n = x^n + \binom{n}{n-1}x^{n-1}y + \binom{n}{n-2}x^{n-2}y^2 + \binom{n}{n-3}x^{n-3}y^3 + \cdots + \binom{n}{n-r}x^{n-r}y^r + \cdots + \binom{n}{1}xy^{n-1} + y^n.$$

The binomial theorem can be proved using mathematical induction (discussed in Section 13.6.) Note that the expansion of $(x + y)^n$ has $n + 1$ terms.

EXAMPLE 3 **EXPANDING A BINOMIAL USING THE BINOMIAL THEOREM**

Write out the binomial expansion of $(x + y)^9$.

Using the binomial theorem,

$$(x + y)^9 = x^9 + \binom{9}{8}x^8y + \binom{9}{7}x^7y^2 + \binom{9}{6}x^6y^3 + \binom{9}{5}x^5y^4 + \binom{9}{4}x^4y^5 + \binom{9}{3}x^3y^6 + \binom{9}{2}x^2y^7 + \binom{9}{1}xy^8 + y^9.$$

Now evaluate each of the binomial coefficients.

$$\begin{aligned}(x + y)^9 &= x^9 + \frac{9!}{8!1!}x^8y + \frac{9!}{7!2!}x^7y^2 + \frac{9!}{6!3!}x^6y^3 + \frac{9!}{5!4!}x^5y^4 \\ &\quad + \frac{9!}{4!5!}x^4y^5 + \frac{9!}{3!6!}x^3y^6 + \frac{9!}{2!7!}x^2y^7 + \frac{9!}{1!8!}xy^8 + y^9 \\ &= x^9 + 9x^8y + 36x^7y^2 + 84x^6y^3 + 126x^5y^4 + 126x^4y^5 \\ &\quad + 84x^3y^6 + 36x^2y^7 + 9xy^8 + y^9 \quad \blacksquare\end{aligned}$$

EXAMPLE 4 **EXPANDING A BINOMIAL USING THE BINOMIAL THEOREM**

Expand $\left(a - \frac{b}{2}\right)^5$.

Again, use the binomial theorem.

$$\begin{aligned}\left(a - \frac{b}{2}\right)^5 &= a^5 + \binom{5}{4}a^4\left(-\frac{b}{2}\right) + \binom{5}{3}a^3\left(-\frac{b}{2}\right)^2 + \binom{5}{2}a^2\left(-\frac{b}{2}\right)^3 \\ &\quad + \binom{5}{1}a\left(-\frac{b}{2}\right)^4 + \left(-\frac{b}{2}\right)^5 \\ &= a^5 + 5a^4\left(-\frac{b}{2}\right) + 10a^3\left(-\frac{b}{2}\right)^2 + 10a^2\left(-\frac{b}{2}\right)^3 \\ &\quad + 5a\left(-\frac{b}{2}\right)^4 + \left(-\frac{b}{2}\right)^5 \\ &= a^5 - \frac{5}{2}a^4b + \frac{5}{2}a^3b^2 - \frac{5}{4}a^2b^3 + \frac{5}{16}ab^4 - \frac{1}{32}b^5 \quad \blacksquare\end{aligned}$$

5 Any single term of a binomial expansion can be determined without writing out the whole expansion. For example, the 7th term of $(x + y)^9$ has y raised to the 6th power (since y has the power 1 in the second term, the power 2 in the third term, and so on). The exponents on x and y in each term must have a sum of 9, so the exponent on x in the 7th term is $9 - 6 = 3$. Thus, writing the coefficient as given in the binomial theorem, the 7th term should be

$$\frac{9!}{3!6!}x^3y^6.$$

This is in fact the 7th term of $(x + y)^9$ found in Example 3 above. This discussion suggests the next theorem.

rTH TERM OF THE BINOMIAL EXPANSION

The rth term of the binomial expansion of $(x + y)^n$, where $n \geq r - 1$, is

$$\binom{n}{n-(r-1)} x^{n-(r-1)} y^{r-1}.$$

EXAMPLE 5

FINDING A SPECIFIC TERM IN A BINOMIAL EXPANSION

Find the fourth term of $(a + 2b)^{10}$.

Start by identifying x, y, n, and r. Since we want the fourth term in the expansion of $(a + 2b)^{10}$, we have

$$x = a, \qquad y = 2b, \qquad n = 10, \qquad \text{and} \qquad r = 4.$$

Substitute these into the formula above.

$$\begin{aligned}
&\binom{10}{10-(4-1)} a^{10-(4-1)}(2b)^{4-1} \\
&= \binom{10}{7} a^7 (2b)^3 \\
&= 120a^7(8b^3) \\
&= 960a^7b^3. \quad \blacksquare
\end{aligned}$$

CAUTION When applying the binomial theorem and the formula for the rth term of $(x + y)^n$, be sure to use parentheses for the factors if necessary. For example, a common error in Example 5 is to forget that the *product* $2b$ must be cubed; that is, we must write $(2b)^3$ rather than $2b^3$ when finding the specified term of the expansion.

13.5 EXERCISES

1. The first seven rows of Pascal's triangle are given in this section. Write the next four rows of the triangle. (See Objective 1.)

2. Suppose that you evaluate $\frac{3!}{0!}$ correctly as 6, but your friend says that it is undefined, since division by zero is undefined. How would you respond? (See Objective 2.)

Evaluate each of the following factorials. Use a calculator for Exercises 9 and 10. See Example 1.

3. $4!$ **4.** $9!$ **5.** $6!$ **6.** $8!$

7. $10!$ **8.** $11!$ **9.** $20!$ **10.** $21!$

Evaluate each binomial coefficient. See Example 2.

11. $\binom{3}{2}$ **12.** $\binom{4}{3}$ **13.** $\binom{5}{2}$ **14.** $\binom{5}{3}$

15. $\binom{7}{3}$ **16.** $\binom{7}{2}$ **17.** $\binom{10}{8}$ **18.** $\binom{10}{7}$

Write out the binomial expansion for each of the following. See Examples 3 and 4.

19. $(x + y)^6$ **20.** $(m + n)^4$ **21.** $(p - q)^5$ **22.** $(a - b)^7$

23. $(r^2 + s)^5$ **24.** $(m + n^2)^4$ **25.** $(p + 2q)^4$ **26.** $(3r - s)^6$

27. $(7p + 2q)^4$ **28.** $(4a - 5b)^5$ **29.** $(3x - 2y)^6$ **30.** $(7k - 9j)^4$

31. $\left(\frac{m}{2} - 1\right)^6$ **32.** $\left(3 + \frac{y}{3}\right)^5$ **33.** $\left(2p + \frac{q}{3}\right)^4$ **34.** $\left(\frac{r}{6} - \frac{m}{2}\right)^3$

For each of the following write the indicated term of the binomial expansion. See Example 5.

35. 5th term of $(m - 2p)^{12}$ **36.** 4th term of $(3x + y)^6$ **37.** 6th term of $(x + y)^9$

38. 12th term of $(a - b)^{15}$ **39.** 9th term of $(2m + n)^{10}$ **40.** 7th term of $(3r - 5s)^{12}$

41. 17th term of $(p^2 + q)^{20}$ **42.** 10th term of $(2x^2 + y)^{14}$ **43.** 8th term of $(x^3 + 2y)^{14}$

44. 13th term of $(a + 2b^3)^{12}$

The binomial theorem can be used to approximate powers of numbers that are close to some integer. For example, we can approximate $(1.01)^8$ *correct to three decimal places as follows.*

$$\begin{aligned}(1.01)^8 &= (1 + .01)^8\\ &= 1^8 + \binom{8}{7}(1)^7(.01) + \binom{8}{6}(1)^6(.01)^2 + \ldots + (.01)^8\\ &= 1 + 8(1)(.01) + 28(1)(.0001) + \ldots + (.0000000000000001)\\ &= 1 + .08 + .0028 + \ldots + .0000000000000001\\ (1.01)^8 &\approx 1.083\end{aligned}$$

Since only three decimal places of accuracy are required, there is no need to evaluate all the terms of the expansion. Additional terms, however, would give the answer to more decimal places of accuracy.

Use the binomial expansion to evaluate each of the following to three decimal places. Check with a calculator.

45. $(1.10)^{10}$ **46.** $(.99)^{15}$ (*Hint:* $.99 = 1 - .01$) **47.** $(1.99)^8$

48. $(2.99)^3$ **49.** $(3.02)^6$ **50.** $(1.01)^9$

In calculus it is shown that

$$(1 + x)^n = 1 + nx + \frac{n(n-1)}{2!}x^2 + \frac{n(n-1)(n-2)}{3!}x^3 + \ldots$$

for any real number n (not just positive integers) and any real number x, where $|x| < 1$. *This result, a generalized binomial theorem, may be used to find approximate values of powers or roots. For example,*

$$\sqrt[4]{630} = (625 + 5)^{1/4} = \left[625\left(1 + \frac{5}{625}\right)\right]^{1/4} = 625^{1/4}\left(1 + \frac{5}{625}\right)^{1/4}$$

In Exercises 51–54, check with a calculator.

51. Use the expression given on page 718 for $(1 + x)^n$ to approximate $(1 + 5/625)^{1/4}$ to the nearest thousandth. Then approximate $\sqrt[4]{630}$.

52. Approximate $\sqrt[3]{9.42}$, using this method.

53. Approximate $(1.02)^{-3}$.

54. Approximate $1/(1.04)^5$.

PREVIEW EXERCISES

Evaluate each expression for $n = 1$. See Section 2.2.

55. $\dfrac{5n(n + 1)}{2}$

56. $\dfrac{3(3^n - 1)}{2}$

57. $\dfrac{n^2(n + 1)^2}{4}$

58. $\dfrac{n(n + 1)(2n + 1)}{6}$

13.6 MATHEMATICAL INDUCTION

OBJECTIVES

1 LEARN THE PRINCIPLE OF MATHEMATICAL INDUCTION.

2 USE THE PRINCIPLE OF MATHEMATICAL INDUCTION TO PROVE A STATEMENT.

Many results in mathematics are claimed true for any positive integer. Any of these results could be checked for $n = 1$, $n = 2$, $n = 3$, and so on, but since the set of positive integers is infinite it would be impossible to check every possible case. For example, let S_n represent the statement that the sum of the first n positive integers is $n(n + 1)/2$,

$$S_n\text{: } 1 + 2 + 3 + \ldots + n = \frac{n(n + 1)}{2}.$$

The truth of this statement can be checked quickly for the first few values of n.

If $n = 1$, S_1 is $\quad 1 = \dfrac{1(1 + 1)}{2}$, a true statement, since $1 = 1$.

If $n = 2$, S_2 is $\quad 1 + 2 = \dfrac{2(2 + 1)}{2}$, a true statement, since $3 = 3$.

If $n = 3$, S_3 is $\quad 1 + 2 + 3 = \dfrac{3(3 + 1)}{2}$, a true statement, since $6 = 6$.

If $n = 4$, S_4 is $\quad 1 + 2 + 3 + 4 = \dfrac{4(4 + 1)}{2}$, a true statement, since $10 = 10$.

Since the statement is true for $n = 1$, 2, 3, and 4, can we conclude that the statement is true for all positive integers by observing this finite number of examples? The answer is no. However, we have an idea that it *may* be true for all positive integers.

1 To prove that such a statement is true for every positive integer, we use the following principle.

PRINCIPLE OF MATHEMATICAL INDUCTION

Let S_n be a statement concerning the positive integer n. Suppose that

(1) S_1 is true;
(2) for any positive integer k, $k \leq n$, if S_k is true, then S_{k+1} is also true.

Then S_n is true for every positive integer value of n.

A proof by mathematical induction can be explained as follows. By assumption (1) above, the statement is true when $n = 1$. By (2) above, the fact that the statement is true for $n = 1$ implies that it is true for $n = 1 + 1 = 2$. Using (2) again, the statement is thus true for $2 + 1 = 3$, for $3 + 1 = 4$, for $4 + 1 = 5$, and so on. Continuing in this way shows that the statement must be true for *every* positive integer, no matter how large.

The situation is similar to that of a number of dominoes lined up as shown in Figure 13.1. If the first domino is pushed over, it pushes the next, which pushes the next, and so on until all are down.

FIGURE 13.1

Another example of the principle of mathematical induction might be an infinite ladder. Suppose the rungs are spaced so that, whenever you are on a rung, you know you can move to the next rung. Then *if* you can get to the first rung, you can go as high up the ladder as you wish.

Two separate steps are required for a proof by mathematical induction:

PROOF BY MATHEMATICAL INDUCTION

Step 1 Prove that the statement is true for $n = 1$.
Step 2 Show that, for any positive integer k, $k \leq n$, if S_k is true, then S_{k+1} is also true.

2 Mathematical induction is used in the next example to prove the statement discussed above: S_n: $1 + 2 + 3 + \ldots + n = \frac{n(n+1)}{2}$.

EXAMPLE 1 PROVING A STATEMENT BY MATHEMATICAL INDUCTION

Let S_n represent the statement

$$1 + 2 + 3 + \ldots + n = \frac{n(n+1)}{2}.$$

Prove that S_n is true for every positive integer n.

Proof

The proof by mathematical induction is as follows.

Step 1 Show that the statement is true when $n = 1$. If $n = 1$, S_1 becomes

$$1 = \frac{1(1 + 1)}{2},$$

which is true.

Step 2 Show that S_k implies S_{k+1}, where S_k is the statement

$$1 + 2 + 3 + \ldots + k = \frac{k(k + 1)}{2},$$

and S_{k+1} is the statement

$$1 + 2 + 3 + \ldots + k + (k + 1) = \frac{(k + 1)[(k + 1) + 1]}{2}.$$

To show that S_k implies S_{k+1}, start with S_k.

$$1 + 2 + 3 + \ldots + k = \frac{k(k + 1)}{2}$$

Add $k + 1$ to both sides of this equation.

$$1 + 2 + 3 + \ldots + k + (k + 1) = \frac{k(k + 1)}{2} + (k + 1)$$

Now factor out the common factor $k + 1$ on the right to get

$$= (k + 1)\left(\frac{k}{2} + 1\right)$$

$$= (k + 1)\left(\frac{k + 2}{2}\right)$$

$$1 + 2 + 3 + \ldots + k + (k + 1) = \frac{(k + 1)[(k + 1) + 1]}{2}.$$

This final result is the statement for $n = k + 1$; it has been shown that if S_k is true, then S_{k+1} is also true. The two steps required for a proof by mathematical induction are now complete, so the statement S_n is true for every positive integer value of n. ■

EXAMPLE 2
PROVING A STATEMENT BY MATHEMATICAL INDUCTION

Prove: $4 + 7 + 10 + \ldots + (3n + 1) = \dfrac{n(3n + 5)}{2}$ for all positive integers n.

Proof

Step 1 Show that the statement is true for S_1. S_1 is

$$4 = \frac{1(3 \cdot 1 + 5)}{2}.$$

Since the right side equals 4, S_1 is a true statement.

Step 2 Show that if S_k is true, then S_{k+1} is true, where S_k is

$$4 + 7 + 10 + \ldots + (3k + 1) = \frac{k(3k + 5)}{2},$$

and S_{k+1} is

$$4 + 7 + 10 + \ldots + (3k + 1) + [3(\mathbf{k + 1}) + 1] = \frac{(\mathbf{k + 1})[3(\mathbf{k + 1}) + 5]}{2}.$$

Start with S_k:

$$4 + 7 + 10 + \ldots + (3k + 1) = \frac{k(3k + 5)}{2}.$$

To get the left side of the equation S_k to be the left side of the equation S_{k+1}, we must add the $(k + 1)$th term. Adding $[3(k + 1) + 1]$ to both sides of S_k gives

$$4 + 7 + 10 + \ldots + (3k + 1) + [3(k + 1) + 1] = \frac{k(3k + 5)}{2} + [3(k + 1) + 1].$$

Clear the parentheses in the new term on the right side of the equals sign and simplify.

$$= \frac{k(3k + 5)}{2} + 3k + 3 + 1$$

$$= \frac{k(3k + 5)}{2} + 3k + 4$$

Now combine the two terms on the right.

$$= \frac{k(3k + 5)}{2} + \frac{2(3k + 4)}{2}$$

$$= \frac{k(3k + 5) + 2(3k + 4)}{2}$$

$$= \frac{3k^2 + 5k + 6k + 8}{2}$$

$$= \frac{3k^2 + 11k + 8}{2}$$

$$= \frac{(k + 1)(3k + 8)}{2}$$

Since $3k + 8$ can be written as $3(k + 1) + 5$,

$$4 + 7 + 10 + \ldots + (3k + 1) + [3(\mathbf{k + 1}) + 1] = \frac{(\mathbf{k + 1})[3(\mathbf{k + 1}) + 5]}{2}.$$

The final result is the statement S_{k+1}. Therefore, if S_k is true, then S_{k+1} is true. The two steps required for a proof by mathematical induction are completed, so the general statement S_n is true for every positive integer value of n. ■

EXAMPLE 3 PROVING A STATEMENT BY MATHEMATICAL INDUCTION

Prove that if x is a real number between 0 and 1, then for every positive integer n,

$$0 < x^n < 1.$$

Proof

Here S_1 is the statement

$$\text{if } 0 < x < 1, \text{ then } 0 < x^1 < 1,$$

which is true. S_k is the statement

$$\text{if } 0 < x < 1, \text{ then } 0 < x^k < 1.$$

To show that this implies that S_{k+1} is true, multiply all members of $0 < x^k < 1$ by x to get

$$x \cdot 0 < x \cdot x^k < x \cdot 1.$$

(Here the fact that $0 < x$ is used.) Simplify to get

$$0 < x^{k+1} < x.$$

Since $x < 1$,

$$x^{k+1} < x < 1$$

and

$$0 < x^{k+1} < 1.$$

By this work, if S_k is true, then S_{k+1} is true, so the given statement is true for every positive integer n. ■

13.6 EXERCISES

Use the method of mathematical induction to prove the following statements. Assume that n is a positive integer. See Examples 1–3.

1. $2 + 4 + 6 + \ldots + 2n = n(n + 1)$

2. $1 + 3 + 5 + \ldots + (2n - 1) = n^2$

3. $3 + 6 + 9 + \ldots + 3n = \frac{3n(n + 1)}{2}$

4. $5 + 10 + 15 + \ldots + 5n = \frac{5n(n + 1)}{2}$

5. $2 + 4 + 8 + \ldots + 2^n = 2^{n+1} - 2$

6. $3 + 3^2 + 3^3 + \ldots + 3^n = \frac{3(3^n - 1)}{2}$

7. $1^2 + 2^2 + 3^2 + \ldots + n^2 = \frac{n(n + 1)(2n + 1)}{6}$

8. $1^3 + 2^3 + 3^3 + \ldots + n^3 = \frac{n^2(n + 1)^2}{4}$

9. $5 \cdot 6 + 5 \cdot 6^2 + 5 \cdot 6^3 + \ldots + 5 \cdot 6^n = 6(6^n - 1)$

10. $7 \cdot 8 + 7 \cdot 8^2 + 7 \cdot 8^3 + \ldots + 7 \cdot 8^n = 8(8^n - 1)$

11. $\dfrac{1}{1 \cdot 2} + \dfrac{1}{2 \cdot 3} + \dfrac{1}{3 \cdot 4} + \ldots + \dfrac{1}{n(n+1)} = \dfrac{n}{n+1}$

12. $\dfrac{1}{1 \cdot 4} + \dfrac{1}{4 \cdot 7} + \dfrac{1}{7 \cdot 10} + \ldots + \dfrac{1}{(3n-2)(3n+1)} = \dfrac{n}{3n+1}$

13. $\dfrac{1}{2} + \dfrac{1}{2^2} + \dfrac{1}{2^3} + \ldots + \dfrac{1}{2^n} = 1 - \dfrac{1}{2^n}$

14. $\dfrac{4}{5} + \dfrac{4}{5^2} + \dfrac{4}{5^3} + \ldots + \dfrac{4}{5^n} = 1 - \dfrac{1}{5^n}$

15. $x^{2n} + x^{2n-1}y + \ldots + xy^{2n-1} + y^{2n} = \dfrac{x^{2n+1} - y^{2n+1}}{x - y}$

16. $x^{2n-1} + x^{2n-2}y + \ldots + xy^{2n-2} + y^{2n-1} = \dfrac{x^{2n} - y^{2n}}{x - y}$

17. $(a^m)^n = a^{mn}$ (Assume that a and m are constant.)

18. $(ab)^n = a^n b^n$ (Assume that a and b are constant.)

19. If $a > 1$, then $a^n > 1$.

20. If $a > 1$, then $a^n > a^{n-1}$.

21. If $0 < a < 1$, then $a^n < a^{n-1}$.

---

22. A proof by mathematical induction allows us to prove that a statement is true for all ______. (See Objective 1.)

23. Suppose that Step 2 in a proof by mathematical induction can be satisfied, but Step 1 cannot. May we conclude that the proof is complete? Explain. (See Objective 2.)

24. What is wrong with the following proof by mathematical induction?

Prove: Any natural number equals the next natural number; that is, $n = n + 1$.

Proof. To begin, we assume the statement true for some natural number $n = k$:

$$k = k + 1.$$

We must now show that the statement is true for $n = k + 1$. If we add 1 to both sides, we have

$$k + 1 = k + 1 + 1$$
$$k + 1 = k + 2.$$

Hence, if the statement is true for $n = k$, it is also true for $n = k + 1$. Thus, the theorem is proved.

25. Suppose that n straight lines (with $n \geq 2$) are drawn in a plane, where no two lines are parallel and no three lines pass through the same point. Show that the number of points of intersection of the lines is $(n^2 - n)/2$.

26. The series of sketches below starts with an equilateral triangle having sides of length 1. In the following steps, equilateral triangles are constructed on each side of the preceding figure. The lengths of the sides of these new triangles are 1/3 the length of the sides of the preceding triangles. Develop a formula for the number of sides of the nth figure. Use mathematical induction to prove your answer.

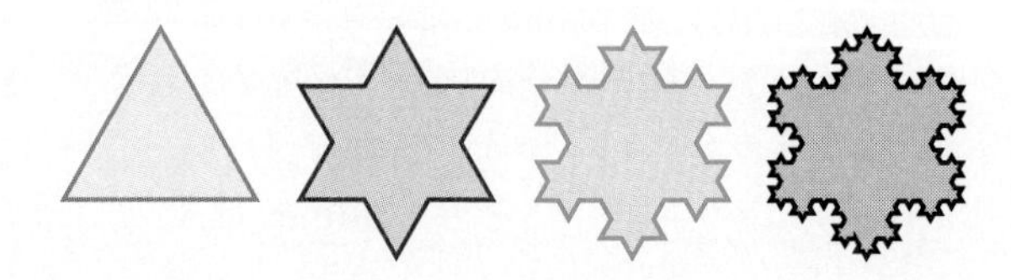

27. Find the perimeter of the nth figure in Exercise 26.

28. Show that the area of the nth figure in Exercise 26 is

$$\sqrt{3}\left[\frac{2}{5} - \frac{3}{20}\left(\frac{4}{9}\right)^{n-1}\right].$$

PREVIEW EXERCISES

Evaluate each of the following. See Section 13.5.

29. $5!$ **30.** $7!$ **31.** $\binom{8}{3}$ **32.** $\binom{9}{7}$

13.7 PERMUTATIONS

OBJECTIVES

1. USE THE FUNDAMENTAL PRINCIPLE OF COUNTING.
2. KNOW THE FORMULA FOR $P(n, r)$.
3. USE PERMUTATIONS TO SOLVE COUNTING PROBLEMS.
4. USE THE FUNDAMENTAL COUNTING PRINCIPLE WITH RESTRICTIONS.

FOCUS ON PROBLEM SOLVING

An automaker produces 7 models, each available in 6 colors, with 4 upholstery fabrics and 5 interior colors. How many varieties of the auto are available?

This problem is an example of a *counting* problem. In this section and the next, we discuss techniques to solve problems of this type. An important principle known as the fundamental principle of counting is introduced in this section, along with a formula for finding the number of permutations, or arrangements, of objects. This problem is Exercise 19 in the exercises for this section. After working through this section, you should be able to solve this problem.

If there are 3 roads from Albany to Baker and 2 roads from Baker to Creswich, in how many ways can one travel from Albany to Creswich by way of Baker? For each of the 3 roads from Albany to Baker, there are 2 different roads from Baker to Creswich, so that there are $3 \cdot 2 = 6$ different ways to make the trip, as shown in the **tree diagram** in Figure 13.2.

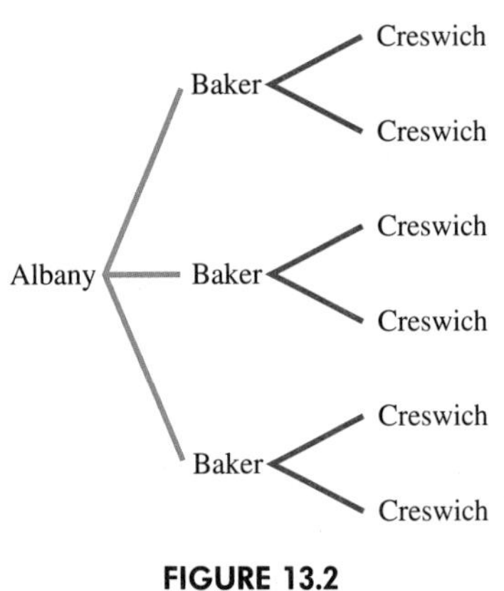

FIGURE 13.2

This example illustrates the following property.

FUNDAMENTAL PRINCIPLE OF COUNTING

If one event can occur in m ways and a second event can occur in n ways, then both events can occur in mn ways, provided the outcome of the first event does not influence the outcome of the second.

1 The fundamental principle of counting can be extended to any number of events, provided the outcome of no one event influences the outcome of another. Such events are called **independent events.**

EXAMPLE 1 USING THE FUNDAMENTAL PRINCIPLE OF COUNTING

A restaurant offers a choice of 3 salads, 5 main dishes, and 2 desserts. Use the fundamental principle of counting to find the number of different 3-course meals that can be selected.

Three independent events are involved: selecting a salad, selecting a main dish, and selecting a dessert. The first event can occur in 3 ways, the second event can occur in 5 ways, and the third event can occur in 2 ways; thus there are

$$3 \cdot 5 \cdot 2 = 30 \text{ possible meals.} \quad \blacksquare$$

EXAMPLE 2 USING THE FUNDAMENTAL PRINCIPLE OF COUNTING

Janet Branson has 5 different books that she wishes to arrange on her desk. How many different arrangements are possible?

Five events are involved: selecting a book for the first spot, selecting a book for the second spot, and so on. Here the outcome of the first event *does* influence the outcome of the other events (since one book has already been chosen). For the first spot Branson has 5 choices, for the second spot 4 choices, for the third spot 3 choices, and so on. Now use the fundamental principle of counting to find that there are

$$5 \cdot 4 \cdot 3 \cdot 2 \cdot 1 \text{ or } 120 \text{ different arrangements.} \quad \blacksquare$$

In using the fundamental principle of counting we often encounter such products as $5 \cdot 4 \cdot 3 \cdot 2 \cdot 1$ from Example 2. For convenience in writing these products, use the symbol $n!$ (read "n factorial"), which was defined in Section 13.5 and is repeated here.

***n*-FACTORIAL ($n!$)**

For any positive integer n:

$$n! = n(n-1)(n-2) \ldots (3)(2)(1)$$

and

$$0! = 1.$$

By the definition, $5 \cdot 4 \cdot 3 \cdot 2 \cdot 1$ is written as $5!$. Also, $3! = 3 \cdot 2 \cdot 1 = 6$. The definition of $n!$ means that $n[(n-1)!] = n!$ for all natural numbers $n \geq 2$. It is useful to have this relation hold also for $n = 1$, so, by definition, $0! = 1$.

EXAMPLE 3 SOLVING A PROBLEM INVOLVING ARRANGEMENTS

Suppose Branson (from Example 2) wishes to place only 3 of the 5 books on her desk. How many arrangements of 3 books are possible?

She still has 5 ways to fill the first spot, 4 ways to fill the second spot, and 3 ways to fill the third. Since she wants to use 3 books, there are only 3 spots to be filled (3 events) instead of 5, so there are $5 \cdot 4 \cdot 3 = 60$ arrangements. $\blacksquare$

2 The number 60 in the example above is called the number of *permutations* of 5 things taken 3 at a time, written $P(5, 3) = 60$. Example 2 showed that the number of ways of arranging 5 elements from a set of 5 elements, written $P(5, 5)$, is 120.

A **permutation** of n elements taken r at a time is one of the ways of arranging r elements taken from a set of n elements ($r \le n$). Generalizing from the examples above, the number of permutations of n elements taken r at a time, denoted by $P(n, r)$, is

$$\begin{aligned} P(n, r) &= n(n-1)(n-2)\cdots(n-r+1) \\ &= \frac{n(n-1)(n-2)\cdots(n-r+1)(n-r)(n-r-1)\cdots(2)(1)}{(n-r)(n-r-1)\cdots(2)(1)} \\ &= \frac{n!}{(n-r)!}. \end{aligned}$$

This proves the following result.

PERMUTATIONS OF n ELEMENTS TAKEN r AT A TIME

If $P(n, r)$ denotes the number of permutations of n elements taken r at a time, $r \le n$, then

$$P(n, r) = \frac{n!}{(n-r)!}.$$

Some other symbols used for permutations of n things taken r at a time are ${}_nP_r$ and P_r^n.

NOTE Some scientific calculators have a key for permutations. Consult the owner's manual for instructions on how to use it.

3 In the next two examples, we use the formula to solve permutation problems.

EXAMPLE 4 USING THE PERMUTATIONS FORMULA

Suppose 8 people enter an event in a swim meet. Assuming there are no ties, in how many ways could the gold, silver, and bronze prizes be awarded?

Using the fundamental principle of counting, there are 3 choices to be made giving $8 \cdot 7 \cdot 6 = 336$. However, we can also use the formula for $P(n, r)$ to get the same result.

$$P(8, 3) = \frac{8!}{(8-3)!} = \frac{8!}{5!} = \frac{8 \cdot 7 \cdot 6 \cdot 5 \cdot 4 \cdot 3 \cdot 2 \cdot 1}{5 \cdot 4 \cdot 3 \cdot 2 \cdot 1} = 8 \cdot 7 \cdot 6 = 336 \quad \blacksquare$$

EXAMPLE 5 USING THE PERMUTATIONS FORMULA

In how many ways can 6 students be seated in a row of 6 desks?

$$P(6, 6) = \frac{6!}{(6-6)!} = \frac{6!}{0!} = 6! = 6 \cdot 5 \cdot 4 \cdot 3 \cdot 2 \cdot 1 = 720 \quad \blacksquare$$

4 The final example involves using the fundamental counting principle with some restrictions.

EXAMPLE 6 USING THE FUNDAMENTAL COUNTING PRINCIPLE WITH RESTRICTIONS

In how many ways can three letters of the alphabet be arranged if a vowel cannot be used in the middle position, and repetitions of the letters are allowed?

We cannot use $P(26, 3)$ here, because of the restriction for the middle position. In the first and third positions, we can use any of the 26 letters of the alphabet, but in the middle position, we can only use one of $26 - 5 = 21$ letters (since there are 5 vowels). Now, using the fundamental counting principle, there are $26 \cdot 21 \cdot 26 = 14{,}196$ ways to arrange the letters according to the problem. ■

13.7 EXERCISES

Evaluate each of the following. See Examples 4 and 5.

1. $P(7, 7)$ **2.** $P(5, 3)$ **3.** $P(6, 5)$ **4.** $P(4, 2)$

5. $P(10, 2)$ **6.** $P(8, 2)$ **7.** $P(8, 3)$ **8.** $P(11, 4)$

9. $P(7, 1)$ **10.** $P(18, 0)$ **11.** $P(9, 0)$ **12.** $P(14, 1)$

13. Explain in your own words what is meant by *permutation*. (See Objective 2.)

14. Explain why the restriction $r \leq n$ is needed in the formula for $P(n, r)$. (See Objective 2.)

Solve each problem. See Examples 1–6.

15. In how many ways can 8 people be seated in a row of 8 seats?

16. In how many ways can 7 out of 10 people be assigned to 7 seats?

17. In how many ways can 5 players be assigned to the 5 positions on a basketball team, assuming that any player can play any position? In how many ways can 10 players be assigned to the 5 positions?

18. A couple has narrowed down their choice of names for a new baby to 3 first names and 5 middle names. How many different first and middle name arrangements are possible?

19. An automaker produces 7 models, each available in 6 colors, with 4 upholstery fabrics and 5 interior colors. How many varieties of the auto are there?

20. How many different homes are available if a builder offers a choice of 5 basic plans, 3 roof styles, and 2 types of siding?

21. A concert is to consist of five works: two modern, two romantic, and one classical. In how many ways can the program be arranged?

22. If the program in Exercise 21 must be shortened to 3 works chosen from the 5, how many arrangements are possible?

23. In Exercise 21, how many different programs are possible if the 2 modern works are to be played first, then the 2 romantic, and then the classical?

24. How many 4-letter radio station call letters can be made if the first letter must be K or W and no letter may be repeated? How many if repeats are allowed?

Photo for Exercise 19

25. How many of the 4-letter call letters in Exercise 24 with no repeats end in K?

26. How many 7-digit telephone numbers are possible if the first digit cannot be zero and
(a) only odd digits can be used;
(b) the number must be a multiple of 10 (that is, must end in zero);
(c) the number must be a multiple of 100;
(d) the first three digits are 481;
(e) no repetitions are allowed?

27. **(a)** How many different license numbers consisting of 3 letters followed by 3 digits are possible?
(b) How many are possible if either the 3 digits or the 3 letters can come first?

28. A business school gives courses in typing, shorthand, transcription, business English, technical writing, and accounting. How many ways can a student arrange his program if he takes 3 courses?

29. How many different license plate numbers can be formed using 3 letters followed by 3 digits if no repeats are allowed? How many if there are no repeats and either letters or numbers come first?

30. Show that $P(n, n-1) = P(n, n)$.

PREVIEW EXERCISES

Evaluate each binomial coefficient. See Section 13.5.

31. $\binom{9}{6}$ **32.** $\binom{10}{8}$ **33.** $\binom{3}{3}$ **34.** $\binom{4}{0}$ **35.** $\binom{n}{n}$ **36.** $\binom{n}{0}$

13.8 COMBINATIONS

OBJECTIVES

1 LEARN THE FORMULA $\binom{n}{r}$ FOR COMBINATIONS.
2 USE THE COMBINATIONS FORMULA TO SOLVE COUNTING PROBLEMS.
3 USE THE COMBINATIONS FORMULA TO FIND PROPERTIES FOR SPECIAL CASES.

FOCUS ON PROBLEM SOLVING

Hal's Hamburger Heaven sells hamburgers with cheese, relish, lettuce, tomato, mustard, or ketchup. How many different kinds of hamburgers can be made using any 3 of the extras?

In the previous section we learned about permutations, or arrangements. In the problem above, the order in which the extras are placed on the hamburger does not matter. When order does not matter, we have a *combination* rather than a permutation. In this section we develop a formula for combinations. This problem is Exercise 19 in the exercises for this section. After working through this section, you should be able to solve this problem.

In the preceding section we discussed a method for finding the number of ways to arrange r elements taken from a set of n elements. Sometimes however, the arrangement (or order) of the elements is not important.

For example, suppose three people (Ms. Opelka, Mr. Adams, and Ms. Jacobs) apply for 2 identical jobs. Ignoring all other factors, in how many ways can the personnel officer select 2 people from the 3 applicants? Here the arrangement or order of the people is unimportant. Selecting Ms. Opelka and Mr. Adams is the same as selecting Mr. Adams and Ms. Opelka. Therefore, there are only 3 ways to select 2 of the 3 applicants:

Ms. Opelka and Mr. Adams;
Ms. Opelka and Ms. Jacobs;
Mr. Adams and Ms. Jacobs.

1 These three choices are called the *combinations* of 3 elements taken 2 at a time. A **combination** of n elements taken r at a time is one of the ways in which r elements can be chosen from n elements.

Each combination of r elements forms $r!$ permutations. Therefore, the number of combinations of n elements taken r at a time is found by dividing the number of permutations, $P(n, r)$, by $r!$ to get

$$\frac{P(n, r)}{r!}$$

combinations. This expression can be rewritten as follows

$$\frac{P(n, r)}{r!} = \frac{\frac{n!}{(n-r)!}}{r!} = \frac{n!}{(n-r)!r!}$$

The symbol $\binom{n}{r}$ is used to represent the number of combinations of n things taken r at a time. (Recall that the quantity expressed by $\binom{n}{r}$ also gives the coefficients in the binomial theorem.) With this symbol the results above are stated as follows.

COMBINATIONS OF n ELEMENTS TAKEN r AT A TIME

If $\binom{n}{r}$ represents the number of combinations of n elements taken r at a time, for $r \leq n$, then

$$\binom{n}{r} = \frac{n!}{(n-r)!r!}.$$

Recall that other symbols used for $\binom{n}{r}$ are $C(n, r)$, ${}_nC_r$, and C^n_r.

2 In the discussion above, it was shown that $\binom{3}{2} = 3$. The same result can be found using the formula.

$$\binom{3}{2} = \frac{3!}{(3-2)!2!} = \frac{3 \cdot 2 \cdot 1}{1 \cdot 2 \cdot 1} = 3$$

The combinations formula is used in the next examples.

EXAMPLE 1 USING THE COMBINATIONS FORMULA

How many different committees of 3 people can be chosen from a group of 8 people?

Since the order in which the members of the committee are chosen does not affect the result, use combinations to get

$$\binom{8}{3} = \frac{8!}{5!3!} = \frac{8 \cdot 7 \cdot 6 \cdot 5 \cdot 4 \cdot 3 \cdot 2 \cdot 1}{5 \cdot 4 \cdot 3 \cdot 2 \cdot 1 \cdot 3 \cdot 2 \cdot 1} = 56. \quad ■$$

EXAMPLE 2 USING THE COMBINATIONS FORMULA

From a group of 30 employees, 3 are to be selected to work on a special project.

(a) In how many different ways can the employees be selected?

The number of 3-element combinations from a set of 30 elements must be found. (Use combinations, not permutations, because order within the group of 3 does not affect the result.) Using the formula gives

$$\binom{30}{3} = \frac{30!}{27!3!} = 4060.$$

There are 4060 ways to select the project group.

(b) In how many different ways can the group of 3 be selected if it has already been decided that a certain employee must work on the project?

Since one employee has already been selected to work on the project, the problem is reduced to selecting 2 more employees from the 29 employees that are left:

$$\binom{29}{2} = \frac{29!}{27!2!} = 406.$$

In this case, the project group can be selected in 406 different ways. ■

EXAMPLE 3 USING THE COMBINATIONS FORMULA AND THE FUNDAMENTAL PRINCIPLE OF COUNTING

A congressional committee consists of four senators and six representatives. A delegation of five members is to be chosen. In how many ways could this delegation include exactly three senators?

"Exactly three senators" implies that there must be $5 - 3 = 2$ representatives as well. The three senators could be chosen in $\binom{4}{3} = 4$ ways, while the two representatives could be chosen in $\binom{6}{2} = 15$ ways. Now apply the fundamental principle of counting to find that there are $4 \cdot 15 = 60$ ways to form the committee. ■

PROBLEM SOLVING

Students often have difficulty determining whether to use permutations or combinations in solving problems. The chart below lists some of the similarities and differences between these two concepts.

PERMUTATIONS	COMBINATIONS
Number of ways of selecting r items out of n items	
Repetitions are not allowed	
Order is important	Order is not important
Arrangements of r items from a set of n items	Subsets of r items from a set of n items
$P(n, r) = \frac{n!}{(n - r)!}$	$\binom{n}{r} = \frac{n!}{(n - r)!r!}$ ■

3 The next examples show how the formula for combinations can be used to find properties that apply to special cases.

EXAMPLE 4 USING THE COMBINATIONS FORMULA FOR A SPECIAL CASE

Find $\binom{n}{n}$.

Use the formula with $r = n$.

$$\binom{n}{n} = \frac{n!}{(n-n)!n!} = \frac{n!}{0!n!} = \frac{n!}{1 \cdot n!} = 1$$

This is reasonable since there is only one way of selecting a group of n objects from a set of n objects when order is disregarded. ■

EXAMPLE 5 USING THE COMBINATIONS FORMULA FOR A SPECIAL CASE

Find $\binom{n}{1}$.

Again use the formula.

$$\binom{n}{1} = \frac{n!}{(n-1)!1!} = \frac{n \cdot (n-1)!}{(n-1)!1!} = n$$

This result agrees with what is already known: there are n distinct one-element subsets of an n-element set. ■

The results of Examples 4 and 5 can be stated as properties and applied directly to special cases. Thus, from Example 4,

$$\binom{5}{5} = 1, \quad \binom{8}{8} = 1,$$

and so on. Also, from Example 5,

$$\binom{6}{1} = 6 \quad \text{and} \quad \binom{9}{1} = 9.$$

EXAMPLE 6 VERIFYING A SPECIAL COMBINATIONS PROPERTY

Show that $\binom{5}{3} = \binom{5}{2}$.

From the formula,

$$\binom{5}{3} = \frac{5!}{2!3!} \quad \text{and} \quad \binom{5}{2} = \frac{5!}{3!2!}$$

Since multiplication is commutative,

$$\binom{5}{3} = \binom{5}{2}. \quad ■$$

Example 6 shows that there are the same number of ways to select three things from a group of five as to select two things from a group of five. This makes sense, because each time you choose a set of two, you are also choosing a set of three, the three not selected.

Example 6 is a special case of the following result.

If $\binom{n}{r}$ represents the number of combinations of n things taken r at a time, then

$$\binom{n}{r} = \binom{n}{n-r}.$$

The proof of this result is left as an exercise.

13.8 EXERCISES

Evaluate each of the following. See Examples 1–6.

1. $\binom{6}{5}$ **2.** $\binom{4}{2}$ **3.** $\binom{8}{5}$ **4.** $\binom{10}{2}$

5. $\binom{15}{4}$ **6.** $\binom{9}{3}$ **7.** $\binom{10}{7}$ **8.** $\binom{10}{3}$

9. $\binom{14}{1}$ **10.** $\binom{20}{2}$ **11.** $\binom{18}{0}$ **12.** $\binom{13}{0}$

13. Explain the difference between a permutation and a combination. (See Objective 1.)

14. Is a telephone number an example of a permutation of digits or a combination of digits? Explain. (See Objective 1 and Objective 2 of Section 13.7.)

15. Padlocks with digit dials are often referred to as "combination locks." According to the mathematical definition of combination, is this an accurate description? Explain. (See Objective 1 and Objective 2 of Section 13.7.)

16. Describe each of the following as either a permutation or a combination. (See Objective 1 and Objective 2 of Section 13.7.)

(a) your social security number

(b) your choice of clothes from your wardrobe that you are wearing today

(c) a committee of school board members

(d) an automobile license plate number

Solve each problem. See Examples 1–3.

17. A club has 30 members. If a committee of 4 is selected at random, how many committees are possible?

18. How many different samples of 3 apples can be drawn from a crate of 25 apples?

19. Hal's Hamburger Heaven sells hamburgers with cheese, relish, lettuce, tomato, mustard, or ketchup. How many different kinds of hamburgers can be made using any 3 of the extras?

20. Three students are to be selected from a group of 12 students to participate in a special class. In how many ways can this be done? In how many ways can the group that will not participate be selected?

21. How many different 2-card hands can be dealt from a deck of 52 cards?

22. How many different 13-card bridge hands can be dealt from a deck of 52 cards?

23. Five cards are marked with the numbers 1, 2, 3, 4, and 5, shuffled, and 2 cards are then drawn. How many different 2-card combinations are possible?

24. If a bag contains 15 marbles, how many samples of 2 marbles can be drawn from it? How many samples of 4 marbles?

25. In Exercise 24, if the bag contains 3 yellow, 4 white, and 8 blue marbles, how many samples of 2 can be drawn in which both marbles are blue?

26. In Exercise 18, if it is known that there are 5 rotten apples in the crate:

(a) How many samples of 3 could be drawn in which all 3 are rotten?

(b) How many samples of 3 could be drawn in which there are 1 rotten apple and 2 good apples?

27. A city council is composed of 5 liberals and 4 conservatives. Three members are to be selected randomly as delegates to a convention.

(a) How many delegations are possible?

(b) How many delegations could have all liberals?

(c) How many delegations could have 2 liberals and 1 conservative?

(d) If 1 member of the council serves as mayor, how many delegations are possible which include the mayor?

28. Seven workers decide to send a delegation of 2 to their supervisor to discuss their grievances.

(a) How many different delegations are possible?

(b) If it is decided that a certain employee must be in the delegation, how many different delegations are possible?

(c) If there are 2 women and 5 men in the group, how many delegations would include at least one woman?

A poker hand is made up of 5 cards drawn at random from a deck of 52 cards. Any 5 cards in one suit are called a flush. The 5 highest cards, that is, the A, K, Q, J, and 10, of any one suit are called a royal flush. Use combinations to set up each of the following. Use a calculator to evaluate.

29. Find the total number of all possible poker hands.

30. How many royal flushes in hearts are possible?

31. How many royal flushes in any of the four suits are possible?

32. How many flushes in hearts are possible?

33. How many flushes in any of the four suits are possible?

34. How many combinations of 3 aces and 2 eights are possible?

Solve the following problems by using either combinations or permutations.

35. How many ways can the letters of the word TOUGH be arranged?

36. If Matthew has 8 courses to choose from, how many ways can he arrange his schedule if he must pick 4 of them?

37. How many samples of 3 pineapples can be drawn from a crate of 12?

38. El-ham specializes in making different vegetable soups with carrots, celery, beans, peas, mushrooms, and potatoes. How many different soups can she make using any 4 ingredients?

For Exercises 39 and 40, see Examples 4–6.

39. Show that $\binom{n}{r} = \binom{n}{n-r}$.

40. Find $\binom{n}{n-1}$.

PREVIEW EXERCISES

Perform the indicated operations and reduce the answers to lowest terms.

41. $\frac{5}{36} + \frac{1}{36} - \frac{2}{36}$

42. $\frac{13}{36} + \frac{2}{36} - \frac{1}{36}$

43. $\frac{12}{52} + \frac{1}{26} - \frac{1}{52}$

44. $\frac{13}{52} + \frac{3}{26} - \frac{1}{52}$

13.9 PROBABILITY

OBJECTIVES

1 LEARN THE TERMINOLOGY OF PROBABILITY THEORY.

2 FIND THE PROBABILITY OF AN EVENT.

3 FIND THE PROBABILITY OF THE COMPLEMENT OF *E*, GIVEN THE PROBABILITY OF *E*.

4 FIND THE ODDS IN FAVOR OF AN EVENT.

5 FIND THE PROBABILITY OF A COMPOUND EVENT.

FOCUS ON PROBLEM SOLVING

Mrs. Elliott invites 10 relatives to a party: her mother, two uncles, three brothers, and four cousins. If the chances of any one guest arriving first are equally likely, find the following probabilities.

(a) The first guest is an uncle or a brother.

(b) The first guest is a brother or cousin.

(c) The first guest is a brother or her mother.

The study of probability has become increasingly popular because it has a wide range of practical applications. Most daily weather forecasts, for example, include the probability of rain for that day. Casinos and gamblers base their decisions on the laws of probability. In this section we introduce some of the basic ideas of probability.

This problem is Exercise 19 in the exercises for this section. After working through this section, you should be able to solve this problem.

1 In probability, each repetition of an experiment is called a **trial.** The possible results of each trial are called **outcomes** of the experiment. In this section, we will be concerned with outcomes that are equally likely to occur. For example, the experiment of tossing a coin has two equally likely possible outcomes: landing heads up (H) or landing tails up (T). Also, the experiment of rolling a die has 6 equally likely outcomes: landing so the face that is up shows 1, 2, 3, 4, 5, or 6 points.

The set S of all possible outcomes of a given experiment is called the **sample space** for the experiment. (In this text all sample spaces are finite.) A sample space for the experiment of tossing a coin consists of the outcomes H and T. This sample space can be written in set notation as

$$S = \{H, T\}.$$

Similarly, a sample space for the experiment of rolling a single die is

$$S = \{1, 2, 3, 4, 5, 6\}.$$

Any subset of the sample space is called an **event.** In the experiment with the die, for example, "the number showing is a three" is an event, say E_1, such that $E_1 = \{3\}$. "The number showing is greater than three" is also an event, say E_2, such that $E_2 = \{4, 5, 6\}$. To represent the number of outcomes that belong to event E, the notation $n(E)$ is used. Then $n(E_1) = 1$ and $n(E_2) = 3$.

2 The notation $P(E)$ is used to define the *probability* of an event.

PROBABILITY OF EVENT E

The **probability** of an event E, written $P(E)$, is the ratio of the number of outcomes in sample space S that belong to event E, $n(E)$, to the total number of outcomes in sample space S, $n(S)$. That is,

$$P(E) = \frac{n(E)}{n(S)}.$$

This definition is used to find the probability of the event E_1 given above, by starting with the sample space for the experiment, $S = \{1, 2, 3, 4, 5, 6\}$, and the desired event, $E_1 = \{3\}$. Since $n(E_1) = 1$ and since there are 6 outcomes in the sample space,

$$P(E_1) = \frac{n(E_1)}{n(S)} = \frac{1}{6}.$$

EXAMPLE 1 FINDING PROBABILITIES OF EVENTS

A single die is rolled. Write the following events in set notation and give the probability for each event.

(a) E_3: the number showing is even

Use the definition above. Since $E_3 = \{2, 4, 6\}$, $n(E_3) = 3$. As shown above, $n(S) = 6$, so

$$P(E_3) = \frac{3}{6} = \frac{1}{2}.$$

(b) E_4: the number showing is greater than 4

Again $n(S) = 6$. Event $E_4 = \{5, 6\}$, with $n(E_4) = 2$. By the definition,

$$P(E_4) = \frac{2}{6} = \frac{1}{3}.$$

(c) E_5: the number showing is less than 7

$$E_5 = \{1, 2, 3, 4, 5, 6\} \text{ and } P(E_5) = \frac{6}{6} = 1$$

(d) E_6: the number showing is 7

$$E_6 = \emptyset \text{ and } P(E_6) = \frac{0}{6} = 0. \quad \blacksquare$$

In Example 1(c), $E_5 = S$. Therefore the event E_5 is certain to occur every time the experiment is performed. An event that is certain to occur, such as E_5, always has a probability of 1. On the other hand, $E_6 = \emptyset$ and $P(E_6)$ is 0. The probability of an impossible event, such as E_6, is always 0, since none of the outcomes in the sample space satisfy the event. For any event E, $P(E)$ is between 0 and 1 inclusive.

3 The set of all outcomes in the sample space that do *not* belong to event E is called the **complement** of E, written E'. For example, in the experiment of drawing a single card from a standard deck of 52 cards, let E be the event "the card is an ace." Then E' is the event "the card is not an ace." From the definition of E', for any event E,

$$E \cup E' = S \quad \text{and} \quad E \cap E' = \emptyset.$$

Probability concepts can be illustrated using **Venn diagrams,** as shown in Figure 13.3. The rectangle in Figure 13.3 represents the sample space in an experiment. The area inside the circle represents event E, while the area inside the rectangle, but outside the circle, represents event E'.

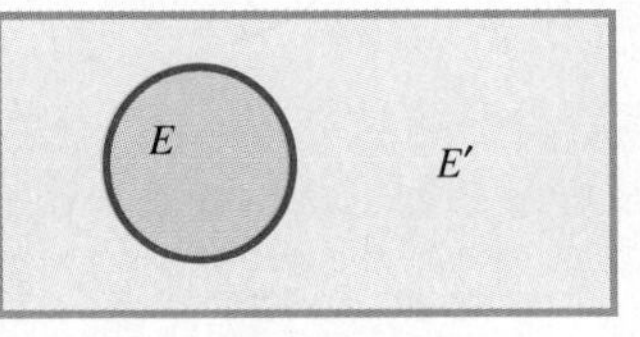

FIGURE 13.3

EXAMPLE 2
USING THE COMPLEMENT IN A PROBABILITY PROBLEM

In the experiment of drawing a card from a well-shuffled deck, find the probability of events E, the card is an ace, and E'.

Since there are four aces in the deck of 52 cards, $n(E) = 4$ and $n(S) = 52$. Therefore, $P(E) = \dfrac{n(E)}{n(S)} = \dfrac{4}{52} = \dfrac{1}{13}$. Of the 52 cards, 48 are not aces, so

$$P(E') = \frac{n(E')}{n(S)} = \frac{48}{52} = \frac{12}{13}. \quad ■$$

In Example 2, $P(E) + P(E') = (1/13) + (12/13) = 1$. This is always true for any event E and its complement E'. That is,

$$P(E) + P(E') = 1.$$

This can be restated as

$$P(E) = 1 - P(E') \qquad \text{or} \qquad P(E') = 1 - P(E).$$

These two equations suggest an alternative way to compute the probability of an event. For example, if it is known that $P(E) = 1/10$, then

$$P(E') = 1 - \frac{1}{10} = \frac{9}{10}.$$

4 Sometimes probability statements are expressed in terms of odds, a comparison of $P(E)$ with $P(E')$. The **odds** in favor of an event E are expressed as the ratio of $P(E)$ to $P(E')$ or as the fraction $P(E)/P(E')$. For example, if the probability of rain can be established as $1/3$, the odds that it will rain are

$$P(\text{rain}) \text{ to } P(\text{no rain}) = \frac{1}{3} \text{ to } \frac{2}{3} = \frac{1/3}{2/3} = \frac{1}{2} \qquad \text{or} \qquad 1 \text{ to } 2.$$

On the other hand, the odds that it will not rain are 2 to 1 (or 2/3 to 1/3). If the odds in favor of an event are, say, 3 to 5, then the probability of the event is 3/8, while the probability of the complement of the event is 5/8. If the odds favoring event E are m to n, then

$$P(E) = \frac{m}{m+n} \qquad \text{and} \qquad P(E') = \frac{n}{m+n}.$$

EXAMPLE 3
FINDING ODDS IN FAVOR OF AN EVENT

A shirt is selected at random from a dark closet containing 6 blue shirts and 4 shirts that are not blue. Find the odds in favor of a blue shirt being selected.

Let E represent "a blue shirt is selected." Then $P(E) = 6/10$ or $3/5$. Also, $P(E') = 1 - (3/5) = 2/5$. Therefore, the odds in favor of a blue shirt being selected are

$$P(E) \text{ to } P(E') = \frac{3}{5} \text{ to } \frac{2}{5} = \frac{3/5}{2/5} = \frac{3}{2} \qquad \text{or} \qquad 3 \text{ to } 2. \quad ■$$

5 A **compound event** involves an *alternative,* such as E or F, where E and F are events. For example, in the experiment of rolling a die, suppose H is the event "the result is a 3," and K is the event "the result is an even number." What is the probability of the compound event "the result is a 3 or an even number"? From the information stated above,

$$H = \{3\} \qquad P(H) = \frac{1}{6}$$

$$K = \{2, 4, 6\} \qquad P(K) = \frac{3}{6} = \frac{1}{2}$$

$$H \text{ or } K = \{2, 3, 4, 6\} \qquad P(H \text{ or } K) = \frac{4}{6} = \frac{2}{3}.$$

Notice that $P(H) + P(K) = P(H \text{ or } K)$.

Before assuming that this relationship is true in general, consider another event for this experiment, "the result is a 2," event G. Now

$$G = \{2\} \qquad P(G) = \frac{1}{6}$$

$$K = \{2, 4, 6\} \qquad P(K) = \frac{3}{6} = \frac{1}{2}$$

$$K \text{ or } G = \{2, 4, 6\} \qquad P(K \text{ or } G) = \frac{3}{6} = \frac{1}{2}.$$

In this case $P(K) + P(G) \neq P(K \text{ or } G)$.

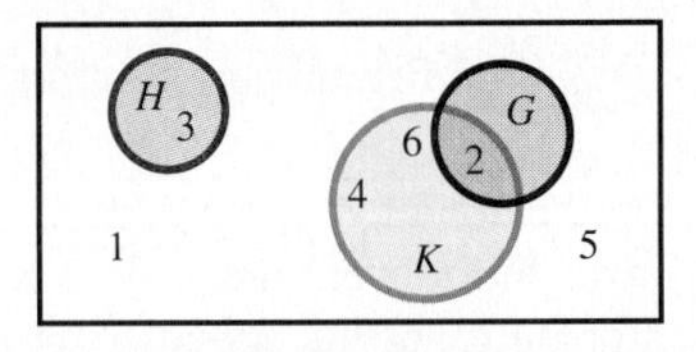

FIGURE 13.4

As Figure 13.4 shows, the difference in the two examples above comes from the fact that events H and K cannot occur simultaneously. Such events are called **mutually exclusive events.** In fact, $H \cap K = \emptyset$, which is true for any two mutually exclusive events. Events K and G, however, can occur simultaneously. Both are satisfied if the result of the roll is a 2, the element in their intersection ($K \cap G = \{2\}$). This example suggests the following property.

PROBABILITY OF ALTERNATIVE EVENTS

For any events E and F,

$$P(E \text{ or } F) = P(E \cup F) = P(E) + P(F) - P(E \cap F).$$

EXAMPLE 4
FINDING THE PROBABILITY OF ALTERNATIVE EVENTS

One card is drawn from a well-shuffled deck of 52 cards. What is the probability of the following events?

(a) The card is an ace or a spade.

The events "drawing an ace" and "drawing a spade" are not mutually exclusive since it is possible to draw the ace of spades, an outcome satisfying both events. The probability is

$$P(\text{ace or spade}) = P(\text{ace}) + \mathbf{P(spade)} - P(\text{ace and spade})$$
$$= \frac{4}{52} + \frac{\mathbf{13}}{\mathbf{52}} - \frac{1}{52} = \frac{16}{52} = \frac{4}{13}.$$

(b) The card is a three or a king.

"Drawing a 3" and "drawing a king" are mutually exclusive events because it is impossible to draw one card that is both a 3 and a king. Using the rule given above,

$$P(3 \text{ or } K) = P(3) + P(K) - P(3 \text{ and } K)$$
$$= \frac{4}{52} + \frac{4}{52} - 0 = \frac{8}{52} = \frac{2}{13}. \quad \blacksquare$$

EXAMPLE 5
FINDING THE PROBABILITY OF ALTERNATIVE EVENTS

For the experiment consisting of one roll of a pair of dice, find the probability that the sum of the points showing is at most 4.

The description "at most 4" can be rewritten as "2 or 3 or 4." (A sum of 1 is meaningless here.) Then

$$P(\text{at most } 4) = P(2 \text{ or } 3 \text{ or } 4)$$
$$= P(2) + P(3) + P(4), \quad \textbf{(1)}$$

since the events represented by "2," "3," and "4" are mutually exclusive.

The sample space for this experiment includes the 36 possible pairs of numbers from 1 to 6: (1, 1), (1, 2), (1, 3), (1, 4), (1, 5), (1, 6), (2, 1), (2, 2), and so on. The pair (1, 1) is the only one with a sum of 2, so $P(2) = 1/36$. Also $P(3) = 2/36$ since both (1, 2) and (2, 1) give a sum of 3. The pairs, (1, 3), (2, 2), and (3, 1) have a sum of 4, so $P(4) = 3/36$. Substituting into equation (1) above gives

$$P(\text{at most } 4) = \frac{1}{36} + \frac{2}{36} + \frac{3}{36} = \frac{6}{36} = \frac{1}{6}. \quad \blacksquare$$

The following box summarizes the properties of probability discussed in this section.

PROPERTIES OF PROBABILITY

For any events E and F:

1. $0 \le P(E) \le 1$
2. $P(\text{a certain event}) = 1$
3. $P(\text{an impossible event}) = 0$
4. $P(E') = 1 - P(E)$
5. $P(E \text{ or } F) = P(E \cup F) = P(E) + P(F) - P(E \cap F)$.

13.9 EXERCISES

Write a sample space with equally likely outcomes for each of the following experiments.

1. A two-headed coin is tossed once.

2. Two ordinary coins are tossed.

3. Three ordinary coins are tossed.

4. Slips of paper marked with the numbers 1, 2, 3, 4, and 5 are placed in a box. After mixing well, two slips are drawn.

5. An unprepared student takes a three-question true/false quiz in which she guesses the answer to all three questions. (Let c = correct, w = wrong.)

6. A die is rolled and then a coin is tossed.

Write the following events in set notation and give the probability of each event. See Example 1.

7. In the experiment described in Exercise 2:
(a) both coins show the same face;
(b) at least one coin turns up heads.

8. In Exercise 1:
(a) the result of the toss is heads;
(b) the result of the toss is tails.

9. In Exercise 4:
(a) both slips are marked with even numbers;
(b) both slips are marked with odd numbers;
(c) both slips are marked with the same number;
(d) one slip is marked with an odd number, the other with an even number.

10. In Exercise 5:
(a) the student gets all three answers correct;
(b) she gets all three answers wrong;
(c) she gets exactly two answers correct;
(d) she gets at least one answer correct.

11. A student gives the answer to a probability problem as 6/5. Explain why this answer must be incorrect. (See Objective 2.)

12. If the probability of an event is .857, what is the probability that the event will not occur? (See Objective 3.)

Work the following problems. See Examples 1–5.

13. A marble is drawn at random from a box containing 3 yellow, 4 white, and 8 blue marbles. Find the following probabilities.
(a) A yellow marble is drawn.
(b) A blue marble is drawn.
(c) A black marble is drawn.
(d) What are the odds in favor of drawing a yellow marble?
(e) What are the odds against drawing a blue marble?

14. A baseball player with a batting average of .300 comes to bat. What are the odds in favor of his getting a hit?

15. In Exercise 4, what are the odds that the sum of the numbers on the two slips of paper is 5?

16. If the odds that it will rain are 4 to 5, what is the probability of rain?

17. If the odds that a candidate will win an election are 3 to 2, what is the probability that the candidate will lose?

18. A card is drawn from a well-shuffled deck of 52 cards. Find the probability that the card is
(a) a 9;
(b) black;
(c) a black 9;
(d) a heart;
(e) the 9 of hearts;
(f) a face card (K, Q, J of any suit).

19. Mrs. Elliott invites ten relatives to a party: her mother, two uncles, three brothers, and four cousins. If the chances of any one guest arriving first are equally likely, find the following probabilities.
(a) The first guest is an uncle or a brother.
(b) The first guest is a brother or cousin.
(c) The first guest is a brother or her mother.

20. One card is drawn from a standard deck of 52 cards. What is the probability that the card is
(a) a 9 or a 10?
(b) red or a 3?
(c) a heart or black?
(d) less than a 4? (Consider aces as 1's.)

21. Two dice are rolled. Find the probability of the following events.
(a) The sum of the points is at least 10.
(b) The sum of the points is either 7 or at least 10.
(c) The sum of the points is 2 or the dice both show the same number.

22. A student estimates that her probability of getting an A in a certain course is .4; a B, .3; a C, .2; and a D, .1.
(a) Assuming that only the grades A, B, C, D, and F are possible, what is the probability that she will fail the course?
(b) What is the probability that she will receive a grade of C or better?
(c) What is the probability that she will receive at least a B in the course?
(d) What is the probability that she will get at most a C in the course?

23. If a marble is drawn from a bag containing 3 yellow, 4 white, and 8 blue marbles, what is the probability of the following?
(a) The marble is either yellow or white.
(b) It is either yellow or blue.
(c) It is either blue or white.

24. The following table shows the probability that a customer of a department store will make a purchase in the indicated price range.

Cost	*Probability*
Below \$5	.25
\$5–\$19.99	.37
\$20–\$39.99	.11
\$40–\$69.99	.09
\$70–\$99.99	.07
\$100–\$149.99	.08
\$150 or more	.03

Find the probability that a customer makes a purchase that is
(a) less than \$20;
(b) \$40 or more;
(c) more than \$99.99;
(d) less than \$100.

CHAPTER 13 GLOSSARY

KEY TERMS

13.1 sequence An ordered list of numbers is called a sequence.

terms of a sequence The elements of a sequence are called the terms of the sequence.

finite sequence A sequence is a finite sequence if the domain is the set $\{1, 2, 3, 4, \ldots, n\}$, where n is a positive integer.

infinite sequence An infinite sequence has the set of all positive integers as its domain.

recursion formula A recursion formula defines the nth term of a sequence in terms of the previous term.

arithmetic sequence (or arithmetic progression) A sequence in which each term after the first is obtained by adding a fixed number to the preceding term is called an arithmetic sequence (or arithmetic progression).

common difference The fixed number that is added in an arithmetic sequence is called the common difference.

13.2 geometric sequence (or geometric progression) A geometric sequence (or geometric progression) is a sequence in which each term after the first is obtained by multiplying the preceding term by a constant nonzero real number.

common ratio The constant nonzero real number in the definition of geometric sequence is called the common ratio of the sequence.

13.3 series The sum of the terms of a sequence is called a series.

index of summation The letter i as used in summation notation $\left(\sum_{i=1}^{k} a_i\right)$ is called the index of summation.

annuity A sequence of equal payments made at equal periods of time is called an annuity.

future value The sum of the payments and the interest on the payments of an annuity is the future value of the annuity.

13.5 Pascal's triangle A triangular array of numbers that gives the values of binomial coefficients is called Pascal's triangle.

13.7 tree diagram A tree diagram is a method of systematically determining the number of different ways that successive events can occur.

independent events Independent events are events such that the outcome of any one does not affect the outcome of any other.

permutation A permutation of n elements taken r at a time is one of the ways of arranging r elements taken from a set of n elements $(r \leq n)$.

13.8 combination A combination of n elements taken r at a time is one of the ways in which r elements can be chosen from n elements $(r \leq n)$.

13.9 trial In probability, each repetition of an experiment is called a trial.

outcomes The possible results of each trial of an experiment are called outcomes of the experiment.

sample space The set S of all possible outcomes of a given experiment is called the sample space for the experiment.

event Any subset of the sample space for an experiment is called an event.

complement The set of all outcomes in the sample space that do not belong to event E is called the complement of E.

Venn diagram A Venn diagram shows relationships among sets.

odds The odds in favor of an event E are expressed as the ratio of the probability of E to the probability of E'.

compound event A compound event involves an alternative, such as E or F, where E and F are events.

mutually exclusive events Two events that cannot occur simultaneously are called mutually exclusive events.

NEW SYMBOLS

$a_1, a_2, a_3, \ldots$ terms of a sequence

a_n general or nth term of a sequence

$\sum_{i=1}^{k} a_i$ summation notation

S_n sum of the first n terms of a sequence

S_∞ sum of an infinite geometric sequence $(-1 < r < 1)$

$n!$ n-factorial

$\binom{n}{r}$ binomial coefficient, or combinations of n elements taken r at a time

$P(n, r)$ permutations of n elements taken r at a time

$n(E)$ number of outcomes of event E

$P(E)$ probability of event E

E' complement of event E

CHAPTER 13 QUICK REVIEW

SECTION	CONCEPTS	EXAMPLES
13.1 INTRODUCTION TO SEQUENCES; ARITHMETIC SEQUENCES	**nth Term of an Arithmetic Sequence** In an arithmetic sequence with first term a_1 and common difference d, the nth term, a_n, is given by $$a_n = a_1 + (n-1)d.$$	Find a_{36} for the arithmetic sequence $$3, 8, 13, 18, \ldots.$$ Here, $a_1 = 3$, $d = 5$, and $n = 36$, so $$\begin{aligned} a_{36} &= 3 + (36-1)5 \\ &= 3 + 35(5) \\ &= 178. \end{aligned}$$
13.2 GEOMETRIC SEQUENCES	**nth Term of a Geometric Sequence** In the geometric sequence with first term a_1 and common ratio r, the nth term is $$a_n = a_1 r^{n-1}.$$	Find a_9 for the geometric sequence $$8, 2, 1/2, 1/8, \ldots.$$ Here, $a_1 = 8$, $r = 1/4$, and $n = 9$, so $$\begin{aligned} a_9 &= 8(1/4)^{9-1} \\ &= 8(1/65536) \\ &= 1/8192. \end{aligned}$$
13.3 SERIES AND APPLICATIONS	**Sum Formulas for an Arithmetic Sequence** If an arithmetic sequence has first term a_1 and common difference d, then the sum of the first n terms is given by $$S_n = \frac{n}{2}(a_1 + a_n)$$ or $S_n = \frac{n}{2}[2a_1 + (n-1)d]$.	For an arithmetic sequence, $a_1 = -7$ and $d = 3$. Find S_{30}. $$\begin{aligned} S_{30} &= \frac{30}{2}[2(-7) + (30-1)(3)] \\ &= 15(-14 + 87) \\ &= 15(73) = 1095 \end{aligned}$$
	Sum Formula for a Geometric Sequence If a geometric sequence has first term a_1 and common ratio r, then the sum of the first n terms is given by $$S_n = \frac{a_1(1 - r^n)}{1 - r},$$ where $r \neq 1$.	For a geometric sequence, $a_1 = 10$ and $r = .1$. Find S_5. $$\begin{aligned} S_5 &= \frac{10(1 - .1^5)}{1 - .1} \\ &= \frac{10(.99999)}{.9} \\ &= 11.111 \end{aligned}$$

SECTION	CONCEPTS	EXAMPLES
13.4 SUMS OF INFINITE GEOMETRIC SEQUENCES	**Sum Formula for an Infinite Geometric Sequence** The sum of an infinite geometric sequence with first term a_1 and common ratio r, where $-1 < r < 1$, is given by $$S_\infty = \frac{a_1}{1 - r}.$$	Find the sum $$\sum_{i=1}^{\infty} 3\left(\frac{1}{2}\right)^i.$$ $a_1 = 3\left(\frac{1}{2}\right)^1 = \frac{3}{2}, \quad r = \frac{1}{2}.$ $$S_\infty = \frac{\frac{3}{2}}{1 - \frac{1}{2}} = \frac{\frac{3}{2}}{\frac{1}{2}} = 3$$
13.5 THE BINOMIAL THEOREM	**Factorials** For any positive integer n: $n! = n(n-1)(n-2) \ldots (3)(2)(1),$ and $0! = 1$	$4! = 4 \cdot 3 \cdot 2 \cdot 1 = 24$
	Binomial Coefficient For nonnegative integers n and r, with $r \le n$, the symbol $\binom{n}{r}$ is defined as $$\binom{n}{r} = \frac{n!}{r!(n-r)!}.$$	$$\binom{5}{2} = \frac{5!}{2!3!} = \frac{5 \cdot 4 \cdot 3 \cdot 2 \cdot 1}{2 \cdot 1 \cdot 3 \cdot 2 \cdot 1} = 10$$
	Binomial Theorem For any positive integer n: $$(x + y)^n = x^n + \binom{n}{n-1} x^{n-1}y + \binom{n}{n-2} x^{n-2}y^2 + \binom{n}{n-3} x^{n-3}y^3 + \ldots + \binom{n}{n-r} x^{n-r}y^r + \ldots + \binom{n}{1} xy^{n-1} + y^n$$	Expand $(a + 2b)^3$. $(a + 2b)^3 = a^3 + 3a^2(2b) + 3a(2b)^2 + (2b)^3$ $= a^3 + 6a^2b + 12ab^2 + 8b^3$

SECTION	CONCEPTS	EXAMPLES
	rth Term of the Expansion of $(x + y)^n$ The rth term of the binomial expansion of $(x + y)^n$, where $n \geq r - 1$, is $$\binom{n}{n-(r-1)} x^{n-(r-1)} y^{r-1}.$$	The third term in the expansion of $(3a + b)^5$ is $$\binom{5}{3}(3a)^3 b^2 = 270a^3b^2.$$
13.6 MATHEMATICAL INDUCTION	**Proof by Mathematical Induction** Let S_n be a statement concerning the positive integer n. Suppose that **(1)** S_1 is true; **(2)** for any positive integer k, $k \leq n$, if S_k is true, then S_{k+1} is also true. Then S_n is true for every positive integer value of n.	See Examples 1–3 in Section 13.6.
13.7 PERMUTATIONS	**Fundamental Principle of Counting** If one event can occur in m ways and a second event can occur in n ways, then both events can occur in mn ways, provided the outcome of the first event does not influence the outcome of the second.	If there are 2 ways to choose a pair of socks and 5 ways to choose a pair of shoes, there are $2 \cdot 5 = 10$ ways to choose socks and shoes.
	Permutations Formula If $P(n, r)$ denotes the number of permutations of n elements taken r at a time, $r \leq n$, then $$P(n, r) = \frac{n!}{(n-r)!}.$$	How many ways are there to arrange the letters of the word *triangle* using 5 letters at a time? Here, $n = 8$ and $r = 5$, so the number of ways is $$P(8, 5) = \frac{8!}{(8-5)!} = \frac{8!}{3!} = 6720.$$

SECTION	CONCEPTS	EXAMPLES
13.8 COMBINATIONS	**Combinations Formula** The binomial coefficient $\binom{n}{r}$, defined in Section 13.5, gives the number of combinations of n elements taken r at a time.	How many committees of 4 senators can be formed from a group of 9 senators? Since the arrangement of senators does not matter, this is a combinations problem. The number of possible committees is $$\binom{9}{4} = \frac{9!}{4!5!} = 126.$$
13.9 PROBABILITY	**Probability of an Event E** In a sample space with equally likely outcomes, the probability of an event E, written $P(E)$, is the ratio of the number of outcomes in sample space S that belong to event E, $n(E)$, to the total number of outcomes in sample space S, $n(S)$. That is, $$P(E) = \frac{n(E)}{n(S)}.$$	A number is chosen at random from the set $$S = \{1, 2, 3, 4, 5, 6\}.$$ What is the probability that the number is less than 3? The event is $E = \{1, 2\}$. $n(S) = 6$ and $n(E) = 2$. Therefore, $P(E) = \frac{2}{6} = \frac{1}{3}$.
	Properties of Probability For any events E and F: **1.** $0 \le P(E) \le 1$ **2.** $P(\text{a certain event}) = 1$ **3.** $P(\text{an impossible event}) = 0$ **4.** $P(E') = 1 - P(E)$ **5.** $P(E \text{ or } F) = P(E \cup F) = P(E) + P(F) - P(E \cap F)$	What is the probability that the number is 3 or more? This event is E' $$P(E') = 1 - \frac{1}{3} = \frac{2}{3}$$

CHAPTER 13 REVIEW EXERCISES

[13.1] *Write the first five terms for each of the following sequences.*

1. $a_n = \dfrac{n}{n+1}$

2. $a_n = (-2)^n$

3. $a_n = 2(n + 3)$

4. $a_n = n(n + 1)$

5. $a_1 = 5$, $a_n = a_{n-1} - 3$

6. $a_1 = -2$, $a_n = 3a_{n-1}$

7. $a_1 = 5$, $a_2 = 3$, $a_n = a_{n-1} - a_{n-2}$ for $n \ge 3$

8. $b_1 = -2$, $b_2 = 2$, $b_3 = -4$, $b_n = -2 \cdot b_{n-2}$ if n is even, and $b_n = 2 \cdot b_{n-2}$ if n is odd.

9. arithmetic, $a_1 = 6$, $d = -4$

10. arithmetic, $a_3 = 9$, $a_4 = 7$

11. arithmetic, $a_1 = 3 - \sqrt{5}$, $a_2 = 4$

12. arithmetic, $a_3 = \pi$, $a_4 = 0$

13. A certain arithmetic sequence has $a_6 = -4$ and $a_{17} = 51$. Find a_1 and a_{20}.

Find a_8 for each of the following arithmetic sequences.

14. $a_1 = 6$, $d = 2$

15. $a_1 = -4$, $d = 3$

16. $a_1 = 6x - 9$, $a_2 = 5x + 1$

17. $a_3 = 11m$, $a_5 = 7m - 4$

[13.2] *Write the first five terms for each of the following sequences.*

18. geometric, $a_1 = 4$, $r = 2$

19. geometric, $a_4 = 8$, $r = 1/2$

20. geometric, $a_1 = -3$, $a_2 = 4$

21. geometric, $a_3 = 8$, $a_5 = 72$

22. For a given geometric sequence, $a_1 = 4$ and $a_5 = 324$. Find a_6.

Find a_5 for each of the following geometric sequences.

23. $a_1 = 3$, $r = 2$

24. $a_2 = 3125$, $r = 1/5$

25. $a_1 = 5x$, $a_2 = x^2$

26. $a_2 = \sqrt{6}$, $a_4 = 6\sqrt{6}$

27. Explain the difference between an arithmetic sequence and a geometric sequence.

[13.3] *Find S_{12} for each of the following arithmetic sequences.*

28. $a_1 = 2$, $d = 3$

29. $a_2 = 6$, $d = 10$

30. $a_1 = -4k$, $d = 2k$

Find S_4 for each of the following geometric sequences.

31. $a_1 = 1$, $r = 2$

32. $a_1 = 3$, $r = 3$

33. $a_1 = 2k$, $a_2 = -4k$

Evaluate each of the following sums.

34. $\sum_{i=1}^{4} \frac{2}{i}$

35. $\sum_{i=1}^{7} (-1)^{i+1} \cdot 6$

36. $\sum_{i=4}^{8} 3i(2i - 5)$

37. $\sum_{i=1}^{6} i(i + 2)$

38. $\sum_{i=1}^{6} 4 \cdot 3^i$

39. $\sum_{i=1}^{4} 8 \cdot 2^i$

Solve each problem.

40. A stack of canned goods in a market display requires 15 cans on the bottom, 13 in the next layer, 11 in the next layer, and so on, to a top layer of 1 can. How many cans are needed for the display?

41. Gale Stockdale borrows \$6000 at simple interest of 12% per year. He will repay the loan and interest in monthly payments of \$260, \$258, \$256, and so on. If he makes 30 payments, what is the total amount required to pay off the loan plus the interest?

42. On a certain production line, new employees during their first week turn out 5/4 as many items each day as on the previous day. If a new employee produces 48 items the first day, how many will she produce on the fifth day of work?

43. The half-life of a radioactive substance is 20 years. If 600 grams are present at the start, how much will be left after 100 years?

[13.4] *Evaluate each of the following sums that exist.*

44. $18 + 9 + 9/2 + 9/4 + \ldots$

45. $20 + 15 + 45/4 + 135/16 + \ldots$

46. $-5/6 + 5/9 - 10/27 + \ldots$

47. $.9 + .09 + .009 + .0009 + \ldots$

48. $\sum_{i=1}^{\infty} \left(\frac{5}{8}\right)^i$

49. $\sum_{i=1}^{\infty} -10\left(\frac{5}{2}\right)^i$

Convert each of the following repeating decimals to a quotient of integers.

50. .6666 . . . **51.** .512512512 . . . **52.** .292929 . . .

[13.5] *Evaluate each binomial coefficient.*

53. $\binom{8}{3}$ **54.** $\binom{10}{5}$ **55.** $\binom{6}{0}$ **56.** $\binom{7}{1}$

Use the binomial theorem to expand each of the following.

57. $(x + 2y)^4$ **58.** $(3z - 5w)^3$ **59.** $\left(\frac{k}{2} - g\right)^5$

Find the indicated term or terms for each of the following expansions.

60. Fifth term of $(3x - 2y)^6$

61. Eighth term of $(2m + n^2)^{12}$

62. First four terms of $(3 + x)^{16}$

63. Last three terms of $(2m - 3n)^{15}$

[13.6] *Use mathematical induction to prove that each of the following is true for every positive integer n.*

64. $2 + 6 + 10 + 14 + \ldots + (4n - 2) = 2n^2$

65. $2^2 + 4^2 + 6^2 + \ldots + (2n)^2 = \dfrac{2n(n + 1)(2n + 1)}{3}$

66. $2 + 2^2 + 2^3 + \ldots + 2^n = 2(2^n - 1)$

67. $1^3 + 3^3 + 5^3 + \ldots + (2n - 1)^3 = n^2(2n^2 - 1)$

[13.7, 13.8] *Find each of the following.*

68. $P(5, 5)$ **69.** $P(9, 2)$ **70.** $P(6, 0)$

Show that each of the following is true for positive integers n and r, where $r \leq n$.

71. $\binom{n}{n} = 1$ **72.** $\binom{n}{0} = 1$ **73.** $\binom{n}{n-1} = n$ **74.** $\binom{n}{n-r} = \binom{n}{r}$

75. $P(n, n - 1) = P(n, n)$ **76.** $P(n, 1) = n$ **77.** $P(n, 0) = 1$

Solve each problem.

78. Two people are planning their wedding. They can select from 2 different chapels, 4 soloists, 3 organists, and 2 ministers. How many different wedding arrangements will be possible?

79. John Jacobs, who is furnishing his apartment, wants to buy a new couch. He can select from 5 different styles, each available in 3 different fabrics, with 6 color choices. How many different couches are available?

80. Four students are to be assigned to 4 different summer jobs. Each student is qualified for all 4 jobs. In how many ways can the jobs be assigned?

81. A student body council consists of a president, vice-president, secretary-treasurer, and 3 representatives at large. Three members are to be selected to attend a conference.

(a) How many different such delegations are possible?

(b) How many if the president must attend?

82. Nine football teams are competing for first-, second-, and third-place titles in a statewide tournament. In how many ways can the winners be determined?

83. How many different license plates can be formed with a letter followed by 3 digits and then 3 letters? How many such license plates have no repeats?

[13.9] *Solve each problem.*

84. A marble is drawn at random from a box containing 4 green, 5 black, and 6 white marbles. Find the following probabilities.

(a) A green marble is drawn.
(b) A marble that is not black is drawn.
(c) A blue marble is drawn.

85. Refer to Exercise 84 and answer each question.

(a) What are the odds in favor of drawing a green marble?
(b) What are the odds against drawing a white marble?

A card is drawn from a standard deck of 52 cards. Find the probability that each of the following is drawn.

86. A black king

87. A face card or an ace

88. An ace or a diamond

89. A card that is not a diamond

90. A card that is not a diamond or not black

A sample shipment of 5 swimming pool filters is chosen. The probability of exactly 0, 1, 2, 3, 4, *or* 5 *filters being defective is given in the following table.*

Number defective	0	1	2	3	4	5
Probability	.31	.25	.18	.12	.08	.06

Find the probability that the following numbers of filters are defective.

91. No more than 3

92. At least 2

CHAPTER 13 TEST

Write the first five terms for each of the following sequences.

1. $a_n = (-1)^n(n^2 + 1)$

2. $a_1 = 5,\ a_n = n + a_{n-1}$ for $n \geq 2$

3. $a_n = n + 1$ if n is odd and $a_n = a_{n-1} + 2$ if n is even.

4. arithmetic, $a_2 = 1,\ a_3 = 25$

---

5. A certain arithmetic sequence has $a_7 = -6$ and $a_{15} = -2$. Find a_{31}.

6. Write the first five terms of the geometric sequence with $a_1 = 81$ and $r = 2/3$.

7. For a given geometric sequence, $a_1 = 12$ and $a_6 = -3/8$. Find a_3.

8. Find a_7 for the geometric sequence with $a_1 = 2x^3$ and $a_3 = 18x^7$.

9. Find S_{10} for the arithmetic sequence with $a_1 = 37$ and $d = 13$.

10. Find S_4 for the geometric sequence with $a_1 = 4$ and $r = 1/2$.

Evaluate each of the following sums that exist.

11. $\sum_{i=1}^{30} (5i + 2)$

12. $\sum_{i=1}^{5} -3 \cdot 2^i$

13. $75 + 30 + 12 + \frac{24}{5} + \ldots$

14. $\sum_{i=1}^{\infty} 54(2/9)^i$

15. Fred Meyers deposited \$50 in a new savings account on February 1. On the first of each month thereafter he deposited \$5 more than the previous month's deposit. Find the total amount of money he deposited after 20 months.

16. The number of bacteria in a certain culture doubles every 30 minutes. If 50 are present at noon, how many will be present at 4:30 P.M.?

17. Use the binomial theorem to expand $(2x - 3y)^4$.

18. Find the fourth term in the expansion of $(w - 2y)^6$.

19. Use mathematical induction to prove that

$$8 + 14 + 20 + 26 + \ldots + (6n + 2) = 3n^2 + 5n$$

for every positive integer n.

Find each of the following.

20. $P(11, 3)$

21. $P(7, 7)$

22. $\binom{10}{2}$

23. A clothing manufacturer makes women's coats in four different styles. Each coat can be made from one of three fabrics. Each fabric comes in five different colors. How many different coats can be made?

24. A club with 30 members is to elect a president, secretary, and treasurer from its membership. If a member can hold at most one position, in how many ways can the offices be filled?

25. In how many ways can a committee of 3 representatives be chosen from a group of 9 representatives?

A card is drawn from a standard deck of 52 cards. Find the probability that each of the following is drawn.

26. A red three

27. A card that is not a face card

28. A king or a spade

29. In the card-drawing experiment above, what are the odds in favor of drawing a face card?

30. A sample of 4 computer chips is chosen. The probability of exactly 0, 1, 2, 3 or 4 chips being defective is given in the following table.

Number defective	0	1	2	3	4
Probability	.19	.43	.30	.07	.01

Find the probability that at most 2 are defective.

APPENDIX A GEOMETRY REVIEW AND FORMULAS

SPECIAL ANGLES

NAME	CHARACTERISTIC	EXAMPLES
RIGHT ANGLE	Measure is 90°.	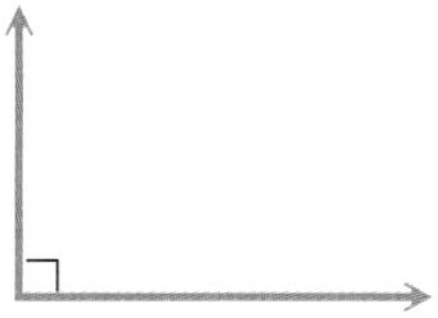
STRAIGHT ANGLE	Measure is 180°.	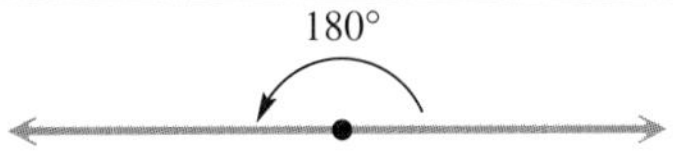 180°
COMPLEMENTARY ANGLES	The sum of the measures of two complementary angles is 90°.	Angle 1 and angle 2 are complementary. 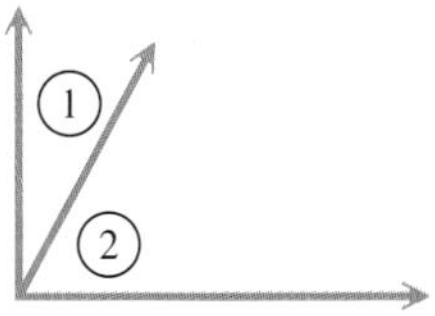
SUPPLEMENTARY ANGLES	The sum of the measures of two supplementary angles is 180°.	Angle 3 and angle 4 are supplementary. 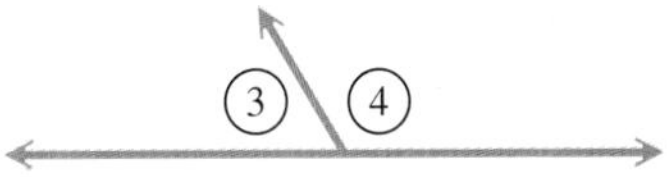
VERTICAL ANGLES	Vertical angles have equal measures.	Angle 2 = Angle 4 Angle 1 = Angle 3 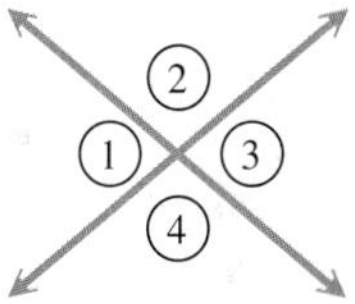
ANGLES FORMED BY PARALLEL LINES AND A TRANSVERSAL		Lines m and n are parallel. 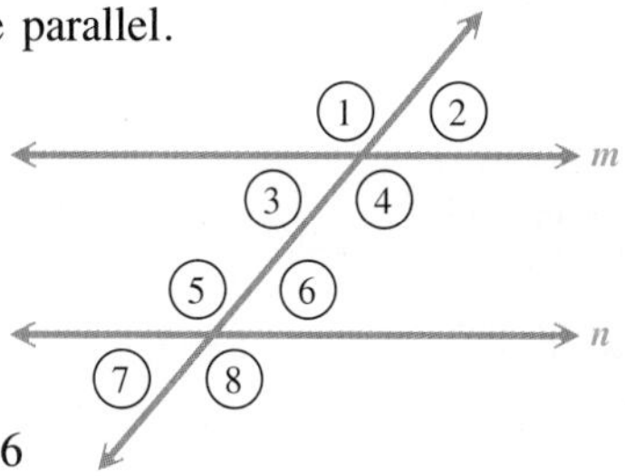
ALTERNATE INTERIOR ANGLES	Measures are equal.	Angle 3 = Angle 6
ALTERNATE EXTERIOR ANGLES	Measures are equal.	Angle 1 = Angle 8
INTERIOR ANGLES ON THE SAME SIDE	Angles are supplements.	Angles 3 and 5 are supplementary.

SPECIAL TRIANGLES

NAME	CHARACTERISTIC	EXAMPLES
RIGHT TRIANGLE	Triangle has a right angle.	90°
ISOSCELES TRIANGLE	Triangle has two equal sides.	$AB = BC$
EQUILATERAL TRIANGLE	Triangle has three equal sides.	$AB = BC = CA$
SIMILAR TRIANGLES	Corresponding angles are equal; corresponding sides are proportional.	$A = D$, $B = E$, $C = F$ $\frac{AB}{DE} = \frac{AC}{DF} = \frac{BC}{EF}$

FORMULAS

FIGURE	FORMULAS	EXAMPLES
SQUARE	Perimeter: $P = 4s$ Area: $A = s^2$	

FIGURE	FORMULAS	EXAMPLES
RECTANGLE	Perimeter: $P = 2L + 2W$ Area: $A = LW$	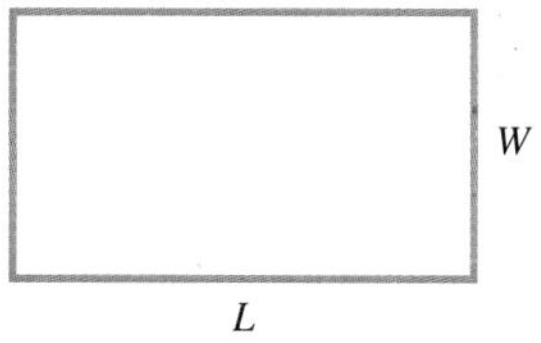 W L
TRIANGLE	Perimeter: $P = a + b + c$ Area: $A = \frac{1}{2}bh$	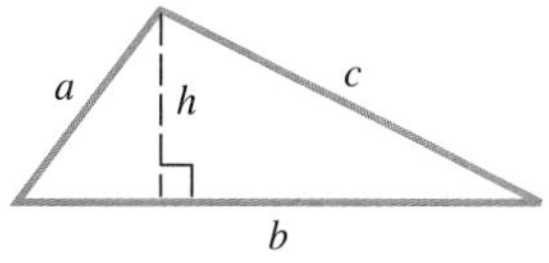 a h c b
PYTHAGOREAN FORMULA (FOR RIGHT TRIANGLES)	For legs a and b and hypotenuse c, $c^2 = a^2 + b^2$.	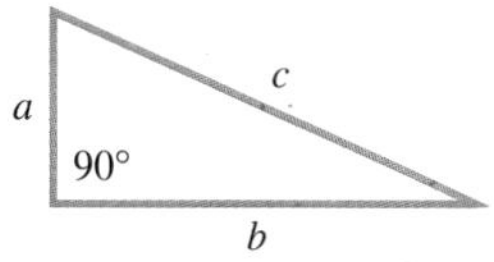 a c 90° b
SUM OF THE ANGLES OF A TRIANGLE	$A + B + C = 180°$	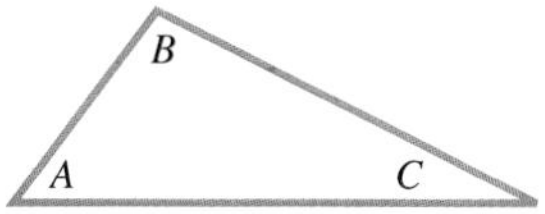 B A C
CIRCLE	Diameter: $d = 2r$ Circumference: $C = 2\pi r = \pi d$ Area: $A = \pi r^2$	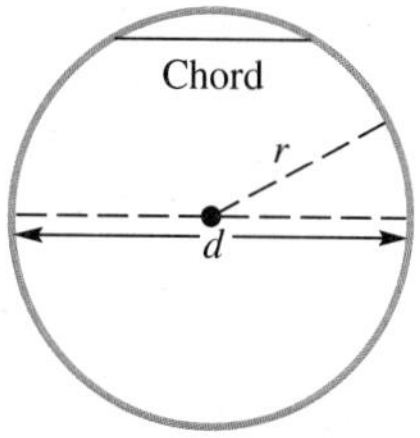 Chord r d
PARALLELOGRAM	Area: $A = bh$ Perimeter: $P = 2a + 2b$	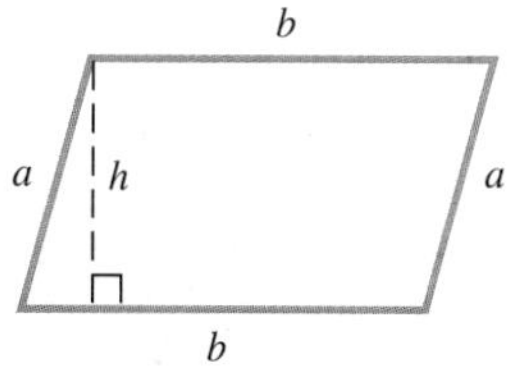 b a h a b
TRAPEZOID	Area: $A = \frac{1}{2}(B + b)h$ Perimeter: $P = a + b + c + B$	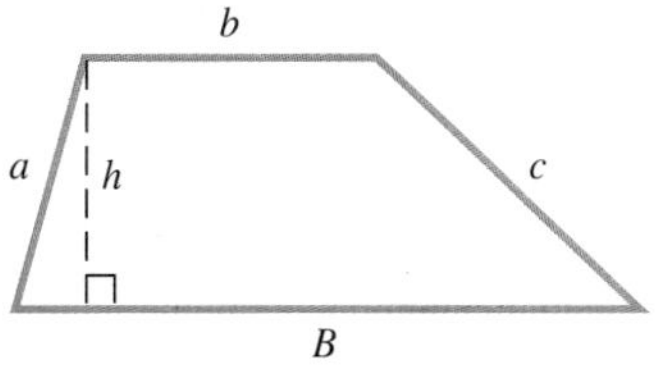 b a h c B

FIGURE	FORMULAS	EXAMPLES
SPHERE	Volume: $V = \frac{4}{3}\pi r^3$ Surface area: $S = 4\pi r^2$	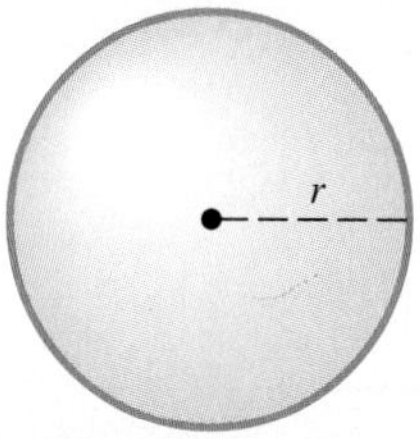 r
CONE	Volume: $V = \frac{1}{3}\pi r^2 h$ Surface area: $S = \pi r\sqrt{r^2 + h^2}$	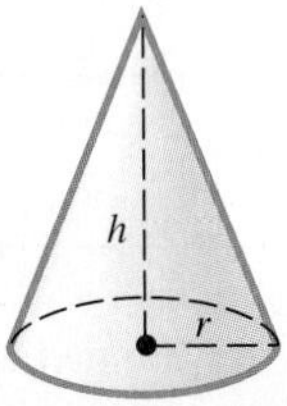 h r
CUBE	Volume: $S = e^3$ Surface area: $A = 6e^2$	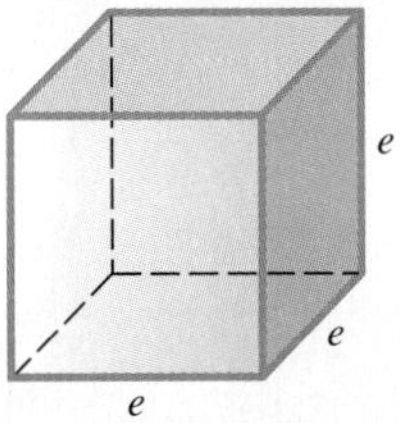 e e e
RECTANGULAR SOLID	Volume: $V = LWH$ Surface area: $A = 2HW + 2LW + 2LH$	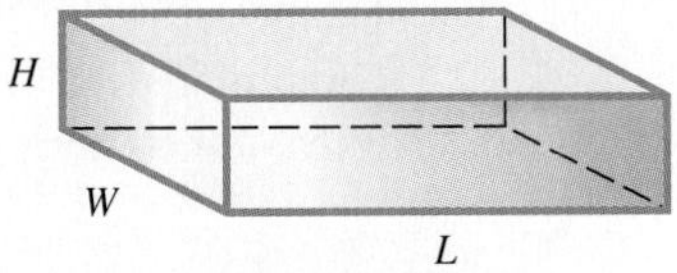 H W L
RIGHT CIRCULAR CYLINDER	Volume: $V = \pi r^2 h$ Surface area: $S = 2\pi rh + 2\pi r^2$	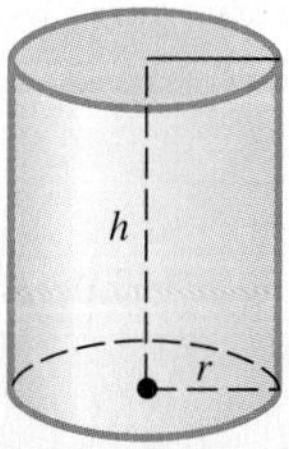 h r
RIGHT PYRAMID	Volume: $V = \frac{1}{3}Bh$, where B = area of the base	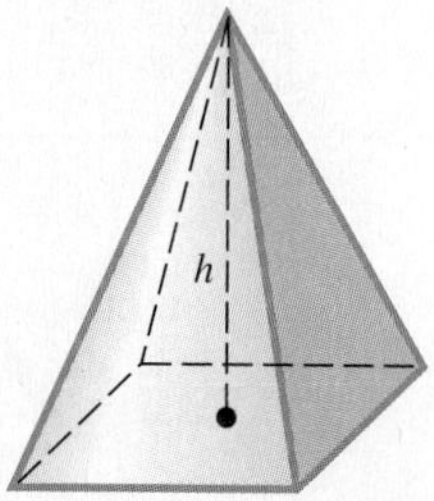 h

APPENDIX B OTHER FORMULAS

Distance:
$d = rt$;
d = distance
r = rate or speed
t = time

Percent:
$p = br$;
p = percentage
b = base
r = rate

Temperature:
$$F = \frac{9}{5}C + 32$$
$$C = \frac{5}{9}(F - 32)$$

FORMULAS FROM CONSUMER MATHEMATICS

Simple Interest:
$I = prt$;
p = principal or amount invested
r = rate or percent
t = time in years

Unearned interest (by the rule of 78):
$$u = f \cdot \frac{k(k + 1)}{n(n + 1)};$$
u = unearned interest
f = total finance charge
k = number of payments remaining
n = total number of payments originally scheduled

Profit = Income − Cost

Approximate annual interest rate for a loan paid off in monthly payments:
$$A = \frac{24f}{b(p + 1)};$$
A = approximate annual interest rate
f = total finance charge
b = amount borrowed
p = total number of payments

APPENDIX C USING THE TABLE OF LOGARITHMS

OBJECTIVES

1 EVALUATE BASE 10 LOGARITHMS BY USING A LOGARITHM TABLE.
2 FIND THE ANTILOGARITHM OF A COMMON LOGARITHM USING A TABLE.
3 INTERPOLATE THE VALUES IN A LOGARITHM TABLE.
4 INTERPOLATE WHEN FINDING ANTILOGARITHMS.

1 This appendix gives a brief explanation of the table of common logarithms. To find logarithms using the table, first write the number in scientific notation (see Section 3.2). For example, to find log 423, first write 423 in scientific notation.

$$\log 423 = \log (4.23 \times 10^2)$$

By the multiplication property for logarithms,

$$\log 423 = \log 4.23 + \log 10^2.$$

The logarithm of 10^2 is 2 (the exponent). To find log 4.23, use the table of common logarithms. (A portion of that table is reproduced here.) Read down on the left to the row headed 4.2. Then read across to the column headed by 3 to get .6263. From the table,

$$\log 423 \approx .6263 + 2 = 2.6263.$$

x	0	1	2	3	4	5	6	7	8	9
4.0	.6021	.6031	.6042	.6053	.6064	.6075	.6085	.6096	.6107	.6117
4.1	.6128	.6138	.6149	.6160	.6170	.6180	.6191	.6201	.6212	.6222
4.2	.6232	.6243	.6253	.6263	.6274	.6284	.6294	.6304	.6314	.6325
4.3	.6335	.6345	.6355	.6365	.6375	.6385	.6395	.6405	.6415	.6425
4.4	.6435	.6444	.6454	.6464	.6474	.6484	.6493	.6503	.6513	.6522

The decimal part of the logarithm, .6263, is called the **mantissa.** The integer part, 2, is called the **characteristic.**

EXAMPLE 1
USING THE TABLE TO FIND COMMON LOGARITHMS

Find the following common logarithms.

(a) log 437,000
Since $437{,}000 = 4.37 \times 10^5$, the characteristic is 5. From the portion of the logarithm table given above, the mantissa of log 437,000 is .6405, and log 437,000 = 5 + .6405 = 5.6405.

(b) log .0415
Express .0415 as 4.15×10^{-2}. The characteristic is -2. The mantissa, from the table, is .6180, and

$$\log .0415 = -2 + .6180 = -1.3820.$$

To retain the mantissa, the answer may be given as $-2 + .6180$, or $.6180 - 2$. A calculator with a logarithm key gives the result as -1.3820. ■

2 In Example 1(a), the number 437,000, whose logarithm is 5.6405, is called the **antilogarithm** (abbreviated *antilog*) of 5.6405. In the same way, in Example 1(b), .0415 is the antilogarithm of the logarithm $.6180 - 2$.

From Example 1,

$$\log 437{,}000 = 5.6405.$$

In exponential form, this is equivalent to

$$437{,}000 = 10^{5.6405}.$$

This shows that the antilogarithm, 437,000, is just an exponential.

EXAMPLE 2 USING THE TABLE TO FIND ANTILOGARITHMS

Find the antilogarithm of each of the following logarithms.

(a) 2.6454

To find the number whose logarithm is 2.6454, first look in the table for a mantissa of .6454. The number whose mantissa is .6454 is 4.42. Since the characteristic is 2, the antilogarithm is $4.42 \times 10^2 = 442$.

(b) $.6314 - 3$

The mantissa is .6314 and the characteristic is -3. In the table, the number with a mantissa of .6314 is 4.28, so the antilogarithm is $4.28 \times 10^{-3} = .00428$. ■

3 The table of logarithms contains logarithms to four decimal places. This table can be used to find the logarithm of any positive number containing three significant digits. **Linear interpolation** is used to find logarithms of numbers containing four significant digits. In Example 3 we show how this is done.

EXAMPLE 3 USING LINEAR INTERPOLATION TO FIND A LOGARITHM

Find log 4238.

Since

$$4230 < 4238 < 4240,$$
$$\log 4230 < \log 4238 < \log 4240.$$

Find log 4230 and log 4240 in the table.

$$10\left\{\begin{array}{l} 8\left\{\begin{array}{l}\log 4230 = 3.6263 \\ \log 4238 = \end{array}\right. \\ \log 4240 = 3.6274 \end{array}\right\}.0011$$

Since 4238 is 8/10 of the way from 4230 to 4240, take 8/10 of .0011, the difference between the two logarithms.

$$\frac{8}{10}(.0011) \approx .0009$$

Now add .0009 to log 4230 to get

$$\begin{aligned}\log 4238 &= 3.6263 + .0009 \\ &= 3.6272.\end{aligned}$$

Check the answer by consulting a more accurate logarithm table or use a calculator to find that $\log 4238 = 3.62716$ (to five places). ■

EXAMPLE 4 USING LINEAR INTERPOLATION TO FIND A LOGARITHM

Find log .02386.

Arrange the work as follows.

$$10\left\{\begin{array}{l} 6\left\{\begin{array}{l}\log .0238 = .3766 - 2 \\ \log .02386 = \end{array}\right. \\ \log .0239 = .3784 - 2 \end{array}\right\}.0018$$

$$\frac{6}{10}(.0018) \approx .0011$$

Now add .3766 − 2 and .0011.

$$\begin{array}{rr} & .3766 - 2 \\ & .0011 \\ \hline \log .02386 = & .3777 - 2 \end{array}$$ ■

4 Interpolation can also be used to find antilogarithms with four significant digits.

EXAMPLE 5 USING LINEAR INTERPOLATION TO FIND AN ANTILOGARITHM

Find the antilogarithm of 3.5894.

From the table, the mantissas closest to .5894 are .5888 and .5899. Find the antilogarithms of 3.5888 and 3.5899.

$$.0011\left\{\begin{array}{l} .0006\left\{\begin{array}{l} 3.5888 = \log 3880 \\ 3.5894 = \end{array}\right. \\ 3.5899 = \log 3890 \end{array}\right\} 10$$

$$\left(\frac{.0006}{.0011}\right)10 \approx .5(10) = 5$$

Add the 5 to 3880, giving

$$\log 3885 = 3.5894.$$

The antilogarithm of 3.5894 is 3885. ■

APPENDIX C EXERCISES

Use Table 1 to find each of the following logarithms. See Example 1.

1. log 2.37 **2.** log 4.69 **3.** log 194
4. log 83 **5.** log 12 **6.** log 1870
7. log 25,000 **8.** log 36,400 **9.** log .6
10. log .05 **11.** log .000211 **12.** log .00432

Use the table to find each of the following antilogarithms. See Example 2.

13. 2.5366 **14.** 1.8407 **15.** .6599 **16.** 3.9258
17. 1.3979 **18.** 4.8716 **19.** .0792 − 3 **20.** .7259 − 1
21. .8727 − 2 **22.** .4843 − 2 **23.** .2041 − 4 **24.** .9586 − 2

Use the table and linear interpolation to find the following logarithms. See Examples 3 and 4.

25. log .8973 **26.** log 2.635 **27.** log 53.89 **28.** log 218.4
29. log 5248 **30.** log 19,040 **31.** log .01253 **32.** log .6529

Use the table and interpolation to find the antilogarithms of the following logarithms to four significant digits. See Example 5.

33. 2.7138 **34.** 1.9146 **35.** .8342 − 1
36. .7138 − 2 **37.** .2008 − 4 **38.** .2413 − 3

TABLE 1 COMMON LOGARITHMS

n	0	1	2	3	4	5	6	7	8	9
1.0	.0000	.0043	.0086	.0128	.0170	.0212	.0253	.0294	.0334	.0374
1.1	.0414	.0453	.0492	.0531	.0569	.0607	.0645	.0682	.0719	.0755
1.2	.0792	.0828	.0864	.0899	.0934	.0969	.1004	.1038	.1072	.1106
1.3	.1139	.1173	.1206	.1239	.1271	.1303	.1335	.1367	.1399	.1430
1.4	.1461	.1492	.1523	.1553	.1584	.1614	.1644	.1673	.1703	.1732
1.5	.1761	.1790	.1818	.1847	.1875	.1903	.1931	.1959	.1987	.2014
1.6	.2041	.2068	.2095	.2122	.2148	.2175	.2201	.2227	.2253	.2279
1.7	.2304	.2330	.2355	.2380	.2405	.2430	.2455	.2480	.2504	.2529
1.8	.2553	.2577	.2601	.2625	.2648	.2672	.2695	.2718	.2742	.2765
1.9	.2788	.2810	.2833	.2856	.2878	.2900	.2923	.2945	.2967	.2989
2.0	.3010	.3032	.3054	.3075	.3096	.3118	.3139	.3160	.3181	.3201
2.1	.3222	.3243	.3263	.3284	.3304	.3324	.3345	.3365	.3385	.3404
2.2	.3424	.3444	.3464	.3483	.3502	.3522	.3541	.3560	.3579	.3598
2.3	.3617	.3636	.3655	.3674	.3692	.3711	.3729	.3747	.3766	.3784
2.4	.3802	.3820	.3838	.3856	.3874	.3892	.3909	.3927	.3945	.3962
2.5	.3979	.3997	.4014	.4031	.4048	.4065	.4082	.4099	.4116	.4133
2.6	.4150	.4166	.4183	.4200	.4216	.4232	.4249	.4265	.4281	.4298
2.7	.4314	.4330	.4346	.4362	.4378	.4393	.4409	.4425	.4440	.4456
2.8	.4472	.4487	.4502	.4518	.4533	.4548	.4564	.4579	.4594	.4609
2.9	.4624	.4639	.4654	.4669	.4683	.4698	.4713	.4728	.4742	.4757
3.0	.4771	.4786	.4800	.4814	.4829	.4843	.4857	.4871	.4886	.4900
3.1	.4914	.4928	.4942	.4955	.4969	.4983	.4997	.5011	.5024	.5038
3.2	.5051	.5065	.5079	.5092	.5105	.5119	.5132	.5145	.5159	.5172
3.3	.5185	.5198	.5211	.5224	.5237	.5250	.5263	.5276	.5289	.5302
3.4	.5315	.5328	.5340	.5353	.5366	.5378	.5391	.5403	.5416	.5428
3.5	.5441	.5453	.5465	.5478	.5490	.5502	.5514	.5527	.5539	.5551
3.6	.5563	.5575	.5587	.5599	.5611	.5623	.5635	.5647	.5658	.5670
3.7	.5682	.5694	.5705	.5717	.5729	.5740	.5752	.5763	.5775	.5786
3.8	.5798	.5809	.5821	.5832	.5843	.5855	.5866	.5877	.5888	.5899
3.9	.5911	.5922	.5933	.5944	.5955	.5966	.5977	.5988	.5999	.6010
4.0	.6021	.6031	.6042	.6053	.6064	.6075	.6085	.6096	.6107	.6117
4.1	.6128	.6138	.6149	.6160	.6170	.6180	.6191	.6201	.6212	.6222
4.2	.6232	.6243	.6253	.6263	.6274	.6284	.6294	.6304	.6314	.6325
4.3	.6335	.6345	.6355	.6365	.6375	.6385	.6395	.6405	.6415	.6425
4.4	.6435	.6444	.6454	.6464	.6474	.6484	.6493	.6503	.6513	.6522
4.5	.6532	.6542	.6551	.6561	.6571	.6580	.6590	.6599	.6609	.6618
4.6	.6628	.6637	.6646	.6656	.6665	.6675	.6684	.6693	.6702	.6712
4.7	.6721	.6730	.6739	.6749	.6758	.6767	.6776	.6785	.6794	.6803
4.8	.6812	.6821	.6830	.6839	.6848	.6857	.6866	.6875	.6884	.6893
4.9	.6902	.6911	.6920	.6928	.6937	.6946	.6955	.6964	.6972	.6981
5.0	.6990	.6998	.7007	.7016	.7024	.7033	.7042	.7050	.7059	.7067
5.1	.7076	.7084	.7093	.7101	.7110	.7118	.7126	.7135	.7143	.7152
5.2	.7160	.7168	.7177	.7185	.7193	.7202	.7210	.7218	.7226	.7235
5.3	.7243	.7251	.7259	.7267	.7275	.7284	.7292	.7300	.7308	.7316
5.4	.7324	.7332	.7340	.7348	.7356	.7364	.7372	.7380	.7388	.7396
n	0	1	2	3	4	5	6	7	8	9

TABLE 1 COMMON LOGARITHMS (CONTINUED)

n	0	1	2	3	4	5	6	7	8	9
5.5	.7404	.7412	.7419	.7427	.7435	.7443	.7451	.7459	.7466	.7474
5.6	.7482	.7490	.7497	.7505	.7513	.7520	.7528	.7536	.7543	.7551
5.7	.7559	.7566	.7574	.7582	.7589	.7597	.7604	.7612	.7619	.7627
5.8	.7634	.7642	.7649	.7657	.7664	.7672	.7679	.7686	.7694	.7701
5.9	.7709	.7716	.7723	.7731	.7738	.7745	.7752	.7760	.7767	.7774
6.0	.7782	.7789	.7796	.7803	.7810	.7818	.7825	.7832	.7839	.7846
6.1	.7853	.7860	.7868	.7875	.7882	.7889	.7896	.7903	.7910	.7917
6.2	.7924	.7931	.7938	.7945	.7952	.7959	.7966	.7973	.7980	.7987
6.3	.7993	.8000	.8007	.8014	.8021	.8028	.8035	.8041	.8048	.8055
6.4	.8062	.8069	.8075	.8082	.8089	.8096	.8102	.8109	.8116	.8122
6.5	.8129	.8136	.8142	.8149	.8156	.8162	.8169	.8176	.8182	.8189
6.6	.8195	.8202	.8209	.8215	.8222	.8228	.8235	.8241	.8248	.8254
6.7	.8261	.8267	.8274	.8280	.8287	.8293	.8299	.8306	.8312	.8319
6.8	.8325	.8331	.8338	.8344	.8351	.8357	.8363	.8370	.8376	.8382
6.9	.8388	.8395	.8401	.8407	.8414	.8420	.8426	.8432	.8439	.8445
7.0	.8451	.8457	.8463	.8470	.8476	.8482	.8488	.8494	.8500	.8506
7.1	.8513	.8519	.8525	.8531	.8537	.8543	.8549	.8555	.8561	.8567
7.2	.8573	.8579	.8585	.8591	.8597	.8603	.8609	.8615	.8621	.8627
7.3	.8633	.8639	.8645	.8651	.8657	.8663	.8669	.8675	.8681	.8686
7.4	.8692	.8698	.8704	.8710	.8716	.8722	.8727	.8733	.8739	.8745
7.5	.8751	.8756	.8762	.8768	.8774	.8779	.8785	.8791	.8797	.8802
7.6	.8808	.8814	.8820	.8825	.8831	.8837	.8842	.8848	.8854	.8859
7.7	.8865	.8871	.8876	.8882	.8887	.8893	.8899	.8904	.8910	.8915
7.8	.8921	.8927	.8932	.8938	.8943	.8949	.8954	.8960	.8965	.8971
7.9	.8976	.8982	.8987	.8993	.8998	.9004	.9009	.9015	.9020	.9025
8.0	.9031	.9036	.9042	.9047	.9053	.9058	.9063	.9069	.9074	.9079
8.1	.9085	.9090	.9096	.9101	.9106	.9112	.9117	.9122	.9128	.9133
8.2	.9138	.9143	.9149	.9154	.9159	.9165	.9170	.9175	.9180	.9186
8.3	.9191	.9196	.9201	.9206	.9212	.9217	.9222	.9227	.9232	.9238
8.4	.9243	.9248	.9253	.9258	.9263	.9269	.9274	.9279	.9284	.9289
8.5	.9294	.9299	.9304	.9309	.9315	.9320	.9325	.9330	.9335	.9340
8.6	.9345	.9350	.9355	.9360	.9365	.9370	.9375	.9380	.9385	.9390
8.7	.9395	.9400	.9405	.9410	.9415	.9420	.9425	.9430	.9435	.9440
8.8	.9445	.9450	.9455	.9460	.9465	.9469	.9474	.9479	.9484	.9489
8.9	.9494	.9499	.9504	.9509	.9513	.9518	.9523	.9528	.9533	.9538
9.0	.9542	.9547	.9552	.9557	.9562	.9566	.9571	.9576	.9581	.9586
9.1	.9590	.9595	.9600	.9605	.9609	.9614	.9619	.9624	.9628	.9633
9.2	.9638	.9643	.9647	.9652	.9657	.9661	.9666	.9671	.9675	.9680
9.3	.9685	.9689	.9694	.9699	.9703	.9708	.9713	.9717	.9722	.9727
9.4	.9731	.9736	.9741	.9745	.9750	.9754	.9759	.9763	.9768	.9773
9.5	.9777	.9782	.9786	.9791	.9795	.9800	.9805	.9809	.9814	.9818
9.6	.9823	.9827	.9832	.9836	.9841	.9845	.9850	.9854	.9859	.9863
9.7	.9868	.9872	.9877	.9881	.9886	.9890	.9894	.9899	.9903	.9908
9.8	.9912	.9917	.9921	.9926	.9930	.9934	.9939	.9943	.9948	.9952
9.9	.9956	.9961	.9965	.9969	.9974	.9978	.9983	.9987	.9991	.9996
n	0	1	2	3	4	5	6	7	8	9

TABLE 2 NATURAL LOGARITHMS AND POWERS OF e

x	e^x	e^{-x}	$\ln x$
0.00	1.00000	1.00000	
0.01	1.01005	0.99004	−4.6052
0.02	1.02020	0.98019	−3.9120
0.03	1.03045	0.97044	−3.5066
0.04	1.04081	0.96078	−3.2189
0.05	1.05127	0.95122	−2.9957
0.06	1.06183	0.94176	−2.8134
0.07	1.07250	0.93239	−2.6593
0.08	1.08328	0.92311	−2.5257
0.09	1.09417	0.91393	−2.4079
0.10	1.10517	0.90483	−2.3026
0.11	1.11628	0.89583	−2.2073
0.12	1.12750	0.88692	−2.1203
0.13	1.13883	0.87810	−2.0402
0.14	1.15027	0.86936	−1.9661
0.15	1.16183	0.86071	−1.8971
0.16	1.17351	0.85214	−1.8326
0.17	1.18530	0.84366	−1.7720
0.18	1.19722	0.83527	−1.7148
0.19	1.20925	0.82696	−1.6607
0.20	1.22140	0.81873	−1.6094
0.30	1.34985	0.74081	−1.2040
0.40	1.49182	0.67032	−0.9163
0.50	1.64872	0.60653	−0.6931
0.60	1.82211	0.54881	−0.5108
0.70	2.01375	0.49658	−0.3567
0.80	2.22554	0.44932	−0.2231
0.90	2.45960	0.40656	−0.1054
1.00	2.71828	0.36787	0.0000
1.10	3.00416	0.33287	0.0953
1.20	3.32011	0.30119	0.1823
1.30	3.66929	0.27253	0.2624
1.40	4.05519	0.24659	0.3365
1.50	4.48168	0.22313	0.4055

x	e^x	e^{-x}	$\ln x$
1.60	4.95302	0.20189	0.4700
1.70	5.47394	0.18268	0.5306
1.80	6.04964	0.16529	0.5878
1.90	6.68589	0.14956	0.6419
2.00	7.38905	0.13533	0.6931
2.10	8.16616	0.12245	0.7419
2.20	9.02500	0.11080	0.7885
2.30	9.97417	0.10025	0.8329
2.40	11.02316	0.09071	0.8755
2.50	12.18248	0.08208	0.9163
2.60	13.46372	0.07427	0.9555
2.70	14.87971	0.06720	0.9933
2.80	16.44463	0.06081	1.0296
2.90	18.17412	0.05502	1.0647
3.00	20.08551	0.04978	1.0986
3.50	33.11545	0.03020	1.2528
4.00	54.59815	0.01832	1.3863
4.50	90.01713	0.01111	1.5041
5.00	148.41316	0.00674	1.6094
5.50	224.69193	0.00409	1.7047
6.00	403.42879	0.00248	1.7918
6.50	665.14163	0.00150	1.8718
7.00	1096.63316	0.00091	1.9459
7.50	1808.04241	0.00055	2.0149
8.00	2980.95799	0.00034	2.0794
8.50	4914.76884	0.00020	2.1401
9.00	8103.08392	0.00012	2.1972
9.50	13359.72683	0.00007	2.2513
10.00	22026.46579	0.00005	2.3026

ADDITIONAL NATURAL LOGARITHMS

x	$\ln x$
11	2.3979
12	2.4849
13	2.5649
14	2.6391
15	2.7081
16	2.7726
17	2.8332
18	2.8904
19	2.9444

x	$\ln x$
20	2.9957
25	3.2189
30	3.4012
35	3.5553
40	3.6889
45	3.8067
50	3.9120
55	4.0073
60	4.0943
65	4.1744

x	$\ln x$
70	4.2485
75	4.3175
80	4.3820
85	4.4427
90	4.4998
95	4.5539
100	4.6052
1000	6.9078

ANSWERS TO SELECTED EXERCISES

TO THE STUDENT

If you need further help with algebra, you may want to obtain a copy of the *Student's Solutions Manual* that goes with this book. It contains solutions to all the odd-numbered exercises and all the chapter test exercises. You also may want the *Student's Study Guide*. It has extra examples and exercises to complete, corresponding to each section of the book. In addition, there is a practice test for each chapter. Your college bookstore either has the *Manual* or *Guide* or can order them for you.

In this section we provide the answers that we think most students will obtain when they work the exercises using the methods explained in the text. If your answer does not look exactly like the one given here, it is not necessarily wrong. In many cases there are equivalent forms of the answer. For example, if the answer section shows 3/4 and your answer is .75, you have obtained the correct answer but written it in a different (yet equivalent) form. Unless the directions specify otherwise, .75 is just as valid an answer as 3/4. In general, if your answer does not agree with the one given in the text, see whether it can be transformed into the other form. If it can, then it is the correct answer. If you still have doubts, talk with your instructor.

CHAPTER 1

1.1 Exercises (page 8)

1. 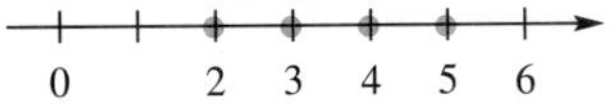

3. 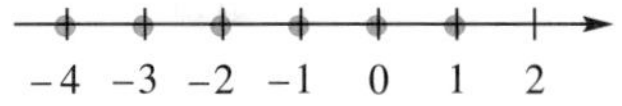

5. 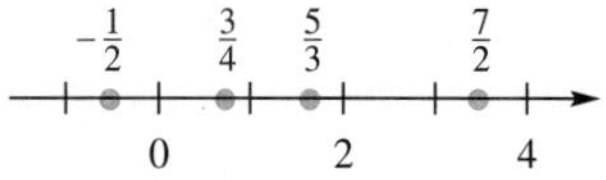

7. {1, 2, 3, 4, 5, 6} **9.** {12, 14, 16, 18, . . .} **11.** ∅ **13.** {3, −3} **15.** {0, 5, 10, 15, . . .} **17.** $\{x \mid x$ is a natural number multiple of 3} **19.** $\{x \mid x$ is one of the first 6 months of the year} **21.** $\{x \mid x$ is an odd natural number less than 10} **23.** 9 **25.** −6 **27.** −3 **29.** 5 **31.** 5 **33.** 4 **35.** −8 **37.** 9 **39.** 0 **41.** −3 **43.** −10 **45.** 8 **47.** 1 **49.** −5 **51.** 2, 3, $\frac{10}{2}$ **53.** −6, 0, 2, 3, $\frac{10}{2}$ **55.** $-\sqrt{3}$, $\sqrt{2}$, π **57.** false **59.** true **61.** false **63.** true **67.** $\subseteq$ **69.** $\subseteq$ **71.** $\subseteq$ **73.** $\subseteq$ **75.** $\not\subseteq$ **77.** $\subseteq$ **79.** $\subseteq$ **81.** $\subseteq$ **83.** $\in$ **85.** $\notin$ **87.** $\subseteq$ **89.** (b), (d) **91.** if $A = B$ **93.** yes **95.** if $x = -1$ **97.** if $x = 0$

1.2 Exercises (page 14)

1. $r \neq 4$ **3.** $2p + 1 < 9$ **5.** $3 \geq 7y$ **7.** $-6 \leq -6$ **9.** $5 < x < 9$ **11.** $-8 \leq 3k \leq 4$ **13.** $1 \leq a < 5$ **15.** $a < b < c$ **17.** false **19.** true **21.** false **23.** true **25.** $4 > -9$ **27.** $2 < 6$ **29.** $4 > x$ **31.** $6 > x > 5$ **33.** false **35.** true **37.** true **39.** false **41.** false

43. $(-1, \infty)$ 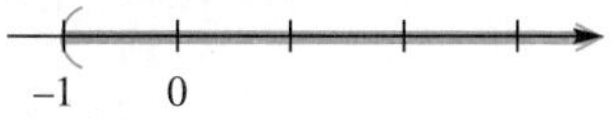

45. $(-\infty, 4]$ 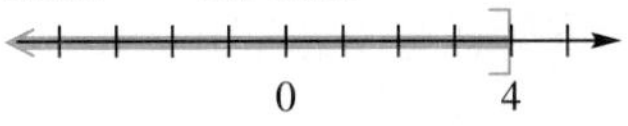

47. $(0, 3)$ 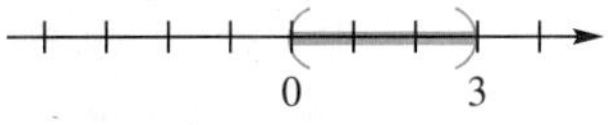

49. $[2, 5]$ 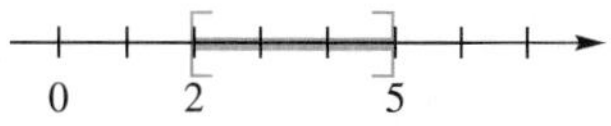

51. $(-4, 2]$ 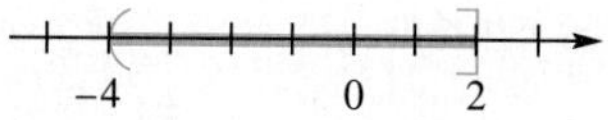

53. transitive **55.** symmetric **57.** transitive or substitution **59.** substitution **61.** substitution or transitive **63.** substitution **65.** Yes, since $-4 \le x \le 4$. **67.** $-1 < x < 0$ or $x > 1$

1.3 Exercises (page 23)

1. -20 **3.** $-\frac{19}{12}$ **5.** -2 **7.** -11 **9.** 9 **11.** 20 **13.** 4 **15.** $-\frac{7}{8}$ **17.** $-\frac{1}{3}$ **19.** $-\frac{5}{4}$ **21.** 6 **23.** 7 **25.** 6 **27.** $\frac{13}{2}$ or $6\frac{1}{2}$ **29.** $|x - a|$ or $|a - x|$ **31.** -102 feet (102 feet below the surface) **33.** 5296 feet **35.** 15 **37.** -1360 **39.** $\frac{5}{4}$ **41.** 2 **43.** undefined **45.** $\frac{8}{69}$ **47.** $-\frac{21}{64}$ **49.** 243 **51.** $-\frac{216}{343}$ **53.** -125 **55.** 16 **57.** -19 **59.** 3 **61.** 3 **65.** 21.838 **67.** .976 **69.** 64 **71.** 29 **73.** 43 **75.** 39 **77.** -2 **79.** 2 **81.** 31 **83.** undefined **85.** 13 **87.** $-\frac{32}{5}$ **89.** 108 **91.** undefined **93.** 200 **95.** False; for example, $10 - (5 - 4) = 10 - 1 = 9$, while $(10 - 5) - 4 = 5 - 4 = 1$. **97.** False; for example, $|6 - (-5)| = |11| = 11$, while $|6| - |-5| = 6 - 5 = 1$. **99.** Yes, a and b must be nonzero numbers with equal absolute values. For example, let $a = 3$ and $b = -3$. Then $\frac{3}{-3} = \frac{-3}{3}$.

1.4 Exercises (page 31)

1. $8k$ **3.** $16r$ **5.** cannot be simplified **7.** $6a$ **9.** $15c$ **11.** $3a + 3b$ **13.** $-10d - 5f$ **15.** $-20k + 12$ **17.** $-3k - 7$ **19.** $7x + 16$ **21.** $-6y - 3$ **23.** $p + 1$ **25.** $-2k + 9$ **27.** $m - 14$ **29.** $-6y + 29$ **31.** $5p + 7$ **33.** $-6z - 39$ **35.** $-4m - 10$ **37.** $-24x - 12$ **39.** $(2 + 3)x$; $5x$ **41.** $(2 \cdot 4)x$; $8x$ **43.** 0 **45.** 7 **47.** $(3 + 5 + 6)a$; $14a$ **49.** -2 **51.** $(8)2 + (8)3$; 40 **53.** 8 **55.** $3 \cdot 4$; 12 **57.** 4 **59.** $2 + 6 \cdot 5 = 2 + 30 = 32$, which does not equal $8 \cdot 5 = 40$. **61.** 1900 **63.** 75 **65.** 431 **67.** No, for example, $7 + (5 \cdot 3) \neq (7 + 5)(7 + 3)$.

Chapter 1 Review Exercises (page 37)

1. 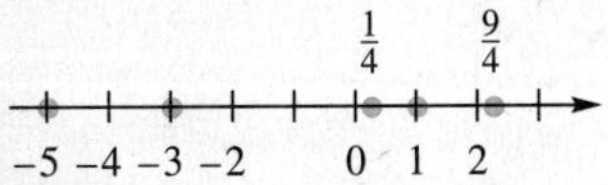

3. 12 **5.** -4 **7.** -15 **9.** -9 **11.** $\frac{12}{3}$ (or 4) **13.** $-9, -1.\overline{3}, -.8, 0, \frac{5}{3}, \frac{12}{3}$ (or 4) **15.** $3 \le 10$ **17.** $-2 < x < 1$ **19.** true **21.** false **23.** true **25.** $(-\infty, -4)$ 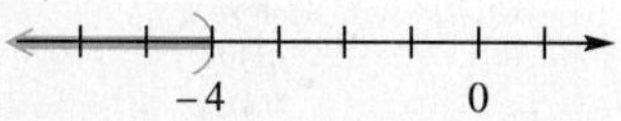

27. $(-2, 5]$ 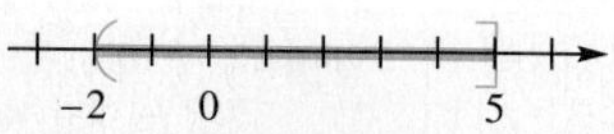

29. $\frac{41}{24}$ **31.** 4 **33.** -39 **35.** $\frac{23}{20}$ **37.** -25 **39.** $\frac{2}{3}$ **41.** 5 **43.** $-\frac{4}{3}$ **45.** 10,000 **47.** -125 **49.** 2.89 **51.** 3 **53.** 3 **55.** -3 **57.** $\frac{7}{3}$ **59.** -29 **61.** -54 **63.** $-\frac{34}{5}$ **65.** $11q$ **67.** $5m$ **69.** $-2k + 6$ **71.** $18m + 27n$ **73.** $p - 6q$ **75.** $-7y^3 + 1$ **77.** $4k^2 - 2$ **79.** $-7p - 10$ **81.** $7(p + r)$ **83.** 11 **85.** 22 **87.** $\frac{256}{625}$ **89.** 43 **91.** 0 **93.** undefined **95.** 10 **97.** $\frac{22}{7}$ **99.** $\frac{107}{9}$

Chapter 1 Test (page 39)

[1.1] **1.** All elements in the set are real numbers. **2.** $-\sqrt{7}, \sqrt{11}$ **3.** $-2, -.5, \frac{0}{2}$ (or 0), $1.\overline{34}, \frac{15}{2}, 3, \frac{24}{2}$ (or 12) **4.** $-2, \frac{0}{2}$ (or 0), $3, \frac{24}{2}$ (or 12) [1.2] **5.** $(-\infty, 2)$ 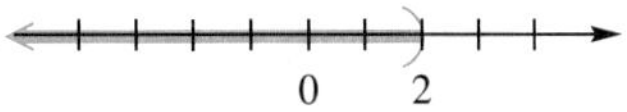

6. $[-2, 5)$ 

[1.3] **7.** -3 **8.** $\frac{3}{5}$ **9.** 0 **10.** -1 **11.** 5 **12.** 1 **13.** $\frac{16}{7}$ **14.** -4 **15.** $-8, 4, y$ **16.** -8 **17.** $\frac{81}{16}$ **18.** 144 **19.** 0 **20.** 4 **21.** 8 **22.** 173 **23.** undefined **24.** -36 **25.** $-\frac{14}{3}$ [1.4] **26.** $6x - 3y$ **27.** $2k - 10$ **28.** $2p + 1$ **29.** $\frac{3}{2}$ **30.** 0

CHAPTER 2

2.1 Exercises (page 48)

1. {3} **3.** {3} **5.** {−3} **7.** {−2} **9.** {−2} **11.** $\left\{-\frac{11}{3}\right\}$ **13.** {−7} **15.** {1} **17.** $\left\{\frac{5}{3}\right\}$ **19.** $\left\{-\frac{10}{3}\right\}$ **21.** $\left\{-\frac{1}{2}\right\}$ **23.** {2} **25.** {4} **27.** {−2} **31.** $\left\{-\frac{18}{5}\right\}$ **33.** $\left\{-\frac{5}{6}\right\}$ **35.** {6} **37.** {−15} **39.** {2} **41.** {2000} **43.** {25} **45.** {40} **47.** identity; contradiction **49.** identity; {all real numbers} **51.** contradiction; ∅ **53.** conditional; {0} **55.** identity; {all real numbers} **57.** contradiction; ∅ **59.** 37 **61.** −14 **63.** $\frac{7}{16}$ **65.** (a) 1940 (b) 1964 **67.** equivalent **69.** not equivalent **71.** not equivalent **73.** 32 **75.** 90 **77.** 100 **79.** 13

2.2 Exercises (page 56)

1. $r = \frac{d}{t}$ **3.** $b = \frac{A}{h}$ **5.** $a = P - b - c$ **7.** $h = \frac{2A}{b}$ **9.** $h = \frac{S - 2\pi r^2}{2\pi r}$ or $h = \frac{S}{2\pi r} - r$ **11.** $F = \frac{9}{5}C + 32$ **13.** $H = \frac{A - 2LW}{2W + 2L}$ **19.** $b = \frac{2A}{h}$ **21.** $p = \frac{I}{rt}$ **23.** $h = \frac{S - 2\pi r^2}{2\pi r}$ or $h = \frac{S}{2\pi r} - r$ **25.** (a) $d = rt$; (b) $r = \frac{d}{t}$; (c) 52 miles per hour **27.** (a) $P = 2L + 2W$; (b) $L = \frac{P - 2W}{2}$; (c) 14.5 meters **29.** (a) $V = LWH$; (b) $H = \frac{V}{LW}$; (c) 10 feet **31.** (a) $I = prt$; (b) $t = \frac{I}{pr}$; (c) 3.5 years **33.** (a) $C = \frac{5}{9}(F - 32)$; (c) 5° C **35.** (a) $F = \frac{9}{5}C + 32$; (c) 104° F **37.** (a) $C = 2\pi r$; (b) $r = \frac{C}{2\pi}$; (c) 15 meters **39.** (a) $S = 2\pi rh + 2\pi r^2$; (b) $h = \frac{S - 2\pi r^2}{2\pi r}$; (c) 6 inches **41.** \$97.59 **43.** 158 days **45.** \$10.51 **47.** \$45.66 **49.** 10% **51.** $x = -8$; 49°; 49° **53.** $x = 27$; 49°; 49° **55.** $x = 47$; 48°; 132° **57.** −7 **59.** −25 **61.** 19

2.3 Exercises (page 66)

1. $x - 4$ **3.** $11 + x$ **5.** $-8x$ **7.** $x - 8$ **9.** $-3 + 4x$ **11.** $\frac{-1}{x}$ **13.** $\frac{x}{-4}$ **15.** $\frac{5}{9}x$ or $\frac{5x}{9}$ **19.** Eisner: 40.1 million dollars; Horrigan: 21.7 million dollars **21.** Ruth: 2873 hits; Hornsby: 2930 hits **23.** 2 **25.** 50 **27.** \$21.20 **29.** $2\frac{1}{2}$ liters **31.** 2 liters **33.** $18\frac{2}{11}$ liters **35.** 5 liters **37.** \$3000 at 8%; \$9000 at 9% **39.** \$2000 at 8%; \$3000 at 14% **41.** \$87,000 **43.** \$5000 **45.** 25° **47.** 20°; 30°; 130°

49. 65°; 115° **51.** ① = 55°; ② = 65°; ③ = 60°; ④ = 65°; ⑤ = 60°; ⑥ = 120°; ⑦ = 60°; ⑧ = 60°; ⑨ = 55°; ⑩ = 55° **53.** 150 **55.** 28 **57.** 17

2.4 Exercises (page 74)

1. 16 nickels; 14 pennies **3.** 25 half-dollars; 20 quarters **5.** 30 quarters; 5 dimes; 5 pennies **7.** 30 three-cent pieces; 10 two-cent pieces **9.** 120 students; 180 non-students **11.** 20 \$2 tickets; 14 \$3 tickets **15.** 3 hours **17.** 5.5 hours **19.** 8 hours **21.** 5/6 hour **23.** 15 miles per hour **25.** 18 miles **27.** 1/2 hour **31.** 16 feet; 16 feet; 29 feet **33.** 23 inches **35.** length: 30 meters; width: 15 meters **37.** Ricardo: 15 years old; Jaime: 30 years old **39.** 65 heads **41.** 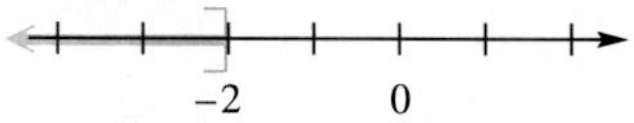

43. 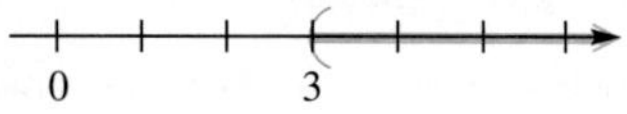

45. 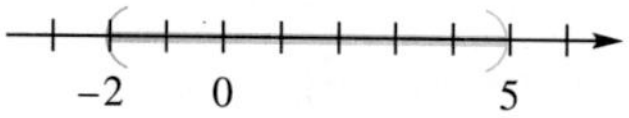

2.5 Exercises (page 84)

1. $(2, \infty)$ 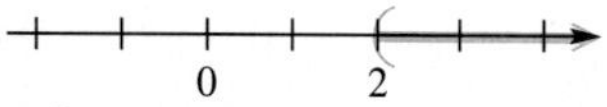

3. $(-\infty, -3]$ 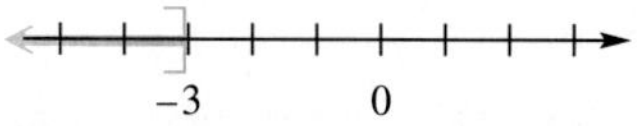

5. $[5, \infty)$ 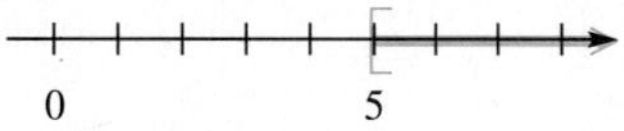

7. $(3, \infty)$ 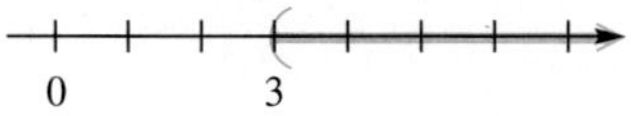

9. $(-\infty, -28]$ 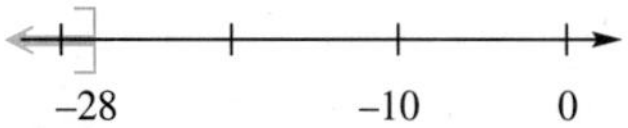

11. $[3, \infty)$ 

13. $(-\infty, -3)$ 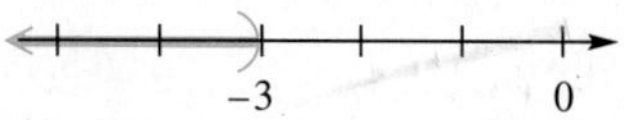

15. $\left(-\infty, \frac{1}{2}\right)$ 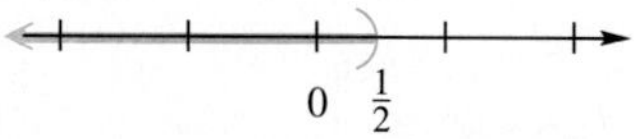

17. $[7, \infty)$ 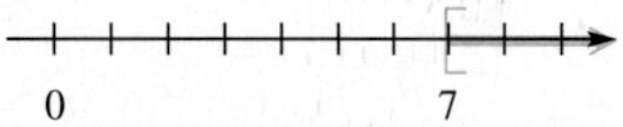

19. $(-3, \infty)$ 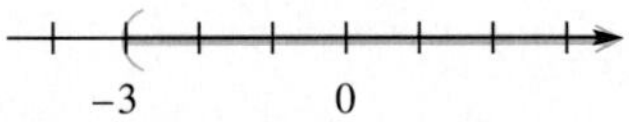

21. $(2, 11)$ 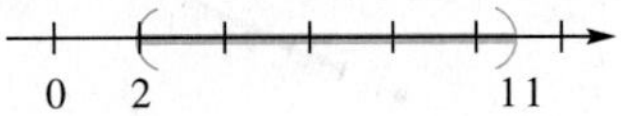

23. $[-5, 4]$ 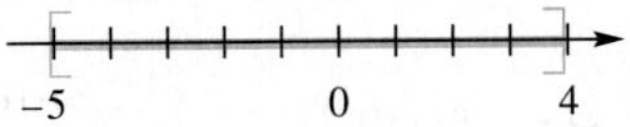

25. $\left[-\frac{14}{3}, -\frac{4}{3}\right]$ 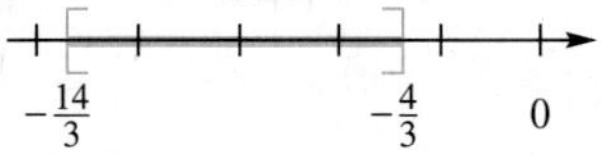

27. $\left[-\frac{19}{2}, \frac{35}{2}\right]$ 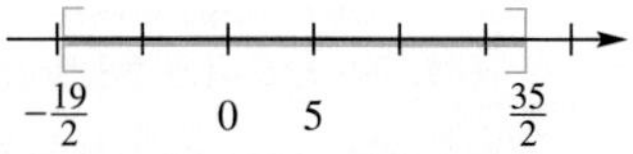

29. $(2, \infty)$ 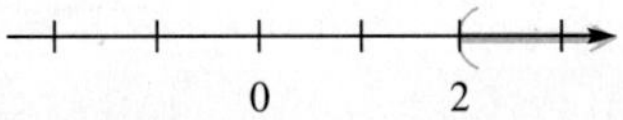

31. $[1, \infty)$ 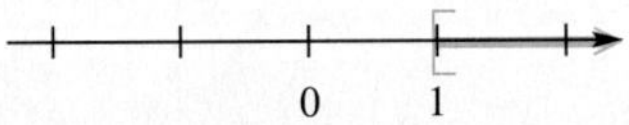

33. $(2, \infty)$ 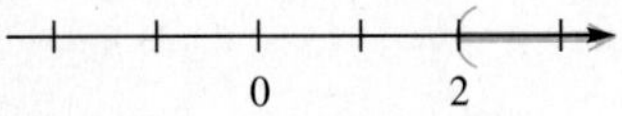

35. $\left(-\infty, \frac{23}{5}\right)$ 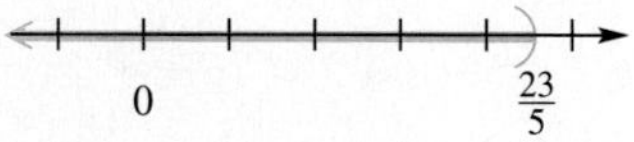

37. $\left[-\frac{5}{6}, \infty\right)$ 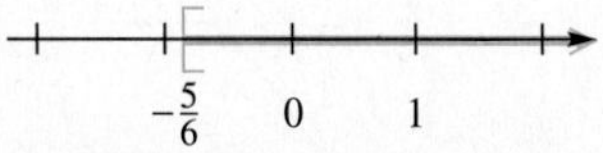

39. $\left(-\infty, \frac{36}{11}\right)$ 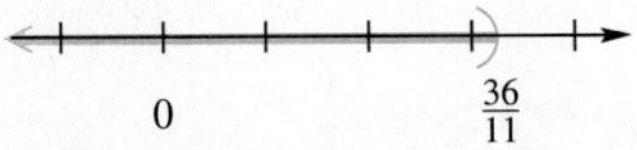

43. at least 74 **45.** 4 green pills **47.** 87 points **49.** 50 miles **51.** 501 units **53.** 46 units **55.** $(-\infty, \infty)$ **57.** $\emptyset$ **59.** all numbers between −2 and 2, that is, $(-2, 2)$ **61.** all numbers greater than or equal to 3, that is, $[3, \infty)$ **63.** all numbers greater than or equal to −9, that is, $[-9, \infty)$ **65.** $1 > -x > -5$, or $-5 < -x < 1$ **69.** $[-1, 2]$ **71.** $(-1, 8)$ **73.** set

2.6 Exercises (page 91)

1. {a, c, e} or B **3.** {d} or D **5.** {a} **7.** {a, b, c, d, e, f} or A **9.** {a, c, e, f} **11.** {a, d, f} **13.** {a, c, e} or B **15.** {a} **17.** {a, c, d, e, f} **19.** {a, c, e, f} **21.** commutative property

25. $(3, 7)$ **27.** $(0, 4)$

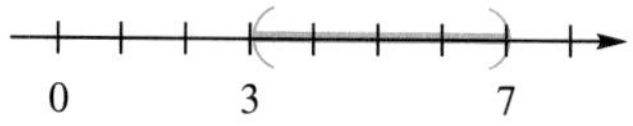

29. $(-\infty, -8]$ 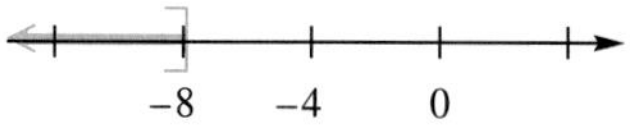 **31.** $[5, \infty)$

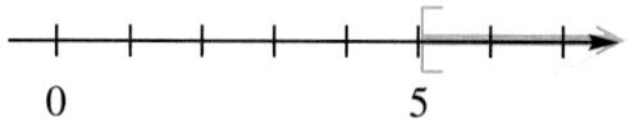

33. 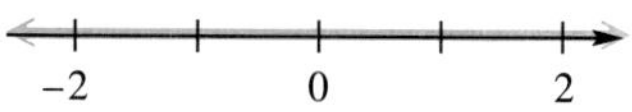 $(-\infty, \infty)$ **35.** $[4, 8]$

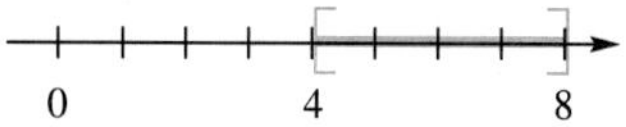

37. $(-2, -1)$ 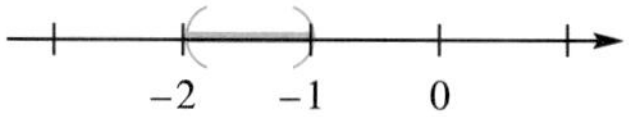 **39.** $\emptyset$ **41.** $(-\infty, 4]$

43. $(-\infty, -4) \cup (4, \infty)$ 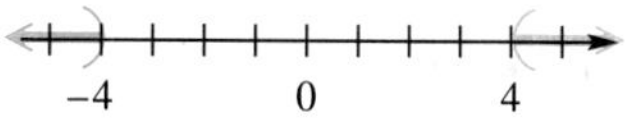 **45.** $(-\infty, 1) \cup (2, \infty)$

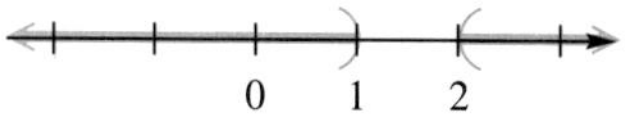

47. $(1, \infty)$ 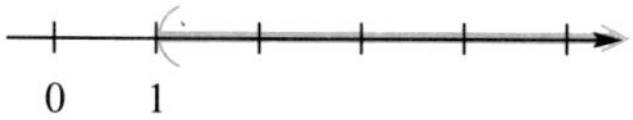 **49.** $(-\infty, \infty)$

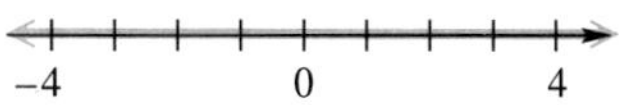

51. $[-4, -1]$ **53.** $[4, 12]$ **55.** $(-\infty, 3)$ **57.** $[6, 11]$ **59.** $(2, 4)$ **61.** $[-1, 5)$ **63.** Each set is {1, 2, 3, 4, 5, 6}. **65.** yes **67.** if A and B are equal **69.** always true **71.** $[-6, \infty)$ **73.** $(-3, 2)$ **75.** -21

2.7 Exercises (page 97)

1. $\{-8, 8\}$ **3.** $\{-3, 3\}$ **5.** $\{-5, 11\}$ **7.** $\{-5, 4\}$ **9.** $\left\{-2, \frac{9}{2}\right\}$ **11.** $\left\{-\frac{17}{2}, \frac{7}{2}\right\}$ **13.** $\{-14, 2\}$ **15.** $\left\{-\frac{8}{3}, \frac{16}{3}\right\}$ **17.** $\left\{-1, \frac{1}{3}\right\}$ **19.** $(-\infty, -2) \cup (2, \infty)$ **21.** $(-\infty, -3] \cup [3, \infty)$ **23.** $(-\infty, -10) \cup (6, \infty)$ **25.** $\left(-\infty, -\frac{4}{3}\right] \cup [2, \infty)$ **27.** $(-\infty, -1) \cup (7, \infty)$ **29.** $[-2, 2]$ **31.** $(-3, 3)$ **33.** $[-10, 6]$ **35.** $\left(-\frac{4}{3}, 2\right)$ **37.** $[-1, 7]$ **39.** $(-\infty, -13) \cup (5, \infty)$ **41.** $\{-5, -2\}$ **43.** $\left[-\frac{7}{3}, 3\right]$ **45.** $\left(-\frac{7}{6}, -\frac{5}{6}\right)$ **47.** $\{-8, 8\}$ **49.** $\{-6, -2\}$ **51.** $\left(-\infty, -\frac{3}{2}\right) \cup \left(\frac{1}{2}, \infty\right)$ **53.** $[-9, -1]$ **55.** $\left\{-8, \frac{6}{5}\right\}$ **57.** $\left\{-3, \frac{5}{3}\right\}$ **59.** $\left\{-\frac{5}{3}, -\frac{1}{3}\right\}$ **61.** $\left\{\frac{1}{4}\right\}$ **63.** $\emptyset$ **65.** $\emptyset$ **67.** $\left\{\frac{1}{4}\right\}$ **69.** $\emptyset$ **71.** $(-\infty, \infty)$ **73.** $\left\{\frac{3}{7}\right\}$ **75.** $(-\infty, \infty)$ **77.** $\left(-\infty, -\frac{9}{10}\right) \cup \left(-\frac{9}{10}, \infty\right)$ **79.** all real numbers **81.** any real number greater than or equal to -6; solution set is $[-6, \infty)$ **83.** $\{2\}$ **85.** $(-\infty, -a]$ **87.** $\{1\}$ **89.** $\{-2\}$ **91.** $\left(\frac{4}{3}, \infty\right)$ **93.** $(-\infty, 1]$ **95.** 8 **97.** 25 **99.** -216 **101.** -216 **103.** $\frac{8}{27}$

Summary on Solving Linear Equations and Inequalities (page 99)

1. $\{13\}$ **3.** $\{-3, 5\}$ **5.** $\left\{\frac{2}{11}\right\}$ **7.** $(-\infty, 1]$ **9.** $\{11\}$ **11.** $\left\{1, \frac{11}{3}\right\}$

13. $(-3, 3)$ 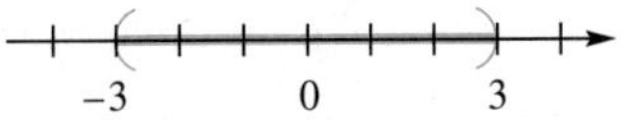

15. $\left\{\frac{8}{5}\right\}$ **17.** $\left\{-1, \frac{1}{5}\right\}$

19. $\left[\frac{8}{5}, \infty\right)$ 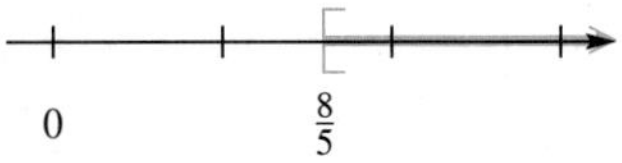

21. $(-\infty, -1) \cup (1, \infty)$ 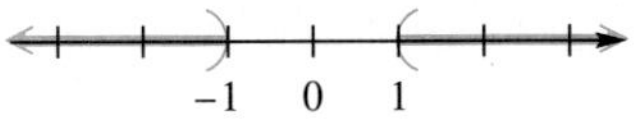

23. $\left\{\frac{96}{5}\right\}$

25. $(-\infty, -12)$ 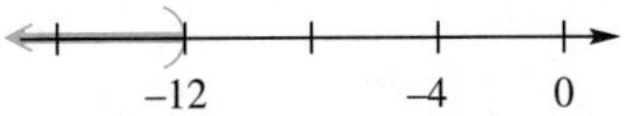

27. $\emptyset$ **29.** $(-4, 6)$ 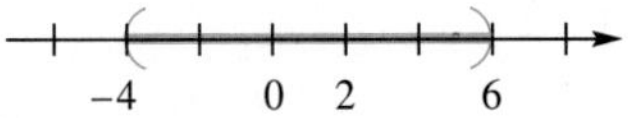

31. $\{0\}$

33. $(-\infty, 1) \cup (2, \infty)$ 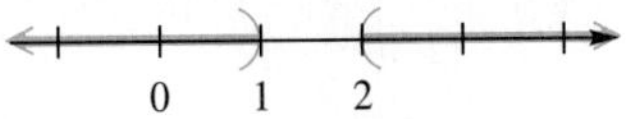

35. $\{12\}$ **37.** $\left(-\infty, -\frac{5}{2}\right]$ 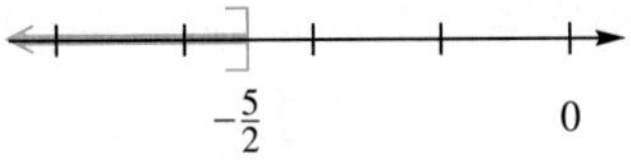

39. $\{3\}$ **41.** $\left[-1, \frac{1}{2}\right]$ 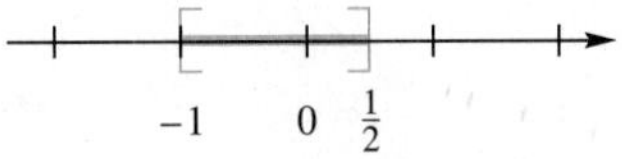

43. $\{0, 1\}$

45. $(-\infty, -1] \cup \left[\frac{5}{3}, \infty\right)$ 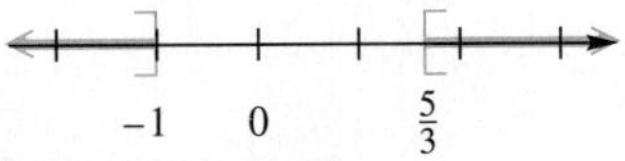

47. $\left\{\frac{7}{2}\right\}$ **49.** $\left[-\frac{1}{2}, \frac{15}{2}\right]$ 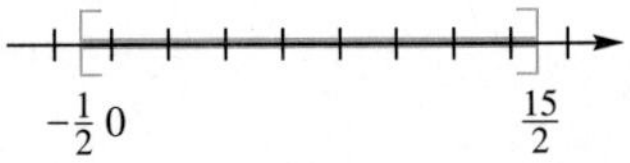

51. $(-\infty, \infty)$ 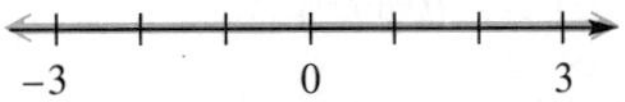

53. $(-\infty, \infty)$ **55.** $\{-1\}$ **57.** $\left(-1, \frac{1}{5}\right)$ 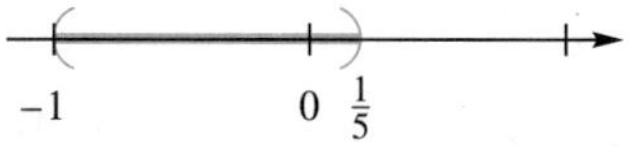

Chapter 2 Review Exercises (page 105)

1. $\{-2\}$ **3.** $\left\{\frac{16}{5}\right\}$ **5.** $\{-1\}$ **7.** $\{2\}$ **9.** $\left\{-\frac{2}{3}\right\}$ **11.** $\{3\}$ **13.** $\left\{\frac{4}{3}\right\}$ **15.** $\{75\}$

17. identity; $(-\infty, \infty)$ **19.** conditional; $\{0\}$ **21.** $H = \frac{V}{LW}$ **23.** $d = \frac{C}{\pi}$ **25.** 20 meters **27.** 6 feet

29. 18 millimeters **31.** 20° C **33.** 105°; 105° **35.** $3x$ **37.** $\frac{1}{2}x + 5$ **39.** 2 **41.** 7 **43.** 400

45. 30 liters **47.** \$25,000 at 15%; \$17,000 at 12% **49.** 40°; 45°; 95° **51.** 850 reserved; 246 general admission **53.** 15 pounds **55.** 50 kilometers per hour; 65 kilometers per hour **57.** 46 miles per hour

59. length: 13 meters; width: 8 meters **61.** $(-\infty, 8)$ **63.** $(-14, \infty)$ **65.** $\left(-\infty, -\frac{14}{5}\right]$ **67.** $[-6, \infty)$

69. $\left[-\frac{9}{7}, \infty\right)$ **71.** $(-\infty, -3)$ **73.** $\left(-\infty, -\frac{11}{9}\right)$ **75.** $[3, 5)$ **77.** $\left(\frac{59}{31}, \infty\right)$ **79.** any grade greater than or equal to 61% **81.** {a, c} **83.** {a, c, e, f, g} **85.** $(-\infty, 3)$ 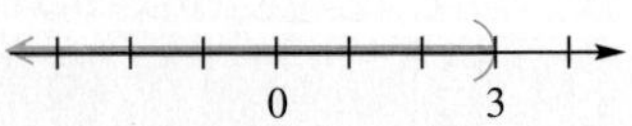

87. $\emptyset$

89. $(-\infty, \infty)$ 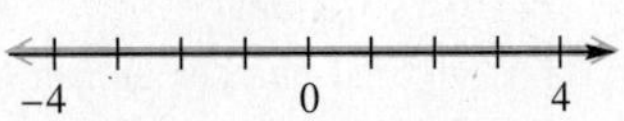

91. $(-3, 4)$ **93.** $(4, \infty)$ **95.** $\{-4, 4\}$ **97.** $\{3, 2\}$

99. $\{6, 1\}$ **101.** $\emptyset$ **103.** $\left\{\frac{36}{17}, 84\right\}$ **105.** $(-4, 4)$ 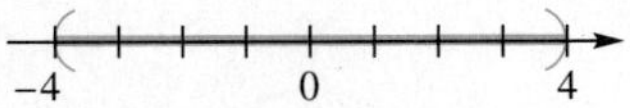

107. $[-4, -1]$ 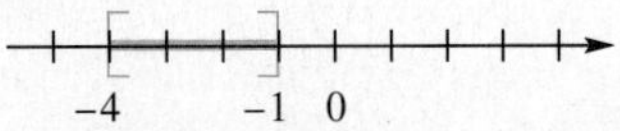

109. $(-\infty, -10] \cup [8, \infty)$ 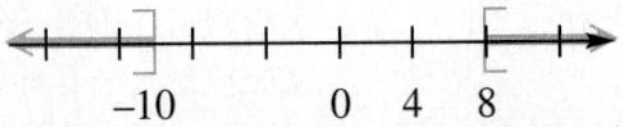

111. $(-\infty, -3) \cup (7, \infty)$ 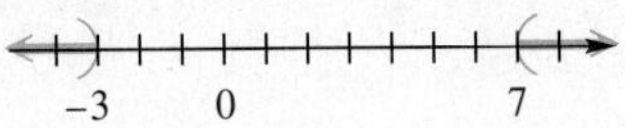

113. $(-2, \infty)$ **115.** $[-2, 3)$ **117.** 15 centimeters

119. $\left(-\infty, -\frac{13}{5}\right) \cup (3, \infty)$ **121.** 5 liters **123.** $\{30\}$ **125.** $\left\{1, \frac{11}{3}\right\}$ **127.** $\emptyset$

Chapter 2 Test (page 109)

[2.1] **1.** {2} **2.** $\left\{\frac{35}{13}\right\}$ **3.** {2} **4.** identity; $(-\infty, \infty)$ [2.2] **5.** $v = \frac{S + 16t^2}{t}$ or $v = \frac{S}{t} + 16t$ **6.** 4 years [2.3] **7.** 4 liters of 90% solution; 16 liters of 75% solution **8.** \$1500 at 7%; \$4500 at 6% **9.** 40°; 40°; 100° [2.4] **10.** slower car: 45 miles per hour; faster car: 60 miles per hour **11.** length: 10 feet; width: 4 feet [2.5] **12.** $(-\infty, 8)$ **13.** $\left(\frac{5}{8}, \infty\right)$

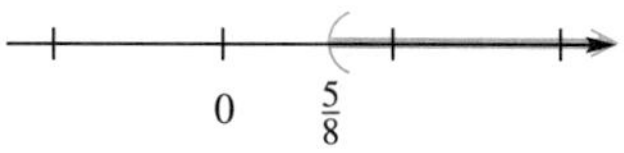

14. $(-\infty, 3]$ 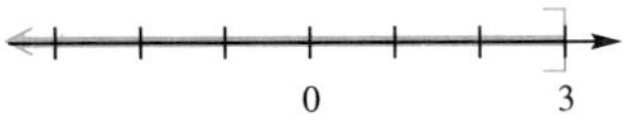 **15.** $\left(\frac{3}{2}, 6\right)$

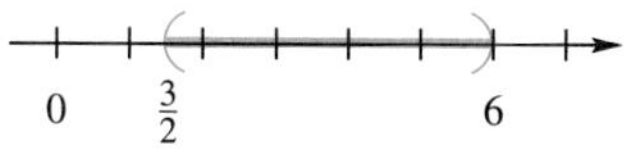

16. (c) **17.** 82% [2.6] **18.** **(a)** {1, 5} **(b)** {1, 2, 5, 7, 9, 12} **19.** $(-5, 4)$ **20.** $(-\infty, 3) \cup [5, \infty)$ [2.7] **21.** {−2, 5} **22.** $\left\{-6, -\frac{4}{3}\right\}$ **23.** $[-7, 4]$

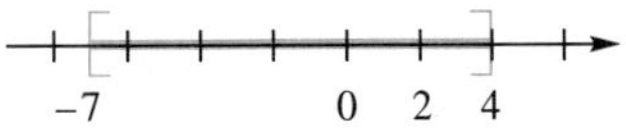

24. $\left(-\infty, -\frac{8}{3}\right] \cup \left[\frac{4}{3}, \infty\right)$ **25.** $\emptyset$

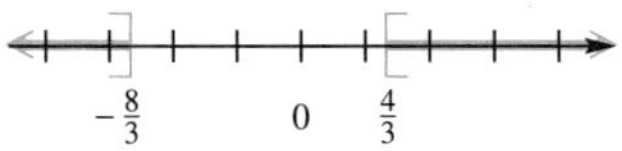

CHAPTER 3

3.1 Exercises (page 119)

In Exercises 1–15, the exponent is given first, followed by the base.

1. 7; 5 **3.** 5; 12 **5.** 4; −9 **7.** 4; 9 **9.** −7; p **11.** 4; y **13.** −4; q **15.** 3; $-m + z$ **17.** 625 **19.** $\frac{25}{9}$ **21.** −32 **23.** −8 **25.** −81 **27.** $\frac{1}{49}$ **29.** $-\frac{1}{49}$ **31.** −128 **33.** $\frac{9}{8}$ **35.** 8 **37.** $\frac{9}{4}$ **39.** $\frac{5}{6}$ **41.** 1 **43.** 1 **45.** −1 **47.** 0 **49.** 2 **53.** 2^{16} **55.** 7^4 **57.** 3^3 or 27 **59.** $\frac{1}{6^2}$ or $\frac{1}{36}$ **61.** $\frac{1}{3^3}$ or $\frac{1}{27}$ **63.** $\frac{1}{9^2}$ or $\frac{1}{81}$ **65.** $\frac{1}{t^7}$ **67.** r^5 **69.** $\frac{1}{a^5}$ **71.** x^{11} **73.** r^6 **75.** $-\frac{56}{k^2}$ **77.** $-24x^4y^2$ **79.** p^{8q} **81.** a^{3r+3} **83.** m^a **85.** $\frac{1}{5}$; $\frac{5}{6}$; $\frac{1}{5} \neq \frac{5}{6}$ **87.** 25; 13; $25 \neq 13$ **89.** 5^6 **91.** -2^6

3.2 Exercises (page 126)

1. 4^{10} **3.** 5^4 **5.** 6^6 **7.** $\frac{16}{9}$ **9.** $\frac{5}{6}$ **11.** $\frac{1}{3^6 \cdot 4^3}$ **13.** $4^6 \cdot 7^{10}$ **15.** $\frac{1}{z^4}$ **17.** $-\frac{3}{r^7}$ **19.** $\frac{27}{a^{18}}$ **21.** $\frac{x^5}{y^2}$ **23.** $\frac{64}{p^2q^4}$ **25.** $\frac{1}{5p^{10}}$ **27.** $\frac{4}{a^2}$ **29.** $\frac{1}{6y^{13}}$ **31.** $\frac{4k^5}{m^2}$ **33.** $\frac{z^5}{4y^5}$ **35.** $\frac{4k^{17}}{125}$ **37.** $\frac{2k^5}{3}$ **39.** $\frac{8}{3pq^{10}}$ **41.** $\frac{y^9}{8}$ **43.** reciprocal; negative (or opposite) **45.** (d) **47.** 2.3×10^2 **49.** 2×10^{-2} **51.** 3.27×10^{-6} **53.** 9.3×10^7 **55.** 6500 **57.** .0152 **59.** .005 **61.** .0469 **63.** 3×10^{-4} or .0003 **65.** 3×10^{-5} or .00003 **67.** 6×10^5 or 600,000 **69.** 2×10^5 or 200,000 **71.** 2×10^5 or 200,000 **73.** 1.711×10^6 or 1,711,000 **75.** 1.198×10^{12} or 1,198,000,000,000 **77.** about $3.2 \times 10^4 = 32{,}000$ hours (about 3.7 years) **79.** 500 seconds, which is about 8.3 minutes **81.** 20,000 hours **83.** **(a)** .7 astronomical unit **(b)** 1 astronomical unit **(c)** 1.6 astronomical units **85.** 2×10^{30} bacteria **87.** x^{3+m} **89.** p^{2y+2} **91.** a^{-4-2k} **93.** r^{p^2+3p-6} **95.** $-\frac{3}{32m^8p^4}$ **97.** $\frac{2}{3y^4}$ **99.** $\frac{3p^8}{16q^{14}}$ **101.** $13x$ **103.** $4y - 6$ **105.** $2x - 2$

3.3 Exercises (page 135)

In Exercises 1–9, the coefficient is given first, followed by the degree.

1. 9; 1 **3.** -11; 7 **5.** π; 3 **7.** 1; 1 **9.** -1; 8 **13.** $3x^3 + 9x^2 + 8x$ **15.** $4y^4 - 6y^3 + 8y^2 - 3$ **17.** $-9y^4 + 13y^3 + 8y^2 - 7y + 2$ **19.** binomial; 1 **21.** trinomial; 2 **23.** trinomial; 4 **25.** monomial; 0 **27.** none of these; 8 **29.** 5 **31.** 2 **33.** 0 **35.** 9 **37.** $12p - 4$ **39.** $-k - 1$ **41.** $3y^3 + 12y - 1$ **43.** $-7z^2 + 7z - 3$ **45.** $10y^3 - 7y^2 + 5y + 8$ **47.** $-4m^3$ **49.** $14p^2 - 32p$ **51.** $m^4 - m^3 + 6m^2$ **53.** $8z^3 + z^2 - 12z - 9$ **55.** -1 **57.** $-8x - 12$ **59.** $5y^3 - 3y^2 + 5y + 1$ **61.** $-4y^2 + 2y - 4$ **63.** $7r$ **65.** $3a^2 + 2a$ **67.** Answers may vary. One example is $7x^5 + 2x^3 - 6x^2 + 9x$. **69.** (a) -11 (b) -8 **71.** (a) 9 (b) 0 **73.** (a) 10 (b) 34 **75.** (a) -11 (b) 10 **77.** $-4y + 11a$ **79.** $3y^2 - 4y + 2$ **81.** $-4m^2 + 4n^2 - 7n$ **83.** $y^4 - 4y^2 - 4$ **85.** $-3xy - 11z^2$ **87.** $10z^2 - 16z$ **89.** $5a^{3x} + 2a^{2x} + 3a^x + 2$ **91.** $-8k^{4p} - 9k^{2p} + 2$ **93.** $3p^{2k} + 8p^k - 7$ **95.** $36p^5$ **97.** $50z^5$ **99.** $12x^2y^7$ **101.** $-6a^9b^2$

3.4 Exercises (page 142)

1. $21p^2 - 12pk$ **3.** $15m^4 + 18m^3 - 9m$ **5.** $12a^7 - 24a^6 + 30a^5$ **7.** $8x^3y^3 + 12x^4y^2 + 16x^3y^2$ **9.** $-30r^3t^3 + 20r^2t^3 - 50r^3t^2$ **11.** $45m^2 - 11m - 4$ **13.** $2g^2 - 13g + 15$ **15.** $-3y^2 + 2y + 1$ **17.** $6a^2 + ab - b^2$ **19.** $-24m^2 + 55mp - 25p^2$ **21.** $k^2 + \frac{1}{6}km - \frac{1}{3}m^2$ **23.** $3w^2 - \frac{23}{4}wz - \frac{1}{2}z^2$ **25.** $15k^2 - 7k - 2$ **27.** $12m^3 + 22m^2 + 4m - 3$ **29.** $8z^4 - 26z^3 + 47z^2 - 28z + 3$ **31.** $-12p^5 + 32p^4 - 43p^3 + 36p^2 - 15p + 2$ **37.** $16z^2 - t^2$ **39.** $4x^2 - 25y^2$ **41.** $25r^2 - 9w^2$ **43.** $x^2 - \frac{1}{9}$ **45.** $\frac{1}{4}y^2 - \frac{4}{9}z^2$ **47.** $y^6 - 25$ **49.** $6t^3 - 24ts^2$ **51.** $r^2 + 12rz + 36z^2$ **53.** $9k^2 - 12kp + 4p^2$ **55.** $100a^2 - 20ab + b^2$ **57.** $9a^2x^2 - 30axby + 25b^2y^2$ **59.** $\frac{9}{64}m^2 - \frac{1}{2}mp + \frac{4}{9}p^2$ **61.** $25x^2 + 10x + 1 + 60xy + 12y + 36y^2$ **63.** $4a^2 + 4ab + b^2 - 12a - 6b + 9$ **65.** $4a^2 + 4ab + b^2 - 9$ **67.** $4h^2 - 4hk + k^2 - 64$ **69.** $25t^2 - 9 - 12s - 4s^2$ **71.** $6a^3 + 7a^2b + 4ab^2 + b^3$ **73.** $4z^4 - 17z^3x + 12z^2x^2 - 6zx^3 + x^4$ **75.** $m^4 - 4m^2p^2 + 4mp^3 - p^4$ **77.** $a^3 - 7ab^2 - 6b^3$ **79.** $y^3 + 6y^2 + 12y + 8$ **81.** $q^4 - 8q^3 + 24q^2 - 32q + 16$ **83.** 25; 13; $25 \neq 13$ **85.** 625; 97; $625 \neq 97$ **87.** $\frac{1}{2}x^2 - 2y^2$ **89.** $15x^2 + 13x - 6$ **91.** $12b^{2p-2} + 6b^{3p}$ **93.** $3z^{2m-2} - 4z^{-4}$ **95.** $12k^{2n} - 16k^n - 3$ **97.** $16k^{3m} + 56k^my^m - 6y^{3m}k^{2m} - 21y^{4m}$ **99.** $y^{4n} - 1$ **101.** $9(6 + r)$ **103.** $7(2x - 3z)$ **105.** $(x + 1)(x + 4)$ **107.** $8z^7w^9$ **109.** $\frac{2p^3}{5q^3}$

3.5 Exercises (page 150)

1. $15(k + 2)$ **3.** $2p(3p + 2)$ **5.** $4mp(2m - p)$ **7.** $xy(5 - 8y)$ **9.** $2m^2(4m^2 + 3m - 2)$ **11.** $2t^3(5t^2 - 4t - 8)$ **13.** $3rs(2r - s - 3r^2)$ **15.** $xy(2x - 3y + 4xy)$ **17.** $12km^2(m - 2k^2 + 3km^2)$ **19.** $16z^3m^4(9z^8 + m - 2zm)$ **21.** $16zn^3(zn^3 + 4n^4 - 2z^2)$ **23.** $3a(5a^2b + 4ac - d^4)$ **25.** $9x^2y^2z(z^3 - 2x)$ **27.** no common factor other than 1 **29.** $7ab(2a^2b + a - 3a^4b^2 + 6b^3)$ **31.** $3mp^3(-5m^2 - 2m^2p - 3 + 10m)$ or $-3mp^3(5m^2 + 2m^2p + 3 - 10m)$ **33.** $(m - 9)(2m + 3)$ **35.** $(3k - 7)(2k + 7)$ **37.** $(r - 6)(r - 2)$ **39.** $m^5(r + s + t + u)$ **41.** $(3 - x)(6 + 2x - x^2)$ **43.** $20z(2z + 1)(3z + 4)$ **45.** $5(m + p)^2(m + p - 2 - 3m^2 - 6mp - 3p^2)$ **49.** $2x^2(-x^3 + 3x + 2)$ or $-2x^2(x^3 - 3x - 2)$ **51.** $16a^2m^3(-2a^2m^2 - 1 - 4a^3m^3)$ or $-16a^2m^3(2a^2m^2 + 1 + 4a^3m^3)$ **53.** $(p + 3r)(q + m)$ **55.** $(b + c)(2 + a)$ **57.** $(p + q)(p - 3y)$ **59.** $(a^2 + 2)(b^2 - 5)$ **61.** $(x - 3)(x + 2)$ **63.** $(3r - 2)(r + 5)$ **65.** $(3y + 2)(7y - 5)$ **67.** $(8p + 3q)(2p - q)$ **69.** $(2m + 3q)(7m - q)$ **71.** $(a^2 + b^2)(3a + 2b)$ **73.** $(4 - 3y^3)(2 - 3y)$ **75.** $(1 - a)(1 - b)$ **77.** (d) **79.** $p^{4m}(p^{2m} - 2)$ **81.** $q^k(q^{2k} + 2q^k + 3)$ **83.** $y^{r+2}(y^3 + y^2 + 1)$ **85.** $(r^p + q^p)(m^p - z^p)$ **87.** $m^{-5}(3 + m^2)$ or $\frac{m^2 + 3}{m^5}$ **89.** $p^{-3}(3 + 2p - 4p^2)$ or $\frac{-4p^2 + 2p + 3}{p^3}$ **91.** $6m^2 + m - 2$ **93.** $16r^2 - 14rs - 15s^2$ **95.** $18z^2 + 31za - 35a^2$ **97.** $14r^2 - 29rs - 15s^2$

3.6 Exercises (page 159)

1. $(c + 5)(c - 1)$ **3.** $(p + 2)(p + 4)$ **5.** $(a - 4)(a + 3)$ **7.** $(r + 4)(r - 5)$ **9.** $(x - 8)(x + 5)$ **11.** $(k - 3n)(k + 2n)$ **13.** $(y - 5x)(y + 2x)$ **15.** $(ab - 3)(ab - 4)$ **17.** $(5y + 6)(y - 1)$ **19.** $(3m + 1)(m + 2)$ **21.** $(8y - 3)(y + 2)$ **23.** $(3x + 2)(6x - 5)$ **25.** $(7p - 5)(5p + 3)$ **27.** $(6a - 5b)(2a + 3b)$ **29.** $(2k - 3a)(2k - 3a)$ or $(2k - 3a)^2$ **31.** $(7x + 3y)(5x - 8y)$ **33.** $(2m + 3p)(4m - 13p)$ **35.** $(3kp + 2)(2kp + 3)$ **37.** $2(3m - 4)(2m + 5)$ **39.** $3(2a - 3)(3a + 2)$ **41.** $6a(a - 3)(a + 5)$ **43.** $13y(y + 4)(y - 1)$ **45.** $2xy^3(x - 12y)(x - 12y)$ or $2xy^3(x - 12y)^2$

In Exercises 47–51, only one of several possible answers is given.

47. $-(x - 9)(x + 2)$ **49.** $-(9a + 5)(2a - 3)$ **51.** $-r(7r + 1)(2r - 3)$ **55.** $-18a^3 + 27a^2b + 315ab^2$ **57.** $(3x^2 + 1)(x^2 - 5)$ **59.** $(z^2 + 3)(z^2 - 10)$ **61.** $(3x^2 - 5)(2x^2 + 5)$ **63.** $(6p^2 - r)(2p^2 + 5r)$ **65.** $(4x^2 - 3a^2)(x^2 + 9a^2)$ **67.** $(3p + 14)(2p + 7)$ **69.** $(3z + 3k - 5)(2z + 2k + 1)$ **71.** $(a + b)^2(a - 3b)(a + 2b)$ **73.** $(p + q)(p + 3q)(p + q)$ or $(p + q)^2(p + 3q)$ **75.** $(z - x)(z - x)(z + 2x)$ or $(z - x)^2(z + 2x)$ **77.** $(p^n - 3)(p^n + 2)$ **79.** $(2z^{2r} + 1)(3z^{2r} - 4)$ **81.** $2k^r(6k^r + 1)(3k^r + 2)$ **83.** -1 **85.** $\frac{1}{m^3p^3}$ **87.** $4m^2 - 25$ **89.** $25a^2 - 30ab + 9b^2$ **91.** $y^3 - 8$

3.7 Exercises (page 164)

1. $(x + 5)(x - 5)$ **3.** $(6m + 5)(6m - 5)$ **5.** $(4y + 9q)(4y - 9q)$ **7.** $(4 + 5ab)(4 - 5ab)$ **9.** $(a^2 + 2b^2)(a^2 - 2b^2)$ **11.** $(x + 2)^2$ **13.** $(a - 5)^2$ **15.** $(3r - s)^2$ **17.** $(5xy - 2)^2$ **19.** $2(6m - 5p)^2$ **21.** $(2a + 1)(4a^2 - 2a + 1)$ **23.** $(3x - 4y)(9x^2 + 12xy + 16y^2)$ **25.** $(4x + 5y)(16x^2 - 20xy + 25y^2)$ **27.** $(5m - 2p)(25m^2 + 10mp + 4p^2)$ **29.** $(10 + 3rs)(100 - 30rs + 9r^2s^2)$ **31.** $(4y^2 + 1)(16y^4 - 4y^2 + 1)$ **33.** $(x - 1)(x^2 + x + 1)(x + 1)(x^2 - x + 1)$; $(x - 1)(x + 1)(x^4 + x^2 + 1)$ **35.** $(5 + r + 3s)(5 - r - 3s)$ **37.** $(m + 3p - 5)(m - 3p + 5)$ **39.** $4ab$ **41.** $(a + b + 1)^2$ **43.** $(m - p + 2)^2$ **45.** $(p - 3 + r)(p - 3 - r)$ **47.** $(3y - 5 + 4x)(3y - 5 - 4x)$ **49.** $(r + 4s - 3)(r - 4s + 3)$ **51.** $(4 - a + b)(16 + 4a - 4b + a^2 - 2ab + b^2)$ **53.** $p(p^2 - 15p + 75)$ **55.** $4(3a^2 - 12a + 16)$ **57.** 9 **59.** 25 **61.** 40 or -40 **63.** $(4m^{2x} + 3)(4m^{2x} - 3)$ **65.** $(8r^{4z} + 1)(8r^{4z} - 1)$ **67.** $(10m^z + 3p^{4z})(10m^z - 3p^{4z})$ **69.** $(3a^{2z} - 5)^2$ **71.** $(x^n - 2)(x^{2n} + 2x^n + 4)$ **73.** $(3z^{4y} + 5x^{2y})(9z^{8y} - 15z^{4y}x^{2y} + 25x^{4y})$ **75.** $8y(2y - 3 + 4y^2)$ **77.** $(x + 2)(y + 4)$ **79.** $(y + 2)(y - 1)$ **81.** $(3r - z)(2r + 7z)$

3.8 Exercises (page 168)

1. $(10a + 3b)(10a - 3b)$ **3.** $3p^2(p - 6)(p + 5)$ **5.** $3pq(a + 6b)(a - 5b)$ **7.** prime **9.** $(6b + 1)(b - 3)$ **11.** $3mn(3m + 2n)(2m - n)$ **13.** $(2p + 5q)(p + 3q)$ **15.** $9m(m - 5 + 2m^2)$ **17.** $2(3m - 10)(9m^2 + 30m + 100)$ **19.** $(2a + 1)(a - 4)$ **21.** $(k - 9)(q + r)$ **23.** prime **25.** $(2p - 5)(4p^2 + 10p + 25)$ **27.** $(x - 5)(x + 5)(x^2 + 25)$ **29.** $(p + 4)(p^2 - 4p + 16)$ **31.** $(8m + 25)(8m - 25)$ **33.** $6z(2z^2 - z + 3)$ **35.** $16(4b + 5c)(4b - 5c)$ **37.** $8(5z + 4)(25z^2 - 20z + 16)$ **39.** $(5r - s)(2r + 5s)$ **41.** $4pq(2p + q)(3p + 5q)$ **43.** $3(4k^2 + 9)(2k + 3)(2k - 3)$ **45.** $2(5p + 9)(5p - 9)$ **47.** $4rx(3m^2 + mn + 10n^2)$ **49.** $(7a - 4b)(3a + b)$ **51.** prime **53.** $(p + 8q - 5)^2$ **55.** $(7m^2 + 1)(3m^2 - 5)$ **57.** $(2r - t)(r^2 - rt + 19t^2)$ **59.** $(x + 3)(x^2 + 1)(x + 1)(x - 1)$ **61.** $(3m^k + 2)(m^k - 3)$ **63.** $3c^y(5c^{3y} + c^y - 2)$ **65.** $(z^{2x} + w^x)(z^{2x} - w^x)$ **67.** $(m + n - 5)(m - n + 1)$ **69.** 9 **71.** 2 **73.** $\{2\}$ **75.** $\{-1\}$ **77.** $\{0\}$ **79.** $\left\{-\frac{4}{3}\right\}$

3.9 Exercises (page 174)

1. $\{2, 3\}$ **3.** $\left\{-\frac{6}{5}, \frac{8}{3}\right\}$ **5.** $\left\{-\frac{5}{2}, 0, 4\right\}$ **7.** $\left\{-\frac{7}{2}, \frac{4}{3}, 5\right\}$ **11.** $\{-3, 4\}$ **13.** $\{-7, -2\}$ **15.** $\left\{-\frac{1}{2}, \frac{1}{6}\right\}$ **17.** $\left\{-\frac{4}{3}, 5\right\}$ **19.** $\{-6, -5\}$ **21.** $\{0, 4\}$ **23.** $\{0, 9\}$ **25.** $\{-2, 2\}$ **27.** $\{-10, 10\}$ **29.** $\{-3, 3\}$ **31.** $\{-2, 4\}$ **33.** $\left\{-\frac{1}{3}, 2\right\}$ **35.** $\{-4, -3\}$ **37.** $\left\{-\frac{1}{3}, 0, 2\right\}$ **39.** $\left\{-1, -\frac{5}{6}, 1\right\}$ **41.** $\{-3, 0, 9\}$ **43.** -2 and 8 **45.** 3 and 9 **47.** -9 and 12 **49.** width: 6 feet; length: 9 feet

51. 4 meters **53.** 8 seconds **55.** 50 feet by 100 feet **57.** $\left\{-\frac{1}{2}, 6\right\}$ **59.** $\left\{-\frac{2}{3}, \frac{4}{15}\right\}$ **61.** $\{2, 4\}$
63. $\left\{-\frac{3}{2}, \frac{1}{2}\right\}$ **65.** $3m^3$ **67.** $-\frac{4}{3ab^3}$ **69.** $\frac{18}{24}$ **71.** $\frac{55}{75}$

Chapter 3 Review Exercises (page 180)

1. exponent: 3; base: 9 **3.** exponent: 2; base: 8 **5.** 243 **7.** -64 **9.** $\frac{64}{27}$ **11.** 1 **13.** -1
15. 10 **17.** 9^8 **19.** m^5 **21.** y^{5b+3} **23.** $\frac{1}{3^8}$ **25.** $\frac{y^6}{x^2}$ **27.** $\frac{25}{m^{18}}$ **29.** $\frac{25}{z^4}$ **31.** $\frac{2025}{8r^4}$ **33.** k^{3r-3}
35. z^{1-6a} **37.** 3.45×10^3 **39.** 1.3×10^{-1} **41.** .0038 **43.** 2×10^{-4} or .0002
45. 4×10^{-5} or .00004 **47.** 14 **49.** 29 **51.** **(a)** $11k^3 - 3k^2 + 9k$ **(b)** trinomial **(c)** 3
53. **(a)** $-5y^4 + 3y^3 + 7y^2 - 2y$ **(b)** none of these **(c)** 4 **55.** **(a)** 31 **(b)** 1 **57.** **(a)** -191 **(b)** 1 **59.** $13p$
61. $m^5 + 3m^4$ **63.** $x^2 + 11x + 6$ **65.** $-4y^3 - 2y^2 + 16y - 14$ **67.** $12m^5 + 11m^3 + 8m^2 - 9m$
69. $6a^3 - 15a^2 + 7a$ **71.** $12y^3$ **73.** $-8a^2 - 22a$ **75.** $2p^3 + 13p^2 + 17p - 12$
77. $6z^4 - 8z^3 + 10z^2 + 9z - 2$ **79.** $m^2 - 4m - 21$ **81.** $6r^2 + 5r - 56$ **83.** $9r^4 - 25$ **85.** $y^2 + 6yz + 9z^2$
87. $18y^{2q-1} - 45$ **89.** $6z^{2r} - 11x^{2r}z^r - 10x^{4r}$ **91.** $5(6p + 5)$ **93.** prime **95.** $6ab(2a + 3b^2 - 4a^2b)$
97. $(x + y)(4 + m)$ **99.** $(m + 3)(2 - a)$ **101.** $(x - 3)(x - 8)$ **103.** $(r + 6)(r + 9)$ **105.** $(4p - 1)(p + 1)$
107. $-(4p - 5)(3p + 4)$ **109.** $(4a - b)(5a - 2b)$ **111.** $(r^2 - 2)(r^2 + 3)$ **113.** $(p + 2)^2(p + 3)(p - 2)$
115. $(3p + 7)(3p - 7)$ **117.** $9(4r + 3s + 7)(4r - 3s + 1)$ **119.** $(3k - 2)^2$ **121.** $(5x - 1)(25x^2 + 5x + 1)$
123. $(11a^{5m} + 12b^{3m})(11a^{5m} - 12b^{3m})$ **125.** $(5r^k + 1)(25r^{2k} - 5r^k + 1)$ **127.** $\left\{\frac{2}{5}, -1\right\}$ **129.** $\{-4, 2\}$
131. $\left\{-\frac{3}{2}, \frac{1}{3}\right\}$ **133.** $\{-3, 3\}$ **135.** $\left\{\frac{1}{2}, 1\right\}$ **137.** $\left\{-\frac{7}{2}, 0, 4\right\}$ **139.** 3 feet **141.** $-\frac{20}{3}$ or 5
143. -9 and 3 **145.** $8x^2 - 10x - 3$ **147.** $\frac{1}{125}$ **149.** $-14 + 16w - 8w^2$ **151.** $-\frac{1}{5z^9}$
153. $\frac{1250z^7x^6}{9}$ **155.** $a(6 - m)(5 + m)$ **157.** $(2 - a)(4 + 2a + a^2)$ **159.** $5y^2(3y + 4)$ **161.** $\left\{-\frac{3}{5}, 4\right\}$

Chapter 3 Test (page 184)

[3.1] **1.** $\frac{27}{8}$ **2.** $-\frac{1}{64}$ **3.** 0 [3.2] **4.** $\frac{1}{8^3}$ or $\frac{1}{512}$ **5.** y^{6r-2} **6.** 1 **7.** $8y^{16}$
8. $\frac{a^8}{3^3 \cdot 2^4}$ or $\frac{a^8}{432}$ **9.** $\frac{r^5}{45}$ **10.** .000046 **11.** 3×10^{-4} or .0003 [3.3] **12.** $-3m^2 - 5m - 3$
13. $-k^3 - 5k^2 - 3k - 9$ **14.** $-10r + 18$ [3.4] **15.** $3k^3 + 20k^2 + 23k - 36$ **16.** $4y^2 - 9z^2 + 6zx - x^2$
[3.3] **17.** 38 [3.5–3.8] **18.** $(x + y)(a + b)$ **19.** $3x(x + 2y^2 + 8y)$ **20.** $(k + 6)(k + 4)$
21. $(2p - 3q)(p - q)$ **22.** prime **23.** $(x^2 - 5)(x + 2)(x - 2)$ **24.** $(10d - 3b^3)(10d + 3b^3)$
25. $(3y - 5z^2)(9y^2 + 15yz^2 + 25z^4)$ **26.** (d) [3.9] **27.** $\left\{-6, \frac{5}{3}\right\}$ **28.** $\left\{\frac{1}{4}, \frac{3}{2}\right\}$ **29.** $\left\{-\frac{1}{4}, 0, 3\right\}$
30. length: 9 meters; width: 7 meters

CHAPTER 4

4.1 Exercises (page 191)

1. $m = 4$ **3.** $z = 0$ **5.** none **7.** $m = -3$ or $m = 4$ **9.** $z = -4$ or $z = 4$ **11.** $x = -2$ or $x = \frac{1}{2}$
13. none **15.** $\frac{3m^5}{7}$ **17.** $\frac{3m}{2n^2}$ **19.** $\frac{2}{3}$ **21.** $\frac{z}{2}$ **23.** $\frac{a - 2}{a + 2}$ **25.** $\frac{r + 4}{r + 6}$ **27.** $\frac{2(c + 6d)}{5(c - d)}$
29. $a^2 + ab + b^2$ **31.** $\frac{z^2 - zx + x^2}{z - x}$ **33.** $\frac{x - w}{x + w}$ **35.** -1 **37.** $-\frac{1}{y + x}$ **39.** $-\frac{4 - n}{4 + n}$ or $\frac{n - 4}{4 + n}$

41. $-x$ **43.** **(a)** numerator: x^2, $4x$; denominator: x, 4 **(b)** numerator: x, $x+2$; denominator: y, $x+2$ **45.** $\frac{6k}{2k^2}$

47. $\frac{30}{5z-10}$ **49.** $\frac{6a(a-1)}{a^2-1}$ **51.** $-\frac{2x}{4-x}$ **53.** $\frac{4(m-1)(m+2)}{16(m+2)}$ or $\frac{4m^2+4m-8}{16(m+2)}$

55. $\frac{4z(z+1)}{2z^2-10z-12}$ or $\frac{4z^2+4z}{2z^2-10z-12}$ **57.** $\frac{-3m(y+m)}{y^2-m^2}$ or $\frac{-3my-3m^2}{y^2-m^2}$ **59.** $\frac{4(m^2-3mp+9p^2)}{m^3+27p^3}$ or $\frac{4m^2-12mp+36p^2}{m^3+27p^3}$ **61.** $\frac{(2r-3s)(3r+s)}{3ar+as+6br+2bs}$ or $\frac{6r^2-7rs-3s^2}{3ar+as+6br+2bs}$ **69.** $\frac{m^{2r}-m^r}{3m^r+3}$ **71.** $\frac{p^y+5}{p^y+2}$

73. -1 **75.** $\frac{1}{m^4}$ **77.** $\frac{5}{16}$ **79.** 6 **81.** $\frac{14}{3}$ **83.** 8

4.2 Exercises (page 197)

1. 2 **3.** $\frac{a^4}{2}$ **5.** $\frac{2y}{x^3}$ **7.** $\frac{4}{3a^3b^2}$ **9.** $\frac{27}{2mn^7}$ **11.** $\frac{4}{3pq}$ **13.** $\frac{7r}{6}$ **15.** $\frac{35}{4}$ **17.** $-(z+1)$ or $-z-1$ **19.** $\frac{-p(p+6)}{p+1}$ **21.** -2 **23.** $m-3$ **25.** $\frac{(r+3)(r-1)}{(r+1)(r-3)}$ **27.** 1 **29.** $\frac{3x-y}{3x+2y}$

31. $\frac{2k+5r}{k+5r}$ **33.** $\frac{a+b}{a-b}$ **35.** $\frac{2m-n^2}{m^2+n^2}$ **37.** $5(x+1)(x-4)$ **39.** $p-q$ **41.** 1 **43.** $\frac{x^2(x-2+y)}{x-2-y}$

45. $\frac{a+5}{3a+1}$ **49.** $\frac{1}{2}$ **51.** $\frac{3}{4}$ **53.** $\frac{11}{48}$ **55.** $-\frac{11}{40}$

4.3 Exercises (page 203)

1. $5k^2$ **3.** $16x^3z^4$ **5.** $4z(z-1)$ **7.** $18(a+3)$ **9.** $(a-4)^2(a+4)$ **11.** $8r(r-1)(r+1)$

13. $(x+5)(x+2)(x-5)$ **15.** $9p^2(p+6)(p+1)$ **19.** $\frac{7m+5}{m(m+1)}$ **21.** $\frac{7m+5}{m(m+1)}$ **23.** $-\frac{7}{p}$ **25.** 1

27. 6 **29.** $a-b$ **31.** $\frac{1}{p-2}$ **33.** $\frac{23}{2r}$ **35.** $\frac{4z+3}{3z^2}$ **37.** $-\frac{17}{18k}$ **39.** $\frac{2m-1}{m(m-1)}$

41. $\frac{2t-16}{(t+2)(t-2)}$ **43.** $\frac{3}{m-4}$ or $\frac{-3}{4-m}$ **45.** $\frac{2y}{x-y}$ or $\frac{-2y}{y-x}$ **47.** $\frac{13}{12(a+3)}$

49. $\frac{x+9}{(x-3)(x-2)(x+1)}$ **51.** $\frac{7b-5a}{(a-3b)(a+2b)(a-b)}$ **53.** $\frac{2r^2-4rs+4s^2}{(r-2s)^2(r+2s)}$ **55.** $\frac{7y^2+11yz-y-2z}{(y+2z)(y+z)(y-z)}$

57. $\frac{4p^2-21p+29}{(p-2)^2}$ **59.** $\frac{13r-16s}{(2r+s)(r-2s)(3r-s)}$ **61.** $\frac{-9q^2+6qs-4s^2-4q-3s}{27q^3+8s^3}$ **63.** $\frac{(m+k)(3m+2k)}{(m-k)^2}$ or $\frac{3m^2+5mk+2k^2}{(m-k)^2}$ **65.** $\frac{p-17}{3p}$ **67.** If $x=4$ and $y=2$, then $\frac{1}{x}+\frac{1}{y}=\frac{1}{4}+\frac{1}{2}=\frac{1}{4}+\frac{2}{4}=\frac{3}{4}$, which does not equal $\frac{1}{x+y}=\frac{1}{4+2}=\frac{1}{6}$. **69.** $\frac{2p-11}{p(p+2)}$ **71.** $\frac{9k-13}{k-3}$ **73.** $\frac{2x-h}{x^2(x-h)^2}$ **75.** $\frac{2a}{(a-b)(a+b)}$

77. $\frac{3-x}{(2x-1)(x+2)}$ **79.** $\frac{15}{4}$ **81.** $\frac{17}{7}$

4.4 Exercises (page 209)

1. $\frac{x-1}{2x}$ **3.** $\frac{8p}{3(p+5)}$ **5.** $\frac{m-1}{4(m+1)}$ **7.** $\frac{5}{27z}$ **9.** $6m^4n^3$ **11.** $\frac{k+m}{4(k-m)}$ **13.** $\frac{y+1}{y-1}$ **15.** $\frac{r+k}{r-k}$

17. $\frac{1}{s+4r}$ **19.** $\frac{3y}{2}$ **21.** $\frac{4x^2+2x}{4x-1}$ **23.** $\frac{a^2-2a-3}{a^2+a-1}$ **25.** $\frac{5a-3}{2a(2a-1)}$ **31.** $\frac{y^2}{y^2-x^2}$ **33.** $\frac{z+m}{z-m}$

35. $\frac{ab}{b+a}$ **37.** $\frac{m+6k}{2km}$ **39.** $\frac{3-4m+8p}{2p-3m}$ **41.** $\frac{3x+2}{2x+1}$ **43.** $\frac{2p}{p+2}$ **45.** $\frac{p+p^2+1}{p^3+p^2+2p+1}$

47. $2p^4$ **49.** $\frac{-4b^6}{3a^2}$ **51.** $-8a^2+a+4$ **53.** $-2x^3-13x^2+11x$

4.5 Exercises (page 215)

1. $a^2 - 2a + 1$ **3.** $m + 2 + \frac{3}{m}$ **5.** $3 + \frac{2}{a} - \frac{5}{2a^2}$ **7.** $\frac{3}{4q} + \frac{3}{2p} - \frac{1}{2pq}$ **9.** $x + 3$ **11.** $r + 3$
13. $m^2 + 2m - 1$ **15.** $3b^2 + b + \frac{1}{2b - 3}$ **17.** $a^2 - 5a - 9 + \frac{-30}{a - 3}$ **19.** $2x - 5 + \frac{-4x + 5}{2x^2 - 4x + 3}$
21. $9t^2 - 4t + 1$ **23.** $4p + 1 + \frac{-11p + 15}{2p^2 + 3}$ **25.** $2y^2 + 3y - 1$ **27.** $p^2 + p + 1$
29. $8r^3 + 4r^2 + 2r + 1 + \frac{2}{2r - 1}$ **31.** $\frac{2}{3}x - 1$ **33.** $\frac{3}{4}a - 2 + \frac{1}{4a + 3}$ **35.** $\frac{2}{3}a^3 - \frac{5}{3}a + \frac{5}{3} + \frac{5a - 13}{3a^2 + 3}$
37. $3k + 2$ **39.** $m^2 + 3m - 7$ kilometers per hour **43.** -6 **45.** -21 **47.** -6

4.6 Exercises (page 220)

1. $x + 7$ **3.** $3a + 4 + \frac{3}{a + 2}$ **5.** $p - 4 + \frac{9}{p + 1}$ **7.** $4a^2 + a + 3 + \frac{4}{a - 1}$
9. $6x^4 + 12x^3 + 22x^2 + 40x + 83 + \frac{164}{x - 2}$ **11.** $-4r^5 - 7r^4 - 10r^3 - 5r^2 - 11r - 8 + \frac{-5}{r - 1}$
13. $y^2 - y + 1 + \frac{-2}{y + 1}$ **15.** $p^3 - 4p^2 + 16p - 64 + \frac{272}{p + 4}$ **17.** 7 **19.** 202 **21.** 0 **23.** 69 **25.** yes
27. no **29.** yes **31.** no **33.** a solution; $(x + 3)(2x^2 - 7x + 8)$ **35.** not a solution **37.** a solution; $\left(y - \frac{3}{2}\right)(2y^3 - 2y^2 + 8y - 2)$ or $(2y - 3)(y^3 - y^2 + 4y - 1)$ **39.** $(x + 2)(x^2 - x - 1) = 0$; -2 is the only rational solution **41.** $(x - 4)(x + 1)^2 = 0$; 4 and -1 are the only rational solutions **43.** $\{3\}$ **45.** $\{3\}$
47. $\{1, 3\}$ **49.** $\left\{-\frac{2}{3}, \frac{1}{2}\right\}$

4.7 Exercises (page 224)

1. $\{16\}$ **3.** $\{-3\}$ **5.** $\left\{\frac{19}{4}\right\}$ **7.** $\{1\}$ **9.** $\{5\}$ **11.** $\{-3\}$ **13.** $\{5\}$ **15.** $\{7\}$ **17.** $\{2\}$ **19.** $\{2\}$
21. $\{-6, 4\}$ **23.** $\left\{-4, \frac{1}{2}\right\}$ **25.** $\{-1\}$ **27.** $\emptyset$ **29.** $\{-3\}$ **31.** $\emptyset$ **33.** $\{-4\}$ **35.** $\emptyset$ **37. (a)** $\left\{\frac{5}{8}\right\}$
(b) $\frac{8a + 9}{2(4a + 1)}$ **39. (a)** $\emptyset$ **(b)** $\frac{4}{a + 5}$ **43.** $\left\{-\frac{1}{4}, 2\right\}$ **45.** $\left\{-\frac{1}{2}\right\}$ **47.** $\left\{\frac{3}{4}\right\}$ **49.** $\left\{-\frac{3}{2}, \frac{9}{5}\right\}$
51. $\frac{1}{2}x + \frac{2}{3}x = 10$; $\frac{60}{7}$ **53.** $\frac{1}{x} + 3 = 4$; 1 **55.** $\frac{1}{2} + \frac{1}{3} = \frac{1}{x}$; $\frac{6}{5}$

Summary of Rational Expressions: Operations and Equations (page 227)

1. $\{10\}$ **3.** $\frac{-2y(y + 2)}{5(y - 2)}$ **5.** $\frac{5}{3m}$ **7.** $\left\{\frac{1}{2}\right\}$ **9.** $\{7\}$ **11.** $\frac{4 - 3q}{2q^2}$ **13.** $\frac{3m + 5}{4m + 1}$ **15.** $\left\{\frac{3}{2}\right\}$
17. $-\frac{36}{5}$ **19.** $\frac{a}{a + 4}$ **21.** $\frac{8a + 10}{a(a - 2)(a + 2)}$ **23.** $\emptyset$ **25.** $\frac{m}{3m + 5k}$ **27.** $\frac{p + 2}{p - 1}$ **29.** $\left\{\frac{5}{4}\right\}$ **31.** $\{9\}$
33. $\{-10\}$ **35.** $\frac{r^2 + 3rp - r - p}{(r + p)(r + 2p)(r + 3p)}$ **37.** $\frac{x - 3y}{x + 3y}$ **39.** $\frac{8m + p}{4m - 3p}$

4.8 Exercises (page 234)

1. $\frac{2}{5}$ **3.** $\frac{25}{4}$ **5.** 24 **7.** $R = \frac{E}{I}$ **9.** $m_1 = \frac{Fd^2}{Gm_2}$ **11.** $c = \frac{ab}{b - a}$ **13.** $T = \frac{PVt}{pv}$ **15.** $V = at + v$
17. $a = \frac{2S}{n} - A$ or $a = \frac{2S - An}{n}$ **19.** $n = \frac{t - a}{d} + 1$ or $n = \frac{t - a + d}{d}$ **21.** Multiply both sides by $a - b$.

23. Let x represent the unknown number. $\frac{15 + x}{17 + x} = \frac{5}{6}$ **25.** Let x represent the smaller integer, and $x + 1$ represent the larger integer. $\frac{1}{x} + \frac{1}{x + 1} = \frac{7}{x(x + 1)}$ **27.** $\frac{6}{5}$ **29.** $\frac{30}{11}$ hours **31.** $\frac{15}{8}$ hours **33.** 9 miles **35.** Mei Lin: 4 meters per second; Sue: 5 meters per second **37.** $x = \frac{ay - z}{y + 1}$ **39.** $R = \frac{Ar}{r - A}$ **41.** $r = \frac{eR}{E - e}$ **43.** $\frac{10}{7}$ hours **45.** $\frac{6}{5}$ hours **47.** $\frac{14}{5}$ hours **49.** 100 miles **51.** 190 miles **53.** $x = 2$; side $AC = 6$, side $DF = 9$ **55.** 2 **57.** 2 **59.** 2

Chapter 4 Review Exercises (page 241)

1. $k = -2$ **3.** none **5.** $\frac{5p^2}{2m}$ **7.** $\frac{y + 5}{y - 3}$ **9.** $r^2 - rs + s^2$ **11.** $\frac{2(r + 3)}{r^2 - 9}$ **13.** $\frac{y(y - 3)}{2y^2 - 5y - 3}$ **15.** $\frac{3y}{2}$ **17.** $\frac{3y^2(2y + 3)}{2y - 3}$ **19.** $\frac{y(y + 5)}{y - 5}$ **21.** $\frac{y(9y + 1)}{3y + 1}$ **23.** $p(p + 6)$ **25.** $9r^2(3r + 1)$ **27.** $(p + 5)(p + 2)(p - 8)$ **29.** $-\frac{3}{k}$ **31.** $\frac{8}{t - 2}$ or $\frac{8}{2 - t}$ **33.** $\frac{15 - m}{(m + 5)(m - 5)}$ **35.** $\frac{13r^2 + 5rs}{(5r + s)(2r - s)(r + s)}$ **37.** $\frac{3(p + 5)}{4p}$ **39.** $\frac{3 - 5x}{6x + 1}$ **41.** $\frac{y^2 - 6}{4y + 2}$ **43.** $\frac{2h^2 + 3g}{gh^2}$ **45.** $\frac{3q + 2x}{q - 4x}$ **47.** $m^2 - \frac{11}{5}m + 2$ **49.** $m - 3$ **51.** $m^2 + 3m - 6 + \frac{-m + 14}{5m^2 - 3}$ **53.** $4y^2 - 2y + 1$ **55.** $10k - 23 + \frac{31}{k + 2}$ **57.** $2k^2 + k + 3 + \frac{21}{k - 3}$ **59.** $9a^4 - a^3 + 4a^2 + 3a + 6 + \frac{-9}{a + 4}$ **61.** $x^5 - x^4 + x^3 - x^2 + x - 1 + \frac{2}{x + 1}$ **63.** -5 **65.** no **67.** $\left\{-\frac{17}{5}\right\}$ **69.** $\{-2\}$ **71.** $\{1, 4\}$ **73.** $\left\{\frac{1}{3}\right\}$ **75.** 6.67×10^{-11} **77.** $M = \frac{Fd^2}{Gm}$ **79.** $v = V - at$ **81.** $d = \frac{A - a}{n - 1}$ **83.** $-\frac{3}{2}$ **85.** 2 meters per second for Ayo, $\frac{100}{52} \approx 1.92$ meters per second for the other swimmer **87.** 60 minutes **89.** 16 kilometers per hour **91.** $\frac{16z - 3}{2z^2}$ **93.** $\frac{2p + 3}{6}$ **95.** $k^2 - 7k + 6$ **97.** $\frac{r - 2}{3r}$ **99.** $\{-3\}$ **101.** $\frac{18}{5}$ hours **103.** 24 **105.** $z = v - \frac{f}{F}(v + w)$ or $z = \frac{Fv - f(v + w)}{F}$

Chapter 4 Test (page 245)

[4.1] **1.** $m = -3, m = \frac{1}{2}$ **2.** $\frac{p}{p - 4}$ [4.2] **3.** $\frac{4}{9k^2}$ **4.** $\frac{x - 5}{x + 4}$ **5.** $\frac{y}{y + 5}$ [4.3] **6.** $36z^3$ **7.** $m(m + 3)(m - 2)$ **8.** $\frac{21}{5k}$ **9.** $\frac{-4}{a - b}$ or $\frac{4}{b - a}$ **10.** $\frac{28 - 5t}{2(t + 4)(t - 4)}$ [4.4] **11.** $\frac{9}{4}$ **12.** $\frac{y + x}{xy}$ [4.5] **13.** $z - 2 + \frac{3}{z}$ **14.** $q - 8 + \frac{8}{2q + 3}$ **15.** $3y^3 - 3y^2 + 4y + 1 + \frac{-10}{2y + 1}$ [4.6] **16.** $9x^4 - 5x^3 + 2x^2 - 2x + 4 + \frac{2}{x + 5}$ **17.** yes [4.7] **18.** $\left\{\frac{1}{2}\right\}$ **19.** $\emptyset$ [4.8] **21.** $m = \frac{5zy}{2y - z}$ **22.** $r = \frac{2k + 2y}{k - y}$ **23.** 6, 7 **24.** $\frac{45}{14}$ hours **25.** 15 miles per hour

CHAPTER 5

5.1 Exercises (page 253)

1. 11 **3.** 21 **5.** 8 **7.** 4 **9.** -9 **11.** $\frac{1}{2}$ **13.** $\frac{4}{5}$ **15.** $\frac{2}{5}$ **17.** 1000 **19.** 64 **21.** -1728 **23.** 144 **25.** $\frac{8}{27}$ **27.** $-\frac{243}{32}$ **29.** $\frac{5}{2}$ **31.** $\frac{216}{1331}$ **35.** 9 **37.** 4 **39.** -6 **41.** 3 **43.** -5 **45.** not a real number **47.** -9 **49.** 2 **51.** 4 **53.** -2 In Exercises 55–69, we give only one form of the answer. There may be others. **55.** $\sqrt{12}$ **57.** $\sqrt[5]{9}$ **59.** $(\sqrt[3]{7y})^2$ **61.** $\sqrt[4]{3py^2}$ **63.** $\frac{1}{(\sqrt{2m})^3}$ **65.** $(\sqrt[3]{2y+x})^2$ **67.** $\sqrt{p^2+q^2}$ **69.** $\frac{1}{(\sqrt[3]{3m^4+2k^2})^2}$ **71.** $\sqrt{a^2+b^2} = \sqrt{3^2+4^2} = 5$; $a + b = 3 + 4 = 7$; $5 \neq 7$ **73.** $13^{1/2}$ **75.** $7^{1/3}$ **77.** $2^{3/2}$ **79.** $6^{4/3}$ **81.** $-8^{5/3}$ **83.** $x^{3/4}$ **85.** $a^{2/7}$ **87.** $|r^3|$ **89.** $|r-2q|$ **91.** $|m-q|$ **93.** $\frac{-5}{|p^2-q^2|}$, if $|p| \neq |q|$ **95.** 73.8 inches **97.** 6 square meters **99.** 30 **101.** $-6x^3y^{13}$ **103.** $27m^9n^6$ **105.** $3xy^4$

5.2 Exercises (page 258)

1. 9^2 or 81 **3.** $81^{1/2}$ or 9 **5.** x **7.** $\frac{1}{k^{2/3}}$ **9.** $4p^6q^4$ **11.** $\frac{1}{r^2}$ **13.** $\frac{z^{5/2}}{x^3}$ **15.** $m^{1/4}h^{3/4}$ **17.** p^2 **19.** $\frac{a^{5/2}}{m^{23/12}}$ **21.** $\frac{q^{5/3}}{9p^{7/2}}$ **23.** $p + 2p^2$ **25.** $k^{7/4} - k^{3/4}$ **27.** $6 + 18a$ **29.** $x^{8/5} - 4x^{2/5}$ **31.** $-5z^{13/6} + 2z^{7/6}$ **33.** $x^{23b/4}$ **35.** $\frac{1}{x^{3b}}$ **37.** $\frac{1}{x^{a/2}}$ **39.** $x^{-1/2}(3-4x)$ **41.** $t^{-1/2}(4+7t^2)$ **43.** $p^{3/4}(4p^{1/4}-1)$ **45.** $k^{-3/4}(9-2k^{1/2})$ **47.** In the definition of $a^{1/n}$, if n is even, then we must have $a \geq 0$. For this reason, the step $(-1)^{2/2} = [(-1)^2]^{1/2}$ is not valid. **49.** For example, let $x = 3$ and $y = 4$. $(x^2+y^2)^{1/2} = (3^2+4^2)^{1/2} = 25^{1/2} = 5$; $x + y = 3 + 4 = 7$; $5 \neq 7$ **51.** $x^{17/20}$ **53.** $\frac{1}{x^{3/2}}$ **55.** $y^{5/6}$ **57.** $m^{1/6}$ **59.** $x^{1/24}$ **61.** 2.646 **63.** -4.359 **65.** -9.055 **67.** 12.247 **69.** -22.583 **71.** 29.833 **73.** 7.507 **75.** 3.162 **77.** 1.885 **79.** 15.155 **81.** .272 **83.** -3.362 **87.** 3.141592653 **89.** p^{11} **91.** $\frac{1}{24x^8}$ **93.** $\frac{s}{16r}$

5.3 Exercises (page 265)

1. $\sqrt{77}$ **3.** $\sqrt{42}$ **5.** $\sqrt[3]{14}$ **7.** $\sqrt[4]{24}$ **9.** $\frac{4}{5}$ **11.** $\frac{\sqrt{7}}{5}$ **13.** $\frac{\sqrt{r}}{10}$ **15.** $\frac{m^4}{5}$ **17.** $-\frac{3}{4}$ **19.** $\frac{\sqrt[3]{k^2}}{5}$ **23.** $3\sqrt{2}$ **25.** $2\sqrt{19}$ **27.** $2\sqrt[3]{2}$ **29.** $4\sqrt[3]{2}$ **31.** $2\sqrt[4]{2}$ **33.** $-5\sqrt[4]{2}$ **35.** $2\sqrt[5]{4}$ **37.** $\frac{6\sqrt{2}}{5}$ **39.** $\frac{2\sqrt[3]{4}}{5}$ **41.** $10y^5$ **43.** $-2k^3$ **45.** $-12m^5z$ **47.** $5m^3b^6c^8$ **49.** $5y\sqrt{3y}$ **51.** $x^2y^3\sqrt{7x}$ **53.** $2z^3r^4$ **55.** $2zx^3\sqrt[3]{3z^2}$ **57.** $2a^2b^3$ **59.** $2km^2\sqrt[4]{2km^2}$ **61.** $\frac{m^4\sqrt{m}}{4}$ **63.** $\frac{y^3\sqrt[3]{y}}{3}$ **65.** $\frac{t^5\sqrt[4]{t^3}}{2}$ **69.** $2\sqrt{3}$ **71.** $m\sqrt{m}$ **73.** $\sqrt{5}$ **75.** $2\sqrt[3]{2}$ **77.** $\sqrt[6]{675}$ **79.** $\sqrt[15]{864}$ **81.** $\sqrt[6]{x^5}$ **83.** 5 **85.** $8\sqrt{2}$ **87.** 6.614 feet **89.** $6x^2$ **91.** $11q - 6q^2$ **93.** $16a^5 - 9a^2 + 4a$ (cannot be simplified further)

5.4 Exercises (page 269)

1. 16 **3.** $-13\sqrt{3}$ **5.** $12\sqrt{2}$ **7.** $5\sqrt{7}$ **9.** $-16\sqrt{5}$ **11.** $10\sqrt{10}$ **13.** $-5\sqrt{2x}$ **15.** $20\sqrt{3r}$ **17.** $27q\sqrt{10}$ **19.** $17m\sqrt{2}$ **21.** $-\sqrt[3]{2}$ **23.** $10\sqrt[3]{x}$ **25.** $-\sqrt[3]{x^2y}$ **27.** $-x\sqrt[3]{xy^2}$ **29.** $19\sqrt[4]{2}$ **31.** $9\sqrt[4]{2a^3}$ **33.** $x\sqrt[4]{xy}$ **35.** $2\sqrt{2} - 2$ **37.** $\frac{5\sqrt{5}}{6}$ **39.** $\frac{7\sqrt{2}}{6}$ **41.** $\frac{5\sqrt{2}}{3}$ **43.** $5\sqrt{2} + 4$

45. $\frac{5-3x}{x^4}$ **47.** $\frac{m\sqrt[3]{m^2}}{2}$ **51.** $12\sqrt{5}+5\sqrt{3}$ centimeters **53.** $37\sqrt{2}+10\sqrt{3}$ yards **55.** $\sqrt{3}\approx 1.732$ and $2\sqrt{3}\approx 3.464$; also, $\sqrt{12}\approx 3.464$, suggesting (but not really proving) that $\sqrt{12}=2\sqrt{3}$. **57.** $-12p^3+6p$ **59.** $15a^2+11ab-14b^2$ **61.** $25x^2-10xy+y^2$ **63.** $49r^2-4s^2$ **65.** $4x-5$

5.5 Exercises (page 276)

1. $3\sqrt{6}+2\sqrt{3}$ **3.** $20\sqrt{2}$ **5.** -2 **7.** -1 **9.** 6 **11.** $\sqrt{6}-\sqrt{2}+\sqrt{3}-1$ **13.** $\sqrt{22}+\sqrt{55}-\sqrt{14}-\sqrt{35}$ **15.** $8-\sqrt{15}$ **17.** $9+4\sqrt{5}$ **19.** $26-2\sqrt{105}$ **21.** $4-\sqrt[3]{36}$ **23.** 10 **25.** $6x+3\sqrt{x}-2\sqrt{5x}-\sqrt{5}$ **27.** $9r-s$ **29.** $4\sqrt[3]{4y^2}-19\sqrt[3]{2y}-5$ **31.** $3\sqrt{5}$ **33.** $\frac{\sqrt{35}}{5}$ **35.** $4\sqrt{15}$ **37.** $-\sqrt{15}$ **39.** $\frac{\sqrt{10}}{6}$ **41.** $\frac{8\sqrt{3k}}{k}$ **43.** $\frac{5\sqrt{2my}}{y^2}$ **45.** $-\frac{4k\sqrt{3z}}{z}$ **47.** $\frac{\sqrt[3]{18}}{4}$ **49.** $\frac{x^2\sqrt[3]{y^2}}{y}$ **51.** $\frac{3(4-\sqrt{5})}{11}$ **53.** $\frac{6\sqrt{2}+4}{7}$ **55.** $\frac{2(3\sqrt{5}-2\sqrt{3})}{33}$ **57.** $2\sqrt{6}+2\sqrt{5}-3\sqrt{2}-\sqrt{15}$ **59.** $\sqrt{m}-2$ **61.** $\frac{3\sqrt{x}(\sqrt{x}+2\sqrt{y})}{x-4y}$ **63.** $3-2\sqrt{6}$ **65.** $1-\sqrt{5}$ **67.** $\frac{4-2\sqrt{2}}{3}$ **69.** $\frac{6-2\sqrt{6p}}{3}$ **71.** $\frac{\sqrt{x+y}}{x+y}$ **73.** $\frac{p\sqrt{p+2}}{p+2}$ **79.** $\frac{319}{6(8\sqrt{5}+1)}$ **81.** $\frac{p-9q}{4q(\sqrt{p}+3\sqrt{q})}$ **85.** $-\frac{3(\sqrt[3]{y^2}-\sqrt[3]{2y}+\sqrt[3]{4})}{y+2}$ **87.** $\{-1\}$ **89.** $\{-2,-1\}$ **91.** $\left\{\frac{2}{5},\frac{1}{3}\right\}$

5.6 Exercises (page 282)

1. $\{11\}$ **3.** $\left\{\frac{1}{3}\right\}$ **5.** $\emptyset$ **7.** $\{5\}$ **9.** $\{18\}$ **11.** $\{5\}$ **13.** $\{4\}$ **15.** $\{17\}$ **17.** $\{5\}$ **19.** $\emptyset$ **21.** $\{4\}$ **23.** $\{0\}$ **25.** $\left\{-\frac{1}{3}\right\}$ **27.** $\emptyset$ **29.** $\{1\}$ **31.** $\{-1\}$ **33.** $\{14\}$ **35.** $\{8\}$ **37.** $\{0\}$ **39.** $\emptyset$ **41.** $\{7\}$ **43.** $\{2\}$ **45.** $\{4, 20\}$ **47.** $\left\{\frac{1}{4}, 1\right\}$ **49.** $\left\{\frac{5}{4}\right\}$ **53.** 7.94 feet **55.** $1+x$ **57.** $2x^2+x-15$ **59.** $\frac{-7(5+\sqrt{2})}{23}$

5.7 Exercises (page 289)

1. $12i$ **3.** $-15i$ **5.** $i\sqrt{3}$ **7.** $5i\sqrt{3}$ **9.** $5i\sqrt{3}$ **11.** $22i\sqrt{5}$ **13.** -5 **15.** -18 **17.** -40 **19.** $\sqrt{2}$ **21.** $3i$ **23.** 6 **27.** $10+5i$ **29.** $-1+3i$ **31.** $0+0i$ **33.** $8-i$ **35.** $7+8i$ **37.** $-2-3i$ **39.** $-6+4i$ **41.** $6+4i$ **43.** $-16+0i$ **45.** $-28+0i$ **47.** $-30+0i$ **49.** $6+6i$ **51.** $-38-8i$ **53.** $-7+i$ **55.** $-8-6i$ **57.** $13+0i$ **59.** $2+0i$ **61.** $145+0i$ **63.** $1-i$ **65.** $\frac{1}{5}+\frac{2}{5}i$ **67.** $0-2i$ **69.** $0-i$ **71.** $\frac{2}{5}-\frac{14}{5}i$ **73.** $0+\frac{9}{2}i$ **75.** -1 **77.** i **79.** i **81.** -1 **83.** $\frac{37}{10}-\frac{19}{10}i$ **85.** $-\frac{13}{10}+\frac{11}{10}i$ **87.** $\frac{1}{2}+\frac{1}{2}i$ **91.** $-5, 5$ **93.** $-3\sqrt{5}, 3\sqrt{5}$ **95.** $\{-8, 5\}$

Chapter 5 Review Exercises (page 295)

1. 1000 **3.** $\frac{1}{100}$ **5.** 42 **7.** -5 **9.** -2 **11.** $\sqrt[3]{5}$ **13.** $(\sqrt[5]{2k})^2$ **15.** $\frac{1}{(\sqrt[3]{3a+b})^5}$ **17.** $9^{1/3}$ **19.** $p^{4/5}$ **21.** 96 **23.** $\frac{1}{m}$ **25.** $\frac{z^{1/2}x^{8/5}}{4}$ **27.** $2y^{2/3}-\frac{10}{y^{5/3}}$ **29.** $r\sqrt{r}$ **31.** $k^2\sqrt[4]{k}$ **33.** $m^4\sqrt[3]{m}$ **35.** $\sqrt[12]{z}$ **37.** -27.495 **39.** 2.783 **41.** 10.903 **43.** -84.202 **45.** $-.189$ **47.** $\sqrt{5r}$ **49.** $5\sqrt{5}$ **51.** $10y^3\sqrt{y}$ **53.** $4p^2q\sqrt[3]{p^2q^2}$ **55.** $\frac{y\sqrt{y}}{12}$ **57.** $\frac{\sqrt[3]{r^2}}{2}$ **59.** $p\sqrt{p}$ **61.** $\sqrt[10]{x^7}$ **63.** $\sqrt{15}$ **65.** $57\sqrt{2}$

67. $19\sqrt[3]{2}$ **69.** $6x\sqrt[3]{y^2}$ **71.** $\frac{16 + 5\sqrt{5}}{20}$ **73.** $1 - \sqrt{3}$ **75.** $9 - 7\sqrt{2}$ **77.** 29 **79.** $\frac{2\sqrt{7}}{7}$ **81.** $\frac{\sqrt{3}}{2}$ **83.** $\frac{4\sqrt{10zk}}{k}$ **85.** $\frac{2\sqrt[3]{4m^2}}{m}$ **87.** $\frac{2 + \sqrt{7}}{-3}$ **89.** $\frac{2\sqrt{z}(\sqrt{z} + 2\sqrt{x})}{z - 4x}$ **91.** $\frac{1 - \sqrt{5}}{4}$ **93.** $2 + \sqrt{3k}$ **97.** $\{6\}$ **99.** $\{25\}$ **101.** $\{3\}$ **103.** $\{6\}$ **105.** $5i$ **107.** $10i\sqrt{2}$ **109.** $-10 - 2i$ **111.** $-8 + 4i$ **113.** $3 + 0i$ **115.** $42 - 32i$ **117.** $-i$ **119.** $7\sqrt{3y}$ **121.** -4 **123.** $26m\sqrt{6m}$ **125.** $13ab^2$ **127.** 6 **129.** $\frac{3\sqrt{7py}}{y}$ **131.** 128 **133.** $49\sqrt[4]{7}$ **135.** $86 + 8\sqrt{55}$ **137.** $1 - i$ **139.** $-5i$

Chapter 5 Test (page 298)

[5.1] **1.** 81 **2.** $\frac{125}{64}$ [5.2] **3.** $\frac{1}{3}$ **4.** $a^{11/4}$ **5.** $r + \frac{1}{r}$ [5.1] **6.** -10 [5.3] **7.** $10y^7$ **8.** $-2mp^3$ [5.2] **9.** 7.905 **10.** $a^3\sqrt[3]{a^2}$ [5.3] **11.** $6x^2\sqrt{3x}$ **12.** $2ab^2\sqrt[4]{ab^3}$ **13.** $3\sqrt[3]{2}$ [5.4] **14.** $8\sqrt{5}$ [5.5] **15.** $12 - 2\sqrt{15}$ **16.** $9\sqrt{5} - 8$ **17.** $-\frac{9\sqrt{10}}{20}$ **18.** $\frac{3\sqrt[3]{2}}{2}$ **19.** $-4(\sqrt{7} + \sqrt{5})$ [5.6] **20.** $\{9\}$ **21.** $\{-1\}$ [5.7] **22.** $-14 + 21i$ **23.** $-2 + 16i$ **24.** $\frac{1}{10} + \frac{7}{10}i$

CHAPTER 6

6.1 Exercises (page 304)

1. (b) and (c) **3.** $\{10, -10\}$ **5.** $\{\sqrt{11}, -\sqrt{11}\}$ **7.** $\{3\sqrt{2}, -3\sqrt{2}\}$ **9.** $\{7, -3\}$ **11.** $\left\{\frac{-4 + \sqrt{5}}{2}, \frac{-4 - \sqrt{5}}{2}\right\}$ **13.** $\left\{\frac{2 + 2\sqrt{6}}{5}, \frac{2 - 2\sqrt{6}}{5}\right\}$ **15.** $\{12i, -12i\}$ **17.** $\{3i\sqrt{2}, -3i\sqrt{2}\}$ **19.** $\{-2 + i\sqrt{2}, -2 - i\sqrt{2}\}$ **21.** $\left\{-\frac{2}{3} + \frac{\sqrt{5}}{3}i, -\frac{2}{3} - \frac{\sqrt{5}}{3}i\right\}$ **25.** 36 **27.** 16 **29.** $\frac{81}{4}$ **31.** $\frac{9}{4}$ **33.** $\{-4, 8\}$ **35.** $\left\{5, -\frac{3}{2}\right\}$ **37.** $\left\{-1, \frac{1}{3}\right\}$ **39.** $\{2 + \sqrt{3}, 2 - \sqrt{3}\}$ **41.** $\{-4 + \sqrt{2}, -4 - \sqrt{2}\}$ **43.** $\left\{\frac{2}{3}, -2\right\}$ **45.** $\{-1 + i, -1 - i\}$ **47.** $\left\{\frac{1 + \sqrt{5}}{2}, \frac{1 - \sqrt{5}}{2}\right\}$ **49.** $\{-2 + 3i, -2 - 3i\}$ **51.** $\left\{\frac{4 + \sqrt{7}}{3}, \frac{4 - \sqrt{7}}{3}\right\}$ **53.** $\left\{\frac{2 + \sqrt{14}}{2}, \frac{2 - \sqrt{14}}{2}\right\}$ **55.** $\left\{1 + \frac{2}{5}i, 1 - \frac{2}{5}i\right\}$ **57.** $\{2\sqrt{b}, -2\sqrt{b}\}$ **59.** $\left\{\frac{5\sqrt{a}}{3}, -\frac{5\sqrt{a}}{3}\right\}$ **61.** $\{\sqrt{a^2 + 36}, -\sqrt{a^2 + 36}\}$ **63.** $\sqrt{13}$ **65.** 29 **67.** $\sqrt{165}$

6.2 Exercises (page 310)

1. $\left\{\frac{2 + \sqrt{3}}{2}, \frac{2 - \sqrt{3}}{2}\right\}$ **3.** $\left\{\frac{-2 + \sqrt{2}}{2}, \frac{-2 - \sqrt{2}}{2}\right\}$ **5.** $\{5 + \sqrt{7}, 5 - \sqrt{7}\}$ **7.** $\left\{\frac{1 + \sqrt{3}}{2}, \frac{1 - \sqrt{3}}{2}\right\}$ **9.** $\left\{\frac{1 + \sqrt{5}}{2}, \frac{1 - \sqrt{5}}{2}\right\}$ **11.** $\left\{\frac{-4 + \sqrt{6}}{5}, \frac{-4 - \sqrt{6}}{5}\right\}$ **13.** $\left\{\frac{1 + \sqrt{6}}{5}, \frac{1 - \sqrt{6}}{5}\right\}$ **15.** $\left\{\frac{2}{3}, -1\right\}$ **17.** $\left\{\frac{-1 + \sqrt{2}}{2}, \frac{-1 - \sqrt{2}}{2}\right\}$ **19.** $\left\{\frac{1 + \sqrt{29}}{2}, \frac{1 - \sqrt{29}}{2}\right\}$ **21.** $\left\{\frac{-1 + \sqrt{7}}{3}, \frac{-1 - \sqrt{7}}{3}\right\}$ **23.** $\left\{\frac{-4 + \sqrt{91}}{3}, \frac{-4 - \sqrt{91}}{3}\right\}$ **29.** $\left\{-\frac{2}{3} + \frac{\sqrt{2}}{3}i, -\frac{2}{3} - \frac{\sqrt{2}}{3}i\right\}$ **31.** $\{3 + i\sqrt{5}, 3 - i\sqrt{5}\}$ **33.** $\left\{\frac{1}{2} + \frac{\sqrt{6}}{2}i, \frac{1}{2} - \frac{\sqrt{6}}{2}i\right\}$ **35.** $\left\{-\frac{1}{2} + \frac{\sqrt{3}}{2}i, -\frac{1}{2} - \frac{\sqrt{3}}{2}i\right\}$ **37.** $\left\{0, \frac{4}{3}i\right\}$ **39.** $\left\{-2i, \frac{1}{2}i\right\}$ **41.** 9.1 hours **43.** Johnson: 4 hours; Tom: 8 hours **45.** Dick: 23.0 hours; Tom: 25.0 hours **47.** $\{-2\sqrt{2}, \sqrt{2}\}$ **49.** $\left\{\frac{-4\sqrt{5}}{5}, \frac{\sqrt{5}}{2}\right\}$ **51.** $\{-i, 2i\}$ **53.** 52; $\frac{1}{2}$; $-\frac{3}{4}$ **55.** 17; $\frac{5\sqrt{2}}{2}$; 1 **57.** 3 **59.** $\frac{1}{4}$

6.3 Exercises (page 317)

1. (b) **3.** (d) **5.** (c) **7.** (c) **9.** (a) **11.** 14 or -14 **13.** 9 **15.** 12 or -12 **17.** 12 or -12 **21.** $(3k+5)(2k-1)$ **23.** $(6z-5)(4z+1)$ **25.** cannot be factored **27.** $(10n-3)(2n+7)$ **29.** cannot be factored In Exercises 31–37, the sum of the roots is listed first and the product of the roots second. **31.** -6; -8 **33.** $\frac{3}{4}$; 0 **35.** 0; $-\frac{25}{4}$ **37.** -10; -12 **39.** yes **41.** yes **43.** no **45.** $10x^2 - x - 3 = 0$ **47.** $x^2 - 20 = 0$ **49.** $x^2 + 16 = 0$ **51.** $15x^2 - 9x - 20 = 0$ **53.** $4x^2 - \sqrt{3}x + 8 = 0$ **55.** $b = \frac{44}{5}$; $x_2 = \frac{3}{10}$ **57.** $\frac{1}{3}$ **59.** -5 **61.** $a^2 + 3a - 5$ **63.** $3m^2 + 4m - 1$ **65.** $\{37\}$ **67.** $\{7\}$

6.4 Exercises (page 325)

1. square root property **3.** quadratic formula **5.** factoring **7.** $\{-7, 4\}$ **9.** $\left\{-\frac{4}{3}, 1\right\}$ **11.** $\left\{-\frac{8}{3}, -1\right\}$ **13.** $\left\{\frac{2+\sqrt{22}}{3}, \frac{2-\sqrt{22}}{3}\right\}$ **15.** $\left\{\frac{11+\sqrt{73}}{4}, \frac{11-\sqrt{73}}{4}\right\}$ **17.** $\left\{\frac{2}{3}\right\}$ **19.** $\left\{\frac{1+\sqrt{13}}{2}\right\}$ **21.** $\{8\}$ **23.** $\emptyset$ **25.** $\{-7\}$ **27.** $\left\{-\frac{4}{3}, \frac{1}{2}\right\}$ **29.** $\left\{\frac{-4+\sqrt{41}}{2}, \frac{-4-\sqrt{41}}{2}\right\}$ **31.** $\left\{-\frac{13}{10}, \frac{1}{5}\right\}$ **33.** $\left\{\frac{4}{3}, -\frac{4}{3}, 1, -1\right\}$ **35.** $\left\{2, -2, \frac{5}{2}, -\frac{5}{2}\right\}$ **37.** $\left\{i, -i, \frac{1}{2}i, -\frac{1}{2}i\right\}$ **39.** 255 miles per hour **41.** 50 miles per hour **43.** 3 hours and 6 hours **45.** $\{1, 2\}$ **47.** $\left\{-\sqrt[3]{5}, -\frac{\sqrt[3]{4}}{2}\right\}$ **49.** $\{2\}$ **51.** $\left\{-\frac{1}{2}, 3\right\}$ **53.** $\{-3, -2, 1, 2\}$ **57.** $h = \frac{2A}{b}$ **59.** $h = \frac{S - 2\pi r^2}{2\pi r}$ **61.** $r^2 = 3q^2 + 6$

6.5 Exercises (page 331)

1. $d = \pm\frac{\sqrt{kR}}{R}$ **3.** $h = \pm\frac{d^2\sqrt{kL}}{L}$ **5.** $r = \pm\frac{\sqrt{3\pi Vh}}{\pi h}$ **7.** $d = \frac{k^2}{F^2}$ **9.** $p = \frac{2L}{M^2}$ **11.** $r = \frac{-\pi h \pm \sqrt{\pi^2h^2 + \pi S}}{\pi}$ **13.** 3 minutes **15.** approximately 3.4 seconds **17.** 5 centimeters, 12 centimeters, 13 centimeters **19.** 16 meters **21.** 5 feet **23.** length: 20 inches; width: 12 inches **25.** 2 feet **27.** \$1.52 **29.** 7.899, 8.899, 11.899 **31.** $R = \frac{E^2 - 2pr \pm E\sqrt{E^2 - 4pr}}{2p}$ **33.** $r = \frac{5pc}{4}$ or $r = -\frac{2pc}{3}$ **35.** $I = \frac{-cR \pm \sqrt{c^2R^2 - 4cL}}{2cL}$ **41.** 4 **43.**

45.

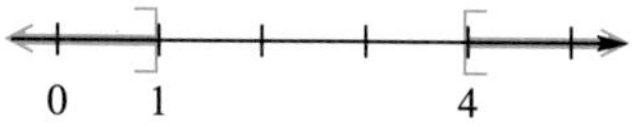

47.

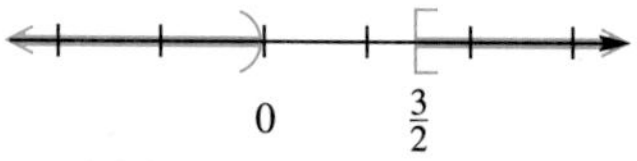

6.6 Exercises (page 338)

1. $\left(-\infty, -\frac{1}{2}\right) \cup (5, \infty)$

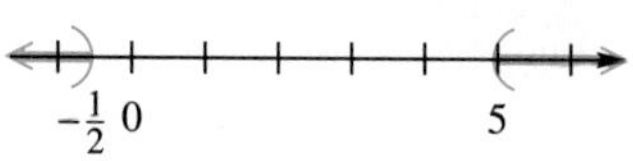

3. $[-4, 6]$

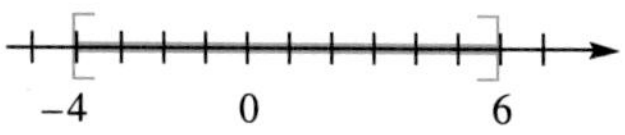

5. $\left[-4, \frac{2}{3}\right]$

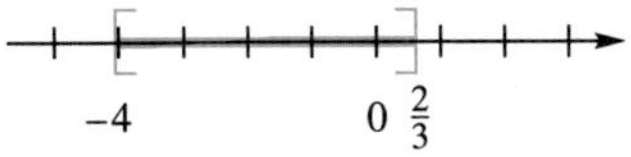

7. $\left(-3, \frac{5}{2}\right)$

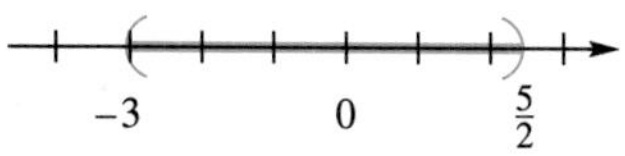

9. $\left(-1, -\frac{3}{4}\right)$ 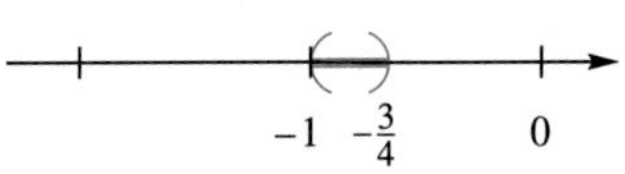 **11.** $(-\infty, 3 - \sqrt{3}] \cup [3 + \sqrt{3}, \infty)$

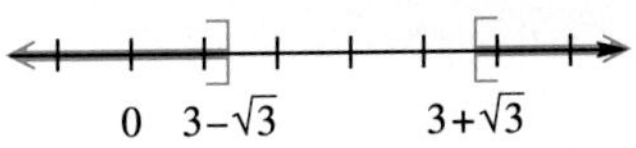

13. $\left(-\infty, \frac{-2 - \sqrt{2}}{2}\right) \cup \left(\frac{-2 + \sqrt{2}}{2}, \infty\right)$

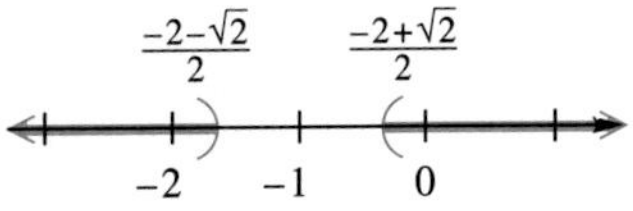

15. $(-\infty, 0] \cup [4, \infty)$ **17.** $\left[0, \frac{5}{3}\right]$

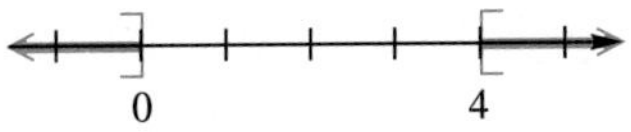

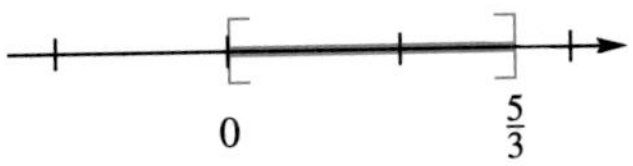

19. $(-\infty, 1) \cup (2, 3)$ 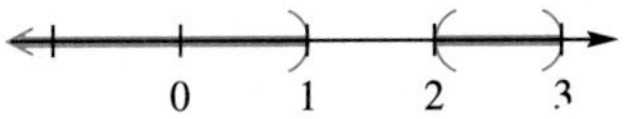 **21.** $\left(-\infty, -\frac{3}{2}\right] \cup \left[\frac{1}{3}, 4\right]$

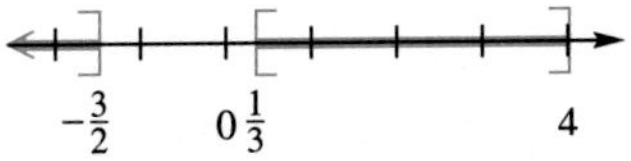

23. $\left(-4, -\frac{1}{2}\right) \cup \left(\frac{1}{3}, \infty\right)$ 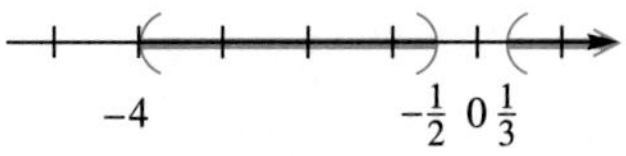 **25.** $(2, 6]$

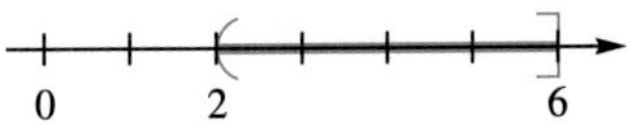

27. $\left[\frac{22}{5}, 5\right)$ **29.** $(-\infty, 1) \cup (4, \infty)$

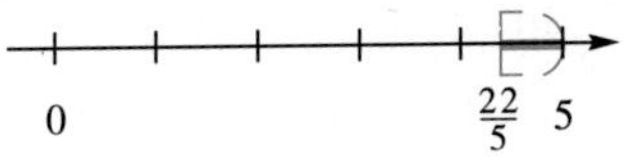

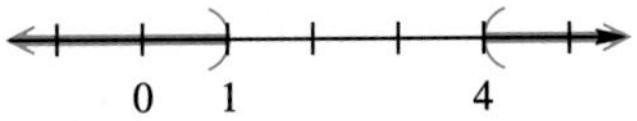

31. $[-4, -2)$ **33.** $\left(1, \frac{3}{2}\right)$

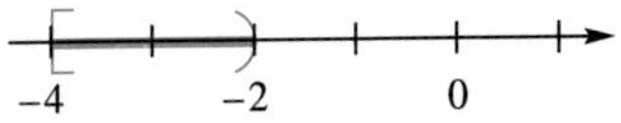

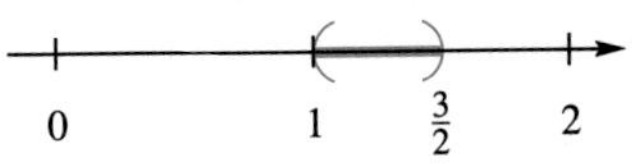

35. $[-1, 0) \cup [4, \infty)$ **37.** $\left[\frac{3}{2}, \infty\right)$

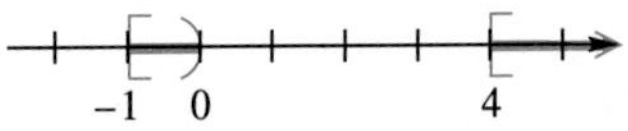

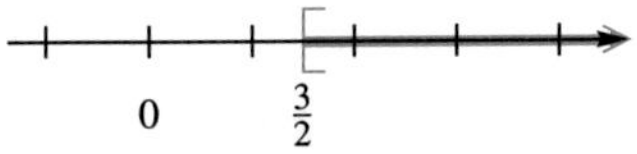

39. $\left(-2, \frac{5}{3}\right) \cup \left(\frac{5}{3}, \infty\right)$ 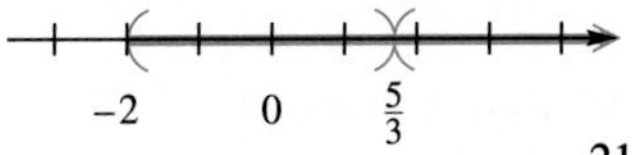 **41.** if x, the number of units sold, is between 0 and 5 or more than $\frac{15}{2}$ (that is, 8 or more) **45.** 0 **47.** $-\frac{21}{2}$

Chapter 6 Review Exercises (page 343)

1. $\{-4, 4\}$ **3.** $\{-6, 1\}$ **5.** $\left\{\frac{2}{3} + \frac{4}{3}i, \frac{2}{3} - \frac{4}{3}i\right\}$ **7.** $\{2 + \sqrt{19}, 2 - \sqrt{19}\}$ **9.** $\left\{\frac{7}{2}, -3\right\}$ **11.** $\left\{\frac{-3 + \sqrt{89}}{4}, \frac{-3 - \sqrt{89}}{4}\right\}$ **13.** $\left\{\frac{-7 + \sqrt{37}}{2}, \frac{-7 - \sqrt{37}}{2}\right\}$ **15.** $\{-2i + i\sqrt{5}, -2i - i\sqrt{5}\}$ **17.** 5.5 hours and 6.5 hours **19.** 6 hours and 8 hours **21.** (a) **23.** (d) **25.** (c) **27.** not factorable **29.** $\frac{5}{2}$; -4 **31.** $3x^2 + 7x + 2 = 0$ **33.** $\left\{-\frac{11}{4}, -1\right\}$ **35.** $\{-2\}$ **37.** $\{-1\}$ **39.** $\left\{1, -1, \frac{\sqrt{15}}{5}i, -\frac{\sqrt{15}}{5}i\right\}$ **41.** $\left\{-\frac{11}{8}, -\frac{7}{4}\right\}$ **43.** 40 kilometers per hour **45.** $v = \pm\frac{\sqrt{rFkw}}{kw}$ **47.** $t = \frac{3m \pm \sqrt{9m^2 + 20m}}{2m}$ **49.** length: 2 centimeters; width: 1 centimeter **51.** 1 inch **53.** 40 miles per hour **55.** $.08 = 8\%$ **57.** $(-4, 3)$

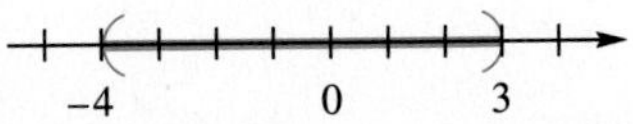

59. $\left(-\infty, -\frac{3}{4}\right] \cup \left[\frac{1}{3}, \infty\right)$ (number line: $-\frac{3}{4}$, 0, $\frac{1}{3}$)

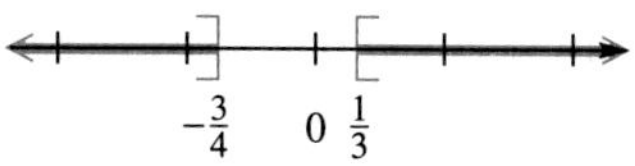

61. $(-\infty, -2) \cup \left(\frac{1}{2}, \frac{4}{5}\right)$ (number line: -2, 0, $\frac{1}{2}$, $\frac{4}{5}$)

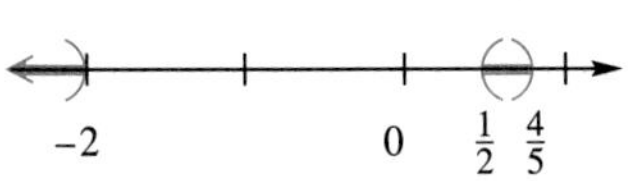

63. $\left(-\infty, \frac{1}{2}\right) \cup \left(\frac{4}{5}, \infty\right)$ (number line: 0, $\frac{1}{2}$, $\frac{4}{5}$)

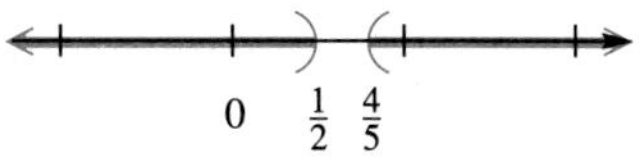

65. $\{5\}$ **67.** $\left\{1 + \frac{\sqrt{3}}{3}i, 1 - \frac{\sqrt{3}}{3}i\right\}$ **69.** $\left\{1, \frac{3}{5}\right\}$

71. 5.2 hours and 7.2 hours **73.** $\left\{\frac{-1 + \sqrt{29}}{2}, \frac{-1 - \sqrt{29}}{2}\right\}$ **75.** $\left\{-\frac{5}{3}, -\frac{3}{2}\right\}$ **77.** $\left\{\frac{2}{3} + \frac{\sqrt{2}}{3}i, \frac{2}{3} - \frac{\sqrt{2}}{3}i\right\}$

Chapter 6 Test (page 346)

[6.1] **1.** $\{-2\sqrt{2}, 2\sqrt{2}\}$ **2.** $\left\{-\frac{4}{3}, 0\right\}$ **3.** $\{1 + \sqrt{2}, 1 - \sqrt{2}\}$ **4.** $\left\{\frac{2}{3}\right\}$ **[6.2]** **5.** $\left\{1, \frac{1}{2}\right\}$

6. $\left\{-\frac{2}{3} + \frac{\sqrt{11}}{3}i, -\frac{2}{3} - \frac{\sqrt{11}}{3}i\right\}$ **[6.3]** **7.** two imaginary solutions **8.** 2; $\frac{5}{2}$ **[6.4]** **9.** $\{-3, 4\}$

10. $\{-3, 3, -i\sqrt{3}, i\sqrt{3}\}$ **11.** $\left\{-\frac{3}{2}, 2\right\}$ **12.** Mario: 11.1 hours; Luis: 9.1 hours **[6.5]** **13.** $r = \pm\frac{\sqrt{\pi S}}{2\pi}$

14. $h = \pm\frac{\sqrt{A^2 - \pi^2 r^4}}{\pi r}$ **15.** $p = \frac{k \pm \sqrt{k^2 - 4kr}}{2k}$ **16.** 2 feet **17.** 120 meters

[6.6] **18.** $\left(-5, \frac{3}{2}\right)$ (number line: -5, 0, $\frac{3}{2}$)

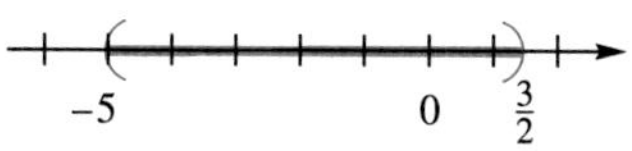

19. $(1, 5]$

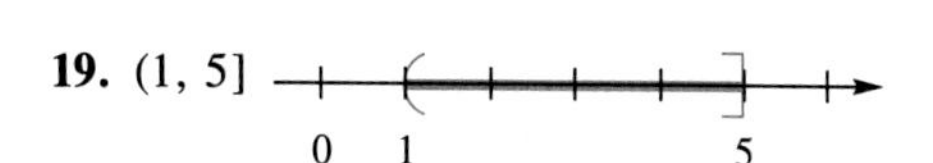

CHAPTER 7

7.1 Exercises (page 355)

1. I
3. III
5. IV
7. II

9.–17.

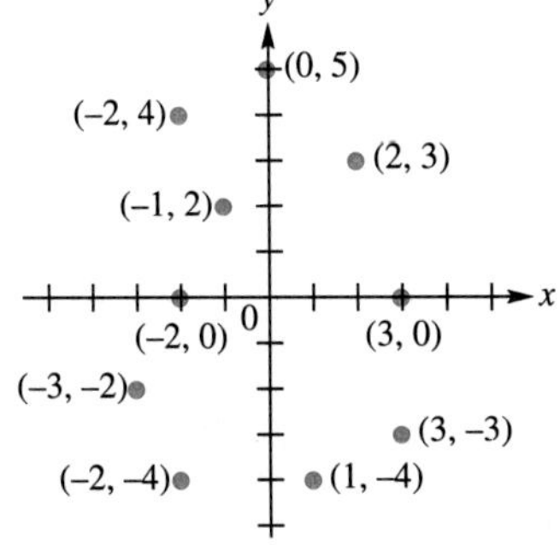

19. $(0, 5)$, $\left(\frac{5}{2}, 0\right)$, $(1, 3)$, $(2, 1)$

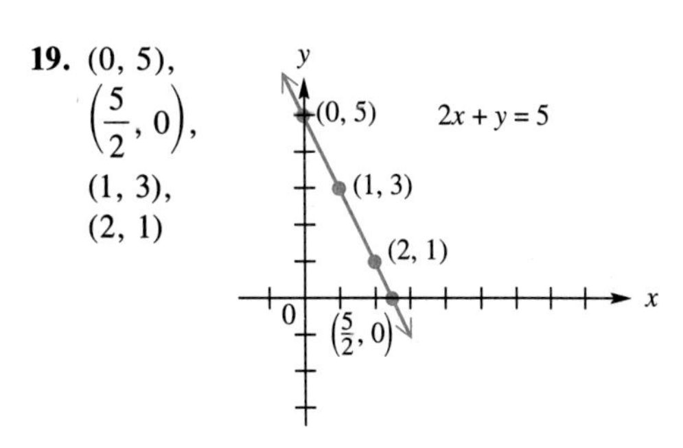

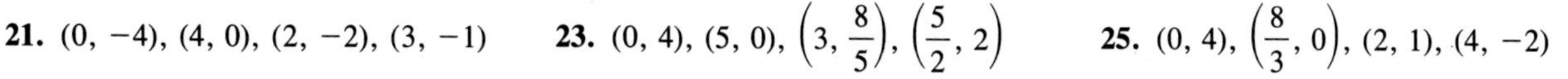

21. $(0, -4)$, $(4, 0)$, $(2, -2)$, $(3, -1)$ **23.** $(0, 4)$, $(5, 0)$, $\left(3, \frac{8}{5}\right)$, $\left(\frac{5}{2}, 2\right)$ **25.** $(0, 4)$, $\left(\frac{8}{3}, 0\right)$, $(2, 1)$, $(4, -2)$

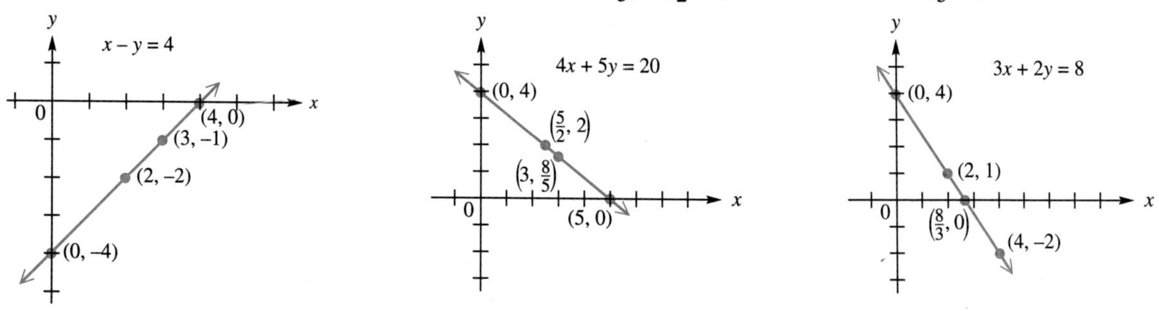

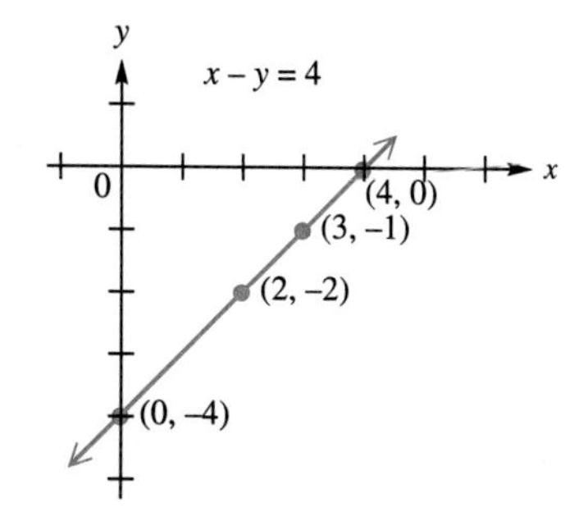

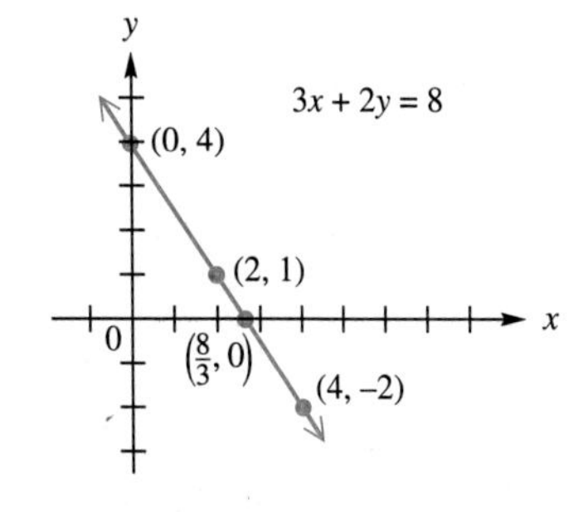

29. (a)

31. (4, 0); (0, 6)

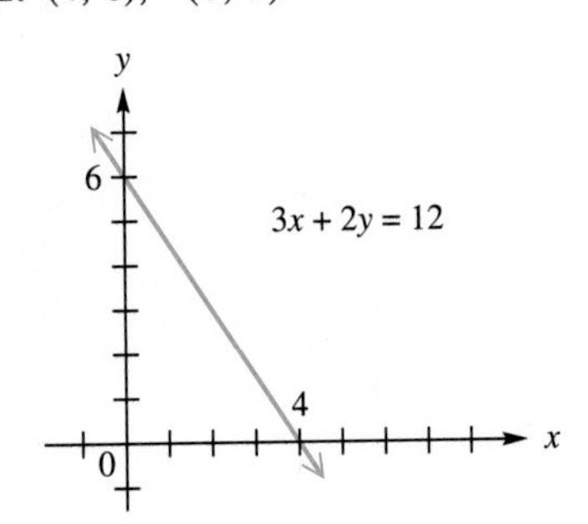

33. (2, 0); $\left(0, \frac{5}{3}\right)$

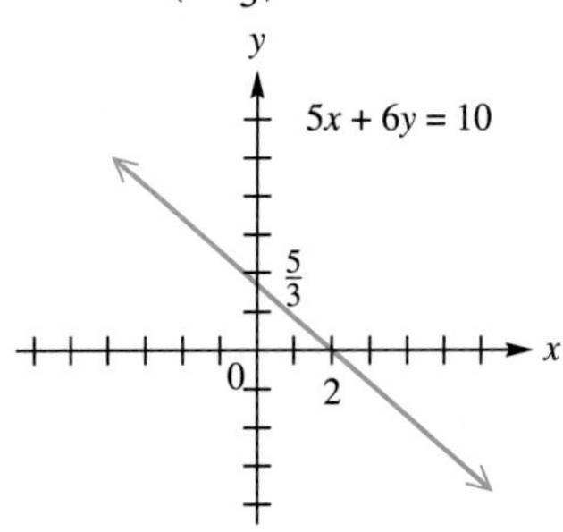

35. $\left(\frac{5}{2}, 0\right)$; (0, −5)

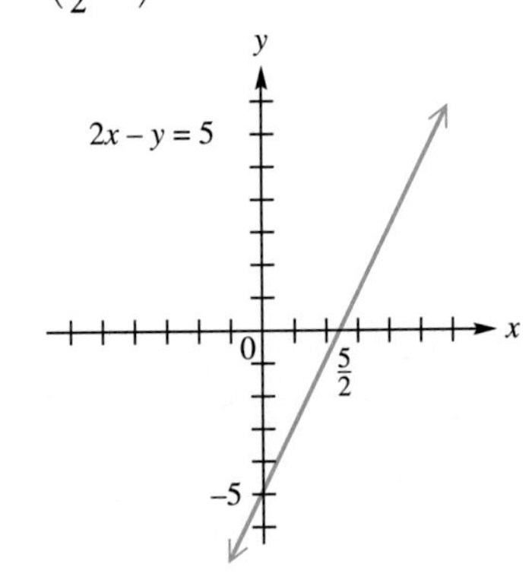

37. (2, 0); $\left(0, -\frac{2}{3}\right)$

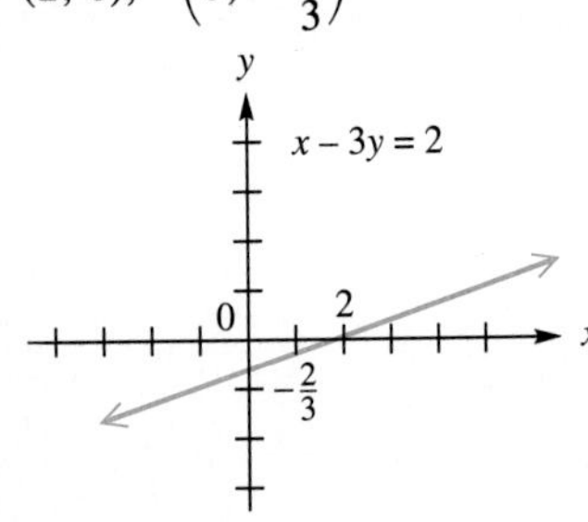

39. (0, 0); (0, 0)

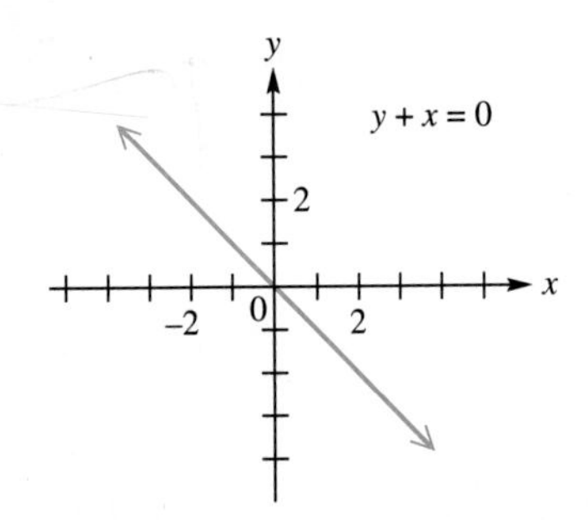

41. (0, 0); (0, 0)

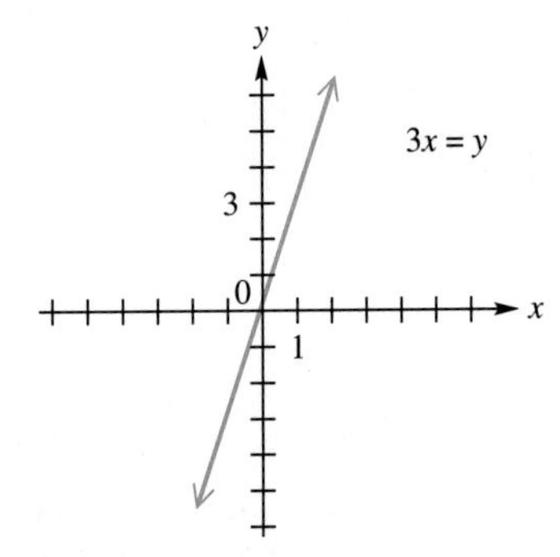

43. (2, 0); none

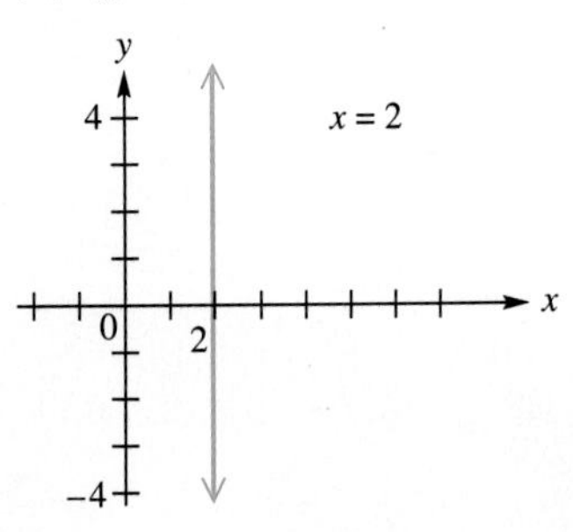

45. none; (0, 4)

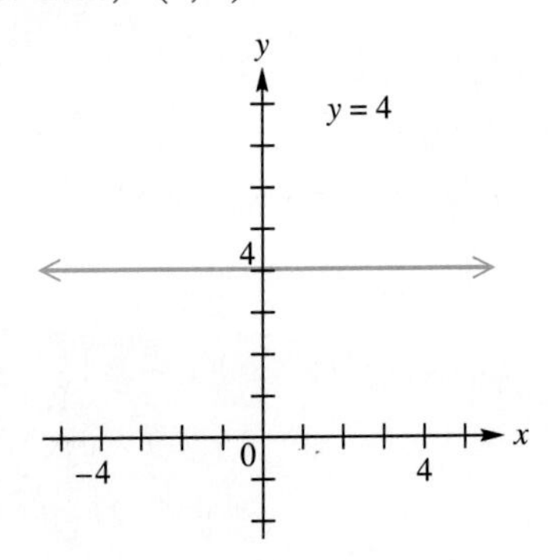

47. $\sqrt{34}$ **49.** $\sqrt{61}$

51. $\sqrt{146}$ **53.** $6\sqrt{2}$

55. $\sqrt{5b^2 - 2ab + a^2}$

57. $[(x_2 - x_1)^2 + (y_2 - y_1)^2]^{1/2}$

59. yes **61.** yes **63.** no

65. 24 **67.** $5 + \sqrt{58} + \sqrt{13}$

69. a right triangle

71. a right triangle

73. $\left(-\frac{1}{2}, 6\right)$ **75.** $\left(-\frac{3}{2}, \frac{3}{2}\right)$

77. (2, −3)

79. **(a)** 233.4 million
(b) 238.5 million
(c) 247 million
(d) 2030
(e)

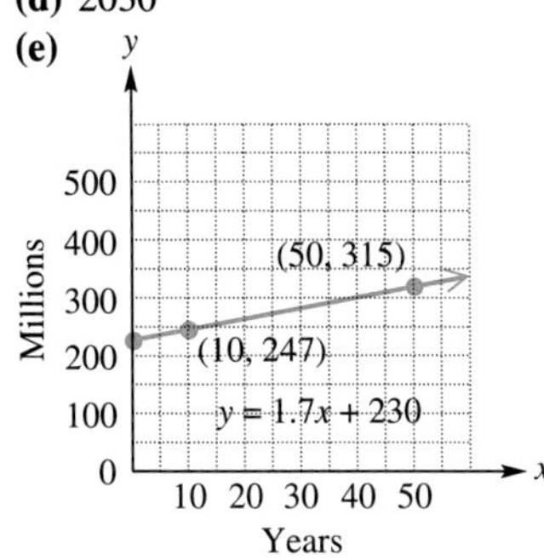

81. $x = 7 + y$

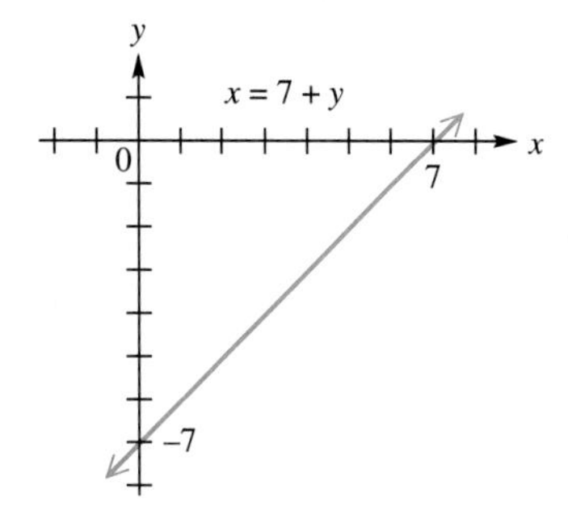

83. 16 feet **85.** $\frac{1}{2}$ **87.** $\frac{2}{5}$ **89.** $-\frac{1}{5}$

7.2 Exercises (page 364)

1. $m = -8$

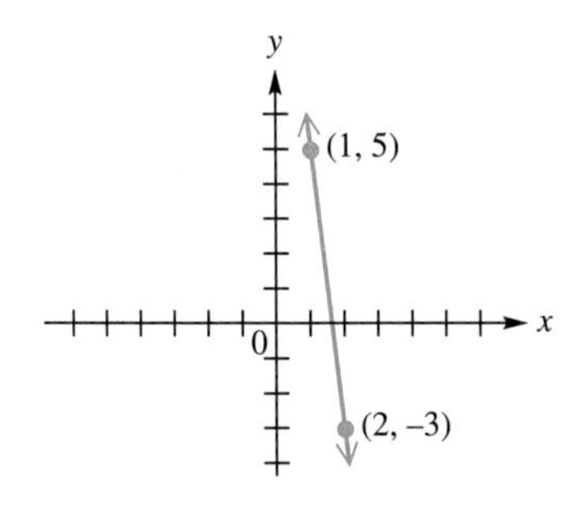

3. $m = -1$

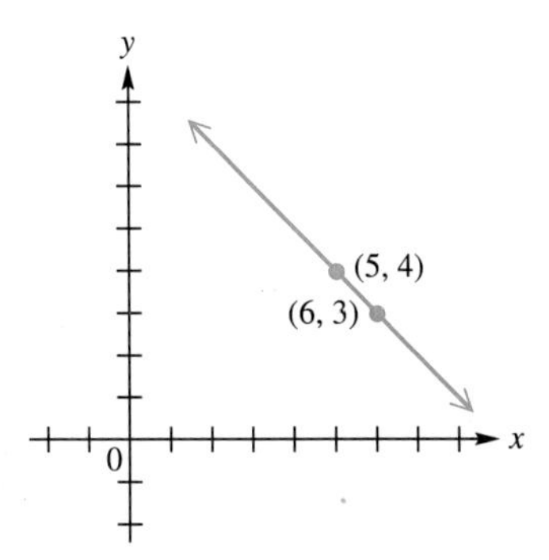

5. $m = \frac{9}{8}$

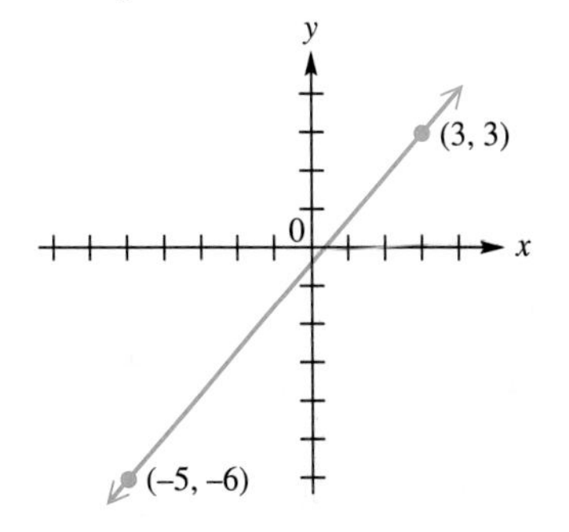

7. $m = 0$

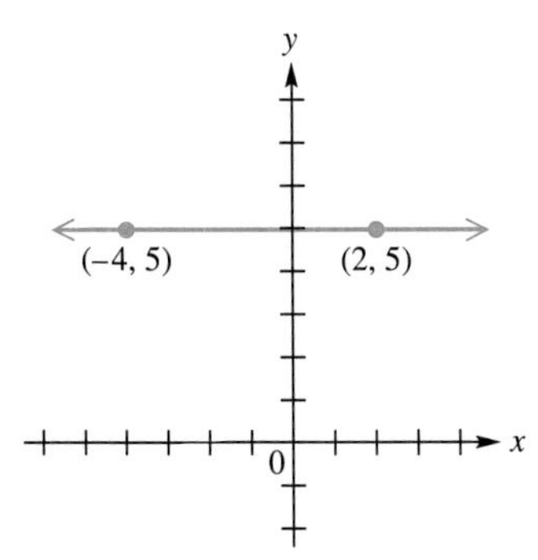

9. positive **11.** negative **13.** 0 **15.** undefined

17. $m = -\frac{1}{2}$

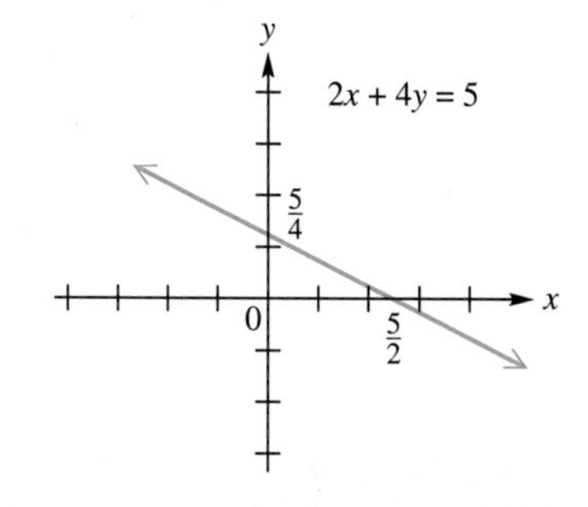

19. $m = 1$

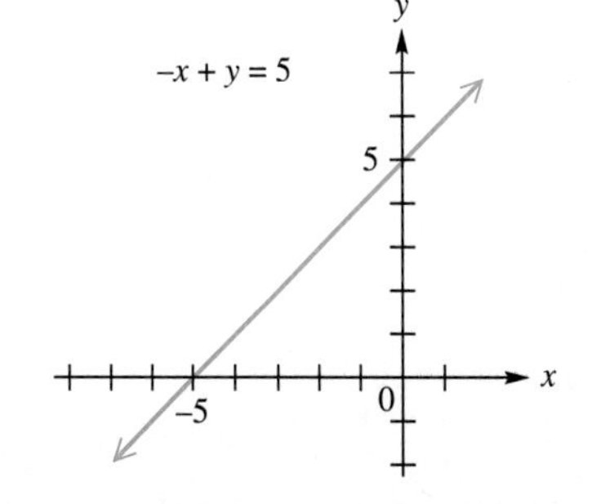

21. $m = \frac{6}{5}$

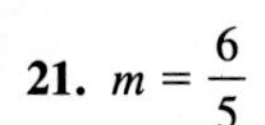

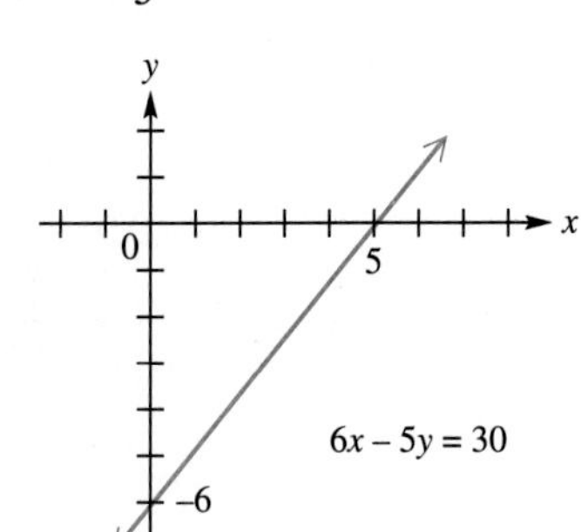

25.

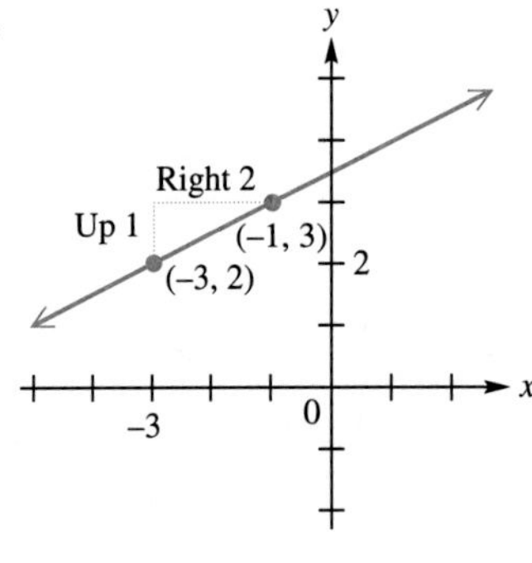

27.

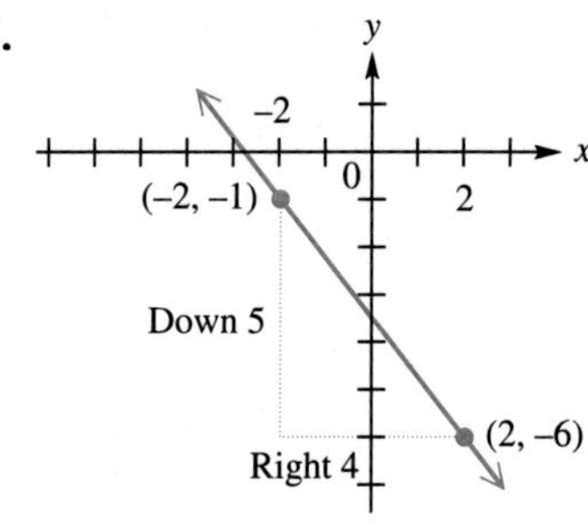

29.

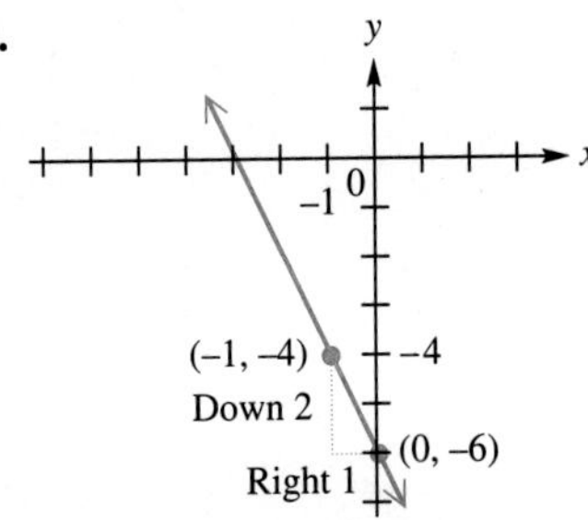

31.

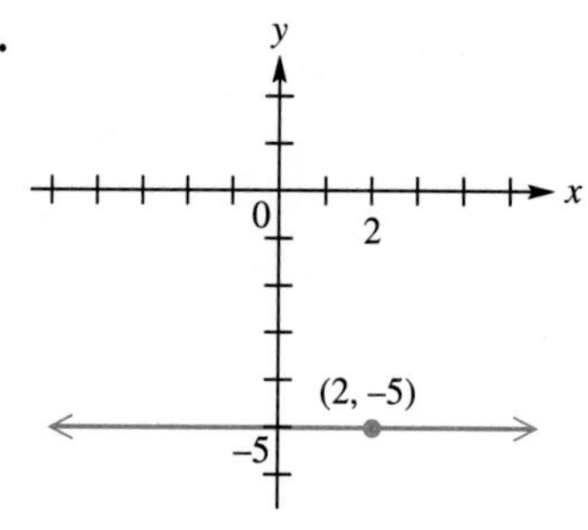

33.

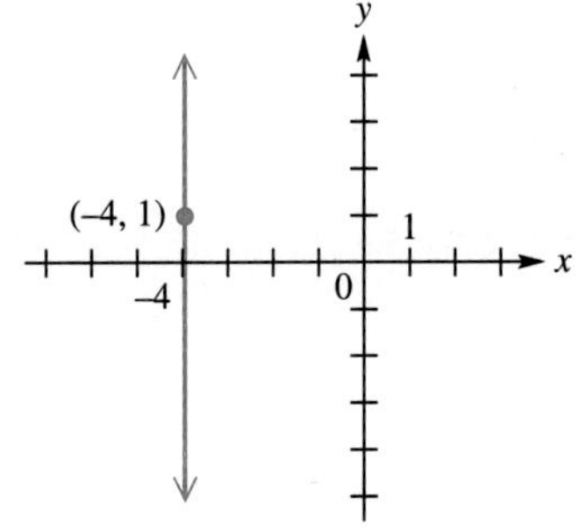

35. $-\frac{4}{9}$; $\frac{9}{4}$ **37.** parallel **39.** not parallel **41.** parallel **43.** perpendicular **45.** not perpendicular **47.** not perpendicular **49.** approximately .92° C per hour **51.** \$1.40 per mile **53.** Since the slopes of both pairs of opposite sides are equal, the figure is a parallelogram. **55.** no **57.** yes **59.** $y = -\frac{2}{3}x + \frac{8}{3}$ **61.** $y = -\frac{3}{4}x + \frac{7}{4}$ **63.** $2x + 3y = 27$ **65.** $5x - 3y = -17$

7.3 Exercises (page 373)

1. $3x + 4y = 14$ **3.** $2x + y = 7$ **5.** $x - 2y = -1$ **7.** $y = 2$ **9.** $4x - y = 12$ **13.** $x = 2$ **15.** $x = -7$ **17.** $2x + y = 10$ **19.** $x + 2y = 8$ **21.** $5x - y = 4$ **23.** $2x - 13y = -6$ **25.** $y = 5$ **27.** $\sqrt{5}x - 2y = -\sqrt{5}$ **29.** $y = 5x + 4$ **31.** $y = -\frac{2}{3}x + \frac{1}{2}$ **33.** $y = \frac{2}{5}x - 1$ **35.** $y = 4$ **37.** $y = -x + 8$; -1; $(0, 8)$ **39.** $y = -\frac{5}{2}x + 5$; $-\frac{5}{2}$; $(0, 5)$ **41.** $y = \frac{2}{3}x - \frac{5}{3}$; $\frac{2}{3}$; $\left(0, -\frac{5}{3}\right)$ **43.** $y = -\frac{5}{3}x - \frac{4}{3}$; $-\frac{5}{3}$; $\left(0, -\frac{4}{3}\right)$ **45.** $y = -\frac{8}{11}x + \frac{9}{11}$; $-\frac{8}{11}$; $\left(0, \frac{9}{11}\right)$ **47.** (a) **49.** (c) **51.** (b) **53.** (d) **55.** (c) **57.** (a) **59.** (d) **61.** (b) **63.** $3x - y = -24$ **65.** $x - 2y = 2$ **67.** $x + 2y = 18$ **69.** $y = 7$ **71.** **(a)** $y = 640x + 1100$ **(b)** \$8780 **73.** **(a)** $y = \frac{4400}{3}x - \frac{94,100}{3}$ **(b)** \$12,633.33 **75.** 45° F **77.** $\left(-\infty, \frac{1}{2}\right)$ **79.** $[-2, \infty)$ **81.** $\left(-\infty, -\frac{8}{5}\right)$

7.4 Exercises (page 380)

1.

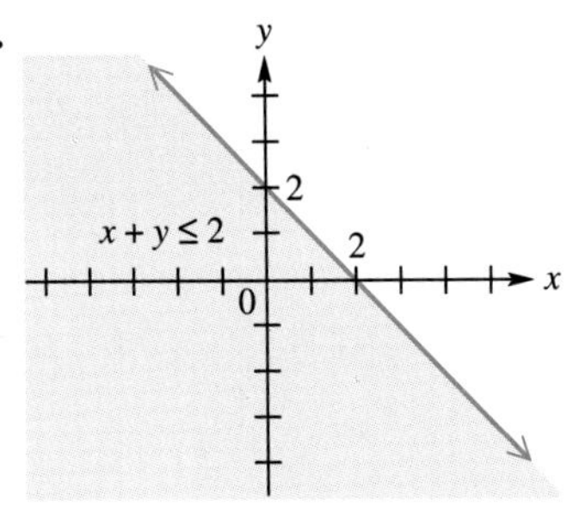

3.

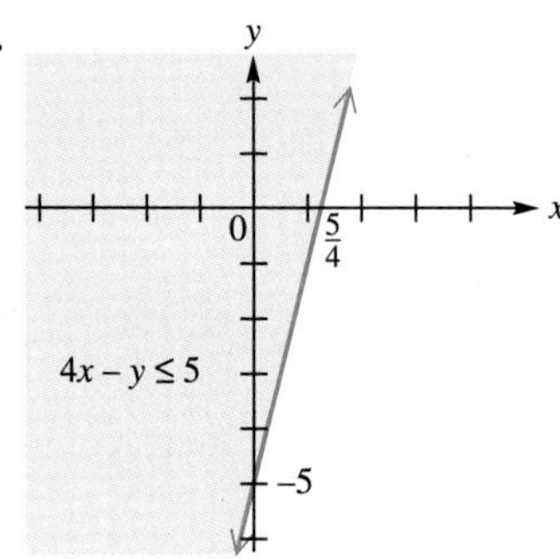

5.

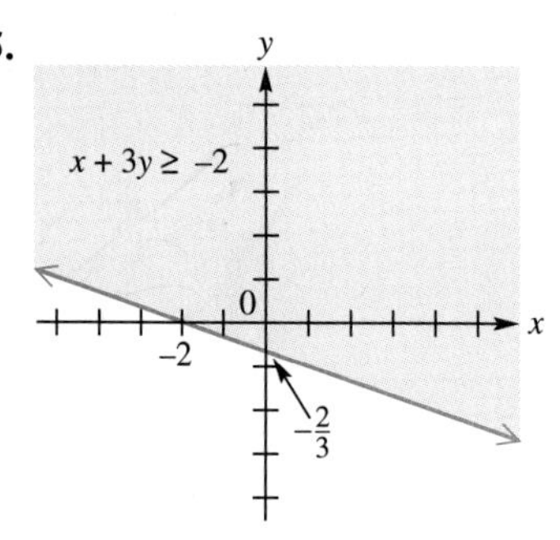

7.

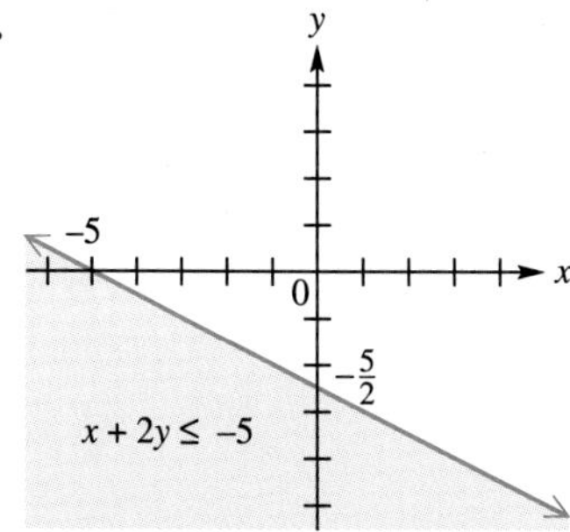

9.

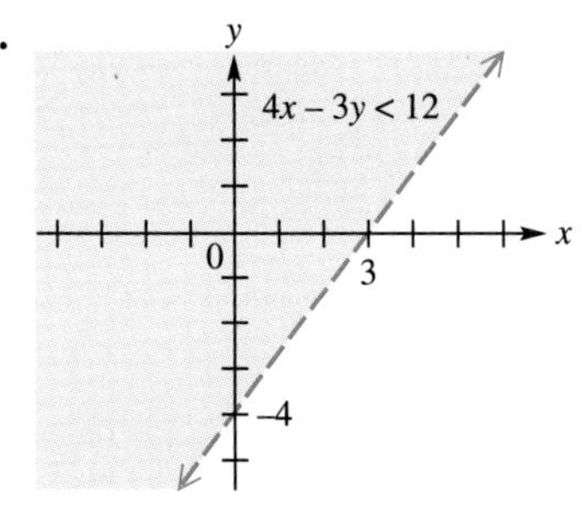

11.

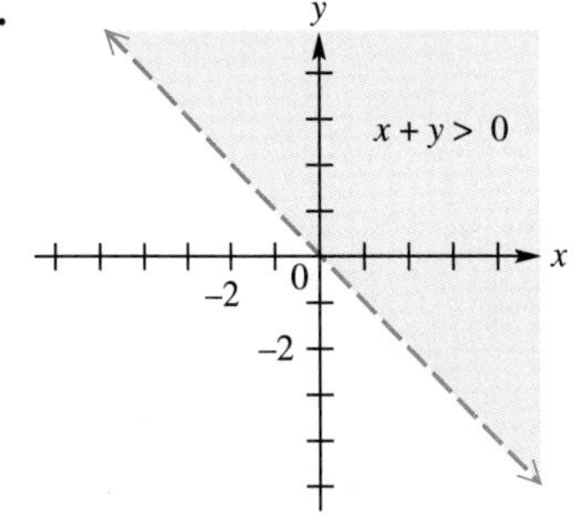

13.

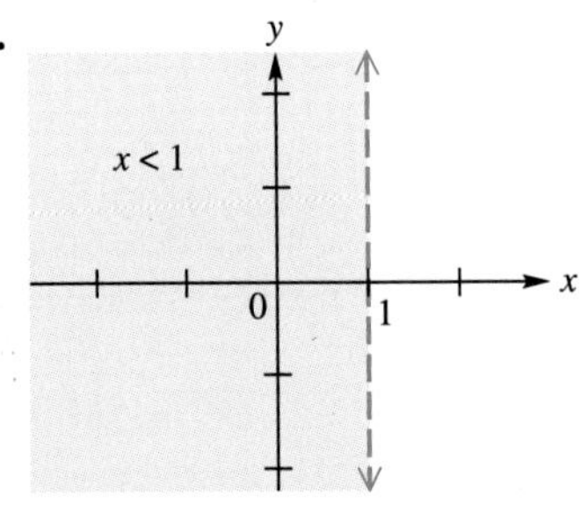

17.

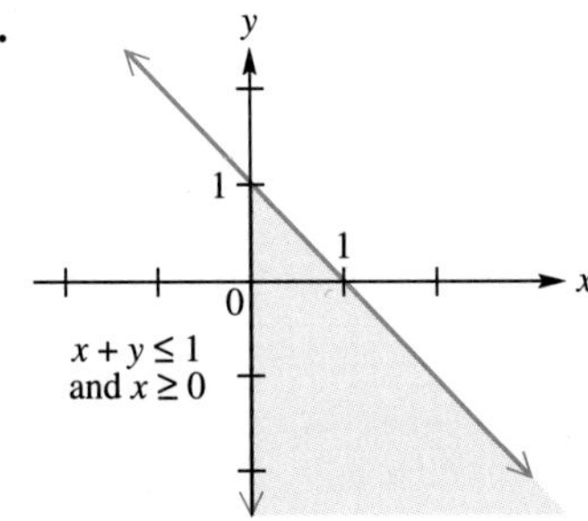

19.

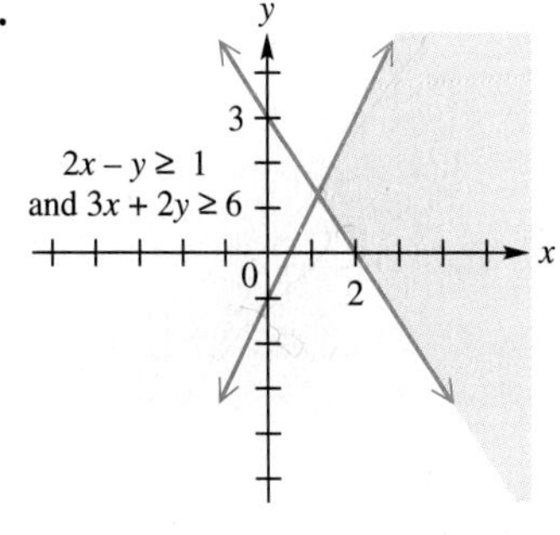

21.

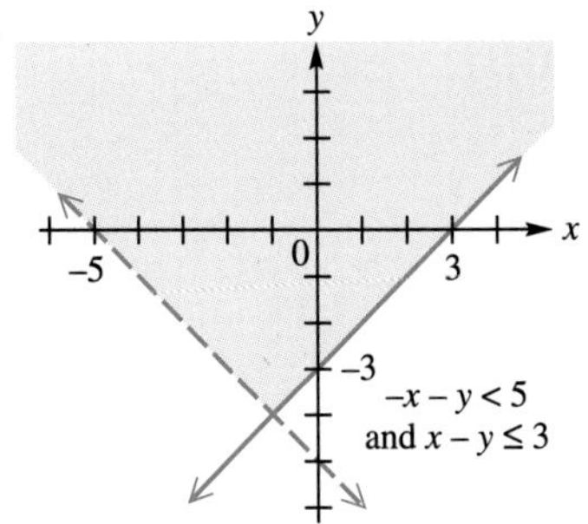

23.

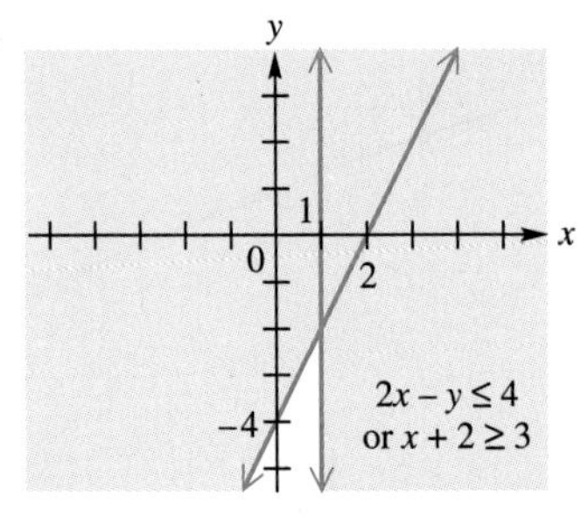

25.

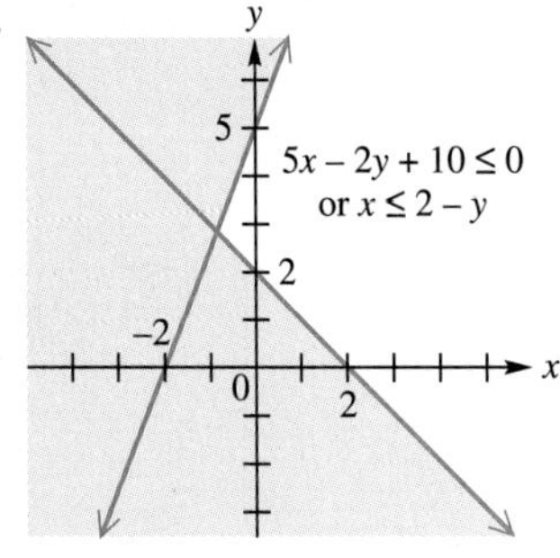

27.

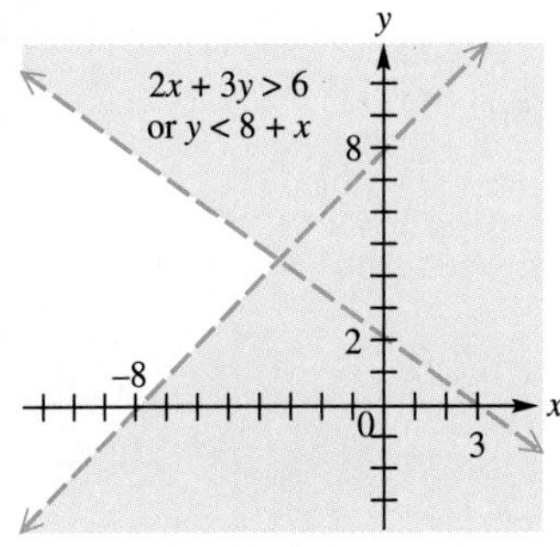

29. $x > 1$ or $x < -1$

31.

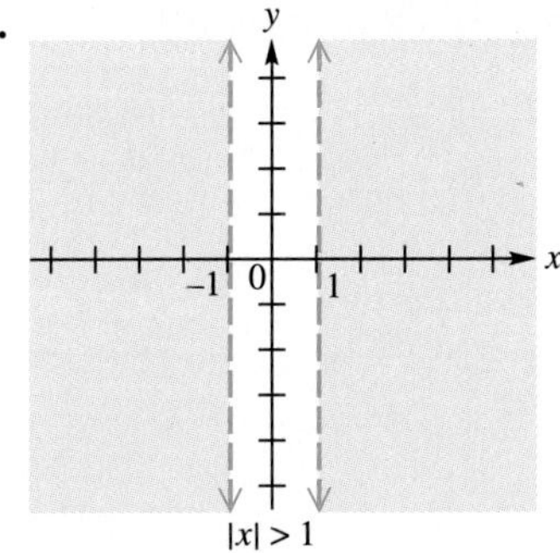

33.

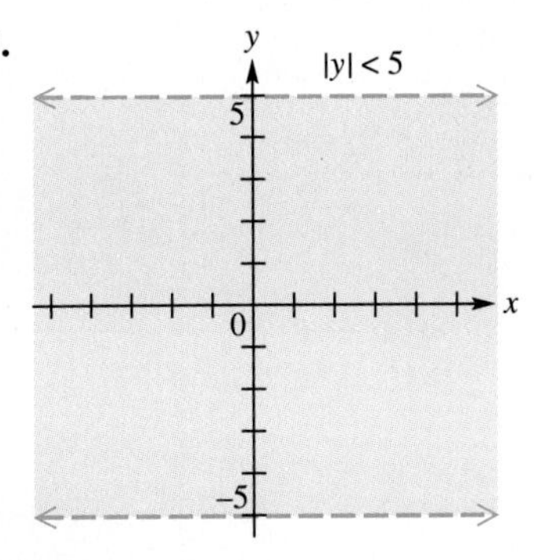

35.

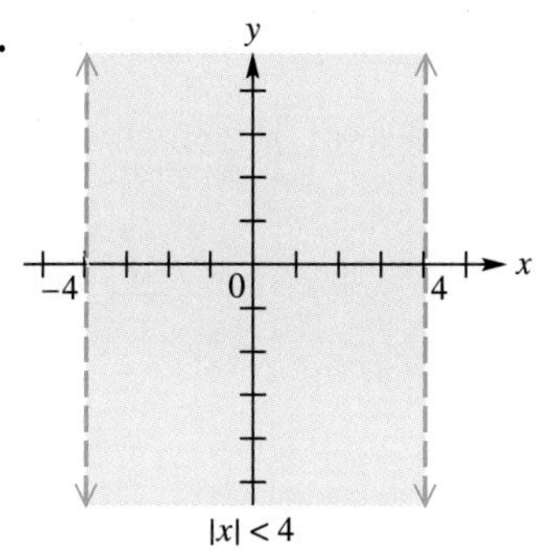

37.

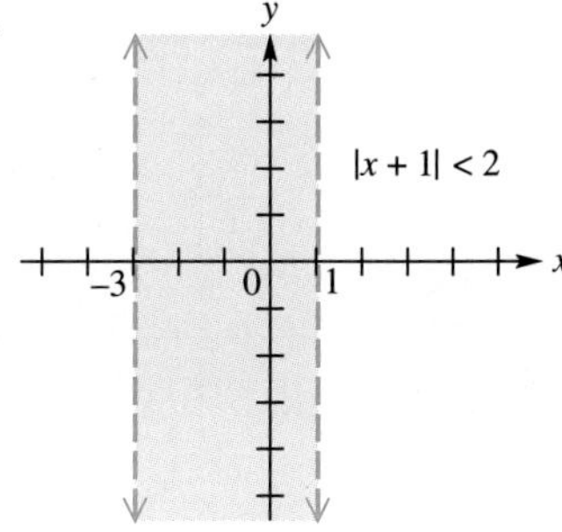

39.

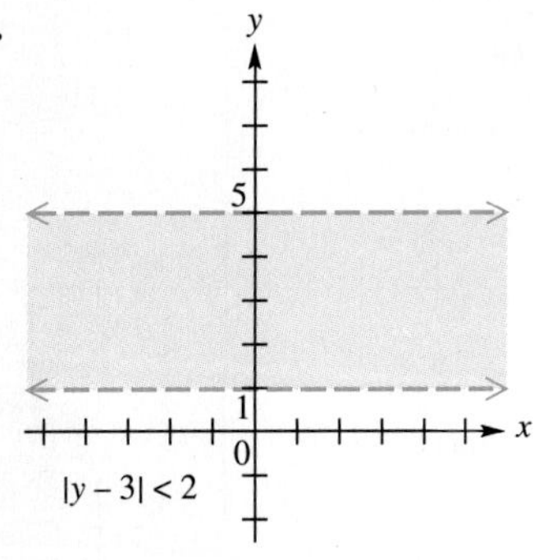

41.

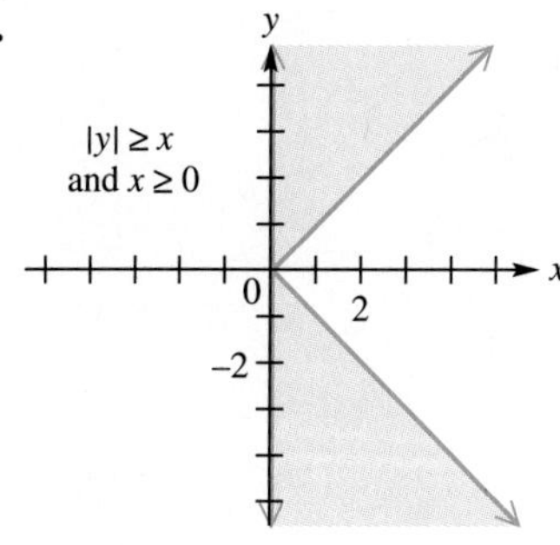

43.

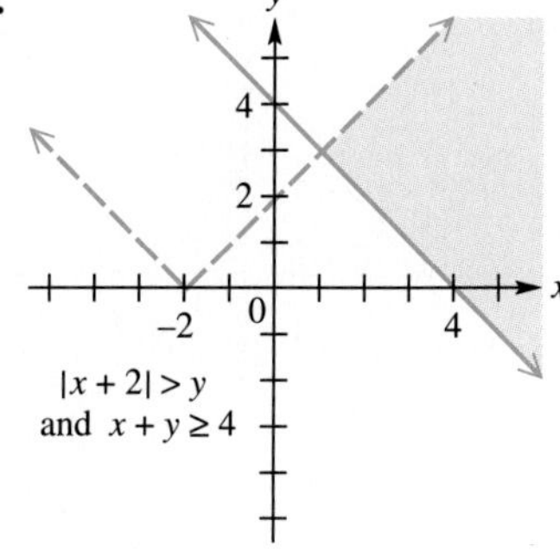

45.

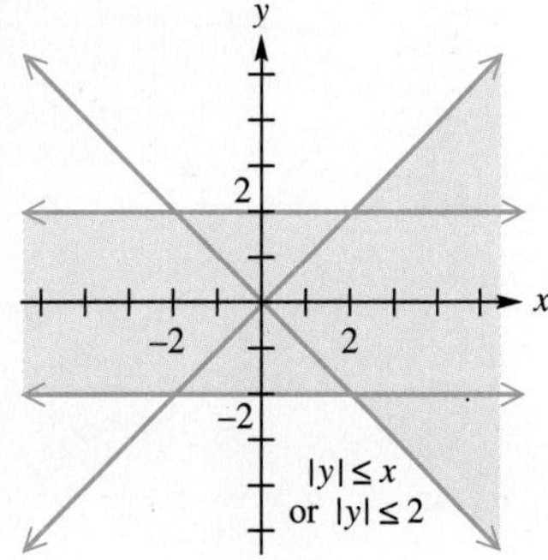

47.

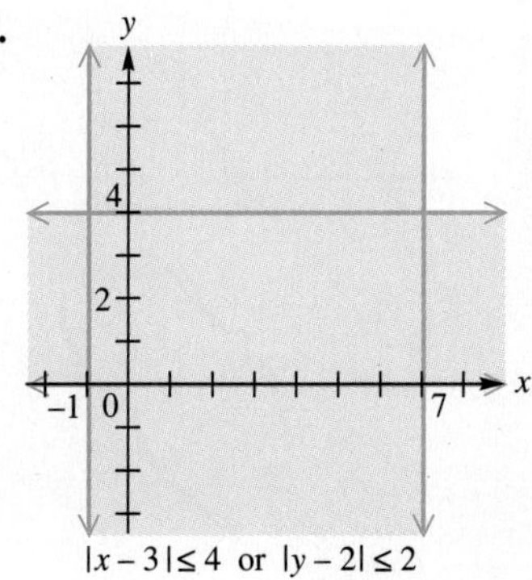

49. $\frac{1}{3}$ **51.** 3

7.5 Exercises (page 386)

1. $\frac{45}{2}$ **3.** $\frac{16}{9}$ **5.** $\frac{5}{16}$ **7.** $\frac{1000}{9}$ **9.** $\frac{1}{6}$ **11.** increases; decreases **13.** inverse **15.** direct **17.** joint **19.** combined **23.** 256 feet **25.** 133.33 newtons per square centimeter **27.** about 126 pounds **29.** 60 kilometers **31.** $\frac{45}{8} \approx 5.63$ kilometers per second **33.** 800 pounds **35.** $\frac{2}{3}(1.06\pi)$ or $.71\pi$ seconds **37.** 480 kilograms **39.** about 8.5 **41.** -9 **43.** 1 **45.** $y = \frac{3}{7}x - \frac{8}{7}$ **47.** $y = \frac{1}{8}x - \frac{5}{4}$

7.6 Exercises (page 397)

1. domain: $\{3, 4, 5, 7\}$; range: $\{1, 2, 6, 9\}$; function **3.** domain: $\{0, 2\}$; range: $\{2, 4, 5\}$; not a function **5.** domain: $\{-3, -2, 4\}$; range: $\{1, 7\}$; function **7.** domain: $\{0, 1, 4, 7\}$; range: $\{2, 3, 6, 7\}$; function **9.** domain: $\{4, 9, 11, 17, 25\}$; range: $\{14, 32, 47, 69\}$; function **11.** domain: $\{14, 17, 23, 75, 91\}$; range: $\{5, 9, 12, 18, 56, 70\}$; not a function **15.** function; domain: $(-\infty, \infty)$; range: $(-\infty, \infty)$ **17.** not a function; domain: $[3, \infty)$; range: $(-\infty, \infty)$ **19.** not a function; domain: $[-4, 4]$; range: $[-3, 3]$ **21.** domain: $(-\infty, \infty)$; function **23.** domain: $(-\infty, \infty)$; function **25.** domain: $[0, \infty)$; function **27.** domain: $(-\infty, \infty)$; not a function **29.** domain: $[3, \infty)$; function **31.** domain: $(-\infty, \infty)$; function **33.** domain: $(-\infty, 7) \cup (7, \infty)$; function **35.** domain: $(-\infty, -4) \cup (-4, 4) \cup (4, \infty)$; function **37.** domain: $(-\infty, \infty)$; function **39.** domain: $(-\infty, 1) \cup (1, 4) \cup (4, \infty)$; function **41.** 5 **43.** 2 **45.** -1 **47.** -13

49. domain: $(-\infty, \infty)$; range: $(-\infty, \infty)$

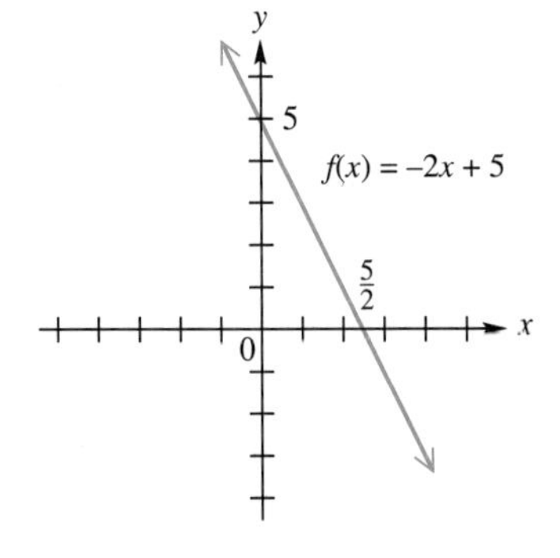

51. domain: $(-\infty, \infty)$; range: $(-\infty, \infty)$

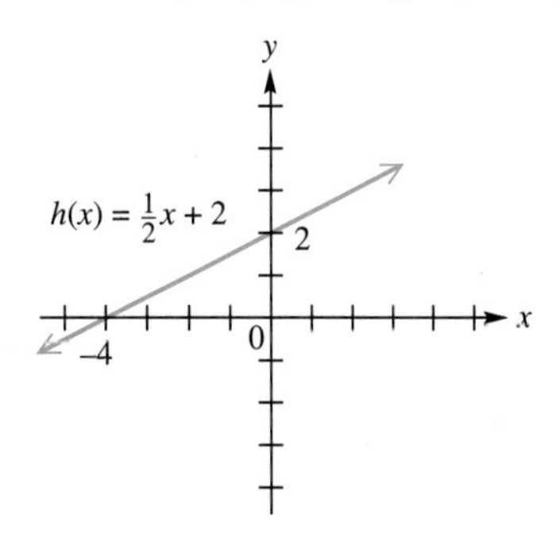

53. domain: $(-\infty, \infty)$; range: $(-\infty, \infty)$

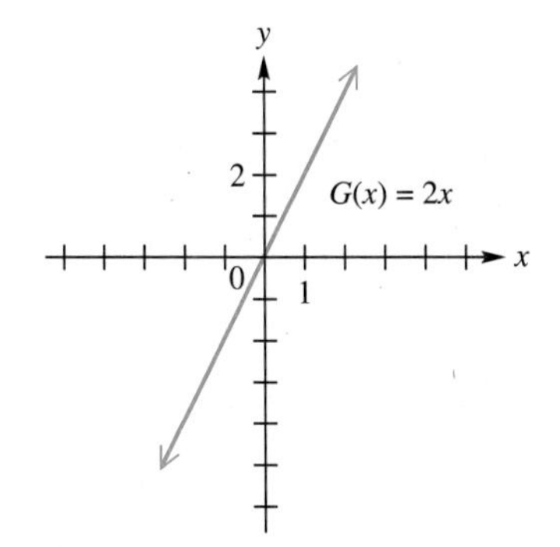

55. domain: $(-\infty, \infty)$; range: $\{5\}$

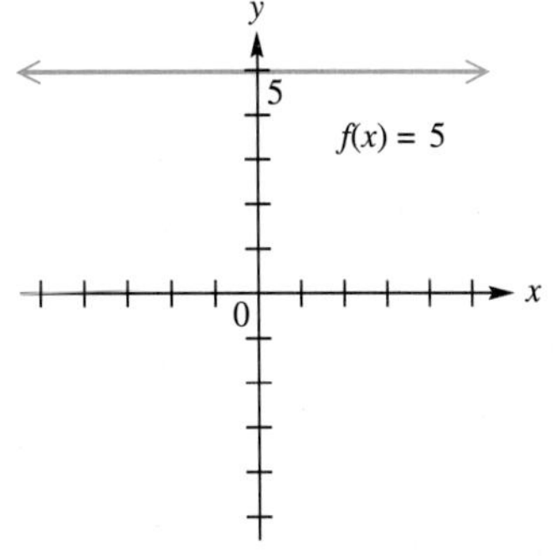

57. $f(x) = 4 - x$; domain: $(-\infty, \infty)$; range: $(-\infty, \infty)$ **59.** $f(x) = -\frac{2}{3}x + 2$; domain: $(-\infty, \infty)$; range: $(-\infty, \infty)$

61. (a) **63.** **(a)** yes **(b)** {Oct. 12, 15, 16, 17, 18, 19, 22, 23, 24, 25} **(c)** {\$39.50, \$38.00, \$39.00, \$37.00, \$34.00, \$28.00, \$29.00, \$30.00} **(d)** \$30.00 **(e)** October 12; October 22 **65.** 194.53 centimeters **67.** 177.41 centimeters **69.** 1.83 cubic meters **71.** 4.11 cubic meters **73.** $\left\{\frac{7}{5}\right\}$ **75.** $\left\{-\frac{25}{11}\right\}$ **77.** $\{-3\}$ **79.** $\{6\}$

Chapter 7 Review Exercises (page 403)

1. $(0, 5)$; $\left(\frac{10}{3}, 0\right)$; $(2, 2)$; $\left(\frac{14}{3}, -2\right)$

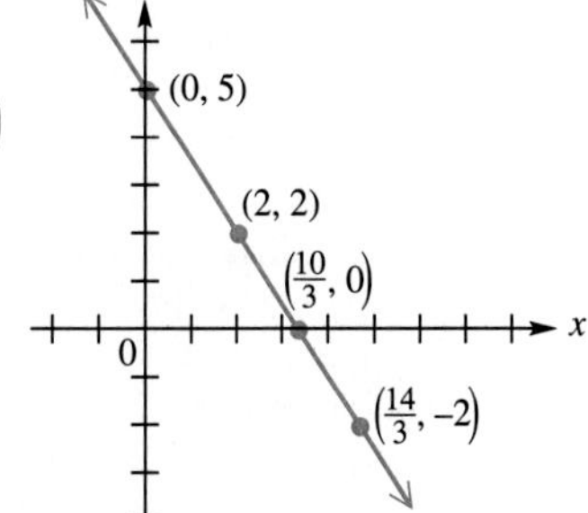

3. $(3, 0)$; $(0, -4)$

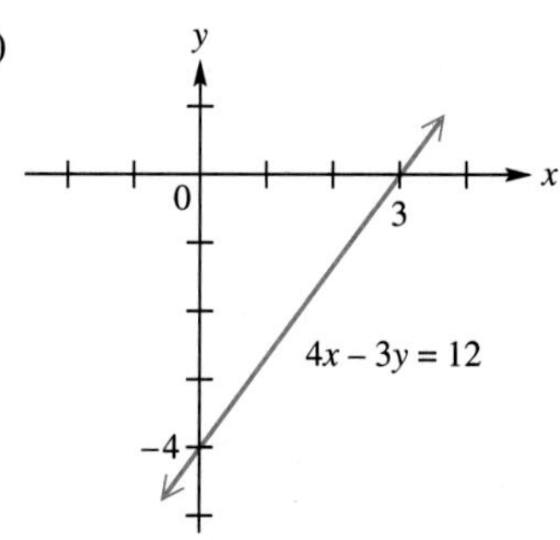

5. $(10, 0)$; $(0, 4)$

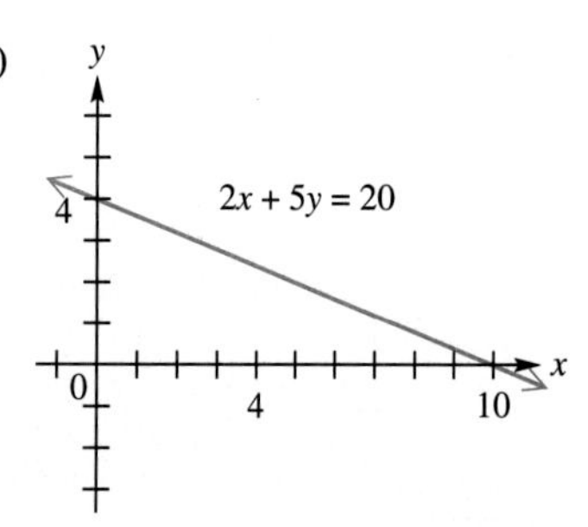

7. $\sqrt{13}$ **9.** $\sqrt{10}$ **11.** $\sqrt{65} + \sqrt{181} + 2\sqrt{17}$
15. $-\frac{1}{2}$ **17.** $\frac{3}{4}$ **19.** $\frac{2}{3}$ **21.** undefined
23. positive **25.** zero **27.** $y = -2$
29. $3x - y = -7$ **31.** $9x + y = 13$
33. $4x - y = 29$ **35.** (b)

37.

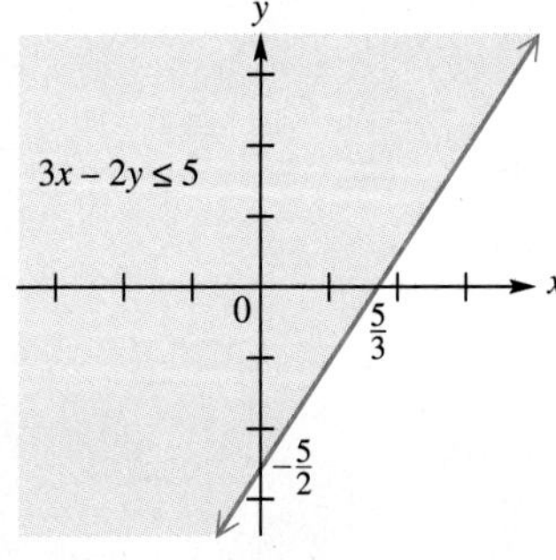

39.

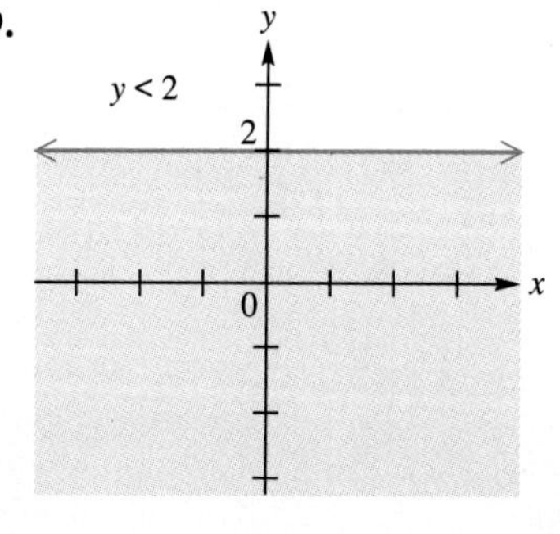

41.

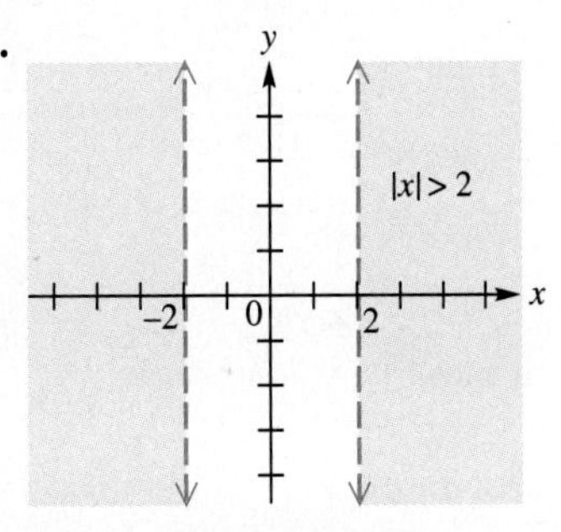

43.

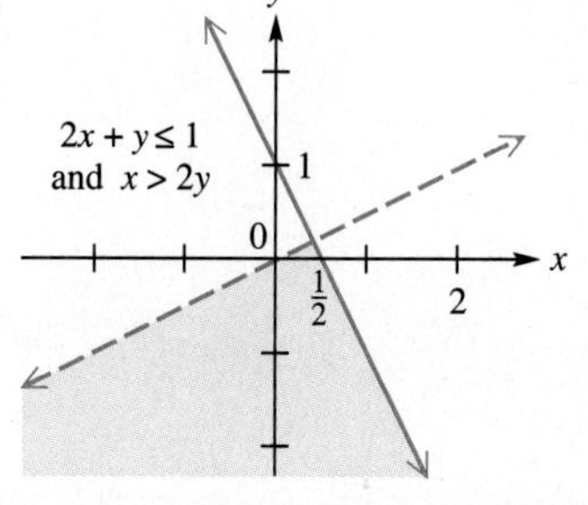

45. 48 **47.** .8 million dollars **49.** domain: {1, 2, 3, 4}; range: {5, 7, 9, 11}; function **51.** domain: {14, 19, 28, 75, 94}; range: {15, 19, 22, 48, 56, 90}; not a function **53.** domain: $(-\infty, -4] \cup [4, \infty)$; range: $(-\infty, \infty)$; not a function **55.** domain: $(-\infty, \infty)$; function **57.** domain: $(-\infty, \infty)$; not a function **59.** -2 **61.** 14 **63.** $k^2 + 2k - 1$ **65.** not a function **67.** function **69.** (c)

Chapter 7 Test (page 406)

[7.2] **1.** $\frac{2}{7}$ [7.1] **2.** $\left(\frac{20}{3}, 0\right)$; $(0, -10)$ **3.** $(8, 0)$; $\left(0, -\frac{8}{5}\right)$

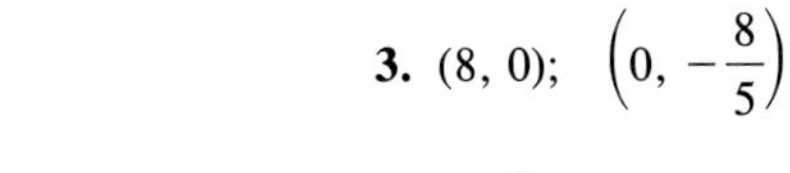

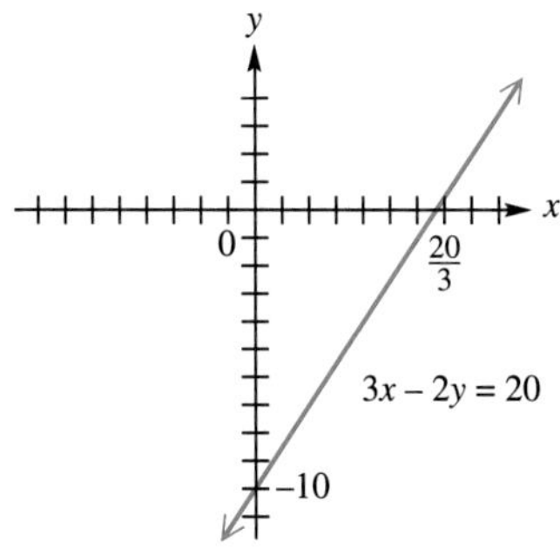

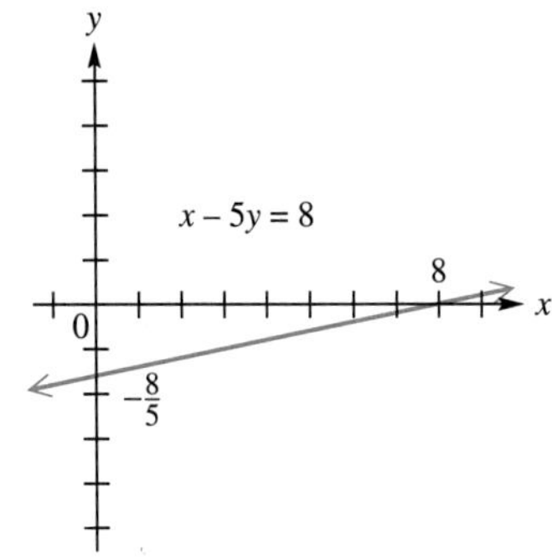

4. none; $(0, 3)$

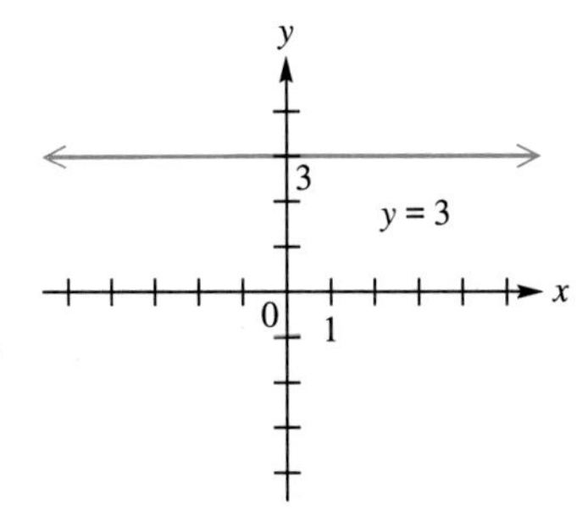

5. $(-2, 0)$; none

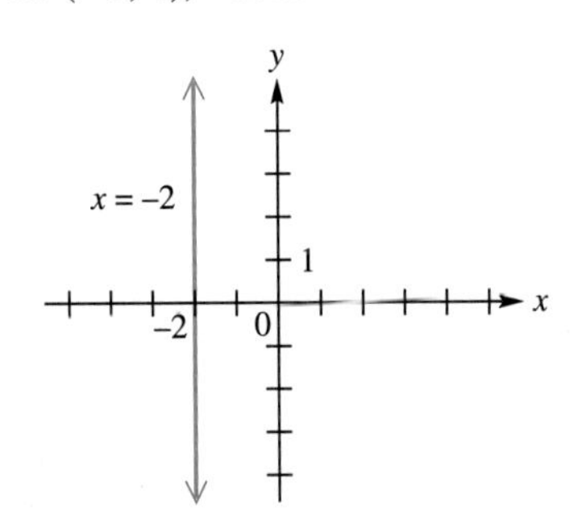

6. $7\sqrt{2}$ **7.** perpendicular **8.** neither [7.3] **9.** $3x + y = 11$ **10.** $5x + 8y = 41$ **11.** $x + 2y = 10$ **12.** $y = 4$ **13.** $5x - 3y = -41$ **14.** (b)

[7.2] **15.**

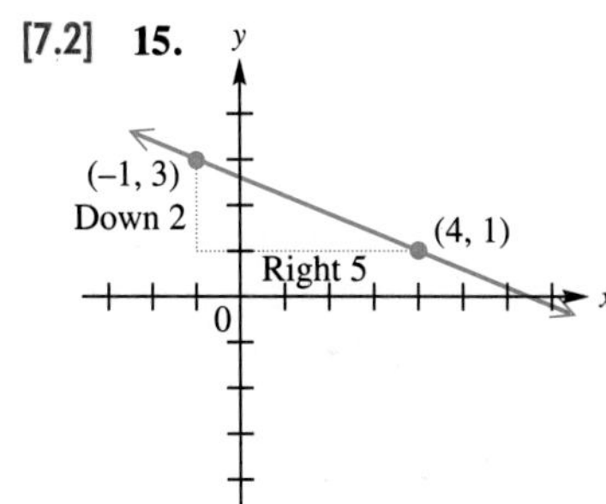

16.

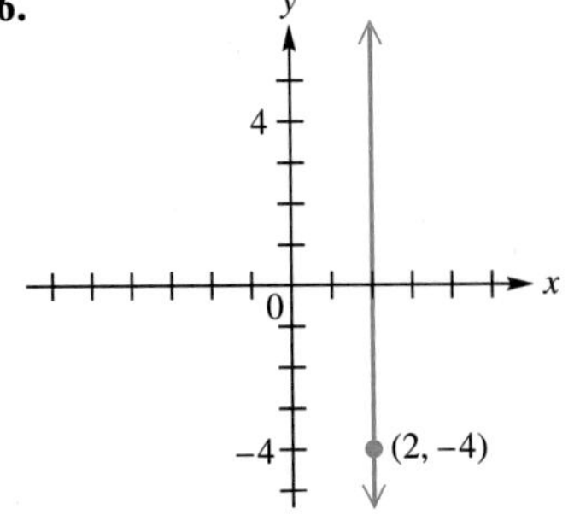

[7.4] **17.**

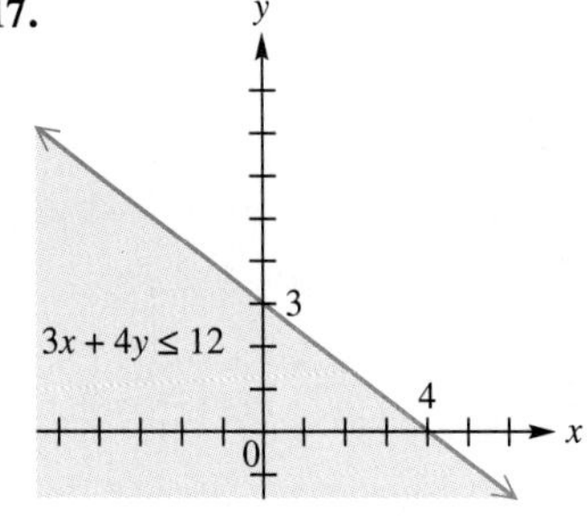

18.

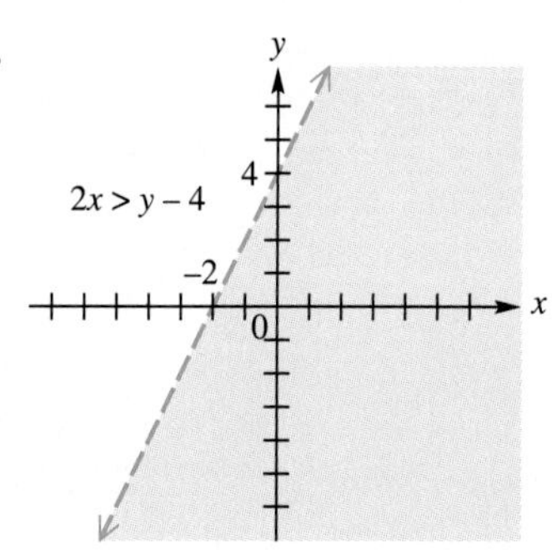

19.

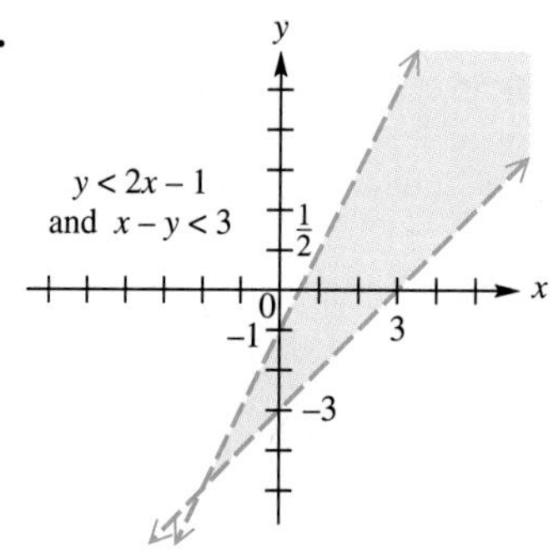

20.

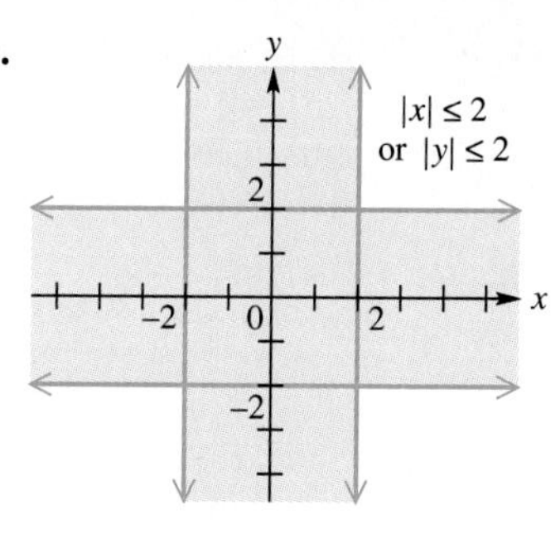

[7.5] **21.** $\dfrac{5}{3}$ **22.** 200 amps [7.6] **23.** (d) **24.** 0 **25.** domain: $(-\infty, \infty)$; range: $(-\infty, \infty)$

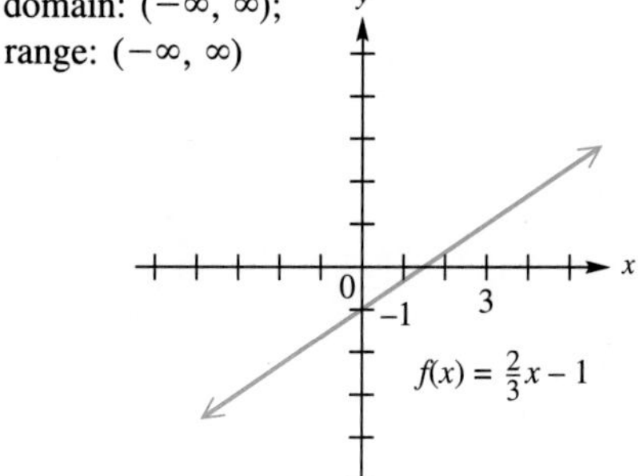

CHAPTER 8

8.1 Exercises (page 413)

1. (a) $10x + 2$ **(b)** $-2x - 4$ **(c)** $24x^2 + 6x - 3$ **(d)** $(4x - 1)/(6x + 3)$; All domains are $(-\infty, \infty)$, except for f/g which is $\left(-\infty, -\dfrac{1}{2}\right) \cup \left(-\dfrac{1}{2}, \infty\right)$. **3. (a)** $4x^2 - 4x + 1$ **(b)** $2x^2 - 1$ **(c)** $(3x^2 - 2x)(x^2 - 2x + 1)$ **(d)** $(3x^2 - 2x)/(x^2 - 2x + 1)$; All domains are $(-\infty, \infty)$, except for f/g which is $(-\infty, 1) \cup (1, \infty)$. **5. (a)** $\sqrt{2x + 5} + \sqrt{4x - 9}$ **(b)** $\sqrt{2x + 5} - \sqrt{4x - 9}$ **(c)** $\sqrt{(2x + 5)(4x - 9)}$ **(d)** $\sqrt{(2x + 5)/(4x - 9)}$; All domains are $[9/4, \infty)$, except f/g which is $(9/4, \infty)$. **7.** 55 **9.** 1848 **11.** $-\dfrac{6}{7}$ **13.** $4m^2 + 6m + 1$ **15.** 1122 **17.** 97 **19.** $256k^2 + 48k + 2$ **21.** $24x + 4$; $24x + 35$; both domains are $(-\infty, \infty)$ **23.** $-5x^2 + 20x + 18$; $-25x^2 - 10x + 6$; both domains are $(-\infty, \infty)$ **25.** $1/x^2$; $1/x^2$; domain of $f \circ g$: $(-\infty, 0) \cup (0, \infty)$; domain of $g \circ f$: $(-\infty, 0) \cup (0, \infty)$ **27.** $2\sqrt{2x - 1}$; $8\sqrt{x + 2} - 6$; domain of $f \circ g$: $\left[\dfrac{1}{2}, \infty\right)$; domain of $g \circ f$: $[-2, \infty)$ **29.** $x/(2 - 5x)$; $2(x - 5)$; domain of $f \circ g$: $(-\infty, 0) \cup \left(0, \dfrac{2}{5}\right) \cup \left(\dfrac{2}{5}, \infty\right)$; domain of $g \circ f$: $(-\infty, 5) \cup (5, \infty)$ **31.** 4 **33.** 0 **35.** 1 **37.** 2 Other correct answers are possible for Exercises 45–49. **45.** $f(x) = x^2$; $g(x) = 6x - 2$ **47.** $f(x) = \sqrt{x}$; $g(x) = x^2 - 1$ **49.** $f(x) = \dfrac{x^2 + 1}{5 - x^2}$; $g(x) = x - 2$ **51.** $D(c) = \dfrac{-c^2 + 10c - 25}{25} + 500$ **53.** $(A \circ r)(t) = 4\pi t^2$

57.

x	-3	0	3
y	-3	0	3

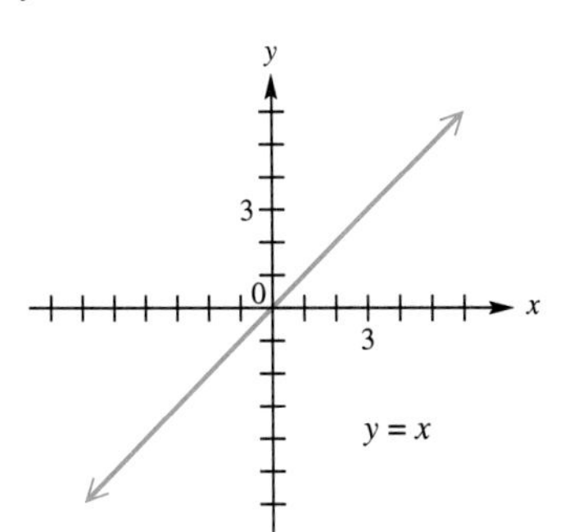

59.

x	-2	0	2
y	-5	-3	-1

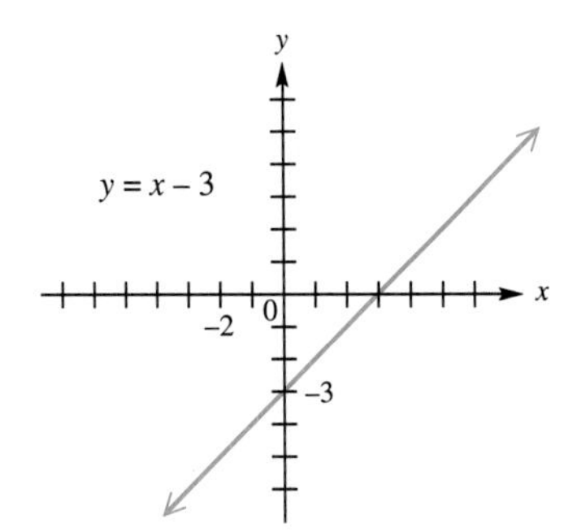

61.

x	-4	0	2
y	7	5	3

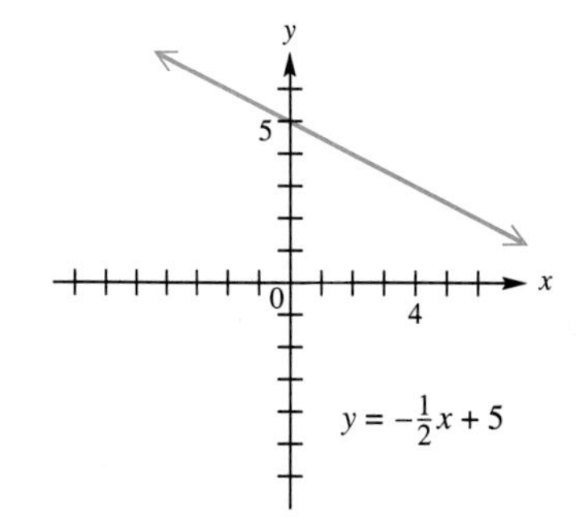

8.2 Exercises (page 420)

1. $(0, -5)$ **3.** $(0, 0)$ **5.** $(-4, 0)$ **7.** $(9, 12)$ **9.** (t, r) **11.** upward; same **13.** downward; narrower **15.** upward; wider **19.** H **21.** B **23.** D **25.** F

27.

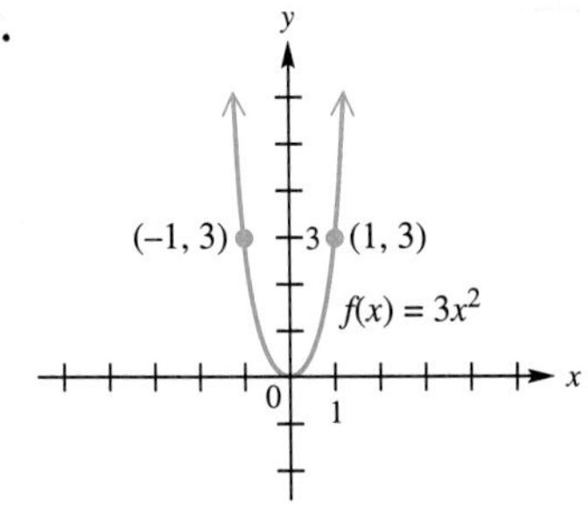

29.

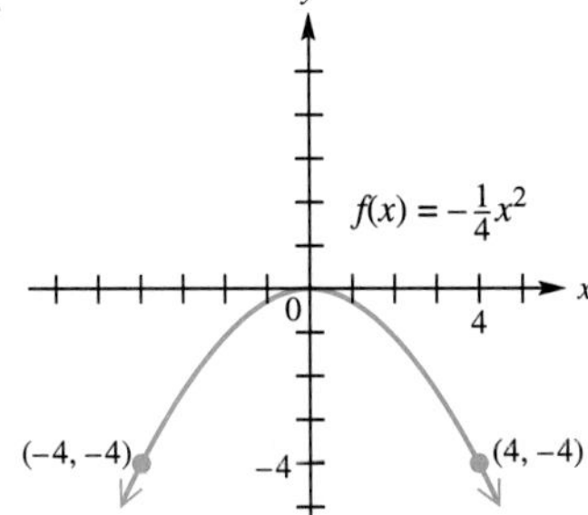

31.

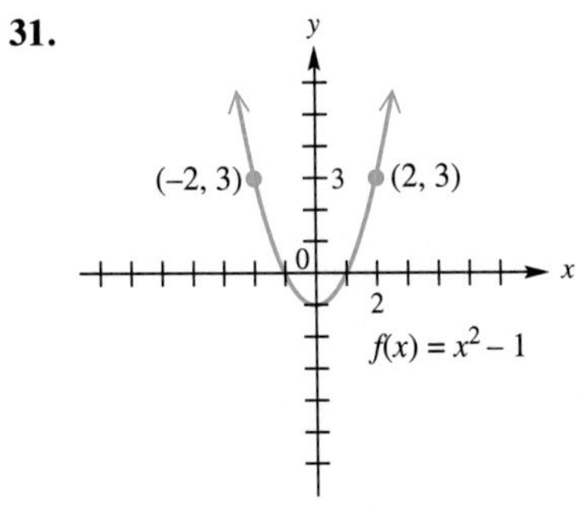

33.

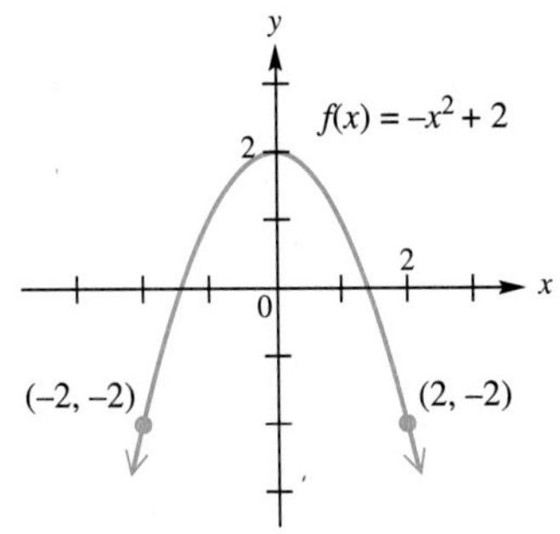

35.

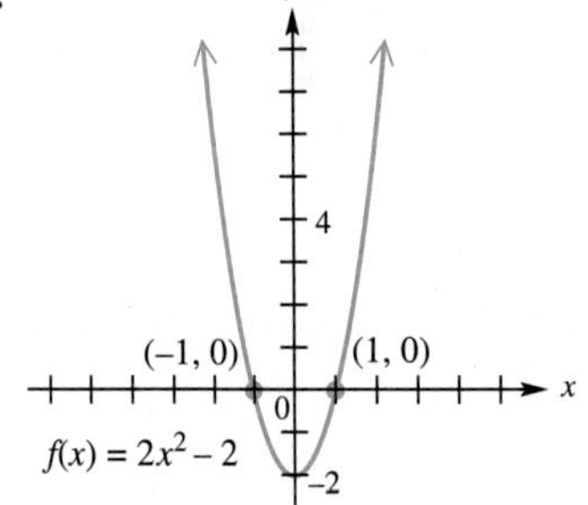

37.

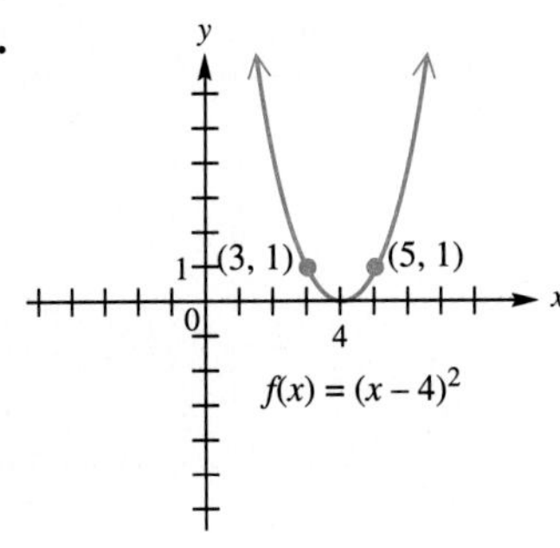

39.

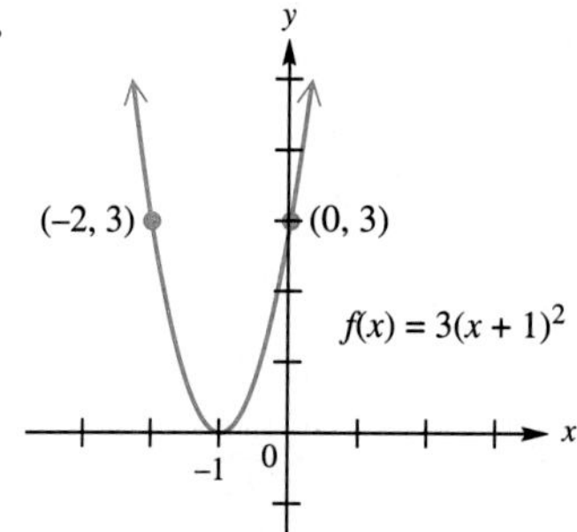

41.

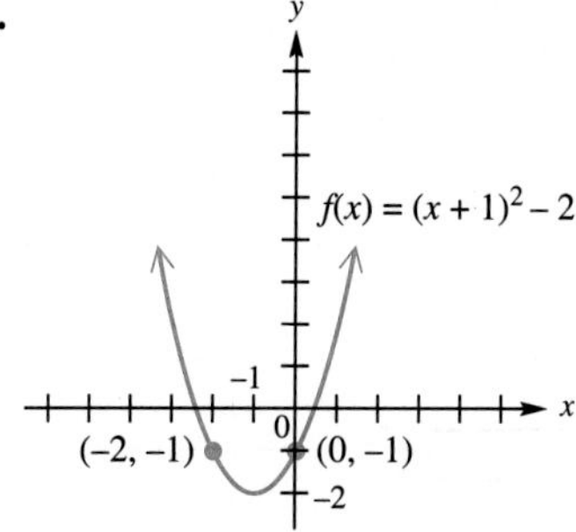

43.

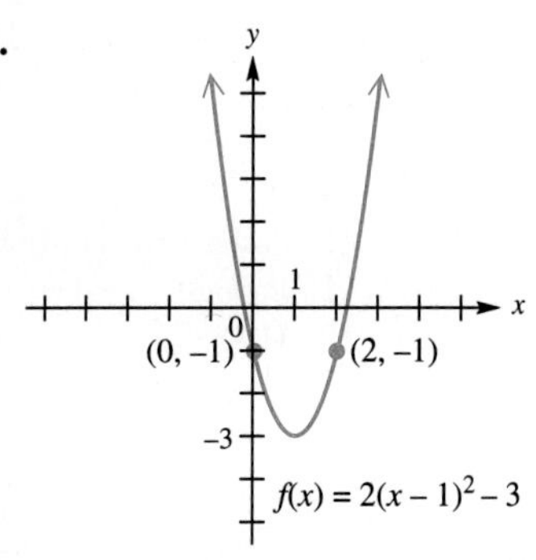

45.

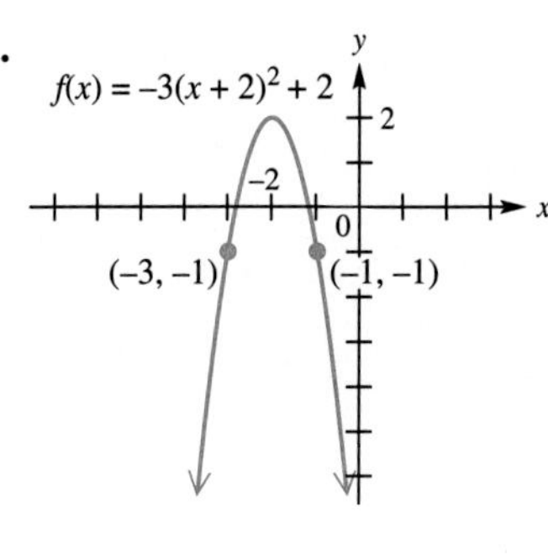

47.

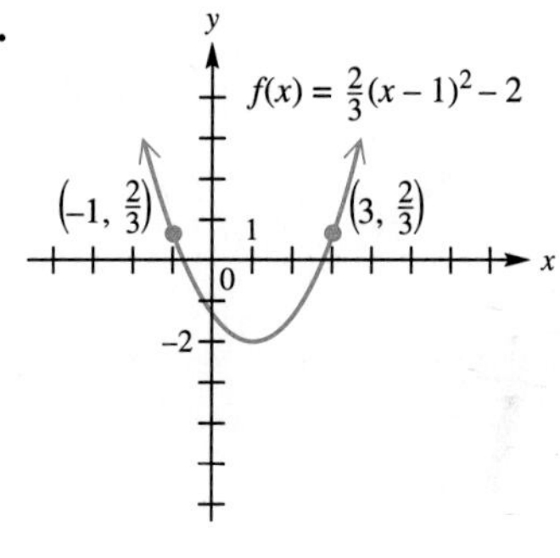

49. $x^2 = 8y$ **51.** $x^2 = -4y$ **53.** $x^2 = 4(y - 3)$ or $x^2 = 4y - 12$ **55.** $\{-3 + 2\sqrt{3}, -3 - 2\sqrt{3}\}$

57. $\left\{\dfrac{6 + \sqrt{46}}{2}, \dfrac{6 - \sqrt{46}}{2}\right\}$ **59.** $\left\{\dfrac{-3 + \sqrt{17}}{2}, \dfrac{-3 - \sqrt{17}}{2}\right\}$

8.3 Exercises (page 429)

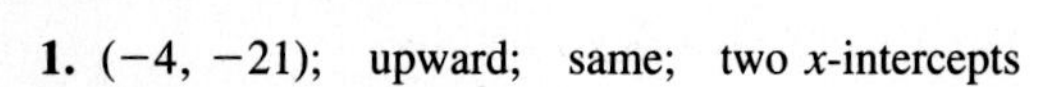

1. $(-4, -21)$; upward; same; two x-intercepts **3.** $\left(\dfrac{3}{2}, \dfrac{21}{4}\right)$; downward; same; two x-intercepts

5. $\left(\dfrac{3}{2}, -\dfrac{3}{2}\right)$; upward; narrower; two x-intercepts **7.** $(-2, -1)$; to the right; narrower **9.** $(1, 2)$; to the right; narrower

15.

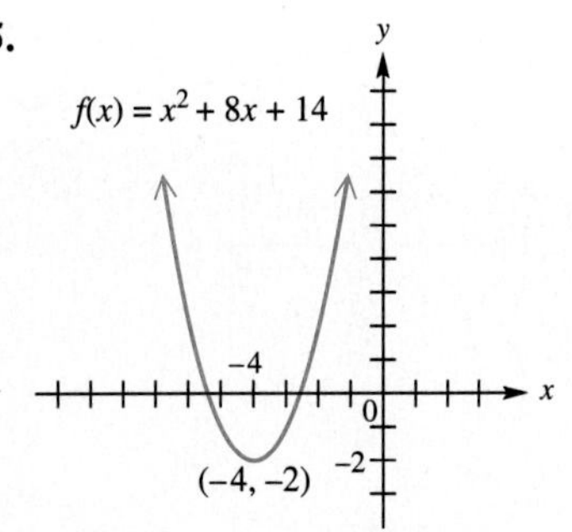

17.

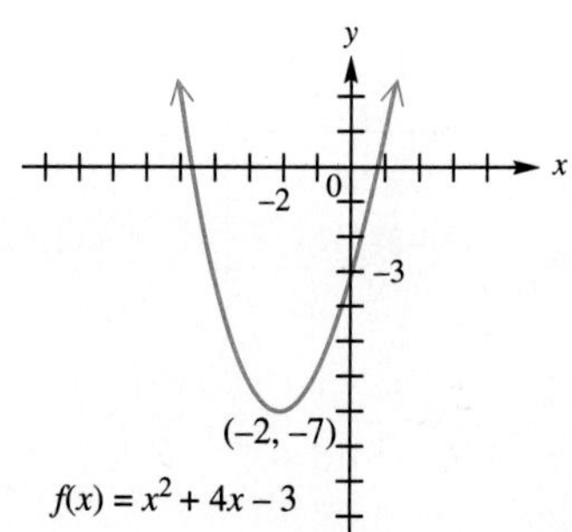

19.

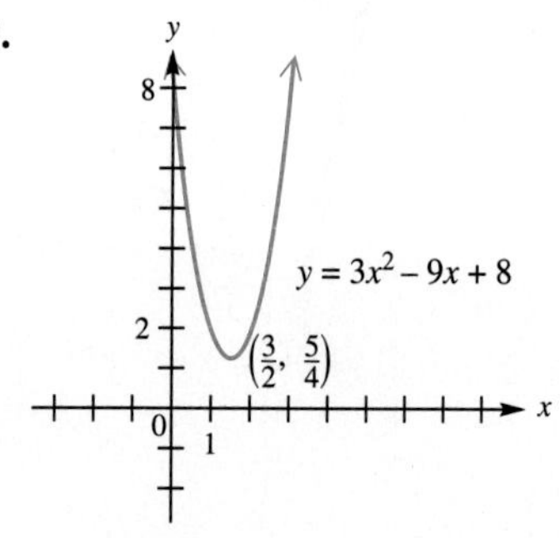

21.

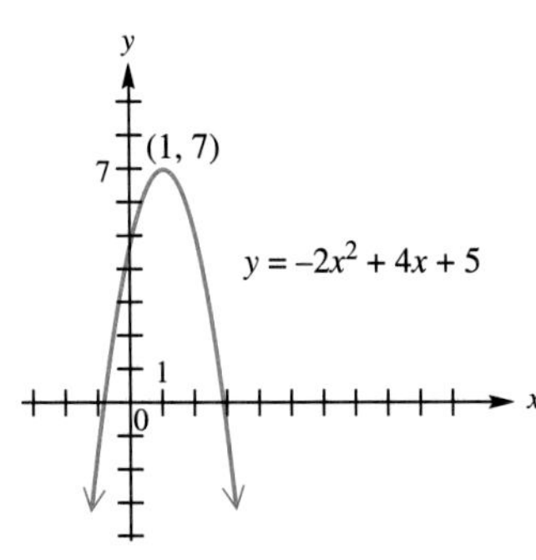

23.

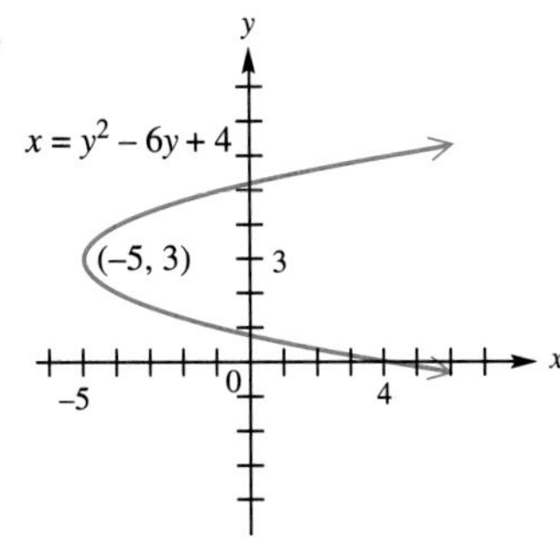

25.

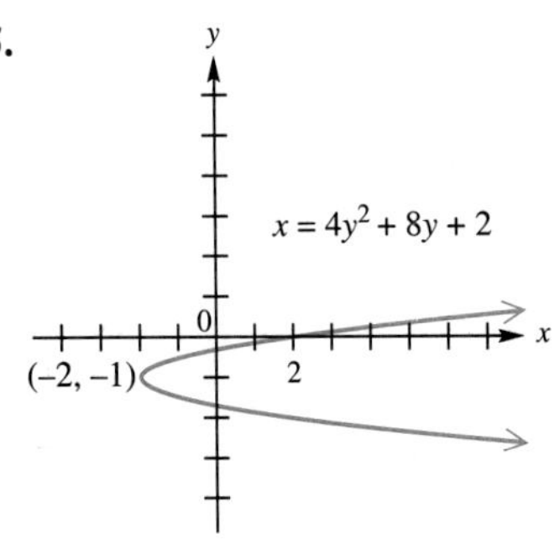

27. 7 units; \$62 **29.** 98; \$48,020 **31.** 2 seconds; 67 feet **33.** 16 feet; 2 seconds **35.** 26 meters **37.** 140 feet by 70 feet; 9800 square feet **39.** $f(-x) = -2x + 3$; $-f(x) = -2x - 3$ **41.** $f(-x) = 4x^2 + 3$; $-f(x) = -4x^2 - 3$ **43.** $f(-x) = -x^2 - 2x$; $-f(x) = x^2 - 2x$

8.4 Exercises (page 435)

1.

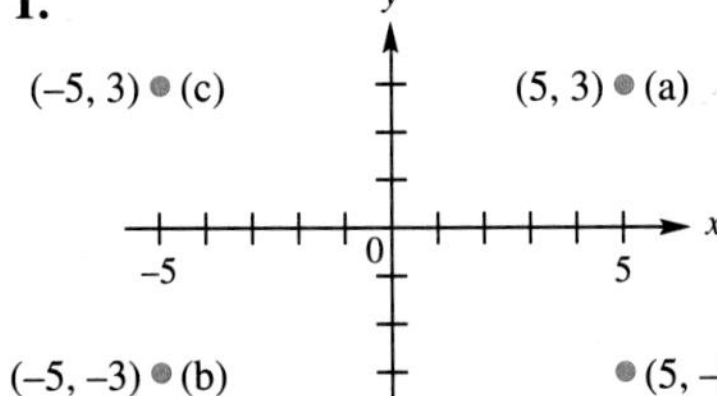

3.

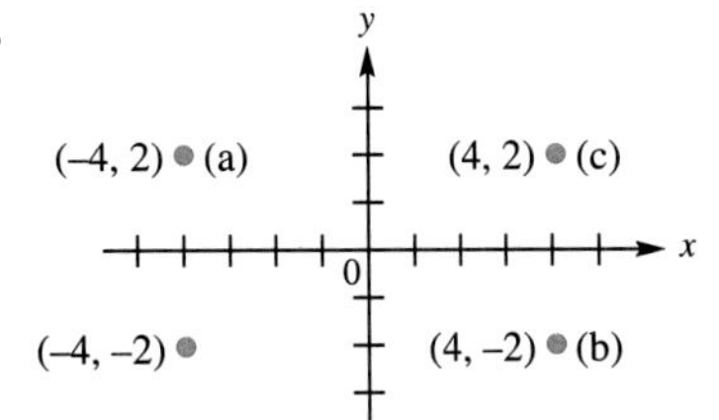

5.

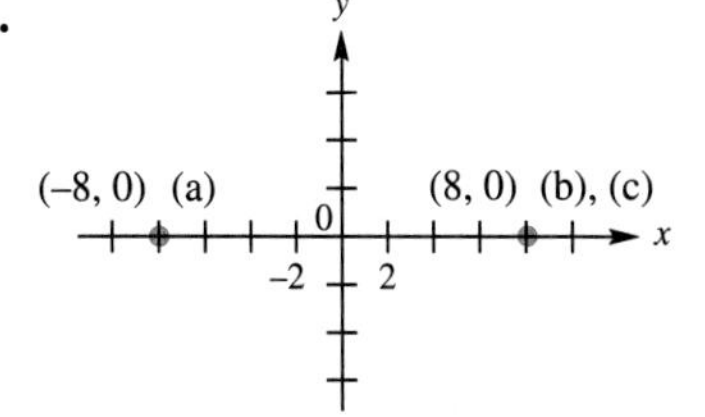

7. symmetric to the x-axis, the y-axis, and the origin **9.** not symmetric to the x-axis, y-axis, or the origin **11.** symmetric to the y-axis **13.** symmetric to the origin **15.** symmetric to the y-axis **17.** symmetric to the origin

19.

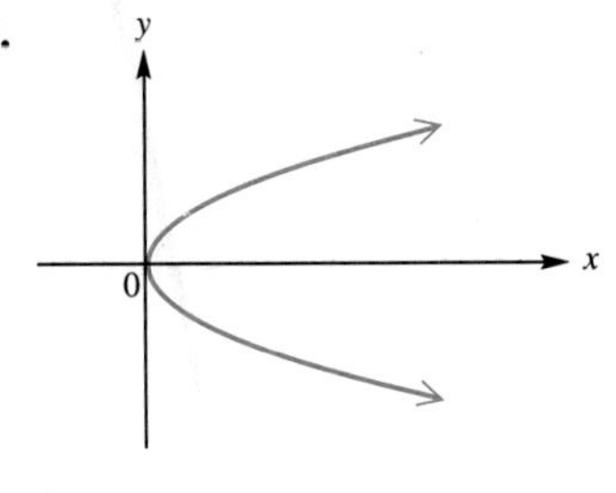

21.

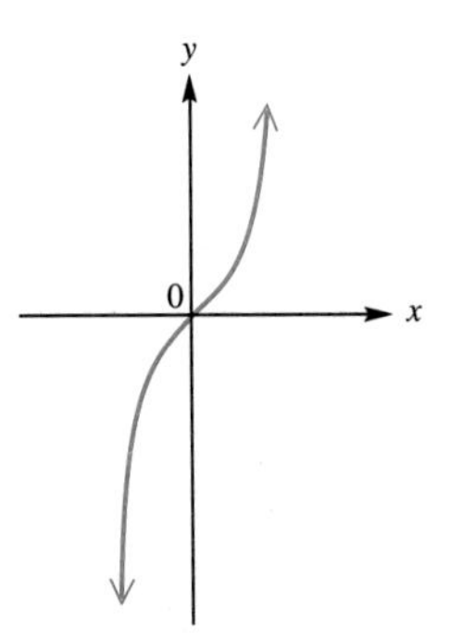

23. increasing on $(-\infty, -3)$; decreasing on $(0, \infty)$
25. increasing on $(-\infty, -2)$ and $(1, \infty)$; decreasing on $(-2, 1)$
27. decreasing on $(-\infty, 0)$; increasing on $(0, \infty)$
29. **(a)** symmetric **(b)** symmetric
31. **(a)** not symmetric **(b)** symmetric
33. **(a)** symmetric **(b)** symmetric
35. $f(4) = 3$

37.

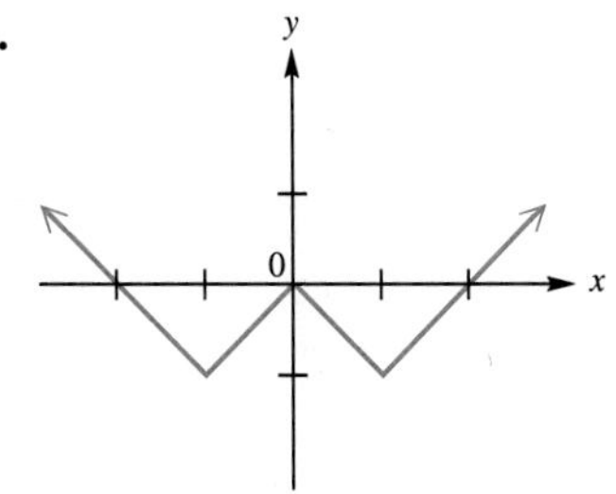

41. $2\sqrt{5}$
43. $\sqrt{234}$
45. $\sqrt{(x + 4)^2 + (y + 3)^2}$

8.5 Exercises (page 442)

1. $(x-2)^2+(y-4)^2=25$ **3.** $x^2+(y-3)^2=2$ **5.** $x^2+(y+1)^2=16$ **7.** $(0, 0)$ **9.** none

11. $(-3, 1)$; 2

13. $(-4, -2)$; 7

15. $(1, -3)$; 3

17.

19.

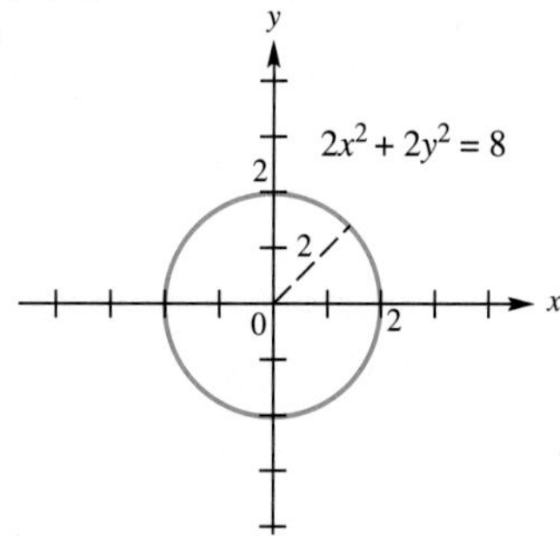

21.

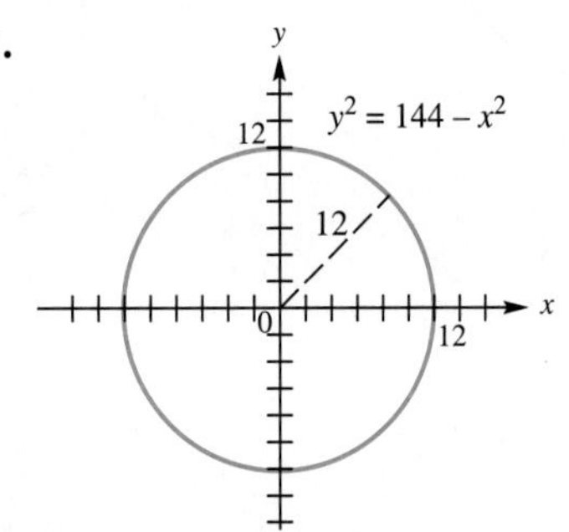

23.

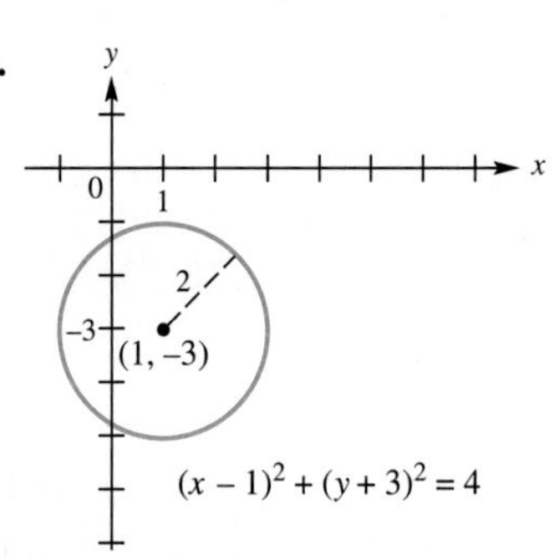

25.

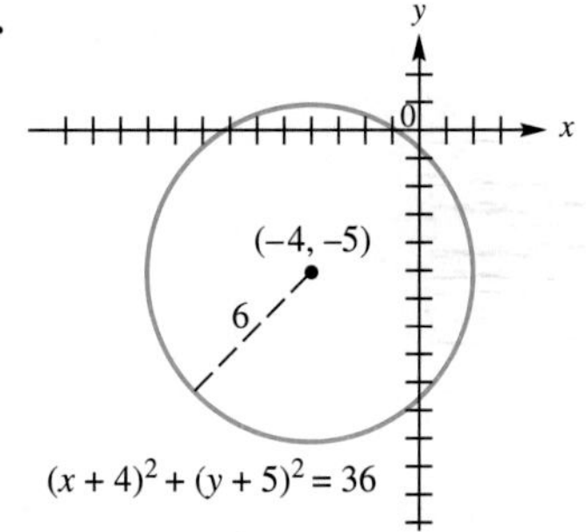

27.

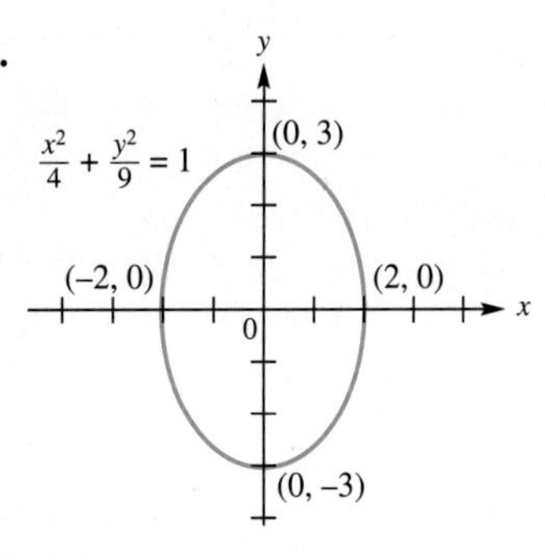

29.

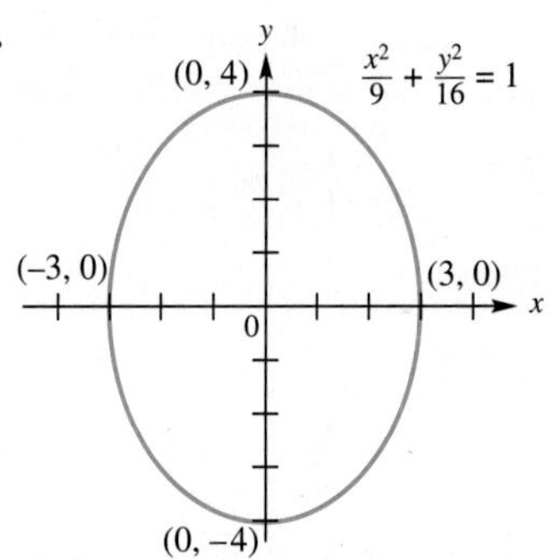

31.

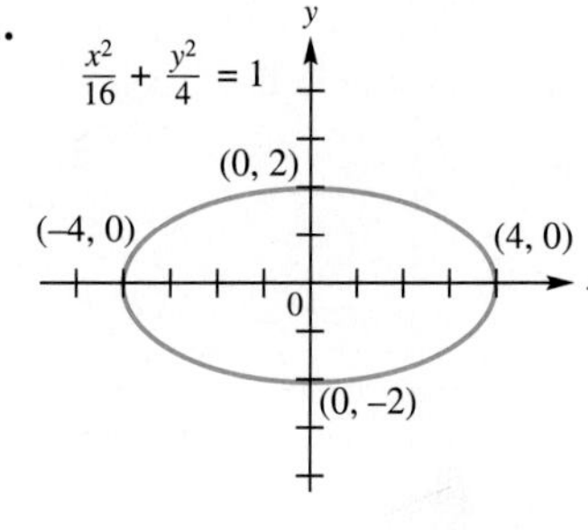

33.

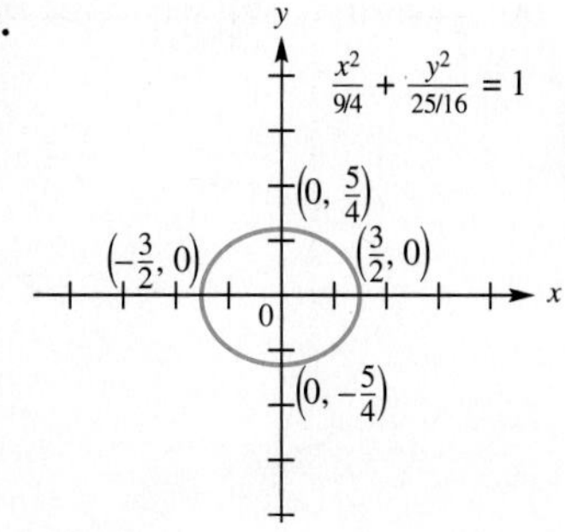

35.

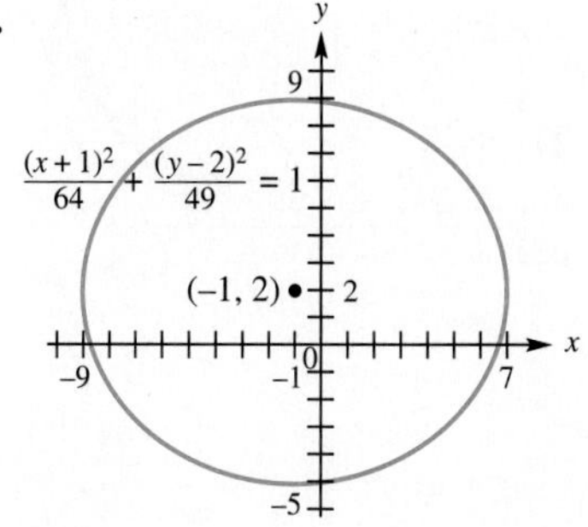

37.

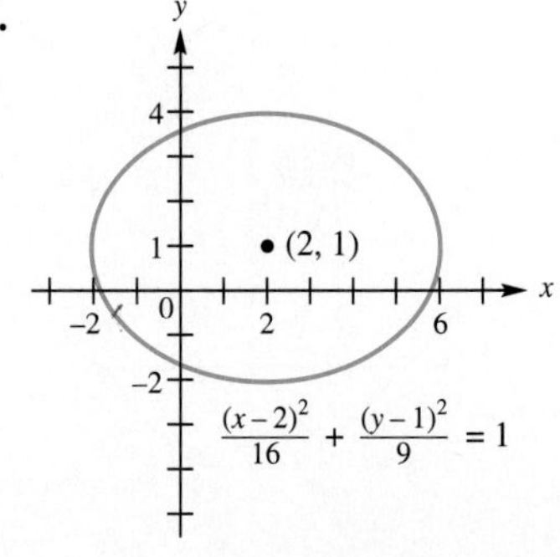

39. $x^2 + y^2 = 8$ **41.** $(x - 4)^2 + (y + 2)^2 = 32$ **43.** $(x - 2)^2 + (y - 4)^2 = 17$ **45.** $(x - 4)^2 + (y - 3)^2 = 16$
49. (a) 77.4 million miles **(b)** 64.2 million miles (Both answers are rounded.) **53.** $(3, 0)$; $(0, 2)$

55. $\left(\frac{5}{4}, 0\right)$; $\left(0, -\frac{5}{3}\right)$ **57.** $\left(\frac{c}{a}, 0\right)$; $\left(0, \frac{c}{b}\right)$ **59.** no x-intercept; $(0, 4)$

8.6 Exercises (page 453)

1.

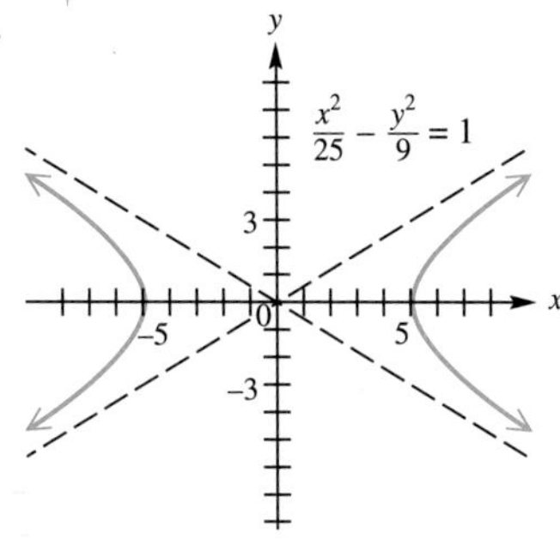

3.

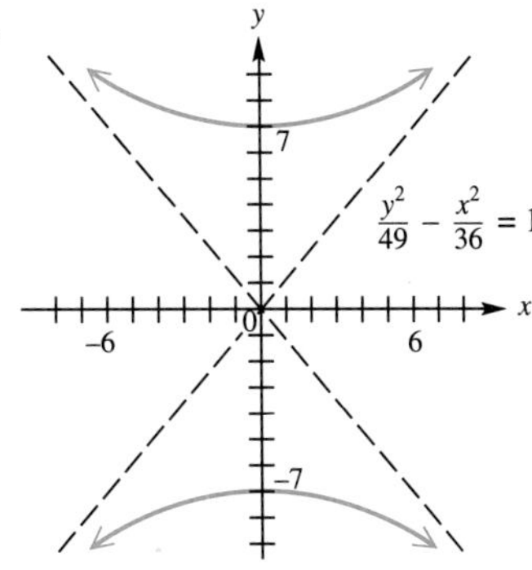

5.

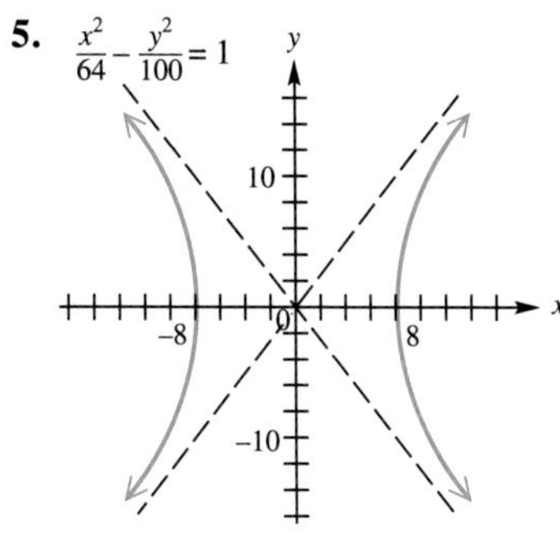

7.

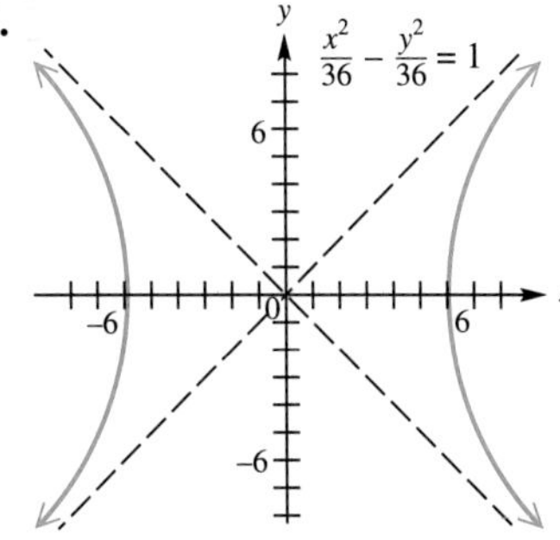

9.

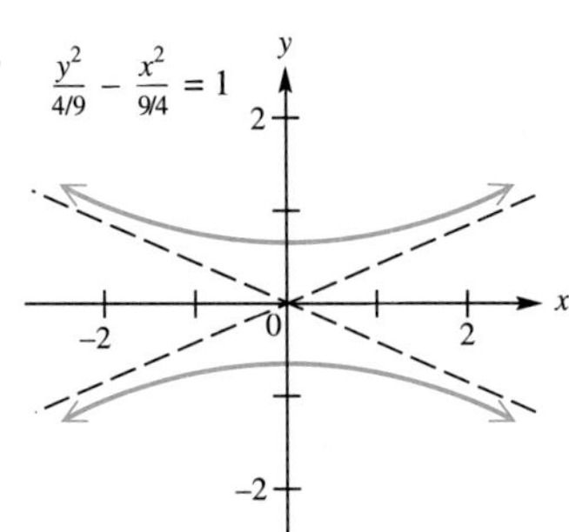

11.

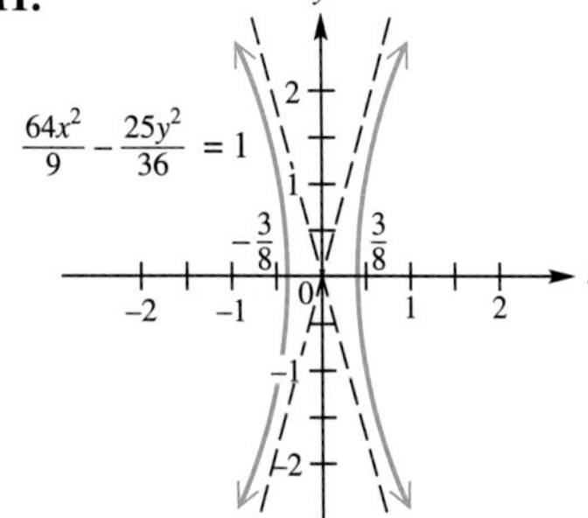

13.

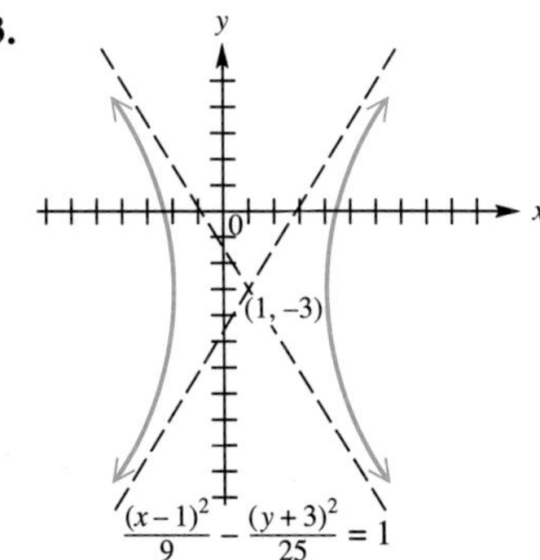

15.

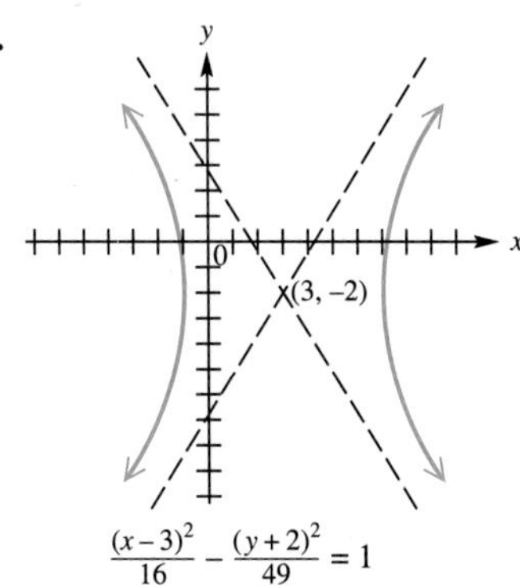

17.

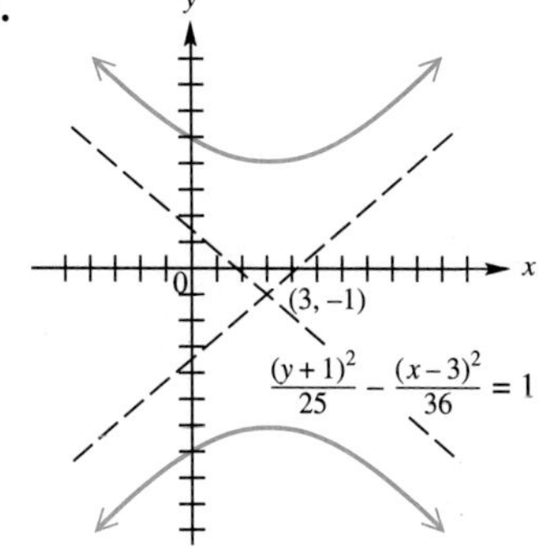

19. circle

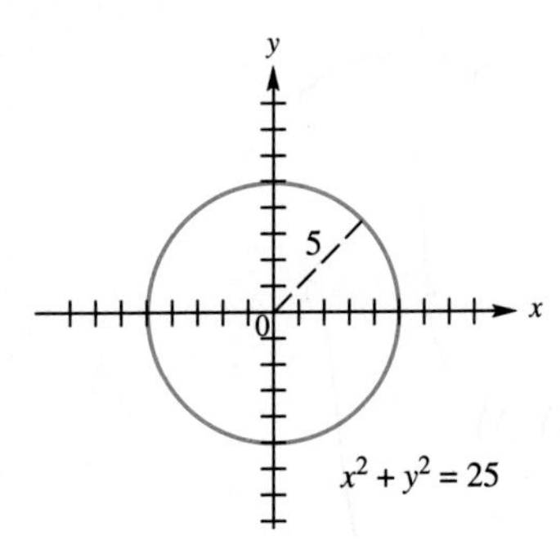

21. ellipse

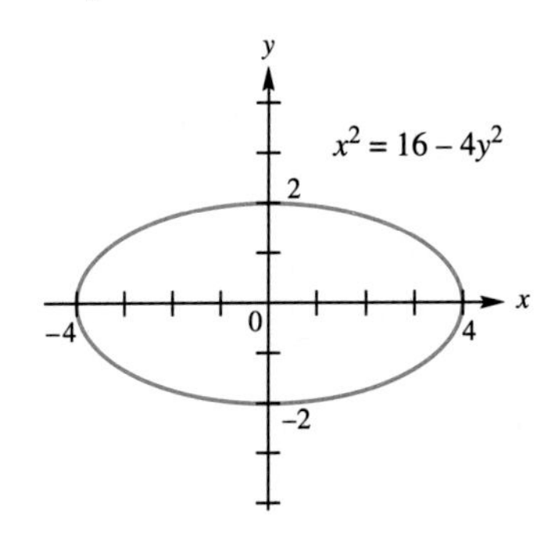

23. parabola

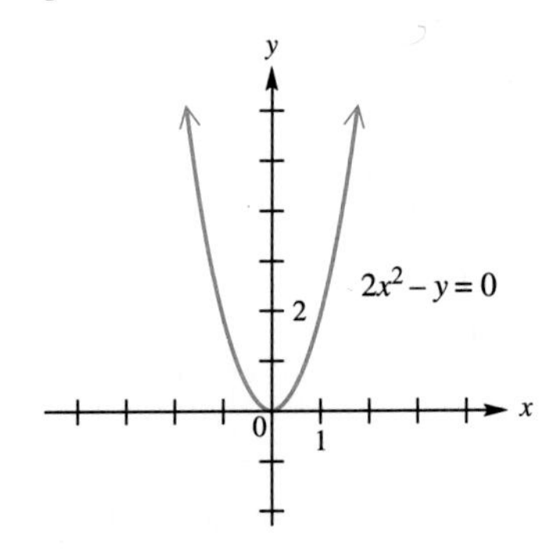

25. circle

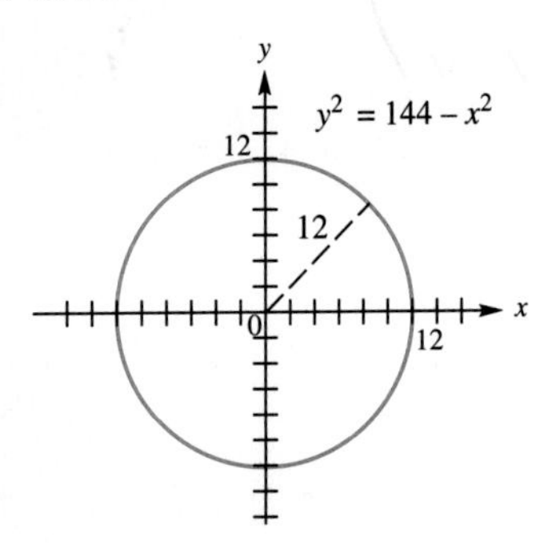

27. ellipse

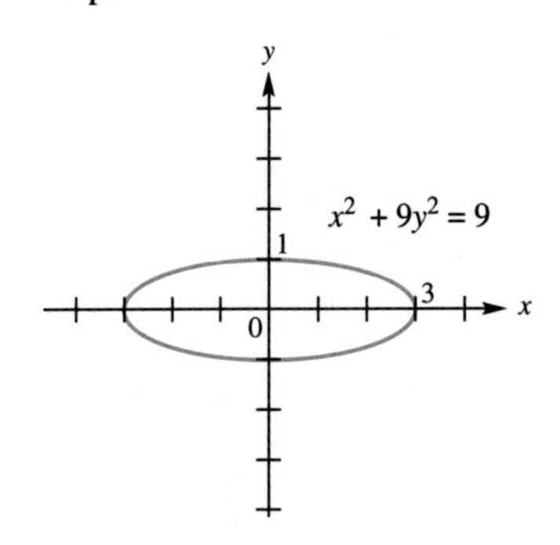

29. hyperbola

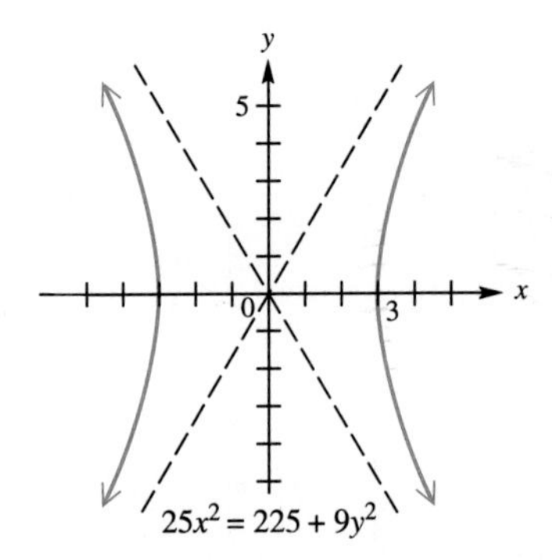

33.

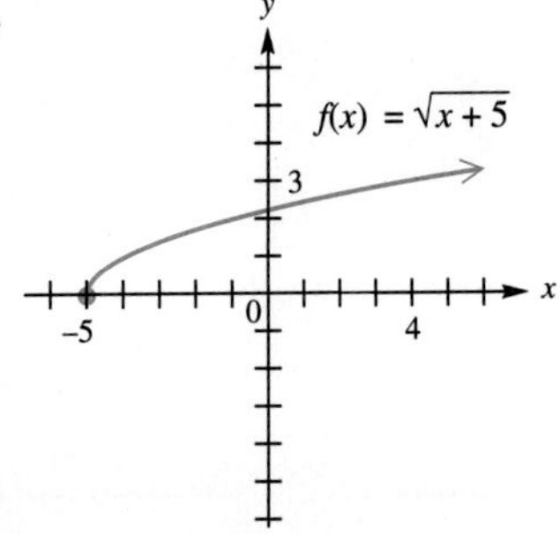

35.

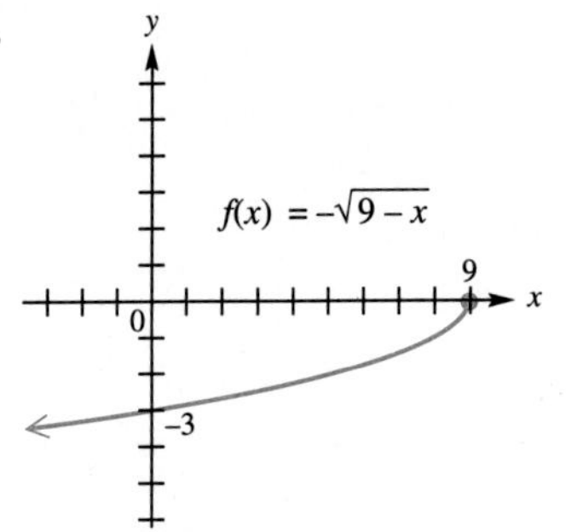

37.

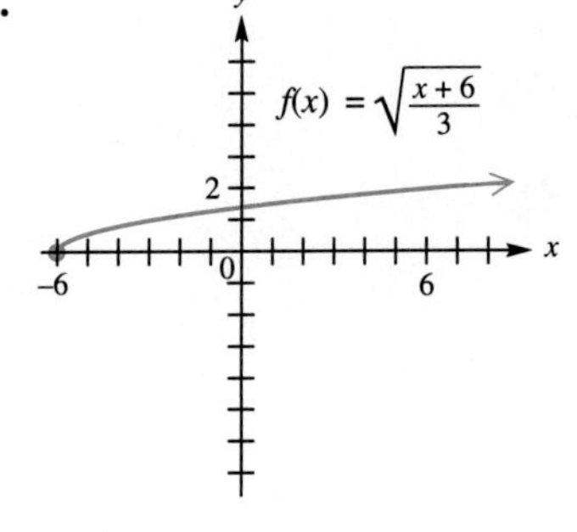

39.

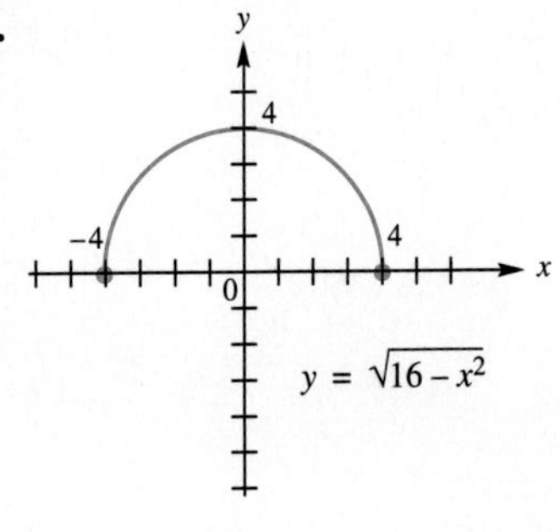

41.

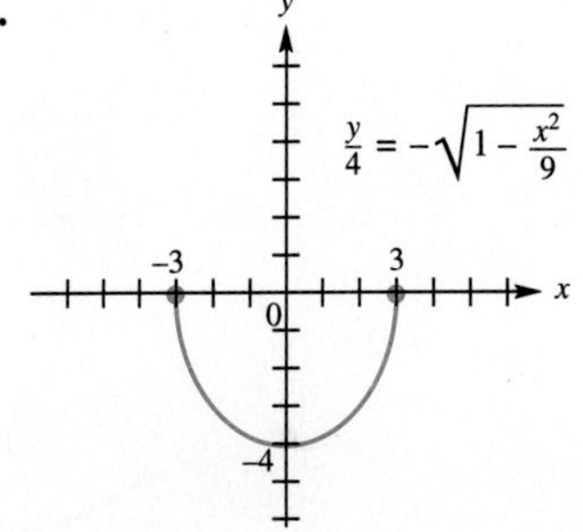

43.

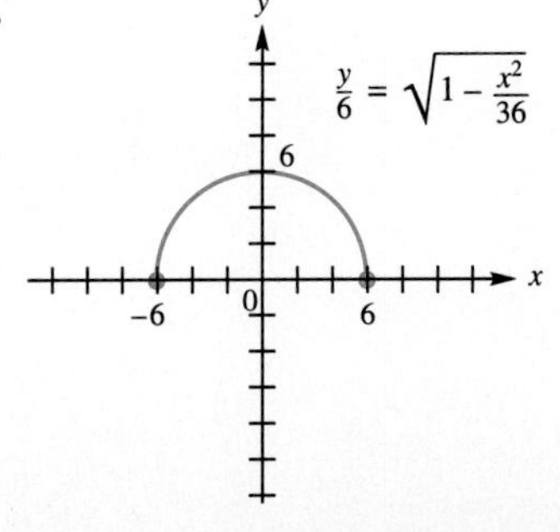

45. **(a)** 10 meters **(b)** 36 meters **47.** **(a)** **(b)**

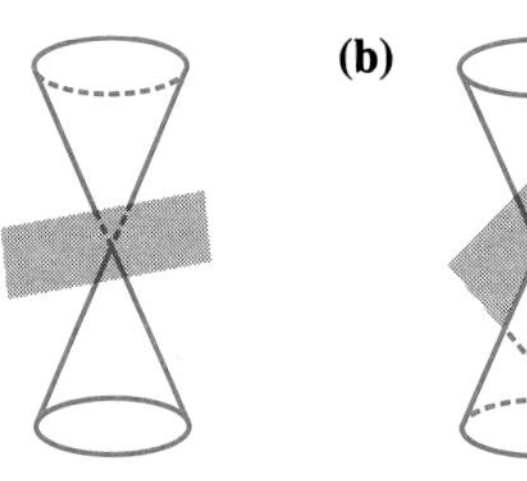

51. 6; 10
53. 2; 0
55. 10; 16

8.7 Exercises (page 459)

1.

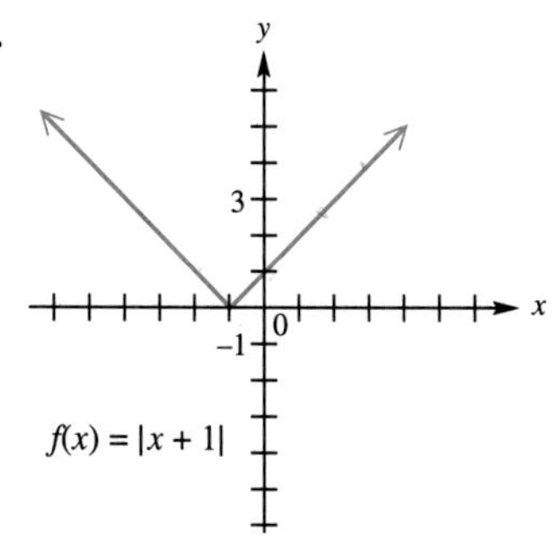

3.

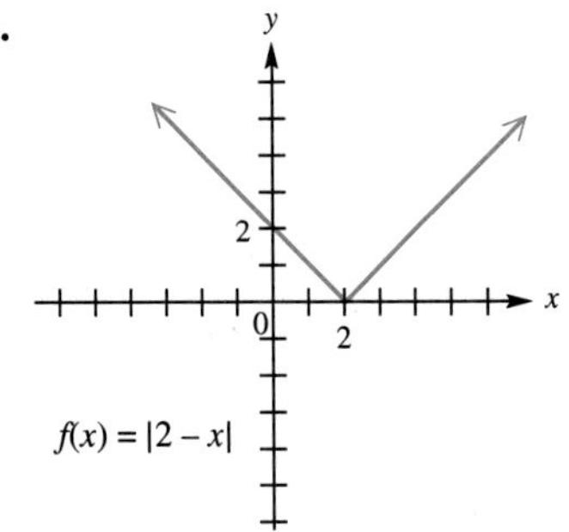

5.

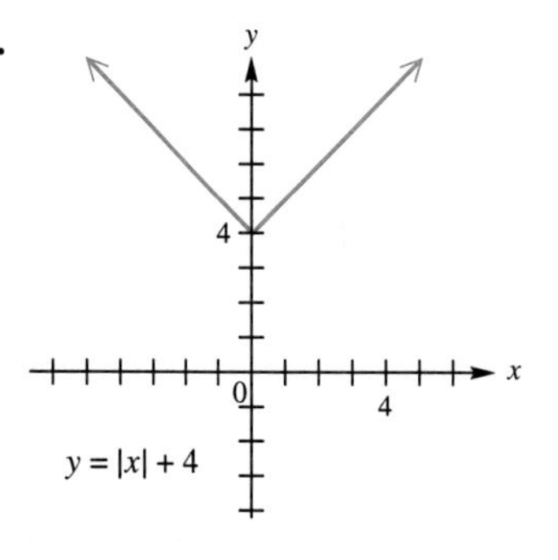

7.

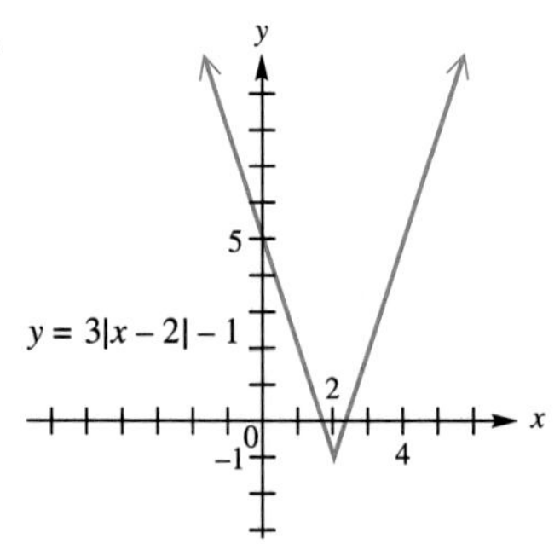

9. **(a)** -10 **(b)** -2 **(c)** -1 **(d)** 2 **(e)** 4
11. **(a)** -10 **(b)** 2 **(c)** 5 **(d)** 3 **(e)** 5

13.

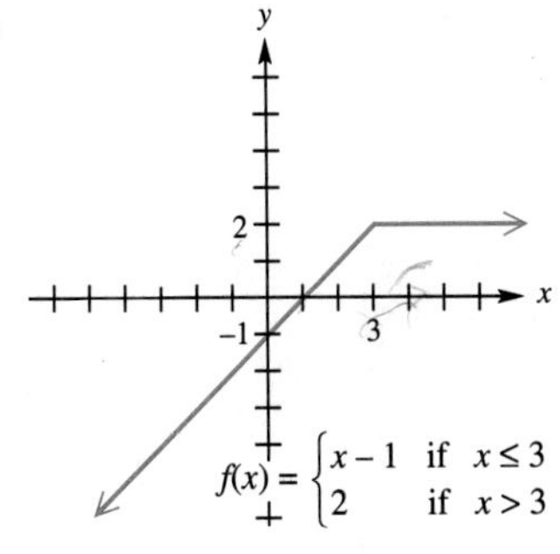

15.

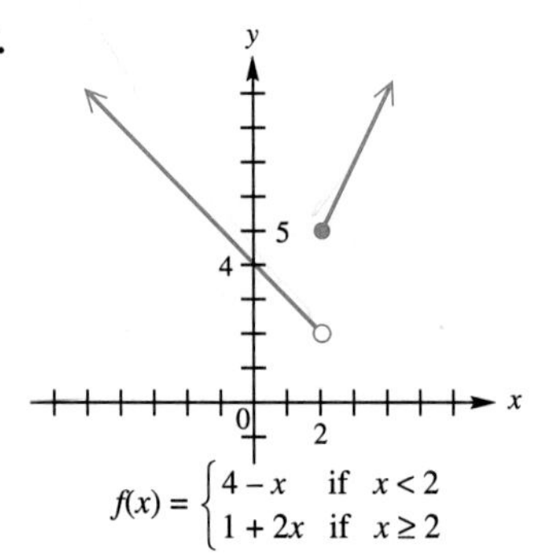

17.

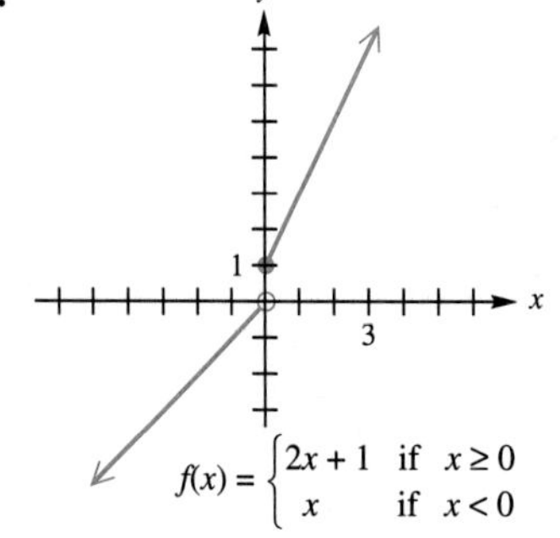

19.

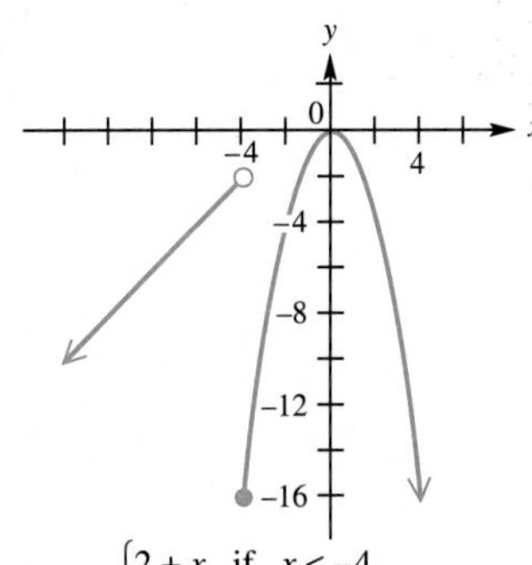

$f(x) = \begin{cases} 2 + x & \text{if } x < -4 \\ -x^2 & \text{if } x \geq -4 \end{cases}$

21.

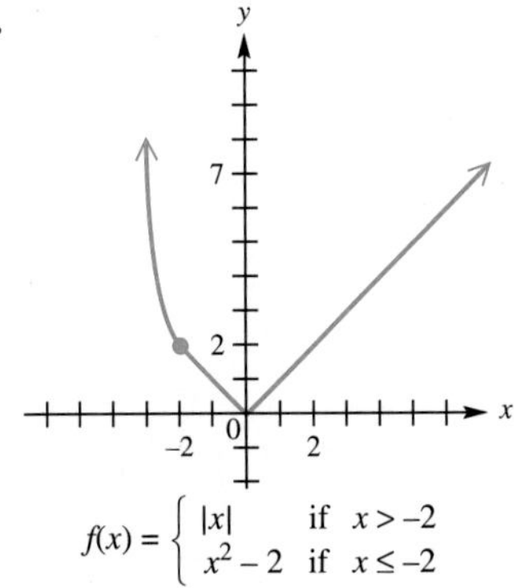

$f(x) = \begin{cases} |x| & \text{if } x > -2 \\ x^2 - 2 & \text{if } x \leq -2 \end{cases}$

23.

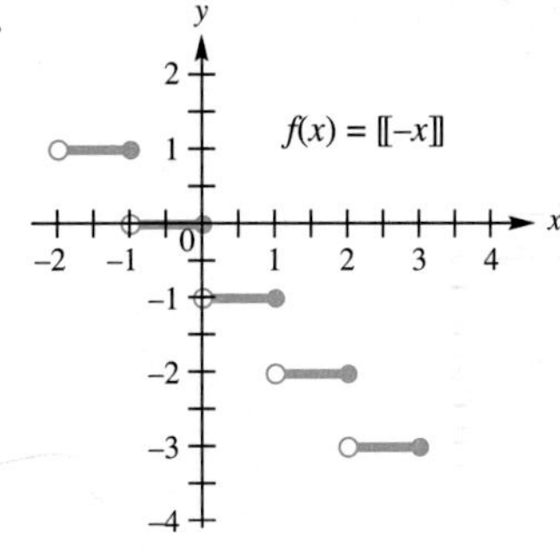

25.

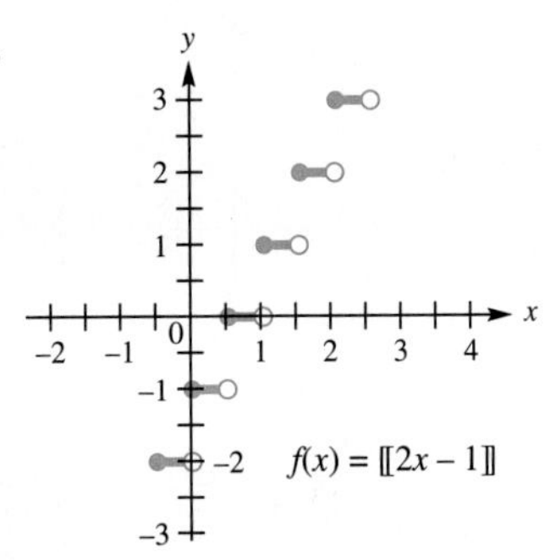

27.

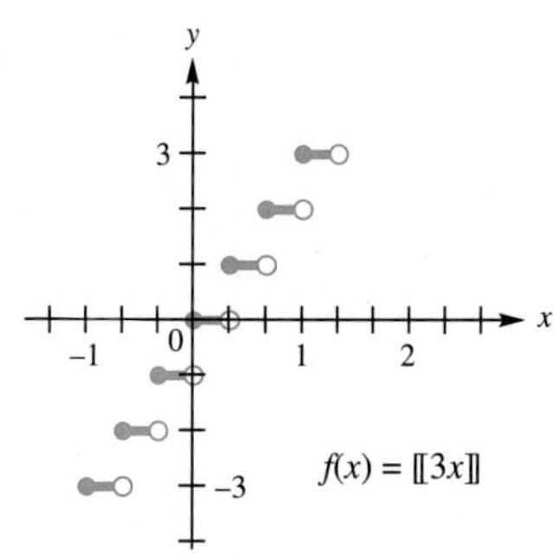

29.

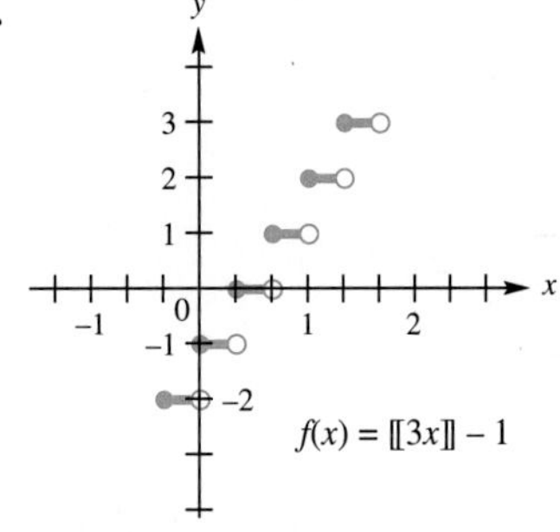

31. **(a)** \$11 **(b)** \$18 **(c)** \$32 **(d)**

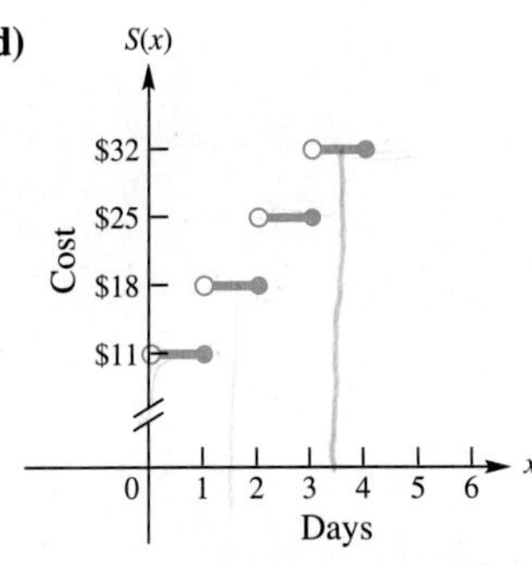

(e) domain: $(0, \infty)$ (at least in theory); range: $\{11, 18, 25, 32, 39, \ldots\}$

33. $-[\![-x]\!]$ ounces

35.

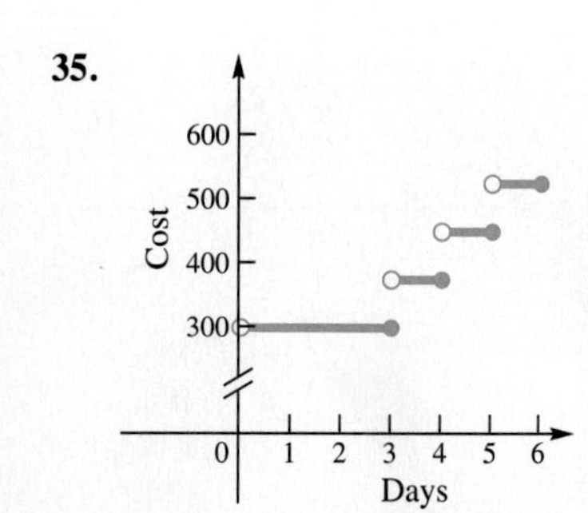

37. **(a)** 140 **(b)** 220 **(c)** 220 **(d)** 220 **(e)** 220 **(f)** 60 **(g)** 60 **(h)**

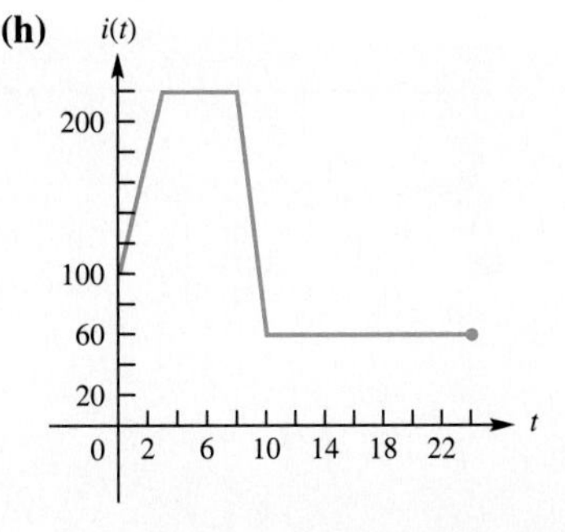

43. $m^2 - m - 2$ **45.** $2p + 3$ **47.** $2z^2 - 2z - 20 - 44/(z - 3)$

Chapter 8 Review Exercises (page 466)

1. $x^2 + 3x + 3$; $(-\infty, \infty)$ **3.** $(x^2 - 2x)(5x + 3)$; $(-\infty, \infty)$ **5.** $5x^2 - 10x + 3$; $(-\infty, \infty)$ **7.** 3 **9.** $7\sqrt{5}$
11. $(4b - 3)(\sqrt{2b})$, $b > 0$ **13.** not real

15.

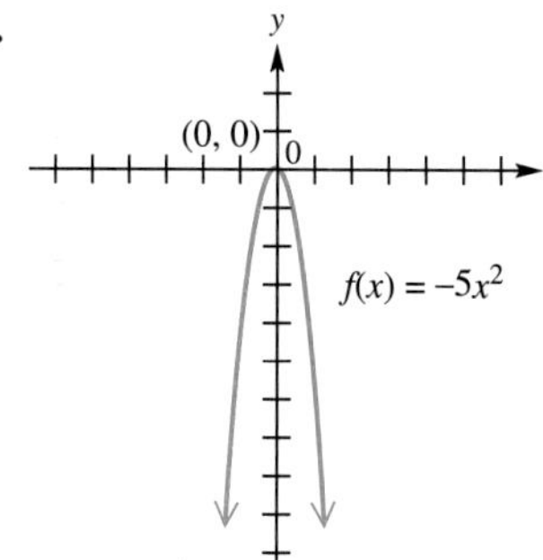

17.

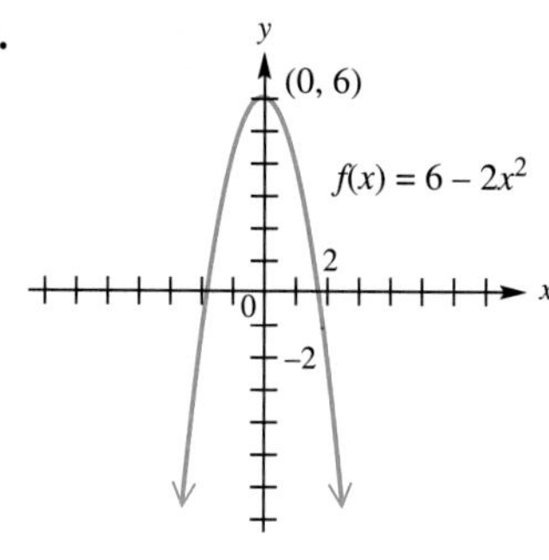

19.

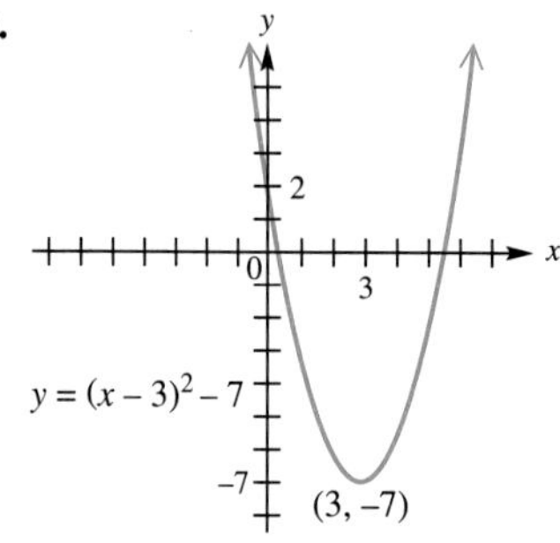

21.

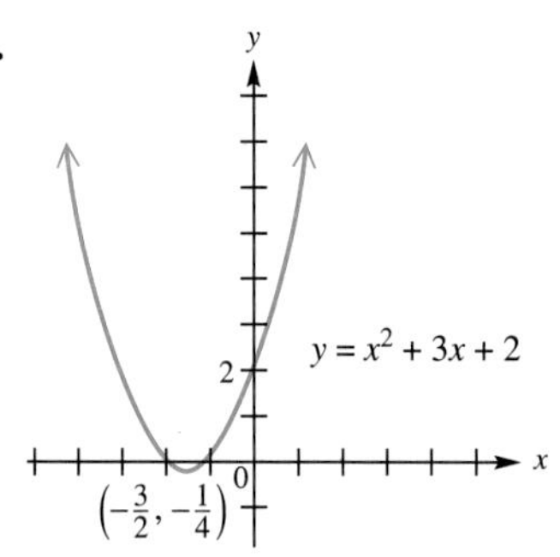

23.

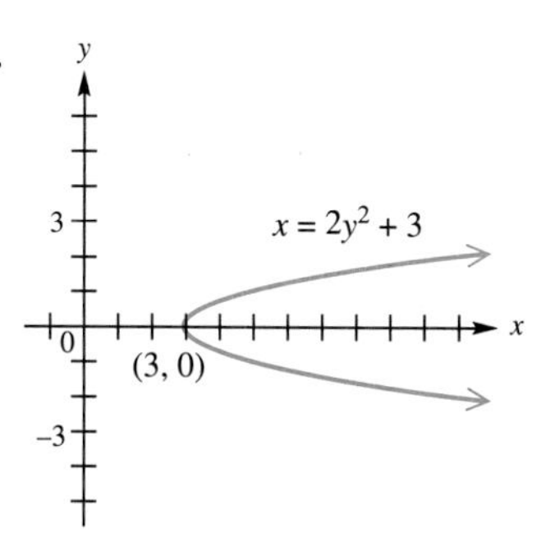

25.

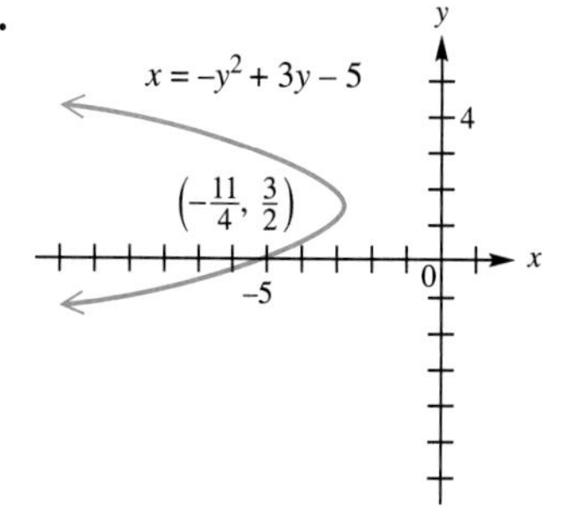

29. 5 seconds **31.** length: 25 meters; width: 25 meters **33.** symmetric to the x-axis, y-axis, and origin
35. symmetric to the x-axis **37.** no symmetry **39.** symmetric to the y-axis **41.** symmetric to the origin
43. increasing on $(0, \infty)$ **45.** increasing on $(1, \infty)$; decreasing on $(-\infty, 1)$ **47.** $(x - 6)^2 + (y + 1)^2 = 9$

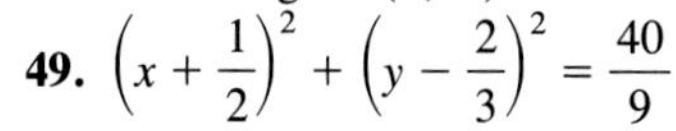

49. $\left(x + \frac{1}{2}\right)^2 + \left(y - \frac{2}{3}\right)^2 = \frac{40}{9}$ **51.** $(5, 2)$; 6 **53.** $\left(\frac{1}{2}, -1\right)$; $\frac{\sqrt{285}}{10}$

55.

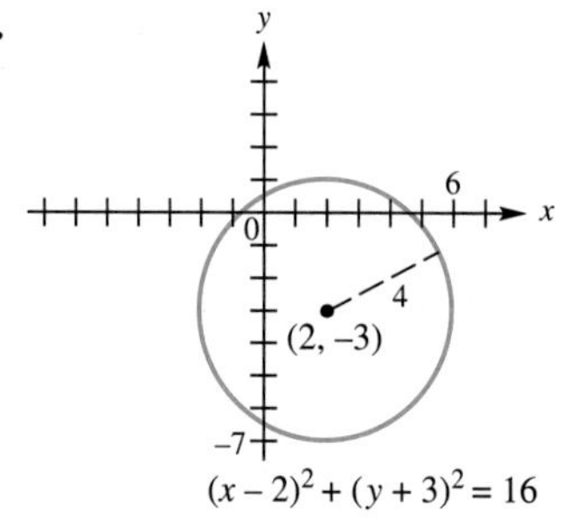

57.

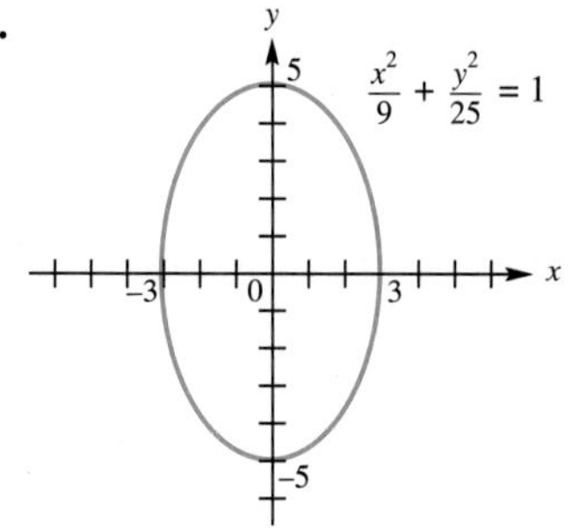

59.

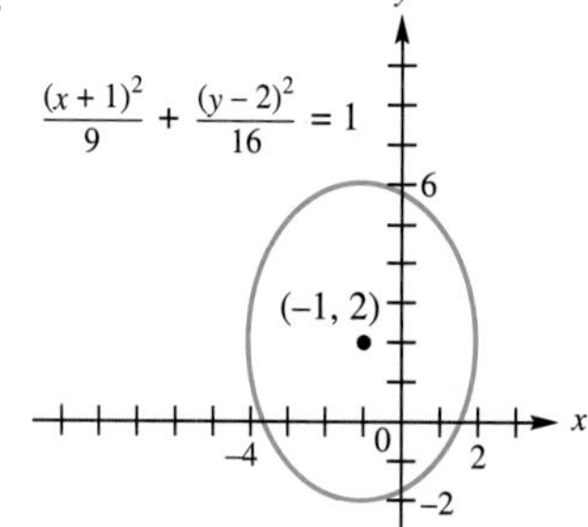

61.

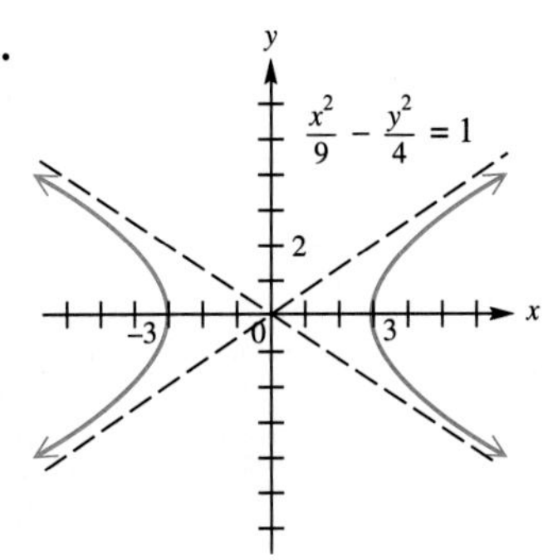

63.

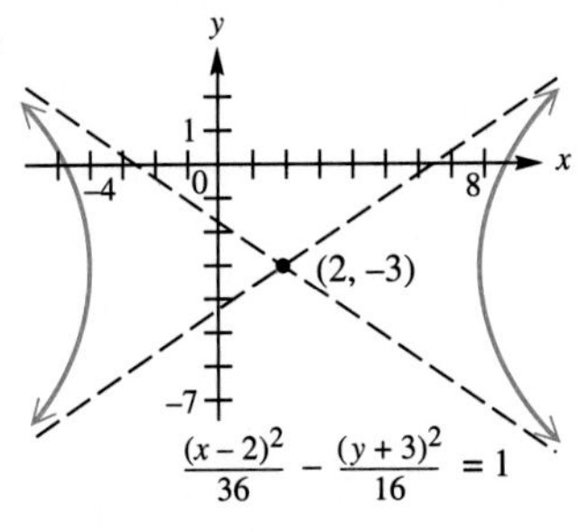

65.

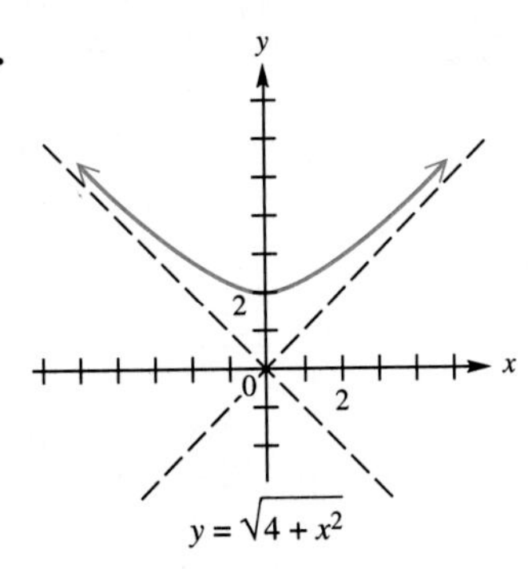

67. circle **69.** ellipse **71.** circle **73.** hyperbola **75.** circle **77.** hyperbola

79.

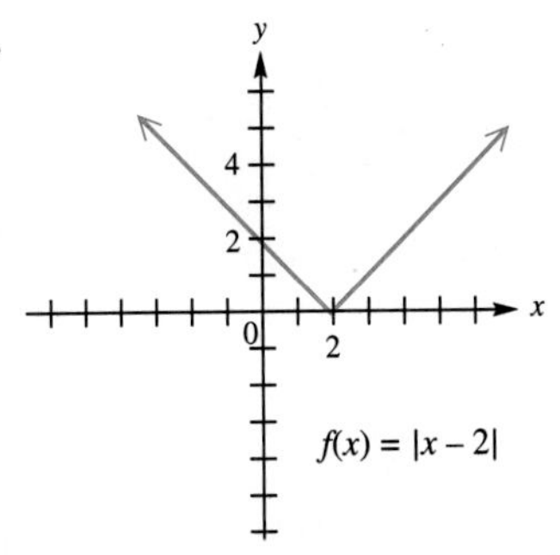

81.

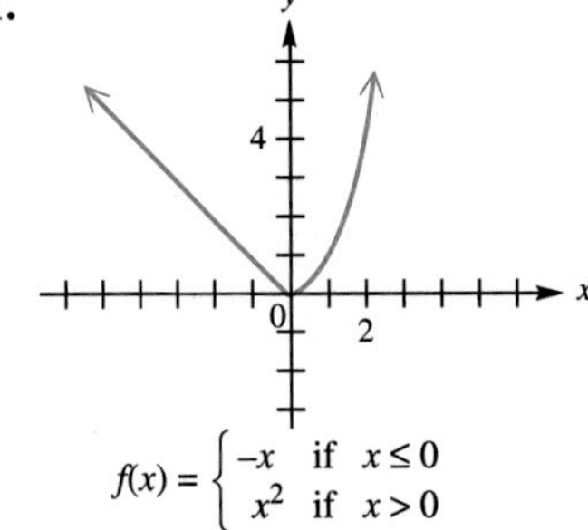

83.

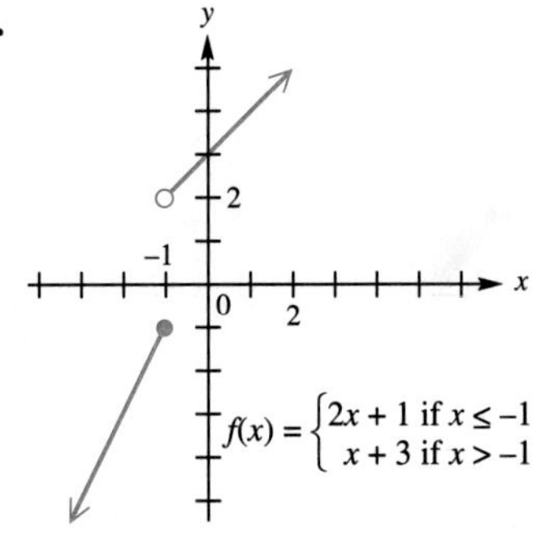

85.

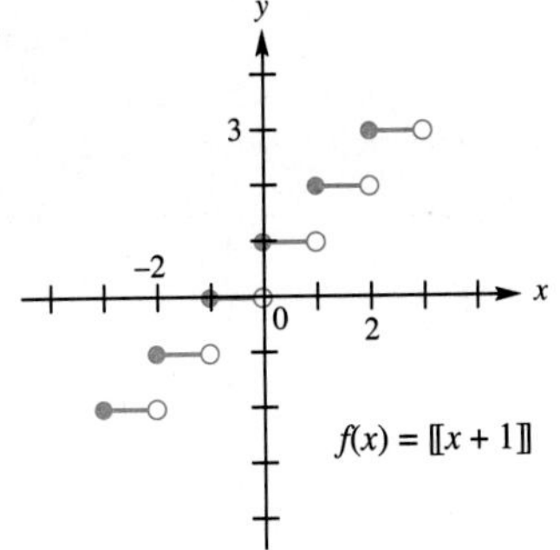

87.

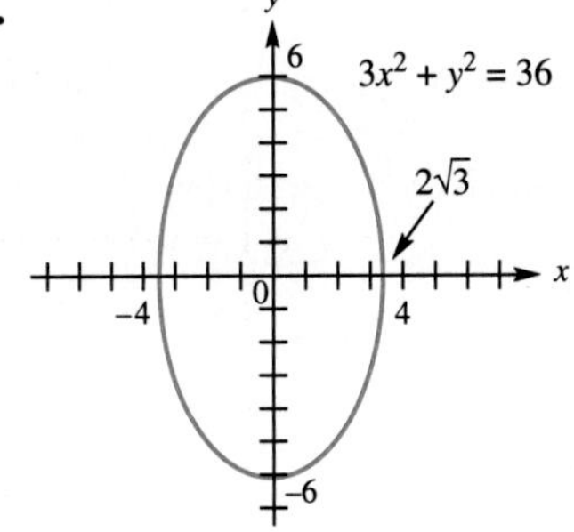

89.

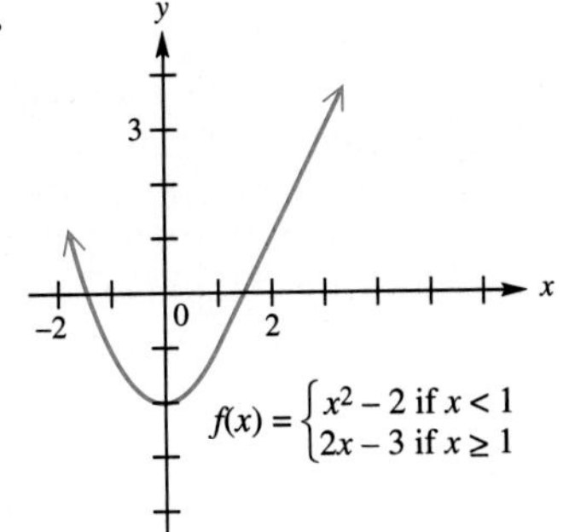

91.

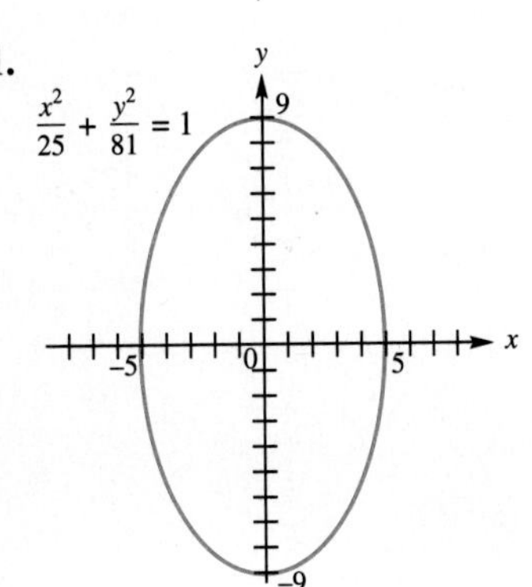

93.

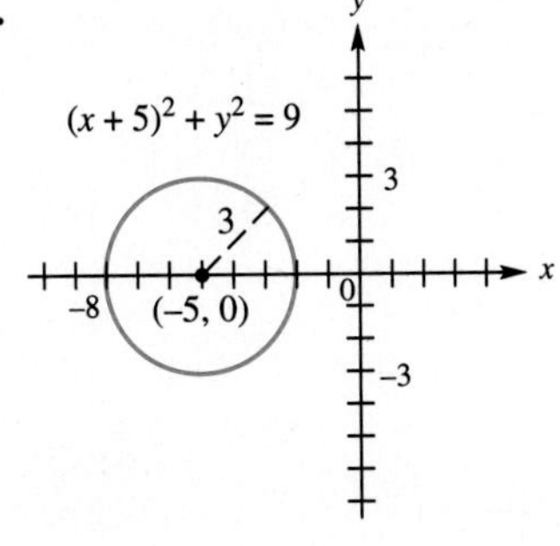

95.

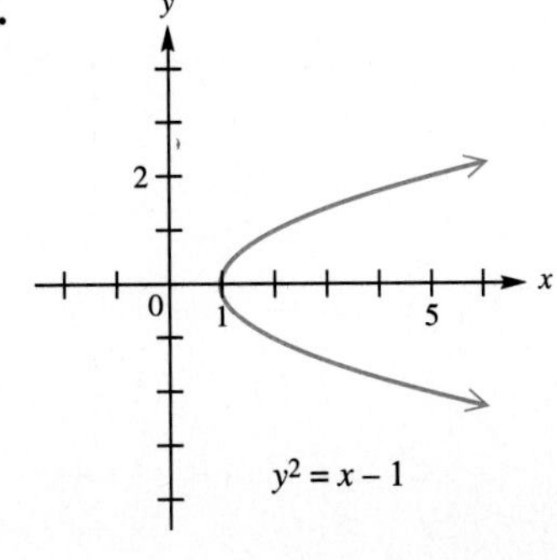

Chapter 8 Test (page 469)

[8.1] **1.** -2 **2.** -12 **3.** $-\frac{4}{5}$ **4.** -16 [8.2] **5.**

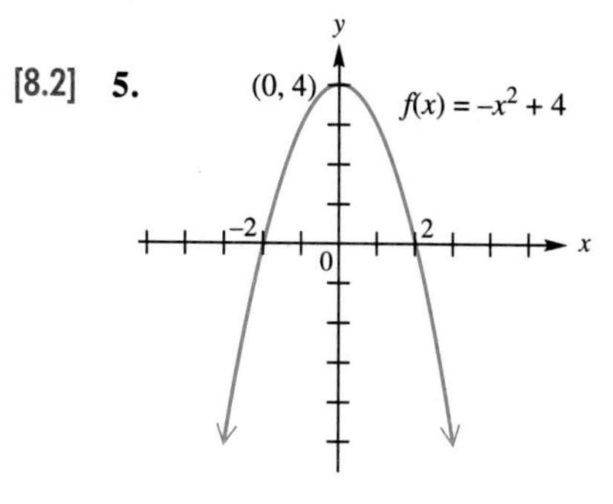

[8.3] **6.** 50 feet; 3 seconds

[8.4] **7.** no symmetry
8. symmetric to the y-axis
9. symmetric to the origin
10. symmetric to the x-axis, y-axis, and the origin
11. increasing on $(-2, 1)$; decreasing on $(-\infty, -2)$; constant on $(1, \infty)$

[8.5] **12.** $(-1, 2)$; 3

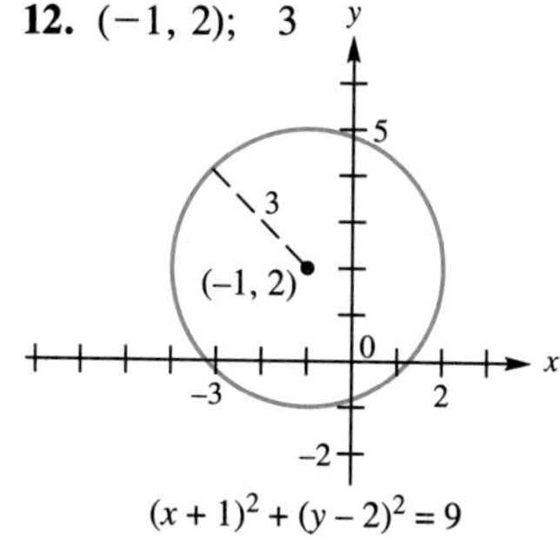

13. $(1, 3)$; 4 [8.6] **14.**

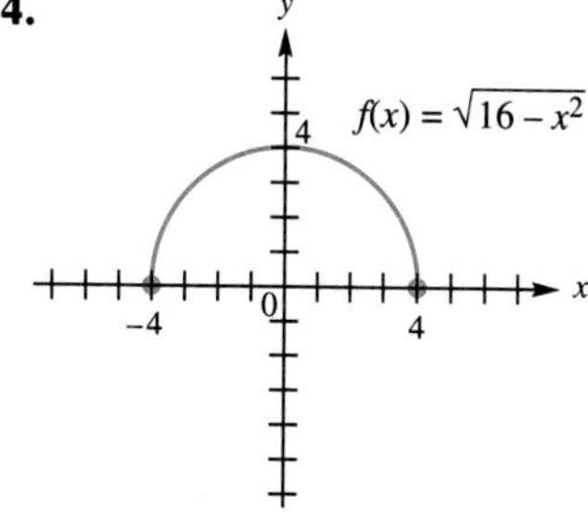

[8.3] **15.**

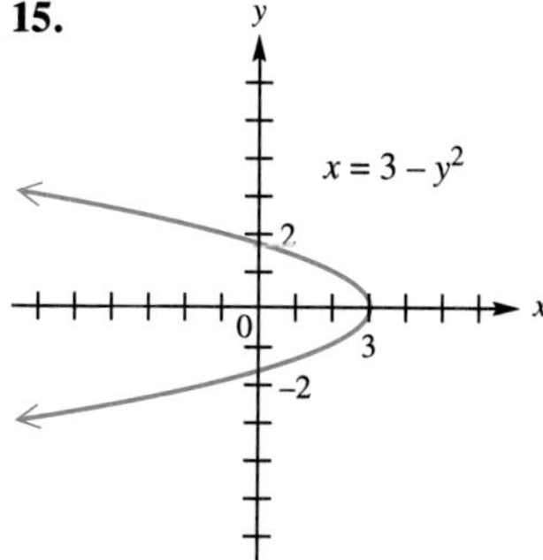

[8.5] **16.**

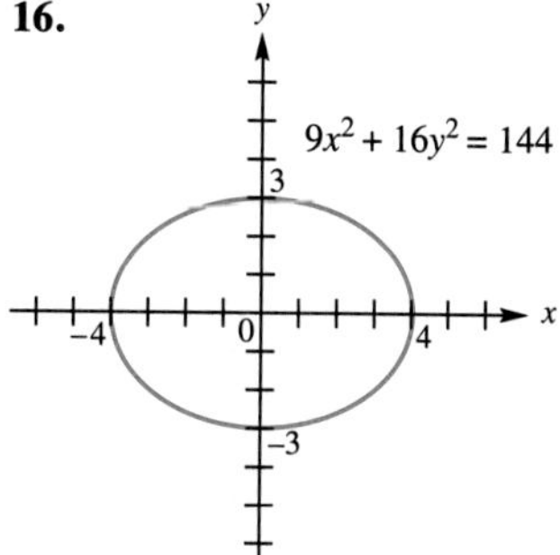

[8.6] **17.**

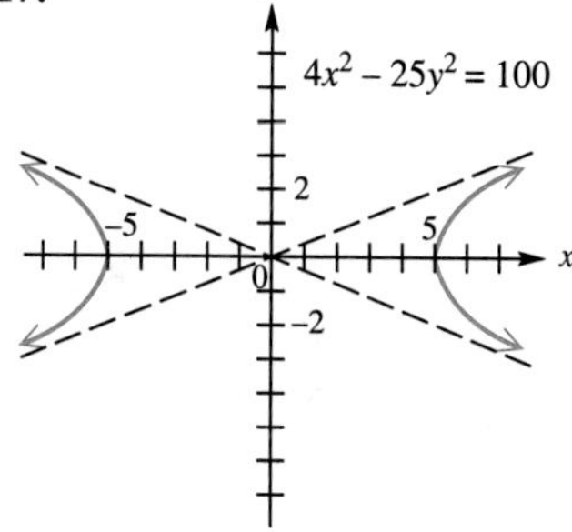

18. circle **19.** hyperbola **20.** parabola **21.** ellipse [8.1] **22.** x; up or down

[8.7] **23.**

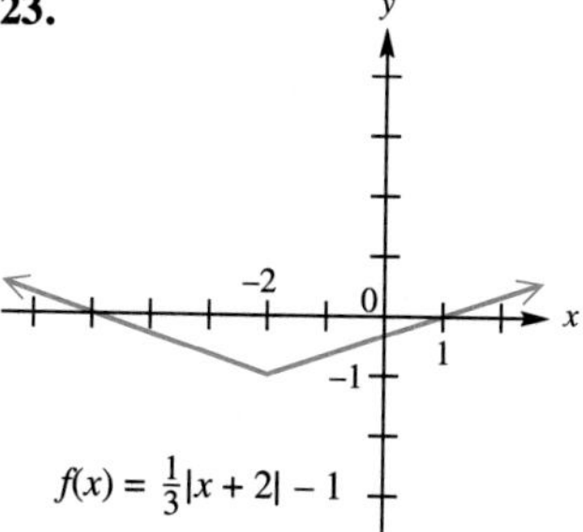

24.

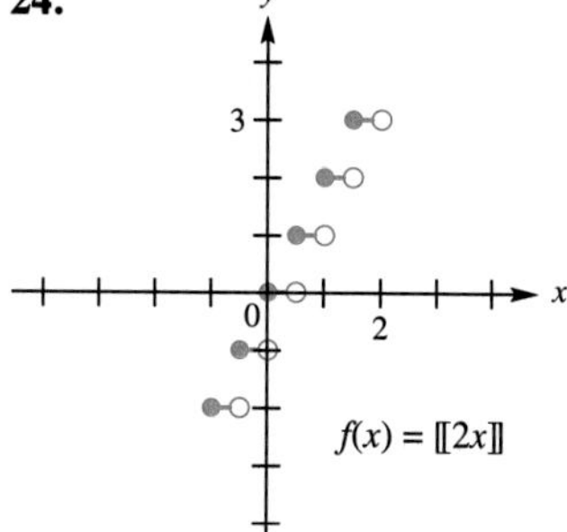

25.

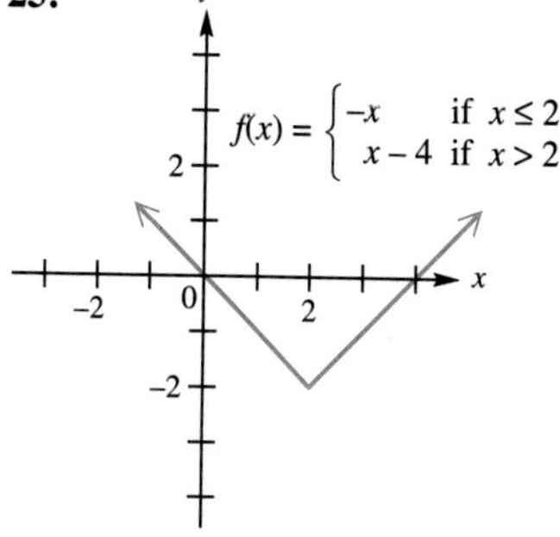

CHAPTER 9

9.1 Exercises (page 479)

1. (a)

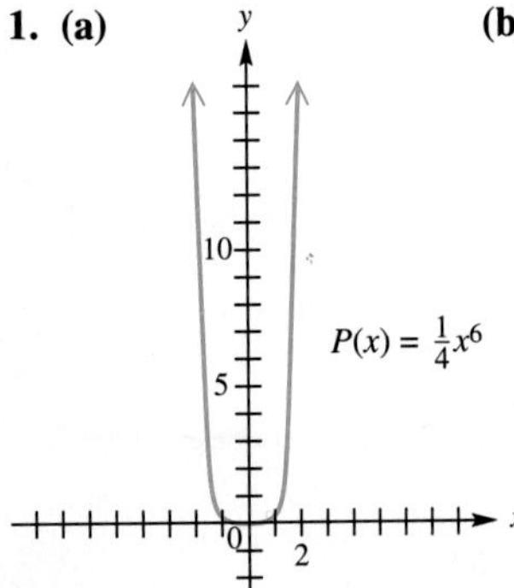

(b) y-axis

3. (a)

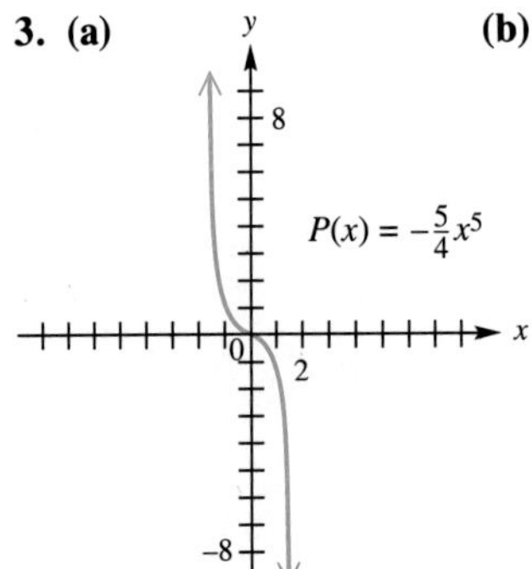

(b) origin

5. (a)

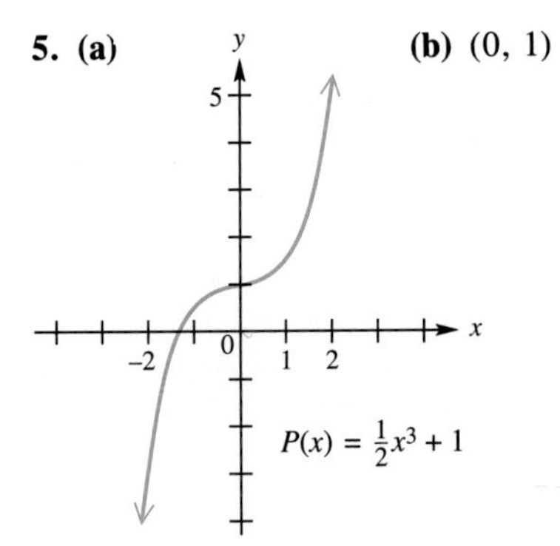

(b) (0, 1)

7. (a)

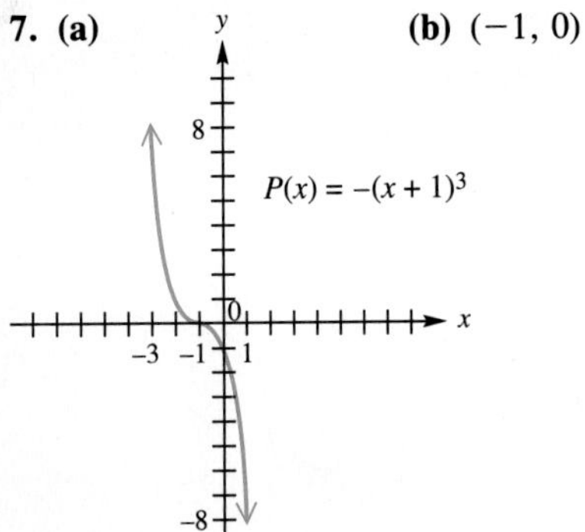

(b) $(-1, 0)$

9. (a)

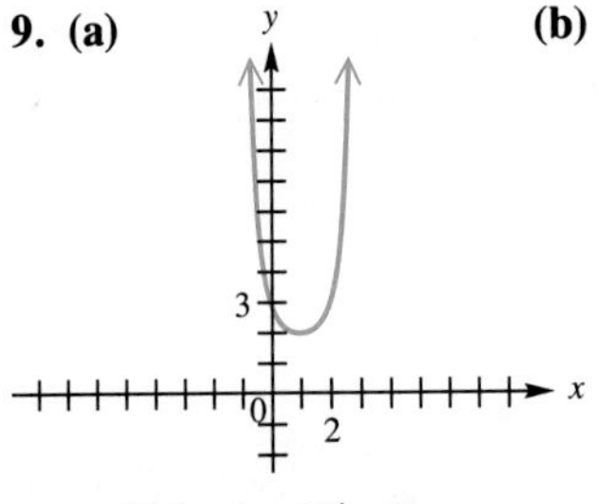

$P(x) = (x - 1)^4 + 2$

(b) $x = 1$

11.

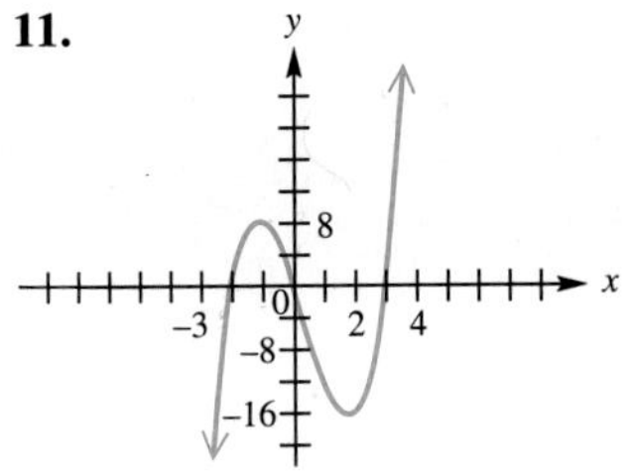

$P(x) = 2x(x - 3)(x + 2)$

13.

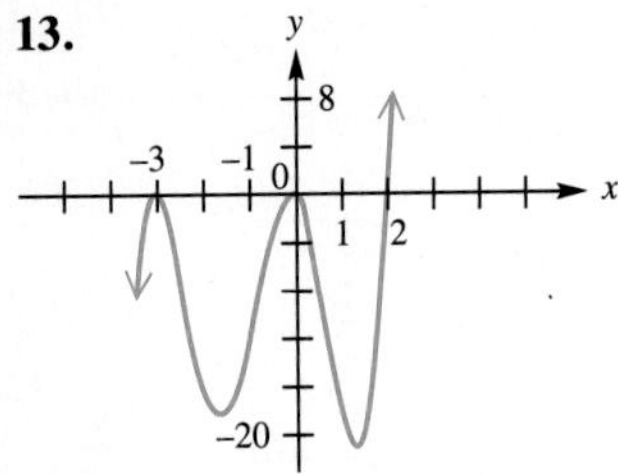

$P(x) = x^2(x - 2)(x + 3)^2$

15.

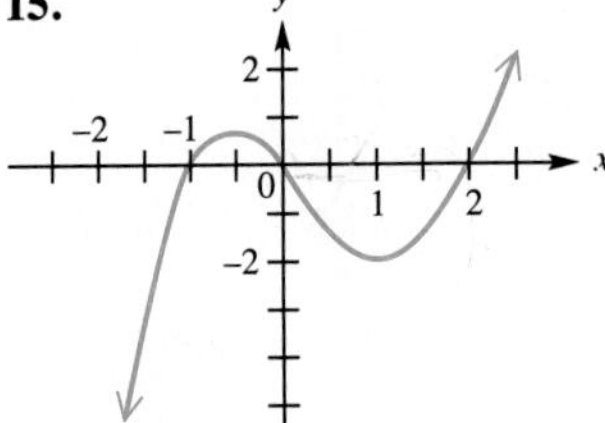

$P(x) = x^3 - x^2 - 2x$

17.

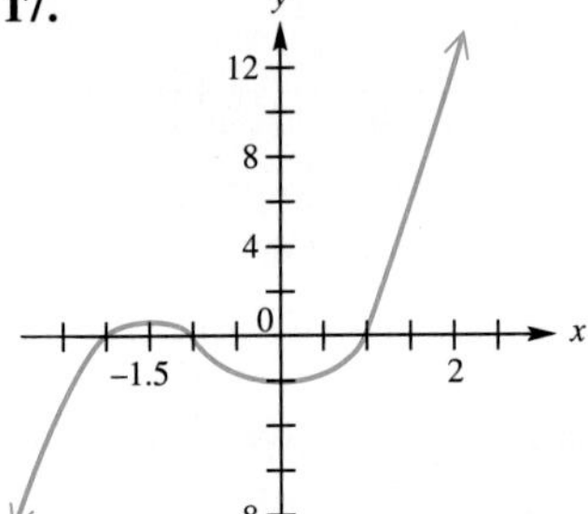

$P(x) = (x + 2)(x - 1)(x + 1)$

19.

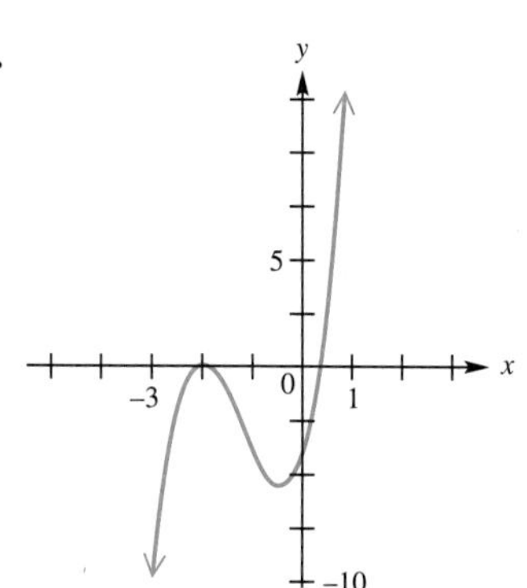

$P(x) = (3x - 1)(x + 2)^2$

21.

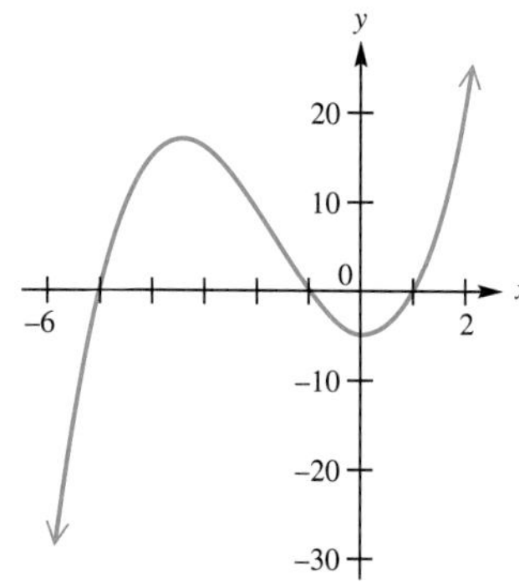

$P(x) = x^3 + 5x^2 - x - 5$

23. (c) **25.** **(a)**

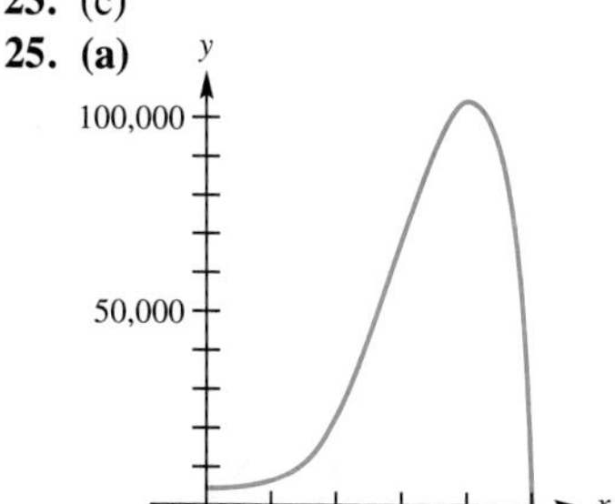

$D(x) = -.125x^5 + 3.125x^4 + 4000$

(b) 1910 to 1925; 1905 to 1910; 1925 to 1930

29. odd **31.** even **33.** odd **35.** neither **37.** even **39.** $P(x) = x^2$ **41.** Maximum is 26.136 when $x = -3.4$; minimum is 25 when $x = -3$. **43.** Maximum is 1.048 when $x = -.1$; minimum is -5 when $x = -1$. **45.** Maximum is 84 when $x = -2$; minimum is -13 when $x = -1$. **47.**

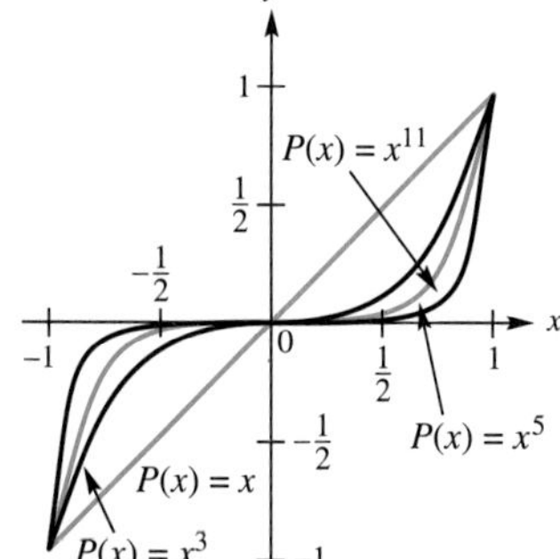

49. $x^2 - 3x - 2$ **51.** $4m^2 - 4m + 1 - \frac{3}{m + 1}$ **53.** no

9.2 Exercises (page 487)

1. no **3.** yes **5.** yes **7.** no **9.** 0: multiplicity 4; -3: multiplicity 5; 8: multiplicity 2 **11.** $\frac{7}{4}$: multiplicity 3; 5: multiplicity 1 **13.** i: multiplicity 4; $-i$: multiplicity 4 **15.** $P(x) = -\frac{1}{6}x^3 + \frac{13}{6}x + 2$ **17.** $P(x) = -\frac{1}{2}x^3 - \frac{1}{2}x^2 + x$ **19.** $P(x) = -10x^3 + 30x^2 - 10x + 30$ **21.** $P(x) = x^2 - 10x + 26$ **23.** $P(x) = x^3 - 4x^2 + 6x - 4$ **25.** $P(x) = x^3 - 3x^2 + x + 1$ **27.** $P(x) = x^4 - 6x^3 + 10x^2 + 2x - 15$ **29.** $P(x) = x^3 - 8x^2 + 22x - 20$ **31.** $P(x) = x^4 - 4x^3 + 5x^2 - 2x - 2$ **33.** $P(x) = x^4 - 10x^3 + 42x^2 - 82x + 65$ **35.** $P(x) = x^5 - 12x^4 + 74x^3 - 248x^2 + 445x - 500$ **37.** $P(x) = x^4 - 6x^3 + 17x^2 - 28x + 20$ **39.** $-1 + i$, $-1 - i$ **41.** $3, -2 - 2i$ **43.** $-\frac{1}{2} + \frac{\sqrt{5}}{2}i, -\frac{1}{2} - \frac{\sqrt{5}}{2}i$ **45.** $i, 2i, -2i$ **47.** $3, -2, 1 - 3i$ **49.** $P(x) = (x - 2)(2x - 5)(x + 3)$ **51.** $P(x) = (x + 4)(3x - 1)(2x + 1)$ **53.** $P(x) = (x - 3i)(x + 4)(x + 3)$ **55.** $P(x) = (x - 1 - i)(2x - 1)(x + 3)$ **57.** zeros are $-2, -1, 3$; $P(x) = (x + 2)^2(x + 1)(x - 3)$ **59.** 1, 3, or 5 **63.** at 3 seconds **73.** $\{-5\}$

9.3 Exercises (page 492)

1. $\pm 1, \pm\frac{1}{2}, \pm\frac{1}{3}, \pm\frac{1}{6}$ **3.** $\pm 1, \pm\frac{1}{2}, \pm\frac{1}{3}, \pm\frac{1}{4}, \pm\frac{1}{6}, \pm\frac{1}{12}, \pm 2, \pm\frac{2}{3}$ **5.** $\pm 1, \pm\frac{1}{2}, \pm 2, \pm 4, \pm 8$ **7.** $\pm 1, \pm 2, \pm 5, \pm 10, \pm 25, \pm 50$ **11.** $-1, -2, 5$ **13.** $2, -3, -5$ **15.** no rational zeros **17.** $1, -2, -3, -5$ **19.** $-4, \frac{3}{2}, -\frac{1}{3}$; $P(x) = (x + 4)(2x - 3)(3x + 1)$ **21.** $-\frac{3}{2}, -\frac{2}{3}, \frac{1}{2}$; $P(x) = (2x - 1)(3x + 2)(2x + 3)$ **23.** $\frac{1}{2}$; $P(x) = (2x - 1)(x^2 + 4x + 8)$ **25.** $-2, -1, \frac{1}{2}, 1$; $P(x) = (x + 2)(x + 1)(2x - 1)(x - 1)$ **27.** $-1, -2, -\frac{2}{3}, 2$; $P(x) = (x + 1)(x + 2)(3x + 2)(x - 2)$ **29.** $-\frac{2}{3}, -1, 3$ **31.** $1, -\frac{5}{4}$ **33.** 1

35. $-1, \frac{3}{2}, 2$; $P(x) = (x + 1)(2x - 3)(x - 2)$

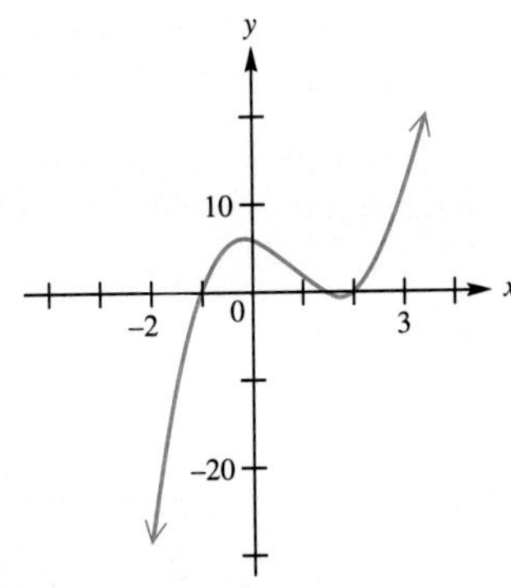

$P(x) = 2x^3 - 5x^2 - x + 6$

37. -2(multiplicity 2), 3; $P(x) = (x + 2)^2(x - 3)$

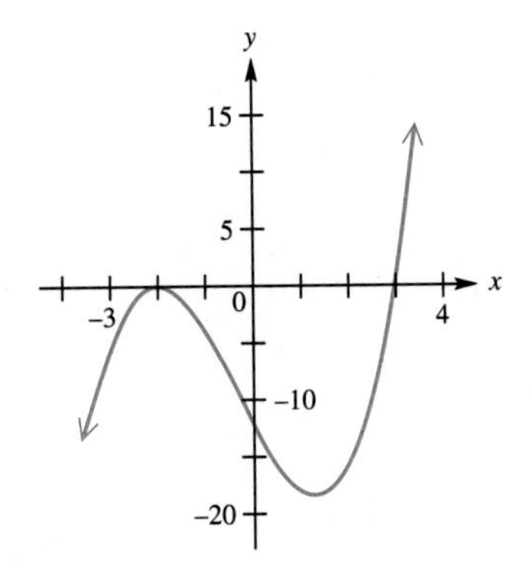

$P(x) = x^3 + x^2 - 8x - 12$

45. **(a)** 18 **(b)** -112 **47.** **(a)** 14 **(b)** 74

9.4 Exercises (page 501)

1. 2 or 0 positive real zeros; 1 negative real zero **3.** 1 positive real zero; 1 negative real zero **5.** 2 or 0 positive real zeros; 3 or 1 negative real zeros **13.** There is a zero between 2 and 2.5. **23.** **(a)** positive: 1; negative: 2 or 0 **(b)** $-3, -1.4, 1.4$ **25.** **(a)** positive: 2 or 0; negative: 1 **(b)** $-2, 1.6, 4.4$ **27.** **(a)** positive: 1; negative: 0 **(b)** 1.5 **29.** **(a)** positive: 3 or 1; negative: 1 **(b)** $1.1, -.7$ **31.** **(a)** positive: 1; negative: 1 **(b)** $-1.5, 3.1$

33.

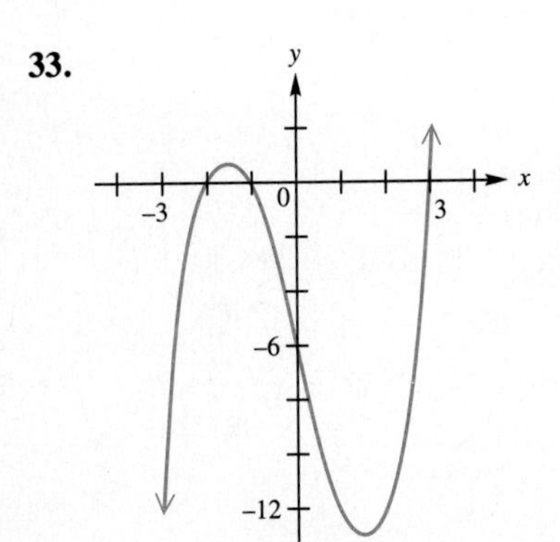

$P(x) = x^3 - 7x - 6$

35.

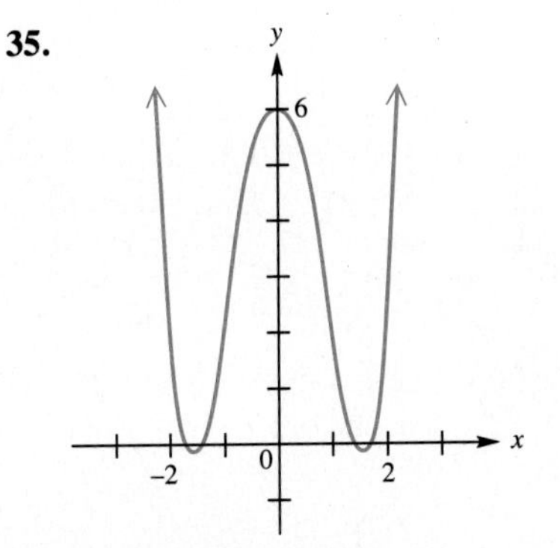

$P(x) = x^4 - 5x^2 + 6$

37.

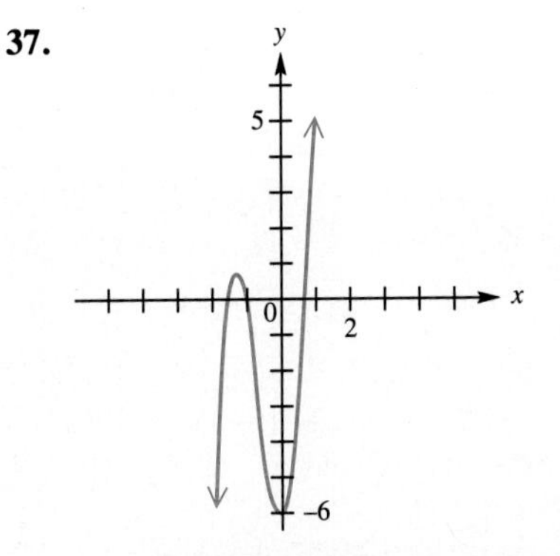

$P(x) = 6x^3 + 11x^2 - x - 6$

39.

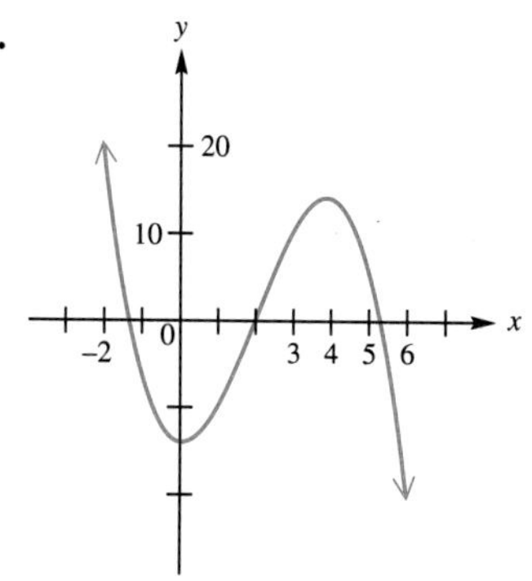

$P(x) = -x^3 + 6x^2 - x - 14$

41. 3.24, −1.24
43. −3.65, −.32, 1.65, 6.32

45. **(a)** 175 (rounded)
(b)

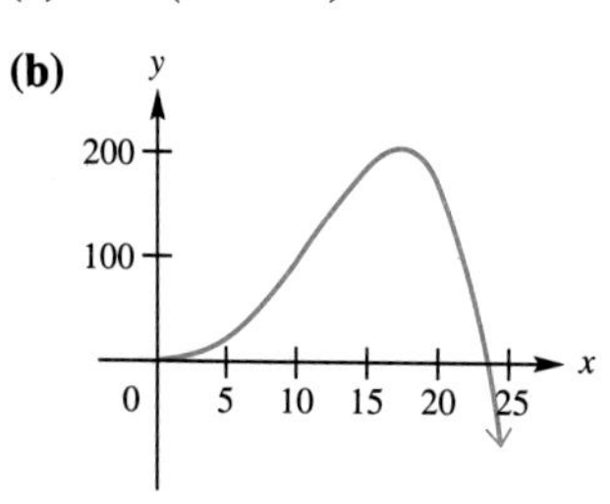

$g(x) = -.006x^4 + .140x^3 - .053x^2 + 1.79x$

47. **(a)** 3 **(b)** −4 **49.** **(a)** −3, 3 **(b)** $-\dfrac{5}{2}$ **51.** **(a)** $-\dfrac{7}{2}$, 3 **(b)** $-\dfrac{1}{3}$, 6

9.5 Exercises (page 513)

1.

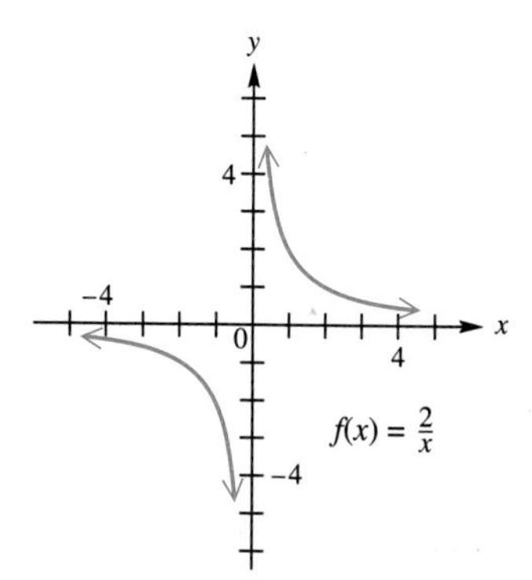

3.

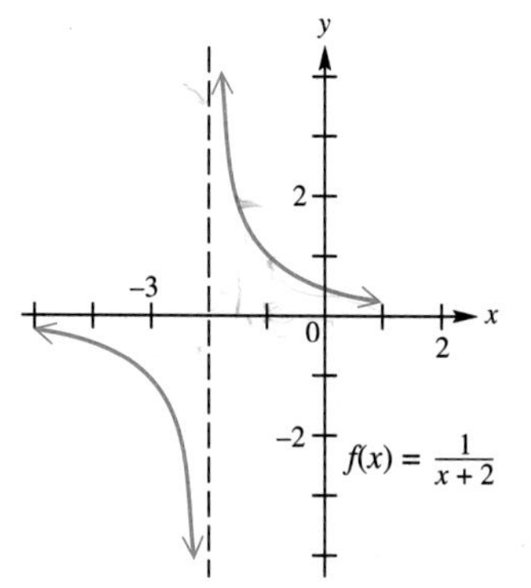

5.

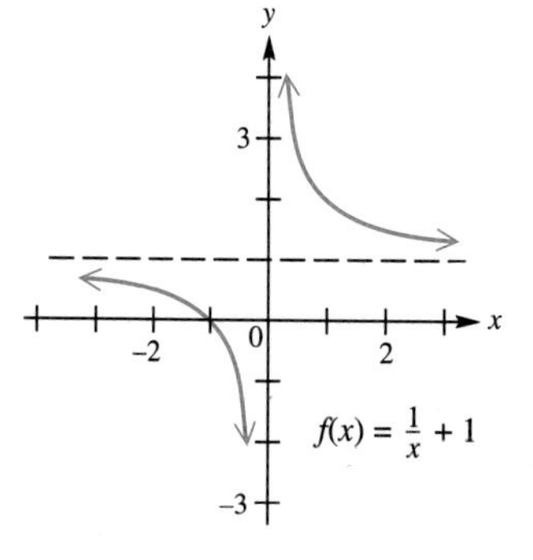

7. **(a)**

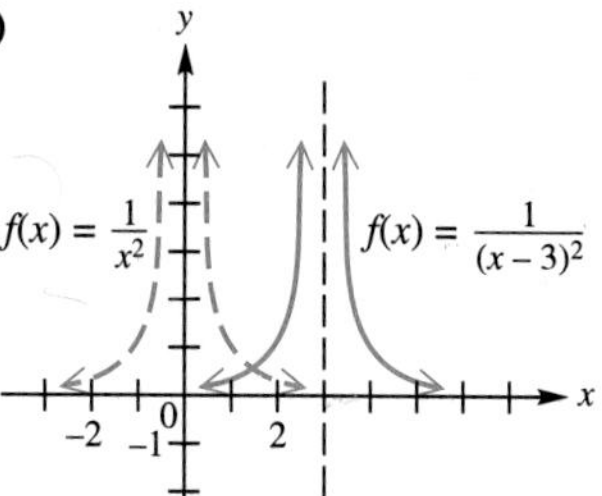

(b)

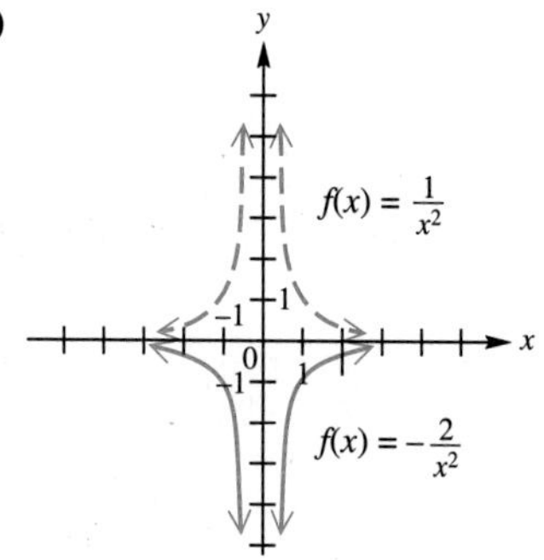

(c)

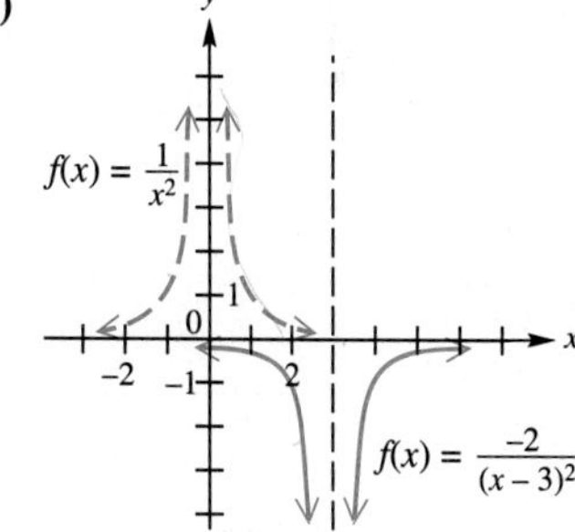

9. vertical asymptote: $x = 5$; horizontal asymptote: $y = 0$ **11.** vertical asymptote: $x = \dfrac{7}{3}$; horizontal asymptote: $y = 0$

13. vertical asymptote: $x = -2$; horizontal asymptote: $y = -1$ **15.** vertical asymptote: $x = -\dfrac{9}{2}$; horizontal asymptote: $y = \dfrac{3}{2}$ **17.** vertical asymptotes: $x = 3$, $x = 1$; horizontal asymptote: $y = 0$

19. vertical asymptote: $x = -3$; oblique asymptote: $y = x - 3$ **21.** vertical asymptotes: $x = -2$, $x = \frac{5}{2}$; horizontal asymptote: $y = \frac{1}{2}$ **23.** (a) **25.**

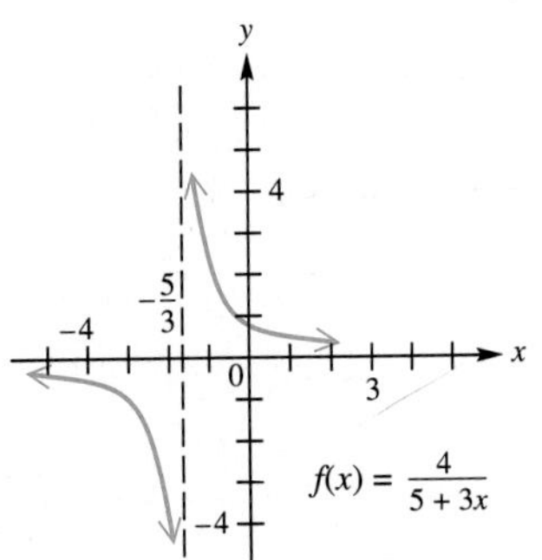

27.

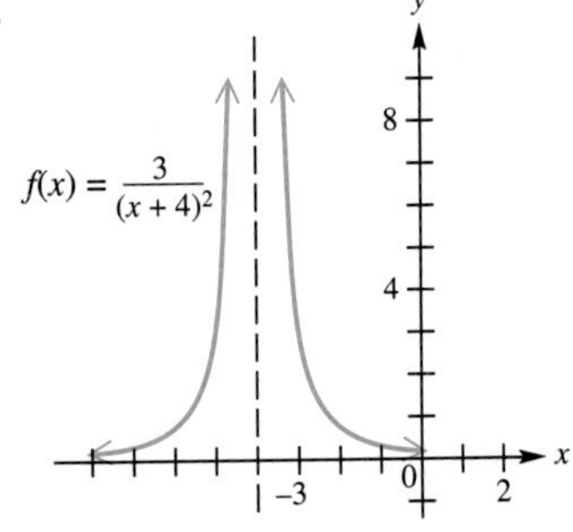

29.

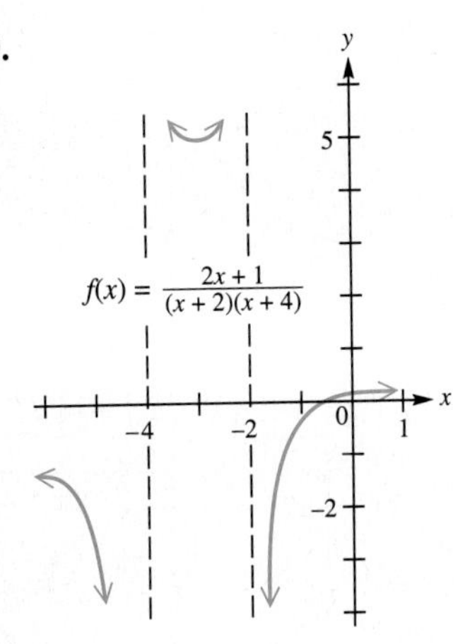

31.

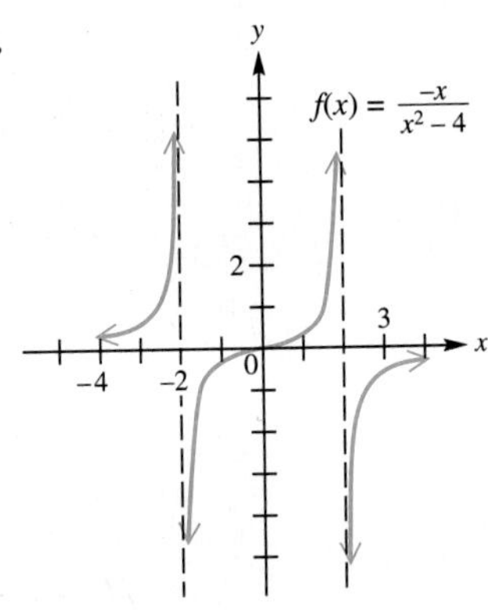

33.

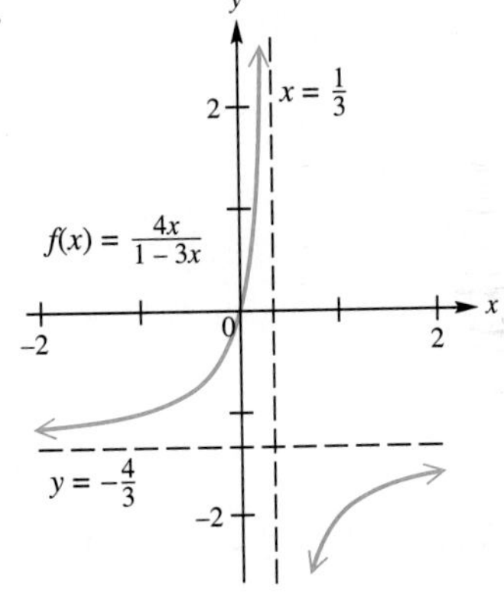

35.

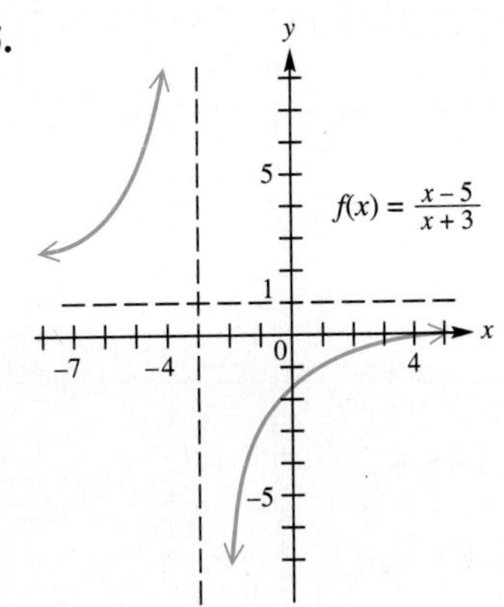

37.

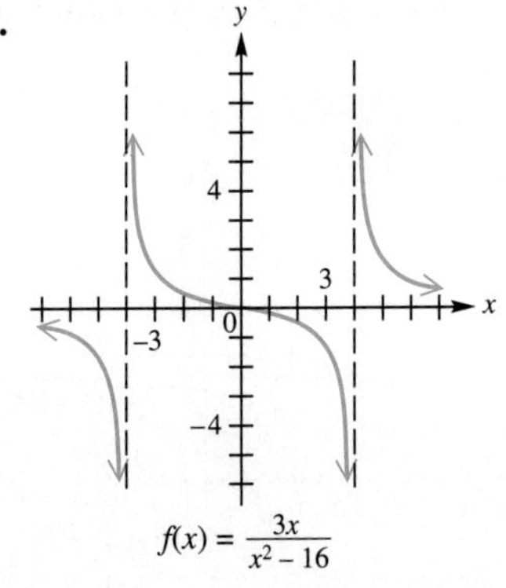

39.

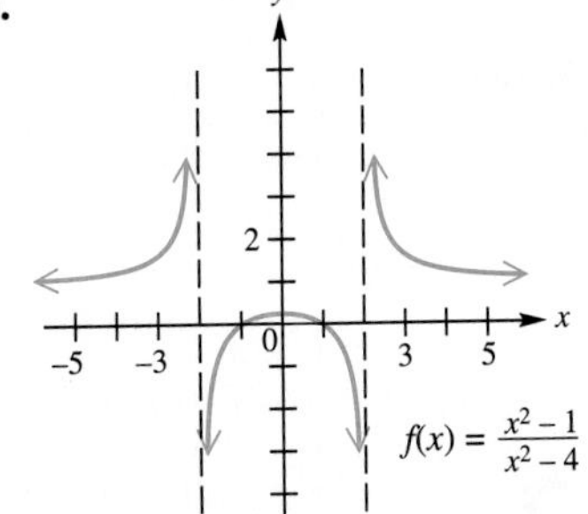

41.

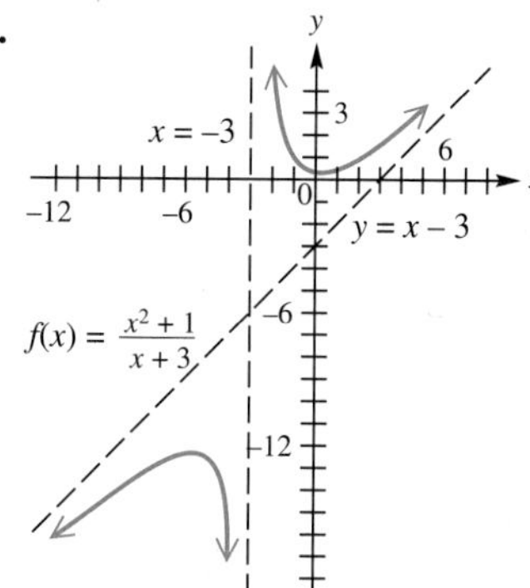

43.

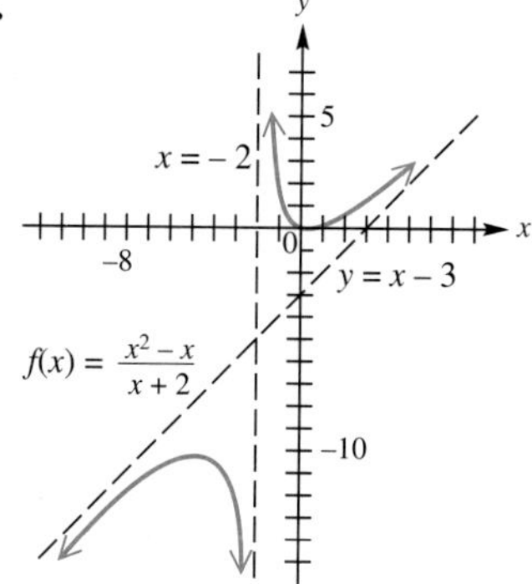

45.

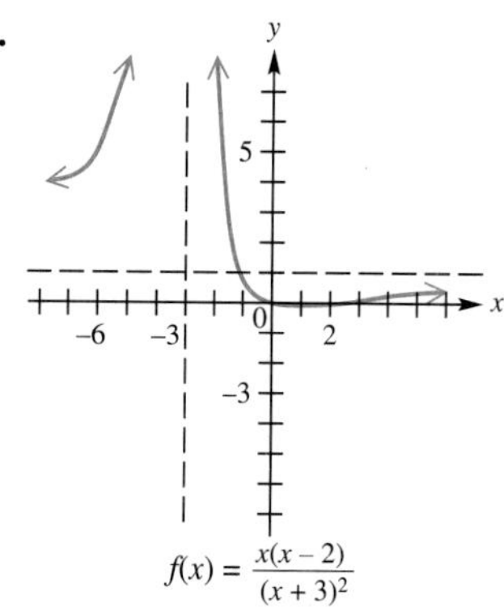

47.

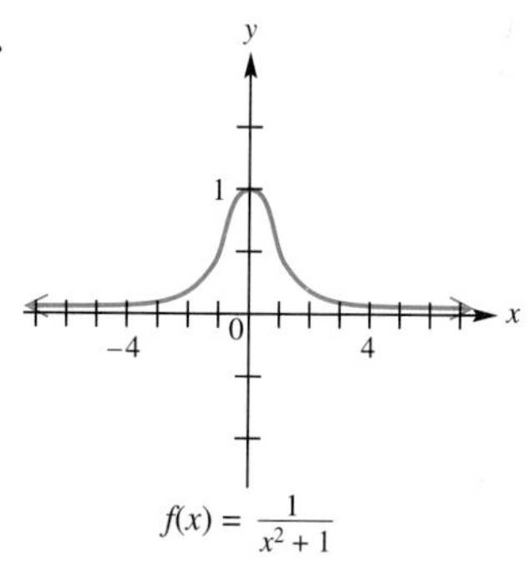

49.

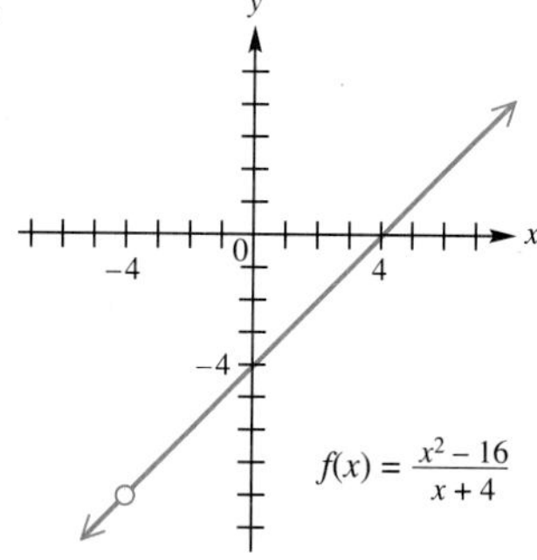

51.

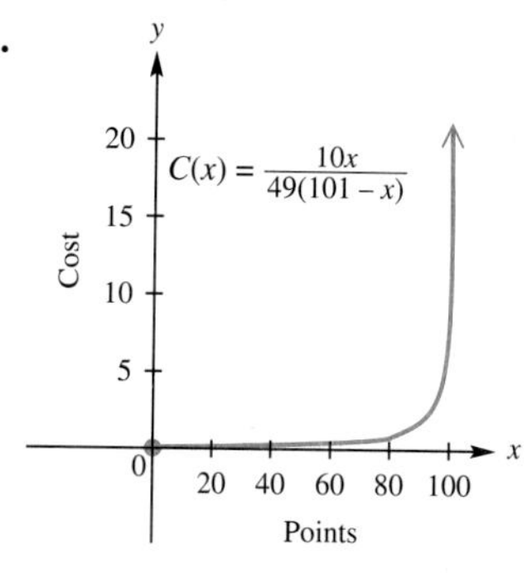

53. (a)

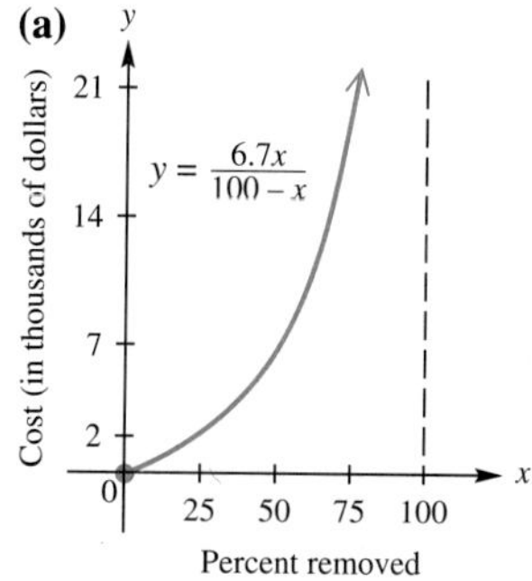

(b) no

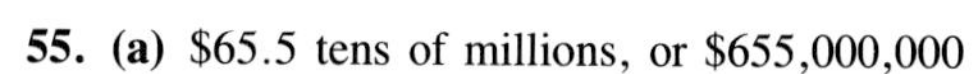

55. (a) \$65.5 tens of millions, or \$655,000,000
(b) \$64 tens of millions, or \$640,000,000
(c) \$60 tens of millions, or \$600,000,000
(d) \$40 tens of millions, or \$400,000,000
(e) \$0

(f)

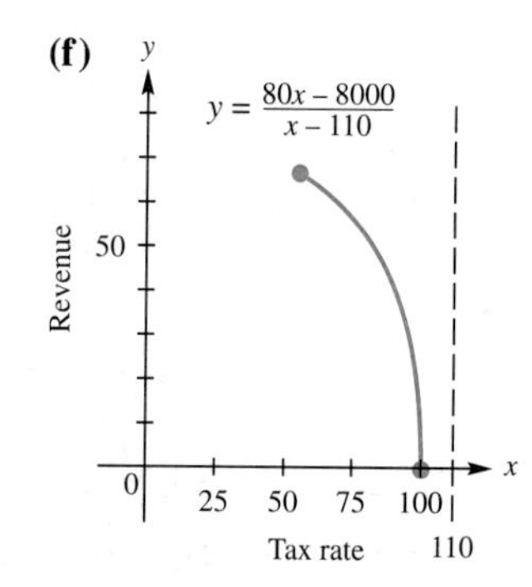

59. 5 **61.** 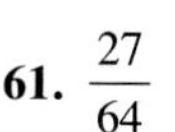 $\frac{27}{64}$ **63.** 1 **65.** $\frac{16}{625}$

Chapter 9 Review Exercises (page 520)

1.

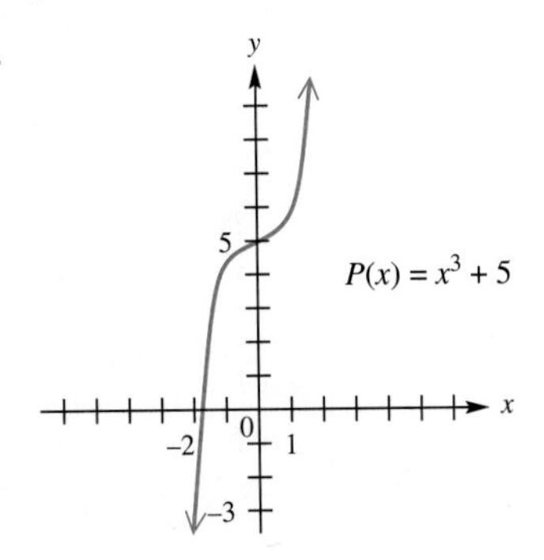

$P(x) = x^3 + 5$

3.

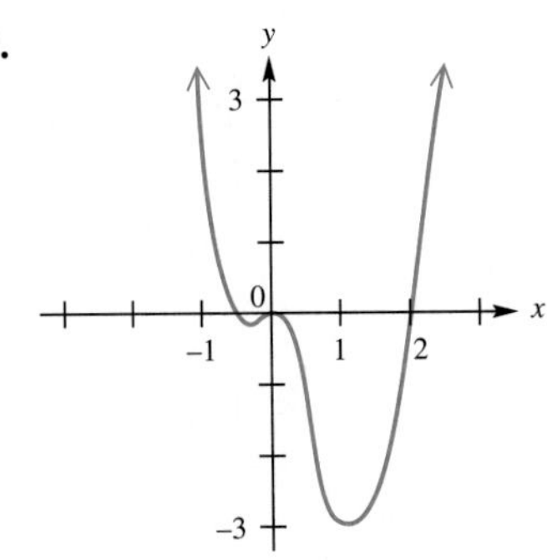

$P(x) = x^2(2x + 1)(x - 2)$

5.

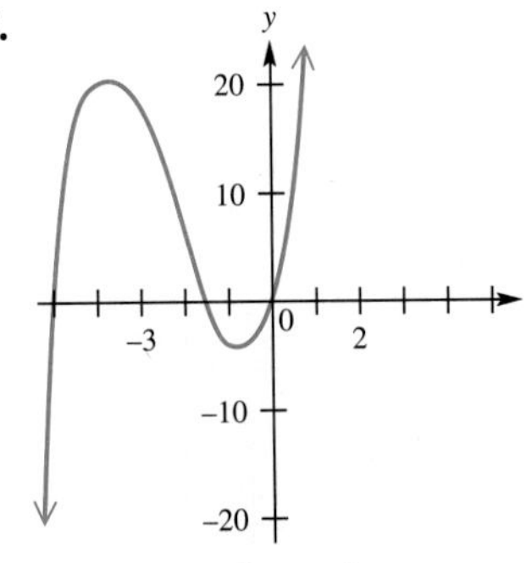

$P(x) = 2x^3 + 13x^2 + 15x$

7.

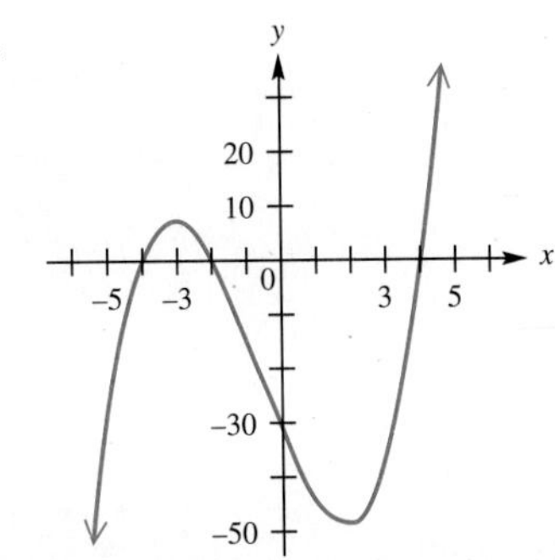

$P(x) = x^3 + 2x^2 - 16x - 32$

9. zero, root, x-intercept **11.** $P(x) = x^3 - 10x^2 + 17x + 28$ **13.** $P(x) = x^4 - x^3 - 9x^2 + 7x + 14$ **15.** no **17.** no **19.** $P(x) = -2x^3 + 6x^2 + 12x - 16$ **21.** $P(x) = x^4 - 3x^2 - 4$ **23.** $P(x) = x^3 + x^2 - 4x + 6$ **25.** $3 + i$, $3 - i$, $2i$, $-2i$; $P(x) = (x - 3 - i)(x - 3 + i)(x - 2i)(x + 2i)$ **27.** $\frac{1}{2}$, -1, 5 **29.** 4, $-\frac{1}{2}$, $-\frac{2}{3}$

31. -1 (multiplicity 2), 7; $P(x) = (x + 1)^2(x - 7)$

$P(x) = x^3 - 5x^2 - 13x - 7$

41. -2.3, 4.6

43.

$P(x) = 2x^3 - 11x^2 - 2x + 2$

45.

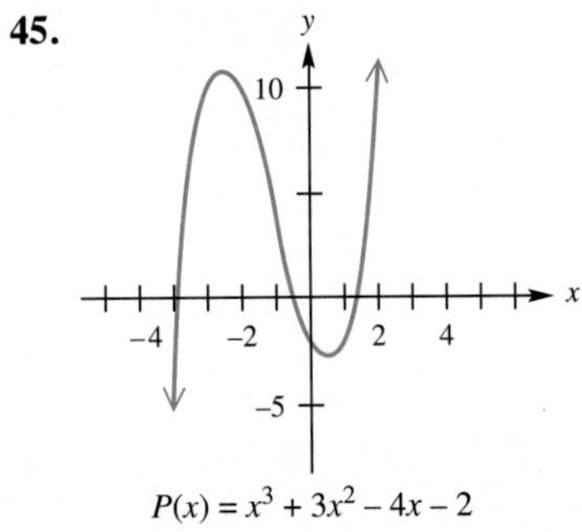

$P(x) = x^3 + 3x^2 - 4x - 2$

47.

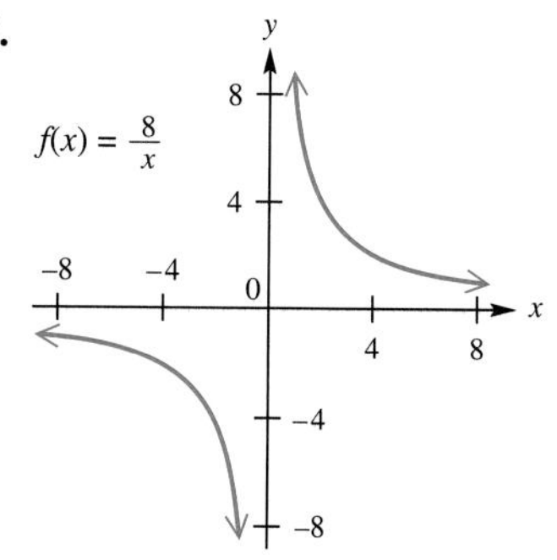

49.

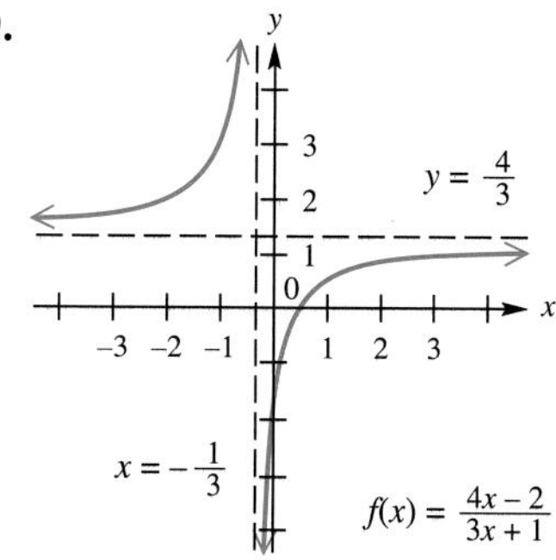

51.

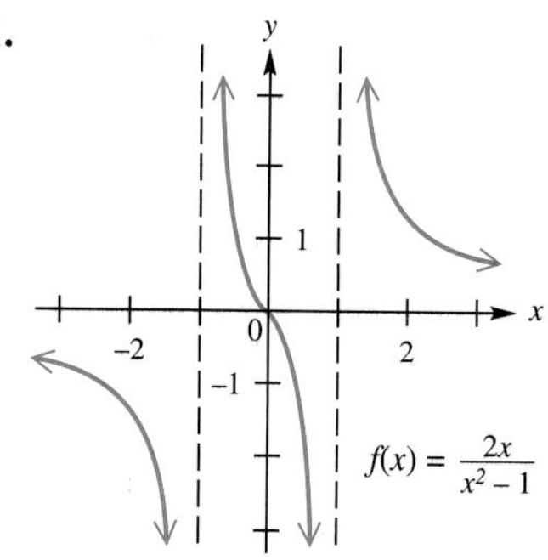

53.

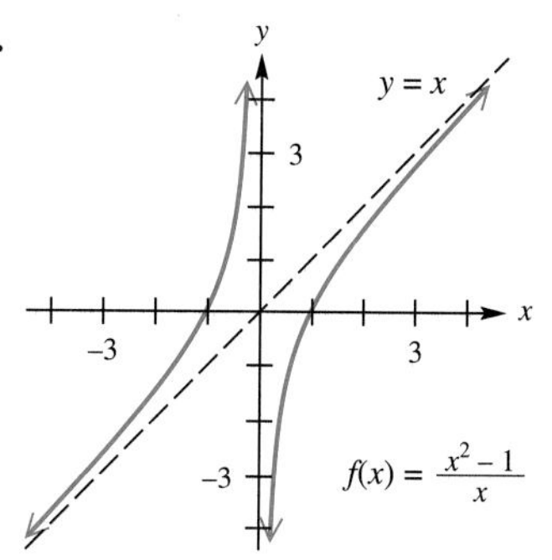

55.

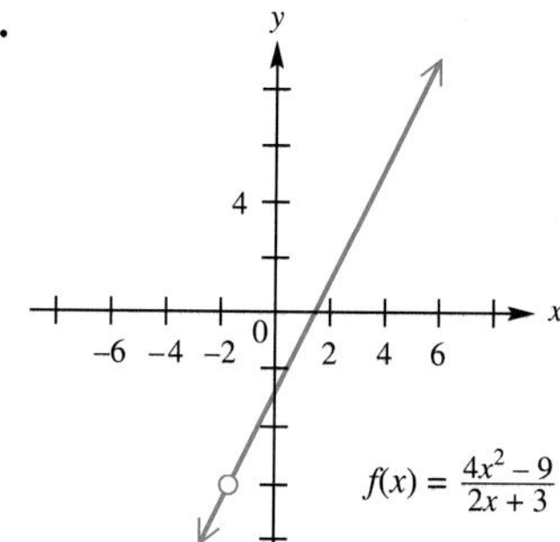

57.

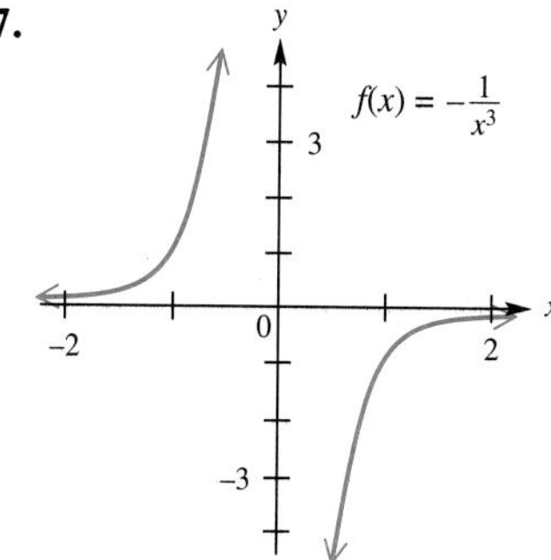

59.

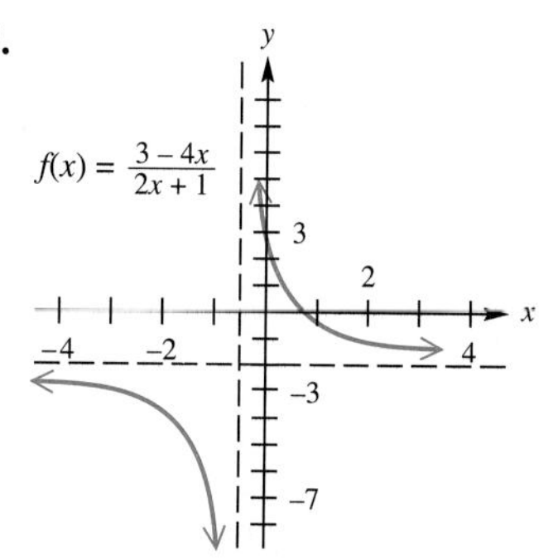

61.

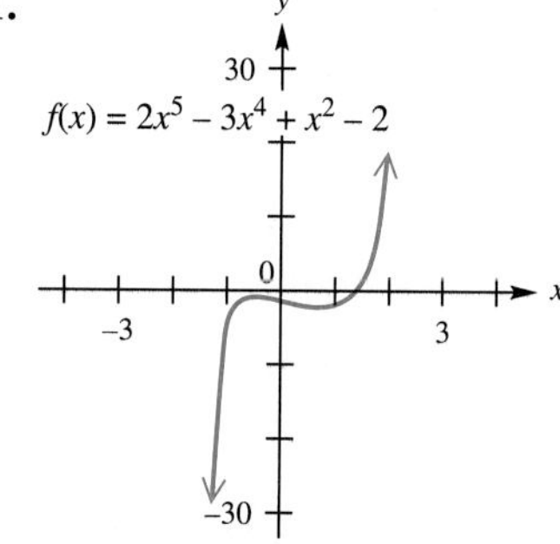

63.

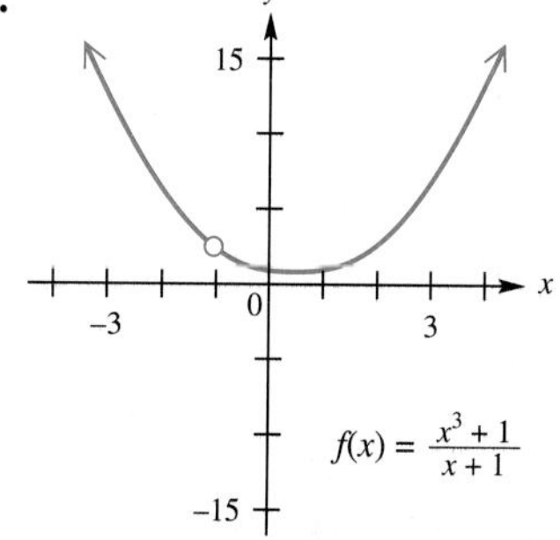

Chapter 9 Test (page 523)

[9.1] 1.

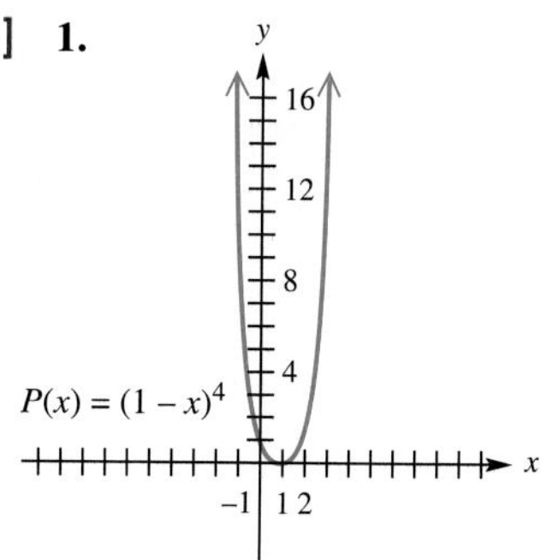

2.

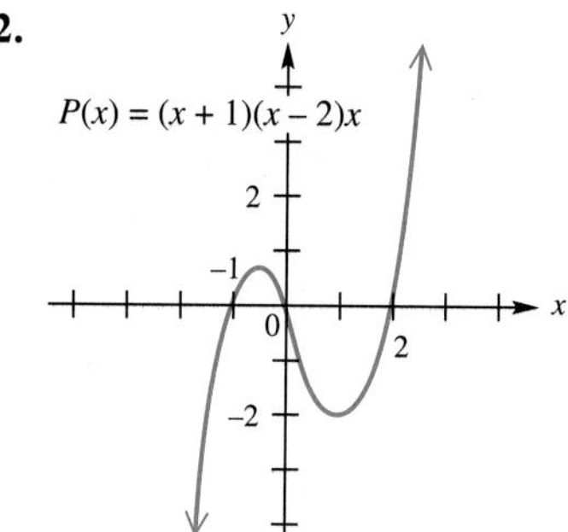

[9.2] 3. $P(x) = x^3 - x^2 - 10x - 8$
4. no **5.** yes
6. $P(x) = 2x^4 - 2x^3 - 2x^2 - 2x - 4$
7. $-3, 4, 2 - i, 2 + i$
8. $P(x) = (x + 2)(2x + 1)(x - 3)$

[9.3] 9. (a) $\pm 1, \pm\frac{1}{2}, \pm\frac{1}{3}, \pm\frac{1}{6}, \pm 7, \pm\frac{7}{2}, \pm\frac{7}{3}, \pm\frac{7}{6}$
(b) $-\frac{1}{3}, 1, \frac{7}{2}$

[9.4] **13.** positive: 2 or 0; negative: 2 or 0

15.

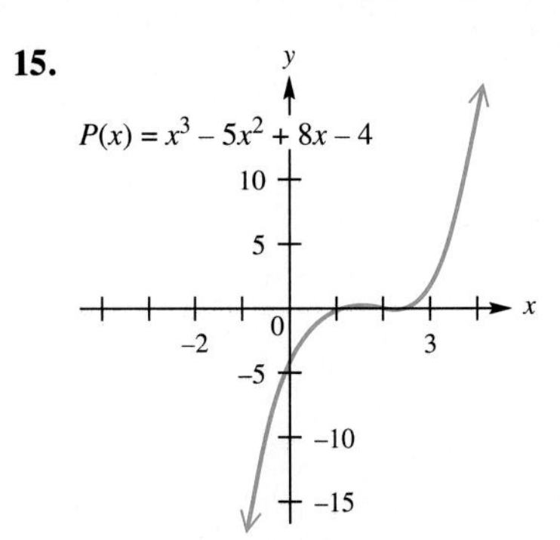

[9.5] **16.**

$f(x) = \frac{-2}{x+3}$

17.

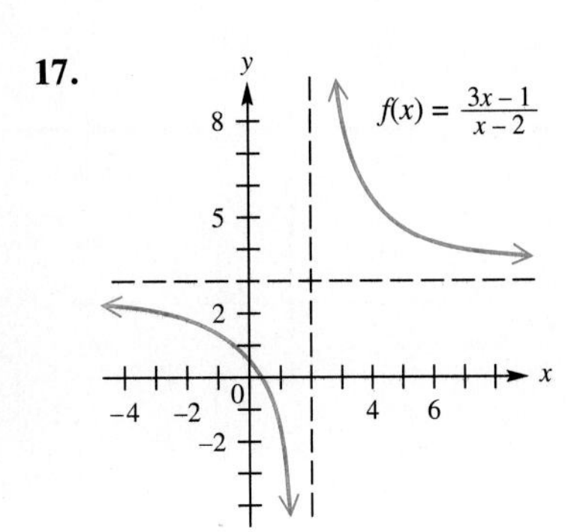

18.

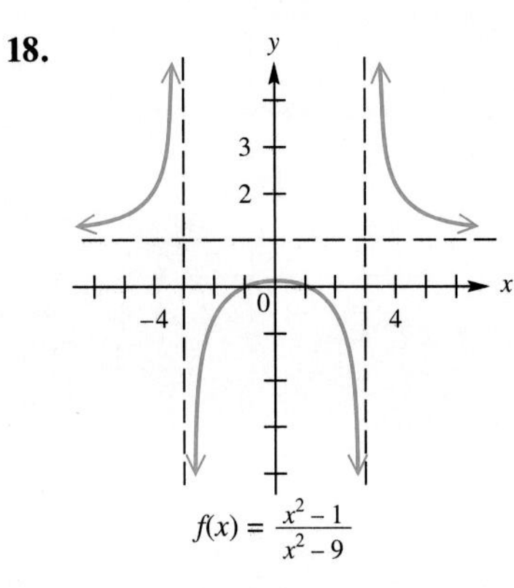

19.

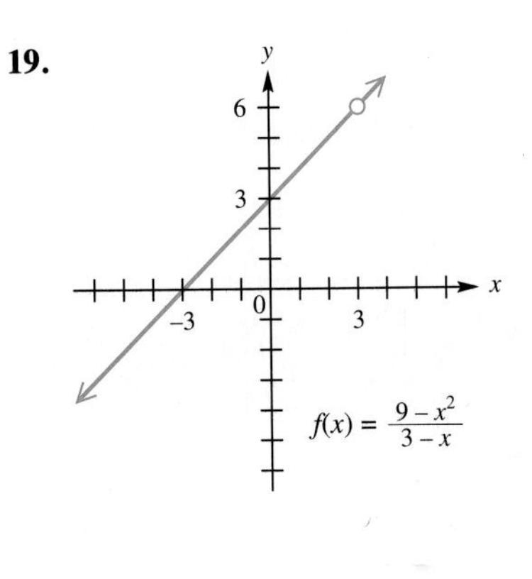

20. (d)

■ CHAPTER 10

10.1 Exercises (page 533)

1. one-to-one **3.** one-to-one **5.** not one-to-one **7.** one-to-one **9.** one-to-one **11.** not one-to-one **13.** not one-to-one **15.** not one-to-one **17.** one-to-one **19.** one-to-one **21.** one-to-one **23.** not one-to-one **25.** taking off your shoes **27.** waking up **29.** spending money **31.** (b) **33.** inverses **35.** not inverses **37.** not inverses **39.** inverses **41.** inverses **43.** inverses **45.** inverses **47.** not inverses

49.

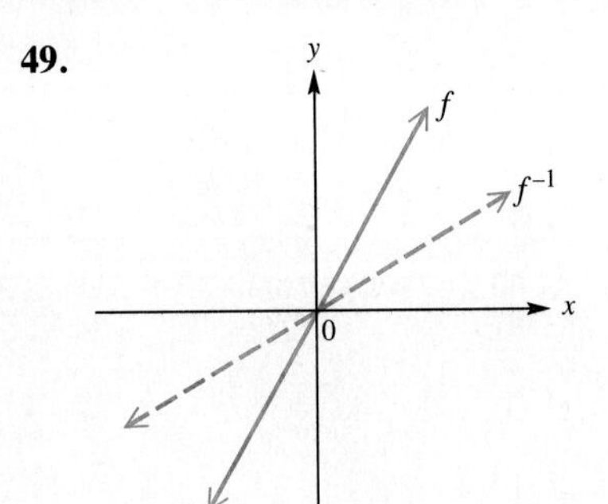

51.

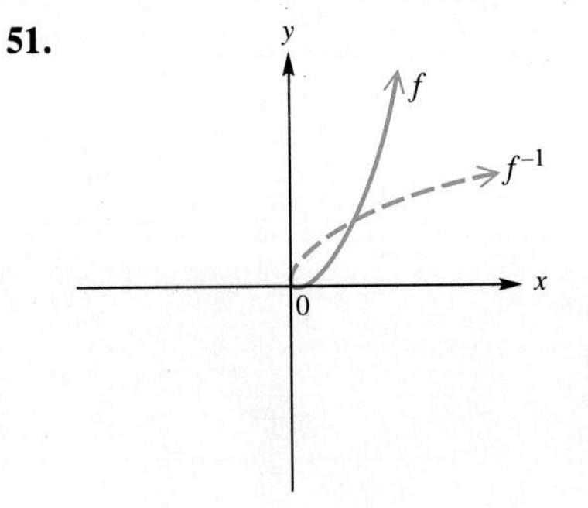

53.

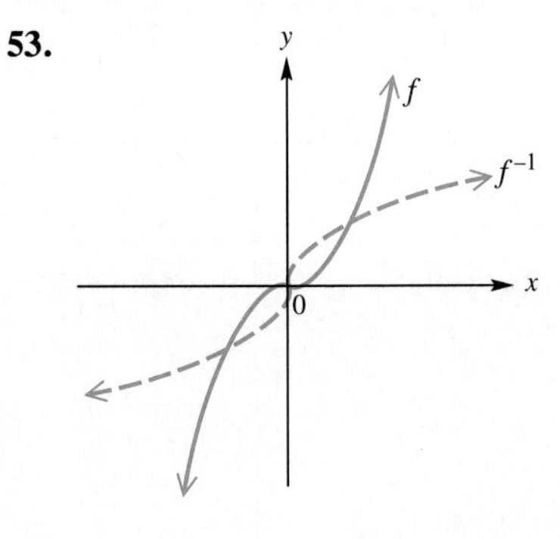

55. $f^{-1}(x) = \dfrac{x+4}{3}$

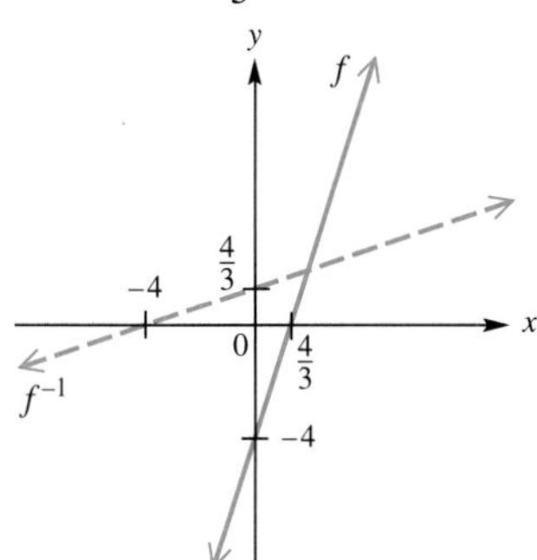

57. $f^{-1}(x) = 3x$

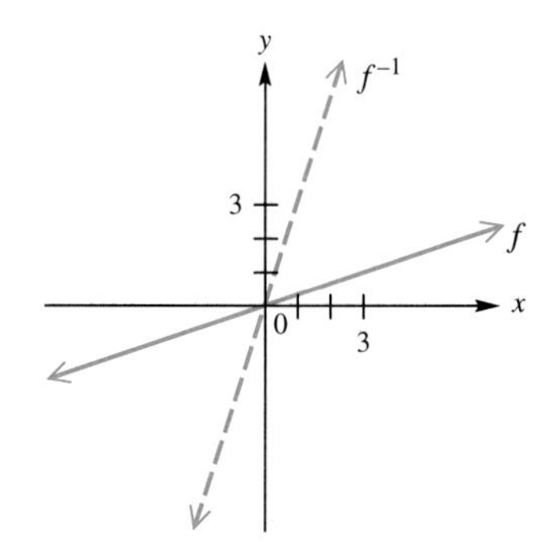

59. $f^{-1}(x) = \sqrt[3]{x-1}$

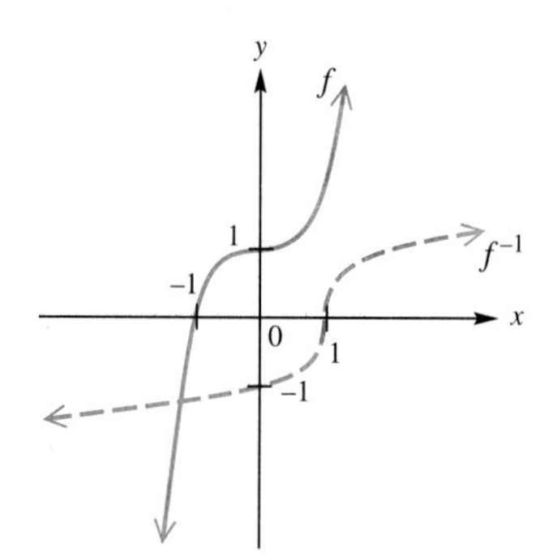

61. $f^{-1}(x) = 12 - 4x$

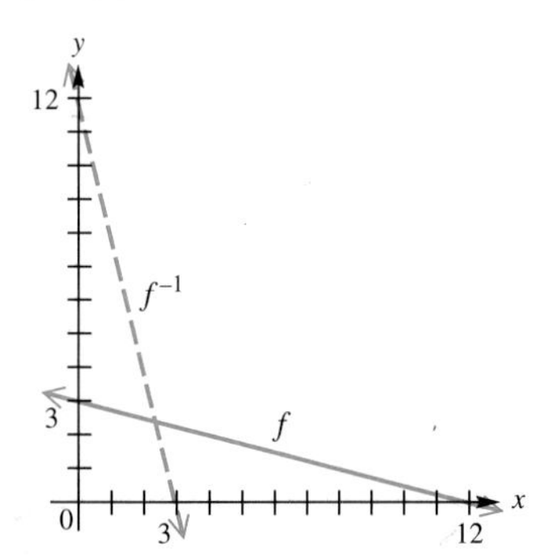

63. not one-to-one

65. $f^{-1}(x) = \dfrac{1}{x}$

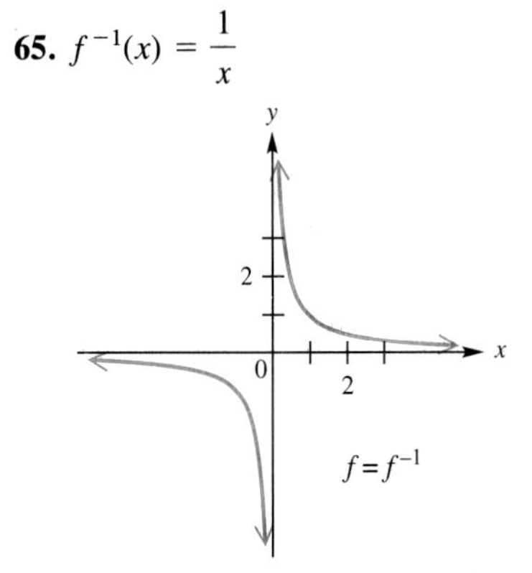

67. $f^{-1}(x) = \dfrac{3}{x}$

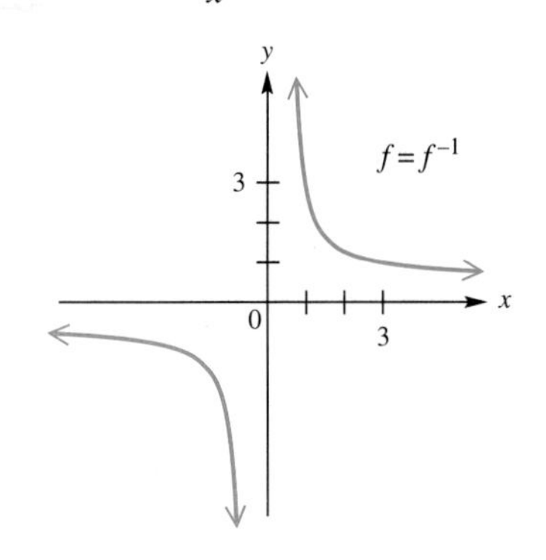

69. $f^{-1}(x) = -\sqrt{4-x}$; domain: $(-\infty, 4]$

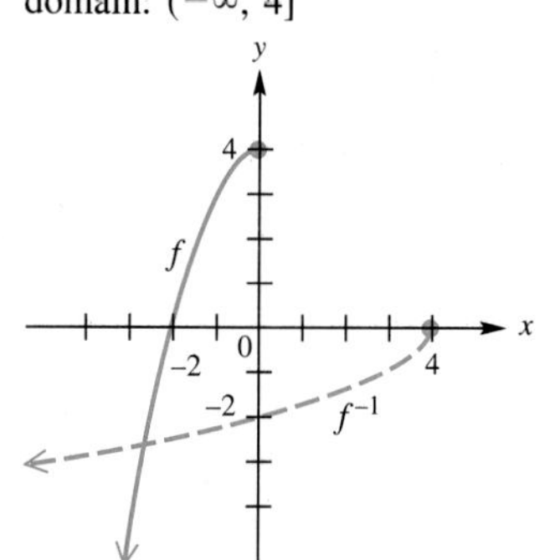

71. $f^{-1}(x) = 9 - x^2$; domain: $[0, \infty)$

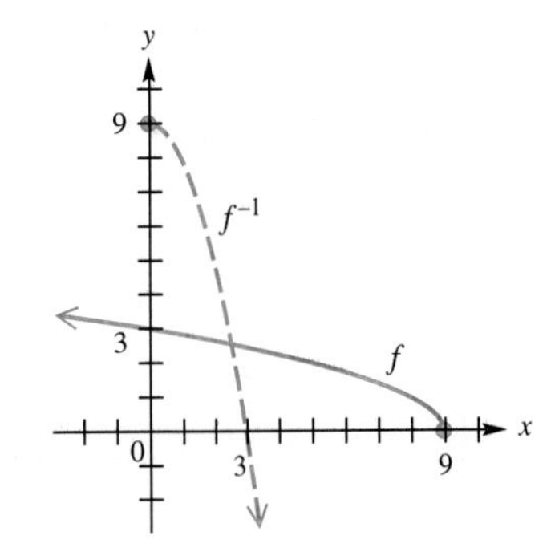

73. $f^{-1}(x) = \sqrt{x^2 + 16}$; domain: $(-\infty, 0]$ **75.** $\dfrac{1}{9}$ **77.** 4 **79.** $\dfrac{1}{243}$

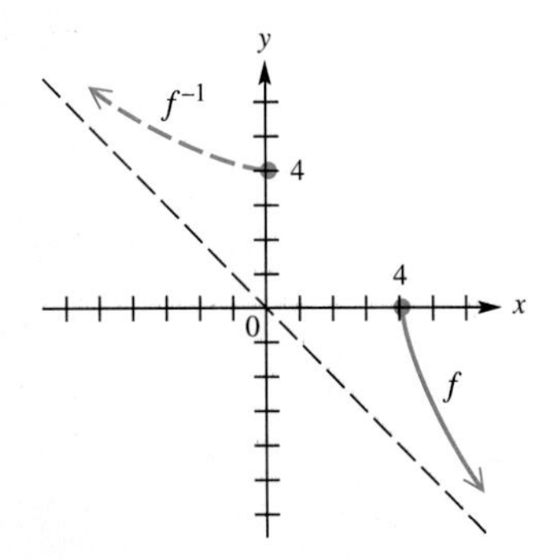

10.2 Exercises (page 545)

1. $\left\{\dfrac{1}{2}\right\}$ **3.** $\{-2\}$ **5.** $\{0\}$ **7.** $\{3\}$ **9.** $\{8\}$ **11.** $\{0\}$ **13.** $\left\{\dfrac{3}{5}\right\}$ **15.** $\left\{-\dfrac{2}{3}\right\}$ **17.** 8.104
19. .134

23. (a)

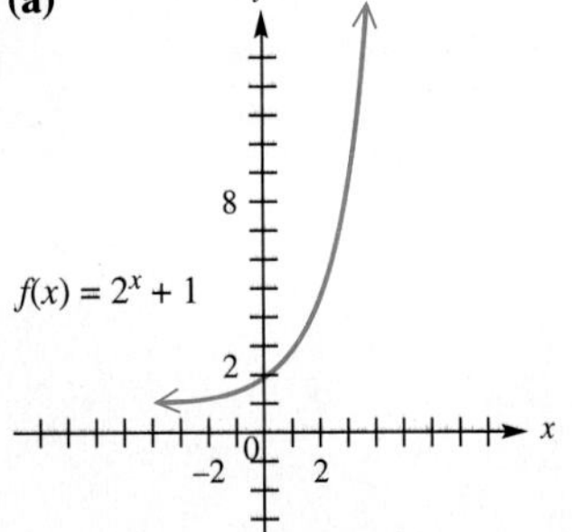

(b)

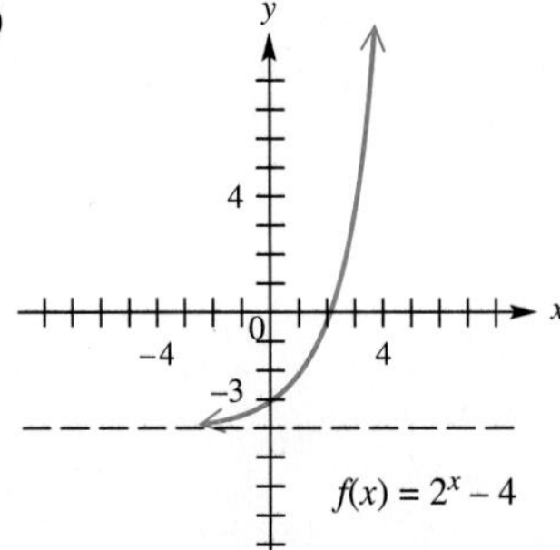

(c)

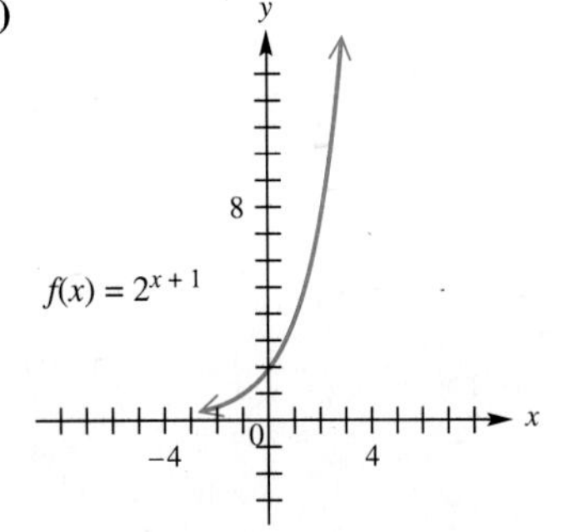

(d)

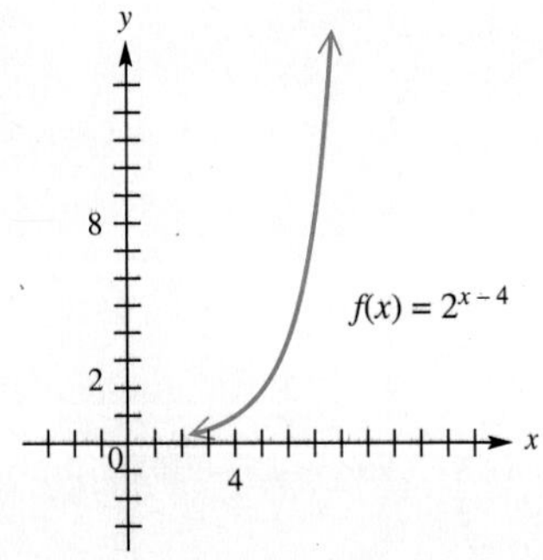

25.

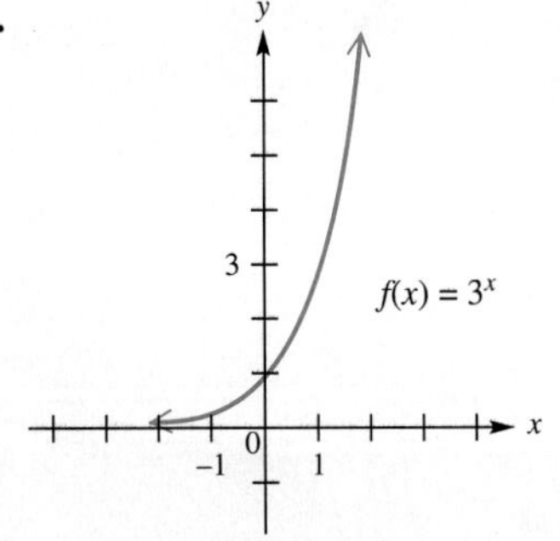

27.

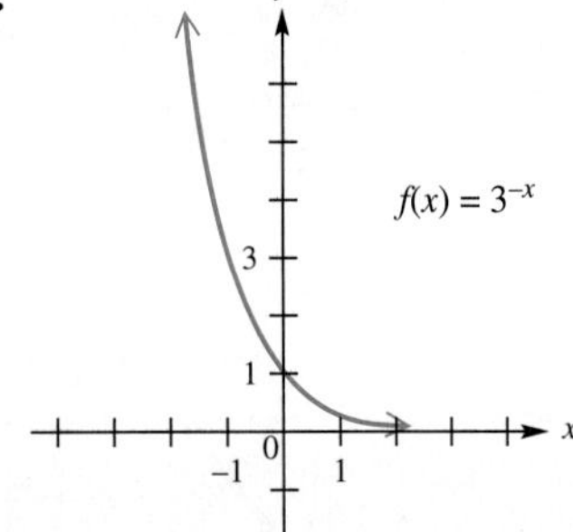

29.

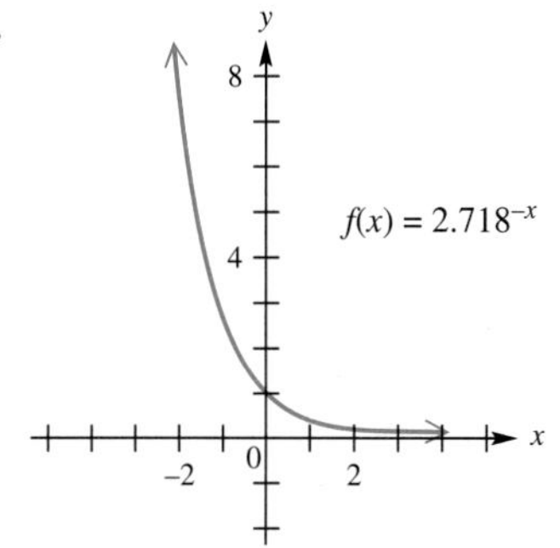

31.

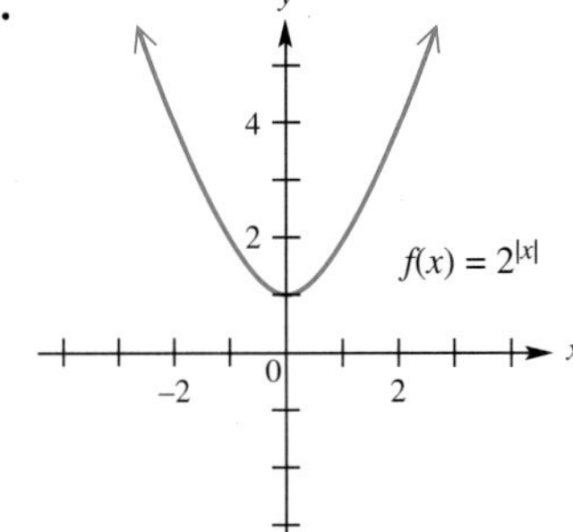

33.

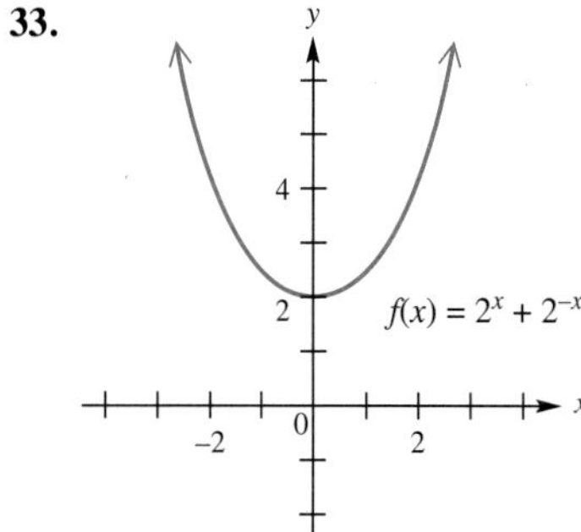

35. $\frac{1}{2}$ **37.** 8 **41.** $(-\infty, \infty)$

43.

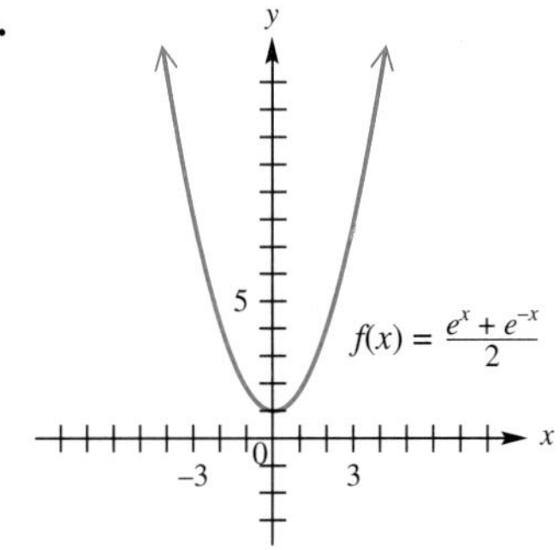

45. **(a)** \$5000 **(b)** \$2973
(c) \$1768

(d)

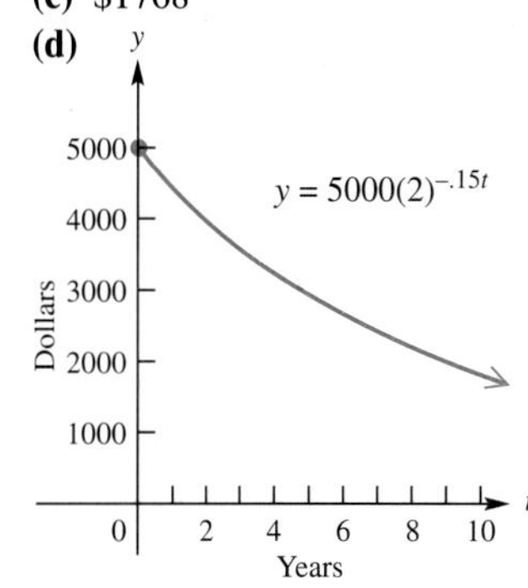

47. **(a)** 1,000,000
(b) about 1,041,000
(c) about 1,083,000
(d) about 1,221,000
(e)

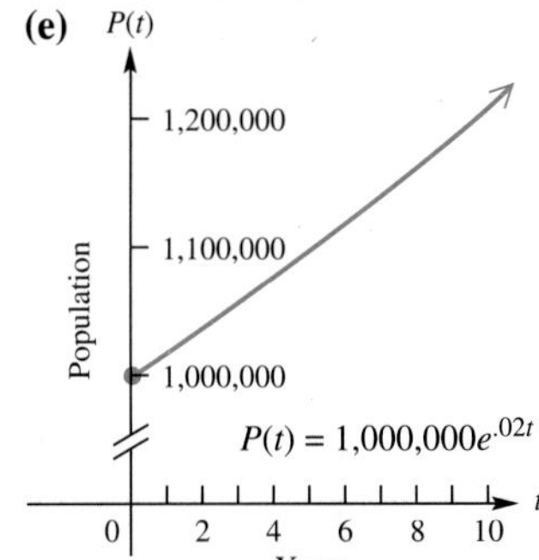

49. **(a)** about 45,200; about 37,000 **(b)** about 72,400; about 48,500
51. \$7686.32 **53.** \$13,891.16 **55.** \$5583.95 **57.** \$41,845.63
59. 12.0% **61.** 8.0% **63.** **(a)** .55° C **(b)** .35° C
65. **(a)** 1.75° C **(b)** .5° C **67.** **(a)** about 2,690,000
(b) about 3,020,000 **(c)** about 9,600,000 **(d)** about 38,400,000
69. 2.717 **71.** $\frac{1}{8}$ **73.** 6 **75.** $9\sqrt{3}$ **77.** $\frac{\sqrt[3]{4}}{4}$

10.3 Exercises (page 554)

1. $\log_3 81 = 4$ **3.** $\log_{10} 10{,}000 = 4$ **5.** $\log_{1/2} 16 = -4$ **7.** $\log_{10} .0001 = -4$ **9.** $6^2 = 36$
11. $(\sqrt{3})^8 = 81$ **13.** $10^{-4} = .0001$ **15.** $m^n = k$ **17.** 2 **19.** 1 **21.** −3 **23.** $-\frac{1}{6}$ **25.** 8
27. 9 **29.** −2 **31.** 4 **37.** {3} **39.** {81} **41.** $\left\{\frac{1}{2}\right\}$ **43.** {2} **45.** {9} **47.** {1} **49.** {12}

51. (a)

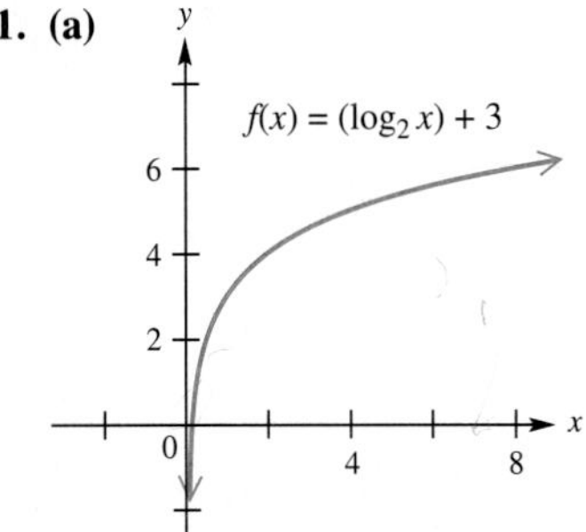

(b)

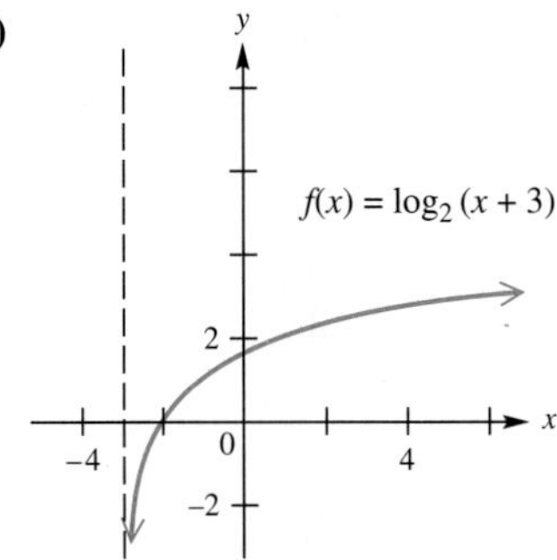

(c)

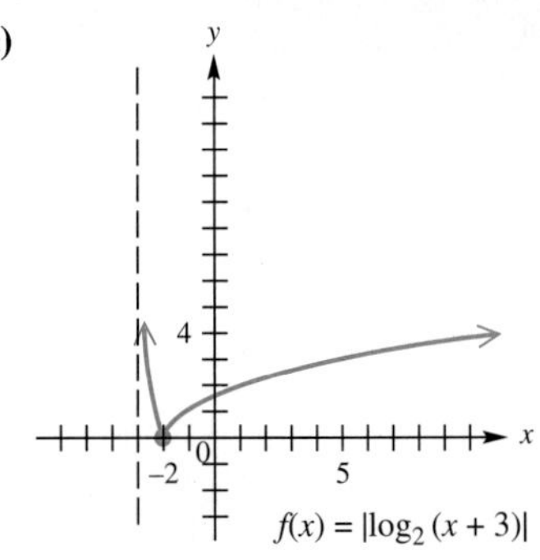

53.

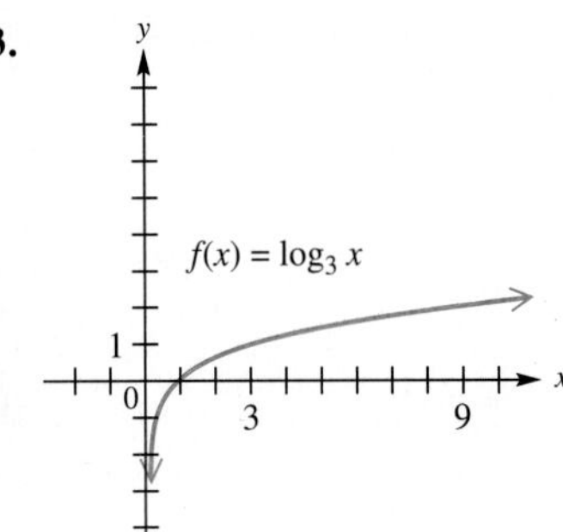

55.

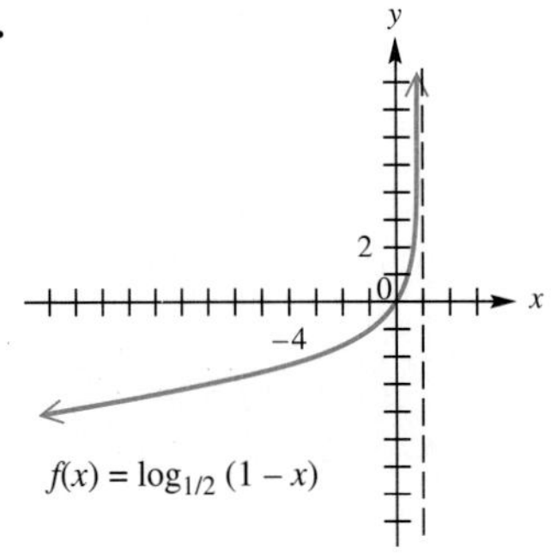

57.

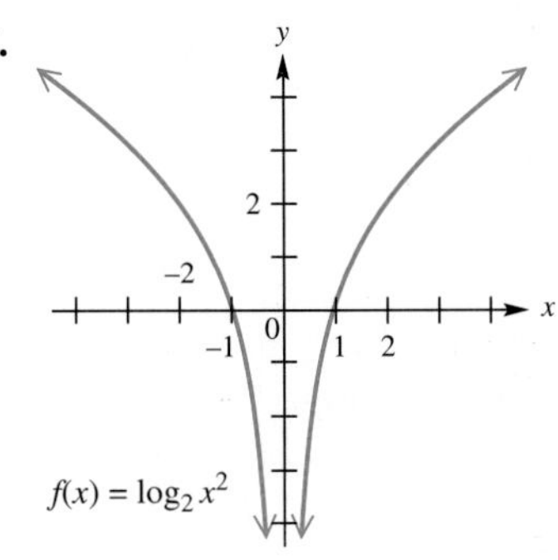

59.

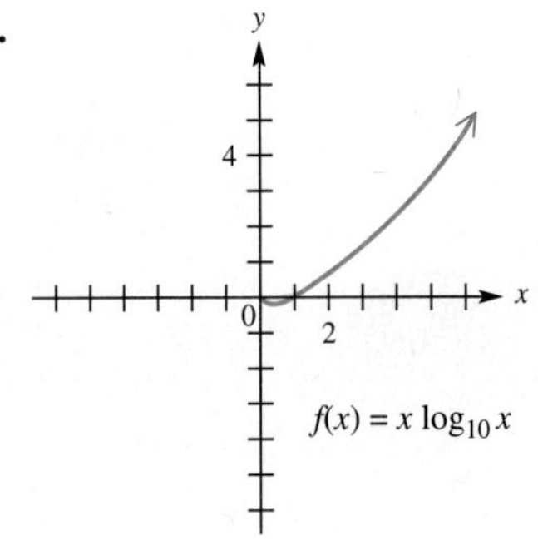

61. (a) 8 **(b)** 16 **(c)** 24 **(d)**

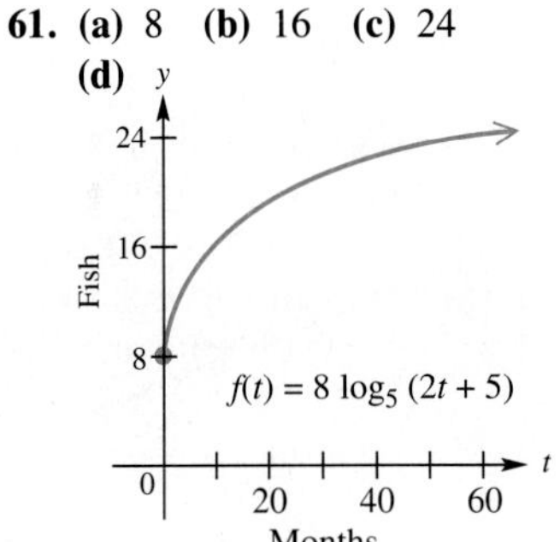

63. (a) about 239 **(b)** about 477 **(c)** about 759 **(d)**

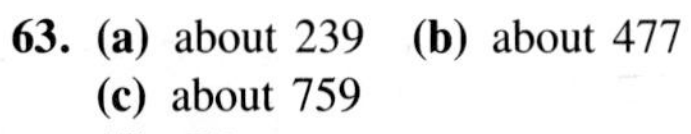

$F(t)$
Population
Months
$F(t) = 500 \log_{10} (2t + 3)$

65. (a) 20 **(b)** 30 **(c)** 50 **(d)** 60 **67. (a)** 3 **(b)** 6 **(c)** 8 **69.** 2^9 **71.** $\frac{1}{4^6}$ **73.** 2^{12} **75.** $\frac{1}{p^4}$

10.4 Exercises (page 562)

1. $\log_6 3 - \log_6 4$ **3.** $\frac{1}{2} \log_2 8 = \frac{3}{2}$ **5.** $2 \log_3 3 = 2$ **7.** $1 + \log_5 x - \log_5 y$

9. $\log_2 5 + \frac{1}{2} \log_2 7 - \log_2 3$ **11.** $\frac{1}{2} \log_3 m + \frac{1}{2} \log_3 a - \frac{1}{2} \log_3 b$ **13.** not possible to rewrite using the properties of this section **17.** $\log_2 8 - \log_2 4 = \log_2 \frac{8}{4} = \log_2 2 = 1$ **19.** $\log_a \frac{xy}{m}$ **21.** $\log_m \frac{a^2}{b^6}$

23. $\log_x (a^{3/2}b^{-4})$ **25.** $\log_b \dfrac{7x(x+2)}{8}$ **27.** $\log_a [(z+1)^2(3z+2)]$ **29.** $\log_5 (5^{1/3}m^{-1/3})$ or $\log_5 \dfrac{5^{1/3}}{m^{1/3}}$ or $\dfrac{1}{3}\log_5 \dfrac{5}{m}$ **31.** .7781 **33.** 1.2040 **35.** 2.2552 **39.** $\log_b AB = -2$; $(\log_b A)\cdot(\log_b B) = -8$; different **41.** $\log_b B^2 = -8$; $(\log_b B)^2 = 16$; different **43.** false **45.** true **47.** false **51.** $\dfrac{2}{3}$ **53.** $-\dfrac{1}{2}$

10.5 Exercises (page 570)

1. 2.4440 **3.** .5159 **5.** .9926 **7.** −.4855 **9.** −3.1726 **11.** 5.8293 **13.** 1.6094 **15.** 6.5468 **17.** .5423 **19.** 4.5809 **21.** 3.1 **23.** 3.3863 **27.** (a) **29.** 3.42 **31.** 454 **33.** .0762 **35.** 47.0 **37.** 4.04 **39.** .659 **41.** 1.4899 **43.** .5794 **45.** 1.0892 **47.** 1.6083 **49.** .8803 **51.** 6.4 **53.** 1.8 **55.** 3.2×10^{-6} **57.** 2.0×10^{-3} **59.** about 530 years **61.** about 82% **63.** 1 **65.** **(a)** 2.03 **(b)** 2.28 **(c)** 2.17 **(d)** 1.21 **67.** **(a)** 11% **(b)** 36% **(c)** 84% **69.** $\left\{-\dfrac{21}{26}\right\}$ **71.** $\left\{\dfrac{1}{2}\right\}$ **73.** {47} **75.** $\{4\sqrt{3}, -4\sqrt{3}\}$

10.6 Exercises (page 579)

1. {1.631} **3.** {−.535} **5.** {−.080} **7.** {2.386} **9.** {−.123} **11.** ∅ **13.** {2} **15.** $\left\{\dfrac{1}{2}\right\}$ **17.** {4} **19.** {−.418} **21.** {−.564} **23.** {17.475} **25.** {11} **27.** {5} **29.** {10} **31.** {4} **33.** ∅ **35.** $\left\{\dfrac{3}{8}\right\}$ **37.** {3} **39.** ∅ **41.** {8} **43.** {−2, 2} **45.** {1, 10} **49.** **(a)** \$9404.63 **(b)** \$9405.31 **(c)** \$9405.34 **51.** \$6549.85 **53.** **(a)** 13.9 years **(b)** 11.6 years **(c)** 8.7 years **55.** **(a)** 6.96 grams **(b)** 4.83 grams **(c)** 5730 years **57.** about 15,000 years **59.** about 34.7 seconds **61.** **(a)** 611 million **(b)** 1007 million **(c)** about 18 days **(d)** January 19 **63.** **(a)** about 961,000 **(b)** about 7.2 years **(c)** about 17.3 years **65.** $\left\{\dfrac{7}{5}\right\}$ **67.** $\left\{-\dfrac{25}{11}\right\}$ **69.** {−3} **71.** {6}

Chapter 10 Review Exercises (page 585)

1. $f^{-1}(x) = 3x + 1$ **3.** not a one-to-one function **5.** $f^{-1}(x) = -2 + \sqrt[3]{x+1}$ **7.** $f^{-1}(x) = \dfrac{x^2-5}{2}$, domain $[0, \infty)$

9.

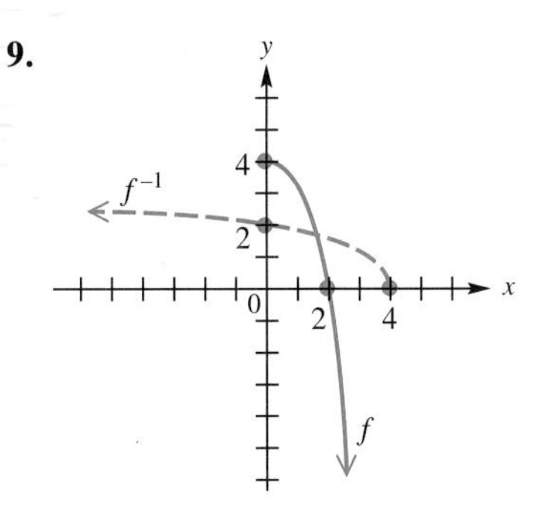

11.

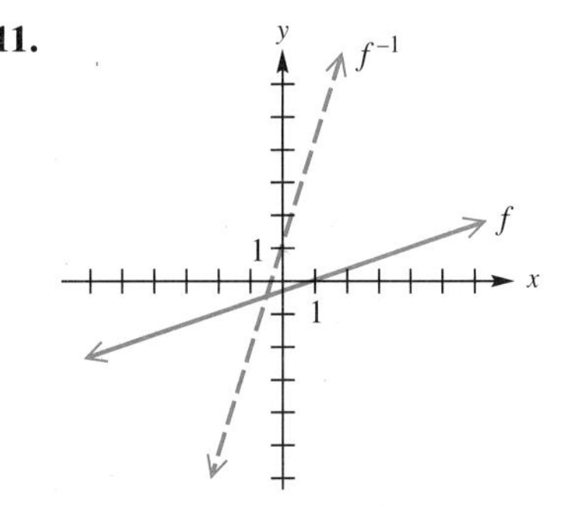

13. true

15.

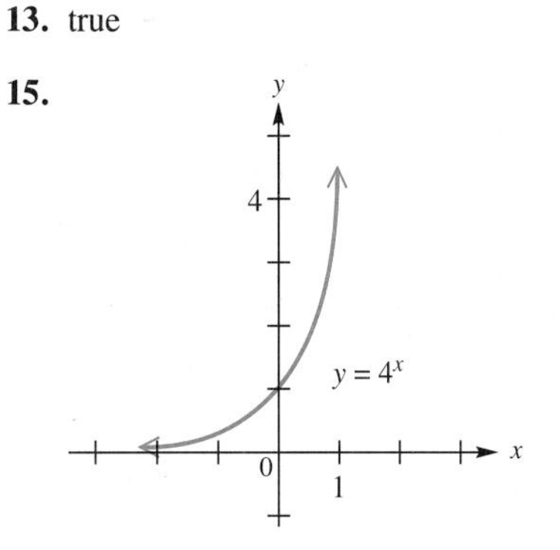

17.

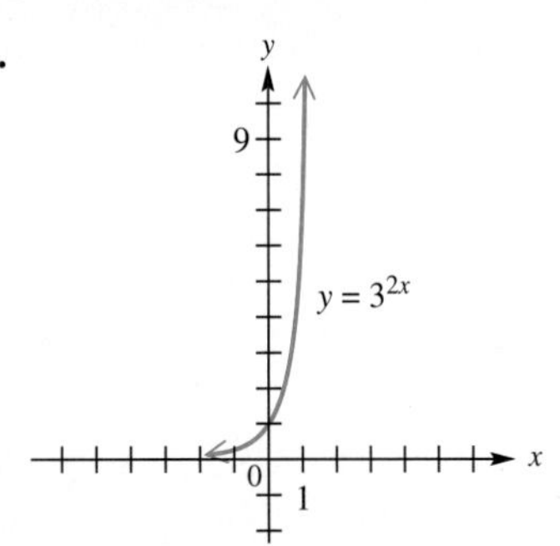

19. \$5375.67

21.

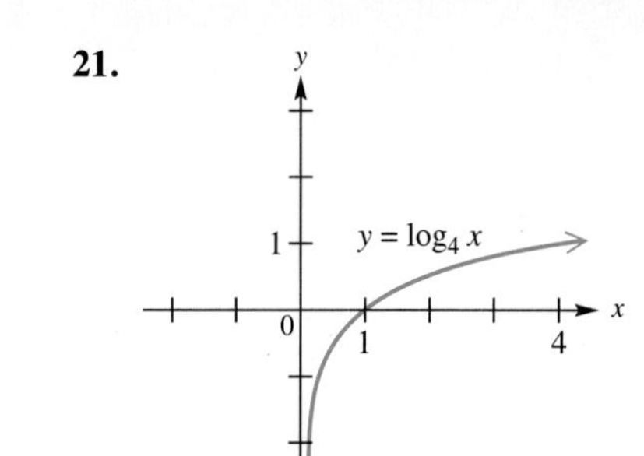

23. $3^4 = 81$ **25.** $8^{1/3} = 2$ **27.** $\log_9 3 = \frac{1}{2}$ **29.** $\log_5 3 + \log_5 x - \log_5 10$ **31.** $\log_2 5 + \log_2 k - \log_2 3 - 3 \log_2 r$ **33.** $\log_a 6x$ **35.** $\log_5 \frac{1}{x+1}$ **37.** $\log_2 x$ **39.** .4698 **41.** −1.1463 **43.** .2 **45.** 2.0794 **47.** 1 **49.** 4 **51.** 1.2851 **53.** 2.1707 **55.** 1.4706 **57.** 2.5863 **59.** {.16} **61.** $\left\{\frac{11}{3}\right\}$ **63.** $\left\{\frac{3}{8}\right\}$ **65.** \$11,972.17 **67.** **(a)** 1105 **(b)** 1284 **(c)** 1649 **69.** \$17,433.92 **71.** $\left\{-\frac{2}{3}\right\}$ **73.** \$28,295.64 **75.** {5}

Chapter 10 Test (page 588)

[10.1] **1.** false **2.** $f^{-1}(x) = 4 - x^3$

3.

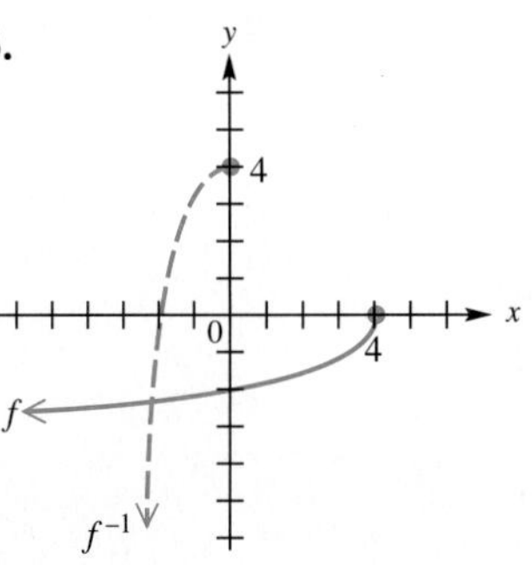

[10.2] **4.**

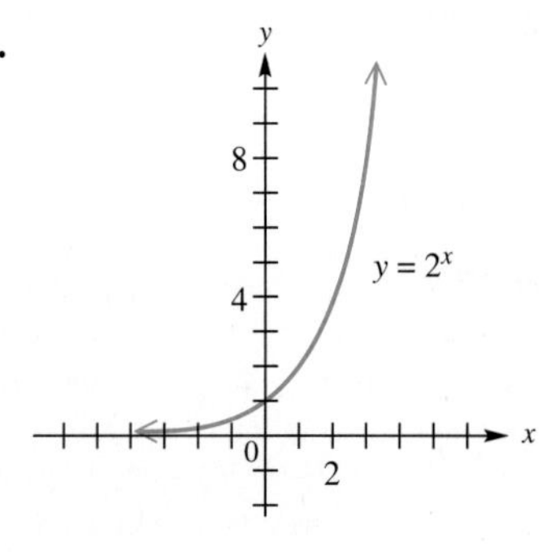

5.

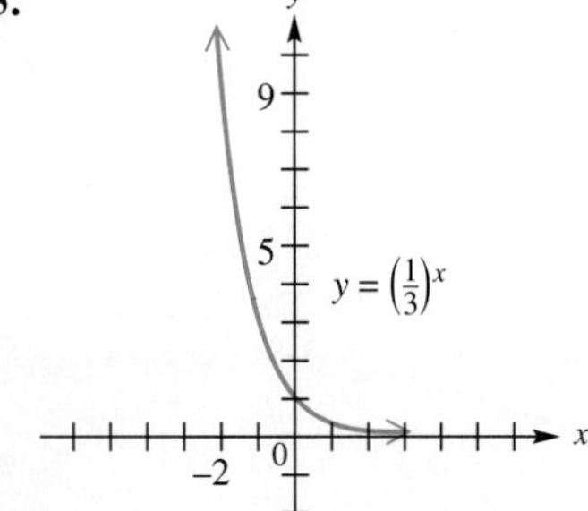

[10.3] **6.**

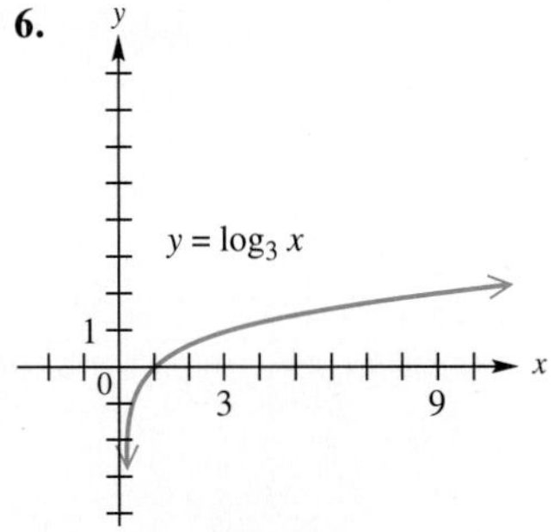

7.

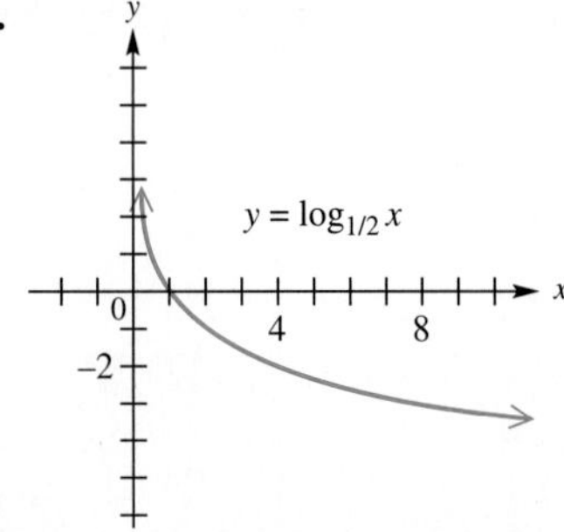

[10.4] **8.** 2.4983 **9.** 2.1329 **10.** 1.1578 **[10.5]** **11.** 2 **12.** −6 **13.** $-\frac{3}{4}$ **14.** 1 **15.** 2.3909 **16.** −3.4989 **17.** 3.3219 **18.** 4.4898 **[10.6]** **19.** {243} **20.** $\left\{\frac{1}{3}\right\}$ **21.** {3.81} (rounded) **22.** {2} **23.** {6} **24.** **(a)** 8000 **(b)** about 59,100 **(c)** at about 1.7 hours **25.** **(a)** \$4804.75 **(b)** \$4819.95

CHAPTER 11

11.1 Exercises (page 598)

1. single solution **3.** infinitely many solutions **5.** no solution **7.** no solution **9.** infinitely many solutions **11.** $\{(5, 5)\}$ **13.** $\{(3, 1)\}$ **15.** $\{(5, 0)\}$ **17.** $\left\{\left(\frac{3-2y}{7}, y\right)\right\}$; dependent **19.** $\{(11, 7)\}$ **21.** $\{(2, -3)\}$ **23.** $\emptyset$; inconsistent **25.** $\{(2, -4)\}$ **27.** $\{(-5y-2, y)\}$; dependent **29.** Multiply (3) by 3 and (4) by 2. **31.** $\{(2, 1)\}$ **33.** $\{(-2, -10)\}$ **35.** $\emptyset$ **37.** $\{(5, 4)\}$ **39.** $\left\{\left(-5, -\frac{10}{3}\right)\right\}$ **41.** $\left\{\left(3, \frac{9}{2}\right)\right\}$ **43.** $\{(2, 1)\}$ **47.** $\{(2, 4)\}$ **49.** $\{(4, -5)\}$ **51.** $\emptyset$ **53.** $\{(4, 0)\}$ **55.** $\left\{\left(\frac{1}{a}, \frac{1}{b}\right)\right\}$ **57.** $\left\{\left(-\frac{3}{5a}, \frac{7}{5}\right)\right\}$ **59. (a)** about \$1000 for A and \$500 for B **(b)** year 3.5: about \$800 **(c)** 7 for A; 1.5 for B **61. (a)** $x = 8$ or 800 items: \$3000 **(b)** about $-\$500$ **(c)** \$1000 **63.** $-6x + 12y - 3z = -24$ **65.** $2x - 6y + 4z = -36$ **67.** $3x + 11y = -3$

11.2 Exercises (page 606)

1. $\{(-1, 2, 3)\}$ **3.** $\{(2, 1, -4)\}$ **5.** $\{(3, -1, 4)\}$ **7.** $\left\{\left(-\frac{7}{3}, \frac{22}{3}, 7\right)\right\}$ **9.** $\{(3, -1, 2)\}$ **11.** $\{(4, 5, 3)\}$ **13.** $\{(2, 2, 2)\}$ **15.** $\{(-1, 2, -2)\}$ **17.** $\emptyset$ **19.** $\left\{\left(\frac{20}{11}, \frac{10}{11}, -\frac{41}{11}\right)\right\}$ **21.** $\{(x, y, x + 4y - 3)\}$ **23.** $\emptyset$ **25.** $\{(0, 0, 0)\}$ **27.** $\{(4y - z - 5, y, z)\}$ **29.** $\{(2, 1, 5, 3)\}$ **31.** $m = 3$, $b = -7$; $y = 3x - 7$ **33.** $a = 3$, $b = 1$, $c = -2$ **37.** 20 centiliters **39.** 25 miles per hour **41.** 40°, 60°, 80°

11.3 Exercises (page 614)

1. 12, 12, and 15 centimeters **3.** square: 10 centimeters; triangle: 6 centimeters **5.** small: \$1.80; medium: \$2 **7.** 6 of yarn; 2 of thread **9.** dark: \$.60 per kilogram; light: \$.40 per kilogram **11.** 5 pounds of oranges; 1 pound of apples **13.** 5 twenties; 30 fifties **15.** \$5000 at 8%; \$10,000 at 7% **17.** \$24,000 at 14%; \$32,000 at 7% **19.** 1 liter of pure acid; 8 liters of 10% acid **21.** 4 liters of pure antifreeze **23.** 28 kilograms of nuts; 32 kilograms of cereal **25.** freight train: 50 kilometers per hour; express train: 80 kilometers per hour **27.** train: 60 kilometers per hour; plane: 160 kilometers per hour **29.** 9, 10, 13 **31.** 84°; 70°; 26° **33.** 16 inches; 18 inches; 22 inches **35.** 7 nickels; 17 dimes; 20 quarters **37.** sub-compact: 9; compact: 5; full size: 6 **39.** "up close": \$10; "in the middle": \$8; "far out": \$7 **41.** \$20,000 in video rental firm; \$30,000 each in the bond and the money market fund **43.** 60 ounces of Orange Pekoe; 30 ounces of Irish Breakfast; 10 ounces of Earl Grey **47.** $\left\{-\sqrt{3}, \sqrt{3}, -\frac{i\sqrt{2}}{2}, \frac{i\sqrt{2}}{2}\right\}$ **49.** $\{-2, 2, -2i, 2i\}$ **51.** $\{-\sqrt{3}, \sqrt{3}, -2, 2\}$

11.4 Exercises (page 624)

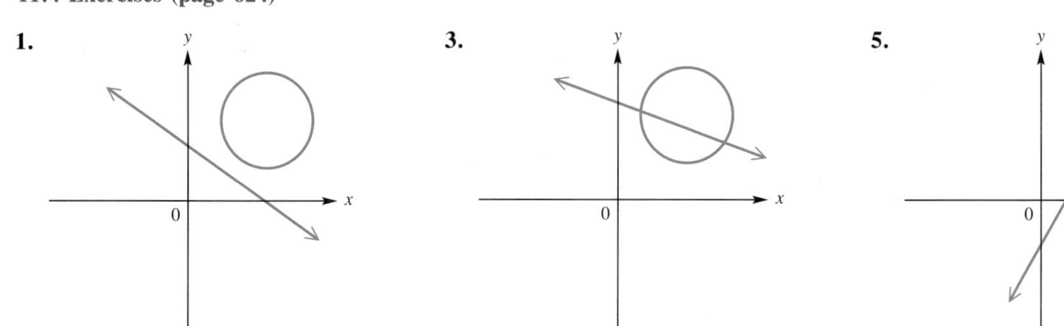

7.

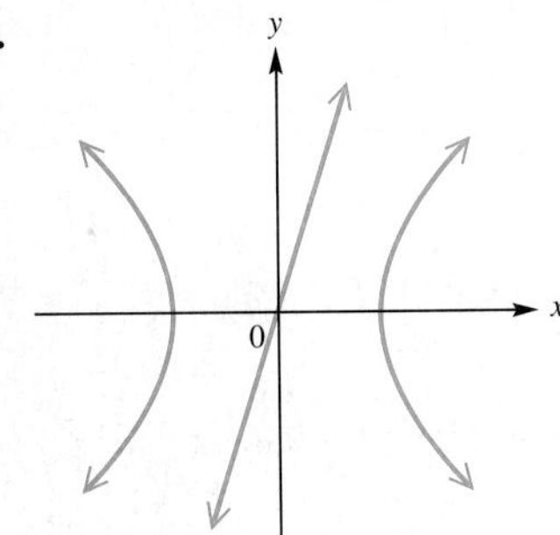

9.

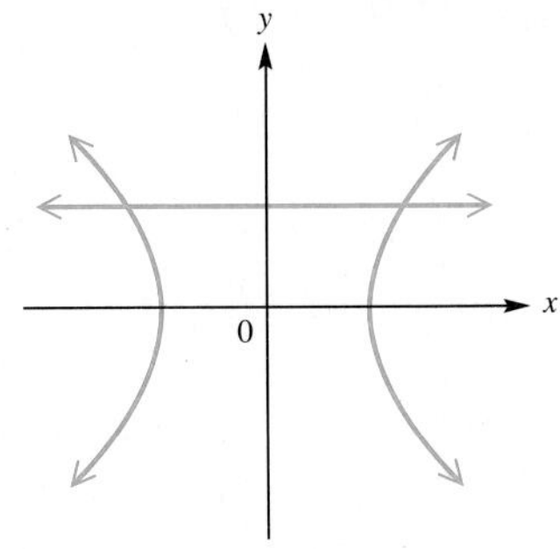

11.

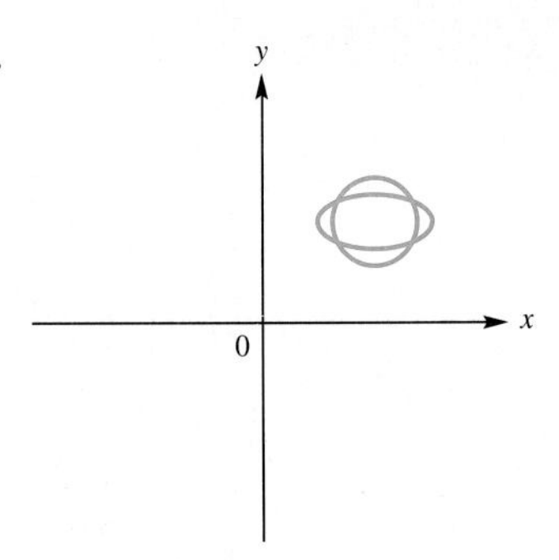

13.

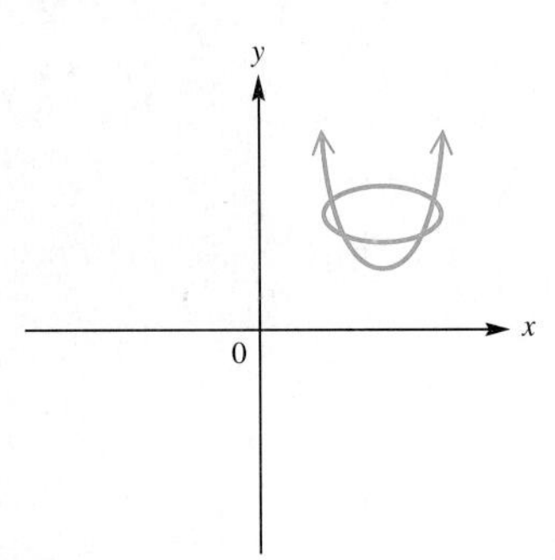

15.

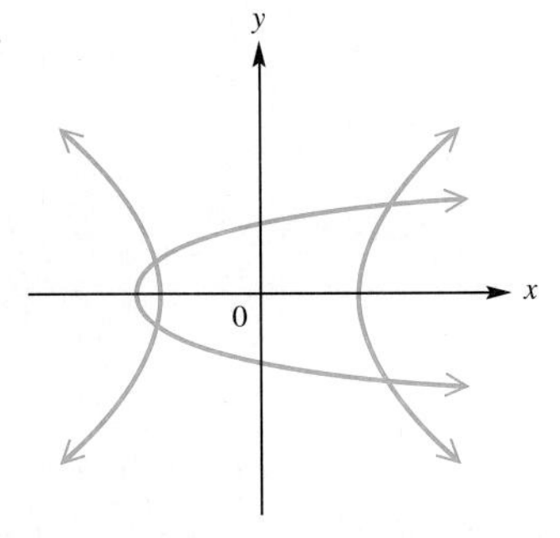

17. $\{(0, 0), (-1, -1)\}$

19. $\{(2, 0), (1, 1)\}$ **21.** $\left\{(1, 0), \left(-\frac{3}{5}, \frac{4}{5}\right)\right\}$ **23.** $\left\{\left(-3, -\frac{2}{3}\right), (-2, -1)\right\}$ **25.** $\{(2, -3), (-3, 2)\}$

27. $\{(-2, 0), (-5, 30)\}$ **29.** $\{(\sqrt{2}, 1), (-\sqrt{2}, 1)\}$ **31.** $\left\{\left(-\frac{\sqrt{3}}{3}i, -\frac{1}{2} - \frac{\sqrt{3}}{6}i\right), \left(\frac{\sqrt{3}}{3}i, -\frac{1}{2} + \frac{\sqrt{3}}{6}i\right)\right\}$

35. $\{(2, 0), (-2, 0)\}$ **37.** $\{(\sqrt{3}, 0), (-\sqrt{3}, 0)\}$ **39.** $\{(1, 3), (1, -3), (-1, 3), (-1, -3)\}$

41. $\{(2, -3), (2, 1), (-2, 3), (-2, -1)\}$ **43.** $\{(2, 1), (-2, -1), (i, -2i), (-i, 2i)\}$

45. $\{(-2, 1), (2, -1), (i, 2i), (-i, -2i)\}$ **47.** $\{(3, -3), (3, 3)\}$

49. $\left\{\left(\frac{1 + \sqrt{13}}{2}, \frac{-1 + \sqrt{13}}{2}\right), \left(\frac{-1 - \sqrt{21}}{2}, \frac{3 + \sqrt{21}}{2}\right)\right\}$ **51.** 2 and 2 or -2 and -2

53. length: 12 feet; width: 7 feet **55.** \$20; 800 calculators

57.

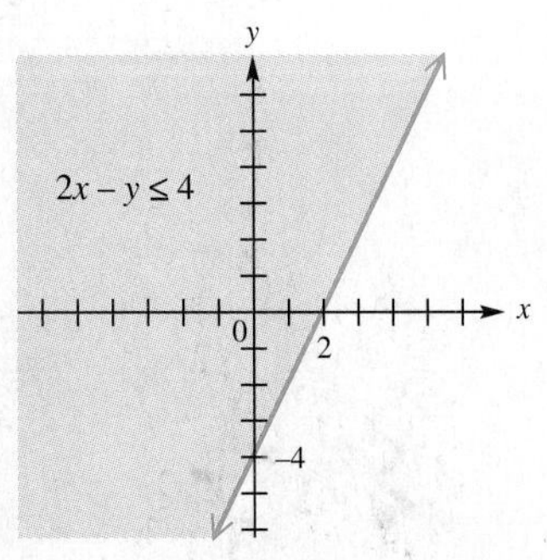

59.

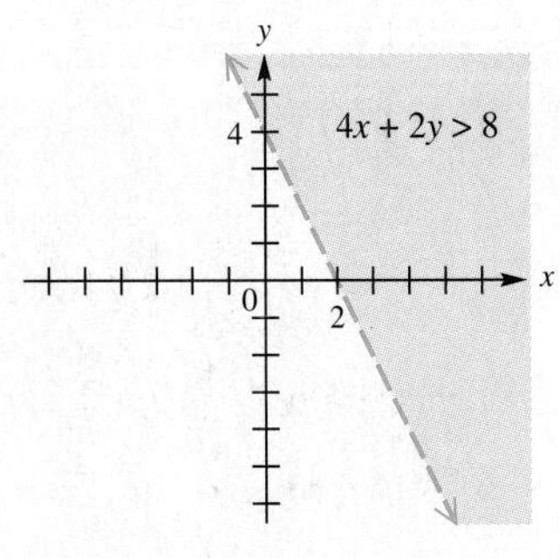

61.

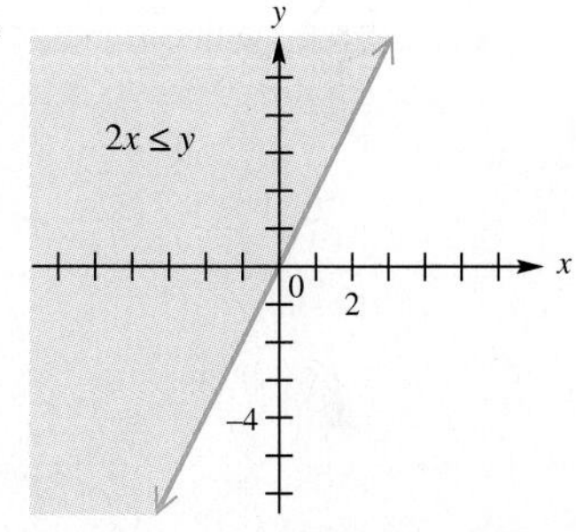

11.5 Exercises (page 632)

1.

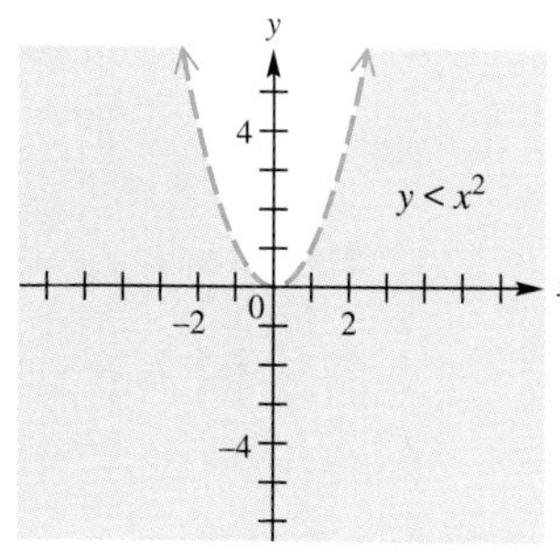

3.

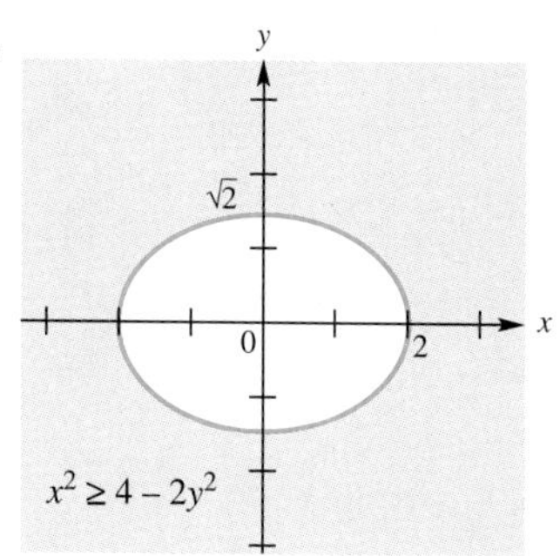

5.

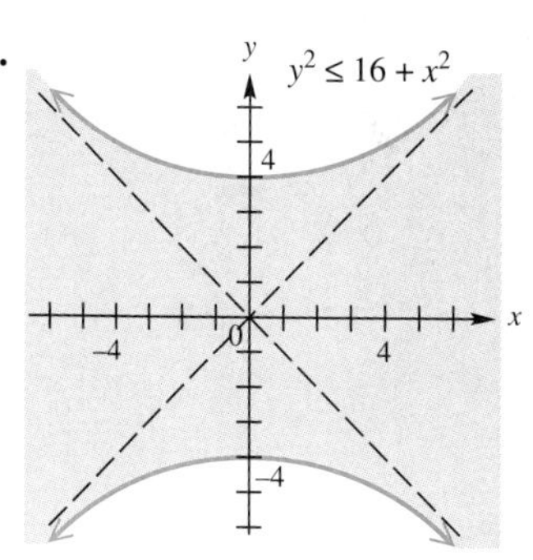

7.

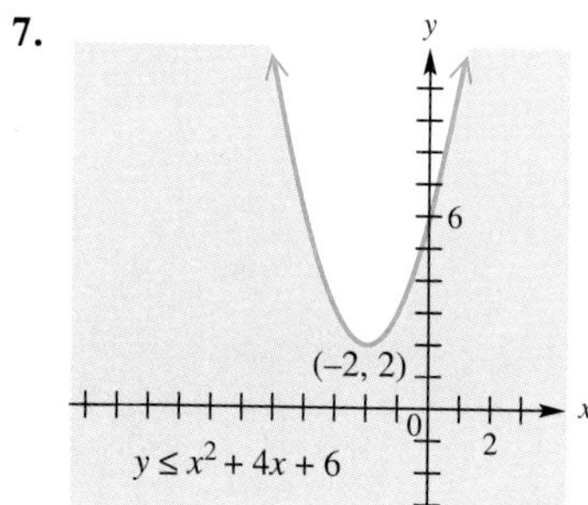

9.

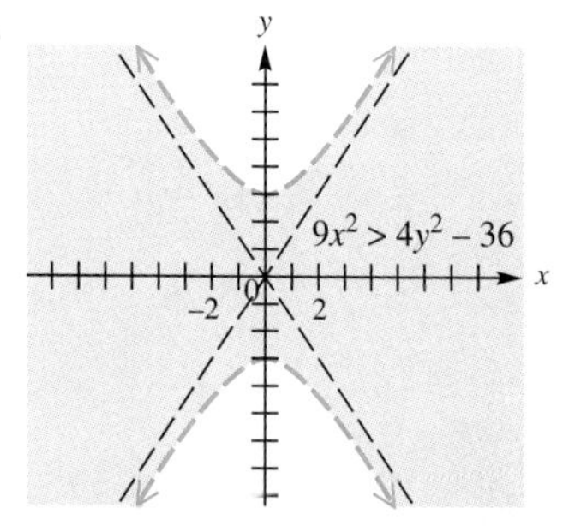

11.

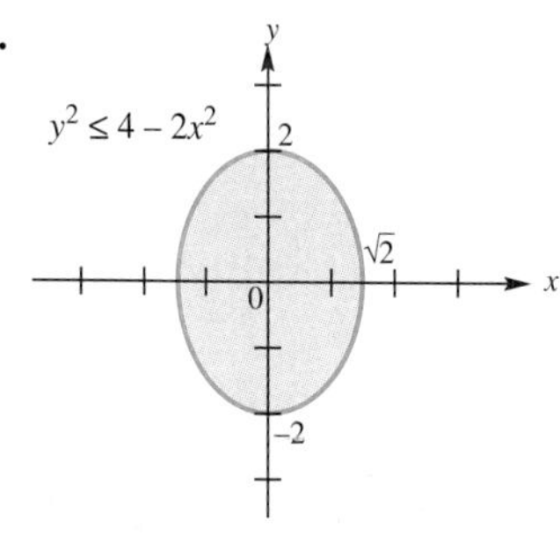

13.

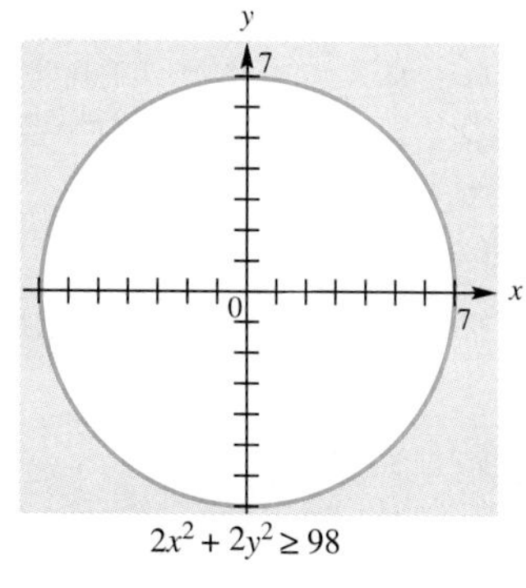

15.

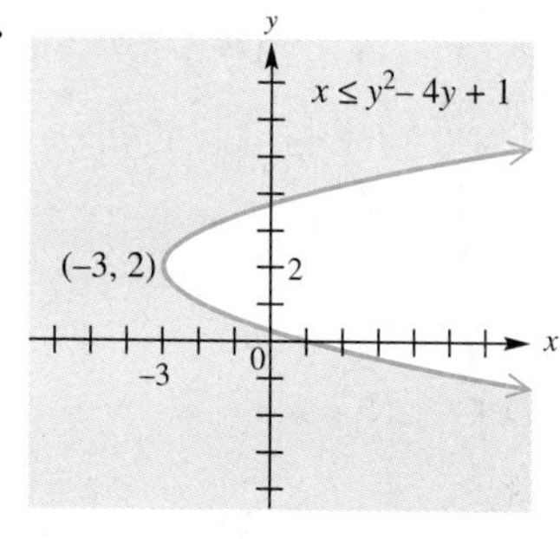

17.

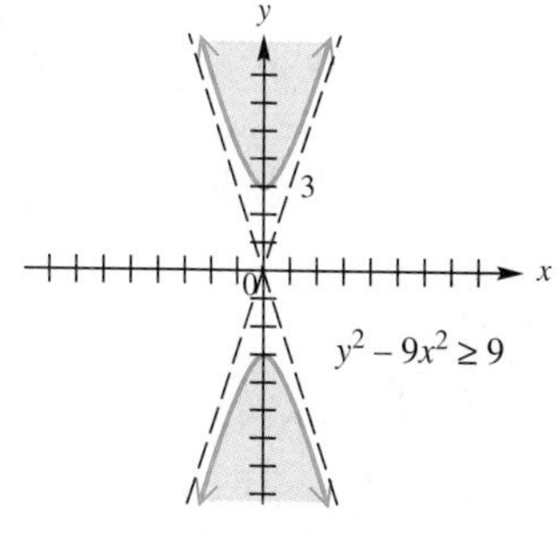

21.

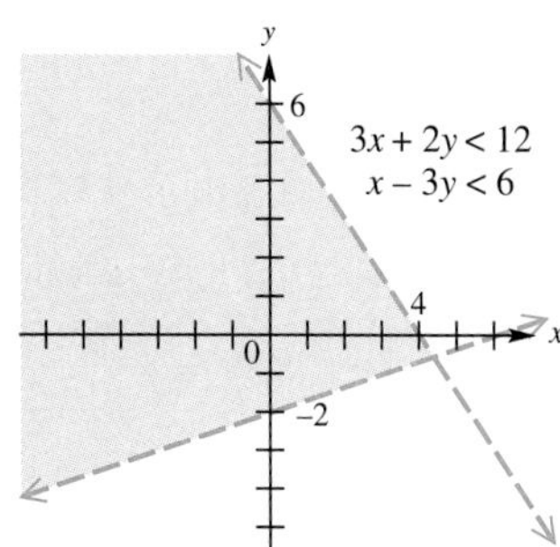

23.

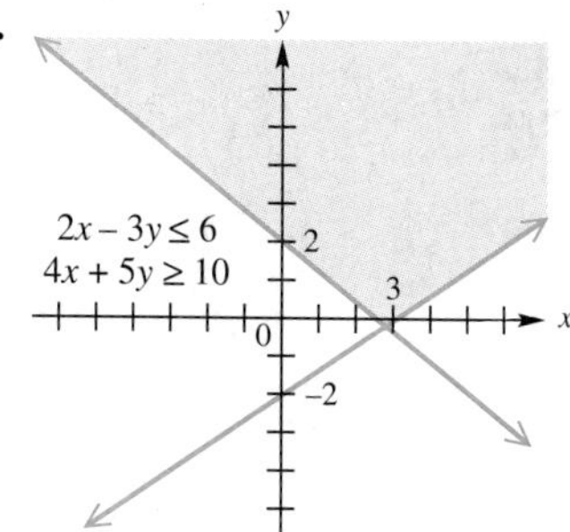

25.

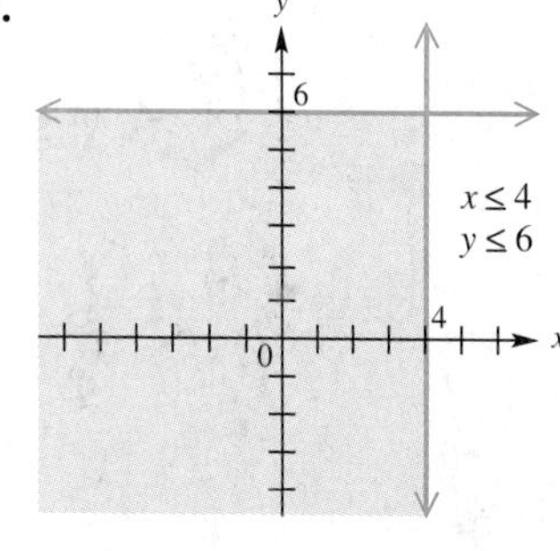

27.

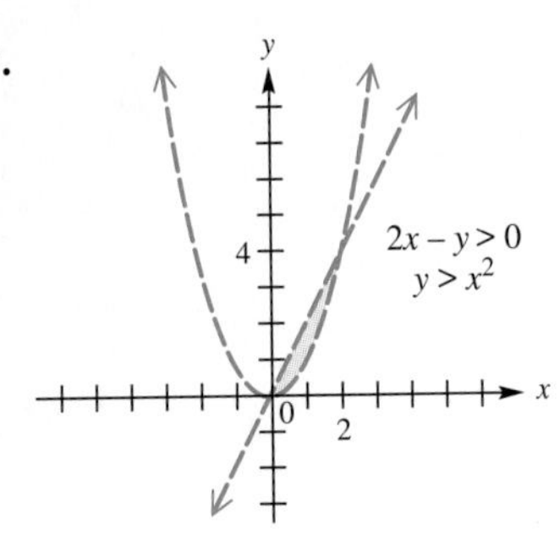

29.

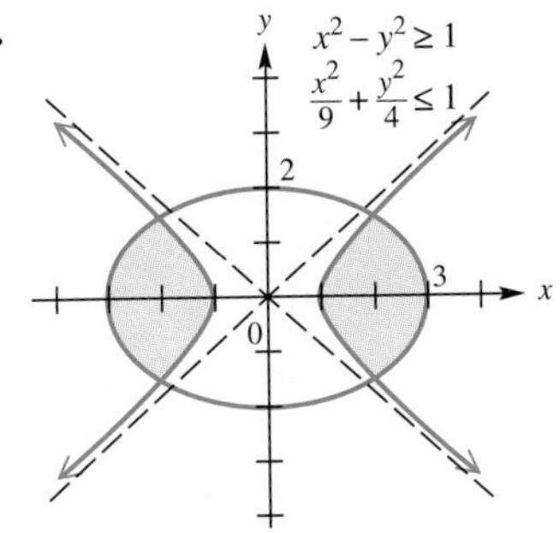

31.

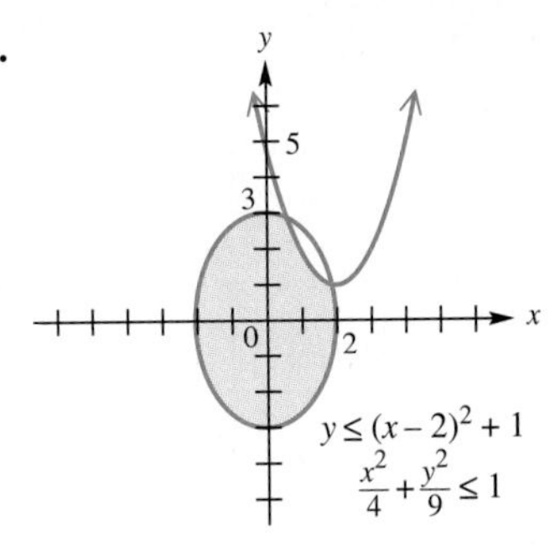

33.

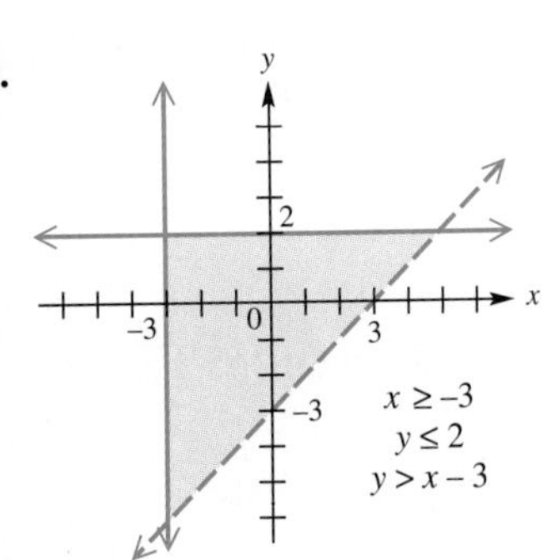

35.

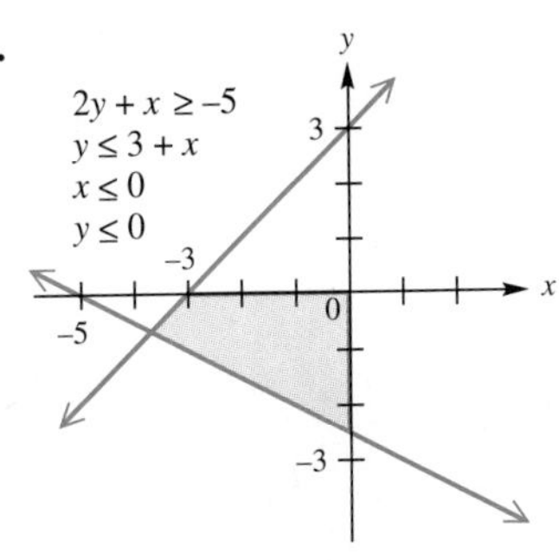

37.

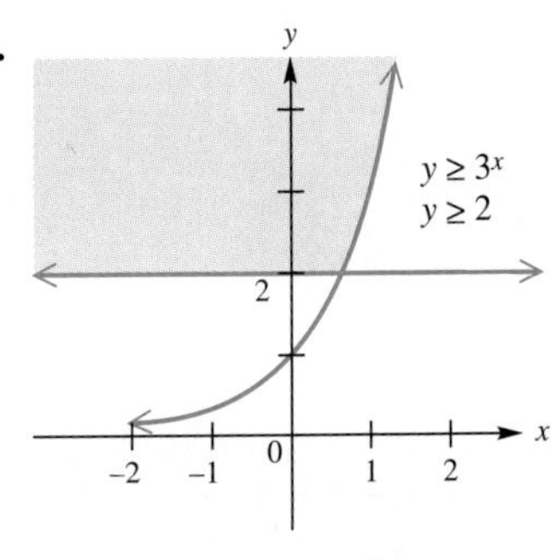

39.

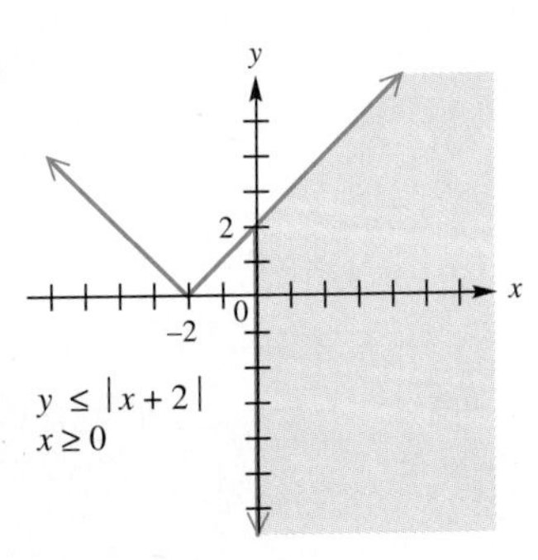

41.

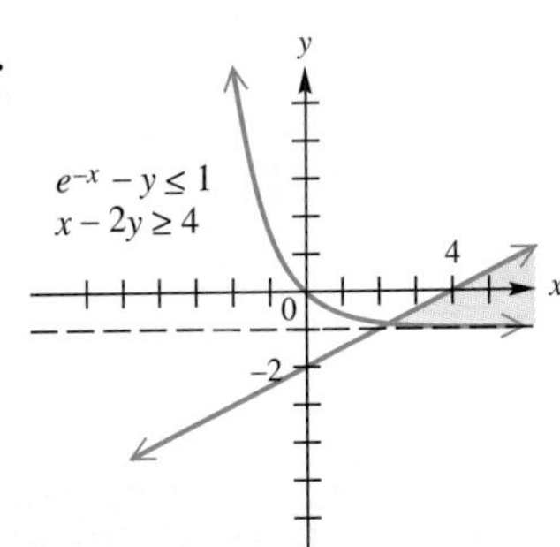

43. (c)
45. maximum: 65; minimum: 8
47. maximum: 900; minimum: 0
49. maximum of $\frac{42}{5}$ at $\left(\frac{6}{5}, \frac{6}{5}\right)$
51. minimum of $\frac{49}{3}$ at $\left(\frac{17}{3}, 5\right)$
53. \$1120 (with 4 pigs, 12 geese)
55. 8 of #1; 3 of #2 (for 100 cubic feet of storage)
57. 6.4 million gallons of gasoline and 3.2 million gallons of fuel oil (for \$16,960,000)
59. −16 **61.** −3 **63.** 7

Chapter 11 Review Exercises (page 641)

1. $\{(2, -3)\}$ **3.** $\{(-1, 2)\}$ **5.** $\left\{\left(-\frac{8}{9}, -\frac{4}{3}\right)\right\}$ **7.** $\{(2, 4)\}$ **9.** $\{(-1, 2)\}$ **11.** $\{(1, -5, 3)\}$ **13.** $\emptyset$

15. 4 days at weekend rates and 6 days at weekday rates **17.** plane: 300 miles per hour; wind: 20 miles per hour

19. 85°, 60°, 35° **21.** $a = 3$, $b = 2$, $c = -4$ **23.** 141.5 pounds of A; 170.3 pounds of B; 48.6 pounds of C

25. $\left\{(2, 3), \left(-\frac{1}{5}, -\frac{18}{5}\right)\right\}$ **27.** $\{(-4, -3), (-2, -6)\}$ **29.** $\left\{\left(\frac{-1 + \sqrt{17}}{2}, \frac{5 + \sqrt{17}}{2}\right), \left(\frac{-1 - \sqrt{17}}{2}, \frac{5 - \sqrt{17}}{2}\right)\right\}$

31. $\left\{\left(-\frac{5}{2}, -6\right), (3, 5)\right\}$ **33.** $\left\{(-1, -3), \left(\frac{3}{2}, 2\right)\right\}$ **35.** 0, 1, 2, 3, 4 **37.** 0, 1, 2, 3, 4

39.

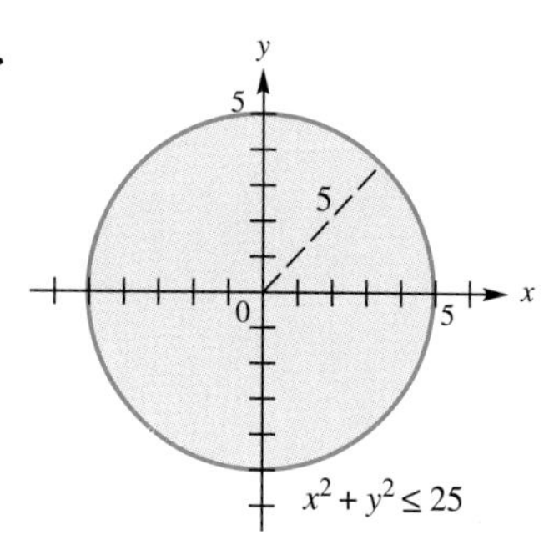

41.

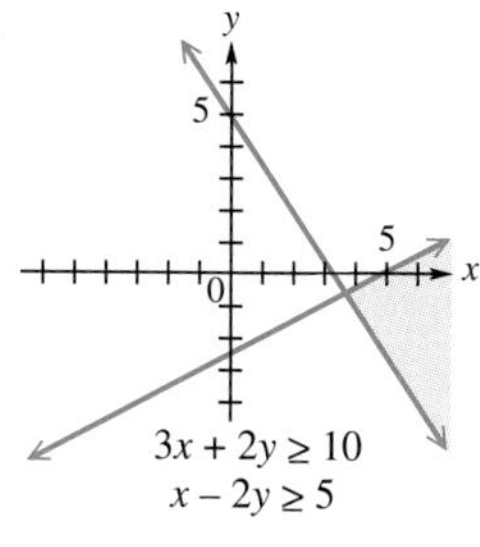

43.

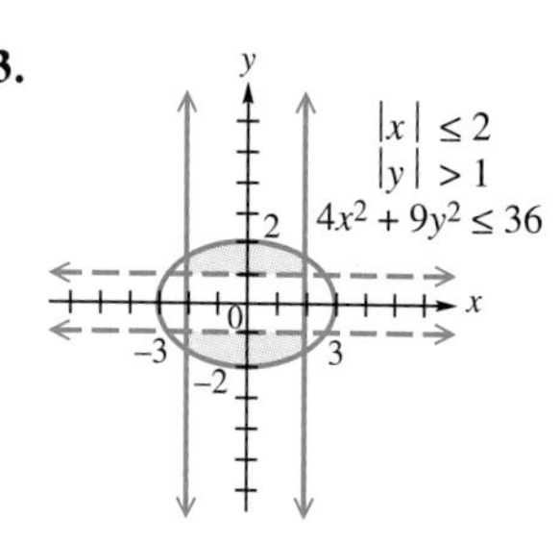

45. Let x = number of batches of cakes and y = number of batches of cookies. Then $x \ge 0$, $y \ge 0$, and $2x + \frac{3}{2}y \le 15$, $3x + \frac{2}{3}y \le 13$.

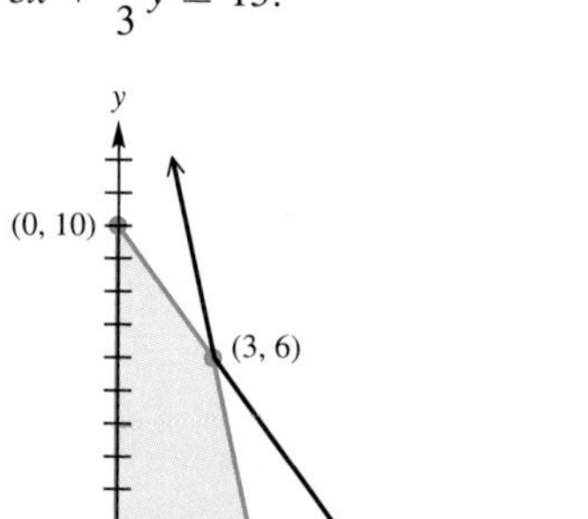

47. 3 batches of cakes and 6 batches of cookies (for maximum profit of $210) **49.** $\{(0, 3, 2)\}$ **51.** $\{(5, -3)\}$ **53.** $4.60 tea: 10 pounds; $5.75 tea: 8 pounds; $6.50 tea: 2 pounds **55.** $\left\{\left(-\frac{5}{4}, -\frac{79}{40}\right)\right\}$ **57.** 12 carat: 17.5 grams; 22 carat: 7.5 grams

Chapter 11 Test (page 643)

[11.1] **1.** $\{(-1, 4)\}$ **2.** $\{(3, -4)\}$ **3.** $\emptyset$ **4.** $\{(-1, 2)\}$ **5.** $\{(4, -3)\}$ **6.** $\{(8, 3)\}$ **[11.2]** **7.** $\{(3, -1, 2)\}$ **8.** $\{(-3, -2, -4)\}$ **[11.3]** **10.** 60 gallons of 30%; 40 gallons of 80% **11.** 40 miles per hour; 60 miles per hour **12.** 6 centimeters; 7 centimeters; 10 centimeters **[11.4]** **13.** 0, 1, 2, 3, 4, or infinitely many (if they are the same parabola) **14.** $\left\{\left(8, \frac{1}{2}\right), (2, 2)\right\}$ **15.** $\left\{(1, 1), \left(\frac{1}{5}, -\frac{7}{5}\right)\right\}$ **16.** $\{(2\sqrt{2}, 2\sqrt{2}), (2\sqrt{2}, -2\sqrt{2}), (-2\sqrt{2}, 2\sqrt{2}), (-2\sqrt{2}, -2\sqrt{2})\}$ **17.** $\{(1 + i\sqrt{6}, -3 + 2i\sqrt{6}), (1 - i\sqrt{6}, -3 - 2i\sqrt{6})\}$

[11.5] **18.**

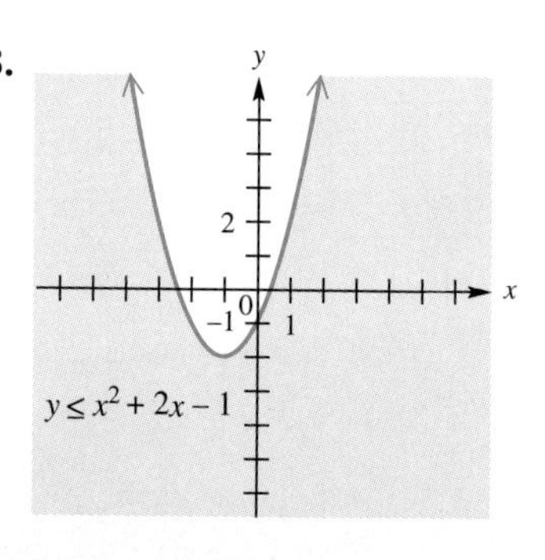

19.

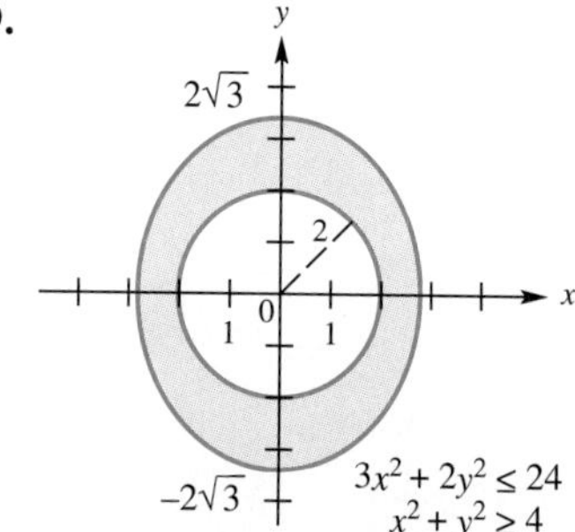

20. 0 VIP rings and 24 SST rings (for maximum profit of $960)

CHAPTER 12

12.1 Exercises (page 653)

1. 2×2; square **3.** 3×4 **5.** 2×1; column **7.** 4×1; column **9.** 3×3; square **11.** -36 **13.** 7 **15.** 0 **17.** -26 **19.** -16 **21.** 0 **23.** $2x - 32$ **25.** $y^2 - 16$ **27.** $x^2 - y^2$ **29.** p^2 **33.** -1 **35.** 8 **37.** 17 **39.** 166 **41.** 0 **43.** 0 **45.** $4x$ **47.** $-10m - 3$ **49.** $\{-4\}$ **51.** $\{13\}$ **55.** -88 **57.** 1 **59.** $\frac{5}{2}$ **61.** 7 **63.** 8 **65.** $x + 3y - 11 = 0$ **67.** commutative property **69.** distributive property

12.2 Exercises (page 662)

1. Property 5 **3.** Property 1 **5.** By Property 4, multiply each element in the second row by $-1/4$. The determinant of the resulting matrix is 0, by Property 5. **7.** Use Property 5 after multiplying each element in the third column by 1/2. **9.** Use Property 5 after multiplying each element in the third column by 1/2. **13.** Property 2 **15.** Property 3 **17.** Property 4 **19.** Property 6 (first row added to second row) **21.** Property 6 (-3 times the elements of the second column added to the elements of the first column) **23.** Property 4 **25.** Property 6 **27.** 0 **29.** 0 **31.** 12 **33.** -49 **35.** -6 **41.** $\{(3, 3)\}$ **43.** $\left\{\left(-\frac{1}{2}, 0, \frac{5}{2}\right)\right\}$

12.3 Exercises (page 668)

1. $\{(2, 2)\}$ **3.** $\{(2, -5)\}$ **5.** $\{(2, 0)\}$ **7.** $\{(2, 3)\}$ **9.** can't use Cramer's rule, $D = 0$; $\left\{\left(\frac{4 - 2y}{3}, y\right)\right\}$ **11.** can't use Cramer's rule, $D = 0$; $\emptyset$ **13.** $\{(-1, 2, 1)\}$ **15.** $\{(-3, 4, 2)\}$ **17.** $\{(0, 0, -1)\}$ **19.** can't use Cramer's rule, $D = 0$; $\emptyset$ **21.** can't use Cramer's rule, $D = 0$; $\left\{\left(\frac{-32 + 19z}{4}, \frac{24 - 13z}{4}, z\right)\right\}$ **23.** $\{(2, 3, 1)\}$ **25.** $\{(0, 4, 2)\}$ **27.** $\left\{\left(\frac{31}{5}, \frac{19}{10}, -\frac{29}{10}\right)\right\}$ **31.** $\{(0, 2, -2, 1)\}$ **33.** Cramer's rule does not apply; $D = 0$ **35.** $\{(-a - b, a^2 + ab + b^2)\}$ **37.** $\{(1, 0)\}$; $b - a \neq 0$ **39.** $2, -\frac{1}{2}$ **41.** $-27, \frac{1}{27}$ **43.** $\frac{4}{5}, -\frac{5}{4}$

12.4 Exercises (page 676)

1. $\begin{bmatrix} 2 & 4 \\ 0 & -1 \end{bmatrix}$ **3.** $\begin{bmatrix} 1 & 1/5 \\ 0 & 1 \end{bmatrix}$ **5.** $\begin{bmatrix} 1 & 5 & 6 \\ 0 & 13 & 11 \\ 4 & 7 & 0 \end{bmatrix}$ **7.** $\begin{bmatrix} -3 & 1 & -4 \\ 2 & 1 & 3 \\ -17 & 0 & -13 \end{bmatrix}$ **9.** $\left[\begin{array}{cc|c} 2 & 3 & 11 \\ 1 & 2 & 8 \end{array}\right]$ **11.** $\left[\begin{array}{cc|c} 1 & 5 & 6 \\ 1 & 2 & 8 \end{array}\right]$ **13.** $\left[\begin{array}{ccc|c} 2 & 1 & 1 & 3 \\ 3 & -4 & 2 & -7 \\ 1 & 1 & 1 & 2 \end{array}\right]$ **15.** $\left[\begin{array}{ccc|c} 1 & 1 & 0 & 2 \\ 0 & 2 & 1 & -4 \\ 0 & 0 & 1 & 2 \end{array}\right]$ **17.** $\left[\begin{array}{ccc|c} 1 & 0 & 0 & 5 \\ 0 & 1 & 0 & -2 \\ 0 & 0 & 1 & 3 \end{array}\right]$ **19.** $x = 2$; $y = 3$ **21.** $2x + y = 1$; $3x - 2y = -9$ **23.** $x = 2$; $y = 3$; $z = -2$ **25.** $3x + 2y + z = 1$; $2y + 4z = 22$; $-x - 2y + 3z = 15$ **27.** $\{(2, 3)\}$ **29.** $\{(-3, 0)\}$ **31.** $\left\{\left(\frac{7}{2}, -1\right)\right\}$ **33.** $\left\{\left(\frac{5}{2}, -1\right)\right\}$ **35.** $\emptyset$ **37.** $\left\{\left(\frac{3y + 1}{6}, y\right)\right\}$ **39.** $\{(-2, 1, 3)\}$ **41.** $\{(-1, 23, 16)\}$ **43.** $\{(3, 2, -4)\}$ **45.** $\{(2, 1, -1)\}$ **49.** $\{(-1, 2, 5, 1)\}$ **51.** \$3000 at 5%, \$1000 at 6%, 6000 at 8% **53.** \$1.40 for premium; \$1.20 for regular; \$1.30 for super **55.** $\left\{\left(\frac{5z + 14}{5}, \frac{5z - 12}{5}, z\right)\right\}$ **57.** $\left\{\left(\frac{12 - z}{7}, \frac{4z - 6}{7}, z\right)\right\}$ **59.** $\emptyset$ **61.** **(a)** $x_3 + x_4 = 600$ **(b)** The system has no unique solution. **(c)** $x_2 - x_4 = 100$ **(d)** $x_4 = 1000 - x_1$; 1000 **(e)** $x_4 = x_2 - 100$; 100 **(f)** $x_4 = 600 - x_3$; 600, 600 **(g)** 600; 600; 700; 1000 **63.** $\left\{\left(-\frac{122}{3}, \frac{17}{3}\right)\right\}$ **65.** $\{(3, 3)\}$

Chapter 12 Review Exercises (page 683)

1. −25 **3.** 0 **5.** −44 **7.** 8 **9.** $\left\{-\frac{7}{3}\right\}$ **11.** {any number} **13.** determinant property 1 **15.** determinant property 4 **17.** determinant property 6 **19.** −24 **21.** $\left\{\left(\frac{37}{11}, \frac{14}{11}\right)\right\}$ **23.** $\left\{\left(\frac{34}{13}, -\frac{20}{39}\right)\right\}$ **25.** {(2, 1, 1)} **27.** $\left\{\left(\frac{172}{67}, -\frac{14}{67}, -\frac{87}{67}\right)\right\}$ **29.** {(−4, 6)} **31.** {(−3, 2)} **33.** {(0, 3, 3)} **35.** {(−1, 2, 3)}

Chapter 12 Test (page 684)

[12.1] **1.** −10 **2.** −6 **3.** 20 **4.** −139 **5.** −5 **6.** −231 [12.2] **7.** true; Property 2 (rows and columns interchanged) **8.** false **9.** true; Property 3 (columns 1 and 2 exchanged) **10.** true; Property 6 (In the first matrix, −3 times the elements of row 2 were added to the elements of row 3.) **11.** 67 **12.** −13 [12.3] **13.** $\left\{\left(-\frac{9}{4}, \frac{5}{4}\right)\right\}$ **14.** $\left\{\left(\frac{2}{5}, \frac{7}{5}\right)\right\}$ **15.** {(−1, 2, 3)} **16.** {(4, 3, −2)} [12.4] **17.** {(2, 2)} **18.** {(2, 1)} **19.** {(−1, 0, 2)} **20.** $\left\{\left(\frac{6z+5}{11}, \frac{31z+2}{11}, z\right)\right\}$

CHAPTER 13

13.1 Exercises (page 690)

1. 10, 16, 22, 28, 34 **3.** 2, 4, 8, 16, 32 **5.** 1, −1, 1, −1, 1 **7.** $\frac{1}{2}, \frac{4}{5}, 1, \frac{8}{7}, \frac{5}{4}$ **9.** $\frac{4}{3}, \frac{12}{5}, \frac{20}{7}, \frac{28}{9}, \frac{36}{11}$ **11.** −2, 8, −24, 64, −160 **13.** $\frac{2}{3}, \frac{5}{6}, \frac{10}{11}, \frac{17}{18}, \frac{26}{27}$ **15.** x, x^2, x^3, x^4, x^5 **17.** 4, 9, 14, 19, 24, 29, 34, 39, 44, 49 **19.** −9, −13, −17, −21, −25, −29, −33, −37, −41, −45 **21.** 2, 4, 8, 16, 32, 64, 128, 256, 512, 1024 **23.** 1, 1, 2, 3, 5, 8, 13, 21, 34, 55 **25.** 4, 6, 8, 10, 12 **27.** 11, 9, 7, 5 **29.** 6, 2, −2, −6 **31.** −6, 0, 6, 12, 18, 24 **33.** $d = 5$; $a_n = 7 + 5n$ **35.** $d = -3$; $a_n = 21 - 3n$ **37.** $d = 7$; $a_n = -26 + 7n$ **39.** $d = m$; $a_n = x + nm - m$ **41.** $d = -m$; $a_n = 2z + 2m - mn$ **43.** $a_8 = 19$; $a_n = 3 + 2n$ **45.** $a_8 = 7$; $a_n = n - 1$ **47.** $a_8 = -6$; $a_n = 10 - 2n$ **49.** $a_8 = -3$; $a_n = -39 + \frac{9n}{2}$ **51.** $a_8 = x + 21$; $a_n = x + 3n - 3$ **53.** $a_8 = 4m$; $a_n = mn - 4m$ **55.** One example is $a_1 = 6$, $d = 2$. Many answers are possible. **57.** (d) **59.** $a_1 = 7$ **61.** $a_1 = 0$ **65.** 8 **67.** 29.0 (rounded) **69.** 18 **71.** $\frac{1}{2}$ **73.** .000001

13.2 Exercises (page 695)

1. 2, 6, 18, 54 **3.** $\frac{1}{2}$, 2, 8, 32 **5.** −2, 6, −18, 54 **7.** 3125, 625, 125, 25, 5, 1, $\frac{1}{5}$ **9.** −1, 1, −1, 1, −1, 1 **11.** $\frac{3}{2}$, 3, 6, 12, 24 **13.** $a_5 = 324$; $a_n = 4 \cdot 3^{n-1}$ **15.** $a_5 = -162$; $a_n = -2 \cdot 3^{n-1}$ **17.** $a_5 = -1875$; $a_n = -3(-5)^{n-1}$ **19.** $a_5 = 24$; $a_n = \left(\frac{3}{2}\right)2^{n-1}$ **21.** $a_5 = -256$; $a_n = -1(-4)^{n-1}$ **23.** $r = 2$; $a_n = 6 \cdot 2^{n-1}$ **25.** $r = 2$; $a_n = \left(\frac{3}{4}\right)2^{n-1}$ **27.** $r = -\frac{1}{2}$; $a_n = -4\left(-\frac{1}{2}\right)^{n-1}$ **29.** not geometric **31.** $a_n = \left(\frac{9}{4}\right)2^{n-1}$ **33.** $a_n = \left(-\frac{2}{9}\right)3^{n-1}$ **35.** One example is $a_1 = \frac{3}{5}$, $r = 5$. Many answers are possible.

37. (d) **39.** $a_1 = 2, r = 3$ or $a_1 = -2, r = -3$ **41.** $a_1 = 128, r = \frac{1}{2}$ or $a_1 = -128, r = -\frac{1}{2}$ **45.** 30 **47.** 10

13.3 Exercises (page 704)

1. 35 **3.** 45 **5.** $\frac{29}{20}$ **7.** $\frac{139}{108}$ **9.** 3 **11.** $\frac{25}{2}$ **13.** 215 **15.** 125 **17.** 230 **19.** $\frac{205}{2}$ **21.** $\frac{155}{2}$ **23.** $a_1 = 7;\ d = 5$ **25.** $a_1 = 1;\ d = -\frac{20}{11}$ **27.** 18 **29.** 140 **31.** 570 **33.** -684 **35.** 500,500 **37.** 400 **39.** 1705 **41.** $\frac{1111}{72}$ **43.** $\frac{1042}{25}$ **45.** 363 **47.** $-\frac{31}{2}$ **49.** 1092 **51.** -120 **53.** $\frac{1330}{9}$ **55.** $-\frac{75}{256}$ **57.** $\frac{7205}{27}$ **59.** 3240 **61.** 181 **63.** 429 **65.** 465 **67.** \$2865 **69.** 54,800 **71.** 713 inches **73.** \$2,147,483,647 **75.** At the end of six weeks, she had 7351 pounds of onions, worth 10.6¢ per pound, for a total value of \$779.21. She lost \$220.79. **77.** $80\left(\frac{4}{5}\right)^8\% \approx 13.42\%$ **79.** \$26,214.40 **81.** \$17,107.43 **83.** \$12,172.78 **85.** $-\frac{12}{5}$ **87.** $-\frac{100}{11}$

13.4 Exercises (page 710)

1. 2 **3.** 10 **5.** $\frac{1}{2}$; sum would exist **7.** $\frac{1}{2}$; sum would exist **11.** $\frac{64}{3}$ **13.** $\frac{1000}{9}$ **15.** 135 **17.** $\frac{512}{3}$ **19.** 81 **21.** $\frac{3}{2}$ **23.** 2 **25.** $\frac{1}{5}$ **27.** The formula of this section does not apply. **29.** $\frac{1}{3}$ **31.** The sum does not exist. **33.** $-\frac{1}{5}$ **35.** $\frac{1}{4}$ **37.** $\frac{1}{9}$ **39.** The sum does not exist. **41.** (b) **43.** $\frac{5}{9}$ **45.** $\frac{4}{33}$ **47.** $\frac{31}{99}$ **49.** $\frac{508}{999}$ **51.** 70 meters **53.** 200 centimeters **55.** 12 meters **57.** $9x^2 + 12xy + 4y^2$ **59.** $a^3 + 3a^2b + 3ab^2 + b^3$

13.5 Exercises (page 717)

1. 8th row: 1, 7, 21, 35, 35, 21, 7, 1; 9th row: 1, 8, 28, 56, 70, 56, 28, 8, 1; 10th row: 1, 9, 36, 84, 126, 126, 84, 36, 9, 1; 11th row: 1, 10, 45, 120, 210, 252, 210, 120, 45, 10, 1 **3.** 24 **5.** 720 **7.** 3,628,800 **9.** $2.432902008 \times 10^{18}$ **11.** 3 **13.** 10 **15.** 35 **17.** 45 **19.** $x^6 + 6x^5y + 15x^4y^2 + 20x^3y^3 + 15x^2y^4 + 6xy^5 + y^6$ **21.** $p^5 - 5p^4q + 10p^3q^2 - 10p^2q^3 + 5pq^4 - q^5$ **23.** $r^{10} + 5r^8s + 10r^6s^2 + 10r^4s^3 + 5r^2s^4 + s^5$ **25.** $p^4 + 8p^3q + 24p^2q^2 + 32pq^3 + 16q^4$ **27.** $2401p^4 + 2744p^3q + 1176p^2q^2 + 224pq^3 + 16q^4$ **29.** $729x^6 - 2916x^5y + 4860x^4y^2 - 4320x^3y^3 + 2160x^2y^4 - 576xy^5 + 64y^6$ **31.** $\frac{m^6}{64} - \frac{3m^5}{16} + \frac{15m^4}{16} - \frac{5m^3}{2} + \frac{15m^2}{4} - 3m + 1$ **33.** $16p^4 + \frac{32p^3q}{3} + \frac{8p^2q^2}{3} + \frac{8pq^3}{27} + \frac{q^4}{81}$ **35.** $7920m^8p^4$ **37.** $126x^4y^5$ **39.** $180m^2n^8$ **41.** $4845p^8q^{16}$ **43.** $439{,}296x^{21}y^7$ **45.** 2.594 **47.** 245.937 **49.** 758.650 **51.** 1.002; 5.010 **53.** .942 **55.** 5 **57.** 1

13.6 Exercises (page 723)

1–21 are proofs, so answers are not supplied. **27.** $\frac{4^{n-1}}{3^{n-2}}$ or $3\left(\frac{4}{3}\right)^{n-1}$ **29.** 120 **31.** 56

13.7 Exercises (page 728)

1. 5040 **3.** 720 **5.** 90 **7.** 336 **9.** 7 **11.** 1 **15.** 40,320 **17.** 120; 30,240 **19.** 840 **21.** 120 **23.** 4 **25.** 552 **27.** **(a)** 17,576,000 **(b)** 35,152,000 **29.** 11,232,000; 22,464,000 **31.** 84 **33.** 1 **35.** 1

13.8 Exercises (page 733)

1. 6 **3.** 56 **5.** 1365 **7.** 120 **9.** 14 **11.** 1 **17.** 27,405 **19.** 20 **21.** 1326 **23.** 10 **25.** 28 **27.** **(a)** 84 **(b)** 10 **(c)** 40 **(d)** 28 **29.** 2,598,960 **31.** 4 **33.** 5148 **35.** 120 **37.** 220 **41.** $\frac{1}{9}$ **43.** $\frac{1}{4}$

13.9 Exercises (page 740)

1. Let H = heads, T = tails; $S = \{H\}$ **3.** $S = \{HHH, HHT, HTH, THH, HTT, THT, TTH, TTT\}$ **5.** Let c = correct, w = wrong; $S = \{ccc, ccw, cwc, wcc, wwc, wcw, cww, www\}$ **7.** **(a)** $\{HH, TT\}$, $\frac{1}{2}$ **(b)** $\{HH, HT, TH\}$, $\frac{3}{4}$ **9.** **(a)** {2 and 4}, $\frac{1}{10}$ **(b)** {1 and 3, 1 and 5, 3 and 5}, $\frac{3}{10}$ **(c)** $\emptyset$, 0 **(d)** {1 and 2, 1 and 4, 2 and 3, 2 and 5, 3 and 4, 4 and 5}, $\frac{3}{5}$ **13.** **(a)** $\frac{1}{5}$ **(b)** $\frac{8}{15}$ **(c)** 0 **(d)** 1 to 4 **(e)** 7 to 8 **15.** 1 to 4 **17.** $\frac{2}{5}$ **19.** **(a)** $\frac{1}{2}$ **(b)** $\frac{7}{10}$ **(c)** $\frac{2}{5}$ **21.** **(a)** $\frac{1}{6}$ **(b)** $\frac{1}{3}$ **(c)** $\frac{1}{6}$ **23.** **(a)** $\frac{7}{15}$ **(b)** $\frac{11}{15}$ **(c)** $\frac{4}{5}$

Chapter 13 Review Exercises (page 746)

1. $\frac{1}{2}, \frac{2}{3}, \frac{3}{4}, \frac{4}{5}, \frac{5}{6}$ **3.** 8, 10, 12, 14, 16 **5.** 5, 2, −1, −4, −7 **7.** 5, 3, −2, −5, −3 **9.** 6, 2, −2, −6, −10 **11.** $3 - \sqrt{5}, 4, 5 + \sqrt{5}, 6 + 2\sqrt{5}, 7 + 3\sqrt{5}$ **13.** $a_1 = -29$; $a_{20} = 66$ **15.** 17 **17.** $m - 10$ **19.** 64, 32, 16, 8, 4 **21.** $\frac{8}{9}, \pm\frac{8}{3}, 8, \pm 24, 72$ **23.** 48 **25.** $\frac{x^5}{125}$ **29.** 612 **31.** 15 **33.** $-10k$ **35.** 6 **37.** 133 **39.** 240 **41.** \$6930 **43.** 18.75 grams **45.** 80 **47.** 1 **49.** $r > 1$ so sum does not exist **51.** $\frac{512}{999}$ **53.** 56 **55.** 1 **57.** $x^4 + 8x^3y + 24x^2y^2 + 32xy^3 + 16y^4$ **59.** $\frac{k^5}{32} - \frac{5k^4g}{16} + \frac{5k^3g^2}{4} - \frac{5k^2g^3}{2} + \frac{5kg^4}{2} - g^5$ **61.** $25{,}344m^5n^{14}$ **63.** $-420 \cdot 3^{13}m^2n^{13} + 30 \cdot 3^{14}mn^{14} - 3^{15}n^{15}$ **69.** 72 **79.** 90 **81.** **(a)** 20 **(b)** 10 **83.** 456,976,000; 258,336,000 **85.** **(a)** 4 to 11 **(b)** 3 to 2 **87.** $\frac{4}{13}$ **89.** $\frac{3}{4}$ **91.** .86

Chapter 13 Test (page 749)

[13.1] **1.** −2, 5, −10, 17, −26 **2.** 5, 7, 10, 14, 19 **3.** 2, 4, 4, 6, 6 **4.** −23, 1, 25, 49, 73 **5.** 6 [13.2] **6.** 81, 54, 36, 24, 16 **7.** 3 **8.** $1458x^{15}$ [13.3] **9.** 955 **10.** $\frac{15}{2}$ **11.** 2385 **12.** −186 **13.** 125 [13.4] **14.** $\frac{108}{7}$ [13.3] **15.** \$1950 **16.** 25,600 [13.5] **17.** $16x^4 - 96x^3y + 216x^2y^2 - 216xy^3 + 81y^4$ **18.** $-160w^3y^3$ [13.7, 13.8] **20.** 990 **21.** 5040 **22.** 45 **23.** 60 **24.** 24,360 **25.** 84 [13.9] **26.** $\frac{1}{26}$ **27.** $\frac{10}{13}$ **28.** $\frac{4}{13}$ **29.** 3 to 10 **30.** .92

GLOSSARY

Absolute value: The absolute value of a number is its distance from 0 or its magnitude without regard to sign. [1.1]

Absolute value equation; absolute value inequality: Absolute value equations and inequalities are equations and inequalities that involve the absolute value of a variable expression. [2.7]

Absolute value function: The absolute value function is defined by $f(x) = |x|$, for all real numbers x. [8.7]

Addition and multiplication properties of equality: These properties state that the same number may be added to (or subtracted from) both sides of an equation to obtain an equivalent equation; and the same nonzero number may be multiplied by or divided into both sides of an equation to obtain an equivalent equation. [2.1]

Addition and multiplication properties of inequality: The same number may be added to (or subtracted from) both sides of an inequality to obtain an equivalent inequality. Both sides of an inequality may be multiplied or divided by the same positive number. If both sides are multiplied by or divided by a negative number, the inequality symbol must be reversed. [2.5]

Additive inverse: The additive inverse (**negative, opposite**) of a number a is $-a$. [1.1]

Algebraic expression: An algebraic expression is an expression indicating any combination of the following operations: addition, subtraction, multiplication, division (except by 0), and taking roots on any collection of variables and numbers. [2.1]

Annuity: A sequence of equal payments made at equal periods of time is called an annuity. [13.3]

Antilogarithm: An antilogarithm is the number that corresponds to a given logarithm. [10.5]

Arithmetic sequence (or arithmetic progression): A sequence in which each term after the first is obtained by adding a fixed number to the preceding term is called an arithmetic sequence (or arithmetic progression). [13.1]

Asymptotes of a hyperbola: The two intersecting lines that the branches of a hyperbola approach are called asymptotes of the hyperbola. [8.6]

Augmented matrix: A matrix with a vertical bar that separates the columns into two groups is called an augmented matrix. [12.4]

Axis of a parabola: The line of symmetry for a parabola is called the axis of the parabola. [8.2]

Base: The base is the repeated factor in an exponential expression. [1.3]

Binomial: A binomial is a polynomial with exactly two terms. [3.3]

Boundary line: In the graph of a linear inequality, the boundary line separates the region that satisfies the inequality from the region that does not satisfy the inequality. [7.4]

Center: The fixed point mentioned in the definition of a circle is the center of the circle. [8.5]

Center of an ellipse: The center of an ellipse is the point halfway between the foci of the ellipse. [8.5]

Circle: A circle is the set of all points in a plane that lie a fixed distance from a fixed point. [8.5]

Coefficient: A coefficient (**numerical coefficient**) is the numerical factor of a term. [1.4]

Collinear: Points that lie on the same straight line are collinear. [7.1]

Column matrix: A column matrix has just one column. [12.1]

Combination: A combination of n elements taken r at a time is one of the ways in which r elements can be chosen from n elements ($r \leq n$). [13.8]

Combining like terms: Combining like terms is a method of adding or subtracting like terms by using the properties of real numbers. [1.4]

Common denominator: A common denominator of several denominators is an expression that is divisible by each of these denominators. [4.3]

Common difference: The fixed number that is added in an arithmetic sequence is called the common difference. [13.1]

Common logarithm: A common logarithm is a logarithm to the base 10. [10.5]

Common ratio: The constant nonzero real number in the definition of geometric sequence is called the common ratio of the sequence. [13.2]

Complement: The set of all outcomes in the sample space that do not belong to event E is called the complement of E. [13.9]

Complementary angles (complements): If the sum of the measures of two angles is 90°, the angles are complementary angles. They are complements of each other. [2.3]

Complex fraction: A fraction that has a fraction in the numerator, the denominator, or both, is called a complex fraction. [4.4]

Complex number: A complex number is a number that can be written in the form $a + bi$, where a and b are real numbers. [5.7]

Components: The components of an ordered pair are the two numbers in the ordered pair. [7.1]

Compound event: A compound event involves an alternative, such as E or F, where E and F are events. [13.9]

Compound inequality: A compound inequality is formed by joining two inequalities with a connective word, such as *and* or *or*. [2.6]

Conditional equation: An equation that has a finite number of elements in its solution set is called a conditional equation. [2.1]

Conic sections: Graphs that result from cutting an infinite cone with a plane are called conic sections. [8.6]

Conjugate: For real numbers a and b, $a + b$ and $a - b$ are conjugates. [5.5]

Constraints: Restrictions that are placed on the variables in a linear programming application are called constraints. [11.5]

Contradiction: An equation that has no solutions (that is, its solution set is $\emptyset$) is called a contradiction. [2.1]

Coordinate: The number that corresponds to a point on the number line is its coordinate. [1.1]

Coordinates: The numbers in an ordered pair are called the coordinates of the corresponding point. [7.1]

Cramer's rule: Cramer's rule is a method of solving a system of equations using determinants. [12.3]

Cube root: The cube root of a number r is the number that can be cubed to get r. [1.3]

Degree: An angle that measures one degree is 1/360 of a complete revolution. [2.2]

Degree of a polynomial: The degree of a polynomial is the largest degree of any of the terms of the polynomial. [3.3]

Degree of a term: The degree of a term with one variable is the exponent on the variable. The degree of a term with more than one variable is the sum of the exponents on the variables. [3.3]

Dependent equations: Dependent equations are equations whose graphs are the same line. [11.1]

Dependent variable: If the quantity y depends on x, then y is called the dependent variable in an equation relating x and y. [7.6]

Descending powers: A polynomial in one variable is written in descending powers of the variable if the exponents on the terms of the polynomial decrease from left to right. [3.3]

Determinant: Every square matrix is associated with a real number called the determinant of the matrix. [12.1]

Difference: The result of subtraction is called the difference. [1.3]

Directrix: The directrix of a parabola is a fixed line that, together with the focus, determines the parabola. [8.2]

Discriminant: The discriminant is the quantity under the radical in the quadratic formula. [6.3]

Domain: The domain of a relation is the set of the first components (x-values) of the ordered pairs of a relation. [7.6]

Element of a matrix: Each number in a matrix is called an element of the matrix. [12.1]

Elements: The elements (or **members**) of a set are the objects in the set. [1.1]

Elimination (or addition) method: The elimination (or addition) method of solving a system of equations involves the elimination of a variable by adding two equations. [11.1]

Ellipse: An ellipse is the set of all points in a plane the sum of whose distances from two fixed points is constant. [8.5]

Empty set: The set with no elements is called the empty set (or **null** set). [1.1]

Equation: An equation is a statement that two algebraic expressions are equal. [2.1]

Equivalent equations: Equivalent equations are equations that have the same solution set. [2.1]

Equivalent inequalities: Equivalent inequalities are inequalities with the same solution set. [2.5]

Even function: A function $y = f(x)$ is an even function if $f(-x) = f(x)$ for all x in the domain of f. [9.1]

Event: Any subset of the sample space for an experiment is called an event. [13.9]

Expansion by minors: Expansion by minors is a method of evaluating a determinant. [12.1]

Exponent: An exponent is a number that shows how many times a factor is repeated in a product. [1.3]

Exponential: A base with an exponent is called an exponential or a **power.** [1.3]

Exponential equation: An equation involving an exponential, where the variable is in the exponent, is an exponential equation. [10.2]

Extraneous solution: An extraneous solution of a radical equation is a solution of $x = a^2$ that is not a solution of $\sqrt{x} = a$. [5.6]

Factors: A factor of a given number is any number that divides evenly into the given number. [1.3]

Factored form: A number is in factored form if it is expressed as a product of two or more numbers. [1.3]

Factoring a polynomial: Writing a polynomial as the product of two or more simpler polynomials is called factoring the polynomial. [3.5]

Finite sequence: A sequence is a finite sequence if the domain is the set $\{1, 2, 3, 4, \ldots, n\}$, where n is a positive integer. [13.1]

Finite set: A finite set has n elements, where n is a counting number. [1.1]

Foci of an ellipse: The foci of an ellipse are two fixed points that determine an ellipse. [8.5]

Focus of a parabola: The focus of a parabola is a point on the axis that determines the curvature of the figure. [8.2]

Formula: A formula is a mathematical statement in which more than one letter is used to express a relationship. [2.2]

Function: A function is a relation in which each value of the first component, x, corresponds to exactly one value of the second component, y. [7.6]

Functions defined piecewise: Functions that have different equations defining different parts of their domains are called functions defined piecewise. [8.7]

Fundamental rectangle: The fundamental rectangle is the rectangle of a hyperbola with vertices at (a, b), $(a, -b)$, $(-a, b)$, $(-a, -b)$. [8.6]

Future value: The sum of the payments and the interest on the payments of an annuity is the future value of the annuity. [13.3]

Gauss-Jordan method: The Gauss-Jordan method is a systematic way of using matrix row transformations to solve a system of equations. [12.4]

Geometric sequence (or geometric progression): A geometric sequence (or geometric progression) is a sequence in which each term after the first is obtained by multiplying the preceding term by a constant nonzero real number. [13.2]

Graph: The point on a number line that corresponds to a number is its graph. [1.1]

Graph of a relation: The graph of a relation is the graph of the ordered pairs of the relation. [7.6]

Greatest common factor: The greatest common factor of a group of integers is the largest integer that will divide without remainder into every integer in the group. [3.5]

Greatest integer function: The greatest integer function defines y as the greatest integer less than or equal to x. [8.7]

Horizontal asymptote: If $f(x)$ is a rational function and if $f(x) \to a$ as $|x| \to \infty$, then the line $y = a$ is a horizontal asymptote. [9.5]

Hyperbola: A hyperbola is the set of all points in a plane such that the absolute value of the difference of the distances from two fixed points is constant. [8.6]

Hypotenuse: The hypotenuse of a right triangle is the longest side, opposite the right angle. [5.3]

Identity: An equation that is satisfied by every number for which both sides are defined is called an identity. [2.1]

Identity element for addition: The identity element for addition is 0. [1.4]

Identity element for multiplication: The identity element for multiplication is 1. [1.4]

Imaginary number: A complex number $a + bi$ with $b \neq 0$ is called an imaginary number. [5.7]

Imaginary part: The imaginary part of $a + bi$ is b. [5.7]

Inconsistent system: A system is inconsistent if it has no solution. [11.1]

Independent events: Independent events are events such that the outcome of any one does not affect the outcome of any other. [13.7]

Independent variable: If y depends on x, then x is the independent variable in an equation relating x and y. [7.6]

Index of summation: The letter i as used in summation notation $(\sum_{i=1}^{k} a_i)$ is called the index of summation. [13.3]

Inequality: An inequality is a mathematical statement that two quantities are not equal. [1.2]

Infinite sequence: An infinite sequence has the set of all positive integers as its domain. [13.1]

Infinite set: The elements of an infinite set are unlimited. [1.1]

Intersection: The intersection of two sets A and B is the set of elements that belong to both A and B. [2.6]

Interval: An interval is a portion of a number line. [1.2]

Interval notation: Interval notation uses symbols to describe an interval on the number line. [1.2]

Inverse of a function f: If f is a one-to-one function, the inverse of f is the set of all ordered pairs of the form (y, x), where (x, y) belongs to f. [10.1]

Leading coefficient: The number a_n in the definition of polynomial function is the leading coefficient of the polynomial. [9.1]

Legs: The legs of a right triangle are the two shorter sides. [5.3]

Like terms: Like terms are terms with the same variables raised to the same powers. [1.4]

Linear equation in two variables: A linear equation in two variables is a first-degree equation with two variables. [7.1]

Linear equation or first-degree equation in one variable: An equation is linear or first-degree in the variable x if it can be written in the form $ax + b = c$, where a, b, and c are real numbers with $a \neq 0$. [2.1]

Linear function: A function that can be written in the form $f(x) = mx + b$ for real numbers m and b is a linear function. [7.6]

Linear inequality in one variable: An inequality is linear in the variable x if it can be written in the form $ax + b < c$, $ax + b \leq c$, $ax + b > c$, or $ax + b \geq c$, where a, b, and c are real numbers, with $a \neq 0$. [2.5]

Linear programming: Linear programming is an application of linear inequalities, used to find such things as minimum cost and maximum profit. [11.5]

Linear system: A linear system is a system of equations that contains only linear equations. [11.1]

Logarithm: A logarithm is an exponent; $\log_a x$ is the exponent on the base a that gives the number x. [10.3]

Logarithmic equation: A logarithmic equation is an equation with a logarithm in at least one term. [10.3]

Major axis: The longer line segment joining the vertices on an ellipse is its major axis. [8.5]

Matrix: A matrix is a rectangular array of numbers enclosed by brackets or parentheses in which the position of each number is meaningful. [12.1]

Minor: The minor of an element of an $n \times n$ matrix is the determinant of an $(n - 1) \times (n - 1)$ matrix, which is a part of the $n \times n$ matrix. [12.1]

Minor axis: The shorter line segment joining the other intercepts of an ellipse is its minor axis. [8.5]

Monomial: A monomial is a polynomial with exactly one term. [3.3]

Multiplicity: The polynomial $P(x) = (x - 3)^2(x + 4)^5$, for example, has a zero of 3 with multiplicity 2 and a zero of -4 with multiplicity 5. [9.2]

Multiplicative inverse: The multiplicative inverse (reciprocal) of a nonzero number b is $1/b$. [1.4]

Mutually exclusive events: Two events that cannot occur simultaneously are called mutually exclusive events. [13.9]

Natural logarithm: A natural logarithm is a logarithm to the base e. [10.5]

Negative of a polynomial: The negative of a polynomial is obtained by changing the sign of each term of the polynomial. [3.3]

Nonlinear equation: An equation that cannot be put into the form $Ax + By = C$ is called a nonlinear equation. [11.4]

Nonlinear system of equations: A nonlinear system of equations is a system with at least one nonlinear equation. [11.4]

Nonlinear system of inequalities: A nonlinear system of inequalities consists of two or more inequalities considered at the same time, where at least one of them is nonlinear. [11.5]

Nonstrict inequality: An inequality that involves $\geq$ or $\leq$ is called a nonstrict inequality. [2.5]

***n*th root:** An nth root of a number a is a number b such that $b^n = a$. [5.1]

Number line: A number line is a line with a scale to indicate the set of real numbers. [1.1]

Numerical coefficient: The number in a product is called the numerical coefficient, or just the coefficient. [3.3]

Oblique asymptote: If $f(x)$ is a rational function and if $f(x) \to mx + b$ as $|x| \to \infty$, then the line $y = mx + b$ is an oblique asymptote. [9.5]

Odds: The odds in favor of an event E are expressed as the ratio of the probability of E to the probability of E'. [13.9]

Odd function: A function $y = f(x)$ is an odd function if $f(-x) = -f(x)$ for all x in the domain of f. [9.1]

One-to-one function: A one-to-one function is a function in which each x-value corresponds to just one y-value and each y-value corresponds to just one x-value. [10.1]

Ordered triple: A solution of an equation in three variables is called an ordered triple. [11.2]

Origin: When two number lines intersect at a right angle, the origin is the common zero point. [7.1]

Outcomes: The possible results of each trial of an experiment are called outcomes of the experiment. [13.9]

Parabola: The graph of a quadratic function is a parabola. A parabola is defined as the set of all points in a plane that are equidistant from a fixed point and a fixed line not containing the point. [8.2]

Parallel lines: Lines that lie in the same plane and do not intersect are called parallel lines. [2.2]

Pascal's triangle: A triangular array of numbers that gives the values of binomial coefficients is called Pascal's triangle. [13.5]

Permutation: A permutation of n elements taken r at a time is one of the ways of arranging r elements taken from a set of n elements ($r \leq n$). [13.7]

Plot: To plot an ordered pair is to locate it on a rectangular coordinate system. [7.1]

Polynomial: A polynomial is a term or a finite sum of terms in which all variables have whole number exponents and no variables appear in denominators. [3.3]

Polynomial function of degree *n*: A polynomial function of degree n is a function defined by

$$P(x) = a_n x^n + a_{n-1} x^{n-1} + \cdots + a_1 x + a_0,$$

for complex numbers $a_n, a_{n-1}, \ldots, a_1, a_0$, where $a_n \neq 0$. [9.1]

Prime factored form: An integer is written in prime factored form when it is written as a product of prime numbers. [3.5]

Prime number: A prime number is a positive integer greater than 1 that can be divided without remainder only by itself and 1. [3.5]

Prime polynomial: If a polynomial cannot be factored, it is called a prime polynomial. [3.6]

Principal *n*th root: For a positive number a and even value of n, the principal nth root of a is the positive nth root of a. [5.1]

Product: The result of multiplication is called the product. [1.3]

Pythagorean formula: The Pythagorean formula states that in a right triangle the square of the length of the hypotenuse equals the sum of the squares of the lengths of the legs. [5.3]

Quadrant: A quadrant is one of the four regions in the plane determined by a rectangular coordinate system. [7.1]

Quadratic equation in standard form: An equation that can be written in the form $ax^2 + bx + c = 0$, where $a \neq 0$, is a quadratic equation. The form is called standard form. [3.9]

Quadratic formula: The quadratic formula is a formula for solving quadratic equations. [6.2]

Quadratic function: A function f is quadratic if $f(x) = ax^2 + bx + c$, where a, b, and c are real numbers with $a \neq 0$. [8.2]

Quadratic in form: An equation that can be written as a quadratic equation is called quadratic in form. [6.4]

Quadratic inequality: A quadratic inequality is an inequality that can be written in the form $ax^2 + bx + c < 0$ or $ax^2 + bx + c > 0$, or with $\leq$ or $\geq$. [6.6]

Quotient: The result of division is called the quotient. [1.3]

Radical: The expression $\sqrt[n]{a}$ is a radical. [5.1]

Radical expression: A radical expression is an algebraic expression that contains radicals. [5.4]

Radicand, index: In the expression $\sqrt[n]{a}$, a is the radicand and n is the index (or *order*). [5.1]

Radius: The radius of a circle is the fixed distance between the center and any point on the circle. [8.5]

Range: The range of a relation is the set of second components (y-values) of the ordered pairs of the relation. [7.6]

Rational expression: A rational expression is the quotient of two polynomials with denominator not 0. [4.1]

Rational function: A function of the form $f(x) = p(x)/q(x)$, where $p(x)$ and $q(x)$ are polynomial functions and $q(x) \neq 0$, is called a rational function. [9.5]

Rational inequality: A rational inequality has a fraction with a variable denominator. [6.6]

Rationalizing the denominator: The process of removing radicals from the denominator so that the denominator contains only rational quantities is called rationalizing the denominator. [5.5]

Real part: The real part of $a + bi$ is a. [5.7]

Reciprocal: The reciprocal of a nonzero number b is $1/b$. [1.3]

Reciprocals: Two rational expressions are reciprocals if their product is 1. [4.2]

Rectangular (Cartesian) coordinate system: Two number lines that intersect at a right angle at their zero points form a rectangular coordinate system. [7.1]

Recursion formula: A recursion formula defines the nth term of a sequence in terms of the previous term. [13.1]

Region of feasible solutions: The shaded region in a linear programming application is called the region of feasible solutions. [11.5]
Relation: A relation is a set of ordered pairs. [7.6]
Right angle: An angle that measures 90° is called a right angle. [2.3]
Row matrix: A row matrix has just one row. [12.1]
Sample space: The set S of all possible outcomes of a given experiment is called the sample space for the experiment. [13.9]
Scientific notation: A number is written in scientific notation when it is expressed in the form $a \times 10^n$, where $1 \leq |a| < 10$, and n is an integer. [3.2]
Second-degree inequality: A second-degree inequality is an inequality with at least one variable of degree two and no variable of degree greater than two. [11.5]
Sequence: An ordered list of numbers is called a sequence. [13.1]
Series: The sum of the terms of a sequence is called a series. [13.3]
Set: A set is a collection of objects. [1.1]
Set-builder notation: Set-builder notation is used to describe a set of numbers without listing them. [1.1]
Signed numbers: Positive and negative numbers are signed numbers. [1.1]
Slope: The ratio of the change in y compared to the change in x along a line is the slope of the line. [7.2]
Solution: A solution of an equation is a number that makes the equation true when substituted for the variable. [2.1]
Solution set: The solution set of an equation is the set of all its solutions. [2.1]
Square matrix: A square matrix has the same number of rows as columns. [12.1]
Square root: A square root of a number r is a number that can be squared to get r. [1.3]
Square root function: A function of the form $f(x) = \sqrt{u}$ for an algebraic expression u, with $u \geq 0$, is called a square root function. [8.6]
Standard form: A complex number written as $a + bi$, where a and b are real numbers, is in standard form. [5.7]
Standard form: A linear equation is in standard form when written as $Ax + By = C$, with $A > 0$, and A, B, and C integers. [7.1]
Step function: A step function has a graph that is a series of horizontal line segments. [8.7]
Straight angle: An angle that measures 180° is called a straight angle. [2.3]
Strict inequality: An inequality that involves $>$ or $<$ is called a strict inequality. [2.5]
Subset: Set A is a subset of set B if every element of A is also an element of B. [1.1]
Substitution method: The substitution method of solving a system involves substituting an expression for one variable into an equation with two variables, in order to get an equation with just one variable. [11.1]
Sum: The result of addition is called the sum. [1.3]
Supplementary angles (supplements): If the sum of the measures of two angles is 180°, the angles are supplementary angles. They are supplements of each other. [2.3]
Synthetic division: Synthetic division is a shortcut division process used when dividing a polynomial by a binomial of the form $x - k$. [4.6]
System of equations: Two or more equations that are to be solved at the same time form a system of equations. [11.1]
Term: A term is a number or the product of a number and one or more variables raised to powers. [1.4]
Terms of a sequence: The elements of a sequence are called the terms of the sequence. [13.1]
Translation: A translation is a horizontal or vertical shift of a graph. [8.2]
Transversal: A line intersecting parallel lines is called a transversal. [2.2]
Tree diagram: A tree diagram is a method of systematically determining the number of different ways that successive events can occur. [13.7]
Trial: In probability, each repetition of an experiment is called a trial. [13.9]
Trinomial: A trinomial is a polynomial with exactly three terms. [3.3]
Union: The union of two sets A and B is the set of elements that belong to either A or B (or both). [2.6]
Variable: A variable is a letter used to represent a number or a set of numbers. [1.1]
Venn diagram: A Venn diagram shows relationships among sets. [13.9]
Vertex of a parabola: The intersection of a parabola with its axis is called the vertex of the parabola. [8.2]
Vertex point: The optimum value in a linear programming application will always occur at a vertex point (or corner point) of the region of feasible solutions. [11.5]
Vertical angles: Vertical angles are formed by intersecting lines, and have the same measure. (See, for example, angles ① and ④ in Figure 2.4.) [2.2]

Vertical asymptote: If $f(x)$ is a rational function and if $|f(x)| \to \infty$ as $x \to a$, then the line $x = a$ is a vertical asymptote. [9.5]

Vertical line test: The vertical line test says that if a vertical line cuts the graph of a relation in more than one point, then the relation is not a function. [7.6]

Vertices of an ellipse: The x- or y-intercepts of the graph of an ellipse with center at the origin are called the vertices. [8.5]

x-axis: The horizontal number line in a rectangular coordinate system is called the x-axis. [7.1]

x-intercept: The point where a line crosses the x-axis is the x-intercept. [7.1]

y-axis: The vertical number line in a rectangular coordinate system is called the y-axis. [7.1]

y-intercept: The point where a line crosses the y-axis is the y-intercept. [7.1]

Zero-factor property: If two numbers have a product of 0, then at least one of the numbers must be 0. [3.9]

Zero polynomial: If all the coefficients of a polynomial are 0, the polynomial equals 0 and is called the zero polynomial. [9.1]

Zeros of a polynomial: The values of x that satisfy $P(x) = 0$ are called the zeros of $P(x)$. [9.1]

ACKNOWLEDGMENTS

ILLUSTRATION CREDITS

All technical line art prepared by Precision Graphics.
All creative illustrations prepared by Rolin Graphics.

PHOTO CREDITS

Unless otherwise acknowledged, all photographs are the property of Scott, Foresman and Company (a subsidiary of HarperCollins Publishers). Page numbers are boldface in the following list. The photos on the title page, in the table of contents, and on chapter opener pages are courtesy of Georg Gerster/Comstock. Text accompanying chapter opener photos supplied by Weldon Owen, Inc.

24: Antonio Bignami/The Image Bank
50: Walter Iooss/The Image Bank
85: Tom Tracy/The Stock Market
128: NASA
160: SUPERSTOCK
216: Harvey Lloyd/The Stock Market
231: Lisl Dennis/The Image Bank
233: Michel Tcherevkof/The Image Bank
236: Craig Hammell/The Stock Market
244: Lew Long/The Stock Market
255: Joe Batar/The Stock Market
266: The Stock Market
311: Jay Maderf/The Image Bank
326: Lew Long/The Stock Market
332: Stephen Green-Armytage/The Stock Market
366: Guido Alberto Rossi/The Image Bank
375: Peter Arnold, Inc.
388: Barry L. Runk from Grant Heilman Photography
399: Robert Essel/The Stock Market
414: Four by Five Inc./SUPERSTOCK
430: Hans Staartjes/Tony Stone Worldwide
460: Magnus Reitz/The Image Bank
479: Giuliano Colliva/The Image Bank
502: Four by Five Inc./SUPERSTOCK
514: The Photo Source/SUPERSTOCK
546: SUPERSTOCK
555: SUPERSTOCK
572: Charles H. Phillips
580: Shostal/SUPERSTOCK
615: Andrew Sacks/Tony Stone Worldwide
625: SUPERSTOCK
635: Tom Mareschal/The Image Bank
677: Four by Five Inc./SUPERSTOCK
678: Four by Five Inc./SUPERSTOCK
679: Four by Five Inc./SUPERSTOCK
706: Four by Five Inc./SUPERSTOCK
728: Peter Poulides/Tony Stone Worldwide
733: Four by Five Inc./SUPERSTOCK
740: Four by Five Inc./SUPERSTOCK

INDEX

INDEX OF APPLICATIONS

5.7 COMPLEX NUMBER	If $i^2 = -1$ and a and b are real numbers, then any number of the form $a + bi$ is called a complex number.
SQUARE ROOT OF A NEGATIVE NUMBER	If b is a positive real number, since $i^2 = -1$, $\sqrt{-b} = i\sqrt{b}$.
IMAGINARY NUMBER	If a and b are real numbers and $b \neq 0$, then $a + bi$ is an imaginary number.
6.1 SQUARE ROOT PROPERTY	If a and b are complex numbers and if $a^2 = b^2$, then $a = b$ or $a = -b$.
6.2 QUADRATIC FORMULA	The solutions of $ax^2 + bx + c = 0$ $(a \neq 0)$ are $$x = \frac{-b \pm \sqrt{b^2 - 4ac}}{2a}.$$
6.3 DISCRIMINANT	The discriminant of $ax^2 + bx + c = 0$ is $b^2 - 4ac$.
7.1 INTERCEPTS	To find the x-intercept, let $y = 0$. To find the y-intercept, let $x = 0$.
DISTANCE FORMULA	The distance between the points (x_1, y_1) and (x_2, y_2) is $$d = \sqrt{(x_2 - x_1)^2 + (y_2 - y_1)^2}.$$
7.2 SLOPE OF A LINE	If $x_1 \neq x_2$, the slope of the line through the distinct points (x_1, y_1) and (x_2, y_2) is $$m = \frac{\text{change in } y}{\text{change in } x} = \frac{y_2 - y_1}{x_2 - x_1}.$$ A vertical line has undefined slope. A horizontal line has 0 slope. Parallel lines have equal slopes. Perpendicular lines have slopes that are negative reciprocals.
7.3 SLOPE-INTERCEPT FORM	The equation of a line with slope m and y-intercept $(0, b)$ is written in slope-intercept form as $y = mx + b$.
POINT-SLOPE FORM	The equation of the line through (x_1, y_1) with slope m is written in point-slope form as $y - y_1 = m(x - x_1)$.
7.5 VARIATION	If $y = kx^n$, then y varies directly as x^n. If $y = \frac{k}{x^n}$, then y varies inversely as x^n.

7.6 FUNCTION A function is a set of ordered pairs in which no first component is repeated.

DOMAIN OF A FUNCTION The set of all first components of the ordered pairs of a function is the domain of the function.

RANGE OF A FUNCTION The set of all second components of the ordered pairs of a function is the range of a function.

8.1 COMPOSITION OF FUNCTIONS If f and g are functions, then the composite function of g and f is defined by

$$(g \circ f)(x) = g[f(x)]$$

for all x in the domain of f such that $f(x)$ is in the domain of g.

8.2–8.6 CONIC SECTIONS

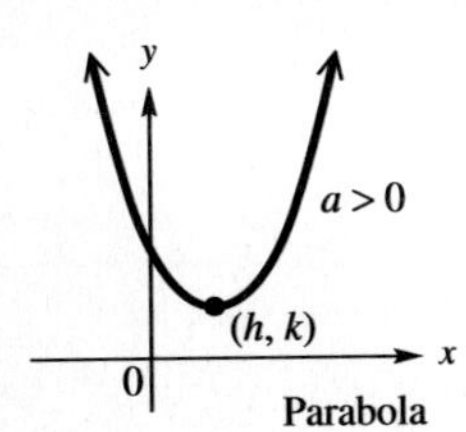

Parabola

$y = a(x - h)^2 + k$

Opens upward if $a > 0$, downward if $a < 0$; vertex is at (h, k)

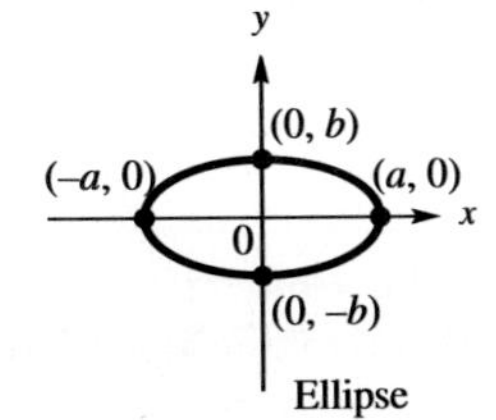

Ellipse

$$\frac{x^2}{a^2} + \frac{y^2}{b^2} = 1$$

x-intercepts are $(a, 0)$ and $(-a, 0)$; y-intercepts are $(0, b)$ and $(0, -b)$

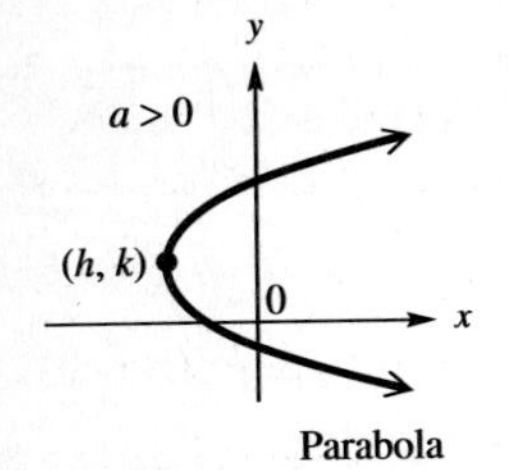

Parabola

$x = a(y - k)^2 + h$

Opens to right if $a > 0$, to left if $a < 0$; vertex is at (h, k)

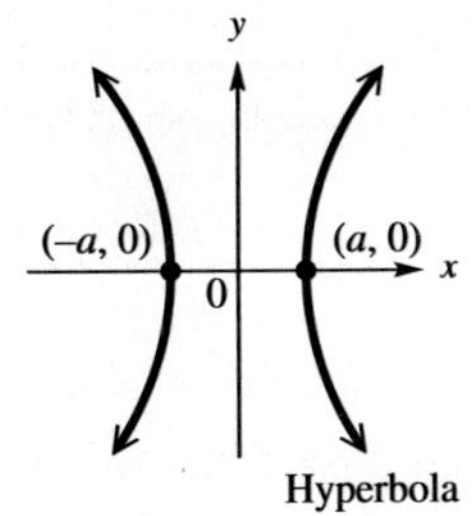

Hyperbola

$$\frac{x^2}{a^2} - \frac{y^2}{b^2} = 1$$

x-intercepts are $(a, 0)$ and $(-a, 0)$; asymptotes are found from (a, b), $(a, -b)$, $(-a, -b)$, and $(-a, b)$

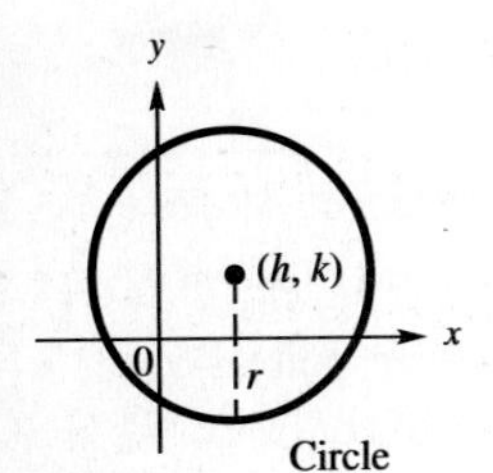

Circle

$(x - h)^2 + (y - k)^2 = r^2$

Center at (h, k), radius r

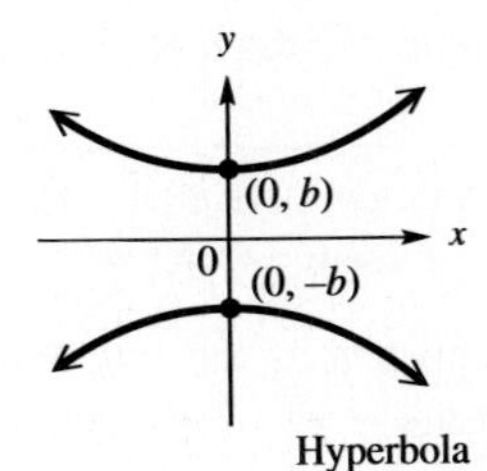

Hyperbola

$$\frac{y^2}{b^2} - \frac{x^2}{a^2} = 1$$

y-intercepts are $(0, b)$ and $(0, -b)$; asymptotes are found from (a, b), $(a, -b)$, $(-a, -b)$, and $(-a, b)$

8.7 ABSOLUTE VALUE FUNCTION

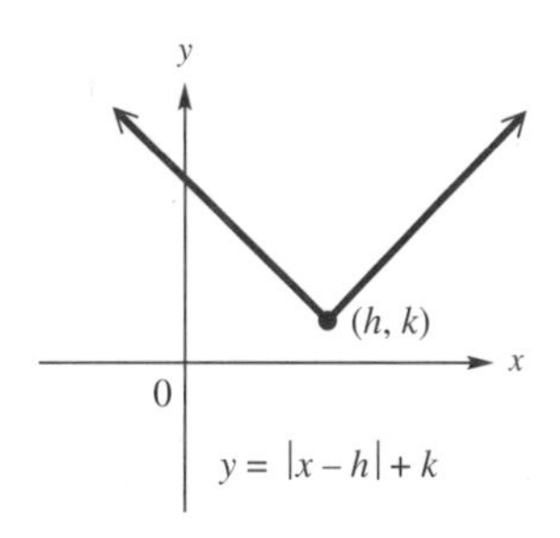

9.1 POLYNOMIAL FUNCTIONS

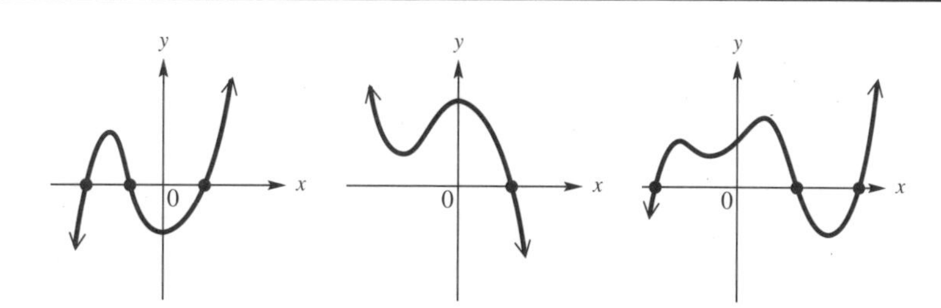

Degree 3; three real zeros

Degree 3; one real zero

Degree 5; three real zeros

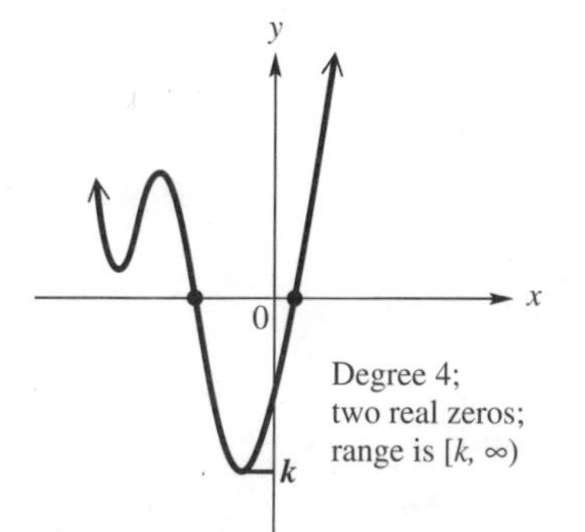

Degree 4; two real zeros; range is $[k, \infty)$

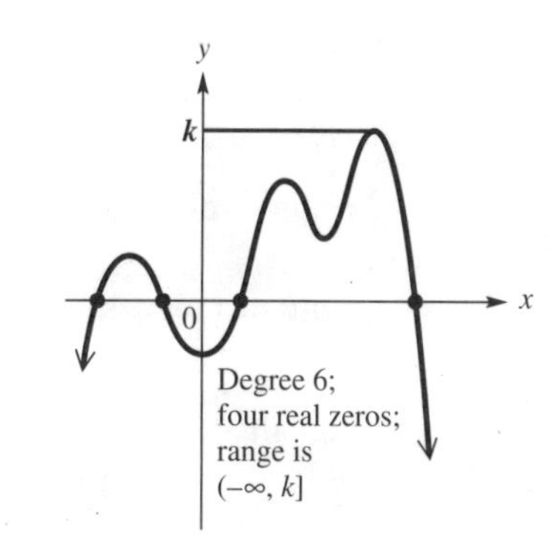

Degree 6; four real zeros; range is $(-\infty, k]$

9.5 RATIONAL FUNCTIONS

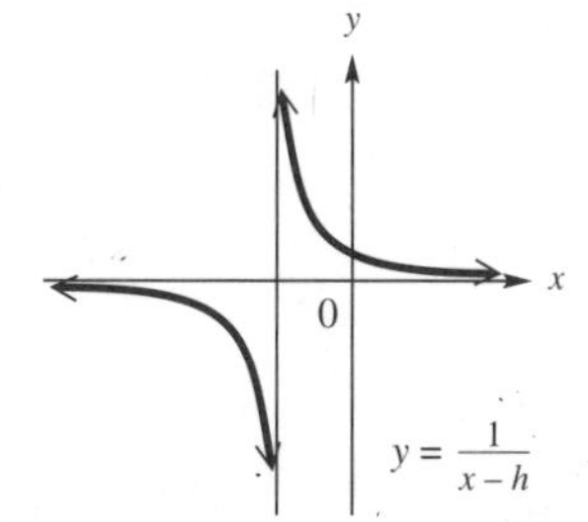

10.1 INVERSE FUNCTIONS

Let f be a one-to-one function. The inverse function, f^{-1}, can be found by exchanging x and y in the equation for f, then solving for y if possible.

10.2–10.3 EXPONENTIAL AND LOGARITHMIC FUNCTIONS

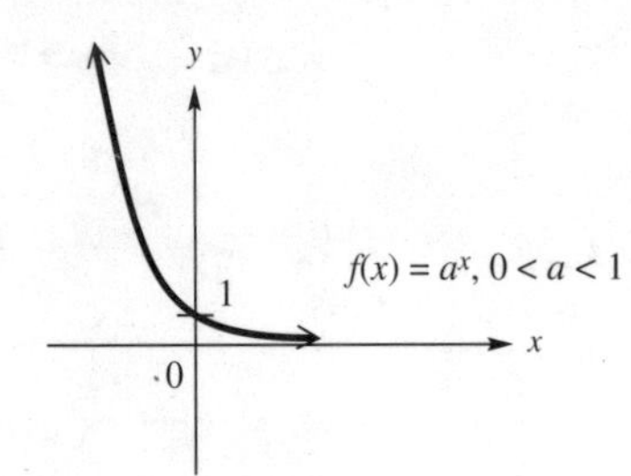

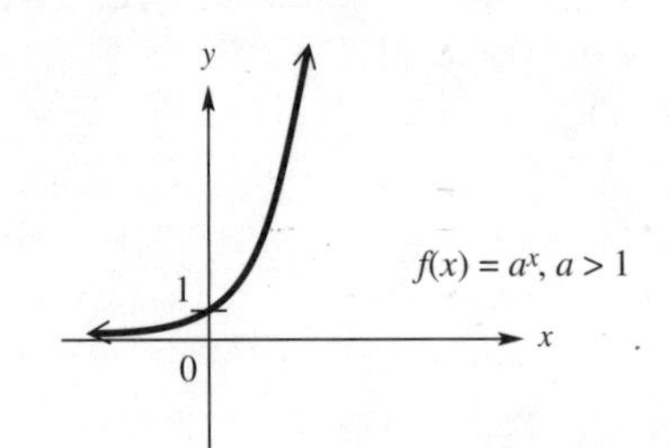

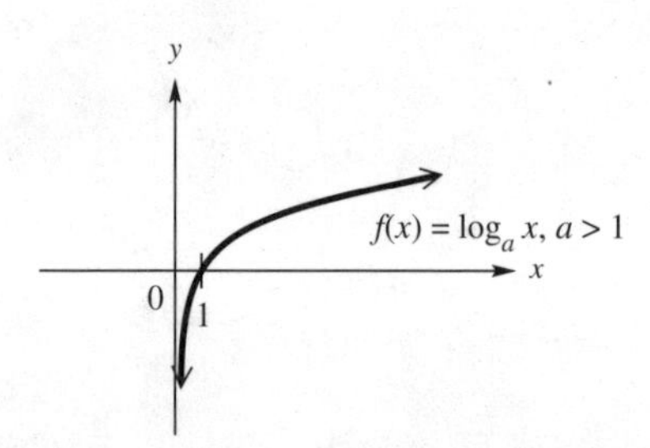

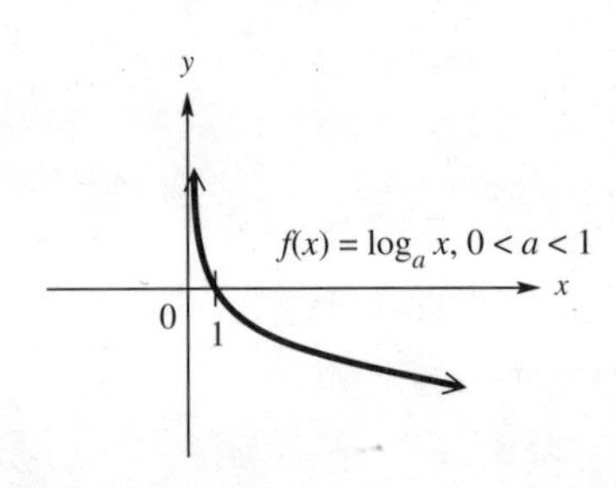

10.3 LOGARITHM

For all positive numbers a, where $a \neq 1$, and all positive numbers x, $y = \log_a x$ means the same as $x = a^y$.

10.4 PROPERTIES OF LOGARITHMS

If x, y, and b are positive numbers, where $b \neq 1$, and if r is any real number, then:

Product rule $\log_b xy = \log_b x + \log_b y$ **Power rule** $\log_b x^r = r(\log_b x)$

Quotient rule $\log_b \dfrac{x}{y} = \log_b x - \log_b y$

12.1 DETERMINANT

The value of a 2×2 determinant is

$$\begin{vmatrix} a & b \\ c & d \end{vmatrix} = ad - bc.$$

Determinants larger than 2×2 are evaluated by expansion by minors about a row or column.

12.3 CRAMER'S RULE FOR 2 × 2 SYSTEMS

Given the system

$$\begin{aligned} a_1x + b_1y &= c_1 \\ a_2x + b_2y &= c_2 \end{aligned}$$

with $a_1b_2 - a_2b_1 = D \neq 0$, then

$$x = \frac{\begin{vmatrix} c_1 & b_1 \\ c_2 & b_2 \end{vmatrix}}{\begin{vmatrix} a_1 & b_1 \\ a_2 & b_2 \end{vmatrix}} = \frac{D_x}{D} \quad \text{and} \quad y = \frac{\begin{vmatrix} a_1 & c_1 \\ a_2 & c_2 \end{vmatrix}}{\begin{vmatrix} a_1 & b_1 \\ a_2 & b_2 \end{vmatrix}} = \frac{D_y}{D}.$$